Cases

CASE Study

Paralyzed Veterans of America

Philanthropic organizations often rely on contributions from individuals to finance the work that they do, and a national veterans' organization is no exception. The Paralyzed Veterans of America (PVA) was founded as a congressionally chartered veterans' service organization more than 60 years ago. It provides a range of services to veterans who have experienced spinal cord injury or dysfunction. Some of the services offered include medical care, research, education, and accessibility and legal consulting. In 2008, this organization had total revenue of more than $135 million, with more than 99 percent of this revenue coming from contributions.

An organization that depends so heavily on contributions needs a multifaceted fundraising program, and PVA solicits donations in a number of ways. From its website (www.pva.org), people can make a one-time donation, donate monthly, donate in honor or in memory of someone, and shop in the PVA online store. People can also support one of the charity events, such as its golf tournament, National Veterans Wheelchair Games, and Charity Ride.

Traditionally, one of PVA's main methods of soliciting funds was the use of return address labels and greeting cards (although still used, this method has declined in recent years). Typically, these gifts were sent to potential donors about every six weeks with a request for a contribution. From its established donors, PVA could expect a response rate of about 5%, which, given the relatively small cost to produce and send the gifts, kept the organization well funded.

But fundraising accounts for 28 percent of expenses, so PVA wanted to know who its donors are, what variables might be useful in predicting whether a donor is likely to give to an upcoming campaign, and what the size of that gift might be. On your DVD is a dataset **Case Study 1**, which includes data designed to be very similar to part of the data that this organization works with. Here is a description of some of the variables. Keep in mind, however, that in the real data set, there would be hundreds more variables given for each donor.

VARIABLE NAME	UNITS (IF APPLICABLE)	DESCRIPTION	REMARKS
Age	Years		
Own Home?	H = Yes U = No or unknown		
Children	Counts		
Income		1 = Lowest ; 7 = Highest	Based on national medians and percentiles
Sex	M = Male; F = Female		
Total Wealth		1 = Lowest ; 9 = Highest	Based on national medians and percentiles
Gifts to Other Orgs	Counts	Number of Gifts (if known) to other philanthropic organizations in the same time period	
Number of Gifts	Counts	Number of Gifts to this organization in this time period	
Time Between Gifts	Months	Time between first and second gifts	

page 187

Case Studies have been added to each Part of the book, giving students experience working through an in-depth application using real data and scenarios from actual organizations.

Brief Cases at the end of each chapter ask students to investigate a question or make a business decision using real data. Students define the objective, plan the process, complete the analysis, and report their conclusion. Data are provided.

Brief CASE

Real Estate

In Chapter 6 we examined the regression of the sales price of a home on its size and saw that larger homes generally fetch a higher price. How much can we learn about a house from the fact that it has a fireplace or more than the average number of bedrooms? Data for a random sample of 1047 homes from the upstate New York area can be found in the file **Saratoga_Real_Estate**. There are 6 quantitative variables: *Price($), Living Area(sq ft), Bathrooms(#), Bedrooms(#), Lot Size(Acres),* and *Age(years)*, and one categorical variable, *Fireplace?(1 = Yes; 0 = No)* denoting whether the house has at least one fireplace. We can use *t*-methods to see, for example, whether homes with fireplaces sell for more, on average, and by how much. For the quantitative variables, create new categorical variables by splitting them at the median or some other splitting point of your choice, and compare home prices above and below this value. For example, the median number of *Bedrooms* of these homes is 2. You might compare the prices of homes with 1 or 2 bedrooms to those with more than 2. Write up a short report summarizing the differences in mean price based on the categorical variables that you created.

Consumer Spending Patterns (Data Analysis)

You are on the financial planning team for monitoring a high spending segment of a credit card. You know that customers tend to spend more during December before the holidays, but you're not sure about the pattern of spending in the months after the holidays. Look at the data set **Consumer_spending_post_holiday**. It contains the monthly credit card spending of 1200 customers during the months December, January, February, March, and April. Report on the spending differences between the months. If you had failed to realize that these are paired data, what difference would that have made in your reported confidence intervals and tests?

page 431

Better Statistical Thinking

Ethics in Action

Lyle Erhart has been working in sales for a leading vendor of Customer Relationship Management (CRM) software for the past three years. He was recently made aware of a published research study that examined factors related to the successful implementation of CRM projects among firms in the financial services industry. Lyle read the research report with interest and was excited to see that his company's CRM software product was included. Among the results were tables reporting the number of projects that were successful based on type of CRM implementation (Operational versus Analytical) for each of the top leading CRM products of 2006. Lyle quickly found the results for his company's product and their major competitor. He summarized the results into one table as follows:

	His Company	Major Competitor
Operational	16 successes out of 20	68 successes out of 80
Analytical	90 successes out of 100	19 successes out of 20

At first he was a bit disappointed, especially since most of their potential clients were interested in Operational CRM. He had hoped to be able to disseminate the findings of this report among the sales force so they could refer to it when visiting potential clients. After some thought, he realized that he could combine the results. His company's overall success rate was 106 out of 120 (over 88%) and was higher than that of its major competitor. Lyle was now happy that he found and read the report.

ETHICAL ISSUE *Lyle, intentionally or not, has benefited from Simpson's Paradox. By combining percentages, he can present the findings in a manner favorable to his company (related to item A, ASA Ethical Guidelines).*

ETHICAL SOLUTION *Lyle should not combine the percentages as the results are misleading. If he decides to disseminate the information to his sales force, he must do so without combining.*

page 69

Ethics in Action vignettes in every chapter teach students to identify challenging issues and propose ethically and statistically sound solutions. Questions are included for study and reflection.

What Can Go Wrong?

- **Nonrespondents.** No survey succeeds in getting responses from everyone. The problem is that those who don't respond may differ from those who do. And if they differ on just the variables we care about, the lack of response will bias the results. Rather than sending out a large number of surveys for which the response rate will be low, it is often better to design a smaller, randomized survey for which you have the resources to ensure a high response rate.

- **Long, dull surveys.** Surveys that are too long are more likely to be refused, reducing the response rate and biasing all the results. Keep it short.

- **Response bias.** Response bias includes the tendency of respondents to tailor their responses to please the interviewer and the consequences of slanted question wording.

- **Push polls.** Push polls, which masquerade as surveys, present one side of an issue before asking a question. For example, a question like

  Would the fact that the new store that just opened by the mall sells mostly goods made overseas by workers in sweatshop conditions influence your decision to shop there rather than in the downtown store that features American-made products?

  is designed not to gather information, but to spread ill-will toward the new store.

THE WIZARD OF ID — parker and hart

What Can Go Wrong? sections near the end of each chapter help students detect common statistical errors and offer practice in debunking misuses of statistics.

page 40

Business Statistics

Second Edition

Norean R. Sharpe
Georgetown University

Richard D. De Veaux
Williams College

Paul F. Velleman
Cornell University

With Contributions by David Bock

Addison Wesley

Boston   Columbus   Indianapolis   New York   San Francisco   Upper Saddle River   Amsterdam   Cape Town
Dubai   London   Madrid   Milan   Munich   Paris   Montreal   Toronto   Delhi   Mexico City
Sao Paulo   Sydney   Hong Kong   Seoul   Singapore   Taipei   Tokyo

Editor in Chief	Deirdre Lynch
Vice President/Executive Director, Development	Carol Trueheart
Senior Development Editor	Elaine Page
Senior Content Editor	Chere Bemelmans
Associate Content Editor	Dana Jones
Associate Managing Editor	Tamela Ambush
Senior Production Project Manager	Peggy McMahon
Design Manager	Andrea Nix
Cover Design	Barbara T. Atkinson
Interior Design	Studio Montage
Cover Photo	*Chisel Carving Wood* © Chris McElcheran/Masterfile
Senior Market Development Manager	Dona Kenly
Executive Marketing Manager	Roxanne McCarley
Senior Marketing Manager	Alex Gay
Marketing Associate	Kathleen DeChavez
Photo Researcher	Diane Austin and Leslie Haimes
Media Producer	Aimee Thorne
Software Development	Edward Chappell and Marty Wright
Senior Author Support/Technology Specialist	Joe Vetere
Rights and Permissions Advisor	Michael Joyce
Senior Manufacturing Buyers	Carol Melville and Ginny Michaud
Production Coordination, Illustration, and Composition	PreMediaGlobal

For permission to use copyrighted material, grateful acknowledgment has been made to the copyright holders listed in Appendix C, which is hereby made part of this copyright page.

Many of the designations used by manufacturers and sellers to distinguish their products are claimed as trademarks. Where those designations appear in this book, and Pearson Education was aware of a trademark claim, the designations have been printed in initial caps or all caps.

Library of Congress Cataloging-in-Publication Data
Sharpe, Norean Radke.
  Business statistics / Norean R. Sharpe, Richard D. De Veaux, Paul F. Velleman; with contributions from Dave Bock. — 2nd ed.
    p. cm.
  ISBN 978-0-321-71609-5
  1. Commercial statistics.   I. De Veaux, Richard D.   II. Velleman, Paul F., 1949–   III. Title.
  HF1017.S467 2012
  650.01'5195—dc22

                                                                        2010001392

Addison-Wesley
is an imprint of

www.pearsonhighered.com

ISBN-13: 978-0-321-71609-5   ISBN-10: 0-321-71609-4
1 2 3 4 5 6 7 8 9 10—WC—13 12 11 10

To my parents, who taught me the importance of education
—Norean

To my family
—Dick

To my father, who taught me about ethical business practice by his constant example as a small businessman and parent
—Paul

Meet the Authors

As a researcher of statistical problems in business and a professor of Statistics at a business school, **Norean Radke Sharpe** (Ph.D. University of Virginia) understands the challenges and specific needs of the business student. She is currently teaching at the McDonough School of Business at Georgetown University, where she is also Associate Dean and Director of Undergraduate Programs. Prior to joining Georgetown, she taught business statistics and operations research courses to both undergraduate and MBA students for fourteen years at Babson College. Before moving into business education, she taught mathematics for several years at Bowdoin College and conducted research at Yale University. Norean is coauthor of the recent text, *A Casebook for Business Statistics: Laboratories for Decision Making*, and she has authored more than 30 articles—primarily in the areas of statistics education and women in science. Norean currently serves as Associate Editor for the journal *Cases in Business, Industry, and Government Statistics*. Her research focuses on business forecasting and statistics education. She is also co-founder of DOME Foundation, Inc., a nonprofit foundation that works to increase Diversity and Outreach in Mathematics and Engineering for the greater Boston area. She has been active in increasing the participation of women and underrepresented students in science and mathematics for several years and has two children of her own.

Richard D. De Veaux (Ph.D. Stanford University) is an internationally known educator, consultant, and lecturer. Dick has taught statistics at a business school (Wharton), an engineering school (Princeton), and a liberal arts college (Williams). While at Princeton, he won a Lifetime Award for Dedication and Excellence in Teaching. Since 1994, he has been a professor of statistics at Williams College, although he returned to Princeton for the academic year 2006–2007 as the William R. Kenan Jr. Visiting Professor of Distinguished Teaching. Dick holds degrees from Princeton University in Civil Engineering and Mathematics and from Stanford University in Dance Education and Statistics, where he studied with Persi Diaconis. His research focuses on the analysis of large data sets and data mining in science and industry. Dick has won both the Wilcoxon and Shewell awards from the American Society for Quality and is a Fellow of the American Statistical Association. Dick is well known in industry, having consulted for such *Fortune* 500 companies as American Express, Hewlett-Packard, Alcoa, DuPont, Pillsbury, General Electric, and Chemical Bank. He was named the "Statistician of the Year" for 2008 by the Boston Chapter of the American Statistical Association for his contributions to teaching, research, and consulting. In his spare time he is an avid cyclist and swimmer. He also is the founder and bass for the doo-wop group, the Diminished Faculty, and is a frequent soloist with various local choirs and orchestras. Dick is the father of four children.

Paul F. Velleman (Ph.D. Princeton University) has an international reputation for innovative statistics education. He designed the Data Desk® software package and is also the author and designer of the award-winning ActivStats® multimedia software, for which he received the EDUCOM Medal for innovative uses of computers in teaching statistics and the ICTCM Award for Innovation in Using Technology in College Mathematics. He is the founder and CEO of Data Description, Inc. (www.datadesk.com), which supports both of these programs. He also developed the Internet site, *Data and Story Library* (DASL; www.dasl.datadesk.com), which provides data sets for teaching Statistics. Paul coauthored (with David Hoaglin) the book *ABCs of Exploratory Data Analysis*. Paul has taught Statistics at Cornell University on the faculty of the School of Industrial and Labor Relations since 1975. His research often focuses on statistical graphics and data analysis methods. Paul is a Fellow of the American Statistical Association and of the American Association for the Advancement of Science. He is also baritone of the barbershop quartet Rowrbazzle! Paul's experience as a professor, entrepreneur, and business leader brings a unique perspective to the book.

Richard De Veaux and Paul Velleman have authored successful books in the introductory college and AP High School market with David Bock, including *Intro Stats*, Third Edition (Pearson, 2009), *Stats: Modeling the World*, Third Edition (Pearson, 2010), and *Stats: Data and Models*, Third Edition (Pearson, 2012).

Contents

---

*Indicates an optional topic

Appendixes

Preface

We set out to write a book for business students that answers the simple question: "How can I make better decisions?" As entrepreneurs and consultants, we know that knowledge of Statistics is essential to survive and thrive in today's competitive environment. As educators, we've seen a disconnect between the way Statistics is taught to business students and the way it is used in making business decisions. In *Business Statistics*, we try to narrow the gap between theory and practice by presenting statistical methods so they are both relevant and interesting.

The data that inform a business decision have a story to tell, and the role of Statistics is to help us hear that story clearly and communicate it to others. Like other textbooks, *Business Statistics* teaches methods and concepts. But, unlike other textbooks, *Business Statistics* also teaches the "why" and insists that results be reported in the context of business decisions. Students will come away knowing how to think statistically to make better business decisions and how to effectively communicate the analysis that led to the decision to others. Our approach requires up-to-date, real-world examples and current data. So, we constantly strive to place our teaching in the context of current business issues and to illustrate concepts with current examples.

What's New in This Edition?

Our overarching goal in the second edition of *Business Statistics* has been to organize the presentation of topics clearly and provide a wealth of examples and exercises so that the story we tell is always tied to the ways Statistics informs sound business practice.

Improved Organization. The Second Edition has been re-designed from the ground up. We have retained our "data first" presentation of topics because we find that it provides students with both motivation and a foundation in real business decisions on which to build an understanding. But we have reorganized the order of topics within chapters, and the order of chapters themselves to tell the story of Statistics in Business more clearly.

- Chapters 1–6 are now devoted entirely to collecting, displaying, summarizing, and understanding data. We find that this gives students a solid foundation to launch their understanding of statistical inference.

- Material on randomness and probability is now grouped together in Chapters 7–10. Material on continuous probability models has been gathered into a single chapter—Chapter 9, which also introduces the Normal model.

- Core material on inference follows in Chapters 11–15. We introduce inference by discussing proportions because most students are better acquainted with proportions reported in surveys and news stories. However, this edition ties in the discussion of means immediately so students can appreciate that the reasoning of inference is the same in a variety of contexts.

- Chapters 16–20 cover regression-based models for decision making.

- Chapters 21–25 discuss special topics that can be selected according to the needs of the course and the preferences of the instructor.

- Chapters 22 (Quality Control) and 23 (Nonparametric Methods) are new in this edition.

Section Examples. Almost every section of every chapter now has a focused example to illustrate and apply the concepts and methods of that section.

Section Exercises. Each chapter's exercises now begin with single-concept exercises that target section topics. This makes it easier to check your understanding of each topic as you learn it.

Recent Data and New Examples. We teach with real data whenever possible. To keep examples and exercises fresh, we've updated data throughout the book. New examples reflect stories in the news and recent economic and business events.

Redesigned Chapter Summaries. Our What Have We Learned chapter summaries have been redesigned to specify learning objectives and place key concepts and skills within those objectives. This makes them even more effective as help for students preparing for exams.

Statistical Case Studies. Each chapter still ends with one or two Brief Cases. Now, in addition, each of the major parts of the book includes a longer case study using larger datasets (found on the CD) and open questions to answer using the data and a computer.

Streamlined Technology Help with additional Excel coverage. Technology Help sections are now in easy-to-follow bulleted lists. Excel screenshots and coverage of Excel 2010 appear throughout the book where appropriate.

What's *Old* in This Edition: Statistical Thinking

For all of our improvements, examples, and updates in this edition of *Business Statistics* we haven't lost sight of our original mission—writing a modern business statistics text that addresses the importance of *statistical thinking* in making business decisions and that acknowledges how Statistics is actually used in business.

Today Statistics is practiced with technology. This insight informs everything from our choice of forms for equations (favoring intuitive forms over calculation forms) to our extensive use of real data. But most important, understanding the value of technology allows us to focus on teaching statistical thinking rather than calculation. The questions that motivate each of our hundreds of examples are not "how do you find the answer?" but "how do you think about the answer, and how does it help you make a better decision?"

Our focus on statistical thinking ties the chapters of the book together. An introductory Business Statistics course covers an overwhelming number of new terms, concepts, and methods. We have organized these to enhance learning. But it is vital that students see their central core: how we can understand more about the world and make better decisions by understanding what the data tell us. From this perspective, it is easy to see that the patterns we look for in graphs are the same as those we think about when we prepare to make inferences. And it is easy to see that the many ways to draw inferences from data are several applications of the same core concepts. And it follows naturally that when we extend these basic ideas into more complex (and even more realistic) situations, the same basic reasoning is still at the core of our analyses.

Our Goal: Read This Book!

The best textbook in the world is of little value if it isn't read. Here are some of the ways we made *Business Statistics* more approachable:

- ***Readability.*** We strive for a conversational, approachable style, and we introduce anecdotes to maintain interest. While using the First Edition, instructors reported (to their amazement) that their students read ahead of their assignments voluntarily. Students write to tell us (to *their* amazement) that they actually enjoy the book.

- ***Focus on assumptions and conditions.*** More than any other textbook, *Business Statistics* emphasizes the need to verify assumptions when using statistical procedures. We reiterate this focus throughout the examples and exercises. We make every effort to provide templates that reinforce the practice of checking these assumptions and conditions, rather than rushing through the computations of a real-life problem.

- ***Emphasis on graphing and exploring data.*** Our consistent emphasis on the importance of displaying data is evident from the first chapters on understanding data to the sophisticated model-building chapters at the end. Examples often illustrate the value of examining data graphically, and the Exercises reinforce this. Good graphics reveal structures, patterns, and occasional anomalies that could otherwise go unnoticed. These patterns often raise new questions and inform both the path of a resulting statistical analysis and the business decisions. The graphics found throughout the book also demonstrate that the simple structures that underlie even the most sophisticated statistical inferences are the same ones we look for in the simplest examples. That helps to tie the concepts of the book together to tell a coherent story.

- ***Consistency.*** We work hard to avoid the "do what we say, not what we do" trap. Having taught the importance of plotting data and checking assumptions and conditions, we are careful to model that behavior throughout the book. (Check the Exercises in the chapters on multiple regression or time series and you'll find us still requiring and demonstrating the plots and checks that were introduced in the early chapters.) This consistency helps reinforce these fundamental principles and provides a familiar foundation for the more sophisticated topics.

- ***The need to read.*** In this book, important concepts, definitions, and sample solutions are not always set aside in boxes. The book needs to be read, so we've tried to make the reading experience enjoyable. The common approach of skimming for definitions or starting with the exercises and looking up examples just won't work here. (It never did work as a way to learn Statistics; we've just made it impractical in our text.)

Coverage

The topics covered in a Business Statistics course are generally mandated by our students' needs in their studies and in their future professions. But the *order* of these topics and the relative emphasis given to each is not well established. *Business Statistics* presents some topics sooner or later than other texts. Although many chapters can be taught in a different order, we urge you to consider the order we have chosen.

We've been guided in the order of topics by the fundamental goal of designing a coherent course in which concepts and methods fit together to provide a new understanding of how reasoning with data can uncover new and important truths. Each new topic should fit into the growing structure of understanding that students develop throughout the course. For example, we teach inference concepts with proportions first and then with means. Most people have a wider experience with proportions, seeing them in polls and advertising. And by starting with proportions, we can teach inference with the Normal model and then introduce inference for means with the Student's *t* distribution.

We introduce the concepts of association, correlation, and regression early in *Business Statistics*. Our experience in the classroom shows that introducing these fundamental ideas early makes Statistics useful and relevant even at the beginning of the course. Later in the semester, when we discuss inference, it is natural and relatively easy to build on the fundamental concepts learned earlier by exploring data with these methods.

We've been guided in our choice of what to emphasize by the GAISE (Guidelines for Assessment and Instruction in Statistics Education) Report, which emerged from extensive studies of how students best learn Statistics (http://www.amstat.org/education/gaise/). Those recommendations, now officially adopted and recommended by the American Statistical Association, urge (among other detailed suggestions) that Statistics education should:

1. Emphasize statistical literacy and develop statistical thinking;
2. Use real data;
3. Stress conceptual understanding rather than mere knowledge of procedures;
4. Foster active learning;
5. Use technology for developing conceptual understanding and analyzing data; and
6. Make assessment a part of the learning process

In this sense, this book is thoroughly modern.

Syllabus Flexibility

But to be effective, a course must fit comfortably with the instructor's preferences. The early chapters—Chapters 1–15—present core material that will be part of any introductory course. Chapters 16–21—multiple regression, time series, model building, and Analysis of Variance—may be included in an introductory course, but our organization provides flexibility in the order and choice of specific topics. Chapters 22–25 may be viewed as "special topics" and selected and sequenced to suit the instructor or the course requirements.

Here are some specific notes:

- Chapter 6, Correlation and Linear Regression, may be postponed until just before covering regression inference in Chapters 16 and 17.
- Chapter 19, Building Multiple Regression Models, must follow the introductory material on multiple regression in Chapter 18.
- Chapter 20, Time Series Analysis, requires material on multiple regression from Chapter 18.
- Chapter 21, Design and Analysis of Experiments and Observational Studies, may be taught before the material on regression—at any point after Chapter 14.

The following topics can be introduced in any order (or omitted):

- Chapter 15, Inference for Counts: Chi-Square Tests
- Chapter 22, Quality Control
- Chapter 23, Nonparametric Methods
- Chapter 24, Decision Making and Risk
- Chapter 25, Introduction to Data Mining

Features

A textbook isn't just words on a page. A textbook is many features that come together to form a big picture. The features in *Business Statistics* provide a real-world context for concepts, help students apply these concepts, promote problem-solving, and integrate technology—all of which help students understand and see the big picture of Business Statistics.

Motivating Vignettes. Each chapter opens with a motivating vignette, often taken from the authors' consulting experiences. These descriptions of companies—such as Amazon.com, Zillow.com, Keen Inc., and Whole Foods Market—enhance and illustrate the story of each chapter and show how and why statistical thinking is so vital to modern business decision-making. We analyze data from or about the companies in the motivating vignettes throughout the chapter.

For Examples. Almost every section of every chapter includes a focused example that illustrates and applies the concepts or methods of that section. The best way to understand and remember a new theoretical concept or method is to see it applied in a real-world business context. That's what these examples do throughout the book.

Step-by-Step Guided Examples. The answer to a statistical question is almost never just a number. Statistics is about understanding the world and making better decisions with data. To that end, some examples in each chapter are presented as Guided Examples. A thorough solution is modeled in the right column while commentary appears in the left column. The overall analysis follows our innovative **Plan, Do, Report** template. That template begins each analysis with a clear question about a business decision and an examination of the data available (**Plan**). It then moves to calculating the selected statistics (**Do**). Finally, it concludes with a **Report** that specifically addresses the question. To emphasize that our goal is to address the motivating question, we present the **Report** step as a business memo that summarizes the results in the context of the example and states a recommendation if the data are able to support one. To preserve the realism of the example, whenever it is appropriate, we include limitations of the analysis or models in the concluding memo, as one should in making such a report.

Brief Cases. Each chapter includes one or two Brief Cases that use real data and ask students to investigate a question or make a decision. Students define the objective, plan the process, complete the analysis, and report a conclusion. Data for the Brief Cases are available on the CD and website, formatted for various technologies.

Case Studies. Each part of the book ends with a Case Study. Students are given realistically large data sets (on the CD) and challenged to respond to open-ended business questions using the data. Students have the opportunity to bring together methods they have learned in the chapters of that part (and indeed, throughout the book) to address the issues raised. Students will have to use a computer to work with the large data sets that accompany these Case Studies.

What Can Go Wrong? Each chapter contains an innovative section called "What Can Go Wrong?" which highlights the most common statistical errors and the misconceptions about Statistics. The most common mistakes for the new user of Statistics involve misusing a method—not miscalculating a statistic. Most of the mistakes we discuss have been experienced by the authors in a business context or a classroom situation. One of our goals is to arm students with the tools to detect statistical errors and to offer practice in debunking misuses of Statistics, whether intentional or not. In this spirit, some of our exercises probe the understanding of such errors.

By Hand

By Hand. Even though we encourage the use of technology to calculate statistical quantities, we recognize the pedagogical benefits of occasionally doing a calculation by hand. The By Hand boxes break apart the calculation of some of the simpler formulas and help the student through the calculation of a worked example.

REALITY CHECK

Reality Check. We regularly offer reminders that Statistics is about understanding the world and making decisions with data. Results that make no sense are probably wrong, no matter how carefully we think we did the calculations. Mistakes are often easy to spot with a little thought, so we ask students to stop for a reality check before interpreting results.

Notation Alert! ——————

Notation Alert. Throughout this book, we emphasize the importance of clear communication. Proper notation is part of the vocabulary of Statistics, but it can be daunting. We all know that in Algebra n can stand for any variable, so it may be surprising to learn that in Statistics n is always and only the sample size. Statisticians dedicate many letters and symbols for specific meanings ($b, e, n, p, q, r, s, t,$ and z, along with many Greek letters, all carry special connotations). To learn Statistics, it is vital to be clear about the letters and symbols statisticians use.

Just Checking

Just Checking. It is easy to start nodding in agreement without really understanding, so we ask questions at points throughout the chapter. These questions are a quick check; most involve very little calculation. The answers are at the end of the exercise sets in each chapter to make them easy to check. The questions can also be used to motivate class discussion.

Math Box

Math Boxes. In many chapters, we present the mathematical underpinnings of the statistical methods and concepts. Different students learn in different ways, and any reader may understand the material best by more than one path. We set proofs, derivations, and justifications apart from the narrative, so the underlying mathematics is there for those who want greater depth, but the text itself presents the logical development of the topic at hand without distractions.

What Have We Learned?

What Have We Learned? These chapter-ending summaries highlight the major learning objectives of the chapter. In that context, we review the concepts, define the terms introduced in the chapter, and list the skills that form the core message of the chapter. These make excellent study guides: the student who understands the concepts in the summary, knows the terms, and has the skills is probably ready for the exam.

Ethics in Action

Ethics in Action. Statistics is not just plugging numbers into formulas; most statistical analyses require a fair amount of judgment. The best guidance for these judgments is that we make an honest and ethical attempt to learn the truth. Anything less than that can lead to poor and even harmful decisions. Our *Ethics in Action* vignettes in each chapter illustrate some of the judgments needed in statistical analyses, identify possible errors, link the issues to the American Statistical Association's Ethical Guidelines, and then propose ethically and statistically sound alternative approaches.

Section Exercises. The Exercises for each chapter begin with straightforward exercises targeted at the topics in each chapter section. This is the place to check understanding of specific topics. Because they are labeled by section, turning back to the right part of the chapter to clarify a concept or review a method is easy.

Chapter Exercises. These exercises are designed to be more realistic than Section Exercises and to lead to conclusions about the real world. They may combine concepts and methods from different sections. We've worked hard to make sure they contain relevant, modern, and real-world questions. Many come from news stories; some come from recent research articles. Whenever possible, the data are on the CD and website (always in a variety of formats) so they can be explored further. The exercises marked with a ❶ indicate that the data are provided on the CD (and on the book's companion website, www.pearsonhighered.com/sharpe). Throughout, we pair the exercises so that each odd-numbered exercise (with answer in the back of the book) is followed by an even-numbered exercise on the same Statistics topic. Exercises are roughly ordered within each chapter by both topic and by level of difficulty.

Data and Sources. Most of the data used in examples and exercises are from real-world sources. Whenever possible, we present the original data as we collected it. Sometimes, due to concerns of confidentiality or privacy, we had to change the values of the data or the names of the variables slightly, always being careful to keep the context as realistic and true to life as possible. Whenever we can, we include references to Internet data sources. As Internet users know well, URLs can break as websites evolve. To minimize the impact of such changes, we point as high in the address tree as is practical, so it may be necessary to search down into a site for the data. Moreover, the data online may change as more recent values become available. The data we use are usually on the CD and on the companion website, www. pearsonhighered.com/sharpe.

Videos with Optional Captioning. Videos, featuring the *Business Statistics* authors, review the high points of each chapter. The presentations feature the same student-friendly style and emphasis on critical thinking as the textbook. In addition, 10 *Business Insight Videos* (concept videos) feature Deckers, Southwest Airlines, Starwood, and other companies and focus on statistical concepts as they pertain to the real world. Videos are available with captioning. They can be viewed from within the online MyStatLab course.

Technology Help

Technology Help. In business, Statistics is practiced with computers using a variety of statistics packages. In Business-school Statistics classes, however, Excel is the software most often used. Throughout the book, we show examples of Excel output and offer occasional tips. At the end of each chapter, we summarize what students can find in the most common software, often with annotated output. We then offer specific guidance for Excel 2007 and 2010, Minitab, SPSS, and JMP, formatted in easy-to-read bulleted lists. (Technology Help for Excel 2003 and Data Desk are on the accompanying CD.) This advice is not intended to replace the documentation for any of the software, but rather to point the way and provide startup assistance.

𝒳LSTAT⁷ An **XLStat Appendix** in the back of the book features chapter-by-chapter guidance for using this new Excel add-in. The XLStat icon in Technology Help sections directs readers to this XLStat-specific guidance in Appendix B.

Supplements

Student Supplements

Business Statistics, for-sale student edition (ISBN-13: 978-0-321-71609-5; ISBN-10: 0-321-71609-4)

Student's Solutions Manual, by Rose Sebastianelli, University of Scranton, and Linda Dawson, University of Washington, provides detailed, worked-out solutions to odd-numbered exercises. (ISBN-13: 978-0-321-68940-5; ISBN-10: 0-321-68940-2)

Excel Manual, by Elaine Newman, Sonoma State University (ISBN-13: 978-0-321-71615-6; ISBN-10: 0-321-71615-9)

Minitab Manual, by Linda Dawson, University of Washington, and Robert H. Carver, Stonehill College (ISBN-13: 978-0-321-71610-1; ISBN-10: 0-321-71610-8)

SPSS Manual (download only), by Rita Akin, Santa Clara University; ISBN-13: 978-0-321-71618-7; ISBN-10: 0-321-71618-3)

Ten **Business Insight Videos** (concept videos) feature Deckers, Southwest Airlines, Starwood, and other companies and focus on statistical concepts as they pertain to the real world. Available with captioning, these 4- to 7-minute videos can be viewed from within the online MyStatLab course or at www.pearsonhighered.com/irc. (ISBN-13: 978-0-321-73874-5; ISBN-10: 0-321-73874-8)

Video Lectures were scripted and presented by the authors themselves, reviewing the important points in each chapter. They can be downloaded from MyStatLab.

Study Cards for *Business Statistics* Software. Technology Study Cards for Business Statistics are a convenient resource for students, with instructions and screenshots for using the most popular technologies. The following Study Cards are available in print (8-page fold-out cards) and within MyStatLab: Excel 2010 with XLStat (0-321-74775-5), Minitab (0-321-64421-2), JMP (0-321-64423-9), SPSS (0-321-64422-0), R (0-321-64469-7), and StatCrunch (0-321-74472-1). A Study Card for the native version of Excel is also available within MyStatLab.

Instructor Supplements

Instructor's Edition contains answers to all exercises. (ISBN-13: 978-0-321-71612-5; ISBN-10: 0-321-71612-4)

Online Test Bank (download only), by Rose Sebastianelli, University of Scranton, includes chapter quizzes and part level tests. The Test Bank is available at www.pearsonhighered.com/irc. (ISBN-13: 978-0-321-68936-8; ISBN-10: 0-321-68936-4)

Instructor's Solutions Manual, by Rose Sebastianelli, University of Scranton, and Linda Dawson, University of Washington, contains detailed solutions to all of the exercises. (ISBN-13: 978-0-321-68935-1; ISBN-10: 0-321-68935-6)

Instructor's Resource Guide contains chapter-by-chapter comments on the major concepts, tips on presenting topics (and what to avoid), teaching examples, suggested assignments, basic exercises, and web links and lists of other resources. Available within MyStatLab or at www.pearsonhighered.com/irc.

Lesson Podcasts for *Business Statistics*. These audio podcasts from the authors focus on the key points of each chapter, helping both new and experienced instructors prepare for class; available in MyStatLab or at www.pearsonhighered.com/irc. (ISBN-13: 978-0-321-74688-7; ISBN-10: 0-321-74688-0)

Business Insight Video Guide to accompany *Business Statistics*. Written to accompany the Business Insight Videos, this guide includes a summary of the video, video-specific questions and answers that can be used for assessment or classroom discussion, a correlation to relevant chapters in *Business Statistics*, concept-centered teaching points, and useful web links. The Video Guide is available for download from MyStatLab or at www.pearsonhighered.com/irc.

PowerPoint® Lecture Slides

PowerPoint Lecture Slides provide an outline to use in a lecture setting, presenting definitions, key concepts, and figures from the text. These slides are available within MyStatLab or at www.pearsonhighered.com/irc.

Active Learning Questions

Prepared in PowerPoint®, these questions are intended for use with classroom response systems. Several multiple-choice questions are available for each chapter of the book, allowing instructors to quickly assess mastery of material in class. The Active Learning Questions are available to download from within MyStatLab and from the Pearson Education online catalog.

Technology Resources

A companion CD is bound in new copies of *Business Statistics*. The CD holds the following supporting materials, including:

- **Data** for exercises marked ⓣ in the text are available on the CD and website formatted for Excel, JMP, Minitab 14 and 15, SPSS, and as text files suitable for these and virtually any other statistics software.

- **XLStat for Pearson.** The CD includes a launch page and instructions for downloading and installing this Excel add-in. Developed in 1993, XLStat is used by leading businesses and universities around the world. It is compatible with all Excel versions from version 97 to version 2010 (except 2008 for Mac), and is compatible with the Windows 9x through Windows 7 systems, as well as with the PowerPC and Intel based Mac systems. For more information, visit www.pearsonhighered.com/xlstat.

ActivStats® *for Business Statistics* (Mac and PC). The award-winning ActivStats multimedia program supports learning chapter by chapter with the book. It complements the book with videos of real-world stories, worked examples, animated expositions of each of the major Statistics topics, and tools for performing simulations, visualizing inference, and learning to use statistics software. ActivStats includes 15 short video clips; 183 animated activities and teaching applets; 260 data sets; interactive graphs, simulations, visualization tools, and much more. ActivStats for *Business Statistics* (Mac and PC) is available in an all-in-one version for Excel, JMP, Minitab, and SPSS. (ISBN-13: 978-0-321-57719-1; ISBN-10: 0-321-57719-1)

MyStatLab™ Online Course (access code required)

MyStatLab™—part of the MyMathLab® product family—is a text-specific, easily customizable online course that integrates interactive multimedia instruction with textbook content. MyStatLab gives you the tools you need to deliver all or a portion of your course online, whether your students are in a lab setting or working from home.

- **Interactive homework exercises,** correlated to your textbook at the objective level, are algorithmically generated for unlimited practice and mastery. Most exercises are free-response and provide guided solutions, sample problems, and learning aids for extra help. StatCrunch, an online data analysis tool, is available with online homework and practice exercises.

- **Personalized homework** assignments that you can design to meet the needs of your class. MyStatLab tailors the assignment for each student based on their test or quiz scores. Each student receives a homework assignment that contains only the problems they still need to master.

- A **Personalized Study Plan,** generated when students complete a test or quiz or homework, indicates which topics have been mastered and links to tutorial exercises for topics students have not mastered. You can customize the Study Plan so that the topics available match your course content.

- **Multimedia learning aids,** such as video lectures and podcasts, animations, and a complete multimedia textbook, help students independently improve their understanding and performance. You can assign these multimedia learning aids as homework to help your students grasp the concepts. In addition, applets are also available to display statistical concepts in a graphical manner for classroom demonstration or independent use.

- **StatCrunch.com** access is now included with MyStatLab. StatCrunch.com is the first web-based data analysis tool designed for teaching statistics. Users can perform complex analyses, share data sets, and generate compelling reports. The vibrant online community offers more than ten thousand data sets for students to analyze.

- **Homework and Test Manager** lets you assign homework, quizzes, and tests that are automatically graded. Select just the right mix of questions from the MyStatLab exercise bank, instructor-created custom exercises, and/or TestGen test items.

- **Gradebook,** designed specifically for mathematics and statistics, automatically tracks students' results, lets you stay on top of student performance, and gives you control over how to calculate final grades. You can also add offline (paper-and-pencil) grades to the gradebook.

- **MathXL Exercise Builder** allows you to create static and algorithmic exercises for your online assignments. You can use the library of sample exercises as an easy starting point or use the Exercise Builder to edit any of the course-related exercises.

- **Pearson Tutor Center** (www.pearsontutorservices.com) access is automatically included with MyStatLab. The Tutor Center is staffed by qualified statistics instructors who provide textbook-specific tutoring for students via toll-free phone, fax, email, and interactive Web sessions.

Students do their assignments in the Flash®-based MathXL Player which is compatible with almost any browser (Firefox®, Safari™, or Internet Explorer®) on almost any platform (Macintosh® or Windows®). MyStatLab is powered by CourseCompass™, Pearson Education's online teaching and learning environment, and by MathXL®, our online homework, tutorial, and assessment system. MyStatLab is available to qualified adopters. For more information, visit www.mystatlab.com or contact your Pearson representative.

MyStatLab™Plus

MyLabsPlus combines effective teaching and learning materials from MyStatLab™ with convenient management tools and a dedicated services team. It is designed to support growing math and statistics programs and includes additional features such as:

- **Batch Enrollment:** Your school can create the login name and password for every student and instructor, so everyone can be ready to start class on the first day. Automation of this process is also possible through integration with your school's Student Information System.

- **Login from your campus portal:** You and your students can link directly from your campus portal into your MyLabsPlus courses. A Pearson service team works with your institution to create a single sign-on experience for instructors and students.

- **Diagnostic Placement:** Students can take a placement exam covering reading, writing, and mathematics to assess their skills. You get the results immediately, and you may customize the exam to meet your department's specific needs.

- **Advanced Reporting:** MyLabsPlus's advanced reporting allows instructors to review and analyze students' strengths and weaknesses by tracking their performance on tests, assignments, and tutorials. Administrators can review grades and assignments across all courses on your MyLabsPlus campus for a broad overview of program performance.

- **24/7 Support:** Students and instructors receive 24/7 support, 365 days a year, by phone, email, or online chat.

MyLabsPlus is available to qualified adopters. For more information, visit our website at www.mylabsplus.com or contact your Pearson representative.

MathXL® for Statistics Online Course (access code required)

MathXL® for Statistics is an online homework, tutorial, and assessment system that accompanies Pearson's textbooks in statistics. MathXL for Statistics is available to qualified adopters. For more information, visit our website at www.mathxl.com, or contact your Pearson representative.

StatCrunch™

StatCrunch™ is web-based statistical software that allows users to perform complex analyses, share data sets, and generate compelling reports of their data. Users can upload their own data to StatCrunch, or search the library of over twelve thousand publicly shared data sets, covering almost any topic of interest. Interactive graphical outputs help users understand statistical concepts, and are available for export to enrich reports with visual representations of data. Additional features include:

- A full range of numerical and graphical methods that allow users to analyze and gain insights from any data set.

- Reporting options that help users create a wide variety of visually appealing representations of their data.

- An online survey tool that allows users to quickly build and administer surveys via a web form.

StatCrunch is available to qualified adopters. For more information, visit our website at www.statcrunch.com, or contact your Pearson representative.

TestGen®

TestGen® (www.pearsoned.com/testgen) enables instructors to build, edit, print, and administer tests using a computerized bank of questions developed to cover all the objectives of the text. TestGen is algorithmically based, allowing instructors to create multiple but equivalent versions of the same question or test with the click of a button. Instructors can also modify test bank questions or add new questions. The software and testbank are available for download from Pearson Education's online catalog.

Pearson Math & Statistics Adjunct Support Center

The Pearson Math & Statistics Adjunct Support Center (http://www.pearsontutorservices.com/math-adjunct.html) is staffed by qualified instructors with more than 100 years of combined experience at both the community college and university levels. Assistance is provided for faculty in the following areas:

- Suggested syllabus consultation
- Tips on using materials packed with your book
- Book-specific content assistance
- Teaching suggestions, including advice on classroom strategies

Companion Website for *Business Statistics*, 2nd edition, includes all of the datasets needed for the book in several formats, tables and selected formulas, and a quick guide to inference. Access this website at www.pearsonhighered.com/sharpe.

Acknowledgements

This book would not have been possible without many contributions from David Bock, our co-author on several other texts. Many of the explanations and exercises in this book benefit from Dave's pedagogical flair and expertise. We are honored to have him as a colleague and friend.

Many people have contributed to this book from the first day of its conception to its publication. *Business Statistics* would have never seen the light of day without the assistance of the incredible team at Pearson. Our Editor in Chief, Deirdre Lynch, was central to the support, development, and realization of the book from day one. Chere Bemelmans, Senior Content Editor, kept us on task as much as humanly possible. Peggy McMahon, Senior Production Project Manager, and Laura Hakala, Senior Project Manager at PreMedia Global, worked miracles to get the book out the door. We are indebted to them. Dana Jones, Associate Content Editor; Alex Gay, Senior Marketing Manager; Kathleen DeChavez, Marketing Associate; and Dona Kenly, Senior Market Development Manager, were essential in managing all of the behind-the-scenes work that needed to be done. Aimee Thorne, Media Producer, put together a top-notch media package for this book. Barbara Atkinson, Senior Designer, and Studio Montage are responsible for the wonderful way the book looks. Evelyn Beaton, Manufacturing Manager, along with Senior Manufacturing Buyers Carol Melville and Ginny Michaud, worked miracles to get this book and CD in your hands, and Greg Tobin, President, was supportive and good-humored throughout all aspects of the project.

Special thanks go out to PreMedia Global, the compositor, for the wonderful work they did on this book and in particular to Laura Hakala, the project manager, for her close attention to detail. We'd also like to thank our accuracy checkers whose monumental task was to make sure we said what we thought we were saying: Jackie Miller, The Ohio State University; Dirk Tempelaar, Maastricht University; and Nicholas Gorgievski, Nichols College.

We wish to thank the following individuals who joined us for a weekend to discuss business statistics education, emerging trends, technology, and business ethics. These individuals made invaluable contributions to *Business Statistics*:

Dr. Taiwo Amoo, CUNY Brooklyn
Dave Bregenzer, Utah State University
Joan Donohue, University of South Carolina
Soheila Fardanesh, Towson University
Chun Jin, Central Connecticut State University
Brad McDonald, Northern Illinois University
Amy Luginbuhl Phelps, Duquesne University
Michael Polomsky, Cleveland State University
Robert Potter, University of Central Florida
Rose Sebastianelli, University of Scranton
Debra Stiver, University of Nevada, Reno
Minghe Sun, University of Texas—San Antonio
Mary Whiteside, University of Texas—Arlington

We also thank those who provided feedback through focus groups, class tests, and reviews (reviewers of the second edition are in boldface):

Alabama: Nancy Freeman, Shelton State Community College; Rich Kern, Montgomery County Community College; Robert Kitahara, Troy University; Tammy Prater, Alabama State University. **Arizona:** Kathryn Kozak, Coconino Community College; Robert Meeks, Pima Community College; Philip J. Mizzi, Arizona State University; **Eugene Round**, Embry-Riddle Aeronautical University; Yvonne Sandoval, Pima Community College; Alex Sugiyama, University of Arizona. **California:** Eugene Allevato, Woodbury University; Randy Anderson, California State University, Fresno; Paul Baum, California State University, Northridge; Giorgio Canarella, California State University, Los Angeles; Natasa Christodoulidou, California State University, Dominguez Hills; Abe Feinberg, California State University, Northridge; Bob Hopfe,

California State University, Sacramento; John Lawrence, California State University, Fullerton; Elaine McDonald-Newman, Sonoma State University; Khosrow Moshirvaziri, California State University; **Sunil Sapra**, California State University, Los Angeles; Carlton Scott, University of California, Irvine; Yeung-Nan Shieh, San Jose State University; Dr. Rafael Solis, California State University, Fresno; T. J. Tabara, Golden Gate University; Dawit Zerom, California State University, Fullerton. **Colorado:** Sally Hay, Western State College; **Austin Lampros**, Colorado State University; Rutilio Martinez, University of Northern Colorado; Gerald Morris, Metropolitan State College of Denver; Charles Trinkel, DeVry University, Colorado. **Connecticut:** Judith Mills, Southern Connecticut State University; William Pan, University of New Haven; Frank Bensics, Central Connecticut State University; Lori Fuller, Tunxis Community College; Chun Jin, Central Connecticut State University; Jason Molitierno, Sacred Heart University. **Florida:** David Afshartous, University of Miami; Dipankar Basu, Miami University; Ali Choudhry, Florida International University; Nirmal Devi, Embry Riddle Aeronautical University; Dr. Chris Johnson, University of North Florida; **Robert Potter**, University of Central Florida; Gary Smith, Florida State University; **Patrick Thompson**, University of Florida; Roman Wong, Barry University. **Georgia: Hope M. Baker**, Kennesaw State University; Dr. Michael Deis, Clayton University; Swarna Dutt, State University of West Georgia; **Kim Gilbert**, University of Georgia; John Grout, Berry College; Michael Parzen, Emory University; Barbara Price, Georgia Southern University; **Dimitry Shishkin**, Georgia Gwinnett College. **Idaho:** Craig Johnson, Brigham Young University; Teri Peterson, Idaho State University; Dan Petrak, Des Moines Area Community College. **Illinois:** Lori Bell, Blackburn College; Jim Choi, DePaul University; David Gordon, Illinois Valley Community College; John Kriz, Joliet Junior College; Constantine Loucopoulos, Northeastern Illinois University; Brad McDonald, Northern Illinois University; Ozgur Orhangazi, Roosevelt University. **Indiana:** H. Lane David, Indiana University South Bend; Ting Liu, Ball State University; Constance McLaren, Indiana State University; Dr. Ceyhun Ozgur, Valparaiso University; Hedayeh Samavati, Indiana University, Purdue; Mary Ann Shifflet, University of Southern Indiana; Cliff Stone, Ball State University; Sandra Strasser, Valparaiso University. **Iowa:** Ann Cannon, Cornell College; Timothy McDaniel, Buena Vista University; Dan Petrack, Des Moines Area Community College; Mount Vernon, Iowa; Osnat Stramer, University of Iowa; Bulent Uyar, University of Northern Iowa; Blake Whitten, University of Iowa. **Kansas:** John E. Boyer, Jr., Kansas State University. **Kentucky: Arnold J. Stromberg**, University of Kentucky. **Louisiana: Jim Van Scotter**, Louisana State University; Zhiwei Zhu, University of Louisiana at Lafayette. **Maryland: John F. Beyers**, University of Maryland University College; Deborah Collins, Anne Arundel Community College; Frederick W. Derrick, Loyola College in Maryland; Soheila Fardanesh, Towson University; Dr. Jeffery Michael, Towson University; Dr. **Timothy Sullivan**, Towson University. **Massachusetts:** Elaine Allen, Babson College; Paul D. Berger, Bentley College; Scott Callan, Bentley College; Ken Callow, Bay Path College; Robert H. Carver, Stonehill College; Richard Cleary, Bentley College; Ismael Dambolena, Babson College; Steve Erikson, Babson College; Elizabeth Haran, Salem State College; David Kopcso, Babson College; Supriya Lahiri, University of Massachusetts, Lowell; John MacKenzie, Babson College; Dennis Mathaisel, Babson College; **Richard McGowan**, Boston College; Abdul Momen, Framingham State University; Ken Parker, Babson College; John Saber, Babson College; Ahmad Saranjam, Bridgewater State College; Daniel G. Shimshak, University of Massachusetts, Boston; Erl Sorensen, Bentley College; Denise Sakai Troxell, Babson College; Janet M. Wagner, University of Massachusetts, Boston; Elizabeth Wark, Worcester State College; Fred Wiseman, Northeastern University. **Michigan:** Sheng-Kai Chang, Wayne State University. **Minnesota:** Daniel G. Brick, University of St. Thomas; Dr. David J. Doorn, University of Minnesota Duluth; Howard Kittleson, Riverland Community College; Craig Miller, Normandale Community College. **Mississippi:** Dal Didia, Jackson State University; **J. H. Sullivan**, Mississippi State University; Wenbin Tang, The University of Mississippi. **Missouri:** Emily Ross, University of Missouri, St. Louis. **Nevada:** Debra K. Stiver, University of Nevada, Reno; Grace Thomson, Nevada State College. **New Hampshire:** Parama Chaudhury, Dartmouth College; Doug Morris, University of New Hampshire. **New Jersey:** Kunle Adamson, DeVry University; Dov Chelst, DeVry University—New Jersey; Leonard

Presby, William Paterson University; Subarna Samanta, The College of New Jersey. **New York:** Dr. Taiwo Amoo, City University of New York, Brooklyn; Bernard Dickman, Hofstra University; Mark Marino, Niagara University. **North Carolina:** Margaret Capen, East Carolina University; Warren Gulko, University of North Carolina, Wilmington; Geetha Vaidyanathan, University of North Carolina. **Ohio:** David Booth, Kent State University, Main Campus; Arlene Eisenman, Kent State University; Michael Herdlick, Tiffin University; Joe Nowakowski, Muskingum College; Jayprakash Patankar, The University of Akron; Michael Polomsky, Cleveland State University; Anirudh Ruhil, Ohio University; Bonnie Schroeder, Ohio State University; Gwen Terwilliger, University of Toledo; Yan Yu, University of Cincinnati. **Oklahoma: Anne M. Davey**, Northeastern State University; Damian Whalen, St. Gregory's University; David Hudgins, University of Oklahoma—Norman; Dr. William D. Warde, Oklahoma State University—Main Campus. **Oregon:** Jodi Fasteen, Portland State University. **Pennsylvania:** Dr. Deborah Gougeon, University of Scranton; Rose Sebastianelli, University of Scranton; Jack Yurkiewicz, Pace University; Rita Akin, Westminster College; H. David Chen, Rosemont College; Laurel Chiappetta, University of Pittsburgh; Burt Holland, Temple University; Ronald K Klimberg, Saint Joseph's University; Amy Luginbuhl Phelps, Duquesne University; Sherryl May, University of Pittsburg—KGSB; Dr. Bruce McCullough, Drexel University; Tracy Miller, Grove City College; Heather O'Neill, Ursinus College; Tom Short, Indiana University of Pennsylvania; Keith Wargo, Philadelphia Biblical University. **Rhode Island:** Paul Boyd, Johnson & Wales University; **Nicholas Gorgievski**, Nichols College; Jeffrey Jarrett, University of Rhode Island. **South Carolina:** Karie Barbour, Lander University; Joan Donohue, University of South Carolina; Woodrow Hughes, Jr., Converse College; Willis Lewis, Lander University; M. Patterson, Midwestern State University; Kathryn A. Szabat, LaSalle University. **Tennessee:** Ferdinand DiFurio, Tennessee Technical University; Farhad Raiszadeh, University of Tennessee—Chattanooga; Scott J. Seipel, Middle Tennessee State University; Han Wu, Austin Peay State University; Jim Zimmer, Chattanooga State University. **Texas:** Raphael Azuaje, Sul Ross State University; Mark Eakin, University of Texas—Arlington; Betsy Greenberg, University of Texas— Austin; Daniel Friesen, Midwestern State University; Erin Hodgess, University of Houston—Downtown; Joyce Keller, St. Edward's University; Gary Kelley, West Texas A&M University; **Monnie McGee**, Southern Methodist University; John M. Miller, Sam Houston State University; Carolyn H. Monroe, Baylor University; Ranga Ramasesh, Texas Christian University; Plamen Simeonov, University of Houston— Downtown; Lynne Stokes, Southern Methodist University; **Minghe Sun**, University of Texas—San Antonio; Rajesh Tahiliani. University of Texas—El Paso; MaryWhiteside, University of Texas—Arlington; Stuart Warnock, Tarleton State University. **Utah:** Dave Bregenzer, Utah State University; Camille Fairbourn, Utah State University. **Virginia:** Sidhartha R. Das, George Mason University; Quinton J. Nottingham, Virginia Polytechnic & State University; Ping Wang, James Madison University. **Washington:** Nancy Birch, Eastern Washington University; Mike Cicero, Highline Community College; Fred DeKay, Seattle University; Stergios Fotopoulous, Washington State University; Teresa Ling, Seattle University; **Motzev Mihail**, Walla Walla University. **West Virginia: Clifford Hawley**, West Virginia University. **Wisconsin:** Daniel G. Brick, University of St. Thomas; Nancy Burnett University of Wisconsin—Oshkosh; Thomas Groleau, Carthage College; Patricia Ann Mullins, University of Wisconsin, Madison. **Canada:** Jianan Peng, Acadia University; Brian E. Smith, McGill University. **The Netherlands:** Dirk Tempelaar, Maastricht University.

Finally, we want to thank our families. This has been a long project, and it has required many nights and weekends. Our families have sacrificed so that we could write the book we envisioned.

Norean Sharpe

Richard De Veaux

Paul Velleman

Index of Applications

BE = Boxed Example; E = Exercises; EIA = Ethics in Action; GE = Guided Example; IE = In-Text Example and For Example; JC = Just Checking; P = Project; TH = Technology Help

Statistics and Variation

Just look at a page from the *Financial Times* website, like the one shown here. It's full of "statistics." Obviously, the writers of the *Financial Times* think all this information is important, but is this what Statistics is all about? Well, yes and no. This page may contain a lot of facts, but as we'll see, the subject is much more interesting and rich than just spreadsheets and tables.

"Why should I learn Statistics?" you might ask. "After all, I don't plan to do this kind of work. In fact, I'm going to hire people to do all of this for me." That's fine. But the decisions you make based on data are too important to delegate. You'll want to be able to interpret the data that surrounds you and to come to your own conclusions. And you'll find that studying Statistics is much more important and enjoyable than you thought.

1.1   So, What Is Statistics?

Q: What is Statistics?

A: Statistics is a way of reasoning, along with a collection of tools and methods, designed to help us understand the world.

Q: What are statistics?

A: Statistics (plural) are quantities calculated from data.

Q: So what is data?

A: You mean, "what *are* data?" Data is the plural form. The singular is datum.

Q: So, what are data?

A: Data are values along with their context.

It seems every time we turn around, someone is collecting data on us, from every purchase we make in the grocery store to every click of our mouse as we surf the Web. The United Parcel Service (UPS) tracks every package it ships from one place to another around the world and stores these records in a giant database. You can access part of it if you send or receive a UPS package. The database is about 17 terabytes—about the same size as a database that contained every book in the Library of Congress would be. (But, we suspect, not quite as interesting.) What can anyone hope to do with all these data?

Statistics plays a role in making sense of our complex world. Statisticians assess the risk of genetically engineered foods or of a new drug being considered by the Food and Drug Administration (FDA). Statisticians predict the number of new cases of AIDS by regions of the country or the number of customers likely to respond to a sale at the supermarket. And statisticians help scientists, social scientists, and business leaders understand how unemployment is related to environmental controls, whether enriched early education affects the later performance of school children, and whether vitamin C really prevents illness. Whenever you have data and a need to understand the world, you need Statistics.

If we want to analyze student perceptions of business ethics (a question we'll come back to in a later chapter), should we administer a survey to every single university student in the United States—or, for that matter, in the world? Well, that wouldn't be very practical or cost effective. What should we do instead? Give up and abandon the survey? Maybe we should try to obtain survey responses from a smaller, representative group of students. Statistics can help us make the leap from the data we have at hand to an understanding of the world at large. We talk about the specifics of sampling in Chapter 3, and the theme of generalizing from the specific to the general is one that we revisit throughout this book. We hope this text will empower *you* to draw conclusions from data and make valid business decisions in response to such questions as:

- Do university students from different parts of the world perceive business ethics differently?
- What is the effect of advertising on sales?
- Do aggressive, "high-growth" mutual funds really have higher returns than more conservative funds?
- Is there a seasonal cycle in your firm's revenues and profits?
- What is the relationship between shelf location and cereal sales?
- How reliable are the quarterly forecasts for your firm?
- Are there common characteristics about your customers and why they choose your products?—and, more importantly, are those characteristics the same among those who aren't your customers?

Our ability to answer questions such as these and draw conclusions from data depends largely on our ability to understand *variation*. That may not be the term you expected to find at the end of that sentence, but it is the essence of Statistics. The key to learning from data is understanding the variation that is all around us.

Data vary. People are different. So are economic conditions from month to month. We can't see everything, let alone measure it all. And even what we do measure, we measure imperfectly. So the data we wind up looking at and basing our decisions on provide, at best, an imperfect picture of the world. Variation lies at the heart of what Statistics is all about. How to make sense of it is the central challenge of Statistics.

1.2   How Will This Book Help?

A fair question. Most likely, this book will not turn out to be what you expect. It emphasizes graphics and understanding rather than computation and formulas. Instead of learning how to plug numbers in formulas you'll learn the process of model development and come to understand the limitations both of the data you analyze and the methods you use. Every chapter uses real data and real business scenarios so you can see how to use data to make decisions.

Graphs

Close your eyes and open the book at random. Is there a graph or table on the page? Do it again, say, ten times. You probably saw data displayed in many ways, even near the back of the book and in the exercises. Graphs and tables help you understand what the data are saying. So, each story and data set and every new statistical technique will come with graphics to help you understand both the methods and the data.

Process

To help you use Statistics to make business decisions, we'll lead you through the entire process of thinking about a problem, finding and showing results, and telling others what you have discovered. The three simple steps to doing Statistics for business right are: **Plan**, **Do**, and **Report**.

PLAN

Plan first. Know where you're headed and why. Clearly defining and understanding your objective will save you a lot of work.

DO

Do is what most students think Statistics is about. The mechanics of calculating statistics and making graphical displays are important, but the computations are usually the least important part of the process. In fact, we usually turn the computations over to technology and get on with understanding what the results tell us.

REPORT

Report what you've learned. Until you've explained your results in a context that someone else can understand, the job is not done.

"Get your facts first, and then you can distort them as much as you please. (Facts are stubborn, but statistics are more pliable.)"

—MARK TWAIN

For Example

At the end of most sections, we present a short example to help you put what you've learned to immediate use. After reading the example, try the corresponding end-of-section exercises at the end of the chapter. These will help prepare you for the other exercises that tend to use all the skills of the chapter.

Guided Example

Each chapter applies the new concepts taught in worked examples called **Guided Examples**. These examples model how you should approach and solve problems using the Plan, Do, Report framework. They illustrate how to plan an analysis, the appropriate techniques to use, and how to report what it all means. These step-by-step examples show you how to produce the kind of solutions and case study reports that instructors and managers or, better yet, clients expect to see. You will find a model solution in the right-hand column and background notes and discussion in the left-hand column.

Just Checking

Sometimes, in the middle of the chapter, you'll find sections called **Just Checking**, which pose a few short questions you can answer without much calculation. Use them to check that you've understood the basic ideas in the chapter. You'll find the answers at the end-of-chapter exercises.

Ethics in Action

Statistics often requires judgment, and the decisions based on statistical analyses may influence people's health and even their lives. Decisions in government can affect policy decisions about how people are treated. In science and industry, interpretations of data can influence consumer safety and the environment. And in business, misunderstanding what the data say can lead to disastrous decisions. The central guiding principle of statistical judgment is the ethical search for a true understanding of the real world. In all spheres of society it is vitally important that a statistical analysis of data be done in an ethical and unbiased way. Allowing preconceived notions, unfair data gathering, or deliberate slanting to affect statistical conclusions is harmful to business and to society.

At various points throughout the book, you will encounter a scenario under the title **Ethics in Action** in which you'll read about an ethical issue. Think about the issue and how you might deal with it. Then read the summary of the issue and one solution to the problem, which follow the scenario. We've related the ethical issues to guidelines that the American Statistical Association has developed.[1] These scenarios can be good topics for discussion. We've presented one solution, but we invite you to think of others.

What Can Go Wrong?

One of the interesting challenges of Statistics is that, unlike some math and science courses, there can be more than one right answer. This is why two statisticians can testify honestly on opposite sides of a court case. And it's why some people think that you can prove anything with statistics. But that's not true. People make mistakes using statistics, and sometimes people misuse statistics to mislead others. Most of the mistakes are avoidable. We're not talking about arithmetic. Mistakes usually involve using a method in the wrong situation or misinterpreting results. So each chapter has a section called **What Can Go Wrong?** to help you avoid some of the most common mistakes that we've seen in our years of consulting and teaching experience.

Brief CASE

At the end of nearly every chapter you'll find a problem or two that use real data sets and ask you to investigate a question or make a decision. These "brief cases" are a good way to test your ability to attack an open-ended (and thus more realistic) problem. You'll be asked to define the objective, plan your process, complete the analysis, and report your conclusion. These are good opportunities to apply the template provided by the **Guided Examples**. And they provide an opportunity to practice reporting your conclusions in written form to refine your communication skills where statistical results are involved. Data sets for these case studies can be found on the disk included with this text.

"Far too many scientists have only a shaky grasp of the statistical techniques they are using. They employ them as an amateur chef employs a cookbook, believing the recipes will work without understanding why. A more cordon bleu attitude . . . might lead to fewer statistical soufflés failing to rise."

—THE ECONOMIST, JUNE 3, 2004, "SLOPPY STATS SHAME SCIENCE."

[1]http://www.amstat.org/about/ethicalguidelines.cfm

CASE
Study

At the end of each section, you'll find a larger project that will help you integrate your knowledge from the entire section you've been studying. These more open-ended projects will help you acquire the skills you'll need to put your knowledge to work in the world of business.

Technology Help: Using the Computer

Although we show you all the formulas you need to understand the calculations, you will most often use a calculator or computer to perform the mechanics of a statistics problem. And the easiest way to calculate statistics with a computer is with a statistics package. Several different statistics packages are used widely. Although they differ in the details of how to use them, they all work from the same basic information and find the same results. Rather than adopt one package for this book, we present generic output and point out common features that you should look for. We also give a table of instructions to get you started on four packages: Excel, Minitab, SPSS, and JMP. Instructions for Excel 2003 and DataDesk can be found on the CD accompanying this textbook.

> From time to time we'll take time out to discuss an interesting or important side issue. We indicate these by setting them apart like this.[2]

What Have We Learned?

At the end of each chapter, you'll see a brief summary of the chapter's learning objectives in a section called **What Have We Learned?** That section also includes a list of the **Terms** you've encountered in the chapter. You won't be able to learn the material from these summaries, but you can use them to check your knowledge of the important ideas in the chapter. If you have the skills, know the terms, and understand the concepts, you should be well prepared—and ready to use Statistics!

Exercises

Beware: No one can learn Statistics just by reading or listening. The only way to learn it is to do it. So, at the end of each chapter (except this one) you'll find **Exercises** designed to help you learn to use the Statistics you've just read about. Some exercises are marked with a red **T**. You'll find the data for these exercises on the book's website, www.aw-bc.com/sharpe or on the book's disk, so you can use technology as you work the exercises.

We've structured the exercises so that the end-of-section exercises are found first. These can be answered after reading each section. After that you'll find end-of-chapter exercises, designed to help you integrate the topics you've learned in the chapter. We've also paired up and grouped the exercises, so if you're having trouble doing an exercise, you'll find a similar exercise either just before or just after it. You'll find answers to the odd-numbered exercises at the back of the book. But these are only "answers" and not complete solutions. What's the difference? The answers are sketches of the complete solutions. For most problems, your solution

You'll find all sorts of stuff in margin notes, such as stories and quotations. For example:

"Computers are useless. They can only give you answers."

—PABLO PICASSO

While Picasso underestimated the value of good statistics software, he did know that creating a solution requires more than just *Doing*—it means you have to *Plan* and *Report*, too!

---

[2]Or in a footnote.

should follow the model of the **Guided Examples**. If your calculations match the numerical parts of the answer and your argument contains the elements shown in the answer, you're on the right track. Your complete solution should explain the context, show your reasoning and calculations, and state your conclusions. Don't worry too much if your numbers don't match the printed answers to every decimal place. Statistics is more than computation—it's about getting the reasoning correct—so pay more attention to how you interpret a result than to what the digit in the third decimal place is.

*Optional Sections and Chapters

Some sections and chapters of this book are marked with an asterisk (*). These are optional in the sense that subsequent material does not depend on them directly. We hope you'll read them anyway, as you did this section.

Getting Started

It's only fair to warn you: You can't get there by just picking out the highlighted sentences and the summaries. This book is different. It's not about memorizing definitions and learning equations. It's deeper than that. And much more interesting. But . . .
    You have to read the book!

Data

Amazon.com

Amazon.com opened for business in July 1995, billing itself even then as "Earth's Biggest Bookstore," with an unusual business plan: They didn't plan to turn a profit for four to five years. Although some shareholders complained when the dotcom bubble burst, Amazon continued its slow, steady growth, becoming profitable for the first time in 2002. Since then, Amazon has remained profitable and has continued to grow. By 2004, they had more than 41 million active customers in over 200 countries and were ranked the 74th most valuable brand by *Business Week*. Their selection of merchandise has expanded to include almost anything you can imagine, from $400,000 necklaces, to yak cheese from Tibet, to the largest book in the world. In 2008, Amazon.com sold nearly $20 billion worth of products online throughout the world.

Amazon R&D is constantly monitoring and evolving their website to best serve their customers and maximize their sales performance. To make changes to the site, they experiment by collecting data and analyzing what works best. As Ronny Kohavi, former director of Data Mining and Personalization, said, "Data trumps intuition. Instead of using our intuition, we experiment on the live site and let our customers tell us what works for them."

Amazon.com has recently stated "many of the important decisions we make at Amazon.com can be made with data. There is a right answer or a wrong answer, a better answer or a worse answer, and math tells us which is which. These are our favorite kinds of decisions."[1] While we might prefer that Amazon refer to these methods as Statistics instead of math, it's clear that data analysis, forecasting, and statistical inference are the core of the decision-making tools of Amazon.com.

"Data is king at Amazon. Clickstream and purchase data are the crown jewels at Amazon. They help us build features to personalize the website experience."

—Ronny Kohavi,
Former Director of Data
Mining and
Personalization,
Amazon.com

Many years ago, stores in small towns knew their customers personally. If you walked into the hobby shop, the owner might tell you about a new bridge that had come in for your Lionel train set. The tailor knew your dad's size, and the hairdresser knew how your mom liked her hair. There are still some stores like that around today, but we're increasingly likely to shop at large stores, by phone, or on the Internet. Even so, when you phone an 800 number to buy new running shoes, customer service representatives may call you by your first name or ask about the socks you bought six weeks ago. Or the company may send an e-mail in October offering new head warmers for winter running. This company has millions of customers, and you called without identifying yourself. How did the sales rep know who you are, where you live, and what you had bought?

The answer to all these questions is data. Collecting data on their customers, transactions, and sales lets companies track inventory and know what their customers prefer. These data can help them predict what their customers may buy in the future so they know how much of each item to stock. The store can use the data and what they learn from the data to improve customer service, mimicking the kind of personal attention a shopper had 50 years ago.

2.1   What *Are* Data?

Businesses have always relied on data for planning and to improve efficiency and quality. Now, more than ever before, businesses rely on the information in data to compete in the global marketplace. Most modern businesses collect information on virtually every transaction performed by the organization, including every item bought or sold. These data are recorded and stored electronically, in vast digital repositories called **data warehouses**.

In the past few decades these data warehouses have grown enormously in size, but with the use of powerful computers, the information contained in them is accessible and used to help make decisions, sometimes almost instantaneously. When you pay with your credit card, for example, the information about the transaction is transmitted to a central computer where it is processed and analyzed. A decision whether to approve or deny your purchase is made and transmitted back to the point of sale, all within a few seconds.

---

[1]Amazon.com 2008 Annual Report

Companies use data to make decisions about other aspects of their business as well. By studying the past behavior of customers and predicting their responses, they hope to better serve their customers and to compete more effectively. This process of using data, especially of **transactional** data (data collected for recording the companies' transactions) to make other decisions and predictions, is sometimes called **data mining** or *predictive analytics*. The more general term **business analytics** (or sometimes simply analytics) describes *any* use of statistical analysis to drive business decisions from data whether the purpose is predictive or simply descriptive.

Leading companies are embracing business analytics. Richard Fairbank, the CEO and founder of Capital One, revolutionized the credit card industry by realizing that credit card transactions hold the key to understanding customer behavior. Reed Hastings, a former computer science major, is the founder and CEO of Netflix. Netflix uses analytics on customer information both to recommend new movies and to adapt the website that customers see to individual tastes. Netflix offered a $1 million prize to anyone who could improve on the accuracy of their recommendations by more than 10%. That prize was won in 2009 by a team of statisticians and computer scientists using predictive analytics and data-mining techniques. The Oakland Athletics use analytics to judge players instead of the traditional methods used by scouts and baseball experts for over a hundred years. The book *Moneyball* documents how business analytics enabled them to put together a team that could compete against the richer teams in spite of the severely limited resources available to the front office.

To understand better what data are, let's look at some hypothetical company records that Amazon might collect:

105-2686834-3759466	B0000010AA	10.99	Chris G.	902	Boston	15.98	Kansas	Illinois
Samuel P.	Orange County	105-9318443-4200264	105-1872500-0198646	N	B000068ZVQ	Bad Blood	Nashville	Katherine H.
Canada	Garbage	16.99	Ohio	N	Chicago	N	11.99	Massachusetts
B000002BK9	312	Monique D.	Y	413	B00000I5Y6	440	103-2628345-9238664	Let Go

Table 2.1 An example of data with no context. It's impossible to say anything about what these values might mean without knowing their context.

THE W'S:
WHO
WHAT
WHEN
WHERE
WHY

Try to guess what these data represent. Why is that hard? Because these data have no *context*. Whether the data are numerical (consisting only of numbers), alphabetic (consisting only of letters), or alphanumerical (mixed numbers and letters), they are useless unless we know what they represent. Newspaper journalists know that the lead paragraph of a good story should establish the "Five W's": *who, what, when, where,* and (if possible) *why*. Often, we add *how* to the list as well. Answering these questions can provide a **context** for data values and make them meaningful. The answers to the first two questions are essential. If you can't answer *who* and *what*, you don't have data, and you don't have any useful information.

We can make the meaning clear if we add the context of *who* the data are about and *what* was measured and organize the values into a **data table** such as this one.

Order Number	Name	State/Country	Price	Area Code	Previous Album Download	Gift?	ASIN	Artist
105-2686834-3759466	Katherine H.	Ohio	10.99	440	Nashville	N	B0000OI5Y6	Kansas
105-9318443-4200264	Samuel P.	Illinois	16.99	312	Orange County	Y	B000002BK9	Boston
105-1872500-0198646	Chris G.	Massachusetts	15.98	413	Bad Blood	N	B000068ZVQ	Chicago
103-2628345-9238664	Monique D.	Canada	11.99	902	Let Go	N	B0000010AA	Garbage
002-1663369-6638649	Katherine H.	Ohio	10.99	440	Best of Kansas	N	B002MXA7Q0	Kansas

Table 2.2 Example of a data table. The variable names are in the top row. Typically, the *Who* of the table are found in the leftmost column.

Look at the rows of Table 2.2. Now we can see that these are five purchase records, relating to album downloads from Amazon. In general, the rows of a data table correspond to individual **cases** about which we've recorded some characteristics called **variables**.

Cases go by different names, depending on the situation. Individuals who answer a survey are referred to as **respondents**. People on whom we experiment are **subjects** or (in an attempt to acknowledge the importance of their role in the experiment) **participants**, but animals, plants, websites, and other inanimate subjects are often called **experimental units**. Often we call cases just what they are: for example, *customers*, *economic quarters*, or *companies*. In a database, rows are called **records**—in this example, purchase records. Perhaps the most generic term is **cases**. In Table 2.2, the cases are the individual orders.

The column titles (variable names) tell *what* has been recorded. What does a row of Table 2.2 represent? Be careful. Even if people are involved, the cases may not correspond to people. For example, in Table 2.2, each row is a different order and not the customer who made the purchases (notice that the same person made two different orders). A common place to find the *who* of the table is the leftmost column. It's often an identifying variable for the cases, in this example, the order number.

If you collect the data yourself, you'll know what the cases are and how the variables are defined. But, often, you'll be looking at data that someone else collected. The information about the data, called the metadata, might have to come from the company's database administrator or from the *information technology* department of a company. **Metadata** typically contains information about *how*, *when*, and *where* (and possibly *why*) the data were collected; *who* each case represents; and the definitions of all the variables.

A general term for a data table like the one shown in Table 2.2 is a **spreadsheet**, a name that comes from bookkeeping ledgers of financial information. The data were typically spread across facing pages of a bound ledger, the book used by an accountant for keeping records of expenditures and sources of income. For the accountant, the columns were the types of expenses and income, and the cases were transactions, typically invoices or receipts. These days, it is common to keep modest-size datasets in a spreadsheet even if no accounting is involved. It is usually easy to move a data table from a spreadsheet program to a program designed for statistical graphics and analysis, either directly or by copying the data table and pasting it into the statistics program.

Although data tables and spreadsheets are great for relatively small data sets, they are cumbersome for the complex data sets that companies must maintain on a day-to-day basis. Try to imagine a spreadsheet from Amazon with customers in the rows and products in the columns. Amazon has tens of millions of customers and hundreds of thousands of products. But very few customers have purchased more than a few dozen items, so almost all the entries would be blank—not a very efficient way to store information. For that reason, various other database architectures are used to store data. The most common is a relational database.

In a **relational database**, two or more separate data tables are linked together so that information can be merged across them. Each data table is a *relation* because it is about a specific set of cases with information about each of these cases for all (or at least most) of the variables ("fields" in database terminology). For example, a table of customers, along with demographic information on each, is such a relation. A data table with information about a different collection of cases is a different relation. For example, a data table of all the items sold by the company, including information on price, inventory, and past history, is a relation as well (for example, as in Table 2.3). Finally, the day-to-day transactions may be held in a third database where each purchase of an item by a customer is listed as a case. In a relational database, these three relations can be linked together. For example, you can look up a customer to see what he or she purchased or look up an item to see which customers purchased it.

In statistics, all analyses are performed on a single data table. But often the data must be retrieved from a relational database. Retrieving data from these databases often requires specific expertise with that software. In the rest of the book, we'll assume that all data have been downloaded to a data table or spreadsheet with variables listed as columns and cases as the rows.

Customers

Customer Number	Name	City	State	Zip Code	Customer since	Gold Member?
473859	R. De Veaux	Williamstown	MA	01267	2007	No
127389	N. Sharpe	Washington	DC	20052	2000	Yes
335682	P. Velleman	Ithaca	NY	14580	2003	No
...						

Items

Product ID	Name	Price	Currently in Stock?
SC5662	Silver Cane	43.50	Yes
TH2839	Top Hat	29.99	No
RS3883	Red Sequined Shoes	35.00	Yes
...			

Transactions

Transaction Number	Date	Customer Number	Product ID	Quantity	Shipping Method	Free Ship?
T23478923	9/15/08	473859	SC5662	1	UPS 2nd Day	N
T23478924	9/15/08	473859	TH2839	1	UPS 2nd Day	N
T63928934	10/20/08	335682	TH2839	3	UPS Ground	N
T72348299	12/22/08	127389	RS3883	1	Fed Ex Ovnt	Y

Table 2.3   A relational database shows all the relevant information for three separate relations linked together by customer and product numbers.

For Example   Identifying variables and the W's

Carly, a marketing manager at a credit card bank, wants to know if an offer mailed 3 months ago has affected customers' use of their cards. To answer that, she asks the information technology department to assemble the following information for each customer: total spending on the card during the 3 months before the offer (*Pre Spending*); spending for 3 months after the offer (*Post Spending*); the customer's *Age* (by category); what kind of expenditure they made (*Segment*); if customers are enrolled in the website (*Enroll?*); what offer they were sent (*Offer*); and the amount each customer has spent on the card in their segment (*Segment Spend*). She gets a spreadsheet whose first six rows look like this:

Account ID	Pre Spending	Post Spending	Age	Segment	Enroll?	Offer	Segment Spend
393371	$2,698.12	$6,261.40	25–34	Travel/Ent	NO	None	$887.36
462715	$2,707.92	$3,397.22	45–54	Retail	NO	Gift Card	$5,062.55
433469	$800.51	$4,196.77	65+	Retail	NO	None	$673.80
462716	$3,459.52	$3,335.00	25–34	Services	YES	Double Miles	$800.75
420605	$2,106.48	$5,576.83	35–44	Leisure	YES	Double Miles	$3,064.81
473703	$2,603.92	$7,397.50	<25	Travel/Ent	YES	Double Miles	$491.29

Question: Identify the cases and the variables. Describe as many of the W's as you can for this data set.

Answer: The cases are individual customers of the credit card bank. The data are from the internal records of the credit card bank from the past 6 months (3 months before and 3 months after an offer was sent to the customers). The variables include the account ID of the customer (*Account ID*) and the amount charged on the card before (*Pre Spending*) and after (*Post Spending*) the offer was sent out. Also included are the customer's *Age*, marketing *Segment*, whether they enrolled on the website (*Enroll?*), what offer they were sent (*Offer*), and how much they charged on the card in their marketing segment (*Segment Spend*).

2.2   Variable Types

Categorical, or Quantitative?
It is wise to be careful. The *what* and *why* of area codes are not as simple as they may first seem.

When area codes were first introduced all phones had dials. To reduce wear and tear on the dials and to speed the most number of calls, the lowest-digit codes (the easiest to dial) were assigned to the largest cities. So, New York City was given 212, Chicago 312, LA 213 and Philadelphia 215, but rural upstate New York was 607, Joliet was 815, and San Diego 619. Back then, the numerical value of an area code could be used to guess something about the population of its region. But after dials gave way to push buttons, new area codes were assigned without regard to population and area codes are now just categories.

Variables play different roles, and knowing the variable's *type* is crucial to knowing what to do with it and what it can tell us. When a variable names categories and answers questions about how cases fall into those categories, we call it a **categorical**, or **qualitative, variable**. When a variable has measured numerical values with *units* and the variable tells us about the quantity of what is measured, we call it a **quantitative variable**. Classifying a variable into categorical or quantitative can help us decide what to do with a variable, but doing so is often more about what we hope to learn from a variable than about the variable itself. It's the questions we ask of a variable (the *why* of our analysis) that shape how we think about it and how we treat it.

Descriptive responses to questions are often categories. For example, the responses to the questions "What type of mutual fund do you invest in?" or "What kind of advertising does your firm use?" yield categorical values. An important special case of categorical variables is one that has only two possible responses (usually "yes" or "no"), which arise naturally from questions like "Do you invest in the stock market?" or "Do you make online purchases from this website?"

If the variable has values that are not numbers, it's clearly categorical (or needs to be recoded). However, if the values are numbers, you need to be careful. It may be considered quantitative if the values actually measure a quantity of something. Otherwise, it's categorical. For example, area codes are numbers, but the numerical values of area codes don't have numerical meaning (see the side bar). The numbers assigned by the area codes are codes that *categorize* the phone number into a geographical area. So, we treat area code as a categorical variable.

For quantitative variables, the **units** tell how each value has been measured. Even more important, units such as yen, cubits, carats, angstroms, nanoseconds, miles per hour, or degrees Celsius tell us the *scale* of measurement, so we know how far apart two values are. Without units, the values of a measured variable have no meaning. It does little good to be promised a raise of 5000 a year if you don't know

Variable Names that Make Sense

One tradition that still hangs on in some places is to name variables with cryptic abbreviations in uppercase letters. This can be traced back to the 1960s, when computer programs were controlled with instructions punched on cards. The earliest punch card equipment used only uppercase letters, and statistics programs limited variable names to six or eight characters, so variables had names like PRSRF3. Modern programs don't have such restrictive limits, so there is no reason not to use names that make sense.

whether it will be paid in euros, dollars, yen, or Estonian krooni. An essential part of a quantitative variable is its units.

Sometimes the type of the variable is clear. But some variables can answer both kinds of questions and how they are classified depends on their use. For example, the variable *Age* would be considered quantitative if the responses were numerical and they had units. A doctor would certainly need *Age* to be quantitative. The units could be years, but for infants, the doctor would want even more precise units, like months, or even days. On the other hand, if Amazon asked your *Age*, it might lump together the values into categories like "Child (12 years or less)," "Teen (13 to 19)," "Adult (20 to 64)," or "Senior (65 or over)." For many purposes, like knowing which CD ad to send you, that might be all the information Amazon might need. In this case, Amazon has made *Age* a categorical variable.

Question	Categories or Responses
Do you invest in the stock market?	__ Yes __ No
What kind of advertising do you use?	__ Newspapers __ Internet __ Direct mailings
What is your class at school?	__ Freshman __ Sophomore __ Junior __ Senior
I would recommend this course to another student.	__ Strongly Disagree __ Slightly Disagree __ Slightly Agree __ Strongly Agree
How satisfied are you with this product?	__ Very Unsatisfied __ Unsatisfied __ Satisfied __ Very Satisfied

Table 2.4   Some examples of categorical variables.

When Amazon considers a special offer of free shipping to customers, they might first analyze how purchases have been shipped in the recent past. They might start by counting the number of purchases shipped in each category: ground transportation, second-day air, and overnight air. Counting is a natural way to summarize a categorical variable like *Shipping Method*. Chapter 4 discusses summaries and displays of categorical variables more fully. Chapter 5 discusses quantitative variables, which require different summaries and displays.

Shipping Method	Number of Purchases
Ground	20,345
Second-day	7890
Overnight	5432

Table 2.5   A summary of the categorical variable *Shipping Method* that shows the counts, or number of cases for each category.

Identifiers

What's your student ID number? It may be numerical, but is it a quantitative variable? No, it doesn't have units. Is it categorical? Yes, but a special kind. Look at how many categories there are and at how many individuals there are in each category. There are exactly as many categories as individuals and only one individual in each category. While it's easy to count the totals for each category, it's not very interesting. This is an **identifier variable**. Amazon wants to know who you are when you sign in again and doesn't want to confuse you with some other customer. So they assign you a unique identifier.

Identifier variables themselves don't tell us anything useful about the categories because we know there is exactly one individual in each. However, they are crucial in this era of large data sets because by uniquely identifying the cases, they make it possible to combine data from different sources, protect confidentiality, and provide unique labels. Most company databases are, in fact, relational databases. The identifier is crucial to linking one data table to another in a relational database. The identifiers in Table 2.3 are the *Customer Number, Product ID*, and *Transaction Number*. Variables like *UPS Tracking Number, Social Security Number*, and Amazon's *ASIN* are other examples of identifiers.

You'll want to recognize when a variable is playing the role of an identifier so you won't be tempted to analyze it. Knowing that Amazon's average ASIN number increased 10% from 2007 to 2008 doesn't really tell you anything—any more than analyzing any categorical variable as if it were quantitative would.

Be careful not to be inflexible in your typing of variables. Variables can play different roles, depending on the question we ask of them, and classifying variables rigidly into types can be misleading. For example, in their annual reports, Amazon refers to its database and looks at the variables *Sales* and *Year*. When analysts ask how many books Amazon sold in 2005, what role does *Year* play? There's only one row for 2005, and *Year* identifies it, so it plays the role of an identifier variable. In its role as an identifier, you might match other data from Amazon, or the economy in general, for the same year. But analysts also track sales growth over time. In this role, *Year* measures time. Now it's being treated as a quantitative variable with units of years.

Other Data Types

A survey might ask:

"How satisfied were you with the service you received?"

1) Not satisfied; 2) Somewhat satisfied; 3) Moderately satisfied; or 4) Extremely satisfied.

Is this variable categorical or quantitative? There is certainly an *order* of perceived worth; higher numbers indicate higher perceived worth. An employee whose customer responses average around 4 seems to be doing a better job than one whose averages are around 2, but are they *twice* as good? Because the values are not strictly numbers, we can't really say and so we should be careful about treating *Customer Satisfaction* as purely quantitative. When, as in this example, the values of a categorical value have an intrinsic order, we can say that the categorical variable is **ordinal**. By contrast, a categorical variable that names categories that don't have order is sometimes called **nominal**. Values can be individually ordered (e.g., the ranks of employees based on the number of days they've worked for the company) or ordered in classes (e.g., Freshman, Sophomore, Junior, Senior). Ordering is not absolute; how the values are ordered depends on the purpose of the ordering. For example, are the categories Infant, Youth, Teen, Adult, and Senior ordinal? Well, if we are ordering on age, they surely are and how to order the categories is clear. But if we are ordering (as Amazon might) on purchase volume, it is likely that either Teen or Adult will be the top group.[2]

Cross-Sectional and Time Series Data

The quantitative variable *Total Revenue* in Table 2.6 is an example of a time series. A **time series** is a single variable measured at regular intervals over time. Time series are common in business. Typical measuring points are months, quarters, or

---

[2]Some people differentiate quantitative variables according to whether their measured values have a defined value for zero. This is a technical distinction and usually not one we'll need to make. (For example, it isn't correct to say that a temperature of 80°F is twice as hot as 40°F because 0° is an arbitrary value. On the Celsius scale those temperatures are 26.67°C and 4.44°C—a ratio of 6.) The term *interval scale* is sometimes applied to data such as these, and the term *ratio scale* is applied to measurements for which such ratios are appropriate.

Year	Total Revenue (in $M)
2002	3288.9
2003	4075.5
2004	5294.2
2005	6369.3
2006	7786.9
2007	9441.5
2008	10,383.0
2009	9774.6

Table 2.6 Starbucks's total revenue (in $M) for the years 2002 to 2009.

years, but virtually any consistently-spaced time interval is possible. Variables collected over time hold special challenges for statistical analysis, and Chapter 20 discusses these in more detail.

By contrast, most of the methods in this book are better suited for **cross-sectional data**, where several variables are measured at the same time point. On the other hand, if we collect data on sales revenue, number of customers, and expenses for last month at *each* Starbucks (more than 16,000 locations as of 2010) at one point in time, this would be cross-sectional data. Cross-sectional data may contain some time information (such as dates), but it isn't a time series because it isn't measured at regular intervals. Because different methods are used to analyze these different types of data, it is important to be able to identify both time series and cross-sectional data sets.

For Example   Identifying the types of variables

Question: Before she can continue with her analysis, Carly (from the example on page 12) must classify each variable as being quantitative or categorical (or possibly both), and whether the data are a time series or cross-sectional. For quantitative variables, what are the units? For categorical variables, are they nominal or ordinal?

Answer:

Account ID – categorical (nominal, identifier)

Pre Spending – quantitative (units $)

Post Spending – quantitative (units $)

Age – categorical (ordinal). Could be quantitative if we had more precise information

Segment – categorical (nominal)

Enroll? – categorical (nominal)

Offer – categorical (nominal)

Segment Spend – quantitative (units $)

The data are cross-sectional. We do not have successive values of a single variable over time.

2.3   Data Sources: Where, How, and When

We must know *who, what,* and *why* to analyze data. Without knowing these three, we don't have enough to start. Of course, we'd always like to know more because the more we know, the more we'll understand. If possible, we'd like to know the *where, how,* and *when* of data as well. Values recorded in 1947 may mean something different than similar values recorded last year. Values measured in Abu Dhabi may differ in meaning from similar measurements made in Mexico.

How the data are collected can make the difference between insight and nonsense. As we'll see later, data that come from a voluntary survey on the Internet are almost always worthless. In a recent Internet poll, 84% of respondents said "no" to the question of whether subprime borrowers should be bailed out. While it may be true that 84% of those 23,418 respondents did say that, it's dangerous to assume that that group is representative of any larger group. To make inferences from the data you have at hand to the world at large, you need to ensure that the data you have are representative of the larger group. Chapter 3 discusses sound methods for *designing* a *survey* or poll to help ensure that the inferences you make are valid.

Another way to collect valid data is by performing an experiment in which you actively manipulate variables (called factors) to see what happens. Most of the "junk mail" credit card offers that you receive are actually experiments done by marketing groups in those companies. They may make different versions of an offer to selected groups of customers to see which one works best before rolling out the winning idea to the entire customer base. Chapter 22 discusses both the design and the analysis of experiments like these.

Sometimes, the answer to the question you have may be found in data that someone, or more typically, some organization has already collected. Internally, companies may analyze data from their own data bases or data warehouse. They may also supplement or rely entirely on data collected by others. Many companies, nonprofit organizations, and government agencies collect vast amounts of data via the Internet. Some organizations may charge a fee for accessing or downloading their data. The U.S. government collects information on nearly every aspect of life in the United States, both social and economic (see for example www.census.gov, or more generally, www.usa.gov), as the European Union does for Europe (see ec.europa.eu/eurostat). International organizations such as the World Health Organization (www.who.org) and polling agencies such as Gallup (www.gallup.com) offer information on a variety of topics as well. Data like these typically do not come from a designed survey or experiment. They are most often collected for different purposes than the analysis you may want to perform. Although they are plentiful, you should be careful when generalizing from data like these. Information about how, when, where, and why the data were collected may not be available. Unless the data were collected in a way that ensures that they are representative of the population in which you are interested, you may be misled if you try to draw conclusions from them. Chapter 24 discusses data mining, which attempts to use large amounts of "found" data to make hypotheses and draw insights. While it can be tempting, interesting, and even useful to analyze such happenstance data, remember that the only way to be sure that a generalization is valid is if the data come from a properly designed survey or experiment.

There's a world of data on the Internet

These days, one of the richest sources of data is the Internet. With a bit of practice, you can learn to find data on almost any subject. We found many of the data sets used in this book by searching on the Internet. The Internet has both advantages and disadvantages as a source of data. Among the advantages are the fact that often you'll be able to find even more current data than we present. One disadvantage is that references to Internet addresses can "break" as sites evolve, move, and die. Another disadvantage is that important metadata—information about the collection, quality, and intent of the data—may be missing.

Our solution to these challenges is to offer the best advice we can to help you search for the data, wherever they may be residing. We usually point you to a website. We'll sometimes suggest search terms and offer other guidance.

Some words of caution, though: Data found on Internet sites may not be formatted in the best way for use in statistics software. Although you may see a data table in standard form, an attempt to copy the data may leave you with a single column of values. You may have to work in your favorite statistics or spreadsheet program to reformat the data into variables. You will also probably want to remove commas from large numbers and such extra symbols as money indicators ($, ¥, £, €); few statistics packages can handle these.

Throughout this book, whenever we introduce data, we will provide a margin note listing some of the W's of the data and, where possible, offer a reference for the source of the data. It's a habit we recommend. The first step of any data analysis is to know why you are examining the data (what you want to know), whom each row of your data table refers to, and what the variables (the columns of the table) record. These are the *Why*, the *Who*, and the *What*. Identifying them is a key part of the *Plan* step of any analysis. Make sure you know all three before you spend time analyzing the data.

For Example   Identifying data sources

On the basis of her initial analysis, Carly asks her colleague Ying Mei to e-mail a sample of customers from the Travel and Entertainment segment and ask about their card use and household demographics. Carly asks another colleague, Gregg, to design a study about their double miles offer. In this study, a random sample of customers receives one of three offers: the standard double miles offer; a double miles offer good on any airline; or no offer.

Question: For each of the three data sets—Carly's original data set and Ying Mei's and Gregg's sets—state whether they come from a designed survey or a designed experiment or are collected in another way.

Answer: Carly's data set was derived from transactional data, not part of a survey or experiment. Ying Mei's data come from a designed survey, and Gregg's data come from a designed experiment.

Just Checking

An insurance company that specializes in commercial property insurance has a separate database for their policies that involve churches and schools. Here is a small portion of that database.

Policy Number	Years Claim Free	Net Property Premium ($)	Net Liability Premium ($)	Total Property Value ($1,000)	Median Age in Zip Code	School?	Territory	Coverage
4000174699	1	3107	503	1036	40	FALSE	AL580	BLANKET
8000571997	2	1036	261	748	42	FALSE	PA192	SPECIFIC
8000623296	1	438	353	344	30	FALSE	ID60	BLANKET
3000495296	1	582	339	270	35	TRUE	NC340	BLANKET
5000291199	4	993	357	218	43	FALSE	OK590	BLANKET
8000470297	2	433	622	108	31	FALSE	NV140	BLANKET
1000042399	4	2461	1016	1544	41	TRUE	NJ20	BLANKET
4000554596	0	7340	1782	5121	44	FALSE	FL530	BLANKET
3000260397	0	1458	261	1037	42	FALSE	NC560	BLANKET
8000333297	2	392	351	177	40	FALSE	OR190	BLANKET
4000174699	1	3107	503	1036	40	FALSE	AL580	BLANKET

1 List as many of the W's as you can for this data set.

2 Classify each variable as to whether you think it should be treated as categorical or quantitative (or both); if quantitative, identify the units.

What Can Go Wrong?

- **Don't label a variable as categorical or quantitative without thinking about the data and what they represent.** The same variable can sometimes take on different roles.

- **Don't assume that a variable is quantitative just because its values are numbers.** Categories are often given numerical labels. Don't let that fool you into thinking they have quantitative meaning. Look at the context.

- **Always be skeptical.** One reason to analyze data is to discover the truth. Even when you are told a context for the data, it may turn out that the truth is a bit (or even a lot) different. The context colors our interpretation of the data, so those who want to influence what you think may slant the context. A survey that seems to be about all students may in fact report just the opinions of those who visited a fan website. The question that respondents answered may be posed in a way that influences responses.

Ethics in Action

Sarah Potterman, a doctoral student in educational psychology, is researching the effectiveness of various interventions recommended to help children with learning disabilities improve their reading skills. Among the approaches examined is an interactive software system that uses analogy-based phonics. Sarah contacted the company that developed this software, RSPT Inc., in order to obtain the system free of charge for use in her research. RSPT Inc. expressed interest in having her compare their product with other intervention strategies and was quite confident that their approach would be the most effective. Not only did the company provide Sarah with free software, but RSPT Inc. also generously offered to fund her research with a grant to cover her data collection and analysis costs.

ETHICAL ISSUE *Both the researcher and company should be careful about the funding source having a vested interest in the research result (related to Item H, ASA Ethical Guidelines).*

ETHICAL SOLUTION *RSPT Inc. should not pressure Sarah Potterman to obtain a particular result. Both parties should agree on paper before the research is begun that the research results can be published even if they show that RSPT's interactive software system is not the most effective.*

Jim Hopler is operations manager for a local office of a top-ranked full-service brokerage firm. With increasing competition from both discount and online brokers, Jim's firm has redirected attention to attaining exceptional customer service through its client-facing staff, namely brokers. In particular, they wish to emphasize the excellent advisory services provided by their brokers. Results from surveying clients about the advice received from brokers at the local office revealed that 20% rated it *poor*, 5% rated it *below average*, 15% rated it *average*, 10% rated

it *above average*, and 50% rated it *outstanding*. With corporate approval, Jim and his management team instituted several changes in an effort to provide the best possible advisory services at the local office. Their goal was to increase the percentage of clients who viewed their advisory services as *outstanding*. Surveys conducted after the changes were implemented showed the following results: 5% *poor*, 5% *below average*, 20% *average*, 40% *above average*, and 30% *outstanding*. In discussing these results, the management team expressed concern that the percentage of clients who considered their advisory services *outstanding* fell from 50% to 30%. One member of the team suggested an alternative way of summarizing the data. By coding the categories on a scale from 1 = poor to 5 = outstanding and computing the average, they found that the average rating increased from 3.65 to 3.85 as a result of the changes implemented. Jim was delighted to see that their changes were successful in improving the level of advisory services offered at the local office. In his report to corporate, he only included average ratings for the client surveys.

ETHICAL ISSUE *By taking an average, Jim is able to show improved customer satisfaction. However, their goal was to increase the percentage of outstanding ratings. Jim redefined his study after the fact to support a position (related to Item A, ASA Ethical Guidelines).*

ETHICAL SOLUTION *Jim should report the percentages for each rating category. He can also report the average. He may wish to include in his report a discussion of what those different ways of looking at the data say and why they appear to differ. He may also want to explore with the survey participants the perceived differences between "above average" and "outstanding."*

What Have We Learned?

Learning Objectives

- Understand that data are values, whether numerical or labels, together with their context.
  - *who, what, why, where, when* (and *how*)—the W's—help nail down the context of the data.
  - We must know *who, what,* and *why* to be able to say anything useful based on the data. The *who* are the cases. The *what* are the variables. A variable gives information about each of the cases. The *why* helps us decide which way to treat the variables.
  - Stop and identify the W's whenever you have data, and be sure you can identify the cases and the variables.

- Identify whether a variable is being used as categorical or quantitative.
  - Categorical variables identify a category for each case. Usually we think about the counts of cases that fall in each category. (An exception is an identifier variable that just names each case.)
  - Quantitative variables record measurements or amounts of something; they must have units.
  - Sometimes we may treat the same variable as categorical or quantitative depending on what we want to learn from it, which means some variables can't be pigeonholed as one type or the other.

- Consider the source of your data and the reasons the data were collected. That can help you understand what you might be able to learn from the data.

Terms

Business analytics	The process of using statistical analysis and modeling to drive business decisions.
Case	A case is an individual about whom or which we have data.
Categorical (or qualitative) variable	A variable that names categories (whether with words or numerals) is called categorical or qualitative.
Context	The context ideally tells *who* was measured, *what* was measured, *how* the data were collected, *where* the data were collected, and *when* and *why* the study was performed.
Cross-sectional data	Data taken from situations that vary over time but measured at a single time instant is said to be a cross-section of the time series.
Data	Recorded values whether numbers or labels, together with their context.
Data mining	The process of using a variety of statistical tools to analyze large data bases or data warehouses.
Data table	An arrangement of data in which each row represents a case and each column represents a variable.
Data warehouse	A large data base of information collected by a company or other organization usually to record transactions that the organization makes, but also used for analysis via data mining.
Experimental unit	An individual in a study for which or for whom data values are recorded. Human experimental units are usually called subjects or participants.
Identifier variable	A categorical variable that records a unique value for each case, used to name or identify it.
Metadata	Auxiliary information about variables in a database, typically including *how, when,* and *where* (and possibly *why*) the data were collected; *who* each case represents; and the definitions of all the variables.
Nominal variable	The term "nominal" can be applied to a variable whose values are used only to name categories.
Ordinal variable	The term "ordinal" can be applied to a variable whose categorical values possess some kind of order.

Participant	A human experimental unit. Also called a subject.
Quantitative variable	A variable in which the numbers are values of measured quantities with units.
Record	Information about an individual in a database.
Relational database	A relational database stores and retrieves information. Within the database, information is kept in data tables that can be "related" to each other.
Respondent	Someone who answers, or responds to, a survey.
Spreadsheet	A spreadsheet is layout designed for accounting that is often used to store and manage data tables. Excel is a common example of a spreadsheet program.
Subject	A human experimental unit. Also called a participant.
Time series	Data measured over time. Usually the time intervals are equally spaced or regularly spaced (e.g., every week, every quarter, or every year).
Transactional Data	Data collected to record the individual transactions of a company or organization.
Units	A quantity or amount adopted as a standard of measurement, such as dollars, hours, or grams.
Variable	A variable holds information about the same characteristic for many cases.

Technology Help: Data on the Computer

Most often we find statistics on a computer using a program, or *package,* designed for that purpose. There are many different statistics packages, but they all do essentially the same things. If you understand what the computer needs to know to do what you want and what it needs to show you in return, you can figure out the specific details of most packages pretty easily.

For example, to get your data into a computer statistics package, you need to tell the computer:

• Where to find the data. This usually means directing the computer to a file stored on your computer's disk or to data on a database. Or it might just mean that you have copied the data

from a spreadsheet program or Internet site and it is currently on your computer's clipboard. Usually, the data should be in the form of a data table. Most computer statistics packages prefer the *delimiter* that marks the division between elements of a data table to be a *tab* character and the delimiter that marks the end of a case to be a *return* character.

• Where to put the data. (Usually this is handled automatically.)

• What to call the variables. Some data tables have variable names as the first row of the data, and often statistics packages can take the variable names from the first row automatically.

Brief CASE

Credit Card Bank

Like all credit and charge card companies, this company makes money on each of its cardholders' transactions. Thus, its profitability is directly linked to card usage. To increase customer spending on its cards, the company sends many different offers to its cardholders, and market researchers analyze the results to see which offers yield the largest increases in the average amount charged.

On your disk (in the file **Credit_Card_Bank**) is part of a database like the one used by the researchers. For each customer, it contains several variables in a spreadsheet.

Examine the data in the data file. List as many of the W's as you can for these data and classify each variable as categorical or quantitative. If quantitative, identify the units.

Exercises

SECTION 2.1

1. A real estate major collected information on some recent local home sales. The first 6 lines of the database appear below. The columns correspond to the house identification number, the community name, the zip code, the number of acres of the property, the year the house was built, the market value, and the size of the living area (in square feet).

HOUSE_ID	NEIGHBORHOOD	MAIL_ZIP	ACRES	YR_BUILT	FULL_MARKET_VALUE	SFLA
413400536	Greenfield Manor	12859	1.00	1967	100400	960
4128001474	Fort Amherst	12801	0.09	1961	132500	906
412800344	Dublin	12309	1.65	1993	140000	1620
4128001552	Granite Springs	10598	0.33	1969	67100	900
412800352	Arcady	10562	2.29	1955	190000	1224
413400322	Ormsbee	12859	9.13	1997	126900	1056

a) What does a row correspond to in this data table? How would you best describe its role: as a participant, subject, case, respondent, or experimental unit?
b) How many variables are measured on each row?

2. A local bookstore is keeping a database of its customers to find out more about their spending habits and so that the store can start to make personal recommendations based on past purchases. Here are the first five rows of their database:

Transaction ID	Customer ID	Date	ISBN Number of Purchase	Price	Coupon?	Gift?	Quantity
29784320912	4J438	11/12/2009	345-23-2355	$29.95	N	N	1
26483589001	3K729	9/30/2009	983-83-2739	$16.99	N	N	1
26483589002	3K729	9/30/2009	102-65-2332	$9.95	Y	N	1
36429489305	3U034	12/5/2009	295-39-5884	$35.00	N	Y	1
36429489306	3U034	12/5/2009	183-38-2957	$79.95	N	Y	1

a) What does a row correspond to in this data table? How would you best describe its role: as a participant, subject, case, respondent, or experimental unit?
b) How many variables are measured on each row?

SECTION 2.2

3. Referring to the real estate data table of Exercise 1,

a) For each variable, would you describe it as primarily categorical, or quantitative? If quantitative, what are the units? If categorical, is it ordinal or simply nominal?
b) Are these data a time series, or are these cross-sectional? Explain briefly.

4. Referring to the bookstore data table of Exercise 2,

a) For each variable, would you describe it as primarily categorical, or quantitative? If quantitative, what are the units? If categorical, is it ordinal or simply nominal?
b) Are these data a time series, or are these cross-sectional? Explain briefly.

SECTION 2.3

5. For the real estate data of Exercise 1, do the data appear to have come from a designed survey or experiment? What concerns might you have about drawing conclusions from this data set?

6. A student finds data on an Internet site that contains financial information about selected companies. He plans to analyze the data and use the results to develop a stock investment strategy. What kind of data source is he using? What concerns might you have about drawing conclusions from this data set?

CHAPTER EXERCISES

For each description of data in Exercises 7 to 26, identify the W's, name the variables, specify for each variable whether its use indicates it should be treated as categorical or quantitative, and for any quantitative variable identify the units in which it was measured (or note that they were not provided). Specify whether the data come from a designed survey or experiment. Are the variables time series or cross-sectional? Report any concerns you have as well.

7. The news. Find a newspaper or magazine article in which some data are reported (e.g., see *The Wall Street Journal*, *Financial Times*, *Business Week*, or *Fortune*). For the data discussed in the article, answer the questions above. Include a copy of the article with your report.

8. The Internet. Find an Internet site on which some data are reported. For the data found on the site, answer as many of the questions above as you can. Include a copy of the URL with your report.

9. Sales. A major U.S. company is interested in seeing how various promotional activities are related to domestic sales. Analysts decide to measure the money spent on different forms of advertising ($ thousand) and sales ($ million) on a monthly basis for three years (2004–2006).

10. Food store. A food retailer that specializes in selling organic food has decided to open a new store. To help determine the best location in the United States for the new store, researchers decide to examine data from existing stores, including weekly sales ($), town population (thousands), median age of town, median income of town ($), and whether or not the store sells wine and beer.

11. Sales II. The company in Exercise 2 is also interested in the impact of national indicators on their sales. It decides to obtain measurements for unemployment rate (%) and inflation rate (%) on a quarterly basis to compare to their quarterly sales ($ million) over the same time period (2004–2006).

12. Arby's menu. A listing posted by the Arby's restaurant chain gives, for each of the sandwiches it sells, the type of meat in the sandwich, number of calories, and serving size in ounces. The data might be used to assess the nutritional value of the different sandwiches.

13. MBA admissions. A school in the northeastern United States is concerned with the recent drop in female students in its MBA program. It decides to collect data from the admissions office on each applicant, including: sex of each applicant, age of each applicant, whether or not they were accepted, whether or not they attended, and the reason for not attending (if they did not attend). The school hopes to find commonalities among the female accepted students who have decided not to attend the business program.

14. MBA admissions II. An internationally recognized MBA program in London intends to also track the GPA of the MBA students and compares MBA performance to standardized test scores over a 5-year period (2000–2005).

15. Pharmaceutical firm. Scientists at a major pharmaceutical firm conducted an experiment to study the effectiveness of an herbal compound to treat the common cold. They exposed volunteers to a cold virus, then gave them either the herbal compound or a sugar solution known to have no effect on colds. Several days later they assessed each patient's condition using a cold severity scale ranging from 0–5. They found no evidence of the benefits of the compound.

16. Start-up company. A start-up company is building a database of customers and sales information. For each customer, it records name, ID number, region of the country (1 = East, 2 = South, 3 = Midwest, 4 = West), date of last purchase, amount of purchase, and item purchased.

17. Vineyards. Business analysts hoping to provide information helpful to grape growers sent out a questionnaire to a sample of growers requesting these data about vineyards: size (acres), number of years in existence, state, varieties of grapes grown, average case price, gross sales, and percent profit.

18. Gallup Poll. The Gallup Poll conducted a representative telephone survey of 1180 American voters. Among the reported results were the voter's region (Northeast, South, etc.), age, political party affiliation, whether the respondent owned any shares of stock, and their attitude (on a scale of 1 to 5) toward unions.

19. EPA. The Environmental Protection Agency (EPA) tracks fuel economy of automobiles. Among the data EPA analysts collect from the manufacturer are the manufacturer (Ford, Toyota, etc.), vehicle type (car, SUV, etc.), weight, horsepower, and gas mileage (mpg) for city and highway driving.

20. Consumer Reports. In 2002, Consumer Reports published an article evaluating refrigerators. It listed 41 models, giving the brand, cost, size (cu ft), type (such as top-freezer), estimated annual energy cost, overall rating (good, excellent, etc.), and repair history for that brand (percentage requiring repairs over the past 5 years).

21. Lotto. A study of state-sponsored Lotto games in the United States (*Chance*, Winter 1998) listed the names of the states and whether or not the state had Lotto. For states that did, the study indicated the number of numbers in the lottery, the number of matches required to win, and the probability of holding a winning ticket.

22. L.L. Bean. L.L. Bean is a large U.S. retailer that depends heavily on its catalog sales. It collects data internally and tracks the number of catalogs mailed out, the number of square inches in each catalog, and the sales ($ thousands) in the 4 weeks following each mailing. The company is interested in learning more about the relationship (if any) among the timing and space of their catalogs and their sales.

23. Stock market. An online survey of students in a large MBA Statistics class at a business school in the northeastern

United States asked them to report their total personal investment in the stock market ($), total number of different stocks currently held, total invested in mutual funds ($), and the name of each mutual fund in which they have invested. The data were used in the aggregate for classroom illustrations.

24. Theme park sites. A study on the potential for developing theme parks in various locations throughout Europe in 2008 collects the following information: the country where the proposed site is located, estimated cost to acquire site (in euros), size of population within a one-hour drive of the site, size of the site (in hectares), mass transportation within 5 minutes of the site. The data will be used to present to prospective developers.

25. Indy 2009. The 2.5-mile Indianapolis Motor Speedway has been the home to a race on Memorial Day nearly every year since 1911. Even during the first race there were controversies. Ralph Mulford was given the checkered flag first but took three extra laps just to make sure he'd completed 500 miles. When he finished, another driver, Ray Harroun, was being presented with the winner's trophy, and Mulford's protests were ignored.

Harroun averaged 74.6 mph for the 500 miles. Here are the data for the first few and three recent Indianapolis 500 races.

Year	Winner	Car	Time (hrs)	Speed (mph)	Car #
1911	Ray Harroun	Marmon Model 32	6.7022	74.602	32
1912	Joe Dawson	National	6.3517	78.719	8
1913	Jules Goux	Peugeot	6.5848	75.933	16
...					
...					
2005	Dan Wheldon	Dallara/Honda	3.1725	157.603	26
2006	Sam Hornish, Jr.	Dallara/Honda	3.1830	157.085	6
2007	Dario Franchitti	Dallara/Honda	2.7343	151.774	27
2008	Scott Dixon	Dallara	3.4826	143.567	9
2009	Hélio Castroneves	Dallara	3.3262	150.318	3

26. Kentucky Derby. The Kentucky Derby is a horse race that has been run every year since 1875 at Churchill Downs, Louisville, Kentucky. The race started as a 1.5-mile race, but in 1896 it was shortened to 1.25 miles because experts felt that 3-year-old horses shouldn't run such a long race that early in the season. (It has been run in May every year but one—1901—when it took place on April 29.) The table at the bottom of the page shows the data for the first few and a few recent races.

Date	Winner	Margin (lengths)	Jockey	Winner's Payoff ($)	Duration (min:sec)	Track Condition
May 17, 1875	Aristides	2	O. Lewis	2850	2:37.75	Fast
May 15, 1876	Vagrant	2	B. Swim	2950	2:38.25	Fast
May 22, 1877	Baden-Baden	2	W. Walker	3300	2:38.00	Fast
May 21, 1878	Day Star	1	J. Carter	4050	2:37.25	Dusty
May 20, 1879	Lord Murphy	1	C. Shauer	3550	2:37.00	Fast
...						
May 5, 2001	Monarchos	4 3/4	J. Chavez	812,000	1:59.97	Fast
May 4, 2002	War Emblem	4	V. Espinoza	1,875,000	2:01.13	Fast
May 3, 2003	Funny Cide	1 3/4	J. Santos	800,200	2:01.19	Fast
May 1, 2004	Smarty Jones	2 3/4	S. Elliott	854,800	2:04.06	Sloppy

When you organize data in a spreadsheet, it is important to lay it out as a data table. For each of these examples in Exercises 27 to 30, show how you would lay out these data. Indicate the headings of columns and what would be found in each row.

27. Mortgages. For a study of mortgage loan performance: amount of the loan, the name of the borrower.

28. Employee performance. Data collected to determine performance-based bonuses: employee ID, average contract closed (in $), supervisor's rating (1–10), years with the company.

29. Company performance. Data collected for financial planning: weekly sales, week (week number of the year), sales predicted by last year's plan, difference between predicted sales and realized sales.

30. Command performance. Data collected on investments in Broadway shows: number of investors, total invested, name of the show, profit/loss after one year.

For the following examples in Exercises 31 to 34, indicate whether the data are a time series or a cross section.

31. Car sales. Number of cars sold by each salesperson in a dealership in September.

32. Motorcycle sales. Number of motorcycles sold by a dealership in each month of 2008.

33. Cross sections. Average diameter of trees brought to a sawmill in each week of a year.

34. Series. Attendance at the third World Series game recording the age of each fan.

Just Checking Answers

1   Who—policies on churches and schools
    What—policy number, years claim free, net property premium ($), net liability premium ($), total property value ($000), median age in zip code, school?, territory, coverage
    How—company records
    When—not given

2   Policy number: identifier (categorical)
    Years claim free: quantitative
    Net property premium: quantitative ($)
    Net liability premium: quantitative ($)
    Total property value: quantitative ($)
    Median age in zip code: quantitative
    School?: categorical (true/false)
    Territory: categorical
    Coverage: categorical

Surveys and Sampling

Roper Polls

Public opinion polls are a relatively new phenomenon. In 1948, as a result of telephone surveys of likely voters, all of the major organizations—Gallup, Roper, and Crossley—consistently predicted, throughout the summer and into the fall, that Thomas Dewey would defeat Harry Truman in the November presidential election. By October the results seemed so clear that *Fortune* magazine declared, "Due to the overwhelming evidence, *Fortune* and Mr. Roper plan no further detailed reports on change of opinion in the forthcoming presidential campaign. . . ."

Of course, Harry Truman went on to win the 1948 election, and the picture of Truman in the early morning after the election holding up the *Chicago Tribune* (printed the night before), with its headline declaring Dewey the winner, has become legend.

The public's faith in opinion polls plummeted after the election, but Elmo Roper vigorously defended the pollsters. Roper was a principal and founder of one of the first market research firms, Cherington, Wood, and Roper, and director of the *Fortune Survey*, which was the first national poll to use scientific sampling techniques. He argued that rather than abandoning polling, business leaders should learn what had gone wrong in the 1948 polls so that market research could be improved. His frank admission of the mistakes made in those polls helped to restore confidence in polling as a business tool.

For the rest of his career, Roper split his efforts between two projects, commercial polling and public opinion. He established the Roper Center for Public Opinion Research at Williams College as a place to house public opinion archives, convincing fellow polling leaders Gallup and Crossley to participate as well. Now located at the University of Connecticut, the Roper Center is one of the world's leading archives of social science data. Roper's market research efforts started as Roper Research Associates and later became the Roper Organization, which was acquired in 2005 by GfK. Founded in Germany in 1934 as the Gesellschaft für Konsumforschung (literally, "Society for Consumption Research"), GfK now stands for "growth from knowledge." It is the fourth largest international market research organization, with over 130 companies in 70 countries and more than 7700 employees worldwide.

GfK Roper Consulting conducts a yearly, global study to examine cultural, economic, and social information that may be crucial to companies doing business worldwide. These companies use the information provided by GfK Roper to help make marketing and advertising decisions in different markets around the world.

How do the researchers at GfK Roper know that the responses they get reflect the real attitudes of consumers? After all, they don't ask everyone, but they don't want to limit their conclusions to just the people they surveyed. Generalizing from the data at hand to the world at large is something that market researchers, investors, and pollsters do every day. To do it wisely, they need three fundamental ideas.

3.1   Three Ideas of Sampling

Idea 1: Sample—Examine a Part of the Whole

We'd like to know about an entire collection of individuals, called a **population**, but examining all of them is usually impractical, if not impossible. So we settle for examining a smaller group of individuals—a **sample**—selected from the population. For the Roper researchers the population of interest is the entire world, but it's not practical, cost-effective, or feasible to survey everyone. So they examine a sample selected from the population.

You take samples every day. For example, if you want to know how the vegetable soup you're cooking for dinner tonight is going to taste, you blow on a spoonful and try it. You certainly don't consume the whole pot. You trust that the taste will *represent* the flavor of the population—the entire pot. The idea of tasting is that a small sample, if selected properly, can represent the larger population.

The GfK Roper Reports® Worldwide poll is an example of a **sample survey**, designed to ask questions of a small group of people in the hope of learning something about the entire population. Most likely, you've never been selected to be part of a national opinion poll. That's true of most people. So how can the pollsters claim that a sample represents the entire population? Selecting a sample to represent the population fairly is easy in theory, but in practice, it's more difficult than it sounds. Polls or surveys most often fail because the sample fails to represent part of the population. For example, a telephone survey may get no responses from people with caller ID and may favor other groups, such as the retired or the homebound, who would be more likely to be near their phones when the interviewer calls. Samples that over- or underemphasize some characteristics of the population are said to be biased. When a sample is **biased**, the summary characteristics of a sample differ from the corresponding characteristics of the population it is trying to represent, so they can produce misleading information. Conclusions based on biased samples are inherently flawed. There is usually no way to fix bias after the sample is drawn and no way to salvage useful information from it.

To make the sample as representative as possible, you might be tempted to handpick the individuals included in the sample. But the best strategy is to do something quite different: We should select individuals for the sample *at random*.

Idea 2: Randomize

Think back to our soup example. Suppose you add some salt to the pot (the population). If you sample it from the top before stirring, you'll get the misleading idea that the whole pot is salty. If you sample from the bottom, you'll get the equally misleading idea that the whole pot is bland. But by stirring the soup, you **randomize** the amount of salt throughout the pot, making each taste more typical of the saltiness of the whole pot. Deliberate randomization is one of the great tools of Statistics.

Randomization can protect against factors that you aren't aware of, as well as those you know are in the data. Suppose, while you aren't looking, a friend adds a handful of peas to the soup. The peas sink to the bottom of the pot, mixing with the other vegetables. If you don't randomize the soup by stirring, your test spoonful from the top won't have any peas. By stirring in the salt, you *also* randomize the peas throughout the pot, making your sample taste more typical of the overall pot *even though you didn't know the peas were there*. So randomizing protects us by giving us a representative sample even for effects we were unaware of.

For a survey, we "stir" the population by selecting the participants at random. Randomizing protects us from the influences of *all* the features of our population by making sure that *on average* the sample looks like the rest of the population.

The essential feature of randomness that we need is that the selection is "fair." We will discuss many facets of randomness in Chapter 7, but what makes the sample fair is that each participant has an equal chance to be selected.

- **Why not match the sample to the population?** Rather than randomizing, we could try to design our sample to include every possible, relevant characteristic: income level, age, political affiliation, marital status, number of children, place of residence, etc. Clearly we can't possibly think of all the things that might be important. Even if we could, we wouldn't be able to match our sample to the population for all these characteristics.

How well can a sample represent the population from which it was selected? Here's an example using the database of the Paralyzed Veterans of America, a philanthropic organization with a donor list of about 3.5 million people. We've taken two samples, each of 8000 individuals at random from the population. Table 3.1 shows how the means and proportions match up on seven variables.

	A	B	C	D	E	F	G	H
1		Age (yr)	White (%)	Female (%)	# of children	Income Bracket (1-7)	Wealth Bracket (1-9)	Homeowner? (% Yes)
2	Sample 1	61.4	85.12	56.2	1.54	3.91	5.29	71.36
3	Sample 2	61.2	84.44	56.4	1.51	3.88	5.33	72.3

Table 3.1   Means and proportions for seven variables from two samples of size 8000 from the Paralyzed Veterans of America data (created in **Excel**). The fact that the summaries of the variables from these two samples are so similar gives us confidence that either one would be representative of the entire population.

The two samples match closely in every category. You can see how well randomizing has stirred the population. We didn't preselect the samples for these variables, but randomizing has matched the results closely. The two samples don't vary much from each other, so we can assume that they don't differ much from the rest of the population either.

Even if a survey is given to multiple random samples, the samples will differ from each other and so, therefore, will the responses. These sample-to-sample differences are referred to as **sampling error** even though no error has occurred.

Idea 3: The Sample Size Is What Matters

You probably weren't surprised by the idea that a sample can represent the whole. And the idea of sampling randomly to make the sample fair makes sense too. But the third important idea of sampling often surprises people. The third idea is that the *size of the sample* determines what we can conclude from the data *regardless of the size of the population*. Many people think that to provide a good representation of the population, the sample must be a large percentage, or *fraction*, of the population, but in fact all that matters is the size of the sample. The size of the *population* doesn't matter at all.[1] A random sample of 100 students in a college represents the student body just about as well as a random sample of 100 voters represents the entire electorate of the United States. This is perhaps the most surprising idea in designing surveys.

Think about the pot of soup again. If you're cooking for a banquet rather than just for a few people, your pot will be bigger, but you don't need a bigger spoon to decide how the soup tastes. You'll get the same information from an ordinary spoonful that you got before, no matter how large the pot—as long as the pot is sufficiently stirred. That's what randomness does for us. What *fraction* of the population you sample doesn't matter. It's the **sample size** itself that's important. This idea is of key importance to the design of any sample survey, because it determines the balance between how well the survey can measure the population and how much the survey costs.

How big a sample do you need? That depends on what you're estimating, but too small a sample won't be representative of the population. To get an idea of what's really in the soup, you need a large enough taste to be a *representative* sample

---

[1]Well, that's not exactly true. If sample is more than about 10% of the whole population, it *can* matter. It doesn't matter whenever, as usual, our sample is a very small fraction of the population.

from the pot, including, say, a selection of the vegetables. For a survey that tries to find the proportion of the population falling into a category, you'll usually need at least several hundred respondents.[2]

- **What do the professionals do?** How do professional polling and market research companies do their work? The most common polling method today is to contact respondents by telephone. Computers generate random telephone numbers for telephone exchanges known to include residential customers; so pollsters can contact people with unlisted phone numbers. The person who answers the phone will be invited to respond to the survey—if that person qualifies. (For example, only adults are usually surveyed, and the respondent usually must live at the residence phoned.) If the person answering doesn't qualify, the caller will ask for an appropriate alternative. When they conduct the interview, the pollsters often list possible responses (such as product names) in randomized orders to avoid biases that might favor the first name on the list.

    Do these methods work? The Pew Research Center for the People and the Press, reporting on one survey, says that

    > *Across five days of interviewing, surveys today are able to make some kind of contact with the vast majority of households (76%), and there is no decline in this contact rate over the past seven years. But because of busy schedules, skepticism and outright refusals, interviews were completed in just 38% of households that were reached using standard polling procedures.*

    Nevertheless, studies indicate that those actually sampled can give a good snapshot of larger populations from which the surveyed households were drawn.

A Census—Does It Make Sense?

Why bother determining the right sample size? If you plan to open a store in a new community, why draw a sample of residents to understand their interests and needs? Wouldn't it be better to just include everyone and make the "sample" be the entire population? Such a special sample is called a **census**. Although a census would appear to provide the best possible information about the population, there are a number of reasons why it might not.

First, it can be difficult to complete a census. There always seem to be some individuals who are hard to locate or hard to measure. Do you really need to contact the folks away on vacation when you collect your data? How about those with no telephone or mail address? The cost of locating the last few cases may far exceed the budget. It can also be just plain impractical to take a census. The quality control manager for Hostess® Twinkies® doesn't want to taste *all* the Twinkies on the production line to determine their quality. Aside from the fact that nobody could eat that many Twinkies, it would defeat their purpose: There would be none left to sell.

Second, the population you're studying may change. For example, in any human population, babies are born, people travel, and folks die during the time it takes to complete the census. News events and advertising campaigns can cause sudden shifts in opinions and preferences. A sample, surveyed in a shorter time frame, may actually generate more accurate information.

Finally, taking a census can be cumbersome. A census usually requires a team of pollsters and the cooperation of the population. Even with both, it's almost impossible to avoid errors. Because it tries to count everyone, the U.S. Census records too many college students. Many are included both by their families and in a report filed by their schools. Errors of this sort, of both under- and overcounting can be found throughout the U.S. Census.

---

[2]Chapter 10 gives the details behind this statement and shows how to decide on a sample size for a survey.

For Example   Identifying sampling terms

A nonprofit organization has taken over the historic State Theater and hopes to preserve it with a combination of attractive shows and fundraising. The organization has asked a team of students to help them design a survey to better understand the customer base likely to purchase tickets. Fortunately, the theater's computerized ticket system records contact and some demographic information for ticket purchasers, and that database of 7345 customers is available.

Questions: What is the population of interest?

What would a census be in this case? Would it be practical?

What is the sampling frame?

Answers: The population is all potential ticket purchasers.

A census would have to reach all potential purchasers. We don't know who they are or have any way to contact them.

The sampling frame is the list of previous ticket purchasers.

3.2   Populations and Parameters

GfK Roper Reports Worldwide reports that 60.5% of people over 50 worry about food safety, but only 43.7% of teens do. What does this claim mean? We can be sure the Roper researchers didn't take a census. So they can't possibly know *exactly* what percentage of teenagers worry about food safety. So what does "43.7%" mean?

To generalize from a sample to the world at large, we need a model of reality. Such a model doesn't need to be complete or perfect. Just as a model of an airplane in a wind tunnel can tell engineers what they need to know about aerodynamics even though it doesn't include every rivet of the actual plane, models of data can give us summaries that we can learn from and use even though they don't fit each data value exactly. It's important to remember that they're only models of reality and not reality itself. But without models, what we can learn about the world at large is limited to only what we can say about the data we have at hand.

Models use mathematics to represent reality. We call the key numbers in those models **parameters**. Sometimes a parameter used in a model for a population is called (redundantly) a **population parameter**.

But let's not forget about the data. We use the data to try to estimate values for the population parameters. Any summary found from the data is a **statistic**. Those statistics that estimate population parameters are particularly interesting. Sometimes—and especially when we match statistics with the parameters they estimate—we use the term **sample statistic**.

We draw samples because we can't work with the entire population. We hope that the statistics we compute from the sample will estimate the corresponding parameters accurately. A sample that does this is said to be **representative**.

> **Statistic**
>
> Any quantity that we calculate from data could be called a "statistic." But in practice, we usually obtain a statistic from a sample and use it to estimate a population parameter.

> **Parameter**
>
> Population model parameters are not just unknown—usually they are *unknowable*. We take a sample and use the sample statistics to estimate them.

Just Checking

1 Various claims are often made for surveys. Why is each of the following claims not correct?

  a) It is always better to take a census than to draw a sample.

  b) Stopping customers as they are leaving a restaurant is a good way to sample opinions about the quality of the food.

  c) We drew a sample of 100 from the 3000 students in a school. To get the same level of precision for a town of 30,000 residents, we'll need a sample of 1000.

  d) A poll taken at a popular website (www.statsisfun.org) garnered 12,357 responses. The majority of respondents said they enjoy doing Statistics. With a sample size that large, we can be sure that most Americans feel this way.

  e) The true percentage of all Americans who enjoy Statistics is called a "population statistic."

3.3   Common Sampling Designs

We've said that every individual in the population should have an equal chance of being selected in a sample. That makes the sample fair, but it's not quite enough to ensure that the sample is representative. Consider, for example, a market analyst who samples customers by drawing at random from product registration forms, half of which arrived by mail and half by online registration. She flips a coin. If it comes up heads, she'll draw 100 mail returns; tails, she'll draw 100 electronic returns. Each customer has an equal chance of being selected, but if tech-savvy customers are different, then the samples are hardly representative.

Simple Random Sample (SRS)

To make the sample representative, we must ensure that our sampling method gives each *combination* of individuals an equal chance as well. A sample drawn in this way is called a **simple random sample**, usually abbreviated **SRS**. An SRS is the sampling method on which the theory of working with sampled data is based and thus the standard against which we measure other sampling methods.

We'd like to select from the population, but often we don't have a list of all the individuals in the population. The list we actually draw from is called a **sampling frame**. A store may want to survey all its regular customers. But it can't draw a sample from the population of all regular customers, because it doesn't have such a list. The store may have a list of customers who have registered as "frequent shoppers." That list can be the sampling frame from which the store can draw its sample.

Of course, whenever the sampling frame and the population differ (as they almost always will), we must deal with the differences. Are the opinions of those who registered as frequent shoppers different from the rest of the regular shoppers? What about customers who used to be regulars but haven't shopped there recently? The answers to questions like these about the sampling frame may depend on the purpose of the survey and may impact the conclusions that one can draw.

Once we have a sampling frame, we need to *randomize* it so we can choose an SRS. Fortunately, random numbers are readily available these days in spreadsheets, statistics programs, and even on the Internet. Before this technology existed, people used to literally draw numbers out of a hat to randomize. But now, the easiest way to randomize your sampling frame is to match it with a parallel list of random numbers and then sort the random numbers, carrying along the cases so that they get "shuffled" into random order. Then you can just pick cases off the top of the randomized list until you have enough for your sample.

Samples drawn at random generally differ one from another. If we were to repeat the sampling process, a new draw of random numbers would select different people for our sample. These differences would lead to different values for the variables we measure. We call these sample-to-sample differences **sampling variability**. Surprisingly, sampling variability isn't a problem; it's an opportunity. If different samples from a population vary little from each other, then most likely the underlying population harbors little variation. If the samples show much sampling variability, the underlying population probably varies a lot. In the coming chapters, we'll spend much time and attention working with sampling variability to better understand what we are trying to measure.

f_x	=RAND()

C	D
0.750306	

A Different Answer Every Time?
The RAND() function in Excel can take you by surprise. Every time the spreadsheet reopens, you get a new column of random numbers. But don't worry. Once you've shuffled the rows, you can ignore the new numbers. The order you got by shuffling won't keep changing.

For Example   Choosing a random sample

Continuing the example on page 30, the student consultants select 200 ticket buyers at random from the database. First, the State Theater database is placed in a spreadsheet. Next, to draw random numbers, the students use the Excel command RAND(). (They type =*RAND()* in the top cell of a column next to the data and then use *Fill Down* to populate the column down to the bottom.) They then sort the spreadsheet to put the random column in order and select ticket buyers from the top of the randomized

(continued)

spreadsheet until they complete 200 interviews. This makes it easy to select more respondents when (as always happens) some of the people they select can't be reached by telephone or decline to participate.

Questions: What is the sampling frame?

If the customer database held 30,000 records instead of 7345, how much larger a sample would we need to get the same information?

If we then draw a different sample of 200 customers and obtain different answers to the questions on the survey, how do we refer to these differences?

Answers: The sampling frame is the customer database.

The size of the sample is all that matters, not the size of the population. We would need a sample of 200.

The differences in the responses are called sampling error, or sampling variability.

Sampling Errors vs. Bias

We referred to sample-to-sample variability earlier in this chapter as *sampling error*, making it sound like it's some kind of mistake. It's not. We understand that samples will vary, so "sampling errors" are to be expected. It's *bias* we must strive to avoid. Bias means our sampling method distorts our view of the population. Of course, bias leads to mistakes. Even more insidious, bias introduces errors that we cannot correct with subsequent analysis.

Simple random sampling is not the only fair way to sample. More complicated designs may save time or money or avert sampling problems. All statistical sampling designs have in common the idea that chance, rather than human choice, is used to select the sample.

Stratified Sampling

Designs that are used to sample from large populations—especially populations residing across large areas—are often more complicated than simple random samples. Sometimes we slice the population into homogeneous groups, called **strata,** and then use simple random sampling within each stratum, combining the results at the end. This is called **stratified random sampling.**

Why would we want to stratify? Suppose we want to survey how shoppers feel about a potential new anchor store at a large suburban mall. The shopper population is 60% women and 40% men, and we suspect that men and women have different views on their choice of anchor stores. If we use simple random sampling to select 100 people for the survey, we could end up with 70 men and 30 women or 35 men and 65 women. Our resulting estimates of the attractiveness of a new anchor store could vary widely. To help reduce this sampling variability, we can force a representative balance, selecting 40 men at random and 60 women at random. This would guarantee that the proportions of men and women within our sample match the proportions in the population, and that should make such samples more accurate in representing population opinion.

You can imagine that stratifying by race, income, age, and other characteristics can be helpful, depending on the purpose of the survey. When we use a sampling method that restricts by strata, additional samples are more like one another, so statistics calculated for the sampled values will vary less from one sample to another. This reduced sampling variability is the most important benefit of stratifying, but the analysis of data sampled with these designs is beyond the scope of our book.

Cluster and Multistage Sampling

Sometimes dividing the sample into homogeneous strata isn't practical, and even simple random sampling may be difficult. For example, suppose we wanted to assess the reading level of a product instruction manual based on the length of the sentences. Simple random sampling could be awkward; we'd have to number each sentence and then find, for example, the 576th sentence or the 2482nd sentence, and so on. Doesn't sound like much fun, does it?

We could make our task much easier by picking a few pages at random and then counting the lengths of the sentences on those pages. That's easier than picking

Strata or Clusters?

We create strata by dividing the population into groups of similar individuals so that each stratum is different from the others. (For example, we often stratify by age, race, or sex.) By contrast, we create clusters that all look pretty much alike, each representing the wide variety of individuals seen in the population.

individual sentences and works if we believe that the pages are all reasonably similar to one another in terms of reading level. Splitting the population in this way into parts or **clusters** that each represent the population can make sampling more practical. We select one or a few clusters at random and perform a census within each of them. This sampling design is called **cluster sampling.** If each cluster fairly represents the population, cluster sampling will generate an unbiased sample.

What's the difference between cluster sampling and stratified sampling? We stratify to ensure that our sample represents different groups in the population, and sample randomly within each stratum. This reduces the sample-to-sample variability. Strata are homogeneous, but differ from one another. By contrast, clusters are more or less alike, each heterogeneous and resembling the overall population. We cluster to save money or even to make the study practical.

Sometimes we use a variety of sampling methods together. In trying to assess the reading level of our instruction manual, we might worry that the "quick start" instructions are easy to read, but the "troubleshooting" chapter is more difficult. If so, we'd want to avoid samples that selected heavily from any one chapter. To guarantee a fair mix of sections, we could randomly choose one section from each chapter of the manual. Then we would randomly select a few pages from each of those sections. If altogether that made too many sentences, we might select a few sentences at random from each of the chosen pages. So, what is our sampling strategy? First we stratify by the chapter of the manual and randomly choose a section to represent each stratum. Within each selected section, we choose pages as clusters. Finally, we consider an SRS of sentences within each cluster. Sampling schemes that combine several methods are called **multistage samples**. Most surveys conducted by professional polling organizations and market research firms use some combination of stratified and cluster sampling as well as simple random samples.

For Example   Identifying more complex designs

The theater board wants to encourage people to come from out of town to attend theater events. They know that, in general, about 40% of ticket buyers are from out of town. These customers often purchase dinner at a local restaurant or stay overnight in a local inn, generating business for the town. The board hopes this information will encourage local businesses to advertise in the theater program, so they want to be sure out-of-town customers are represented in the samples. The database includes zip codes. The student consultants decide to sample 80 ticket buyers from zip codes outside the town and 120 from the town's zip code.

Questions: What kind of sampling scheme are they using to replace the simple random sample?

What are the advantages of selecting 80 out of town and 120 local customers?

Answers: A stratified sample, consisting of a sample of 80 out-of-town customers and a sample of 120 local customers.

By stratifying, they can guarantee that 40% of the sample is from out of town, reflecting the overall proportions among ticket buyers. If out-of-town customers differ in important ways from local ticket buyers, a stratified sample will reduce the variation in the estimates for each group so that the combined estimates can be more precise.

Systematic Samples

Sometimes we draw a sample by selecting individuals systematically. For example, a **systematic sample** might select every tenth person on an alphabetical list of employees. To make sure our sample is random, we still must start the systematic selection with a randomly selected individual—not necessarily the first person on the list. When there is no reason to believe that the order of the list could be associated in any way with the responses measured, systematic sampling can give a

Just Checking

2 We need to survey a random sample of the 300 passengers on a flight from San Francisco to Tokyo. Name each sampling method described below.

 a) Pick every tenth passenger as people board the plane.

 b) From the boarding list, randomly choose five people flying first class and 25 of the other passengers.

 c) Randomly generate 30 seat numbers and survey the passengers who sit there.

 d) Randomly select a seat position (right window, right center, right aisle, etc.) And survey all the passengers sitting in those seats.

representative sample. Systematic sampling can be much less expensive than true random sampling. When you use a systematic sample, you should justify the assumption that the systematic method is not associated with any of the measured variables.

Think about the reading level sampling example again. Suppose we have chosen a section of the manual at random, then three pages at random from that section, and now we want to select a sample of 10 sentences from the 73 sentences found on those pages. Instead of numbering each sentence so we can pick a simple random sample, it would be easier to sample systematically. A quick calculation shows $73/10 = 7.3$, so we can get our sample by picking every seventh sentence on the page. But where should you start? At random, of course. We've accounted for $10 \times 7 = 70$ of the sentences, so we'll throw the extra three into the starting group and choose a sentence at random from the first 10. Then we pick every seventh sentence after that and record its length.

Guided Example   Market Demand Survey

In a course at a business school in the United States, the students form business teams, propose a new product, and use seed money to launch a business to sell the product on campus.

Before committing funds for the business, each team must complete the following assignment: "Conduct a survey to determine the potential market demand on campus for the product you are proposing to sell." Suppose your team's product is a 500-piece jigsaw puzzle of the map of your college campus. Design a marketing survey and discuss the important issues to consider.

PLAN

Setup State the goals and objectives of the survey.

Our team designed a study to find out how likely students at our school are to buy our proposed product—a 500-piece jigsaw puzzle of the map of our college campus.

Population and Parameters Identify the population to be studied and the associated sampling frame. What are the parameters of interest?

The population studied will be students at our school. We have obtained a list of all students currently enrolled to use as the sampling frame. The parameter of interest is the proportion of students likely to buy this product. We'll also collect some demographic information about the respondents.

Sampling Plan Specify the sampling method and the sample size, n. Specify how the sample was actually drawn. What is the sampling frame?

The description should, if possible, be complete enough to allow someone to replicate the procedure, drawing another sample from the same population in the same manner. A good description of the procedure is essential, even if it could never practically be repeated. The question you ask is important, so state the wording of the question clearly. Be sure that the question is useful in helping you with the overall goal of the survey.

We will select a simple random sample of students. We decided against stratifying by sex or class because we thought that students were all more or less alike in their likely interest in our product.

We will ask the students we contact:

Do you solve jigsaw puzzles for fun?

Then we will show them a prototype puzzle and ask:

If this puzzle sold for $10, would you purchase one?

We will also record the respondent's sex and class.

DO

Sampling Practice Specify *when*, *where*, and *how* the sampling will be performed. Specify any other details of your survey, such as how respondents were contacted, any incentives that were offered to encourage them to respond, how nonrespondents were treated, and so on.

The survey will be administered in the middle of the fall semester during October. We have a master list of registered students, which we will randomize by matching it with random numbers from www.random.org and sorting on the random numbers, carrying the names. We will contact selected students by phone or e-mail and arrange to meet with them. If a student is unwilling to participate, the next name from the randomized list will be substituted until a sample of 200 participants is found.

We will meet with students in an office set aside for this purpose so that each will see the puzzle under similar conditions.

REPORT

Summary and Conclusion This report should include a discussion of all the elements needed to design the study. It's good practice to discuss any special circumstances or other issues that may need attention.

MEMO

Re: Survey Plans

Our team's plans for the puzzle market survey call for a simple random sample of students. Because subjects need to be shown the prototype puzzle, we must arrange to meet with selected participants. We have arranged an office for that purpose.

We will also collect demographic information so we can determine whether there is in fact a difference in interest level among classes or between men and women.

The Real Sample

We have been discussing sampling in a somewhat idealized setting. In the real word, things can be a bit messier. Here are some things to consider.

The population may not be as well-defined as it seems. For example, if a company wants the opinions of a typical mall "shopper," who should they sample? Should they only ask shoppers carrying a purchase? Should they include people eating at the food court? How about teenagers just hanging out by the mall's video store? Even when the population is clear, it may not be possible to establish an appropriate sampling frame.

Usually, the practical sampling frame is not the group you *really* want to know about. For example, election polls want to sample from those who will actually vote in the next election—a group that is particularly tricky to identify before election day. The sampling frame limits what your survey can find out.

Then there's your target sample. These are the individuals selected according to your sample design for whom you *intend* to measure responses. You're not likely to get responses from all of them. ("I know it's dinner time, but I'm sure you wouldn't mind answering a few questions. It'll only take 20 minutes or so. Oh, you're busy?") Nonresponse is a problem in many surveys.

Sample designs are usually about the target sample. But in the real world, you won't get responses from everyone your design selects. So in reality, your sample consists of the actual respondents. These are the individuals about whom you *do* get data and can draw conclusions. Unfortunately, they might not be representative of either the sampling frame or the population.

At each step, the group we can study may be constrained further. The *who* of our study keeps changing, and each constraint can introduce biases. A careful study should address the question of how well each group matches the population of interest. The *who* in an SRS is the population of interest from which we've drawn a representative sample. That's not always true for other kinds of samples.

When people (or committees!) decide on a survey, they often fail to think through the important questions about who are the *who* of the study and whether they are the individuals about whom the answers would be interesting or have meaningful business consequences. This is a key step in performing a survey and should not be overlooked.

3.4 The Valid Survey

It isn't sufficient to draw a sample and start asking questions. You want to feel confident your survey can yield the information you need about the population you are interested in. We want a *valid survey*.

To help ensure a valid survey, you need to ask four questions:

- What do I want to know?
- Who are the right respondents?
- What are the right questions?
- What will be done with the results?

These questions may seem obvious, but there are a number of specific pitfalls to avoid:

Know what you want to know. Far too often, decisionmakers decide to perform a survey without any clear idea of what they hope to learn. Before considering a survey, you must be clear about what you hope to learn and what population you want to learn about. If you don't know that, you can't even judge whether you have a valid survey. The survey *instrument*—the questionnaire itself—can be a source of errors. Perhaps the most common error is to ask unnecessary questions. The longer the survey, the fewer people will complete it, leading to greater nonresponse bias. For each question on your survey, you should ask yourself whether you really want to know this and what you would do with the responses if you had them. If you don't have a good use for the answer to a question, don't ask it.

Use the right sampling frame. A valid survey obtains responses from appropriate respondents. Be sure you have a suitable sampling frame. Have you identified the population of interest and sampled from it appropriately? A company looking to expand its base might survey customers who returned warrantee registration cards—after all, that's a readily available sampling frame—but if the company wants to know how to make its product more attractive, it needs to survey customers who rejected its product in favor of a competitor's product. This is the population that can tell the company what about its product needs to change to capture a larger market share. The errors in the presidential election polls of 1948 were likely due to the use of telephone samples in an era when telephones were not affordable by the less affluent—who were the folks most likely to vote for Truman.

It is equally important to be sure that your respondents actually know the information you hope to discover. Your customers may not know much about the competing products, so asking them to compare your product with others may not yield useful information.

Ask specific rather than general questions. It is better to be specific. "Do you usually recall TV commercials?" won't be as useful as "How many TV commercials can you recall from last night?" or better, yet, "Please describe for me all the TV commercials you can recall from your viewing last night."

Watch for biases. Even with the right sampling frame, you must beware of bias in your sample. If customers who purchase more expensive items are less likely to respond to your survey, this can lead to **nonresponse bias**. Although you can't expect all mailed surveys to be returned, if those individuals who don't respond have common characteristics, your sample will no longer represent the population you hope to learn about. Surveys in which respondents volunteer to participate, such as online surveys, suffer from **voluntary response bias**. Individuals with the strongest feelings on either side of an issue are more likely to respond; those who don't care may not bother.

Be careful with question phrasing. Questions must be carefully worded. A respondent may not understand the question—or may not understand the question the way the researcher intended it. For example, "Does anyone in your family own a Ford truck?" leaves the term "family" unclear. Does it include only spouses and children or parents and siblings, or do in-laws and second cousins count too? A question like "Was your Twinkie fresh?" might be interpreted quite differently by different people.

Be careful with answer phrasing. Respondents and survey-takers may also provide inaccurate responses, especially when questions are politically or sociologically sensitive. This also applies when the question does not take into account all possible answers, such as a true-false or multiple-choice question to which there may be other answers. Or the respondent may not know the correct answer to the question on the survey. In 1948, there were four major candidates for President,[3] but some

---

[3] Harry Truman, Thomas Dewey, Strom Thurmond, and Henry Wallace.

survey respondents might not have been able to name them all. A survey question that just asked "Who do you plan to vote for?" might have underrepresented the less prominent candidates. And one that just asked "What do you think of Wallace?" might yield inaccurate results from voters who simply didn't know who he was. We refer to inaccurate responses (intentional or unintentional) as **measurement errors**. One way to cut down on measurement errors is to provide a range of possible responses. But be sure to phrase them in neutral terms.

The best way to protect a survey from measurement errors is to perform a pilot test. In a **pilot test,** a small sample is drawn from the sampling frame, and a draft form of the survey instrument is administered. A pilot test can point out flaws in the instrument. For example, during a staff cutback at one of our schools, a researcher surveyed faculty members to ask how they felt about the reduction in staff support. The scale ran from "It's a good idea" to "I'm very unhappy." Fortunately, a pilot study showed that everyone was very unhappy or worse. The scale was retuned to run from "unhappy" to "ready to quit."

For Example   Survey design

A nonprofit organization has enlisted some student consultants to help design a fundraising survey. The student consultants suggest to the board of directors that they may want to rethink their survey plans. They point out that there are differences between the population, the sampling frame, the target sample contacted by telephone, and the actual sample.

Question: How do the population, sampling frame, target sample, and sample differ?

Answer: The population is all potential ticket buyers.

The sampling frame is only those who have previously purchased tickets. Anyone who wasn't attracted to previous productions wouldn't be surveyed. That could keep the board from learning of ways to make the theater's offering more attractive to those who hadn't purchased tickets before.

The target sample is those selected from the database who can be contacted by telephone. Those with unlisted numbers or who had declined to give their phone number can't be contacted. It may be more difficult to contact those with caller ID.

The actual sample will be those previous customers selected at random from the database who can be reached by telephone and who agree to complete the survey.

3.5   How to Sample Badly

Bad sample designs yield worthless data. Many of the most convenient forms of sampling can be seriously biased. And there is no way to correct for the bias from a bad sample. So it's wise to pay attention to sample design—and to beware of reports based on poor samples.

Voluntary Response Sample

One of the most common dangerous sampling methods is the voluntary response sample. In a **voluntary response sample,** a large group of individuals is invited to respond, and all who do respond are counted. This method is used by call-in shows, 900 numbers, Internet polls, and letters written to members of Congress. Voluntary response samples are almost always biased, and so conclusions drawn from them are almost always wrong.

It's often hard to define the sampling frame of a voluntary response study. Practically, the frames are groups such as Internet users who frequent a particular website or viewers of a particular TV show. But those sampling frames don't correspond to the population you are likely to be interested in.

Even if the sampling frame is of interest, voluntary response samples are often biased toward those with strong opinions or those who are strongly motivated—and especially from those with strong negative opinions. A request that travelers who have used the local airport visit a survey site to report on their experiences is much more likely to hear from those who had long waits, cancelled flights, and lost luggage than from those whose flights were on time and carefree. The resulting voluntary response bias invalidates the survey.

Convenience Sampling

Another sampling method that doesn't work is convenience sampling. As the name suggests, in **convenience sampling** we simply include the individuals who are convenient. Unfortunately, this group may not be representative of the population. A survey of 437 potential home buyers in Orange County, California, found, among other things, that

> *all but 2 percent of the buyers have at least one computer at home, and 62 percent have two or more. Of those with a computer, 99 percent are connected to the Internet (Jennifer Hieger, "Portrait of Homebuyer Household: 2 Kids and a PC," Orange County Register, July 27, 2001).*

Later in the article, we learn that the survey was conducted via the Internet. That was a convenient way to collect data and surely easier than drawing a simple random sample, but perhaps home builders shouldn't conclude from this study that *every* family has a computer and an Internet connection.

Many surveys conducted at shopping malls suffer from the same problem. People in shopping malls are not necessarily representative of the population of interest. Mall shoppers tend to be more affluent and include a larger percentage of teenagers and retirees than the population at large. To make matters worse, survey interviewers tend to select individuals who look "safe," or easy to interview.

Convenience sampling is not just a problem for beginners. In fact, convenience sampling is a widespread problem in the business world. When a company wants to find out what people think about its products or services, it may turn to the easiest people to sample: its own customers. But the company will never learn how those who *don't* buy its product feel about it.

Bad Sampling Frame?

An SRS from an incomplete sampling frame introduces bias because the individuals included may differ from the ones not in the frame. It may be easier to sample workers from a single site, but if a company has many sites and they differ in worker satisfaction, training, or job descriptions, the resulting sample can be biased. There is serious concern among professional pollsters that the increasing numbers of people who can be reached only by cell phone may bias telephone-based market research and polling.

Undercoverage

Many survey designs suffer from **undercoverage,** in which some portion of the population is not sampled at all or has a smaller representation in the sample than it has in the population. Undercoverage can arise for a number of reasons, but it's always a potential source of bias. Are people who use answering machines to screen callers (and are thus less available to blind calls from market researchers) different from other customers in their purchasing preferences?

Do you use the Internet?
Click here ○ for yes
Click here ○ for no

Internet Surveys

Internet convenience surveys are often worthless. As voluntary response surveys, they have no well-defined sampling frame (all those who use the Internet and visit their site?) and thus report no useful information. Do not use them.

| For Example | Common mistakes in survey design |

A board member proposes that rather than telephoning past customers, they simply post someone at the door to ask theater goers their opinions. Another suggests that it would be even easier to post a questionnaire on the theater website and invite responses there. A third suggests that rather than working with random numbers, they simply phone every 200th person on the list of past customers.

Question: Identify the three methods proposed and explain what strengths and weaknesses they have.

Answer: Questioning customers at the door would be a convenience sample. It would be cheap and fast but is likely to be biased by the nature and quality of the particular performance where the survey takes place.

Inviting responses on the website would be a voluntary response sample. Only customers who frequented the website and decided to respond would be surveyed. This might, for example, underrepresent older customers or those without home Internet access.

Sampling every 200th name from the customer list would be a systematic sample. It is slightly easier than randomizing. If the order of names on the list is unrelated to any questions asked, then this might be an acceptable method. But if, for example, the list is kept in the order of first purchases (when a customer's name and information were added to the database), then there might be a relationship between opinions and location on the list.

What Can Go Wrong?

- **Nonrespondents.** No survey succeeds in getting responses from everyone. The problem is that those who don't respond may differ from those who do. And if they differ on just the variables we care about, the lack of response will bias the results. Rather than sending out a large number of surveys for which the response rate will be low, it is often better to design a smaller, randomized survey for which you have the resources to ensure a high response rate.

- **Long, dull surveys.** Surveys that are too long are more likely to be refused, reducing the response rate and biasing all the results. Keep it short.

- **Response bias.** Response bias includes the tendency of respondents to tailor their responses to please the interviewer and the consequences of slanted question wording.

- **Push polls.** Push polls, which masquerade as surveys, present one side of an issue before asking a question. For example, a question like

  Would the fact that the new store that just opened by the mall sells mostly goods made overseas by workers in sweatshop conditions influence your decision to shop there rather than in the downtown store that features American-made products?

  is designed not to gather information, but to spread ill-will toward the new store.

THE WIZARD OF ID   parker and hart

How to Think about Biases

- **Look for biases in any survey.** If you design a survey of your own, ask someone else to help look for biases that may not be obvious to you. Do this *before* you collect your data. There's no way to recover from a biased sample or a survey that asks biased questions.

  A bigger sample size for a biased study just gives you a bigger useless study. A really big sample gives you a really big useless study.

- **Spend your time and resources reducing biases.** No other use of resources is as worthwhile as reducing the biases.

- **If you possibly can, pretest or pilot your survey.** Administer the survey in the exact form that you intend to use it to a small sample drawn from the population you intend to sample. Look for misunderstandings, misinterpretation, confusion, or other possible biases. Then redesign your survey instrument.

- **Always report your sampling methods in detail.** Others may be able to detect biases where you did not expect to find them.

Ethics in Action

The Lackawax River Group is interested in applying for state funds in order to continue their restoration and conservation of the Lackawax River, a river that has been polluted from years of industry and agricultural discharge. While they have managed to gain significant support for their cause through education and community involvement, the executive committee is now interested in presenting the state with more compelling evidence. They decided to survey local residents regarding their attitudes toward the proposed expansion of the river restoration and conservation project. With limited time and money (the deadline for the grant application was fast approaching), the executive committee was delighted that one of its members, Harry Greentree, volunteered to undertake the project. Harry owned a local organic food store and agreed to have a sample of his shoppers interviewed during the next one-week period. The only concern that the committee had was that the shoppers be selected in a systematic fashion, for instance, by interviewing every fifth person who entered the store. Harry had no problem with this request and was eager to help the Lackawax River Group.

ETHICAL ISSUE *Introducing bias into the results (even if not intentional). One might expect consumers of organic food to be more concerned about the environment than the general population (related to Item C, ASA Ethical Guidelines).*

ETHICAL SOLUTION *Harry is using a convenience sample from which results cannot be generalized. If the Lackawax River Group cannot improve their sampling scheme and survey design (for example, for lack of expertise or time), they should openly discuss the weaknesses of their sampling method when they disclose details of their study. When reporting the results, they should note that their findings are from a convenience sample and include an appropriate disclaimer.*

What Have We Learned?

Learning Objectives

- Know the three ideas of sampling.
  - Examine a part of the whole: A sample can give information about the population.
  - Randomize to make the sample representative.
  - The sample size is what matters. It's the size of the sample—and not its fraction of the larger population—that determines the precision of the statistics it yields.

- Be able to draw a Simple Random Sample (SRS) using a table of random digits or a list of random numbers from technology or an Internet site.
  - In a **simple random sample** (SRS), every possible group of *n* individuals has an equal chance of being our sample.

- Know the definitions of other sampling methods:
  - **Stratified samples** can reduce sampling variability by identifying homogeneous subgroups and then randomly sampling within each.
  - **Cluster samples** randomly select among heterogeneous subgroups that each resemble the population at large, making our sampling tasks more manageable.
  - **Systematic samples** can work in some situations and are often the least expensive method of sampling. But we still want to start them randomly.
  - **Multistage samples** combine several random sampling methods.

- Identify and avoid causes of bias.
  - **Nonresponse bias** can arise when sampled individuals will not or cannot respond.
  - **Response bias** arises when respondents' answers might be affected by external influences, such as question wording or interviewer behavior.
  - **Voluntary response samples** are almost always biased and should be avoided and distrusted.
  - **Convenience samples** are likely to be flawed for similar reasons.
  - **Undercoverage** occurs when individuals from a subgroup of the population are selected less often than they should be.

Terms

Bias	Any systematic failure of a sampling method to represent its population.
Census	An attempt to collect data on the entire population of interest.
Cluster	A representative subset of a population chosen for reasons of convenience, cost, or practicality.
Cluster sampling	A sampling design in which groups, or clusters, representative of the population are chosen at random and a census is then taken of each.
Convenience sampling	A sample that consists of individuals who are conveniently available.
Measurement error	Any inaccuracy in a response, from any source, whether intentional or unintentional.
Multistage sample	A sampling scheme that combines several sampling methods.
Nonresponse bias	Bias introduced to a sample when a large fraction of those sampled fails to respond.
Parameter	A numerically valued attribute of a model for a population. We rarely expect to know the value of a parameter, but we do hope to estimate it from sampled data.
Pilot test	A small trial run of a study to check that the methods of the study are sound.
Population	The entire group of individuals or instances about whom we hope to learn.

Population parameter	A numerically valued attribute of a model for a population.
Randomization	A defense against bias in the sample selection process, in which each individual is given a fair, random chance of selection.
Representative sample	A sample from which the statistics computed accurately reflect the corresponding population parameters.
Response bias	Anything in a survey design that influences responses.
Sample	A subset of a population, examined in hope of learning about the population.
Sample size	The number of individuals in a sample.
Sample survey	A study that asks questions of a sample drawn from some population in the hope of learning something about the entire population.
Sampling frame	A list of individuals from which the sample is drawn. Individuals in the population of interest but who are not in the sampling frame cannot be included in any sample.
Sampling variability (or sampling error)	The natural tendency of randomly drawn samples to differ, one from another.
Simple random sample (SRS)	A sample in which each set of n elements in the population has an equal chance of selection.
Statistic, sample statistic	A value calculated for sampled data, particularly one that corresponds to, and thus estimates, a population parameter. The term "sample statistic" is sometimes used, usually to parallel the corresponding term "population parameter."
Strata	Subsets of a population that are internally homogeneous but may differ one from another.
Stratified random sample	A sampling design in which the population is divided into several homogeneous subpopulations, or strata, and random samples are then drawn from each stratum.
Systematic sample	A sample drawn by selecting individuals systematically from a sampling frame.
Voluntary response bias	Bias introduced to a sample when individuals can choose on their own whether to participate in the sample.
Voluntary response sample	A sample in which a large group of individuals are invited to respond and decide individually whether or not to participate. Voluntary response samples are generally worthless.
Undercoverage	A sampling scheme that biases the sample in a way that gives a part of the population less representation than it has in the population.

Technology Help: Random Sampling XLSTAT

Computer-generated pseudorandom numbers are usually quite good enough for drawing random samples. But there is little reason not to use the truly random values available on the Internet. Here's a convenient way to draw an SRS of a specified size using a computer-based sampling frame. The sampling frame can be a list of names or of identification numbers arrayed, for example, as a column in a spreadsheet, statistics program, or database:

1. Generate random numbers of enough digits so that each exceeds the size of the sampling frame list by several digits. This makes duplication unlikely. (For example, in Excel, use the RAND function to fill a column with random numbers between 0 and 1. With many digits they will almost surely be unique.)

2. Assign the random numbers arbitrarily to individuals in the sampling frame list. For example, put them in an adjacent column.

3. Sort the list of random numbers, *carrying* along the sampling frame list.

4. Now the first n values in the sorted sampling frame column are an SRS of n values from the entire sampling frame.

Brief CASE

Market Survey Research

You are part of a marketing team that needs to research the potential of a new product. Your team decides to e-mail an interactive survey to a random sample of consumers. Write a short questionnaire that will generate the information you need about the new product. Select a sample of 200 using an SRS from your sampling frame. Discuss how you will collect the data and how the responses will help your market research.

The GfK Roper Reports Worldwide Survey

GfK Roper Consulting conducts market research for multinational companies who want to understand attitudes in different countries so they can market and advertise more effectively to different cultures. Every year they conduct a poll worldwide, which asks hundreds of questions of people in approximately 30 different countries. Respondents are asked a variety of questions about food. Some of the questions are simply yes/no (agree/disagree) questions: Please tell me whether you agree or disagree with each of these statements about your appearance: (Agree = 1; Disagree = 2; Don't know = 9).

The way you look affects the way you feel.

I am very interested in new skin care breakthroughs.

People who don't care about their appearance don't care about themselves.

Other questions are asked on a 5-point scale (Please tell me the extent to which you disagree or agree with it using the following scale: Disagree completely = 1; Disagree somewhat = 2; Neither disagree nor agree = 3; Agree somewhat = 4; Agree completely = 5; Don't know = 9).

Examples of such questions include:

I read labels carefully to find out about ingredients, fat content, and/or calories.

I try to avoid eating fast food.

When it comes to food I'm always on the lookout for something new.

Think about designing a survey on such a global scale:

- What is the population of interest?
- Why might it be difficult to select an SRS from this sampling frame?
- What are some potential sources of bias?
- Why might it be difficult to ensure a representative number of men and women and all age groups in some countries?
- What might be a reasonable sampling frame?

Exercises

SECTION 3.1

1. Indicate whether each statement below is true or false. If false, explain why.

a) We can eliminate sampling error by selecting an unbiased sample.

b) Randomization helps to ensure that our sample is representative.

c) Sampling error refers to sample-to-sample differences and is also known as sampling variability.

d) It is better to try to match the characteristics of the sample to the population rather than relying on randomization.

2. Indicate whether each statement below is true or false. If false, explain why.

a) To get a representative sample, you must sample a large fraction of the population.

b) Using modern methods, it is best to select a representative subset of a population systematically.

c) A census is the only true representative sample.

d) A random sample of 100 students from a school with 2000 students has the same precision as a random sample of 100 from a school with 20,000 students.

SECTION 3.2

3. A consumer advocacy group is interested in gauging perceptions about food safety among professionals in the food industry. Specifically, they wish to determine the percentage of professional food preparers in the United States that believe food safety has improved. They use an alphabetized list of members of the organization *Chef's Collaborative* and use Excel to generate a randomly shuffled list of the members. They then select members to contact from this list until they have succeeded in contacting 150 members.

a) What is the population?

b) What is the sampling frame?

c) What is the population parameter of interest?

d) What sampling method is used?

4. An airline company is interested in the opinions of their frequent flyer customers about their proposed new routes. Specifically they want to know what proportion of them plan to use one of their new hubs in the next 6 months. They take a random sample of 10,000 from the database of all frequent flyers and send them an e-mail message with a request to fill out a survey in exchange for 1500 miles.

a) What is the population?

b) What is the sampling frame?

c) What is the population parameter of interest?

d) What sampling method is used?

SECTION 3.3

5. As discussed in the chapter, GfK Roper Consulting conducts a global consumer survey to help multinational companies understand different consumer attitudes throughout the world. In India, the researchers interviewed 1000 people aged 13–65 (www.gfkamerica.com). Their sample is designed so that they get 500 males and 500 females.

a) Are they using a simple random sample? How do you know?

b) What kind of design do you think they are using?

6. For their class project, a group of Business students decide to survey the student body to assess opinions about a proposed new student coffee shop to judge how successful it might be. Their sample of 200 contained 50 first-year students, 50 sophomores, 50 juniors, and 50 seniors.

a) Do you think the group was using an SRS? Why?

b) What kind of sampling design do you think they used?

7. The consumer advocacy group from Exercise 3 that was interested in gauging perceptions about food safety among professionals in the food industry has decided to use a different method to sample. Instead of randomly selecting members from a shuffled list, they listed the members in alphabetical order and took every tenth member until they succeeded in contacting 150 members. What kind of sampling method have they used?

8. The airline company from Exercise 4, interested in the opinions of their frequent flyer customers about their proposed new routes, has decided that different types of customers might have different opinions. Of their customers, 50% are silver-level, 30% are blue, and 20% are red. They first compile separate lists of silver, blue, and red members and then randomly select 5000 silver members, 3000 blue members, and 2000 red members to e-mail. What kind of sampling method have they used?

For Exercises 9 and 10, identify the following if possible. (If not, say why.)

a) The population

b) The population parameter of interest

c) The sampling frame

d) The sample

e) The sampling method, including whether or not randomization was employed

f) Any potential sources of bias you can detect and any problems you see in generalizing to the population of interest

9. A business magazine mailed a questionnaire to the human resources directors of all Fortune 500 companies, and received responses from 23% of them. Those responding reported that they did not find that such surveys intruded significantly on their workday.

10. A question posted on the Lycos website asked visitors to the site to say whether they thought that businesses should be required to pay for their employees' health insurance.

SECTION 3.4

11. An intern for the consumer advocacy group in Exercise 3 has decided to make the survey process simpler by calling 150 of the members who attended the recent symposium on "Food Safety in the 21st century" that was recently held in Las Vegas. He has all the phone numbers, so it will be easy to contact them. He will start calling members from the top of the list, which was generated as the members enrolled for the symposium. He has written a script to read to them that follows,

"As we learned in Las Vegas, food safety is of utmost importance in the restaurant business today. Given the enormous effort of the Food Safety Institute in developing proper guidelines and education for food professionals, do you agree that food safety has improved in the U.S.?"

a) What is the population of interest?
b) What is the sampling frame?
c) Point out any problems you see either with the sampling procedure and/or the survey itself. What are the potential impacts of these problems?

12. The airline company in Exercise 4 has realized that some of its customers don't have e-mail or don't read it regularly. They decide to restrict the mailing only to customers who have recently registered for a "Win a trip to Miami" contest, figuring that those with Internet access are more likely to read and to respond to their e-mail. They send an e-mail with the following message:

"Did you know that National Airlines has just spent over $3 million refurbishing our brand new hub in Miami? By answering the following question, you may be eligible to win $1000 worth of coupons that can be spent in any of the fabulous restaurants or shops in the Miami airport. Might you possibly think of traveling to Miami in the next six months on your way to one of your destinations?"

a) What is the population?
b) What is the sampling frame?
c) Point out any problems you see either with the sampling procedure and/or the survey itself. What are the potential impacts of these problems?

13. An intern is working for Pacific TV (PTV), a small cable and Internet provider, and has proposed some questions that might be used in the survey to assess whether customers are willing to pay $50 for a new service.

Question 1: If PTV offered state-of-the-art high-speed Internet service for $50 per month, would you subscribe to that service?

Question 2: Would you find $50 per month—less than the cost of a daily cappuccino—an appropriate price for high-speed Internet service?

a) Do you think these are appropriately worded questions? Why or why not?
b) Suggest a question with better wording.

14. Here are more proposed survey questions for the survey in Exercise 13:

Question 3: Do you find that the slow speed of dial-up Internet access reduces your enjoyment of web services?

Question 4: Given the growing importance of high-speed Internet access for your children's education, would you subscribe to such a service if it were offered?

a) Do you think these are appropriately worded questions? Why or why not?
b) Which one has more neutral wording? Explain.

SECTION 3.5

15. Indicate whether each statement below is true or false. If false, explain why.

a) A local television news program that asks viewers to call in and give their opinion on an issue typically results in a biased voluntary response sample.
b) Convenience samples are generally not representative of the population.
c) Measurement error is the same as sampling error.
d) A pilot test can be useful for identifying poorly worded questions on a survey.

16. Indicate whether each statement below is true or false. If false, explain why.

a) Asking viewers to call into a 900 number (for which there is a toll charge) is a good way to produce a representative sample.
b) When writing a survey, it's a good idea to include as many questions as possible to ensure efficiency and to lower costs.
c) A recent poll on a website was valid because the sample size was over 1,000,000 respondents.
d) Malls are not necessarily good places to conduct surveys because people who frequent malls may not be representative of the population at large.

17. For your marketing class, you'd like to take a survey from a sample of all the Catholic Church members in your city to assess the market for a DVD about the pope's visit to the United States. A list of churches shows 17 Catholic churches within the city limits. Rather than try to obtain a list of all members of all these churches, you decide to pick 3 churches at random. For those churches, you'll ask to get a list of all current members and contact 100 members at random.

a) What kind of design have you used?
b) What could go wrong with the design that you have proposed?

18. The U.S. Fish and Wildlife Service plans to study the fishing industry around Saginaw Bay. To do that, they decide to randomly select 5 fishing boats at the end of a randomly chosen fishing day and to count the numbers and types of all the fish on those boats.

a) What kind of design have they used?
b) What could go wrong with the design that they have proposed?

CHAPTER EXERCISES

19. Software licenses. The website www.gamefaqs.com asked, as their question of the day to which visitors to the site were invited to respond, *"Do you ever read the end-user license agreements when installing software or games?"* Of the 98,574 respondents, 63.47% said they never read those agreements—a fact that software manufacturers might find important.

a) What kind of sample was this?
b) How much confidence would you place in using 63.47% as an estimate of the fraction of people who don't read software licenses?

20. Drugs in baseball. Major League Baseball, responding to concerns about their "brand," tests players to see whether they are using performance-enhancing drugs. Officials select a team at random, and a drug-testing crew shows up unannounced to test all 40 players on the team. Each testing day can be considered a study of drug use in Major League Baseball.

a) What kind of sample is this?
b) Is that choice appropriate?

21. Gallup. At its website (www.galluppoll.com) the Gallup Poll publishes results of a new survey each day. Scroll down to the end, and you'll find a statement that includes an explanation such as this:

Results are based on telephone interviews with 1,008 national adults, aged 18 and older, conducted April 2–5, 2007. . . . In addition to sampling error, question wording and practical difficulties in conducting surveys can introduce error or bias into the findings of public opinion polls.

a) For this survey, identify the population of interest.
b) Gallup performs its surveys by phoning numbers generated at random by a computer program. What is the sampling frame?
c) What problems, if any, would you be concerned about in matching the sampling frame with the population?

22. Defining the survey. At its website (www.gallupworldpoll .com) the Gallup World Poll reports results of surveys conducted in various places around the world. At the end of one of these reports, they describe their methods, including explanations such as the following:

Results are based on face-to-face interviews with randomly selected national samples of approximately 1000 adults, aged 15 and older, who live permanently in each of the 21 sub-Saharan African nations surveyed. Those countries include Angola (areas where land mines might be expected were excluded), Benin, Botswana, Burkina Faso, Cameroon, Ethiopia, Ghana, Kenya, Madagascar (areas where interviewers had to walk more than 20 kilometers from a road were excluded), Mali, Mozambique, Niger, Nigeria, Senegal, Sierra Leone, South Africa, Tanzania, Togo, Uganda (the area of activity of the Lord's Resistance Army was excluded from the survey), Zambia, and Zimbabwe. . . . In all countries except Angola, Madagascar, and Uganda, the sample is representative of the entire population.

a) Gallup is interested in sub-Saharan Africa. What kind of survey design are they using?
b) Some of the countries surveyed have large populations. (Nigeria is estimated to have about 130 million people.) Some are quite small. (Togo's population is estimated at 5.4 million.) Nonetheless, Gallup sampled 1000 adults in each country. How does this affect the precision of its estimates for these countries?

23–30. Survey details. For the following reports about statistical studies, identify the following items (if possible). If you can't tell, then say so—this often happens when we read about a survey.

a) The population
b) The population parameter of interest
c) The sampling frame
d) The sample
e) The sampling method, including whether or not randomization was employed
f) Any potential sources of bias you can detect and any problems you see in generalizing to the population of interest

23. Alternative medicine. Consumers Union asked all subscribers whether they had used alternative medical treatments and, if so, whether they had benefited from them. For almost all of the treatments, approximately 20% of those responding reported cures or substantial improvement in their condition.

24. Global warming. The Gallup Poll interviewed 1012 randomly selected U.S. adults aged 18 and older, March 6–9, 2008. Gallup reports that when asked when (if ever) the effects of global warming will begin to happen, 61% of respondents said the effects had already begun. Only 11% thought that they would never happen.

25. At the bar. Researchers waited outside a bar they had randomly selected from a list of such establishments. They stopped every 10th person who came out of the bar and asked whether he or she thought drinking and driving was a serious problem.

26. Election poll. Hoping to learn what issues may resonate with voters in the coming election, the campaign director for a mayoral candidate selects one block from each of the city's election districts. Staff members go there and interview all the residents they can find.

27. Toxic waste. The Environmental Protection Agency took soil samples at 16 locations near a former industrial waste dump and checked each for evidence of toxic chemicals. They found no elevated levels of any harmful substances.

28. Housing discrimination. Inspectors send trained "renters" of various races and ethnic backgrounds, and of both sexes to inquire about renting randomly assigned advertised apartments. They look for evidence that landlords deny access illegally based on race, sex, or ethnic background.

29. Quality control. A company packaging snack foods maintains quality control by randomly selecting 10 cases from each day's production and weighing the bags. Then they open one bag from each case and inspect the contents.

30. Contaminated milk. Dairy inspectors visit farms unannounced and take samples of the milk to test for contamination. If the milk is found to contain dirt, antibiotics, or other foreign matter, the milk will be destroyed and the farm is considered to be contaminated pending further testing.

31. Instant poll. A local TV station conducted an "Instant Poll" to predict the winner in the upcoming mayoral election. Evening news viewers were invited to phone in their votes, with the results to be announced on the late-night news. Based on the phone calls, the station predicted that Amabo would win the election with 52% of the vote. They were wrong: Amabo lost, getting only 46% of the vote. Do you think the station's faulty prediction is more likely to be a result of bias or sampling error? Explain.

32. Paper poll. Prior to the mayoral election discussed in Exercise 31, the newspaper also conducted a poll. The paper surveyed a random sample of registered voters stratified by political party, age, sex, and area of residence. This poll predicted that Amabo would win the election with 52% of the vote. The newspaper was wrong: Amabo lost, getting only 46% of the vote. Do you think the newspaper's faulty prediction is more likely to be a result of bias or sampling error? Explain.

33. Cable company market research. A local cable TV company, Pacific TV, with customers in 15 towns is considering offering high-speed Internet service on its cable lines. Before launching the new service they want to find out whether customers would pay the $50 per month that they plan to charge. An intern has prepared several alternative plans for assessing customer demand. For each, indicate what kind of sampling strategy is involved and what (if any) biases might result.

a) Put a big ad in the newspaper asking people to log their opinions on the PTV website.
b) Randomly select one of the towns and contact every cable subscriber by phone.

c) Send a survey to each customer and ask them to fill it out and return it.
d) Randomly select 20 customers from each town. Send them a survey, and follow up with a phone call if they do not return the survey within a week.

34. Cable company market research, part 2. Four new sampling strategies have been proposed to help PTV determine whether enough cable subscribers are likely to purchase high-speed Internet service. For each, indicate what kind of sampling strategy is involved and what (if any) biases might result.

a) Run a poll on the local TV news, asking people to dial one of two phone numbers to indicate whether they would be interested.
b) Hold a meeting in each of the 15 towns, and tally the opinions expressed by those who attend the meetings.
c) Randomly select one street in each town and contact each of the households on that street.
d) Go through the company's customer records, selecting every 40th subscriber. Send employees to those homes to interview the people chosen.

35. Amusement park riders. An amusement park has opened a new roller coaster. It is so popular that people are waiting for up to three hours for a two-minute ride. Concerned about how patrons (who paid a large amount to enter the park and ride on the rides) feel about this, they survey every 10th person on the line for the roller coaster, starting from a randomly selected individual.

a) What kind of sample is this?
b) Is it likely to be representative?
c) What is the sampling frame?

36. Playground. Some people have been complaining that the children's playground at a municipal park is too small and is in need of repair. Managers of the park decide to survey city residents to see if they believe the playground should be rebuilt. They hand out questionnaires to parents who bring children to the park. Describe possible biases in this sample.

37. Another ride. The survey of patrons waiting in line for the roller coaster in Exercise 35 asks whether they think it is worthwhile to wait a long time for the ride and whether they'd like the amusement park to install still more roller coasters. What biases might cause a problem for this survey?

38. Playground bias. The survey described in Exercise 36 asked,

Many people believe this playground is too small and in need of repair. Do you think the playground should be repaired and expanded even if that means raising the entrance fee to the park?

Describe two ways this question may lead to response bias.

39. (Possibly) Biased questions. Examine each of the following questions for possible bias. If you think the question is biased, indicate how and propose a better question.

a) *Should companies that pollute the environment be compelled to pay the costs of cleanup?*
b) *Should a company enforce a strict dress code?*

40. More possibly biased questions. Examine each of the following questions for possible bias. If you think the question is biased, indicate how and propose a better question.

a) *Do you think that price or quality is more important in selecting an MP3 player?*
b) *Given humanity's great tradition of exploration, do you favor continued funding for space flights?*

41. Phone surveys. Anytime we conduct a survey, we must take care to avoid undercoverage. Suppose we plan to select 500 names from the city phone book, call their homes between noon and 4 p.m., and interview whoever answers, anticipating contacts with at least 200 people.

a) Why is it difficult to use a simple random sample here?
b) Describe a more convenient, but still random, sampling strategy.
c) What kinds of households are likely to be included in the eventual sample of opinion? Who will be excluded?
d) Suppose, instead, that we continue calling each number, perhaps in the morning or evening, until an adult is contacted and interviewed. How does this improve the sampling design?
e) Random-digit dialing machines can generate the phone calls for us. How would this improve our design? Is anyone still excluded?

42. Cell phone survey. What about drawing a random sample only from cell phone exchanges? Discuss the advantages and disadvantages of such a sampling method compared with surveying randomly generated telephone numbers from non–cell phone exchanges. Do you think these advantages and disadvantages have changed over time? How do you expect they'll change in the future?

43. Change. How much change do you have on you right now? Go ahead, count it.

a) How much change do you have?
b) Suppose you check on your change every day for a week as you head for lunch and average the results. What parameter would this average estimate?
c) Suppose you ask 10 friends to average *their* change every day for a week, and you average those 10 measurements. What is the population now? What parameter would this average estimate?
d) Do you think these 10 average change amounts are likely to be representative of the population of change amounts in your class? In your college? In the country? Why or why not?

44. Fuel economy. Occasionally, when I fill my car with gas, I figure out how many miles per gallon my car got. I wrote down those results after six fill-ups in the past few

months. Overall, it appears my car gets 28.8 miles per gallon.

a) What statistic have I calculated?
b) What is the parameter I'm trying to estimate?
c) How might my results be biased?
d) When the Environmental Protection Agency (EPA) checks a car like mine to predict its fuel economy, what parameter is it trying to estimate?

45. Accounting. Between quarterly audits, a company likes to check on its accounting procedures to address any problems before they become serious. The accounting staff processes payments on about 120 orders each day. The next day, the supervisor rechecks 10 of the transactions to be sure they were processed properly.

a) Propose a sampling strategy for the supervisor.
b) How would you modify that strategy if the company makes both wholesale and retail sales, requiring different bookkeeping procedures?

46. Happy workers? A manufacturing company employs 14 project managers, 48 foremen, and 377 laborers. In an effort to keep informed about any possible sources of employee discontent, management wants to conduct job satisfaction interviews with a simple random sample of employees every month.

a) Do you see any danger of bias in the company's plan? Explain.
b) How might you select a simple random sample?
c) Why do you think a simple random sample might not provide the best estimate of the parameters the company wants to estimate?
d) Propose a better sampling strategy.
e) Listed below are the last names of the project managers. Use random numbers to select two people to be interviewed. Be sure to explain your method carefully.

Barrett	Bowman	Chen
DeLara	DeRoos	Grigorov
Maceli	Mulvaney	Pagliarulo
Rosica	Smithson	Tadros
Williams	Yamamoto	

47. Quality control. Sammy's Salsa, a small local company, produces 20 cases of salsa a day. Each case contains 12 jars and is imprinted with a code indicating the date and batch number. To help maintain consistency, at the end of each day, Sammy selects three bottles of salsa, weighs the contents, and tastes the product. Help Sammy select the sample jars. Today's cases are coded 07N61 through 07N80.

a) Carefully explain your sampling strategy.
b) Show how to use random numbers to pick the three jars for testing.
c) Did you use a simple random sample? Explain.

48. Fish quality. Concerned about reports of discolored scales on fish caught downstream from a newly sited chemical plant, scientists set up a field station in a shoreline public park. For one week they asked fishermen there to bring any fish they caught to the field station for a brief inspection. At the end of the week, the scientists said that 18% of the 234 fish that were submitted for inspection displayed the discoloration. From this information, can the researchers estimate what proportion of fish in the river have discolored scales? Explain.

49. Sampling methods. Consider each of these situations. Do you think the proposed sampling method is appropriate? Explain.

a) We want to know what percentage of local doctors accept Medicaid patients. We call the offices of 50 doctors randomly selected from local Yellow Page listings.

b) We want to know what percentage of local businesses anticipate hiring additional employees in the upcoming month. We randomly select a page in the Yellow Pages and call every business listed there.

50. More sampling methods. Consider each of these situations. Do you think the proposed sampling method is appropriate? Explain.

a) We want to know if business leaders in the community support the development of an "incubator" site at a vacant lot on the edge of town. We spend a day phoning local businesses in the phone book to ask whether they'd sign a petition.

b) We want to know if travelers at the local airport are satisfied with the food available there. We go to the airport on a busy day and interview every 10th person in line in the food court.

Just Checking Answers

1 a) It can be hard to reach all members of a population, and it can take so long that circumstances change, affecting the responses. A well-designed sample is often a better choice.
   b) This sample is probably biased—people who didn't like the food at the restaurant might not choose to eat there.
   c) No, only the sample size matters, not the fraction of the overall population.
   d) Students who frequent this website might be more enthusiastic about Statistics than the overall population of Statistics students. A large sample cannot compensate for bias.
   e) It's the population "parameter." "Statistics" describe samples.

2 a) systematic
   b) stratified
   c) simple
   d) cluster

Displaying and Describing Categorical Data

Keen, Inc.

KEEN, Inc. was started to create a sandal designed for a variety of water activities. The sandals quickly became popular due to their unique patented toe protection—a black bumper to protect the toes when adventuring out on rivers and trails. Today the KEEN brand offers over 300 different outdoor performance and outdoor inspired casual footwear styles as well as bags and socks.

Few companies experience the kind of growth that KEEN did in less than seven years. Amazingly, they've done this with relatively little advertising and by selling primarily to specialty footwear and outdoor stores, in addition to online outlets.

After the 2004 Tsunami disaster, KEEN cut its advertising budget almost completely and donated over $1 million to help the victims and establish the KEEN Foundation to support environmental and social causes. Philanthropy and community projects continue to play an integral part of the KEEN brand values. In fact, KEEN has established a giving program with a philanthropic effort devoted to helping the environment, conservation, and social movements involving the outdoors.

WHO	Visits to the KEEN, Inc. website
WHAT	Search Engine that led to KEEN's website
WHEN	September 2006
WHERE	Worldwide
HOW	Data compiled via *Google®Analytics* from KEEN website
WHY	To understand customer use of the website and how they got there

KEEN, Inc., like most companies, collects data on visits to its website. Each visit to the site and each subsequent action the visitor takes (changing the page, entering data, etc.) is recorded in a file called a usage, or access weblog. These logs contain a lot of potentially worthwhile information, but they are not easy to use. Here's one line from a log:

```
245.240.221.71 -- [03/Jan/2007:15:20:06-0800]" GET
http://www.keenfootwear.com/pdp_page.cfm?productID=148"
200 8788 "http://www.google.com/" "Mozilla/3.0WebTV/1.2
(compatible; MSIE 2.0)"
```

Unless the company has the analytic resources to deal with these files, it must rely on a third party to summarize the data. KEEN, like many other small and midsized companies, uses *Google Analytics* to collect and summarize its log data.

Imagine a whole table of data like the one above—with a line corresponding to every visit. In September 2006 there were 93,173 visits to the KEEN site, which would be a table with 93,173 rows. The problem with a file like this—and in fact even with data tables—is that we can't see what's going on. And seeing is exactly what we want to do. We need ways to show the data so that we can see patterns, relationships, trends, and exceptions.

4.1   Summarizing a Categorical Variable

The Three Rules of Data Analysis

There are three things you should always do with data:

1. **Make a picture.** A display of your data will reveal things you are not likely to see in a table of numbers and will help you to *plan* your approach to the analysis and think clearly about the patterns and relationships that may be hiding in your data.
2. **Make a picture.** A well-designed display will *do* much of the work of analyzing your data. It can show the important features and patterns. A picture will also reveal things you did not expect to see: extraordinary (possibly wrong) data values or unexpected patterns.
3. **Make a picture.** The best way to *report* to others what you find in your data is with a well-chosen picture.

These are the three rules of data analysis. These days, technology makes drawing pictures of data easy, so there is no reason not to follow the three rules. Here are some displays showing various aspects of traffic on one of the authors' websites.

Some displays communicate information better than others. We'll discuss some general principles for displaying information honestly in this chapter.

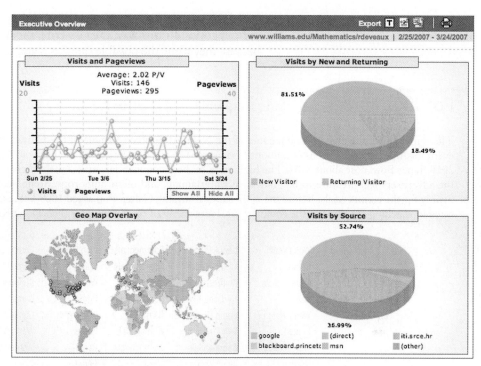

Figure 4.1   Part of the output from *Google Analytics* (www.google.com/analytics) for the period Feb. 25 to March 24, 2007 displaying website traffic.

Search Engine	Visits
Google	50,629
Direct	22,173
Yahoo	7272
MSN	3166
SnapLink	946
All Others	8987
Total	**93,167**

Table 4.1   A frequency table of the Search Engine used by visitors to the KEEN, Inc. website.

Search Engine	Visits by %
Google	54.34%
Direct	23.80%
Yahoo	7.80%
MSN	3.40%
SnapLink	1.02%
All Others	9.65%
Total	**100.00%**

Table 4.2   A relative frequency table for the same data.

Frequency Tables

KEEN might be interested to know how people find their website. They might use the information to allocate their advertising revenue to various search engines, putting ads where they'll be seen by the most potential customers. The variable *Search Engine* records, for each visit to KEEN's website, where the visit came from. The categories are all the search engines used, plus the label "Direct," which indicates that the customer typed in KEEN's web address (or URL) directly into the browser. In order to make sense of the 93,167 visits for which they have data, they'd like to summarize the variable and display the information in a way that can easily communicate the results to others.

In order to make a picture of any variable, we first need to organize its values. For a categorical variable, like *Search Engine*, this is easy—we just count the number of cases corresponding to each category. A **frequency table** (Table 4.1) records the counts for each of the categories of the variable and lists the counts under the category name. By ordering the categories by number of counts, we can easily see, for example, that the most popular source was Google.

The names of the categories label each row in the frequency table. For *Search Engine* these are "Google," "Direct," "Yahoo," and so on. Even with thousands of cases, a variable that doesn't have too many categories produces a frequency table that is easy to read. A frequency table with dozens or hundreds of categories would be much harder to read. Notice the label of the last line of the table—"All Others." When the number of categories gets too large, we often lump together values of the variable into "Other." When to do that is a judgment call, but it's a good idea to have fewer than about a dozen categories.

Counts are useful, but sometimes we want to know the fraction or **proportion** of the data in each category, so we divide the counts by the total number of cases. Usually we multiply by 100 to express these proportions as **percentages**. A **relative frequency table** (Table 4.2) displays the *percentages*, rather than the counts, of the values in each category. Both types of table show how the cases are distributed across the categories. In this way, they describe the **distribution** of a categorical variable because they name the possible categories and tell how frequently each occurs.

For Example   Making frequency and relative frequency tables

The Super Bowl, the championship game of the National Football League of the United States, is an important annual social event for Americans, with tens of millions of viewers. The ads that air during the game are expensive: a 30-second ad during the 2010 Super Bowl cost about $3M. The high price of these commercials makes them high-profile and much anticipated, and so the advertisers feel pressure to be innovative, entertaining, and often humorous. Some people, in fact, watch the Super Bowl mainly for the commercials. Before the 2007 Super Bowl, the Gallup Poll asked 1008 U.S. adults whether they were more interested in watching the game or the commercials. Here are 40 of those responses (NA/Don't Know = No Answer or Don't Know):

Won't Watch	Game	Commercials	Won't Watch	Game
Game	Won't Watch	Commercials	Game	Game
Commercials	Commercials	Game	Won't Watch	Commercials
Game	NA/Don't Know	Commercials	Game	Game
Won't Watch	Game	Game	Won't Watch	Game
Game	Won't Watch	Won't Watch	Game	Won't Watch
Won't Watch	Commercials	Commercials	Game	Won't Watch
NA/Don't Know	Won't Watch	Game	Game	Game

Question: Make a frequency table for this variable. Include the percentages to display both a frequency and relative frequency table at the same time.

Answer: There were four different responses to the question about watching the Super Bowl. Counting the number of participants who responded to each of these gives the following table:

Response	Counts	Percentage
Commercials	8	20.0%
Game	18	45.0%
Won't Watch	12	30.0%
No Answer/Don't Know	2	5.0%
Total	**40**	**100.0%**

4.2   Displaying a Categorical Variable

The Area Principle

Now that we have a frequency table, we're ready to follow the three rules of data analysis. But we can't make just any display; a bad picture can distort our understanding rather than help it. For example, Figure 4.2 is a graph of the frequencies of Table 4.1. What impression do you get of the relative frequencies of visits from each source?

While it's true that the majority of people came to KEEN's website from Google, in Figure 4.2 it looks like nearly all did. That doesn't seem right. What's wrong? The lengths of the sandals *do* match the totals in the table. But our eyes tend to be more impressed by the *area* (or perhaps even the *volume*) than by other aspects of each sandal image, and it's that aspect of the image that we notice. Since there were about twice as many people who came from Google as those who typed

100.01%?

If you are careful to add the percentages in Table 4.2, you will notice the total is 100.01%. Of course the real total has to be 100.00%. The discrepancy is due to individual percentages being rounded. You'll often see this in tables of percents, sometimes with explanatory footnotes.

Figure 4.2   Although the length of each sandal corresponds to the correct number, the impression we get is all wrong because we perceive the entire area of the sandal. In fact, only a little more than 50% of all visitors used Google to get to the website.

the URL in directly, the sandal depicting the number of Google visitors is about two times longer than the sandal below it, but it occupies about four times the area. As you can see from the frequency table, that just isn't a correct impression.

The best data displays observe a fundamental principle of graphing data called the **area principle**, which says that the area occupied by a part of the graph should correspond to the magnitude of the value it represents.

Bar Charts

Figure 4.3 gives us a chart that obeys the area principle. It's not as visually entertaining as the sandals, but it does give a more *accurate* visual impression of the distribution. The height of each bar shows the count for its category. The bars are the

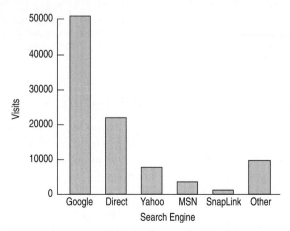

Figure 4.3   Visits to the KEEN, Inc. website by *Search Engine* choice. With the area principle satisfied, the true distribution is clear.

same width, so their heights determine their areas, and the areas are proportional to the counts in each class. Now it's easy to see that nearly half the site hits came from places other than Google—not the impression that the sandals in Figure 4.2 conveyed. We can also see that there were a little more than twice as many visits that originated with a Google search as there were visits that came directly. Bar charts make these kinds of comparisons easy and natural.

A **bar chart** displays the distribution of a categorical variable, showing the counts for each category next to each other for easy comparison. Bar charts should have small spaces between the bars to indicate that these are freestanding bars that could be rearranged into any order. The bars are lined up along a common base with labels for each category. The variable name is often used as a subtitle for the x-axis.

Bar charts are usually drawn vertically in columns, , but sometimes they are drawn with horizontal bars, like this.[1]

If we want to draw attention to the relative *proportion* of visits from each *Search Engine*, we could replace the counts with percentages and use a **relative frequency bar chart**, like the one shown in Figure 4.4.

Figure 4.4   The relative frequency bar chart, created in **Excel**, looks the same as the bar chart (Figure 4.3) but shows the proportion of visits in each category rather than the counts.

Pie Charts

Another common display that shows how a whole group breaks into several categories is a pie chart. **Pie charts** show the whole group of cases as a circle. They slice the circle into pieces whose size is proportional to the fraction of the whole in each category.

----

[1]Excel refers to this display as a column chart when the bars are vertical and a bar chart when they are horizontal.

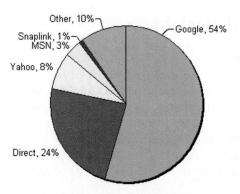

Figure 4.5   A pie chart shows the proportion of visits by Search Engine. Pie chart created in **Excel**.

Pie charts give a quick impression of how a whole group is partitioned into smaller groups. Because we're used to cutting up pies into 2, 4, or 8 pieces, pie charts are good for seeing relative frequencies near 1/2, 1/4, or 1/8. For example, in Figure 4.5, you can easily see that the slice representing Google is just slightly more than half the total. Unfortunately, other comparisons are harder to make with pie charts. Were there more visits from Yahoo, or from All Others? It's hard to tell since the two slices look about the same. Comparisons such as these are usually easier in a bar chart. (Compare to Figure 4.4.)

• **Think before you draw.** Our first rule of data analysis is *Make a picture*. But what kind of picture? We don't have a lot of options—yet. There's more to Statistics than pie charts and bar charts, and knowing when to use every type of display we'll discuss is a critical first step in data analysis. That decision depends in part on what type of data you have and on what you hope to communicate.

We always have to check that the data are appropriate for whatever method of analysis we choose. Before you make a bar chart or a pie chart, always check the **Categorical Data Condition:** that the data are counts or percentages of individuals in categories.

If you want to make a pie chart or relative frequency bar chart, you'll need to also make sure that the categories don't overlap, so that no individual is counted in two categories. If the categories do overlap, it's misleading to make a pie chart, since the percentages won't add up to 100%. For the *Search Engine* data, either kind of display is appropriate because the categories don't overlap—each visit comes from a unique source.

Throughout this course, you'll see that doing Statistics right means selecting the proper methods. That means you have to think about the situation at hand. An important first step is to check that the type of analysis you plan is appropriate. The Categorical Data Condition is just the first of many such checks.

For Example   Making a bar chart

Question: Make a bar chart for the 40 Super Bowl responses of the example on page 54.

Answer: Use the frequencies in the table in the example on page 54 to produce the heights of the bars:

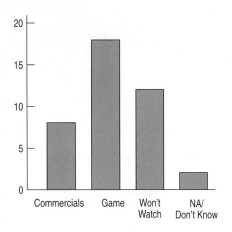

4.3   Exploring Two Categorical Variables: Contingency Tables

WHO	Respondents in the GfK Roper Reports Worldwide Survey
WHAT	Responses to questions relating to perceptions of food and health
WHEN	Fall 2005; published in 2006
WHERE	Worldwide
HOW	Data collected by GfK Roper Consulting using a multistage design
WHY	To understand cultural differences in the perception of the food and beauty products we buy and how they affect our health

In Chapter 3 we saw how GfK Roper Consulting gathered information on consumers attitudes about health, food, and health care products. In order to effectively market food products across different cultures, it's essential to know how strongly people in different cultures feel about their food. One question in the Roper survey asked respondents whether they agreed with the following statement: "I have a strong preference for regional or traditional products and dishes from where I come from." Here is a frequency table (Table 4.3) of the responses.

Response to *Regional Food Preference Question*	Counts	Relative Frequency
Agree Completely	2346	30.51%
Agree Somewhat	2217	28.83%
Neither Disagree Nor Agree	1738	22.60%
Disagree Somewhat	811	10.55%
Disagree Completely	498	6.48%
Don't Know	80	1.04%
Total	**7690**	**100.00%**

Table 4.3   A combined frequency and relative frequency table for the responses (from all 5 countries represented: China, France, India, the U.K., and the U.S.) to the statement "I have a strong preference for regional or traditional products and dishes from where I come from."

The pie chart (Figure 4.6) shows clearly that more than half of all the respondents agreed with the statement.

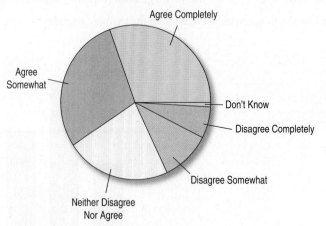

Figure 4.6   It's clear from the pie chart that the majority of respondents identify with their local foods.

But if we want to target our marketing differently in different countries, wouldn't it be more interesting to know how opinions vary from country to country?

To find out, we need to look at the two categorical variables *Regional Preference* and *Country* together, which we do by arranging the data in a two-way table. Table 4.4 is a two-way table of *Regional Preference* by *Country*. Because the table shows how the individuals are distributed along each variable, depending on, or *contingent on*, the value of the other variable, such a table is called a **contingency table**.

		Regional Preference						
		Agree Completely	Agree Somewhat	Neither Disagree Nor Agree	Disagree Somewhat	Disagree Completely	Don't Know	Total
Country	China	518	576	251	117	33	7	**1502**
	France	347	475	400	208	94	15	**1539**
	India	960	282	129	65	95	4	**1535**
	U.K.	214	407	504	229	175	28	**1557**
	U.S.	307	477	454	192	101	26	**1557**
	Total	**2346**	**2217**	**1738**	**811**	**498**	**80**	**7690**

Table 4.4   Contingency table of *Regional Preference* and *Country*. The bottom line "Totals" are the values that were in Table 4.3.

The margins of a contingency table give totals. In the case of Table 4.4, these are shown in both the right-hand column (in bold) and the bottom row (also in bold). The totals in the bottom row of the table show the frequency distribution of the variable *Regional Preference*. The totals in the right-hand column of the table show the frequency distribution of the variable *Country*. When presented like this, at the margins of a contingency table, the frequency distribution of either one of the variables is called its **marginal distribution**. The marginal distribution for a variable in a contingency table is the same frequency distribution we found by considering each variable separately.

Each **cell** of a contingency table (any intersection of a row and column of the table) gives the count for a combination of values of the two variables. For example, in Table 4.4 you can see that 504 people in the United Kingdom neither agreed nor disagreed. Looking down the Agree Completely column, you can see that the largest number of responses in that column (960) are from India. Are Britons less likely to agree with the statement than Indians or Chinese? Questions like this are more naturally addressed using percentages.

We know that 960 people from India agreed completely with the statement. We could display this number as a percentage, but as a percentage of what? The total number of people in the survey? (960 is 12.5% of the total.) The number of Indians in the survey? (960 is 62.5% of the row total.) The number of people who agree completely? (960 is 40.9% of the column total.) All of these are possibilities, and all are potentially useful or interesting. You'll probably wind up calculating (or letting your technology calculate) lots of percentages. Most statistics programs offer a choice of **total percent**, **row percent**, or **column percent** for contingency tables. Unfortunately, they often put them all together with several numbers in each cell of the table. The resulting table (Table 4.5) holds lots of information but is hard to understand.

Country	Regional Preference						
	Agree Completely	**Agree Somewhat**	**Neither Disagree Nor Agree**	**Disagree Somewhat**	**Disagree Completely**	**Don't Know**	**Total**
China	518	576	251	117	33	7	**1502**
% of Row	*34.49*	*38.35*	*16.71*	*7.79*	*2.20*	*0.47*	**100.00%**
% of Column	*22.08*	*25.98*	*14.44*	*14.43*	*6.63*	*8.75*	**19.53%**
% of Total	*6.74*	*7.49*	*3.26*	*1.52*	*0.43*	*0.09*	**19.53%**
France	347	475	400	208	94	15	**1539**
% of Row	*22.55*	*30.86*	*25.99*	*13.52*	*6.11*	*0.97*	**100.00%**
% of Column	*14.79*	*21.43*	*23.01*	*25.65*	*18.88*	*18.75*	**20.01%**
% of Total	*4.51*	*6.18*	*5.20*	*2.70*	*1.22*	*0.20*	**20.01%**
India	960	282	129	65	95	4	**1535**
% of Row	*62.54*	*18.37*	*8.40*	*4.23*	*6.19*	*0.26*	**100.00%**
% of Column	*40.92*	*12.72*	*7.42*	*8.01*	*19.08*	*5.00*	**19.96%**
% of Total	*12.48*	*3.67*	*1.68*	*0.85*	*1.24*	*0.05*	**19.96%**
U.K.	214	407	504	229	175	28	**1557**
% of Row	*13.74*	*26.14*	*32.37*	*14.71*	*11.24*	*1.80*	**100.00%**
% of Column	*9.12*	*18.36*	*29.00*	*28.24*	*35.14*	*35.00*	**20.24%**
% of Total	*2.78*	*5.29*	*6.55*	*2.98*	*2.28*	*0.36*	**20.24%**
U.S.	307	477	454	192	101	26	**1557**
% of Row	*19.72*	*30.64*	*29.16*	*12.33*	*6.49*	*1.67*	**100.00%**
% of Column	*13.09*	*21.52*	*26.12*	*23.67*	*20.28*	*32.50*	**20.24%**
% of Total	*3.99*	*6.20*	*5.90*	*2.50*	*1.31*	*0.34*	**20.24%**
Total	**2346**	**2217**	**1738**	**811**	**498**	**80**	**7690**
% of Row	**30.51%**	**28.83%**	**22.60%**	**10.55%**	**6.48%**	**1.04%**	**100.00%**
% of Column	**100.00%**	**100.00%**	**100.00%**	**100.00%**	**100.00%**	**100.00%**	**100.00%**
% of Total	**30.51%**	**28.83%**	**22.60%**	**10.55%**	**6.48%**	**1.04%**	**100.00%**

Table 4.5   Another contingency table of *Regional Preference* and *Country*. This time we see not only the counts for each combination of the two variables, but also the percentages these counts represent. For each count, there are three choices for the percentage: by row, by column, and by table total. There's probably too much information here for this table to be useful.

To simplify the table, let's pull out the values corresponding to the percentages of the total.

Country	Regional Preference						
	Agree Completely	**Agree Somewhat**	**Neither Disagree Nor Agree**	**Disagree Somewhat**	**Disagree Completely**	**Don't Know**	**Total**
China	*6.74*	*7.49*	*3.26*	*1.52*	*0.43*	*0.09*	**19.53**
France	*4.51*	*6.18*	*5.20*	*2.70*	*1.22*	*0.20*	**20.01**
India	*12.48*	*3.67*	*1.68*	*0.85*	*1.24*	*0.05*	**19.96**
U.K.	*2.78*	*5.29*	*6.55*	*2.98*	*2.28*	*0.36*	**20.25**
U.S.	*3.99*	*6.20*	*5.90*	*2.50*	*1.31*	*0.34*	**20.25**
Total	**30.51**	**28.83**	**22.60**	**10.55**	**6.48**	**1.04**	**100.00**

Table 4.6   A contingency table of *Regional Preference* and *Country* showing only the total percentages.

These percentages tell us what percent of *all* respondents belong to each combination of column and row category. For example, we see that 3.99% of the respondents were Americans who agreed completely with the question, which is slightly more than the percentage of Indians who agreed somewhat. Is this fact useful? Is that really what we want to know?

> ### Percent of what?
>
> The English language can be tricky when we talk about percentages. If asked, "What percent of those answering 'I Don't Know' were from India?" it's pretty clear that you should focus only on the *Don't Know* column. The question itself seems to restrict the *who* in the question to that column, so you should look at the number of those in each country among the 80 people who replied "I don't know." You'd find that in the column percentages, and the answer would be 4 out of 80 or 5.00%.
>
> But if you're asked, "What percent were Indians who replied 'I don't know?'" you'd have a different question. Be careful. The question really means "what percent of the entire sample were both from India and replied 'I don't know'?" So the *who* is all respondents. The denominator should be 7690, and the answer is the table percent $4/7690 = 0.05\%$.
>
> Finally, if you're asked, "What percent of the Indians replied 'I don't know'?" you'd have a third question. Now the *who* is Indians. So the denominator is the 1535 Indians, and the answer is the row percent, $4/1535 = 0.26\%$.

Always be sure to ask "percent of what?" That will help define the *who* and will help you decide whether you want *row*, *column*, or *table* percentages.

Conditional Distributions

The more interesting questions are contingent on something. We'd like to know, for example, what percentage *of Indians* agreed completely with the statement and how that compares to the percentage *of Britons* who also agreed. Equivalently, we might ask whether the chance of agreeing with the statement depended on the *Country* of the respondent. We can look at this question in two ways. First, we could ask how the distribution of *Regional Preference* changes across *Country*. To do that we look at the row percentages.

		Agree Completely	Agree Somewhat	Neither Disagree Nor Agree	Disagree Somewhat	Disagree Completely	Don't Know	Total
Country	India	960	282	129	65	95	4	1535
		62.54	18.37	8.40	4.23	6.19	0.26	100%
	U.K.	214	407	504	229	175	28	1557
		13.74	26.14	32.37	14.71	11.24	1.80	100%

Regional Preference

Table 4.7 The conditional distribution of *Regional Preference* conditioned on two values of *Country:* India and the United Kingdom. This table shows the row percentages.

By focusing on each row separately, we see the distribution of *Regional Preference* under the condition of being in the selected *Country*. The sum of the percentages in each row is 100%, and we divide that up by the responses to the question. In effect, we can temporarily restrict the *who* first to Indians and look at how their responses are distributed. A distribution like this is called a **conditional distribution** because it shows the distribution of one variable for just those cases that satisfy a condition on another. We can compare the two conditional distributions with pie

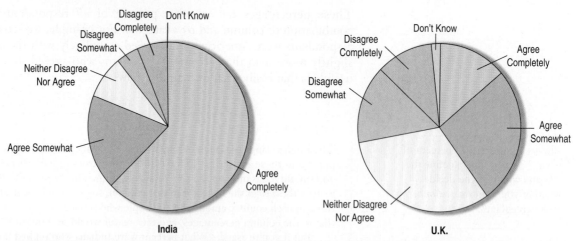

| Figure 4.7   Pie charts of the conditional distributions of *Regional Food Preference* importance for India and the United Kingdom. The percentage of people who agree is much higher in India than in the United Kingdom.

charts (Figure 4.7). Of course, we could also turn the question around. We could look at the distribution of *Country* for each category of *Regional Preference*. To do this, we would look at the column percentages.

Looking at how the percentages change across each row, it sure looks like the distribution of responses to the question is different in each *Country*. To make the differences more vivid, we could also display the conditional distributions. Figure 4.8 shows an example of a side-by-side bar chart, displaying the responses to the questions for India and the United Kingdom.

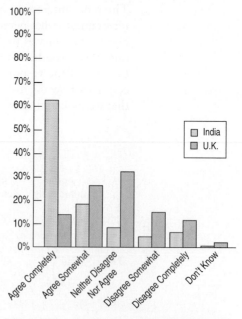

Figure 4.8   Side-by-side bar charts showing the conditional distribution of *Regional Food Preference* for both India and the United Kingdom. It's easier to compare percentages within each country with side-by-side bar charts than pie charts.

From Figure 4.8, it is clear that Indians have a stronger preference for their own cuisine than Britons have for theirs. For food companies, including GfK Roper's clients, that means Indians are less likely to accept a food product they perceive as foreign, and people in Great Britain are more accepting of "foreign" foods. This could be invaluable information for marketing products.

Variables can be associated in many ways and to different degrees. The best way to tell whether two variables are associated is to ask whether they are *not*.[2] In a contingency table, when the distribution of one variable is the same for all categories of another variable, we say that the two variables are **independent**. That tells us there's no association between these variables. We'll see a way to check for independence formally later in the book. For now, we'll just compare the distributions.

For Example   Contingency tables and side-by-side bar charts

Here is a contingency table of the responses to the question Gallup asked about the Super Bowl by sex:

	Sex		
	Female	Male	**Total**
Game	198	277	**475**
Commercials	154	79	**233**
NA/Don't Know	4	4	**8**
Won't Watch	160	132	**292**
Total	**516**	**492**	**1008**

Question: Does it seem that there is an association between what viewers are interested in watching and their sex?

Answer: First, find the conditional distributions of the four responses for each sex:

For Men:

Game = 277/492 = 56.3%

Commercials = 79/492 = 16.1%

Won't Watch = 132/492 = 26.8%

NA/Don't Know = 4/492 = 0.8%

For Women:

Game = 198/516 = 38.4%

Commercials = 154/516 = 29.8%

Won't Watch = 160/516 = 31.0%

NA/Don't Know = 4/516 = 0.8%

Now display the two distributions with side-by-side bar charts:

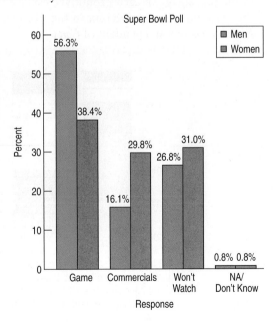

(continued)

---

[2]This kind of "backwards" reasoning shows up surprisingly often in science—and in statistics.

Based on this poll it appears that women were only slightly less interested than men in watching the Super Bowl telecast: 31% of the women said they didn't plan to watch, compared to just under 27% of men. Among those who planned to watch, however, there appears to be an association between the viewer's sex and what the viewer is most looking forward to. While more women are interested in the game (38%) than the commercials (30%), the margin among men is much wider: 56% of men said they were looking forward to seeing the game, compared to only 16% who cited the commercials.

Just Checking

So that they can balance their inventory, an optometry shop collects the following data for customers in the shop.

		Eye Condition			
		Nearsighted	Farsighted	Need Bifocals	Total
Sex	Males	6	20	6	32
	Females	4	16	12	32
	Total	**10**	**36**	**18**	**64**

1 What percent of females are farsighted?
2 What percent of nearsighted customers are female?
3 What percent of all customers are farsighted females?
4 What's the distribution of *Eye Condition*?
5 What's the conditional distribution of *Eye Condition* for males?
6 Compare the percent who are female among nearsighted customers to the percent of all customers who are female.
7 Does it seem that *Eye Condition* and *Sex* might be dependent? Explain.

Segmented Bar Charts

We could display the Roper survey information by dividing up bars rather than circles as we did when making pie charts. The resulting **segmented bar chart** treats each bar as the "whole" and divides it proportionally into segments corresponding to the percentage in each group. We can see that the distributions of responses to the question are very different in the two countries, indicating again that *Regional Preference* is not independent of *Country*.

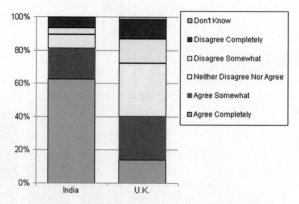

Figure 4.9   Although the totals for India and the United Kingdom are different, the bars are the same height because we have converted the numbers to percentages. Compare this display (created in **Excel**) with the side-by-side pie charts of the same data in Figure 4.7.

Guided Example   Food Safety

Food storage and food safety are major issues for multinational food companies. A client wants to know if people of all age groups have the same degree of concern so GfK Roper Consulting asked 1500 people in five countries whether they agree with the following statement: "I worry about how safe the food I buy is." We might want to report to a client who was interested in how concerns about food safety were related to age.

PLAN

Setup
- State the objectives and goals of the study.
- Identify and define the variables.
- Provide the time frame of the data collection process.

Determine the appropriate analysis for data type.

The client wants to examine the distribution of responses to the food safety question and see whether they are related to the age of the respondent. GfK Roper Consulting collected data on this question in the fall of 2005 for their 2006 Worldwide report. We will use the data from that study.

The variable is *Food Safety*. The responses are in nonoverlapping categories of agreement, from Agree Completely to Disagree Completely (and Don't Know). There were originally 12 Age groups, which we can combine into five:

Teen	13–19
Young Adult	20–29
Adult	30–39
Middle Aged	40–49
Mature	50 and older

Both variables, *Food Safety* and *Age*, are ordered categorical variables. To examine any differences in responses across age groups, it is appropriate to create a contingency table and a side-by-side bar chart.
Here is a contingency table of "Food Safety" by "Age."

DO

Mechanics For a large data set like this, we rely on technology to make table and displays.

	Food Safety						
	Agree Completely	Agree Somewhat	Neither Disagree Nor Agree	Disagree Somewhat	Disagree Completely	Don't Know	Total
Teen	16.19	27.50	24.32	19.30	10.58	2.12	100%
Young Adult	20.55	32.68	23.81	14.94	6.98	1.04	100%
Adult	22.23	34.89	23.28	12.26	6.75	0.59	100%
Middle Aged	24.79	35.31	22.02	12.43	5.06	0.39	100%
Mature	26.60	33.85	21.21	11.89	5.82	0.63	100%

Age (row label for the five age groups)

(continued)

A side-by-side bar chart is particularly helpful when comparing multiple groups.

A side-by-side bar chart shows the percent of each response to the question by Age group.

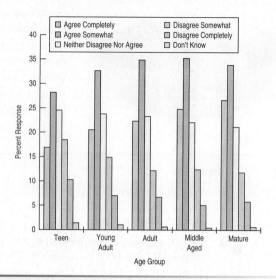

REPORT

Summary and Conclusions Summarize the charts and analysis in context. Make recommendations if possible and discuss further analysis that is needed.

MEMO

Re: Food safety concerns by age

Our analysis of the GfK Roper Reports™ Worldwide survey data for 2006 shows a pattern of concern about food safety that generally increases from youngest to oldest.

Our analysis thus far has not considered whether this trend is consistent across countries. If it were of interest to your group, we could perform a similar analysis for each of the countries.

The enclosed tables and plots provide support for these conclusions.

What Can Go Wrong?

- **Don't violate the area principle.** This is probably the most common mistake in a graphical display. Violations of the area principle are often made for the sake of artistic presentation. Here, for example, are two versions of the same pie chart for the *Regional Preference* data.

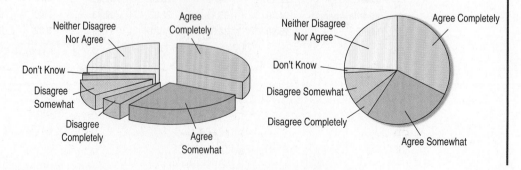

The one on the left looks interesting, doesn't it? But showing the pie three dimensionally on a slant violates the area principle and makes it much more difficult to compare fractions of the whole made up of each category of the response—the principal feature that a pie chart ought to show.

- **Keep it honest.** Here's a pie chart that displays data on the percentage of high school students who engage in specified dangerous behaviors as reported by the Centers for Disease Control. What's wrong with this plot?

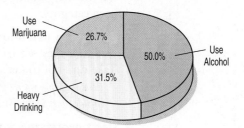

Try adding up the percentages. Or look at the 50% slice. Does it look right? Then think: What are these percentages of? Is there a "whole" that has been sliced up? In a pie chart, the proportions shown by each slice of the pie must add up to 100%, and each individual must fall into only one category. Of course, showing the pie on a slant makes it even harder to detect the error.

Here's another example. This bar chart shows the number of airline passengers searched by security screening.

Looks like things didn't change much in the final years of the 20th century—until you read the bar labels and see that the last three bars represent single years, while all the others are for *pairs* of years. The false depth makes it even harder to see the problem.

- **Don't confuse percentages.** Many percentages based on a conditional and joint distributions sound similar, but are different (see Table 4.5):
  - The percentage of French who answered "Agree Completely": This is 347/1539 or 22.55%.
  - The percentage of those who answered "Don't Know" who were French: This is 15/80 or 18.75%.
  - The percentage of those who were French *and* answered "Agree Completely": This is 347/7690 or 4.51%.

(continued)

In each instance, pay attention to the wording that makes a restriction to a smaller group (those who are French, those who answered "Don't Know," and all respondents, respectively) before a percentage is found. This restricts the *who* of the problem and the associated denominator for the percentage. Your discussion of results must make these differences clear.

- **Don't forget to look at the variables separately, too.** When you make a contingency table or display a conditional distribution, be sure to also examine the marginal distributions. It's important to know how many cases are in each category.

- **Be sure to use enough individuals.** When you consider percentages, take care that they are based on a large enough number of individuals (or cases). Take care not to make a report such as this one:

  We found that 66.67% of the companies surveyed improved their performance by hiring outside consultants. The other company went bankrupt.

- **Don't overstate your case.** Independence is an important concept, but it is rare for two variables to be *entirely* independent. We can't conclude that one variable has no effect whatsoever on another. Usually, all we know is that little effect was observed in our study. Other studies of other groups under other circumstances could find different results.

- **Don't use unfair or inappropriate percentages.** Sometimes percentages can be misleading. Sometimes they don't make sense at all. Be careful when finding percentages across different categories not to combine percentages inappropriately. The next section gives an example.

Simpson's Paradox

Here's an example showing that combining percentages across very different values or groups can give absurd results. Suppose there are two sales representatives, Peter and Katrina. Peter argues that he's the better salesperson, since he managed to close 83% of his last 120 prospects compared with Katrina's 78%. But let's look at the data a little more closely. Here (Table 4.8) are the results for each of their last 120 sales calls, broken down by the product they were selling.

Sales Rep	Product Printer Paper	Product USB Flash Drive	Product Overall
Peter	90 out of 100 90%	10 out of 20 50%	100 out of 120 83%
Katrina	19 out of 20 95%	75 out of 100 75%	94 out of 120 78%

Table 4.8   Look at the percentages within each Product category. Who has a better success rate closing sales of paper? Who has the better success rate closing sales of Flash Drives? Who has the better performance overall?

Look at the sales of the two products separately. For printer paper sales, Katrina had a 95% success rate, and Peter only had a 90% rate. When selling flash drives, Katrina closed her sales 75% of the time, but Peter only 50%. So Peter has better "overall" performance, but Katrina is better selling each product. How can this be?

This problem is known as **Simpson's Paradox**, named for the statistician who described it in the 1960s. Although it is rare, there have been a few well-publicized cases of it. As we can see from the example, the problem results from inappropriately combining percentages of different groups. Katrina concentrates on selling flash drives, which is more difficult, so her *overall* percentage is heavily influenced by her flash drive average. Peter sells more printer paper, which appears to be easier to sell. With their different patterns of selling, taking an overall percentage is misleading. Their manager should be careful not to conclude rashly that Peter is the better salesperson.

The lesson of Simpson's Paradox is to be sure to combine comparable measurements for comparable individuals. Be especially careful when combining across different levels of a second variable. It's usually better to compare percentages *within* each level, rather than across levels.

Discrimination?

One famous example of Simpson's Paradox arose during an investigation of admission rates for men and women at the University of California at Berkeley's graduate schools. As reported in an article in *Science*, about 45% of male applicants were admitted, but only about 30% of female applicants got in. It looked like a clear case of discrimination. However, when the data were broken down by school (Engineering, Law, Medicine, etc.), it turned out that within each school, the women were admitted at nearly the same or, in some cases, much *higher* rates than the men. How could this be? Women applied in large numbers to schools with very low admission rates. (Law and Medicine, for example, admitted fewer than 10%.) Men tended to apply to Engineering and Science. Those schools have admission rates above 50%. When the total applicant pool was combined and the percentages were computed, the women had a much lower *overall* rate, but the combined percentage didn't really make sense.

Ethics in Action

Lyle Erhart has been working in sales for a leading vendor of Customer Relationship Management (CRM) software for the past three years. He was recently made aware of a published research study that examined factors related to the successful implementation of CRM projects among firms in the financial services industry. Lyle read the research report with interest and was excited to see that his company's CRM software product was included. Among the results were tables reporting the number of projects that were successful based on type of CRM implementation (Operational versus Analytical) for each of the top leading CRM products of 2006. Lyle quickly found the results for his company's product and their major competitor. He summarized the results into one table as follows:

	His Company	Major Competitor
Operational	16 successes out of 20	68 successes out of 80
Analytical	90 successes out of 100	19 successes out of 20

At first he was a bit disappointed, especially since most of their potential clients were interested in Operational CRM. He had hoped to be able to disseminate the findings of this report among the sales force so they could refer to it when visiting potential clients. After some thought, he realized that he could combine the results. His company's overall success rate was 106 out of 120 (over 88%) and was higher than that of its major competitor. Lyle was now happy that he found and read the report.

ETHICAL ISSUE *Lyle, intentionally or not, has benefited from Simpson's Paradox. By combining percentages, he can present the findings in a manner favorable to his company (related to item A, ASA Ethical Guidelines).*

ETHICAL SOLUTION *Lyle should not combine the percentages as the results are misleading. If he decides to disseminate the information to his sales force, he must do so without combining.*

What Have We Learned?

Learning Objectives	▪ Make and interpret a frequency table for a categorical variable.
	• We can summarize categorical data by counting the number of cases in each category, sometimes expressing the resulting distribution as percentages.
	▪ Make and interpret a bar chart or pie chart.
	• We display categorical data using the area principle in either a **bar chart** or a **pie chart**.
	▪ Make and interpret a contingency table.
	• When we want to see how two categorical variables are related, we put the counts (and/or percentages) in a two-way table called a **contingency table**.
	▪ Make and interpret bar charts and pie charts of marginal distributions.
	• We look at the **marginal distribution** of each variable (found in the margins of the table). We also look at the **conditional distribution** of a variable within each category of the other variable.
	• Comparing conditional distributions of one variable across categories of another tells us about the association between variables. If the conditional distributions of one variable are (roughly) the same for every category of the other, the variables are **independent**.

Terms	
Area principle	In a statistical display, each data value is represented by the same amount of area.
Bar chart (relative frequency bar chart)	A chart that represents the count (or percentage) of each category in a categorical variable as a bar, allowing easy visual comparisons across categories.
Cell	Each location in a contingency table, representing the values of two categorical variables, is called a cell.
Segmented bar chart	A segmented bar chart displays the conditional distribution of a categorical variable within each category of another variable.
Column percent	The proportion of each column contained in the cell of a frequency table.
Conditional distribution	The distribution of a variable restricting the *who* to consider only a smaller group of individuals.
Contingency table	A table displaying the frequencies (sometimes percentages) for each combination of two or more variables.
Distribution	The distribution of a variable is a list of:
	• all the possible values of the variable
	• the relative frequency of each value
Frequency table (relative frequency table)	A table that lists the categories in a categorical variable and gives the number (the percentage) of observations for each category. The row percent is the proportion of each row contained in the cell of a frequency table, while the column percent is the proportion of each column contained in the cell of a frequency table.
Independent variables	Variables for which the conditional distribution of one variable is the same for each category of the other.
Marginal distribution	In a contingency table, the distribution of either variable alone. The counts or percentages are the totals found in the margins (usually the right-most column or bottom row) of the table.
Pie chart	Pie charts show how a "whole" divides into categories by showing a wedge of a circle whose area corresponds to the proportion in each category.

Row percent	The proportion of each row contained in the cell of a frequency table.
Simpson's paradox	A phenomenon that arises when averages, or percentages, are taken across different groups, and these group averages appear to contradict the overall averages.
Total percent	The proportion of the total contained in the cell of a frequency table.

Technology Help: Displaying Categorical Data on the Computer

Although every package makes a slightly different bar chart, they all have similar features:

May have a box around it or not

You may be able to add color later on in some programs

Counts or relative frequencies on this axis

Bar order may be arbitrary, alphabetical, or by first occurrence of the category

Bar charts should have spaces between the bars

Sometimes the count or a percentage is printed above or on top of each bar to give some additional information. You may find that your statistics package sorts category names in annoying orders by default. For example, many packages sort categories alphabetically or by the order the categories are seen in the data set. Often, neither of these is the best choice.

EXCEL 2007 XLSTAT

To make a bar chart:

- Select the variable in Excel you want to work with.
- Choose the **Column** command from the Insert tab in the Ribbon.
- Select the appropriate chart from the drop down dialog.

To change the bar chart into a pie chart:

- Right-click the chart and select **Change Chart Type**. . . from the menu. The Chart type dialog opens.
- Select a pie chart type.
- Click the **OK** button. Excel changes your bar chart into a pie chart.

(continued)

JMP

JMP makes a bar chart and frequency table together.

- From the **Analyze** menu, choose **Distribution.**
- In the Distribution dialog, drag the name of the variable into the empty variable window beside the label "Y, Columns"; click **OK.**
- To make a pie chart, choose **Chart** from the **Graph** menu.
- In the Chart dialog, select the variable name from the Columns list, click on the button labeled "Statistics," and select "N" from the drop-down menu.
- Click the "**Categories, X, Levels**" button to assign the same variable name to the X-axis.
- Under Options, click on the **second** button—labeled "**Bar Chart**"—and select "Pie" from the drop-down menu.

MINITAB

To make a bar chart,

- Choose **Bar Chart** from the **Graph** menu.
- Then select a Simple, Cluster, or Stack chart from the options and click **OK.**

- To make a **Simple** bar chart, enter the name of the variable to graph in the dialog box.
- To make a relative frequency chart, click **Chart Options**, and choose **Show Y as Percent.**
- In the Chart dialog, enter the name of the variable that you wish to display in the box labeled "Categorical variables."
- Click **OK.**

SPSS

To make a bar chart,

- Open the **Chart Builder** from the **Graphs** menu.
- Click the **Gallery** tab.
- Choose **Bar Chart** from the list of chart types.
- Drag the appropriate bar chart onto the canvas.
- Drag a categorical variable onto the x-axis drop zone.
- Click **OK.**

Comments

A similar path makes a pie chart by choosing **Pie chart** from the list of chart types.

Brief CASE   KEEN

More of the data that KEEN, Inc. obtained from *Google Analytics* are in the file **KEEN**. Open the data file using a statistics package and find data on *Country of Origin, Top Keywords, Online Retailers, User Statistics,* and *Page Visits.* Create frequency tables, bar charts, and pie charts using your software. What might KEEN want to know about their Web traffic? Which of these tables and charts is most useful to address the question of where they should advertise and how they should position their products? Write a brief case report summarizing your analysis and results.

Exercises

SECTION 4.1

1. As part of the human resource group of your company you are asked to summarize the educational levels of the 512 employees in your division. From company records, you find that 164 have no college degree (None), 42 have an associate's degree (AA), 225 have a bachelor's degree (BA), 52 have a master's degree (MA), and 29 have PhDs. For the educational level of your division:

a) Make a frequency table.
b) Make a relative frequency table.

2. As part of the marketing group at Pixar, you are asked to find out the age distribution of the audience of Pixar's latest film. With the help of 10 of your colleagues, you conduct exit interviews by randomly selecting people to question at 20 different movie theatres. You ask them to tell you if they are younger than 6 years old, 6 to 9 years old, 10 to 14 years old, 15 to 21 years old, or older than 21. From 470 responses, you find out that 45 are younger than 6, 83 are 6 to 9 years old, 154 are 10 to 14, 18 are 15 to 21, and 170 are older than 21. For the age distribution:

a) Make a frequency table.
b) Make a relative frequency table.

SECTION 4.2

3. From the educational level data described in Exercise 1:

a) Make a bar chart using counts on the *y*-axis.
b) Make a relative frequency bar chart using percentages on the *y*-axis.
c) Make a pie chart.

4. From the age distribution data described in Exercise 2:

a) Make a bar chart using counts on the *y*-axis.
b) Make a relative frequency bar chart using percentages on the *y*-axis.
c) Make a pie chart.

5. For the educational levels described in Exercise 1:

a) Write two to four sentences summarizing the distribution.
b) What conclusions, if any, could you make about the educational level at other companies?

6. For the ages described in Exercise 2:

a) Write two to four sentences summarizing the distribution.
b) What possible problems do you see in concluding that the age distribution from these surveys accurately represents the ages of the national audience for this film?

SECTION 4.3

7. From Exercise 1, we also have data on how long each person has been with the company (tenure) categorized into three levels: less than 1 year, between 1 and 5 years, and more than 5 years. A table of the two variables together looks like:

	None	AA	BA	MA	PhD
<1 year	10	3	50	20	12
1–5 years	42	9	112	27	15
more than 5 years	112	30	63	5	2

a) Find the marginal distribution of the tenure. (*Hint*: find the row totals.)
b) Verify that the marginal distribution of the education level is the same as that given in Exercise 1.

8. In addition to their age levels, the movie audiences in Exercise 2 were also asked if they had seen the movie before (Never, Once, More than Once). Here is a table showing the responses by age group:

	Under 6	6 to 9	10 to 14	15 to 21	Over 21
Never	39	60	84	16	151
Once	3	20	38	2	15
More than Once	3	3	32	0	4

a) Find the marginal distribution of their previous viewing of the movie. (Hint: find the row totals.)
b) Verify that the marginal distribution of the ages is the same as that given in Exercise 2.

9. For the table in Exercise 7,

a) Find the column percentages.
b) Looking at the column percentages in part a, does the *tenure* distribution (how long the employee has been with the company) for each educational level look the same? Comment briefly.
c) Make a stacked bar chart showing the *tenure* distribution for each educational level.
d) Is it easier to see the differences in the distributions using the column percentages or the stacked bar chart?

10. For the table in Exercise 8,

a) Find the column percentages.
b) Looking at the column percentages in part a, does the distribution of how many times someone has seen the movie look the same for each age group? Comment briefly.
c) Make a stacked bar chart, showing the distribution of viewings for each age level.

d) Is it easier to see the differences in the distributions using the column percentages or the stacked bar chart?

CHAPTER EXERCISES

11. Graphs in the news. Find a bar graph of categorical data from a business publication (e.g., *Business Week, Fortune, The Wall Street Journal*, etc.).

a) Is the graph clearly labeled?
b) Does it violate the area principle?
c) Does the accompanying article tell the W's of the variable?
d) Do you think the article correctly interprets the data? Explain.

12. Graphs in the news, part 2. Find a pie chart of categorical data from a business publication (e.g., *Business Week, Fortune, The Wall Street Journal*, etc.).

a) Is the graph clearly labeled?
b) Does it violate the area principle?
c) Does the accompanying article tell the W's of the variable?
d) Do you think the article correctly interprets the data? Explain.

13. Tables in the news. Find a frequency table of categorical data from a business publication (e.g., *Business Week, Fortune, The Wall Street Journal*, etc.).

a) Is it clearly labeled?
b) Does it display percentages or counts?
c) Does the accompanying article tell the W's of the variable?
d) Do you think the article correctly interprets the data? Explain.

14. Tables in the news, part 2. Find a contingency table of categorical data from a business publication (e.g., *Business Week, Fortune, The Wall Street Journal*, etc.).

a) Is it clearly labeled?
b) Does it display percentages or counts?
c) Does the accompanying article tell the W's of the variable?
d) Do you think the article correctly interprets the data? Explain.

15. U.S. market share. An article in the *The Wall Street Journal* (March 16, 2007) reported the 2006 U.S. market share of leading sellers of carbonated drinks, summarized in the following pie chart:

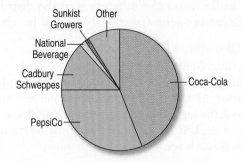

a) Is this an appropriate display for these data? Explain.
b) Which company had the largest share of the market?

16. World market share. *The Wall Street Journal* article described in Exercise 15 also indicated the 2005 world market share for leading distributors of total confectionery products. The following bar chart displays the values:

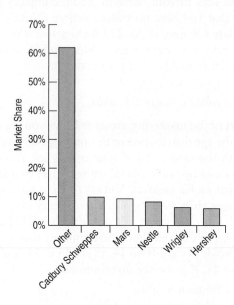

a) Is this an appropriate display for these data? Explain.
b) Which company had the largest share of the candy market?

17. Market share again. Here's a bar chart of the data in Exercise 15.

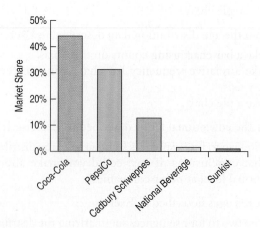

a) Compared to the pie chart in Exercise 15, which is better for displaying the relative portions of market share? Explain.
b) What is missing from this display that might make it misleading?

18. World market share again. Here's a pie chart of the data in Exercise 16.

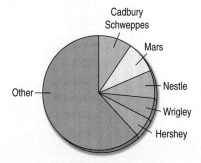

a) Which display of these data is best for comparing the market shares of these companies? Explain.
b) Does Cadbury Schweppes or Mars have a bigger market share?

19. Insurance company. An insurance company is updating its payouts and cost structure for their insurance policies. Of particular interest to them is the risk analysis for customers currently on heart or blood pressure medication. The Centers for Disease Control lists causes of death in the United States during one year as follows.

Cause of Death	Percent
Heart disease	30.3
Cancer	23.0
Circulatory diseases and stroke	8.4
Respiratory diseases	7.9
Accidents	4.1

a) Is it reasonable to conclude that heart or respiratory diseases were the cause of approximately 38% of U.S. deaths during this year?
b) What percent of deaths were from causes not listed here?
c) Create an appropriate display for these data.

20. Revenue growth. A 2005 study by Babson College and The Commonwealth Institute surveyed the top women-led businesses in the state of Massachusetts in 2003 and 2004. The study reported the following results for continuing participants with a 9% response rate. (Does not add up to 100% due to rounding.)

2003–2004 Revenue Growth	
Decline	7%
Modest Decline	9%
Steady State	10%
Modest Growth	18%
Growth	54%

a) Describe the distribution of companies with respect to revenue growth.
b) Is it reasonable to conclude that 72% of all women-led businesses in the U.S. reported some level of revenue growth? Explain.

21. Web conferencing. Cisco Systems Inc. announced plans in March 2007 to buy WebEx Communications, Inc. for $3.2 billion, demonstrating their faith in the future of Web conferencing. The leaders in market share for the vendors in the area of Web conferencing in 2006 are as follows: WebEx 58.4% and Microsoft 26.3%. Create an appropriate graphical display of this information and write a sentence or two that might appear in a newspaper article about the market share.

22. Mattel. In their 2006 annual report, Mattel Inc. reported that their domestic market sales were broken down as follows: 44.1% Mattel Girls and Boys brand, 43.0% Fisher-Price brand and the rest of the nearly $3.5 billion revenues were due to their American Girl brand. Create an appropriate graphical display of this information and write a sentence or two that might appear in a newspaper article about their revenue breakdown.

23. Small business productivity. The Wells Fargo/Gallup Small Business Index asked 592 small business owners in March 2004 what steps they had taken in the past year to increase productivity. They found that 60% of small business owners had updated their computers, 52% had made other (noncomputer) capital investments, 37% hired part-time instead of full-time workers, 24% had not replaced workers who left voluntarily, 15% had laid off workers, and 10% had lowered employee salaries.

a) What do you notice about the percentages listed? How could this be?
b) Make a bar chart to display the results and label it clearly.
c) Would a pie chart be an effective way of communicating this information? Why or why not?
d) Write a couple of sentences on the steps taken by small businesses to increase productivity.

24. Small business hiring. In 2004, the Wells Fargo/Gallup Small Business Index found that 86% of the 592 small business owners they surveyed said their productivity for the previous year had stayed the same or increased and most had substituted productivity gains for labor. (See Exercise 23.) As a follow-up question, the survey gave them a list of possible economic outcomes and asked if that would make them hire more employees. Here are the percentages of owners saying that they would "definitely or probably hire more employees" for each scenario: a substantial increase in sales—79%, a major backlog of sales orders—71%, a general improvement in the economy—57%, a gain in productivity—50%, a reduction in overhead costs—43%, and more qualified employees available—39%.

a) What do you notice about the percentages listed?

b) Make a bar chart to display the results and label it clearly.

c) Would a pie chart be an effective way of communicating this information? Why or why not?

d) Write a couple of sentences on the responses to small business owners about hiring given the scenarios listed.

25. Environmental hazard. Data from the International Tanker Owners Pollution Federation Limited (www.itopf .com) give the cause of spillage for 312 large oil tanker accidents from 1974–2006. Here are the displays. Write a brief report interpreting what the displays show. Is a pie chart an appropriate display for these data? Why or why not?

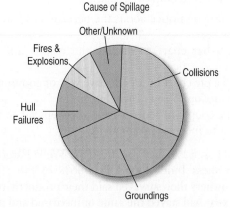

26. Winter Olympics. Twenty-six countries won medals in the 2010 Winter Olympics. The following table lists them, along with the total number of medals each won.

a) Try to make a display of these data. What problems do you encounter?

b) Can you find a way to organize the data so that the graph is more successful?

Country	Medals	Country	Medals
United States	37	Poland	6
Germany	30	Italy	5
Canada	26	Japan	5
Norway	23	Finland	5
Austria	16	Australia	3
Russia	15	Belarus	3
South Korea	14	Slovakia	3
China	11	Croatia	3
Sweden	11	Slovenia	3
France	11	Latvia	2
Switzerland	9	Great Britain	1
Netherlands	8	Estonia	1
Czech Republic	6	Kazakhstan	1

27. Importance of wealth. GfK Roper Reports Worldwide surveyed people in 2004, asking them "How important is acquiring wealth to you?" The percent who responded that it was of more than average importance were: 71.9% China, 59.6% France, 76.1% India, 45.5% UK, and 45.3% USA. There were about 1500 respondents per country. A report showed the following bar chart of these percentages.

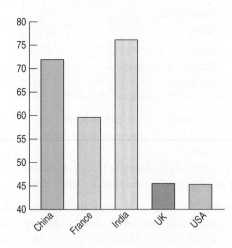

a) How much larger is the proportion of those who said acquiring wealth was important in India than in the United States?

b) Is that the impression given by the display? Explain.

c) How would you improve this display?

d) Make an appropriate display for the percentages.

e) Write a few sentences describing what you have learned about attitudes toward acquiring wealth.

28. Importance of power. In the same survey as that discussed in Exercise 27, GfK Roper Consulting also asked

"How important is having control over people and resources to you?" The percent who responded that it was of more than average importance are given in the following table:

China	49.1%
France	44.1%
India	74.2%
UK	27.8%
USA	36.0%

Here's a pie chart of the data:

a) List the errors you see in this display.
b) Make an appropriate display for the percentages.
c) Write a few sentences describing what you have learned about attitudes toward acquiring power.

29. Google financials. Google Inc. derives revenue from three major sources: advertising revenue from their websites, advertising revenue from the thousands of third-party websites that comprise the Google Network, and licensing and miscellaneous revenue. The following table shows the percentage of all revenue derived from these sources for the period 2002 to 2006.

		Year				
		2002	2003	2004	2005	2006
Revenue Source	Google websites	70%	54%	50%	55%	60%
	Google network websites	24%	43%	49%	44%	39%
	Licensing & other revenue	6%	3%	1%	1%	1%

a) Are these row or column percentages?
b) Make an appropriate display of these data.
c) Write a brief summary of this information.

30. Real estate pricing. A study of a sample of 1057 houses in upstate New York reports the following percentages of houses falling into different Price and Size categories.

		Price			
		Low	Med Low	Med High	High
Size	Small	61.5%	35.2%	5.2%	2.4%
	Med Small	30.4%	45.3%	26.4%	4.7%
	Med Large	5.4%	17.6%	47.6%	21.7%
	Large	2.7%	1.9%	20.8%	71.2%

a) Are these column, row, or total percentages? How do you know?
b) What percent of the highest priced houses were small?
c) From this table, can you determine what percent of all houses were in the low price category?
d) Among the lowest prices houses, what percent were small or medium small?
e) Write a few sentences describing the association between *Price* and *Size*.

31. Stock performance. The following table displays information for 40 widely held U.S. stocks, on how their one-day change on March 15, 2007 compared with their previous 52-week change.

		Over prior 52 weeks	
		Positive Change	Negative Change
MARCH 15, 2007	Positive Change	14	9
	Negative Change	11	6

a) What percent of the companies reported a positive change in their stock price over the prior 52 weeks?
b) What percent of the companies reported a positive change in their stock price over both time periods?
c) What percent of the companies reported a negative change in their stock price over both time periods?
d) What percent of the companies reported a positive change in their stock price over one period and then a negative change in the other period?
e) Among those companies reporting a positive change in their stock price over the prior day what percentage also reported a positive change over the prior year?
f) Among those companies reporting a negative change in their stock price over the prior day what percentage also reported a positive change over the prior year?
g) What relationship, if any, do you see between the performance of a stock on a single day and its 52-week performance?

32. New product. A company started and managed by business students is selling campus calendars. The students have conducted a market survey with the various campus constituents to determine sales potential and identify which market segments should be targeted. (Should they advertise in the Alumni Magazine and/or the local newspaper?) The following table shows the results of the market survey.

Buying Likelihood

Campus Group	Unlikely	Moderately Likely	Very Likely	Total
Students	197	388	320	905
Faculty/Staff	103	137	98	338
Alumni	20	18	18	56
Town Residents	13	58	45	116
Total	333	601	481	1415

a) What percent of all these respondents are alumni?
b) What percent of these respondents are very likely to buy the calendar?
c) What percent of the respondents who are very likely to buy the calendar are alumni?
d) Of the alumni, what percent are very likely to buy the calendar?
e) What is the marginal distribution of the campus constituents?
f) What is the conditional distribution of the campus constituents among those very likely to buy the calendar?
g) Does this study present any evidence that this company should focus on selling to certain campus constituents?

33. Real estate. *The Wall Street Journal* reported in March 2007 that the real estate market in Nashville, Tennessee, slowed slightly from 2006 to 2007. The supporting data are summarized in the following table.

Type of Sale

Year	Condos	Farms/Land	Residential	Multi-family	Total
2006	266	177	2119	48	2610
2007	341	190	2006	38	2575
Total	607	367	4125	86	5185

a) What percent of all sales in February 2006 were condominiums (condos)? In February 2007?
b) What percent of the sales in February 2006 were multifamily? In February 2007?
c) What was the change in the percent of residential real estate sales in Nashville, Tennessee, from February 2006 to February 2007?

34. Google financials, part 2. Google Inc. divides their total costs and expenses into five categories: cost of revenues, research and development, sales and marketing, general administrative, and miscellaneous. See the table at the bottom of the page.

a) What percent of all costs and expenses were cost of revenues in 2005? In 2006?
b) What percent of all costs and expenses were due to research and development in 2005? In 2006?
c) Have general administrative costs grown as a percentage of all costs and expenses over this time period?

35. Movie ratings. The movie ratings system is a voluntary system operated jointly by the Motion Picture Association of America (MPAA) and the National Association of Theatre Owners (NATO). The ratings themselves are given by a board of parents who are members of the Classification and Ratings Administration (CARA). The board was created in response to outcries from parents in the 1960s for some kind of regulation of film content, and the first ratings were introduced in 1968. Here is information on

Cost and Expenses	2002	2003	2004	2005	2006
Cost of revenues	$132,575	$634,411	$1,468,967	$2,577,088	$4,225,027
Research and development	$40,494	$229,605	$385,164	$599,510	$1,228,589
Sales and marketing	$48,783	$164,935	$295,749	$468,152	$849,518
General administrative	$31,190	$94,519	$188,151	$386,532	$751,787
Miscellaneous	$0	$0	$201,000	$90,000	$0
Total Costs and Expenses	$253,042	$1,123,470	$2,539,031	$4,121,282	$7,054,921

Table for Exercise 34

the ratings of 120 movies that came out in 2005, also classified by their genre.

	Rating				
Genre	**G**	**PG**	**PG-13**	**R**	**Total**
Action/Adventure	4	5	17	9	35
Comedy	2	12	20	4	38
Drama	0	3	8	17	28
Thriller/Horror	0	0	11	8	19
Total	**6**	**20**	**56**	**38**	**120**

a) Find the conditional distribution (in percentages) of movie ratings for action/adventure films.
b) Find the conditional distribution (in percentages) of movie ratings for thriller/horror films.
c) Create a graph comparing the ratings for the four genres.
d) Are *Genre* and *Rating* independent? Write a brief summary of what these data show about movie ratings and the relationship to the genre of the film.

36. Wireless access. The Pew Internet and American Life Project has monitored access to the Internet since the 1990s. Here is an income breakdown of 798 Internet users surveyed in December 2006, asking whether they have logged on to the Internet using a wireless device or not.

	Wireless Users	**Other Internet Users**	**Total**
Income			
Under $30K	34	128	162
$30K–50K	31	133	164
$50K–$75K	44	72	116
Over $75K	83	111	194
Don't know/refused	51	111	162
Total	**243**	**555**	**798**

a) Find the conditional distribution (in percentages) of income distribution for the wireless users.
b) Find the conditional distribution (in percentages) of income distribution for other Internet users.
c) Create a graph comparing the income distributions of the two groups.
d) Do you see any differences between the conditional distributions? Write a brief summary of what these data show about wireless use and its relationship to income.

37. MBAs. A survey of the entering MBA students at a university in the United States classified the country of origin of the students, as seen in the table.

	MBA Program		
Origin	**Two-Year MBA**	**Evening MBA**	**Total**
Asia/Pacific Rim	31	33	64
Europe	5	0	5
Latin America	20	1	21
Middle East/Africa	5	5	10
North America	103	65	168
Total	**164**	**104**	**268**

a) What percent of all MBA students were from North America?
b) What percent of the Two-Year MBAs were from North America?
c) What percent of the Evening MBAs were from North America?
d) What is the marginal distribution of origin?
e) Obtain the column percentages and show the conditional distributions of origin by MBA Program.
f) Do you think that origin of the MBA student is independent of the MBA program? Explain.

38. MBAs, part 2. The same university as in Exercise 37 reported the following data on the gender of their students in their two MBA programs.

	Type		
Sex	**Two-Year**	**Evening**	**Total**
Men	116	66	182
Women	48	38	86
Total	**164**	**104**	**268**

a) What percent of all MBA students are women?
b) What percent of Two-Year MBAs are women?
c) What percent of Evening MBAs are women?
d) Do you see evidence of an association between the *Type* of MBA program and the percentage of women students? If so, why do you believe this might be true?

T **39.** Top producing movies. The following table shows the Motion Picture Association of America (MPA) (www.mpaa.org) ratings for the top 20 grossing films in the United States for each of the 10 years from 1999 to 2008. (Data are number of films.)

a) What percent of all these top 20 films are G rated?
b) What percent of all top 20 films in 2005 were G rated?
c) What percent of all top 20 films were PG-13 and came out in 1999?
d) What percent of all top 20 films produced in 2006 or later were PG-13?

	Rating				
	G	PG	PG-13	R	Total
2008	2	4	10	4	20
2007	1	5	11	3	20
2006	1	4	13	2	20
2005	1	4	13	2	20
2004	1	6	10	3	20
2003	1	3	11	5	20
2002	1	6	13	0	20
2001	2	4	10	4	20
2000	0	3	12	5	20
1999	2	3	7	8	20
Total	**12**	**42**	**110**	**36**	**200**

(left column, *Year* labels the rows)

e) What percent of all top 20 films produced from 1999 to 2002 were rated PG-13 or R?

f) Compare the conditional distributions of the ratings for films produced in 2004 or later to those produced from 1999 to 2003. Write a couple of sentences summarizing what you see.

T 40. Movie admissions. The following table shows attendance data collected by the Motion Picture Association of America during the period 2002 to 2006. Figures are in millions of movie admissions.

	Patron Age						
	12 to 24	25 to 29	30 to 39	40 to 49	50 to 59	60 and over	Total
2006	485	136	246	219	124	124	1334
2005	489	135	194	216	125	122	1281
2004	567	132	265	236	145	132	1477
2003	567	124	269	193	152	118	1423
2002	551	158	237	211	119	130	1406
Total	**2659**	**685**	**1211**	**1075**	**665**	**626**	**6921**

(left column, *Year* labels the rows)

a) What percent of all admissions during this period were bought by people between the ages of 12 and 24?

b) What percent of admissions in 2003 were bought by people between the ages of 12 and 24?

c) What percent of the admission were bought by people between the ages of 12 and 24 in 2006?

d) What percent of admissions in 2006 were bought by people 60 years old and older?

e) What percent of the admissions bought by people 60 and over were in 2002?

f) Compare the conditional distributions of the age groups across years. Write a couple of sentences summarizing what you see.

41. Tattoos. A study by the University of Texas Southwestern Medical Center examined 626 people to see if there was an increased risk of contracting hepatitis C associated with having a tattoo. If the subject had a tattoo, researchers asked whether it had been done in a commercial tattoo parlor or elsewhere. Write a brief description of the association between tattooing and hepatitis C, including an appropriate graphical display.

	Tatto done in commercial parlor	Tattoo done elsewhere	No tattoo
Has hepatitis C	17	8	18
No hepatitis C	35	53	495

42. Working parents. In July 1991 and again in April 2001, the Gallup Poll asked random samples of 1015 adults about their opinions on working parents. The following table summarizes responses to this question: *"Considering the needs of both parents and children, which of the following do you see as the ideal family in today's society?"* Based upon these results, do you think there was a change in people's attitudes during the 10 years between these polls? Explain.

		Year	
		1991	2001
	Both work full-time	142	131
	One works full-time, other part-time	274	244
	One works, other works at home	152	173
	One works, other stays home for kids	396	416
	No opinion	51	51

(left column, *Response* labels the rows)

43. Revenue growth, last one. The study completed in 2005 and described in Exercise 20 also reported on education levels of the women chief executives. The column percentages for CEO education for each level of revenue are summarized in the following table. (Revenue is in $ million.)

	Graduate Education and Firm Revenue Size		
	< $10 M revenue	$10–$49.999 M revenue	≥ $50 M revenue
% with High School Education only	8%	4%	8%
% with College Education, but no Graduate Education	48%	42%	33%
% with Graduate Education	44%	54%	59%
Total	**100%**	**100%**	**100%**

a) What percent of these CEOs in the highest revenue category had only a high school education?
b) From this table, can you determine what percent of all these CEOs had graduate education? Explain.
c) Among the CEOs in the lowest revenue category, what percent had more than a high school education?
d) Write a few sentences describing the association between *Revenue* and *Education*.

T 44. Minimum wage workers. The U.S. Department of Labor (www.bls.gov) collects data on the number of U.S. workers who are employed at or below the minimum wage. Here is a table showing the number of hourly workers by *Age* and *Sex* and the number who were paid at or below the prevailing minimum wage:

Age	Hourly Workers (in thousands)		At or Below Minimum Wage (in thousands)	
	Men	Women	Men	Women
16–24	7978	7701	384	738
25–34	9029	7864	150	332
35–44	7696	7783	71	170
45–54	7365	8260	68	134
55–64	4092	4895	35	72
65+	1174	1469	22	50

a) What percent of the women were ages 16 to 24?
b) Using side-by-side bar graphs, compare the proportions of the men and women who worked at or below minimum wage at each *Age* group. Write a couple of sentences summarizing what you see.

45. Moviegoers and ethnicity. The Motion Picture Association of America studies the ethnicity of moviegoers to understand changes in the demographics of moviegoers over time. Here are the numbers of moviegoers (in millions) classified as to whether they were Hispanic, African-American, or Caucasian for the years 2002 to 2006.

	Year					
Ethnicity	2002	2003	2004	2005	2006	Total
Hispanic	21	23	25	25	26	120
African-American	21	20	22	21	20	104
Caucasian	118	127	127	113	120	605
Total	160	170	174	159	166	829

a) Find the marginal distribution *Ethnicity* of moviegoers.
b) Find the conditional distribution of *Ethnicity* for the year 2006.
c) Compare the conditional distribution of *Ethnicity* for all 5 years with a segmented bar graph.
d) Write a brief description of the association between *Year* and *Ethnicity* among these respondents.

46. Department store. A department store is planning its next advertising campaign. Since different publications are read by different market segments, they would like to know if they should be targeting specific age segments. The results of a marketing survey are summarized in the following table by *Age* and *Shopping Frequency* at their store.

	Age			
Shopping	Under 30	30–49	50 and Over	Total
Low	27	37	31	95
Moderate	48	91	93	232
High	23	51	73	147
Total	98	179	197	474

a) Find the marginal distribution of *Shopping Frequency*.
b) Find the conditional distribution of *Shopping Frequency* within each age group.
c) Compare these distributions with a segmented bar graph.
d) Write a brief description of the association between *Age* and *Shopping Frequency* among these respondents.
e) Does this prove that customers ages 50 and over are more likely to shop at this department store? Explain.

47. Women's business centers. A study conducted in 2002 by Babson College and the Association of Women's Centers surveyed women's business centers in the United States. The data showing the location of established centers (at least 5 years old) and less established centers are summarized in the following table.

	Location	
	Urban	Nonurban
Less Established	74%	26%
Established	80%	20%

a) Are these percentages column percentages, row percentages, or table percentages?
b) Use graphical displays to compare these percentages of women's business centers by location.

48. Advertising. A company that distributes a variety of pet foods is planning their next advertising campaign. Since different publications are read by different market segments, they would like to know how pet ownership is distributed across different income segments. The U.S. Census Bureau reports the number of households owning various types of pets. Specifically, they keep track of dogs, cats, birds, and horses.

INCOME DISTRIBUTION OF HOUSEHOLDS OWING PETS (PERCENT)

		Dog	Cat	Bird	Horse
Income	Under $12,500	14	15	16	9
	$12,500 to $24,999	20	20	21	21
	$25,000 to $39,999	24	23	24	25
	$40,000 to $59,999	22	22	21	22
	$60,000 and over	20	20	18	23
	Total	**100**	**100**	**100**	**100**

Column header spanning Dog Cat Bird Horse: **Pet**

a) Do you think the income distributions of the households who own these different animals would be roughly the same? Why or why not?
b) The table shows the percentages of income levels for each type of animal owned. Are these row percentages, column percentages, or table percentages?
c) Do the data support that the pet food company should not target specific market segments based on household income? Explain.

49. Worldwide toy sales. Around the world, toys are sold through different channels. For example, in some parts of the world toys are sold primarily through large toy store chains, while in other countries department stores sell more toys. The following table shows the percentages by region of the distribution of toys sold through various channels in Europe and America in 2003, accumulated by the International Council of Toy Industries (www.toy-icti.org).

a) Are these row percentages, column percentages, or table percentages?
b) Can you tell what percent of toys sold by mail order in both Europe and America are sold in Europe? Why or why not?
c) Use a graphical display to compare the distribution of channels between Europe and America.

d) Summarize the distribution of toy sales by channel in a few sentences. What are the biggest differences between these two continents?

50. Internet piracy. Illegal downloading of copyrighted movies is an international problem estimated to have cost the international movie industry more than $18 billion in 2005. The typical pirate worldwide is a 16 to 24-year old male living in an urban area, according to a study by the international strategy consulting firm LEK (www.mpaa.org/researchStatistics.asp). The following table compares the age distribution of the U.S. pirate to the rest of the world.

		Age			
		16–24	25–29	30–39	Over 40
Region	United States	71	11	7	11
	Rest of World	58	15	18	9

a) Are these row percentages, column percentages, or table percentages?
b) Can you tell what percent of pirates worldwide are in the 16 to 24 age group?
c) Use a graphical display to compare the age distribution of pirates in the United States to the distribution in the rest of the world.
d) Summarize the distribution of *Age* by *Region* in a few sentences. What are the biggest differences between these two regions?

51. Insurance company, part 2. An insurance company that provides medical insurance is concerned with recent data. They suspect that patients who undergo surgery at large hospitals have their discharges delayed for various reasons—which results in increased medical costs to the insurance company. The recent data for area hospitals and two types of surgery (major and minor) are shown in the following table.

		Discharge Delayed	
		Large Hospital	Small Hospital
Procedure	Major surgery	120 of 800	10 of 50
	Minor surgery	10 of 200	20 of 250

		Channel					
		General Merchandise	Toy Specialists	Department Stores	Mass Merchant Discounters & Food Hypermarkets	Mail Order	Other
Location	America	9%	25%	3%	51%	4%	8%
	Europe	13%	36%	7%	24%	5%	15%

Table for Exercise 49

a) Overall, for what percent of patients was discharge delayed?

b) Were the percentages different for major and minor surgery?

c) Overall, what were the discharge delay rates at each hospital?

d) What were the delay rates at each hospital for each kind of surgery?

e) The insurance company is considering advising their clients to use large hospitals for surgery to avoid postsurgical complications. Do you think they should do this?

f) Explain, in your own words, why this confusion occurs.

52. Delivery service. A company must decide which of two delivery services they will contract with. During a recent trial period, they shipped numerous packages with each service and have kept track of how often deliveries did not arrive on time. Here are the data.

Delivery Service	Type of Service	Number of Deliveries	Number of Late Packages
Pack Rats	Regular	400	12
	Overnight	100	16
Boxes R Us	Regular	100	2
	Overnight	400	28

a) Compare the two services' overall percentage of late deliveries.

b) Based on the results in part a, the company has decided to hire Pack Rats. Do you agree they deliver on time more often? Why or why not? Be specific.

c) The results here are an instance of what phenomenon?

53. Graduate admissions. A 1975 article in the magazine *Science* examined the graduate admissions process at Berkeley for evidence of gender bias. The following table shows the number of applicants accepted to each of four graduate programs.

Program	Males Accepted (of Applicants)	Females Accepted (of Applicants)
1	511 of 825	89 of 108
2	352 of 560	17 of 25
3	137 of 407	132 of 375
4	22 of 373	24 of 341
Total	**1022 of 2165**	**262 of 849**

a) What percent of total applicants were admitted?

b) Overall, were a higher percentage of males or females admitted?

c) Compare the percentage of males and females admitted in each program.

d) Which of the comparisons you made do you consider to be the most valid? Why?

54. Simpson's Paradox. Develop your own table of data that is a business example of Simpson's Paradox. Explain the conflict between the conclusions made from the conditional and marginal distributions.

Just Checking Answers

1 50.0%
2 40.0%
3 25.0%
4 15.6% Nearsighted, 56.3% Farsighted, 28.1% Need Bifocals
5 18.8% Nearsighted, 62.5% Farsighted, 18.8% Need Bifocals
6 40% of the nearsighted customers are female, while 50% of customers are female.
7 Since nearsighted customers appear less likely to be female, it seems that they may not be independent. (But the numbers are small.)

Displaying and Describing Quantitative Data

AIG

The American International Group (AIG) was once the 18th largest corporation in the world. AIG was founded nearly 100 years ago by Cornelius Vander Starr who opened an insurance agency in Shanghai, China. As the first Westerner to sell insurance to the Chinese, Starr grew his business rapidly until 1949 when Mao Zedong and the People's Liberation Army took over Shanghai. Starr moved the company to New York City, where it continued to grow, expanding its markets worldwide. In 2004, AIG stock hit an all-time high of $76.77, putting its market value at nearly $300 billion.

According to its own website, "By early 2007 AIG had assets of $1 trillion, $110 billion in revenues, 74 million customers and 116,000 employees in 130 countries and jurisdictions. Yet just 18 months later, AIG found itself on the brink of failure and in need of emergency government assistance." AIG was one of the largest beneficiaries of the U.S. government's Troubled Asset Relief Program (TARP), established in 2008 during the financial crisis to purchase assets and equity from financial institutions. TARP was an attempt to strengthen the financial sector and avoid a repeat of a depression as severe as the 1930s. Many banks quickly repaid the government part or all of the money given to them under the TARP program, but AIG, which received $170 billion, did not.

By 2009, AIG stock had lost more than 99% of its value, hitting $0.35 in early March. That same month AIG became embroiled in controversy when it disclosed that it had paid $218 million in bonuses to employees of its financial services division. AIG's drop in stock price represented a loss of nearly $300 billion for investors. Portfolio managers typically examine stock prices and volumes to determine stock volatility and to help them decide which stocks to buy and sell. Were there early warning signs in AIG's data?

To learn more about the behavior and volatility of AIG's stock, let's start by looking at Table 5.1, which gives the monthly average stock price (in dollars) for the six years leading up to the company's crisis.

	Jan.	Feb.	Mar.	Apr.	May	June	July	Aug.	Sept.	Oct.	Nov.	Dec.
2002	77.26	72.95	73.72	71.57	68.42	65.99	61.22	64.10	58.04	60.26	65.03	59.96
2003	59.74	49.57	49.41	54.38	56.52	57.88	59.80	61.51	59.39	60.93	58.73	62.37
2004	69.02	73.25	72.06	74.21	70.93	72.61	69.85	69.58	70.67	62.31	62.17	65.33
2005	66.74	68.96	61.55	51.77	53.81	55.66	60.27	60.86	60.54	62.64	67.06	66.72
2006	68.33	67.02	67.15	64.29	63.14	59.74	59.40	62.00	65.25	67.02	69.86	71.35
2007	70.45	68.99	68.14	68.25	71.78	71.75	68.64	65.21	66.02	66.12	56.86	58.13

| Table 5.1   Monthly stock price in dollars of AIG stock for the period 2002 through 2007.

It's hard to tell very much from tables of values like this. You might get a rough idea of how much the stock cost—usually somewhere around $60 or so, but that's about it.

5.1   Displaying Quantitative Variables

Instead, let's follow the first rule of data analysis and make a picture. What kind of picture should we make? It can't be a bar chart or a pie chart. Those are only for categorical variables, and AIG's stock price is a *quantitative* variable, whose units are dollars.

WHO	Months
WHAT	Monthly average price for AIG's stock (in dollars)
WHEN	2002 through 2007
WHERE	New York Stock Exchange
WHY	To examine AIG stock volatility

Histograms

Here are the monthly prices of AIG stock displayed in a histogram.

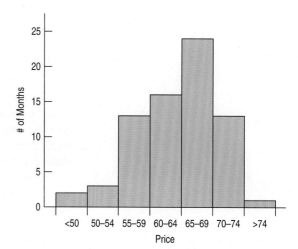

Figure 5.1   Monthly average prices of AIG stock. The histogram displays the distribution of prices by showing for each "bin" of prices, the number of months having prices in that bin.

Like a bar chart, a **histogram** plots the bin counts as the heights of bars. It counts the number of cases that fall into each bin, and displays that count as the height of the corresponding bar. In this histogram of monthly average prices, each bin has a width of $5, so, for example, the height of the tallest bar says that there were 24 months whose average price of AIG stock was between $65 and $70. In this way, the histogram displays the entire distribution of prices at a glance. Unlike a bar chart, which puts gaps between bars to separate the categories, there are no gaps between the bars of a histogram unless there are actual gaps in the data. **Gaps** indicate a region where there are no values. Gaps can be important features of the distribution so watch out for them and point them out.

For categorical variables, each category got its own bar. The only choice was whether to combine categories for ease of display. For quantitative variables, we have to choose the width of the bins.

- **Making a Histogram by Hand** Although you'll rarely make a histogram by hand, it can be instructive to see how it might be done. Some of the same choices that you have to make by hand will either be made by software automatically, or with input from you.

  Step 1. Organize your data into a table. Divide the data into equal intervals or bins so that all values are covered. The number of bins depends on how many data values you have. For small (fewer than 25 or so values) data sets, 5 bins is fine. For large data sets you may need 20 or more.

  You will probably want the width of intervals to be aesthetically pleasing (bins that have widths that end with 5's or 0's are popular—for example, 35–40,40–45, 45–50, etc). Now create two columns—one for the bins and the other for the frequencies. This creates a frequency distribution much like that for a categorical variable, but here instead of categories, we have bins of equal width. You'll have to decide whether to put values that lie at the end points of the bin into the left or right bin. Most software histogram programs put values into the bin on the right with the larger values, so 40 would go into the bin 40–45, not in the bin 35–40, but

Price Bin	# of Months
45–50	2
50–55	3
55–60	13
60–65	16
65–70	24
70–75	13
75–80	1

either choice is possible. In fact, Excel chooses to put the values to the left, so that 5 goes into the bin 0–5 not the bin 5–10.

Step 2. On a sheet of paper, mark the bins on the *x*-axis (horizontal axis) with no spaces between the bins. Mark the frequencies on the *y*-axis (vertical axis) over the center of each bin.

Step 3. Plot your data. For each bin, draw a horizontal line at the appropriate frequency over the bin. Then, draw vertical bars on the sides of each bin, reaching up to the corresponding frequency.

From the histogram, we can see that in these months a typical AIG stock price was near $60 or so. We can see that although they vary, most of the monthly prices were between $55 and $75. Only in a very few months was the average price below $55. It's important to note that the histogram is a static picture. We have treated these prices simply as a collection of months, with no sense of time, and shown their distribution. Later in the chapter we will add time to the story.

Does the distribution look as you expected? It's often a good idea to imagine what the distribution might look like before making the display. That way you're less likely to be fooled by errors either in your display or in the data themselves.

If our focus is on the overall pattern of how the values are distributed rather than on the counts themselves, it can be useful to make a relative frequency histogram, replacing the counts on the vertical axis with the percentage of the total number of cases falling in each bin (simply divide the counts in each bin by the total number of data values). The shape of the histogram is exactly the same; only the labels are different. **A relative frequency histogram** is faithful to the area principle by displaying the *percentage* of cases in each bin instead of the count.

Figure 5.2 A relative frequency histogram looks just like a frequency histogram except that the *y*-axis now shows the percentage of months in each bin.

For Example   Creating a histogram

1. As the chief financial officer of a music download site, you've just secured the rights to offer downloads of a new album. You'd like to see how well it's selling, so you collect the number of downloads per hour for the past 24 hours:

Hour	Downloads	Hour	Downloads
12:00 a.m.	36	12:00 p.m.	25
1:00 a.m.	28	1:00 p.m.	22
2:00 a.m.	19	2:00 p.m.	17
3:00 a.m.	10	3:00 p.m.	18
4:00 a.m.	5	4:00 p.m.	20
5:00 a.m.	3	5:00 p.m.	23
6:00 a.m.	2	6:00 p.m.	21
7:00 a.m.	6	7:00 p.m.	18
8:00 a.m.	12	8:00 p.m.	24
9:00 a.m.	14	9:00 p.m.	30
10:00 a.m.	20	10:00 p.m.	27
11:00 a.m.	18	11:00 p.m.	30

Question: Make a histogram for this variable.

Answer: Create a frequency table of bins of width five from 0 to 40 and put values at the ends of bins into the right bin:

Downloads	Number OF Hours
0–5	2
5–10	2
10–15	3
15–20	5
20–25	6
25–30	3
30–35	2
35–40	1
Total	**24**

The histogram looks like this:

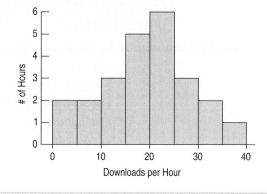

*Stem-and-Leaf Displays

Histograms provide an easy-to-understand summary of the distribution of a quantitative variable, but they don't show the data values themselves. **Stem-and-leaf displays** are like histograms, but they also give the individual values. They are easy to make by hand for data sets that aren't too large, so they're a great way to look at a small batch of values quickly.[1] Here's a stem-and-leaf display for the AIG stock data, alongside a histogram of the same data.

Figure 5.3   The AIG monthly average stock prices displayed both by a histogram (left) and stem-and-leaf display (right). Stem-and-leaf displays are typically made by hand, so we are most likely to use them for small data sets. For much larger data sets, we use a histogram.

- **How do stem-and-leaf displays work?** A stem-and-leaf display breaks each number into two parts: the stem shown to the left of the solid line and the leaf, to the right. For the AIG data, each price change, for example $67.02 is first truncated to two digits, $67. Then it is split into its two components: 6|7. The line 5|134 therefore, shows the values $51, $53, and $54 and corresponds to the histogram bin from $50 to $55. The stem-and-leaf in Figure 5.3 uses a bin width of 5. Another choice would be to increase the bin size and put all the prices from $50 to $60 on one line:

$$5\,|\,1345667888999999$$

That would decrease the number of bins to 4, but makes the bin from $60 to $70 too crowded:

```
4|99
5|1345667888999999
6|00000111222223445555566667777788888889999
7|00011112223347
```

Sometimes the stem-and-leaf display puts the higher numbers on top:

```
7|7
7|00001111222334
6|55555666677778888889999
6|0000011122222344
5|5667888999999
5|134
4|99
```

Either choice is possible, although putting the lower numbers on top makes the correspondence between histogram and stem-and-leaf easier to see.

---

[1]The authors like to make stem-and-leaf displays whenever data are presented (without a suitable display) at committee meetings or working groups. The insights from just that quick look at the distribution are often quite valuable.

In Chapter 4, you learned to check the Categorical Data Condition before making a pie chart or a bar chart. Now, by contrast, before making a stem-and-leaf display, or a histogram, you need to check the **Quantitative Data Condition**: The data are values of a quantitative variable whose units are known.

Although a bar chart and a histogram may look similar, they're not the same display. You can't display categorical data in a histogram or quantitative data in a bar chart. Always check the condition that confirms what type of data you have before making your display.

5.2   Shape

Once you've displayed the distribution in a histogram or stem-and-leaf display, what can you say about it? When you describe a distribution, you should pay attention to three things: its **shape**, its **center**, and its **spread**.

We describe the shape of a distribution in terms of its modes, its symmetry, and whether it has any gaps or outlying values.

Mode

Does the histogram have a single, central hump (or peak) or several, separated humps? These humps are called **modes**.[2] Formally, the mode is the single, most frequent value, but we rarely use the term that way. Sometimes we talk about the mode as being the value of the variable at the center of this hump. The AIG stock prices have a single mode around $65 (Figure 5.1). We often use modes to describe the shape of the distribution. A distribution whose histogram has one main hump, such as the one for the AIG stock prices, is called **unimodal**; distributions whose histograms have two humps are **bimodal**, and those with three or more are called **multimodal**. For example, here's a bimodal distribution.

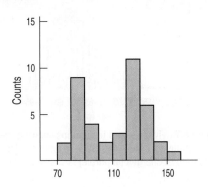

Figure 5.4   A bimodal distribution has two apparent modes.

A bimodal histogram is often an indication that there are two groups in the data. It's a good idea to investigate when you see bimodality.

A distribution whose histogram doesn't appear to have any mode and in which all the bars are approximately the same height is called **uniform**. (Chapter 9 gives a more formal definition.)

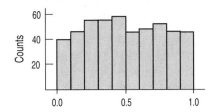

Figure 5.5   In a uniform distribution, bars are all about the same height. The histogram doesn't appear to have a mode.

---

[2]Technically, the mode is the value on the *x*-axis of the histogram below the highest peak, but informally we often refer to the peak or hump itself as a mode.

Where's the Mode?

The **mode** is typically defined as the single value that appears most often. That definition is fine for categorical variables because we need only to count the number of cases for each category. For quantitative variables, the meaning of *mode* is more ambiguous. For example, what's the mode of the AIG data? No two prices were exactly the same, but 7 months had prices between $68 and $69. Should that be the mode? Probably not—that seems a little arbitrary. For quantitative data, it makes more sense to use the word *mode* in the more general sense of "peak in a histogram," rather than as a single summary value.

Pie à la Mode

Is there a connection between pie and the mode of a distribution? Actually, there is! The mode of a distribution is a *popular* value near which a lot of the data values gather. And à la mode means "in style"—*not* "with ice cream." That just happened to be a *popular* way to have pie in Paris around 1900.

Symmetry

Could you fold the histogram along a vertical line through the middle and have the edges match pretty closely, as in Figure 5.6, or are more of the values on one side, as in the histograms in Figure 5.7? A distribution is **symmetric** if the halves on either side of the center look, at least approximately, like mirror images.

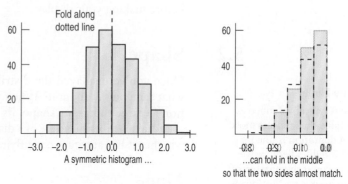

Figure 5.6   A symmetric histogram can fold in the middle so that the two sides almost match.

The (usually) thinner ends of a distribution are called the **tails**. If one tail stretches out farther than the other, the distribution is said to be **skewed** to the side of the longer tail.

> Amounts of things (dollars, employees, waiting times) can't be negative and have no natural upper limit. So, they often have distributions that are skewed to the right.

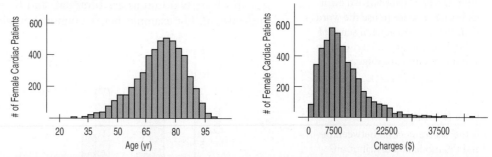

Figure 5.7   Two skewed histograms showing the age (left) and hospital charges (right) for all female heart attack patients in New York State in one year. The histogram of Age (in blue) is skewed to the left, while the histogram of Charges (in purple) is skewed to the right.

Outliers

Do any features appear to stick out? Often such features tell us something interesting or exciting about the data. You should always point out any stragglers or **outliers** that stand off away from the body of the distribution. For example, if you're studying the personal wealth of Americans and Bill Gates is in your sample, he would certainly be an outlier. Because his wealth would be so obviously atypical, you'd want to point it out as a special feature.

Outliers can affect almost every method we discuss in this book, so we'll always be on the lookout for them. An outlier can be the most informative part of your data, or it might just be an error. Either way, you shouldn't throw it away without comment. Treat it specially and discuss it when you report your conclusions about your data. (Or find the error and fix it if you can.) We'll soon learn a rule of thumb for how we can decide if and when a value might be considered to be an outlier and some advice for what to do when you encounter them.

- **Using Your Judgment.** How you characterize a distribution is often a judgment call. Does the gap you see in the histogram really reveal that you have two subgroups, or will it go away if you change the bin width slightly? Are those observations at the high end of the histogram truly unusual, or are they just the largest ones at the end of a long tail? These are matters of judgment on which different people can legitimately disagree. There's no automatic calculation or rule of thumb that can make the decision for you. Understanding your data and how they arose can help. What should guide your decisions is an honest desire to understand what is happening in the data.

Looking at a histogram at several different bin widths can help you to see how persistent some of the features are. Some technologies offer ways to change the bin width interactively to get multiple views of the histogram. If the number of observations in each bin is small enough so that moving a couple of values to the next bin changes your assessment of how many modes there are, be careful. Be sure to think about the data, where they came from, and what kinds of questions you hope to answer from them.

For Example   Describing the shape of a distribution

Question: Describe the shape of the distribution of downloads from the example on page 90.

Answer: It is symmetric and unimodal with no outliers.

5.3   Center

Look again at the AIG prices in Figure 5.1. If you had to pick one number to describe a *typical* price, what would you pick? When a histogram is unimodal and fairly symmetric, most people would point to the center of the distribution, where the histogram peaks. The typical price is around $65.00.

If we want to be more precise and *calculate* a number, we can *average* the data. In the AIG example, the average monthly prices is $64.48, about what we might expect from the histogram. You already know how to average values, but this is a good place to introduce notation that we'll use throughout the book. We'll call the generic variable y, and use the Greek capital letter sigma, Σ, to mean "sum" (sigma is "S" in Greek), and write[3]:

$$\bar{y} = \frac{Total}{n} = \frac{\Sigma y}{n}.$$

According to this formula, we add up all the values of the variable, y, and divide that sum (*Total*, or Σy) by the number of data values, n. We call the resulting value the **mean** of y.[4]

Notation Alert! ⎯⎯⎯⎯

A bar over any symbol indicates the mean of that quantity.

⎯⎯⎯⎯⎯⎯⎯⎯⎯⎯

[3]You may also see the variable called x and the equation written $\bar{x} = \frac{Total}{n} = \frac{\Sigma x}{n}$. We prefer to call a single variable y instead of x, because x will later be used to name a variable that predicts another (which we'll call y), but when you have only one variable either name is common. Most calculators call a single variable x.

[4]Once you've averaged the data, you might logically expect the result to be called the *average*. But average is used too colloquially as in the "average" home buyer, where we don't sum up anything. Even though average *is* sometimes used in the way we intend, as in the Dow Jones Industrial Average (which is actually a weighted average) or a batting average, we'll usually use the term *mean* throughout the book.

Although the mean is a natural summary for unimodal, symmetric distributions, it can be misleading for skewed data or for distributions with gaps or outliers. The histogram of AIG monthly prices in Figure 5.1 is unimodal, and nearly symmetric, with a slight left skew. A look at the total volume of stocks sold each month for the same 6 years tells a very different story. Figure 5.8 shows a unimodal but strongly right-skewed distribution with two gaps. The mean monthly volume was 170.1 million shares. Locate that value on the histogram. Does it seem a little high as a summary of a typical month's volume? In fact, more than two out of three months have volumes that are less than that value. It might be better to use the **median**—the value that splits the histogram into two equal *areas*. We find the median by counting in from the ends of the data until we reach the middle value. The median is commonly used for variables such as cost or income, which are likely to be skewed. That's because the median is *resistant* to unusual observations and to the shape of the distribution. For the AIG monthly trading volumes, the median is 135.9 million shares, which seems like a more appropriate summary.

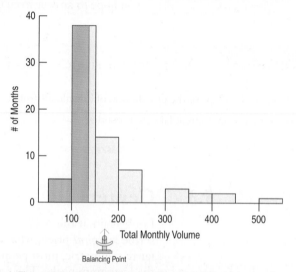

Figure 5.8   The median splits the area of the histogram in half at 135.9 million shares. Because the distribution is skewed to the right, the mean 170.1 million shares is *higher* than the median. The points at the right have pulled the mean toward them, away from the median.

Does it really make a difference whether we choose a mean or a median? The mean monthly price for the AIG stock is $64.48. Because the distribution of the prices is roughly symmetric, we'd expect the mean and median to be close. In fact, we compute the median to be $65.23. But for variables with skewed distributions, the story is quite different. For a right-skewed distribution like the monthly volumes in Figure 5.8, the mean is larger than the median: 170.1 compared to 135.9. The two give quite different summaries. The difference is due to the overall shape of the distributions.

By Hand

Finding the Median

Finding the median of a batch of n numbers is easy as long as you remember to order the values first. If n is odd, the median is the middle value.

Counting in from the ends, we find this value in the $\frac{n+1}{2}$ position.

When n is even, there are two middle values. So, in this case, the median is the average of the two values in positions $\frac{n}{2}$ and $\frac{n}{2} + 1$.

Here are two examples:

Suppose the batch has the values 14.1, 3.2, 25.3, 2.8, −17.5, 13.9, and 45.8. First we order the values: −17.5, 2.8, 3.2, 13.9, 14.1, 25.3, and 45.8. There are 7 values, so the median is the $(7 + 1)/2 = $ 4th value counting from the top or bottom: 13.9.

Suppose we had the same batch with another value at 35.7. Then the ordered values are −17.5, 2.8, 3.2, 13.9, 14.1, 25.3, 35.7, and 45.8. The median is the average of the 8/2, or 4th, and the $(8/2) + 1$, or 5th, values. So the median is $(13.9 + 14.1)/2 = $ 14.0.

The mean is the point at which the histogram would balance. Just like a child who moves away from the center of a see-saw, a bar of the histogram far from the center has more leverage, pulling the mean in its direction. It's hard to argue that a summary that's been pulled aside by only a few outlying values or by a long tail is what we mean by the center of the distribution. That's why the median is usually a better choice for skewed data.

However, when the distribution is unimodal and symmetric, the mean offers better opportunities to calculate useful quantities and draw more interesting conclusions. It will be the summary value we work with much more throughout the rest of the book.

For Example   Finding the mean and median

Question: From the data on page 90, what is a typical number of downloads per hour?

Answer: The mean number is 18.7 downloads per hour. The median is 19.5 downloads per hour. Because the distribution is unimodal and roughly symmetric, we shouldn't be surprised that the two are close. There are a few more hours (in the middle of the night) with small numbers of downloads that pull the mean lower than the median, but either one seems like a reasonable summary to report.

5.4   Spread of the Distribution

We know that the typical price of the AIG stock is around $65, but knowing the mean or median alone doesn't tell us about the entire distribution. A stock whose price doesn't move away from its center isn't very interesting.[5] The more the data vary, the less a measure of center can tell us. We need to know how spread out the data are as well.

One simple measure of spread is the **range**, defined as the difference between the extremes:

$$\text{Range} = max - min.$$

For the AIG price data, the range is $77.26 − $49.41 = $27.85. Notice that the range is *a single number* that describes the spread of the data, not an interval of values—as you might think from its use in common speech. If there are any unusual

---

[5]And not much of an investment, either.

observations in the data, the range is not resistant and will be influenced by them. Concentrating on the middle of the data avoids this problem.

The **quartiles** are the values that frame the middle 50% of the data. One quarter of the data lies below the lower quartile, Q1, and one quarter of the data lies above the upper quartile, Q3. The **interquartile range (IQR)** summarizes the spread by focusing on the middle half of the data. It's defined as the difference between the two quartiles:

$$IQR = Q3 - Q1.$$

By Hand

Finding Quartiles

Quartiles are easy to find in theory, but more difficult in practice. The three quarties, Q1 (lower quartile), Q2 (the median) and Q3 (the upper quartile) split the sorted data values into quarters. So, for example, 25% of the data values will lie at or below Q1. The problem lies in the fact that unless your sample size divides nicely by 4, there isn't just one way to split the data into quarters. The statistical software package SAS offers at least five different ways to compute quartiles. The differences are usually small, but can be annoying. Here are two of the most common methods for finding quartiles by hand or with a calculator:

1. The Tukey Method

   Split the sorted data at the median. (If n is odd, include the median with each half). Then find the median of each of these halves—use these as the quartiles.

   Example: The data set {14.1, 3.2, 25.3, 2.8, −17.5, 13.9, 45.8}

   First we order the values: {−17.5, 2.8, 3.2, 13.9, 14.1, 25.3, 45.8}. We found the median to be 13.9, so form two data sets: {−17.5, 2.8, 3.2, 13.9} and { 13.9, 14.1, 25.3, 45.8}. The medians of these are $3.0 = (2.8 + 3.2)/2$ and $19.7 = (14.1 + 25.3)/2$. So we let Q1 = 3.0 and Q3 = 19.7.

2. The TI calculator method

   The same as the Tukey method, except we *don't* include the median with each half. So for {14.1, 3.2, 25.3, 2.8, −17.5, 13.9, and 45.8} we find the two data sets:

   {−17.5, 2.8, 3.2} and {14.1, 25.3, 45.8} by not including the median in either.

   Now the medians of these are Q1 = 2.8 and Q3 = 25.3.

   Notice the effect on the IQR. For Tukey:
   $IQR = Q3 - Q1 = 19.7 - 3.0 = 16.7$, but for TI,
   $IQR = 25.3 - 2.8 = 22.5$.

For both of these methods, notice that the quartiles are either data values, or the average of two adjacent values. In Excel, and other software, the quartiles are *interpolated*, so they may not be simple averages of two values. Be aware that there may be differences, but the idea is the same: the quartiles Q1, Q2, and Q3 split the data roughly into quarters.

For the AIG data, there are 36 values on either side of the median. After ordering the data, we average the 18th and 19th values to find $Q1 = (59.96 + 60.26)/2 = \$60.11$. We average the 54th and 55th values to find Q3 = (68.99 + 69.02)/2 = \$69.01. So the IQR = Q3 − Q1 = \$69.01 − \$60.11 = \$8.90.

The IQR is usually a reasonable summary of spread, but because it uses only the two quartiles of the data, it ignores much of the information about how individual values vary.

A more powerful measure of spread—and the one we'll use most often—is the **standard deviation**, which, as we'll see, takes into account how far each value is from the mean. Like the mean, the standard deviation is appropriate only for symmetric data and can be influenced by outlying observations.

As the name implies, the standard deviation uses the *deviations* of each data value from the mean. If we tried to average these deviations, the positive and negative differences would cancel each other out, giving an average deviation of 0—not very useful. Instead, we square each deviation. The average[6] of the *squared* deviations is called the **variance** and is denoted by s^2:

$$s^2 = \frac{\sum(y - \bar{y})^2}{n - 1}.$$

The variance plays an important role in statistics, but as a measure of spread, it has a problem. Whatever the units of the original data, the variance is in *squared* units. We want measures of spread to have the same units as the data, so we usually take the square root of the variance. That gives the **standard deviation**.

$$s = \sqrt{\frac{\sum(y - \bar{y})^2}{n - 1}}.$$

For the AIG stock prices, $s = \$6.12$.

For Example   Describing the spread

Question: For the data on page 90, describe the spread of the number of downloads per hour.

Answer: The range of downloads is 36 − 2 = 34 downloads per hour.
The quartiles are 13 and 24.5, so the IQR is 24.5 − 13 = 11.5 downloads per hour. The standard deviation is 8.94 downloads per hour.

By Hand

Finding the Standard Deviation

To find the standard deviation, start with the mean, $\bar{y}$. Then find the *deviations* by taking $\bar{y}$ from each value: $(y - \bar{y})$. Square each deviation: $(y - \bar{y})^2$.

Now you're nearly home. Just add these up and divide by $n - 1$. That gives you the variance, s^2. To find the standard deviation, s, take the square root.

Suppose the batch of values is 4, 3, 10, 12, 8, 9, and 3.

(continued)

---

[6]For technical reasons, we divide by $n - 1$ instead of n to take this average.

The mean is $\bar{y} = 7$. So find the deviations by subtracting 7 from each value:

Original Values	Deviations	Squared Deviations
4	$4 - 7 = -3$	$(-3)^2 = 9$
3	$3 - 7 = -4$	$(-4)^2 = 16$
10	$10 - 7 = 3$	9
12	$12 - 7 = 5$	25
8	$8 - 7 = 1$	1
9	$9 - 7 = 2$	4
3	$3 - 7 = -4$	16

Add up the squared deviations:
$9 + 16 + 9 + 25 + 1 + 4 + 16 = 80$.
Now, divide by $n - 1$: $80/6 = 13.33$.
Finally, take the square root: $s = \sqrt{13.33} = 3.65$

Just Checking

Thinking About Variation

1 The U.S. Census Bureau reports the median family income in its summary of census data. Why do you suppose they use the median instead of the mean? What might be the disadvantages of reporting the mean?

2 You've just bought a new car that claims to get a highway fuel efficiency of 31 miles per gallon. Of course, your mileage will "vary." If you had to guess, would you expect the IQR of gas mileage attained by all cars like yours to be 30 mpg, 3 mpg, or 0.3 mpg? Why?

3 A company selling a new MP3 player advertises that the player has a mean lifetime of 5 years. If you were in charge of quality control at the factory, would you prefer that the standard deviation of life spans of the players you produce be 2 years or 2 months? Why?

5.5 Shape, Center, and Spread—A Summary

What should you report about a quantitative variable? Report the shape of its distribution, and include a center and a spread. But which measure of center and which measure of spread? The guidelines are pretty easy.

- If the shape is skewed, point that out and report the median and IQR. You may want to include the mean and standard deviation as well, explaining why the mean and median differ. The fact that the mean and median do not agree is a sign that the distribution may be skewed. A histogram will help you make the point.

- If the shape is unimodal and symmetric, report the mean and standard deviation and possibly the median and IQR as well. For unimodal symmetric data, the IQR is usually a bit larger than the standard deviation. If that's not true for your data set, look again to make sure the distribution isn't skewed or mutimodal and that there are no outliers.

- If there are multiple modes, try to understand why. If you can identify a reason for separate modes, it may be a good idea to split the data into separate groups.

- If there are any clearly unusual observations, point them out. If you are reporting the mean and standard deviation, report them computed with and without the unusual observations. The differences may be revealing.
- Always pair the median with the IQR and the mean with the standard deviation. It's not useful to report one without the other. Reporting a center without a spread can lead you to think you know more about the distribution than you do. Reporting only the spread omits important information.

For Example   **Summarizing data**

Question: Report on the shape, center, and spread of the downloads data; see page 90.

Answer: The distribution of downloads per hour over the past 24 hours is unimodal and roughly symmetric. The mean number of downloads per hour is 18.7 and the standard deviation is 8.94. There are several hours in the middle of the night with very few downloads, but none seem to be so unusual as to be considered outliers.

5.6   Five-Number Summary and Boxplots

One good way to summarize a distribution with just a few values is with a five-number summary. The **five-number summary** of a distribution reports its median, quartiles, and extremes (maximum and minimum). For example, the five-number summary of the monthly trading volumes of AIG stock for the period 2002 to 2007 looks like this (in millions of shares).

Max	515.62
Q3	182.32
Median	135.87
Q1	121.04
Min	83.91

Table 5.2   The five-number summary of monthly trading volume of AIG shares (in millions of shares) for the period 2002 to 2007.

The five-number summary provides a good overall look at the distribution. For example, because the quartiles frame the middle half of the data, we can see that on half of the days the volume was between 121.04 and 182.32 million shares. We can also see the extremes of over 500 million shares on the high end and 83.91 million shares on the low end. Were those days extraordinary for some reason or just the busiest and quietest days? To answer that, we'll need to work with the summaries a bit more.

Once we have a five-number summary of a (quantitative) variable, we can display that information in a **boxplot** (see Figure 5.9).

A boxplot highlights several features of the distribution of a variable. The central box shows the middle half of the data, between the quartiles. Because the top of the box is at the third quartile (Q3) and the bottom is at Q1, the height of the box

Figure 5.9   Boxplot of monthly volumes of AIG stock traded in the period 2002–2007 (in millions of shares).

is equal to Q3 − Q1 which is the IQR. (For the AIG data, it's 61.28.) The median is displayed as a horizontal line. If the median is roughly centered between the quartiles, then the middle half of the data is roughly symmetric. If it is not centered, the distribution is skewed. In extreme cases, the median can coincide with one of the quartiles.

The whiskers reach out from the box to the most extreme values that are not considered outliers. The boxplot nominates points as outliers if they fall farther than 1.5 IQRs beyond either quartile (for the AIG data, 1.5 IQR = 1.5 × 61.28 = 91.92). Outliers are displayed individually, both to keep them out of the way for judging skewness and to encourage you to give them special attention. They may be mistakes or they may be the most interesting cases in your data. This rule is not a definition of what makes a point an outlier. It just nominates cases for special attention. But it is not a substitute for careful analysis and thought about whether a value is special.

It's easy to make a boxplot. First locate the median and quartiles on an axis and draw three short lines. For the AIG data, those are at approximately 121 (Q1), 136 (median), and 182 (Q3). The axis is usually vertical (as in Figure 5.9), but it can be horizontal. Connect the quartile lines to make a box. Identify the "fences" at 1.5 IQR beyond each quartile. For the AIG data, that's 121.04 − 1.5 × 61.28 = 29.12 (lower fence) and 182.32 + 1.5 × 61.28 = 274.24 (upper fence). These fences are not drawn on the final boxplot. They are used to decide which points to display as outliers. Draw whiskers to the most extreme data value not outside the fences. In the AIG data, there are no values below the lower fence since the minimum 83.91 is greater than 29.12, but there are 7 points above the upper fence at 274.24. Finally, draw any outliers individually. Some boxplots use a special symbol for "far" outliers that lie more than 3 IQR's from the fences, as shown in Figure 5.9.

Some features of the distribution are lost in a boxplot, but as we'll soon see, they are especially useful when comparing several distributions side by side.

From the shape of the box in Figure 5.9, it looks like the central part of the distribution of volume is skewed to the right (upward here) and the dissimilar length of the two whiskers shows the outer parts of the distribution to be skewed as well. We also see several high volume and some extremely high volume days. Boxplots are particularly good at exhibiting outliers. These extreme days may deserve more attention. (When and why did they occur?)

> **The 1.5 IQR Rule for Nomination Outliers**
>
> Designate a point as an outlier if it lies farther than 1.5 IQRs from either the first (Q1) or third (Q3) quartile. Some boxplots also designate points as "far" outliers if they lie more than 3 IQRs from the quartiles. The prominent statistician John W. Tukey, the originator of the boxplot, was asked (by one of the authors) why the outlier nomination rule cut at 1.5 IQRs beyond each quartile. He answered that the reason was that 1 IQR would be too small and 2 IQRs would be too large.

For Example   The boxplot rule for nominating outliers

Question: From the histogram on page 90, we saw that no points seemed to be so far from the center as to be considered outliers. Use the 1.5 IQR rule to see if it nominates any points as outliers.

Answer: The quartiles are 13 and 24.5 and the IQR is 11.5. 1.5 *IQR = 17.25. A point would have to be larger than 24.5 + 17.25 = 41.25 downloads/hr or smaller than 13 − 17.25 = −4.25. The largest value was 36 downloads/hr and all values must be nonnegative, so there are no points nominated as outliers.

Guided Example | Credit Card Bank Customers

To focus on the needs of particular customers, companies often segment their customers into groups with similar needs or spending patterns. A major credit card bank wanted to see how much a particular group of cardholders charged per month on their cards in order to understand the potential growth in their card use. The data for each customer was the amount he or she spent using the card during a three-month period in 2008. Boxplots are especially useful for one variable when combined with a histogram and numerical summaries. Let's summarize the spending of this market segment.

PLAN	**Setup** Identify the *variable*, the time frame of the data, and the objective of the analysis.	We want to summarize the average monthly charges (in dollars) made by 500 cardholders from a market segment of interest during a three-month period in 2008. The data are quantitative, so we'll use histograms and boxplots, as well as numerical summaries.
DO	**Mechanics** Select an appropriate display based on the nature of the data and what you want to know about it.	
REALITY CHECK ▶	It is always a good idea to think about what you expected to see and to check whether the histogram is close to what you expected. Are the data about what you might expect for customers to charge on their cards in a month? A typical value is a few hundred dollars. That seems like the right ballpark. Note that outliers are often easier to see with boxplots than with histograms, but the histogram provides more details about the shape of the distribution. This computer program "jitters" the outliers in the boxplot so they don't lie on top of each other, making them easier to see.	Both graphs show a distribution that is highly skewed to the right with several outliers and an extreme outlier near $7000.

Summary of Monthly Charges	
Count	500
Mean	544.749
Median	370.65
StdDev	661.244
IQR	624.125
Q1	114.54
Q3	738.665

The mean is much larger than the median. The data do not have a symmetric distribution.

(continued)

REPORT

Interpretation Describe the shape, center, and spread of the distribution. Be sure to report on the symmetry, number of modes, and any gaps or outliers.

Recommendation State a conclusion and any recommended actions or analysis.

MEMO

Re: Report on segment spending.

The distribution of charges for this segment during this time period is unimodal and skewed to the right. For that reason, we have summarized the data with the median and interquartile range (IQR).

The median amount charged was $370.65. Half of the card-holders charged between $114.54 and $738.67.

In addition, there are several high outliers, with one extreme value at $6745.

There are also a few negative values. We suspect that these are people who returned more than they charged in a month, but because the values might be data errors, we suggest that they be checked.

Future analyses should look at whether charges during these three months in 2008 were similar to charges in the rest of the year. We would also like to investigate if there is a seasonal pattern and, if so, whether it can be explained by our advertising campaigns or by other factors.

5.7 Comparing Groups

As we saw earlier, the volume of a stock can vary greatly from month to month or even day to day, but if we step back a bit, we may be able to find patterns that can help us understand, model, and predict it. We started the chapter by looking at monthly summaries of the price and volume of AIG stock. If, instead, we consider the individual daily values, we can group them into periods such as weeks, months, seasons, or years. The picture can change depending on what grouping we use. Comparing the distributions can reveal patterns, differences, and trends.

Let's start with the "big picture." Instead of taking monthly averages, let's look at the daily closing prices for the first two years of our data, 2002 and 2003:

| Figure 5.10   Daily closing prices of AIG on the NYSE for the two years 2002 and 2003. How do the two distributions differ?

It's not hard to see that prices were generally lower in 2003 than 2002. The price distribution for 2002 appears to be symmetric with a center in the high $60s while the 2003 distribution is left skewed with a center below $60. We were able to make the comparison easily because we displayed the two histograms on the same scale. Histograms with very different centers and spreads may appear similar unless you do that.

Histograms work well for comparing two groups, but what if we want to compare the prices across several years? Histograms are best at displaying one or two distributions. When we compare several groups, boxplots usually do a better job. Boxplots offer an ideal balance of information and simplicity, hiding the details while displaying the overall summary information. And we can plot them side by side, making it easy to compare multiple groups or categories.

When we place boxplots side by side, we can easily see which group has the higher median, which has the greater IQR, where the central 50% of the data is located, and which has the greater overall range. We can also get a general idea of symmetry from whether the medians are centered within their boxes and whether the whiskers extend roughly the same distance on either side of the boxes. Equally important, we can see past any outliers in making these comparisons because they've been displayed separately. We can also begin to look for trends in the medians and in the IQRs.

Guided Example   AIG Stock Price and Volume

What really happened to the AIG stock price from the beginning of the period we've been studying through the financial crisis of 2008/2009? Boxplots of the number of shares traded by month are a good way to see such patterns. We're interested not only in the centers, but also in the spreads. Are volumes equally variable from year to year or are they more spread out in some years?

| PLAN | **Setup** Identify the variables, report the time frame of the data, and state the objective. | We want to compare the daily price of shares traded from year to year on the NYSE from 2002 through 2009. |
| | | The daily price is quantitative and measured in dollars. We can partition the values by year and use side-by-side boxplots to compare the daily prices across years. |

| DO | **Mechanics** Plot the side-by-side boxplots of the data. | 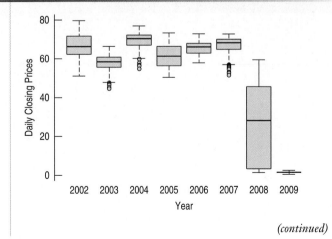 |

(continued)

Display any other plots suggested by the previous.

What happened in 2008? We'd better look there with a finer partition. Here are boxplots by month for 2008.

REPORT

Conclusion Report what you've learned about the data and any recommended action or analysis.

MEMO

Re: Research on price of AIG stock

We have examined the daily closing prices of AIG stock on the NYSE for the period 2002 through 2009. As the displays show, prices were relatively stable for the period 2002 through 2007. Prices lowered in 2003 but recovered and stayed generally above $60 for 2004 through 2007. Then in 2008, prices dropped dramatically, and throughout 2009 AIG's stock price was a small fraction of what it had once been. A boxplot by month during 2008 shows that the decline in price was constant throughout the entire year but most noticeable in September 2008. Most analysts point to that month as the beginning of the financial meltdown, but clearly there were signs in the price of AIG that trouble had been brewing for much longer. By October and for the rest of the year, the price was very low with almost no variation.

For Example   Comparing boxplots

Question: For the data on page 90, compare the AM downloads to the PM downloads by displaying the two distributions side-by-side with boxplots.

Answer: There are generally more downloads in the afternoon than in the morning. The median number of afternoon downloads is around 22 as compared with 14 for the morning hours. The PM downloads are also much more consistent. The entire range of the PM hours, 15, is about the size of the IQR for AM hours. Both distributions appear to be fairly symmetric, although the AM hour distribution has some high points which seem to give some asymmetry.

5.8 Identifying Outliers

We've just seen that the price of AIG shares dropped precipitously during the year 2008. Let's look at a boxplot by month of the daily volumes to see if a similar pattern appears.

Figure 5.11 In January, there was a high volume day of 38 million shares that is nominated as an outlier for that month. In February there were three outliers with a maximum of over 100 million shares. In most months one or more high volume days are identified as outliers for their month. But none of these high volume days would have been considered unusual during September, when the median daily volume of AIG stock was 170 million shares. Days that may have seemed ordinary for September if placed in another month would have seemed extraordinary and *vice versa*. That high volume day in January certainly wouldn't stand out in September or even October or November, but for January it was remarkable.

Cases that stand out from the rest of the data deserve our attention. Boxplots have a rule for nominating extreme cases to display as outliers, but that's just a rule of thumb—not a definition. The rule doesn't tell you what to do with them. It's never a substitute for careful thinking about the data and their context.

So, what *should* we do with outliers? The first thing to do is to try to understand them in the context of the data. Once you've identified likely outliers, you should always investigate them. Some outliers are unbelievable and may simply be errors. A decimal point may have been misplaced, digits transposed, or digits repeated or omitted. Sometimes a number is transcribed incorrectly, perhaps copying an adjacent value on the original data sheet. Or, the units may be wrong. If you saw the number of AIG shares traded on the NYSE listed as 2 shares for a particular day, you'd know something was wrong. It could be that it was meant as 2 million shares, but you'd have to check to be sure. If you can identify the error, then you should certainly correct it.

Many outliers are not wrong; they're just different. These are the cases that often repay your efforts to understand them. You may learn more from the extraordinary cases than from summaries of the overall dataset.

What about those two days in September that stand out as extreme even during that volatile month? Those were September 15 and 16, 2008. On the 15th, 740 million shares of AIG stock were traded. That was followed by an incredible volume of over one billion shares of stock from a single company traded the following day. Here's how Barron's described the trading of September 16:

Record Volume for NYSE Stocks, Nasdaq Trades Surge
Beats Its July Record

Yesterday's record-setting volume of 8.14 billion shares traded of all stocks listed on the New York Stock Exchange was pushed aside today by 9.31 billion shares in NYSE Composite volume. The biggest among those trades was the buying and selling of American International Group, with 1.11 billion shares traded as of 4 p.m. today. The AIG trades were 12% of all NYSE Composite volume.

For Example   Identifying outliers and summarizing data

Question: A real estate report lists the following prices for sales of single family homes in a small town in Virginia (rounded to the nearest thousand). Write a couple of sentences describing house prices in this town.

155,000	329,000	172,000	122,000	260,000
139,000	178,000	339,435,000	136,000	330,000
158,000	194,000	279,000	167,000	159,000
149,000	160,000	231,000	136,000	128,000

Answer: A box plot shows an extreme outlier:

That extreme point is a home whose sale price is listed at $339.4 M.

A check on the Internet shows that the most expensive homes ever sold are less than $200 M. This is clearly a mistake.

Setting aside this point, we find the following histogram and summary statistics:

The distribution of prices is strongly skewed to the right. The median price is $160,000. The minimum is $122,000 and the maximum (without the outlier) is $330,000. The middle 50% of house prices lie between $144,000 and $212,500 with an IQR of $68,500.

5.9 Standardizing

Forbes magazine lists the 258 largest privately held companies in the United States. What do we mean by large? *Forbes* provides two measures: *Revenue* (measured in $B) and number of *Employees*. *Forbes* uses only revenue in its rankings, but couldn't the size of the workforce be a way to measure size as well? How can we compare the two? How does having revenue of $20B compare to having a workforce of 50,000 employees? Which is "larger"? They don't have the same units, so we can't compare them directly. The trick is to standardize each variable first and then compare them. By doing this, we avoid comparing apples to oranges. Over and over during this course (and in many other courses you may take), questions such as "How does this value compare to a typical value?" or "How different are these two values?" will be answered by measuring the distance or difference in standard deviations from the mean.

Here are two companies listed by *Forbes*:

US Foodservice (A diversified food company, ranked #11) with $19.81B revenue and 26,000 employees

Toys "R" Us (the toy chain, ranked #21) with revenues of only $13.72B but 69,000 employees

It's easy to see which company earns more and which has more employees, but which company stands out more relative to others in the Forbes list?

How does standardizing work?

We first need to find the mean and standard deviation of each variable for all 258 companies in the *Forbes* list:

	Mean (all companies)	SD (all companies)
Revenue ($B)	6.23	10.56
Employees	19,629	32,055

Next we measure how *far* each of our values are by subtracting the mean and then dividing by the standard deviation:

$$z = (y - \bar{y})/s$$

We call the resulting value a **standardized value** and denote it with the letter *z*. Usually, we just call it a **z-score**. The *z*-score tells us how many standard deviations the value is from its mean.

Let's look at revenues first.

To compute the *z*-score for US Foodservice, take its value (19.81), subtract the mean (6.23) and divide by 10.56:

$$z = (19.81 - 6.23)/10.56 = 1.29$$

That means that US Foodservice's revenue is 1.29 standard deviations *above* the mean. How about employees?

$$z = (26,000 - 19,629)/32,055 = 0.20$$

So US Foodservice's workforce is not nearly as large (relative to the rest of the companies) as their revenue. The number of employees is only 0.20 standard deviations larger than the mean.

What about Toys "R" Us?

For revenue, $z = (13.72 - 6.23)/10.56 = 0.71$ and for employees, $z = (69,000 - 19,629)/32,055 = 1.54$

So who's bigger? If we use revenue, US Foodservice is the winner. If we use workforce, it's Toys "R" Us.

It's not clear which one we should use, but standardizing gives us a way to compare variables even when they're measured in different units. In this case, one could argue that Toys "R" Us is the bigger company. Its revenue *z*-score is 0.71 compared to US Foodservice's 1.29 but its employee size is 1.54 compared to 0.20 for US Foodservice.

It's not clear how to combine these two variables, although people do this sort of thing all the time. *Fortune* magazine with the help of the Great Places to Work Institute ranks the best companies to work for. In 2009 the software company SAS won. How did they get that honor? Overall, the analysts measured 50 different aspects of the companies. Was SAS better on all 50 variables? Certainly not, but it's almost certain that to combine the variables the analysts had to standardize the variables before combining them, no matter what their methodology.

Standardizing into *z*-Scores:

- Shifts the mean to 0.
- Changes the standard deviation to 1.
- Does not change the shape.
- Removes the units.

For Example   Comparing values by standardizing

Question: A real estate analyst finds more data from home sales as discussed in the example on page 106. Of 350 recent sales, the average price was $175,000 with a standard deviation of $55,000. The size of the houses (in square feet) averaged 2100 sq. ft. with a standard deviation of 650 sq. ft. Which is more unusual, a house in this town that costs $340,000, or a 5000 sq. ft. house?

Answer: Compute the *z*-scores to compare. For the $340,000 house:

$$z = \frac{y - \bar{y}}{s} = \frac{(340,000 - 175,000)}{55,000} = 3.0$$

The house price is 3 standard deviations above the mean.
For the 5000 sq. ft. house:

$$z = \frac{y - \bar{y}}{s} = \frac{(5,000 - 2,100)}{650} = 4.46$$

This house is 4.46 standard deviations above the mean in size. That's more unusual than the house that costs $340,000.

5.10 Time Series Plots

The price and volume of stocks traded on the NYSE are reported daily. Earlier, we grouped the days into months and years, but we could simply look at the price day by day. A histogram can provide information about the distribution of a variable, but it can't show any pattern over time. Whenever we have time series data, it is a good idea to look for patterns by plotting the data in time order. Figure 5.12 shows the *daily prices* plotted over time for 2007.

A display of values against time is called a **time series plot**. This plot reflects the pattern that we were unable to see by displaying the entire year's prices in either a histogram or a boxplot. Now we can see that although the price rallied in the spring of 2007, after July there were already signs that the price might not stay above $60. By October, that pattern was clear.

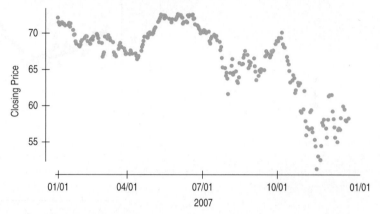

Figure 5.12   A time series plot of daily closing *prices* of AIG stock shows the overall pattern and changes in variation.

Time series plots often show a great deal of point-to-point variation, as Figure 5.12 does, and you'll often see time series plots drawn with all the points connected, especially in financial publications.

Figure 5.13   The *daily prices* of Figure 5.12, drawn by connecting all the points. Sometimes this can help us see the underlying pattern.

Often it is better to try to smooth out the local point-to-point variability. After all, we usually want to see past this variation to understand any underlying trend and think about how the values vary around that trend—the time series version of center and spread. There are many ways for computers to run a smooth trace through a time series plot. Some follow local bumps, others emphasize long-term trends. Some provide an equation that gives a typical value for any given time point, others just offer a smooth trace.

A smooth trace can highlight long-term patterns and help us see them through the more local variation. Figure 5.14 shows the daily prices of Figures 5.12 and 5.13 with a typical smoothing function, available in many statistics programs. With the smooth trace, it's a bit easier to see a pattern. The trace helps our eye follow the main trend and alerts us to points that don't fit the overall pattern.

Figure 5.14   The daily volumes of Figure 5.12, with a smooth trace added to help your eye see the long-term pattern.

It is always tempting to try to extend what we see in a timeplot into the future. Sometimes that makes sense. Most likely, the NYSE volume follows some regular patterns throughout the year. It's probably safe to predict more volume on triple witching days (when contracts expire) and less activity in the week between Christmas and New Year's Day.

Other patterns are riskier to extend into the future. If a stock's price has been rising, how long will it continue to go up? No stock has ever increased in value indefinitely, and no stock analyst has consistently been able to forecast when a stock's value will turn around. Stock prices, unemployment rates, and other economic, social, or psychological measures are much harder to predict than physical quantities. The path a ball will follow when thrown from a certain height at a given speed and direction is well understood. The path interest rates will take is much less clear.

Unless we have strong (nonstatistical) reasons for doing otherwise, we should resist the temptation to think that any trend we see will continue indefinitely. Statistical models often tempt those who use them to think beyond the data. We'll pay close attention later in this book to understanding when, how, and how much we can justify doing that.

Look at the prices in Figures 5.12 through 5.14 and try to guess what happened in the subsequent months. Was that drop from October to December a sign of trouble ahead, or was the increase in December back to around $60 where the stock had comfortably traded for several years a sign that stability had returned to AIG's stock price? Perhaps those who picked up the stock for $51 in early November really got a bargain. Let's look ahead to 2008:

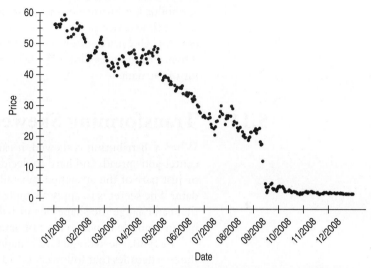

Figure 5.15 A time series plot of daily AIG *prices* shows what happened to the company in 2008.

Even through the spring of 2008, although the price was gently falling, nothing prepared traders following only the time series plot for what was to follow. In September the stock lost 99% of its value and as of 2010 was still trading below $2 an original share.

For Example   Plotting time series data

Question: The downloads from the example on page 90 are a time series. Plot the data by hour of the day and describe any patterns you see.

Answer: For this day, downloads were highest at midnight with about 36 downloads/hr, then dropped sharply until about 5–6 AM when they reached their minimum at 2–3 per hour. They gradually increased to about 20/hr by noon, and then stayed in the twenties until midnight, with a slight increase during the evening hours. When we split the data at midnight and noon, as we did earlier, we missed this pattern entirely.

The histogram we saw the beginning of the chapter (Figure 5.1) summarized the distribution of prices fairly well because during that period the prices were fairly stable. When a time series is **stationary**[7] (without a strong trend or change in variability), a histogram can provide a useful summary, especially in conjunction with a time series plot. However, when the time series is not stationary as was the case for AIG prices after 2007, a histogram is unlikely to capture much of interest. Then, a time series plot is the best graphical display to use in describing the behavior of the data.

5.11 Transforming Skewed Data

When a distribution is skewed, it can be hard to summarize the data simply with a center and spread, and hard to decide whether the most extreme values are outliers or just part of the stretched-out tail. How can we say anything useful about such data? The secret is to apply a simple function to each data value. One such function that can change the shape of a distribution is the logarithmic function. Let's examine an example in which a set of data is severely skewed.

In 1980, the average CEO made about 42 times the average worker's salary. In the two decades that followed, CEO compensation soared when compared with the average worker's pay; by 2000, that multiple had jumped to 525.[8] What does the distribution of the Fortune 500 companies' CEOs look like? Figure 5.16 shows a boxplot and a histogram of the 2005 compensation.

Figure 5.16 The total compensation for CEOs (in $000) of the 500 largest companies is skewed and includes some extraordinarily large values.

These values are reported in *thousands* of dollars. The boxplot indicates that some of the 500 CEOs received extraordinarily high compensation. The first bin of the histogram, containing about half the CEOs, covers the range $0 to $5,000,000. The reason that the histogram seems to leave so much of the area blank is that the largest observations are so far from the bulk of the data, as we can see from the boxplot. Both the histogram and boxplot make it clear that this distribution is very skewed to the right.

---

[7]Sometimes we separate out the properties and say the series is stationary with respect to the mean (if there is no trend) or stationary with respect to the variance (if the spread doesn't change), but unless otherwise noted, we'll assume that *all the statistical properties* of a stationary series are constant over time.

[8]Sources: United for a Fair Economy, *Business Week* annual CEO pay surveys, Bureau of Labor Statistics, "Average Weekly Earnings of Production Workers, Total Private Sector." Series ID: EEU00500004.

Total compensation for CEOs consists of their base salaries, bonuses, and extra compensation, usually in the form of stock or stock options. Data that add together several variables, such as the compensation data, can easily have skewed distributions. It's often a good idea to separate the component variables and examine them individually, but we don't have that information for the CEOs.

Skewed distributions are difficult to summarize. It's hard to know what we mean by the "center" of a skewed distribution, so it's not obvious what value to use to summarize the distribution. What would you say was a typical CEO total compensation? The mean value is $10,307,000, while the median is "only" $4,700,000. Each tells something different about how the data are distributed.

One way to make a skewed distribution more symmetric is to **re-express**, or **transform**, the data by applying a simple function to all the data values. Variables with a distribution that is skewed to the right often benefit from a re-expression by logarithms or square roots. Those skewed to the left may benefit from squaring the data values. It doesn't matter what base you use for a logarithm.

- **Dealing with logarithms** You probably don't encounter logarithms every day. In this book, we use them to make data behave better by making model assumptions more reasonable. Base 10 logs are the easiest to understand, but natural logs are often used as well. (Either one is fine.) You can think of base 10 logs as roughly one less than the number of digits you need to write the number. So 100, which is the smallest number to require 3 digits, has a $\log_{10}$ of 2. And 1000 has a $\log_{10}$ of 3. The $\log_{10}$ of 500 is between 2 and 3, but you'd need a calculator to find that it's approximately 2.7. All salaries of "six figures" have $\log_{10}$ between 5 and 6. Logs are incredibly useful for making skewed data more symmetric. Fortunately, with technology, remaking a histogram or other display of the data is as easy as pushing a button.

The histogram of the logs of the total CEO compensations in Figure 5.17 is much more symmetric, so we can see that a typical *log compensation* is between 6.0 and 7.0, which means that it lies between $1 million and $10 million. To be more precise, the mean log_{10} value is 6.73, while the median is 6.67 (that's $5,370,317 and $4,677,351, respectively). Note that nearly all the values are between 6.0 and 8.0—in other words, between $1,000,000 and $100,000,000 per year. Logarithmic transformations are common, and because computers and calculators are available to do the calculating, you should consider transformation as a helpful tool whenever you have skewed data.

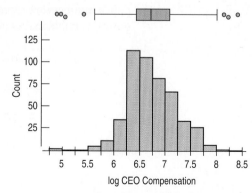

Figure 5.17 Taking logs makes the histogram of CEO total compensation nearly symmetric.

For Example   Transforming skewed data

Question: Every year *Fortune* magazine publishes a list of the 100 best companies to work for (http://money.cnn.com/magazines/fortune/bestcompanies/2010/). One statistic often looked at is the average annual pay for the most common job title at the company. Can we characterize those pay values? Here is a histogram of the average annual pay values and a histogram of the logarithm of the pay values. Which would provide the better basis for summarizing pay?

Answer: The pay values are skewed to the high end. The logarithm transformation makes the distribution more nearly symmetric. A symmetric distribution is more appropriate to summarize with a mean and standard deviation.

What Can Go Wrong?

A data display should tell a story about the data. To do that it must speak in a clear language, making plain what variable is displayed, what any axis shows, and what the values of the data are. And it must be consistent in those decisions.

The task of summarizing a quantitative variable requires that we follow a set of rules. We need to watch out for certain features of the data that make summarizing them with a number dangerous. Here's some advice:

- **Don't make a histogram of a categorical variable.** Just because the variable contains numbers doesn't mean it's quantitative. Here's a histogram of the insurance policy numbers of some workers. It's not very informative because the policy numbers are categorical. A histogram or stem-and-leaf display of a categorical variable makes no sense. A bar chart or pie chart may do better.

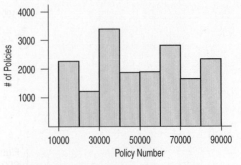

Figure 5.18   It's not appropriate to display categorical data like policy numbers with a histogram.

- **Choose a scale appropriate to the data.** Computer programs usually do a pretty good job of choosing histogram bin widths. Often, there's an easy way to adjust the width, sometimes interactively. Figure 15.19 shows the AIG price change histogram with two other choices for the bin size.

- **Avoid inconsistent scales.** Parts of displays should be mutually consistent—no fair changing scales in the middle or plotting two variables on different scales but on the same display. When comparing two groups, be sure to draw them on the same scale.

- **Label clearly.** Variables should be identified clearly and axes labeled so a reader knows what the plot displays.

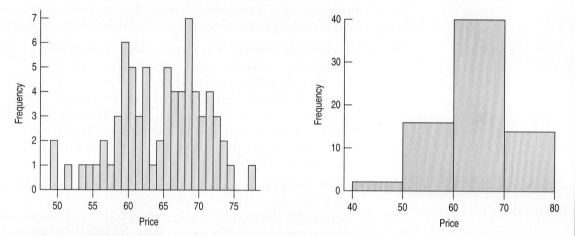

Figure 5.19 Changing the bin width changes how the histogram looks. The AIG stock prices look very different with these two choices.

Here's a remarkable example of a plot gone wrong. It illustrated a news story about rising college costs. It uses time series plots, but it gives a misleading impression. First, think about the story you're being told by this display. Then try to figure out what has gone wrong.

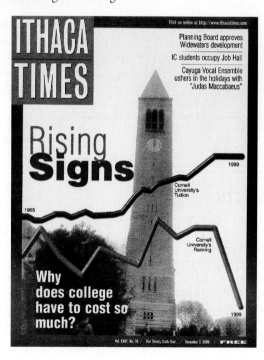

(continued)

What's wrong? Just about everything.

- The horizontal scales are inconsistent. Both lines show trends over time, but for what years? The tuition sequence starts in 1965, but rankings are graphed from 1989. Plotting them on the same (invisible) scale makes it seem that they're for the same years.

- The vertical axis isn't labeled. That hides the fact that it's using two different scales. Does it graph dollars (of tuition) or ranking (of Cornell University)?

This display violates three of the rules. And it's even worse than that. It violates a rule that we didn't even bother to mention. The two inconsistent scales for the vertical axis don't point in the same direction! The line for Cornell's rank shows that it has "plummeted" from 15th place to 6th place in academic rank. Most of us think that's an *improvement*, but that's not the message of this graph.

- **Do a reality check.** Don't let the computer (or calculator) do your thinking for you. Make sure the calculated summaries make sense. For example, does the mean look like it is in the center of the histogram? Think about the spread. An IQR of 50 mpg would clearly be wrong for a family car. And no measure of spread can be negative. The standard deviation can take the value 0, but only in the very unusual case that all the data values equal the same number. If you see the IQR or standard deviation equal to 0, it's probably a sign that something's wrong with the data.

- **Don't compute numerical summaries of a categorical variable.** The mean zip code or the standard deviation of Social Security numbers is not meaningful. If the variable is categorical, you should instead report summaries such as percentages. It is easy to make this mistake when you let technology do the summaries for you. After all, the computer doesn't care what the numbers mean.

- **Watch out for multiple modes.** If the distribution—as seen in a histogram, for example—has multiple modes, consider separating the data into groups. If you cannot separate the data in a meaningful way, you should not summarize the center and spread of the variable.

- **Beware of outliers.** If the data have outliers but are otherwise unimodal, consider holding the outliers out of the further calculations and reporting them individually. If you can find a simple reason for the outlier (for instance, a data transcription error), you should remove or correct it. If you cannot do either of these, then choose the median and IQR to summarize the center and spread.

Ethics in Action

Beth Tully owns Zenna's Café, an independent coffee shop located in a small midwestern city. Since opening Zenna's in 2002, she has been steadily growing her business and now distributes her custom coffee blends to a number of regional restaurants and markets. She operates a microroaster that offers specialty grade Arabica coffees recognized as some as the best in the area. In addition to providing the highest quality coffees, Beth also wants her business to be socially responsible. Toward that end, she pays fair prices to coffee farmers and donates funds to help charitable causes in Panama, Costa Rica, and Guatemala. In addition, she encourages her employees to get involved in the local community. Recently, one of the well-known multinational coffeehouse chains announced plans to locate shops in her area. This chain is one of the few to offer Certified Free Trade coffee products and work toward social justice in the global community. Consequently, Beth thought it might be a good idea for her to begin communicating Zenna's socially responsible efforts to the public, but with an emphasis on their commitment to the local community. Three months ago she began collecting data on the number of volunteer hours donated by her employees per week. She has a total of 12 employees, of whom 10 are full time. Most employees volunteered less than 2 hours per week, but Beth noticed that one part-time employee volunteered more than 20 hours per week. She discovered that her employees collectively volunteered an average of 15 hours per month (with a median of 8 hours). She planned to report the average number and believed most people would be impressed with Zenna's level of commitment to the local community.

ETHICAL ISSUE *The outlier in the data affects the average in a direction that benefits Beth Tully and Zenna's Café (related to Item C, ASA Ethical Guidelines).*

ETHICAL SOLUTION *Beth's data are highly skewed. There is an outlier value (for a part-time employee) that pulls the average number of volunteer hours up. Reporting the average is misleading. In addition, there may be justification to eliminate the value since it belongs to a part-time employee (10 of the 12 employees are full time). It would be more ethical for Beth to: (1) report the average but discuss the outlier value, (2) report the average for only full-time employees, or (3) report the median instead of the average.*

What Have We Learned?

Learning Objectives

- Make and interpret histograms to display the distribution of a variable.
  - We understand distributions in terms of their shape, center, and spread.

- Describe the shape of a distribution.
  - A **symmetric** distribution has roughly the same shape reflected around the center
  - A **skewed** distribution extends farther on one side than on the other.
  - A **unimodal** distribution has a single major hump or mode; a bimodal distribution has two; multimodal distributions have more.
  - **Outliers** are values that lie far from the rest of the data.

- Compute the mean and median of a distribution, and know when it is best to use each to summarize the center.
  - The **mean** is the sum of the values divided by the count. It is a suitable summary for unimodal, symmetric distributions.
  - The **median** is the middle value; half the values are above and half are below the median. It is a better summary when the distribution is skewed or has outliers.

- Compute the standard deviation and interquartile range (IQR), and know when it is best to use each to summarize the spread.
  - The **standard deviation** is roughly the square root of the average squared difference between each data value and the mean. It is the summary of choice for the spread of unimodal, symmetric variables.
  - The **IQR** is the difference between the quartiles. It is often a better summary of spread for skewed distributions or data with outliers.

- Find a five-number summary and, using it, make a boxplot. Use the boxplot's outlier nomination rule to identify cases that may deserve special attention.
  - A **five-number summary** consists of the median, the quartiles, and the extremes of the data.
  - A **boxplot** shows the quartiles as the upper and lower ends of a central box, the median as a line across the box, and "whiskers" that extend to the most extreme values that are not nominated as outliers.

(continued)

- Boxplots display separately any case that is more than 1.5 IQRs beyond each quartile. These cases should be considered as possible outliers.

■ **Use boxplots to compare distributions.**
  - Boxplots facilitate comparisons of several groups. It is easy to compare centers (medians) and spreads (IQRs).
  - Because boxplots show possible outliers separately, any outliers don't affect comparisons.

■ **Standardize values and use them for comparisons of otherwise disparate variables.**
  - We standardize by finding **z-scores**. To convert a data value to its z-score, subtract the mean and divide by the standard deviation.
  - z-scores have no units, so they can be compared to z-scores of other variables.
  - The idea of measuring the distance of a value from the mean in terms of standard deviations is a basic concept in Statistics and will return many times later in the course.

■ **Make and interpret time plots for time series data.**
  - Look for the trend and any changes in the spread of the data over time.

Terms

Bimodal	Distributions with two modes.
Boxplot	A boxplot displays the 5-number summary as a central box with whiskers that extend to the nonoutlying values. Boxplots are particularly effective for comparing groups.
Center	The middle of the distribution, usually summarized numerically by the mean or the median.
Distribution	The distribution of a variable gives: • possible values of the variable • frequency or relative frequency of each value
Five-number summary	A five-number summary for a variable consists of: • The minimum and maximum • The quartiles Q1 and Q3 • The median
Histogram (relative frequency histogram)	A histogram uses adjacent bars to show the distribution of values in a quantitative variable. Each bar represents the frequency (relative frequency) of values falling in an interval of values.
Interquartile range (IQR)	The difference between the first and third quartiles. $IQR = Q3 - Q1$.
Mean	A measure of center found as $\bar{y} = \Sigma y/n$.
Median	The middle value with half of the data above it and half below it.
Mode	A peak or local high point in the shape of the distribution of a variable. The apparent location of modes can change as the scale of a histogram is changed.
Multimodal	Distributions with more than two modes.
Outliers	Extreme values that don't appear to belong with the rest of the data. They may be unusual values that deserve further investigation or just mistakes; there's no obvious way to tell.
Quartile	The lower quartile (Q1) is the value with a quarter of the data below it. The upper quartile (Q3) has a quarter of the data above it. The median and quartiles divide the data into four equal parts.
Range	The difference between the lowest and highest values in a data set: $Range = max - min$.
Re-express or transform	To re-express or transform data, take the logarithm, square root, reciprocal, or some other mathematical operation on all values of the data set. Re-expression can make the distribution of a variable more nearly symmetric and the spread of groups more nearly alike.

Shape	The visual appearance of the distribution. To describe the shape, look for:
	• single vs. multiple modes
	• symmetry vs. skewness
Skewed	A distribution is skewed if one tail stretches out farther than the other.
Spread	The description of how tightly clustered the distribution is around its center. Measures of spread include the IQR and the standard deviation.
Standard deviation	A measure of spread found as $s = \sqrt{\dfrac{\Sigma(y - \bar{y})^2}{n-1}}$.
Standardized value	We standardize a value by subtracting the mean and dividing by the standard deviation for the variable. These values, called z-scores, have no units.
Stationary	A time series is said to be stationary if its statistical properties don't change over time.
Stem-and-leaf display	A stem-and-leaf display shows quantitative data values in a way that sketches the distribution of the data. It's best described in detail by example.
Symmetric	A distribution is symmetric if the two halves on either side of the center look approximately like mirror images of each other.
Tail	The tails of a distribution are the parts that typically trail off on either side.
Time series plot	Displays data that change over time. Often, successive values are connected with lines to show trends more clearly.
Uniform	A distribution that's roughly flat is said to be uniform.
Unimodal	Having one mode. This is a useful term for describing the shape of a histogram when it's generally mound-shaped.
Variance	The standard deviation squared.
z-score	A standardized value that tells how many standard deviations a value is from the mean; z-scores have a mean of 0 and a standard deviation of 1.

Technology Help: Displaying and Summarizing Quantitative Variables

Almost any program that displays data can make a histogram, but some will do a better job of determining where the bars should start and how they should partition the span of the data (see the art on the next page).

Many statistics packages offer a prepackaged collection of summary measures. The result might look like this:

```
Variable: Weight
N = 234
Mean =143.3      Median = 139
St. Dev = 11.1   IQR = 14
```

Alternatively, a package might make a table for several variables and summary measures:

```
Variable   N     mean    median   stdev   IQR
Weight     234   143.3   139      11.1    14
Height     234   68.3    68.1     4.3     5
Score      234   86      88       9       5
```

It is usually easy to read the results and identify each computed summary. You should be able to read the summary statistics produced by any computer package.

Packages often provide many more summary statistics than you need. Of course, some of these may not be appropriate when the data are skewed or have outliers. It is your responsibility to check a histogram or stem-and-leaf display and decide which summary statistics to use.

It is common for packages to report summary statistics to many decimal places of "accuracy." Of course, it is rare to find data that have such accuracy in the original measurements. The ability to calculate to six or seven digits beyond the decimal point doesn't mean that those digits have any meaning. Generally, it's a good idea to round these values, allowing perhaps one more digit of precision than was given in the original data.

Displays and summaries of quantitative variables are among the simplest things you can do in most statistics packages.

(continued)

The vertical scale may be counts or proportions. Sometimes it isn't clear which. But the shape of the histogram is the same either way.

Most packages choose the number of bars for you automatically. Often you can adjust that choice.

The axis should be clearly labeled so you can tell what "pile" each bar represents. You should be able to tell the lower and upper bounds of each bar.

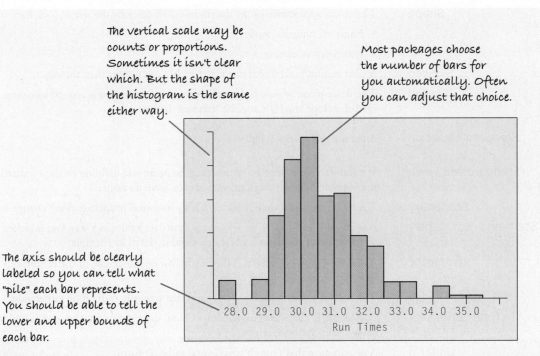

Run Times

EXCEL   XLSTAT

To make a histogram in Excel 2007 or 2010, use the Data Analysis add-in. If you have not installed that, you must do that first.

- From the Data ribbon, select the Data Analysis add-in.
- From its menu, select Histograms.
- Indicate the range of the data whose histogram you wish to draw.
- Indicate the bin ranges that are up to and including the right end points of each bin.
- Check **Labels** if your columns have names in the first cell.
- Check **Chart output** and click **OK**.
- Right-click on any bar of the resulting graph and, from the menu that drops down, select **Format Data Series . . .**
- In the dialog box that opens, select **Series Options** from the sidebar.
- Slide the Gap Width slider to **No Gap**, and click **Close**.
- In the pivot table on the left, use your pointing tool to slide the bottom of the table up to get rid of the "more" bin.
- Edit the bin names in Column A to properly identify the contents of each bin.

- You can right click on the legend or axis names to edit or remove them.
- Following these instructions, you can reproduce Figure 5.1 using the data set AIG.

Alternatively, you can set up your own bin boundaries and count the observations falling within each bin using an Excel function such as FREQUENCY (Data array, Bins array). Consult your Excel manual or help files for details of how to do this.

JMP

To make a histogram and find summary statistics:

- Choose **Distribution** from the **Analyze** menu.
- In the **Distribution** dialog, drag the name of the variable that you wish to analyze into the empty window beside the label "**Y, Columns**."
- Click **OK**. JMP computes standard summary statistics along with displays of the variables.

To make boxplots:

- Choose **Fit y By x**. Assign a continuous response variable to **Y, Response** and a nominal group variable holding the group names to **X, Factor,** and click **OK**. JMP will offer (among other things) dotplots of the data. click the red triangle and, under **Display Options**, select Boxplots. Note: If the variables are of the wrong type, the display options might not offer boxplots.

MINITAB

To make a histogram:

- Choose **Histogram** from the **Graph** menu.
- Select "Simple" for the type of graph and click **OK**.
- Enter the name of the quantitative variable you wish to display in the box labeled "Graph variables." Click **OK**.

To make a boxplot:

- Choose **Boxplot** from the **Graph** menu and specify your data format.

To calculate summary statistics:

- Choose **Basic Statistics** from the **Stat** menu. From the **Basic Statistics** submenu, choose **Display Descriptive Statistics**.
- Assign variables from the variable list box to the Variables box. MINITAB makes a Descriptive Statistics table.

SPSS

To make a histogram or boxplot in SPSS open the Chart Builder from the Graphs menu.

- Click the **Gallery** tab.

- Choose **Histogram** or **Boxplot** from the list of chart types.
- Drag the icon of the plot you want onto the canvas.
- Drag a scale variable to the y-axis drop zone.
- Click **OK**.

   To make side-by-side boxplots, drag a categorical variable to the x-axis drop zone and click **OK**.

   To calculate summary statistics:

- Choose **Explore** from the **Descriptive Statistics** submenu of the **Analyze** menu. In the Explore dialog, assign one or more variables from the source list to the Dependent List and click the **OK** button.

Brief CASE

Hotel Occupancy Rates

Many properties in the hospitality industry experience strong seasonal fluctuations in demand. To be successful in this industry it is important to anticipate such fluctuations and to understand demand patterns. The file **Occupancy_Rates** contains data on monthly *Hotel Occupancy Rates* (in % capacity) for Honolulu, Hawaii, from January 2000 to December 2007.

Examine the data and prepare a report for the manager of a hotel chain in Honolulu on patterns in *Hotel Occupancy* during this period. Include both numerical summaries and graphical displays and summarize the patterns that you see. Discuss any unusual features of the data and explain them if you can, including a discussion of whether the manager should take these features into account for future planning.

Value and Growth Stock Returns

Investors in the stock market have choices of how aggressive they would like to be with their investments. To help investors, stocks are classified as "growth" or "value" stocks. Growth stocks are generally shares in high quality companies that have demonstrated consistent performance and are expected to continue to do well. Value stocks on the other hand are stocks whose prices seem low compared to their inherent worth (as measured by the book to price ratio). Managers invest in these hoping that their low price is simply an overreaction to recent negative events.

In the data set **Returns**[9] are the monthly returns of 2500 stocks classified as Growth and Value for the time period January 1975 to June 1997. Examine the distributions of the two types of stocks and discuss the advantages and disadvantages of each. Is it clear which type of stock offers the best investment? Discuss briefly.

---

[9]Source: Independence International Associates, Inc. maintains a family of international style indexes covering 22 equity markets. The highest book-to-price stocks are selected one by one from the top of the list. The top half of these stocks become the constituents of the "value index," and the remaining stocks become the "growth index."

Exercises

SECTION 5.1

1. As part of the marketing team at an Internet music site, you want to understand who your customers are. You send out a survey to 25 customers (you use an incentive of $50 worth of downloads to guarantee a high response rate) asking for demographic information. One of the variables is the customer's age. For the 25 customers the ages are:

20	32	34	29	30
30	30	14	29	11
38	22	44	48	26
25	22	32	35	32
35	42	44	44	48

a) Make a histogram of the data using a bar width of 10 years.
b) Make a histogram of the data using a bar width of 5 years.
c) Make a relative frequency histogram of the data using a bar width of 5 years.
d) *Make a stem-and-leaf plot of the data using 10s as the stems and putting the youngest customers on the top of the plot.

2. As the new manager of a small convenience store, you want to understand the shopping patterns of your customers. You randomly sample 20 purchases from yesterday's records (all purchases in U.S. dollars):

39.05	2.73	32.92	47.51
37.91	34.35	64.48	51.96
56.95	81.58	47.80	11.72
21.57	40.83	38.24	32.98
75.16	74.30	47.54	65.62

a) Make a histogram of the data using a bar width of $20.
b) Make a histogram of the data using a bar width of $10.
c) Make a relative frequency histogram of the data using a bar width of $10.
d) *Make a stem-and-leaf plot of the data using $10 as the stems and putting the smallest amounts on top.

SECTION 5.2

3. For the histogram you made in Exercise 1a,

a) Is the distribution unimodal or multimodal?
b) Where is (are) the mode(s)?
c) Is the distribution symmetric?
d) Are there any outliers?

4. For the histogram you made in Exercise 2a:

a) Is the distribution unimodal or multimodal?
b) Where is (are) the mode(s)?
c) Is the distribution symmetric?
d) Are there any outliers?

SECTION 5.3

5. For the data in Exercise 1:

a) Would you expect the mean age to be smaller than, bigger than, or about the same size as the median? Explain.
b) Find the mean age.
c) Find the median age.

6. For the data in Exercise 2:

a) Would you expect the mean purchase to be smaller than, bigger than, or about the same size as the median? Explain.
b) Find the mean purchase.
c) Find the median purchase.

SECTION 5.4

7. For the data in Exercise 1:

a) Find the quartiles using your calculator.
b) Find the quartiles using the method on page 96.
c) Find the IQR using the quartiles from part b.
d) Find the standard deviation.

8. For the data in Exercise 2:

a) Find the quartiles using your calculator.
b) Find the quartiles using the method on page 96.
c) Find the IQR using the quartiles from part b.
d) Find the standard deviation.

SECTION 5.5

9. The histogram shows the December charges (in $) for 5000 customers from one marketing segment from a credit card company. (Negative values indicate customers who received more credits than charges during the month.)

a) Write a short description of this distribution (shape, center, spread, unusual features).
b) Would you expect the mean or the median to be larger? Explain.
c) Which would be a more appropriate summary of the center, the mean or the median? Explain.

10. Adair Vineyard is a 10-acre vineyard in New Paltz, New York. The winery itself is housed in a 200-year-old historic Dutch barn, with the wine cellar on the first floor and the tasting room and gift shop on the second. Since they are relatively small and considering an expansion, they are curious about how their size compares to that of other vineyards. The histogram shows the sizes (in acres) of 36 wineries in upstate New York.

a) Write a short description of this distribution (shape, center, spread, unusual features).

b) Would you expect the mean or the median to be larger? Explain.

c) Which would be a more appropriate summary of the center, the mean or the median? Explain.

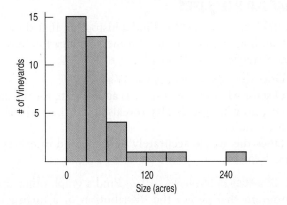

SECTION 5.6

11. For the data in Exercise 1:

a) Draw a boxplot using the quartiles from Exercise 7b.

b) Does the boxplot nominate any outliers?

c) What age would be considered a high outlier?

12. For the data in Exercise 2:

a) Draw a boxplot using the quartiles from Exercise 8b.

b) Does the boxplot nominate any outliers?

c) What purchase amount would be considered a high outlier?

13. Here are summary statistics for the sizes (in acres) of upstate New York vineyards from Exercise 10.

Variable	N	Mean	StDev	Minimum	Q1	Median	Q3	Maximum
Acres	36	46.50	47.76	6	18.50	33.50	55	250

a) From the summary statistics, would you describe this distribution as symmetric or skewed? Explain.

b) From the summary statistics, are there any outliers? Explain.

c) Using these summary statistics, sketch a boxplot. What additional information would you need to complete the boxplot?

14. A survey of major universities asked what percentage of incoming freshmen usually graduate "on time" in 4 years. Use the summary statistics given to answer these questions.

	% on time
Count	48
Mean	68.35
Median	69.90
StdDev	10.20
Min	43.20
Max	87.40
Range	44.20
25th %tile	59.15
75th %tile	74.75

a) Would you describe this distribution as symmetric or skewed?

b) Are there any outliers? Explain.

c) Create a boxplot of these data.

SECTION 5.7

15. The survey from Exercise 1 had also asked the customers to say whether they were male or female. Here are the data:

Age	Sex	Age	Sex	Age	Sex	Age	Sex	Age	Sex
20	M	32	F	34	F	29	M	30	M
30	F	30	M	14	M	29	M	11	M
38	F	22	M	44	F	48	F	26	F
25	M	22	M	32	F	35	F	32	F
35	F	42	F	44	F	44	F	48	F

Construct boxplots to compare the ages of men and women and write a sentence summarizing what you find.

16. The store manager from Exercise 2 has collected data on purchases from weekdays and weekends. Here are some summary statistics (rounded to the nearest dollar):

Weekdays n = 230

Min = 4, Q1 = 28, Median = 40, Q3 = 68, Max = 95

Weekend n = 150

Min = 10, Q1 = 35, Median = 55, Q3 = 70, Max = 100

From these statistics, construct side-by-side boxplots and write a sentence comparing the two distributions.

17. Here are boxplots of the weekly sales (in $ U.S.) over a two-year period for a regional food store for two locations. Location #1 is a metropolitan area that is known to be residential where shoppers walk to the store. Location #2 is a suburban area where shoppers drive to the store. Assume that the two towns have similar populations and that the two stores are similar in square footage. Write a brief report discussing what these data show.

18. Recall the distributions of the weekly sales for the regional stores in Exercise 17. Following are boxplots of weekly sales for this same food store chain for three stores of similar size and location for two different states: Massachusetts (MA) and Connecticut (CT). Compare the distribution of sales for the two states and describe in a report.

SECTION 5.9

19. Using the ages from Exercise 1:

a) Standardize the minimum and maximum ages using the mean from Exercise 5b and the standard deviation from Exercise 7d.
b) Which has the more extreme z-score, the min or the max?
c) How old would someone with a z-score of 3 be?

20. Using the purchases from Exercise 2:

a) Standardize the minimum and maximum purchase using the mean from Exercise 6b and the standard deviation from Exercise 8d.
b) Which has the more extreme z-score, the min or the max?
c) How large a purchase would a purchase with a z-score of 3.5 be?

SECTION 5.11

21. When analyzing data on the number of employees in small companies in one town, a researcher took square roots of the counts. Some of the resulting values, which are reasonably symmetric, were:

4, 4, 6, 7, 7, 8, 10

What were the original values, and how are they distributed?

22. You wish to explain to your boss what effect taking the base-10 logarithm of the salary values in the company's database will have on the data. As simple, example values, you compare a salary of $10,000 earned by a part-time shipping clerk, a salary of $100,000 earned by a manager, and the CEO's $1,000,000 compensation package. Why might the average of these values be a misleading summary? What would the logarithms of these three values be?

CHAPTER EXERCISES

23. Statistics in business. Find a histogram that shows the distribution of a variable in a business publication (e.g., *The Wall Street Journal, Business Week*, etc.).

a) Does the article identify the W's?
b) Discuss whether the display is appropriate for the data.
c) Discuss what the display reveals about the variable and its distribution.
d) Does the article accurately describe and interpret the data? Explain.

24. Statistics in business, part 2. Find a graph other than a histogram that shows the distribution of a quantitative variable in a business publication (e.g., *The Wall Street Journal, Business Week*, etc.).

a) Does the article identify the W's?
b) Discuss whether the display is appropriate for the data.
c) Discuss what the display reveals about the variable and its distribution.
d) Does the article accurately describe and interpret the data? Explain.

25. Two-year college tuition. The histogram shows the distribution of average tuitions charged by each of the 50 U.S. states for public two-year colleges in the 2007–2008

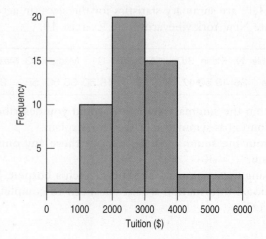

academic year. Write a short description of this distribution (shape, center, spread, unusual features).

26. Gas prices. The website MSN auto (www.autos.msn .com) provides prices of gasoline at stations all around the United States. This histogram shows the price of regular gas (in $/gallon) for 57 stations in the Los Angeles area during the week before Christmas 2007. Describe the shape of this distribution (shape, center, spread, unusual features).

27. Mutual funds. The histogram displays the 12-month returns (in percent) for a collection of mutual funds in 2007. Give a short summary of this distribution (shape, center, spread, unusual features).

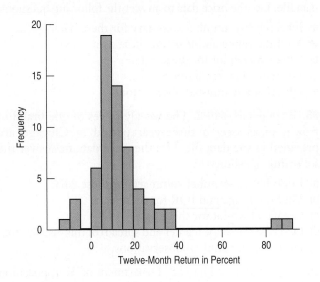

28. Car discounts. A researcher, interested in studying gender differences in negotiations, collects data on the prices that men and women pay for new cars. Here is a histogram of the discounts (the amount in $ below the list price) that men and women received at one car dealership for the last 100 transactions (54 men and 46 women). Give a short summary of this distribution (shape, center, spread, unusual features). What do you think might account for this particular shape?

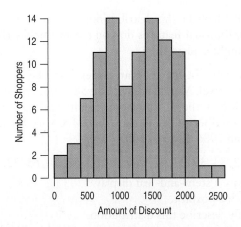

29. Mutual funds, part 2. Use the data set of Exercise 27 to answer the following questions.
a) Find the five-number summary for these data.
b) Find appropriate measures of center and spread for these data.
c) Create a boxplot for these data.
d) What can you see, if anything, in the histogram that isn't clear in the boxplot?

30. Car discounts, part 2. Use the data set of Exercise 28 to answer the following questions.
a) Find the five-number summary for these data.
b) Create a boxplot for these data.
c) What can you see, if anything, in the histogram of Exercise 28 that isn't clear in the boxplot?

*31. Vineyards. The data set provided contains the data from Exercises 10 and 13. Create a stem-and-leaf display of the sizes of the vineyards in acres. Point out any unusual features of the data that you can see from the stem-and-leaf.

*32. Gas prices, again. The data set provided contains the data from Exercise 26 on the price of gas for 57 stations around Los Angeles in December 2007. Round the data to the nearest penny (e.g., 3.459 becomes 3.46) and create a stem-and-leaf display of the data. Point out any unusual features of the data that you can see from the stem-and-leaf.

33. Gretzky. During his 20 seasons in the National Hockey League, Wayne Gretzky scored 50% more points than anyone else who ever played professional hockey. He accomplished this amazing feat while playing in 280 fewer games than Gordie Howe, the previous record holder. Here are the number of games Gretzky played during each season:

79, 80, 80, 80, 74, 80, 80, 79, 64, 78, 73, 78, 74, 45, 81, 48, 80, 82, 82, 70

*a) Create a stem-and-leaf display.
b) Sketch a boxplot.

c) Briefly describe this distribution.

d) What unusual features do you see in this distribution? What might explain this?

34. McGwire. In his 16-year career as a player in major league baseball, Mark McGwire hit 583 home runs, placing him eighth on the all-time home run list (as of 2008). Here are the number of home runs that McGwire hit for each year from 1986 through 2001:

3, 49, 32, 33, 39, 22, 42, 9, 9, 39, 52, 58, 70, 65, 32, 29

a) *Create a stem-and-leaf display.

b) Sketch a boxplot.

c) Briefly describe this distribution.

d) What unusual features do you see in this distribution? What might explain this?

35. Gretzky returns. Look once more at data of hockey games played each season by Wayne Gretzky, seen in Exercise 33.

a) Would you use the mean or the median to summarize the center of this distribution? Why?

b) Without actually finding the mean, would you expect it to be lower or higher than the median? Explain.

c) A student was asked to make a histogram of the data in Exercise 33 and produced the following. Comment.

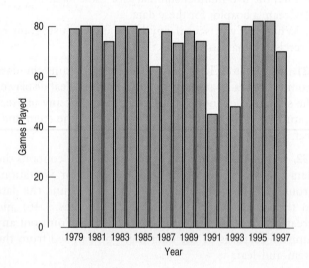

36. McGwire, again. Look once more at data of home runs hit by Mark McGwire during his 16-year career as seen in Exercise 34.

a) Would you use the mean or the median to summarize the center of this distribution? Why?

b) Find the median.

c) Without actually finding the mean, would you expect it to be lower or higher than the median? Explain.

d) A student was asked to make a histogram of the data in Exercise 34 and produced the following. Comment.

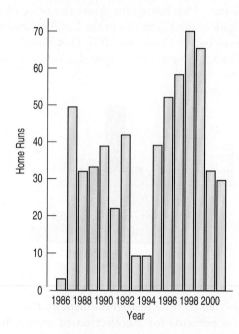

37. Pizza prices. The weekly prices of one brand of frozen pizza over a three-year period in Dallas are provided in the data file. Use the price data to answer the following questions.

a) Find the five-number summary for these data.

b) Find the range and IQR for these data.

c) Create a boxplot for these data.

d) Describe this distribution.

e) Describe any unusual observations.

38. Pizza prices, part 2. The weekly prices of one brand of frozen pizza over a three-year period in Chicago are provided in the data file. Use the price data to answer the following questions.

a) Find the five-number summary for these data.

b) Find the range and IQR for these data.

c) Create a boxplot for these data.

d) Describe the shape (center and spread) of this distribution.

e) Describe any unusual observations.

39. Gasoline usage. The U.S. Department of Transportation collects data on the amount of gasoline sold in each state and the District of Columbia. The following data show the per capita (gallons used per person) consumption in the year 2005. Write a report on the gasoline usage by state in the year 2005, being sure to include appropriate graphical displays and summary statistics.

State	Gasoline Usage	State	Gasoline Usage
Alabama	556.91	Montana	486.15
Alaska	398.99	Nebraska	439.46
Arizona	487.52	Nevada	484.26
Arkansas	491.85	New Hampshire	521.45
California	434.11	New Jersey	481.79
Colorado	448.33	New Mexico	482.33
Connecticut	441.39	New York	283.73
Delaware	514.78	North Carolina	491.07
District of Columbia	209.47	North Dakota	513.16
Florida	485.73	Ohio	434.65
Georgia	560.90	Oklahoma	501.12
Hawaii	352.02	Oregon	415.67
Idaho	414.17	Pennsylvania	402.85
Illinois	392.13	Rhode Island	341.67
Indiana	497.35	South Carolina	570.24
Iowa	509.13	South Dakota	498.36
Kansas	399.72	Tennessee	509.77
Kentucky	511.30	Texas	505.39
Louisiana	489.84	Utah	409.93
Maine	531.77	Vermont	537.94
Maryland	471.52	Virginia	518.06
Massachusetts	427.52	Washington	423.32
Michigan	470.89	West Virginia	444.22
Minnesota	504.03	Wisconsin	440.45
Mississippi	539.39	Wyoming	589.18
Missouri	530.72		

40. OECD Established in Paris in 1961, the Organisation for Economic Co-operation and Development (OECD) (www.oced.org) collects information on many economic and social aspects of countries around the world. Here are the 2005 gross domestic product (GDP) growth rates (in percentages) of 30 industrialized countries. Write a brief report on the 2005 GDP growth rates of these countries being sure to include appropriate graphical displays and summary statistics.

Country	Growth Rate
Turkey	0.074
Czech Republic	0.061
Slovakia	0.061
Iceland	0.055
Ireland	0.055
Hungary	0.041
Korea, Republic of (South Korea)	0.040
Luxembourg	0.040
Greece	0.037

Country	Growth Rate
Poland	0.034
Spain	0.034
Denmark	0.032
United States	0.032
Mexico	0.030
Canada	0.029
Finland	0.029
Sweden	0.027
Japan	0.026
Australia	0.025
New Zealand	0.023
Norway	0.023
Austria	0.020
Switzerland	0.019
United Kingdom	0.019
Belgium	0.015
The Netherlands	0.015
France	0.012
Germany	0.009
Portugal	0.004
Italy	0.000

41. Golf courses. A start-up company is planning to build a new golf course. For marketing purposes, the company would like to be able to advertise the new course as one of the more difficult courses in the state of Vermont. One measure of the difficulty of a golf course is its length: the total distance (in yards) from tee to hole for all 18 holes. Here are the histogram and summary statistics for the lengths of all the golf courses in Vermont.

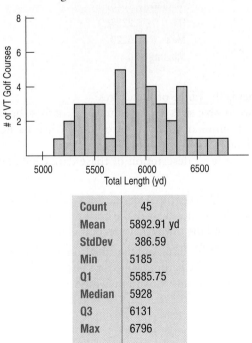

Count	45
Mean	5892.91 yd
StdDev	386.59
Min	5185
Q1	5585.75
Median	5928
Q3	6131
Max	6796

a) What is the range of these lengths?

b) Between what lengths do the central 50% of these courses lie?

c) What summary statistics would you use to describe these data?

d) Write a brief description of these data (shape, center, and spread).

42. Real estate. A real estate agent has surveyed houses in 20 nearby zip codes in an attempt to put together a comparison for a new property that she would like to put on the market. She knows that the size of the living area of a house is a strong factor in the price, and she'd like to market this house as being one of the biggest in the area. Here is a histogram and summary statistics for the sizes of all the houses in the area.

Count	1057
Mean	1819.498 sq. ft
Std Dev	662.9414
Min	672
Q1	1342
Median	1675
Q3	2223
Max	5228
Missing	0

a) What is the range of these sizes?

b) Between what sizes do the central 50% of these houses lie?

c) What summary statistics would you use to describe these data?

d) Write a brief description of these data (shape, center, and spread).

43. Food sales. Sales (in $) for one week were collected for 18 stores in a food store chain in the northeastern United States. The stores and the towns they are located in vary in size.

a) Make a suitable display of the sales from the data provided.

b) Summarize the central value for sales for this week with a median and mean. Why do they differ?

c) Given what you know about the distribution, which of these measures does the better job of summarizing the stores' sales? Why?

d) Summarize the spread of the sales distribution with a standard deviation and with an IQR.

e) Given what you know about the distribution, which of these measures does the better job of summarizing the spread of stores' sales? Why?

f) If we were to remove the outliers from the data, how would you expect the mean, median, standard deviation, and IQR to change?

44. Insurance profits. Insurance companies don't know whether a policy they've written is profitable until the policy matures (expires). To see how they've performed recently, an analyst looked at mature policies and investigated the net profit to the company (in $).

a) Make a suitable display of the profits from the data provided.

b) Summarize the central value for the profits with a median and mean. Why do they differ?

c) Given what you know about the distribution, which of these measures might do a better job of summarizing the company's profits? Why?

d) Summarize the spread of the profit distribution with a standard deviation and with an IQR.

e) Given what you know about the distribution, which of these measures might do a better job of summarizing the spread in the company's profits? Why?

f) If we were to remove the outliers from the data, how would you expect the mean, median, standard deviation, and IQR to change?

45. iPod failures. MacInTouch (www.macintouch.com/reliability/ipodfailures.html) surveyed readers about the reliability of their iPods. Of the 8926 iPods owned, 7510 were problem-free while the other 1416 failed. From the data on the CD, compute the failure rate for each of the 17 iPod models. Produce an appropriate graphical display of the failure rates and briefly describe the distribution. (To calculate the failure rate, divide the number failed by the sum of the number failed and the number OK for each model and then multiply by 100.)

46. Unemployment. The data set provided contains 2008 unemployment rates for 23 developed countries (www.oecd.org). Produce an appropriate graphical display and briefly describe the distribution of unemployment rates.

47. Gas prices, part 2. Below are boxplots of weekly gas prices at a service station in the Midwest United States (prices in $ per gallon).

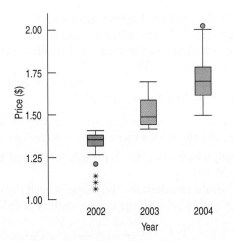

a) Compare the distribution of prices over the three years.
b) In which year were the prices least stable (most volatile)? Explain.

48. Fuel economy. American automobile companies are becoming more motivated to improve the fuel efficiency of the automobiles they produce. It is well known that fuel efficiency is impacted by many characteristics of the car. Describe what these boxplots tell you about the relationship between the number of cylinders a car's engine has and the car's fuel economy (mpg).

49. Wine prices. The boxplots display case prices (in dollars) of wines produced by vineyards along three of the Finger Lakes in upstate New York.

a) Which lake region produces the most expensive wine?
b) Which lake region produces the cheapest wine?
c) In which region are the wines generally more expensive?
d) Write a few sentences describing these prices.

50. Ozone. Ozone levels (in parts per billion, ppb) were recorded at sites in New Jersey monthly between 1926 and 1971. Here are boxplots of the data for each month (over the 46 years) lined up in order (January = 1).

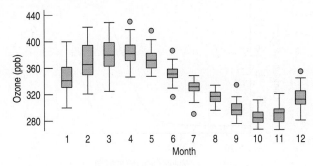

a) In what month was the highest ozone level ever recorded?
b) Which month has the largest IQR?
c) Which month has the smallest range?
d) Write a brief comparison of the ozone levels in January and June.
e) Write a report on the annual patterns you see in the ozone levels.

51. Derby speeds. How fast do horses run? Kentucky Derby winners top 30 miles per hour, as shown in the graph. This graph shows the percentage of Kentucky Derby winners that have run *slower* than a given speed. Note that few have won running less than 33 miles per hour, but about 95% of the winning horses have run less than 37 miles per hour. (A cumulative frequency graph like this is called an **ogive**.)

a) Estimate the median winning speed.
b) Estimate the quartiles.
c) Estimate the range and the IQR.
d) Create a boxplot of these speeds.
e) Write a few sentences about the speeds of the Kentucky Derby winners.

52. Mutual fund, part 3. Here is an ogive of the distribution of monthly returns for a group of aggressive (or high growth) mutual funds over a period of 25 years from 1975 to 1999. (Recall from Exercise 51 that an ogive, or cumulative relative frequency graph, shows the percent of cases at or below a certain value. Thus this graph always begins at 0% and ends at 100%.)

a) Estimate the median.
b) Estimate the quartiles.
c) Estimate the range and the IQR.
d) Create a boxplot of these returns.

53. Test scores. Three Statistics classes all took the same test. Here are histograms of the scores for each class.

a) Which class had the highest mean score?
b) Which class had the highest median score?
c) For which class are the mean and median most different? Which is higher? Why?
d) Which class had the smallest standard deviation?
e) Which class had the smallest IQR?

54. Test scores, again. Look again at the histograms of test scores for the three Statistics classes in Exercise 53.

a) Overall, which class do you think performed better on the test? Why?
b) How would you describe the shape of each distribution?
c) Match each class with the corresponding boxplot.

T 55. Quality control holes. Engineers at a computer production plant tested two methods for accuracy in drilling holes into a PC board. They tested how fast they could set the drilling machine by running 10 boards at each of two different speeds. To assess the results, they measured the distance (in inches) from the center of a target on the board to the center of the hole. The data and summary statistics are shown in the table.

	Fast	Slow
	0.000101	0.000098
	0.000102	0.000096
	0.000100	0.000097
	0.000102	0.000095
	0.000101	0.000094
	0.000103	0.000098
	0.000104	0.000096
	0.000102	0.975600
	0.000102	0.000097
	0.000100	0.000096
Mean	0.000102	0.097647
StdDev	0.000001	0.308481

Write a report summarizing the findings of the experiment. Include appropriate visual and verbal displays of the distributions, and make a recommendation to the engineers if they are most interested in the accuracy of the method.

T 56. Fire sale. A real estate agent notices that houses with fireplaces often fetch a premium in the market and wants to assess the difference in sales price of 60 homes that recently sold. The data and summary are shown in the table.

	No Fireplace	Fireplace
	142,212	134,865
	206,512	118,007
	50,709	138,297
	108,794	129,470
	68,353	309,808
	123,266	157,946
	80,248	173,723
	135,708	140,510
	122,221	151,917
	128,440	235,105,000
	221,925	259,999
	65,325	211,517
	87,588	102,068
	88,207	115,659
	148,246	145,583
	205,073	116,289
	185,323	238,792
	71,904	310,696
	199,684	139,079
	81,762	109,578
	45,004	89,893
	62,105	132,311
	79,893	131,411
	88,770	158,863
	115,312	130,490
	118,952	178,767
		82,556
		122,221
		84,291
		206,512
		105,363
		103,508
		157,513
		103,861
Mean	116,597.54	7,061,657.74
Median	112,053	136,581

Write a report summarizing the findings of the investigation. Include appropriate visual and verbal displays of the distributions, and make a recommendation to the agent about the average premium that a fireplace is worth in this market.

57. Customer database. A philanthropic organization has a database of millions of donors that they contact by mail to raise money for charities. One of the variables in the database, *Title*, contains the title of the person or persons printed on the address label. The most common are Mr., Ms., Miss, and Mrs., but there are also Ambassador and Mrs., Your Imperial Majesty, and Cardinal, to name a few

others. In all there are over 100 different titles, each with a corresponding numeric code. Here are a few of them.

Code	Title
000	MR.
001	MRS.
1002	MR. and MRS.
003	MISS
004	DR.
005	MADAME
006	SERGEANT
009	RABBI
010	PROFESSOR
126	PRINCE
127	PRINCESS
128	CHIEF
129	BARON
130	SHEIK
131	PRINCE AND PRINCESS
132	YOUR IMPERIAL MAJESTY
135	M. ET MME.
210	PROF.
⋮	⋮

An intern who was asked to analyze the organization's fundraising efforts presented these summary statistics for the variable *Title*.

Mean	54.41
StdDev	957.62
Median	1
IQR	2
n	94649

a) What does the mean of 54.41 mean?
b) What are the typical reasons that cause measures of center and spread to be as different as those in this table?
c) Is that why these are so different?

58. CEOs. For each CEO, a code is listed that corresponds to the industry of the CEO's company. Here are a few of the codes and the industries to which they correspond.

Industry	Industry Code	Industry	Industry Code
Financial services	1	Energy	12
Food/drink/tobacco	2	Capital goods	14
Health	3	Computers/communications	16
Insurance	4	Entertainment/information	17
Retailing	6	Consumer non-durables	18
Forest products	9	Electric utilities	19
Aerospace/defense	11		

A recently hired investment analyst has been assigned to examine the industries and the compensations of the CEOs. To start the analysis, he produces the following histogram of industry codes.

a) What might account for the gaps seen in the histogram?
b) What advice might you give the analyst about the appropriateness of this display?

T 59. Mutual funds types. The 64 mutual funds of Exercise 27 are classified into three types: U.S. Domestic Large Cap Funds, U.S. Domestic Small/Mid Cap Funds, and International Funds. Compare the 3-month return of the three types of funds using an appropriate display and write a brief summary of the differences.

T 60. Car discounts, part 3. The discounts negotiated by the car buyers in Exercise 28 are classified by whether the buyer was Male (code = 0) or Female (code = 1). Compare the discounts of men vs. women using an appropriate display and write a brief summary of the differences.

61. Houses for sale. Each house listed on the multiple listing service (MLS) is assigned a sequential ID number. A recently hired real estate agent decided to examine the MLS numbers in a recent random sample of homes for sale by one real estate agency in nearby towns. To begin the analysis, the agent produces the following histogram of ID numbers.

a) What might account for the distribution seen in the histogram?
b) What advice might you give the analyst about the appropriateness of this display?

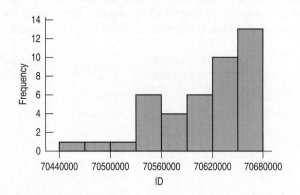

62. Zip codes. Holes-R-Us, an Internet company that sells piercing jewelry, keeps transaction records on its sales. At a

recent sales meeting, one of the staff presented the following histogram and summary statistics of the zip codes of the last 500 customers, so that the staff might understand where sales are coming from. Comment on the usefulness and appropriateness of this display.

T *63. Hurricanes. Buying insurance for property loss from hurricanes has become increasingly difficult since hurricane Katrina caused record property loss damage. Many companies have refused to renew policies or write new ones. The data set provided contains the total number of hurricanes by every full decade from 1851 to 2000 (from the National Hurricane Center). Some scientists claim that there has been an increase in the number of hurricanes in recent years.

a) Create a histogram of these data.
b) Describe the distribution.
c) Create a time series plot of these data.
d) Discuss the time series plot. Does this graph support the claim of these scientists, at least up to the year 2000?

T *64. Hurricanes, part 2. Using the hurricanes data set, examine the number of major hurricanes (category 3, 4, or 5) by every full decade from 1851 to 2000.

a) Create a histogram of these data.
b) Describe the distribution.
c) Create a timeplot of these data.
d) Discuss the timeplot. Does this graph support the claim of scientists that the number of major hurricanes has been increasing (at least up through the year 2000)?

65. Productivity study. The National Center for Productivity releases information on the efficiency of workers. In a recent report, they included the following graph showing a rapid rise in productivity. What questions do you have about this?

66. Productivity study revisited. A second report by the National Center for Productivity analyzed the relationship between productivity and wages. Comment on the graph they used.

67. Real estate, part 2. The 1057 houses described in Exercise 42 have a mean price of $167,900, with a standard deviation of $77,158. The mean living area is 1819 sq. ft., with a standard deviation of 663 sq. ft. Which is more unusual, a house in that market that sells for $400,000 or a house that has 4000 sq. ft of living area? Explain.

T 68. Tuition, 2008. The data set provided contains the average tuition of private four-year colleges and universities as well as the average 2007–2008 tuitions for each state seen in Exercise 25. The mean tuition charged by a public two-year college was $2763, with a standard deviation of $988. For private four-year colleges the mean was $21,259, with a standard deviation of $6241. Which would be more unusual: a state whose average public two-year college is $700 or a state whose average private four-year college tuition was $10,000? Explain.

T 69. Food consumption. FAOSTAT, the Food and Agriculture Organization of the United Nations, collects information on the production and consumption of more than

Country	Alcohol	Meat	Country	Alcohol	Meat
Australia	29.56	242.22	Luxembourg	34.32	197.34
Austria	40.46	242.22	Mexico	13.52	126.50
Belgium	34.32	197.34	Netherlands	23.87	201.08
Canada	26.62	219.56	New Zealand	25.22	228.58
Czech Republic	43.81	166.98	Norway	17.58	129.80
Denmark	40.59	256.96	Poland	20.70	155.10
Finland	25.01	146.08	Portugal	33.02	194.92
France	24.88	225.28	Slovakia	26.49	121.88
Germany	37.44	182.82	South Korea	17.60	93.06
Greece	17.68	201.30	Spain	28.05	259.82
Hungary	29.25	179.52	Sweden	20.07	155.32
Iceland	15.94	178.20	Switzerland	25.32	159.72
Ireland	55.80	194.26	Turkey	3.28	42.68
Italy	21.68	200.64	United Kingdom	30.32	171.16
Japan	14.59	93.28	United States	26.36	267.30

200 food and agricultural products for 200 countries around the world. Here are two tables, one for meat consumption (per capita in kg per year) and one for alcohol consumption (per capita in gallons per year). The United States leads in meat consumption with 267.30 pounds, while Ireland is the largest alcohol consumer at 55.80 gallons.

Using z-scores, find which country is the larger consumer of both meat and alcohol together.

70. World Bank. The World Bank, through their Doing Business project (www.doingbusiness.org), ranks nearly 200 economies on the ease of doing business. One of their rankings measures the ease of starting a business and is made up (in part) of the following variables: number of required start-up procedures, average start-up time (in days), and average start-up cost (in % of per capita income). The following table gives the mean and standard deviations of these variables for 95 economies.

	Procedures (#)	Time (Days)	Cost (%)
Mean	7.9	27.9	14.2
SD	2.9	19.6	12.9

Here are the data for three countries.

	Procedures	Time	Cost
Spain	10	47	15.1
Guatemala	11	26	47.3
Fiji	8	46	25.3

a) Use z-scores to combine the three measures.
b) Which country has the best environment after combining the three measures? Be careful—a lower rank indicates a better environment to start up a business.

T *71. Regular gas. The data set provided contains U.S. regular retail gasoline prices (cents/gallon) from August 20, 1990 to May 28, 2007, from a national sample of gasoline stations obtained from the U.S. Department of Energy.

a) Create a histogram of the data and describe the distribution.
b) Create a time series plot of the data and describe the trend.
c) Which graphical display seems the more appropriate for these data? Explain.

T *72. Home price index. Standard and Poor's Case-Shiller® Home Price Index measures the residential housing market in 20 metropolitan regions across the United States. The national index is a composite of the 20 regions and can be found in the data set provided.

a) Create a histogram of the data and describe the distribution.
b) Create a time series plot of the data and describe the trend.
c) Which graphical display seems the more appropriate for these data? Explain.

*73. Unemployment rate, 2010. The histogram shows the monthly U.S. unemployment rate from January 2001 to January 2010.

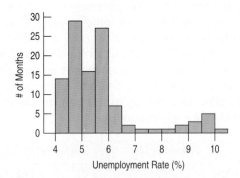

Here is the time series plot for the same data.

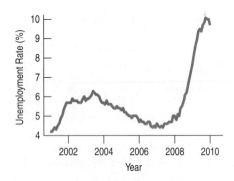

a) What features of the data can you see in the histogram that aren't clear in the time series plot?
b) What features of the data can you see in the time series plot that aren't clear in the histogram?
c) Which graphical display seems the more appropriate for these data? Explain.
d) Write a brief description of unemployment rates over this time period in the United States.

*74. Mutual fund performance. The following histogram displays the monthly returns for a group of mutual funds considered aggressive (or high growth) over a period of 22 years from 1975 to 1997.

Here is the time series plot for the same data.

a) What features of the data can you see in the histogram that aren't clear from the time series plot?
b) What features of the data can you see in the time series plot that aren't clear in the histogram?
c) Which graphical display seems the more appropriate for these data? Explain.
d) Write a brief description of monthly returns over this time period.

75. Assets. Here is a histrogram of the assets (in millions of dollars) of 79 companies chosen from the *Forbes* list of the nation's top corporations.

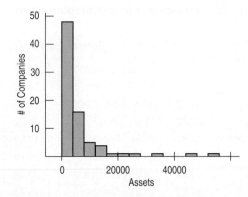

a) What aspect of this distribution makes it difficult to summarize, or to discuss, center and spread?
b) What would you suggest doing with these data if we want to understand them better?

76. Assets, again. Here are the same data you saw in Exercise 75 after re-expressions as the square root of assets and the logarithm of assets.

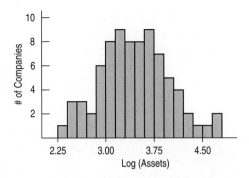

a) Which re-expression do you prefer? Why?
b) In the square root re-expression, what does the value 50 actually indicate about the company's assets?

1 Incomes are probably skewed to the right and not symmetric, making the median the more appropriate measure of center. The mean will be influenced by the high end of family incomes and not reflect the "typical" family income as well as the median would. It will give the impression that the typical income is higher than it is.

2 An IQR of 30 mpg would mean that only 50% of the cars get gas mileages in an interval 30 mpg wide. Fuel economy doesn't vary that much. 3 mpg is reasonable. It seems plausible that 50% of the cars will be within about 3 mpg of each other. An IQR of 0.3 mpg would mean that the gas mileage of half the cars varies little from the estimate. It's unlikely that cars, drivers, and driving conditions are that consistent.

3 We'd prefer a standard deviation of 2 months. Making a consistent product is important for quality. Customers want to be able to count on the MP3 player lasting somewhere close to 5 years, and a standard deviation of 2 years would mean that life spans were highly variable.

Correlation and Linear Regression

Lowe's

In 1921 Lucius S. Lowe opened a hardware store in North Wilkesboro, North Carolina. After his death, his son Jim and son-in-law Carl Buchan took over the store. After World War II, the company expanded under Buchan's leadership. By purchasing materials directly from manufacturers, Lowe's was able to offer lower prices to its customers, most of whom were contractors. By 1955 Lowe's had six stores. Most had a small retail floor with limited inventory and a lumberyard out back near the railroad tracks. By the late 1960s, Lowe's had grown to more than 50 stores and sales of about $100 million. When new home construction almost stopped in the later part of the 1970s, Lowe's researched the market and found that stores that served do-it-yourself homeowners did well even during home-building slumps. So they began to shift their focus and expand their stores.

By the late 1980s Lowe's had more than 300 stores, but those stores still averaged barely more than 20,000 square feet, and The Home Depot had shot past Lowe's, pioneering a new big-box era. Lowe's studied that market and, in 1989, committed to developing big-box stores, taking a $71.3 million restructuring charge in 1991 to cover the costs of closing, relocating, and remodeling about half of the company's stores. By 1996 there were more than 400 Lowe's stores, now averaging more than 75,000 square feet per unit. Sales grew rapidly after the restructuring, increasing from $3.1 billion to $8.6 billion. Net earnings reached $292.2 million in 1996. The company has continued to grow rapidly, working to bolster its number two position and to cut into The Home Depot's lead.

In 2000, Bob Tillman, Chairman and CEO of Lowe's, released a policy promising that all wood products sold would not be sourced from rainforests. Lowe's was awarded the Energy Star retail partner of the year in 2004 for its outstanding contribution to reducing greenhouse gas emissions and in 2007 Lowe's won an Environmental Excellence Award from the U.S. Environmental Protection Agency SmartWay Transport Partnership.

WHO	Years
WHAT	Lowe's Net *Sales* and U.S. Expenditures on Residential *Improvements* and Repairs
UNITS	Both in $M
WHEN	1985–2007
WHERE	United States
WHY	To assess Lowe's sales relative to the home improvement market

Lowe's sells to both contractors and homeowners. Perhaps knowing how much Americans spend on home improvement nationally can help us predict Lowe's sales. Here's a plot showing Lowe's annual net sales against the U.S. Census Bureau's measure of the amount spent by homeowners on residential improvement and repairs.[1]

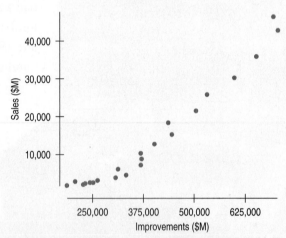

Figure 6.1 Lowe's annual net *Sales* ($M) and the amount spent annually on residential *Improvements* and Repairs ($M) from 1985–2007.

If you were asked to summarize this relationship, what would you say? Clearly Lowe's sales grew when home improvement expenses grew. This plot is an example of a **scatterplot**, which plots one quantitative variable against another. Just by looking at a scatterplot, you can see patterns, trends, relationships, and even the occasional unusual values standing apart from the others. Scatterplots are the best way to start observing the relationship between two *quantitative* variables.

---

[1]www.census.gov/const/C50/histtab1.pdf. The census bureau gives quarterly values. We've combined them to obtain annual totals.

Relationships between variables are often at the heart of what we'd like to learn from data.

- Is consumer confidence related to oil prices?
- What happens to customer satisfaction as sales increase?
- Is an increase in money spent on advertising related to sales?
- What is the relationship between a stock's sales volume and its price?

Questions such as these relate two quantitative variables and ask whether there is an association between them. Scatterplots are the ideal way to picture such associations.

6.1   Looking at Scatterplots

WHO	Cities in the United States
WHAT	*Congestion Cost* Per Person and Peak Period *Freeway Speed*
UNITS	*Congestion Cost* Per Person ($ per person per year); Peak Period *Freeway Speed* (mph)
WHEN	2000
WHERE	United States
WHY	To examine the relationship between congestion on the highways and its impact on society and business

The Texas Transportation Institute, which studies the mobility provided by the nation's transportation system, issues an annual report on traffic congestion and its costs to society and business. Figure 6.2 shows a scatterplot of the annual *Congestion Cost* Per Person of traffic delays (in dollars) in 65 cities in the United States against the Peak Period *Freeway Speed* (mph).

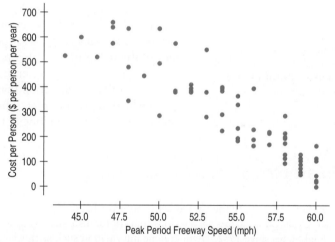

Figure 6.2   Congestion Cost Per Person ($ per year) of traffic delays against Peak Period Freeway Speed (mph) for 65 U.S. cities.

Everyone looks at scatterplots. But, if asked, many people would find it hard to say what to look for in a scatterplot. What do *you* see? Try to describe the scatterplot of *Congestion Cost* against *Freeway Speed*.

You might say that the **direction** of the association is important. As the peak freeway speed goes up, the cost of congestion goes down. A pattern that runs from

Look for **Direction**: What's the sign—positive, negative, or neither?

the upper left to the lower right is said to be **negative**. A pattern running the other way is called **positive**.

The second thing to look for in a scatterplot is its **form**. If there is a straight line relationship, it will appear as a cloud or swarm of points stretched out in a generally consistent, straight form. For example, the scatterplot of traffic congestion has an underlying **linear** form, although some points stray away from it.

Scatterplots can reveal many different kinds of patterns. Often they will not be straight, but straight line patterns are both the most common and the most useful for statistics.

If the relationship isn't straight, but curves gently, while still increasing or decreasing steadily, we can often find ways to straighten it out. But if

> **Look for Form:** Straight, curved, something exotic, or no pattern?

it curves sharply—up and then down, for example, —then you'll need more advanced methods.

The third feature to look for in a scatterplot is the **strength** of the relationship.

At one extreme, do the points appear tightly clustered in a single stream (whether straight, curved, or bending all over the place)? Or, at the other extreme, do the points seem to be so variable and spread out that we can barely discern any

> **Look for Strength:** How much scatter?

trend or pattern? The traffic congestion plot shows moderate scatter around a generally straight form. That indicates that there's a moderately strong linear relationship between cost and speed.

Finally, always look for the unexpected. Often the most interesting discovery in a scatterplot is something you never thought to look for. One example of such a surprise is an unusual observation, or **outlier**, standing away from the overall pattern of the scatterplot. Such a point is almost always interesting and deserves special attention. You may see entire clusters or subgroups that stand away or show a trend in a different direction than the rest of the plot. That should raise questions about why they are different. They may be a clue that you should split the data into subgroups instead of looking at them all together.

> **Look for Unusual Features:** Are there unusual observations or subgroups?

For Example   Creating a scatterplot

The first automobile crash in the United States occurred in New York City in 1896, when a motor vehicle collided with a "pedal-cycle" rider. Cycle/car accidents are a serious concern for insurance companies. About 53,000 cyclists have died in traffic crashes in the United States since 1932. Demographic information such as this is often available from government agencies. It can be useful to insurers, who use it to set appropriate rates, and to retailers, who must plan what safety equipment to stock and how to present it to their customers. This becomes a more pressing concern when the demographic profiles change over time.

Here's data on the mean age of cyclists killed each year during the decade from 1998 to 2008. (Source: National Highway Transportation Safety Agency, http://www-nrd.nhtsa.dot.gov/Pubs/811156.PDF)

Year	Mean Age
1998	32
1999	33
2000	35
2001	36
2002	37
2003	36
2004	39
2005	39
2006	41
2007	40
2008	41

Question: Make a scatterplot and summarize what it says.

Answer:

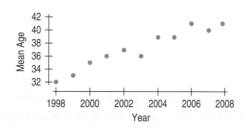

The mean age of cyclist traffic deaths has been increasing almost linearly during this period. The trend is a strong one.

6.2   Assigning Roles to Variables in Scatterplots

Descartes was a philosopher, famous for his statement *cogito, ergo sum*: I think, therefore I am.

Scatterplots were among the first modern mathematical displays. The idea of using two axes at right angles to define a field on which to display values can be traced back to René Descartes (1596–1650), and the playing field he defined in this way is formally called a *Cartesian plane*, in his honor.

The two axes Descartes specified characterize the scatterplot. The axis that runs up and down is, by convention, called the y-axis, and the one that runs from side to side is called the x-axis. These terms are standard.[2]

To make a scatterplot of two quantitative variables, assign one to the y-axis and the other to the x-axis. As with any graph, be sure to label the axes clearly, and indicate the scales of the axes with numbers. Scatterplots display *quantitative* variables. Each variable has units, and these should appear with the display—usually near each axis. Each point is placed on a scatterplot at a position that corresponds to values of these two variables. Its horizontal location is specified by its x-value, and its vertical location is specified by its y-value variable. Together, these are known as *coordinates* and written (x, y).

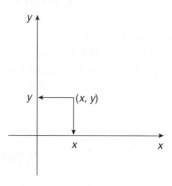

Scatterplots made by computer programs (such as the two we've seen in this chapter) often do not—and usually should not—show the *origin*, the point at $x = 0$, $y = 0$ where the axes meet. If both variables have values near or on both sides of zero, then the origin will be part of the display. If the values are far from zero, though, there's no reason to include the origin. In fact, it's far better to focus on the part of the Cartesian plane that contains the data. In our example about freeways,

---

[2]The axes are also called the "ordinate" and the "abscissa"—but we can never remember which is which because statisticians don't generally use these terms. In Statistics (and in all statistics computer programs) the axes are generally called "x" (abscissa) and "y" (ordinate) and are usually labeled with the names of the corresponding variables.

none of the speeds was anywhere near 0 mph, so the computer drew the scatterplot in Figure 6.2 with axes that don't quite meet.

Which variable should go on the *x*-axis and which on the *y*-axis? What we want to know about the relationship can tell us how to make the plot. We often have questions such as:

- Is Lowe's employee satisfaction related to productivity?
- Are increased sales at Lowe's reflected in the stock price?
- What other factors besides residential improvements are related to Lowe's sales?

In all of these examples, one variable plays the role of the **explanatory** or **predictor variable**, while the other takes on the role of the **response variable**. We place the explanatory variable on the *x*-axis and the response variable on the *y*-axis. When you make a scatterplot, you can assume that those who view it will think this way, so choose which variables to assign to which axes carefully.

The roles that we choose for variables have more to do with how we *think* about them than with the variables themselves. Just placing a variable on the *x*-axis doesn't necessarily mean that it explains or predicts *anything*, and the variable on the *y*-axis may not respond to it in any way. We plotted *Congestion Cost* Per Person against peak *Freeway Speed*, thinking that the slower traffic moves, the more it costs in delays. But maybe *spending* $500 per person in freeway improvement would increase speed. If we were examining that option, we might choose to plot *Congestion Cost* Per Person as the explanatory variable and *Freeway Speed* as the response.

The *x*- and *y*-variables are sometimes referred to as the **independent** and **dependent** variables, respectively. The idea is that the *y*-variable *depends* on the *x*-variable and the *x*-variable acts *independently* to make *y* respond. These names, however, conflict with other uses of the same terms in Statistics. Instead, we'll sometimes use the terms "explanatory" or "predictor variable" and "response variable" when we're discussing roles, but we'll often just say *x-variable* and *y-variable*.

Notation Alert! ───────

So *x* and *y* are reserved letters as well, but not just for labeling the axes of a scatterplot. In Statistics, the assignment of variables to the *x*- and *y*-axes (and choice of notation for them in formulas) often conveys information about their roles as predictor or response.

For Example   Assigning roles to variables

Question: When examining the ages of victims in cycle/car accidents, why does it make the most sense to plot *year* on the *x*-axis and *mean age* on the *y*-axis? (See the example on page 140.)

Answer: We are interested in how the age of accident victims might change over time, so we think of the year as the basis for prediction and the mean age of victims as the variable that is predicted.

6.3   Understanding Correlation

WHO	Quarters
WHAT	Expenditures for *Improvement* and *Replacement* of residences
UNITS	Both in $M
WHEN	1985–2004
WHERE	United States
WHY	To understand components of the U.S. Census Bureau's total expenditures for residential maintenance

The U.S. Census Bureau reports separate components of their quarterly home improvement expenditure data. For example, they categorize some expenditures as *Improvement* and others as *Replacement*. How are these related to each other? Figure 6.3 shows the scatterplot.

As you might expect, expenses for both improvement and replacement tend to rise and fall together. There is a clear positive association, and the scatterplot looks linear. But how strong is the association? If you had to put a number (say, between 0 and 1) on the strength of the association, what would it be? Your measure shouldn't depend on the choice of units for the variables. After all, if sales had been recorded in euros instead of dollars or maintenance expenditures in billions of dollars rather than millions, the scatterplot would look the same. The direction, form, and strength won't change, so neither should our measure of the association's strength.

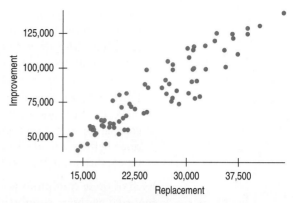

Figure 6.3 Quarterly expenditures on *Improvement* and *Replacement* in residential maintenance and repairs from 1985 to 2004, both in $M.

We saw a way to remove the units in the previous chapter. We can standardize each of the variables, finding $z_x = \left(\dfrac{x - \bar{x}}{s_x}\right)$ and $z_y = \left(\dfrac{y - \bar{y}}{s_y}\right)$. With these, we can compute a measure of strength that you've probably heard of: **the correlation coefficient**:

$$r = \frac{\sum z_x z_y}{n - 1}.$$

Keep in mind that the x's and y's are paired. For each quarter, we have a replacement expenditure and an improvement expenditure. To find the correlation we multiply each standardized value by the standardized value it is paired with and add up those *crossproducts*. Then we divide the total by the number of pairs minus one, $n - 1$.[3]

Keep in mind that the x's and y's are paired. For each quarter, we have a replacement expenditure and an improvement expenditure. To find the correlation we multiply each standardized value by the standardized value it is paired with and add up those *crossproducts*. Then we divide the total by the number of pairs minus one, $n - 1$.[3]

For residential *Improvement* and *Replacement*, the correlation coefficient is 0.92.

There are alternative formulas for the correlation in terms of the variables x and y. Here are two of the more common:

$$r = \frac{\sum (x - \bar{x})(y - \bar{y})}{\sqrt{\sum (x - \bar{x})^2 \sum (y - \bar{y})^2}} = \frac{\sum (x - \bar{x})(y - \bar{y})}{(n - 1)s_x s_y}.$$

These formulas can be more convenient for calculating correlation by hand, but the form given using z-scores is best for understanding what correlation means.

Correlation Conditions

Correlation measures the strength of the *linear* association between two *quantitative* variables. Before you use correlation, you must check three *conditions*:

- **Quantitative Variables Condition:** Correlation applies only to quantitative variables. Don't apply correlation to categorical data masquerading as quantitative. Check that you know the variables' units and what they measure.

- **Linearity Condition:** Sure, you can *calculate* a correlation coefficient for any pair of variables. But correlation measures the strength only of the *linear* association and will be misleading if the relationship is not straight enough. What is "straight enough"? This question may sound too informal for a statistical condition, but that's really the point. We can't verify whether a relationship is linear or not. Very

Notation Alert! ——————

The letter r is always used for correlation, so you can't use it for anything else in Statistics. Whenever you see an "r," it's safe to assume it's a correlation.

---
[3]The same $n - 1$ we used for calculating the standard deviation.

few relationships between variables are perfectly linear, even in theory, and scatterplots of real data are never perfectly straight. How nonlinear looking would the scatterplot have to be to fail the condition? This is a judgment call that you just have to think about. Do you think that the underlying relationship is curved? If so, then summarizing its strength with a correlation would be misleading.

- **Outlier Condition:** Unusual observations can distort the correlation and can make an otherwise small correlation look big or, on the other hand, hide a large correlation. It can even give an otherwise positive association a negative correlation coefficient (and vice versa). When you see an outlier, it's often a good idea to report the correlation both with and without the point.

Each of these conditions is easy to check with a scatterplot. Many correlations are reported without supporting data or plots. You should still think about the conditions. You should be cautious in interpreting (or accepting others' interpretations of) the correlation when you can't check the conditions for yourself.

By Hand

Finding the correlation coefficient

To find the correlation coefficient by hand, we'll use a formula in original units, rather than z-scores. This will save us the work of having to standardize each individual data value first. Start with the summary statistics for both variables: $\bar{x}, \bar{y}, s_x,$ and s_y. Then find the deviations as we did for the standard deviation, but now in both x and y: $(x - \bar{x})$ and $(y - \bar{y})$. For each data pair, multiply these deviations together: $(x - \bar{x}) \times (y - \bar{y})$. Add the products up for all data pairs. Finally, divide the sum by the product of $(n - 1) \times s_x \times s_y$ to get the correlation coefficient.

Here we go.

Suppose the data pairs are:

x	6	10	14	19	21
y	5	3	7	8	12

Then $\bar{x} = 14, \bar{y} = 7, s_x = -6.20,$ and $s_y = 3.39$

Deviations in x	Deviations in y	Product
6 − 14 = −8	5 − 7 = −2	−8 × −2 = 16
10 − 14 = −4	3 − 7 = −4	16
14 − 14 = 0	7 − 7 = 0	0
19 − 14 = 5	8 − 7 = 1	5
21 − 14 = 7	12 − 7 = 5	35

Add up the products: $16 + 16 + 0 + 5 + 35 = 72$
Finally, we divide by $(n - 1) \times s_x \times s_y = (5 - 1) \times 6.20 \times 3.39 = 84.07$
The ratio is the correlation coefficient:

$$r = 72/84.07 = 0.856$$

Just Checking

For the years 1992 to 2002, the quarterly stock price of the semiconductor companies Cypress and Intel have a correlation of 0.86.

1 Before drawing any conclusions from the correlation, what would you like to see? Why?

2 If your coworker tracks the same prices in euros, how will this change the correlation? Will you need to know the exchange rate between euros and U.S. dollars to draw conclusions?

3 If you standardize both prices, how will this affect the correlation?

4 In general, if on a given day the price of Intel is relatively low, is the price of Cypress likely to be relatively low as well?

5 If on a given day the price of Intel stock is high, is the price of Cypress stock definitely high as well?

Guided Example   Customer Spending

A major credit card company sends an incentive to its best customers in hope that the customers will use the card more. They wonder how often they can offer the incentive. Will repeated offerings of the incentive result in repeated increased credit card use? To examine this question, an analyst took a random sample of 184 customers from their highest use segment and investigated the charges in the two months in which the customers had received the incentive.

PLAN

Setup State the objective. Identify the quantitative variables to examine. Report the time frame over which the data have been collected and define each variable. (State the W's.)

Our objective is to investigate the association between the amount that a customer charges in the two months in which they received an incentive. The customers have been randomly selected from among the highest use segment of customers. The variables measured are the total credit card charges (in $) in the two months of interest.

✓ **Quantitative Variable Condition.** Both variables are quantitative. Both charges are measured in dollars.

Make the scatterplot and clearly label the axes to identify the scale and units.

Because we have two quantitative variables measured on the same cases, we can make a scatterplot.

Check the conditions.

✓ **Linearity Condition.** The scatterplot is straight enough.

✓ **Outlier Condition.** There are no obvious outliers.

(continued)

DO	**Mechanics** Once the conditions are satisfied, calculate the correlation with technology.	The correlation is −0.391. The negative correlation coefficient confirms the impression from the scatterplot.
REPORT	**Conclusion** Describe the direction, form, and the strength of the plot, along with any unusual points or features. Be sure to state your interpretation in the proper context.	MEMO **Re: Credit Card Spending** We have examined some of the data from the incentive program. In particular, we looked at the charges made in the first two months of the program. We noted that there was a negative association between charges in the second month and charges in the first month. The correlation was −0.391, which is only moderately strong, and indicates substantial variation. We've concluded that although the observed pattern is negative, these data do not allow us to find the causes of this behavior. It is likely that some customers were encouraged by the offer to increase their spending in the first month, but then returned to former spending patterns. It is possible that others didn't change their behavior until the second month of the program, increasing their spending at that time. Without data on the customers' pre-incentive spending patterns it would be hard to say more. We suggest further research, and we suggest that the next trial extend for a longer period of time to help determine whether the patterns seen here persist.

Correlation Properties

Because correlation is so widely used as a measure of association it's a good idea to remember some of its basic properties. Here's a useful list of facts about the correlation coefficient:

- **The sign of a correlation coefficient gives the direction of the association.**
- **Correlation is always between −1 and +1.** Correlation *can* be exactly equal to −1.0 or +1.0, but watch out. These values are unusual in real data because they mean that all the data points fall *exactly* on a single straight line.
- **Correlation treats *x* and *y* symmetrically.** The correlation of *x* with *y* is the same as the correlation of *y* with *x*.
- **Correlation has no units.** This fact can be especially important when the data's units are somewhat vague to begin with (customer satisfaction, worker efficiency, productivity, and so on).
- **Correlation is not affected by changes in the center or scale of either variable.** Changing the units or baseline of either variable has no effect on the correlation coefficient because the correlation depends only on the *z*-scores.
- **Correlation measures the strength of the *linear* association between the two variables.** Variables can be strongly associated but still have a small correlation if the association is not linear.
- **Correlation is sensitive to unusual observations.** A single outlier can make a small correlation large or make a large one small.

How Strong Is Strong?

There's little agreement on what the terms "weak," "moderate," and "strong" mean. The same correlation might be strong in one context and weak in another. A correlation of 0.7 between an economic index and stock market prices would be exciting, but finding "only" a correlation of 0.7 between a drug dose and blood pressure might be seen as a failure by a pharmaceutical company. Use these terms cautiously and be sure to report the correlation and show a scatterplot so others can judge the strength for themselves.

Correlation Tables

Sometimes you'll see the correlations between each pair of variables in a data set arranged in a table. The rows and columns of the table name the variables, and the cells hold the correlations.

	Volume	Close	Net earnings
Volume	1.000		
Close	0.396	1.000	
Net earnings	0.477	0.464	1.000

Table 6.1   A correlation table for some other variables measured quarterly during the period 1985 to 2007. *Volume* = number of shares of Lowe's traded, *Close* = closing price of Lowe's stock, *Net Earnings* = Lowe's reported net earnings for the quarter.

Correlation tables are compact and give a lot of summary information at a glance. They can be an efficient way to start to look at a large data set. The diagonal cells of a correlation table always show correlations of exactly 1.000, and the upper half of the table is symmetrically the same as the lower half (can you see why?), so by convention, only the lower half is shown. A table like this can be convenient, but be sure to check for linearity and unusual observations or the correlations in the table may be misleading or meaningless. Can you be sure, looking at Table 6.1, that the variables are linearly associated? Correlation tables are often produced by statistical software packages. Fortunately, these same packages often offer simple ways to make all the scatterplots you need to look at.[4]

For Example   Finding the correlation coefficient

Question: What is the correlation of *mean age* and *year* for the cyclist accident data on page 140?

Answer: Working by hand following the method in the sidebar:

$$\bar{x} = 2003, s_x = 3.32$$
$$\bar{y} = 37.18, s_y = 3.09$$

The sum of the cross-product of the deviations is found as follows:

$$\sum(x - \bar{x})(y - \bar{y}) = 99$$

Putting the sum of the cross-products in the numerator and $(n - 1) \times s_x \times s_y$ in the denominator, we get

$$\frac{99}{(11 - 1) \times 3.32 \times 3.09} = 0.965$$

For *mean age* and *year*, the correlation coefficient is 0.965. That indicates a strong linear association. Because this is a time series, it indicates a strong trend.

[4]A table of scatterplots arranged just like a correlation table is sometimes called a *scatterplot matrix*, or SPLOM, and is easily created using a statistics package.

6.4   Lurking Variables and Causation

An educational researcher finds a strong association between height and reading ability among elementary school students in a nationwide survey. Taller children tend to have higher reading scores. Does that mean that students' height *causes* their reading scores to go up? No matter how strong the correlation is between two variables, there's no simple way to show from observational data that one variable causes the other. A high correlation just increases the temptation to think and to say that the *x*-variable *causes* the *y*-variable. Just to make sure, let's repeat the point again.

No matter how strong the association, no matter how large the *r* value, no matter how straight the form, there is no way to conclude from a high correlation *alone* that one variable causes the other. There's always the possibility that some third variable—a **lurking variable**—is affecting both of the variables you have observed. In the reading score example, you may have already guessed that the lurking variable is the age of the child. Older children tend to be taller and have stronger reading skills. But even when the lurking variable isn't as obvious, resist the temptation to think that a high correlation implies causation. Here's another example.

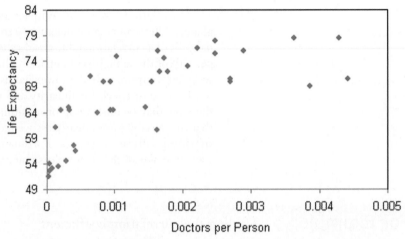

Figure 6.4   *Life Expectancy* and numbers of *Doctors per Person* in 40 countries shows a fairly strong, positive linear relationship with a correlation of 0.705. (Scatterplot created in **Excel**.)

The scatterplot shows the *Life Expectancy* (average of men and women, in years) for each of 40 countries of the world, plotted against the number of *Doctors per Person* in each country. The strong positive association ($r = 0.705$) seems to confirm our expectation that more *Doctors per Person* improves health care, leading to longer lifetimes and a higher *Life Expectancy*. Perhaps we should send more doctors to developing countries to increase life expectancy.

If we increase the number of doctors, will the life expectancy increase? That is, would adding more doctors *cause* greater life expectancy? Could there be another explanation of the association? Figure 6.5 shows another scatterplot. *Life Expectancy* is still the response, but this time the predictor variable is not the number of doctors, but the number of *Televisions per Person* in each country. The positive association in this scatterplot looks even *stronger* than the association in the previous plot. If we wanted to calculate a correlation, we should straighten the plot first, but even from this plot, it's clear that higher life expectancies are associated with more televisions per person. Should we conclude that increasing the number of televisions extends lifetimes? If so, we should send televisions instead of doctors to developing

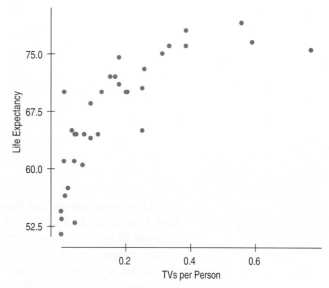

Figure 6.5 *Life Expectancy* and number of *Televisions per Person* shows a strong, positive (although clearly not linear) relationship.

countries. Not only is the association with life expectancy stronger, but televisions are cheaper than doctors.

What's wrong with this reasoning? Maybe we were a bit hasty earlier when we concluded that doctors *cause* greater life expectancy. Maybe there's a lurking variable here. Countries with higher standards of living have both longer life expectancies *and* more doctors. Could higher living standards cause changes in the other variables? If so, then improving living standards might be expected to prolong lives, increase the number of doctors, and increase the number of televisions. From this example, you can see how easy it is to fall into the trap of mistakenly inferring causality from a correlation. For all we know, doctors (or televisions) *do* increase life expectancy. But we can't tell that from data like these no matter how much we'd like to. Resist the temptation to conclude that *x* causes *y* from a correlation, no matter how obvious that conclusion seems to you.

For Example   Understanding causation

Question: An insurance company analyst suggests that the data on ages of cyclist accident deaths are actually due to the entire population of cyclists getting older and not to a change in the safe riding habits of older cyclists (see page 140). What would we call the *mean cyclist age* if we had that variable available?

Answer: It would be a lurking variable. If the entire population of cyclists is aging then that would lead to the average age of cyclists in accidents increasing.

6.5   The Linear Model

Let's return to the relationship between Lowe's sales and home improvement expenditures between 1985 and 2007. In Figure 6.1 (repeated here) we saw a strong, positive, linear relationship, so we can summarize its strength with a correlation. For this relationship, the correlation is 0.976.

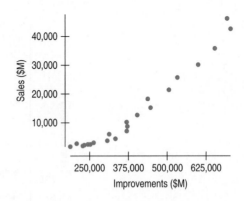

That's quite strong, but the strength of the relationship is only part of the picture. Lowe's management might want to predict sales based on the census bureau's estimate of residential improvement expenditures for the next year. That's a reasonable business question, but to answer it we'll need a model for the trend. The correlation says that there seems to be a strong linear association between the variables, but it doesn't tell us what that association is.

Of course, we can say more. We can model the relationship with a line and give the equation. For Lowe's, we can find a linear model to describe the relationship we saw in Figure 6.1 between Lowe's *Sales* and residential *Improvements*. A **linear model** is just an equation of a straight line through the data. The points in the scatterplot don't all line up, but a straight line can summarize the general pattern with only a few parameters. This model can help us understand how the variables are associated.

Residuals

We know the model won't be perfect. No matter what line we draw, it won't go through many of the points. The best line might not even hit any of the points. Then how can it be the "best" line? We want to find the line that somehow comes *closer* to all the points than any other line. Some of the points will be above the line and some below. A linear model can be written as $\hat{y} = b_0 + b_1 x$, where b_0 and b_1 are numbers estimated from the data and $\hat{y}$ (pronounced *y*-hat) is the **predicted value**. We use the *hat* to distinguish the predicted value from the observed value *y*. The difference between these two is called the **residual**:

$$e = y - \hat{y}.$$

The residual value tells us how far the model's prediction is from the observed value at that point. To find the residuals, we always subtract the predicted values from the observed ones.

Our question now is how to find the right line.

The Line of "Best Fit"

When we draw a line through a scatterplot, some residuals are positive, and some are negative. We can't assess how well the line fits by adding up all the residuals—the positive and negative ones would just cancel each other out. We need to find the line that's closest to all the points, and to do that, we need to make all the distances positive. We faced the same issue when we calculated a standard deviation to measure spread. And we deal with it the same way here: by squaring the residuals to make them positive. The sum of all the squared residuals tells us how well the line we drew fits the data—the smaller the sum, the better the fit. A different line will produce a different sum, maybe bigger, maybe smaller. The **line of best fit** is the line for which the sum of the squared residuals is smallest—often called the **least squares line**.

─────────────

[5]Stigler, Steven M., "Gauss and the Invention of Least Squares," **Annals of Statistics, 9, (3)**, 1981, pp. 465-474.

This line has the special property that the variation of the data around the model, as seen in the residuals, is the smallest it can be for any straight line model for these data. No other line has this property. Speaking mathematically, we say that this line minimizes the sum of the squared residuals. You might think that finding this "least squares line" would be difficult. Surprisingly, it's not, although it was an exciting mathematical discovery when Legendre published it in 1805.

For Example   **Interpreting the equation of a linear model**

Question: The data on cyclist accident deaths show a linear pattern. Find and interpret the equation of a linear model for that pattern. Refer to the values given in the answer to the example on page 147.

Answer:

$$b = 0.965 \times \frac{3.09}{3.32} = 0.90$$

$$a = 37.18 - 0.90 \times 2003 = -1765.52$$

$$MeanAge = -1765.52 + 0.90\ Year$$

The mean age of cyclists killed in vehicular accidents has increased by about 0.9 years of age (about 11 months) per year during the decade observed by these data.

6.6   Correlation and the Line

Any straight line can be written as:

$$y = b_0 + b_1 x.$$

If we were to plot all the (x, y) pairs that satisfy this equation, they'd fall exactly on a straight line. We'll use this form for our linear model. Of course, with real data, the points won't all fall on the line. So, we write our model as $\hat{y} = b_0 + b_1 x$, using $\hat{y}$ for the predicted values, because it's the predicted values (not the data values) that fall on the line. If the model is a good one, the data values will scatter closely around it.

For the Lowe's sales data, the line is:

$$\widehat{Sales} = -19{,}679 + 0.346\ Improvements.$$

What does this mean? The **slope,** 0.346, says that we can expect a year in which residential improvement spending is 1 million dollars higher to be one in which Lowe's sales will be about 0.346 \$M (\$346,000) higher. Slopes are always expressed in y-units per x-units. They tell you how the response variable changes for a one unit step in the predictor variable. So we'd say that the slope is 0.346 million dollars of *Sales* per million dollars of *Improvements*.

The **intercept,** −19,679, is the value of the line when the x-variable is zero. What does it mean here? The intercept often serves just as a starting value for our predictions. We don't interpret it unless a 0 value for the predictor variable would really mean something under the circumstances. The Lowe's model is based on years in which annual spending on residential improvements is between 50 and 100 billion dollars. It's unlikely to be appropriate if there were no such spending at all. In this case, we wouldn't interpret the intercept.

How do we find the slope and intercept of the least squares line? The formulas are simple. The model is built from the summary statistics we've used before. We'll need the correlation (to tell us the strength of the linear association), the standard deviations (to give us the units), and the means (to tell us where to locate the line).

Who Was First?

One of history's most famous disputes of authorship was between Gauss and Legendre over the method of "least squares." Legendre was the first to publish the solution to finding the best fit line through data in 1805, at which time Gauss claimed to have known it for years. There is some evidence that, in fact, Gauss may have been right, but he hadn't bothered to publish it, and had been unable to communicate its importance to other scientists[5]. Gauss later referred to the solution as "*our* method" (*principium nostrum*), which certainly didn't help his relationship with Legendre.

Just Checking

A scatterplot of sales per month (in thousands of dollars) vs. number of employees for all the outlets of a large computer chain shows a relationship that is straight, with only moderate scatter and no outliers. The correlation between *Sales* and *Employees* is 0.85, and the equation of the least squares model is:

$$\widehat{Sales} = 9.564 + 122.74 \, Employees$$

6   What does the slope of 122.74 mean?

7   What are the units of the slope?

8   The outlet in Dallas, Texas, has 10 more employees than the outlet in Cincinnati. How much more *Sales* do you expect it to have?

The slope of the line is computed as:

$$b_1 = r \frac{s_y}{s_x}.$$

We've already seen that the correlation tells us the sign and the strength of the relationship, so it should be no surprise to see that the slope inherits this sign as well. If the correlation is positive, the scatterplot runs from lower left to upper right, and the slope of the line is positive.

Correlations don't have units, but slopes do. How *x* and *y* are measured—what units they have—doesn't affect their correlation, but does change the slope. The slope gets its units from the ratio of the two standard deviations. Each standard deviation has the units of its respective variable. So, the units of the slope are a ratio, too, and are always expressed in units of *y* per unit of *x*.

How do we find the intercept? If you had to predict the *y*-value for a data point whose *x*-value was average, what would you say? The best fit line predicts $\bar{y}$ for points whose *x*-value is $\bar{x}$. Putting that into our equation and using the slope we just found gives:

$$\bar{y} = b_0 + b_1\bar{x}$$

and we can rearrange the terms to find:

$$b_0 = \bar{y} - b_1\bar{x}.$$

Finding the Regression Coefficients for the Lowe's Data
Summary statistics:

$$Sales: \bar{y} = 13{,}564.17; \, s_y = 14{,}089.61$$

$$Improvements: \bar{x} = 96{,}009.8; \, s_x = 39{,}036.8$$

$$Correlation = 0.976$$

So, $b_1 = r\dfrac{s_y}{s_x} = (0.976)\dfrac{14{,}089.61}{39{,}036.8}$

$= 0.352$ ($M Sales per $M Improvement expenditures)

And

$$b_0 = \bar{y} - b_1\bar{x} = 13{,}564.17 - (0.352)96{,}009.8 = -20{,}231.3$$

The equation from the computer output has slope 0.346 and intercept −19,679. The differences are due to rounding error. We've shown the calculation using rounded summary statistics, but if you are doing this by hand, you should always keep all digits in intermediate steps.

It's easy to use the estimated linear model to predict Lowe's *Sales* for any amount of national spending on residential *Improvements*. For example, in 2007 the total was $172,150(M). To estimate Lowe's *Sales*, we substitute this value for x in the model:

$$\widehat{Sales} = -19{,}679 + 0.346 \times 172{,}150 = 39{,}885$$

Sales actually were 46,927 ($M), so the residual of $46{,}927 - 39{,}885 = 7{,}042$ ($M) tells us how much better Lowe's did than the model predicted.

Least squares lines are commonly called **regression** lines. Although this name is an accident of history (as we'll soon see), "regression" almost always means "the linear model fit by least squares." Clearly, regression and correlation are closely related. We'll need to check the same conditions for regression as we did for correlation:

1. **Quantitative Variables Condition**
2. **Linearity Condition**
3. **Outlier Condition**

A little later in the chapter we'll add two more.

Understanding Regression from Correlation

The slope of a regression line depends on the units of both x and y. Its units are the units of y per unit of x. The units are expressed in the slope because $b_1 = r\dfrac{s_y}{s_x}$. The correlation has no units, but each standard deviation is measured in the units of its respective variable. For our regression of Lowe's *Sales* on home *Improvements*, the slope was millions of dollars of sales *per* million dollars of improvement expenditure.

It can be useful to see what happens to the regression equation if we were to standardize both the predictor and response variables and regress z_y on z_x. For both these standardized variables, the standard deviation is 1 and the means are zero. That means that the slope is just r, and the intercept is 0 (because both $\bar{y}$ and $\bar{x}$ are now 0).

This gives us the simple equation for the regression of standardized variables:

$$\hat{z}_y = rz_x.$$

Although we don't usually standardize variables for regression, it can be useful to think about what this means. Thinking in z-scores is a good way to understand what the regression equation is doing. The equation says that for every standard deviation we deviate from the mean in x, we predict that y will be r standard deviations away from the mean in y.

Let's be more specific. For the Lowe's example, the correlation is 0.976. So, we know immediately that:

$$\hat{z}_{Sales} = 0.976\, z_{Improvements}.$$

That means that a change of one standard deviation in expenditures on *Improvements* corresponds in our model to a 0.976 standard deviation change in *Sales*.

6.7 Regression to the Mean

Suppose you were told that a new male student was about to join the class and you were asked to guess his height in inches. What would be your guess? A good guess would be the mean height of male students. Now suppose you are also told that this student had a grade point average (GPA) of 3.9—about 2 SDs above the mean

GPA. Would that change your guess? Probably not. The correlation between GPA and height is near 0, so knowing the GPA value doesn't tell you anything and doesn't move your guess. (And the standardized regression equation, $\hat{z}_y = rz_x$, tells us that as well, since it says that we should move 0×2 SDs from the mean.)

On the other hand, if you were told that, measured in centimeters, the student's height was 2 SDs above the mean, you'd know his height in inches. There's a perfect correlation between *Height* in inches and *Height* in centimeters ($r = 1$), so you know he's 2 SDs above mean height in inches as well.

What if you were told that the student was 2 SDs above the mean in shoe size? Would you still guess that he's of average height? You might guess that he's taller than average, since there's a positive correlation between height and shoe size. But would you guess that he's 2 SDs above the mean? When there was no correlation, we didn't move away from the mean at all. With a perfect correlation, we moved our guess the full 2 SDs. Any correlation between these extremes should lead us to move somewhere between 0 and 2 SDs above the mean. (To be exact, our best guess would be to move $r \times 2$ standard deviations away from the mean.)

Notice that if x is 2 SDs above its mean, we won't ever move more than 2 SDs away for y, since r can't be bigger than 1.0. So each predicted y tends to be closer to its mean (in standard deviations) than its corresponding x was. This property of the linear model is called **regression to the mean**. This is why the line is called the regression line.

Sir Francis Galton was the first to speak of "regression," although others had fit lines to data by the same method.

The First Regression

Sir Francis Galton related the heights of sons to the heights of their fathers with a regression line. The slope of his line was less than 1. That is, sons of tall fathers were tall, but not as much above the average height as their fathers had been above their mean. Sons of short fathers were short, but generally not as far from their mean as their fathers. Galton interpreted the slope correctly as indicating a "regression" toward the mean height—and "regression" stuck as a description of the method he had used to find the line.

Math Box

Equation of the line of best fit

Where does the equation of the line of best fit come from? To write the equation of any line, we need to know a point on the line and the slope. It's logical to expect that an average x will correspond to an average y, and, in fact, the line does pass through the point $(\bar{x}, \bar{y})$. (This is not hard to show as well.)

To think about the slope, we look once again at the z-scores. We need to remember a few things.

1. The mean of any set of z-scores is 0. This tells us that the line that best fits the z-scores passes through the origin (0, 0).

2. The standard deviation of a set of z-scores is 1, so the variance is also 1. This means that $\dfrac{\Sigma(z_y - \bar{z}_y)^2}{n-1} = \dfrac{\Sigma(z_y - 0)^2}{n-1} = \dfrac{\Sigma z_y^2}{n-1} = 1$, a fact that will be important soon.

3. The correlation is $r = \dfrac{\Sigma z_x z_y}{n-1}$, also important soon.

Remember that our objective is to find the slope of the best fit line. Because it passes through the origin, the equation of the best fit line will be of the form $\hat{z}_y = mz_x$. We want to find the value for m that will minimize the sum of the squared errors. Actually we'll divide that sum by $n-1$ and minimize this mean squared error (MSE). Here goes:

Minimize: $$MSE = \frac{\Sigma(z_y - \hat{z}_y)^2}{n-1}$$

Since $\hat{z}_y = mz_x$: $$MSE = \frac{\Sigma(z_y - mz_x)^2}{n-1}$$

Square the binomial:
$$= \frac{\sum(z_y^2 - 2mz_xz_y + m^2z_x^2)}{n-1}$$

Rewrite the summation:
$$= \frac{\sum z_y^2}{n-1} - 2m\frac{\sum z_xz_y}{n-1} + m^2\frac{\sum z_x^2}{n-1}$$

4. Substitute from (2) and (3): $= 1 - 2mr + m^2$

This last expression is a quadratic. A parabola in the form $y = ax^2 + bx + c$ reaches its minimum at its turning point, which occurs when $x = \frac{-b}{2a}$. We can minimize the mean of squared errors by choosing
$$m = \frac{-(-2r)}{2(1)} = r.$$

The slope of the best fit line for z-scores is the correlation, r. This fact leads us immediately to two important additional results:

A slope with value r for z-scores means that a difference of 1 standard deviation in z_x corresponds to a difference of r standard deviations in $\hat{z}_y$. Translate that back to the original x and y values: "Over one standard deviation in x, up r standard deviations in $\hat{y}$."

The slope of the regression line is $b = \frac{rs_y}{s_x}$.

We know choosing $m = r$ minimizes the sum of the squared errors (SSE), but how small does that sum get? Equation (4) told us that the mean of the squared errors is $1 - 2mr + m^2$. When $m = r$, $1 - 2mr + m^2 = 1 - 2r^2 + r^2 = 1 - r^2$. This is the percentage of variability *not* explained by the regression line. Since $1 - r^2$ of the variability is *not* explained, the percentage of variability in y that *is* explained by x is r^2. This important fact will help us assess the strength of our models.

And there's still another bonus. Because r^2 is the percent of variability explained by our model, r^2 is at most 100%. If $r^2 \leq 1$, then $-1 \leq r \leq 1$, proving that correlations are always between -1 and $+1$.

Why *r* for *Correlation*?

In his original paper on correlation, Galton used r for the "index of correlation"—what we now call the correlation coefficient. He calculated it from the regression of y on x or of x on y after standardizing the variables, just as we have done. It's fairly clear from the text that he used r to stand for (standardized) regression.

6.8 Checking the Model

The linear regression model is perhaps the most widely used model in all of Statistics. It has everything we could want in a model: two easily estimated parameters, a meaningful measure of how well the model fits the data, and the ability to predict new values. It even provides a self-check in plots of the residuals to help us avoid all kinds of mistakes. Most models are useful only when specific **assumptions** are true. Of course, assumptions are hard—often impossible—to check. That's why we *assume* them. But we should check to see whether the assumptions are *reasonable*. Fortunately, we can often check *conditions* that provide information about the assumptions. For the linear model, we start by checking the same ones we checked earlier in this chapter for using correlation.

Linear models only make sense for quantitative data. The **Quantitative Data Condition** is pretty easy to check, but don't be fooled by categorical data recorded as numbers. You probably don't want to predict zip codes from credit card account numbers.

The regression model *assumes* that the relationship between the variables is, in fact, linear. If you try to model a curved relationship with a straight line, you'll usually get what you deserve. We can't ever verify that the underlying relationship between two variables is truly linear, but an examination of the scatterplot will let you decide whether the **Linearity Assumption** is reasonable. The **Linearity Condition** we used for correlation is designed to do precisely that and is satisfied if the scatterplot looks reasonably straight. If the scatterplot is not straight enough, stop. You can't use a linear model for just *any* two variables, even if they are related. The two variables must have a *linear* association, or the model won't mean a thing. Some nonlinear relationships can be saved by re-expressing the data to make the scatterplot more linear.

Watch out for outliers. The linearity assumption also requires that no points lie far enough away to distort the line of best fit. Check the **Outlier Condition** to make sure no point needs special attention. Outlying values may have large residuals, and squaring makes their influence that much greater. Outlying points can dramatically change a regression model. Unusual observations can even change the sign of the slope, misleading us about the direction of the underlying relationship between the variables.

Another assumption that is usually made when fitting a linear regression is that the residuals are independent of each other. We don't strictly need this assumption to fit the line, but to generalize from the data it's a crucial assumption and one that we'll come back to when we discuss inference. As with all assumptions, there's no way to be sure that **Independence Assumption** is true. However, we could check that the cases are a random sample from the population.

We can also check displays of the regression residuals for evidence of patterns, trends, or clumping, any of which would suggest a failure of independence. In the special case when we have a time series, a common violation of the Independence Assumption is for the errors to be correlated with each other (autocorrelation). The error our model makes today may be similar to the one it made yesterday. We can check this violation by plotting the residuals against time (usually *x* for a time series) and looking for patterns.

When our goal is just to explore and describe the relationship, independence isn't essential (and so we won't insist that the conditions relating to it be formally checked). However, when we want to go beyond the data at hand and make inferences for other situations (in Chapter 16) this will be a crucial assumption, so it's good practice to think about it even now, especially for time series.

We always check conditions with a scatterplot of the data, but we can learn even more after we've fit the regression model. There's extra information in the residuals that we can use to help us decide how reasonable our model is and how well the model fits. So, we plot the residuals and check the conditions again.

The residuals are the part of the data that *hasn't* been modeled. We can write

$$Data = Predicted + Residual$$

or, equivalently,

$$Residual = Data - Predicted.$$

Or, as we showed earlier, in symbols:

$$e = y - \hat{y}.$$

A scatterplot of the residuals versus the *x*-values should be a plot without patterns. It shouldn't have any interesting features—no direction, no shape. It should stretch horizontally, showing no bends, and it should have no outliers. If you see nonlinearities, outliers, or clusters in the residuals, find out what the regression model missed.

Make a Picture

Check the scatterplot. The shape must be linear, or you can't use regression for the variables in their current form. And watch out for outliers.

Why *e* for *Residual?*

The easy answer is that *r* is already taken for correlation, but the truth is that *e* stands for "error." It's not that the data point is a mistake but that statisticians often refer to variability not explained by a model as error.

Let's examine the residuals from our regression of Lowe's *Sales* on residential *Improvement* expenditures.[6]

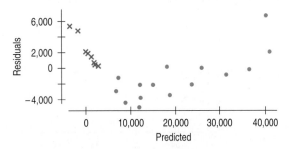

Figure 6.6   Residuals of the regression model predicting Lowe's *Sales* from Residential *Improvement* expenses 1985–2007. Now we can see that there are really two groups. Can we identify them?

These residuals hold a surprise. They seem to fall into two groups, one with a steeply declining trend (indicating sales that did not increase with residential improvements as fast as the regression model predicts) and one with an increasing trend. If we identify the two groups, we find that the red **x**'s are for the years 1985–1992 and the green dots are for the years 1993–2007. Recall that in 1991 Lowe's started a re-structuring plan to convert to big-box stores. This analysis suggests that it was a successful business decision but that we ought to fit a separate linear model to the years 1993–2007 where the linearity condition holds.

Not only can the residuals help check the conditions, but they can also tell us how well the model performs. The better the model fits the data, the less the residuals will vary around the line. The standard deviation of the residuals, s_e, gives us a measure of how much the points spread around the regression line. Of course, for this summary to make sense, the residuals should all share the same underlying spread. So we must *assume* that the standard deviation around the line is the same wherever we want the model to apply.

This new assumption about the standard deviation around the line gives us a new condition, called the **Equal Spread Condition**. The associated question to ask is does the plot have a consistent spread or does it fan out? We check to make sure that the spread of the residuals is about the same everywhere. We can check that either in the original scatterplot of *y* against *x* or in the scatterplot of residuals (or, preferably, in both plots). We estimate the **standard deviation of the residuals** in almost the way you'd expect:

$$s_e = \sqrt{\frac{\sum e^2}{n-2}}.$$

We don't need to subtract the mean of the residuals because $\bar{e} = 0$. Why divide by $n - 2$ rather than $n - 1$? We used $n - 1$ for *s* when we estimated the mean. Now we're estimating both a slope and an intercept. Looks like a pattern—and it is. We subtract one more for each parameter we estimate.

If we predict Lowe's *Sales* in 1999 when home *Improvements* totaled 100,250 $M, the regression model gives a predicted value of 15,032 $M. The actual value was about 12,946 $M. So our residual is 12,946 − 15,032 = −2,086. The value of s_e from the regression is 3170, so our residual is only 2086/3170 = 0.66 standard deviations away from the actual value. That's a fairly typical size for a residual because it's within two standard deviations.

Equal Spread Condition

This condition requires that the scatter is about equal for all *x*-values. It's often checked using a plot of residuals against predicted values. The underlying assumption of equal variance is also called *homoscedasticity*.

---

[6]Most computer statistics packages plot the residuals as we did in Figure 6.6, against the predicted values, rather than against *x*. When the slope is positive, the scatterplots are virtually identical except for the axes labels. When the slope is negative, the two versions are mirror images. Since all we care about is the patterns (or, better, lack of patterns) in the plot, either plot is useful.

For Example   Examining the residuals

Here is a scatterplot of the residuals for the linear model found in the example on page 151 plotted against the predicted values:

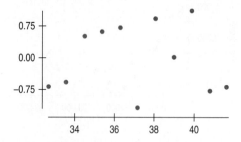

Question: Show how the plotted values were calculated. What does the plot suggest about the model?

Answer: The predicted values are the values of $\widehat{MeanAge}$ found for each year by substituting the year value in the linear model. The residuals are the differences between the actual mean ages and the predicted values for each year.

The plot shows some remaining pattern in the form of three nearly parallel trends. A further analysis may want to determine the reason for this pattern.

6.9   Variation in the Model and R^2

The variation in the residuals is the key to assessing how well the model fits. Let's compare the variation of the response variable with the variation of the residuals. Lowe's *Sales* has a standard deviation of 14,090 ($M). The standard deviation of the residuals is only 3,097 ($M). If the correlation were 1.0 and the model predicted the *Sales* values perfectly, the residuals would all be zero and have no variation. We couldn't possibly do any better than that.

On the other hand, if the correlation were zero, the model would simply predict 13,564 ($M) (the mean) for all menu items. The residuals from that prediction would just be the observed *Sales* values minus their mean. These residuals would

Figure 6.7   Compare the variability of *Sales* with the variability of the residuals from the regression. The means have been subtracted to make it easier to compare spreads. The variation left in the residuals is unaccounted for by the model, but it's less than the variation in the original data.

r and R^2

Is a correlation of 0.80 twice as strong as a correlation of 0.40? Not if you think in terms of R^2. A correlation of 0.80 means an R^2 of $0.80^2 = 64\%$. A correlation of 0.40 means an R^2 of $0.40^2 = 16\%$—only a quarter as much of the variability accounted for. A correlation of 0.80 gives an R^2 *four* times as strong as a correlation of 0.40 and accounts for four times as much of the variability.

Some Extreme Tales

One major company developed a method to differentiate between proteins. To do so, they had to distinguish between regressions with R^2 of 99.99% and 99.98%. For this application, 99.98% was not high enough.

The president of a financial services company reports that although his regressions give R^2 below 2%, they are highly successful because those used by his competition are even lower.

have the same variability as the original data because, as we know, just subtracting the mean doesn't change the spread.

How well does the regression model do? Look at the boxplots in Figure 6.7. The variation in the residuals is smaller than in the data, but bigger than zero. That's nice to know, but how much of the variation is still left in the residuals? If you had to put a number between 0% and 100% on the fraction of the variation left in the residuals, what would you say?

All regression models fall somewhere between the two extremes of zero correlation and perfect correlation. We'd like to gauge where our model falls. Can we use the correlation to do that? Well, a regression model with correlation -0.5 is doing as well as one with correlation $+0.5$. They just have different directions. But if we *square* the correlation coefficient, we'll get a value between 0 and 1, and the direction won't matter. The squared correlation, r^2, gives the fraction of the data's variation accounted for by the model, and $1 - r^2$ is the fraction of the original variation left in the residuals. For the Lowe's *Sales* model, $r^2 = 0.976^2 = 0.952$ and $1 - r^2$ is 0.048, so only 4.8% of the variability in *Sales* has been left in the residuals.

All regression analyses include this statistic, although by tradition, it is written with a capital letter, R^2, and pronounced "R-squared." An R^2 of 0 means that none of the variance in the data is in the model; all of it is still in the residuals. It would be hard to imagine using that model for anything.

Because R^2 is a fraction of a whole, it is often given as a percentage.[7] For the Lowe's *Sales* data, R^2 is 95.2%.

When interpreting a regression model, you need to report what R^2 means. According to our linear model, 95.2% of the variability in Lowe's *Sales* is accounted for by variation in residential *Improvement* expenditures.

- **How can we see that R^2 is really the fraction of variance accounted for by the model?** It's a simple calculation. The variance of *Sales* is $14089.6^2 = 198,516,828$. If we treat the residuals as data, the variance of the residuals is $9,592,072$.[8] As a fraction of the variance of *Sales*, that's 0.483 or 4.83%. That's the fraction of the variance that is *not* accounted for by the model. The fraction that *is* accounted for is $100\% - 4.83\% = 95.2$, just the value we got for R^2.

How Big Should R^2 Be?

The value of R^2 is always between 0% and 100%. But what is a "good" R^2 value? The answer depends on the kind of data you are analyzing and on what you want to do with it. Just as with correlation, there is no value for R^2 that automatically

Just Checking

Let's go back to our regression of sales ($000) on number of employees again.

$$\widehat{Sales} = 9.564 + 122.74\ Employees$$

The R^2 value is reported as 71.4%.

9 What does the R^2 value mean about the relationship of *Sales* and *Employees*?

10 Is the correlation of *Sales* and *Employees* positive or negative? How do you know?

11 If we measured the *Sales* in thousands of euros instead of thousands of dollars, would the R^2 value change? How about the slope?

[7]By contrast, we usually give correlation coefficients as decimal values between -1.0 and 1.0.
[8]This isn't quite the same as squaring s_e which we discussed previously, but it's very close.

Sums of Squares

The sum of the squared residuals $\Sigma(y - \hat{y})^2$ is sometimes written as SSE (sum of squared errors). If we call $\Sigma(y - \bar{y})^2$ SST (for total sum of squares) then

$$R^2 = 1 - \frac{SSE}{SST}.$$

determines that the regression is "good." Data from scientific experiments often have R^2 in the 80% to 90% range and even higher. Data from observational studies and surveys, though, often show relatively weak associations because it's so difficult to measure reliable responses. An R^2 of 30% to 50% or even lower might be taken as evidence of a useful regression. The standard deviation of the residuals can give us more information about the usefulness of the regression by telling us how much scatter there is around the line.

As we've seen, an R^2 of 100% is a perfect fit, with no scatter around the line. The s_e would be zero. All of the variance would be accounted for by the model with none left in the residuals. This sounds great, but it's too good to be true for real data.[9]

For Example   Understanding R^2

Question: Find and interpret the R^2 for the regression of cyclist death ages vs. time found in the example on page 151. (Hint: The calculation is a simple one.)

Answer: We are given the correlation, $r = 0.965$. R^2 is the square of this, or 0.9312. It tells us that 93.1% of the variation in the mean age of cyclist deaths can be accounted for by the trend of increasing age over time.

6.10   Reality Check: Is the Regression Reasonable?

Statistics don't come out of nowhere. They are based on data. The results of a statistical analysis should reinforce common sense. If the results are surprising, then either you've learned something new about the world or your analysis is wrong.

Whenever you perform a regression, think about the coefficients and ask whether they make sense. Is the slope reasonable? Does the direction of the slope seem right? The small effort of asking whether the regression equation is plausible will be repaid whenever you catch errors or avoid saying something silly or absurd about the data. It's too easy to take something that comes out of a computer at face value and assume that it makes sense.

Always be skeptical and ask yourself if the answer is reasonable.

Guided Example   Home Size and Price

Real estate agents know the three most important factors in determining the price of a house are *location*, *location*, and *location*. But what other factors help determine the price at which a house should be listed? Number of bathrooms? Size of the yard? A student amassed publicly available data on thousands of homes in upstate New York. We've drawn a random sample of 1057 homes to examine house pricing. Among the variables she collected were the total living area (in square feet), number of bathrooms, number of bedrooms, size of lot (in acres), and age of house (in years). We will investigate how well the size of the house, as measured by living area, can predict the selling price.

---

[9]If you see an R^2 of 100%, it's a good idea to investigate what happened. You may have accidentally regressed two variables that measure the same thing.

PLAN

Setup State the objective of the study.

Identify the variables and their context.

We want to find out how well the living area of a house in upstate NY can predict its selling price.

We have two quantitative variables: the living area (in square feet) and the selling price ($). These data come from public records in upstate New York in 2006.

✓ **Quantitative Variables Condition**

Model We need to check the same conditions for regression as we did for correlation. To do that, make a picture. Never fit a regression without looking at the scatterplot first.

Check the Linearity, Equal Spread, and Outlier Conditions.

✓ **Linearity Condition** The scatterplot shows two variables that appear to have a fairly strong positive association. The plot appears to be fairly linear.

✓ **Equal Spread Condition** The scatterplot shows a consistent spread across the x-values.

✓ **Outlier Condition** There appear to be a few possible outliers, especially among large, relatively expensive houses. A few smaller houses are expensive for their size. We will check their influence on the model later.

We have two quantitative variables that appear to satisfy the conditions, so we will model this relationship with a regression line.

DO

Mechanics Find the equation of the regression line using a statistics package. Remember to write the equation of the model using meaningful variable names.

Our software produces the following output.

```
Dependent variable is: Price
1057 total cases
R squared = 62.43%
s = 57930 with 1000 − 2 = 998 df
Variable          Coefficient
Intercept         6378.08
Living Area       115.13
```

Once you have the model, plot the residuals and check the Equal Spread Condition again.

(continued)

The residual plot appears generally patternless. The few relatively expensive small houses are evident, but setting them aside and refitting the model did not change either the slope or intercept very much so we left them in. There is a slight tendency for cheaper houses to have less variation, but the spread is roughly the same throughout.

REPORT

Conclusion Interpret what you have found in the proper context.

MEMO

Re: Report on housing prices.

We examined how well the size of a house could predict its selling price. Data were obtained from recent sales of 1057 homes in upstate New York. The model is:

$$\widehat{Price} = \$6376.08 + 115.13 \times Living\ Area$$

In other words, from a base of $6376.08, houses cost about $115.13 per square foot in upstate NY.

This model appears reasonable from both a statistical and real estate perspective. Although we know that size is not the only factor in pricing a house, the model accounts for 62.4% of the variation in selling price.

As a reality check, we checked with several real estate pricing sites (www.realestateabc.com, www.zillow.com) and found that houses in this region were averaging $100 to $150 per square foot, so our model is plausible.

Of course, not all house prices are predicted well by the model. We computed the model without several of these houses, but their impact on the regression model was small. We believe that this is a reasonable place to start to assess whether a house is priced correctly for this market. Future analysis might benefit by considering other factors.

6.11   Nonlinear Relationships

Everything we've discussed in this chapter requires that the underlying relationship between two variables be linear. But what should we do when the relationship is nonlinear and we can't use the correlation coefficient or a linear model? There are three basic approaches, each with its advantages and disadvantages.

Let's consider an example. The Human Development Index (HDI) was developed by the United Nations as a general measure of quality of life in countries around the world. It combines economic information (GDP), life expectancy, and education. The growth of cell phone usage has been phenomenal worldwide. Is cell phone usage related to the developmental state of a country? Figure 6.8 shows a scatterplot of number of cell phones vs. HDI for 152 countries of the world.

We can look at the scatterplot and see that cell phone usage increases with increasing HDI. But the relationship is not straight. In Figure 6.8, we can easily see the bend in the form. But that doesn't help us summarize or model the relationship.

You might think that we should just fit some curved function such as an exponential or quadratic to a shape like this. But using curved functions is complicated, and the resulting model can be difficult to interpret. And many of the convenient

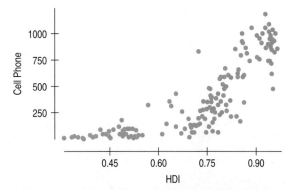

Figure 6.8   The scatterplot of number of cell phones (000s) vs. HDI for countries shows a bent relationship not suitable for correlation or regression.

associated statistics (which we'll see in Chapters 18 and 19) are not appropriate for such models. So this approach isn't often used.

Another approach allows us to summarize the strength of the association between the variables even when we don't have a linear relationship. The **Spearman rank correlation**[10] works with the *ranks* of the data rather than their values. To find the ranks we simply count from the lowest value to the highest so that rank 1 is assigned to the lowest value, rank 2 to the next lowest, and so on. Using ranks for both variables generally straightens out the relationship, as Figure 6.9 shows.

Figure 6.9   Plotting the ranks results in a plot with a straight relationship.

Now we can calculate a correlation on the ranks. The resulting correlation[11] summarizes the degree of relationship between two variables—but not, of course, of the degree of *linear* relationship. The Spearman correlation for these variables is 0.876. That says there's a reasonably strong relationship between cell phones and HDI. We don't usually fit a linear model to the ranks because that would be difficult to interpret and because the supporting statistics (as we said, see Chapters 18 and 19) wouldn't be appropriate.

A third approach to a nonlinear relationship is to transform or re-express one or both of the variables by a function such as the square root, logarithm, or reciprocal. We saw in Chapter 5 that a transformation can improve the symmetry of the distribution of a single variable. In the same way—and often with the same transforming function—transformations can make a relationship more nearly linear.

---

[10]Due to Charles Spearman, a psychologist who did pioneering work in intelligence testing.

[11]Spearman rank correlation is a *nonparametric* statistical method. You'll find other nonparametric methods in Chapter 17.

Figure 6.10   Taking the logarithm of Cell Phones results in a more nearly linear relationship.

Figure 6.10, for example, shows the relationship between the log of the number of cell phones and the HDI for the same countries.

The advantage of re-expressing variables is that we *can* use regression models, along with all the supporting statistics still to come. The disadvantage is that we must interpret our results in terms of the re-expressed data, and it can be difficult to explain what we mean by the logarithm of the number of cell phones in a country. We can, of course, reverse the transformation to transform a predicted value or residual back to the original units. (In the case of a logarithmic transformation, calculate 10^y to get back to the original units.)

Which approach you choose is likely to depend on the situation and your needs. Statisticians, economists, and scientists generally prefer to transform their data, and many of their laws and theories include transforming functions.[12] But for just understanding the shape of a relationship, a scatterplot does a fine job, and as a summary of the strength of a relationship, a Spearman correlation is a good general-purpose tool.

For Example   Re-expressing for linearity

Consider the relationship between a company's *Assets* and its *Sales* as reported in annual financial statements. Here's a scatterplot of those variables for 79 of the largest companies:

---

[12]In fact, the HDI itself includes such transformed variables in its construction.

The Pearson correlation is 0.746, and the Spearman rank correlation is 0.50. Taking the logarithm of both variables produces the following scatterplot:

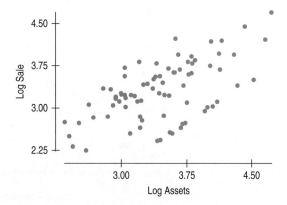

Question: What should we say about the relationship between Assets and Sales?

Answer: The Pearson correlation is not appropriate because the scatterplot of the data is not linear. The Spearman correlation is a more appropriate summary. The scatterplot of the log transformed variables is linear and shows a strong pattern. We could find a linear model for this relationship, but we'd have to interpret it in terms of log Sales and log Assets.

What Can Go Wrong?

- **Don't say "correlation" when you mean "association."** How often have you heard the word "correlation"? Chances are pretty good that when you've heard the term, it's been misused. It's one of the most widely misused Statistics terms, and given how often Statistics are misused, that's saying a lot. One of the problems is that many people use the specific term *correlation* when they really mean the more general term *association*. Association is a deliberately vague term used to describe the relationship between two variables.

  Correlation is a precise term used to describe the strength and direction of a linear relationship between quantitative variables.

- **Don't correlate categorical variables.** Be sure to check the Quantitative Variables Condition. It makes no sense to compute a correlation of categorical variables.

- **Make sure the association is linear.** Not all associations between quantitative variables are linear. Correlation can miss even a strong nonlinear association. And linear regression models are never appropriate for relationships that are not linear. A company, concerned that customers might use ovens with imperfect temperature controls, performed a series of experiments[13] to assess the effect of baking temperature on the quality of brownies made from their freeze-dried reconstituted brownies. The company wants to understand the sensitivity of brownie quality to variation in oven temperatures around the recommended baking temperature of 325°F. The lab reported a correlation of −0.05 between the scores awarded by a panel of trained taste-testers and baking

(continued)

---

[13]Experiments designed to assess the impact of environmental variables outside the control of the company on the quality of the company's products were advocated by the Japanese quality expert Dr. Genichi Taguchi starting in the 1980s in the United States.

temperature and a regression slope of −0.02, so they told management that there is no relationship. Before printing directions on the box telling customers not to worry about the temperature, a savvy intern asks to see the scatterplot.

Figure 6.11   The relationship between brownie taste score and baking temperature is strong, but not linear.

The plot actually shows a strong association—but not a linear one. Don't forget to check the Linearity Condition.

- **Beware of outliers.** You can't interpret a correlation coefficient or a regression model safely without a background check for unusual observations. Here's an example. The relationship between IQ and shoe size among comedians shows a surprisingly strong positive correlation of 0.50. To check assumptions, we look at the scatterplot.

| Figure 6.12   IQ vs. Shoe Size.

From this "study," what can we say about the relationship between the two? The correlation is 0.50. But who *does* that point in the upper right-hand corner belong to? The outlier is Bozo the Clown, known for his large shoes and widely acknowledged to be a comic "genius." Without Bozo the correlation is near zero.

Even a single unusual observation can dominate the correlation value. That's why you need to check the Unusual Observations Condition.

- **Don't confuse correlation with causation.** Once we have a strong correlation, it's tempting to try to explain it by imagining that the predictor variable has *caused* the response to change. Putting a regression line on a scatterplot tempts us even further. Humans are like that; we tend to see causes and effects in everything. Just because two variables are related does not mean that one *causes* the other.

Does cancer cause smoking?

Even if the correlation of two variables is due to a causal relationship, the correlation itself cannot tell us what causes what.

Sir Ronald Aylmer Fisher (1890–1962) was one of the greatest statisticians of the 20th century. Fisher testified in court (paid by the tobacco companies) that a causal relationship might underlie the correlation of smoking and cancer:

> "Is it possible, then, that lung cancer . . . is one of the causes of smoking cigarettes? I don't think it can be excluded . . . the pre-cancerous condition is one involving a certain amount of slight chronic inflammation . . .
>
> A slight cause of irritation . . . is commonly accompanied by pulling out a cigarette, and getting a little compensation for life's minor ills in that way. And . . . is not unlikely to be associated with smoking more frequently."

Ironically, the proof that smoking indeed is the cause of many cancers came from experiments conducted following the principles of experiment design and analysis that Fisher himself developed.

Scatterplots, correlation coefficients, and regression models *never* prove causation. This is, for example, partly why it took so long for the U.S. Surgeon General to get warning labels on cigarettes. Although there was plenty of evidence that increased smoking was *associated* with increased levels of lung cancer, it took years to provide evidence that smoking actually *causes* lung cancer. (The tobacco companies used this to great advantage.)

- **Watch out for lurking variables.** A scatterplot of the damage (in dollars) caused to a house by fire would show a strong correlation with the number of firefighters at the scene. Surely the damage doesn't cause firefighters. And firefighters actually do cause damage, spraying water all around and chopping holes, but does that mean we shouldn't call the fire department? Of course not. There is an underlying variable that leads to both more damage and more firefighters—the size of the blaze. A hidden variable that stands behind a relationship and determines it by simultaneously affecting the other two variables is called a **lurking variable**. You can often debunk claims made about data by finding a lurking variable behind the scenes.

- **Don't fit a straight line to a nonlinear relationship.** Linear regression is suited only to relationships that are, in fact, linear.

- **Beware of extraordinary points.** Data values can be extraordinary or unusual in a regression in two ways. They can have y-values that stand off from the linear pattern suggested by the bulk of the data. These are what we have been calling outliers; although with regression, a point can be an outlier by being far from the linear pattern even if it is not the largest or smallest y-value. Points can also be extraordinary in their x-values. Such points can exert a strong influence on the line. Both kinds of extraordinary points require attention.

- **Don't extrapolate far beyond the data. A linear model will often do a reasonable job of summarizing a relationship in the range of observed x-values.** Once we have a working model for the relationship, it's tempting to use it. But beware of predicting y-values for x-values that lie too far outside the range of the original data. The model may no longer hold there, so such extrapolations too far from the data are dangerous.

(continued)

- **Don't choose a model based on R^2 alone.** Although R^2 measures the *strength* of the linear association, a high R^2 does not demonstrate the *appropriateness* of the regression. A single unusual observation, or data that separate into two groups, can make the R^2 seem quite large when, in fact, the linear regression model is simply inappropriate. Conversely, a low R^2 value may be due to a single outlier. It may be that most of the data fall roughly along a straight line, with the exception of a single point. Always look at the scatterplot.

Ethics in Action

An ad agency hired by a well-known manufacturer of dental hygiene products (electric toothbrushes, oral irrigators, etc.) put together a creative team to brainstorm ideas for a new ad campaign. Trisha Simes was chosen to lead the team as she has had the most experience with this client to date. At their first meeting, Trisha communicated to her team the client's desire to differentiate themselves from their competitors by not focusing their message on the cosmetic benefits of good dental care. As they brainstormed ideas, one member of the team, Brad Jonns, recalled a recent CNN broadcast that reported a "correlation" between flossing teeth and reduced risk of heart disease. Seeing potential in promoting the health benefits of proper dental care, the team agreed to pursue this idea further. At their next meeting several team members commented on how surprised they were to find so many articles, medical, scientific, and popular, that seemed to claim good dental hygiene resulted in good health. One member noted that he found articles that linked gum disease not only to heart attacks and strokes but to diabetes and even cancer. Although Trisha puzzled over why their client's competitors had not yet capitalized on these research findings, her team was on a roll and had already begun to focus on designing the campaign around this core message.

ETHICAL ISSUE *Correlation does not imply causation. The possibility of lurking variables is not explored. For example, it is likely that those who take better care of themselves would floss regularly and also have less risk of heart disease (related to Item C, ASA Ethical Guidelines).*

ETHICAL SOLUTION *Refrain from implying cause and effect from correlation results.*

Jill Hathway is looking for a career change and is interested in starting a franchise. After spending the last 20 years working as a mid-level manager for a major corporation, Jill wants to indulge her entrepreneurial spirit and strike out on her own. She currently lives in a small southwestern city and is considering a franchise in the health and fitness industry. She is considering several possibilities including *Pilates One*, for which she requested a franchise packet. Included in the packet information were data showing how various regional demographics (age, gender, income) related to franchise success (revenue, profit, return on investment). *Pilates One* is a relatively new franchise with only a few scattered locations. Nonetheless, the company reported various graphs and data analysis results to help prospective franchisers in their decision-making process. Jill was particularly interested in the graph and the regression analysis that related the proportion of women over the age of 40 within a 20-mile radius of a *Pilates One* location to return on investment for the franchise. She noticed that there was a positive relationship. With a little research, she discovered that the proportion of women over the age of 40 in her city was higher than for any other *Pilates One* location (attributable, in part, to the large number of retirees relocating to the southwest). She then used the regression equation to project return on investment for a *Pilates One* located in her city and was very pleased with the result. With such objective data, she felt confident that *Pilates One* was the franchise for her.

ETHICAL ISSUE *Pilates One is reporting analysis based on only a few observations. Jill is extrapolating beyond the range of x-values (related to Item C, ASA Ethical Guidelines).*

ETHICAL SOLUTION *Pilates One should include a disclaimer that the analysis was based on very few observations and that the equation should not be used to predict success at other locations or beyond the range of x-values used in the analysis.*

What Have We Learned?

Learning Objectives	■ Make a scatterplot to display the relationship between two quantitative variables.

Learning Objectives

■ Make a scatterplot to display the relationship between two quantitative variables.

- Look at the direction, form, and strength of the relationship, and any outliers that stand away from the overall pattern.

■ Provided the form of the relationship is linear, summarize its strength with a correlation, r.

- The sign of the correlation gives the direction of the relationship.

- $-1 \leq r \leq 1$; A correlation of 1 or -1 is a perfect linear relationship. A correlation of 0 is a lack of linear relationship.

- Correlation has no units, so shifting or scaling the data, standardizing, or even swapping the variables has no effect on the numerical value.

- A large correlation is not a sign of a causal relationship

■ Model a linear relationship with a least squares regression model.

- The regression (best fit) line doesn't pass through all the points, but it is the best compromise in the sense that the sum of squares of the residuals is the smallest possible.

- The slope tells us the change in y per unit change in x.

- The R^2 gives the fraction of the variation in y accounted for by the linear regression model.

■ Recognize regression to the mean when it occurs in data.

- A deviation of one standard deviation from the mean in one variable is predicted to correspond to a deviation of r standard deviations from the mean in the other. Because r is never more than 1, we predict a change toward the mean.

■ Examine the residuals from a linear model to assess the quality of the model.

- When plotted against the predicted values, the residuals should show no pattern and no change in spread.

Terms

Association

- **Direction:** A positive direction or association means that, in general, as one variable increases, so does the other. When increases in one variable generally correspond to decreases in the other, the association is negative.

- **Form:** The form we care about most is straight, but you should certainly describe other patterns you see in scatterplots.

- **Strength:** A scatterplot is said to show a strong association if there is little scatter around the underlying relationship.

Correlation coefficient

A numerical measure of the direction and strength of a linear association.

$$r = \frac{\sum z_x z_y}{n - 1}$$

Explanatory or independent variable (*x*-variable)

The variable that accounts for, explains, predicts, or is otherwise responsible for the y-variable.

Intercept

The intercept, b_0, gives a starting value in y-units. It's the $\hat{y}$ value when x is 0.

$$b_0 = \bar{y} - b_1 \bar{x}$$

(continued)

Least squares	A criterion that specifies the unique line that minimizes the variance of the residuals or, equivalently, the sum of the squared residuals.
Linear model (Line of best fit)	The linear model of the form $\hat{y} = b_0 + b_1 x$ fit by least squares. Also called the regression line. To interpret a linear model, we need to know the variables and their units.
Lurking variable	A variable other than x and y that simultaneously affects both variables, accounting for the correlation between the two.
Outlier	A point that does not fit the overall pattern seen in the scatterplot.
Predicted value	The prediction for y found for each x-value in the data. A predicted value, $\hat{y}$, is found by substituting the x-value in the regression equation. The predicted values are the values on the fitted line; the points $(x, \hat{y})$ lie exactly on the fitted line.
Re-expression or transformation	Re-expressing one or both variables using functions such as log, square root, or reciprocal can improve the straightness of the relationship between them.
Residual	The difference between the actual data value and the corresponding value predicted by the regression model—or, more generally, predicted by any model.
Spearman rank correlation	The correlation between the ranks of two variables may be an appropriate measure of the strength of a relationship when the form isn't straight.
Regression line	The particular linear equation that satisfies the least squares criterion, often called the line of best fit.
Regression to the mean	Because the correlation is always less than 1.0 in magnitude, each predicted y tends to be fewer standard deviations from its mean than its corresponding x is from its mean.
Response or dependent variable (y-variable)	The variable that the scatterplot is meant to explain or predict.
R^2	• The square of the correlation between y and x • The fraction of the variability of y accounted for by the least squares linear regression on x • An overall measure of how successful the regression is in linearly relating y to x
Scatterplot	A graph that shows the relationship between two quantitative variables measured on the same cases.
Standard deviation of the residuals	s_e is found by: $$s_e = \sqrt{\frac{\Sigma e^2}{n-2}}.$$
Slope	The slope, b_1, is given in y-units per x-unit. Differences of one unit in x are associated with differences of b_1 units in predicted values of y: $$b_1 = r\frac{s_y}{s_x}$$

Technology Help: Correlation and Regression

All statistics packages make a table of results for a regression. These tables may differ slightly from one package to another, but all are essentially the same—and all include much more than we need to know for now. Every computer regression table includes a section that looks something like this:

Standard dev of residuals (s_e)

R squared

The "dependent," response, or y-variable

```
Dependent variable is: Sales
R squared = 69.0%
s = 9.277

Variable      Coefficient   SE(Coeff)   t-ratio   P-value
Intercept     6.83077       2.664       2.56      0.0158
Shelf Space   0.971381      0.1209      8.04      ≤0.0001
```

The "independent," predictor, or x-variable

The slope

The intercept

We'll deal with all of these later in the book. You may ignore them for now.

The slope and intercept coefficient are given in a table such as this one. Usually the slope is labeled with the name of the *x*-variable, and the intercept is labeled "Intercept" or "Constant." So the regression equation shown here is

$$\widehat{Sales} = 6.83077 + 0.97138 \text{ } Shelf\ Space.$$

EXCEL

XLSTAT

To make a scatterplot in Excel 2007 or Excel 2010,

- From the Data ribbon, select the Data Analysis add-in.
- From its menu, select **Regression**.
- Indicate the range of the data whose scatterplot you wish to draw.
- Check the Labels box if your data columns have names in the first cell.
- Check the **Line Fit Plots** box, and click **OK**.

- Excel will place regression output and the scatterplot on a new sheet.
- The correlation is in cell B4.
- The slope and *y*-intercept are in cells B18 and B17 respectively.
- You can edit or remove any part of the scatterplot by right clicking on the part you want to edit.
- For example, to remove the Predicted Values, right click on one of the points and **Delete**.
- To add the Least Squares Regression Line, right click on the data and **Add Trendline . . .**

(continued)

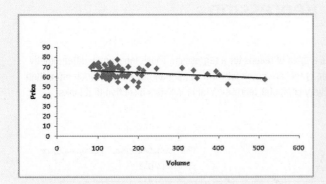

But we aren't quite done yet. Excel always scales the axes of a scatterplot to show the origin (0, 0). But most data are not near the origin, so you may get a plot that, like this one, is bunched up in one corner.

- Right-click on the *y*-axis labels. From the menu that drops down, choose Format axis. . .
- Choose Scale.
- Set the *y*-axis minimum value.

One useful trick is to use the dialog box itself as a straightedge to read over to the *y*-axis so you can estimate a good minimum value. Here 75 seems appropriate.

- Repeat the process with the *x*-axis.

JMP

To make a scatterplot and compute correlation

- Choose Fit Y by X from the Analyze menu.
- In the Fit Y by X dialog, drag the Y variable into the "Y, Response" box, and drag the X variable into the "X, Factor" box.
- Click the OK button.

Once JMP has made the scatterplot, click on the red triangle next to the plot title to reveal a menu of options.

- Select Density Ellipse and select .95. JMP draws an ellipse around the data and reveals the Correlation tab.
- Click the blue triangle next to Correlation to reveal a table containing the correlation coefficient.

To compute a regression,

- Choose Fit Y by X from the Analyze menu. Specify the *y*-variable in the — Select Columns box and click the "Y, Response" button.
- Specify the *x*-variable and click the "X, Factor" button.
- Click OK to make a scatterplot.

- In the scatterplot window, click on the red triangle beside the heading labeled "Bivariate Fit. . ." and choose "Fit Line." JMP draws the least squares regression line on the scatterplot and displays the results of the regression in tables below the plot.

MINITAB

To make a scatterplot,

- Choose Scatterplot from the Graph menu.
- Choose "Simple" for the type of graph. Click OK.
- Enter variable names for the *Y*-variable and *X*-variable into the table. Click OK.

To compute a correlation coefficient,

- Choose Basic Statistics from the Stat menu.
- From the Basic Statistics submenu, choose Correlation. Specify the names of at least two quantitative variables in the "Variables" box.
- Click OK to compute the correlation table.

SPSS

To make a scatterplot in SPSS, open the Chart Builder from the Graphs menu. Then

- Click the Gallery tab.
- Choose Scatterplot from the list of chart types.
- Drag the scatterplot onto the canvas.
- Drag a scale variable you want as the response variable to the *y*-axis drop zone.
- Drag a scale variable you want as the factor or predictor to the *x*-axis drop zone.
- Click OK.

To compute a correlation coefficient,

- Choose Correlate from the Analyze menu.
- From the Correlate submenu, choose Bivariate.
- In the Bivariate Correlations dialog, use the arrow button to move variables between the source and target lists. Make sure the Pearson option is selected in the Correlation Coefficients field.

To compute a regression, from the Analyze menu, choose

- Regression > Linear . . . In the Linear Regression dialog, specify the Dependent (*y*), and Independent (*x*) variables.
- Click the Plots button to specify plots and Normal Probability Plots of the residuals. Click OK.

BriefCASE

Fuel Efficiency

With the ever increasing price of gasoline, both drivers and auto companies are motivated to raise the fuel efficiency of cars. Recent information posted by the U.S. government proposes some simple ways to increase fuel efficiency (see www.fueleconomy.gov): avoid rapid acceleration, avoid driving over 60 mph, reduce idling, and reduce the vehicle's weight. An extra 100 pounds can reduce fuel efficiency (mpg) by up to 2%. A marketing executive is studying the relationship between the fuel efficiency of cars (as measured in miles per gallon) and their weight to design a new compact car campaign. In the data set **Fuel_Efficiency** you'll find data on the variables below.[14]

- Make of Car
- Model of Car
- Engine Size (L)
- Cylinders
- MSRP (Manufacturer's Suggested Retail Price in $)
- City (mpg)
- Highway (mpg)
- Weight (pounds)
- Type and Country of manufacturer

Describe the relationship of Weight, MSRP, and Engine Size with fuel efficiency (both City and Highway) in a written report. Only in the U.S. is fuel efficiency measured in miles per gallon. The rest of the world uses liters per 100 kilometers. To convert mpg to l/100km, compute 235.215/mpg. Try that form of the variable and compare the resulting models. Be sure to plot the residuals.

The U.S. Economy and the Home Depot Stock Prices

The file **Home_Depot** contains economic variables, as well as stock market data for The Home Depot. Economists, investors, and corporate executives use measures of the U.S. economy to evaluate the impact of inflationary pressures and employment fluctuations on the stock market. Inflation is often tracked through interest rates. Although there are many different types of interest rates, here we include the monthly values for the bank prime loan rate, a rate posted by a majority of the top 25 (based on assets) insured U.S.-chartered commercial banks. The prime rate is often used by banks to price short-term business loans. In addition, we provide the interest rates on 6-month CDs, the unemployment rates (seasonally adjusted), and the rate on Treasury bills. Investigate the relationships between *Closing Price* for The Home Depot stock and the following variables from 2006 to 2008:[15]

- Unemployment Rate (%)
- Bank Prime Rate (Interest Rate in %)
- CD Rate (%)
- Treasury Bill Rate (%)

[14]Data are from the 2004 model year and were compiled from www.Edmonds.com.
[15]Sources: Unemployment rate—U.S. Bureau of Labor Statistics. See unemployment page at www.bls.gov/cps/home.htm#data. Interest rates—Federal Reserve. See www.federalreserve.gov/releases/H15/update/. Home Depot stock prices on HD/Investor Relations website. See ir.homedepot.com/quote.cfm.

(continued)

Describe the relationship of each of these variables with The Home Depot *Closing Price* in a written report. Be sure to use scatterplots and correlation tables in your analysis and transform variables, if necessary.

Cost of Living

The Mercer Human Resource Consulting website (www.mercerhr.com) lists prices of certain items in selected cities around the world. They also report an overall cost-of-living index for each city compared to the costs of hundreds of items in New York City. For example, London at 110.6 is 10.6% more expensive than New York. You'll find the 2006 data for 16 cities in the data set **Cost_of_living_vs_cost_of_items**. Included are the 2006 cost of living index, cost of a luxury apartment (per month), price of a bus or subway ride, price of a compact disc, price of an international newspaper, price of a cup of coffee (including service), and price of a fast-food hamburger meal. All prices are in U.S. dollars.

Examine the relationship between the overall cost of living and the cost of each of these individual items. Verify the necessary conditions and describe the relationship in as much detail as possible. (Remember to look at direction, form, and strength.) Identify any unusual observations.

Based on the correlations and linear regressions, which item would be the best predictor of overall cost in these cities? Which would be the worst? Are there any surprising relationships? Write a short report detailing your conclusions.

Mutual Funds

According to the U.S. Securities and Exchange Commission (SEC), a mutual fund is a professionally-managed collection of investments for a group of investors in stocks, bonds, and other securities. The fund manager manages the investment portfolio and tracks the wins and losses. Eventually the dividends are passed along to the individual investors in the mutual fund. The first group fund was founded in 1924, but the spread of these types of funds was slowed by the stock market crash in 1929. Congress passed the Securities Act in 1933 and the Securities Exchange Act in 1934 to require that investors be provided disclosures about the fund, the securities, and the fund manager. The SEC drafted the Investment Company Act, which provided guidelines for registering all funds with the SEC. By the end of the 1960s, funds reported $48 billion in assets and, by October 2007 there were over 8,000 mutual funds with combined assets under management of over $12 trillion.

Investors often choose mutual funds on the basis of past performance, and many brokers, mutual fund companies, and other websites offer such data. In the file **Mutual_funds_returns**, you'll find the 3-month return, the annualized 1 yr, 5 yr, and 10 yr returns, and the return since inception of 64 funds of various types. Which data from the past provides the best predictions of the recent 3 months? Examine the scatterplots and regression models for predicting 3-month returns and write a short report containing your conclusions.

The calculations for correlation and regression models can be very sensitive to how intermediate results are rounded. If you find your answers using a calculator and writing down intermediate results, you may obtain slightly different answers that you would have had you used statistics software. Different programs can also yield different results. So your answers may differ in the trailing digits from those in the Appendix. That should not concern you. The meaningful digits are the first few; the trailing digits may be essentially random results of the rounding of intermediate results.

Exercises

SECTION 6.1

1. Consider the following data from a small bookstore.

Number of sales people working	Sales (in $1000)
2	10
3	11
7	13
9	14
10	18
10	20
12	20
15	22
16	22
20	26
$\bar{x} = 10.4$	$\bar{y} = 17.6$
$SD(x) = 5.64$	$SD(y) = 5.34$

a) Prepare a scatterplot of *Sales* against *Number of sales people* working.
b) What can you say about the direction of the association?
c) What can you say about the form of the relationship?
d) What can you say about the strength of the relationship?
e) Does the scatterplot show any outliers?

2. Disk drives have been getting larger. Their capacity is now often given in *terabytes* (TB) where 1 TB = 1000 gigabytes, or about a trillion bytes. A survey of prices for external disk drives found the following data:

Capacity (in TB)	Price (in $)
.080	29.95
.120	35.00
.200	299.00
.250	49.95
.320	69.95
1.0	99.00
2.0	205.00
4.0	449.00

a) Prepare a scatterplot of *Price* against *Capacity*.
b) What can you say about the direction of the association?
c) What can you say about the form of the relationship?

d) What can you say about the strength of the relationship?
e) Does the scatterplot show any outliers?

SECTION 6.2

3. The human resources department at a large multinational corporation wants to be able to predict average salary for a given number of years experience. Data on salary (in $1000's) and years of experience were collected for a sample of employees.

a) Which variable is the explanatory or predictor variable?
b) Which variable is the response variable?
c) Which variable would you plot on the *y* axis?

4. A company that relies on Internet-based advertising linked to key search terms wants to understand the relationship between the amount it spends on this advertising and revenue (in $).

a) Which variable is the explanatory or predictor variable?
b) Which variable is the response variable?
c) Which variable would you plot on the *x* axis?

SECTION 6.3

5. If we assume that the conditions for correlation are met, which of the following are true? If false, explain briefly.

a) A correlation of −0.98 indicates a strong, negative association.
b) Multiplying every value of *x* by 2 will double the correlation.
c) The units of the correlation are the same as the units of *y*.

6. If we assume that the conditions for correlation are met, which of the following are true? If false, explain briefly.

a) A correlation of 0.02 indicates a strong positive association.
b) Standardizing the variables will make the correlation 0.
c) Adding an outlier can dramatically change the correlation.

SECTION 6.4

7. A larger firm is considering acquiring the bookstore of Exercise 1. An analyst for the firm, noting the relationship seen in Exercise 1, suggests that when they acquire the store they should hire more people because that will drive higher sales. Is his conclusion justified? What alternative explanations can you offer? Use appropriate statistics terminology.

8. A study finds that during blizzards, online sales are highly associated with the number of snow plows on the road; the more plows, the more online purchases. The director of an association of online merchants suggests that the organization should encourage municipalities to send out more plows whenever it snows because, he says, that will increase business. Comment.

SECTION 6.5

9. True or False. If False, explain briefly.

a) We choose the linear model that passes through the most data points on the scatterplot.
b) The residuals are the observed *y*-values minus the *y*-values predicted by the linear model.
c) Least squares means that the square of the largest residual is as small as it could possibly be.

10. True or False. If False, explain briefly.

a) Some of the residuals from a least squares linear model will be positive and some will be negative.
b) Least Squares means that some of the squares of the residuals are minimized.
c) We write $\hat{y}$ to denote the predicted values and *y* to denote the observed values.

SECTION 6.6

11. For the bookstore sales data in Exercise 1, the correlation is 0.965.

a) If the number of people working is 2 standard deviations above the mean, how many standard deviations above or below the mean do you expect sales to be?
b) What value of sales does that correspond to?
c) If the number of people working is 1 standard deviation below the mean, how many standard deviations above or below the mean do you expect sales to be?
d) What value of sales does that correspond to?

12. For the hard drive data in Exercise 2, some research on the prices discovered that the 200 GB hard drive was a special "hardened" drive designed to resist physical shocks and work under water. Because it is completely different from the other drives, it was removed from the data. The correlation is now 0.994 and other summary statistics are:

Capacity (in TB)	Price (in $)
$\bar{x} = 1.110$	$\bar{y} = 133.98$
$SD(x) = 1.4469$	$SD(y) = 151.26$

a) If a drive has a capacity of 2.55669 TB (or 1 SD above the mean of 1.100 TB), how many standard deviations above or below the mean price of $133.98 do you expect the drive to cost?
b) What price does that correspond to?

13. For the bookstore of Exercise 1, the manager wants to predict *Sales* from *Number of Sales People Working*.
a) Find the slope estimate, b_1.
b) What does it mean, in this context?
c) Find the intercept, b_0.
d) What does it mean, in this context? Is it meaningful?
e) Write down the equation that predicts *Sales* from *Number of Sales People Working*.
f) If 18 people are working, what *Sales* do you predict?
g) If sales are actually $25,000, what is the value of the residual?
h) Have we overestimated or underestimated the sales?

14. For the disk drives in Exercise 2 (as corrected in Exercise 12), we want to predict *Price* from *Capacity*.
a) Find the slope estimate, b_1.
b) What does it mean, in this context?
c) Find the intercept, b_0.
d) What does it mean, in this context? Is it meaningful?
e) Write down the equation that predicts *Price* from *Capacity*.
f) What would you predict for the price of a 3.0 TB disk?
g) You have found a 3.0 TB drive for $300. Is this a good buy? How much would you save compared to what you expected to pay?
h) Does the model overestimate or underestimate the price?

SECTION 6.7

15. A CEO complains that the winners of his "rookie junior executive of the year" award often turn out to have less impressive performance the following year. He wonders whether the award actually encourages them to slack off. Can you offer a better explanation?

16. An online investment blogger advises investing in mutual funds that have performed badly the past year because "regression to the mean tells us that they will do well next year." Is he correct?

SECTION 6.8

17. Here are the residuals for a regression of *Sales* on *Number of Sales People Working* for the bookstore of Exercise 1:

Sales People Working	Residual
2	0.07
3	0.16
7	−1.49
9	−2.32
10	0.77
10	2.77
12	0.94
15	0.20
16	−0.72
20	−0.37

a) What are the units of the residuals?
b) Which residual contributes the most to the sum that was minimized according to the Least Squares Criterion to find this regression?
c) Which residual contributes least to that sum?

18. Here are residual plots (residuals plotted against predicted values) for three linear regression models. Indicate which condition appears to be violated (linearity, outlier or equal spread) in each case.

a)

b)

c)

SECTION 6.9

19. For the regression model for the bookstore of Exercise 1, what is the value of R^2 and what does it mean?

20. For the disk drive data of Exercise 2 (as corrected in Exercise 12), find and interpret the value of R^2.

SECTION 6.11

21. When analyzing data on the number of employees in small companies in one town, a researcher took square roots of the counts. Some of the resulting values, which are reasonably symmetric were:

$$4, 4, 6, 7, 7, 8, 10$$

What were the original values, and how are they distributed?

22. You wish to explain to your boss what effect taking the base-10 logarithm of the salary values in the company's database will have on the data. As simple, example values, you compare a salary of $10,000 earned by a part-time shipping clerk, a salary of $100,000 earned by a manager, and the CEO's $1,000,000 compensation package. Why might the average of these values be a misleading summary? What would the logarithms of these three values be?

CHAPTER EXERCISES

23. Association. Suppose you were to collect data for each pair of variables. You want to make a scatterplot. Which variable would you use as the explanatory variable and which as the response variable? Why? What would you expect to see in the scatterplot? Discuss the likely direction and form.

a) Cell phone bills: number of text messages, cost.
b) Automobiles: Fuel efficiency (mpg), sales volume (number of autos).
c) For each week: Ice cream cone sales, air conditioner sales.
d) Product: Price ($), demand (number sold per day).

24. Association, part 2. Suppose you were to collect data for each pair of variables. You want to make a scatterplot. Which variable would you use as the explanatory variable and which as the response variable? Why? What would you expect to see in the scatterplot? Discuss the likely direction and form.

a) T- shirts at a store: price each, number sold.
b) Real estate: house price, house size (square footage).
c) Economics: Interest rates, number of mortgage applications.
d) Employees: Salary, years of experience.

25. Scatterplots. Which of the scatterplots show:

a) Little or no association?
b) A negative association?
c) A linear association?
d) A moderately strong association?
e) A very strong association?

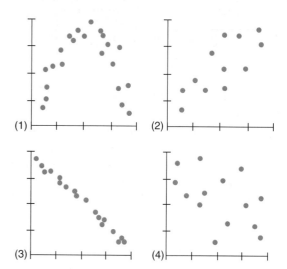

26. Scatterplots, part 2. Which of the scatterplots show:

a) Little or no association?
b) A negative association?
c) A linear association?
d) A moderately strong association?
e) A very strong association?

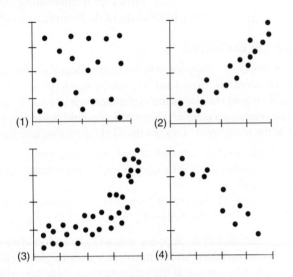

(1)   (2)

(3)   (4)

27. Manufacturing. A ceramics factory can fire eight large batches of pottery a day. Sometimes a few of the pieces break in the process. In order to understand the problem better, the factory records the number of broken pieces in each batch for three days and then creates the scatterplot shown.

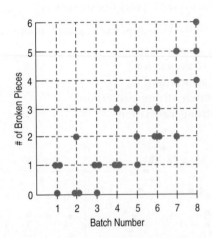

a) Make a histogram showing the distribution of the number of broken pieces in the 24 batches of pottery examined.
b) Describe the distribution as shown in the histogram. What feature of the problem is more apparent in the histogram than in the scatterplot?

c) What aspect of the company's problem is more apparent in the scatterplot?

28. Coffee sales. Owners of a new coffee shop tracked sales for the first 20 days and displayed the data in a scatterplot (by day).

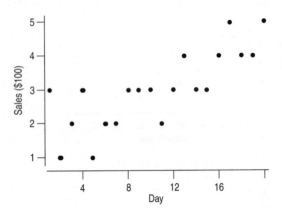

a) Make a histogram of the daily sales since the shop has been in business.
b) State one fact that is obvious from the scatterplot, but not from the histogram.
c) State one fact that is obvious from the histogram, but not from the scatterplot.

29. Matching. Here are several scatterplots. The calculated correlations are −0.923, −0.487, 0.006, and 0.777. Which is which?

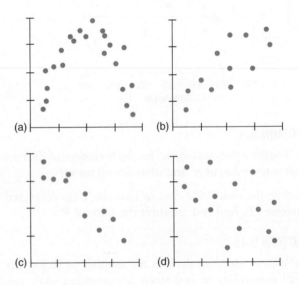

(a)   (b)

(c)   (d)

30. Matching, part 2. Here are several scatterplots. The calculated correlations are −0.977, −0.021, 0.736, and 0.951. Which is which?

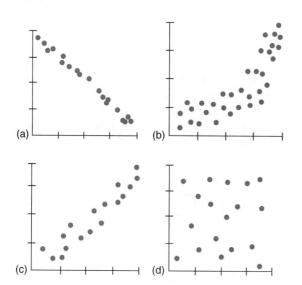

(a) (b) (c) (d)

31. Pizza sales and price. A linear model fit to predict weekly *Sales* of frozen pizza (in pounds) from the average *Price* ($/unit) charged by a sample of stores in the city of Dallas in 39 recent weeks is:

$$\widehat{Sales} = 141{,}865.53 - 24{,}369.49\,Price.$$

a) What is the explanatory variable?
b) What is the response variable?
c) What does the slope mean in this context?
d) What does the *y*-intercept mean in this context? Is it meaningful?
e) What do you predict the sales to be if the average price charged was $3.50 for a pizza?
f) If the sales for a price of $3.50 turned out to be 60,000 pounds, what would the residual be?

32. Used Saab prices. A linear model to predict the *Price* of a 2004 Saab 9-3 (in $) from its *Mileage* (in miles) was fit to 38 cars that were available during the week of January 11, 2008 (Kelly's Blue Book, www.kbb.com). The model was:

$$\widehat{Price} = 24{,}356.15 - 0.0151\,Mileage.$$

a) What is the explanatory variable?
b) What is the response variable?
c) What does the slope mean in this context?
d) What does the *y*-intercept mean in this context? Is it meaningful?
e) What do you predict the price to be for a car with 100,000 miles on it?
f) If the price for a car with 100,000 miles on it was $24,000, what would the residual be?

33. Football salaries. Is there a relationship between total team salary and the performance of teams in the National Football League (NFL)? For the 2006 season, a linear model

predicting *Wins* (out of 16 regular season games) from the total team *Salary* ($M) for the 32 teams in the league is:

$$\widehat{Wins} = 1.783 + 0.062\,Salary.$$

a) What is the explanatory variable?
b) What is the response variable?
c) What does the slope mean in this context?
d) What does the *y*-intercept mean in this context? Is it meaningful?
e) If one team spends $10 million more than another on salary, how many more games on average would you predict them to win?
f) If a team spent $50 million on salaries and won 8 games, would they have done better or worse than predicted?
g) What would the residual of the team in part f be?

34. Baseball salaries. In 2007, the Boston Red Sox won the World Series and spent $143 million on salaries for their players (benfry.com/salaryper). Is there a relationship between salary and team performance in Major League Baseball? For the 2007 season, a linear model fit to the number of *Wins* (out of 162 regular season games) from the team *Salary* ($M) for the 30 teams in the league is:

$$\widehat{Wins} = 70.097 + 0.132\,Salary.$$

a) What is the explanatory variable?
b) What is the response variable?
c) What does the slope mean in this context?
d) What does the *y*-intercept mean in this context? Is it meaningful?
e) If one team spends $10 million more than another on salaries, how many more games on average would you predict them to win?
f) If a team spent $110 million on salaries and won half (81) of their games, would they have done better or worse than predicted?
g) What would the residual of the team in part f be?

35. Pizza sales and price, part 2. For the data in Exercise 31, the average *Sales* was 52,697 pounds (SD = 10,261 pounds), and the correlation between *Price* and *Sales* was = −0.547.

If the *Price* in a particular week was one SD higher than the mean *Price*, how much pizza would you predict was sold that week?

36. Used Saab prices, part 2. The 38 cars in Exercise 32 had an average *Price* of $23,847 (SD = $923), and the correlation between *Price* and *Mileage* was = −0.169.

If the *Mileage* of a 2004 Saab was 1 SD below the average number of miles, what *Price* would you predict for it?

37. Packaging. A CEO announces at the annual shareholders meeting that the new see-through packaging for the company's flagship product has been a success. In fact, he says, "There is a strong correlation between packaging and sales." Criticize this statement on statistical grounds.

38. Insurance. Insurance companies carefully track claims histories so that they can assess risk and set rates appropriately. The National Insurance Crime Bureau reports that Honda Accords, Honda Civics, and Toyota Camrys are the cars most frequently reported stolen, while Ford Tauruses, Pontiac Vibes, and Buick LeSabres are stolen least often. Is it reasonable to say that there's a correlation between the type of car you own and the risk that it will be stolen?

39. Sales by region. A sales manager for a major pharmaceutical company analyzes last year's sales data for her 96 sales representatives, grouping them by region (1 = East Coast U.S.; 2 = Mid West U.S.; 3 = West U.S.; 4 = South U.S.; 5 = Canada; 6 = Rest of World). She plots *Sales* (in $1000) against *Region* (1–6) and sees a strong negative correlation.

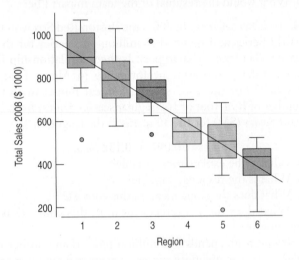

She fits a regression to the data and finds:

$$\widehat{Sales} = 1002.5 - 102.7 \, Region.$$

The R^2 is 70.5%.

Write a few sentences interpreting this model and describing what she can conclude from this analysis.

40. Salary by job type. At a small company, the head of human resources wants to examine salary to prepare annual reviews. He selects 28 employees at random with job types ranging from 01 = Stocking clerk to 99 = President. He plots *Salary* ($) against *Job Type* and finds a strong linear relationship with a correlation of 0.96.

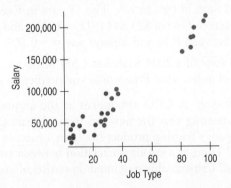

The regression output gives:

$$\widehat{Salary} = 15827.9 + 1939.1 \, Job \, Type$$

Write a few sentences interpreting this model and describing what he can conclude from this analysis.

41. Carbon footprint. The scatterplot shows, for 2008 cars, the carbon footprint (tons of CO_2 per year) vs. the new Environmental Protection Agency (EPA) highway mileage for 82 family sedans as reported by the U.S. government (www.fueleconomy.gov/feg/byclass.htm). The car with the highest highway mpg and lowest carbon footprint is the Toyota Prius.

a) The correlation is −0.947. Describe the association.
b) Are the assumptions and conditions met for computing correlation?
c) Using technology, find the correlation of the data when the Prius is not included with the others. Can you explain why it changes in that way?

42. EPA mpg. In 2008, the EPA revised their methods for estimating the fuel efficiency (mpg) of cars—a factor that plays an increasingly important role in car sales. How do the new highway and city estimated mpg values relate to each other? Here's a scatterplot for 83 family sedans as reported by the U.S. government. These are the same cars as in Exercise 41 except that the Toyota Prius has been removed from the data and two other hybrids, the Nissan Altima and Toyota Camry, are included in the data (and are the cars with highest city mpg).

a) The correlation of these two variables is 0.823. Describe the association.

b) If the two hybrids were removed from the data, would you expect the correlation to increase, decrease, or stay the same? Try it using technology. Report and discuss what you find.

🅣 43. Real estate. Is the number of total rooms in the house associated with the price of a house? Here is the scatterplot of a random sample of homes for sale:

Homes for Sale

a) Is there an association?

b) Check the assumptions and conditions for correlation.

🅣 44. Economic analysis. An economics student is studying the American economy and finds that the correlation between the inflation adjusted Dow Jones Industrial Average and the Gross Domestic Product (GDP) (also inflation adjusted) is 0.77 (www.measuringworth.com). From that he concludes that there is a strong linear relationship between the two series and predicts that a drop in the GDP will make the stock market go down. Here is a scatterplot of the adjusted DJIA against the GDP (in year 2000 $). Describe the relationship and comment on the student's conclusions.

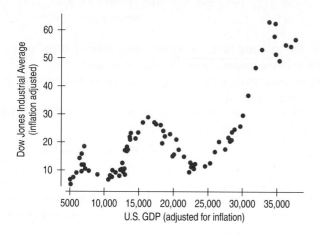

🅣 45. GDP growth. Is economic growth in the developing world related to growth in the industrialized countries? Here's a scatterplot of the growth (in % of Gross Domestic

Product) of the developing countries vs. the growth of developed countries for 180 countries as grouped by the World Bank (www.ers.usda.gov/data/macroeconomics). Each point represents one of the years from 1970 to 2007. The output of a regression analysis follows.

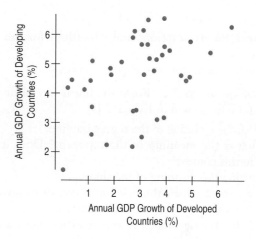

Dependent variable: GDP Growth Developing Countries
$R^2 = 20.81\%$
$s = 1.244$

Variable	Coefficient
Intercept	3.46
GDP Growth Developed Countries	0.433

a) Check the assumptions and conditions for the linear model.

b) Explain the meaning of R^2 in this context.

c) What are the cases in this model?

🅣 46. European GDP growth. Is economic growth in Europe related to growth in the United States? Here's a scatterplot of the average growth in 25 European countries (in % of Gross Domestic Product) vs. the growth in the United States. Each point represents one of years from 1970 to 2007.

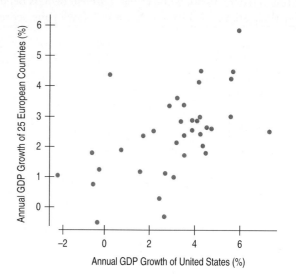

```
Dependent variable: 25 European Countries GDP
Growth
R² = 29.65%
s = 1.156
Variable                Coefficient
Intercept               1.330
U.S. GDP Growth         0.3616
```

a) Check the assumptions and conditions for the linear model.

b) Explain the meaning of R^2 in this context.

47. GDP growth, part 2. From the linear model fit to the data on GDP growth in Exercise 45.

a) Write the equation of the regression line.

b) What is the meaning of the intercept? Does it make sense in this context?

c) Interpret the meaning of the slope.

d) In a year in which the developed countries grow 4%, what do you predict for the developing world?

e) In 2007, the developed countries experienced a 2.65% growth, while the developing countries grew at a rate of 6.09%. Is this more or less than you would have predicted?

f) What is the residual for this year?

48. European GDP growth, part 2. From the linear model fit to the data on GDP growth of Exercise 46.

a) Write the equation of the regression line.

b) What is the meaning of the intercept? Does it make sense in this context?

c) Interpret the meaning of the slope.

d) In a year in which the United States grows at 0%, what do you predict for European growth?

e) In 2007, the United States experienced a 3.20% growth, while Europe grew at a rate of 2.16%. Is this more or less than you would have predicted?

f) What is the residual for this year?

49. Attendance 2006. American League baseball games are played under the designated hitter rule, meaning that weak-hitting pitchers do not come to bat. Baseball owners believe that the designated hitter rule means more runs scored,

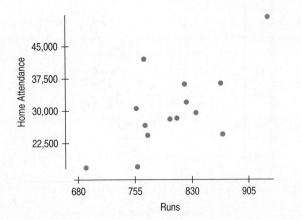

which in turn means higher attendance. Is there evidence that more fans attend games if the teams score more runs? Data collected from American League games during the 2006 season have a correlation of 0.667 between *Runs Scored* and the number of people at the game (www.mlb.com).

a) Does the scatterplot indicate that it's appropriate to calculate a correlation? Explain.

b) Describe the association between attendance and runs scored.

c) Does this association prove that the owners are right that more fans will come to games if the teams score more runs?

50. Attendance 2006, part 2. Perhaps fans are just more interested in teams that win. Here are displays of other variables in the dataset of exercise 49 (espn.go.com). Are the teams that win necessarily those that score the most runs?

	Correlation		
	Wins	**Runs**	**Attend**
Wins	1.000		
Runs	0.605	1.000	
Attend	0.697	0.667	1.000

a) Do winning teams generally enjoy greater attendance at their home games? Describe the association.

b) Is attendance more strongly associated with winning or scoring runs? Explain.

c) How strongly is scoring more runs associated with winning more games?

51. Tuition 2008. All 50 states offer public higher education through four-year colleges and universities and two-year colleges (often called community colleges). Tuition charges by different states vary widely for both types. Would you expect to find a relationship between the tuition states charge for the two types?

a) Using the data on the CD, make a scatterplot of the average tuition for four-year colleges against the tuition charged for two-year colleges. Describe the relationship.

b) Is the direction of the relationship what you expected?

c) What is the regression equation for predicting the tuition at a four-year college from the tuition at a two-year college in the same state?

d) Is a linear model appropriate?

e) How much more do states charge on average in yearly tuition for four-year colleges compared to two-year colleges according to this model?

f) What is the R^2 value for this model? Explain what it says.

52. Tuition 2008, part 2. Exercise 51 examined the relationship between the tuition charged by states for four-year colleges and universities compared to the tuition for two-year colleges. Now, examine the relationship between private and public four-year colleges and universities in the states.

a) Would you expect the relationship between tuition ($ per year) charged by private and public four-year colleges and universities to be as strong as the relationship between public four-year and two-year institutions?

b) Using the data on the CD, examine a scatterplot of the average tuition for four-year private institutions against the tuition charged for four-year public institutions. Describe the relationship.

c) What is the regression equation for predicting the tuition at a four-year private institution from the tuition at a four-year public institution in the same state?

d) Is a linear model appropriate?

e) Interpret the regression equation. How much more is the tuition for four-year private institutions compared to four-year public institutions in the same state according to this model?

f) What is the R^2 value for this model? Explain what it says.

53. Mutual fund flows. As the nature of investing shifted in the 1990s (more day traders and faster flow of information using technology), the relationship between mutual fund monthly performance (*Return*) in percent and money flowing (*Flow*) into mutual funds ($ million) shifted. Using only the values for the 1990s (we'll examine later years in later chapters), answer the following questions. (You may assume that the assumptions and conditions for regression are met.)

The least squares linear regression is:

$$\widehat{Flow} = 9747 + 771 \, Return.$$

a) Interpret the intercept in the linear model.

b) Interpret the slope in the linear model.

c) What is the predicted fund *Flow* for a month that had a market *Return* of 0%?

d) If during this month, the recorded fund *Flow* was $5 billion, what is the residual using this linear model? Did the model provide an underestimate or overestimate for this month?

54. Online clothing purchases. An online clothing retailer examined their transactional database to see if total yearly *Purchases* ($) were related to customers' *Incomes* ($). (You may assume that the assumptions and conditions for regression are met.)

The least squares linear regression is:

$$\widehat{Purchases} = -31.6 + 0.012 \, Income.$$

a) Interpret the intercept in the linear model.

b) Interpret the slope in the linear model.

c) If a customer has an *Income* of $20,000, what is his predicted total yearly *Purchases*?

d) This customer's yearly *Purchases* were actually $100. What is the residual using this linear model? Did the model provide an underestimate or overestimate for this customer?

55. Residual plots. Tell what each of the following residual plots indicates about the appropriateness of the linear model that was fit to the data.

56. Residual plots, again. Tell what each of the following residual plots indicates about the appropriateness of the linear model that was fit to the data.

57. Consumer spending. An analyst at a large credit card bank is looking at the relationship between customers' charges to the bank's card in two successive months. He selects 150 customers at random, regresses charges in *March* ($) on charges in *February* ($), and finds an R^2 of 79%. The intercept is $730.20, and the slope is 0.79. After verifying all the data with the company's CPA, he concludes that the model is a useful one for predicting one month's charges from the other. Examine the data on the CD and comment on his conclusions.

58. Insurance policies. An actuary at a mid-sized insurance company is examining the sales performance of the company's sales force. She has data on the average size of the policy ($) written in two consecutive years by 200 salespeople. She fits a linear model and finds the slope to be 3.00 and the R^2 is 99.92%. She concludes that the predictions for next year's policy size will be very accurate. Examine the data on the CD and comment on her conclusions.

59. What slope? If you create a regression model for predicting the sales ($ million) from money spent on advertising the prior month ($ thousand), is the slope most likely to be 0.03, 300 or 3000? Explain.

60. What slope, part 2? If you create a regression model for estimating a student's business school GPA (on a scale of 1–5) based on his math SAT (on a scale of 200–800), is the slope most likely to be 0.01, 1, or 10? Explain.

61. Misinterpretations. An advertising agent who created a regression model using amount spent on *Advertising* to predict annual *Sales* for a company made these two statements. Assuming the calculations were done correctly, explain what is wrong with each interpretation.

a) My R^2 of 93% shows that this linear model is appropriate.
b) If this company spends $1.5 million on advertising, then annual sales will be $10 million.

62. More misinterpretations. An economist investigated the association between a country's *Literacy Rate* and *Gross Domestic Product (GDP)* and used the association to draw the following conclusions. Explain why each statement is incorrect. (Assume that all the calculations were done properly.)

a) The *Literacy Rate* determines 64% of the *GDP* for a country.
b) The slope of the line shows that an increase of 5% in *Literacy Rate* will produce a $1 billion improvement in *GDP*.

63. Business admissions. An analyst at a business school's admissions office claims to have developed a valid linear model predicting success (measured by starting salary ($) at time of graduation) from a student's undergraduate performance (measured by GPA). Describe how you would check each of the four regression conditions in this context.

64. School rankings. A popular magazine annually publishes rankings of both U.S. business programs and international business programs. The latest issue claims to have developed a linear model predicting the school's ranking (with "1" being the highest ranked school) from its financial resources (as measured by size of the school's endowment). Describe how you would apply each of the four regression conditions in this context.

Ⓣ 65. Used BMW prices. A business student needs cash, so he decides to sell his car. The car is a valuable BMW 840 that was only made over the course of a few years in the late 1990s. He would like to sell it on his own, rather than through a dealer so he'd like to predict the price he'll get for his car's model year.

a) Make a scatterplot for the data on used BMW 840s provided.
b) Describe the association between year and price.
c) Do you think a linear model is appropriate?
d) Computer software says that $R^2 = 57.4\%$. What is the correlation between year and price?
e) Explain the meaning of R^2 in this context.
f) Why doesn't this model explain 100% of the variability in the price of a used BMW 840?

Ⓣ 66. Used BMW prices, part 2. Use the advertised prices for BMW 840s given in Exercise 65 to create a linear model for the relationship between a car's *Year* and its *Price*.

a) Find the equation of the regression line.
b) Explain the meaning of the slope of the line.
c) Explain the meaning of the intercept of the line.
d) If you want to sell a 1997 BMW 840, what price seems appropriate?
e) You have a chance to buy one of two cars. They are about the same age and appear to be in equally good condition. Would you rather buy the one with a positive residual or the one with a negative residual? Explain.

Ⓣ 67. Cost of living index. The *Worldwide Cost of Living Survey City Rankings* determine the cost of living in the most expensive cities in the world as an index. This index scales New York City as 100 and expresses the cost of living in other cities as a percentage of the New York cost. For example, in 2007, the cost of living index in Tokyo was 122.1, which means that it was 22% higher than New York. The scatterplot shows the index for 2007 plotted against the 2006 index for the 15 most expensive cities of 2007.

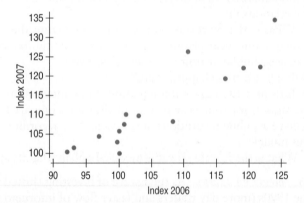

a) Describe the association between cost of living indices in 2007 and 2006.
b) The R^2 for the regression equation is 0.837. Interpret the value of R^2.
c) Using the data provided, find the correlation.
d) Predict the 2007 cost of living of Moscow and find its residual.

Ⓣ 68. Lobster prices. Over the past few decades both the demand for lobster and the price of lobster have continued to increase. The scatterplot shows this increase in the *Price* of Maine lobster (*Price*/pound) since 1990.

a) Describe the increase in the *Price* of lobster since 1990.
b) The R^2 for the regression equation is 88.5%. Interpret the value of R^2.
c) Find the correlation.
d) Find the linear model and examine the plot of residuals versus predicted values. Is the Equal Spread Condition satisfied? (Use time starting at 1990 so that 1990 = 0.)

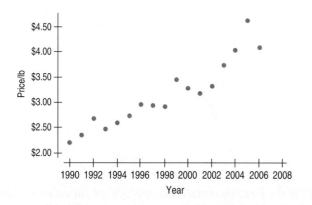

g) CO_2 levels may reach 364 *ppm* in the near future. What *Mean Temperature* does the model predict for that value?

69. El Niño. Concern over the weather associated with El Niño has increased interest in the possibility that the climate on Earth is getting warmer. The most common theory relates an increase in atmospheric levels of carbon dioxide (CO_2), a greenhouse gas, to increases in temperature. Here is a scatterplot showing the mean annual CO_2 concentration in the atmosphere, measured in parts per million (ppm) at the top of Mauna Loa in Hawaii, and the mean annual air temperature over both land and sea across the globe, in degrees Celsius (C).

A regression predicting *Mean Temperature* from CO_2 produces the following output table (in part).

```
Dependent variable: Temperature
R-squared = 33.4%

Variable              Coefficient
Intercept             15.3066
CO2                   0.004
```

a) What is the correlation between CO_2 and *Mean Temperature*?
b) Explain the meaning of *R*-squared in this context.
c) Give the regression equation.
d) What is the meaning of the slope in this equation?
e) What is the meaning of the intercept of this equation?
f) Here is a scatterplot of the residuals vs. CO_2. Does this plot show evidence of the violations of any of the assumptions of the regression model? If so, which ones?

T 70. U.S. birthrates. The table shows the number of live births per 1000 women aged 15–44 years in the United States, starting in 1965. (National Center for Health Statistics, www.cdc.gov/nchs/)

Year	1965	1970	1975	1980	1985
Rate	19.4	18.4	14.8	15.9	15.6

Year	1990	1995	2000	2005
Rate	16.4	14.8	14.4	14.0

a) Make a scatterplot and describe the general trend in *Birthrates*. (Enter *Year* as years since 1900: 65, 70, 75, etc.)
b) Find the equation of the regression line.
c) Check to see if the line is an appropriate model. Explain.
d) Interpret the slope of the line.
e) The table gives rates only at 5-year intervals. Estimate what the rate was in 1978.
f) In 1978, the birthrate was actually 15.0. How close did your model come?
g) Predict what the *Birthrate* will be in 2010. Comment on your faith in this prediction.
h) Predict the *Birthrate* for 2025. Comment on your faith in this prediction.

T 71. Dirt bikes. Off-road motorcycles, commonly called "dirt bikes," are engineered for a particular kind of performance. One measure of the power of the engine is its displacement, measured in cubic centimeters. The first scatterplot shows the relationship between *Displacement* and *Total Weight* for a selection of 4-stroke off-road bikes. The other scatterplot plots $\sqrt{Displacement}$ instead.

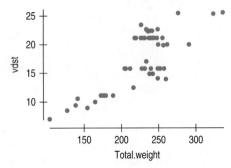

a) What statistic would be appropriate to summarize the strength of the association between *Displacement* and *Total Weight*? Explain.

b) Which form would be the best choice if you wanted to fit a linear model to the relationship? Explain.

T 72. Lobsters. The Maine lobster fishery is closely monitored, and statistics about the lobster business are published annually. Here are plots relating the total value ($M) of lobsters harvested between 1950 and 2006 to the price of lobster ($ per pound) and the Log*Value* offered as an alternative transformation.

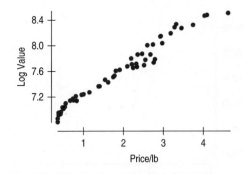

a) Is the Pearson correlation appropriate for either of these relationships? If not, what would you use? Explain.

b) Would it be appropriate to fit a linear model by least squares to either of these relationships? Explain.

Just Checking Answers

1   We know the scores are quantitative. We should check to see if the *Linearity Condition* and the *Outlier Condition* are satisfied by looking at a scatterplot of the two scores.

2   It won't change.

3   It won't change.

4   They are more likely to do poorly. The positive correlation means that low closing prices for Intel are associated with low closing prices for Cypress.

5   No, the general association is positive, but daily closing prices may vary.

6   For each additional employee, monthly sales increase, on average, $122,740.

7   Thousands of $ per employee.

8    $1,227,400 per month.

9   Differences in the number of employees account for about 71.4% of the variation in the monthly sales.

10   It's positive. The correlation and the slope have the same sign.

11   R^2, No. Slope, Yes.

CASE Study

Paralyzed Veterans of America

Philanthropic organizations often rely on contributions from individuals to finance the work that they do, and a national veterans' organization is no exception. The Paralyzed Veterans of America (PVA) was founded as a congressionally chartered veterans' service organization more than 60 years ago. It provides a range of services to veterans who have experienced spinal cord injury or dysfunction. Some of the services offered include medical care, research, education, and accessibility and legal consulting. In 2008, this organization had total revenue of more than $135 million, with more than 99% of this revenue coming from contributions.

An organization that depends so heavily on contributions needs a multifaceted fundraising program, and PVA solicits donations in a number of ways. From its website (www.pva.org), people can make a one-time donation, donate monthly, donate in honor or in memory of someone, and shop in the PVA online store. People can also support one of the charity events, such as its golf tournament, National Veterans Wheelchair Games, and Charity Ride.

Traditionally, one of PVA's main methods of soliciting funds was the use of return address labels and greeting cards (although still used, this method has declined in recent years). Typically, these gifts were sent to potential donors about every six weeks with a request for a contribution. From its established donors, PVA could expect a response rate of about 5%, which, given the relatively small cost to produce and send the gifts, kept the organization well funded.

But fundraising accounts for 28% of expenses, so PVA wanted to know who its donors are, what variables might be useful in predicting whether a donor is likely to give to an upcoming campaign, and what the size of that gift might be. On your DVD is a dataset **Case Study 1**, which includes data designed to be very similar to part of the data that this organization works with. Here is a description of some of the variables. Keep in mind, however, that in the real data set, there would be hundreds more variables given for each donor.

VARIABLE NAME	UNITS (IF APPLICABLE)	DESCRIPTION	REMARKS
Age	Years		
Own Home?	H = Yes; U = No or unknown		
Children	Counts		
Income		1 = Lowest ; 7 = Highest	Based on national medians and percentiles
Sex	M = Male; F = Female		
Total Wealth		1 = Lowest; 9 = Highest	Based on national medians and percentiles
Gifts to Other Orgs	Counts	Number of Gifts (if known) to other philanthropic organizations in the same time period	
Number of Gifts	Counts	Number of Gifts to this organization in this time period	
Time Between Gifts	Months	Time between first and second gifts	
Smallest Gift	$	Smallest Gift (in $) in the time period	See also Sqrt(Smallest Gift)
Largest Gift	$	Largest Gift (in $) in the time period	See also Sqrt(Largest Gift)
Previous Gift	$	Gift (in $) for previous campaign	See also Sqrt(Previous Gift)
Average Gift	$	Total amount donated divided by total number of gifts	See also Sqrt(Average Gift)
Current Gift	$	Gift (in $) to organization this campaign	See also Sqrt(Current Gift)
Sqrt(Smallest Gift)	Sqrt($)	Square Root of Smallest Gift in $	
Sqrt(Largest Gift)	Sqrt($)	Square Root of Largest Gift in $	
Sqrt(Previous Gift)	Sqrt($)	Square Root of Previous Gift in $	
Sqrt(Average Gift)	Sqrt($)	Square Root of Average Gift in $	
Sqrt(Current Gift)	Sqrt($)	Square Root of Current Gift in $	

Let's see what the data can tell us. Are there any interesting relationships between the current gift and other variables? Is it possible to use the data to predict who is going to respond to the next direct-mail campaign?

> Recall that when variables are highly skewed or the relationship between variables is not linear, reporting a correlation coefficient is not appropriate. You may want to consider a transformed version of those variables (square roots are provided for all the variables concerning gifts) or a correlation based on the ranks of the values rather than the values themselves.

Suggested Study Plan and Questions

Write a report of what you discover about the donors to this organization. Be sure to follow the Plan, Do, Report outline for your report. Include a basic description of each variable (shape, center, and spread), point out any interesting features, and explore the relationships between the variables. In particular you should describe any interesting relationships between the current gift and other variables. Use these questions as a guide:

- Is the age distribution of the clients a typical one found in most businesses?
- Do people who give more often make smaller gifts on average?
- Do people who give to other organizations tend to give to this organization?

Describe the relationship between the Income and Wealth rankings. How do you explain this relationship (or lack of one)? (*Hint*: Look at the age distribution.)

What variables (if any) seem to have an association with the Current Gift? Do you think the organization can use any of these variables to predict the gift for the next campaign?

*This file includes people who did not give to the current campaign. Do your answers to any of the questions above change if you consider only those who gave to this campaign?

Randomness and Probability

Credit Reports and the Fair Isaacs Corporation

You've probably never heard of the Fair Isaacs Corporation, but they probably know you. Whenever you apply for a loan, a credit card, or even a job, your credit "score" will be used to determine whether you are a good risk. And because the most widely used credit scores are Fair Isaacs' FICO® scores, the company may well be involved in the decision. The Fair Isaacs Corporation (FICO) was founded in 1956, with the idea that data, used intelligently, could improve business decision-making. Today, Fair Isaacs claims that their services provide companies around the world with information for more than 180 billion business decisions a year.

Your credit score is a number between 350 and 850 that summarizes your credit "worthiness." It's a snapshot of credit risk today based on your credit history and past behavior. Lenders of all kinds use credit scores to predict behavior, such as how likely you are to make your loan payments on time or to default on a loan. Lenders use the score to determine not only whether to give credit, but also the cost of the credit that they'll offer. There are no established boundaries, but generally scores over 750 are considered excellent, and applicants with those scores get the best rates. An applicant with a score below 620 is generally considered to be a poor risk. Those with very low scores may be denied credit outright or only offered "subprime" loans at substantially higher rates.

It's important that you be able to verify the information that your score is based on, but until recently, you could only hope that your score was based on correct information. That changed in 2000, when a California law gave mortgage applicants the right to see their credit scores. Today, the credit industry is more open about giving consumers access to their scores and the U.S. government, through the Fair and Accurate Credit Transaction Act (FACTA), now guarantees that you can access your credit report at no cost, at least once a year.[1]

Companies have to manage risk to survive, but by its nature, risk carries uncertainty. A bank can't know for certain that you'll pay your mortgage on time—or at all. What can they do with events they can't predict? They start with the fact that, although individual outcomes cannot be anticipated with certainty, random phenomena do, in the long run, settle into patterns that are consistent and predictable. It's this property of random events that makes Statistics practical.

7.1   Random Phenomena and Probability

When a customer calls the 800 number of a credit card company, he or she is asked for a card number before being connected with an operator. As the connection is made, the purchase records of that card and the demographic information of the customer are retrieved and displayed on the operator's screen. If the customer's FICO score is high enough, the operator may be prompted to "cross-sell" another service—perhaps a new "platinum" card for customers with a credit score of at least 750.

Of course, the company doesn't know which customers are going to call. Call arrivals are an example of a random phenomenon. With **random phenomena**, we can't predict the individual outcomes, but we can hope to understand characteristics of their long-run behavior. We don't know whether the *next* caller will qualify for the platinum card, but as calls come into the call center, the company will find that the percentage of platinum-qualified callers who qualify for cross-selling will settle into a pattern, like that shown in the graph in Figure 7.1.

As calls come into the call center, the company might record whether each caller qualifies. The first caller today qualified. Then the next five callers' qualifications were no, yes, yes, no, and no. If we plot the percentage who qualify against the call number, the graph would start at 100% because the first caller qualified (1 out of 1, for 100%). The next caller didn't qualify, so the accumulated percentage dropped to 50% (1 out of 2). The third caller qualified (2 out of 3, or 67%), then yes again (3 out of 4, or 75%), then no twice in a row (3 out of 5, for 60%, and then 3 out of 6, for 50%), and so on (Table 7.1). With each new call, the new datum is a smaller fraction of the accumulated experience, so, in the long run, the graph settles down. As it settles down, it appears that, in fact, the fraction of customers who qualify is about 35%.

When talking about long-run behavior, it helps to define our terms. For any random phenomenon, each attempt, or **trial**, generates an **outcome**. For the call center, each call is a trial. Something happens on each trial, and we call whatever happens the outcome. Here the outcome is whether the caller qualifies or not.

[1]However, the score you see in your report will be an "educational" score intended to show consumers how scoring works. You still have to pay a "reasonable fee" to see your FICO score.

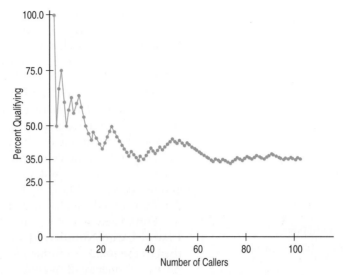

Figure 7.1   The percentage of credit card customers who qualify for the premium card.

Call	FICO Score	Qualify?	% Qualify
1	750	Yes	100
2	640	No	50
3	765	Yes	66.7
4	780	Yes	75
5	680	No	60
6	630	No	50
⋮	⋮		⋮

Table 7.1   Data on the first six callers showing their FICO score, whether they qualified for the platinum card offer, and a running percentage of number of callers who qualified.

> A **phenomenon** consists of **trials**. Each trial has an **outcome**. Outcomes combine to make **events**.

We use the more general term **event** to refer to outcomes or combinations of outcomes. For example, suppose we categorize callers into 6 risk categories and number these outcomes from 1 to 6 (of increasing credit worthiness). The three outcomes 4, 5, or 6 could make up the event "caller is at least a category 4."

We sometimes talk about the collection of *all possible outcomes*, a special event that we'll refer to as the **sample space**. We denote the sample space **S**; you may also see the Greek letter Ω used. But whatever symbol we use, the sample space is the set that contains all the possible outcomes. For the calls, if we let Q = qualified and N = not qualified, the sample space is simple: **S** = {Q, N}. If we look at two calls together, the sample space has four outcomes: **S** = {QQ, QN, NQ, NN}. If we were interested in at least one qualified caller from the two calls, we would be interested in the event (call it **A**) consisting of the three outcomes QQ, QN, and NQ, and we'd write **A** = {QQ, QN, NQ}.

Although we may not be able to predict a *particular* individual outcome, such as which incoming call represents a potential upgrade sale, we can say a lot about the long-run behavior. Look back at Figure 7.1. If you were asked for the probability that a random caller will qualify, you might say that it was 35% because, in the *long run*, the percentage of the callers who qualify is about 35%. And, that's exactly what we mean by **probability**.

> The **probability** of an event is its long-run relative frequency. A relative frequency is a fraction, so we can write it as $\frac{35}{100}$, as a decimal, 0.35, or as a percentage, 35%.

That seems simple enough, but do random phenomena always behave this well? Couldn't it happen that the frequency of qualified callers never settles down, but just bounces back and forth between two numbers? Maybe it hovers around 45% for awhile, then goes down to 25%, and then back and forth forever. When we think about what happens with a series of trials, it really simplifies things if the individual trials are independent. Roughly speaking, **independence** means that the outcome of one trial doesn't influence or change the outcome of another. Recall, that in Chapter 4, we called two variables *independent* if the value of one categorical variable did not influence the value of another categorical variable. (We checked for independence by comparing relative frequency distributions across variables.) There's no reason to think that whether the one caller qualifies influences whether another caller qualifies, so these are independent trials. We'll see a more formal definition of independence later in the chapter.

Fortunately, for independent events, we can depend on a principle called the **Law of Large Numbers (LLN)**, which states that if the events are independent, then as the number of calls increases, over days or months or years, the long-run relative frequency of qualified calls gets closer and closer to a single value. This gives us the guarantee we need and makes probability a useful concept.

Because the LLN guarantees that relative frequencies settle down in the long run, we know that the value we called the probability is legitimate and the number it settles down to is called the probability of that event. For the call center, we can write $P(\text{qualified}) = 0.35$. Because it is based on repeatedly observing the event's outcome, this definition of probability is often called **empirical probability**.

> **Law of Large Numbers**
>
> The *long-run relative frequency* of repeated, independent events eventually produces the *true relative frequency* as the number of trials increases.

7.2 The Nonexistent Law of Averages

> *"Slump? I ain't in no slump. I just ain't hittin'."*
>
> —YOGI BERRA

You may think it's obvious that the frequency of repeated events settles down in the long run to a single number. The discoverer of the Law of Large Numbers thought so, too. The way he put it was: *"For even the most stupid of men is convinced that the more observations have been made, the less danger there is of wandering from one's goal."*

—JACOB BERNOULLI, 1713

The Law of Large Numbers says that the relative frequency of a random event settles down to a single number in the long run. But, it is often misunderstood to be a "law of averages," perhaps because the concept of "long run" is hard to grasp. Many people believe, for example, that an outcome of a random event that hasn't occurred in many trials is "due" to occur. The original "dogs of the Dow" strategy for buying stocks recommended buying the 10 worst performing stocks of the 30 that make up the Dow Jones Industrial Average, figuring that these "dogs" were bound to do better next year. After all, we know that in the long run, the relative frequency will settle down to the probability of that outcome, so now we have some "catching up" to do, right? Wrong. In fact, Louis Rukeyser (the former host of *Wall Street Week*) said of the "dogs of the Dow" strategy, "that theory didn't work as promised."

Actually, we know very little about the behavior of random events in the short run. The fact that we are seeing independent random events makes each individual result impossible to predict. Relative frequencies even out *only* in the long run. And, according to the LLN, the long run is really long (infinitely long, in fact). The "Large" in the law's name means *infinitely* large. Sequences of random events don't compensate in the short run and don't need to do so to get back to the right long-run probability. Any short-run deviations will be overwhelmed in the long run. If the probability of an outcome doesn't change and the events are independent, the probability of any outcome in another trial is always what it was, no matter what has happened in other trials.

Many people confuse the Law of Large numbers with the so-called Law of Averages that would say that things have to even out in the short run. But even though the Law of Averages doesn't exist at all, you'll hear people talk about it as if it does. Is a good hitter in baseball who has struck out the last six times due for a hit his next time up? If the stock market has been down for the last three sessions, is it due to increase today? No. This isn't the way random phenomena work. There is no Law of Averages for short runs—no "Law of Small Numbers." A belief in such a "law" can lead to poor business decisions.

Keno and the Law of Averages

Of course, sometimes an apparent drift from what we expect means that the probabilities are, in fact, *not* what we thought. If you get 10 heads in a row, maybe the coin has heads on both sides!

Keno is a simple casino game in which numbers from 1 to 80 are chosen. The numbers, as in most lottery games, are supposed to be equally likely. Payoffs are made depending on how many of those numbers you match on your card. A group of graduate students from a Statistics department decided to take a field trip to Reno. They (*very* discreetly) wrote down the outcomes of the games for a couple of days, then drove back to test whether the numbers were, in fact, equally likely. It turned out that some numbers were *more likely* to come up than others. Rather than bet on the Law of Averages and put their money on the numbers that were "due," the students put their faith in the LLN—and all their (and their friends') money on the numbers that had come up before. After they pocketed more than $50,000, they were escorted off the premises and invited never to show their faces in that casino again. Not coincidentally, the ringleader of that group currently makes his living on Wall Street.

"In addition, in time, if the roulette-betting fool keeps playing the game, the bad histories [outcomes] will tend to catch up with him."

—NASSIM NICHOLAS TALEB IN *FOOLED BY RANDOMNESS*

You've just flipped a fair coin and seen six heads in a row. Does the coin "owe" you some tails? Suppose you spend that coin and your friend gets it in change. When she starts flipping the coin, should we expect a run of tails? Of course not. Each flip is a new event. The coin can't "remember" what it did in the past, so it can't "owe" any particular outcomes in the future. Just to see how this works in practice, we simulated 100,000 flips of a fair coin on a computer. In our 100,000 "flips," there were 2981 streaks of at least 5 heads. The "Law of Averages" suggests that the next flip after a run of 5 heads should be tails more often to even things out. Actually, the next flip was heads more often than tails: 1550 times to 1431 times. That's 51.9% heads. You can perform a similar simulation easily.

Just Checking

1 It has been shown that the stock market fluctuates randomly. Nevertheless, some investors believe that they should buy right after a day when the market goes down because it is bound to go up soon. Explain why this is faulty reasoning.

7.3 Different Types of Probability

Model-Based (Theoretical) Probability

We've discussed *empirical probability*—the relative frequency of an event's occurrence as the probability of an event. There are other ways to define probability as well. Probability was first studied extensively by a group of French mathematicians who were interested in games of chance. Rather than experiment with the games and risk losing their money, they developed mathematical models of probability. To make things simple (as we usually do when we build models), they started by looking at games in which the different outcomes were equally likely. Fortunately, many games of chance are like that. Any of 52 cards is equally likely to be the next one dealt from a well-shuffled deck. Each face of a die is equally likely to land up (or at least it should be).

We can write:

$$P(A) = \frac{\text{\# of outcomes in } A}{\text{total \# of outcomes}}$$

and call this the **(theoretical) probability** of the event.

When outcomes are equally likely, their probability is easy to compute—it's just 1 divided by the number of possible outcomes. So the probability of rolling a 3 with a fair die is one in six, which we write as 1/6. The probability of picking the ace of spades from the top of a well-shuffled deck is 1/52.

It's almost as simple to find probabilities for events that are made up of several equally likely outcomes. We just count all the outcomes that the event contains. The probability of the event is the number of outcomes in the event divided by the total number of possible outcomes.

For example, Pew Research[2] reports that of 10,190 randomly generated working phone numbers called for a survey, the initial results of the calls were as follows:

> **Equally Likely?**
>
> In an attempt to understand why, an interviewer asked someone who had just purchased a lottery ticket, "What do you think your chances are of winning the lottery?" The reply was, "Oh, about 50–50." The shocked interviewer asked, "How do you get that?" to which the response was, "Well, the way I figure it, either I win or I don't!" The moral of this story is that events are not always equally likely.

Result	Number of Calls
No Answer	311
Busy	61
Answering Machine	1336
Callbacks	189
Other Non-Contacts	893
Contacted Numbers	7400

The phone numbers were generated randomly, so each was equally likely. To find the probability of a contact, we just divide the number of contacts by the number of calls: 7400/10,190 = 0.7262.

But don't get trapped into thinking that random events are always equally likely. The chance of winning a lottery—especially lotteries with very large payoffs—is small. Regardless, people continue to buy tickets.

Personal Probability

What's the probability that gold will sell for more than $1000 an ounce at the end of next year? You may be able to come up with a number that seems reasonable. Of course, no matter how confident you feel about your prediction, your probability should be between 0 and 1. How did you come up with this probability? In our discussion of probability, we've defined probability in two ways: 1) in terms of the relative frequency—or the fraction of times—that an event occurs in the long run or 2) as the number of outcomes in the event divided by the total number of outcomes. Neither situation applies to your assessment of gold's chances of selling for more than $1000.

We use the *language* of probability in everyday speech to express a degree of uncertainty without basing it on long-run relative frequencies. Your personal assessment of an event expresses your uncertainty about the outcome. That uncertainty may be based on your knowledge of commodities markets, but it can't be based on long-run behavior. We call this kind of probability a subjective, or **personal probability**.

Although personal probabilities may be based on experience, they are not based either on long-run relative frequencies or on equally likely events. Like the two other probabilities we defined, they need to satisfy the same rules as both empirical and theoretical probabilities that we'll discuss in the next section.

---

[2]www.pewinternet.org/pdfs/PIP_Digital_Footprints.pdf.

7.4   Probability Rules

For some people, the phrase "50/50" means something vague like "I don't know" or "whatever." But when we discuss probabilities, 50/50 has the precise meaning that two outcomes are *equally likely*. Speaking vaguely about probabilities can get you into trouble, so it's wise to develop some formal rules about how probability works. These rules apply to probability whether we're dealing with empirical, theoretical, or personal probability.

"Baseball is 90% mental. The other half is physical."

—YOGI BERRA

Rule 1. If the probability of an event occurring is 0, the event can't occur; likewise if the probability is 1, the event *always* occurs. Even if you think an event is very unlikely, its probability can't be negative, and even if you're sure it will happen, its probability can't be greater than 1. So we require that:

A probability is a number between 0 and 1.

For any event A, $0 \leq P(A) \leq 1$.

Rule 2. If a random phenomenon has only one possible outcome, it's not very interesting (or very random). So we need to distribute the probabilities among all the outcomes a trial can have. How can we do that so that it makes sense? For example, consider the behavior of a certain stock. The possible daily outcomes might be:

A: The stock price goes up.
B: The stock price goes down.
C: The stock price remains the same.

When we assign probabilities to these outcomes, we should be sure to distribute all of the available probability. Something always occurs, so the probability of *something* happening is 1. This is called the **Probability Assignment Rule**:

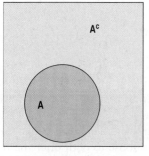

The set **A** and its complement **A**^C. Together, they make up the entire sample space **S**.

The probability of the set of all possible outcomes must be 1.

$$P(S) = 1$$

where **S** represents the set of all possible outcomes and is called the **sample space**.

Rule 3. Suppose the probability that you get to class on time is 0.8. What's the probability that you don't get to class on time? Yes, it's 0.2. The set of outcomes that are *not* in the event **A** is called the "complement" of **A**, and is denoted **A**^C. This leads to the **Complement Rule**:

The probability of an event occurring is 1 minus the probability that it doesn't occur.

$$P(A) = 1 - P(A^C)$$

For Example   Applying the complement rule

Lee's Lights sells lighting fixtures. Some customers are there only to browse, so Lee records the behavior of all customers for a week to assess how likely it is that a customer will make a purchase. Lee finds that of 1000 customers entering the store during the week, 300 make purchases. Lee concludes that the probability of a customer making a purchase is 0.30.

Question: If *P*(purchase) = 0.30, what is the probability that a customer *doesn't* make a purchase?

Answer: Because "no purchase" is the complement of "purchase,"

$$P(\text{no purchase}) = 1 - P(\text{purchase})$$
$$= 1 - 0.30 = 0.70$$

There is a 70% chance a customer won't make a purchase.

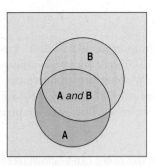

Two sets **A** and **B** that are not disjoint. The event (**A** and **B**) is their intersection.

Rule 4. Whether or not a caller qualifies for a platinum card is a random outcome. Suppose the probability of qualifying is 0.35. What's the chance that the next two callers qualify? The **Multiplication Rule** says that to find the probability that two independent events occur, we multiply the probabilities:

> For two independent events A and B, the probability that both
> A *and* B occur is the product of the probabilities of the two events.
> $P(\text{A and B}) = P(\text{A}) \times P(\text{B})$, provided that A and B are independent.

Thus if **A** = {customer 1 qualifies} and **B** = {customer 2 qualifies}, the chance that both qualify is:

$$0.35 \times 0.35 = 0.1225$$

Of course, to calculate this probability, we have used the assumption that the two events are independent. We'll expand the multiplication rule to be more general later in this chapter.

For Example   Using the multiplication rule

Lee knows that the probability that a customer will make a purchase is 30%.

Question: If we can assume that customers behave independently, what is the probability that the next two customers entering Lee's Lights make purchases?

Answer: Because the events are independent, we can use the multiplication rule.

$P(\text{first customer makes a purchase } and \text{ second customer makes a purchase})$

$= P(\text{purchase}) \times P(\text{purchase})$

$= 0.30 \times 0.30 = 0.09$

There's about a 9% chance that the next two customers will both make purchases.

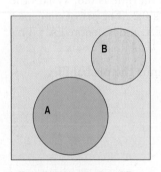

Two disjoint sets, **A** and **B**.

Rule 5. Suppose the card center operator has more options. She can **A**: offer a special travel deal, **B**: offer a platinum card, or **C**: decide to send information about a new affinity card. If she can do one, but only one, of these, then these outcomes are **disjoint** (or **mutually exclusive**). To see whether two events are disjoint, we separate them into their component outcomes and check whether they have any outcomes in common. For example, if the operator can choose to both offer the travel deal and send the affinity card information, those would not be disjoint. The **Addition Rule** allows us to add the probabilities of disjoint events to get the probability that *either* event occurs:

$$P(\text{A or B}) = P(\text{A}) + P(\text{B}).$$

Thus the probability that the caller *either* is offered a platinum card *or* is sent the affinity card information is the sum of the two probabilities, since the events are disjoint.

For Example   Using the addition rule

Some customers prefer to see the merchandise but then make their purchase later using Lee's Lights' new Internet site. Tracking customer behavior, Lee determines that there's an 8% chance of a customer making a purchase in this way. We know that about 30% of customers make purchases when they enter the store.

Question: What is the probability that a customer who enters the store makes no purchase at all?

Answer: We can use the Addition Rule because the alternatives "no purchase," "purchase in the store," and "purchase online" are disjoint events.

$$P(\text{purchase in the store } or \text{ online}) = P(\text{purchase in store}) + P(\text{purchase online})$$

$$= 0.30 + 0.09 = 0.39$$

$$P(\text{no purchase}) = P(\text{not } P(\text{purchase in the store } or \text{ purchase online}))$$

$$= 1 - P(\text{in store } or \text{ online})$$

$$= 1 - 0.39 = 0.61$$

Notation Alert! ─────────

You may see the event (**A** or **B**) written as (**A** ∪ **B**). The symbol ∪ means "union" and represents the outcomes in event **A** or event **B**. Similarly the symbol ∩ means intersection and represents outcomes that are in *both* event **A** and event **B**. You may see the event (**A** and **B**) written as (**A** ∩ **B**).

───────────────────

Rule 6. Suppose we would like to know the probability that either of the next two callers is qualified for a platinum card? We know $P(\mathbf{A}) = P(\mathbf{B}) = 0.35$, but $P(\mathbf{A} \text{ or } \mathbf{B})$ is not simply the sum $P(\mathbf{A}) + P(\mathbf{B})$ because the events **A** and **B** are not disjoint in this case. Both customers could qualify. So we need a new probability rule.

We can't simply add the probabilities of **A** and **B** because that would count the outcome of *both* customers qualifying twice. So, if we started by adding the two probabilities, we could compensate by subtracting out the probability of that outcome. In other words,

$P(\text{customer A } or \text{ customer B qualifies}) =$

$P(\text{customer A qualifies}) + P(\text{customer B qualifies}) - P(\text{both customers qualify})$

$= (0.35) + (0.35) - (0.35 \times 0.35)$ (since events are independent)

$= (0.35) + (0.35) - (0.1225)$

$= 0.5775$

It turns out that this method works in general. We add the probabilities of two events and then subtract out the probability of their intersection. This gives us the **General Addition Rule**, which does not require disjoint events:

$$P(\mathbf{A} \text{ or } \mathbf{B}) = P(\mathbf{A}) + P(\mathbf{B}) - P(\mathbf{A} \text{ and } \mathbf{B})$$

For Example   *Using the general addition rule*

Lee notices that when two customers enter the store together, their behavior isn't independent. In fact, there's a 20% chance they'll *both* make a purchase.

Question: When two customers enter the store together, what is the probability that *at least one* of them will make a purchase?

Answer: Now we know that the events are not independent, so we must use the General Addition Rule

$$P(\text{Both purchase}) = P(\text{A purchases } or \text{ B purchases})$$

$$= P(\text{A purchases}) + P(\text{B purchases}) - P(\text{A and B both purchase})$$

$$= 0.30 + 0.30 - 0.20 = 0.40$$

Just Checking

2 MP3 players have relatively high failure rates for a consumer product, especially those models that contain a disk drive as opposed to those that have less storage but no drive. The worst failure rate for all iPod models was the 40GB Click wheel (as reported by MacIntouch.com) at 30%. If a store sells this model and failures are independent,

  a) What is the probability that the next one they sell will have a failure?

  b) What is the probability that there will be failures on *both* of the next two?

  c) What is the probability that the store's first failure problem will be with the third one they sell?

  d) What is the probability the store will have a failure problem with at least one of the next five that they sell?

Guided Example   M&M's Modern Market Research

In 1941, when M&M's® milk chocolate candies were introduced to American GIs in World War II, there were six colors: brown, yellow, orange, red, green, and violet. Mars®, the company that manufactures M&M's, has used the introduction of a new color as a marketing and advertising event several times in the years since then. In 1980, the candy went international adding 16 countries to their markets. In 1995, the company conducted a "worldwide survey" to vote on a new color. Over 10 million people voted to add blue. They even got the lights of the Empire State Building in New York City to glow blue to help announce the addition. In 2002, they used the Internet to help pick a new color. Children from over 200 countries were invited to respond via the Internet, telephone, or mail. Millions of voters chose among purple, pink, and teal. The global winner was purple, and for a brief time, purple M&M's could be found in packages worldwide (although in 2010, the colors were brown, yellow, red, blue, orange, and green). In the United States, 42% of those who voted said purple, 37% said teal, and only 19% said pink. But in Japan the percentages were 38% pink, 36% teal, and only 16% purple. Let's use Japan's percentages to ask some questions.

1. What's the probability that a Japanese M&M's survey respondent selected at random preferred either pink or teal?

2. If we pick two respondents at random, what's the probability that they *both* selected purple?

3. If we pick three respondents at random, what's the probability that *at least one* preferred purple?

PLAN	Setup The probability of an event is its long-term relative frequency. This can be determined in several ways: by looking at many replications of an event, by deducing it from equally likely events, or by using some other information. Here, we are told the relative frequencies of the three responses.	The M&M's website reports the proportions of Japanese votes by color. These give the probability of selecting a voter who preferred each of the colors: $P(pink) = 0.38$ $P(teal) = 0.36$ $P(purple) = 0.16$
	Make sure the probabilities are legitimate. Here, they're not. Either there was a mistake or the other voters must have chosen a color other than the three given. A check of other countries shows a similar deficit, so probably we're seeing those who had no preference or who wrote in another color.	Each is between 0 and 1, but these don't add up to 1. The remaining 10% of the voters must have not expressed a preference or written in another color. We'll put them together into "other" and add $P(other) = 0.10$. With this addition, we have a legitimate assignment of probabilities.

Question 1. What's the probability that a Japanese M&M's survey respondent selected at random preferred either pink or teal?

PLAN	Setup Decide which rules to use and check the conditions they require.	The events "pink" and "teal" are individual outcomes (a respondent can't choose both colors), so they are disjoint. We can apply the General Addition Rule anyway.

DO	Mechanics Show your work.	$P(\text{pink or teal}) = P(\text{pink}) + P(\text{teal})$ $\qquad - P(\text{pink and teal})$ $= 0.38 + 0.36 - 0 = 0.74$ The probability that both pink and teal were chosen is zero, since respondents were limited to one choice.
REPORT	Conclusion Interpret your results in the proper context.	The probability that the respondent said pink or teal is 0.74.

Question 2. If we pick two respondents at random, what's the probability that they both said purple?

PLAN	Setup The word "both" suggests we want $P(\mathbf{A} \text{ and } \mathbf{B})$, which calls for the Multiplication Rule. Check the required condition.	**Independence** It's unlikely that the choice made by one respondent affected the choice of the other, so the events seem to be independent. We can use the Multiplication Rule.
DO	Mechanics Show your work. For both respondents to pick purple, each one has to pick purple.	$P(\text{both purple})$ $= P(\text{first respondent picks purple and second respondent picks purple})$ $= P(\text{first respondent picks purple})$ $\times P(\text{second respondent picks purple})$ $= 0.16 \times 0.16 = 0.0256$
REPORT	Conclusion Interpret your results in the proper context.	The probability that both respondents pick purple is 0.0256.

Question 3. If we pick three respondents at random, what's the probability that at least one preferred purple?

PLAN	Setup The phrase "at least one" often flags a question best answered by looking at the complement, and that's the best approach here. The complement of "at least one preferred purple" is "none of them preferred purple." Check the conditions.	$P(\text{at least one picked purple})$ $= P(\{\text{none picked purple}\}^c)$ $= 1 - P(\text{none picked purple})$. **Independence.** These are independent events because they are choices by three random respondents. We can use the Multiplication Rule.
DO	Mechanics We calculate $P(\text{none purple})$ by using the Multiplication Rule.	$P(\text{none picked purple}) = P(\text{first not purple})$ $\times P(\text{second not purple})$ $\times P(\text{third not purple})$ $= [P(\text{not purple})]^3$. $P(\text{not purple}) = 1 - P(\text{purple})$ $= 1 - 0.16 = 0.84$.

(continued)

Then we can use the Complement Rule to get the probability we want.	So P(none picked purple) $= (0.84)^3 = 0.5927$. P(at least 1 picked purple) $= 1 - P$(none picked purple) $= 1 - 0.5927 = 0.4073$.
REPORT **Conclusion** Interpret your results in the proper context.	There's about a 40.7% chance that at least one of the respondents picked purple.

7.5   Joint Probability and Contingency Tables

As part of a Pick Your Prize Promotion, a chain store invited customers to choose which of three prizes they'd like to win (while providing name, address, phone number, and e-mail address). At one store, the responses could be placed in the contingency table in Table 7.2.

		Prize preference			
		MP3	**Camera**	**Bike**	**Total**
Sex	**Man**	117	50	60	**227**
	Woman	130	91	30	**251**
	Total	**247**	**141**	**90**	**478**

| Table 7.2   Prize preference for 478 customers.

> A **marginal probability** uses a marginal frequency (from either the Total row or Total column) to compute the probability.

If the winner is chosen at random from these customers, the probability we select a woman is just the corresponding relative frequency (since we're equally likely to select any of the 478 customers). There are 251 women in the data out of a total of 478, giving a probability of:

$$P(\text{woman}) = 251/478 = 0.525$$

This is called a **marginal probability** because it depends only on totals found in the margins of the table. The same method works for more complicated events. For example, what's the probability of selecting a woman whose preferred prize is the camera? Well, 91 women named the camera as their preference, so the probability is:

$$P(\text{woman } and \text{ camera}) = 91/478 = 0.190$$

Probabilities such as these are called **joint probabilities** because they give the probability of two events occurring together.

The probability of selecting a customer whose preferred prize is a bike is:

$$P(\text{bike}) = 90/478 = 0.188$$

For Example — **Marginal probabilities**

Lee suspects that men and women make different kinds of purchases at Lee's Lights (see the example on page 195). The table shows the purchases made by the last 100 customers.

	Utility Lighting	Fashion Lighting	Total
Men	40	20	60
Women	10	30	40
Total	50	50	100

Question: What's the probability that one of Lee's customers is a woman? What is the probability that a random customer is a man who purchases fashion lighting?

Answer: From the marginal totals we can see that 40% of Lee's customers are women, so the probability that a customer is a woman is 0.40. The cell of the table for Men who purchase Fashion lighting has 20 of the 100 customers, so the probability of that event is 0.20.

Women

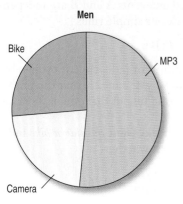

Men

Figure 7.2 Conditional distributions of *Prize Preference* for *Women* and for *Men*.

Notation Alert!

$P(\mathbf{B}|\mathbf{A})$ is the conditional probability of **B** *given* **A**.

7.6 Conditional Probability

Since our sample space is these 478 customers, we can recognize the relative frequencies as probabilities. What if we are given the information that the selected customer is a woman? Would that change the probability that the selected customer's preferred prize is a bike? You bet it would! The pie charts show that women are much less likely to say their preferred prize is a bike than are men. When we restrict our focus to women, we look only at the women's row of the table, which gives the conditional distribution of preferred prizes given "woman." Of the 251 women, only 30 of them said their preferred prize was a bike. We write the probability that a selected customer wants a bike *given* that we have selected a woman as:

$$P(\text{bike}|\text{woman}) = 30/251 = 0.120$$

For men, we look at the conditional distribution of preferred prizes given "man" shown in the top row of the table. There, of the 227 men, 60 said their preferred prize was a bike. So, $P(\text{bike}|\text{man}) = 60/227 = 0.264$, more than twice the women's probability (see Figure 7.2).

In general, when we want the probability of an event from a *conditional* distribution, we write $P(\mathbf{B}|\mathbf{A})$ and pronounce it "the probability of **B** *given* **A**." A probability that takes into account a given *condition* such as this is called a **conditional probability**.

Let's look at what we did. We worked with the counts, but we could work with the probabilities just as well. There were 30 women who selected a bike as a prize, and there were 251 women customers. So we found the probability to be 30/251. To find the probability of the event **B** *given* the event **A**, we restrict our attention to the outcomes in **A**. We then find in what fraction of *those* outcomes **B** also occurred. Formally, we write:

$$P(\mathbf{B}|\mathbf{A}) = \frac{P(\mathbf{A} \text{ and } \mathbf{B})}{P(\mathbf{A})}$$

We can use the formula directly with the probabilities derived from the contingency table (Table 7.2) to find:

$$P(\text{bike}|\text{woman}) = \frac{P(\text{bike and woman})}{P(\text{woman})} = \frac{30/478}{251/478} = \frac{0.063}{0.525} = 0.120 \text{ as before.}$$

The formula for conditional probability requires one restriction. The formula works only when the event that's given has probability greater than 0. The formula doesn't work if $P(A)$ is 0 because that would mean we had been "given" the fact that **A** was true even though the probability of **A** is 0, which would be a contradiction.

Rule 7. Remember the Multiplication Rule for the probability of **A** *and* **B**? It said

$$P(A \text{ and } B) = P(A) \times P(B)$$

when **A** and **B** are independent. Now we can write a more general rule that doesn't require independence. In fact, we've already written it. We just need to rearrange the equation a bit.

The equation in the definition for conditional probability contains the probability of **A** *and* **B**. Rearranging the equation gives the **General Multiplication Rule** for compound events that does not require the events to be independent:

$$P(A \text{ and } B) = P(A) \times P(B|A)$$

The probability that two events, **A** and **B**, both occur is the probability that event **A** occurs multiplied by the probability that event **B** *also* occurs—that is, by the probability that event **B** occurs given that event **A** occurs.

Of course, there's nothing special about which event we call **A** and which one we call **B**. We should be able to state this the other way around. Indeed we can. It is equally true that:

$$P(A \text{ and } B) = P(B) \times P(A|B).$$

Let's return to the question of just what it means for events to be independent. We said informally in Chapter 4 that what we mean by independence is that the outcome of one event does not influence the probability of the other. With our new notation for conditional probabilities, we can write a formal definition. Events **A** and **B** are **independent** whenever:

$$P(B|A) = P(B).$$

Now we can see that the Multiplication Rule for independent events is just a special case of the General Multiplication Rule. The general rule says

$$P(A \text{ and } B) = P(A) \times P(B|A)$$

whether the events are independent or not. But when events **A** and **B** are independent, we can write $P(B)$ for $P(B|A)$ and we get back our simple rule:

$$P(A \text{ and } B) = P(A) \times P(B).$$

> **Independence**
>
> If we had to pick one key idea in this chapter that you should understand and remember, it's the definition and meaning of independence.

Sometimes people use this statement as the definition of independent events, but we find the other definition more intuitive. Either way, the idea is that the probabilities of independent events don't change when you find out that one of them has occurred.

Using our earlier example, is the probability of the event *choosing a bike* independent of the sex of the customer? We need to check whether

$$P(\text{bike}|\text{man}) = \frac{P(\text{bike } and \text{ man})}{P(\text{man})} = \frac{0.126}{0.475} = 0.265$$

is the same as $P(\text{bike}) = 0.189$.

Because these probabilities aren't equal, we can say that prize preference is *not* independent of the sex of the customer. Whenever at least one of the joint probabilities in the table is *not* equal to the product of the marginal probabilities, we say that the variables are not independent.

- **Independent *vs.* Disjoint.** Are disjoint events independent? Both concepts seem to have similar ideas of separation and distinctness about them, but in fact disjoint events *cannot* be independent.[3] Let's see why. Consider the two disjoint events {you get an A in this course} and {you get a B in this course}. They're disjoint because they have no outcomes in common. Suppose you learn that you *did* get an A in the course. Now what is the probability that you got a B? You can't get both grades, so it must be 0.

  Think about what that means. Knowing that the first event (getting an A) occurred changed your probability for the second event (down to 0). So these events aren't independent.

  Mutually exclusive events can't be independent. They have no outcomes in common, so knowing that one occurred means the other didn't. A common error is to treat disjoint events as if they were independent and apply the Multiplication Rule for independent events. Don't make that mistake.

For Example   Conditional probability

Question: Using the table from the example on page 201, if customer purchases a Fashion light, what is the probability that customer is a woman?

Answer: $P(\text{woman} \mid \text{Fashion}) = P(\text{Woman and Fashion})/P(\text{Fashion})$

$\qquad\qquad\qquad = 0.30/0.50 = 0.60$

7.7   Constructing Contingency Tables

Sometimes we're given probabilities without a contingency table. You can often construct a simple table to correspond to the probabilities.

A survey of real estate in upstate New York classified homes into two price categories (Low—less than $175,000 and High—over $175,000). It also noted whether the houses had at least 2 bathrooms or not (True or False). We are told that 56% of the houses had at least 2 bathrooms, 62% of the houses were Low priced, and 22% of the houses were both. That's enough information to fill out the table. Translating the percentages to probabilities, we have:

		At least 2 Bathrooms		
		True	False	Total
Price	Low	0.22		0.62
	High			
	Total	0.56		1.00

The 0.56 and 0.62 are marginal probabilities, so they go in the margins. What about the 22% of houses that were both Low priced and had at least 2 bathrooms? That's a *joint* probability, so it belongs in the interior of the table.

Because the cells of the table show disjoint events, the probabilities always add to the marginal totals going across rows or down columns.

---

[3]Technically two disjoint events *can* be independent, but only if the probability of one of the events is 0. For practical purposes, we can ignore this case, since we don't anticipate collecting data about things that can't possibly happen.

		At least 2 Bathrooms		
		True	False	Total
Price	Low	0.22	0.40	**0.62**
	High	0.64	0.04	**0.38**
	Total	**0.56**	**0.44**	**1.00**

Now, finding any other probability is straightforward. For example, what's the probability that a high-priced house has at least 2 bathrooms?

P(at least 2 bathrooms | high-priced)

= P(at least 2 bathrooms *and* high-priced)/P(high-priced)

= 0.34/0.38 = 0.895 or 89.5%.

Just Checking

3 Suppose a supermarket is conducting a survey to find out the busiest time and day for shoppers. Survey respondents are asked 1) whether they shopped at the store on a weekday or on the weekend and 2) whether they shopped at the store before or after 5 p.m. The survey revealed that:

• 48% of shoppers visited the store before 5 p.m.
• 27% of shoppers visited the store on a weekday (Mon.–Fri.)

• 7% of shoppers visited the store before 5 p.m. on a weekday.

a) Make a contingency table for the variables *time of day* and *day of week*.
b) What is the probability that a randomly selected shopper who shops before 5 p.m. also shops on a weekday?
c) Are time and day of the week disjoint events?
d) Are time and day of the week independent events?

What Can Go Wrong?

• **Beware of probabilities that don't add up to 1.** To be a legitimate assignment of probability, the sum of the probabilities for all possible outcomes must total 1. If the sum is less than 1, you may need to add another category ("other") and assign the remaining probability to that outcome. If the sum is more than 1, check that the outcomes are disjoint. If they're not, then you can't assign probabilities by counting relative frequencies.

• **Don't add probabilities of events if they're not disjoint.** Events must be disjoint to use the Addition Rule. The probability of being under 80 *or* a female is not the probability of being under 80 *plus* the probability of being female. That sum may be more than 1.

• **Don't multiply probabilities of events if they're not independent.** The probability of selecting a customer at random who is over 70 years old *and* retired is not the probability the customer is over 70 years old *times* the probability the customer is retired. Knowing that the customer is over 70 changes the probability of his or her being retired. You can't multiply these probabilities. The multiplication of probabilities of events that are not independent is one of the most common errors people make in dealing with probabilities.

• **Don't confuse disjoint and independent.** Disjoint events *can't* be independent. If **A** = {you get a promotion} and **B** = {you don't get a promotion}, **A** and **B** are disjoint. Are they independent? If you find out that **A** is true, does that change the probability of **B**? You bet it does! So they can't be independent.

Ethics in Action

A national chain of hair salons is considering the inclusion of some spa services. A management team is organized to investigate the possibility of entering the spa market via two offerings: facials or massages. One member of the team, Sherrie Trapper, found some results published by a spa industry trade journal regarding the probability of salon customers purchasing these types of services: There is an 80% chance that a customer visiting a hair salon that offers spa services is there for hair styling services. Of those, 50% will also purchase facials. On the other hand, 90% of customers visiting salons that offer spa services will be there for hair styling services or massages. Sherry wasn't quite sure how to interpret all the numbers, but argued in favor of offering massages rather than facials on their initial spa menu since 90% is greater than 50%.

ETHICAL ISSUE *Sherrie does not understand what she is reporting and consequently should not use this information to persuade others on the team (related to Item A, ASA Ethical Guidelines).*

ETHICAL SOLUTION *Sherrie should share all details of the published results with the management team. The probabilities she is reporting are not comparable (one is conditional and the other is the probability of a union).*

What Have We Learned?

Learning Objectives

- Apply the facts about probability to determine whether an assignment of probabilities is legitimate.
  - Probability is long-run relative frequency.
  - Individual probabilities must be between 0 and 1.
  - The sum of probabilities assigned to all outcomes must be 1.

- Understand the Law of Large Numbers and that the common understanding of the "Law of Averages" is false.
- Know the rules of probability and how to apply them.
  - The **Complement Rule** says that $P(\text{not } A) = P(A^C) = 1 - P(A)$.
  - The **Multiplication Rule** for independent events says that $P(A \text{ and } B) = P(A) \times P(B)$ provided events A and B are independent.
  - The **General Multiplication Rule** says that $P(A \text{ and } B) = P(A) \times P(B|A)$.
  - The **Addition Rule** for disjoint events says that $P(A \text{ or } B) = P(A) + P(B)$ provided events A and B are disjoint.
  - The **General Addition Rule** says that $P(A \text{ or } B) = P(A) + P(B) - P(A \text{ and } B)$.

- Know how to construct and read a contingency table.
- Know how to define and use independence.
  - Events A and B are independent if $P(A|B) = P(A)$.

Terms

Addition Rule

If A and B are disjoint events, then the probability of A or B is

$$P(A \text{ or } B) = P(A) + P(B).$$

Complement Rule

The probability of an event occurring is 1 minus the probability that it doesn't occur:

$$P(A) = 1 - P(A^C).$$

Conditional probability	$$P(\mathbf{B}|\mathbf{A}) = \frac{P(\mathbf{A} \text{ and } \mathbf{B})}{P(\mathbf{A})}.$$
	$P(\mathbf{B}|\mathbf{A})$ is read "the probability of $\mathbf{B}$ *given* $\mathbf{A}$."
Disjoint (or Mutually Exclusive) Events	Two events are disjoint if they share no outcomes in common. If $\mathbf{A}$ and $\mathbf{B}$ are disjoint, then knowing that $\mathbf{A}$ occurs tells us that $\mathbf{B}$ cannot occur. Disjoint events are also called "mutually exclusive."
Empirical probability	When the probability comes from the long-run relative frequency of the event's occurrence, it is an empirical probability.
Event	A collection of outcomes. Usually, we identify events so that we can attach probabilities to them. We denote events with bold capital letters such as $\mathbf{A}$, $\mathbf{B}$, or $\mathbf{C}$.
General Addition Rule	For any two events, $\mathbf{A}$ and $\mathbf{B}$, the probability of $\mathbf{A}$ *or* $\mathbf{B}$ is: $$P(\mathbf{A} \text{ or } \mathbf{B}) = P(\mathbf{A}) + P(\mathbf{B}) - P(\mathbf{A} \text{ and } \mathbf{B}).$$
General Multiplication Rule	For any two events, $\mathbf{A}$ and $\mathbf{B}$, the probability of $\mathbf{A}$ and $\mathbf{B}$ is: $$P(\mathbf{A} \text{ and } \mathbf{B}) = P(\mathbf{A}) \times P(\mathbf{B}|\mathbf{A}).$$
Independence (informally)	Two events are *independent* if the fact that one event occurs does not change the probability of the other.
Independence (used formally)	Events $\mathbf{A}$ and $\mathbf{B}$ are independent when $P(\mathbf{B}|\mathbf{A}) = P(\mathbf{B})$.
Joint probabilities	The probability that two events both occur.
Law of Large Numbers (LLN)	The Law of Large Numbers states that the *long-run relative frequency* of repeated, independent events settles down to the *true relative frequency* as the number of trials increases.
Marginal probability	In a joint probability table a marginal probability is the probability distribution of either variable separately, usually found in the rightmost column or bottom row of the table.
Multiplication Rule	If $\mathbf{A}$ and $\mathbf{B}$ are independent events, then the probability of $\mathbf{A}$ *and* $\mathbf{B}$ is: $$P(\mathbf{A} \text{ and } \mathbf{B}) = P(\mathbf{A}) \times P(\mathbf{B}).$$
Outcome	The outcome of a trial is the value measured, observed, or reported for an individual instance of that trial.
Personal probability	When the probability is subjective and represents your personal degree of belief, it is called a personal probability.
Probability	The probability of an event is a number between 0 and 1 that reports the likelihood of the event's occurrence. A probability can be derived from a model (such as equally likely outcomes), from the long-run relative frequency of the event's occurrence, or from subjective degrees of belief. We write $P(\mathbf{A})$ for the probability of the event $\mathbf{A}$.
Probability Assignment Rule	The probability of the entire sample space must be 1: $$P(S) = 1.$$
Random phenomenon	A phenomenon is random if we know what outcomes could happen, but not which particular values will happen.
Sample space	The collection of all possible outcome values. The sample space has a probability of 1.
Theoretical probability	When the probability comes from a mathematical model (such as, but not limited to, equally likely outcomes), it is called a theoretical probability.
Trial	A single attempt or realization of a random phenomenon.

Brief CASE

Market Segmentation

The data from the "Chicago Female Fashion Study"[4] were collected using a self-administered survey of a sample of homes in the greater Chicago metropolitan area. Marketing managers want to know how important quality is to their customers. A consultant reports that based on past research, 30% of all consumers nationwide are more interested in quantity than quality. The marketing manager of a particular department store suspects that customers from her store are different, and that customers of different ages might have different views as well. Using conditional probabilities, marginal probabilities, and joint probabilities constructed from the data in the file **Market_Segmentation**,[5] write a report to the manager on what you find.

Keep in mind: The manager may be more interested in the opinions of "frequent" customers than those who never or hardly ever shop at her store. These "frequent" customers contribute a disproportionate amount of profit to the store. Keep that in mind as you do your analysis and write up your report.

VARIABLE AND QUESTION	CATEGORIES
Age *Into which of the following age categories do you belong?*	18–24 yrs old 25–34 35–44 45–54 55–64 65 or over
Frequency *How often do you shop for women's clothing at [department store X]?*	0. Never–hardly ever 1–2 times per year 3–4 times per year 5 times or more
Quality *For the same amount of money, I will generally buy one good item than several of lower price and quality.*	1. Definitely Disagree 2. Generally Disagree 3. Moderately Disagree 4. Moderately Agree 5. Generally Agree 6. Definitely Agree

Exercises

SECTION 7.1

1. Indicate which of the following represent independent events. Explain briefly.

a) The gender of customers using an ATM machine.

b) The last digit of the social security numbers of students in a class.

c) The scores you receive on the first midterm, second midterm, and the final exam of a course.

2. Indicate which of the following represent independent events. Explain briefly.

a) Prices of houses on the same block.

b) Successive measurements of your heart rate as you exercise on a treadmill.

c) Measurements of the heart rates of all students in the gym.

---

[4]Original *Market Segmentation Exercise* prepared by K. Matsuno, D. Kopcso, and D. Tigert, Babson College in 1997 (Babson Case Series #133-C97A-U).

[5]For a version with the categories coded as integers see **Market_Segmentation_Coded**.

SECTION 7.2

3. In many state lotteries you can choose which numbers to play. Consider a common form in which you choose 5 numbers. Which of the following strategies can improve your chance of winning? If the method works, explain why. If not, explain why using appropriate statistics terms.

a) Always play 1, 2, 3, 4, 5.
b) Generate random numbers using a computer or calculator and play those.

4. For the same kind of lottery as in Exercise 3, which of the following strategies can improve your chance of winning? If the method works, explain why. If not, explain why using appropriate statistics terms.

a) Choose randomly from among the numbers that have *not* come up in the last 3 lottery drawings.
b) Choose the numbers that did come up in the most recent lottery drawing.

SECTION 7.4

5. You and your friend decide to get your cars inspected. You are informed that 75% of cars pass inspection. If the event of your car's passing is independent of your friend's car,

a) What is the probability that your car passes inspection?
b) What is the probability that your car doesn't pass inspection?
c) What is the probability that both of the cars pass?
d) What is the probability that at least one of the two cars passes?

6. At your school, 10% of the class are marketing majors. If you are randomly assigned to two partners in your statistics class,

a) What is the probability that the first partner will be a marketing major?
b) What is the probability that the first partner won't be a marketing major?
c) What is the probability that both will be marketing majors?
d) What is the probability that one or the other will be marketing majors?

SECTION 7.5

7. The following contingency table shows opinion about global warming (nonissue vs. serious concern) among registered voters, broken down by political party affiliation (Democratic, Republican, and Independent).

		Opinion on Global Warming		
		Nonissue	Serious Concern	Total
Political Party	Democratic	60	440	500
	Republican	290	210	500
	Independent	90	110	200
	Total	440	760	1200

a) What is the probability that a registered voter selected at random believes that global warming is a serious issue?
b) What type of probability did you find in part a?
c) What is the probability that a registered voter selected at random is a Republican and believes that global warming is a serious issue?
d) What type of probability did you find in part c?

8. Multigenerational families can be categorized as having two adult generations such as parents living with adult children, "skip" generation families, such as grandparents living with grandchildren, and three or more generations living in the household. Pew Research surveyed multigenerational households. This table is based on their reported results.

	2 Adult Gens	2 Skip Gens	3 or More Gens	
White	509	55	222	786
Hispanic	139	11	142	292
Black	119	32	99	250
Asian	61	1	48	110
	828	99	511	1438

a) What is the probability that a multigenerational family is Hispanic?
b) What is the probability that a multigenerational family selected at random is a Black, two-adult-generation family?
c) What type of probability did you find in parts a and b?

SECTION 7.6

9. Using the table from Exercise 7,

a) What is the probability that a randomly selected registered voter who is a Republican believes that global warming is a serious issue?
b) What is the probability that a randomly selected registered voter is a Republican given that he or she believes global warming is a serious issue?
c) What is $P(\text{Serious Concern} \mid \text{Democratic})$?

10. Using the table from Exercise 8,

a) What is the probability that a randomly selected Black multigenerational family is a two-generation family?
b) What is the probability that a randomly selected multigenerational family is White, given that it is a "skip" generation family?
c) What is $P(3 \text{ or more Generations} \mid \text{Asian})$?

SECTION 7.7

11. A national survey indicated that 30% of adults conduct their banking online. It also found that 40% are under the age of 50, and that 25% are under the age of 50 and conduct their banking online.

a) What percentage of adults do not conduct their banking online?
b) What type of probability is the 25% mentioned above?
c) Construct a contingency table showing all joint and marginal probabilities.
d) What is the probability that an individual conducts banking online given that the individual is under the age of 50?
e) Are *Banking online* and *Age* independent? Explain.

12. Facebook reports that 70% of their users are from outside the United States and that 50% of their users log on to Facebook every day. Suppose that 20% of their users are United States users who log on every day.

a) What percentage of Facebook's users are from the United States?
b) What type of probability is the 20% mentioned above?
c) Construct a contingency table showing all the joint and marginal probabilities.
d) What is the probability that a user is from the United States given that he or she logs on every day?
e) Are *From United States* and *Log on Every Day* independent? Explain.

CHAPTER EXERCISES

13. What does it mean? part 1. Respond to the following questions:

a) A casino claims that its roulette wheel is truly random. What should that claim mean?
b) A reporter on *Market Place* says that there is a 50% chance that the Federal Reserve Bank will cut interest rates by a quarter point at their next meeting. What is the meaning of such a phrase?

14. What does it mean? part 2. Respond to the following questions:

a) After an unusually dry autumn, a radio announcer is heard to say, "Watch out! We'll pay for these sunny days later on this winter." Explain what he's trying to say, and comment on the validity of his reasoning.
b) A batter who had failed to get a hit in seven consecutive times at bat then hits a game-winning home run. When talking to reporters afterward, he says he was very confident that last time at bat because he knew he was "due for a hit." Comment on his reasoning.

15. Airline safety. Even though commercial airlines have excellent safety records, in the weeks following a crash, airlines often report a drop in the number of passengers, probably because people are afraid to risk flying.

a) A travel agent suggests that since the law of averages makes it highly unlikely to have two plane crashes within a few weeks of each other, flying soon after a crash is the safest time. What do you think?
b) If the airline industry proudly announces that it has set a new record for the longest period of safe flights, would you be reluctant to fly? Are the airlines due to have a crash?

16. Economic predictions. An investment newsletter makes general predictions about the economy to help their clients make sound investment decisions.

a) Recently they said that because the stock market had been up for the past three months in a row that it was "due for a correction" and advised their client to reduce their holdings. What "law" are they applying? Comment.
b) They advised buying a stock that had gone down in the past four sessions because they said that it was clearly "due to bounce back." What "law" are they applying? Comment.

17. Fire insurance. Insurance companies collect annual payments from homeowners in exchange for paying to rebuild houses that burn down.

a) Why should you be reluctant to accept a $300 payment from your neighbor to replace his house should it burn down during the coming year?
b) Why can the insurance company make that offer?

18. Casino gambling. Recently, the International Gaming Technology company issued the following press release:

(LAS VEGAS, Nev.)—Cynthia Jay was smiling ear to ear as she walked into the news conference at the Desert Inn Resort in Las Vegas today, and well she should. Last night, the 37-year-old cocktail waitress won the world's largest slot jackpot—$34,959,458—on a Megabucks machine. She said she had played $27 in the machine when the jackpot hit. Nevada Megabucks has produced 49 major winners in its 14-year history. The top jackpot builds from a base amount of $7 million and can be won with a 3-coin ($3) bet.

a) How can the Desert Inn afford to give away millions of dollars on a $3 bet?
b) Why did the company issue a press release? Wouldn't most businesses want to keep such a huge loss quiet?

19. Toy company. A toy company manufactures a spinning game and needs to decide what probabilities are involved in the game. The plastic arrow on the spinner stops rotating to point at a color that will determine what happens next. Knowing these probabilities will help determine how easy or difficult it is for a person to win the game and helps to determine how long the average game will last. Are each of the following probability assignments possible? Why or why not?

	Probabilities of ...			
	Red	**Yellow**	**Green**	**Blue**
a)	0.25	0.25	0.25	0.25
b)	0.10	0.20	0.30	0.40
c)	0.20	0.30	0.40	0.50
d)	0	0	1.00	0
e)	0.10	0.20	1.20	−1.50

20. Store discounts. Many stores run "secret sales": Shoppers receive cards that determine how large a discount they get, but the percentage is revealed by scratching off that black stuff (what *is* that?) only after the purchase has been totaled at the cash register. The store is required to reveal (in the fine print) the distribution of discounts available. Are each of these probability assignments plausible? Why or why not?

	Probabilities of ...			
	10% off	**20% off**	**30% off**	**50% off**
a)	0.20	0.20	0.20	0.20
b)	0.50	0.30	0.20	0.10
c)	0.80	0.10	0.05	0.05
d)	0.75	0.25	0.25	−0.25
e)	1.00	0	0	0

21. Quality control. A tire manufacturer recently announced a recall because 2% of its tires are defective. If you just bought a new set of four tires from this manufacturer, what is the probability that at least one of your new tires is defective?

22. Pepsi promotion. For a sales promotion, the manufacturer places winning symbols under the caps of 10% of all Pepsi bottles. If you buy a six-pack of Pepsi, what is the probability that you win something?

23. Auto warranty. In developing their warranty policy, an automobile company estimates that over a 1-year period 17% of their new cars will need to be repaired once, 7% will need repairs twice, and 4% will require three or more repairs. If you buy a new car from them, what is the probability that your car will need:

a) No repairs?
b) No more than one repair?
c) Some repairs?

24. Consulting team. You work for a large global management consulting company. Of the entire work force of analysts, 55% have had no experience in the telecommunications

industry, 32% have had limited experience (less than 5 years), and the rest have had extensive experience (5 years or more). On a recent project, you and two other analysts were chosen at random to constitute a team. It turns out that part of the project involves telecommunications. What is the probability that the first teammate you meet has:

a) Extensive telecommunications experience?
b) Some telecommunications experience?
c) No more than limited telecommunications experience?

25. Auto warranty, part 2. Consider again the auto repair rates described in Exercise 23. If you bought two new cars, what is the probability that:

a) Neither will need repair?
b) Both will need repair?
c) At least one car will need repair?

26. Consulting team, part 2. You are assigned to be part of a team of three analysts of a global management consulting company as described in Exercise 24. What is the probability that of your other two teammates:

a) Neither has any telecommunications experience?
b) Both have some telecommunications experience?
c) At least one has had extensive telecommunications experience?

27. Auto warranty, again. You used the Multiplication Rule to calculate repair probabilities for your cars in Exercise 23.

a) What must be true about your cars in order to make that approach valid?
b) Do you think this assumption is reasonable? Explain.

28. Final consulting team project. You used the Multiplication Rule to calculate probabilities about the telecommunications experience of your consulting teammates in Exercise 24.

a) What must be true about the groups in order to make that approach valid?
b) Do you think this assumption is reasonable? Explain.

29. Real estate. Real estate ads suggest that 64% of homes for sale have garages, 21% have swimming pools, and 17% have both features. What is the probability that a home for sale has:

a) A pool or a garage?
b) Neither a pool nor a garage?
c) A pool but no garage?

30. Human resource data. Employment data at a large company reveal that 72% of the workers are married, 44% are college graduates, and half of the college grads are married. What's the probability that a randomly chosen worker is:

a) Neither married nor a college graduate?
b) Married but not a college graduate?
c) Married or a college graduate?

31. Market research on energy. A Gallup Poll in March 2007 asked 1005 U.S. adults whether increasing domestic energy production or protecting the environment should be given higher priority. Here are the results.

Response	Number
Increase Production	342
Protect the Environment	583
Equally Important	50
No Opinion	30
Total	**1005**

If we select a person at random from this sample of 1005 adults:

a) What is the probability that the person responded "Increase Production"?

b) What is the probability that the person responded "Equally Important" or had "No Opinion"?

32. More market research on energy. Exercise 31 shows the results of a Gallup Poll about energy. Suppose we select three people at random from this sample.

a) What is the probability that all three responded "Protect the Environment"?

b) What is the probability that none responded "Equally Important"?

c) What assumption did you make in computing these probabilities?

d) Explain why you think that assumption is reasonable.

33. Telemarketing contact rates. Marketing research firms often contact their respondents by sampling random telephone numbers. Although interviewers currently reach about 76% of selected U.S. households, the percentage of those contacted who agree to cooperate with the survey has fallen. Assume that the percentage of those who agree to cooperate in telemarketing surveys is now only 38%. Each household is assumed to be independent of the others.

a) What is the probability that the next household on the list will be contacted but will refuse to cooperate?

b) What is the probability of failing to contact a household or of contacting the household but not getting them to agree to the interview?

c) Show another way to calculate the probability in part b.

34. Telemarketing contact rates, part 2. According to Pew Research, the contact rate (probability of contacting a selected household) in 1997 was 69%, and in 2003, it was 76%. However, the cooperation rate (probability of someone at the contacted household agreeing to be interviewed) was 58% in 1997 and dropped to 38% in 2003.

a) What is the probability (in 2003) of obtaining an interview with the next household on the sample list? (To obtain an interview, an interviewer must both contact the household and then get agreement for the interview.)

b) Was it more likely to obtain an interview from a randomly selected household in 1997 or in 2003?

35. Mars product information. The Mars company says that before the introduction of purple, yellow made up 20% of their plain M&M candies, red made up another 20%, and orange, blue, and green each made up 10%. The rest were brown.

a) If you picked an M&M at random from a pre-purple bag of candies, what is the probability that it was:

    i) Brown?

    ii) Yellow or orange?

    iii) Not green?

    iv) Striped?

b) Assuming you had an infinite supply of M&M's with the older color distribution, if you picked three M&M's in a row, what is the probability that:

    i) They are all brown?

    ii) The third one is the first one that's red?

    iii) None are yellow?

    iv) At least one is green?

36. American Red Cross. The American Red Cross must track their supply and demand for various blood types. They estimate that about 45% of the U.S. population has Type O blood, 40% Type A, 11% Type B, and the rest Type AB.

a) If someone volunteers to give blood, what is the probability that this donor:

    i) Has Type AB blood?

    ii) Has Type A or Type B blood?

    iii) Is not Type O?

b) Among four potential donors, what is the probability that:

    i) All are Type O?

    ii) None have Type AB blood?

    iii) Not all are Type A?

    iv) At least one person is Type B?

37. More Mars product information. In Exercise 35, you calculated probabilities of getting various colors of M&M's.

a) If you draw one M&M, are the events of getting a red one and getting an orange one disjoint or independent or neither?

b) If you draw two M&M's one after the other, are the events of getting a red on the first and a red on the second disjoint or independent or neither?

c) Can disjoint events ever be independent? Explain.

38. American Red Cross, part 2. In Exercise 36, you calculated probabilities involving various blood types.

a) If you examine one donor, are the events of the donor being Type A and the donor being Type B disjoint or independent or neither? Explain your answer.

b) If you examine two donors, are the events that the first donor is Type A and the second donor is Type B disjoint or independent or neither?

c) Can disjoint events ever be independent? Explain.

39. Tax accountant. A recent study of IRS audits showed that, for estates worth less than $5 million, about 1 out of 7 of all estate tax returns are audited, but that probability increases to 50% for estates worth over $5 million. Suppose a tax accountant has three clients who have recently filed returns for estates worth more than $5 million. What are the probabilities that:

a) All three will be audited?

b) None will be audited?

c) At least one will be audited?

d) What did you assume in calculating these probabilities?

40. Casinos. Because gambling is big business, calculating the odds of a gambler winning or losing in every game is crucial to the financial forecasting for a casino. A standard slot machine has three wheels that spin independently. Each has 10 equally likely symbols: 4 bars, 3 lemons, 2 cherries, and a bell. If you play once, what is the probability that you will get:

a) 3 lemons?

b) No fruit symbols?

c) 3 bells (the jackpot)?

d) No bells?

e) At least one bar (an automatic loser)?

41. Information technology. A company has recently replaced their e-mail server because previously mail was interrupted on about 15% of workdays. To see how bad the situation was, calculate the probability that during a 5-day work week, there would be an e-mail interruption.

a) On Monday and again on Tuesday?

b) For the first time on Thursday?

c) Every day?

d) At least once during the week?

42. Information technology, part 2. At a mid-sized Web design and maintenance company, 57% of the computers are PCs, 29% are Macs, and the rest are Unix-based machines. Assuming that users of each of the machines are equally likely to call in to the information technology help line, what is the probability that of the next three calls:

a) All are Macs?

b) None are PCs?

c) At least one is a Unix machine?

d) All are Unix machines?

43. Casinos, part 2. In addition to slot machines, casinos must understand the probabilities involved in card games. Suppose you are playing at the blackjack table, and the dealer shuffles a deck of cards. The first card shown is red. So is the second and the third. In fact, you are surprised to see 5 red cards in a row. You start thinking, "The next one is due to be black!"

a) Are you correct in thinking that there's a higher probability that the next card will be black than red? Explain.

b) Is this an example of the Law of Large Numbers? Explain.

44. Inventory. A shipment of road bikes has just arrived at The Spoke, a small bicycle shop, and all the boxes have been placed in the back room. The owner asks her assistant to start bringing in the boxes. The assistant sees 20 identical-looking boxes and starts bringing them into the shop at random. The owner knows that she ordered 10 women's and 10 men's bicycles, and so she's surprised to find that the first six are all women's bikes. As the seventh box is brought in, she starts thinking, "This one is bound to be a men's bike."

a) Is she correct in thinking that there's a higher probability that the next box will contain a men's bike? Explain.

b) Is this an example of the Law of Large Numbers? Explain.

45. International food survey. A GfK Roper Worldwide survey in 2005 asked consumers in five countries whether they agreed with the statement "I am worried about the safety of the food I eat." Here are the responses classified by the age of the respondent.

		Agree	Neither Agree nor Disagree	Disagree	Don't Know/ No Response	Total
Age	13–19	661	368	452	32	**1513**
	20–29	816	365	336	16	**1533**
	30–39	871	355	290	9	**1525**
	40–49	914	335	266	6	**1521**
	50 +	966	339	283	10	**1598**
	Total	**4228**	**1762**	**1627**	**73**	**7690**

If we select a person at random from this sample:

a) What is the probability that the person agreed with the statement?

b) What is the probability that the person is younger than 50 years old?

c) What is the probability that the person is younger than 50 *and* agrees with the statement?

d) What is the probability that the person is younger than 50 *or* agrees with the statement?

46. Cosmetics marketing. A GfK Roper Worldwide survey asked consumers in five countries whether they agreed with the statement "I follow a skin care routine every day." Here are the responses classified by the country of the respondent.

	Response			
Country	**Agree**	**Disagree**	**Don't know**	**Total**
China	361	988	153	1502
France	695	763	81	1539
India	828	689	18	1535
U.K.	597	898	62	1557
USA	668	841	48	1557
Total	**3149**	**4179**	**362**	**7690**

If we select a person at random from this sample:

a) What is the probability that the person agreed with the statement?
b) What is the probability that the person is from China?
c) What is the probability that the person is from China *and* agrees with the statement?
d) What is the probability that the person is from China *or* agrees with the statement?

47. E-commerce. Suppose an online business organizes an e-mail survey to find out if online shoppers are concerned with the security of business transactions on the Web. Of the 42 individuals who respond, 24 are concerned, and 18 are not concerned. Eight of those concerned about security are male and 6 of those not concerned are male. If a respondent is selected at random, find each of the following conditional probabilities:

a) The respondent is male, given that the respondent is not concerned about security.
b) The respondent is not concerned about security, given that she is female.
c) The respondent is female, given that the respondent is concerned about security.

48. Automobile inspection. Twenty percent of cars that are inspected have faulty pollution control systems. The cost of repairing a pollution control system exceeds $100 about 40% of the time. When a driver takes her car in for inspection, what's the probability that she will end up paying more than $100 to repair the pollution control system?

49. Pharmaceutical company. A U.S. pharmaceutical company is considering manufacturing and marketing a pill that will help to lower both an individual's blood pressure and cholesterol. The company is interested in understanding the demand for such a product. The joint probabilities that an adult American man has high blood pressure and/or high cholesterol are shown in the table.

	Blood Pressure	
Cholesterol	**High**	**OK**
High	0.11	0.21
OK	0.16	0.52

a) What's the probability that an adult American male has both conditions?
b) What's the probability that an adult American male has high blood pressure?
c) What's the probability that an adult American male with high blood pressure also has high cholesterol?
d) What's the probability that an adult American male has high blood pressure if it's known that he has high cholesterol?

50. International relocation. A European department store is developing a new advertising campaign for their new U.S. location, and their marketing managers need to better understand their target market. Based on survey responses, a joint probability table that an adult shops at their new U.S. store classified by their age is shown below.

	Shop		
Age	**Yes**	**No**	**Total**
< 20	0.26	0.04	0.30
20–40	0.24	0.10	0.34
> 40	0.12	0.24	0.36
Total	**0.62**	**0.38**	**1.00**

a) What's the probability that a survey respondent will shop at the U.S. store?
b) What is the probability that a survey respondent will shop at the store given that they are younger than 20 years old?
c) What is the probability that a survey respondent who is older than 40 shops at the store?
d) What is the probability that a survey respondent is younger than 20 or will shop at the store?

51. Pharmaceutical company, again. Given the table of probabilities compiled for marketing managers in Exercise 49, are high blood pressure and high cholesterol independent? Explain.

52. International relocation, again. Given the table of probabilities compiled for a department store chain in Exercise 50, are age and shopping at the department store independent? Explain.

53. International food survey, part 2. Look again at the data from the GfK Roper Worldwide survey on food attitudes in Exercise 45.

a) If we select a respondent at random, what's the probability we choose a person between 13 and 19 years old who agreed with the statement?
b) Among the 13- to 19-year-olds, what is the probability that a person responded "Agree"?
c) What's the probability that a person who agreed was between 13 and 19?
d) If the person responded "Disagree," what is the probability that they are at least 50 years old?
e) What's the probability that a person 50 years or older disagreed?
f) Are response to the question and age independent?

54. Cosmetics marketing, part 2. Look again at the data from the GfK Roper Worldwide survey on skin care in Exercise 46.

a) If we select a respondent at random, what's the probability we choose a person from the U.S.A. who agreed with the statement?
b) Among those from the U.S.A., what is the probability that a person responded "Agree"?
c) What's the probability that a person who agreed was from the U.S.A.?
d) If the person responded "Disagree," what is the probability that they are from the U.S.A.?
e) What's the probability that a person from the U.S.A. disagreed?
f) Are response to the question and Country independent?

55. Real estate, part 2. In the real estate research described in Exercise 29, 64% of homes for sale have garages, 21% have swimming pools, and 17% have both features.

a) What is the probability that a home for sale has a garage, but not a pool?
b) If a home for sale has a garage, what's the probability that it has a pool, too?
c) Are having a garage and a pool independent events? Explain.
d) Are having a garage and a pool mutually exclusive? Explain.

56. Employee benefits. Fifty-six percent of all American workers have a workplace retirement plan, 68% have health insurance, and 49% have both benefits. If we select a worker at random:

a) What's the probability that the worker has neither employer-sponsored health insurance nor a retirement plan?
b) What's the probability that the worker has health insurance if they have a retirement plan?
c) Are having health insurance and a retirement plan independent? Explain.
d) Are having these two benefits mutually exclusive? Explain.

57. Telemarketing. Telemarketers continue to attempt to reach consumers by calling landline phone numbers. According to estimates from a national 2003 survey, based on face-to-face interviews in 16,677 households, approximately 58.2% of U.S. adults have both a landline in their residence and a cell phone, 2.8% have only cell phone service but no land line, and 1.6% have no telephone service at all.

a) Polling agencies won't phone cell phone numbers because customers object to paying for such calls. What proportion of U.S. households can be reached by a landline call?
b) Are having a cell phone and having a landline independent? Explain.

58. Snoring. According to the British United Provident Association (BUPA), a major health care provider in the U.K., snoring can be an indication of sleep apnea which can cause chronic illness if left untreated. In the U.S.A., the National Sleep Foundation reports that 36.8% of the 995 adults they surveyed snored. Of the respondents, 81.5% were over the age of 30, and 32% were both over the age of 30 and snorers.

a) What percent of the respondents were 30 years old or younger and did not snore?
b) Is snoring independent of age? Explain.

59. Selling cars. A recent ad campaign for a major automobile manufacturer is clearly geared toward an older demographic. You are surprised, so you decide to conduct a quick survey of your own. A random survey of autos parked in the student and staff lots at your university classified the brands by country of origin, as seen in the table. Is country of origin independent of type of driver?

		Driver	
		Student	Staff
Origin	American	107	105
	European	33	12
	Asian	55	47

60. Fire sale. A survey of 1056 houses in the Saratoga Springs, New York, area found the following relationship between price (in $) and whether the house had a fireplace in 2006. Is the price of the house independent of whether it has a fireplace?

		Fireplace	
		No	Yes
House Price	Low—less than $112,000	198	66
	Med Low ($112 to $152K)	133	131
	Med High ($152 to $207K)	65	199
	High—over $207,000	31	233

61. Used cars. A business student is searching for a used car to purchase, so she posts an ad to a website saying she wants to buy a used Jeep between $18,000 and $20,000. From Kelly's BlueBook.com, she learns that there are 149 cars matching that description within a 30-mile radius of her home. If we assume that those are the people who will call her and that they are equally likely to call her:

a) What is the probability that the first caller will be a Jeep Liberty owner?

b) What is the probability that the first caller will own a Jeep Liberty that costs between $18,000 and $18,999?

c) If the first call offers her a Jeep Liberty, what is the probability that it costs less than $19,000?

d) Suppose she decides to ignore calls with cars whose cost is ≥ $19,000. What is the probability that the first call she takes will offer to sell her a Jeep Liberty?

		Price		
Make		$18,000–$18,999	$19,000–$19,999	**Total**
	Commander	3	6	9
	Compass	6	1	7
Car	Grand Cherokee	33	33	66
	Liberty	17	6	23
	Wrangler	33	11	44
	Total	**92**	**57**	**149**

62. CEO relocation. The CEO of a mid-sized company has to relocate to another part of the country. To make it easier, the company has hired a relocation agency to help purchase a house. The CEO has 5 children and so has specified that the house have at least 5 bedrooms, but hasn't put any other constraints on the search. The relocation agency has narrowed the search down to the houses in the table and has selected one house to showcase to the CEO and family on their trip out to the new site. The agency doesn't know it, but the family has its heart set on a Cape Cod house with a fireplace. If the agency selected the house at random, without regard to this:

		Fireplace?		
		No	Yes	Total
House Type	Cape Cod	7	2	9
	Colonial	8	14	22
	Other	6	5	11
	Total	**21**	**21**	**42**

a) What is the probability that the selected house is a Cape Cod?

b) What is the probability that the house is a Colonial with a fireplace?

c) If the house is a Cape Cod, what is the probability that it has a fireplace?

d) What is the probability that the selected house is what the family wants?

Just Checking Answers

1 The probability of going up on the next day is not affected by the previous day's outcome.

2 a) 0.30
 b) $0.30(0.30) = 0.09$
 c) $(1 - 0.30)^2(0.30) = 0.147$
 d) $1 - (1 - 0.30)^5 = 0.832$

3 a)

		Weekday		
		Yes	No	Total
Before Five	Yes	0.07	0.41	**0.48**
	No	0.20	0.32	**0.52**
	Total	**0.27**	**0.73**	**1.00**

b) $P(\mathbf{BF}|\mathbf{WD}) = P(\mathbf{BF}\ and\ \mathbf{WD})/P(\mathbf{WD}) = 0.07/0.27 = .259$

c) No, shoppers can do both (and 7% do).

d) To be independent, we'd need $P(\mathbf{BF}|\mathbf{WD})= P(\mathbf{BF})$. $P(\mathbf{BF}|\mathbf{WD}) = 0.259$, but $P(\mathbf{BF}) = \mathbf{0.48}$. They do not appear to be independent.

Random Variables and Probability Models

Metropolitan Life Insurance Company

In 1863, at the height of the U.S. Civil War, a group of businessmen in New York City decided to form a new company to insure Civil War soldiers against disabilities and injuries suffered from the war. After the war ended, they changed direction and decided to focus on selling life insurance. The new company was named Metropolitan Life (MetLife) because the bulk of the company's clients were in the "metropolitan" area of New York City.

Although an economic depression in the 1870s put many life insurance companies out of business, MetLife survived, modeling their business on similar successful programs in England. Taking advantage of spreading industrialism and the selling methods of British insurance agents, the company soon was enrolling as many as 700 new policies per day. By 1909, MetLife was the nation's largest life insurer in the United States.

During the Great Depression of the 1930s, MetLife expanded their public service by promoting public health campaigns, focusing on educating the urban poor in major U.S. cities about the risk of tuberculosis. Because the company invested primarily in urban and farm mortgages, as opposed to the stock market, they survived the crash of 1929 and ended up investing heavily in the post-war U.S. housing boom. They were the principal investors in both the Empire State

Building (1929) and Rockefeller Center (1931). During World War II, the company was the single largest contributor to the Allied cause, investing more than half of their total assets in war bonds.

Today, in addition to life insurance, MetLife manages pensions and investments. In 2000, the company held an initial public offering and entered the retail banking business in 2001 with the launch of MetLife Bank. The company's public face is well known because of their use of Snoopy, the dog from the cartoon strip "Peanuts."

Insurance companies make bets all the time. For example, they bet that you're going to live a long life. Ironically, you bet that you're going to die sooner. Both you and the insurance company want the company to stay in business, so it's important to find a "fair price" for your bet. Of course, the right price for *you* depends on many factors, and nobody can predict exactly how long you'll live. But when the company averages its bets over enough customers, it can make reasonably accurate estimates of the amount it can expect to collect on a policy before it has to pay out the benefit. To do that effectively, it must model the situation with a probability model. Using the resulting probabilities, the company can find the fair price of almost any situation involving risk and uncertainty.

Here's a simple example. An insurance company offers a "death and disability" policy that pays $100,000 when a client dies or $50,000 if the client is permanently disabled. It charges a premium of only $500 per year for this benefit. Is the company likely to make a profit selling such a plan? To answer this question, the company needs to know the *probability* that a client will die or become disabled in any year. From actuarial information such as this and the appropriate model, the company can calculate the expected value of this policy.

8.1   Expected Value of a Random Variable

To model the insurance company's risk, we need to define a few terms. The amount the company pays out on an individual policy is an example of a **random variable**, called that because its value is based on the outcome of a random event. We use a capital letter, in this case, X, to denote a random variable. We'll denote a particular *value* that it can have by the corresponding lowercase letter, in this case, x. For the insurance company, x can be $100,000 (if you die that year), $50,000 (if you are disabled), or $0 (if neither occurs). Because we can list all the outcomes, we call this random variable a **discrete random variable**. A random variable that can take on any value between two values is called a **continuous random variable**. Continuous random variables are common in business applications for modeling physical quantities like heights and weights, and monetary quantities such as profits, revenues, and spending.

Sometimes it is obvious whether to treat a random variable as discrete or continuous, but at other times the choice is more subtle. Age, for example, might be viewed as discrete if it is measured only to the nearest decade with possible values 10, 20, 30, ... In a scientific context, however, it might be measured more precisely and treated as continuous.

For both discrete and continuous variables, the collection of all the possible values and the probabilities associated with them is called the **probability model** for the random variable. For a discrete random variable, we can list the probability of all possible values in a table, or describe it by a formula. For example, to model the possible outcomes of a fair die, we can let X be the number showing on the face. The probability model for X is simply:

$$P(X = x) = \begin{cases} 1/6 & \text{if } x = 1, 2, 3, 4, 5, \text{ or } 6 \\ 0 & \text{otherwise} \end{cases}$$

Suppose in our insurance risk example that the death rate in any year is 1 out of every 1000 people and that another 2 out of 1000 suffer some kind of disability. The loss, which we'll denote as X, is a discrete random variable because it takes on only 3 possible values. We can display the probability model for X in a table, as in Table 8.1.

Notation Alert!

The most common letters for random variables are X, Y, and Z, but any capital letter might be used.

Policyholder Outcome	Payout x (cost)	Probability $P(X = x)$
Death	100,000	$\frac{1}{1000}$
Disability	50,000	$\frac{2}{1000}$
Neither	0	$\frac{997}{1000}$

Table 8.1 Probability model for an insurance policy.

Of course, we can't predict exactly what *will* happen during any given year, but we can say what we *expect* to happen—in this case, what we expect the profit of a policy will be. The expected value of a policy is a **parameter** of the probability model. In fact, it's the mean. We'll signify this with the notation $E(X)$, for expected value (or sometimes μ to indicate that it is a mean). This isn't an average of data values, so we won't estimate it. Instead, we calculate it directly from the probability model for the random variable. Because it comes from a model and not data, we use the parameter μ to denote it (and *not $\bar{y}$ or $\bar{x}$.*)

To see what the insurance company can expect, think about some convenient number of outcomes. For example, imagine that they have exactly 1000 clients and that the outcomes in one year followed the probability model exactly: 1 died, 2 were disabled, and 997 survived unscathed. Then our expected payout would be:

$$\mu = E(X) = \frac{100{,}000(1) + 50{,}000(2) + 0(997)}{1000} = 200$$

Notation Alert!

The expected value (or mean) of a random variable is written $E(X)$ or μ. (Be sure not to confuse the mean of a random variable, calculated from probabilities, with the mean of a collection of data values which is denoted by $\bar{y}$ or $\bar{x}$.)

So our expected payout comes to $200 per policy.

Instead of writing the expected value as one big fraction, we can rewrite it as separate terms, each divided by 1000.

$$\mu = E(X) = \$100{,}000\left(\frac{1}{1000}\right) + \$50{,}000\left(\frac{2}{1000}\right) + \$0\left(\frac{997}{1000}\right)$$
$$= \$200$$

Writing it this way, we can see that for each policy, there's a 1/1000 chance that we'll have to pay $100,000 for a death and a 2/1000 chance that we'll have to pay $50,000 for a disability. Of course, there's a 997/1000 chance that we won't have to pay anything.

So the **expected value** of a (discrete) random variable is found by multiplying each possible value of the random variable by the probability that it occurs and then summing all those products. This gives the general formula for the expected value of a discrete random variable:[1]

$$E(X) = \sum x \, P(x).$$

Be sure that *every* possible outcome is included in the sum. Verify that you have a valid probability model to start with—the probabilities should each be between 0 and 1 and should sum to one. (Recall the rules of probability in Chapter 7.)

For Example   Calculating expected value of a random variable

Question: A fund-raising lottery offers 500 tickets for $3 each. If the grand prize is $250 and 4 second prizes are $50 each, what is the expected value of a single ticket? (Don't count the price of the ticket in this yet). Now, including the price, what is the expected value of the ticket? (Knowing this value, does it make any "sense" to buy a lottery ticket?) The fund-raising group has a target of $1000 to be raised by the lottery. Can they expect to make this much?

Answer: Each ticket has a 1/500 chance of winning the grand prize of $250, a 4/500 chance of winning $50 and a 495/500 chance of winning nothing. So $E(X) = (1/500) \times \$250 + (4/500) \times \$50 + (495/500) \times \$0 = \$0.50 + \$0.40 + \$0.00 = \$0.90$. Including the price the expected value is $\$0.90 - \$3 = -\$2.10$. The expected value of a ticket is $-\$2.10$. Although no single person will lose $2.10 (they either lose $3 or win $50 or $250), $2.10 is the amount, on average, that the lottery gains per ticket. Therefore, they can expect to make $500 \times \$2.10 = \$1050$.

8.2   Standard Deviation of a Random Variable

Of course, this expected value (or mean) is not what actually happens to any *particular* policyholder. No individual policy actually costs the company $200. We are dealing with random events, so some policyholders receive big payouts and others nothing. Because the insurance company must anticipate this variability, it needs to know the standard deviation of the random variable.

For data, we calculate the standard deviation by first computing the deviation of each data value from the mean and squaring it. We perform a similar calculation when we compute the **standard deviation** of a (discrete) random variable as well. First, we find the deviation of each payout from the mean (expected value). (See Table 8.2.)

Policyholder Outcome	Payout x (cost)	Probability $P(X = x)$	Deviation $(x - EV)$
Death	100,000	$\dfrac{1}{1000}$	$(100{,}000 - 200) = 99{,}800$
Disability	50,000	$\dfrac{2}{1000}$	$(50{,}000 - 200) = 49{,}800$
Neither	0	$\dfrac{997}{1000}$	$(0 - 200) = -200$

| Table 8.2   Deviations between the expected value and each payout (cost).

Next, we square each deviation. The **variance** is the expected value of those squared deviations. To find it, we multiply each by the appropriate probability and sum those products:

$$Var(X) = 99{,}800^2\left(\frac{1}{1000}\right) + 49{,}800^2\left(\frac{2}{1000}\right) + (-200)^2\left(\frac{997}{1000}\right)$$

$$= 14{,}960{,}000.$$

---

[1]The concept of expected values for continuous random variables is similar, but the calculation requires calculus and is beyond the scope of this text.

Finally, we take the square root to get the standard deviation:

$$SD(X) = \sqrt{14{,}960{,}000} \approx \$3867.82$$

The insurance company can expect an average payout of $200 per policy, with a standard deviation of $3867.82.

Think about that. The company charges $500 for each policy and expects to pay out $200 per policy. Sounds like an easy way to make $300. (In fact, most of the time—probability 997/1000—the company pockets the entire $500.) But would you be willing to take on this risk yourself and sell all your friends policies like this? The problem is that occasionally the company loses big. With a probability of 1/1000, it will pay out $100,000, and with a probability of 2/1000, it will pay out $50,000. That may be more risk than you're willing to take on. The standard deviation of $3867.82 gives an indication of the uncertainty of the profit, and that seems like a pretty big spread (and risk) for an average profit of $300.

Here are the formulas for these arguments. Because these are parameters of our probability model, the variance and standard deviation can also be written as σ^2 and σ, respectively (sometimes with the name of the random variable as a subscript). You should recognize both kinds of notation:

$$\sigma^2 = Var(X) = \sum(x - \mu)^2 P(x), \text{ and}$$
$$\sigma = SD(X) = \sqrt{Var(X)}.$$

Guided Example Computer Inventory

 As the head of inventory for a computer company, you've had a challenging couple of weeks. One of your warehouses recently had a fire, and you had to flag all the computers stored there to be recycled. On the positive side, you were thrilled that you had managed to ship two computers to your biggest client last week. But then you discovered that your assistant hadn't heard about the fire and had mistakenly transported a whole truckload of computers from the damaged warehouse into the shipping center. It turns out that 30% of all the computers shipped last week were damaged. You don't know whether your biggest client received two damaged computers, two undamaged ones, or one of each. Computers were selected at random from the shipping center for delivery.

If your client received two undamaged computers, everything is fine. If the client gets one damaged computer, it will be returned at your expense—$100—and you can replace it. However, if both computers are damaged, the client will cancel all other orders this month, and you'll lose $10,000. What is the expected value and the standard deviation of your loss under this scenario?

PLAN	Setup State the problem.	We want to analyze the potential consequences of shipping damaged computers to a large client. We'll look at the expected value and standard deviation of the amount we'll lose.
		Let X = amount of loss. We'll denote the receipt of an undamaged computer by **U** and the receipt of a damaged computer by **D**. The three possibilities are: two undamaged computers (**U** and **U**), two damaged computers (**D** and **D**), and one of each (**UD** or **DU**). Because the computers were selected randomly and the number in the warehouse is large, we can assume independence.

(continued)

DO

Model List the possible values of the random variable, and compute all the values you'll need to determine the probability model.

Because the events are independent, we can use the multiplication rule (Chapter 7) and find:

$$P(\mathbf{UU}) = P(\mathbf{U}) \times P(\mathbf{U})$$
$$= 0.7 \times 0.7 = 0.49$$
$$P(\mathbf{DD}) = P(\mathbf{D}) \times P(\mathbf{D})$$
$$= 0.3 \times 0.3 = 0.09$$

So, $P(\mathbf{UD} \text{ or } \mathbf{DU}) = 1 - (0.49 + 0.09) = 0.42$

We have the following model for all possible values of X.

Outcome	x	$P(X=x)$
Two damaged	10000	$P(\mathbf{DD}) = 0.09$
One damaged	100	$P(\mathbf{UD} \text{ or } \mathbf{DU}) = 0.42$
Neither damaged	0	$P(\mathbf{UU}) = 0.49$

Mechanics Find the expected value.

$$E(X) = 0(0.49) + 100(0.42) + 10000(0.09)$$
$$= \$942.00$$

Find the variance.

$$Var(X) = (0 - 942)^2 \times (0.49)$$
$$+ (100 - 942)^2 \times (0.42)$$
$$+ (10000 - 942)^2 \times (0.09)$$
$$= 8,116,836$$

Find the standard deviation.

$$SD(X) = \sqrt{8,116,836} = \$2849.01$$

REPORT

Conclusion Interpret your results in context.

REALITY CHECK

MEMO

Re: Damaged Computers

The recent shipment of two computers to our large client may have some serious negative impact. Even though there is about a 50% chance that they will receive two perfectly good computers, there is a 9% chance that they will receive two damaged computers and will cancel the rest of their monthly order. We have analyzed the expected loss to the firm as $942 with a standard deviation of $2849.01. The large standard deviation reflects the fact that there is a real possibility of losing $10,000 from the mistake.

Both numbers seem reasonable. The expected value of $942 is between the extremes of $0 and $10,000, and there's great variability in the outcome values.

For Example   Calculating standard deviation of a random variable

Question: In the lottery on page 220, we found the expected gain per ticket to be $2.10. What is the standard deviation? What does it say about your chances in the lottery? Comment.

Answer:

$$\sigma^2 = Var(X) = \sum (x - E(X))^2 P(X) = \sum (x - 2.10)^2 P(x)$$

$$= \sum (250 - 2.10)^2 \frac{1}{500} + (50 - 2.10)^2 \frac{4}{500} + (0 - 2.10)^2 \frac{495}{500}$$

$$= \sum 61,454.41 \times \frac{1}{500} + 2,294.41 \times \frac{4}{500} + 4.41 \times \frac{495}{500}$$

$$= 145.63$$

so $\sigma = \sqrt{145.63} = \$12.07$

That's a lot of variation for a mean of $2.10, which reflects the fact that there is a small chance that you'll win a lot but a large chance you'll win nothing.

8.3   Properties of Expected Values and Variances

Our example insurance company expected to pay out an average of $200 per policy, with a standard deviation of about $3868. The expected profit then was $500 − $200 = $300 per policy. Suppose that the company decides to lower the price of the premium by $50 to $450. It's pretty clear that the expected profit would drop an average of $50 per policy, to $450 − $200 = $250.

What about the standard deviation? We know that adding or subtracting a constant from data shifts the mean but doesn't change the variance or standard deviation. The same is true of random variables:[2]

$$E(X \pm c) = E(X) \pm c,$$
$$Var(X \pm c) = Var(X), \text{ and}$$
$$SD(X \pm c) = SD(X).$$

What if the company decides to *double* all the payouts—that is, pay $200,000 for death and $100,000 for disability? This would double the average payout per policy and also increase the variability in payouts. In general, multiplying each value of a random variable by a constant multiplies the mean by that constant and multiplies the variance by the *square* of the constant:

$$E(aX) = aE(X), \text{ and}$$
$$Var(aX) = a^2 Var(X).$$

Taking square roots of the last equation shows that the standard deviation is multiplied by the absolute value of the constant:

$$SD(aX) = |a|SD(X).$$

This insurance company sells policies to more than just one person. We've just seen how to compute means and variances for one person at a time. What happens to the mean and variance when we have a collection of customers? The profit on a group of customers is the *sum* of the individual profits, so we'll need to know how to find expected values and variances for sums. To start, consider a simple case with

---

[2]The rules in this section are true for both discrete *and* continuous random variables.

just two customers who we'll call Mr. Ecks and Ms. Wye. With an expected payout of $200 on each policy, we might expect a total of $200 + $200 = $400 to be paid out on the two policies—nothing surprising there. In other words, we have the **Addition Rule for Expected Values of Random Variables**: *The expected value of the sum (or difference) of random variables is the sum (or difference) of their expected values:*

$$E(X \pm Y) = E(X) \pm E(Y).$$

The variability is another matter. Is the risk of insuring two people the same as the risk of insuring one person for twice as much? We wouldn't expect both clients to die or become disabled in the same year. In fact, because we've spread the risk, the standard deviation should be smaller. Indeed, this is the fundamental principle behind insurance. By spreading the risk among many policies, a company can keep the standard deviation quite small and predict costs more accurately. It's much less risky to insure thousands of customers than one customer when the total expected payout is the same, assuming that the events are independent. Catastrophic events such as hurricanes or earthquakes that affect large numbers of customers at the same time destroy the independence assumption, and often the insurance company along with it.

But how much smaller is the standard deviation of the sum? It turns out that, if the random variables are independent, we have the **Addition Rule for Variances of (Independent) Random Variables**: *The variance of the sum or difference of two independent random variables is the sum of their individual variances:*

$$Var(X \pm Y) = Var(X) + Var(Y)$$

if X and Y are independent.

Math Box

Pythagorean Theorem of Statistics

We often use the standard deviation to measure variability, but when we add independent random variables, we use their variances. Think of the Pythagorean Theorem. In a right triangle (only), the *square* of the length of the hypotenuse is the sum of the *squares* of the lengths of the other two sides:

$$c^2 = a^2 + b^2.$$

For independent random variables (only), the *square* of the standard deviation of their sum is the sum of the *squares* of their standard deviations:

$$SD^2(X + Y) = SD^2(X) + SD^2(Y).$$

It's simpler to write this with *variances*:

$$Var(X + Y) = Var(X) + Var(Y),$$

but we'll use the standard deviation formula often as well:

$$SD(X + Y) = \sqrt{Var(X) + Var(Y)}.$$

For Mr. Ecks and Ms. Wye, the insurance company can expect their outcomes to be independent, so (using X for Mr. Ecks's payout and Y for Ms. Wye's):

$$Var(X + Y) = Var(X) + Var(Y)$$
$$= 14,960,000 + 14,960,000$$
$$= 29,920,000.$$

Let's compare the variance of writing two independent policies to the variance of writing only one for twice the size. If the company had insured only Mr. Ecks for twice as much, the variance would have been

$$Var(2X) = 2^2 Var(X) = 4 \times 14,960,000 = 59,840,000, \text{or}$$

twice as big as with two independent policies, even though the expected payout is the same.

Of course, variances are in squared units. The company would prefer to know standard deviations, which are in dollars. The standard deviation of the payout for two independent policies is $SD(X + Y) = \sqrt{Var(X + Y)} = \sqrt{29,920,000} = \$5469.92$. But the standard deviation of the payout for a single policy of twice the size is twice the standard deviation of a single policy: $SD(2X) = 2SD(X) = 2(\$3867.82) = \$7735.64$, or about 40% more than the standard deviation of the sum of the two independent policies.

If the company has two customers, then it will have an expected annual total payout (cost) of $400 with a standard deviation of about $5470. If they write one policy with an expected annual payout of $400, they increase the standard deviation by about 40%. Spreading risk by insuring many independent customers is one of the fundamental principles in insurance and finance.

Let's review the rules of expected values and variances for sums and differences.

- *The expected value of the sum of two random variables is the sum of the expected values.*

- *The expected value of the difference of two random variables is the difference of the expected values:*

$$E(X \pm Y) = E(X) \pm E(Y).$$

- *If the random variables are independent, the variance of their sum or difference is always the sum of the variances:*

$$Var(X \pm Y) = Var(X) + Var(Y).$$

Do we always *add* variances? Even when we take the *difference* of two random quantities? Yes! Think about the two insurance policies. Suppose we want to know the mean and standard deviation of the *difference* in payouts to the two clients. Since each policy has an expected payout of $200, the expected difference is $200 − $200 = $0. If we computed the variance of the difference by subtracting variances, we would get $0 for the variance. But that doesn't make sense. Their difference won't always be exactly $0. In fact, the difference in payouts could range from $100,000 to −$100,000, a spread of $200,000. The variability in differences *increases* as much as the variability in sums. If the company has two customers, the difference in payouts has a mean of $0 and a standard deviation of about $5470.

- **For random variables, does $X + X + X = 3X$?** Maybe, but be careful. As we've just seen, insuring one person for $300,000 is not the same risk as insuring three people for $100,000 each. When each instance represents a different outcome for the same random variable, though, it's easy to fall into the trap of writing all of them with the same symbol. Don't make this common mistake. Make sure you write each instance as a *different* random variable. Just because each random variable describes a similar situation doesn't mean that each random outcome will be the same. What you really mean is $X_1 + X_2 + X_3$. Written this way, it's clear that the sum shouldn't necessarily equal 3 times *anything*.

For Example   Sums of random variables

You are considering investing $1000 into one or possibly two different investment funds. Historically, each has delivered 5% a year in profit with a standard deviation of 3%. So, a $1000 investment would produce $50 with a standard deviation of $30.

Question: Assuming the two funds are independent, what are the relative advantages and disadvantages of putting $1000 into one, or splitting the $1000 and putting $500 into each? Compare the means and SDs of the profit from the two strategies.

Answer: Let X = amount gained by putting $1000 into one

$$E(X) = 0.05 \times 1000 = \$50 \text{ and } SD(X) = 0.03 \times 1000 = \$30.$$

Let W = amount gained by putting $500 into each. W_1 and W_2 are the amounts from each fund respectively. $E(W_1) = E(W_2) = 0.05 \times 500 = \$25$. So $E(W) = E(W_1) + E(W_2) = \$25 + \$25 = \$50$. The expected values of the two strategies are the same. You expect on average to earn $50 on $1000.

$$SD(W) = \sqrt{SD^2(W_1) + SD^2(W_2)}$$
$$= \sqrt{(0.03 \times 500)^2 + (0.03 \times 500)^2}$$
$$= \sqrt{15^2 + 15^2}$$
$$= \$21.213$$

The standard deviation of the amount earned is $21.213 by splitting the investment amount compared to $30 for investing in one. The expected values are the same. Spreading the investment into more than one vehicle *reduces* the variation. On the other hand, keeping it all in one vehicle increases the chances of both extremely good and extremely bad returns. Which one is better depends on an individual's appetite for risk.[3]

Just Checking

1 Suppose that the time it takes a customer to get and pay for seats at the ticket window of a baseball park is a random variable with a mean of 100 seconds and a standard deviation of 50 seconds. When you get there, you find only two people in line in front of you.

a) How long do you expect to wait for your turn to get tickets?
b) What's the standard deviation of your wait time?
c) What assumption did you make about the two customers in finding the standard deviation?

8.4   Discrete Probability Distributions

We've seen how to compute means and standard deviations of random variables. But plans based just on averages are, on average, wrong. At least that's what Sam Savage, Professor at Stanford University says in his book, *The Flaw of Averages*. Unfortunately, many business owners make decisions based solely on averages—the average amount sold last year, the average number of customers seen last month, etc. Instead of relying on averages, the business decisionmaker can incorporate much more by modeling the situation with a probability model. Probability models can play an important and pivotal role in helping decisionmakers better predict both the outcome and the consequences of their decisions. In this section we'll see that some fairly simple models provide a framework for thinking about how to model a wide variety of business phenomena.

---

[3]The assumption of independence is crucial, but not always (or ever) reasonable. As a March 3, 2010, article on *CNN Money* stated:

"It's only when economic conditions start to return to normal . . . that investors, and investments, move independently again. That's when diversification reasserts its case. . . ."

http://money.cnn.com/2010/03/03/pf/funds/diversification.moneymag/index.htm

The Uniform Distribution

When we first studied probability in Chapter 7, we saw that equally likely events were the simplest case. For example, a single die can turn up 1, 2, . . . , 6 on one toss. A probability model for the toss is Uniform because each of the outcomes has the same probability (1/6) of occurring. Similarly if X is a random variable with possible outcomes 1, 2, . . . , n and $P(X = i) = 1/n$ for each value of i, then we say X has a **discrete Uniform distribution, U[1, . . . , *n*]**.

Bernoulli Trials

When Google Inc. designed their web browser *Chrome*, they worked hard to minimize the probability that their browser would have trouble displaying a website. Before releasing the product, they had to test many websites to discover those that might fail. Although web browsers are relatively new, *quality control inspection* such as this is common throughout manufacturing worldwide and has been in use in industry for nearly 100 years.

The developers of *Chrome* sampled websites, recording whether the browser displayed the website correctly or had a problem. We call the act of inspecting a website a trial. There are two possible outcomes—either the website is displayed correctly or it isn't. The developers thought that whether any particular website displayed correctly was independent from other sites. Situations like this occur often and are called **Bernoulli trials.** To summarize, trials are Bernoulli if:

- There are only two possible outcomes (called *success* and *failure*) for each **trial**.
- The probability of success, denoted p, is the same on every trial. (The probability of failure, $1 - p$ is often denoted q.)
- The trials are independent. Finding that one website does not display correctly does not change what might happen with the next website.

Common examples of Bernoulli trials include tossing a coin, collecting responses on Yes/No questions from surveys or even shooting free throws in a basketball game. Bernoulli trials are remarkably versatile and can be used to model a wide variety of real-life situations. The specific question you might ask in different situations will give rise to different random variables that, in turn, have different probability models.

Of course, the *Chrome* developers wanted to find websites that wouldn't display so they could fix any problems in the browser. So for them a "success" was finding a failed website. The labels "success" and "failure" are often applied arbitrarily, so be sure you know what they mean in any particular situation.

The Geometric Distribution

What's the probability that the first website that fails to display is the second one that we test? Let X denote the number of trials (websites) until the first such "success." For X to be 2, the first website must have displayed correctly (which has probability $1 - p$), and then the second one must have failed to display correctly—a success, with probability p. Since the trials are independent, these probabilities can be multiplied, and so $P(X = 2) = (1 - p)(p)$ or qp. Maybe you won't find a success until the fifth trial. What are the chances of that? You'd have to fail 4 times in a row and then succeed, so $P(X = 5) = (1 - p)^4(p) = q^4p$. See the Math Box for an extension and more explanation.

Whenever we want to know how long (how many trials) it will take us to achieve the first success, the model that tells us this probability is called the **geometric probability distribution**. Geometric distributions are completely specified by one parameter, p, the probability of success. We denote them Geom(p).

Daniel Bernoulli (1700–1782) was the nephew of Jakob, whom you saw in Chapter 7. He was the first to work out the mathematics for what we now call Bernoulli trials.

The geometric distribution can tell Google something important about its software. No large complex program is entirely free of bugs. So before releasing a program or upgrade, developers typically ask not whether it is free of bugs, but how long it is likely to be until the next bug is discovered. If the expected number of pages displayed until the next failure is high enough, then the program is ready to ship.

Notation Alert! ——————

Now we have two more reserved letters. Whenever we deal with Bernoulli trials, p represents the probability of success, and q represents the probability of failure. (Of course, $q = 1 - p$.)

Geometric probability model for Bernoulli trials: Geom(p)

p = probability of success (and $q = 1 - p$ = probability of failure)
X = number of trials until the first success occurs

$$P(X = x) = q^{x-1}p$$

Expected value: $\mu = \dfrac{1}{p}$

Standard deviation: $\sigma = \sqrt{\dfrac{q}{p^2}}$

Math Box

Finding the Expected Value of a Geometric Distribution

We want to find the mean (expected value) of random variable X, using a geometric distribution with probability of success p.

First write the probabilities:

x	1	2	3	4	...
$P(X = x)$	p	qp	q^2p	q^3p	...

The expected value is: $E(X) = 1p + 2qp + 3q^2p + 4q^3p + \cdots$

Let $p = 1 - q$: $= (1 - q) + 2q(1 - q) + 3q^2(1 - q) + 4q^3(1 - q) + \cdots$

Simplify: $= 1 - q + 2q - 2q^2 + 3q^2 - 3q^3 + 4q^3 - 4q^4 + \cdots$

That's an infinite geometric series, with first term 1 and common ratio q: $= 1 + q + q^2 + q^3 + \cdots$

$= \dfrac{1}{1 - q}$

So, finally $E(X) = \dfrac{1}{p}$.

Independence

One of the important requirements for Bernoulli trials is that the trials be independent. Sometimes that's a reasonable assumption. Is it true for our example? It's easy to imagine that related sites might have similar problems, but if the sites are selected at random, whether one has a problem should be independent of others.

The 10% Condition: Bernoulli trials must be independent. In theory, we need to sample from a population that's infinitely big. However, if the population is finite, it's still okay to proceed as long as the sample is smaller than 10% of the population. In Google's case, they just happened to have a directory of millions of websites, so most samples would easily satisfy the 10% condition.

The Binomial Distribution

Suppose Google tests 5 websites. What's the probability that *exactly* 2 of them have problems (2 "successes")? When we studied the geometric distribution we asked how long it would take until our first success. Now we want to find the probability of getting exactly 2 successes among the 5 trials. We are still talking about Bernoulli trials, but we're asking a different question.

This time we're interested in the *number of successes* in the 5 trials, which we'll denote by X. We want to find $P(X = 2)$. Whenever the random variable of interest is the number of successes in a series of Bernoulli trials, it's called a **Binomial random variable.** It takes two parameters to define this **Binomial probability distribution:** the number of trials, n, and the probability of success, p. We denote this distribution $\text{Binom}(n, p)$.

Suppose that in this phase of development, 10% of the sites exhibited some sort of problem so that $p = 0.10$. (Early in the development phase of a product, it is not uncommon for the number of defects to be much higher than it is when the product is released.) Exactly 2 successes in 5 trials means 2 successes and 3 failures. It seems logical that the probability should be $(p)^2(1 - p)^3$. Unfortunately, it's not *quite* that easy. That calculation would give you the probability of finding two successes and then three failures—*in that order*. But you could find the two successes in a lot of other ways, for example in the 2nd and 4th website you test. The probability of that sequence is $(1 - p)p(1 - p)p(1 - p)$ which is also $p^2(1 - p)^3$. In fact, as long as there are two successes and three failures, the probability will always be the same, regardless of the order of the sequence of successes and failures. The probability will be $(p)^2(1 - p)^3$. To find the probability of getting 2 successes in 5 trials in any order, we just need to know how many ways that outcome can occur.

Fortunately, the possible sequences that lead to the same number of successes are *disjoint*. (For example, if your successes came on the first two trials, they couldn't come on the last two.) So once we find all the different sequences, we can add up their probabilities. And since the probabilities are all the same, we just need to find how many sequences there are and multiply $(p)^2(1 - p)^3$ by that number.

Each different order in which we can have k successes in n trials is called a "combination." The total number of ways this can happen is written $\binom{n}{k}$ or $_nC_k$ and pronounced "n choose k:"

$$\binom{n}{k} = {_nC_k} = \frac{n!}{k!(n - k)!} \text{ where } n! = n \times (n - 1) \times \cdots \times 1.$$

For 2 successes in 5 trials,

$$\binom{5}{2} = \frac{5!}{2!(5 - 2)!} = \frac{(5 \times 4 \times 3 \times 2 \times 1)}{(2 \times 1 \times 3 \times 2 \times 1)} = \frac{(5 \times 4)}{(2 \times 1)} = 10.$$

So there are 10 ways to get 2 successes in 5 websites, and the probability of each is $(p)^2(1 - p)^3$. To find the probability of exactly 2 successes in 5 trials, we multiply the probability of any particular order by this number:

$P(\textit{exactly 2 successes in 5 trials}) = 10p^2(1 - p)^3 = 10(0.10)^2(0.90)^3 = 0.0729$

In general, we can write the probability of exactly k successes in n trials as $P(X = k) = \binom{n}{k}p^k q^{n-k}$.

If the probability that any single website has a display problem is 0.10, what's the expected number of websites with problems if we test 100 sites? You probably said 10. We suspect you didn't use the formula for expected value that involves multiplying each value times its probability and adding them up. In fact, there is an easier way to find the expected value for a Binomial random variable. You just multiply the probability of success by n. In other words, $E(X) = np$. We prove this in the next Math Box.

The standard deviation is less obvious and you can't just rely on your intuition. Fortunately, the formula for the standard deviation also boils down to something simple: $SD(X) = \sqrt{npq}$. If you're curious to know where that comes from, it's in the Math Box, too.

In our website example, with $n = 100$, $E(X) = np = 100(0.10) = 10$ so we expect to find 10 successes out of the 100 trials. The standard deviation is

$$\sqrt{100 \times 0.10 \times 0.90} = 3 \text{ websites.}$$

To summarize, a Binomial probability model describes the distribution of the number of successes in a specified number of trials.

Binomial model for Bernoulli trials: Binom(n, p)

n = number of trials
p = probability of success (and $q = 1 - p$ = probability of failure)
X = number of successes in n trials

$$P(X = x) = \binom{n}{x} p^x q^{n-x}, \text{where } \binom{n}{x} = \frac{n!}{x!(n-x)!}$$

Mean: $\mu = np$
Standard deviation: $\sigma = \sqrt{npq}$

Math Box

Mean and Standard Deviation of the Binomial Model

To derive the formulas for the mean and standard deviation of the Binomial distribution we start with the most basic situation.

Consider a single Bernoulli trial with probability of success p. Let's find the mean and variance of the number of successes.

Here's the probability model for the number of successes:

x	0	1
$P(X = x)$	q	p

Find the expected value:

$$E(X) = 0q + 1p$$
$$E(X) = p$$

Now the variance:

$$Var(X) = (0 - p)^2 q + (1 - p)^2 p$$
$$= p^2 q + q^2 p$$
$$= pq(p + q)$$
$$= pq(1)$$
$$Var(X) = pq$$

What happens when there is more than one trial? A Binomial distribution simply counts the number of successes in a series of n independent Bernoulli trials. That makes it easy to find the mean and standard deviation of a binomial random variable, Y.

$$\text{Let } Y = X_1 + X_2 + X_3 + \cdots + X_n$$
$$E(Y) = E(X_1 + X_2 + X_3 + \cdots + X_n)$$
$$= E(X_1) + E(X_2) + E(X_3) + \cdots + E(X_n)$$
$$= p + p + p + \cdots + p \,(\text{There are } n \text{ terms.})$$

So, as we thought, the mean is $E(Y) = np$.

And since the trials are independent, the variances add:

$$Var(Y) = Var(X_1 + X_2 + X_3 + \cdots + X_n)$$
$$= Var(X_1) + Var(X_2) + Var(X_3) + \cdots + Var(X_n)$$
$$= pq + pq + pq + \cdots + pq \,(\text{Again, } n \text{ terms.})$$
$$Var(Y) = npq$$

Voila! The standard deviation is $SD(Y) = \sqrt{npq}$.

Guided Example   The American Red Cross

Every two seconds someone in America needs blood.

The American Red Cross is a non-profit organization that runs like a large business. It serves over 3000 hospitals around the United States, providing a wide range of high quality blood products and blood donor and patient testing services. It collects blood from over 4 million donors and provides blood to millions of patients with a dedication to meeting customer needs.[4]

The balancing of supply and demand is complicated not only by the logistics of finding donors that meet health criteria, but by the fact that the blood type of donor and patient must be matched. People with O-negative blood are called "universal donors" because O-negative blood can be given to patients with any blood type. Only about 6% of people have O-negative blood, which presents a challenge in managing and planning. This is especially true, since, unlike a manufacturer who can balance supply by planning to produce or to purchase more or less of a key item, the Red Cross gets its supply from volunteer donors who show up more-or-less at random (at least in terms of blood type). Modeling the arrival of samples with various blood types helps Red Cross managers to plan their blood allocations.

Here's a small example of the kind of planning required. Of the next 20 donors to arrive at a blood donation center, how many universal donors can be expected? Specifically, what are the mean and standard deviation of the number of universal donors? What is the probability that there are 2 or 3 universal donors?

Question 1: What are the mean and standard deviation of the number of universal donors?
Question 2: What is the probability that there are exactly 2 or 3 universal donors out of the 20 donors?

(continued)

---

[4]Source: www.redcross.org

PLAN	**Setup** State the question.	We want to know the mean and standard deviation of the number of universal donors among 20 people and the probability that there are 2 or 3 of them.
	Check to see that these are Bernoulli trials.	✓ There are two outcomes: success = O-negative failure = other blood types
	Variable Define the random variable.	✓ $p = 0.06$
	Model Specify the model.	✓ **10% Condition:** Fewer than 10% of all possible donors have shown up.
		Let X = number of O-negative donors among $n = 20$ people.
		We can model X with a Binom(20, 0.06).

DO	**Mechanics** Find the expected value and standard deviation.	$E(X) = np = 20(0.06) = 1.2$ $SD(X) = \sqrt{npq} = \sqrt{20(0.06)(0.94)} \approx 1.06$
	Calculate the probability of 2 or 3 successes.	$P(X = 2 \text{ or } 3) = P(X = 2) + P(X = 3)$ $= \binom{20}{2}(0.06)^2(0.94)^{18}$ $+ \binom{20}{3}(0.06)^3(0.94)^{17}$ $\approx 0.2246 + 0.0860$ $= 0.3106$

REPORT	**Conclusion** Interpret your results in context.	**MEMO** **Re: Blood Drive** In groups of 20 randomly selected blood donors, we'd expect to find an average of 1.2 universal donors, with a standard deviation of 1.06. About 31% of the time, we'd expect to find exactly 2 or 3 universal donors among the 20 people.

Simeon Denis Poisson was a French mathematician interested in rare events. He originally derived his model to approximate the Binomial model when the probability of a success, p, is very small and the number of trials, n, is very large. Poisson's contribution was providing a simple approximation to find that probability. When you see the formula, however, you won't necessarily see the connection to the Binomial.

The Poisson Distribution

Not all discrete events can be modeled as Bernoulli trials. Sometimes we're interested simply in the number of events that occur over a given interval of time or space. For example, we might want to model the number of customers arriving in our store in the next ten minutes, the number of visitors to our website in the next minute, or the number of defects that occur in a computer monitor of a certain size. In cases like these, the number of occurrences can be modeled by a **Poisson random variable.** The Poisson's parameter, the mean of the distribution, is usually denoted by λ.

W. S. Gosset, the quality control chemist at the Guinness brewery in the early 20th century who developed the methods of Chapters 13 and 14, was one of the first to use the Poisson in industry. He used it to model and predict the number of yeast cells so he'd know how much to add to the stock. The Poisson is a good model to consider whenever your data consist of counts of occurrences. It requires only that the events be independent and that the mean number of occurrences stays constant.

Poisson probability model for occurrences: Poisson (λ)

λ = mean number of occurrences
X = number of occurrences

$$P(X = x) = \frac{e^{-\lambda}\lambda^x}{x!}$$

Expected value: $\qquad E(X) = \lambda$
Standard deviation: $\qquad SD(X) = \sqrt{\lambda}$

Where Does *e* Come From?

The constant *e* equals 2.7182818 . . . (to 7 decimal places). One of the places *e* originally turned up was in calculating how much money you'd earn if you could get interest compounded more often. If you earn 100% per year simple interest, at the end of the year, you'd have twice as much money as when you started. But if the interest were compounded and paid at the end of every month, each month you'd earn 1/12 of 100% interest. At the year's end you'd have $(1 + 1/12)^{12} = 2.613$ times as much instead of 2. If the interest were paid every day, you'd get $(1 + 1/365)^{365} = 2.714$ times as much. If the interest were paid every second, you'd get $(1 + 1/3153600)^{3153600} = 2.7182818$ times as much. This is where *e* shows up. If you could get the interest compounded continually, you'd get *e* times as much. In other words, as *n* gets large, the limit of $(1 + 1/n)^n = e$. This unexpected result was discovered by Jacob Bernoulli in 1683.

For example, data show an average of about 4 hits per minute to a small business website during the afternoon hours from 1:00 to 5:00 P.M. We can use the Poisson distribution to find the probability that any number of hits will arrive. For example, if we let X be the number of hits arriving in the next minute, then $P(X = x) = \frac{e^{-\lambda}\lambda^x}{x!} = \frac{e^{-4}4^x}{x!}$, using the given average rate of 4 per minute. So, the probability of no hits during the next minute would be $P(X = 0) = \frac{e^{-4}4^0}{0!} = e^{-4} = 0.0183$ (The constant *e* is the base of the natural logarithms and is approximately 2.71828).

One interesting and useful feature of the Poisson distribution is that it scales according to the interval size. For example, suppose we want to know the probability of no hits to our website in the next 30 seconds. Since the mean rate is 4 hits per minute, it's 2 hits per 30 seconds, so we can use the model with $\lambda = 2$ instead. If we let Y be the number of hits arriving in the next 30 seconds, then:

$$P(Y = 0) = \frac{e^{-2}2^0}{0!} = e^{-2} = 0.1353.$$

(Recall that 0! = 1.) The Poisson distribution has been used to model phenomena such as customer arrivals, hot streaks in sports, and disease clusters.

Whenever or wherever rare events happen closely together, people want to know whether the occurrence happened by chance or whether an underlying change caused the unusual occurrence. The Poisson distribution can be used to find the probability of the occurrence and can be the basis for making the judgment.

Just Checking

Roper Worldwide reports that they are able to contact 76% of the randomly selected households drawn for a telephone survey.

2 Explain why these phone calls can be considered Bernoulli trials.

3 Which of the models of this chapter (Geometric, Binomial, or Poisson) would you use to model the number of successful contracts from a list of 1000 sampled households?

4 Roper also reports that even after they contacted a household, only 38% of the contacts agreed to he interviewed. So the probability of getting a completed interview from a randomly selected household is only 0.29. Which of the models of this chapter would you use to model the number of households Roper has to call before they get the first completed interview?

For Example   Probability models

A venture capital firm has a list of potential investors who have previously invested in new technologies. On average, these investors invest about 5% of the time. A new client of the firm is interested in finding investors for a mobile phone application that enables financial transactions, an application that is finding increasing acceptance in much of the developing world. An analyst at the firm starts calling potential investors.

Questions:

1. What is the probability that the first person she calls will want to invest?

2. What is the probability that none of the first five people she calls will be interested?

3. How many people will she have to call until the probability of finding someone interested is at least 0.50?

4. How many investors will she have to call, on average, to find someone interested?

5. If she calls 10 investors, what is the probability that exactly 2 of them will be interested?

6. What assumptions are you making to answer these questions?

Answers:

1. Each investor has a 5% or 1/20 chance of wanting to invest, so the chance that the first person she calls is interested is 1/20.

2. P(first one not interested) $= 1 - 1/20 = 19/20$. Assuming the trials are independent, P(none are interested) $= P$(1st not interested) $\times P$(2nd not interested) $\times \cdots \times P$(5th not interested) $= (19/20)^5 = 0.774$.

3. By trial and error, $(19/20)^{13} = 0.513$ and $(19/20)^{14} = 0.488$, so she would need to call 14 people to have the probability of *no one* interested drop below 0.50, therefore making the probability that someone is interested greater than 0.50.

4. This uses a geometric model. Let $X =$ number of people she calls until 1st person is interested. $E(X) = 1/p = 1/(1/20) = 20$ people.

5. Using the Binomial model, let $Y =$ number of people interested in 10 calls, then

$$P(Y = 2) = \binom{10}{2}p^2(1 - p)^8 = \frac{10 \times 9}{2}(1/20)^2(19/20)^8 = 0.0746$$

6. We are assuming that the trials are independent and that the probability of being interested in investing is the same for all potential investors.

What Can Go Wrong?

- **Probability distributions are still just models.** Models can be useful, but they are not reality. Think about the assumptions behind your models. Question probabilities as you would data.

- **If the model is wrong, so is everything else.** Before you try to find the mean or standard deviation of a random variable, check to make sure the probability distribution is reasonable. As a start, the probabilities should all be between 0 and 1 and they should add up to 1. If not, you may have calculated a probability incorrectly or left out a value of the random variable.

- **Watch out for variables that aren't independent.** You can add expected values of *any* two random variables, but you can only add variances of independent random variables. Suppose a survey includes questions about the number of hours of sleep people get each night and also the number of hours they are awake each day. From their answers, we find the mean and standard deviation of hours asleep and hours awake. The expected total must be 24 hours; after all, people are either asleep or awake. The means still add just fine. Since all the totals are exactly 24 hours, however, the standard deviation of the total will be 0. We can't add variances here because the number of hours you're awake depends on the number of hours you're asleep. Be sure to check for independence before adding variances.

- **Don't write independent instances of a random variable with notation that looks like they are the same variables.** Make sure you write each instance as a different random variable. Just because each random variable describes a similar situation doesn't mean that each random outcome will be the same. These are *random* variables, not the variables you saw in Algebra. Write $X_1 + X_2 + X_3$ rather than $X + X + X$.

- **Don't forget:** Variances of independent random variables add. Standard deviations don't.

- **Don't forget:** Variances of independent random variables add, even when you're looking at the difference between them.

- **Be sure you have Bernoulli trials.** Be sure to check the requirements first: two possible outcomes per trial ("success" and "failure"), a constant probability of success, and independence. Remember that the 10% Condition provides a reasonable substitute for independence.

Ethics in Action

Kurt Williams was about to open a new SEP IRA account and was interested in exploring various investment options. Although he had some ideas about how to invest his money, Kurt thought it best to seek the advice of a professional, so he made an appointment with Keith Klingman, a financial advisor at James, Morgan, and Edwards, LLC. Prior to their first meeting, Kurt told Keith that he preferred to keep his investments simple and wished to allocate his money equally among only two funds. Also, he mentioned that while he was willing to take on some risk to yield higher returns, he was concerned about taking on too much risk given the increased volatility in the markets. After their conversation, Keith began to prepare for their first meeting. Based on historical performance, the firm expected annual returns and standard deviations for the various funds it offers investors. Because Kurt was interested in investing his SEP IRA money in only two funds, Keith decided to compile figures on the expected annual return and standard deviation (a measure of risk) for potential SEP IRA account consisting of different combinations of two funds. If X and Y represent the annual returns for two different funds, Keith knew he could represent the annual return for a specific SEP IRA account as $\frac{1}{2}X + \frac{1}{2}Y$. While calculating the expected annual, $E\left(\frac{1}{2}X + \frac{1}{2}Y\right)$, was straightforward, Keith seemed to recall a more complicated formula for finding the standard deviation. He remembered that he would first need to compute the variance, and after doing some research, decided to use the expression $Var\left(\frac{1}{2}X + \frac{1}{2}Y\right) =$ $\frac{1}{2}^2 Var(X) + \frac{1}{2}^2 Var(Y)$. After completing his computations, he noticed that various combinations of two different equity funds offered some of the highest expected annual returns with relatively low standard deviations. He had anticipated lower standard deviations for accounts that involved mixed assets, such as an equity fund and a bond fund. He was pleasantly surprised with the results, since his firm made more money with investments in equity funds. Keith was confident that these figures would help Kurt realize that he would be best served by investing his SEP IRA money in equity funds only.

ETHICAL ISSUE *Keith incorrectly assumed that funds are independent. It is likely that similar funds are positively correlated (e.g., two equity funds would tend to move in the same direction) while different types of funds are likely negatively correlated (e.g., equity and bond funds tend to move in opposite directions). By not taking the covariance into account, Keith's computations for the variance (and standard deviation) are incorrect. He would have underestimated the variance (and therefore the volatility) for an account consisting of two equity funds (related to Items A and B, ASA Ethical Guidelines).*

ETHICAL SOLUTION *Keith should have recognized that these funds are not independent. His initial uncertainty about how to compute the variance should have made him cautious about his computation results; instead he let his bias toward wanting to sell equity funds affect his judgment. He should not present these figures as fact to Kurt.*

What Have We Learned?

Learning Objectives

■ Understand how probability models relate values to probabilities.

• For discrete random variables, probability models assign a probability to each possible outcome.

■ Know how to find the mean, or expected value, of a discrete probability model from $\mu = \Sigma x P(X = x)$ and the standard deviation from $\sigma = \sqrt{\Sigma (x - \mu)^2 P(x)}$.

■ Foresee the consequences of shifting and scaling random variables, specifically

$$E(X \pm c) = E(X) \pm c \qquad E(aX) = aE(X)$$
$$Var(X \pm c) = Var(X) \qquad Var(aX) = a^2 Var(X)$$
$$SD(X \pm c) = SD(X) \qquad SD(aX) = |a|SD(X)$$

■ Understand that when adding or subtracting random variables the expected values add or subtract well: $E(X \pm y) = E(X) \pm E(Y)$. However, when adding or subtracting independent random variables, the variances *add*:

$$Var(X \pm Y) = Var(X) + Var(Y)$$

■ Be able to explain the properties and parameters of the Uniform, the Binomial, the Geometric, and the Poisson distributions.

Terms

Addition Rule for Expected Values of Random Variables
$E(X \pm Y) = E(X) \pm E(Y)$

Addition Rule for Variances of Random Variables
(Pythagorean Theorem of Statistics)

If X and Y are *independent*: $Var(X \pm Y) = Var(X) + Var(Y)$,

and $SD(X \pm Y) = \sqrt{Var(X) + Var(Y)}$.

Bernoulli trials
A sequence of trials are called Bernoulli trials if:

1. There are exactly two possible outcomes (usually denoted *success* and *failure*).

2. The probability of success is constant.

3. The trials are independent.

Binomial probability distribution
A Binomial distribution is appropriate for a random variable that counts the number of Bernoulli trials.

Changing a random variable by a constant
$E(X \pm c) = E(X) \pm c \qquad Var(X \pm c) = Var(X) \qquad SD(X \pm c) = SD(X)$
$E(aX) = aE(X) \qquad Var(aX) = a^2 Var(X) \qquad SD(aX) = |a|SD(X)$

Discrete random variable
A random variable that can take one of a finite number[5] of distinct outcomes.

Expected value
The expected value of a random variable is its theoretical long-run average value, the center of its model. Denoted μ or $E(X)$, it is found (if the random variable is discrete) by summing the products of variable values and probabilities:

$$\mu = E(X) = \Sigma x P(x)$$

Geometric probability distribution
A model appropriate for a random variable that counts the number of Bernoulli trials until the first success.

---

[5]Technically, there could be an infinite number of outcomes as long as they're *countable*. Essentially, that means we can imagine listing them all in order, like the counting numbers 1, 2, 3, 4, 5, ...

Parameter	A numerically valued attribute of a model, such as the values of μ and σ representing the mean and standard deviation.
Poisson model	A discrete model often used to model the number of arrivals of events such as customers arriving in a queue or calls arriving into a call center.
Probability density function (pdf)	A function $f(x)$ that represents the probability distribution of a continuous random variable X. The probability that X is in an interval A is the area under the curve $f(x)$ over A.
Probability model	A function that associates a probability P with each value of a discrete random variable X, denoted $P(X = x)$, or with any interval of values of a continuous random variable.
Random variable	Assumes any of several different values as a result of some random event. Random variables are denoted by a capital letter, such as X.
Standard deviation of a random variable	Describes the spread in the model and is the square root of the variance.
Uniform distribution	For a discrete uniform distribution over a set of n values, each value has probability $1/n$.
Variance	The variance of a random variable is the expected value of the squared deviations from the mean. For discrete random variables, it can be calculated as:

$$\sigma^2 = Var(X) = \sum (x - \mu)^2 P(x).$$

Brief CASE

Investment Options

A young entrepreneur has just raised $30,000 from investors, and she would like to invest it while she continues her fund-raising in hopes of starting her company one year from now. She wants to do due diligence and understand the risk of each of her investment options. After speaking with her colleagues in finance, she believes that she has three choices: (1) she can purchase a $30,000 certificate of deposit (CD); (2) she can invest in a mutual fund with a balanced portfolio; or (3) she can invest in a growth stock that has a greater potential payback but also has greater volatility. Each of her options will yield a different payback on her $30,000, depending on the state of the economy.

During the next year, she knows that the CD yields a constant annual percentage rate, regardless of the state of the economy. If she invests in a balanced mutual fund, she estimates that she will earn as much as 12% if the economy remains strong, but could possibly lose as much as 4% if the economy takes a downturn. Finally, if she invests all $30,000 in a growth stock, experienced investors tell her that she can earn as much as 40% in a strong economy, but may lose as much as 40% in a poor economy.

Estimating these returns, along with the likelihood of a strong economy, is challenging. Therefore, often a "sensitivity analysis" is conducted, where figures are computed using a range of values for each of the uncertain parameters in the problem. Following this advice, this investor decides to compute measures for a range of interest rates for CDs, a range of returns for the mutual fund, and a range of returns for the growth stock. In addition, the likelihood of a strong economy is unknown, so she will vary these probabilities as well.

Assume that the probability of a strong economy over the next year is 0.3, 0.5, or 0.7. To help this investor make an informed decision, evaluate the expected value and volatility of each of her investments using the following ranges of rates of growth:

CD: Look up the current annual rate for the return on a 3-year CD and use this value $\pm\, 0.5\%$

(continued)

Mutual Fund: Use values of 8%, 10%, and 12% for a strong economy and values of 0%, −2%, and −4% for a weak economy.

Growth Stock: Use values of 10%, 25%, and 40% in a strong economy and values of −10%, −25%, and −40% in a weak economy.

Discuss the expected returns and uncertainty of each of the alternative investment options for this investor in each of the scenarios you analyzed. Be sure to compare the volatility of each of her options.

Exercises

SECTION 8.1

1. A company's employee database includes data on whether or not the employee includes a dependent child in his or her health insurance.

a) Is this variable discrete or continuous?
b) What are the possible values it can take on?

2. The database also, of course, includes each employee's compensation.

a) Is this variable discrete or continuous?
b) What are the possible values it can take on?

3. Suppose that the probabilities of a customer purchasing 0, 1, or 2 books at a book store are 0.2, 0.4, and 0.4, respectively. What is the expected number of books a customer will purchase?

4. A day trader buys an option on a stock that will return $100 profit if the stock goes up today and lose $400 if it goes down. If the trader thinks there is a 75% chance that the stock will go up,

a) What is her expected value of the option's profit?
b) What do you think of this option?

SECTION 8.2

5. Find the standard deviation of the book purchases in exercise 3.

6. Find the standard deviation of the day trader's option value in exercise 4.

7. An orthodontist has three financing packages, and each has a different service charge. He estimates that 30% of patients use the first plan, which has a $10 finance charge; 50% use the second plan, which has a $20 finance charge; and 20% use the third plan, which has a $30 finance charge.

a) Find the expected value of the service charge.
b) Find the standard deviation of the service charge.

8. A marketing agency has developed three vacation packages to promote a timeshare plan at a new resort. They estimate that 20% of potential customers will choose the Day Plan, which does not include overnight accommodations; 40% will choose the Overnight Plan, which includes one night at the resort; and 40% will choose the Weekend Plan, which includes two nights.

a) Find the expected value of the number of nights potential customers will need.
b) Find the standard deviation of the number of nights potential customers will need.

SECTION 8.3

9. Given independent random variables, X and Y, with means and standard deviations as shown, find the mean and standard deviation of each of the variables in parts a to d.

a) $3X$
b) $Y + 6$
c) $X + Y$
d) $X - Y$

	Mean	SD
X	10	2
Y	20	5

10. Given independent random variables, X and Y, with means and standard deviations as shown, find the mean and standard deviation of each of the variables in parts a to d.

a) $X - 20$
b) $0.5Y$
c) $X + Y$
d) $X - Y$

	Mean	SD
X	80	12
Y	12	3

11. A broker has calculated the expected values of two different financial instruments X and Y. Suppose that $E(X) =$ $100, E(Y) = $90, SD(X) = $12 and SD(Y) = $8. Find each of the following.

a) $E(X + 10)$ and $SD(X + 10)$
b) $E(5Y)$ and $SD(5Y)$
c) $E(X + Y)$ and $SD(X + Y)$
d) What assumption must you make in part c?

12. A company selling glass ornaments by mail-order expects, from previous history, that 6% of the ornaments it ships will break in shipping. You purchase two ornaments as gifts and have them shipped separately to two different addresses. What is the probability that both arrive safely? What did you assume?

SECTION 8.4

13. At many airports, a traveler entering the U.S. is sent randomly to one of several stations where his passport and visa are checked. If each of the 6 stations is equally likely, can the probabilities of which station a traveler will be sent be modeled with a Uniform model?

14. At the airport entry sites, a computer is used to randomly decide whether a traveler's baggage should be opened for inspection. If the chance of being selected is 12%, can you model your chance of having your baggage opened with a Bernoulli model? Check each of the conditions specifically.

15. The 2000 Census showed that 26% of all firms in the United States are owned by women. You are phoning local businesses, assuming that the national percentage is true in your area. You wonder how many calls you will have to make before you find one owned by a woman. What probability model should you use? (Specify the parameters as well.)

16. As in Exercise 15, you are phoning local businesses. You call 3 firms. What is the probability that all three are owned by women?

17. A manufacturer of clothing knows that the probability of a button flaw (broken, sewed on incorrectly, or missing) is 0.002. An inspector examines 50 shirts in an hour, each with 6 buttons. Using a Poisson probability model:
a) What is the probability that she finds no button flaws?
b) What is the probability that she finds at least one?

18. Replacing the buttons with snaps increases the probability of a flaw to 0.003, but the inspector can check 70 shirts an hour (still with 6 snaps each). Now what is the probability she finds no snap flaws?

CHAPTER EXERCISES

19. New website. You have just launched the website for your company that sells nutritional products online. Suppose X = the number of different pages that a customer hits during a visit to the website.
a) Assuming that there are n different pages in total on your website, what are the possible values that this random variable may take on?
b) Is the random variable discrete or continuous?

20. New website, part 2. For the website described in Exercise 1, let Y = the total time (in minutes) that a customer spends during a visit to the website.
a) What are the possible values of this random variable?
b) Is the random variable discrete or continuous?

21. Job interviews. Through the career services office, you have arranged preliminary interviews at four companies for summer jobs. Each company will either ask you to come to their site for a follow-up interview or not. Let X be the

random variable equal to the total number of follow-up interviews that you might have.
a) List all the possible values of X.
b) Is the random variable discrete or continuous?
c) Do you think a uniform distribution might be appropriate as a model for this random variable? Explain briefly.

22. Help desk. The computer help desk is staffed by students during the 7:00 P.M. to 11:00 P.M. shift. Let Y denote the random variable that represents the number of students seeking help during the 15-minute time slot 10:00 to 10:15 P.M.
a) What are the possible values of Y?
b) Is the random variable discrete or continuous?

23. Lottery. Iowa has a lottery game called Pick 3 in which customers buy a ticket for $1 and choose three numbers, each from zero to nine. They also must select the play type, which determines what combinations are winners. In one type of play, called the "Straight/Box," they win if they match the three numbers in any order, but the payout is greater if the order is exact. For the case where all three of the numbers selected are different, the probabilities and payouts are:

	Probability	Payout
Straight/Box Exact	1 in 1000	$350
Straight/Box Any	5 in 1000	$50

a) Find the amount a Straight/Box player can expect to win.
b) Find the standard deviation of the player's winnings.
c) Tickets to play this game cost $1 each. If you subtract $1 from the result in part a, what is the expected result of playing this game?

24. Software company. A small software company will bid on a major contract. It anticipates a profit of $50,000 if it gets it, but thinks there is only a 30% chance of that happening.
a) What's the expected profit?
b) Find the standard deviation for the profit.

25. Commuting to work. A commuter must pass through five traffic lights on her way to work and will have to stop at each one that is red. After keeping record for several months, she developed the following probability model for the number of red lights she hits:

X = # of red	0	1	2	3	4	5
P(X = x)	0.05	0.25	0.35	0.15	0.15	0.05

a) How many red lights should she expect to hit each day?
b) What's the standard deviation?

26. Defects. A consumer organization inspecting new cars found that many had appearance defects (dents, scratches, paint chips, etc.). While none had more than three of these defects, 7% had three, 11% had two, and 21% had one defect.

a) Find the expected number of appearance defects in a new car.
b) What is the standard deviation?

27. Fishing tournament. A sporting goods manufacturer was asked to sponsor a local boy in two fishing tournaments. They claim the probability that he will win the first tournament is 0.4. If he wins the first tournament, they estimate the probability that he will also win the second is 0.2. They guess that if he loses the first tournament, the probability that he will win the second is 0.3.

a) According to their estimates, are the two tournaments independent? Explain your answer.
b) What's the probability that he loses both tournaments?
c) What's the probability he wins both tournaments?
d) Let random variable X be the number of tournaments he wins. Find the probability model for X.
e) What are the expected value and standard deviation of X?

28. Contracts. Your company bids for two contracts. You believe the probability that you get contract #1 is 0.8. If you get contract #1, the probability that you also get contract #2 will be 0.2, and if you do not get contract #1, the probability that you get contract #2 will be 0.3.

a) Are the outcomes of the two contract bids independent? Explain.
b) Find the probability you get both contracts.
c) Find the probability you get neither contract.
d) Let X be the number of contracts you get. Find the probability model for X.
e) Find the expected value and standard deviation of X.

29. Battery recall. A company has discovered that a recent batch of batteries had manufacturing flaws, and has issued a recall. You have 10 batteries covered by the recall, and 3 are dead. You choose 2 batteries at random from your package of 10.

a) Has the assumption of independence been met? Explain.
b) Create a probability model for the number of good batteries chosen.
c) What's the expected number of good batteries?
d) What's the standard deviation?

30. Grocery supplier. A grocery supplier believes that the mean number of broken eggs per dozen is 0.6, with a standard deviation of 0.5. You buy 3 dozen eggs without checking them.

a) How many broken eggs do you expect to get?
b) What's the standard deviation?
c) Is it necessary to assume the cartons of eggs are independent? Why?

31. Commuting, part 2. A commuter finds that she waits an average of 14.8 seconds at each of five stoplights, with a standard deviation of 9.2 seconds. Find the mean and the standard deviation of the total amount of time she waits at all five lights. What, if anything, did you assume?

32. Repair calls. A small engine shop receives an average of 1.7 repair calls per hour, with a standard deviation of 0.6. What is the mean and standard deviation of the number of calls they receive for an 8-hour day? What, if anything, did you assume?

33. Insurance company. An insurance company estimates that it should make an annual profit of $150 on each homeowner's policy written, with a standard deviation of $6000.

a) Why is the standard deviation so large?
b) If the company writes only two of these policies, what are the mean and standard deviation of the annual profit?
c) If the company writes 1000 of these policies, what are the mean and standard deviation of the annual profit?
d) What circumstances could violate the assumption of independence of the policies?

34. Casino. At a casino, people play the slot machines in hopes of hitting the jackpot, but most of the time, they lose their money. A certain machine pays out an average of $0.92 (for every dollar played), with a standard deviation of $120.

a) Why is the standard deviation so large?
b) If a gambler plays 5 times, what are the mean and standard deviation of the casino's profit?
c) If gamblers play this machine 1000 times in a day, what are the mean and standard deviation of the casino's profit?

35. Bike sale. A bicycle shop plans to offer 2 specially priced children's models at a sidewalk sale. The basic model will return a profit of $120 and the deluxe model $150. Past experience indicates that sales of the basic model will have a mean of 5.4 bikes with a standard deviation of 1.2, and sales of the deluxe model will have a mean of 3.2 bikes with a standard deviation of 0.8 bikes. The cost of setting up for the sidewalk sale is $200.

a) Define random variables and use them to express the bicycle shop's net profit.
b) What's the mean of the net profit?
c) What's the standard deviation of the net profit?
d) Do you need to make any assumptions in calculating the mean? How about the standard deviation?

36. Farmers' market. A farmer has 100 lbs of apples and 50 lbs of potatoes for sale. The market price for apples (per pound) each day is a random variable with a mean of 0.5 dollars and a standard deviation of 0.2 dollars. Similarly, for a pound of potatoes, the mean price is 0.3 dollars and the standard deviation is 0.1 dollars. It also costs him 2 dollars to bring all the apples and potatoes to the market. The market is busy with eager shoppers, so we can assume that he'll be able to sell all of each type of produce at that day's price.

a) Define your random variables, and use them to express the farmer's net income.
b) Find the mean of the net income.
c) Find the standard deviation of the net income.
d) Do you need to make any assumptions in calculating the mean? How about the standard deviation?

37. Movie rentals. To compete with Netflix, the owner of a movie rental shop decided to try sending DVDs through the mail. In order to determine how many copies of newly released titles he should purchase, he carefully observed turnaround times. Since nearly all of his customers were in his local community, he tested delivery times by sending DVDs to his friends. He found the mean delivery time was 1.3 days, with a standard deviation of 0.5 days. He also noted that the times were the same whether going to the customer or coming back to the shop.

a) Find the mean and standard deviation of the round-trip delivery times for a DVD (mailed to the customer and then mailed back to the shop).
b) The shop owner tries to process a DVD that is returned to him and get it back in the mail in one day, but circumstances sometimes prevent it. His mean turnaround time is 1.1 days, with a standard deviation of 0.3 days. Find the mean and standard deviation of the turnaround times combined with the round-trip times in part a.

38. Online applications. Researchers for an online marketing company suggest that new customers who have to become a member before they can check out on the website are very intolerant of long applications. One way to rate an application is by the total number of keystrokes required to fill it out.

a) One common frustration is having to enter an e-mail address twice. If the mean length of e-mail addresses is 13.3 characters, with a standard deviation of 2.8 characters, what is the mean and standard deviation of total characters typed if entered twice?
b) The company found the mean and standard deviation of the length of customers' names (including spaces) were 13.4 and 2.4 characters, respectively, and for addresses, 30.8 and 6.3 characters. What is the mean and standard deviation of the combined lengths of entering the e-mail addresses twice and then the name and the address?

39. eBay. A collector purchased a quantity of action figures and is going to sell them on eBay. He has 19 Hulk figures. In recent auctions, the mean selling price of similar figures has been $12.11, with a standard deviation of $1.38. He also has 13 Iron Man figures which have had a mean selling price of $10.19, with a standard deviation of $0.77.

His insertion fee will be $0.55 on each item, and the closing fee will be 8.75% of the selling price. He assumes all will sell without having to be relisted.

a) Define your random variables, and use them to create a random variable for the collector's net income.
b) Find the mean (expected value) of the net income.
c) Find the standard deviation of the net income.
d) Do you have to assume independence for the sales on eBay? Explain.

40. Real estate. A real estate broker purchased 3 two-bedroom houses in a depressed market for a combined cost of $71,000. He expects the cleaning and repair costs on each house to average $3700, with a standard deviation of $1450. When he sells them, after subtracting taxes and other closing costs, he expects to realize an average of $39,000 per house, with a standard deviation of $1100.

a) Define your random variables, and use them to create a random variable for the broker's net profit.
b) Find the mean (expected value) of the net profit.
c) Find the standard deviation of the net profit.
d) Do you have to assume independence for the repairs and sale prices of the houses? Explain.

41. Bernoulli. Can we use probability models based on Bernoulli trials to investigate the following situations? Explain.

a) Each week a doctor rolls a single die to determine which of his six office staff members gets the preferred parking space.
b) A medical research lab has samples of blood collected from 120 different individuals. How likely is it that the majority of them are Type A blood, given that Type A is found in 43% of the population?
c) From a workforce of 13 men and 23 women, all five promotions go to men. How likely is that, if promotions are based on qualifications rather than gender?
d) We poll 500 of the 3000 stockholders to see how likely it is that the proposed budget will pass.
e) A company realizes that about 10% of its packages are not being sealed properly. In a case of 24 packages, how likely is it that more than 3 are unsealed?

42. Bernoulli, part 2. Can we use probability models based on Bernoulli trials to investigate the following situations? Explain.

a) You are rolling 5 dice. How likely is it to get at least two 6's to win the game?
b) You survey 500 potential customers to determine their color preference.
c) A manufacturer recalls a doll because about 3% have buttons that are not properly attached. Customers return 37 of these dolls to the local toy store. How likely are they to find any buttons not properly attached?

d) A city council of 11 Republicans and 8 Democrats picks a committee of 4 at random. How likely are they to choose all Democrats?
e) An executive reads that 74% of employees in his industry are dissatisfied with their jobs. How many dissatisfied employees can he expect to find among the 481 employees in his company?

43. Closing sales. A salesman normally makes a sale (closes) on 80% of his presentations. Assuming the presentations are independent, find the probability of each of the following.

a) He fails to close for the first time on his fifth attempt.
b) He closes his first presentation on his fourth attempt.
c) The first presentation he closes will be on his second attempt.
d) The first presentation he closes will be on one of his first three attempts.

44. Computer chip manufacturer. Suppose a computer chip manufacturer rejects 2% of the chips produced because they fail presale testing. Assuming the bad chips are independent, find the probability of each of the following.

a) The fifth chip they test is the first bad one they find.
b) They find a bad one within the first 10 they examine.
c) The first bad chip they find will be the fourth one they test.
d) The first bad chip they find will be one of the first three they test.

45. Side effects. Researchers testing a new medication find that 7% of users have side effects. To how many patients would a doctor expect to prescribe the medication before finding the first one who has side effects?

46. Credit cards. College students are a major target for advertisements for credit cards. At a university, 65% of students surveyed said they had opened a new credit card account within the past year. If that percentage is accurate, how many students would you expect to survey before finding one who had not opened a new account in the past year?

47. Missing pixels. A company that manufactures large LCD screens knows that not all pixels on their screen light, even if they spend great care when making them. In a sheet 6 ft by 10 ft (72 in. by 120 in.) that will be cut into smaller screens, they find an average of 4.7 blank pixels. They believe that the occurrences of blank pixels are independent. Their warranty policy states that they will replace any screen sold that shows more than 2 blank pixels.

a) What is the mean number of blank pixels per square foot?
b) What is the standard deviation of blank pixels per square foot?
c) What is the probability that a 2 ft by 3 ft screen will have at least one defect?
d) What is the probability that a 2 ft by 3 ft screen will be replaced because it has too many defects?

48. Bean bags. Cellophane that is going to be formed into bags for items such as dried beans or bird seed is passed over a light sensor to test if the alignment is correct before it passes through the heating units that seal the edges. Small adjustments can be made by the machine automatically. But if the alignment is too bad, the process is stopped and an operator has to manually adjust it. These misalignment stops occur randomly and independently. On one line, the average number of stops is 52 per 8-hour shift.

a) What is the mean number of stops per hour?
b) What is the standard deviation of stops per hour?

49. Hurricane insurance. An insurance company needs to assess the risks associated with providing hurricane insurance. Between 1990 and 2006, Florida was hit by 22 tropical storms or hurricanes. If tropical storms and hurricanes are independent and the mean has not changed, what is the probability of having a year in Florida with each of the following. (Note that 1990 to 2006 is 17 years.)

a) No hits?
b) Exactly one hit?
c) More than three hits?

50. Hurricane insurance, part 2. Between 1965 and 2007, there were 95 major hurricanes (category 3 or more) in the Atlantic basin. Assume that hurricanes are independent and the mean has not changed.

a) What is the mean number of major hurricanes per year? (There are 43 years from 1965 to 2007.)
b) What is the standard deviation of the frequency of major hurricanes?
c) What is the probability of having a year with no major hurricanes?
d) What is the probability of going three years in a row without a major hurricane?

51. Professional tennis. Serena Williams made a successful first serve 67% of the time in a Wimbledon finals match against her sister Venus. If she continues to serve at the same rate the next time they play and serves 6 times in the first game, determine the following probabilities. (Assume that each serve is independent of the others.)

a) All 6 first serves will be in.
b) Exactly 4 first serves will be in.
c) At least 4 first serves will be in.

52. American Red Cross. Only 4% of people have Type AB blood. A bloodmobile has 12 vials of blood on a rack. If the distribution of blood types at this location is consistent with the general population, what's the probability they find AB blood in:

a) None of the 12 samples?
b) At least 2 samples?
c) 3 or 4 samples?

53. Satisfaction survey. A cable provider wants to contact customers in a particular telephone exchange to see how satisfied they are with the new digital TV service the company has provided. All numbers are in the 452 exchange, so there are 10,000 possible numbers from 452-0000 to 452-9999. If they select the numbers with equal probability:

a) What distribution would they use to model the selection?

b) What is the probability the number selected will be an even number?

c) What is the probability the number selected will end in 000?

54. Manufacturing quality. In an effort to check the quality of their cell phones, a manufacturing manager decides to take a random sample of 10 cell phones from yesterday's production run, which produced cell phones with serial numbers ranging (according to when they were produced) from 43005000 to 43005999. If each of the 1000 phones is equally likely to be selected:

a) What distribution would they use to model the selection?

b) What is the probability that a randomly selected cell phone will be one of the last 100 to be produced?

c) What is the probability that the first cell phone selected is either from the last 200 to be produced or from the first 50 to be produced?

d) What is the probability that the first two cell phones are both from the last 100 to be produced?

55. Web visitors. A website manager has noticed that during the evening hours, about 3 people per minute check out from their shopping cart and make an online purchase. She believes that each purchase is independent of the others and wants to model the number of purchases per minute.

a) What model might you suggest to model the number of purchases per minute?

b) What is the probability that in any one minute at least one purchase is made?

c) What is the probability that no one makes a purchase in the next 2 minutes?

56. Quality control. The manufacturer in Exercise 54 has noticed that the number of faulty cell phones in a production run of cell phones is usually small and that the quality of one day's run seems to have no bearing on the next day.

a) What model might you use to model the number of faulty cell phones produced in one day?

b) If the mean number of faulty cell phones is 2 per day, what is the probability that no faulty cell phones will be produced tomorrow?

c) If the mean number of faulty cell phones is 2 per day, what is the probability that 3 or more faulty cell phones were produced in today's run?

Just Checking Answers

1 a) $100 + 100 = 200$ seconds
   b) $\sqrt{50^2 + 50^2} = 70.7$ seconds
   c) The times for the two customers are independent.

2 There are two outcomes (contact, no contact), the probability of contact stays constant at 0.76, and random calls should be independent.

3 Binomial

4 Geometric

The Normal Distribution

The NYSE

The New York Stock Exchange (NYSE) was founded in 1792 by 24 stockbrokers who signed an agreement under a buttonwood tree on Wall Street in New York. The first offices were in a rented room at 40 Wall Street. In the 1830s traders who were not part of the Exchange did business in the street. They were called "curbstone brokers." It was the curbstone brokers who first made markets in gold and oil stocks and, after the Civil War, in small industrial companies such as the emerging steel, textile, and chemical industries.

By 1903 the New York Stock Exchange was established at its current home at 18 Broad Street. The curbstone brokers finally moved indoors in 1921 to a building on Greenwich street in lower Manhattan. In 1953 the curb market changed its name to the American Stock Exchange. In 1993 the American Stock Exchange pioneered the market for derivatives by introducing the first exchange-traded fund, Standard & Poor's Depositary Receipts (SPDRs).

The NYSE Euronext holding company was created in 2007 as a combination of the NYSE Group, Inc., and Euronext N.V. And in 2008, NYSE Euronext merged with the American Stock Exchange. The combined exchange is the world's largest and most liquid exchange group.

9.1   The Standard Deviation as a Ruler

Investors have always sought ways to help them decide when to buy and when to sell. Such measures have become increasingly sophisticated. But all rely on identifying when the stock market is in an unusual state—either unusually undervalued (buy!) or unusually overvalued (sell!). One such measure is the Cyclically Adjusted Price/Earnings Ratio (CAPE10) developed by Yale professor Robert Shiller. The CAPE10 is based on the standard Price/Earnings (P/E) ratio of stocks, but designed to smooth out short-term fluctuations by "cyclically adjusting" them. The CAPE10 has been as low as 4.78, in 1920, and as high as 44.20, in late 1999. The long-term average CAPE10 (since year 1881) is 16.34.

Investors who follow the CAPE10 use the metric to signal times to buy and sell. One mutual fund strategy buys only when the CAPE10 is 33% lower than the long-term average and sells (or "goes into cash") when the CAPE10 is 50% higher than the long-term average. Between January 1, 1971, and October 23, 2009, this strategy would have outperformed such standard measures as the Wiltshire 5000 in both average return and volatility, but it is important to note that the strategy would have been completely in cash from just before the stock market crash of 1987 all the way to March of 2009! Shiller popularized the strategy in his book *Irrational Exuberance*. Figure 9.1 shows a time series plot of the CAPE10 values for the New York Stock Exchange from 1880 until the middle of 2010. Generally, the CAPE10 hovers around 15. But occasionally, it can take a large excursion. One such time was in 1999 and 2000, when the CAPE10 exceeded 40. But was this just a random peak or were these values really extraordinary?

| Figure 9.1   CAPE10 values for the NYSE from 1880 to 2010.

We can look at the overall distribution of CAPE10 values. Figure 9.2 shows a histogram of the same values. Now we don't see patterns over time, but we may be able to make a better judgment of whether values are extraordinary.

Overall, the main body of the distribution looks unimodal and reasonably symmetric. But then there's a tail of values that trails off to the high end. How can we assess how extraordinary they are?

Investors follow a wide variety of measures that record various aspects of stocks, bonds, and other investments. They are usually particularly interested in identifying times when these measures are extraordinary because those often represent times of increased risk or opportunity. But these are quantitative values, not categories.

How can we characterize the usual behavior of a random variable that can take on any value in a range of values? The distributions of Chapter 8 won't provide the tools we need, but many of the basic concepts still work. The random variables we need are *continuous*.

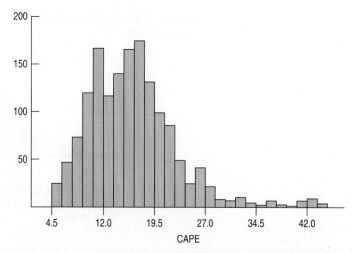

| Figure 9.2   The distribution of the CAPE10 values shown in Figure 9.1.

We saw in Chapter 5 that z-scores provide a standard way to compare values. In a sense, we use the standard deviation as a ruler, asking how many standard deviations a value is from the mean. That's what a z-score reports; the number of standard deviations away from the mean. We can convert the CAPE10 values to z-scores by subtracting their mean (16.3559) and dividing by their standard deviation (6.58). Figure 9.3 shows the resulting distribution.

| Figure 9.3   The CAPE10 values as z-scores.

It's easy to see that the z-scores have the same distribution as the original values, but now we can also see that the largest of them is above 4. How extraordinary is it for a value to be four standard deviations away from the mean? Fortunately, there's a fact about unimodal, symmetric distributions that can guide us.[1]

The 68–95–99.7 Rule

In a unimodal, symmetric distribution, about 68% of the values fall within 1 standard deviation of the mean, about 95% fall within 2 standard deviations of the mean, and about 99.7%—almost all—fall within 3 standard deviations of the

---

[1]All of the CAPE10 values in the right tail occurred after 1993. Until that time the distribution of CAPE10 values was quite symmetric and clearly unimodal.

mean. Calling this rule the **68–95–99.7 Rule** provides a mnemonic for these three values.[2]

Figure 9.4 The 68–95–99.7 Rule tells us how much of most unimodal, symmetric models is found within one, two, or three standard deviations of the mean.

For Example   An extraordinary day for the Dow?

On May 6, 2010, the Dow Jones Industrial Average (DJIA) lost 404.7 points. Although that wasn't the most ever lost in a day, it was a large amount for that period. During the previous year, the mean change in the DJIA was −9.767 with a standard deviation of 98.325 points. A histogram of day-to-day changes in the DJIA looks like this:

Figure 9.5 Day-to-day changes in the Dow Jones Industrial Average for the year ending June 2010.

Question: Use the 68–95–99.7 Rule to characterize how extraordinary the May 6 changes were. Is the rule appropriate?

Answer: The histogram is unimodal and symmetric, so the 68–95–99.7 rule is an appropriate model. The z-score corresponding to the May 6 change is

$$\frac{-404.7 - (-9.767)}{98.325} = -4.017$$

A z-score bigger than 3 in magnitude will occur with a probability of less than 0.015. A z-score of 4 is even less likely. This was a truly extraordinary event.

9.2   The Normal Distribution

The 68–95–99.7 Rule is useful in describing how unusual a z-score is. But often in business we want a more precise answer than one of these three values. To say more about how big we expect a z-score to be, we need to *model* the data's distribution.

"All models are wrong—but some are useful."

—GEORGE BOX, FAMOUS STATISTICIAN

[2]This rule is also called the "Empirical Rule" because it originally was observed without any proof. It was first published by Abraham de Moivre in 1733, 75 years before the underlying reason for it—which we're about to see—was known.

Notation Alert! ─────────

$N(\mu, \sigma)$ always denotes a Normal. The μ, pronounced "mew," is the Greek letter for "m," and always represents the mean in a model. The σ, sigma, is the lowercase Greek letter for "s," and always represents the standard deviation in a model.

─────────────────────

Is the Standard Normal a Standard?

Yes. We call it the "Standard Normal" because it models standardized values. It is also a "standard" because this is the particular Normal model that we almost always use.

A model will let us say much more precisely how often we'd be likely to see z-scores of different sizes. Of course, like all models of the real world, the model will be wrong—wrong in the sense that it can't match reality exactly. But it can still be useful. Like a physical model, it's something we can look at and manipulate to learn more about the real world.

Models help our understanding in many ways. Just as a model of an airplane in a wind tunnel can give insights even though it doesn't show every rivet,[3] models of data give us summaries that we can learn from and use, even though they don't fit each data value exactly. It's important to remember that they're only *models* of reality and not reality itself. But without models, what we can learn about the world at large is limited to only what we can say about the data we have at hand.

There is no universal standard for z-scores, but there is a model that shows up over and over in Statistics. You may have heard of "bell-shaped curves." Statisticians call them Normal distributions. **Normal distributions** are appropriate models for distributions whose shapes are unimodal and roughly symmetric. There is a Normal distribution for every possible combination of mean and standard deviation. We write $N(\mu, \sigma)$ to represent a Normal distribution with a mean of μ and a standard deviation of σ. We use Greek symbols here because *this* mean and standard deviation are not numerical summaries of data. They are part of the model. They don't come from the data. Rather, they are numbers that we choose to help specify the distribution. Such numbers are called **parameters**.

If we model data with a Normal distribution and standardize them using the corresponding μ and σ, we still call the standardized values **z-scores,** and we write

$$z = \frac{y - \mu}{\sigma}.$$

If we standardize the data first (using its mean and standard deviation) it will have mean 0 and standard deviation 1. Then, to model it with a Normal, we'll need only the model $N(0,1)$. The Normal distribution with mean 0 and standard deviation 1 is called the **standard Normal distribution** (or the **standard Normal model**).

But be careful. You shouldn't use a Normal model for just any data set. Remember that standardizing won't change the shape of the distribution. If the distribution is not unimodal and symmetric to begin with, standardizing won't make it Normal.

Just Checking

1 Your Accounting teacher has announced that the lower of your two tests will be dropped. You got a 90 on test 1 and an 80 on test 2. You're all set to drop the 80 until she announces that she grades "on a curve." She standardized the scores in order to decide which is the lower one. If the mean on the first test was 88 with a standard deviation of 4 and the mean on the second was 75 with a standard deviation of 5,

   a) Which one will be dropped?

   b) Does this seem "fair"?

Finding Normal Percentiles

Finding the probability that a proportion is at least 1 SD above the mean is easy. We know that 68% of the values lie within 1 SD of the mean, so 32% lie farther away. Since the Normal distribution is symmetric, half of those 32% (or 16%) are

─────────────────────

[3]In fact, the model is useful *because* it doesn't have every rivet. It is because models offer a simpler view of reality that they are so useful as we try to understand reality.

more than 1 SD above the mean. But what if we want to know the percentage of observations that fall more than 1.8 SD above the mean? We already know that no more than 16% of observations have z-scores above 1. By similar reasoning, no more than 2.5% of the observations have a z-score above 2. Can we be more precise with our answer than "between 16% and 2.5%"?

Figure 9.6   A table of Normal percentiles (Table Z in Appendix D) lets us find the percentage of individuals in a standard Normal distribution falling below any specified z-score value.

When the value doesn't fall exactly 0, 1, 2, or 3 standard deviations from the mean, we can look it up in a table of **Normal percentiles**.[4] Tables use the standard Normal distribution, so we'll have to convert our data to z-scores before using the table. If our data value was 1.8 standard deviations above the mean, we would standardize it to a z-score of 1.80, and then find the value associated with a z-score of 1.80. If we use a table, as shown in Figure 9.6, we find the z-score by looking down the left column for the first two digits (1.8) and across the top row for the third digit, 0. The table gives the percentile as 0.9641. That means that 96.4% of the z-scores are less than 1.80. Since the total area is always 1, and $1 - 0.9641 = 0.0359$ we know that only 3.6% of all observations from a Normal distribution have z-scores higher than 1.80. We can also find the probabilities associated with z-scores using technology such as calculators, statistical software, and various websites.

Finding Normal Percentiles

These days, finding percentiles from a Normal table is rarely necessary. Most of the time, we can use a calculator, a computer, or a website.

For Example   GMAT scores and the Normal model

The Graduate Management Admission Test (GMAT) has scores from 200 to 800. Scores are supposed to follow a distribution that is roughly unimodal and symmetric and is designed to have an overall mean of 500 and a standard deviation of 100. In any one year, the mean and standard deviation may differ from these target values by a small amount, but we can use these values as good overall approximations.

Question: Suppose you earned a 600 on your GMAT test. From that information and the 68–95–99.7 Rule, where do you stand among all students who took the GMAT?

Answer: Because we're told that the distribution is unimodal and symmetric, we can approximate the distribution with a Normal model. We are also told the scores have a mean of 500 and an SD of 100. So, we'll use a $N(500,100)$. It's good practice at this point to draw the distribution. Find the score whose percentile you want to know and locate it on the picture. When you finish the calculation, you should check to make sure that it's a reasonable percentile from the picture.

A score of 600 is 1 SD above the mean. That corresponds to one of the points in the 68–95–99.7% Rule. About 32% (100% − 68%) of those who took the test were more than one standard deviation from the mean, but only half of those were on the high side. So about 16% (half of 32%) of the test scores were better than 600.

[4]See Table Z in Appendix D. Many calculators and statistics computer packages do this as well.

For Example   More GMAT scores

Question: Assuming the GMAT scores are nearly Normal with $N(500,100)$, what proportion of GMAT scores falls between 450 and 600?

Answer: *The first step is to find the z-scores associated with each value.* Standardizing the scores we are given, we find that for 600, $z = (600 - 500)/100 = 1.0$ and for 450, $z = (450 - 500)/100 = -0.50$. We can label the axis below the picture either in the original values or the z-scores or even use both scales as the following picture shows.

From Table Z, we find the area $z \le 1.0 = 0.8413$, which means that 84.13% of scores fall below 1.0, and the area $z \le -0.50 = 0.3085$, which means that 30.85% of the values fall below -0.5, so the proportion of z-scores *between* them is 84.13% − 30.85% = 53.28%. So, the Normal model estimates that about 53.3% of GMAT scores fall between 450 and 600.

Finding areas from z-scores is the simplest way to work with the Normal distribution. But sometimes we start with areas and are asked to work backward to find the corresponding z-score or even the original data value. For instance, what z-score represents the first quartile, Q1, in a Normal distribution? In our first set of examples, we knew the z-score and used the table or technology to find the percentile. Now we want to find the cut point for the 25th percentile. Make a picture, shading the leftmost 25% of the area. Look in Table Z for an area of 0.2500. The exact area is not there, but 0.2514 is the closest number. That shows up in the table with -0.6 in the left margin and .07 in the top margin. The z-score for Q1, then, is approximately $z = -0.67$. Computers and calculators can determine the cut point more precisely (and more easily).[5]

For Example   An exclusive MBA program

Question: Suppose an MBA program says it admits only people with GMAT scores among the top 10%. How high a GMAT score does it take to be eligible?

Answer: The program takes the top 10%, so their cutoff score is the 90th percentile. Draw an approximate picture like this one.

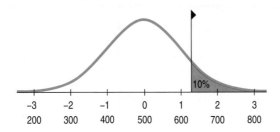

	0.07	0.08	0.09
1.0	0.8577	0.8599	0.8621
1.1	0.8790	0.8810	0.8830
1.2	0.8980	0.8997	0.9015
1.3	0.9147	0.9162	0.9177
1.4	0.9292	0.9306	0.9319

(continued)

---

[5]We'll often use those more precise values in our examples. If you're finding the values from the table you may not get *exactly* the same number to all decimal places as your classmate who's using a computer package.

From our picture we can see that the z-value is between 1 and 1.5 (if we've judged 10% of the area correctly), and so the cut-off score is between 600 and 650 or so. Using technology, you may be able to select the 10% area and find the z-value directly. Using a table, such as Table Z, locate 0.90 (or as close to it as you can; here 0.8997 is closer than 0.9015) in the *interior* of the table and find the corresponding z-score (see table above). Here the 1.2 is in the left margin, and the .08 is in the margin above the entry. Putting them together gives 1.28. Now, convert the z-score back to the original units. From Table Z, the cut point is $z = 1.28$. A z-score of 1.28 is 1.28 standard deviations above the mean. Since the standard deviation is 100, that's 128 GMAT points. The cutoff is 128 points above the mean of 500, or 628. Because the program wants GMAT scores in the top 10%, the cutoff is 628. (Actually since GMAT scores are reported only in multiples of 10, you'd have to score at least a 630.)

Guided Example   Cereal Company

A cereal manufacturer has a machine that fills the boxes. Boxes are labeled "16 oz," so the company wants to have that much cereal in each box. But since no packaging process is perfect, there will be minor variations. If the machine is set at exactly 16 oz and the Normal distribution applies (or at least the distribution is roughly symmetric), then about half of the boxes will be underweight, making consumers unhappy and exposing the company to bad publicity and possible lawsuits. To prevent underweight boxes, the manufacturer has to set the mean a little higher than 16.0 oz. Based on their experience with the packaging machine, the company believes that the amount of cereal in the boxes fits a Normal distribution with a standard deviation of 0.2 oz. The manufacturer decides to set the machine to put an average of 16.3 oz in each box. Let's use that model to answer a series of questions about these cereal boxes.

Question 1: What fraction of the boxes will be underweight?

PLAN	Setup State the variable and the objective.	The variable is weight of cereal in a box. We want to determine what fraction of the boxes risk being underweight.
	Model Check to see if a Normal distribution is appropriate.	We have no data, so we cannot make a histogram. But we are told that the company believes the distribution of weights from the machine is Normal.
	Specify which Normal distribution to use.	We use an N(16.3, 0.2) model.
DO	Mechanics Make a graph of this Normal distribution. Locate the value you're interested in on the picture, label it, and shade the appropriate region.	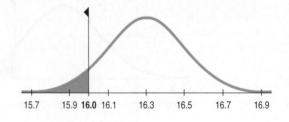

REALITY CHECK	Estimate from the picture the percentage of boxes that are underweight. (This will be useful later to check that your answer makes sense.)	(It looks like a low percentage—maybe less than 10%.)

We want to know what fraction of the boxes will weigh less than 16 oz.

Convert your cutoff value into a *z*-score.

$$z = \frac{y - \mu}{\sigma} = \frac{16 - 16.3}{0.2} = -1.50.$$

Look up the area in the Normal table, or use technology.

Area $(y < 16) =$ Area $(Z < -1.50) = 0.0668$.

REPORT **Conclusion** State your conclusion in the context of the problem.

We estimate that approximately 6.7% of the boxes will contain less than 16 oz of cereal.

Question 2: The company's lawyers say that 6.7% is too high. They insist that no more than 4% of the boxes can be underweight. So the company needs to set the machine to put a little more cereal in each box. What mean setting do they need?

PLAN **Setup** State the variable and the objective.

The variable is weight of cereal in a box. We want to determine a setting for the machine.

Model Check to see if a Normal model is appropriate.

We have no data, so we cannot make a histogram. But we are told that a Normal model applies.

Specify which Normal distribution to use. This time you are not given a value for the mean!

We don't know μ, the mean amount of cereal. The standard deviation for this machine is 0.2 oz. The model, then, is $N(\mu, 0.2)$.

REALITY CHECK We found out earlier that setting the machine to $\mu = 16.3$ oz made 6.7% of the boxes too light. We'll need to raise the mean a bit to reduce this fraction.

We are told that no more than 4% of the boxes can be below 16 oz.

DO **Mechanics** Make a graph of this Normal distribution. Center it at μ (since you don't know the mean) and shade the region below 16 oz.

Using the Normal table, a calculator, or software, find the *z*-score that cuts off the lowest 4%.

The z-score that has 0.04 area to the left of it is $z = -1.75$.

Use this information to find μ. It's located 1.75 standard deviations to the right of 16.

Since 16 must be 1.75 standard deviations below the mean, we need to set the mean at $16 + 1.75 \cdot 0.2 = 16.35$.

REPORT **Conclusion** State your conclusion in the context of the problem.

The company must set the machine to average 16.35 oz of cereal per box.

(continued)

Question 3: The company president vetoes that plan, saying the company should give away less free cereal, not more. Her goal is to set the machine no higher than 16.2 oz and still have only 4% underweight boxes. The only way to accomplish this is to reduce the standard deviation. What standard deviation must the company achieve, and what does that mean about the machine?

PLAN	**Setup** State the variable and the objective.	The variable is weight of cereal in a box. We want to determine the necessary standard deviation to have only 4% of boxes underweight.
	Model Check that a Normal model is appropriate. Specify which Normal distribution to use. This time you don't know σ.	The company believes that the weights are described by a Normal distribution. Now we know the mean, but we don't know the standard deviation. The model is therefore $N(16.2, \sigma)$.
REALITY CHECK	We know the new standard deviation must be less than 0.2 oz.	
DO	**Mechanics** Make a graph of this Normal distribution. Center it at 16.2, and shade the area you're interested in. We want 4% of the area to the left of 16 oz.	
	Find the z-score that cuts off the lowest 4%.	We already know that the z-score with 4% below it is $z = -1.75$.
	Solve for σ. (Note that we need 16 to be 1.75 σ's below 16.2, so 1.75σ must be 0.2 oz. You could just start with that equation.)	$$z = \frac{y - \mu}{\sigma}$$ $$-1.75 = \frac{16 - 16.2}{\sigma}$$ $$1.75\sigma = 0.2$$ $$\sigma = 0.114.$$
REPORT	**Conclusion** State your conclusion in the context of the problem. As we expected, the standard deviation is lower than before—actually, quite a bit lower.	The company must get the machine to box cereal with a standard deviation of only 0.114 oz. This means the machine must be more consistent (by nearly a factor of 2) in filling the boxes.

9.3   Normal Probability Plots

A specialized graphical display can help you to decide whether the Normal model is appropriate: the **Normal probability plot.** If the distribution of the data is roughly Normal, the plot is roughly a diagonal straight line. Deviations from a straight line indicate that the distribution is not Normal. This plot is usually able to show deviations from Normality more clearly than the corresponding histogram, but it's usually easier to understand *how* a distribution fails to be Normal by looking at its histogram. Normal probability plots are difficult to make by hand, but are provided by most statistics software.

Some data on a car's fuel efficiency provide an example of data that are nearly Normal. The overall pattern of the Normal probability plot is straight. The two trailing low values correspond to the values in the histogram that trail off the low end. They're not quite in line with the rest of the data set. The Normal probability plot shows us that they're a bit lower than we'd expect of the lowest two values in a Normal distribution.

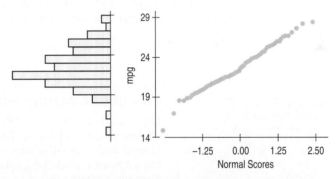

Figure 9.7   Histogram and Normal probability plot for gas mileage (mpg) recorded for a Nissan Maxima. The vertical axes are the same, so each dot on the probability plot would fall into the bar on the histogram immediately to its left.

By contrast, the Normal probability plot of a sample of men's *Weights* in Figure 9.8 from a study of lifestyle and health is far from straight. The weights are skewed to the high end, and the plot is curved. We'd conclude from these pictures that approximations using the Normal model for these data would not be very accurate.

Figure 9.8   Histogram and Normal probability plot for men's weights. Note how a skewed distribution corresponds to a bent probability plot.

For Example   Using a normal probability plot

A normal probability plot of the CAPE10 prices from page 247 looks like this:

Question: What does this plot say about the distribution of the CAPE10 scores?

Answer: The bent shape of the probability plot—and in particular, the sharp bend on the right—indicates a deviation from Normality—in this case the CAPE scores in the right tail do not stretch out as far as we'd expect for a Normal model.

How does a normal probability plot work?

Why does the Normal probability plot work like that? We looked at 100 fuel efficiency measures for a car. The smallest of these has a z-score of -3.16. The Normal model can tell us what value to expect for the smallest z-score in a batch of 100 if a Normal model were appropriate. That turns out to be -2.58. So our first data value is smaller than we would expect from the Normal.

We can continue this and ask a similar question for each value. For example, the 14th-smallest fuel efficiency has a z-score of almost exactly -1, and that's just what we should expect (-1.1 to be exact). We can continue in this way, comparing each observed value with the value we'd expect from a Normal model. The easiest way to make the comparison, of course, is to graph it.[6] If our observed values look like a sample from a Normal model, then the probability plot stretches out in a straight line from lower left to upper right. But if our values deviate from what we'd expect, the plot will bend or have jumps in it. The values we'd expect from a Normal model are called Normal scores, or sometimes nscores. You can't easily look them up in the table, so probability plots are best made with technology and not by hand.

The best advice on using Normal probability plots is to see whether they are straight. If so, then your data look like data from a Normal model. If not, make a histogram to understand how they differ from the model.

9.4   The Distribution of Sums of Normals

Another reason normal models show up so often is that they have some special properties. An important one is that the sum or difference of two independent Normal random variables is also Normal.

A company manufactures small stereo systems. At the end of the production line, the stereos are packaged and prepared for shipping. Stage 1 of this process is called "packing." Workers must collect all the system components (a main unit, two speakers, a power cord, an antenna, and some wires), put each in plastic bags, and then place everything inside a protective form. The packed form then moves on to

---

[6]Sometimes the Normal probability plot switches the two axes, putting the data on the *x*-axis and the *z*-scores on the *y*-axis.

Stage 2, called "boxing," in which workers place the form and a packet of instructions in a cardboard box and then close, seal, and label the box for shipping.

The company says that times required for the packing stage are unimodal and symmetric and can be described by a Normal distribution with a mean of 9 minutes and standard deviation of 1.5 minutes. (See Figure 9.9.) The times for the boxing stage can also be modeled as Normal, with a mean of 6 minutes and standard deviation of 1 minute.

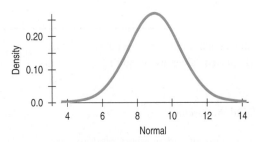

Figure 9.9   The Normal model for the packing stage with a mean of 9 minutes and standard deviation of 1.5 minutes.

The company is interested in the total time that it takes to get a system through both packing and boxing, so they want to model the sum of the two random variables. Fortunately, the special property that adding independent Normals yields another Normal allows us to apply our knowledge of Normal probabilities to questions about the sum or difference of independent random variables. To use this property of Normals, we'll need to check two assumptions: that the variables are Independent and that they can be modeled by the Normal distribution.

Guided Example   Packaging Stereos

Consider the company that manufactures and ships small stereo systems that we discussed previously.

If the time required to pack the stereos can be described by a Normal distribution, with a mean of 9 minutes and standard deviation of 1.5 minutes, and the times for the boxing stage can also be modeled as Normal, with a mean of 6 minutes and standard deviation of 1 minute, what is the probability that packing an order of two systems takes over 20 minutes? What percentage of the stereo systems takes longer to pack than to box?

Question 1: What is the probability that packing an order of two systems takes more than 20 minutes?

PLAN	**Setup** State the problem.	We want to estimate the probability that packing an order of two systems takes more than 20 minutes.
	Variables Define your random variables.	Let P_1 = time for packing the first system $\quad P_2$ = time for packing the second system $\quad T$ = total time to pack two systems
	Write an appropriate equation for the variables you need.	$T = P_1 + P_2$

(continued)

	Think about the model assumptions.	✓ **Normal Model Assumption.** We are told that packing times are well modeled by a Normal model, and we know that the sum of two Normal random variables is also Normal.
		✓ **Independence Assumption.** There is no reason to think that the packing time for one system would affect the packing time for the next, so we can reasonably assume the two are independent.

DO   Mechanics Find the expected value. (Expected values always add.)

$$E(T) = E(P_1 + P_2)$$
$$= E(P_1) + E(P_2)$$
$$= 9 + 9 = 18 \text{ minutes}$$

Find the variance.

For sums of independent random variables, variances add. (In general, we don't need the variables to be Normal for this to be true—just independent.)

Since the times are independent,

$$Var(T) = Var(P_1 + P_2)$$
$$= Var(P_1) + Var(P_2)$$
$$= 1.5^2 + 1.5^2$$
$$Var(T) = 4.50$$

Find the standard deviation.

$$SD(T) = \sqrt{4.50} \approx 2.12 \text{ minutes}$$

Now we use the fact that both random variables follow Normal distributions to say that their sum is also Normal.

We can model the time, T, with a $N(18, 2.12)$ model.

Sketch a picture of the Normal distribution for the total time, shading the region representing over 20 minutes.

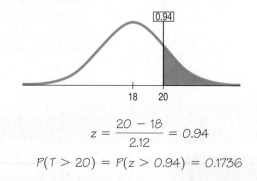

Find the z-score for 20 minutes.

$$z = \frac{20 - 18}{2.12} = 0.94$$

Use technology or a table to find the probability.

$$P(T > 20) = P(z > 0.94) = 0.1736$$

REPORT   Conclusion Interpret your result in context.

MEMO

Re: Computer Systems Packing

Using past history to build a model, we find slightly more than a 17% chance that it will take more than 20 minutes to pack an order of two stereo systems.

Question 2: What percentage of stereo systems take longer to pack than to box?

PLAN   Setup State the question.

We want to estimate the percentage of the stereo systems that takes longer to pack than to box.

Variables Define your random variables.

Let P = time for packing a system
B = time for boxing a system
D = difference in times to pack and box a system

Write an appropriate equation.

$$D = P - B$$

What are we trying to find? Notice that we can tell which of two quantities is greater by subtracting and asking whether the difference is positive or negative.

A system that takes longer to pack than to box will have $P > B$, and so D will be positive. We want to find $P(D > 0)$.

Remember to think about the assumptions.

✓ **Normal Model Assumption.** We are told that both random variables are well modeled by Normal distributions, and we know that the difference of two Normal random variables is also Normal.

✓ **Independence Assumption.** There is no reason to think that the packing time for a system will affect its boxing time, so we can reasonably assume the two are independent.

DO

Mechanics Find the expected value.

$$
\begin{aligned}
E(D) &= E(P - B) \\
&= E(P) - E(B) \\
&= 9 - 6 = 3 \text{ minutes}
\end{aligned}
$$

For the difference of independent random variables, the variance is the sum of the individual variances.

Since the times are independent,

$$
\begin{aligned}
Var(D) &= Var(P - B) \\
&= Var(P) + Var(B) \\
&= 1.5^2 + 1^2
\end{aligned}
$$

$$Var(D) = 3.25$$

Find the standard deviation.

$$SD(D) = \sqrt{3.25} \approx 1.80 \text{ minutes}$$

State what model you will use.

We can model D with $N(3, 1.80)$.

Sketch a picture of the Normal distribution for the difference in times and shade the region representing a difference greater than zero.

Find the z-score. Then use a table or technology to find the probability.

$$z = \frac{0 - 3}{1.80} = -1.67$$

$$P(D > 0) = P(z > -1.67) = 0.9525$$

REPORT

Conclusion Interpret your result in context.

MEMO

Re: Computer Systems Packing

In our second analysis, we found that just over 95% of all the stereo systems will require more time for packing than for boxing.

9.5   The Normal Approximation for the Binomial

The Normal distribution can approximate discrete events when the number of possible events is large. In particular, it is a good model for sums of independent random variables of which a Binomial random variable is a special case. Here's an example of how the Normal can be used to calculate binomial probabilities. Suppose that the Tennessee Red Cross anticipates the need for at least 1850 units of O-negative blood this year. It estimates that it will collect blood from 32,000 donors. How likely is the Tennessee Red Cross to meet its need? We learned how to calculate such probabilities in Chapter 8. We could use the binomial model with $n = 32{,}000$ and $p = 0.06$. The probability of getting *exactly* 1850 units of O-negative blood from 32,000 donors is $\binom{32000}{1850} \times 0.06^{1850} \times 0.94^{30150}$. No calculator on earth can calculate 32000 choose 1850 (it has more than 100,000 digits).[7] And that's just the beginning. The problem said *at least* 1850, so we would have to calculate it again for 1851, for 1852, and all the way up to 32,000. When we're dealing with a large number of trials like this, making direct calculations of the probabilities becomes tedious (or outright impossible).

The Binomial model has mean $np = 1920$ and standard deviation $\sqrt{npq} \approx 42.48$. We can approximate its distribution with a Normal distribution using the same mean and standard deviation. Remarkably enough, that turns out to be a very good approximation. Using that mean and standard deviation, we can find the *probability*:

$$P(X \geq 1850) = P\left(z \geq \frac{1850 - 1920}{42.48}\right) \approx P(z \geq -1.65) \approx 0.95$$

There seems to be about a 95% chance that this Red Cross chapter will have enough O-negative blood.

We can't always use a Normal distribution to make estimates of Binomial probabilities. The success of the approximation depends on the sample size. Suppose we are searching for a prize in cereal boxes, where the probability of finding a prize is 20%. If we buy five boxes, the actual Binomial probabilities that we get 0, 1, 2, 3, 4, or 5 prizes are 33%, 41%, 20%, 5%, 1%, and 0.03%, respectively. The histogram just below shows that this probability model is skewed. We shouldn't try to estimate these probabilities by using a Normal model.

But if we open 50 boxes of this cereal and count the number of prizes we find, we'll get the histogram below. It is centered at $np = 50(0.2) = 10$ prizes, as expected, and it appears to be fairly symmetric around that center.

---

[7] If your calculator *can* find Binom(32000, 0.06), then apparently it's smart enough to use an approximation.

The third histogram (shown just below) shows the same distribution, still centered at the expected value of 10 prizes. It looks close to Normal for sure. With this larger sample size, it appears that a Normal distribution might be a useful approximation.

---

*The continuity correction

When we use a continuous model to model a set of discrete events, we may need to make an adjustment called the **continuity correction**. We approximated the Binomial distribution (50, 0.2) with a Normal distribution. But what does the Normal distribution say about the probability that $X = 10$? Every specific value in the Normal probability model has probability 0. That's not the answer we want.

Because X is really discrete, it takes on the exact values 0, 1, 2, . . . , 50, each with positive probability. The histogram holds the secret to the correction. Look at the bin corresponding to $X = 10$ in the histogram. It goes from 9.5 to 10.5. What we really want is to find the area under the normal curve *between* 9.5 and 10.5. So when we use the Normal distribution to approximate discrete events, we go halfway to the next value on the left and/or the right. We approximate $P(X = 10)$ by finding $P(9.5 \leq X \leq 10.5)$. For a Binomial (50, 0.2), $\mu = 10$ and $\sigma = 2.83$.

$$\text{So } P(9.5 \leq X \leq 10.5) \approx P\left(\frac{9.5 - 10}{2.83} \leq z \leq \frac{10.5 - 10}{2.83} \right)$$

$$= P(-0.177 \leq z \leq 0.177)$$

$$= 0.1405$$

By comparison, the *exact* Binomial probability is 0.1398.

A Normal distribution is a close enough approximation to the Binomial only for a large enough number of trials. And what we mean by "large enough" depends on the probability of success. We'd need a larger sample if the probability of success were very low (or very high). It turns out that a Normal distribution works pretty well if we expect to see at least 10 successes and 10 failures. That is, we check the Success/Failure Condition.

Success/Failure Condition: A Binomial model is approximately Normal if we expect at least 10 successes and 10 failures:

$$np \geq 10 \text{ and } nq \geq 10.$$

Why 10? Well, actually it's 9, as revealed in the following Math Box.

Math Box

Why Check $np > 10$?

It's easy to see where the magic number 10 comes from. You just need to remember how Normal models work. The problem is that a Normal model extends infinitely in both directions. But a Binomial model must have between 0 and n successes, so if we use a Normal to approximate a Binomial, we have to cut off its tails. That's not very important if the center of the Normal model is so far from 0 and n that the lost tails have only a negligible area. More than three standard deviations should do it because a Normal model has little probability past that.

So the mean needs to be at least 3 standard deviations from 0 and at least 3 standard deviations from n. Let's look at the 0 end.

We require:	$\mu - 3\sigma > 0$
Or, in other words:	$\mu > 3\sigma$
For a Binomial that's:	$np > 3\sqrt{npq}$
Squaring yields:	$n^2p^2 > 9npq$
Now simplify:	$np > 9q$
Since $q \leq 1$, we require:	$np > 9$

For simplicity we usually demand that np (and nq for the other tail) be at least 10 to use the Normal approximation which gives the Success/Failure Condition.[8]

For Example   Using the Normal distribution

Some LCD panels have stuck or "dead" pixels that have defective transistors and are permanently unlit. If a panel has too many dead pixels, it must be rejected. A manufacturer knows that, when the production line is working correctly, the probability of rejecting a panel is .07.

Questions:

a) How many screens do they expect to reject in a day's production run of 500 screens? What is the standard deviation?

b) If they reject 40 screens today, is that a large enough number that they should be concerned that something may have gone wrong with the production line?

c) In the past week of 5 days of production, they've rejected 200 screens—an average of 40 per day. Should that raise concerns?

Answers:

a) $\mu = 0.07 \times 500 = 35$ is the expected number of rejects

$\sigma = \sqrt{npq} = \sqrt{500 \times 0.07 \times 0.93} = 5.7$

b) $P(X \geq 45) = P\left(z \geq \dfrac{40 - 35}{5.7}\right) = P(z \geq 0.877) \approx 0.29$, not an extraordinarily large number of rejects

c) Using the Normal approximation:

$\mu = 0.07 \times 2500 = 1.75$

$\sigma = \sqrt{2500 \times 0.07 \times 0.93} = 12.757$

$P(X \geq 200) = P\left(z \geq \dfrac{200 - 175}{12.757}\right) = P(z \geq 1.96) \approx 0.025$

---

[8]Looking at the final step, we see that we need $np > 9$ in the worst case, when q (or p) is near 1, making the Binomial model quite skewed. When q and p are near 0.5—for example, between 0.4 and 0.6—the Binomial model is nearly symmetric, and $np > 5$ ought to be safe enough. Although we'll always check for 10 expected successes and failures, keep in mind that for values of p near 0.5, we can be somewhat more forgiving.

9.6   Other Continuous Random Variables

The Normal distribution differs from the probability distributions we saw in Chapter 8 because it doesn't specify probabilities for individual values, but rather, for intervals of values. When a random variable can take on any value in an interval, we can't model it using a discrete probability model and must use a continuous probability model instead. For any continuous random variable, the distribution of its probability can be shown with a curve. That curve is called the **probability density function (pdf),** usually denoted as $f(x)$. Technically, the curve we've been using to work with the Normal distribution is known as the Normal probability density function.

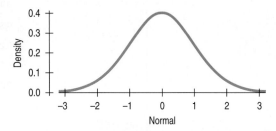

Figure 9.10   The standard Normal density function (a normal with mean 0 and standard deviation 1). The probability of finding a *z*-score in any interval is the area over that interval under the curve. For example, the probability that the *z*-score falls between −1 and 1 is about 68%, which can be seen approximately from the density function or found more precisely from a table or technology.

Density functions must satisfy two requirements. They must stay nonnegative for every possible value, and the total area under the curve must be exactly 1.0. This last requirement corresponds to the Probability Assignment Rule of Chapter 7, which said that the total probability (equal to 1.0) must be assigned somewhere.

Any density function can give the probability that the random variable lies in an interval. But remember, the probability that X lies in the interval from a to b is the *area* under the density function, $f(x)$, between the values a and b and not the value $f(a)$ or $f(b)$. In general, finding that area requires calculus or numerical analysis, and is beyond the scope of this text. But for the models we'll discuss, the probabilities are found either from tables (the Normal) or simple computations (Uniform).

There are many (in fact, there are an infinite number of) possible continuous distributions, but we'll explore only three of the most commonly used to model business phenomena. In addition to the Normal distribution, we'll look at the Uniform distribution and the Exponential distribution.

How can *every* value have probability 0?

At first it may seem illogical that each value of a continuous random variable has probability 0. Let's look at the standard Normal random variable, Z. We could find (from a table, website, or computer program) that the probability that Z lies between 0 and 1 is 0.3413.

(continued)

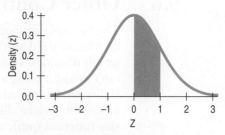

That's the area under the Normal pdf (in red) between the values 0 and 1.
So, what's the probability that Z is between 0 and 1/10?

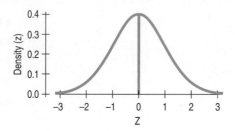

That area is only 0.0398. What is the chance then that Z will fall between 0 and
1/100? There's not much area—the probability is only 0.0040. If we kept going, the
probability would keep getting smaller. The probability that Z is between 0 and
1/100,000 is less than 0.0001.

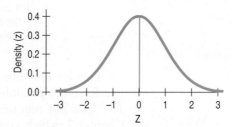

So, what's the probability that Z is *exactly* 0? Well, there's *no* area under the curve right at
$x = 0$, so the probability is 0. It's only intervals that have positive probability, but that's
OK. In real life we never mean exactly 0.0000000000 or any other value. If you say "exactly 164 pounds," you might really mean between 163.5 and 164.5 pounds or even between 163.99 and 164.01 pounds, but realistically not 164.000000000 . . . pounds.

The Uniform Distribution

We've already seen the discrete version of the uniform probability model. A continuous uniform shares the principle that all events should be equally likely, but with a
continuous distribution we can't talk about the probability of a particular value because each value has probability zero. Instead, for a continuous random variable X,
we say that the probability that X lies in any interval depends only on the length of
that interval. Not surprisingly the density function of a continuous uniform random variable looks flat (see Figure 9.11).

The density function of a continuous uniform random variable defined on the
interval a to b can be defined by the formula (see Figure 9.11)

$$f(x) = \begin{cases} \dfrac{1}{b-a} & if \quad a \le x \le b \\ 0 & otherwise \end{cases}$$

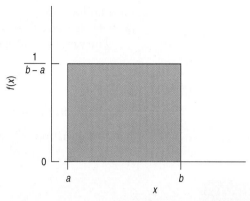

Figure 9.11 The density function of a continuous uniform random variable on the interval from *a* to *b*.

From Figure 9.11, it's easy to see that the probability that *X* lies in any interval between *a* and *b* is the same as any other interval of the same length. In fact, the probability is just the ratio of the length of the interval to the total length: $b - a$. In other words:

For values c and d ($c \leq d$) *both within the interval* $[a, b]$:

$$P(c \leq X \leq d) = \frac{(d - c)}{(b - a)}$$

As an example, suppose you arrive at a bus stop and want to model how long you'll wait for the next bus. The sign says that busses arrive about every 20 minutes, but no other information is given. You might assume that the arrival is equally likely to be anywhere in the next 20 minutes, and so the density function would be

$$f(x) = \begin{cases} \dfrac{1}{20} & if \quad 0 \leq x \leq 20 \\ 0 & otherwise \end{cases}$$

and would look as shown in Figure 9.12.

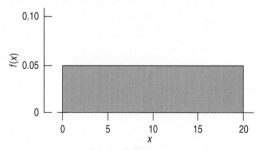

Figure 9.12 The density function of a continuous uniform random variable on the interval [0,20]. Notice that the mean (the balancing point) of the distribution is at 10 minutes and that the area of the box is 1.

Just as the mean of a data distribution is the balancing point of a histogram, the mean of any continuous random variable is the balancing point of the density function. Looking at Figure 9.12, we can see that the balancing point is halfway between the end points at 10 minutes. In general, the expected value is:

$$E(X) = \frac{a + b}{2}$$

for a uniform distribution on the interval (a, b). With $a = 0$ and $b = 20$, the expected value would be 10 minutes.

The variance and standard deviation are less intuitive:

$$Var(X) = \frac{(b - a)^2}{12}; SD(X) = \sqrt{\frac{(b - a)^2}{12}}.$$

Using these formulas, our bus wait will have an expected value of 10 minutes with a standard deviation of $\sqrt{\frac{(20 - 0)^2}{12}} = 5.77$ minutes.

Just Checking

2 As a group, the Dutch are among the tallest people in the world. The average Dutch man is 184 cm tall—just over 6 feet (and the average Dutch woman is 170.8 cm tall—just over 5'7"). If a Normal model is appropriate and the standard deviation for men is about 8 cm, what percentage of all Dutch men will be over 2 meters (6'6") tall?

3 Suppose it takes you 20 minutes, on average, to drive to work, with a standard deviation of 2 minutes. Suppose a

Normal model is appropriate for the distributions of driving times.

a) How often will you arrive at work in less than 22 minutes?

b) How often will it take you more than 24 minutes?

c) Do you think the distribution of your driving times is unimodal and symmetric?

d) What does this say about the accuracy of your prediction? Explain.

The Exponential Model

We saw in Chapter 8 that the Poisson distribution is a good model for the arrival of, or occurrence, of events. We found, for example, the probability that x visits to our website will occur within the next minute. The exponential distribution with parameter λ can be used to model the time *between* those events. Its density function has the form:

$$f(x) = \lambda e^{-\lambda x} \quad for \; x \geq 0 \; and \; \lambda > 0$$

The use of the parameter λ again is not coincidental. It highlights the relationship between the exponential and the Poisson.

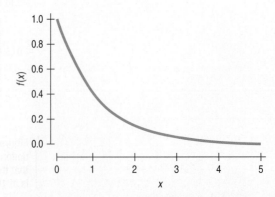

| Figure 9.13 The exponential density function with $\lambda = 1$.

If a discrete random variable can be modeled by a Poisson model with rate λ, then the times between events can be modeled by an exponential model with the same parameter λ. The mean of the exponential is $1/\lambda$. The inverse relationship between the two means makes intuitive sense. If λ increases and we expect *more* hits per minute, then the expected time between hits should go down. The standard deviation of an exponential random variable is $1/\lambda$.

Like any continuous random variable, probabilities of an exponential random variable can be found only through the density function. Fortunately, the area under the exponential density between any two values, s and t ($s \leq t$), has a particularly easy form:

$$P(s \leq X \leq t) = e^{-\lambda s} - e^{-\lambda t}.$$

In particular, by setting s to be 0, we can find the probability that the waiting time will be less than t from

$$P(X \leq t) = P(0 \leq X \leq t) = e^{-\lambda 0} - e^{-\lambda t} = 1 - e^{-\lambda t}.$$

The function $P(X \leq t) = F(t)$ is called the **cumulative distribution function (cdf)** of the random variable X. If arrivals of hits to our website can be well modeled by a Poisson with $\lambda = 4$/minute, then the probability that we'll have to wait less than 20 seconds (1/3 of a minute) is $F(1/3) = P(0 \leq X \leq 1/3) = 1 - e^{-4/3} = 0.736$. That seems about right. Arrivals are coming about every 15 seconds on average, so we shouldn't be surprised that nearly 75% of the time we won't have to wait more than 20 seconds for the next hit.

What Can Go Wrong?

- **Probability models are still just models.** Models can be useful, but they are not reality. Think about the assumptions behind your models. Question probabilities as you would data.

- **Don't assume everything's Normal.** Just because a random variable is continuous or you happen to know a mean and standard deviation doesn't mean that a Normal model will be useful. You must think about whether the **Normality Assumption** is justified. Using a Normal model when it really does not apply will lead to wrong answers and misleading conclusions.

  A sample of CEOs has a mean total compensation of \$10,307,311.87 with a standard deviation of \$17,964,615.16. Using the Normal model rule, we should expect about 68% of the CEOs to have compensations between −\$7,657,303.29 and \$28,271,927.03. In fact, more than 90% of the CEOs have annual compensations in this range. What went wrong? The distribution is skewed, not symmetric. Using the 68–95–99.7 Rule for data like these will lead to silly results.

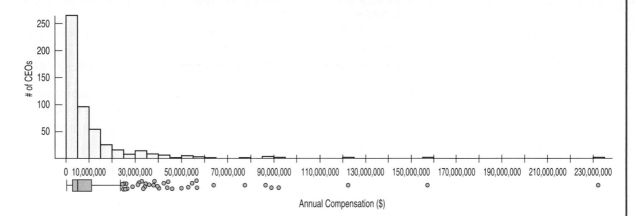

- **Don't use the Normal approximation with small _n_.** To use a Normal approximation in place of a Binomial model, there must be at least 10 expected successes and 10 expected failures.

Ethics in Action

Although e-government services are available online, many Americans, especially those who are older, prefer to deal with government agencies in person. For this reason, the U.S. Social Security Administration (SSA) has local offices distributed across the country. Pat Mennoza is the office manager for one of the larger SSA offices in Phoenix. Since the initiation of the SSA website, his staff has been severely reduced. Yet, because of the number of retirees in the area, his office is one of the busiest. Although there have been no formal complaints, Pat expects that customer waiting times have increased. He decides to keep track of customer wait times for a one-month period in the hopes of making a case for hiring additional staff. He finds that the average wait time is 5 minutes with a standard deviation of 6 minutes. He reasons that 50% of customers who visit his office wait longer than 5 minutes for service. The target wait time is 10 minutes or less. Applying the Normal probability model, Pat finds that more than 20% of customers will have to wait longer than 10 minutes! He has uncovered what he suspected. His next step is to request additional staff based on his findings.

ETHICAL ISSUE *Waiting times are generally skewed and therefore not usually modeled using the Normal distribution. Pat should have checked the data to see if a Normal model was appropriate. Using the Normal for data that are highly skewed to the right will inflate the probability a customer will have to wait longer than 10 minutes. Related to Item A, ASA Ethical Guidelines.*

ETHICAL SOLUTION *Check reasonableness of applying the Normal probability model.*

What Have We Learned?

Learning Objectives

■ Recognize normally distributed data by making a histogram and checking whether it is unimodal, symmetric, and bell-shaped, or by making a normal probability plot using technology and checking whether the plot is roughly a straight line.

- The Normal model is a distribution that will be important for much of the rest of this course.
- Before using a Normal model, we should check that our data are plausibly from a normally distributed population.
- A Normal probability plot provides evidence that the data are Normally distributed if it is linear.

■ Understand how to use the Normal model to judge whether a value is extreme.

- Standardize values to make z-scores and obtain a standard scale. Then refer to a standard Normal distribution.
- Use the 68–95–99.7 Rule as a rule-of-thumb to judge whether a value is extreme.

■ Know how to refer to tables or technology to find the probability of a value randomly selected from a Normal model falling in any interval.

- Know how to perform calculations about Normally distributed values and probabilities.

■ Recognize when independent random Normal quantities are being added or subtracted.

- The sum or difference will also follow a Normal model
- The *variance* of the sum or difference will be the sum of the individual variances.
- The mean of the sum or difference will be the sum or difference, respectively, of the means.

■ Recognize when other continuous probability distributions are appropriate models.

Terms

68–95–99.7 Rule (or Empirical Rule)	In a Normal model, 68% of values fall within one standard deviation of the mean, 95% fall within two standard deviations of the mean, and 99.7% fall within three standard deviations of the mean. This is also approximately true for most unimodal, symmetric distributions.
Cumulative distribution function (cdf)	A function for a continuous probability model that gives the probability of all values below a given value.
Exponential Distribution	A continuous distribution appropriate for modeling the times between events whose occurrences follow a Poisson model.
Normal Distribution	A unimodal, symmetric, "bell-shaped" distribution that appears throughout Statistics.
Normal probability plot	A display to help assess whether a distribution of data is approximately Normal. If the plot is nearly straight, the data satisfy the Nearly Normal Condition.
Probability Density Function (pdf)	A function for any continuous probability model that gives the probability of a random value falling between any two values as the area under the pdf between those two values.
Standard Normal model or Standard Normal distribution	A Normal model, $N(\mu, \sigma)$ with mean $\mu = 0$ and standard deviation $\sigma = 1$.
Uniform Distribution	A continuous distribution that assigns a probability to any range of values (between 0 and 1) proportional to the difference between the values.

Brief CASE

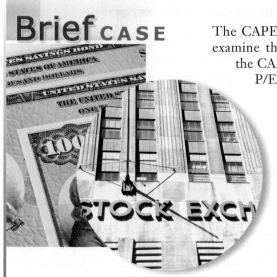

The CAPE10 index is based on the Price/Earnings (P/E) ratios of stocks. We can examine the P/E ratios without applying the smoothing techniques used to find the CAPE10. The file **CAPE10** holds the data, giving dates, CAPE10 values, and P/E values.

Examine the P/E values. Would you judge that a Normal model would be appropriate for those values from the 1880s through the 1980s? Explain (and show the plots you made.)

Now consider the more recent P/E values in this context. Do you think they have been extreme? Explain.

Technology Help: Making Normal Probability Plots

The best way to tell whether your data can be modeled well by a Normal model is to make a picture or two. We've already talked about making histograms. Normal probability plots are almost never made by hand because the values of the Normal scores are tricky to find. But most statistics software make Normal plots, though various packages call the same plot by different names and array the information differently.

EXCEL XLSTAT

Excel offers a "Normal probability plot" as part of the Regression command in the Data Analysis extension, but (as of this writing) it is not a correct Normal probability plot and should not be used.

JMP

To make a "Normal Quantile Plot" in JMP,

- Make a histogram using **Distributions** from the **Analyze** menu.
- Click on the drop-down menu next to the variable name.
- Choose **Normal Quantile Plot** from the drop-down menu.
- JMP opens the plot next to the histogram.

Comments
JMP places the ordered data on the vertical axis and the Normal scores on the horizontal axis. The vertical axis aligns with the histogram's axis, a useful feature.

MINITAB

To make a "Normal Probability Plot" in MINITAB,

- Choose **Probability Plot** from the **Graph** menu.
- Select "Single" for the type of plot. Click **OK**.
- Enter the name of the variable in the "Graph variables" box. Click **OK**.

Comments
MINITAB places the ordered data on the horizontal axis and the Normal scores on the vertical axis.

SPSS

To make a Normal "P-P plot" in SPSS,

- Choose **P-P** from the **Graphs** menu.
- Select the variable to be displayed in the source list.
- Click the arrow button to move the variable into the target list.
- Click the **OK** button.

Comments
SPSS places the ordered data on the horizontal axis and the Normal scores on the vertical axis. You may safely ignore the options in the P-P dialog.

Exercises

SECTION 9.1

1. An incoming MBA student took placement exams in economics and mathematics. In economics, she scored 82 and in math 86. The overall results on the economics exam had a mean of 72 and a standard deviation of 8, while the mean math score was 68, with a standard deviation of 12. On which exam did she do better compared with the other students?

2. The first Statistics exam had a mean of 65 and a standard deviation of 10 points; the second had a mean of 80 and a standard deviation of 5 points. Derrick scored an 80 on both tests. Julie scored a 70 on the first test and a 90 on the second. They both totaled 160 points on the two exams, but Julie claims that her total is better. Explain.

3. Your company's Human Resources department administers a test of "Executive Aptitude." They report test grades as z-scores, and you got a score of 2.20. What does this mean?

4. After examining a child at his 2-year checkup, the boy's pediatrician said that the z-score for his height relative to American 2-year-olds was -1.88. Write a sentence to explain to the parents what that means.

5. Your company will admit to the executive training program only people who score in the top 3% on the executive aptitude test discussed in Exercise 1.

a) With your z-score of 2.20, did you make the cut?
b) What do you need to assume about test scores to find your answer in part a?

6. The pediatrician in Exercise 4 explains to the parents that the most extreme 5% of cases often require special treatment or attention.

a) Does this child fall into that group?
b) What do you need to assume about the heights of 2-year-olds to find your answer to part a?

SECTION 9.2

7. The Environmental Protection Agency (EPA) fuel economy estimates for automobiles suggest a mean of 24.8 mpg and a standard deviation of 6.2 mpg for highway driving. Assume that a Normal model can be applied.

a) Draw the model for auto fuel economy. Clearly label it, showing what the 68–95–99.7 Rule predicts about miles per gallon.
b) In what interval would you expect the central 68% of autos to be found?
c) About what percent of autos should get more than 31 mpg?
d) About what percent of cars should get between 31 and 37.2 mpg?
e) Describe the gas mileage of the worst 2.5% of all cars.

8. Some IQ tests are standardized to a Normal model with a mean of 100 and a standard deviation of 16.

a) Draw the model for these IQ scores. Clearly label it, showing what the 68–95–99.7 Rule predicts about the scores.
b) In what interval would you expect the central 95% of IQ scores to be found?
c) About what percent of people should have IQ scores above 116?
d) About what percent of people should have IQ scores between 68 and 84?
e) About what percent of people should have IQ scores above 132?

9. What percent of a standard Normal model is found in each region? Be sure to draw a picture first.

a) $z > 1.5$
b) $z < 2.25$
c) $-1 < z < 1.15$
d) $|z| > 0.5$

10. What percent of a standard Normal model is found in each region? Draw a picture first.

a) $z > -2.05$
b) $z < -0.33$
c) $1.2 < z < 1.8$
d) $|z| < 1.28$

11. In a standard Normal model, what value(s) of z cut(s) off the region described? Don't forget to draw a picture.

a) the highest 20%
b) the highest 75%

c) the lowest 3%
d) the middle 90%

12. In a standard Normal model, what value(s) of z cut(s) off the region described? Remember to draw a picture first.

a) the lowest 12%
b) the highest 30%
c) the highest 7%
d) the middle 50%

SECTION 9.3

13. Speeds of cars were measured as they passed one point on a road to study whether traffic speed controls were needed. Here's a histogram and normal probability plot of the measured speeds. Is a Normal model appropriate for these data? Explain.

14. Has the Consumer Price Index (CPI) fluctuated around its mean according to a Normal model? Here are some displays. Is a Normal model appropriate for these data? Explain.

SECTION 9.4

15. For a new type of tire, a NASCAR team found the average distance a set of tires would run during a race is 168 miles, with a standard deviation of 14 miles. Assume that tire mileage is independent and follows a Normal model.

a) If the team plans to change tires twice during a 500-mile race, what is the expected value and standard deviation of miles remaining after two changes?
b) What is the probability they won't have to change tires a third time before the end of a 500-mile race?

16. In the 4×100 medley relay event, four swimmers swim 100 yards, each using a different stroke. A college team preparing for the conference championship looks at the times their swimmers have posted and creates a model based on the following assumptions:

- The swimmers' performances are independent.
- Each swimmer's times follow a Normal model.
- The means and standard deviations of the times (in seconds) are as shown here.

Swimmer	Mean	SD
1 (backstroke)	50.72	0.24
2 (breaststroke)	55.51	0.22
3 (butterfly)	49.43	0.25
4 (freestyle)	44.91	0.21

a) What are the mean and standard deviation for the relay team's total time in this event?
b) The team's best time so far this season was 3:19.48. (That's 199.48 seconds.) What is the probability that they will beat that time in the next event?

SECTION 9.5

17. Because many passengers who make reservations do not show up, airlines often overbook flights (sell more tickets than there are seats). A Boeing 767-400ER holds 245 passengers. If the airline believes the rate of passenger no-shows is 5% and sells 255 tickets, is it likely they will not have enough seats and someone will get bumped?

a) Use the Normal model to approximate the Binomial to determine the probability of at least 246 passengers showing up.
b) Should the airline change the number of tickets they sell for this flight? Explain.

18. Shortly after the introduction of the Belgian euro coin, newspapers around the world published articles claiming the coin is biased. The stories were based on reports that someone had spun the coin 250 times and gotten 140 heads—that's 56% heads.

a) Use the Normal model to approximate the Binomial to determine the probability of spinning a fair coin 250 times and getting at least 140 heads.

b) Do you think this is evidence that spinning a Belgian euro is unfair? Would you be willing to use it at the beginning of a sports event? Explain.

SECTION 9.6

19. A cable provider wants to contact customers in a particular telephone exchange to see how satisfied they are with the new digital TV service the company has provided. All numbers are in the 452 exchange, so there are 10,000 possible numbers from 452-0000 to 452-9999. If they select the numbers with equal probability:

a) What distribution would they use to model the selection?
b) The new business "incubator" was assigned the 200 numbers between 452-2500 and 452-2699, but these businesses don't subscribe to digital TV. What is the probability that the randomly selected number will be for an incubator business?
c) Numbers above 9000 were only released for domestic use last year, so they went to newly constructed residences. What is the probability that a randomly selected number will be one of these?

20. In an effort to check the quality of their cell phones, a manufacturing manager decides to take a random sample of 10 cell phones from yesterday's production run, which produced cell phones with serial numbers ranging (according to when they were produced) from 43005000 to 43005999. If each of the 1000 phones is equally likely to be selected:

a) What distribution would they use to model the selection?
b) What is the probability that a randomly selected cell phone will be one of the last 100 to be produced?
c) What is the probability that the first cell phone selected is either from the last 200 to be produced or from the first 50 to be produced?

CHAPTER EXERCISES

For Exercises 21–28, use the 68–95–99.7 Rule to approximate the probabilities rather than using technology to find the values more precisely. Answers given for probabilities or percentages from Exercise 29 and on assume that a calculator or software has been used. Answers found from using *Z*-tables may vary slightly.

T 21. Mutual fund returns. In the last quarter of 2007, a group of 64 mutual funds had a mean return of 2.4% with a standard deviation of 5.6%. If a Normal model can be used to model them, what percent of the funds would you expect to be in each region?

Be sure to draw a picture first.

a) Returns of 8.0% or more
b) Returns of 2.4% or less
c) Returns between −8.8% and 13.6%
d) Returns of more than 19.2%

22. Human resource testing. Although controversial and the subject of some recent law suits (e.g., *Satchell et al. vs. FedEx Express*), some human resource departments administer standard IQ tests to all employees. The Stanford-Binet test scores are well modeled by a Normal model with mean 100 and standard deviation 16. If the applicant pool is well modeled by this distribution, a randomly selected applicant would have what probability of scoring in the following regions?

a) 100 or below
b) Above 148
c) Between 84 and 116
d) Above 132

23. Mutual funds, again. From the 64 mutual funds in Exercise 21 with quarterly returns that are well modeled by a Normal model with a mean of 2.4% and a standard deviation of 5.6%, find the cutoff return value(s) that would separate the

a) highest 50%.
b) highest 16%.
c) lowest 2.5%.
d) middle 68%.

24. Human resource testing, again. For the IQ test administered by human resources and discussed in Exercise 22, what cutoff value would separate the

a) lowest 0.15% of all applicants?
b) lowest 16%?
c) middle 95%?
d) highest 2.5%?

25. Currency exchange rates. The daily exchange rates for the five-year period 2003 to 2008 between the euro (EUR) and the British pound (GBP) are well modeled by a Normal distribution with mean 1.459 euros (to pounds) and standard deviation 0.033 euros. Given this model, what is the probability that on a randomly selected day during this period, the pound was worth

a) less than 1.459 euros?
b) more than 1.492 euros?
c) less than 1.393 euros?
d) Which would be more unusual, a day on which the pound was worth less than 1.410 euros or more than 1.542 euros?

26. Stock prices. For the 900 trading days from January 2003 through July 2006, the daily closing price of IBM stock (in $) is well modeled by a Normal model with mean $85.60 and standard deviation $6.20. According to this model, what is the probability that on a randomly selected day in this period the stock price closed

a) above $91.80?
b) below $98.00?

c) between $73.20 and $98.00?
d) Which would be more unusual, a day on which the stock price closed above $93 or below $70?

27. Currency exchange rates, again. For the model of the EUR/GBP exchange rate discussed in Exercise 25, what would the cutoff rates be that would separate the

a) highest 16% of EUR/GBP rates?
b) lowest 50%?
c) middle 95%?
d) lowest 2.5%?

28. Stock prices, again. According to the model in Exercise 26, what cutoff value of price would separate the

a) lowest 16% of the days?
b) highest 0.15%?
c) middle 68%?
d) highest 50%?

29. Mutual fund probabilities. According to the Normal model $N(0.024, 0.056)$ describing mutual fund returns in the 4th quarter of 2007 in Exercise 21, what percent of this group of funds would you expect to have return

a) over 6.8%?
b) between 0% and 7.6%?
c) more than 1%?
d) less than 0%?

30. Normal IQs. Based on the Normal model $N(100, 16)$ describing IQ scores from Exercise 22, what percent of applicants would you expect to have scores

a) over 80?
b) under 90?
c) between 112 and 132?
d) over 125?

31. Mutual funds, once more. Based on the model $N(0.024, 0.056)$ for quarterly returns from Exercise 21, what are the cutoff values for the

a) highest 10% of these funds?
b) lowest 20%?
c) middle 40%?
d) highest 80%?

32. More IQs. In the Normal model $N(100, 16)$ for IQ scores from Exercise 22, what cutoff value bounds the

a) highest 5% of all IQs?
b) lowest 30% of the IQs?
c) middle 80% of the IQs?
d) lowest 90% of all IQs?

33. Mutual funds, finis. Consider the Normal model $N(0.024, 0.056)$ for returns of mutual funds in Exercise 21 one last time.

a) What value represents the 40th percentile of these returns?
b) What value represents the 99th percentile?
c) What's the IQR of the quarterly returns for this group of funds?

34. IQs, finis. Consider the IQ model $N(100, 16)$ one last time.

a) What IQ represents the 15th percentile?
b) What IQ represents the 98th percentile?
c) What's the IQR of the IQs?

35. Parameters. Every Normal model is defined by its parameters, the mean and the standard deviation. For each model described here, find the missing parameter. As always, start by drawing a picture.

a) $\mu = 20$, 45% above 30; $\sigma = ?$
b) $\mu = 88$, 2% below 50; $\sigma = ?$
c) $\sigma = 5$, 80% below 100; $\mu = ?$
d) $\sigma = 15.6$, 10% above 17.2; $\mu = ?$

36. Parameters, again. Every Normal model is defined by its parameters, the mean and the standard deviation. For each model described here, find the missing parameter. Don't forget to draw a picture.

a) $\mu = 1250$, 35% below 1200; $\sigma = ?$
b) $\mu = 0.64$, 12% above 0.70; $\sigma = ?$
c) $\sigma = 0.5$, 90% above 10.0; $\mu = ?$
d) $\sigma = 220$, 3% below 202; $\mu = ?$

37. SAT or ACT? Each year thousands of high school students take either the SAT or ACT, standardized tests used in the college admissions process. Combined SAT scores can go as high as 1600, while the maximum ACT composite score is 36. Since the two exams use very different scales, comparisons of performance are difficult. (A convenient rule of thumb is $SAT = 40 \times ACT + 150$; that is, multiply an ACT score by 40 and add 150 points to estimate the equivalent SAT score.) Assume that one year the combined SAT can be modeled by $N(1000, 200)$ and the ACT can be modeled by $N(27, 3)$. If an applicant to a university has taken the SAT and scored 1260 and another student has taken the ACT and scored 33, compare these students scores using z-values. Which one has a higher relative score? Explain.

38. Economics. Anna, a business major, took final exams in both Microeconomics and Macroeconomics and scored 83 on both. Her roommate Megan, also taking both courses, scored 77 on the Micro exam and 95 on the Macro exam. Overall, student scores on the Micro exam had a mean of 81 and a standard deviation of 5, and the Macro scores had a mean of 74 and a standard deviation of 15. Which student's overall performance was better? Explain.

39. Claims. Two companies make batteries for cell phone manufacturers. One company claims a mean life span of 2 years, while the other company claims a mean life span of 2.5 years (assuming average use of minutes/month for the cell phone).

a) Explain why you would also like to know the standard deviations of the battery life spans before deciding which brand to buy.
b) Suppose those standard deviations are 1.5 months for the first company and 9 months for the second company. Does this change your opinion of the batteries? Explain.

40. Car speeds. The police department of a major city needs to update its budget. For this purpose, they need to understand the variation in their fines collected from motorists for speeding. As a sample, they recorded the speeds of cars driving past a location with a 20 mph speed limit, a place that in the past has been known for producing fines. The mean of 100 readings was 23.84 mph, with a standard deviation of 3.56 mph. (The police actually recorded every car for a two-month period. These are 100 representative readings.)

a) How many standard deviations from the mean would a car going the speed limit be?
b) Which would be more unusual, a car traveling 34 mph or one going 10 mph?

41. CEOs. A business publication recently released a study on the total number of years of experience in industry among CEOs. The mean is provided in the article, but not the standard deviation. Is the standard deviation most likely to be 6 months, 6 years, or 16 years? Explain which standard deviation is correct and why.

42. Stocks. A newsletter for investors recently reported that the average stock price for a blue chip stock over the past 12 months was $72. No standard deviation was given. Is the standard deviation more likely to be $6, $16, or $60? Explain.

43. Web visitors. A website manager has noticed that during the evening hours, about 3 people per minute check out from their shopping cart and make an online purchase. She believes that each purchase is independent of the others.

a) What model might you suggest to model the number of purchases per minute?
b) What model would you use to model the time between events?
c) What is the mean time between purchases?
d) What is the probability that the time to the next purchase will be between 1 and 2 minutes?

44. Monitoring quality. A cell phone manufacturer samples cell phones from the assembly to test. She noticed that the number of faulty cell phones in a production run of cell phones is usually small and that the quality of one day's run seems to have no bearing on the next day.

a) What model might you use to model the number of faulty cell phones produced in one day?
She wants to model the time between the events of producing a faulty phone. The mean number of defective cell phones is 2 per day.
b) What model would you use to model the time between events?
c) What would the probability be that the time to the next failure is 1 day or less?
d) What is the mean time between failures?

45. Lefties. A lecture hall has 200 seats with folding arm tablets, 30 of which are designed for left-handers. The typical size of classes that meet there is 188, and we can assume that about 13% of students are left-handed. Use a Normal approximation to find the probability that a right-handed student in one of these classes is forced to use a lefty arm tablet.

46. Seatbelts. Police estimate that 80% of drivers wear their seatbelts. They set up a safety roadblock, stopping cars to check for seatbelt use. If they stop 120 cars, what's the probability they find at least 20 drivers not wearing their seatbelt? Use a Normal approximation.

47. Rickets. Vitamin D is essential for strong, healthy bones. Although the bone disease rickets was largely eliminated in England during the 1950s, some people there are concerned that this generation of children is at increased risk because they are more likely to watch TV or play computer games than spend time outdoors. Recent research indicated that about 20% of British children are deficient in vitamin D. A company that sells vitamin D supplements tests 320 elementary school children in one area of the country. Use a Normal approximation to find the probability that no more than 50 of them have vitamin D deficiency.

48. Tennis. A tennis player has taken a special course to improve her serving. She thinks that individual serves are independent of each other. She has been able to make a successful first serve 70% of the time. Use a Normal approximation to find the probability she'll make at least 65 of her first serves out of the 80 she serves in her next match if her success percentage has not changed.

49. Low job satisfaction. Suppose that job satisfaction scores can be modeled with $N(100, 12)$. Human resource departments of corporations are generally concerned if the job satisfaction drops below a certain score. What score would you consider to be unusually low? Explain.

50. Low return. Exercise 21 proposes modeling quarterly returns of a group of mutual funds with $N(0.024, 0.056)$. The manager of this group of funds would like to flag any fund whose return is unusually low for a quarter. What level of return would you consider to be unusually low? Explain.

51. Management survey. A survey of 200 middle managers showed a distribution of the number of hours of exercise they participated in per week with a mean of 3.66 hours and a standard deviation of 4.93 hours.

a) According to the Normal model, what percent of managers will exercise fewer than one standard deviation below the mean number of hours?
b) For these data, what does that mean? Explain.
c) Explain the problem in using the Normal model for these data.

52. Customer database. A large philanthropic organization keeps records on the people who have contributed to their cause. In addition to keeping records of past giving, the organization buys demographic data on neighborhoods from the U.S. Census Bureau. Eighteen of these variables concern the ethnicity of the neighborhood of the donor. Here is a histogram and summary statistics for the percentage of whites in the neighborhoods of 500 donors.

Count	500
Mean	83.59
Median	93
StdDev	22.26
IQR	17
Q1	80
Q3	97

a) Which is a better summary of the percentage of white residents in the neighborhoods, the mean or the median? Explain.
b) Which is a better summary of the spread, the IQR or the standard deviation? Explain.
c) From a Normal model, about what percentage of neighborhoods should have a percent white residents within one standard deviation of the mean?
d) What percentage of neighborhoods actually have a percent white within one standard deviation of the mean?
e) Explain the problem in using the Normal model for these data.

53. Drug company. Manufacturing and selling drugs that claim to reduce an individual's cholesterol level is big business. A company would like to market their drug to women if their cholesterol is in the top 15%. Assume the cholesterol levels of adult American women can be described by a Normal model with a mean of 188 mg/dL and a standard deviation of 24.

a) Draw and label the Normal model.
b) What percent of adult women do you expect to have cholesterol levels over 200 mg/dL?
c) What percent of adult women do you expect to have cholesterol levels between 150 and 170 mg/dL?
d) Estimate the interquartile range of the cholesterol levels.
e) Above what value are the highest 15% of women's cholesterol levels?

54. Tire company. A tire manufacturer believes that the tread life of its snow tires can be described by a Normal model with a mean of 32,000 miles and a standard deviation of 2500 miles.

a) If you buy a set of these tires, would it be reasonable for you to hope that they'll last 40,000 miles? Explain.
b) Approximately what fraction of these tires can be expected to last less than 30,000 miles?
c) Approximately what fraction of these tires can be expected to last between 30,000 and 35,000 miles?
d) Estimate the IQR for these data.
e) In planning a marketing strategy, a local tire dealer wants to offer a refund to any customer whose tires fail to last a certain number of miles. However, the dealer does not want to take too big a risk. If the dealer is willing to give refunds to no more than 1 of every 25 customers, for what mileage can he guarantee these tires to last?

Just Checking Answers

1 a) On the first test, the mean is 88 and the SD is 4, so $z = (90 - 88)/4 = 0.5$. On the second test, the mean is 75 and the SD is 5, so $z = (80 - 75)/5 = 1.0$. The first test has the lower z-score, so it is the one that will be dropped.

   b) The second test is 1 standard deviation above the mean, farther away than the first test, so it's the better score relative to the class.

2 The mean is 184 centimeters, with a standard deviation of 8 centimeters. 2 meters is 200 centimeters, which is 2 standard deviations above the mean. We expect 5% of the men to be more than 2 standard deviations below or above the mean, so half of those, 2.5%, are likely to be above 2 meters.

3 a) We know that 68% of the time we'll be within 1 standard deviation (2 min) of 20. So 32% of the time we'll arrive in less than 18 or more than 22 minutes. Half of those times (16%) will be greater than 22 minutes, so 84% will be less than 22 minutes.

   b) 24 minutes is 2 standard deviations above the mean. Because of the 95% rule, we know 2.5% of the times will be more than 24 minutes.

   c) Traffic incidents may occasionally increase the time it takes to get to school, so the driving times may be skewed to the right, and there may be outliers.

   d) If so, the Normal model would not be appropriate and the percentages we predict would not be accurate.

Sampling Distributions

Marketing Credit Cards: The MBNA Story

When Delaware substantially raised its interest rate ceiling in 1981, banks and other lending institutions rushed to establish corporate headquarters there. One of these was the Maryland Bank National Association, which established a credit card branch in Delaware using the acronym MBNA. Starting in 1982 with 250 employees in a vacant supermarket in Ogletown, Delaware, MBNA grew explosively in the next two decades.

One of the reasons for this growth was MBNA's use of affinity groups—issuing cards endorsed by alumni associations, sports teams, interest groups, and labor unions, among others. MBNA sold the idea to these groups by letting them share a small percentage of the profit. By 2006, MBNA had become Delaware's largest private employer. At its peak, MBNA had more than 50 million cardholders and had outstanding credit card loans of $82.1 billion, making MBNA the third-largest U.S. credit card bank.

"In American corporate history, I doubt there are many companies that burned as brightly, for such a short period of time, as MBNA," said Rep. Mike Castle, R-Del.[1] MBNA was bought by Bank of America in 2005 for $35 billion. Bank of America kept the brand briefly before issuing all cards under its own name in 2007.

Unlike the early days of the credit card industry when MBNA established itself, the environment today is intensely competitive, with companies constantly looking for ways to attract new customers and to maximize the profitability of the customers they already have. Many of the large companies have millions of customers, so instead of trying out a new idea with all their customers, they almost always conduct a pilot study or trial first, conducting a survey or an experiment on a sample of their customers.

Credit card companies make money on their cards in three ways: they earn a percentage of every transaction, they charge interest on balances that are not paid in full, and they collect fees (yearly fees, late fees, etc.). To generate all three types of revenue, the marketing departments of credit card banks constantly seek ways to encourage customers to increase the use of their cards.

A marketing specialist at one company had an idea of offering double air miles to their customers with an airline-affiliated card if they increased their spending by at least $800 in the month following the offer. To forecast the cost and revenue of the offer, the finance department needed to know what percent of customers would actually qualify for the double miles. The marketer decided to send the offer to a random sample of 1000 customers to find out. In that sample, she found that 211 (21.1%) of the cardholders increased their spending by more than the required $800. But, another analyst drew a different sample of 1000 customers of whom 202 (20.2%) of the cardholders exceeded $800.

The two samples don't agree. We know that observations vary, but how much variability among samples should we expect to see?

Why do sample proportions vary at all? How can two samples of the same population measuring the same quantity get different results? The answer is fundamental to statistical inference. Each proportion is based on a *different* sample of cardholders. The proportions vary from sample to sample because the samples are comprised of different people.

WHO	Cardholders of a bank's credit card
WHAT	Whether cardholders increased their spending by at least $800 in the subsequent month
WHEN	February 2008
WHERE	United States
WHY	To predict costs and benefits of a program offer

10.1   The Distribution of Sample Proportions

We'd like to know how much proportions can vary from sample to sample. We've talked about *Plan, Do,* and *Report,* but to learn more about the variability, we have to add *Imagine.* When we sample, we see only the results from the actual sample that we draw, but we can *imagine* what we might have seen had we drawn *all* other possible random samples. What would the histogram of all those sample proportions look like?

If we could take many random samples of 1000 cardholders, we would find the proportion of each sample who spent more than $800 and collect all of those proportions into a histogram. Where would you expect the center of that histogram to be? Of course, we don't *know* the answer, but it is reasonable to think that it will be at the true proportion in the population. We probably will never know the value of the true proportion. But it is important to us, so we'll give it a label, *p* for "true proportion."

Imagine
We see only the sample we actually drew, but if we *imagine* the results of all the other possible samples we could have drawn (by modeling or simulating them), we can learn more.

---
[1]Delaware *News Online,* January 1, 2006.

In fact, we can do better than just imagining. We can *simulate*. We can't really take all those different random samples of size 1000, but we can use a computer to pretend to draw random samples of 1000 individuals from some population of values over and over. In this way, we can study the process of drawing many samples from a real population. A *simulation* can help us understand how sample proportions vary due to random sampling.

When we have only two possible outcomes for an event, the convention in Statistics is to arbitrarily label one of them "success" and the other "failure." Here, a "success" would be that a customer increases card charges by at least $800, and a "failure" would be that the customer didn't. In the simulation, we'll set the true proportion of successes to a known value, draw random samples, and then record the sample proportion of successes, which we'll denote by $\hat{p}$, for each sample.

The proportion of successes in each of our simulated samples will vary from one sample to the next, but the *way* in which the proportions vary shows us how the proportions of real samples would vary. Because we can specify the true proportion of successes, we can see how close each sample comes to estimating that true value. Here's a histogram of the proportions of cardholders who increased spending by at least $800 in 2000 independent samples of 1000 cardholders, when the true proportion is $p = 0.21$. (We know this is the true value of p because in a simulation we can control it.)

Figure 10.1   The distribution of 2000 sample values of $\hat{p}$, from simulated samples of size 1000 drawn from a population in which the true p is 0.21.

It should be no surprise that we don't get the same proportion for each sample we draw, even though the underlying true value, p, stays the same at $p = 0.21$. Since each $\hat{p}$ comes from a random sample, we don't expect them to all be equal to p. And since each comes from a *different* independent random sample, we don't expect them to be equal to each other, either. The remarkable thing is that even though the $\hat{p}$'s vary from sample to sample, they do so in a way that we can model and understand.

For Example   The distribution of a sample proportion

A supermarket has installed "self-checkout" stations that allow customers to scan and bag their own groceries. These are popular, but because customers occasionally encounter a problem, a staff member must be available to help out. The manager wants to estimate what proportion of customers need help so that he can optimize the number of self-check stations per staff monitor. He collects data from the stations for 30 days, recording the proportion of customers on each day that need help and makes a histogram of the observed proportions.

Questions:

1. If the proportion needing help is independent from day to day, what shape would you expect his histogram to follow?

2. Is the assumption of independence reasonable?

Answers:

1. Normal, centered at the true proportion.

2. Possibly not. For example, shoppers on weekends might be less experienced than regular weekday shoppers and would then need more help.

10.2   Sampling Distribution for Proportions

The collection of $\hat{p}$'s may be better behaved than you expected. The histogram in Figure 10.1 is unimodal and symmetric. It is also bell-shaped—and that means that the Normal may be an appropriate model. It is one of the early discoveries and successes of statistics that this distribution of sample proportions can be modeled by the Normal.

The distribution that we displayed in Figure 10.1 is just one simulation. The more proportions we simulate, the more the distribution settles down to a smooth bell shape. The distribution of proportions over all possible independent samples from the same population is called the **sampling distribution** of the proportions.

In fact, we can use the Normal model to describe the behavior of proportions. With the Normal model, we can find the percentage of values falling between any two values. But to make that work, we need to know the mean and standard deviation.

We know already that a sampling distribution of a sample proportion is centered around the true proportion, p. An amazing fact about proportions gives us the appropriate standard deviation to use as well. It turns out that once we know the mean, p, and the sample size, n, we also know the standard deviation of the sampling distribution as you can see from its formula:

$$SD(\hat{p}) = \sqrt{\frac{p(1-p)}{n}} = \sqrt{\frac{pq}{n}}.$$

If the true proportion of credit cardholders who increased their spending by more than \$800 is 0.21, then for samples of size 1000, we expect the distribution of sample proportions to have a standard deviation of:

$$SD(\hat{p}) = \sqrt{\frac{p(1-p)}{n}} = \sqrt{\frac{0.21(1-0.21)}{1000}} = 0.0129, \text{ or about } 1.3\%.$$

Remember that the two samples of size 1000 had proportions of 21.1% and 20.2%. Since the standard deviation of proportions is 1.3%, these two proportions are not even a full standard deviation apart. In other words, the two samples don't really disagree. Proportions of 21.1% and 20.2% from samples of 1000 are both *consistent* with a true proportion of 21%. We know from Chapter 3 that this difference between sample proportions is referred to as **sampling error**. But it's not really an *error*. It's just the *variability* you'd expect to see from one sample to another. A better term might be *sampling variability*.

Look back at Figure 10.1 to see how well the model worked in our simulation. If $p = 0.21$, we now know that the standard deviation should be about 0.013. The 68–95–99.7 Rule from the Normal model says that 68% of the samples will have proportions within 1 SD of the mean of 0.21. How closely does our simulation match the predictions? The actual standard deviation of our 2000 *sample* proportions is 0.0129 or 1.29%. And, of the 2000 simulated samples, 1346 of them had proportions between 0.197 and .223 (one standard deviation on either side of 0.21). The 68–95–99.7 Rule predicts 68%—the actual number is 1346/2000 or 67.3%.

> We have now answered the question raised at the start of the chapter. To discover how variable a sample proportion is, we need to know the proportion and the size of the sample. That's all.

Now we know everything we need to know to model the sampling distribution. We know the mean and standard deviation of the sampling distribution of proportions: they're p, the true population proportion, and $\sqrt{\frac{pq}{n}}$. So the particular Normal model, $N\left(p, \sqrt{\frac{pq}{n}}\right)$, is a **sampling distribution model for the sample proportion**.

> **Effect of Sample Size**
>
> Because n is in the denominator of $SD(\hat{p})$, the larger the sample, the smaller the standard deviation. We need a small standard deviation to make sound business decisions, but larger samples cost more. That tension is a fundamental issue in statistics.

We saw this worked well in a simulation, but can we rely on it in all situations? It turns out that this model can be justified theoretically with just a little mathematics.

It won't work for *all* situations, but it works for most situations that you'll encounter in practice. We'll provide conditions to check so you'll know when the model is useful.

> ### The sampling distribution model for a proportion
>
> Provided that the sampled values are independent and the sample size is large enough, the sampling distribution of $\hat{p}$ is modeled by a Normal model with mean $\mu(\hat{p}) = p$ and standard deviation $SD(\hat{p}) = \sqrt{\dfrac{pq}{n}}$.

Just Checking

1 You want to poll a random sample of 100 shopping mall customers about whether they like the proposed location for the new coffee shop on the third floor, with a panoramic view of the food court. Of course, you'll get just one number, your sample proportion, $\hat{p}$. But if you imagined all the possible samples of 100 customers you could draw and imagined the histogram of all the sample proportions from these samples, what shape would it have?

2 Where would the center of that histogram be?

3 If you think that about half the customers are in favor of the plan, what would the standard deviation of the sample proportions be?

The sampling distribution model for $\hat{p}$ is valuable for a number of reasons. First, because it is known from mathematics to be a good model (and one that gets better and better as the sample size gets larger), we don't need to actually draw many samples and accumulate all those sample proportions, or even to simulate them. The Normal sampling distribution model tells us what the distribution of sample proportions would look like. Second, because the Normal model is a mathematical model, we can calculate what fraction of the distribution will be found in any region. You can find the fraction of the distribution in *any* interval of values using Table Z at the back of the book or with technology.

How Good Is the Normal Model?

We've seen that the simulated proportions follow the 68–95–99.7 Rule well. But do all sample proportions really work like this? Stop and think for a minute about what we're claiming. We've said that if we draw repeated random samples of the same size, n, from some population and measure the proportion, $\hat{p}$, we get for each sample, then the collection of these proportions will pile up around the underlying population proportion, p, in such a way that a histogram of the sample proportions can be modeled well by a Normal model.

There must be a catch. Suppose the samples were of size 2, for example. Then the only possible numbers of successes could be 0, 1, or 2, and the proportion values would be 0, 0.5, and 1. There's no way the histogram could ever look like a Normal model with only three possible values for the variable (Figure 10.2).

Well, there *is* a catch. The claim is only approximately true. (But, that's fine. Models are *supposed* to be only approximately true.) And the model becomes a better and better representation of the distribution of the sample proportions as the sample size gets bigger.[2] Samples of size 1 or 2 just aren't going to work very well, but the distributions of proportions of many larger samples do have histograms that are remarkably close to a Normal model.

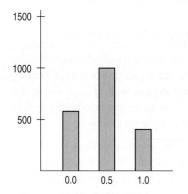

Figure 10.2 Proportions from samples of size 2 can take on only three possible values. A Normal model does not work well here.

---

[2]Formally, we say the claim is true in the limit as the sample size (n) grows.

For Example   Sampling distribution for proportions

Time-Warner provides cable, phone, and Internet services to customers, some of whom subscribe to "packages" including several services. Nationwide, suppose that 30% of their customers are "package subscribers" and subscribe to all three types of service. A local representative in Phoenix, Arizona, wonders if the proportion in his region is the same as the national proportion.

Questions: If the same proportion holds in his region and he takes a survey of 100 customers at random from his subscriber list:

1. What proportion of customers would you expect to be package subscribers?

2. What is the standard deviation of the sample proportion?

3. What shape would you expect the sampling distribution of the proportion to have?

4. Would you be surprised to find out that in a sample of 100, 49 of the customers are package subscribers? Explain. What might account for this high percentage?

Answers:

1. Because 30% of customers nationwide are package subscribers, we would expect the same for the sample proportion.

2. The standard deviation is $SD(\hat{p}) = \sqrt{\dfrac{pq}{n}} = \sqrt{\dfrac{(0.3)(0.7)}{100}} = 0.046$.

3. Normal.

4. 49 customers results in a sample proportion of 0.49. The mean is 0.30 with a standard deviation of 0.046. This sample proportion is more than 4 standard deviations higher than the mean: $\dfrac{(0.49 - 0.30)}{0.046} = 4.13$. It would be very unusual to find such a large proportion in a random sample. Either it is a very unusual sample, or the proportion in his region is not the same as the national average.

Assumptions and Conditions

Most models are useful only when specific assumptions are true. In the case of the model for the distribution of sample proportions, there are two assumptions:

> **Independence Assumption:** The sampled values must be *independent* of each other.
>
> **Sample Size Assumption:** The sample size, n, must be *large* enough.

Of course, the best we can do with assumptions is to think about whether they are likely to be true, and we should do so. However, we often can check corresponding *conditions* that provide information about the assumptions as well. Think about the Independence Assumption and check the following corresponding conditions before using the Normal model to model the distribution of sample proportions:

> **Randomization Condition:** If your data come from an experiment, subjects should have been randomly assigned to treatments. If you have a survey, your sample should be a simple random sample of the population. If some other sampling design was used, be sure the sampling method was not biased and that the data are representative of the population.
>
> **10% Condition:** If sampling has not been made with replacement (that is, returning each sampled individual to the population before drawing the next individual), then the sample size, n, must be no larger than 10% of the population. If it is, you must adjust the size of the confidence interval with methods more advanced than those found in this book.
>
> **Success/Failure Condition:** The Success/Failure condition says that the sample size must be big enough so that both the number of "successes," np, and the number of "failures," nq, are expected to be at least 10.[3] Expressed

---

[3]We saw where the 10 came from in the Math Box on page 262.

without the symbols, this condition just says that we need to expect at least 10 successes and at least 10 failures to have enough data for sound conclusions. For the bank's credit card promotion example, we labeled as a "success" a cardholder who increases monthly spending by at least $800 during the trial. The bank observed 211 successes and 789 failures. Both are at least 10, so there are certainly enough successes and enough failures for the condition to be satisfied.[4]

These two conditions seem to contradict each other. The Success/Failure condition wants a big sample size. How big depends on p. If p is near 0.5, we need a sample of only 20 or so. If p is only 0.01, however, we'd need 1000. But the 10% condition says that the sample size can't be too large a fraction of the population. Fortunately, the tension between them isn't usually a problem in practice. Often, as in polls that sample from all U.S. adults, or industrial samples from a day's production, the populations are much larger than 10 times the sample size.

For Example   Assumptions and conditions for sample proportions

The analyst conducting the Time-Warner survey says that, unfortunately, only 20 of the customers he tried to contact actually responded, but that of those 20, 8 are package subscribers.

Questions:

1. If the proportion of package subscribers in his region is 0.30, how many package subscribers, on average, would you expect in a sample of 20?

2. Would you expect the shape of the sampling distribution of the proportion to be Normal? Explain.

Answers:

1. You would expect $0.30 \cdot 20 = 6$ package subscribers.

2. No. Because 6 is less than 10, we should be cautious in using the Normal as a model for the sampling distribution of proportions. (The number of *observed* successes, 8, is also less than 10.)

Guided Example   Foreclosures

An analyst at a home loan lender was looking at a package of 90 mortgages that the company had recently purchased in central California. The analyst was aware that in that region about 13% of the homeowners with current mortgages will default on their loans in the next year and the house will go into foreclosure. In deciding to buy the collection of mortgages, the finance department assumed that no more than 15 of the mortgages would go into default. Any amount above that will result in losses for the company. In the package of 90 mortgages, what's the probability that there will be more than 15 foreclosures?

| **PLAN** | Setup State the objective of the study. | We want to find the probability that in a group of 90 mortgages, more than 15 will default. Since 15 out of 90 is 16.7%, we need the probability of finding more than 16.7% defaults out of a sample of 90, if the proportion of defaults is 13%. |

(continued)

---

[4]The Success/Failure condition is about the number of successes and failures we *expect*, but if the number of successes and failures that *occurred* is ≥ 10, then you can use that.

Model Check the conditions.

✓ **Independence Assumption** If the mortgages come from a wide geographical area, one homeowner defaulting should not affect the probability that another does. However, if the mortgages come from the same neighborhood(s), the independence assumption may fail and our estimates of the default probabilities may be wrong.

✓ **Randomization Condition.** The 90 mortgages in the package can be considered as a random sample of mortgages in the region.

✓ **10% Condition.** The 90 mortgages are less than 10% of the population.

✓ **Success/Failure Condition**

$$np = 90(0.13) = 11.7 \geq 10$$
$$nq = 90(0.87) = 78.3 \geq 10$$

State the parameters and the sampling distribution model.

The population proportion is $p = 0.13$. The conditions are satisfied, so we'll model the sampling distribution of $\hat{p}$ with a Normal model, with mean 0.13 and standard deviation

$$SD(\hat{p}) = \sqrt{\frac{pq}{n}} = \sqrt{\frac{(0.13)(0.87)}{90}} \approx 0.035.$$

Our model for $\hat{p}$ is $N(0.13, 0.035)$. We want to find $P(\hat{p} > 0.167)$.

Plot Make a picture. Sketch the model and shade the area we're interested in, in this case the area to the right of 16.7%.

| 0.025 | 0.06 | 0.095 | 0.130 | 0.165 | 0.2 | 0.235 |
| -3σ | -2σ | -1σ | p | 1σ | 2σ | 3σ |

DO

Mechanics Use the standard deviation as a ruler to find the z-score of the cutoff proportion. Find the resulting probability from a table, a computer program, or a calculator.

$$z = \frac{\hat{p} - p}{SD(\hat{p})} = \frac{0.167 - 0.13}{0.035} = 1.06$$
$$P(\hat{p} > 0.167) = P(z > 1.06) = 0.1446$$

REPORT

Conclusion Interpret the probability in the context of the question.

MEMO

Re: Mortgage Defaults

Assuming that the 90 mortgages we recently purchased are a random sample of mortgages in this region, there is about a 14.5% chance that we will exceed the 15 foreclosures that Finance has determined as the break-even point.

10.3 The Central Limit Theorem

Proportions summarize categorical variables. When we sample at random, the results we get will vary from sample to sample. The Normal model seems an incredibly simple way to summarize all that variation. Could something that simple work for means? We won't keep you in suspense. It turns out that means also have a sampling distribution that we can model with a Normal model. And it turns out that there's a theoretical result that proves it to be so. As we did with proportions, we can get some insight from a simulation.

Simulating the Sampling Distribution of a Mean

Here's a simple simulation with a quantitative variable. Let's start with one fair die. If we toss this die 10,000 times, what should the histogram of the numbers on the face of the die look like? Here are the results of a simulated 10,000 tosses:

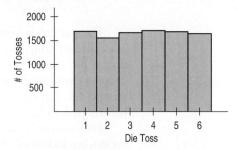

That's called the *uniform distribution*, and it's certainly not Normal. Now let's toss a *pair* of dice and record the average of the two. If we repeat this (or at least simulate repeating it) 10,000 times, recording the average of each pair, what will the histogram of these 10,000 averages look like? Before you look, think a minute. Is getting an average of 1 on *two* dice as likely as getting an average of 3 or 3.5? Let's see:

We're much more likely to get an average near 3.5 than we are to get one near 1 or 6. Without calculating those probabilities exactly, it's fairly easy to see that the *only* way to get an average of 1 is to get two 1s. To get a total of 7 (for an average of 3.5), though, there are many more possibilities. This distribution even has a name—the *triangular distribution*.

What if we average three dice? We'll simulate 10,000 tosses of three dice and take their average.

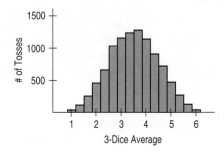

What's happening? First notice that it's getting harder to have averages near the ends. Getting an average of 1 or 6 with three dice requires all three to come up 1 or 6, respectively. That's less likely than for two dice to come up both 1 or both 6. The distribution is being pushed toward the middle. But what's happening to the shape?

Let's continue this simulation to see what happens with larger samples. Here's a histogram of the averages for 10,000 tosses of five dice.

The pattern is becoming clearer. Two things are happening. The first fact we knew already from the Law of Large Numbers, which we saw in Chapter 7. It says that as the sample size (number of dice) gets larger, each sample average tends to become closer to the population mean. So we see the shape continuing to tighten around 3.5. But the shape of the distribution is the surprising part. It's becoming bell-shaped. In fact, it's approaching the Normal model.

Are you convinced? Let's skip ahead and try 20 dice. The histogram of averages for throws 10,000 of 20 dice looks like this.

Now we see the Normal shape again (and notice how much smaller the spread is). But can we count on this happening for situations other than dice throws? What kinds of sample means have sampling distributions that we can model with a Normal model? It turns out that Normal models work well amazingly often.

The Central Limit Theorem

The dice simulation may look like a special situation. But it turns out that what we saw with dice is true for means of repeated samples for almost every situation. When we looked at the sampling distribution of a proportion, we had to check only a few conditions. For means, the result is even more remarkable. There are almost no conditions at all.

Let's say that again: The sampling distribution of *any* mean becomes Normal as the sample size grows. All we need is for the observations to be independent and collected with randomization. We don't even care about the shape of the population distribution![5] This surprising fact was proved in a fairly general form in 1810 by Pierre-Simon Laplace, and caused quite a stir (at least in mathematics circles)

| Pierre-Simon Laplace, 1749–1827.

"The theory of probabilities is at bottom nothing but common sense reduced to calculus."

—LAPLACE, IN *THÉORIE ANALYTIQUE DES PROBABILITIÉS*, 1812

---

[5]Technically, the data must come from a population with a finite variance.

because it is so unintuitive. Laplace's result is called the **Central Limit Theorem**[6] (CLT).

Not only does the distribution of means of many random samples get closer and closer to a Normal model as the sample size grows, but *this is true regardless of the shape of the population distribution!* Even if we sample from a skewed or bimodal population, the Central Limit Theorem tells us that means of repeated random samples will tend to follow a Normal model as the sample size grows. Of course, you won't be surprised to learn that it works better and faster the closer the population distribution is to a Normal model. And it works better for larger samples. If the data come from a population that's exactly Normal to start with, then the observations themselves are Normal. If we take samples of size 1, their "means" are just the observations—so, of course, they have a Normal sampling distribution. But now suppose the population distribution is very skewed (like the CEO data from Chapter 5, for example). The CLT works, although it may take a sample size of dozens or even hundreds of observations for the Normal model to work well.

For example, think about a real bimodal population, one that consists of only 0s and 1s. The CLT says that even means of samples from this population will follow a Normal sampling distribution model. But wait. Suppose we have a categorical variable and we assign a 1 to each individual in the category and a 0 to each individual not in the category. Then we find the mean of these 0s and 1s. That's the same as counting the number of individuals who are in the category and dividing by n. That mean will be the *sample proportion*, $\hat{p}$, of individuals who are in the category (a "success"). So maybe it wasn't so surprising after all that proportions, like means, have Normal sampling distribution models; proportions are actually just a special case of Laplace's remarkable theorem. Of course, for such an extremely bimodal population, we need a reasonably large sample size—and that's where the Success/Failure condition for proportions comes in.

> ## The Central Limit Theorem (CLT)
> The mean of a random sample has a sampling distribution whose shape can be approximated by a Normal model. The larger the sample, the better the approximation will be.

Be careful. We have been slipping smoothly between the real world, in which we draw random samples of data, and a magical mathematical-model world, in which we describe how the sample means and proportions we observe in the real world might behave if we could see the results from every random sample that we might have drawn. Now we have *two* distributions to deal with. The first is the real-world distribution of the sample, which we might display with a histogram (for quantitative data) or with a bar chart or table (for categorical data). The second is the math-world *sampling distribution* of the statistic, which we model with a Normal model based on the Central Limit Theorem. Don't confuse the two.

For example, don't mistakenly think the CLT says that the *data* are Normally distributed as long as the sample is large enough. In fact, as samples get larger, we expect the distribution of the data to look more and more like the distribution of the population from which it is drawn—skewed, bimodal, whatever—but not necessarily Normal. You can collect a sample of CEO salaries for the next 1000 years, but the histogram will never look Normal. It will be skewed to the right.

[6]The word "central" in the name of the theorem means "fundamental." It doesn't refer to the center of a distribution.

The Central Limit Theorem doesn't talk about the distribution of the data from the sample. It talks about the sample *means* and sample *proportions* of many different random samples drawn from the same population. Of course, we never actually draw all those samples, so the CLT is talking about an imaginary distribution—the sampling distribution model.

The CLT does require that the sample be big enough when the population shape is not unimodal and symmetric. But it is still a very surprising and powerful result.

For Example   **The Central Limit Theorem**

The supermarket manager in the example on page 279 also examines the amount spent by customers using the self-checkout stations. He finds that the distribution of these amounts is unimodal but skewed to the high end because some customers make unusually expensive purchases. He finds the mean spent on each of the 30 days studied and makes a histogram of those values.

Questions:

1. What shape would you expect for this histogram?

2. If, instead of averaging all customers on each day, he selects the first 10 for each day and just averages those, how would you expect his histogram of the means to differ from the one in (1)?

Answers:

1. Normal. It doesn't matter that the sample is drawn from a skewed distribution; the CLT tells us that the means will follow a Normal model.

2. The CLT requires large samples. Samples of 10 are not large enough.

10.4   The Sampling Distribution of the Mean

The CLT says that the sampling distribution of any mean or proportion is approximately Normal. But which Normal? We know that any Normal model is specified by its mean and standard deviation. For proportions, the sampling distribution is centered at the population proportion. For means, it's centered at the population mean. What else would we expect?

What about the standard deviations? We noticed in our dice simulation that the histograms got narrower as the number of dice we averaged increased. This shouldn't be surprising. Means vary less than the individual observations. Think about it for a minute. Which would be more surprising, having *one* person in your Statistics class who is over 6'9" tall or having the *mean* of 100 students taking the course be over 6'9"? The first event is fairly rare.[7] You may have seen somebody this tall in one of your classes sometime. But finding a class of 100 whose mean height is over 6'9" tall just won't happen. Why? *Means have smaller standard deviations than individuals.*

"The n's justify the means."

—APOCRYPHAL STATISTICAL
SAYING

That is, the Normal model for the sampling distribution of the mean has a standard deviation equal to $SD(\bar{y}) = \dfrac{\sigma}{\sqrt{n}}$ where σ is the standard deviation of the population. To emphasize that this is a standard deviation *parameter* of the sampling distribution model for the sample mean, $\bar{y}$, we write $SD(\bar{y})$ or $\sigma(\bar{y})$.

---

[7]If students are a random sample of adults, fewer than 1 out of 10,000 should be taller than 6'9". Why might college students not really be a random sample with respect to height? Even if they're not a perfectly random sample, a college student over 6'9" tall is still rare.

> ### The sampling distribution model for a mean
>
> When a random sample is drawn from any population with mean μ and standard deviation σ, its sample mean, $\bar{y}$, has a sampling distribution with the same mean μ but whose standard deviation is $\dfrac{\sigma}{\sqrt{n}}$, and we write $\sigma(\bar{y}) = SD(\bar{y}) = \dfrac{\sigma}{\sqrt{n}}$. No matter what population the random sample comes from, the shape of the sampling distribution is approximately Normal as long as the sample size is large enough. The larger the sample used, the more closely the Normal approximates the sampling distribution model for the mean.

We now have two closely related sampling distribution models. Which one we use depends on which kind of data we have.

- When we have categorical data, we calculate a sample proportion, $\hat{p}$. Its sampling distribution follows a Normal model with a mean at the population proportion, p, and a standard deviation $SD(\hat{p}) = \sqrt{\dfrac{pq}{n}} = \dfrac{\sqrt{pq}}{\sqrt{n}}$.

- When we have quantitative data, we calculate a sample mean, $\bar{y}$. Its sampling distribution has a Normal model with a mean at the population mean, μ, and a standard deviation $SD(\bar{y}) = \dfrac{\sigma}{\sqrt{n}}$.

The means of these models are easy to remember, so all you need to be careful about is the standard deviations. Remember that these are standard deviations of the *statistics* $\hat{p}$ and $\bar{y}$. They both have a square root of n in the denominator. That tells us that the larger the sample, the less either statistic will vary. The only difference is in the numerator. If you just start by writing $SD(\bar{y})$ for quantitative data and $SD(\hat{p})$ for categorical data, you'll be able to remember which formula to use.

Assumptions and Conditions for the Sampling Distribution of the Mean

The CLT requires essentially the same assumptions as we saw for modeling proportions:

Independence Assumption: The sampled values must be independent of each other.

Randomization Condition: The data values must be sampled randomly, or the concept of a sampling distribution makes no sense.

Sample Size Assumption: The sample size must be sufficiently large. We can't check these directly, but we can think about whether the Independence Assumption is plausible. We can also check some related conditions:

10% Condition: When the sample is drawn without replacement (as is usually the case), the sample size, n, should be no more than 10% of the population.

Large Enough Sample Condition: The CLT doesn't tell us how large a sample we need. The truth is, it depends; there's no one-size-fits-all rule. If the population is unimodal and symmetric, even a fairly small sample is okay. You may hear that 30 or 50 observations is always enough to guarantee Normality, but in truth, it depends on the shape of the original data distribution. For highly skewed distributions, it may require samples of several hundred for the sampling distribution of means to be approximately Normal. Always plot the data to check.

Sample Size—Diminishing Returns

The standard deviation of the sampling distribution declines only with the square root of the sample size. The mean of a random sample of 4 has half $\left(\frac{1}{\sqrt{4}} = \frac{1}{2}\right)$ the standard deviation of an individual data value. To cut it in half again, we'd need a sample of 16, and a sample of 64 to halve it once more. In practice, random sampling works well, and means have smaller standard deviations than the individual data values that were averaged. This is the power of averaging.

If only we could afford a much larger sample, we could get the standard deviation of the sampling distribution *really* under control so that the sample mean could tell us still more about the unknown population mean. As we shall see, that square root limits how much we can make a sample tell about the population. This is an example of something that's known as the **Law of Diminishing Returns**.

For Example   Working with the sampling distribution of the mean

Suppose that the weights of boxes shipped by a company follow a unimodal, symmetric distribution with a mean of 12 lbs and a standard deviation of 4 lbs. Boxes are shipped in palettes of 10 boxes. The shipper has a limit of 150 lbs for such shipments.

Question: What's the probability that a palette will exceed that limit?

Answer: Asking the probability that the total weight of a sample of 10 boxes exceeds 150 lbs is the same as asking the probability that the *mean* weight exceeds 15 lbs. First we'll check the conditions. We will assume that the 10 boxes on the palette are a random sample from the population of boxes and that their weights are mutually independent. We are told that the underlying distribution of weights is unimodal and symmetric, so a sample of 10 boxes should be large enough. And 10 boxes is surely less than 10% of the population of boxes shipped by the company.

Under these conditions, the CLT says that the sampling distribution of $\bar{y}$ has a Normal model with mean 12 and standard deviation

$$SD(\bar{y}) = \frac{\sigma}{\sqrt{n}} = \frac{4}{\sqrt{10}} = 1.26 \text{ and } z = \frac{\bar{y} - \mu}{SD(\bar{y})} = \frac{15 - 12}{1.26} = 2.38$$

$$P(\bar{y} > 150) = P(z > 2.38) = 0.0087$$

So the chance that the shipper will reject a palette is only .0087—less than 1%.

10.5 How Sampling Distribution Models Work

Both of the sampling distributions we've looked at are Normal. We know for proportions, $SD(\hat{p}) = \sqrt{\frac{pq}{n}}$, and for means, $SD(\bar{y}) = \frac{\sigma}{\sqrt{n}}$. These are great if we know, or can pretend that we know, p or σ, and sometimes we'll do that.

Often we know only the observed proportion, $\hat{p}$, or the sample standard deviation, s. So of course we just use what we know, and we estimate. That may not seem like a big deal, but it gets a special name. Whenever we estimate the standard deviation of a sampling distribution, we call it a **standard error (SE)**.

For a sample proportion, $\hat{p}$, the standard error is:

$$SE(\hat{p}) = \sqrt{\frac{\hat{p}\hat{q}}{n}}.$$

For the sample mean, $\bar{y}$, the standard error is:

$$SE(\bar{y}) = \frac{s}{\sqrt{n}}.$$

You may see a "standard error" reported by a computer program in a summary or offered by a calculator. It's safe to assume that if no statistic is specified, what was meant is $SE(\bar{y})$, the standard error of the mean.

Just Checking

4 The entrance exam for business schools, the GMAT, given to 100 students had a mean of 520 and a standard deviation of 120. What was the standard error for the mean of this sample of students?

5 As the sample size increases, what happens to the standard error, assuming the standard deviation remains constant?

6 If the sample size is doubled, what is the impact on the standard error?

To keep track of how the concepts we've seen combine, we can draw a diagram relating them. At the heart is the idea that *the statistic itself (the proportion or the mean) is a random quantity*. We can't know what our statistic will be because it comes from a random sample. A different random sample would have given a different result. This sample-to-sample variability is what generates the sampling distribution, the distribution of all the possible values that the statistic could have had.

We could simulate that distribution by pretending to take lots of samples. Fortunately, for the mean and the proportion, the CLT tells us that we can model their sampling distribution directly with a Normal model.

The two basic truths about sampling distributions are:

1. Sampling distributions arise because samples vary. Each random sample will contain different cases and, so, a different value of the statistic.

2. Although we can always simulate a sampling distribution, the Central Limit Theorem saves us the trouble for means and proportions.

When we don't know σ, we estimate it with the standard deviation of the one real sample. That gives us the standard error, $SE(\bar{y}) = \dfrac{s}{\sqrt{n}}$.

Figure 10.3 diagrams the process.

Figure 10.3 We start with a population model, which can have any shape. It can even be bimodal or skewed (as this one is). We label the mean of this model μ and its standard deviation, σ.

We draw one real sample (solid line) of size n and show its histogram and summary statistics. We *imagine* (or simulate) drawing many other samples (dotted lines), which have their own histograms and summary statistics.

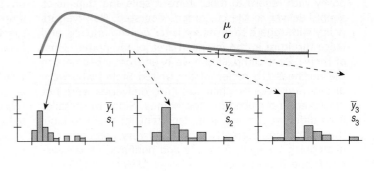

We (imagine) gathering all the means into a histogram.

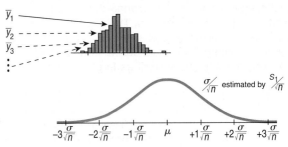

The CLT tells us we can model the shape of this histogram with a Normal model. The mean of this Normal is μ, and the standard deviation is $SD(\bar{y}) = \dfrac{\sigma}{\sqrt{n}}$.

What Can Go Wrong?

- **Don't confuse the sampling distribution with the distribution of the sample.** When you take a sample, you always look at the distribution of the values, usually with a histogram, and you may calculate summary statistics. Examining the distribution of the sample like this is wise. But that's not the sampling distribution. The sampling distribution is an imaginary collection of the values that a statistic might have taken for all the random samples—the one you got and the ones that you didn't get. Use the sampling distribution model to make statements about how the statistic varies.

- **Beware of observations that are not independent.** The CLT depends crucially on the assumption of independence. Unfortunately, this isn't something you can check in your data. You have to think about how the data were gathered. Good sampling practice and well-designed randomized experiments ensure independence.

- **Watch out for small samples from skewed populations.** The CLT assures us that the sampling distribution model is Normal if n is large enough. If the population is nearly Normal, even small samples may work. If the population is very skewed, then n will have to be large before the Normal model will work well. If we sampled 15 or even 20 CEOs and used $\bar{y}$ to make a statement about the mean of all CEOs' compensation, we'd likely get into trouble because the underlying data distribution is so skewed. Unfortunately, there's no good rule to handle this.[8] It just depends on how skewed the data distribution is. Always plot the data to check.

Ethics in Action

Home Illusions, a national retailer of contemporary furniture and home décor has recently experienced customer complaints about the delivery of its products. This retailer uses different carriers depending on the order destination. Its policy with regard to most items it sells and ships is to simply deliver to the customer's doorstep. However, its policy with regard to furniture is to "deliver, unpack, and place furniture in the intended area of the home." Most of their recent complaints have been from customers in the northeastern region of the United States who were dissatisfied because their furniture deliveries were not unpacked and placed in their homes. Since the retailer uses different carriers, it is important for them to label their packages correctly so the delivery company can distinguish between furniture and nonfurniture deliveries. Home Illusions sets as a target "1% or less" for incorrect labeling of packages. Joe Zangard, V.P. Logistics, was asked to look into the problem. The retailer's largest warehouse in the northeast prepares about 1000 items per week for shipping. Joe's initial attention was directed at this facility, not only because of its large volume, but also because he had some reservations about the newly hired warehouse manager, Brent Mossir. Packages at the warehouse were randomly selected and examined over a period of several weeks. Out of 1000 packages, 13 were labeled incorrectly. Since Joe had expected the count to be 10 or fewer, he was confident that he had now pinpointed the problem. His next step was to set up a meeting with Brent in order to discuss the ways in which he can improve the labeling process at his warehouse.

ETHICAL ISSUE *Joe is treating the sample proportion as if it were the true fixed value. By not recognizing that this sample proportion varies from sample to sample, he has unfairly judged the labeling process at Brent's warehouse. This is consistent with his initial misgivings about Brent being hired as warehouse manager (related to Item A, ASA Ethical Guidelines).*

ETHICAL SOLUTION *Joe Zangard needs to use the normal distribution to model the sampling distribution for the sample proportion. In this way, he would realize that the sample proportion observed is less than one standard deviation away from 1% (the upper limit of the target) and thus not conclusively larger than the limit.*

---

[8] For proportions, there is a rule: the Success/Failure condition. That works for proportions because the standard deviation of a proportion is linked to its mean. You may hear that 30 or 50 observations is enough to guarantee Normality, but it really depends on the skewness of the original data distribution.

What Have We Learned?

Learning Objectives

■ Model the variation in statistics from sample to sample with a sampling distribution.

• The Central Limit Theorem tells us that the sampling distribution of both the sample proportion and the sample mean are Normal.

■ Understand that, usually, the mean of a sampling distribution is the value of the parameter estimated.

• For the sampling distribution of $\hat{p}$, the mean is p.

• For the sampling distribution of $\bar{y}$ the mean is μ.

■ Interpret the standard deviation of a sampling distribution.

• The standard deviation of a sampling model is the most important information about it.

• The standard deviation of the sampling distribution of a proportion is $\sqrt{\dfrac{pq}{n}}$, where $q = 1 - p$.

• The standard deviation of the sampling distribution of a mean is $\dfrac{\sigma}{\sqrt{n}}$, where σ is the population standard deviation.

■ Understand that the Central Limit Theorem is a *limit* theorem.

• The sampling distribution of the mean is Normal, *no matter what the underlying distribution of the data is.*

• The CLT says that this happens in the limit, as the sample size grows. The Normal model applies sooner when sampling from a unimodal, symmetric population and more gradually when the population is very non-Normal.

Terms

Central Limit Theorem
:   The Central Limit Theorem (CLT) states that the sampling distribution model of the sample mean (and proportion) is approximately Normal for large n, regardless of the distribution of the population, as long as the observations are independent.

Sampling distribution
:   The distribution of a statistic over many independent samples of the same size from the same population.

Sampling distribution model for a mean
:   If the independence assumption and randomization condition are met and the sample size is large enough, the sampling distribution of the sample mean is well modeled by a Normal model with a mean equal to the population mean, μ, and a standard deviation equal to $\dfrac{\sigma}{\sqrt{n}}$.

Sampling distribution model for a proportion
:   If the independence assumption and randomization condition are met and we expect at least 10 successes and 10 failures, then the sampling distribution of a proportion is well modeled by a Normal model with a mean equal to the true proportion value, p, and a standard deviation equal to $\sqrt{\dfrac{pq}{n}}$.

Sampling error
:   The variability we expect to see from sample to sample is often called the sampling error, although sampling variability is a better term.

Standard error
:   When the standard deviation of the sampling distribution of a statistic is estimated from the data, the resulting statistic is called a standard error (SE).

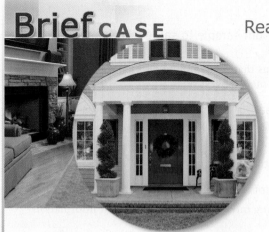

Brief CASE

Real Estate Simulation

Many variables important to the real estate market are skewed, limited to only a few values or considered as categorical variables. Yet, marketing and business decisions are often made based on means and proportions calculated over many homes. One reason these statistics are useful is the Central Limit Theorem.

Data on 1063 houses sold recently in the Saratoga, New York area, are in the file **Saratoga_Real_Estate** on your disk. Let's investigate how the CLT guarantees that the sampling distribution of proportions approaches the Normal and that the same is true for means of a quantitative variable even when samples are drawn from populations that are far from Normal.

Part 1: Proportions

The variable *Fireplace* is a dichotomous variable where 1 = *has a fireplace* and 0 = *does not have a fireplace*.

• Calculate the proportion of homes that have fireplaces for all 1063 homes. Using this value, calculate what the standard error of the sample proportion would be for a sample of size 50.

• Using the software of your choice, draw 100 samples of size 50 from this population of homes, find the proportion of homes with fireplaces in each of these samples, and make a histogram of these proportions.

• Compare the mean and standard deviation of this (sampling) distribution to what you previously calculated.

Part 2: Means

• Select one of the quantitative variables and make a histogram of the entire population of 1063 homes. Describe the distribution (including its mean and SD).

• Using the software of your choice, draw 100 samples of size 50 from this population of homes, find the means of these samples, and make a histogram of these means.

• Compare the (sampling) distribution of the means to the distribution of the population.

• Repeat the exercise with samples of sizes 10 and of 30. What do you notice about the effect of the sample size?

Some statistics packages make it easier than others to draw many samples and find means. Your instructor can provide advice on the path to follow for your package.

An alternative approach is to have each member of the class draw one sample to find the proportion and mean and then combine the statistics for the entire class.

Exercises

SECTION 10.1

1. An investment website can tell what devices are used to access the site. The site managers wonder whether they should enhance the facilities for trading via "smart phones" so they want to estimate the proportion of users who access the site that way (even if they also use their computers sometimes). They draw a random sample of 200 investors from their customers. Suppose that the true proportion of smart phone users is 36%.

a) What would you expect the shape of the sampling distribution for the sample proportion to be?
b) What would be the mean of this sampling distribution?
c) If the sample size were increased to 500, would your answers change? Explain.

2. The proportion of adult women in the United States is approximately 51%. A marketing survey telephones 400 people at random.

a) What proportion of women in the sample of 400 would you expect to see?
b) How many women, on average, would you expect to find in a sample of that size? (*Hint:* Multiply the expected proportion by the sample size.)

SECTION 10.2

3. The investment website of Exercise 1 draws a random sample of 200 investors from their customers. Suppose that the true proportion of smart phone users is 36%.

a) What would the standard deviation of the sampling distribution of the proportion of smart phone users be?
b) What is the probability that the sample proportion of smart phone users is greater than 0.36?
c) What is the probability that the sample proportion is between 0.30 and 0.40?
d) What is the probability that the sample proportion is less than 0.28?
e) What is the probability that the sample proportion is greater than 0.42?

4. The proportion of adult women in the United States is approximately 51%. A marketing survey telephones 400 people at random.

a) What is the sampling distribution of the observed proportion that are women?
b) What is the standard deviation of that proportion?
c) Would you be surprised to find 53% women in a sample of size 400? Explain.
d) Would you be surprised to find 41% women in a sample of size 400? Explain.
e) Would you be surprised to find that there were fewer than 160 women in the sample? Explain.

5. A real estate agent wants to know how many owners of homes worth over $1,000,000 might be considering putting their home on the market in the next 12 months. He surveys 40 of them and finds that 10 of them are considering such a move. Are all the assumptions and conditions for finding the sampling distribution of the proportion satisfied? Explain briefly.

6. A tourist agency wants to know what proportion of vistors to the Eiffel Tower are from the Far East. To find out they survey 100 people in the line to purchase tickets to the top of the tower one Sunday afternoon in May. Are all the assumptions and conditions for finding the sampling distribution of the proportion satisfied? Explain briefly.

SECTION 10.3

7. A sample of 40 games sold for the iPad has prices that have a distribution that is skewed to the high end with a mean of $3.48 and a standard deviation of $2.23. Teens who own iPads typically own about 20 games. Using the 68–95–99.7 Rule, draw and label an appropriate sampling model for the average amount a teen would spend per game if they had 20 games, assuming those games were a representative sample of available games.

8. Statistics for the closing price of the USAA Aggressive Growth Fund for the year 2009 indicate that the average closing price was $23.90, with a standard deviation of $3.00. Using the 68–95–99.7 Rule, draw and label an appropriate sampling model for the mean closing price of 36 days' closing prices selected at random. What (if anything) do you need to assume about the distribution of prices? Are those assumptions reasonable?

SECTION 10.4

9. According to the Gallup poll, 27% of U.S. adults have high levels of cholesterol. They report that such elevated levels "could be financially devastating to the U.S. health-care system" and are a major concern to health insurance providers. According to recent studies, cholesterol levels in healthy U.S. adults average about 215 mg/dL with a standard deviation of about 30 mg/dL and are roughly Normally distributed. If the cholesterol levels of a sample of 42 healthy U.S. adults is taken,

a) What shape should the sampling distribution of the mean have?
b) What would the mean of the sampling distribution be?
c) What would its standard deviation be?
d) If the sample size were increased to 100, how would your answers to parts a–c change?

10. As in Exercise 9, cholesterol levels in healthy U.S. adults average about 215 mg/dL with a standard deviation of about 30 mg/dL and are roughly Normally distributed. If the cholesterol levels of a sample of 42 healthy US adults is taken, what is the probability that the mean cholesterol level of the sample

a) Will be no more than 215?
b) Will be between 205 and 225?
c) Will be less than 200?
d) Will be greater than 220?

SECTION 10.5

11. A marketing researcher for a phone company surveys 100 people and finds that that proportion of clients who are likely to switch providers when their contract expires is 0.15.

a) What is the standard deviation of the sampling distribution of the proportion?
b) If she wants to reduce the standard deviation by half, how large a sample would she need?

12. A market researcher for a provider of iPod accessories wants to know the proportion of customers who own cars to assess the market for a new iPod car charger. A survey of 500 customers indicates that 76% own cars.

a) What is the standard deviation of the sampling distribution of the proportion?
b) How large would the standard deviation have been if he had surveyed only 125 customers (assuming the proportion is about the same)?

13. Organizers of a fishing tournament believe that the lake holds a sizable population of largemouth bass. They assume that the weights of these fish have a model that is skewed to the right with a mean of 3.5 pounds and a standard deviation of 2.32 pounds.

a) Explain why a skewed model makes sense here.
b) Explain why you cannot determine the probability that a largemouth bass randomly selected ("caught") from the lake weighs over 3 pounds.
c) Each contestant catches 5 fish each day. Can you determine the probability that someone's catch averages over 3 pounds? Explain.
d) The 12 contestants competing each caught the limit of 5 fish. What's the standard deviation of the mean weight of the 60 fish caught?
e) Would you be surprised if the mean weight of the 60 fish caught in the competition was more than 4.5 pounds? Use the 68–95–99.7 Rule.

14. In 2008 and 2009, Systemax bought two failing electronics stores, Circuit City and CompUSA. They have kept both the names active and customers can purchase products from either website. If they take a random sample of a mixture of recent purchases from the two websites, the distribution of the amounts purchased will be bimodal.

a) As their sample size increases, what's the expected shape of the distribution of amounts purchased in the sample?
b) As the sample size increases, what's the expected shape of the sampling model for the mean amount purchased of the sample?

CHAPTER EXERCISES

15. Send money. When they send out their fundraising letter, a philanthropic organization typically gets a return from about 5% of the people on their mailing list. To see what the response rate might be for future appeals, they did a simulation using samples of size 20, 50, 100, and 200. For each sample size, they simulated 1000 mailings with success rate $p = 0.05$ and constructed the histogram of the 1000 sample proportions, shown below. Explain how these histograms demonstrate what the Central Limit Theorem says about the sampling distribution model for sample proportions. Be sure to talk about shape, center, and spread.

16. Character recognition. An automatic character recognition device can successfully read about 85% of handwritten credit card applications. To estimate what might happen when this device reads a stack of applications, the company did a simulation using samples of size 20, 50, 75, and 100. For each sample size, they simulated 1000 samples with success rate $p = 0.85$ and constructed the histogram of the 1000 sample proportions, shown here. Explain how these histograms demonstrate what the Central Limit Theorem says about the sampling distribution model for sample proportions. Be sure to talk about shape, center, and spread.

Samples of Size 20

Samples of Size 50

Samples of Size 75

Samples of Size 100

17. Send money, again. The philanthropic organization in Exercise 15 expects about a 5% success rate when they send fundraising letters to the people on their mailing list. In Exercise 15 you looked at the histograms showing distributions of sample proportions from 1000 simulated mailings for samples of size 20, 50, 100, and 200. The sample statistics from each simulation were as follows:

n	mean	st. dev.
20	0.0497	0.0479
50	0.0516	0.0309
100	0.0497	0.0215
200	0.0501	0.0152

a) According to the Central Limit Theorem, what should the theoretical mean and standard deviations be for these sample sizes?
b) How close are those theoretical values to what was observed in these simulations?
c) Looking at the histograms in Exercise 15, at what sample size would you be comfortable using the Normal model as an approximation for the sampling distribution?
d) What does the Success/Failure Condition say about the choice you made in part c?

18. Character recognition, again. The automatic character recognition device discussed in Exercise 16 successfully reads about 85% of handwritten credit card applications. In Exercise 16 you looked at the histograms showing distributions of sample proportions from 1000 simulated samples of size 20, 50, 75, and 100. The sample statistics from each simulation were as follows:

n	mean	st. dev.
20	0.8481	0.0803
50	0.8507	0.0509
75	0.8481	0.0406
100	0.8488	0.0354

a) According to the Central Limit Theorem, what should the theoretical mean and standard deviations be for these sample sizes?
b) How close are those theoretical values to what was observed in these simulations?
c) Looking at the histograms in Exercise 16, at what sample size would you be comfortable using the Normal model as an approximation for the sampling distribution?
d) What does the Success/Failure Condition say about the choice you made in part c?

19. Stock picking. In a large Business Statistics class, the professor has each person select stocks by throwing 16 darts at pages of the *Wall Street Journal*. They then check to see whether their stock picks rose or fell the next day and report their proportion of "successes." As a lesson, the professor has selected pages of the *Journal* for which exactly half the publicly traded stocks went up and half went down. The professor then makes a histogram of the reported proportions.

a) What shape would you expect this histogram to be? Why?
b) Where do you expect the histogram to be centered?
c) How much variability would you expect among these proportions?
d) Explain why a Normal model should not be used here.

20. Quality management. Manufacturing companies strive to maintain production consistency, but it is often difficult for outsiders to tell whether they have succeeded. Sometimes, however, we can find a simple example. The candy company that makes M&M's candies claims that 10% of the candies it produces are green and that bags are packed randomly. We can check on their production controls by sampling bags of candies. Suppose we open bags containing about 50 M&M's and record the proportion of green candies.

a) If we plot a histogram showing the proportions of green candies in the various bags, what shape would you expect it to have?
b) Can that histogram be approximated by a Normal model? Explain.

c) Where should the center of the histogram be?
d) What should the standard deviation of the proportion be?

21. Bigger portfolio. The class in Exercise 19 expands its stock-picking experiment.

a) The students use computer-generated random numbers to choose 25 stocks each. Use the 68–95–99.7 Rule to describe the sampling distribution model.
b) Confirm that you can use a Normal model here.
c) They increase the number of stocks picked to 64 each. Draw and label the appropriate sampling distribution model. Check the appropriate conditions to justify your model.
d) Explain how the sampling distribution model changes as the number of stocks picked increases.

22. More quality. Would a bigger sample help us to assess manufacturing consistency? Suppose instead of the 50-candy bags of Exercise 20, we work with bags that contain 200 M&M's each. Again we calculate the proportion of green candies found.

a) Explain why it's appropriate to use a Normal model to describe the distribution of the proportion of green M&M's they might expect.
b) Use the 68–95–99.7 Rule to describe how this proportion might vary from bag to bag.
c) How would this model change if the bags contained even more candies?

23. A winning investment strategy? One student in the class of Exercise 19 claims to have found a winning strategy. He watches a cable news show about investing and *during the show* throws his darts at the pages of the *Journal*. He claims that of 200 stocks picked in this manner, 58% were winners.

a) What do you think of his claim? Explain.
b) If there are 100 students in the class, are you surprised that one was this successful? Explain.

24. Even more quality. In a really large bag of M&M's, we found 12% of 500 candies were green. Is this evidence that the manufacturing process is out of control and has made too many greens? Explain.

25. Speeding. State police believe that 70% of the drivers traveling on a major interstate highway exceed the speed limit. They plan to set up a radar trap and check the speeds of 80 cars.

a) Using the 68–95–99.7 Rule, draw and label the distribution of the proportion of these cars the police will observe speeding.
b) Do you think the appropriate conditions necessary for your analysis are met? Explain.

26. Smoking, 2008. Public health statistics indicate that 20.6% of American adults smoke cigarettes. Using the 68–95–99.7 Rule, describe the sampling distribution model

for the proportion of smokers among a randomly selected group of 50 adults. Be sure to discuss your assumptions and conditions.

27. Vision. It is generally believed that nearsightedness affects about 12% of all children. A school district has registered 170 incoming kindergarten children.

a) Can you apply the Central Limit Theorem to describe the sampling distribution model for the sample proportion of children who are nearsighted? Check the conditions and discuss any assumptions you need to make.
b) Sketch and clearly label the sampling model, based on the 68–95–99.7 Rule.
c) How many of the incoming students might the school expect to be nearsighted? Explain.

28. Mortgages. In early 2007 the Mortgage Lenders Association reported that homeowners, hit hard by rising interest rates on adjustable-rate mortgages, were defaulting in record numbers. The foreclosure rate of 1.6% meant that millions of families were in jeopardy of losing their homes. Suppose a large bank holds 1731 adjustable-rate mortgages.

a) Can you use the Normal model to describe the sampling distribution model for the sample proportion of foreclosures? Check the conditions and discuss any assumptions you need to make.
b) Sketch and clearly label the sampling model, based on the 68–95–99.7 Rule.
c) How many of these homeowners might the bank expect will default on their mortgages? Explain.

29. Loans. Based on past experience, a bank believes that 7% of the people who receive loans will not make payments on time. The bank has recently approved 200 loans.

a) What are the mean and standard deviation of the proportion of clients in this group who may not make timely payments?
b) What assumptions underlie your model? Are the conditions met? Explain.
c) What's the probability that over 10% of these clients will not make timely payments?

30. Contacts. The campus representative for Lens.com wants to know what percentage of students at a university currently wear contact lens. Suppose the true proportion is 30%.

a) We randomly pick 100 students. Let $\hat{p}$ represent the proportion of students in this sample who wear contacts. What's the appropriate model for the distribution of $\hat{p}$? Specify the name of the distribution, the mean, and the standard deviation. Be sure to verify that the conditions are met.
b) What's the approximate probability that more than one third of this sample wear contacts?

31. Back to school? Best known for its testing program, ACT, Inc., also compiles data on a variety of issues in education. In 2004 the company reported that the national college freshman-to-sophomore retention rate held steady at 74% over the previous four years. Consider colleges with freshman classes of 400 students. Use the 68–95–99.7 Rule to describe the sampling distribution model for the percentage of those students we expect to return to that school for their sophomore years. Do you think the appropriate conditions are met?

32. Binge drinking. A national study found that 44% of college students engage in binge drinking (5 drinks at a sitting for men, 4 for women). Use the 68–95–99.7 Rule to describe the sampling distribution model for the proportion of students in a randomly selected group of 200 college students who engage in binge drinking. Do you think the appropriate conditions are met?

33. Back to school, again. Based on the 74% national retention rate described in Exercise 31, does a college where 522 of the 603 freshman returned the next year as sophomores have a right to brag that it has an unusually high retention rate? Explain.

34. Binge sample. After hearing of the national result that 44% of students engage in binge drinking (5 drinks at a sitting for men, 4 for women), a professor surveyed a random sample of 244 students at his college and found that 96 of them admitted to binge drinking in the past week. Should he be surprised at this result? Explain.

35. Polling. Just before a referendum on a school budget, a local newspaper polls 400 voters in an attempt to predict whether the budget will pass. Suppose that the budget actually has the support of 52% of the voters. What's the probability the newspaper's sample will lead them to predict defeat? Be sure to verify that the assumptions and conditions necessary for your analysis are met.

36. Seeds. Information on a packet of seeds claims that the germination rate is 92%. What's the probability that more than 95% of the 160 seeds in the packet will germinate? Be sure to discuss your assumptions and check the conditions that support your model.

37. Apples. When a truckload of apples arrives at a packing plant, a random sample of 150 is selected and examined for bruises, discoloration, and other defects. The whole truckload will be rejected if more than 5% of the sample is unsatisfactory. Suppose that in fact 8% of the apples on the truck do not meet the desired standard. What's the probability that the shipment will be accepted anyway?

38. Genetic defect. It's believed that 4% of children have a gene that may be linked to juvenile diabetes. Researchers hoping to track 20 of these children for several years test

732 newborns for the presence of this gene. What's the probability that they find enough subjects for their study?

39. Nonsmokers. While some nonsmokers do not mind being seated in a smoking section of a restaurant, about 60% of the customers demand a smoke-free area. A new restaurant with 120 seats is being planned. How many seats should be in the nonsmoking area in order to be very sure of having enough seating there? Comment on the assumptions and conditions that support your model, and explain what "very sure" means to you.

40. Meals. A restaurateur anticipates serving about 180 people on a Friday evening, and believes that about 20% of the patrons will order the chef's steak special. How many of those meals should he plan on serving in order to be pretty sure of having enough steaks on hand to meet customer demand? Justify your answer, including an explanation of what "pretty sure" means to you.

41. Sampling. A sample is chosen randomly from a population that can be described by a Normal model.

a) What's the sampling distribution model for the sample mean? Describe shape, center, and spread.
b) If we choose a larger sample, what's the effect on this sampling distribution model?

42. Sampling, part II. A sample is chosen randomly from a population that was strongly skewed to the left.

a) Describe the sampling distribution model for the sample mean if the sample size is small.
b) If we make the sample larger, what happens to the sampling distribution model's shape, center, and spread?
c) As we make the sample larger, what happens to the expected distribution of the data in the sample?

43. Waist size. A study commissioned by a clothing manufacturer measured the *Waist Size* of 250 men, finding a mean of 36.33 inches and a standard deviation of 4.02 inches. Here is a histogram of these measurements:

a) Describe the histogram of *Waist Size*.
b) To explore how the mean might vary from sample to sample, they simulated by drawing many samples of size 2, 5, 10, and 20, with replacement, from the 250 measurements.

Here are histograms of the sample means for each simulation. Explain how these histograms demonstrate what the Central Limit Theorem says about the sampling distribution model for sample means.

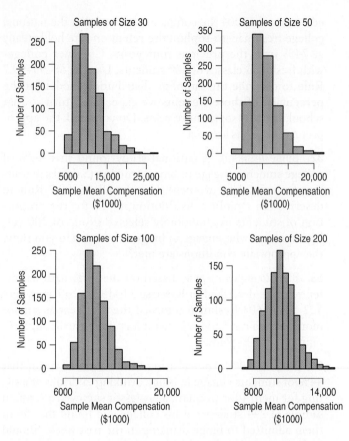

44.   CEO compensation. The average total annual compensation for CEOs of the 800 largest U.S. companies (in $1000) is 10,307.31 and the standard deviation is 17964.62. Here is a histogram of their annual compensations (in $1000):

a) Describe the histogram of *Total Compensation*. A research organization simulated sample means by drawing samples of 30, 50, 100, and 200, with replacement, from the 800 CEOs. The histograms show the distributions of means for many samples of each size.
b) Explain how these histograms demonstrate what the Central Limit Theorem says about the sampling distribution model for sample means. Be sure to talk about shape, center, and spread.

c) Comment on the "rule of thumb" that "With a sample size of at least 30, the sampling distribution of the mean is Normal."

45.   Waist size revisited. A study commissioned by a clothing manufacturer measured the *Waist Sizes* of a random sample of 250 men. The mean and standard deviation of the *Waist Sizes* for all 250 men are 36.33 in and 4.019 inches, respectively. In Exercise 43 you looked at the histograms of simulations that drew samples of sizes 2, 5, 10, and 20 (with replacement). The summary statistics for these simulations were as follows:

n	mean	st. dev.
2	36.314	2.855
5	36.314	1.805
10	36.341	1.276
20	36.339	0.895

a) According to the Central Limit Theorem, what should the theoretical mean and standard deviation be for each of these sample sizes?
b) How close are the theoretical values to what was observed in the simulation?
c) Looking at the histograms in Exercise 43, at what sample size would you be comfortable using the Normal model as an approximation for the sampling distribution?
d) What about the shape of the distribution of *Waist Size* explains your choice of sample size in part c?

46. CEOs revisited. In Exercise 44 you looked at the annual compensation for 800 CEOs, for which the true mean and standard deviation were (in thousands of dollars) 10,307.31 and 17,964.62, respectively. A simulation drew samples of sizes 30, 50, 100, and 200 (with replacement) from the total annual compensations of the Fortune 800 CEOs. The summary statistics for these simulations were as follows:

n	mean	st. dev.
30	10,251.73	3359.64
50	10,343.93	2483.84
100	10,329.94	1779.18
200	10,340.37	1230.79

a) According to the Central Limit Theorem, what should the theoretical mean and standard deviation be for each of these sample sizes?
b) How close are the theoretical values to what was observed from the simulation?
c) Looking at the histograms in Exercise 44, at what sample size would you be comfortable using the Normal model as an approximation for the sampling distribution?
d) What about the shape of the distribution of *Total Compensation* explains your answer in part c?

47. GPAs. A college's data about the incoming freshmen indicates that the mean of their high school GPAs was 3.4, with a standard deviation of 0.35; the distribution was roughly mound-shaped and only slightly skewed. The students are randomly assigned to freshman writing seminars in groups of 25. What might the mean GPA of one of these seminar groups be? Describe the appropriate sampling distribution model—shape, center, and spread—with attention to assumptions and conditions. Make a sketch using the 68–95–99.7 Rule.

48. Home values. Assessment records indicate that the value of homes in a small city is skewed right, with a mean of $140,000 and standard deviation of $60,000. To check the accuracy of the assessment data, officials plan to conduct a detailed appraisal of 100 homes selected at random. Using the 68–95–99.7 Rule, draw and label an appropriate sampling model for the mean value of the homes selected.

49. The trial of the pyx. In 1150, it was recognized in England that coins should have a standard weight of precious metal as the basis for their value. A guinea, for example, was supposed to contain 128 grains of gold. (There are 360 grains in an ounce.) In the "trial of the pyx," coins minted under contract to the crown were weighed and compared to standard coins (which were kept in a wooden box called the pyx). Coins were allowed to deviate by no more than 0.28 grains—roughly equivalent to specifying that the standard deviation should be no greater than 0.09 grains (although they didn't know what a standard deviation was in 1150). In fact, the trial was performed by weighing 100 coins at a time and requiring the *sum* to deviate by no more than $100 \times 0.28 = 28$ or 28 grains—equivalent to the sum having a standard deviation of about 9 grains.

a) In effect, the trial of the pyx required that the mean weight of the sample of 100 coins have a standard deviation of 0.09 grains. Explain what was wrong with performing the trial in this manner.
b) What should the limit have been on the standard deviation of the mean?

Note: Because of this error, the crown was exposed to being cheated by private mints that could mint coins with greater variation and then, after their coins passed the trial, select out the heaviest ones and recast them at the proper weight, retaining the excess gold for themselves. The error persisted for over 600 years, until sampling distributions became better understood.

50. Safe cities. Allstate Insurance Company identified the 10 safest and 10 least-safe U.S. cities from among the 200 largest cities in the United States, based on the mean number of years drivers went between automobile accidents. The cities on both lists were all smaller than the 10 largest cities. Using facts about the sampling distribution model of the mean, explain why this is not surprising.

Just Checking Answers

1. A Normal model (approximately).
2. At the actual proportion of all customers who like the new location.
3. $SD(\hat{p}) = \sqrt{\dfrac{(0.5)(0.5)}{100}} = 0.05$
4. $SE(\bar{y}) = 120/\sqrt{100} = 12$
5. Decreases.
6. The standard error decreases by $1/\sqrt{2}$.

CASE
Study

Investigating the Central Limit Theorem

In the data set you investigated for Case Study Part I, were four variables: Age, Own Home?, Time Between Gifts, and Largest Gift. (The square root of Largest Gift was also included.) The Central Limit Theorem (CLT) says that for distributions that are nearly Normal, the sampling distribution of the mean will be approximately Normal even for small samples. For moderately skewed distributions, one rule of thumb says that sample sizes of 30 are large enough for the CLT to be approximately true. After looking at the distribution of each quantitative variable, think about how large a sample you think you would need for the mean (or proportion) to have a sampling distribution that is approximately Normal. The rest of this case study will investigate how soon the Central Limit Theorem actually starts to work for a variety of distributions.

Before starting, suppose that the organization tells us that no valid donor is under 21 years old. Exclude any such cases from the rest of the analysis (exclude the entire row for all donors under 21 years old).

For each quantitative variable, investigate the relationship between skewness and sample size by following the steps below.

a) Describe the distribution of each quantitative variable for all valid cases, paying special attention to the shape.

b) What is the mean of this variable? From that, what is the standard error of the mean?

c) How large do you think a sample would have to be for the sampling distribution of the mean to be approximately Normal?

d) Shuffle the valid cases, randomizing the order of the rows. (Note: if you are unable to generate different random samples from each variable, the means of a simulation are available on the book's website. In that case, skip steps d and e, use those files, and proceed to step f.)

e) Split 2500 of the valid cases into 100 samples of size 25, find the mean of each sample, and save those values.

f) Plot the histogram of the 100 means. Describe the distribution. Does the CLT seem to be working?

g) From the standard error in step b, according to the Normal distribution, what percent of the sample means should be farther from the mean Age than two standard errors?

h) From the 100 means that you found, what percent are *actually* more than two standard errors from the mean? What does that tell you about the CLT for samples of size 25 for this variable?

i) If the distribution is not sufficiently Normal for this sample size, increase the sample size to 50 and repeat steps e–h.

j) If the distribution is still not Normal enough for $n = 50$, try $n = 100$ or even $n = 250$ and repeat steps e–h until the CLT seems to be in effect (or use the sample means from various sample sizes in the files on the book's website).

For the categorical variable Own Home?, a histogram is inappropriate. Instead, follow these steps:

k) What proportion of the valid cases own their own home?

l) How large do you think a sample would have to be for the sampling distribution of the proportion to be approximately Normal?

m) Shuffle the valid cases, randomizing the order of the rows.

n) Split 2500 of the valid cases into 100 samples of size 25, find the proportion that own their own homes in each sample, and save those values. Given your answer to step a, what should the standard error of the sample proportions be?

o) Plot the histogram of the 100 proportions. Describe the distribution. Is the distribution approximately Normal?

p) From the standard error in step b, if the distribution is Normal, what percent of the sample proportions should be farther from the proportion you found in step a than two standard errors?

q) From the 100 proportions that you found, what percent are actually more than two standard errors from the true proportion? What does that tell you about the CLT for samples of size 25 for this variable?

r) If the distribution is not sufficiently Normal for this sample size, increase the sample size to 50 and repeat steps e–h.

s) If the distribution is still not sufficiently Normal for *n = 50*, try *n = 100* or even *n = 250* and repeat steps d–g until the CLT seems to be in effect.

Write a short report including a summary of what you found out about the relationship between the skewness of the distribution of a quantitative variable and the sample size needed for the sampling distribution of the mean to be approximately Normal. For the categorical variable, how large a sample seems to be needed for the CLT to work? For Largest Gift, what does this say about the wisdom of using the square root of Largest Gift instead of the original variable?

Confidence Intervals for Proportions

The Gallup Organization

Dr. George Gallup was working as a market research director at an advertising agency in the 1930s when he founded the Gallup Organization to measure and track the public's attitudes toward political, social, and economic issues. He gained notoriety a few years later when he defied common wisdom and predicted that Franklin Roosevelt would win the U.S. presidential election in 1936. Today, the Gallup Poll is a household name. During the late 1930s, he founded the Gallup International Research Institute to conduct polls across the globe. International businesses use the Gallup polls to track how consumers think and feel about such issues as corporate behavior, government policies, and executive compensation.

During the late twentieth century, the Gallup Organization partnered with CNN and *USA Today* to conduct and publish public opinion polls. As Gallup once said, "If politicians and special interests have polls to guide them in pursuing their interests, the voters should have polls as well."[1]

[1]Source: The Gallup Organization, Princeton, NJ, www.gallup.com.

Gallup's Web-based data storage system now holds data from polls taken over the last 65 years on a variety of topics, including consumer confidence, household savings, stock market investment, and unemployment.

WHO	U.S. adults
WHAT	Proportion who think economy is getting better
WHEN	March 2010
WHY	To measure expectations about the economy

To plan their inventory and production needs, businesses use a variety of forecasts about the economy. One important attribute is consumer confidence in the overall economy. Tracking changes in consumer confidence over time can help businesses gauge whether the demand for their products is on an upswing or about to experience a downturn. The Gallup Poll periodically asks a random sample of U.S. adults whether they think economic conditions are getting better, getting worse, or staying about the same. When they polled 2976 respondents in March 2010, only 1012 thought economic conditions in the United States were getting better—a sample proportion of $\hat{p} = 1012/2976 = 34.0\%$.[2] We (and Gallup) hope that this observed proportion is close to the population proportion, p, but we know that a second sample of 2976 adults wouldn't have a sample proportion of exactly 34.0%. In fact, Gallup did sample another group of adults just a few days later and found a sample proportion of 38.0%.

From Chapter 10, we know it isn't surprising that two random samples give slightly different results. We'd like to say something, not about different random *samples*, but about the proportion of *all* adults who thought that economic conditions in the United States were getting better in March 2010. The sampling distribution will be the key to our ability to generalize from our sample to the population.

11.1   A Confidence Interval

Notation Alert! ————

Remember that $\hat{p}$ is our sample estimate of the true proportion p. Recall also that q is just shorthand for $1 - p$, and $\hat{q} = 1 - \hat{p}$.

What do we know about our sampling distribution model? We know that it's centered at the true proportion, p, of all U.S. adults who think the economy is improving. But we don't know p. It isn't 34.0%. That's the $\hat{p}$ from our sample. What we do know is that the sampling distribution model of $\hat{p}$ is centered at p, and we know that the standard deviation of the sampling distribution is $\sqrt{\dfrac{pq}{n}}$. We also know, from the Central Limit Theorem, that the shape of the sampling distribution is approximately Normal, when the sample is large enough.

We don't know p, so we can't find the true standard deviation of the sampling distribution model. But we'll use $\hat{p}$ and find the standard error:

$$SE(\hat{p}) = \sqrt{\frac{\hat{p}\hat{q}}{n}} = \sqrt{\frac{(0.34)(1 - 0.34)}{2976}} = 0.009$$

Because the Gallup sample of 2976 is large, we know that the sampling distribution model for $\hat{p}$ should look approximately like the one shown in Figure 11.1.

---

[2]A proportion is a *number* between 0 and 1. In business it's usually reported as a percentage. You may see it written either way.

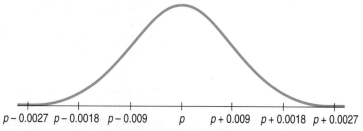

$$p-0.0027 \quad p-0.0018 \quad p-0.009 \qquad p \qquad p+0.009 \quad p+0.0018 \quad p+0.0027$$

> **Figure 11.1** The sampling distribution of sample proportions is centered at the true proportion, *p*, with a standard deviation of 0.009.

The sampling distribution model for $\hat{p}$ is Normal with a mean of *p* and a standard deviation we estimate to be $\sqrt{\dfrac{\hat{p}\hat{q}}{n}}$. Because the distribution is Normal, we'd expect that about 68% of all samples of 2976 U.S. adults taken in March 2010 would have had sample proportions within 1 standard deviation of *p*. And about 95% of all these samples will have proportions within $p \pm 2$ SEs. But where is *our* sample proportion in this picture? And what value does *p* have? We still don't know!

We do know that for 95% of random samples, $\hat{p}$ will be no more than 2 SEs away from *p*. So let's reverse it and look at it from $\hat{p}$'s point of view. If I'm $\hat{p}$, there's a 95% chance that *p* is no more than 2 SEs away from me. If I reach out 2 SEs, or 2×0.009, away from me on both sides, I'm 95% sure that *p* will be within my grasp. Of course, I won't know, and even if my interval does catch *p*, I still don't know its true value. The best I can do is state a probability that I've covered the true value in the interval.

> **Figure 11.2** Reaching out 2 *SEs* on either side of $\hat{p}$ makes us 95% confident we'll trap the true proportion, *p*.

What Can We Say about a Proportion?

So what can we really say about *p*? Here's a list of things we'd like to be able to say and the reasons we can't say most of them:

1. **"34.0% of *all* U.S. adults thought the economy was improving."** It would be nice to be able to make absolute statements about population values with certainty, but we just don't have enough information to do that. There's no way to be sure that the population proportion is the same as the sample proportion; in fact, it almost certainly isn't. Observations vary. Another sample would yield a different sample proportion.

2. **"It is *probably* true that 34.0% of all U.S. adults thought the economy was improving."** No. In fact, we can be pretty sure that whatever the true proportion is, it's not exactly 34.0%, so the statement is not true.

3. **"We don't know exactly what proportion of U.S. adults thought the economy was improving, but we know that it's within the interval 34.0% ± 2 × 0.9%. That is, it's between 32.2% and 35.8%."** This is getting closer, but we still can't be certain. We can't know for sure that the true proportion is in this interval—or in any particular range.

4. **"We don't know exactly what proportion of U.S. adults thought the economy was improving, but the interval from 32.2% to 35.8% *probably* contains the true proportion."** We've now fudged twice—first by giving an interval and second by admitting that we only think the interval "probably" contains the true value.

"Far better an approximate answer to the right question, . . . than an exact answer to the wrong question."

—JOHN W. TUKEY

That last statement is true, but it's a bit wishy-washy. We can tighten it up by quantifying what we mean by "probably." We saw that 95% of the time when we reach out 2 SEs from $\hat{p}$, we capture p, *so we can be 95% confident that this is one of those times.* After putting a number on the probability that this interval covers the true proportion, we've given our best guess of where the parameter is and how certain we are that it's within some range.

5. **"We are 95% confident that between 32.2% and 35.8% of U.S. adults thought the economy was improving."** This is now an appropriate interpretation of our confidence intervals. It's not perfect, but it's about the best we can do.

Each confidence interval discussed in the book has a name. You'll see many different kinds of confidence intervals in the following chapters. Some will be about more than *one* sample, some will be about statistics other than *proportions*, and some will use models other than the Normal. The interval calculated and interpreted here is an example of a **one-proportion z-interval**.[3] We'll lay out the formal definition in the next few pages.

What Does "95% Confidence" Really Mean?

What do we mean when we say we have 95% confidence that our interval contains the true proportion? Formally, what we mean is that "95% of samples of this size will produce confidence intervals that capture the true proportion." This is correct but a little long-winded, so we sometimes say "we are 95% confident that the true proportion lies in our interval." Our uncertainty is about whether the particular sample we have at hand is one of the successful ones or one of the 5% that fail to produce an interval that captures the true value. In Chapter 10, we saw how proportions vary from sample to sample. If other pollsters had selected their own samples of adults, they would have found some who thought the economy was getting better, but each sample proportion would almost certainly differ from ours. When they each tried to estimate the true proportion, they'd center their confidence intervals at the proportions they observed in their own samples. Each would have ended up with a different interval.

Figure 11.3 shows the confidence intervals produced by simulating 20 samples. The purple dots are the simulated proportions of adults in each sample who thought the economy was improving, and the orange segments show the confidence intervals found for each simulated sample. The green line represents the true percentage of adults who thought the economy was improving. You can see that most of the simulated confidence intervals include the true value—but one missed. (Note that it is the *intervals* that vary from sample to sample; the green line doesn't move.)

---

[3] In fact, this confidence interval is so standard for a single proportion that you may see it simply called a "confidence interval for the proportion."

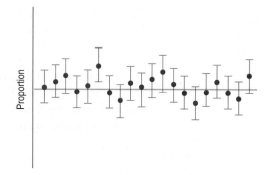

Figure 11.3 The horizontal green line shows the true proportion of people in March 2010 who thought the economy was improving. Most of the 20 simulated samples shown here produced 95% confidence intervals that captured the true value, but one missed.

Of course, a huge number of possible samples *could* be drawn, each with its own sample proportion. This simulation approximates just some of them. Each sample can be used to make a confidence interval. That's a large pile of possible confidence intervals, and ours is just one of those in the pile. Did *our* confidence interval "work"? We can never be sure because we'll never know the true proportion of all U.S. adults who thought in March 2010 that the economy was improving. However, the Central Limit Theorem assures us that 95% of the intervals in the pile are winners, covering the true value, and only 5%, on average, miss the target. That's why we're 95% *confident* that our interval is a winner.

For Example   **Finding a 95% confidence interval for a proportion**

The Chamber of Commerce of a mid-sized city has supported a proposal to change the zoning laws for a new part of town. The new regulations would allow for mixed commercial and residential development. The vote on the measure is scheduled for three weeks from today, and the president of the Chamber of Commerce is concerned that they may not have the majority of votes that they will need to pass the measure. She commissions a survey that asks likely voters if they plan to vote for the measure. Of the 516 people selected at random from likely voters, 289 said they would likely vote for the measure.

Questions:

a. Find a 95% confidence interval for the true proportion of voters who will vote for the measure. (Use the 68–95–99.7% Rule.)

b. What would you report to the president of the Chamber of Commerce?

Answer:

a. $\hat{p} = \dfrac{289}{516} = 0.56$ So, $SE(\hat{p}) = \sqrt{\dfrac{\hat{p}\hat{q}}{n}} = \sqrt{\dfrac{(0.56)(0.44)}{516}} = 0.022$

A 95% confidence interval for p can be found from $\hat{p} \pm 2\,SE(\hat{p}) = 0.56 \pm 2(0.022) = (0.516, 0.604)$ or 51.6% to 60.4%.

b. We are 95% confident that the true proportion of voters who plan to vote for the measure is between 51.6% and 60.4%. This assumes that the sample we have is representative of all likely voters.

11.2   Margin of Error: Certainty vs. Precision

We've just claimed that at a certain confidence level we've captured the true proportion of all U.S. adults who thought the economy was improving in March 2010. Our confidence interval stretched out the same distance on either side of the estimated proportion with the form:

$$\hat{p} \pm 2\,SE(\hat{p}).$$

The *extent* of that interval on either side of $\hat{p}$ is called the **margin of error (ME)**. In general, confidence intervals look like this:

$$estimate \pm ME.$$

The margin of error for our 95% confidence interval was 2 SEs. What if we wanted to be more confident? To be more confident, we'd need to capture p more often, and to do that, we'd need to make the interval wider. For example, if we want to be 99.7% confident, the margin of error will have to be 3 SEs.

$\hat{p} - 3\ SE$ $\hat{p}$ $\hat{p} + 3\ SE$

Figure 11.4   Reaching out 3 SEs on either side of $\hat{p}$ makes us 99.7% confident we'll trap the true proportion p. Compare the width of this interval with the interval in Figure 11.2.

The more confident we want to be, the larger the margin of error must be. We can be 100% confident that any proportion is between 0% and 100%, but that's not very useful. Or we could give a narrow confidence interval, say, from 33.98% to 34.02%. But we couldn't be very confident about a statement this precise. Every confidence interval is a balance between certainty and precision.

The tension between certainty and precision is always there. There is no simple answer to the conflict. Fortunately, in most cases we can be both sufficiently certain and sufficiently precise to make useful statements. The choice of confidence level is somewhat arbitrary, but you must choose the level yourself. The data can't do it for you. The most commonly chosen confidence levels are 90%, 95%, and 99%, but any percentage can be used. (In practice, though, using something like 92.9% or 97.2% might be viewed with suspicion.)

Garfield © 1999 Paws, Inc. Reprinted with permission of UNIVERSAL PRESS SYNDICATE. All rights reserved.

Critical Values

In our opening example, our margin of error was 2 SEs, which produced a 95% confidence interval. To change the confidence level, we'll need to change the *number* of SEs to correspond to the new level. A wider confidence interval means more confidence. For any confidence level the number of SEs we must stretch out on either side of $\hat{p}$ is called the **critical value**. Because it is based on the Normal

Notation Alert!

We put an asterisk on a letter to indicate a critical value. We usually use "z" when we talk about Normal models, so z^* is always a critical value from a Normal model.

Some common confidence levels and their associated critical values:

CI	z^*
90%	1.645
95%	1.960
99%	2.576

model, we denote it z^*. For any confidence level, we can find the corresponding critical value from a computer, a calculator, or a Normal probability table, such as Table Z in the back of the book.

For a 95% confidence interval, the precise critical value is $z^* = 1.96$. That is, 95% of a Normal model is found within ± 1.96 standard deviations of the mean. We've been using $z^* = 2$ from the 68–95–99.7 Rule because 2 is very close to 1.96 and is easier to remember. Usually, the difference is negligible, but if you want to be precise, use 1.96.[4]

Suppose we could be satisfied with 90% confidence. What critical value would we need? We can use a smaller margin of error. Our greater precision is offset by our acceptance of being wrong more often (that is, having a confidence interval that misses the true value). Specifically, for a 90% confidence interval, the critical value is only 1.645 because for a Normal model, 90% of the values are within 1.645 standard deviations from the mean. By contrast, suppose your boss demands more confidence. If she wants an interval in which she can have 99% confidence, she'll need to include values within 2.576 standard deviations, creating a wider confidence interval.

Figure 11.5 For a 90% confidence interval, the critical value is 1.645 because for a Normal model, 90% of the values fall within 1.645 standard deviations of the mean.

For Example   Finding confidence intervals for proportions with different levels of confidence

The president of the Chamber of Commerce is worried that 95% confidence is too low and wants a 99% confidence interval.

Question: Find a 99% confidence interval. Would you reassure her that the measure will pass? Explain.

Answer: In the example on page 309, we used 2 as the value of z^* for 95% confidence. A more precise value would be 1.96 for 95% confidence. For 99% confidence, the critical z-value is 2.756. So, a 99% confidence interval for the true proportion is

$$\hat{p} \pm 2.576 \, SE(\hat{p}) = 0.56 \pm 2.576(0.022) = (0.503, 0.617)$$

The confidence interval is now wider: 50.3% to 61.7%.

The Chamber of Commerce needs at least 50% for the vote to pass. At a 99% confidence level, it looks now as if the measure will pass. However, we must assume that the sample is representative of the voters in the actual election and that people vote in the election as they said they will when they took the survey.

11.3   Assumptions and Conditions

The statements we made about what all U.S. adults thought about the economy were possible because we used a Normal model for the sampling distribution. But is that model appropriate?

As we've seen, all statistical models make assumptions. If those assumptions are not true, the model might be inappropriate, and our conclusions based on it may be

[4]It's been suggested that since 1.96 is both an unusual value and so important in Statistics, you can recognize someone who's had a Statistics course by just saying "1.96" and seeing whether they react.

wrong. Because the confidence interval is built on the Normal model for the sampling distribution, the assumptions and conditions are the same as those we discussed in Chapter 10. But, because they are so important, we'll go over them again.

You can never be certain that an assumption is true, but you can decide intelligently whether it is reasonable. When you have data, you can often decide whether an assumption is plausible by checking a related condition in the data. However, you'll want to make a statement about the world at large, not just about the data. So the assumptions you make are not just about how the data look, but about how representative they are.

Here are the assumptions and the corresponding conditions to check before creating (or believing) a confidence interval about a proportion.

Independence Assumption

You first need to think about whether the independence assumption is plausible. You can look for reasons to suspect that it fails. You might wonder whether there is any reason to believe that the data values somehow affect each other. (For example, might any of the adults in the sample be related?) This condition depends on your knowledge of the situation. It's not one you can check by looking at the data. However, now that you have data, there are two conditions that you can check:

- **Randomization Condition:** Were the data sampled at random or generated from a properly randomized experiment? Proper randomization can help ensure independence.

- **10% Condition:** Samples are almost always drawn without replacement. Usually, you'd like to have as large a sample as you can. But if you sample from a small population, the probability of success may be different for the last few individuals you draw than it was for the first few. For example, if most of the women have already been sampled, the chance of drawing a woman from the remaining population is lower. If the sample exceeds 10% of the population, the probability of a success changes so much during the sampling that a Normal model may no longer be appropriate. But if less than 10% of the population is sampled, it is safe to assume to have independence.

Sample Size Assumption

The model we use for inference is based on the Central Limit Theorem. So, the sample must be large enough for the Normal sampling model to be appropriate. It turns out that we need more data when the proportion is close to either extreme (0 or 1). This requirement is easy to check with the following condition:

- **Success/Failure Condition:** We must expect our sample to contain at least 10 "successes" and at least 10 "failures." Recall that by tradition we arbitrarily label one alternative (usually the outcome being counted) as a "success" even if it's something bad. The other alternative is then a "failure." So we check that both $n\hat{p} \geq 10$ and $n\hat{q} \geq 10$.

One-proportion *z*-interval

When the conditions are met, we are ready to find the confidence interval for the population proportion, p. The confidence interval is $\hat{p} \pm z^* \times SE(\hat{p})$, where the standard deviation of the proportion is estimated by $SE(\hat{p}) = \sqrt{\dfrac{\hat{p}\hat{q}}{n}}$.

For Example   Assumptions and conditions for a confidence interval for proportions

We previously reported a confidence interval to the president of the Chamber of Commerce.

Question: Were the assumptions and conditions for making this interval satisfied?

Answer: Because the sample was randomized, we assume that the responses of the people surveyed were independent so the randomization condition is met. We assume that 516 people represent fewer than 10% of the likely voters in the town so the 10% condition is met. Because 289 people said they were likely to vote for the measure and thus 227 said they were not, both are much larger than 10 so the Success/Failure condition is also met.

All the conditions to make a confidence interval for the proportion appear to have been satisfied.

Guided Example   Public Opinion

In the film *Up in the Air*, George Clooney portrays a man whose job it is to tell workers that they have been fired. The reactions to such news took a somewhat odd turn in France in the spring of 2009 when workers at Sony France took the boss hostage for a night and barricaded their factory entrance with a tree trunk. He was freed only after he agreed to reopen talks on their severance packages. Similar incidents occurred at 3M and Caterpillar plants in France. A poll taken by *Le Parisien* in April 2009 found 45% of the French "supportive" of such action. A similar poll taken by *Paris Match*, April 2–3, 2009, found 30% "approving" and 63% were "understanding" or "sympathetic" of the action. Only 7% condemned the practice of "bossnapping."

The *Paris Match* poll was based on a random representative sample of 1010 adults. What can we conclude about the proportion of all French adults who sympathize with (without supporting outright) the practice of bossnapping?

To answer this question, we'll build a confidence interval for the proportion of all French adults who sympathize with the practice of bossnapping. As with other procedures, there are three steps to building and summarizing a confidence interval for proportions: Plan, Do, and Report.

WHO	Adults in France
WHAT	Proportion who sympathize with the practice of bossnapping
WHEN	April 2–3, 2009
WHERE	France
HOW	1010 adults were randomly sampled by the French Institute of Public Opinion (l'Ifop) for the magazine *Paris Match*
WHY	To investigate public opinion of bossnapping

PLAN

Setup State the context of the question.

Identify the *parameter* you wish to estimate. Identify the *population* about which you wish to make statements.

Choose and state a confidence level.

Model Think about the assumptions and check the conditions to decide whether we can use the Normal model.

We want to find an interval that is likely with 95% confidence to contain the true proportion, *p*, of French adults who sympathize with the practice of bossnapping. We have a random sample of 1010 French adults, with a sample proportion of 63%.

✓ **Independence Assumption:** A French polling agency, l'Ifop, phoned a random sample of French adults. It is unlikely that any respondent influenced another.

✓ **Randomization Condition:** l'Ifop drew a random sample from all French adults. We don't have details of their randomization but assume that we can trust it.

✓ **10% Condition:** Although sampling was necessarily without replacement, there are many more French adults than were sampled. The sample is certainly less than 10% of the population.

(continued)

	✓ **Success/Failure Condition:** $n\hat{p} = 1010 \times 0.63 = 636 \geq 10$ and $n\hat{q} = 1010 \times 0.37 = 374 \geq 10$, so the sample is large enough.
State the sampling distribution model for the statistic. Choose your method.	The conditions are satisfied, so I can use a Normal model to find a one-proportion z-interval.

DO

Mechanics Construct the confidence interval. First, find the standard error. (Remember: It's called the "standard error" because we don't know p and have to use $\hat{p}$ instead.)

$n = 1010$, $\hat{p} = 0.63$, so

$$SE(\hat{p}) = \sqrt{\frac{0.63 \times 0.37}{1010}} = 0.015$$

Next, find the margin of error. We could informally use 2 for our critical value, but 1.96 is more accurate.[5]

Because the sampling model is Normal, for a 95% confidence interval, the critical value $z^* = 1.96$. The margin of error is:

$$ME = z^* \times SE(\hat{p}) = 1.96 \times 0.015 = 0.029$$

So the 95% confidence interval is:

$$0.63 \pm 0.029 \text{ or } (0.601, 0.659).$$

Write the confidence interval.

REALITY CHECK ▶ Check that the interval is plausible. We may not have a strong expectation for the center, but the width of the interval depends primarily on the sample size—especially when the estimated proportion is near 0.5.

The confidence interval covers a range of about plus or minus 3%. That's about the width we might expect for a sample size of about 1000 (when $\hat{p}$ is reasonably close to 0.5).

REPORT

Conclusion Interpret the confidence interval in the proper context. We're 95% confident that our interval captured the true proportion.

MEMO

Re: Bossnapping Survey

The polling agency l'Ifop surveyed 1010 French adults and asked whether they approved, were sympathetic to or disapproved of recent bossnapping actions. Although we can't know the true proportion of French adults who were sympathetic (without supporting outright), based on this survey we can be 95% confident that between 60.1% and 65.9% of all French adults were. Because this is an ongoing concern for public safety, we may want to repeat the survey to obtain more current data. We may also want to keep these results in mind for future corporate public relations.

Just Checking

Think some more about the 95% confidence interval we just created for the proportion of French adults who were sympathetic to bossnapping.

1 If we wanted to be 98% confident, would our confidence interval need to be wider or narrower?

2 Our margin of error was about ±3% If we wanted to reduce it to ±2% without increasing the sample size, would our level of confidence be higher or lower?

3 If the organization had polled more people, would the interval's margin of error have likely been larger or smaller?

---

[5]If you are following along on your calculator and not rounding off (as we have done for this example), you'll get $SE = 0.0151918$ and a ME of 0.029776.

315

11.4 Choosing the Sample Size

Every confidence interval must balance precision—the width of the interval—against confidence. Although it is good to be precise and comforting to be confident, there is a trade-off between the two. A confidence interval that says that the percentage is between 10% and 90% wouldn't be of much use, although you could be quite confident that it covered the true proportion. An interval from 43% to 44% is reassuringly precise, but not if it carries a confidence level of 35%. It's a rare study that reports confidence levels lower than 80%. Levels of 95% or 99% are more common.

The time to decide whether the margin of error is small enough to be useful is when you design your study. Don't wait until you compute your confidence interval. To get a narrower interval without giving up confidence, you need to have less variability in your sample proportion. How can you do that? Choose a larger sample.

Consider a company planning to offer a new service to their customers. Product managers want to estimate the proportion of customers who are likely to purchase this new service to within 3% with 95% confidence. How large a sample do they need?

Let's look at the margin of error:

$$ME = z^* \sqrt{\frac{\hat{p}\hat{q}}{n}}$$

$$0.03 = 1.96 \sqrt{\frac{\hat{p}\hat{q}}{n}}.$$

They want to find n, the sample size. To find n, they need a value for $\hat{p}$. They don't know $\hat{p}$ because they don't have a sample yet, but they can probably guess a value. The worst case—the value that makes the SD (and therefore n) largest—is 0.50, so if they use that value for $\hat{p}$, they'll certainly be safe.

The company's equation, then, is:

$$0.03 = 1.96 \sqrt{\frac{(0.5)(0.5)}{n}}.$$

To solve for n, just multiply both sides of the equation by $\sqrt{n}$ and divide by 0.03:

$$0.03\sqrt{n} = 1.96\sqrt{(0.5)(0.5)}$$

$$\sqrt{n} = \frac{1.96\sqrt{(0.5)(0.5)}}{0.03} \approx 32.67$$

Then square the result to find n:

$$n \approx (32.67)^2 \approx 1067.1$$

That method will probably give a value with a fraction. To be safe, always round up. The company will need at least 1068 respondents to keep the margin of error as small as 3% with a confidence level of 95%.

Unfortunately, bigger samples cost more money and require more effort. Because the standard error declines only with the *square root* of the sample size, to cut the standard error (and thus the ME) in half, you must *quadruple* the sample size.

Generally a margin of error of 5% or less is acceptable, but different circumstances call for different standards. The size of the margin of error may be a marketing decision or one determined by the amount of financial risk you (or the company) are willing to accept. Drawing a large sample to get a smaller ME,

> **What $\hat{p}$ Should We Use?**
>
> Often you'll have an estimate of the population proportion based on experience or perhaps on a previous study. If so, use that value as $\hat{p}$ in calculating what size sample you need. If not, the cautious approach is to use $\hat{p} = 0.5$. That will determine the largest sample necessary regardless of the true proportion. It's the *worst case* scenario.

however, can run into trouble. It takes time to survey 2400 people, and a survey that extends over a week or more may be trying to hit a target that moves during the time of the survey. A news event or new product announcement can change opinions in the middle of the survey process.

Keep in mind that the sample size for a survey is the number of respondents, not the number of people to whom questionnaires were sent or whose phone numbers were dialed. Also keep in mind that a low response rate turns any study essentially into a voluntary response study, which is of little value for inferring population values. It's almost always better to spend resources on increasing the response rate than on surveying a larger group. A complete or nearly complete response by a modest-size sample can yield useful results.

Surveys are not the only place where proportions pop up. Credit card banks sample huge mailing lists to estimate what proportion of people will accept a credit card offer. Even pilot studies may be mailed to 50,000 customers or more. Most of these customers don't respond. But in this case, that doesn't make the sample smaller. In fact, they did respond in a way—they just said "No thanks." To the bank, the response rate[6] is $\hat{p}$. With a typical success rate below 1%, the bank needs a very small margin of error—often as low as 0.1%—to make a sound business decision. That calls for a large sample, and the bank should take care when estimating the size needed. For our election poll example, we used $p = 0.5$, both because it's safe and because we honestly believed p to be near 0.5. If the bank used 0.5, they'd get an absurd answer. Instead they base their calculation on a value of p that they expect to find from their experience.

> **Why 1000?**
>
> Public opinion polls often use a sample size of 1000, which gives an ME of about 3% (at 95% confidence) when p is near 0.5. But businesses and nonprofit organizations often use much larger samples to estimate the response to a direct mail campaign. Why? Because the proportion of people who respond to these mailings is very low, often 5% or even less. An ME of 3% may not be precise enough if the response rate is that low. Instead, an ME like 0.1% would be more useful, and that requires a very large sample size.

How much of a difference can it make?

A credit card company is about to send out a mailing to test the market for a new credit card. From that sample, they want to estimate the true proportion of people who will sign up for the card nationwide. To be within a tenth of a percentage point, or 0.001 of the true acquisition rate with 95% confidence, how big does the test mailing have to be? Similar mailings in the past lead them to expect that about 0.5% of the people receiving the offer will accept it. Using those values, they find:

$$ME = 0.001 = z^* \sqrt{\frac{pq}{n}} = 1.96 \sqrt{\frac{(0.005)(0.995)}{n}}$$

$$(0.001)^2 = 1.96^2 \frac{(0.005)(0.995)}{n} \Rightarrow n = \frac{1.96^2(0.005)(0.995)}{(0.001)^2}$$

$$= 19{,}111.96 \ or \ 19{,}112$$

That's a perfectly reasonable size for a trial mailing. But if they had used 0.50 for their estimate of p they would have found:

$$ME = 0.001 = z^* \sqrt{\frac{pq}{n}} = 1.96 \sqrt{\frac{(0.5)(0.5)}{n}}$$

$$(0.001)^2 = 1.96^2 \frac{(0.5)(0.5)}{n} \Rightarrow n = \frac{1.96^2(0.5)(0.5)}{(0.001)^2} = 960{,}400.$$

Quite a different result!

---

[6]Be careful. In marketing studies like this *every* mailing yields a response—"yes" or "no"—and response rate means the success rate, the proportion of customers who accept the offer. That's a different use of the term response rate from the one used in survey response.

For Example Sample size calculations for a confidence interval for a proportion

The President of the Chamber of Commerce in the previous examples is worried that the 99% confidence interval is too wide. Recall that it was (0.503, 0.617), which has a width of 0.114.

Question: How large a sample would she need to take to have a 99% interval half as wide? One quarter as wide? What if she wanted a 95% confidence interval that was plus or minus 3 percentage points? How large a sample would she need?

Answer: Because the formula for the confidence interval is dependent on the inverse of the square root of the sample size:

$$\hat{p} \pm z^* \sqrt{\frac{\hat{p}\hat{q}}{n}},$$

a sample size four times as large will produce a confidence interval *half* as wide. The original 99% confidence interval had a sample size of 516. If she wants it half as wide, she will need about $4 \times 516 = 2064$ respondents. To get it a quarter as wide she'd need $4^2 \times 516 = 8192$ respondents!

If she wants a 99% confidence interval that's plus or minus 3 percentage points, she must calculate

$$\hat{p} \pm z^* \sqrt{\frac{\hat{p}\hat{y}}{n}} = \hat{p} \pm 0.03$$

so

$$2.576\sqrt{\frac{(0.5)(0.5)}{n}} = 0.03$$

which means that

$$n \approx \left(\frac{2.576}{0.03}\right)^2 (0.5)(0.5) = 1843.27$$

Rounding up, she'd need 1844 respondents. We used 0.5 because we didn't have any information about the election before taking the survey. Using $p = 0.56$ instead would give $n = 1817$.

*11.5 A Confidence Interval for Small Samples

When the Success/Failure condition fails, all is not lost. A simple adjustment to the calculation lets us make a confidence interval anyway. All we do is add four *synthetic* observations, two to the successes and two to the failures. So instead of the proportion $\hat{p} = \dfrac{y}{n}$, we use the adjusted proportion $\widetilde{p} = \dfrac{y + 2}{n + 4}$, and for convenience, we write $\widetilde{n} = n + 4$. We modify the interval by using these adjusted values for both the center of the interval *and* the margin of error. Now the adjusted interval is:

$$\widetilde{p} \pm z^* \sqrt{\frac{\widetilde{p}(1 - \widetilde{p})}{\widetilde{n}}}.$$

This adjusted form gives better performance overall[7] and works much better for proportions near 0 or 1. It has the additional advantage that we don't need to check the Success/Failure condition that $n\hat{p}$ and $n\hat{q}$ are greater than 10.

Suppose a student in an advertising class is studying the impact of ads placed during the Super Bowl, and wants to know what proportion of students

---

[7]By "better performance" we mean that the 95% confidence interval's actual chance of covering the true population proportion is closer to 95%. Simulation studies have shown that our original, simpler confidence interval covers the true population proportion less than 95% of the time when the sample size is small or the proportion is very close to 0 or 1. The original idea was E. B. Wilson's, but the simplified approach we suggest here appeared in A. Agresti and B. A. Coull, "Approximate Is Better Than 'Exact' for Interval Estimation of Binomial Proportions," *The American Statistician*, 52 (1998): 119–126.

on campus watched it. She takes a random sample of 25 students and find that all 25 watched the Super Bowl for a $\hat{p}$ of 100%. A 95% confidence interval is $\hat{p} \pm 1.96\sqrt{\dfrac{\hat{p}\hat{q}}{n}} = 1.0 \pm 1.96\sqrt{\dfrac{1.0(0.0)}{25}} = (1.0, 1.0)$. Does she really believe that *every* one of the 30,000 students on her campus watched the Super Bowl? Probably not. And she realizes that the Success/Failure condition is severely violated because there are *no* failures.

Using the pseudo observation method described above, she adds two successes and two failures to the sample to get 27/29 successes, for $\tilde{p} = \dfrac{27}{29} = 0.931$. The standard error is no longer 0, but $SE(\tilde{p}) = \sqrt{\dfrac{\tilde{p}\tilde{q}}{\tilde{n}}} = \sqrt{\dfrac{(0.931)(0.069)}{29}} = 0.047$.

Now, a 95% confidence interval is $0.931 \pm 1.96(0.047) = (0.839, 1.023)$. In other words, she's 95% confident that between 83.9% and 102.3% of all students on campus watched the Super Bowl. Because any number greater than 100% makes no sense, she will report simply that with 95% confidence the proportion is at least 83.9%.

What Can Go Wrong?

Confidence intervals are powerful tools. Not only do they tell us what is known about the parameter value, but—more important—they also tell us what we *don't* know. In order to use confidence intervals effectively, you must be clear about what you say about them.

* **Be sure to use the right language to describe your confidence intervals.** Technically, you should say "I am 95% confident that the interval from 32.2% to 35.8% captures the true proportion of U.S. adults who thought the economy was improving in March 2010." That formal phrasing emphasizes that *your confidence (and your uncertainty) is about the interval, not the true proportion.* But you may choose a more casual phrasing like "I am 95% confident that between 32.2% and 35.8% of U.S. adults thought the economy was improving in March 2010." Because you've made it clear that the uncertainty is yours and you didn't suggest that the randomness is in the true proportion, this is OK. Keep in mind that it's the interval that's random. It's the focus of both our confidence and our doubt.

* **Don't suggest that the parameter varies.** A statement like "there is a 95% chance that the true proportion is between 32.2% and 35.8%" sounds as though you think the population proportion wanders around and sometimes happens to fall between 32.2% and 35.8%. When you interpret a confidence interval, make it clear that *you* know that the population parameter is fixed and that it is the interval that varies from sample to sample.

* **Don't claim that other samples will agree with yours.** Keep in mind that the confidence interval makes a statement about the true population proportion. An interpretation such as "in 95% of samples of U.S. adults the proportion who thought the economy was improving in March 2010 will be between 32.2% and 35.8%" is just wrong. The interval isn't about sample proportions but about the population proportion. There is nothing special about the sample we happen to have; it doesn't establish a standard for other samples.

* **Don't be certain about the parameter.** Saying "between 32.2% and 35.8% of U.S. adults thought the economy was improving in March 2010" asserts that the population proportion cannot be outside that interval. Of course, you can't be absolutely certain of that (just pretty sure).

- **Don't forget: It's about the parameter.** Don't say "I'm 95% confident that $\hat{p}$ is between 32.2% and 35.8%." Of course, you are—in fact, we calculated that our sample proportion was 34.0%. So we already *know* the sample proportion. The confidence interval is about the (unknown) population parameter, p.

- **Don't claim to know too much.** Don't say "I'm 95% confident that between 32.2% and 35.8% of all U.S. adults think the economy is improving." Gallup sampled adults during March 2010, and public opinion shifts over time.

- **Do take responsibility.** Confidence intervals are about *uncertainty. You* are the one who is uncertain, not the parameter. You have to accept the responsibility and consequences of the fact that not all the intervals you compute will capture the true value. In fact, about 5% of the 95% confidence intervals you find will fail to capture the true value of the parameter. You *can* say "I am 95% confident that between 32.2% and 35.8% of U.S. adults thought the economy was improving in March 2010."

Confidence intervals and margins of error depend crucially on the assumption and conditions. When they're not true the results may be invalid. For your own surveys, follow the survey designs from Chapter 3. For surveys you read about, so be sure to:

- **Watch out for biased sampling.** Just because we have more statistical machinery now doesn't mean we can forget what we've already learned. A questionnaire that finds that 85% of people enjoy filling out surveys still suffers from nonresponse bias even though now we're able to put confidence intervals around this (biased) estimate.

- **Think about independence.** The assumption that the values in a sample are mutually independent is one that you usually cannot check. It always pays to think about it, though.

- **Be careful of sample size.** The validity of the confidence interval for proportions may be affected by sample size. Avoid using the confidence interval on "small" samples.

Ethics in Action

One of Tim Solsby's major responsibilities at MassEast Federal Credit Union is managing online services and website content. In an effort to better serve MassEast members, Tim routinely visits the sites of other financial institutions to get ideas on how he can improve MassEast's online presence. One of the features that caught his attention was a "teen network" that focused on educating teenagers about personal finances. He thought that this was a novel idea and one that could help build a stronger online community among MassEast's members. The executive board of MassEast was meeting next month to consider proposals for improving credit union services, and Tim was eager to present his idea for adding an online teen network. To strengthen his proposal, he decided to poll current credit union members. On the MassEast Federal Credit Union website, he posted an online survey. Among the questions he asked are "Do you have teenage children in your household?" and "Would you encourage your teenage children to learn more about managing personal finances?" Based on 850 responses, Tim constructed a 95% confidence interval and was able to estimate (with 95% confidence) that between 69% and 75% of MassEast members had teenage children at home and that between 62% and 68% would encourage their teenagers to learn more about managing personal finances. Tim believed these results would help convince the executive board that MassEast should add this feature to its website.

ETHICAL ISSUE *The sampling method introduces bias because it is a voluntary response sample and not a random sample. Customers who do have teenagers are more likely to respond than those that do not (related to Item A, ASA Ethical Guidelines).*

ETHICAL SOLUTION *Tim should revise his sampling methods. He might draw a simple random sample of credit union customers and try and contact them by mail or telephone. Whatever method he uses, Tim needs to disclose the sampling procedure to the Board and discuss possible sources of bias.*

What Have We Learned?

Learning Objectives

- Construct a confidence interval for a proportion, p, as the statistic, $\hat{p}$, plus and minus a **margin of error**.
  - The margin of error consists of a **critical value** based on the sampling model times a **standard error** based on the sample.
  - The critical value is found from the Normal model.
  - The standard error of a sample proportion is calculated as $\sqrt{\dfrac{\hat{p}\hat{q}}{n}}$.

- Interpret a confidence interval correctly.
  - You can claim to have the specified level of confidence that the interval you have computed actually covers the true value.

- Understand the importance of the sample size, n, in improving both the certainty (confidence level) and precision (margin of error).
  - For the same sample size and proportion, more certainty requires less precision and more precision requires less certainty.

- Know and check the assumptions and conditions for finding and interpreting confidence intervals.
  - Independence Assumption or Randomization Condition
  - 10% Condition
  - Success/Failure Condition

- Be able to invert the calculation of the margin of error to find the sample size required, given a proportion, a confidence level, and a desired margin of error.

Terms

Confidence interval
: An interval of values usually of the form

$$estimate \pm margin\ of\ error$$

found from data in such a way that a percentage of all random samples can be expected to yield intervals that capture the true parameter value.

Critical value
: The number of standard errors to move away from the mean of the sampling distribution to correspond to the specified level of confidence. The critical value, denoted z^*, is usually found from a table or with technology.

Margin of error (ME)
: In a confidence interval, the extent of the interval on either side of the observed statistic value. A margin of error is typically the product of a critical value from the sampling distribution and a standard error from the data. A small margin of error corresponds to a confidence interval that pins down the parameter precisely. A large margin of error corresponds to a confidence interval that gives relatively little information about the estimated parameter.

One-proportion z-interval
: A confidence interval for the true value of a proportion. The confidence interval is

$$\hat{p} \pm z^*SE(\hat{p})$$

where z^* is a critical value from the Standard Normal model corresponding to the specified confidence level.

Technology Help: Confidence Intervals for Proportions

Confidence intervals for proportions are so easy and natural that many statistics packages don't offer special commands for them. Most statistics programs want the "raw data" for computations. For proportions, the raw data are the "success" and "failure" status for each case. Usually, these are given as 1 or 0, but they might be category names like "yes" and "no." Often we just know the proportion of successes, $\hat{p}$, and the total count, n. Computer packages don't usually deal with summary data like this easily, but the statistics routines found on many graphing calculators allow you to create confidence intervals from summaries of the data—usually all you need to enter are the number of successes and the sample size.

In some programs you can reconstruct variables of 0's and 1's with the given proportions. But even when you have (or can reconstruct) the raw data values, you may not get *exactly* the same margin of error from a computer package as you would find working by hand. The reason is that some packages make approximations or use other methods. The result is very close but not exactly the same. Fortunately, Statistics means never having to say you're certain, so the approximate result is good enough.

EXCEL   XLSTAT

Inference methods for proportions are not part of the standard Excel tool set, but you can compute a confidence interval using Excel's equations:

For example, suppose you have 100 observations in cells A1:A100 and each cell is "yes" or "no."

- In cell B2, enter =countif(A1:A100,"yes")/100 to compute the proportion of "yes" responses. (The 100 is because you have 100 observations. Replace it with the number of observations you actually have.)
- In cell B3, enter =sqrt(B2*(1-B2)/100) to compute the standard error.
- In cell B4, enter =normsinv(.975) for a 95% confidence interval.
- In cell B5, enter =B2-B4*B3 as the lower end of the CI.
- In cell B6, enter =B2+B4*B3 as the upper end of the CI.

Comments

For summarized data, compute the proportion in cell B2 according to whether your summaries are counts, percentages, or already

proportions, and continue with the example, using total count in place of the "100" in the second step.

JMP

For a **categorical** variable that holds category labels, the **Distribution** platform includes tests and intervals for proportions. For summarized data,

- Put the category names in one variable and the frequencies in an adjacent variable.
- Designate the frequency column to have the role of frequency.
- Then use the Distribution platform.

Comments

JMP uses slightly different methods for proportion inferences than those discussed in this text. Your answers are likely to be slightly different, especially for small samples.

MINITAB

Choose **Basic Statistics** from the **Stat** menu.

- Choose **1Proportion** from the Basic Statistics submenu.
- If the data are category names in a variable, assign the variable from the variable list box to the **Samples in columns** box. If you have summarized data, click the **Summarized Data** button and fill in the number of trials and the number of successes.
- Click the **Options** button and specify the remaining details.
- If you have a large sample, check **Use test and interval based on normal distribution.** Click the **OK** button.

Comments

When working from a variable that names categories, MINITAB treats the last category as the "success" category. You can specify how the categories should be ordered.

SPSS

SPSS does not find confidence intervals for proportions.

Brief CASE

Investment

During the period from June 27–29, 2003, the Gallup Organization asked stock market investors questions about the amount and type of their investments. The questions asked the investors were:

1. Is the total amount of your investments right now $10,000 or more, or is it less than $10,000?

2. If you had $1000 to invest, would you be more likely to invest it in stocks or bonds?

In response to the first question, 65% of the 692 investors reported that they currently have at least $10,000 invested in the stock market. In response to the second question, 48% of the 692 investors reported that they would be more likely to invest in stocks (over bonds). Compute the standard error for each sample proportion. Compute and describe the 95% confidence intervals in the context of the question. What would the size of the sample need to be for the margin of error to be 3%?

Find a recent survey about investment practices or opinions and write up a short report on your findings.

Forecasting Demand

Utilities must forecast the demand for energy use far into the future because it takes decades to plan and build new power plants. Ron Bears, who worked for a northeast utility company, had the job of predicting the proportion of homes that would choose to use electricity to heat their homes. Although he was prepared to report a confidence interval for the true proportion, after seeing his preliminary report, his management demanded a single number as his prediction.

Help Ron explain to his management why a confidence interval for the desired proportion would be more useful for planning purposes. Explain how the precision of the interval and the confidence we can have in it are related to each other. Discuss the business consequences of an interval that is too narrow and the consequences of an interval with too low a confidence level.

Exercises

SECTION 11.1

1. For each situation below identify the population and the sample and identify p and $\hat{p}$ if appropriate and what the value of $\hat{p}$ is. Would you trust a confidence interval for the true proportion based on these data? Explain briefly why or why not.

a) As concertgoers enter a stadium, a security guard randomly inspects their backpacks for alcoholic beverages. Of the 130 backpacks checked so far, 17 contained alcoholic beverages of some kind. The guards want to estimate the percentage of all backpacks of concertgoers at this concert that contain alcoholic beverages.

b) The website of the English newspaper *The Guardian* asked visitors to the site to say whether they approved of recent "bossnapping" actions by British workers who were outraged over being fired. Of those who responded, 49.2% said "Yes. Desperate times, desperate measures."

c) An airline wants to know the weight of carry-on baggage that customers take on their international routes, so they take a random sample of 50 bags and find that the average weight is 17.3 pounds.

2. For each situation below identify the population and the sample and explain what p and $\hat{p}$ represent and what the value of $\hat{p}$ is. Would you trust a confidence interval for the true proportion based on these data? Explain briefly why or why not.

a) A marketing analyst conducts a large survey of her customers to find out how much money they plan to spend at the company website in the next 6 months. The average amount reported from the 534 respondents is $145.34.

b) A campus survey on a large campus (40,000 students) is trying to find out whether students approve of a new parking policy allowing students to park in previously inaccessible parking lots, but for a small fee. Surveys are sent out by mail and e-mail. Of the 243 surveys returned, 134 are in favor of the change.

c) The human resources department of a large Fortune 100 company wants to find out how many employees would take advantage of an on-site day care facility. They send out an e-mail to 500 employees and receive responses from 450 of them. Of those responding, 75 say that they would take advantage of such a facility.

3. A survey of 200 students is selected randomly on a large university campus. They are asked if they use a laptop in class to take notes. Suppose that based on the survey, 70 of the 200 students responded "yes."

a) What is the value of the sample proportion $\hat{p}$?
b) What is the standard error of the sample proportion?
c) Construct an approximate 95% confidence interval for the true proportion p by taking ±2 SEs from the sample proportion.

4. From a survey of 250 coworkers you find that 155 would like the company to provide on-site day care.

a) What is the value of the sample proportion $\hat{p}$?
b) What is the standard error of the sample proportion?
c) Construct an approximate 95% confidence interval for the true proportion p by taking ±2 SEs from the sample proportion.

SECTION 11.2

5. From a survey of coworkers you find that 48% of 200 have already received this year's flu vaccine. An approximate 95% confidence interval is (0.409, 0.551). Which of the following are true? If not, explain briefly.

a) 95% of the coworkers fall in the interval (0.409, 0.551).
b) We are 95% confident that the proportion of coworkers who have received this year's flu vaccine is between 40.9% and 55.1%.
c) There is a 95% chance that a random selected coworker has received the vaccine.
d) There is a 48% chance that a random selected coworker has received the vaccine.
e) We are 95% confident that between 40.9% and 55.1% of the samples will have a proportion near 48%.

6. As in Exercise 5, from a survey of coworkers you find that 48% of 200 have already received this year's flu

vaccine. An approximate 95% confidence interval is (0.409, 0.551).

a) How would the confidence interval change if the sample size had been 800 instead of 200?
b) How would the confidence interval change if the confidence level had been 90% instead of 95%?
c) How would the confidence interval change if the confidence level had been 99% instead of 95%?

SECTION 11.3

7. Consider each situation described. Identify the population and the sample, explain what p and $\hat{p}$ represent, and tell whether the methods of this chapter can be used to create a confidence interval.

a) A consumer group hoping to assess customer experiences with auto dealers surveys 167 people who recently bought new cars; 3% of them expressed dissatisfaction with the salesperson.
b) A cell phone service provider wants to know what percent of U.S. college students have cell phones. A total of 2883 students were asked as they entered a football stadium, and 2243 indicated they had phones with them.

8. Consider each situation described. Identify the population and the sample, explain what p and $\hat{p}$ represent, and tell whether the methods of this chapter can be used to create a confidence interval.

a) A total of 240 potato plants in a field in Maine are randomly checked, and only 7 show signs of blight. How severe is the blight problem for the U.S. potato industry?
b) Concerned about workers' compensation costs, a small company decided to investigate on-the-job injuries. The company reported that 12 of their 309 employees suffered an injury on the job last year. What can the company expect in future years?

SECTION 11.4

9. Suppose you want to estimate the proportion of traditional college students on your campus who own their own car. You have no preconceived idea of what that proportion might be.

a) What sample size is needed if you wish to be 95% confident that your estimate is within 0.02 of the true proportion?
b) What sample size is needed if you wish to be 99% confident that your estimate is within 0.02 of the true proportion?
c) What sample size is needed if you wish to be 95% confident that your estimate is within 0.05 of the true proportion?

10. As in Exercise 9, you want to estimate the proportion of traditional college students on your campus who own their own car. However, from some research on other college campuses, you believe the proportion will be near 20%.

a) What sample size is needed if you wish to be 95% confident that your estimate is within 0.02 of the true proportion?

b) What sample size is needed if you wish to be 99% confident that your estimate is within 0.02 of the true proportion?

c) What sample size is needed if you wish to be 95% confident that your estimate is within 0.05 of the true proportion?

11. It's believed that as many as 25% of adults over age 50 never graduated from high school. We wish to see if this percentage is the same among the 25 to 30 age group.

a) How many of this younger age group must we survey in order to estimate the proportion of nongrads to within 6% with 90% confidence?

b) Suppose we want to cut the margin of error to 4%. What's the necessary sample size?

c) What sample size would produce a margin of error of 3%?

12. In preparing a report on the economy, we need to estimate the percentage of businesses that plan to hire additional employees in the next 60 days.

a) How many randomly selected employers must we contact in order to create an estimate in which we are 98% confident with a margin of error of 5%?

b) Suppose we want to reduce the margin of error to 3%. What sample size will suffice?

c) Why might it not be worth the effort to try to get an interval with a margin of error of 1%?

CHAPTER EXERCISES

13. Margin of error. A corporate executive reports the results of an employee satisfaction survey, stating that 52% of employees say they are either "satisfied" or "extremely satisfied" with their jobs, and then says "the margin of error is plus or minus 4%." Explain carefully what that means.

14. Margin of error. A market researcher estimates the percentage of adults between the ages of 21 and 39 who will see their television ad is 15%, adding that he believes his estimate has a margin of error of about 3%. Explain what the margin of error means.

15. Conditions. Consider each situation described below. Identify the population and the sample, explain what p and $\hat{p}$ represent, and tell whether the methods of this chapter can be used to create a confidence interval.

a) Police set up an auto checkpoint at which drivers are stopped and their cars inspected for safety problems. They find that 14 of the 134 cars stopped have at least one safety violation. They want to estimate the proportion of all cars in this area that may be unsafe.

b) A CNN show asks viewers to register their opinions on corporate corruption by logging onto a website. Of the 602 people who voted, 488 thought corporate corruption was "worse" this year than last year. The show wants to estimate the level of support among the general public.

16. More conditions. Consider each situation described below. Identify the population and the sample, explain what p and $\hat{p}$ represent, and tell whether the methods of this chapter can be used to create a confidence interval.

a) A large company with 10,000 employees at their main research site is considering moving its day care center off-site to save money. Human resources gathers employees' opinions by sending a questionnaire home with all employees; 380 surveys are returned, with 228 employees in favor of the change.

b) A company sold 1632 MP3 players last month, and within a week, 1388 of the customers had registered their products online at the company website. The company wants to estimate the percentage of all their customers who enroll their products.

17. Catalog sales. A catalog sales company promises to deliver orders placed on the Internet within 3 days. Follow-up calls to a few randomly selected customers show that a 95% confidence interval for the proportion of all orders that arrive on time is 88% ± 6%. What does this mean? Are the conclusions in parts a–e correct? Explain.

a) Between 82% and 94% of all orders arrive on time.

b) 95% of all random samples of customers will show that 88% of orders arrive on time.

c) 95% of all random samples of customers will show that 82% to 94% of orders arrive on time.

d) The company is 95% sure that between 82% and 94% of the orders placed by the customers in this sample arrived on time.

e) On 95% of the days, between 82% and 94% of the orders will arrive on time.

18. Belgian euro. Recently, two students made worldwide headlines by spinning a Belgian euro 250 times and getting 140 heads—that's 56%. That makes the 90% confidence interval (51%, 61%). What does this mean? Are the conclusions in parts a–e correct? Explain your answers.

a) Between 51% and 61% of all euros are unfair.

b) We are 90% sure that in this experiment this euro landed heads between 51% and 61% of the spins.

c) We are 90% sure that spun euros will land heads between 51% and 61% of the time.

d) If you spin a euro many times, you can be 90% sure of getting between 51% and 61% heads.

e) 90% of all spun euros will land heads between 51% and 61% of the time.

19. Confidence intervals. Several factors are involved in the creation of a confidence interval. Among them are the sample size, the level of confidence, and the margin of error. Which statements are true?

a) For a given sample size, higher confidence means a smaller margin of error.

b) For a specified confidence level, larger samples provide smaller margins of error.

c) For a fixed margin of error, larger samples provide greater confidence.
d) For a given confidence level, halving the margin of error requires a sample twice as large.

20. Confidence intervals, again. Several factors are involved in the creation of a confidence interval. Among them are the sample size, the level of confidence, and the margin of error. Which statements are true?

a) For a given sample size, reducing the margin of error will mean lower confidence.
b) For a certain confidence level, you can get a smaller margin of error by selecting a bigger sample.
c) For a fixed margin of error, smaller samples will mean lower confidence.
d) For a given confidence level, a sample 9 times as large will make a margin of error one third as big.

21. Cars. A student is considering publishing a new magazine aimed directly at owners of Japanese automobiles. He wanted to estimate the fraction of cars in the United States that are made in Japan. The computer output summarizes the results of a random sample of 50 autos. Explain carefully what it tells you.

```
z-interval for proportion
With 90.00% confidence
0.29938661 < p(japan) < 0.46984416
```

22. Quality control. For quality control purposes, 900 ceramic tiles were inspected to determine the proportion of defective (e.g., cracked, uneven finish, etc.) tiles. Assuming that these tiles are representative of all tiles manufactured by an Italian tile company, what can you conclude based on the computer output?

```
z-interval for proportion
With 95.00% confidence
0.025 < p(defective) < 0.035
```

23. E-mail. A small company involved in e-commerce is interested in statistics concerning the use of e-mail. A poll found that 38% of a random sample of 1012 adults, who use a computer at their home, work, or school, said that they do not send or receive e-mail.

a) Find the margin of error for this poll if we want 90% confidence in our estimate of the percent of American adults who do not use e-mail.
b) Explain what that margin of error means.
c) If we want to be 99% confident, will the margin of error be larger or smaller? Explain.
d) Find that margin of error.
e) In general, if all other aspects of the situation remain the same, will smaller margins of error involve greater or less confidence in the interval?

24. Biotechnology. A biotechnology firm in Boston is planning its investment strategy for future products and research labs. A poll found that only 8% of a random sample of 1012 U.S. adults approved of attempts to clone a human.

a) Find the margin of error for this poll if we want 95% confidence in our estimate of the percent of American adults who approve of cloning humans.
b) Explain what that margin of error means.
c) If we only need to be 90% confident, will the margin of error be larger or smaller? Explain.
d) Find that margin of error.
e) In general, if all other aspects of the situation remain the same, would smaller samples produce smaller or larger margins of error?

25. Teenage drivers. An insurance company checks police records on 582 accidents selected at random and notes that teenagers were at the wheel in 91 of them.

a) Create a 95% confidence interval for the percentage of all auto accidents that involve teenage drivers.
b) Explain what your interval means.
c) Explain what "95% confidence" means.
d) A politician urging tighter restrictions on drivers' licenses issued to teens says, "In one of every five auto accidents, a teenager is behind the wheel." Does your confidence interval support or contradict this statement? Explain.

26. Advertisers. Direct mail advertisers send solicitations ("junk mail") to thousands of potential customers in the hope that some will buy the company's product. The response rate is usually quite low. Suppose a company wants to test the response to a new flyer and sends it to 1000 people randomly selected from their mailing list of over 200,000 people. They get orders from 123 of the recipients.

a) Create a 90% confidence interval for the percentage of people the company contacts who may buy something.
b) Explain what this interval means.
c) Explain what "90% confidence" means.
d) The company must decide whether to now do a mass mailing. The mailing won't be cost-effective unless it produces at least a 5% return. What does your confidence interval suggest? Explain.

27. Retailers. Some food retailers propose subjecting food to a low level of radiation in order to improve safety, but sale of such "irradiated" food is opposed by many people. Suppose a grocer wants to find out what his customers think. He has cashiers distribute surveys at checkout and ask customers to fill them out and drop them in a box near the front door. He gets responses from 122 customers, of whom 78 oppose the radiation treatments. What can the grocer conclude about the opinions of all his customers?

28. Local news. The mayor of a small city has suggested that the state locate a new prison there, arguing that the construction project and resulting jobs will be good for the local economy. A total of 183 residents show up for a public hearing on the proposal, and a show of hands finds 31 in favor of the prison project. What can the city council conclude about public support for the mayor's initiative?

29. Internet music. In a survey on downloading music, the Gallup Poll asked 703 Internet users if they "ever downloaded music from an Internet site that was not authorized by a record company, or not," and 18% responded "yes." Construct a 95% confidence interval for the true proportion of Internet users who have downloaded music from an Internet site that was not authorized.

30. Economy worries. In 2008, a Gallup Poll asked 2335 U.S. adults, aged 18 or over, how they rated economic conditions. In a poll conducted from January 27–February 1, 2008, only 24% rated the economy as Excellent/Good. Construct a 95% confidence interval for the true proportion of Americans who rated the U.S. economy as Excellent/Good.

31. International business. In Canada, the vast majority (90%) of companies in the chemical industry are ISO 14001 certified. The ISO 14001 is an international standard for environmental management systems. An environmental group wished to estimate the percentage of U.S. chemical companies that are ISO 14001 certified. Of the 550 chemical companies sampled, 385 are certified.

a) What proportion of the sample reported being certified?
b) Create a 95% confidence interval for the proportion of U.S. chemical companies with ISO 14001 certification. (Be sure to check conditions.) Compare to the Canadian proportion.

32. Worldwide survey. In Chapter 4, Exercise 27, we learned that GfK Roper surveyed people worldwide asking them "how important is acquiring wealth to you." Of 1535 respondents in India, 1168 said that it was of more than average importance. In the United States of 1317 respondents, 596 said it was of more than average importance.

a) What proportion thought acquiring wealth was of more than average importance in each country's sample?
b) Create a 95% confidence interval for the proportion who thought it was of more than average importance in India. (Be sure to test conditions.) Compare that to a confidence interval for the U.S. population.

33. Business ethics. In a survey on corporate ethics, a poll split a sample at random, asking 538 faculty and corporate recruiters the question: "Generally speaking, do you believe that MBAs are more or less aware of ethical issues in business today than five years ago?" The other half were asked: "Generally speaking, do you believe that MBAs are less or more aware of ethical issues in business today than five years ago?" These may seem like the same questions, but sometimes the order of the choices matters. In response to the first question, 53% thought MBA graduates were more aware of

ethical issues, but when the question was phrased differently, this proportion dropped to 44%.

a) What kind of bias may be present here?
b) Each group consisted of 538 respondents. If we combine them, considering the overall group to be one larger random sample, what is a 95% confidence interval for the proportion of the faculty and corporate recruiters that believe MBAs are more aware of ethical issues today?
c) How does the margin of error based on this pooled sample compare with the margins of error from the separate groups? Why?

34. Media survey. In 2007, a Gallup Poll conducted face-to-face interviews with 1006 adults in Saudi Arabia, aged 15 and older, asking them questions about how they get information. Among them was the question: "Is international television very important in keeping you well-informed about events in your country?" Gallup reported that 82% answered "yes" and noted that at 95% confidence there was a 3% margin of error and that "in addition to sampling error, question wording and practical difficulties in conducting surveys can introduce error or bias into the findings of public opinion polls."

a) What kinds of bias might they be referring to?
b) Do you agree with their margin of error? Explain.

35. Gambling. A city ballot includes a local initiative that would legalize gambling. The issue is hotly contested, and two groups decide to conduct polls to predict the outcome. The local newspaper finds that 53% of 1200 randomly selected voters plan to vote "yes," while a college Statistics class finds 54% of 450 randomly selected voters in support. Both groups will create 95% confidence intervals.

a) Without finding the confidence intervals, explain which one will have the larger margin of error.
b) Find both confidence intervals.
c) Which group concludes that the outcome is too close to call? Why?

36. Casinos. Governor Deval Patrick of Massachusetts proposed legalizing casinos in Massachusetts although they are not currently legal, and he included the revenue from them in his latest state budget. The website www.boston .com conducted an Internet poll on the question: "Do you agree with the casino plan the governor is expected to unveil?" As of the end of 2007, there were 8663 votes cast, of which 63.5% of respondents said: "No. Raising revenues by allowing gambling is shortsighted."

a) Find a 95% confidence interval for the proportion of voters in Massachusetts who would respond this way.
b) Are the assumptions and conditions satisfied? Explain.

37. Pharmaceutical company. A pharmaceutical company is considering investing in a "new and improved" vitamin D supplement for children. Vitamin D, whether ingested as a dietary supplement or produced naturally when sunlight falls upon the skin, is essential for strong, healthy bones. The bone disease rickets was largely eliminated in England during the 1950s, but now there is concern that a generation of children more likely to watch TV or play computer games than spend time outdoors is at increased risk. A recent study of 2700 children randomly selected from all parts of England found 20% of them deficient in vitamin D.

a) Find a 98% confidence interval for the proportion of children in England who are deficient in vitamin D.
b) Explain carefully what your interval means.
c) Explain what "98% confidence" means.
d) Does the study show that computer games are a likely cause of rickets? Explain.

38. Wireless access. In Chapter 4, Exercise 36, we saw that the Pew Internet and American Life Project polled 798 Internet users in December 2006, asking whether they have logged on to the Internet using a wireless device or not and 243 responded "Yes."

a) Find a 98% confidence interval for the proportion of all U.S. Internet users who have logged in using a wireless device.
b) Explain carefully what your interval means.
c) Explain what "98% confidence" means.

*39. Funding. In 2005, a survey developed by Babson College and the Association of Women's Business Centers (WBCs) was distributed to WBCs in the United States. Of a representative sample of 20 WBCs, 40% reported that they had received funding from the national Small Business Association (SBA).

a) Check the assumptions and conditions for inference on proportions.
b) If it's appropriate, find a 90% confidence interval for the proportion of WBCs that receive SBA funding. If it's not appropriate, explain and/or recommend an alternative action.

*40. Real estate survey. A real estate agent looks over the 15 listings she has in a particular zip code in California and finds that 80% of them have swimming pools.

a) Check the assumptions and conditions for inference on proportions.
b) If it's appropriate, find a 90% confidence interval for the proportion of houses in this zip code that have swimming pools. If it's not appropriate, explain and/or recommend an alternative action.

*41. Benefits survey. A paralegal at the Vermont State Attorney General's office wants to know how many companies in Vermont provide health insurance benefits to all employees. She chooses 12 companies at random and finds that all 12 offer benefits.

a) Check the assumptions and conditions for inference on proportions.
b) Find a 95% confidence interval for the true proportion of companies that provide health insurance benefits to all their employees.

*42. Awareness survey. A telemarketer at a credit card company is instructed to ask the next 18 customers that call into the 800 number whether they are aware of the new Platinum card that the company is offering. Of the 18, 17 said they were aware of the program.

a) Check the assumptions and conditions for inference on proportions.
b) Find a 95% confidence interval for the true proportion of customers who are aware of the new card.

43. IRS. In a random survey of 226 self-employed individuals, 20 reported having had their tax returns audited by the IRS in the past year. Estimate the proportion of self-employed individuals nationwide who've been audited by the IRS in the past year.

a) Check the assumptions and conditions (to the extent you can) for constructing a confidence interval.
b) Construct a 95% confidence interval.
c) Interpret your interval.
d) Explain what "95% confidence" means in this context.

44. ACT, Inc. In 2004, ACT, Inc. reported that 74% of 1644 randomly selected college freshmen returned to college the next year. Estimate the national freshman-to-sophomore retention rate.

a) Check that the assumptions and conditions are met for inference on proportions.
b) Construct a 98% confidence interval.
c) Interpret your interval.
d) Explain what "98% confidence" means in this context.

45. Internet music, again. A Gallup Poll (Exercise 29) asked Americans if the fact that they can make copies of songs on the Internet for free made them more likely—or less likely—to buy a performer's CD. Only 13% responded that it made them "less likely." The poll was based on a random sample of 703 Internet users.

a) Check that the assumptions and conditions are met for inference on proportions.
b) Find the 95% confidence interval for the true proportion of all U.S. Internet users who are "less likely" to buy CDs.

46. ACT, Inc., again. The ACT, Inc. study described in Exercise 44 was actually stratified by type of college—public or private. The retention rates were 71.9% among 505 students enrolled in public colleges and 74.9% among 1139 students enrolled in private colleges.

a) Will the 95% confidence interval for the true national retention rate in private colleges be wider or narrower than the 95% confidence interval for the retention rate in public colleges? Explain.
b) Find the 95% confidence interval for the public college retention rate.
c) Should a public college whose retention rate is 75% proclaim that they do a better job than other public colleges of keeping freshmen in school? Explain.

47. Politics. A poll of 1005 U.S. adults split the sample into four age groups: ages 18–29, 30–49, 50–64, and 65+. In the youngest age group, 62% said that they thought the U.S. was ready for a woman president, as opposed to 35% who said "no, the country was not ready" (3% were undecided). The sample included 250 18- to 29-year-olds.

a) Do you expect the 95% confidence interval for the true proportion of all 18- to 29-year-olds who think the U.S. is ready for a woman president to be wider or narrower than the 95% confidence interval for the true proportion of all U.S. adults? Explain.

b) Find the 95% confidence interval for the true proportion of all 18- to 29-year-olds who believe the U.S. is ready for a woman president.

48. Wireless access, again. The survey in Exercise 38 asking about wireless Internet access also classified the 798 respondents by income.

	A	B	C	D
1		Wireless Users	Other Internet Users	Total
2	Under $30K	34	128	162
3	$30K--$50K	31	133	164
4	$50K--$75K	44	72	116
5	Over $75K	83	111	194
6	Don't know/refused	51	111	162
7	Total	243	555	798

a) Do you expect the 95% confidence interval for the true proportion of all those making more than $75K who are wireless users to be wider or narrower than the 95% confidence interval for the true proportion among those who make between $50K and $75K? Explain briefly.

b) Find the 95% confidence interval for the true proportion of those making more than $75K who are wireless users.

49. More Internet music. A random sample of 168 students was asked how many songs were in their digital music library and what fraction of them was legally purchased. Overall, they reported having a total of 117,079 songs, of which 23.1% were legal. The music industry would like a good estimate of the proportion of songs in students' digital music libraries that are legal.

a) Think carefully. What is the parameter being estimated? What is the population? What is the sample size?

b) Check the conditions for making a confidence interval.

c) Construct a 95% confidence interval for the fraction of legal digital music.

d) Explain what this interval means. Do you believe that you can be this confident about your result? Why or why not?

50. Trade agreement. Results from a January 2008 telephone survey conducted by Gallup showed that 57% of urban Colombian adults support a free trade agreement (FTA) with the United States. Gallup used a sample of 1000 urban Colombians aged 15 and older.

a) What is the parameter being estimated? What is the population? What is the sample size?

b) Check the conditions for making a confidence interval.

c) Construct a 95% confidence interval for the fraction of Colombians in agreement with the FTA.

d) Explain what this interval means. Do you believe that you can be this confident about your result? Why or why not?

51. CDs. A company manufacturing CDs is working on a new technology. A random sample of 703 Internet users were asked: "As you may know, some CDs are being manufactured so that you can only make one copy of the CD after you purchase it. Would you buy a CD with this technology, or would you refuse to buy it even if it was one you would normally buy?" Of these users, 64% responded that they would buy the CD.

a) Create a 90% confidence interval for this percentage.

b) If the company wants to cut the margin of error in half, how many users must they survey?

52. Internet music, last time. The research group that conducted the survey in Exercise 49 wants to provide the music industry with definitive information, but they believe that they could use a smaller sample next time. If the group is willing to have twice as big a margin of error, how many songs must be included?

53. Graduation. As in Exercise 11, we hope to estimate the percentage of adults aged 25 to 30 who never graduated from high school. What sample size would allow us to increase our confidence level to 95% while reducing the margin of error to only 2%?

54. Better hiring info. Editors of the business report in Exercise 12 are willing to accept a margin of error of 4% but want 99% confidence. How many randomly selected employers will they need to contact?

55. Pilot study. A state's environmental agency worries that a large percentage of cars may be violating clean air emissions standards. The agency hopes to check a sample of vehicles in order to estimate that percentage with a margin of error of 3% and 90% confidence. To gauge the size of the problem, the agency first picks 60 cars and finds 9 with faulty emissions systems. How many should be sampled for a full investigation?

56. Another pilot study. During routine conversations, the CEO of a new start-up reports that 22% of adults between the ages of 21 and 39 will purchase her new product. Hearing this, some investors decide to conduct a large-scale study, hoping to estimate the proportion to within 4% with 98% confidence. How many randomly selected adults between the ages of 21 and 39 must they survey?

57. Approval rating. A newspaper reports that the governor's approval rating stands at 65%. The article adds that the poll is based on a random sample of 972 adults and has a margin of error of 2.5%. What level of confidence did the pollsters use?

58. Amendment. The Board of Directors of a publicly traded company says that a proposed amendment to their bylaws is likely to win approval in the upcoming election because a poll of 1505 stock owners indicated that 52% would vote in favor. The Board goes on to say that the margin of error for this poll was 3%.
a) Explain why the poll is actually inconclusive.
b) What confidence level did the pollsters use?

T 59. Customer spending. The data set provided contains last month's credit card purchases of 500 customers randomly chosen from a segment of a major credit card issuer. The marketing department is considering a special offer for customers who spend more than $1000 per month on their card. From these data construct a 95% confidence interval for the proportion of customers in this segment who will qualify.

T 60. Advertising. A philanthropic organization knows that its donors have an average age near 60 and is considering taking out an ad in the *American Association of Retired People (AARP)* magazine. An analyst wonders what proportion of their donors are actually 50 years old or older. He takes a

random sample of the records of 500 donors. From the data provided, construct a 95% confidence interval for the proportion of donors who are 50 years old or older.

61. Health insurance. Based on a 2007 survey of U.S. households (see www.census.gov), 87% (out of 3060) of males in Massachusetts (MA) have health insurance.
a) Examine the conditions for constructing a confidence interval for the proportion males in MA who had health insurance.
b) Find the 95% confidence interval for the percent of males who have health insurance.
c) Interpret your confidence interval.

62. Health insurance, part 2. Using the same survey and data as in Exercise 61, we find that 84% of those respondents in Massachusetts who identified themselves as Black/African-Americans (out of 440) had health insurance.
a) Examine the conditions for constructing a confidence interval for the proportion of Black/African-Americans in MA who had health insurance.
b) Find the 95% confidence interval.
c) Interpret your confidence interval.

Just Checking Answers

1 Wider
2 Lower
3 Smaller

Confidence Intervals for Means

CHAPTER

12

Guinness & Co.

In 1759, when Arthur Guinness was 34 years old, he took an incredible gamble, signing a 9000-year lease on a run-down, abandoned brewery in Dublin. The brewery covered four acres and consisted of a mill, two malt houses, stabling for 12 horses, and a loft that could hold 200 tons of hay. At the time, brewing was a difficult and competitive market. Gin, whiskey, and the traditional London porter were the drinks of choice.

In addition to the lighter ales that Dublin was known for, Guinness began to brew dark porters to compete directly with those of the English brewers. Forty years later, Guinness stopped brewing light Dublin ales altogether to concentrate on his stouts and porters. Upon his death in 1803, his son Arthur Guinness II took over the business, and a few years later the company began to export Guinness stout to other parts of Europe. By the 1830s, the Guinness St. James's Gate Brewery had become the largest in Ireland. In 1886, the Guinness Brewery, with an annual production of 1.2 million barrels, was the first major brewery to be incorporated as a public company on the London Stock Exchange. During the 1890s, the company began to employ scientists. One of those, William S. Gosset, was hired as a chemist to test the quality of the brewing process. Gosset was not only an early pioneer of quality control methods in industry but a statistician whose work made modern statistical inference possible.[1]

As a chemist at the Guinness Brewery in Dublin, William S. Gosset was in charge of quality control. His job was to make sure that the stout (a thick, dark beer) leaving the brewery was of high enough quality to meet the standards of the brewery's many discerning customers. It's easy to imagine, when testing stout, why testing a large amount of stout might be undesirable, not to mention dangerous to one's health. So to test for quality Gosset often used a sample of only 3 or 4 observations per batch. But he noticed that with samples of this size, his tests for quality weren't quite right. He knew this because when the batches that he rejected were sent back to the laboratory for more extensive testing, too often the test results turned out to be wrong. As a practicing statistician, Gosset knew he had to be wrong *some* of the time, but he hated being wrong more often than the theory predicted. One result of Gosset's frustrations was the development of a test to handle small samples, the main subject of this chapter.

12.1   The Sampling Distribution for the Mean

You've learned how to create confidence intervals for proportions. Now we want to do the same thing for means. For proportions we found the confidence interval as

$$\hat{p} \pm ME.$$

The *ME* was equal to a critical value, z^*, times $SE(\hat{p})$. Our confidence interval for means will look very similar:

$$\bar{y} \pm ME.$$

And our *ME* will be a critical value times $SE(\bar{y})$. So let's put the pieces together. What the Central Limit Theorem told us back in Chapter 10 is exactly what we need.

> ### The Central Limit Theorem
> When a random sample is drawn from *any* population with mean μ and standard deviation σ, its sample mean, $\bar{y}$, has a sampling distribution whose *shape* is approximately Normal as long as the sample size is large enough. The larger the sample used, the more closely the Normal approximates the sampling distribution for the mean. The mean of the sampling distribution is μ, and its standard deviation is $SD(\bar{y}) = \dfrac{\sigma}{\sqrt{n}}$.

---

[1]Source: Guinness & Co., www.guinness.com/global/story/history.

This gives us a sampling distribution and a standard deviation for the mean. All we need is a random sample of quantitative data and the true value of the population standard deviation σ.

But wait. That could be a problem. To compute $\sigma/\sqrt{n}$ we need to know σ. How are we supposed to know σ? Suppose we told you that for 25 young executives the mean value of their stock portfolios is \$125,672. Would that tell you the value of σ? No, the standard deviation depends on how similarly the executives invest, not on how well they invested (the mean tells us that). But we need σ because it's the numerator of the standard deviation of the sample mean: $SD(\bar{y}) = \dfrac{\sigma}{\sqrt{n}}$. So what can we do? The obvious answer is to use the sample standard deviation, s, from the data instead of σ. The result is the standard error: $SE(\bar{y}) = \dfrac{s}{\sqrt{n}}$.

A century ago, people just plugged the standard error into the Normal model, assuming it would work. And for large sample sizes it *did* work pretty well. But they began to notice problems with smaller samples. The extra variation in the standard error was wreaking havoc with the margins of error.

Gosset was the first to investigate this phenomenon. He realized that not only do we need to allow for the extra variation with larger margins of error, but we also need a new sampling distribution model. In fact, we need a whole *family* of models, depending on the sample size, n. These models are unimodal, symmetric, and bell-shaped, but the smaller our sample, the more we must stretch out the tails. Gosset's work transformed Statistics, but most people who use his work don't even know his name.

Gosset's *t*

Gosset made decisions about the stout's quality by using statistical inference. He knew that if he used a 95% confidence interval, he would fail to capture the true quality of the batch about 5% of the time. However, the lab told him that he was in fact rejecting about 15% of the good batches. Gosset knew something was wrong, and it bugged him.

Gosset took time off from his job to study the problem and earn a graduate degree in the emerging field of Statistics. He figured out that when he used the standard error $\dfrac{s}{\sqrt{n}}$, the shape of the sampling model was no longer Normal. He even figured out what the new model was and called it a *t*-distribution.

The Guinness Company didn't give Gosset a lot of support for his work. In fact, it had a policy against publishing results. Gosset had to convince the company that he was not publishing an industrial secret and (as part of getting permission to publish) had to use a pseudonym. The pseudonym he chose was "Student," and ever since, the model he found has been known as **Student's *t*.**

Gosset's model is always bell-shaped, but the details change with the sample sizes (Figure 12.1). So the Student's *t*-models form a family of related distributions that depend on a parameter known as **degrees of freedom**. We often denote degrees of freedom as df and the model as t_{df}, with the numerical value of the degrees of freedom as a subscript.

Student's *t*-models are unimodal, symmetric, and bell-shaped, just like the Normal model. But *t*-models with only a few degrees of freedom have a narrower peak than the Normal model and have much fatter tails. (That's what makes the margin of error bigger.) As the degrees of freedom increase, the *t*-models look more and more like the Normal model. In fact, the *t*-model with infinite degrees of freedom is exactly Normal.[2] This is great news if you happen to have an infinite

To find the sampling distribution of $\dfrac{\bar{y}}{s/\sqrt{n}}$, Gosset simulated it *by hand.* He drew paper slips of small samples from a hat *hundreds of times* and computed the means and standard deviations with a mechanically cranked calculator. Today you could repeat in seconds on a computer the experiment that took him over a year. Gosset's work was so meticulous that not only did he get the shape of the new histogram approximately right, but he even figured out the exact *formula* for it from his sample. The formula was not confirmed mathematically until years later by Sir Ronald Aylmer Fisher.

---

[2]Formally, in the limit as the number of degrees of freedom goes to infinity.

Figure 12.1 The *t*-model (solid curve) with 2 degrees of freedom has fatter tails than the Normal model (dashed curve). So the 68–95–99.7 Rule doesn't work for *t*-models with only a few degrees of freedom.

z or t?

If you know σ, use *z*. (That's rare!) Whenever you use *s* to estimate σ, use *t*.

number of data values. Unfortunately, that's not practical. Fortunately, above a few hundred degrees of freedom it's very hard to tell the difference. Of course, in the rare situation that we *know* σ, it would be foolish not to use that information. If we don't have to estimate σ, we can use the Normal model. Typically that value of σ would be based on (lots of) experience, or on a theoretical model. Usually, however, we estimate σ by *s* from the data and use the *t*-model.

> ### Using a known standard deviation
>
> Variation is inherent in manufacturing, even under the most tightly controlled processes. To ensure that parts do not vary too much, however, quality professionals monitor the processes by selecting samples at regular intervals (see Chapter 22 for more details). The mean performance of these samples is measured, and if it lies too far from the desired target mean, the process may be stopped until the underlying cause of the problem can be determined. In silicon wafer manufacturing, the thickness of the film is a crucial measurement. To assess a sample of wafers, quality engineers compare the mean thickness of the sample to the target mean. But, they don't estimate the standard deviation of the mean by using the standard error derived from the same sample. Instead they base the standard deviation of the mean on the historical process standard deviation, estimated from a vast collection of similar parts. In this case, the standard deviation can be treated as "known" and the normal model can be used for the sampling distribution instead of the *t* distribution.

12.2   A Confidence Interval for Means

Notation Alert! ————

Ever since Gosset, the letter *t* has been reserved in Statistics for his distribution.

To make confidence intervals, we need to use Gosset's model. Which one? Well, for means, it turns out the right value for degrees of freedom is df = $n - 1$.

> ### Practical sampling distribution model for means
>
> When certain conditions are met, the standardized sample mean,
>
> $$t = \frac{\bar{y} - \mu}{SE(\bar{y})}$$
>
> follows a Student's *t*-model with $n - 1$ degrees of freedom. We find the standard error from:
>
> $$SE(\bar{y}) = \frac{s}{\sqrt{n}}.$$

When Gosset corrected the Normal model for the extra uncertainty, the margin of error got bigger, as you might have guessed. When you use Gosset's model instead of the Normal model, your confidence intervals will be just a bit wider. That's just the correction you need. By using the *t*-model, you've compensated for the extra variability in precisely the right way.

One-sample *t*-interval

When the assumptions and conditions are met, we are ready to find the **confidence interval for the population mean, μ.** The confidence interval is:

$$\bar{y} \pm t^*_{n-1} \times SE(\bar{y})$$

where the standard error of the mean is:

$$SE(\bar{y}) = \frac{s}{\sqrt{n}}.$$

The critical value t^*_{n-1} depends on the particular confidence level, *C*, that you specify and on the number of degrees of freedom, $n - 1$, which we get from the sample size.

Finding t^*-Values

The Student's *t*-model is different for each value of degrees of freedom. We might print a table like Table Z (in Appendix D) for each degrees of freedom value, but that's a lot of pages and not likely to be a bestseller. One way to shorten the book is to limit ourselves to 80%, 90%, 95% and 99% confidence levels. So Statistics books usually have one table of *t*-model critical values for a selected set of confidence levels. This one does too; see Table T in Appendix D. (You can also find tables on the Internet.)

The *t*-tables run down the page for as many degrees of freedom as can fit, and, as you can see from Figure 12.2, they are much easier to use than the Normal tables. Then they get to the bottom of the page and run out of room. Of course, for *enough* degrees of freedom, the *t*-model gets closer and closer to the Normal, so the tables give a final row with the critical values from the Normal model and label it "∞ df."

		Two tail probability	0.20	0.10	0.05
		One tail probability	0.10	0.05	0.025
Table T	**df**				
Values of t_α	*1*		3.078	6.314	12.706
	2		1.886	2.920	4.303
	3		1.638	2.353	3.182
	4		1.533	2.132	2.776
	5		1.476	2.015	2.571
	6		1.440	1.943	2.447
	7		1.415	1.895	2.365
	8		1.397	1.860	2.306
	9		1.383	1.833	2.262
	10		1.372	1.812	2.228
	11		1.363	1.796	2.201
	12		1.356	1.782	2.179
	13		1.350	1.771	2.160
	14		1.345	1.761	2.145
	15		1.341	1.753	2.131
	16		1.337	1.746	2.120
	17		1.333	1.740	2.110
	18		1.330	1.734	2.101
	19		1.328	1.729	2.093
	⋮		⋮	⋮	⋮
	∞		1.282	1.645	1.960
	Confidence levels		80%	90%	95%

(Two tails diagram: $\frac{\alpha}{2}$ and $\frac{\alpha}{2}$ with $-t_{\alpha/2}$, 0, $t_{\alpha/2}$ — "Two tails")

(One tail diagram: α with 0, t_α — "One tail")

| Figure 12.2   Part of Table T in Appendix D.

| For Example | Finding a confidence interval for the mean |

According to the Environmental Defense Fund, "Americans are eating more and more salmon, drawn to its rich taste and health benefits. Increasingly they are choosing *farmed* salmon because of its wide availability and low price. But in the last few years, farmed salmon has been surrounded by controversy over its health risks and the ecological impacts of salmon aquaculture operations. Studies have shown that some farmed salmon is relatively higher in contaminants like PCBs than wild salmon, and there is mounting concern over the industry's impact on wild salmon populations."

In a widely cited study of contaminants in farmed salmon, fish from many sources were analyzed for 14 organic contaminants.[3] One of those was the insecticide mirex, which has been shown to be carcinogenic and is suspected of being toxic to the liver, kidneys, and endocrine system. Summaries for 150 mirex concentrations (in parts per million) from a variety of farmed salmon sources were reported as:

$$n = 150; \quad \bar{y} = 0.0913 \text{ ppm}; \quad s = 0.0495 \text{ ppm}$$

Question: The Environmental Protection Agency (EPA) recommends to recreational fishers as a "screening value" that mirex concentrations be no larger than 0.08 ppm. What does the 95% confidence interval say about that value?

Answer: Because $n = 150$, there are 149 df. From Table T in Appendix D, we find $t^*_{140, 0.025} = 1.977$ (from technology, $t^*_{149, 0.025} = 1.976$), so a 95% confidence interval can be found from:

$$\bar{y} \pm t^* \times SE(\bar{y}) = \bar{y} \pm 1.977 \times \frac{s}{\sqrt{n}} = 0.0913 \pm 1.977\frac{0.0495}{\sqrt{150}} = (0.0833, 0.0993)$$

If this sample is representative (as the authors claim it is), we can be 95% confident that it contains the true value of the mean mirex concentration. Because the interval from 0.0834 to 0.0992 ppm is entirely above the recommended value set by the EPA, we have reason to believe that the true mirex concentration exceeds the EPA guidelines.

12.3   Assumptions and Conditions

Gosset found the *t*-model by simulation. Years later, when Fisher showed mathematically that Gosset was right, he needed to make some assumptions to make the proof work. These are the assumptions we need in order to use the Student's *t*-models.

Independence Assumption

Independence Assumption: The data values should be independent. There's really no way to check independence of the data by looking at the sample, but we should think about whether the assumption is reasonable.

Randomization Condition: The data arise from a random sample or suitably randomized experiment. Randomly sampled data—and especially data from a Simple Random Sample (SRS)—are ideal.

When a sample is drawn without replacement, technically we ought to confirm that we haven't sampled a large fraction of the population, which would threaten the independence of our selections.

10% Condition: The sample size should be no more than 10% of the population. In practice, though, we often don't mention the 10% Condition when estimating means. Why not? When we made inferences about proportions, this condition was crucial because we usually had large samples. But for means our samples are

We Don't *Want* to Stop
We check conditions hoping that we can make a meaningful analysis of our data. The conditions serve as *disqualifiers*—we keep going unless there's a serious problem. If we find minor issues, we note them and express caution about our results. If the sample is not an SRS, but we believe it's representative of some populations, we limit our conclusions accordingly. If there are outliers, rather than stop, we perform the analysis both with and without them. If the sample looks bimodal, we try to analyze subgroups separately. Only when there's major trouble—like a strongly skewed small sample or an obviously non-representative sample—are we unable to proceed at all.

[3]Ronald A. Hites, Jeffery A. Foran, David O. Carpenter, M. Coreen Hamilton, Barbara A. Knuth, and Steven J. Schwager, "Global Assessment of Organic Contaminants in Farmed Salmon," *Science* 9 January 2004: Vol. 303, no. 5655, pp. 226–229.

generally smaller, so this problem arises only if we're sampling from a small population (and then there's a correction formula we could use).

Normal Population Assumption

Student's *t*-models won't work for data that are badly skewed. How skewed is too skewed? Well, formally, we assume that the data are from a population that follows a Normal model. Practically speaking, there's no way to be certain this is true.

And it's almost certainly *not* true. Models are idealized; real data are, well, real. The good news, however, is that even for small samples, it's sufficient to check a condition.

Nearly Normal Condition. The data come from a distribution that is unimodal and symmetric. This is a much more practical condition and one we can check by making a histogram.[4] For small samples, it can be hard to see any distribution shape in the histogram. Unfortunately, the condition matters most when it's hardest to check.

For very small samples (*n* < 15 or so), the data should follow a Normal model pretty closely. Of course, with so little data, it's rather hard to tell. But if you do find outliers or strong skewness, don't use these methods.

For moderate sample sizes (*n* between 15 and 40 or so), the *t* methods will work well as long as the data are unimodal and reasonably symmetric. Make a histogram to check.

When the sample size is larger than 40 or 50, the *t* methods are safe to use unless the data are extremely skewed. Make a histogram anyway. If you find outliers in the data and they aren't errors that are easy to fix, it's always a good idea to perform the analysis twice, once with and once without the outliers, even for large samples. The outliers may well hold additional information about the data, so they deserve special attention. If you find multiple modes, you may well have different groups that should be analyzed and understood separately.

If the data are extremely skewed, the mean may not be the most appropriate summary. But when our data consist of a collection of instances whose *total* is the business consequence—as when we add up the profits (or losses) from many transactions or the costs of many supplies—then the mean is just that total divided by *n*. And that's the value with a business consequence. Fortunately, in this instance, the Central Limit Theorem comes to our rescue. Even when we must sample from a very skewed distribution, the sampling distribution of our sample mean will be close to Normal, so we can use Student's *t* methods without much worry as long as the sample size is *large enough*.

How large is large enough? Here's the histogram of CEO compensations ($000) for Fortune 500 companies.

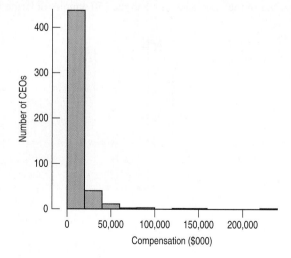

Figure 12.3 It's hard to imagine a distribution more skewed than these annual compensations from the Fortune 500 CEOs.

---

[4]Or we could check a normal probability plot.

Although this distribution is very skewed, the Central Limit Theorem will make the sampling distribution of the means of samples from this distribution more and more Normal as the sample size grows. Here's a histogram of the means of many samples of 100 CEOs:

Figure 12.4 Even samples as small as 100 from the CEO data set produce means whose sampling distribution is nearly normal. Larger samples will have sampling distributions even more Normal.

Often, in modern business applications, we have samples of many hundreds, or thousands. We should still be on guard for outliers and multiple modes and we should be sure that the observations are independent. But if the mean is of interest, the Central Limit Theorem works quite well in ensuring that the sampling distribution of the mean will be close to the Normal for samples of this size.

For Example   Checking the assumptions and conditions for a confidence interval for means

Researchers purchased whole farmed salmon from 51 farms in eight regions in six countries (see page 336). The histogram shows the concentrations of the insecticide mirex in the 150 samples of farmed salmon we examined in the previous example.

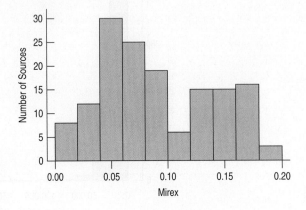

Question: Are the assumptions and conditions for making a confidence interval for the mean mirex concentration satisfied?

Answer:

✓ **Independence Assumption:** The fish were raised in many different places, and samples were purchased independently from several sources.

✓ **Randomization Condition:** The fish were selected randomly from those available for sale.

✓ **10% Condition:** There are lots of fish in the sea (and at the fish farms); 150 is certainly far fewer than 10% of the population.

✓ **Nearly Normal Condition:** The histogram of the data looks bimodal. While it might be interesting to learn the reason for that and possibly identify the subsets, we can proceed because the sample size is large.

It's okay to use these data about farm-raised salmon to make a confidence interval for the mean.

Just Checking

Every 10 years, the United States takes a census that tries to count every resident. In addition, the census collects information on a variety of economic and social questions. Businesses of all types use the census data to plan sales and marketing strategies and to understand the underlying demographics of the areas that they serve.

There are two census forms: the "short form," answered by most people, and the "long form," sent only to about one in six or seven households chosen at random. According to the Census Bureau (factfinder.census.gov), ". . . each estimate based on the long form responses has an associated confidence interval."

1 Why does the Census Bureau need a confidence interval for long-form information, but not for the questions that appear on both the long and short forms?

2 Why must the Census Bureau base these confidence intervals on *t*-models?

The Census Bureau goes on to say, "These confidence intervals are wider . . . for geographic areas with smaller populations and for characteristics that occur less frequently in the area being examined (such as the proportion of people in poverty in a middle-income neighborhood)."

3 Why is this so? For example, why should a confidence interval for the mean amount families spend monthly on housing be wider for a sparsely populated area of farms in the Midwest than for a densely populated area of an urban center? How does the formula for the one-sample *t*-interval show this will happen?

To deal with this problem, the Census Bureau reports long-form data only for ". . . geographic areas from which about two hundred or more long forms were completed—which are large enough to produce good quality estimates. If smaller weighting areas had been used, the confidence intervals around the estimates would have been significantly wider, rendering many estimates less useful."

4 Suppose the Census Bureau decided to report on areas from which only 50 long forms were completed. What effect would that have on a 95% confidence interval for, say, the mean cost of housing? Specifically, which values used in the formula for the margin of error would change? Which values would change a lot, and which values would change only slightly? Approximately how much wider would that confidence interval based on 50 forms be than the one based on 200 forms?

Guided Example   Insurance Profits

Insurance companies take risks. When they insure a property or a life, they must price the policy in such a way that their expected profit enables them to survive. They can base their projections on actuarial tables, but the reality of the insurance business often demands that they discount policies to a variety of customers and situations. Managing this risk is made even more difficult by the fact that until the policy expires, the company won't know if they've made a profit, no matter what premium they charge.

A manager wanted to see how well one of her sales representatives was doing, so she selected 30 matured policies that had been sold by the sales

rep and computed the (net) profit (premium charged minus paid claims), for each of the 30 policies.

The manager would like you, as a consultant, to construct a 95% confidence interval for the mean profit of the policies sold by this sales rep.

Profit (in $) from 30 policies		
222.80	463.35	2089.40
1756.23	−66.20	2692.75
1100.85	57.90	2495.70
3340.66	833.95	2172.70
1006.50	1390.70	3249.65
445.50	2447.50	−397.10
3255.60	1847.50	−397.31
3701.85	865.40	186.25
−803.35	1415.65	590.85
3865.90	2756.94	578.95

PLAN	**Setup** State what we want to know. Identify the variables and their context.	We wish to find a 95% confidence interval for the mean profit of policies sold by this sales rep. We have data for 30 matured policies.
	Make a picture. Check the distribution shape and look for skewness, multiple modes, and outliers.	Here's a boxplot and histogram of these values.

The sample appears to be unimodal and fairly symmetric with profit values between −$1000 and $4000 and no outliers.

Model Think about the assumptions and check the conditions.

✓ **Independence Assumption**

This is a random sample so observations should be independent.

✓ **Randomization Condition**

This sample was selected randomly from the matured policies sold by the sales representative of the company.

✓ **Nearly Normal Condition**

The distribution of profits is unimodal and fairly symmetric without strong skewness.

State the sampling distribution model for the statistic.

We will use a Student's t-model with $n - 1 = 30 - 1 = 29$ degrees of freedom and find a one-sample t-interval for the mean.

DO

Mechanics Compute basic statistics and construct the confidence interval.

Using software, we obtain the following basic statistics:

$$n = 30$$
$$\bar{y} = \$1438.90$$
$$s = \$1329.60$$

Remember that the standard error of the mean is equal to the standard deviation divided by the square root of *n*.

The standard error of the mean is:

$$SE(\bar{y}) = \frac{s}{\sqrt{n}} = \frac{1329.60}{\sqrt{30}} = \$242.75$$

The critical value we need to make a 95% confidence interval comes from a Student's *t* table, a computer program, or a calculator. We have 30 − 1 = 29 degrees of freedom. The selected confidence level says that we want 95% of the probability to be caught in the middle, so we exclude 2.5% in *each* tail, for a total of 5%. The degrees of freedom and 2.5% tail probability are all we need to know to find the critical value. Here it's 2.045.

There are $30 - 1 = 29$ degrees of freedom. The manager has specified a 95% level of confidence, so the critical value (from table T) is 2.045.

The margin of error is:

$$ME = 2.045 \times SE(\bar{y})$$
$$= 2.045 \times 242.75$$
$$= \$496.42$$

The 95% confidence interval for the mean profit is:

$$\$1438.90 \pm \$496.42$$
$$= (\$942.48, \$1935.32)$$

REPORT

Conclusion Interpret the confidence interval in the proper context.

MEMO

Re: Profit from Policies

From our analysis of the selected policies, we are 95% confident that the true mean profit of policies sold by this sales rep is contained in the interval from $942.48 to $1935.32.

Caveat: Insurance losses are notoriously subject to outliers. One very large loss could influence the average profit substantially. However, there were no such cases in this data set.

When we construct confidence intervals in this way, we expect 95% of them to cover the true mean and 5% to miss the true value. That's what "95% confident" means.

Finding Student's *t* Critical Values

The critical value in the Guided Example was found in the Student's *t* Table in Appendix D. To find the critical value, locate the row of the table corresponding to the degrees of freedom and the column corresponding to the probability you want. Since a 95% confidence interval leaves 2.5% of the values on either side, we look for 0.025 at the top of the column or look for 95% confidence directly in the bottom row of the table. The value in the table at that intersection is the critical value we need. In the Guided Example, the number of degrees of freedom was 30 − 1 = 29, so we located the value of 2.045.

Figure 12.5   Using Table T to look up the critical value *t** for a 95% confidence level with 29 degrees of freedom.

12.4 Cautions about Interpreting Confidence Intervals

Confidence intervals for means offer new, tempting, wrong interpretations. Here are some ways to keep from going astray:

- ***Don't say,*** "*95% of all the policies* sold by this sales rep have profits between $942.48 and $1935.32." The confidence interval is about the *mean*, not about the measurements of individual policies.

- ***Don't say,*** "We are 95% confident that *a randomly selected policy* will have a net profit between $942.48 and $1935.32." This false interpretation is also about individual policies rather than about the *mean* of the policies. We are 95% confident that the *mean* profit of all (similar) policies sold by this sales rep is between $942.48 and $1935.32.

- ***Don't say,*** "The mean profit is $1438.90 95% *of the time*." That's about means, but still wrong. It implies that the true mean varies, when in fact it is the confidence interval that would have been different had we gotten a different sample.

- Finally, ***don't say,*** "95% *of all samples* will have mean profits between $942.48 and $1935.32." That statement suggests that *this* interval somehow sets a standard for every other interval. In fact, this interval is no more (or less) likely to be correct than any other. You could say that 95% of all possible samples would produce intervals that contain the true mean profit. (The problem is that because we'll never know what the true mean profit is, we can't know if our sample was one of those 95%.)

Just Checking

In discussing estimates based on the long-form samples, the Census Bureau notes, "The disadvantage . . . is that . . . estimates of characteristics that are also reported on the short form will not match the [long-form estimates]."

The short-form estimates are values from a complete census, so they are the "true" values—something we don't usually have when we do inference.

5 Suppose we use long-form data to make 100 95% confidence intervals for the mean age of residents, one for each of 100 of the census-defined areas. How many of these 100 intervals should we expect will *fail* to include the true mean age (as determined from the complete short-form census data)?

12.5   Sample Size

How large a sample do we need? The simple answer is always "larger." But more data cost money, effort, and time. So how much is enough? Suppose your computer took an hour to download a movie you wanted to watch. You wouldn't be happy. Then you hear about a program that claims to download movies in under a half hour. You're interested enough to spend $29.95 for it, but only if it really delivers. So you get the free evaluation copy and test it by downloading a movie 10 times. Of course, the mean download time is not exactly 30 minutes as claimed. Observations vary. If the margin of error were 8 minutes, though, you'd probably be able to decide whether the software was worth the money. Doubling the sample size would require another 5 or so hours of testing and would reduce your margin of error to a bit under 6 minutes. You'd need to decide whether that's worth the effort.

As we make plans to collect data, we should have some idea of how small a margin of error is required to be able to draw a conclusion or detect a difference we want to see. If the size of the effect we're studying is large, then we may be able to tolerate a larger ME. If we need great precision, however, we'll want a smaller ME, and, of course, that means a larger sample size.

Armed with the ME and confidence level, we can find the sample size we'll need. Almost.

We know that for a mean, $ME = t^*_{n-1} \times SE(\bar{y})$ and that $SE(\bar{y}) = \dfrac{s}{\sqrt{n}}$, so we can determine the sample size by solving this equation for n:

$$ME = t^*_{n-1} \times \frac{s}{\sqrt{n}}.$$

The good news is that we have an equation; the bad news is that we won't know most of the values we need to compute it. When we thought about sample size for proportions, we ran into a similar problem. There we had to guess a working value for p to compute a sample size. Here, we need to know s. We don't know s until we get some data, but we want to calculate the sample size *before* collecting the data. We might be able to make a good guess, and that is often good enough for this purpose. If we have no idea what the standard deviation might be or if the sample size really matters (for example, because each additional individual is very expensive to sample or experiment on), it might be a good idea to run a small *pilot study* to get some feeling for the size of the standard deviation.

That's not all. Without knowing n, we don't know the degrees of freedom, and we can't find the critical value, t^*_{n-1}. One common approach is to use the corresponding z^* value from the Normal model. If you've chosen a 95% confidence level, then just use 2, following the 68–95–99.7 Rule, or 1.96 to be more precise. If your estimated sample size is 60 or more, it's probably okay—z^* was a good guess. If it's smaller than that, you may want to add a step, using z^* at first, finding n, and then replacing z^* with the corresponding t^*_{n-1} and calculating the sample size once more.

Sample size calculations are *never* exact. The margin of error you find *after* collecting the data won't match exactly the one you used to find n. The sample size formula depends on quantities that you won't have until you collect the data, but using it is an important first step. Before you collect data, it's always a good idea to know whether the sample size is large enough to give you a good chance of being able to tell you what you want to know.

By Hand

Sample size calculations

Let's give the sample size formula a spin. Suppose we want an ME of 8 minutes and we think the standard deviation of download times is about 10 minutes. Using a 95% confidence interval and $z^* = 1.96$, we solve for n:

$$8 = 1.96\frac{10}{\sqrt{n}}$$

$$\sqrt{n} = \frac{1.96 \times 10}{8} = 2.45$$

$$n = (2.45)^2 = 6.0025$$

That's a small sample size, so we use $(6 - 1) = 5$ degrees of freedom to substitute an appropriate t^* value. At 95%, $t_5^* = 2.571$. Now we can solve the equation one more time:

$$8 = 2.571\frac{10}{\sqrt{n}}$$

$$\sqrt{n} = \frac{2.571 \times 10}{8} \approx 3.214$$

$$n = (3.214)^2 \approx 10.33$$

To make sure the ME is no larger than you want, you should always round *up*, which gives $n = 11$ runs. So, to get an ME of 8 minutes, we should find the downloading times for $n = 11$ movies.

For Example   Finding the sample size for a confidence interval for means

In the 150 samples of farmed salmon (see page 336), the mean concentration of mirex was 0.0913 ppm with a standard deviation of 0.0495 ppm. A 95% confidence interval for the mean mirex concentration was found to be: (0.0833, 0.0993).

Question: How large a sample would be needed to produce a 95% confidence interval with a margin of error of 0.004?

Answer: We will assume that the standard deviation is 0.0495 ppm. The margin of error is equal to the critical value times the standard error. Using z^*, we find:

$$0.004 = 1.96 \times \frac{0.0495}{\sqrt{n}}$$

Solving for n, we find:

$$\sqrt{n} = 1.96 \times \frac{0.0495}{0.004}$$

or

$$n = \left(1.96 \times \frac{0.0495}{0.004}\right)^2 = 588.3$$

The t^* critical value with 400 df is 1.966 instead of 1.960. Using that value, the margin of error is:

$$1.966 \times \frac{0.0495}{\sqrt{589}} = 0.00401$$

You could go back and use 1.966 instead of 1.960 in the equation for n, above, and you would find that n should be 592. That will give a margin of error of 0.004, but the uncertainty in the standard deviation is likely to make such differences unimportant.

12.6 Degrees of Freedom—Why *n* − 1?

The number of degrees of freedom (*n* − 1) might have reminded you of the value we divide by to find the standard deviation of the data (since, after all, it's the same number). We promised back when we introduced that formula to say a bit more about why we divide by *n* − 1 rather than by *n*. The reason is closely tied to the reasoning of the *t*-distribution.

If only we knew the true population mean, μ, we would find the sample standard deviation using *n* instead of *n* − 1 as:

$$s = \sqrt{\frac{\sum (y - \mu)^2}{n}} \text{ and we'd call it } s.$$

We have to use $\bar{y}$ instead of μ, though, and that causes a problem. For any sample, $\bar{y}$ is as close to the data values as possible. Generally the population mean, μ, will be farther away. Think about it. GMAT scores have a population mean of 525. If you took a random sample of 5 students who took the test, their sample mean wouldn't be 525. The five data values will be closer to their own $\bar{y}$ than to 525. So if we use $\sum (y - \bar{y})^2$ instead of $\sum (y - \mu)^2$ in the equation to calculate *s*, our standard deviation estimate will be too small. The amazing mathematical fact is that we can compensate for the fact that $\sum (y - \bar{y})^2$ is too small just by dividing by *n* − 1 instead of by *n*. So that's all the *n* − 1 is doing in the denominator of *s*. We call *n* − 1 the degrees of freedom.

What Can Go Wrong?

First, you must decide when to use Student's *t* methods.

- **Don't confuse proportions and means.** When you treat your data as categorical, counting successes and summarizing with a sample proportion, make inferences using the Normal model methods. When you treat your data as quantitative, summarizing with a sample mean, make your inferences using Student's *t* methods.

Student's *t* methods work only when the Normal Population Assumption is true. Naturally, many of the ways things can go wrong turn out to be ways that the Normal Population Assumption can fail. It's always a good idea to look for the most common kinds of failure. It turns out that you can even fix some of them.

- **Beware of multimodality.** The Nearly Normal Condition clearly fails if a histogram of the data has two or more modes. When you see this, look for the possibility that your data come from two groups. If so, your best bet is to try to separate the data into groups. (Use the variables to help distinguish the modes, if possible. For example, if the modes seem to be composed mostly of men in one and women in the other, split the data according to the person's sex.) Then you can analyze each group separately.

- **Beware of skewed data.** Make a histogram of the data. If the data are severely skewed, you might try re-expressing the variable. Re-expressing may yield a distribution that is unimodal and symmetric, making it more appropriate for the inference methods for means. Re-expression cannot help if the sample distribution is not unimodal.

(continued)

• **Investigate outliers.** The Nearly Normal Condition also fails if the data have outliers. If you find outliers in the data, you need to investigate them. Sometimes, it's obvious that a data value is wrong and the justification for removing or correcting it is clear. When there's no clear justification for removing an outlier, you might want to run the analysis both with and without the outlier and note any differences in your conclusions. Any time data values are set aside, you *must* report on them individually. Often they will turn out to be the most informative part of your report on the data.[5]

Of course, Normality issues aren't the only risks you face when doing inferences about means.

• **Watch out for bias.** Measurements of all kinds can be biased. If your observations differ from the true mean in a systematic way, your confidence interval may not capture the true mean. And there is no sample size that will save you. A bathroom scale that's 5 pounds off will be 5 pounds off even if you weigh yourself 100 times and take the average. We've seen several sources of bias in surveys, but measurements can be biased, too. Be sure to think about possible sources of bias in your measurements.

• **Make sure data are independent.** Student's *t* methods also require the sampled values to be mutually independent. We check for random sampling and the 10% Condition. You should also think hard about whether there are likely violations of independence in the data collection method. If there are, be very cautious about using these methods.

Ethics in Action

Recent reports have indicated that waiting times in hospital emergency rooms (ERs) across the United States are getting longer, with the average reported as 30 minutes in January 2008 (WashingtonPost.com). Several reasons have been cited for this rise in average ER waiting time including the closing of hospital emergency rooms in urban areas and problems with managing hospital flow. Tyler Hospital, located in rural Ohio, joined the Joint Commission's Continuous Service Readiness program and consequently agreed to monitor its ER waiting times. After collecting data for a random sample of 30 ER patients arriving at Tyler's ER during the last month, they found an average waiting time of 26 minutes with a standard deviation of 8.25 minutes. Further statistical analysis yielded a 95% confidence interval of 22.92 to 29.08 minutes, clear indication that Tyler's ER patients wait less than 30 minutes to see a doctor.

Tyler's administration was not only pleased with the findings, but also sure that the Joint Commission would also be impressed. Their next step was to consider ways of including this message, "95% of Tyler's ER patients can expect to wait less than the national average to see a doctor," in their advertising and promotional materials.

ETHICAL ISSUE *Interpretation of the confidence interval is incorrect and misleading (related to Item C, ASA Ethical Guidelines). The confidence interval does not provide results for individual patients. So, it is incorrect to state that 95% of individual ER patients wait less (or can expect to wait less) than 30 minutes to see a doctor.*

ETHICAL SOLUTION *Interpret the results of the confidence interval correctly, in terms of the mean waiting time and not individual patients.*

---

[5]This suggestion may be controversial in some disciplines. Setting aside outliers is seen by some as unethical because the result is likely to be a narrower confidence interval or a smaller P-value. But an analysis of data with outliers left in place is *always* wrong. The outliers violate the Nearly Normal Condition and also the implicit assumption of a homogeneous population, so they invalidate inference procedures. An analysis of the nonoutlying points, along with a separate discussion of the outliers, is often much more informative, and can reveal important aspects of the data.

What Have We Learned?

Learning Objectives

■ Know the sampling distribution of the mean.

• To apply the Central Limit Theorem for the mean in practical applications, we must estimate the standard deviation. This *standard error* is

$$SE(\bar{y}) = \frac{s}{\sqrt{n}}$$

• When we use the SE, the sampling distribution that allows for the additional uncertainty is Student's *t*.

■ Construct confidence intervals for the true mean, μ.

• A confidence interval for the mean has the form $\bar{y} \pm ME$.

• The Margin of Error is $ME = t^*_{df} SE(\bar{y})$.

■ Find *t** values by technology or from tables.

• When constructing confidence intervals for means, the correct degrees of freedom is $n - 1$.

■ Check the Assumptions and Conditions before using any sampling distribution for inference.

■ Write clear summaries to interpret a confidence interval.

Terms

Degrees of freedom (df)   A parameter of the Student's *t*-distribution that depends upon the sample size. Typically, more degrees of freedom reflects increasing information from the sample.

One-sample *t*-interval for the mean   A one-sample *t*-interval for the population mean is:

$$\bar{y} \pm t^*_{n-1} \times SE(\bar{y}) \text{ where } SE(\bar{y}) = \frac{s}{\sqrt{n}}.$$

The critical value t^*_{n-1} depends on the particular confidence level, *C*, that you specify and on the number of degrees of freedom, $n - 1$.

Student's *t*   A family of distributions indexed by its degrees of freedom. The *t*-models are unimodal, symmetric, and bell-shaped, but generally have fatter tails and a narrower center than the Normal model. As the degrees of freedom increase, *t*-distributions approach the Normal model.

Technology Help: Inference for Means

Statistics packages offer convenient ways to make histograms of the data. That means you have no excuse for skipping the check that the data are nearly Normal.

Any standard statistics package can compute a confidence interval.

Inference results are sometimes reported in a table. You may have to read carefully to find the values you need. Often, confidence interval bounds are given together with related results for hypothesis tests (which we'll learn about in the next chapter). Here is an example of that kind of output (although no package we know gives results in exactly this form).

The commands to do inference for means on common statistics programs and calculators are not always obvious. (By contrast, the resulting output is usually clearly labeled and easy to read.) The guides for each program can help you start navigating.

Calculated mean, $\bar{y}$

Hypothesized value	30		
Estimated mean	31.043478		
DF	22		
Std Error	0.886		
Alpha	0.05		

	tTest		tinterval
Statistic	1.178		
Prob > |t|	0.2513	Upper 95%	32.880348
Prob > t	0.1257	Lower 95%	29.206608
Prob < t	0.8743		

Confidence interval bounds

(continued)

EXCEL   *XLSTAT*›

To find a confidence interval for a mean in Excel, you can set up the calculations using Excel's functions. For example, suppose you have 100 observations in cells A1:A100.

- In cell B2, enter "=AVERAGE(A1:A100)" to compute the sample mean.
- In cell B3, enter "=STDEV(A1:A11)/SQRT(100)" to compute the standard error of the mean.
- In cell B4, enter "=TINV(.05,99)" to compute t*.
- In cell B5, enter "=B2-B4*B3" as the lower end of the CI.
- In cell B6, enter "=B2+B4*B3" as the upper end of the CI.

JMP

- From the **Analyze** menu, select **Distribution**.
- For a confidence interval, scroll down to the "Moments" section to find the interval limits. (Be sure that your variables are "Continuous" type so that this section will be available.)

Comments

"Moment" is a fancy statistical term for means, standard deviations, and other related statistics.

MINITAB

- From the **Stat** menu, choose the **Basic Statistics** submenu.
- From that menu, choose 1-sample t . . . .
- Then fill in the dialog.

Comments

The dialog offers a clear choice between confidence interval and hypothesis test.

SPSS

- From the **Analyze** menu, choose the **Compare Means** submenu.
- From that, choose **One-Sample t-test** command.

Comments

The commands suggest neither a single mean nor an interval. But the results provide both.

Brief CASE   Real Estate

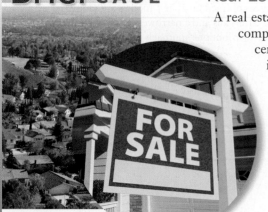

A real estate agent is trying to understand the pricing of homes in her area, a region comprised of small to midsize towns and a small city. For each of 1200 homes recently sold in the region, the file **Real_Estate_sample1200** holds the following variables:

- *Sale Price* (in $)
- *Lot size* (size of the lot in acres)
- *Waterfront* (Yes, No)
- *Age* (in years)
- *Central Air* (Yes, No)
- *Fuel Type* (Wood, Oil, Gas, Electric, Propane, Solar, Other)
- *Condition* (1 to 5, 1 = Poor, 5 = Excellent)
- *Living Area* (living area in square feet)
- *Pct College* (% in zip code who attend a four-year college)

- *Full Baths* (number of full bathrooms)
- *Half Baths* (number of half bathrooms)
- *Bedrooms* (number of bedrooms)
- *Fireplaces* (number of fireplaces)

The agent has a family interested in a four bedroom house. Using confidence intervals, how should she advise the family on what the average price of a four bedroom house might be in this area? Compare that to a confidence interval for two bedroom homes. How does the presence of central air conditioning affect the mean price of houses in this area? Use confidence intervals and graphics to help answer that question.

Explore other questions that might be useful for the real estate agent in knowing how different categorical factors affect the sale price and write up a short report on your findings.

Donor Profiles

A philanthropic organization collects and buys data on their donor base. The full database contains about 4.5 million donors and over 400 variables collected on each, but the data set **Donor_Profiles** is a sample of 916 donors and includes the variables:

- *Age* (in years)
- *Homeowner* (H = Yes, U = Unknown)
- *Gender* (F = Female, M = Male, U = Unknown)
- *Wealth* (Ordered categories of total household wealth from 1 = Lowest to 9 = Highest)
- *Children* (Number of children)
- *Donated Last* (0 = Did not donate to last campaign, 1 = Did donate to last campaign)
- *Amt Donated Last* ($ amount of contribution to last campaign)

The analysts at the organization want to know how much people donate on average to campaigns, and what factors might influence that amount. Compare the confidence intervals for the mean *Amt Donated Last* by those known to own their homes with those whose homeowner status is unknown. Perform similar comparisons for *Gender* and two of the *Wealth* categories. Write up a short report using graphics and confidence intervals for what you have found. (Be careful not to make inferences directly about the differences between groups. We'll discuss that in Chapter 14. Your inference should be about single groups.)

(The distribution of *Amt Donated Last* is highly skewed to the right, and so the median might be thought to be the appropriate summary. But the median is $0.00 so the analysts must use the mean. From simulations, they have ascertained that the sampling distribution for the mean is unimodal and symmetric for samples larger than 250 or so. Note that small differences in the mean could result in millions of dollars of added revenue nationwide. The average cost of their solicitation is $0.67 per person to produce and mail.)

Exercises

SECTION 12.1

1. From Chapter 5 a survey of 25 randomly selected customers found the following ages (in years):

20	32	34	29	30
30	30	14	29	11
38	22	44	48	26
25	22	32	35	32
35	42	44	44	48

Recall that the mean was 31.84 years and the standard deviation was 9.84 years.

a) What is the standard error of the mean?
b) How would the standard error change if the same size had been 100 instead of 25? (Assume that the sample standard deviation didn't change.)

2. From Chapter 5, a random sample of 20 purchases showed the following amounts (in $):

39.05	2.73	32.92	47.51
37.91	34.35	64.48	51.96
56.95	81.58	47.80	11.72
21.57	40.83	38.24	32.98
75.16	74.30	47.54	65.62

Recall that the mean was $45.26 and the standard deviation was $20.67.

a) What is the standard error of the mean?
b) How would the standard error change if the same size had been 5 instead of 20? (Assume that the sample standard deviation didn't change.)

3. For the data in Exercise 1:

a) How many degrees of freedom does the *t*-statistic have?
b) How many degrees of freedom would the *t*-statistic have if the sample size had been 100?

4. For the data in Exercise 2:

a) How many degrees of freedom does the *t*-statistic have?
b) How many degrees of freedom would the *t*-statistic have if the sample size had been 5?

SECTION 12.2

5. Find the critical value *t** for:

a) a 95% confidence interval based on 24 df.
b) a 95% confidence interval based on 99 df.

6. Find the critical value *t** for:

a) a 90% confidence interval based on 19 df.
b) a 90% confidence interval based on 4 df.

7. For the ages in Exercise 1:

a) Construct a 95% confidence interval for the mean age of all customers, assuming that the assumptions and conditions for the confidence interval have been met.
b) How large is the margin of error?
c) How would the confidence interval change if you had assumed that the standard deviation was known to be 10.0 years?

8. For the purchase amounts in Exercise 2:

a) Construct a 90% confidence interval for the mean purchases of all customers, assuming that the assumptions and conditions for the confidence interval have been met.
b) How large is the margin of error?
c) How would the confidence interval change if you had assumed that the standard deviation was known to be $20?

SECTION 12.3

9. For the confidence intervals of Exercise 7, a histogram of the data looks like this:

Check the assumptions and conditions for your inference.

10. For confidence intervals of Exercise 8, a histogram of the data looks like this:

Check the assumptions and conditions for your inference.

SECTION 12.5

11. For the confidence interval in Exercise 7:

a) How large would the sample size have to be to cut the margin of error in half?
b) About how large would the sample size have to be to cut the margin of error by a factor of 10?

12. For the confidence interval in Exercise 8:

a) To reduce the margin of error to about $4, how large would the sample size have to be?

b) How large would the sample size have to be to reduce the margin of error to $0.80?

CHAPTER EXERCISES

13. *t*-models. Using the *t* tables, software, or a calculator, estimate:

a) the critical value of *t* for a 90% confidence interval with df = 17.

b) the critical value of *t* for a 98% confidence interval with df = 88.

14. *t*-models, part 2. Using the *t* tables, software, or a calculator, estimate:

a) the critical value of *t* for a 95% confidence interval with df = 7.

b) the critical value of *t* for a 99% confidence interval with df = 102.

15. Confidence intervals. Describe how the width of a 95% confidence interval for a mean changes as the standard deviation (*s*) of a sample increases, assuming sample size remains the same.

16. Confidence intervals, part 2. Describe how the width of a 95% confidence interval for a mean changes as the sample size (*n*) increases, assuming the standard deviation remains the same.

17. Confidence intervals and sample size. A confidence interval for the price of gasoline from a random sample of 30 gas stations in a region gives the following statistics:

$$\bar{y} = \$4.49 \quad s = \$0.29$$

a) Find a 95% confidence interval for the mean price of regular gasoline in that region.

b) Find the 90% confidence interval for the mean.

c) If we had the same statistics from a sample of 60 stations, what would the 95% confidence interval be now?

18. Confidence intervals and sample size, part 2. A confidence interval for the price of gasoline from a random sample of 30 gas stations in a region gives the following statistics:

$$\bar{y} = \$4.49 \quad SE(\bar{y}) = \$0.06$$

a) Find a 95% confidence interval for the mean price of regular gasoline in that region.

b) Find the 90% confidence interval for the mean.

c) If we had the same statistics from a sample of 60 stations, what would the 95% confidence interval be now?

19. Marketing livestock feed. A feed supply company has developed a special feed supplement to see if it will promote weight gain in livestock. Their researchers report that the 77 cows studied gained an average of 56 pounds and that a 95% confidence interval for the mean weight gain this supplement produces has a margin of error of ±11 pounds. Staff in their marketing department wrote the following conclusions. Did anyone interpret the interval correctly? Explain any misinterpretations.

a) 95% of the cows studied gained between 45 and 67 pounds.

b) We're 95% sure that a cow fed this supplement will gain between 45 and 67 pounds.

c) We're 95% sure that the average weight gain among the cows in this study was between 45 and 67 pounds.

d) The average weight gain of cows fed this supplement is between 45 and 67 pounds 95% of the time.

e) If this supplement is tested on another sample of cows, there is a 95% chance that their average weight gain will be between 45 and 67 pounds.

20. Meal costs. A company is interested in estimating the costs of lunch in their cafeteria. After surveying employees, the staff calculated that a 95% confidence interval for the mean amount of money spent for lunch over a period of six months is ($780, $920). Now the organization is trying to write its report and considering the following interpretations. Comment on each.

a) 95% of all employees pay between $780 and $920 for lunch.

b) 95% of the sampled employees paid between $780 and $920 for lunch.

c) We're 95% sure that employees in this sample averaged between $780 and $920 for lunch.

d) 95% of all samples of employees will have average lunch costs between $780 and $920.

e) We're 95% sure that the average amount all employees pay for lunch is between $780 and $920.

21. CEO compensation. A sample of 20 CEOs from the Forbes 500 shows total annual compensations ranging from a minimum of $0.1 to $62.24 million. The average for these 20 CEOs is $7.946 million. The histogram and boxplot are as follows:

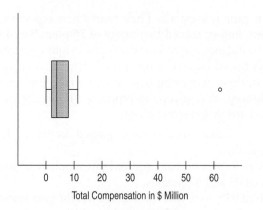

Total Compensation in $ Million

Based on these data, a computer program found that a confidence interval for the mean annual compensation of all Forbes 500 CEOs is (1.69, 14.20) $M. Why should you be hesitant to trust this confidence interval?

22. Credit card charges. A credit card company takes a random sample of 100 cardholders to see how much they charged on their card last month. A histogram and boxplot are as follows:

March 2005 Charges

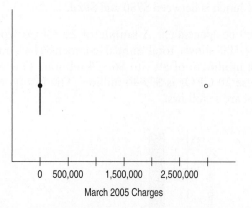

March 2005 Charges

A computer program found that the 95% confidence interval for the mean amount spent in March 2005 is (−$28,366.84, $90,691.49). Explain why the analysts didn't find the confidence interval useful, and explain what went wrong.

23. Parking. Hoping to lure more shoppers downtown, a city builds a new public parking garage in the central

business district. The city plans to pay for the structure through parking fees. For a random sample of 44 weekdays, daily fees collected averaged $126, with a standard deviation of $15.

a) What assumptions must you make in order to use these statistics for inference?
b) Find a 90% confidence interval for the mean daily income this parking garage will generate.
c) Explain in context what this confidence interval means.
d) Explain what 90% confidence means in this context.
e) The consultant who advised the city on this project predicted that parking revenues would average $128 per day. Based on your confidence interval, what do you think of the consultant's prediction? Why?

24. Housing 2008 was a difficult year for the economy. There were a large number of foreclosures of family homes. In one large community, realtors randomly sampled 36 bids from potential buyers to determine the average loss in home value. The sample showed the average loss was $11,560 with a standard deviation of $1500.

a) What assumptions and conditions must be checked before finding a confidence interval? How would you check them?
b) Find a 95% confidence interval for the mean loss in value per home.
c) Interpret this interval and explain what 95% confidence means.
d) Suppose nationally, the average loss in home values at this time was $10,000. Do you think the loss in the sampled community differs significantly from the national average? Explain.

25. Parking, part 2. Suppose that for budget planning purposes the city in Exercise 23 needs a better estimate of the mean daily income from parking fees.

a) Someone suggests that the city use its data to create a 95% confidence interval instead of the 90% interval first created. How would this interval be better for the city? (You need not actually create the new interval.)
b) How would the 95% confidence interval be worse for the planners?
c) How could they achieve a confidence interval estimate that would better serve their planning needs?

26. Housing, part 2. In Exercise 24, we found a 95% confidence interval to estimate the loss in home values.

a) Suppose the standard deviation of the losses was $3000 instead of the $1500 used for that interval. What would the larger standard deviation do to the width of the confidence interval (assuming the same level of confidence)?
b) Your classmate suggests that the margin of error in the interval could be reduced if the confidence level were changed to 90% instead of 95%. Do you agree with this statement? Why or why not?
c) Instead of changing the level of confidence, would it be more statistically appropriate to draw a bigger sample?

27. State budgets. States that rely on sales tax for revenue to fund education, public safety, and other programs often end up with budget surpluses during economic growth periods (when people spend more on consumer goods) and budget deficits during recessions (when people spend less on consumer goods). Fifty-one small retailers in a state with a growing economy were recently sampled. The sample showed a mean increase of $2350 in additional sales tax revenue collected per retailer compared to the previous quarter. The sample standard deviation = $425.

a) Find a 95% confidence interval for the mean increase in sales tax revenue.
b) What assumptions have you made in this inference? Do you think the appropriate conditions have been satisfied?
c) Explain what your interval means and provide an example of what it does not mean.

28. State budgets, part 2. Suppose the state in Exercise 27 sampled 16 small retailers instead of 51, and for the sample of 16, the sample mean increase again equaled $2350 in additional sales tax revenue collected per retailer compared to the previous quarter. Also assume the sample standard deviation = $425.

a) What is the standard error of the mean increase in sales tax revenue collected?
b) What happens to the accuracy of the estimate when the interval is constructed using the smaller sample size?
c) Find and interpret a 95% confidence interval.
d) How does the margin of error for the interval constructed in Exercise 27 compare with the margin of error constructed in this exercise? Explain statistically how sample size changes the accuracy of the constructed interval. Which sample would you prefer if you were a state budget planner? Why?

29. Departures. What are the chances your flight will leave on time? The U.S. Bureau of Transportation Statistics of the Department of Transportation publishes information about airline performance. Here are a histogram and summary statistics for the percentage of flights departing on time each month from 1995 through 2006.

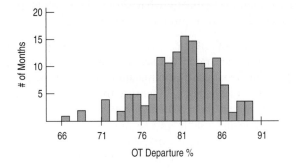

n	144
$\bar{y}$	81.1838
s	4.47094

There is no evidence of a trend over time. (The correlation of On Time Departure% with time is $r = -0.016$.)

a) Check the assumptions and conditions for inference.
b) Find a 90% confidence interval for the true percentage of flights that depart on time.
c) Interpret this interval for a traveler planning to fly.

30. Late arrivals. Will your flight get you to your destination on time? The U.S. Bureau of Transportation Statistics reported the percentage of flights that were late each month from 1995 through 2006. Here's a histogram, along with some summary statistics:

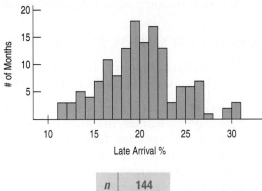

n	144
$\bar{y}$	20.0757
s	4.08837

We can consider these data to be a representative sample of all months. There is no evidence of a time trend.

a) Check the assumptions and conditions for inference about the mean.
b) Find a 99% confidence interval for the true percentage of flights that arrive late.
c) Interpret this interval for a traveler planning to fly.

31. Computer lab fees. The technology committee has stated that the average time spent by students per lab visit has increased, and the increase supports the need for increased lab fees. To substantiate this claim, the committee randomly samples 12 student lab visits and notes the amount of time spent using the computer. The times in minutes are as follows:

Time	Time
52	74
57	53
54	136
76	73
62	8
52	62

a) Plot the data. Are any of the observations outliers? Explain.

b) The previous mean amount of time spent using the lab computer was 55 minutes. Find a 95% confidence interval for the true mean. What do you conclude about the claim? If there are outliers, find intervals with and without the outliers present.

T 32. Cell phone batteries. A company that produces cell phones claims its standard phone battery lasts longer on average than other batteries in the market. To support this claim, the company publishes an ad reporting the results of a recent experiment showing that under normal usage, their batteries last at least 35 hours. To investigate this claim, a consumer advocacy group asked the company for the raw data. The company sends the group the following results:

35, 34, 32, 31, 34, 34, 32, 33, 35, 55, 32, 31

Find a 95% confidence interval and state your conclusion. Explain how you dealt with the outlier, and why.

33. Growth and air pollution. Government officials have difficulty attracting new business to communities with troubled reputations. Nevada has been one of the fastest growing states in the country for a number of years. Accompanying the rapid growth are massive new construction projects. Since Nevada has a dry climate, the construction creates visible dust pollution. High pollution levels may paint a less than attractive picture of the area, and can also result in fines levied by the federal government. As required by government regulation, researchers continually monitor pollution levels. In the most recent test of pollution levels, 121 air samples were collected. The dust particulate levels must be reported to the federal regulatory agencies. In the report sent to the federal agency, it was noted that the mean particulate level = 57.6 micrograms/cubic liter of air, and the 95% confidence interval estimate is (52.06 mg to 63.07 mg). A graph of the distribution of the particulate amounts was also included and is shown below.

a) Discuss the assumptions and conditions for using Student's *t* inference methods with these data.
b) Do you think the confidence interval noted in the report is valid? Briefly explain why or why not.

34. Convention revenues. At one time, Nevada was the only U.S. state that allowed gambling. Although gambling continues to be one of the major industries in Nevada, the proliferation of legalized gambling in other areas of the country has required state and local governments to look at other growth possibilities. The convention and visitor's authorities in many Nevada cities actively recruit national conventions that bring thousands of visitors to the state. Various demographic and economic data are collected from surveys given to convention attendees. One statistic of interest is the amount visitors spend on slot machine gambling. Nevada often reports the slot machine expenditure as amount spent per hotel guest room. A recent survey of 500 visitors asked how much they spent on gambling. The average expenditure per room was $180.

Casinos will use the information reported in the survey to estimate slot machine expenditure per hotel room. Do you think the estimates produced by the survey will accurately represent expenditures? Explain using the statistics reported and graph shown.

35. Traffic speed. Police departments often try to control traffic speed by placing speed-measuring machines on roads that tell motorists how fast they are driving. Traffic safety experts must determine where machines should be placed. In one recent test, police recorded the average speed clocked by cars driving on one busy street close to an elementary school. For a sample of 25 speeds, it was determined that the average amount over the speed limit for the 25 clocked speeds was 11.6 mph with a standard deviation of 8 mph. The 95% confidence interval estimate for this sample is 8.30 mph to 14.90 mph.

a) What is the margin of error for this problem?
b) The researchers commented that the interval was too wide. Explain specifically what should be done to reduce the margin of error to no more than ±2 mph.

36. Traffic speed, part 2. The speed-measuring machines must measure accurately to maximize effectiveness in slowing traffic. The accuracy of the machines will be tested before placement on city streets. To ensure that error rates are estimated accurately, the researchers want to take a large enough sample to ensure usable and accurate interval estimates of how much the machines may be off in measuring actual speeds. Specially, the researchers want the margin of error for a single speed measurement to be no more than ±1.5 mph.

a) Discuss how the researchers may obtain a reasonable estimate of the standard deviation of error in the measured speeds.

b) Suppose the standard deviation for the error in the measured speeds equals 4 mph. At 95% confidence, what sample size should be taken to ensure that the margin of error is no larger than ±1.0 mph?

37. Tax audits. Certified public accountants are often required to appear with clients if the IRS audits the client's tax return. Some accounting firms give the client an option to pay a fee when the tax return is completed that guarantees tax advice and support from the accountant if the client were audited. The fee is charged up front like an insurance premium and is less than the amount that would be charged if the client were later audited and then decided to ask the firm for assistance during the audit. A large accounting firm is trying to determine what fee to charge for next year's returns. In previous years, the actual mean cost to the firm for attending a client audit session was $650. To determine if this cost has changed, the firm randomly samples 32 client audit fees. The sample mean audit cost was $680 with a standard deviation of $75.

a) Develop a 95% confidence interval estimate for the mean audit cost.
b) Based on your confidence interval, what do you think of the claim that the mean cost has changed?

T 38. Tax audits, part 2. While reviewing the sample of audit fees, a senior accountant for the firm notes that the fee charged by the firm's accountants depends on the complexity of the return. A comparison of actual charges therefore might not provide the information needed to set next year's fees. To better understand the fee structure, the senior accountant requests a new sample that measures the time the accountants spent on the audit. Last year, the average hours charged per client audit was 3.25 hours. A new sample of 10 audit times shows the following times in hours:

4.2, 3.7, 4.8, 2.9, 3.1, 4.5, 4.2, 4.1, 5.0, 3.4

a) Assume the conditions necessary for inference are met. Find a 90% confidence interval estimate for the mean audit time.
b) Based on your answer to part a, comment on the claim that the mean fees have increased.

T 39. Wind power. Should you generate electricity with your own personal wind turbine? That depends on whether you have enough wind on your site. To produce enough energy, your site should have an annual average wind speed of at least 8 miles per hour, according to the Wind Energy Association. One candidate site was monitored for a year, with wind speeds recorded every 6 hours. A total of 1114 readings of wind speed averaged 8.019 mph with a standard deviation of 3.813 mph. You've been asked to make a statistical report to help the landowner decide whether to place a wind turbine at this site.

a) Discuss the assumptions and conditions for using Student's *t* inference methods with these data. Here are some plots that may help you decide whether the methods can be used:

b) What would you tell the landowner about whether this site is suitable for a small wind turbine? Explain

40. Real estate crash? After the sub-prime crisis of late 2007, real estate prices fell almost everywhere in the U.S. In 2006–2007 before the crisis, the average selling price of homes in a region in upstate New York was $191,300. A real estate agency wants to know how much the prices have fallen since then. They collect a sample of 1231 homes in the region and find the average asking price to be $178,613.50 with a standard deviation of $92,701.56. You have been retained by the real estate agency to report on the current situation.

a) Discuss the assumptions and conditions for using *t*-methods for inference with these data. Here are some plots that may help you decide what to do.

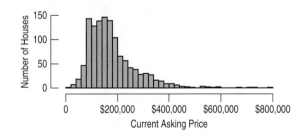

b) What would you report to the real estate agency about the current situation?

Just Checking Answers

1   Questions on the short form are answered by every-
    one in the population. This is a census, so means or
    proportions *are* the true population values. The long
    forms are just given to a sample of the population.
    When we estimate parameters from a sample, we use
    a confidence interval to take sample-to-sample vari-
    ability into account.

2   They don't know the population standard deviation,
    so they must use the sample SD as an estimate.
    The additional uncertainty is taken into account by
    t-models.

3   The margin of error for a confidence interval for a
    mean depends, in part, on the standard error:

$$SE(\bar{y}) = \frac{s}{\sqrt{n}}$$

Since *n* is in the denominator, smaller sample sizes
generally lead to larger SEs and correspondingly
wider intervals. Because long forms are sampled at
the same rate of one in every six or seven households
throughout the country, samples will be smaller in
less populous areas and result in wider confidence
intervals.

4   The critical values for *t* with fewer degrees of freedom
    would be slightly larger. The $\sqrt{n}$ part of the standard
    error changes a lot, making the SE much larger. Both
    would increase the margin of error. The smaller sam-
    ple is one fourth as large, so the confidence interval
    would be roughly twice as wide.

5   We expect 95% of such intervals to cover the true
    value, so 5 of the 100 intervals might be expected to
    miss.

Testing Hypotheses

Dow Jones Industrial Average

More than a hundred years ago Charles Dow changed the way people look at the stock market. Surprisingly, he wasn't an investment wizard or a venture capitalist. He was a journalist who wanted to make investing understandable to ordinary people. Although he died at the relatively young age of 51 in 1902, his impact on how we track the stock market has been both long-lasting and far-reaching.

In the late 1800s, when Charles Dow reported on Wall Street, investors preferred bonds, not stocks. Bonds were reliable, backed by the real machinery and other hard assets the company owned. What's more, bonds were predictable; the bond owner knew when the bond would mature and so, knew when and how much the bond would pay. Stocks simply represented "shares" of ownership, which were risky and erratic. In May 1896, Dow and Edward Jones, whom he had known since their days as reporters for the *Providence Evening Press*, launched the now-famous Dow Jones Industrial Average (DJIA) to help the public understand stock market trends.

The original DJIA averaged 11 stock prices. Of those original industrial stocks, only General Electric is still in the DJIA.

Since then, the DJIA has become synonymous with overall market performance and is often referred to simply as the Dow. The index was expanded to 20 stocks in 1916 and to 30 in 1928 at the height of the roaring 20's bull market. That bull market peaked on September 3, 1929, when the Dow reached 381.17. On October 28 and 29, 1929, the Dow lost nearly 25% of its value. Then things got worse. Within four years, on July 8, 1932, the 30 industrials reached an all-time low of 40.65. The highs of September 1929 were not reached again until 1954.

Today the Dow is a weighted average of 30 industrial stocks, with weights used to account for splits and other adjustments. The "Industrial" part of the name is largely historic. Today's DJIA includes the service industry and financial companies and is much broader than just heavy industry. And it is still one of the most watched indicators of the state of the U.S. stock market and the global economy.

WHO	Days on which the stock market was open ("trading days")
WHAT	Closing price of the Dow Jones Industrial Average (*Close*)
UNITS	Points
WHEN	August 1982 to December 1986
WHY	To test theory of stock market behavior

How does the stock market move? Here are the DJIA closing prices for the bull market that ran from mid 1982 to the end of 1986.

Figure 13.1   Daily closing prices of the Dow Jones Industrials from mid 1982 to the end of 1986.

The DJIA clearly increased during this famous bull market, more than doubling in value in less than five years. One common theory of market behavior says that on a given day, the market is just as likely to move up as down. Another way of phrasing this is that the daily behavior of the stock market is random. Can that be true during such periods of obvious increase? Let's investigate if the Dow is just as likely to move higher or lower on any given day. Out of the 1112 trading days in that period, the average increased on 573 days, a sample proportion of 0.5153 or 51.53%. That *is* more "up" days than "down" days, but is it far enough from 50% to cast doubt on the assumption of equally likely up or down movement?

13.1   Hypotheses

Hypothesis n.;

pl. Hypotheses.

A supposition; a proposition or principle which is supposed or taken for granted, in order to draw a conclusion or inference for proof of the point in question; something not proved, but assumed for the purpose of argument.

—*Webster's Unabridged Dictionary, 1913*

We've learned how to create confidence intervals for both means and proportions, but now we are not estimating values. We have a specific value in mind and a question to go with it. We wonder whether the market moves randomly up and down with equal probability regardless of its long-term trend. A confidence interval provides plausible values for the parameter, but now we seek a more direct test. Tests like this are useful, for example, if we want to know whether our customers are really more satisfied since the launch of our new website, whether the mean income of our preferred customers is higher than those of our regular customers, or whether our recent ad campaign really reached our target of 20% of adults in our region.

A confidence interval starts with the sample statistic (mean or proportion) and builds an interval around it. A hypothesis test turns that idea on its head. How can we state and test a hypothesis about daily changes in the DJIA? Hypotheses are working models that we adopt temporarily. To test whether the daily fluctuations are equally likely to be up as down, we assume that they are, and that any apparent difference from 50% is just random fluctuation. So, our starting hypothesis, called the null hypothesis, is that the proportion of days on which the DJIA increases is 50%. The **null hypothesis**, which we denote H_0, specifies a population model parameter and proposes a value for that parameter. We usually write down a null hypothesis about a proportion in the form $H_0: p = p_0$. (For a mean, we would write $H_0: \mu = \mu_0$.) This is a concise way to specify the two things we need most: the identity of the parameter we hope to learn about (the true proportion) and a specific hypothesized value for that parameter (in this case, 50%). We need a hypothesized value so we can compare our observed statistic to it. Which value to use for the hypothesis is not a statistical question. It may be obvious from the context of the data, but sometimes it takes a bit of thinking to translate the question we hope to answer into a hypothesis about a parameter. For our hypothesis about whether the DJIA moves up or down with equal likelihood, it's pretty clear that we need to test $H_0: p = 0.5$.

The **alternative hypothesis**, which we denote H_A, contains the values of the parameter that we consider plausible if we reject the null hypothesis. In our example, our null hypothesis is that the proportion, p, of "up" days is 0.5. What's the alternative? During a bull market, you might expect more up days than down, but for now we'll assume that we're interested in a deviation in either direction from the null hypothesis, so our alternative is $H_A: p \neq 0.5$.

What would convince you that the proportion of up days was not 50%? If on 95% of the days, the DJIA closed up, you'd probably be convinced that up and down days were not equally likely. But if the sample proportion of up days were only slightly higher than 50%, you might not be sure. After all, observations do vary, so you wouldn't be surprised to see some difference. How different from 50% must the sample proportion be before you would be convinced that the true proportion wasn't 50%? Whenever we ask about the size of a statistical difference, we naturally think of the standard deviation. So let's start by finding the standard deviation of the sample proportion of days on which the DJIA increased.

Notation Alert! ─────

Capital H is the standard letter for hypotheses. H_0 labels the null hypothesis, and H_A labels the alternative.

We've seen 51.53% up days out of 1112 trading days. Is 51.53% far enough from 50% to be convincing evidence that the true proportion of up days is greater than 50? To be formal, we'll need a probability. And to find a probability we'd like to model the behavior of the sample proportion with the Normal model, so we'll check the assumptions and conditions. The sample size of 1112 is certainly big enough to satisfy the Success/Failure condition. (We expect $0.50 \times 1112 = 556$ daily increases.) It is reasonable to assume that the daily price changes are random and independent. And we know what hypothesis we are testing. To test a hypothesis we (temporarily) *assume* that it is true so we can see whether that description of the world is plausible. If we assume that the Dow increases or decreases with equal likelihood, we'll need to center our Normal sampling model at a mean of 0.5. Then, we can find the standard deviation of the sampling model as

$$SD(\hat{p}) = \sqrt{\frac{pq}{n}} = \sqrt{\frac{(0.5)(1 - 0.5)}{1112}} = 0.015$$

> ### Why is this a standard deviation and not a standard error?
>
> This is a standard deviation because we are using the model (hypothesized) value for p and *not* the estimated value, $\hat{p}$. Once we assume that the null hypothesis is true, it gives us a value for the model parameter p. With proportions, if we know p then we also automatically know its standard deviation. Because we find the standard deviation from the model parameter, this is a standard deviation and not a standard error. When we found a confidence interval for p, we could not assume that we knew its value, so we estimated the standard deviation from the sample value, $\hat{p}$.

To remind us that the parameter value comes from the null hypothesis, it is sometimes written as p_0 and the standard deviation as

$$SD(\hat{p}) = \sqrt{\frac{p_0 q_0}{n}}.$$

Now we know both parameters of the Normal sampling distribution model for our null hypothesis. For the mean of the Normal we use $p = 0.50$, and for the standard deviation we use the standard deviation of the sample proportions found using the null hypothesis value, $SD(\hat{p}) = 0.015$. We want to know how likely it would be to see the observed value $\hat{p}$ as far away from 50% as the value of 51.53% that we actually have observed. Looking first at a picture (Figure 13.2), we can see that 51.53% doesn't look very surprising. The more exact answer (from a calculator, computer program, or the Normal table) is that the probability is about 0.308. This is the probability of observing more than 51.53% up days or more than 51.53% down days if the null model were true. In other words, if the chance of an up day for the Dow is 50%, we'd expect to see stretches of 1112 trading days with as many as 51.53% up days about 15.4% of the time and with as many as 51.53% down days about 15.4% of the time. That's not terribly unusual, so there's really no convincing evidence that the market did not act randomly.

0.455   0.47   0.485   0.5   0.515   0.53   0.55

Figure 13.2   How likely is a proportion of more than 51.5% or less than 48.5% when the true mean is 50%? This is what it looks like. Each red area is 0.154 of the total area under the curve.

It may surprise you that even during a bull market, the direction of daily movements is random. But, the probability that any given day will end up or down appears to be about 0.5 regardless of the longer-term trends. It may be that when the stock market has a long run up (or possibly down, although we haven't checked that), it does so not by having more days of increasing or decreasing value, but by the actual amounts of the increases or decreases being unequal.

For Example | Framing hypotheses

Summit Projects is a full-service interactive agency, based in Hood River, OR, that offers companies a variety of website services. One of Summit's clients is SmartWool®, which produces and sells wool apparel including the famous SmartWool socks. Summit recently re-designed SmartWool's apparel website, and analysts at SmartWool wondered whether traffic has changed since the new website went live. In particular, an analyst might want to know if the proportion of visits resulting in a sale has changed since the new site went online. She might also wonder if the average sale amount has changed.

Questions: If the old site's proportion was 20%, frame appropriate null and alternative hypotheses for the proportion. If last year's average sale was $24.85, frame appropriate null and alternative hypotheses for the mean.[1]

Answers: For the proportion, let p = proportion of visits that result in a sale.

$$H_0: p = 0.2 \ vs. \ H_A: p \neq 0.2$$

For the average amount purchased, let μ = mean amount purchased per visit. Then

$$H_0: \mu = \$24.85 \ vs. \ H_A: \mu \neq \$24.85$$

13.2 A Trial as a Hypothesis Test

"If the People fail to satisfy their burden of proof, you must find the defendant not guilty."

—NEW YORK STATE JURY INSTRUCTIONS

We started by assuming that the probability of an up day was 50%. Then we looked at the data and concluded that we couldn't say otherwise because the proportion that we actually observed wasn't far enough from 50%. Does this reasoning of hypothesis tests seem backwards? That could be because we usually prefer to think about getting things right rather than getting them wrong. But, you've seen this reasoning before in a different context. This is the logic of jury trials.

Let's suppose a defendant has been accused of robbery. In British common law and those systems derived from it (including U.S. law), the null hypothesis is that the defendant is innocent. Instructions to juries are quite explicit about this.

The evidence takes the form of facts that seem to contradict the presumption of innocence. For us, this means collecting data. In the trial, the prosecutor presents evidence. ("If the defendant were innocent, wouldn't it be remarkable that the police found him at the scene of the crime with a bag full of money in his hand, a mask on his face, and a getaway car parked outside?") The next step is to judge the evidence. Evaluating the evidence is the responsibility of the jury in a trial, but it falls on your shoulders in hypothesis testing. The jury considers the evidence in light of the *presumption* of innocence and judges whether the evidence against the defendant would be plausible *if the defendant were in fact innocent*.

Like the jury, we ask: "Could these data plausibly have happened by chance if the null hypothesis were true?" If they are very unlikely to have occurred, then the evidence raises a reasonable doubt about the null hypothesis. Ultimately, *you* must make a decision. The standard of "beyond a reasonable doubt" is purposefully ambiguous because it leaves the jury to decide the degree to which the evidence contradicts the hypothesis of innocence. Juries don't explicitly use probability to help them decide whether to reject that hypothesis. But when you ask the same question of your null hypothesis, you have the advantage of being able to quantify exactly how surprising the evidence would be if the null hypothesis were true.

How unlikely is unlikely? Some people set rigid standards. Levels like 1 time out of 20 (0.05) or 1 time out of 100 (0.01) are common. But if *you* have to make the decision, you must judge for yourself in each situation whether the probability of observing your data is small enough to constitute "reasonable doubt."

---

[1] These numbers are hypothetical, but typical of the values that might have occurred.

13.3   P-Values

The fundamental step in our reasoning is the question: "Are the data surprising, given the null hypothesis?" And the key calculation is to determine exactly how likely the data we observed would be if the null hypothesis were the true model of the world. So we need a *probability*. Specifically, we want to find the probability of seeing data like these (or something even less likely) *given* the null hypothesis. This probability is the value on which we base our decision, so statisticians give this probability a special name. It's called the **P-value**.

A low enough P-value says that the data we have observed would be very unlikely if our null hypothesis were true. We started with a model, and now that same model tells us that the data we have are unlikely to have happened. That's surprising. In this case, the model and data are at odds with each other, so we have to make a choice. Either the null hypothesis is correct and we've just seen something remarkable, or the null hypothesis is wrong, (and, in fact, we were wrong to use it as the basis for computing our P-value). When you see a low P-value, you should reject the null hypothesis. There is no hard and fast rule about how low the P-value has to be. In fact, that decision is the subject of much of the rest of this chapter. Almost everyone would agree, however, that a P-value less than 0.001 indicates very strong evidence *against* the null hypothesis but a P-value greater than 0.05 provides very weak evidence.

When the P-value is *high* (or just not low *enough*), what do we conclude? In that case, we haven't seen anything unlikely or surprising at all. The data are consistent with the model from the null hypothesis, and we have no reason to reject the null hypothesis. Events that have a high probability of happening happen all the time. So, when the P-value is high does that mean we've proved the null hypothesis is true? No! We realize that many other similar hypotheses could also account for the data we've seen. The most we can say is that it doesn't appear to be false. Formally, we say that we "fail to reject" the null hypothesis. That may seem to be a pretty weak conclusion, but it's all we can say when the P-value is not low enough. All that means is that the data are consistent with the model that we started with.

What to Do with an "Innocent" Defendant

Let's see what that last statement means in a jury trial. If the evidence is not strong enough to reject the defendant's presumption of innocence, what verdict does the jury return? They do not say that the defendant is innocent. They say "not guilty." All they are saying is that they have not seen sufficient evidence to reject innocence and convict the defendant. The defendant may, in fact, be innocent, but the jury has no way to be sure.

Said statistically, the jury's null hypothesis is: innocent defendant. If the evidence is too unlikely (the P-value is low) then, given the assumption of innocence, the jury rejects the null hypothesis and finds the defendant guilty. But—and this is an important distinction—if there is *insufficient evidence* to convict the defendant (if the P-value is *not* low), the jury does not conclude that the null hypothesis is true and declare that the defendant is innocent. Juries can only *fail to reject* the null hypothesis and declare the defendant "not guilty."

In the same way, if the data are not particularly unlikely under the assumption that the null hypothesis is true, then the most we can do is to "fail to reject" our null hypothesis. We never declare the null hypothesis to be true. In fact, we simply do not know whether it's true or not. (After all, more evidence may come along later.)

Imagine a test of whether a company's new website design encourages a higher percentage of visitors to make a purchase (as compared to the site they've used for years). The null hypothesis is that the new site is no more effective at stimulating purchases than the old one. The test sends visitors randomly to one version of the

> **Beyond a Reasonable Doubt**
>
> We ask whether the data were unlikely beyond a reasonable doubt. We've just calculated how unlikely the data are if the null hypothesis is true. The probability that the observed statistic value (or an even more extreme value) could occur if the null model is true—in this case, 0.308—is the P-value That probability is certainly not *beyond a reasonable doubt*, so we fail to reject the null hypothesis here.

> **Don't We Want to Reject the Null?**
>
> Often, people who collect data or perform an experiment hope to reject the null. They hope the new ad campaign is *better* than the old one, or they hope their candidate is *ahead* of the opponent. But, when we test a hypothesis, we must stay neutral. We can't let our hope bias our decision. As in a jury trial, we must stay with the null hypothesis until we are convinced otherwise. The burden of proof rests with the alternative hypothesis—innocent until proven guilty. When you test a hypothesis, you must act as judge and jury, but not as prosecutor.

Conclusion

If the P-value is "low," reject H_0 and conclude H_A.

If the P-value is not "low enough," then fail to reject H_0 and the test is inconclusive.

website or the other. Of course, some will make a purchase, and others won't. If we compare the two websites on only 10 customers each, the results are likely *not to be clear*, and we'll be unable to reject the hypothesis. Does this mean the new design is a complete bust? Not necessarily. It simply means that we don't have enough evidence to reject our null hypothesis. That's why we don't start by assuming that the new design is *more* effective. If we were to do that, then we could test just a few customers, find that the results aren't clear, and claim that since we've been unable to reject our original assumption the redesign must be effective. The Board of Directors is unlikely to be impressed by that argument.

For Example   Conclusions from P-values

Question: The SmartWool analyst (see page 361) collects a representative sample of visits since the new website has gone online and finds that the P-value for the test of proportion is 0.0015 and the P-value for the test of the mean is 0.3740. What conclusions can she draw?

Answer: The proportion of visits that resulted in a sale since the new website went online is very unlikely to still be 0.20. There is strong evidence to suggest that the proportion has changed. She should reject the null hypothesis. However, the mean amount spent is consistent with the null hypothesis and therefore she is unable to reject the null hypothesis that the mean is still $24.85 against the alternative that it increased.

Just Checking

1 A pharmaceutical firm wants to know whether aspirin helps to thin blood. The null hypothesis says that it doesn't. The firm's researchers test 12 patients, observe the proportion with thinner blood, and get a P-value of 0.32. They proclaim that aspirin doesn't work. What would you say?

2 An allergy drug has been tested and found to give relief to 75% of the patients in a large clinical trial. Now the scientists want to see whether a new, "improved" version works even better. What would the null hypothesis be?

3 The new allergy drug is tested, and the P-value is 0.0001. What would you conclude about the new drug?

13.4   The Reasoning of Hypothesis Testing

Hypothesis tests follow a carefully structured path. To avoid getting lost, it helps to divide that path into four distinct sections: hypothesis, model, mechanics, and conclusion.

Hypotheses

First, state the null hypothesis. That's usually the skeptical claim that nothing's different. The null hypothesis assumes the default (often the status quo) is true (the defendant is innocent, the new method is no better than the old, customer preferences haven't changed since last year, etc.).

In statistical hypothesis testing, hypotheses are almost always about model parameters. To assess how unlikely our data may be, we need a null model. The null hypothesis specifies a particular parameter value to use in our model. In the usual notation, we write H_0: *parameter = hypothesized value*. The alternative hypothesis, H_A, contains the values of the parameter we consider plausible when we reject the null.

"The null hypothesis is never proved or established, but is possibly disproved, in the course of experimentation. Every experiment may be said to exist only in order to give the facts a chance of disproving the null hypothesis."

—SIR RONALD FISHER,
THE DESIGN OF EXPERIMENTS, 1931

Model

To plan a statistical hypothesis test, specify the *model* for the sampling distribution of the statistic you will use to test the null hypothesis and the parameter of interest. For proportions, we use the Normal model for the sampling distribution. Of course, all models require assumptions, so you will need to state them and check any corresponding conditions. For a test of a proportion, the assumptions and conditions are the same as for a one-proportion *z*-interval.

Your model step should end with a statement such as: *Because the conditions are satisfied, we can model the sampling distribution of the proportion with a Normal model.* Watch out, though. Your Model step could end with: *Because the conditions are not satisfied, we can't proceed with the test.* (If that's the case, stop and reconsider.)

Each test we discuss in this book has a name that you should include in your report. We'll see many tests in the following chapters. Some will be about more than one sample, some will involve statistics other than proportions, and some will use models other than the Normal (and so will not use *z*-scores). The test about proportions is called a **one-proportion *z*-test**.[2]

> ### One-proportion *z*-test
>
> The conditions for the one-proportion *z*-test are the same as for the one-proportion *z*-interval (except that we use the hypothesized values, p_0 and q_0, to check the Success/Failure condition). We test the hypothesis $H_0: p = p_0$ using the statistic
>
> $$z = \frac{(\hat{p} - p_0)}{SD(\hat{p})}.$$
>
> We also use p_0 to find the standard deviation: $SD(\hat{p}) = \sqrt{\dfrac{p_0 q_0}{n}}$. When the conditions are met and the null hypothesis is true, this statistic follows the standard Normal model, so we can use that model to obtain a P-value.

Mechanics

Under "Mechanics" we perform the actual calculation of our test statistic from the data. Different tests we encounter will have different formulas and different test statistics. Usually, the mechanics are handled by a statistics program or calculator. The ultimate goal of the calculation is to obtain a P-value—the probability that the observed statistic value (or an even more extreme value) could occur if the null model were correct. If the P-value is small enough, we'll reject the null hypothesis.

Conclusions and Decisions

The primary conclusion in a formal hypothesis test is only a statement about the null hypothesis. It simply states whether we reject or fail to reject that hypothesis. As always, the conclusion should be stated in context, but your conclusion about the null hypothesis should never be the end of the process. You can't make a decision based solely on a P-value. Business decisions have consequences, with actions to take or policies to change. The conclusions of a hypothesis test can help *inform* your decision, but they shouldn't be the only basis for it.

Business decisions should always take into consideration three things: the statistical significance of the test, the *cost* of the proposed action, and the **effect size** (the difference between the hypothesized and observed value) of the statistic. For

---

[2]It's also called the "one-sample test for a proportion."

example, a cellular telephone provider finds that 30% of their customers switch providers (or *churn*) when their two-year subscription contract expires. They try a small experiment and offer a random sample of customers a free $350 top-of-the-line phone if they renew their contracts for another two years. Not surprisingly, they find that the new switching rate is lower by a statistically significant amount. Should they offer these free phones to all their customers? Obviously, the answer depends on more than the P-value of the hypothesis test. Even if the P-value is statistically significant, the correct business decision also depends on the cost of the free phones and by how much the churn rate is lowered (the effect size). It's rare that a hypothesis test alone is enough to make a sound business decision.

For Example   The reasoning of hypothesis tests

Question: The analyst at SmartWool (see page 361) selects 200 recent weblogs at random and finds that 58 of them have resulted in a sale. The null hypothesis is that $p = 0.20$. Would this be a surprising proportion of sales if the true proportion of sales were 20%?

Answer: To judge whether 58 is a surprising number of sales given the null hypothesis, we use the Normal model based on the null hypothesis. That is, use 0.20 as the mean and $\sqrt{\dfrac{p_0 q_0}{n}} = \sqrt{\dfrac{(0.2)(0.8)}{200}} = 0.02828$ as the standard deviation.

58 sales is a sample proportion of $\hat{p} = \dfrac{58}{200} = 0.29$ or 29%.

The z-value for 0.29 is then $z = \dfrac{\hat{p} - p_0}{SD(\hat{p})} = \dfrac{0.29 - 0.20}{0.02828} = 3.182$.

In other words, given that the null hypothesis is true, our sample proportion is 3.182 standard deviations higher than the mean. That seems like a surprisingly large value since the probability of being farther than 3 standard deviations from the mean is (from the 68–95–99.7 Rule) only 0.3%.

13.5   Alternative Hypotheses

"They make things admirably plain,
But one hard question will remain:
If one hypothesis you lose,
Another in its place you choose . . ."

—James Russell Lowell,
Credidimus Jovem Regnare

In our example about the DJIA, we were equally interested in proportions that deviate from 50% in *either* direction. So we wrote our alternative hypothesis as H_A: $p \neq 0.5$. Such an alternative hypothesis is known as a **two-sided alternative** because we are equally interested in deviations on either side of the null hypothesis value. For two-sided alternatives, the P-value is the probability of deviating in *either* direction from the null hypothesis value.

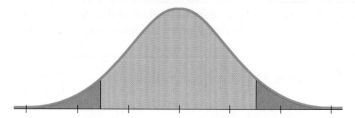

Figure 13.3   The P-value for a two-sided alternative adds the probabilities in both tails of the sampling distribution model outside the value that corresponds to the test statistic.

Suppose we want to test whether the proportion of customers returning merchandise has decreased under our new quality monitoring program. We know the quality has improved, so we can be pretty sure things haven't gotten worse.

Alternative Hypotheses

Proportions:

Two-sided

$H_0: p = p_0$

$H_A: p \neq p_0$

One-sided

$H_0: p = p_0$

$H_A: p < p_0$ or $p > p_0$

Means:

Two-sided

$H_0: \mu = \mu_0$

$H_A: \mu \neq \mu_0$

One-sided

$H_0: \mu = \mu_0$

$H_A: \mu < \mu_0$ or $\mu > \mu_0$

But have the customers noticed? We would only be interested in a sample proportion *smaller* than the null hypothesis value. We'd write our alternative hypothesis as $H_A: p < p_0$. An alternative hypothesis that focuses on deviations from the null hypothesis value in only one direction is called a **one-sided alternative**.

Figure 13.4 The P-value for a one-sided alternative considers only the probability of values beyond the test statistic value in the specified direction.

For a hypothesis test with a one-sided alternative, the P-value is the probability of deviating *only in the direction of the alternative* away from the null hypothesis value.

Guided Example   Home Field Advantage

Major league sports are big business. And the fans are more likely to come out to root for the team if the home team has a good chance of winning. Anyone who follows or plays sports has heard of the "home field advantage." It is said that teams are more likely to win when they play at home. That *would* be good for encouraging the fans to come to the games. But is it true?

In the 2009 Major League Baseball (MLB) season, there were 2430 regular season games. (Tied at the end of the season the Colorado Rockies and San Diego Padres played an extra game to determine

who won the Wild Card playoff spot.) It turns out that the home team won 1332 of the 2430 games, or 54.81% of the time. If there were no home field advantage, the home teams would win about half of all games played. Could this deviation from 50% be explained just from natural sampling variability, or does this evidence suggest that there really is a home field advantage, at least in professional baseball?

To test the hypothesis, we will ask whether the observed rate of home team victories, 54.81%, is so much greater than 50% that we cannot explain it away as just chance variation.

Remember the four main steps to performing a hypothesis test—hypotheses, model, mechanics, and conclusion? Let's put them to work and see what this will tell us about the home team's chances of winning a baseball game.

PLAN   Setup State what we want to know.

Define the variables and discuss their context.

We want to know whether the home team in professional baseball is more likely to win. The data are all 2430 games from the 2009 Major League Baseball season. The variable is whether or not the home team won. The parameter of interest is the proportion of home team wins. If there is an advantage, we'd expect that proportion to be greater than 0.50. The observed statistic value is $\hat{p} = 0.5481$.

Hypotheses The null hypothesis makes the claim of no home field advantage.

$$H_0: p = 0.50$$

We are interested only in a home field *advantage*, so the alternative hypothesis is one-sided.

$$H_A: p > 0.50$$

Model Think about the assumptions and check the appropriate conditions.

✓ **Independence Assumption.** Generally, the outcome of one game has no effect on the outcome of another game. But this may not always be strictly true. For example, if a key player is injured, the probability that the team will win in the next couple of games may decrease slightly, but independence is still roughly true.

Consider the time frame carefully.

✓ **Randomization Condition.** We have results for all 2430 games of the 2009 season. But we're not just interested in 2009. While these games were not randomly selected, they may be reasonably representative of all recent professional baseball games.

✓ **10% Condition.** This is not a random sample, but these 2430 games are fewer than 10% of all games played over the years.

✓ **Success/Failure Condition.** Both
$np_0 = 2430(0.50) = 1215$ and
$nq_0 = 2430(0.50) = 1215$ are at least 10.

Specify the sampling distribution model.

Tell what test you plan to use.

Because the conditions are satisfied, we'll use a Normal model for the sampling distribution of the proportion and do a one-proportion z-test.

DO

Mechanics The null model gives us the mean, and (because we are working with proportions) the mean gives us the standard deviation.

The null model is a Normal distribution with a mean of 0.50 and a standard deviation of:

$$SD(\hat{p}) = \sqrt{\frac{p_0 q_0}{n}} = \sqrt{\frac{(0.5)(1 - 0.5)}{2430}} = 0.01014$$

The observed proportion $\hat{p}$ is 0.5481.

$$z = \frac{(\hat{p} - p_0)}{SD(\hat{p})}$$

$$= \frac{0.0481}{0.01014} = 4.74.$$

The observed proportion is 4.74 standard deviations above the hypothesized proportion.

From technology, we can find the P-value, which tells us the probability of observing a value that extreme (or more).

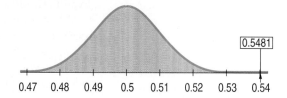

The corresponding P-value is <0.001.

The probability of observing a $\hat{p}$ of 0.5481 or more in our Normal model can be found by computer, calculator, or table to be <0.001.

(continued)

REPORT

Conclusion State your conclusion about the parameter—in context.

MEMO

Re: Home field advantage

Our analysis of outcomes during the 2009 Major League Baseball season showed a statistically significant advantage to the home team ($P < 0.001$). We can be quite confident that playing at home gives a baseball team an advantage.

For Example — Finding P-values for a test of proportion

Question: From the fact that of 200 randomly selected access logs, 58 resulted in sales (see page 365), find the P-value for testing the hypothesis that $p = 0.20$.

Answer: The alternative is *two-sided*, and the z-value is 3.182. So, we find the probability that z is greater than 3.182 or less than -3.182. Because the Normal is symmetric, that's the same as $2 \times P(z > 3.182) = 2 \times .00073 = 0.00146$. This is strong evidence against the null hypothesis so we *reject* it and conclude that the proportion of sales is no longer 0.20.

13.6 Testing Hypothesis about Means— the One-Sample *t*-Test

When we made confidence intervals about proportions, we could base the interval on the z statistic because proportions have a natural link between their value ($\hat{p}$) and the standard error ($\sqrt{(\hat{p} \times \hat{q}/n)}$). For testing a hypothesis about a proportion, we used z as the reference for the same reason. But, for means, we saw that sample means have no link between their value (μ) and the value of the standard error ($s/\sqrt{(n)}$). That seemingly small extra variation due to estimating the standard error rocked Gosset's world in 1906 and led to the discovery of the *t*-distribution. It should come as no surprise then, that for testing a hypothesis about a mean, we base the test on the t distribution. In fact, other than that, the test looks just like the test for a proportion.

In Chapter 12, we built a confidence interval for the mean profit of policies sold by an insurance agent whose average profit from a sample of 30 policies was $1438.90. Now the manager has a more specific concern. Company policy states that if a sales rep's mean profit is below $1500, the sales rep has been discounting too much and will have to adjust his pricing strategy. Is there evidence from this sample that the mean is really less than $1500? This question calls for a hypothesis test called the **one-sample *t*-test for the mean**.

When testing a hypothesis, it's natural to compare the difference between the observed statistic and a hypothesized value to the standard error. For means that looks like: $\dfrac{\bar{y} - \mu_0}{SE(\bar{y})}$. We already know from Chapter 12 that the appropriate probability model to use is Student's t with $n - 1$ degrees of freedom.

One-sample *t*-test for the mean

The conditions for the one-sample *t*-test for the mean are the same as for the one-sample *t*-interval (see page 336 in Chapter 12). We test the hypothesis $H_0: \mu = \mu_0$ using the statistic

$$t_{n-1} = \frac{\bar{y} - \mu_0}{SE(\bar{y})},$$

where the standard error of $\bar{y}$ is: $SE(\bar{y}) = \frac{s}{\sqrt{n}}$.

When the conditions are met and the null hypothesis is true, this statistic follows a Student's *t*-model with $n - 1$ degrees of freedom. We use that model to obtain a P-value.

Guided Example   **Insurance Profits Revisited**

Let's apply the one-sample *t*-test to the 30 mature policies sampled by the manager in Chapter 12. From these 30 policies, the management would like to know if there's evidence that the mean profit of policies sold by this sales rep is less than $1500.

PLAN

Setup State what we want to know. Make clear what the population and parameter are.

Identify the variables and context.

We want to test whether the mean profit of the sales rep's policies is less than $1500. We have a random sample of 30 mature policies from which to judge.

Hypotheses We give benefit of the doubt to the sales rep. The null hypothesis is that the true mean profit is equal to $1500. Because we're interested in whether the profit is less, the alternative is one-sided.

$$H_O: \mu = \$1500$$
$$H_A: \mu < \$1500$$

Make a graph. Check the distribution for skewness, multiple modes, and outliers.

We checked the histogram of these data in a previous Guided Example (Chapter 12) and saw that it had a unimodal, symmetric distribution.

Model Check the conditions.

We checked the Randomization and Nearly Normal Conditions in the previous Guided Example in Chapter 12.

State the sampling distribution model.

Choose your method.

The conditions are satisfied, so we'll use a Student's *t*-model with $n - 1 = 29$ degrees of freedom and a one-sample *t*-test for the mean.

DO

Mechanics Compute the sample statistics. Be sure to include the units when you write down what you know from the data.

Using software, we obtain the following basic statistics:

$$n = 30$$
Mean = $1438.90
StDev = $1329.60

(continued)

The *t*-statistic calculation is just a standardized value. We subtract the hypothesized mean and divide by the standard error.

We assume the null model is true to find the P-value. Make a picture of the *t*-model, centered at μ_0. Since this is a lower-tail test, shade the region to the left of the observed average profit.

The P-value is the probability of observing a sample mean as small as $1438.90 (or smaller) *if* the true mean were $1500, as the null hypothesis states. We can find this P-value from a table, calculator, or computer program.

$$t = \frac{1438.90 - 1500}{1329.60/\sqrt{30}} = -0.2517$$

(The observed mean is less than one standard error below the hypothesized value.)

P-value $= P(t_{29} < -0.2517) = 0.4015$ (or from a table $0.1 < P < 0.5$)

REPORT

Conclusion Link the P-value to your decision about H_0, and state your conclusion in context.

MEMO

Re: Sales Performance

The mean profit on 30 sampled contracts closed by the sales rep in question has fallen below our standard of $1500, but there is not enough evidence in this sample of policies to indicate that the true mean is below $1500. If the mean were $1500, we would expect a sample of size 30 to have a mean this low about 40.15% of the time.

Notice that the way this hypothesis was set up, the sales rep's mean profit would have to be well below $1500 to reject the null hypothesis. Because the null hypothesis was that the mean was $1500 and the alternative was that it was less, this setup gave some benefit of the doubt to the sales rep. There's nothing intrinsically wrong with that, but keep in mind that it's always a good idea to make sure that the hypotheses are stated in ways that will guide you to make the right business decision.

For Example   Testing a mean

Question: From the 58 sales recorded in the sample of 200 access logs (see page 365), the analyst finds that the mean amount spent is $26.05 with a standard deviation of $10.20. Test the hypothesis that the mean is still $24.85 against the alternative that it has increased.

Answer: We can write: $H_0: \mu = \$24.85$ *vs.* $H_A: \mu > \$24.85$.

Then $t = \dfrac{(26.05 - 24.85)}{10.2/\sqrt{58}} = 0.896$.

Because the alternative is *one-sided*, we find $P(t > 0.896)$ with 57 degrees of freedom. From technology, $P(t > 0.896) = 0.1870$, a large P-value. This is not a surprising value given the hypothesized mean of $24.85. Therefore we *fail to reject* the null hypothesis and conclude that there is not sufficient evidence to suggest that the mean has increased from 24.85. Had we used a two-sided alternative, the P-value would have been twice 0.1870 or 0.3740.

13.7   Alpha Levels and Significance

Sometimes we need to make a firm decision about whether or not to reject the null hypothesis. A jury must *decide* whether the evidence reaches the level of "beyond a reasonable doubt." A business must *select* a Web design. You need to decide which section of a Statistics course to enroll in.

When the P-value is small, it tells us that our data are rare *given the null hypothesis.* As humans, we are suspicious of rare events. If the data are "rare enough," we just don't think that could have happened due to chance. Since the data *did* happen, something must be wrong. All we can do now is to reject the null hypothesis.

But how rare is "rare"? How low does the P-value have to be?

We can define "rare event" arbitrarily by setting a threshold for our P-value. If our P-value falls below that point, we'll reject the null hypothesis. We call such results *statistically significant*. The threshold is called an **alpha level**. Not surprisingly, it's labeled with the Greek letter α. Common α-levels are 0.10, 0.05, 0.01, and 0.001. You have the option—almost the *obligation*—to consider your alpha level carefully and choose an appropriate one for the situation. If you're assessing the safety of air bags, you'll want a low alpha level; even 0.01 might not be low enough. If you're just wondering whether folks prefer their pizza with or without pepperoni, you might be happy with $\alpha = 0.10$. It can be hard to justify your choice of α, though, so often we arbitrarily choose 0.05.

Sir Ronald Fisher (1890–1962) was one of the founders of modern Statistics.

Where did the value 0.05 come from?

In 1931, in a famous book called *The Design of Experiments*, Sir Ronald Fisher discussed the amount of evidence needed to reject a null hypothesis. He said that it was *situation dependent*, but remarked, somewhat casually, that for many scientific applications, 1 out of 20 *might be* a reasonable value, especially in a *first* experiment—one that will be followed by confirmation. Since then, some people—indeed some entire disciplines—have acted as if the number 0.05 were sacrosanct.

Notation Alert! —————

The first Greek letter, α, is used in Statistics for the threshold value of a hypothesis test. You'll hear it referred to as the alpha level. Common values are 0.10, 0.05, 0.01, and 0.001.

The alpha level is also called the **significance level**. When we reject the null hypothesis, we say that the test is "significant at that level." For example, we might say that we reject the null hypothesis "at the 5% level of significance." You must select the alpha level *before* you look at the data. Otherwise you can be accused of finagling the conclusions by tuning the alpha level to the results after you've seen the data.

What can you say if the P-value does not fall below α? When you have not found sufficient evidence to reject the null according to the standard you have established, you should say: "The data have failed to provide sufficient evidence to reject the null hypothesis." Don't say: "We accept the null hypothesis." You certainly haven't proven or established the null hypothesis; it was assumed to begin with. You *could* say that you have *retained* the null hypothesis, but it's better to say that you've failed to reject it.

It could happen to you!

Of course, if the null hypothesis *is* true, no matter what alpha level you choose, you still have a probability α of rejecting the null hypothesis by mistake. When we do reject the null hypothesis, no one ever thinks that *this* is one of those rare times. As statistician Stu Hunter notes, "The statistician says 'rare events do happen—but not to me!'"

Look again at the home field advantage example. The P-value was <0.001. This is so much smaller than any reasonable alpha level that we can reject H_0. We concluded: "We reject the null hypothesis. There is sufficient evidence to conclude that there is a home field advantage over and above what we expect with random variation." On the other hand, when testing the mean in the insurance example, the

Conclusion

If the P-value $< \alpha$, then reject H_0.

If the P-value $\geq \alpha$, then fail to reject H_0.

Practical vs. Statistical Significance

A large insurance company mined its data and found a statistically significant ($P = 0.04$) difference between the mean value of policies sold in 2001 and those sold in 2002. The difference in the mean values was $0.98. Even though it was statistically significant, management did not see this as an important difference when a typical policy sold for more than $1000. On the other hand, a marketable improvement of 10% in relief rate for a new pain medicine may not be statistically significant unless a large number of people are tested. The effect, which is economically significant, might not be statistically significant.

P-value was 0.4015, a very high P-value. In this case we can say only that we have failed to reject the null hypothesis that $\mu = \$1500$. We certainly can't say that we've proved it, or even that we've accepted it.

The automatic nature of the reject/fail-to-reject decision when we use an alpha level may make you uncomfortable. If your P-value falls just slightly above your alpha level, you're not allowed to reject the null. Yet a P-value just barely below the alpha level leads to rejection. If this bothers you, you're in good company. Many statisticians think it better to report the P-value than to choose an alpha level and carry the decision through to a final reject/fail-to-reject verdict. So when you declare your decision, it's always a good idea to report the P-value as an indication of the strength of the evidence.

It's in the stars

Some disciplines carry the idea further and code P-values by their size. In this scheme, a P-value between 0.05 and 0.01 gets highlighted by a single asterisk (*). A P-value between 0.01 and 0.001 gets two asterisks (**), and a P-value less than 0.001 gets three (***). This can be a convenient summary of the weight of evidence against the null hypothesis, but it isn't wise to take the distinctions too seriously and make black-and-white decisions near the boundaries. The boundaries are a matter of tradition, not science; there is nothing special about 0.05. A P-value of 0.051 should be looked at seriously and not casually thrown away just because it's larger than 0.05, and one that's 0.009 is not very different from one that's 0.011.

Sometimes it's best to report that the conclusion is not yet clear and to suggest that more data be gathered. (In a trial, a jury may "hang" and be unable to return a verdict.) In such cases, it's an especially good idea to report the P-value, since it's the best summary we have of what the data say or fail to say about the null hypothesis.

What do we mean when we say that a test is statistically significant? All we mean is that the test statistic had a P-value lower than our alpha level. Don't be lulled into thinking that "statistical significance" necessarily carries with it any practical importance or impact.

For large samples, even small, unimportant ("insignificant") deviations from the null hypothesis can be statistically significant. On the other hand, if the sample is not large enough, even large, financially or scientifically important differences may not be statistically significant.

When you report your decision about the null hypothesis, it's good practice to report the effect size (the magnitude of the difference between the observed statistic value and the null hypothesis value in the data units) along with the P-value.

For Example   Setting the α level

Question: The manager of the analyst at SmartWool (see pages 368 and 370) wants her to use an α level of 0.05 for all her hypothesis tests. Would her conclusions for the two hypothesis tests have changed if she used an α level of 0.05?

Answer: Using $\alpha = 0.05$, we reject the null hypothesis when the P-value is less than 0.05 and fail to reject when the P-value is greater than or equal to 0.05. For the test of proportion, $P = 0.00146$, which is much less than 0.05 and so we reject. For the test of means, the P-value was 0.1870, which is greater than 0.05, so we fail to reject. Our conclusions are the same as before using this α level.

13.8   Critical Values

When building a confidence interval, we calculated the margin of error as the product of an estimated standard error for the statistic and a critical value. For proportions, we found a **critical value**, z^*, to correspond to our selected confidence level. For means, we found the critical value t^* based on both the confidence level and the degrees of freedom. Critical values can also be used as a shortcut for hypothesis tests. Before computers and calculators were common, P-values were hard to find. It was easier to select a few common alpha levels (0.05, 0.01, 0.001, for example) and learn the corresponding critical values for the Normal model (that is, the critical values corresponding to confidence levels 0.95, 0.99, and 0.999, respectively). Rather than find the probability that corresponded to your observed statistic, you'd just calculate how many standard deviations it was away from the hypothesized value and compare that value directly against these z^* values. (Remember that whenever we measure the distance of a value from the mean in standard deviations, we are finding a z-score.) Any z-score larger in magnitude (that is, more extreme) than a particular critical value has to be less likely, so it will have a P-value smaller than the corresponding alpha.

If we were willing to settle for a flat reject/fail-to-reject decision, comparing an observed z-score with the critical value for a specified alpha level would give a shortcut path to that decision. For the home field advantage example, if we choose $\alpha = 0.05$, then in order to reject H_0, our z-score has to be larger than the one-sided critical value of 1.645. The observed proportion was 4.74 standard deviations above 0.5, so we clearly reject the null hypothesis. This is perfectly correct and does give us a yes/no decision, but it gives us less information about the hypothesis because we don't have the P-value to think about. With technology, P-values are easy to find. And since they give more information about the strength of the evidence, you should report them.

Here are the traditional z^* critical values from the Normal model[3]:

α	1-sided	2-sided
0.05	1.645	1.96
0.01	2.33	2.576
0.001	3.09	3.29

Quick Decisions

If you need to make a decision on the fly with no technology, remember "2." That's our old friend from the 68–95–99.7 Rule. It's roughly the critical value for testing a hypothesis against a two-sided alternative at $\alpha = 0.05$ using z. In fact, it's just about the t^* critical value with 60 degrees of freedom. The exact critical value for z^* is 1.96, but 2 is close enough for most decisions.

Figure 13.5 When the alternative is one-sided, the critical value puts all of α on one side.

Figure 13.6 When the alternative is two-sided, the critical value splits α equally into two tails.

When testing means, you'll need to know both the α level and the degrees of freedom to find the t^* critical value. With large n, the t^* critical values will be close to the z^* critical values (above) that you use for testing proportions.

For Example   Testing using critical values

Question: Find the critical z and t values for the SmartWool hypotheses (see pages 368 and 370) using $\alpha = 0.05$ and show that the same decisions would have been made using critical values.

Answer: For the two-sided test of proportions, the critical z values at $\alpha = 0.05$ are ± 1.96. Because the z value was 3.182, much larger than 1.96, we reject the null hypothesis.

For the one-sided test of means, with $n = 58$, the critical t value at $\alpha = 0.05$ is 1.676 (using df = 50 from a table) or 1.672 (using df = 57 from technology). In either case, the t value of 0.896 is smaller than the critical value, so we fail to reject the null hypothesis.

---

[3]In a sense, these are the flip side of the 68–95–99.7 Rule. There we chose simple statistical distances from the mean and recalled the areas of the tails. Here we select convenient tail areas (0.05, 0.01, and 0.001, either on one side or adding both together), and record the corresponding statistical distances.

13.9   Confidence Intervals and Hypothesis Tests

Confidence intervals and hypothesis tests are built from the same calculations. They have the same assumptions and conditions. As we have just seen, you can approximate a hypothesis test by examining the confidence interval. Just ask whether the null hypothesis value is consistent with a confidence interval for the parameter at the corresponding confidence level. Because confidence intervals are naturally two-sided, they correspond to two-sided tests. For example, a 95% confidence interval corresponds to a two-sided hypothesis test at $\alpha = 5\%$. In general, a confidence interval with a confidence level of $C\%$ corresponds to a two-sided hypothesis test with an α level of $100 - C\%$.

The relationship between confidence intervals and one-sided hypothesis tests gives us a choice. For a one-sided test with $\alpha = 5\%$, you could construct a one-sided confidence level of 95%, leaving 5% in one tail.

A one-sided confidence interval leaves one side unbounded. For example, in the home field example, we wondered whether the home field gave the home team an *advantage*, so our test was naturally one-sided. A 95% one-sided confidence interval for a proportion would be constructed from one side of the associated two-sided confidence interval:

$$0.5481 - 1.645 \times 0.01014 = 0.531.$$

In order to leave 5% on one side, we used the z^* value 1.645 that leaves 5% in one tail. Writing the one-sided interval as $(0.531, \infty)$ allows us to say with 95% confidence that we know the home team will win, on average, at least 53.1% of the time. To test the hypothesis H_0: $p = 0.50$ we note that the value 0.50 is not in this interval. The lower bound of 0.531 is clearly above 0.50, showing the connection between hypothesis testing and confidence intervals.

For convenience, and to provide more information, however, we sometimes report a two-sided confidence interval even though we are interested in a one-sided test. For the home field example, we could report a 90% two-sided confidence interval:

$$0.5481 \pm 1.645 \times 0.01014 = (0.531, 0.565).$$

Notice that we *matched* the left end point by leaving α in *both* sides, which made the corresponding confidence level 90%. We can still see the correspondence. Because the 90% (two-sided) confidence interval for $\hat{p}$ doesn't contain 0.50, we reject the null hypothesis. Using the two-sided interval also tells us that the home team winning percentage is unlikely to be greater than 56.5%, an added benefit to understanding. You can see the relationship between the two confidence intervals in Figure 13.7.

There's another good reason for finding a confidence interval along with a hypothesis test. Although the test can tell us whether the observed statistic differs from the hypothesized value, it doesn't say by how much. Often, business decisions depend not only on whether there is a statistically significant difference, but also on whether the difference is meaningful. For the home field advantage, the corresponding confidence interval shows that over a full season, home field advantage adds an average of about two to six extra victories for a team. That could make a meaningful difference in both the team's standing and in the size of the crowd.

The story is similar for means. In Chapter 12, we found a 95% confidence interval for the mean profit on a sales representative's policies to be ($942.48, $1935.32). Intervals are naturally two-sided and this 95% confidence interval leaves 0.025 on each side of the mean not covered. In the current chapter, we tested the

"Extraordinary claims require extraordinary proof."

—CARL SAGAN

hypothesis that the mean value was $1500. But unlike the confidence interval, our test is one-sided because our alternative is $\mu < 1500$. So checking to see if $1500 is too high to be in the interval is equivalent to performing the hypothesis test at an α of 0.025—one side of the confidence interval. (Or, you could use a 90% confidence interval to test the hypothesis at $\alpha = 0.05$.) In this example, $1500 is clearly within the interval, so we fail to reject the null hypothesis. There is no evidence to suggest that the true mean is less than $1500.

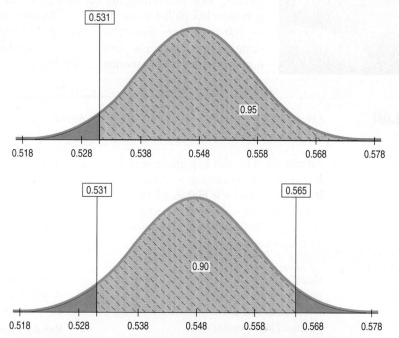

Figure 13.7 The one-sided 95% confidence interval (top) leaves 5% on one side (in this case the left), but leaves the other side unbounded. The 90% confidence interval is symmetric and matches the one-sided interval on the side of interest. Both intervals indicate that a one-sided test of $p = 0.50$ would be rejected at $\alpha = 0.05$ for any value of $\hat{p}$ greater than 0.531.

Just Checking

4 A bank is testing a new method for getting delinquent customers to pay their past-due credit card bills. The standard way was to send a letter (costing about $0.60 each) asking the customer to pay. That worked 30% of the time. The bank wants to test a new method that involves sending a DVD to the customer encouraging them to contact the bank and set up a payment plan. Developing and sending the DVD costs about $10.00 per customer. What is the parameter of interest? What are the null and alternative hypotheses?

5 The bank sets up an experiment to test the effectiveness of the DVD. The DVD is mailed to several randomly selected delinquent customers, and employees keep track of how many customers then contact the bank to arrange payments. The bank just got back the results on their test of the DVD strategy. A 90% confidence interval for the

success rate is (0.29, 0.45). Their old send-a-letter method had worked 30% of the time. Can you reject the null hypothesis and conclude that the method increases the proportion at $\alpha = 0.05$? Explain.

6 Given the confidence interval the bank found in the trial of the DVD mailing, what would you recommend be done? Should the bank scrap the DVD strategy?

7 The mileage rewards program at a major airline company has just completed a test market of a new arrangement with a hotel partner to try to increase engagement of valued customers with the program. The cost to the rewards program of the hotel offer is $35.00 per customer. From the test market, a 95% confidence interval for the revenue generated is ($37.95, $55.05). What does that say about the viability of the new arrangement?

A credit card company plans to offer a special incentive program to customers who charge at least $500 next month. The marketing department has pulled a sample of 500 customers from the same month last year and noted that the mean amount charged was $478.19 and the median amount was $216.48. The finance department says that the only relevant quantity is the proportion of customers who spend more than $500. If that proportion is not more than 25%, the program will lose money.

Among the 500 customers, 148 or 29.6% of them charged $500 or more. Can we use a confidence interval to test whether the goal of 25% for all customers was met?

PLAN	

Setup State the problem and discuss the variables and the context.

Hypotheses The null hypothesis is that the proportion qualifying is 25%. The alternative is that it is higher. It's clearly a one-sided test, so if we use a confidence interval, we'll have to be careful about what level we use.

Model Check the conditions. (Because this is a confidence interval, we use the observed successes and failures to check the Success/Failure condition.)

We want to know whether 25% or more of the customers will spend $500 or more in the next month and qualify for the special program. We will use the data from the same month a year ago to estimate the proportion and see whether the proportion was at least 25%.

The statistic is $\hat{p} = 0.296$, the proportion of customers who charged $500 or more.

$$H_0: p = 0.25$$
$$H_A: p > 0.25$$

✓ **Independence Assumption.** Customers are not likely to influence one another when it comes to spending on their credit cards.

✓ **Randomization Condition.** This is a random sample from the company's database.

✓ **10% Condition.** The sample is less than 10% of all customers.

✓ **Success/Failure Condition.** There were 148 successes and 352 failures, both at least 10. The sample is large enough.

State your method. Here we are using a confidence interval to test a hypothesis.

Under these conditions, the sampling model is Normal. We'll create a one-proportion z-interval.

DO	

Mechanics Write down the given information and determine the sample proportion.

To use a confidence interval, we need a confidence level that corresponds to the alpha level of the test. If we use $\alpha = 0.05$, we should construct a 90% confidence interval because this is a one-sided test. That will leave 5% on *each* side of the observed proportion. Determine the standard error of the sample proportion and the margin of error. The critical value is $z^* = 1.645$.

The confidence interval is estimate ± margin of error.

$n = 500$, so

$$\hat{p} = \frac{148}{500} = 0.296 \text{ and}$$

$$SE(\hat{p}) = \sqrt{\frac{\hat{p}\hat{q}}{n}} = \sqrt{\frac{(0.296)(0.704)}{500}} = 0.020$$

$$ME = z^* \times SE(\hat{p})$$

$$= 1.645(0.020) = 0.033$$

The 90% confidence interval is 0.296 ± 0.033 or $(0.263, 0.329)$.

REPORT   **Conclusion** Link the confidence interval to your decision about the null hypothesis, then state your conclusion in context.

MEMO

Re: Credit card promotion

Our study of a sample of customer records indicates that between 26.3% and 32.9% of customers charge $500 or more. We are 90% confident that this interval includes the true value. Because the minimum suitable value of 25% is below this interval, we conclude that it is not a plausible value, and so we reject the null hypothesis that only 25% of the customers charge more than $500 a month. The goal appears to have been met assuming that the month we studied is typical.

For Example   Confidence intervals and hypothesis tests

Question: Construct appropriate confidence intervals for testing the two hypotheses (see page 361) and show how we could have reached the same conclusions from these intervals.

Answer: The test of proportion was two-sided, so we construct a 95% confidence interval for the true proportion:

$\hat{p} \pm 1.96 SE(\hat{p}) = 0.29 \pm 1.96 \times \sqrt{\dfrac{(0.29)(0.71)}{200}} = (0.227, 0.353)$. Since 0.20 is not a plausible value, we reject the null hypothesis.

The test of means is one-sided so we construct a one-sided 95% confidence interval, using the t critical value of 1.672:

$$\left(\bar{y} - t^*SE(\bar{y}), \infty\right) = \left(26.05 - 1.672 \times \frac{10.2}{\sqrt{58}}, \infty\right) = (23.81, \infty)$$

We can see that the hypothesized value of $24.85 is in this interval, so we fail to reject the null hypothesis.

13.10   Two Types of Errors

Nobody's perfect. Even with lots of evidence, we can still make the wrong decision. In fact, when we perform a hypothesis test, we can make mistakes in *two* ways:

I. The null hypothesis is true, but we mistakenly reject it.
II. The null hypothesis is false, but we fail to reject it.

These two types of errors are known as **Type I** and **Type II errors**, respectively. One way to keep the names straight is to remember that we start by assuming the null hypothesis is true, so a Type I error is the first kind of error we could make.

In medical disease testing, the null hypothesis is usually the assumption that a person is healthy. The alternative is that he or she has the disease we're testing for. So a Type I error is a *false positive*—a healthy person is diagnosed with the disease. A Type II error, in which an infected person is diagnosed as disease free, is a *false negative*. These errors have other names, depending on the particular discipline and context.

Which type of error is more serious depends on the situation. In a jury trial, a Type I error occurs if the jury convicts an innocent person. A Type II error occurs if the jury fails to convict a guilty person. Which seems more serious? In medical diagnosis, a false negative could mean that a sick patient goes untreated. A false positive might mean that the person receives unnecessary treatments or even surgery.

In business planning, a false positive result could mean that money will be invested in a project that turns out not to be profitable. A false negative result might mean that money won't be invested in a project that would have been profitable. Which error is worse, the lost investment or the lost opportunity? The answer always depends on the situation, the cost, and your point of view.

Here's an illustration of the situations:

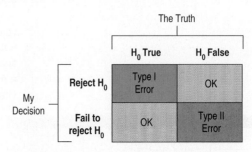

Figure 13.8   The two types of errors occur on the diagonal where the truth and decision don't match. Remember that we *start* by assuming H_0 to be true, so an error made (rejecting it) when H_0 is true is called a Type I error. A Type II error is made when H_0 is false (and we fail to reject it).

How often will a Type I error occur? It happens when the null hypothesis is true but we've had the bad luck to draw an unusual sample. To reject H_0, the P-value must fall below α. When H_0 is true, that happens *exactly* with probability α. So when you choose level α, you're setting the probability of a Type I error to α.

What if H_0 is not true? Then we can't possibly make a Type I error. You can't get a false positive from a sick person. A Type I error can happen only when H_0 is true.

When H_0 is false and we reject it, we have done the right thing. A test's ability to detect a false hypothesis is called the **power** of the test. In a jury trial, power is a measure of the ability of the criminal justice system to convict people who are guilty. We'll have a lot more to say about power soon.

When H_0 is false but we fail to reject it, we have made a Type II error. We assign the letter β to the probability of this mistake. What's the value of β? That's harder to assess than α because we don't know what the value of the parameter really is. When H_0 is true, it specifies a single parameter value. But when H_0 is false, we don't have a specific one; we have many possible values. We can compute the probability β for any parameter value in H_A, but the choice of which one to pick is not always clear.

One way to focus our attention is by thinking about the *effect size*. That is, ask: "How big a difference would matter?" Suppose a charity wants to test whether placing personalized address labels in the envelope along with a request for a donation increases the response rate above the baseline of 5%. If the minimum response that would pay for the address labels is 6%, they would calculate β for the alternative $p = 0.06$.

Of course, we could reduce β for *all* alternative parameter values by increasing α. By making it easier to reject the null, we'd be more likely to reject it whether it's true or not. The only way to reduce *both* types of error is to collect more evidence or, in statistical terms, to collect more data. Otherwise, we just wind up trading off one kind of error against the other. Whenever you design a survey or experiment, it's a good idea to calculate β (for a reasonable α level). Use a parameter value in the alternative that corresponds to an effect size that you want to be able to detect. Too often, studies fail because their sample sizes are too small to detect the change they are looking for.

Notation Alert!

In Statistics, α is the probability of a Type I error and β is the probability of a Type II error.

The *Probabilities* of a Type II Error

The null hypothesis specifies a single value for the parameter. So it's easy to calculate the probability of a Type I error. But the alternative gives a whole range of possible values, and we may want to find a β for several of them.

Sample Size and Power

We have seen ways to find a sample size by specifying the margin of error. Choosing the sample size to achieve a specified β (for a particular alternative value) is sometimes more appropriate, but the calculation is more complex and lies beyond the scope of this book.

8 Remember our bank that's sending out DVDs to try to get customers to make payments on delinquent loans? It is looking for evidence that the costlier DVD strategy produces a higher success rate than the letters it has been sending. Explain what a Type I error is in this context and what the consequences would be to the bank.

9 What's a Type II error in the bank experiment context and what would the consequences be?

10 If the DVD strategy *really* works well—actually getting 60% of the people to pay off their balances—would the power of the test be higher or lower compared to a 32% payoff rate? Explain briefly.

11 Recall the mileage program test market of Question 7. Suppose after completing the hotel partnership, the mean revenue per customer is $40.26. Has a Type I or Type II error been made? Explain.

For Example   Type I and Type II errors

Question: Suppose that a year later, a full accounting of all the transactions (see page 368) finds that 26.5% of visits resulted in sales with an average purchase amount of $26.25. Have any errors been made?

Answer: We rejected the null hypothesis that $p = 0.20$ and in fact $p = 0.265$, so we did not make a Type I error (the only error we could have made when rejecting the null hypothesis). For the mean amount, we failed to reject that the mean had increased from $24.85 but actually the mean had increased to $26.25 so we made a Type II error—we failed to reject the null hypothesis when it was false.

13.11   Power

Power and Effect Size

When planning a study, it's wise to think about the size of the effect we're looking for. We've called the effect size the difference between the null hypothesis and the observed statistic. In planning, it's the difference between the null hypothesis and a particular alternative we're interested in. It's easier to see a larger effect size, so the power of the study will increase with the effect size.

Once the study has been completed we'll base our business decision on the observed effect size, the difference between the null hypothesis and the observed value.

Remember, we can never prove a null hypothesis true. We can only fail to reject it. But when we fail to reject a null hypothesis, it's natural to wonder whether we looked hard enough. Might the null hypothesis actually be false and our test too weak to tell?

When the null hypothesis actually *is* false, we hope our test is strong enough to reject it. We'd like to know how likely we are to succeed. The power of the test gives us a way to think about that. The power of a test is the probability that it correctly rejects a false null hypothesis. When the power is high, we can be confident that we've looked hard enough. We know that β is the probability that a test *fails* to reject a false null hypothesis, so the power of the test is the complement, $1 - \beta$. We might have just written $1 - \beta$, but power is such an important concept that it gets its own name.

Whenever a study fails to reject its null hypothesis, the test's power comes into question. Was the sample size big enough to detect an effect had there been one? Might we have missed an effect large enough to be interesting just because we failed to gather sufficient data or because there was too much variability in the data we could gather? Might the problem be that the experiment simply lacked adequate power to detect their ability?

When we calculate power, we base our calculation on the smallest effect that might influence our business decision. The value of the power depends on how large this effect is. For proportions, that effect size is $p - p_0$; for means it's $\mu - \mu_0$. The power depends directly on the effect size. It's easier to see larger effects, so the further p_0 is from p (or μ is from μ_0), the greater the power.

How can we decide what power we need? Choice of power is more a financial or scientific decision than a statistical one because to calculate the power, we need to specify the alternative parameter value we're interested in. In other words, power is calculated for a particular effect size, and it changes depending on the size of the effect we want to detect.

Graph It!

It makes intuitive sense that the larger the effect size, the easier it should be to see it. Obtaining a larger sample size decreases the probability of a Type II error, so it increases the power. It also makes sense that the more we're willing to accept a Type I error, the less likely we will be to make a Type II error.

Figure 13.9 may help you visualize the relationships among these concepts. Although we'll use proportions to show the ideas, a similar picture and similar statements also hold true for means. Suppose we are testing $H_0: p = p_0$ against the alternative $H_A: p > p_0$. We'll reject the null if the observed proportion, $\hat{p}$, is big enough. By *big enough*, we mean $\hat{p} > p^*$ for some critical value p^* (shown as the red region in the right tail of the upper curve). The upper model shows a picture of the sampling distribution model for the proportion when the null hypothesis is true. If the null were true, then this would be a picture of that truth. We'd make a Type I error whenever the sample gave us $\hat{p} > p^*$ because we would reject the (true) null hypothesis. Unusual samples like that would happen only with probability α.

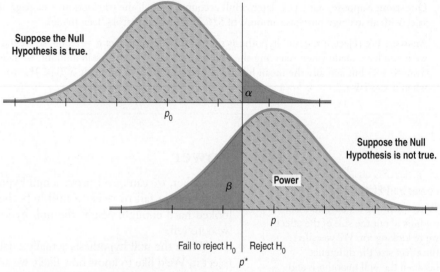

Figure 13.9   The power of a test is the probability that it rejects a false null hypothesis. The upper figure shows the null hypothesis model. We'd reject the null in a one-sided test if we observed a value in the red region to the right of the critical value, p^*. The lower figure shows the model if we assume that the true value is p. If the true value of p is greater than p_0, then we're more likely to observe a value that exceeds the critical value and make the correct decision to reject the null hypothesis. The power of the test is the green region on the right of the lower figure. Of course, even drawing samples whose observed proportions are distributed around p, we'll sometimes get a value in the red region on the left and make a Type II error of failing to reject the null.

In reality, though, the null hypothesis is rarely *exactly* true. The lower probability model supposes that H_0 is not true. In particular, it supposes that the true value is p, not p_0. It shows a distribution of possible observed $\hat{p}$ values around this true value. Because of sampling variability, sometimes $\hat{p} < p^*$ and we fail to reject the (false) null hypothesis. Then we'd make a Type II error. The area under the curve to the left of p^* in the bottom model represents how often this happens. The probability is β. In this picture, β is less than half, so most of the time we *do* make the right decision. The *power* of the test—the probability that we make the right decision—is shown as the region to the right of p^*. It's $1 - \beta$.

We calculate p^* based on the upper model because p^* depends only on the null model and the alpha level. No matter what the true proportion, p^* doesn't change. After all, we don't *know* the truth, so we can't use it to determine the critical value. But we always reject H_0 when $\hat{p} > p^*$.

How often we reject H_0 when it's *false* depends on the effect size. We can see from the picture that if the true proportion were further from the hypothesized value, the bottom curve would shift to the right, making the power greater.

We can see several important relationships from this figure:

- Power $= 1 - \beta$.
- Moving the critical value (p^* in the case of proportions) to the right, reduces α, the probability of a Type I error, but increases β, the probability of a Type II error. It correspondingly reduces the power.
- The larger the true effect size, the real difference between the hypothesized value and the true population value, the smaller the chance of making a Type II error and the greater the power of the test.

If the two proportions (or means) are very far apart, the two models will barely overlap, and we would not be likely to make any Type II errors at all—but then, we are unlikely to really need a formal hypothesis testing procedure to see such an obvious difference.

Reducing Both Type I and Type II Errors

Figure 13.9 seems to show that if we reduce Type I errors, we automatically must increase Type II errors. But there is a way to reduce both. Can you think of it?

If we can make both curves narrower, as shown in Figure 13.10, then the probability of both Type I errors and Type II errors will decrease, and the power of the test will increase.

How can we do that? The only way is to reduce the standard deviations by increasing the sample size. (Remember, these are pictures of sampling distribution models, not of data.) Increasing the sample size works regardless of the true population parameters. But recall the curse of diminishing returns. The standard deviation of the sampling distribution model decreases only as the *square root* of the sample size, so to halve the standard deviations, we must *quadruple* the sample size.

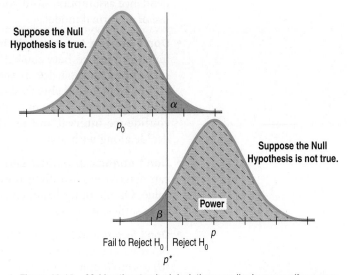

Figure 13.10 Making the standard deviations smaller increases the power without changing the alpha level or the corresponding z-critical value. The proportions are just as far apart as in Figure 13.9, but the error rates are reduced. A similar picture could be drawn for testing means.

What Can Go Wrong?

- **Don't confuse proportions and means.** When you treat your data as categorical, counting successes and summarizing with a sample proportion, make inferences using the Normal model methods. When you treat your data as quantitative, summarizing with a sample mean, make your inferences using Student's t methods.

- **Don't base your null hypotheses on what you see in the data.** You are not allowed to look at the data first and then adjust your null hypothesis so that it will be rejected. If the mean spending of 10 customers turns out to be $\bar{y} = \$24.94$ with a standard error of $5, don't form a null hypothesis just big enough so you'll be able to reject it like $H_0: \mu = \$26.97$. The null hypothesis should not be based on the data you collect. It should describe the "nothing interesting" or "nothing has changed" situation.

- **Don't base your alternative hypothesis on the data either.** You should always think about the situation you are investigating and base your alternative hypothesis on that. Are you interested only in knowing whether something has *increased*? Then write a one-tail (upper tail) alternative. Or would you be equally interested in a change in either direction? Then you want a two-tailed alternative. You should decide whether to do a one- or two-tailed test based on what results would be of interest to you, not on what you might see in the data.

- **Don't make your null hypothesis what you want to show to be true.** Remember, the null hypothesis is the status quo, the nothing-is-strange-here position a skeptic would take. You wonder whether the data cast doubt on that. You can reject the null hypothesis, but you can never "accept" or "prove" the null.

- **Don't forget to check the conditions.** The reasoning of inference depends on randomization. No amount of care in calculating a test result can save you from a biased sample. The probabilities you compute depend on the independence assumption. And your sample must be large enough to justify your use of a Normal model.

- **Don't believe too strongly in arbitrary alpha levels.** There's not really much difference between a P-value of 0.051 and a P-value of 0.049, but sometimes it's regarded as the difference between night (having to retain H_0) and day (being able to shout to the world that your results are "statistically significant"). It may just be better to report the P-value and a confidence interval and let the world (perhaps your manager or client) decide along with you.

- **Don't confuse practical and statistical significance.** A large sample size can make it easy to discern even a trivial change from the null hypothesis value. On the other hand, you could miss an important difference if your test lacks sufficient power.

- **Don't forget that in spite of all your care, you might make a wrong decision.** No one can ever reduce the probability of a Type I error (α) or of a Type II error (β) to zero (but increasing the sample size helps).

Ethics in Action

Many retailers have recognized the importance of staying connected to their in-store customers via the Internet. Retailers not only use the Internet to inform their customers about specials and promotions, but also to send them e-coupons redeemable for discounts. Shellie Cooper, longtime owner of a small organic food store, specializes in locally produced organic foods and products. Over the years Shellie's customer base has been quite stable, consisting mainly of health-conscious individuals who tend not to be very price sensitive, opting to pay higher prices for better-quality local, organic products. However, faced with increasing competition from grocery chains offering more organic choices, Shellie is now thinking of offering coupons. She needs to decide between the newspaper and the Internet. She recently read that the percentage of consumers who use printable Internet coupons is on the rise but, at 15%, is much less than the 40% who clip and redeem newspaper coupons. Nonetheless, she's interested in learning more about the Internet and sets up a meeting with Jack Kasor, a Web consultant. She discovers that for an initial investment and continuing monthly fee, Jack would design Shellie's website, host it on his server, and broadcast e-coupons to her customers at regular intervals. While she was concerned about the difference in redemption rates for e-coupons vs. newspaper coupons, Jack assured her that e-coupon redemptions are continuing to rise and that she should expect between 15% and 40% of her customers to redeem them. Shellie agreed to give it a try. After the first six months, Jack informed Shellie that the proportion of her customers who redeemed e-coupons was significantly greater than 15%. He determined this by selecting several broadcasts at random and found the number redeemed (483) out of the total number sent (3000). Shellie thought that this was positive and made up her mind to continue the use of e-coupons.

ETHICAL ISSUE *Statistical vs. practical significance. While it is true that the percentage of Shellie's customers redeeming e-coupons is significantly greater than 15% statistically, in fact, the percentage is just over 16%. This difference amounts to about 33 customers more than 15%, which may not be of practical significance to Shellie (related to Item A, ASA Ethical Guidelines). Mentioning a range of 15% to 40% may mislead Shellie into expecting a value somewhere in the middle.*

ETHICAL SOLUTION *Jack should report the difference between the observed value and the hypothesized value to Shellie, especially since there are costs associated with continuing e-coupons. Perhaps he should recommend that she reconsider using the newspaper.*

What Have We Learned?

Learning Objectives

- Know how to formulate a null and alternative hypothesis for a question of interest.
  - The null hypothesis specifies a parameter and a (null) value for that parameter.
  - The alternative hypothesis specifies a range of plausible values should we fail to reject the null.

- Be able to perform a hypothesis test for a proportion.
  - The null hypothesis has the form $H_0: p = p_0$.
  - We estimate the standard deviation of the sampling distribution of the sample proportion by assuming that the null hypothesis is true:

$$SD(\hat{p}) = \sqrt{\frac{p_0 q_0}{n}}$$

  - We refer the statistic $z = \dfrac{\hat{p} - p_0}{SD(\hat{p})}$ to the standard Normal model.

(continued)

- Be able to perform a hypothesis test for a mean.
  - The null hypothesis has the form $H_0: \mu = \mu_0$.
  - We estimate the standard error of the sampling distribution as

  $$SE(\bar{y}) = \frac{s}{\sqrt{n}}.$$

  - We refer the test statistic $t = \dfrac{\bar{y} - \mu_0}{SE(\bar{y})}$ to the Student's t distribution with $n - 1$ degrees of freedom.

- Understand P-values.
  - A P-value is the estimated probability of observing a statistic value at least as far from the (null) hypothesized value as the one we have actually observed.
  - A small P-value indicates that the statistic we have observed would be unlikely were the null hypothesis true. That leads us to doubt the null.
  - A large P-value just tells us that we have insufficient evidence to doubt the null hypothesis. In particular, it does not prove the null to be true.

- Know the reasoning of hypothesis testing.
  - State the **hypotheses**.
  - Determine (and check assumptions for) the sampling distribution **model**.
  - Calculate the test statistic—the **mechanics**.
  - State your **conclusions and decisions**.

- Be able to decide on a two-sided or one-sided alternative hypothesis, and justify your decision.

- Compare P-values to a pre-determined α-level to decide whether to reject the null hypothesis.

- Know the value of estimating and reporting the effect size.
  - A test may be statistically significant, but practically meaningless if the estimated effect is of trivial importance.

- Be aware of the risks of making errors when testing hypotheses.
  - A Type I error can occur when rejecting a null hypothesis if that hypothesis is, in fact, true.
  - A Type II error can occur when failing to reject a null hypothesis if that hypothesis is, in fact, false.

- Understand the concept of the power of a test.
  - We are particularly concerned with power when we fail to reject a null hypothesis.
  - The power of a test reports, for a specified effect size, the probability that the test would reject a false null hypothesis.
  - Remember that increasing the sample size will generally improve the power of any test.

Terms

Alpha level   The threshold P-value that determines when we reject a null hypothesis. Using an alpha level of α, if we observe a statistic whose P-value based on the null hypothesis is less than α, we reject that null hypothesis.

Alternative hypothesis   The hypothesis that proposes what we should conclude if we find the null hypothesis to be unlikely.

Critical value	The value in the sampling distribution model of the statistic whose P-value is equal to the alpha level. Any statistic value further from the null hypothesis value than the critical value will have a smaller P-value than α and will lead to rejecting the null hypothesis. The critical value is often denoted with an asterisk, as z^* or t^*, for example.
Effect size	The difference between the null hypothesis and an observed (or proposed) value.
Null hypothesis	The claim being assessed in a hypothesis test. Usually, the null hypothesis is a statement of "no change from the traditional value," "no effect," "no difference," or "no relationship." For a claim to be a testable null hypothesis, it must specify a value for some population parameter that can form the basis for assuming a sampling distribution for a test statistic.
One-sample t-test for the mean	The one-sample-t-test for the mean tests the hypothesis $H_0\colon \mu = \mu_0$ using the statistic $t_{n-1} = \dfrac{\bar{y} - \mu_0}{SE(\bar{y})}$, where $SE(\bar{y}) = \dfrac{s}{\sqrt{n}}$.
One-proportion z-test	A test of the null hypothesis that the proportion of a single sample equals a specified value ($H_0\colon p = p_0$) by comparing the statistic $z = \dfrac{\hat{p} - p_0}{SD(\hat{p})}$ to a standard Normal model.
One-sided alternative	An alternative hypothesis is one-sided (e.g., $H_A\colon p > p_0$ or $H_A\colon p < p_0$) when we are interested in deviations in *only one* direction away from the hypothesized parameter value.
P-value	The probability of observing a value for a test statistic at least as far from the hypothesized value as the statistic value actually observed if the null hypothesis is true. A small P-value indicates that the observation obtained is improbable given the null hypothesis and thus provides evidence against the null hypothesis.
Power	The probability that a hypothesis test will correctly reject a false null hypothesis. To find the power of a test, we must specify a particular alternative parameter value as the "true" value. For any specific value in the alternative, the power is $1 - \beta$.
Significance level	Another term for the alpha level, used most often in a phrase such as "at the 5% significance level."
Two-sided alternative	An alternative hypothesis is two-sided (e.g. $H_A\colon \mu \neq \mu_0$) when we are interested in deviations in *either* direction away from the hypothesized parameter value.
Type I error	The error of rejecting a null hypothesis when in fact it is true (also called a "false positive"). The probability of a Type I error is α.
Type II error	The error of failing to reject a null hypothesis when in fact it is false (also called a "false negative"). The probability of a Type II error is commonly denoted β and depends on the effect size.

Technology Help: Hypothesis Tests

Although most statistics packages offer ways to test hypotheses for means, many do not offer special commands for proportions. You can always report the "success" and "failure" status for each case as 1 or 0, but packages are likely to then treat the proportion as a mean (which, of course, it is for such data) and base inferences on Student's *t*. Some programs will allow you to specify the standard deviation and some will accept summary proportions.

In some programs you can reconstruct the original values from the summary proportions. Because of the different approaches, if you use a computer package, you may notice slight discrepancies between your answers and the answers in the back of the book, but they're not important.

Reports about hypothesis tests generated by technologies don't follow a standard form. Most will name the test and provide the test statistic value, its standard deviation, and the P-value. But these elements may not be labeled clearly. For example, the expressions "Prob > $|z|$" or "Prob > $|t|$" denote a two-sided alternative by use of the absolute value symbol. They give the probability of being as far or farther on either side from the null hypothesis than the observed data. That is a fancy (and not very clear) way of saying P-value. In some packages, you can specify that the test be one-sided. Others might report three P-values, covering the ground for both one-sided tests and the two-sided test.

Sometimes a confidence interval and hypothesis test are automatically given together. The confidence interval ought to be for the corresponding confidence level: $1 - \alpha$.

Often, the standard deviation of the statistic is called the "standard error," and usually that's appropriate because we've had to estimate its value from the data. That's not the case for

(continued)

proportions, however: We get the standard deviation for a proportion from the null hypothesis value. Nevertheless, you may see the standard deviation called a "standard error" even for tests with proportions.

It's common for statistics packages and calculators to report more digits of "precision" than could possibly have been found from the data. You can safely ignore them. Round values such as the standard deviation to one digit more than the number of digits reported in your data.

Here are the kind of results you might see in typical computer output.

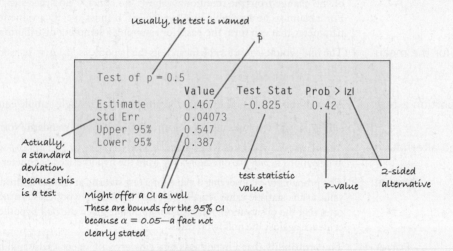

Statistics packages offer convenient ways to make histograms of the data. That means you have no excuse for skipping the check that the data are nearly Normal.

Here's what the package output might look like for a hypothesis test of a mean (although no package we know gives the results in exactly this form):

The package computes the sample mean and sample standard deviation of the variable and finds the P-value from the *t*-distribution based on the appropriate number of degrees of freedom. All modern statistics packages report P-values. The package may also provide additional information such as the sample mean, sample standard deviation, *t*-statistic value, and degrees of freedom. These are useful for interpreting the resulting P-value and telling the difference between a meaningful result

and one that is merely statistically significant. Statistics packages that report the estimated standard deviation of the sampling distribution usually label it "standard error" or "SE."

Inference results are also sometimes reported in a table. You may have to read carefully to find the values you need. Often, test results and the corresponding confidence interval bounds are given together. And often you must read carefully to find the alternative hypotheses. On the next page is an example of that kind of output.

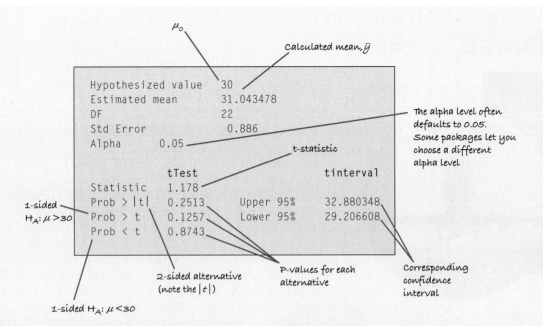

μ_0 — Hypothesized value

Calculated mean, $\bar{y}$ — Estimated mean

```
Hypothesized value    30
Estimated mean        31.043478
DF                    22
Std Error             0.886
Alpha    0.05

                  tTest                    tinterval
Statistic         1.178
Prob > |t|        0.2513     Upper 95%     32.880348
Prob > t          0.1257     Lower 95%     29.206608
Prob < t          0.8743
```

The alpha level often defaults to 0.05. Some packages let you choose a different alpha level

t-statistic

1-sided $H_A: \mu > 30$

1-sided $H_A: \mu < 30$

2-sided alternative (note the $|t|$)

P-values for each alternative

Corresponding confidence interval

The commands to do inference for means on common statistics programs and calculators are not always obvious. (By contrast, the resulting output is usually clearly labeled and easy to read.) The guides for each program can help you start navigating.

EXCEL   XLSTAT

Inference methods in Excel can be handled by simply typing in the appropriate formulas. For example, suppose your data consist of 100 observations in cells A1:A100 and you want to do a two sided test of the null hypothesis that the mean is equal to 0.5.

- In cell B2, enter "=AVERAGE(A1:A100)" to compute the sample mean.
- In cell B3, enter "=STDEV(A1:A100)/SQRT(100)" to compute the standard error of the mean.
- In cell B4, enter "=(B2 – 0.5)/B3" to compute the test statistic.
- In cell B5, enter "=TDIST(B4,99,2)" to compute the p-value.

JMP

For a **categorical** variable that holds category labels, the **Distribution** platform includes tests and intervals of proportions.

- For summarized data, put the category names in one variable and the frequencies in an adjacent variable.
- Designate the frequency column to have the role of frequency.
- Then use the Distribution platform test.

For a hypothesis test,

- Click the red triangle next to the variable's name and choose **Test Mean** from the menu. Then fill in the resulting dialog.

Comments

JMP uses slightly different methods for proportion inferences than those discussed in this text. Your answers are likely to be slightly different. "Moment" is a fancy statistical term for means, standard deviations, and other related statistics.

MINITAB

For proportions, choose **Basic Statistics** from the **Stat** menu.

- Choose **1 Proportion** from the Basic Statistics submenu.
- If the data are category names in a variable, assign the variable from the variable list box to the **Samples in columns** box.
- If you have summarized data, click the **Summarized Data** button and fill in the number of trials and the number of successes.
- Click the **Options** button and specify the remaining details.
- If you have a large sample, check **Use test and interval based on Normal distribution.**
- Click the **OK** button.

For means,

- From the **Stat** menu, choose the **Basic Statistics** submenu.
- From that menu, choose **1-sample t**. . . . Then fill in the dialog.

Comments

When working from a variable that names categories, MINITAB treats the last category as the "success" category. You can specify how the categories should be ordered.

SPSS

SPSS does not find hypothesis tests for proportions. For means,

- From the **Analyze** menu, choose the **Compare Means** submenu.
- From that, choose **One-Sample t-test** command.

Comments

The commands suggest neither a single mean nor an interval. But the results provide both a test and an interval.

Brief CASE

Metal Production

Ingots are huge pieces of metal, often weighing in excess of 20,000 pounds, made in a giant mold. They must be cast in one large piece for use in fabricating large structural parts for cars and planes. If they crack while being made, the crack may propagate into the zone required for the part, compromising its integrity. Airplane manufacturers insist that metal for their planes be defect-free, so the ingot must be made over if any cracking is detected.

Even though the metal from the cracked ingot is recycled, the scrap cost runs into the tens of thousands of dollars. Metal manufacturers would like to avoid cracking if at all possible, but only about 75% of the ingots have been free of cracks. The data from 5000 ingots produced after some changes were made to the process are found in the file **Ingots**. The variable *Crack* indicates whether a crack was found (1) or not (0). The variable *Impurities* shows the amount (in ppm) of impurities found in a sample from each ingot. Select a random sample of 100 ingots and test the claim that the cracking rate has decreased from 25%. Also test whether the mean impurity level for the cracked ingots is greater than 500 ppm. Find confidence intervals for the cracking rate and for the mean impurity of the cracked ingots. Now select a random sample of 1000 ingots and test the claims and find the confidence intervals again. Prepare a short report about your findings including any differences you see between the conclusions from two samples.

Loyalty Program

A marketing manager has sent out 10,000 mail pieces to a random sample of customers to test a new web-based loyalty program and see its impact on customer spending. The customers either received nothing (No Offer), a free companion airline ticket (Free Flight), or free flight insurance on their next flight (Free Insurance). The analyst who designed the test used a stratified random sample to ensure that all the market segments are represented, but the manager has several concerns. First, she worries that the *Travel* segment (which comprises 25% of all customers) is underrepresented (variable *Segment*). In addition, she worries that fewer than 1/3 of the customers in that segment were held out as controls and received no offer. Finally, she wants to know the mean amount of customers in the *Travel* segment for each of the three possible *Offers*. Using the data found in **Loyalty_Program** write a short report to the manager testing the appropriate hypotheses and summarizing your findings. Include in your report 95% confidence intervals for the proportion of customers who responded to the offer by signing up for the loyalty program in the various segments. (The variable *Response* indicates a 1 for responders and 0 for nonresponders.)

Exercises

SECTION 13.1

1. For each of the following situations, define the parameter (proportion or mean) and write the null and alternative hypotheses in terms of parameter values. Example: We want to know if the proportion of up days in the stock market is 50%. Answer: Let p = the proportion of up days. H_0: $p = 0.5$ vs. H_A: $p \neq 0.5$.

a) A casino wants to know if their slot machine really delivers the 1 in 100 win rate that it claims.

b) Last year customers spent an average of $35.32 per visit to the company's website. Based on a random sample of purchases this year, the company wants to know if the mean this year has changed.

c) A pharmaceutical company wonders if their new drug has a cure rate different from the 30% reported by the placebo.

d) A bank wants to know if the percentage of customers using their website has changed from the 40% that used it before their system crashed last week.

2. As in Exercise 1, for each of the following situations, define the parameter and write the null and alternative hypotheses in terms of parameter values.

a) Seat-belt compliance in Massachusetts was 65% in 2008. The state wants to know if it has changed.

b) Last year, a survey found that 45% of the employees were willing to pay for on-site day care. The company wants to know if that has changed.

c) Regular card customers have a default rate of 6.7%. A credit card bank wants to know if that rate is different for their Gold card customers.

d) Regular card customers have been with the company for an average of 17.3 months. The credit card bank wants to know if their Gold card customers have been with the company on average the same amount of time.

SECTION 13.3

3. Which of the following are true? If false, explain briefly.

a) A very high P-value is strong evidence that the null hypothesis is false.

b) A very low P-value proves that the null hypothesis is false.

c) A high P-value shows that the null hypothesis is true.

d) A P-value below 0.05 is always considered sufficient evidence to reject a null hypothesis.

4. Which of the following are true? If false, explain briefly.

a) A very low P-value provides evidence against the null hypothesis.

b) A high P-value is strong evidence in favor of the null hypothesis.

c) A P-value above 0.10 shows that the null hypothesis is true.

d) If the null hypothesis is true, you can't get a P-value below 0.01.

SECTION 13.4

5. A consulting firm had predicted that 35% of the employees at a large firm would take advantage of a new company Credit Union, but management is skeptical. They doubt the rate is that high. A survey of 300 employees shows that 138 of them are currently taking advantage of the Credit Union. From the sample proportion

a) Find the standard deviation of the sample proportion based on the null hypothesis.

b) Find the z-statistic.

c) Does the z-statistic seem like a particularly large or small value?

6. A survey of 100 CEOs finds that 60 think the economy will improve next year. Is there evidence that the rate is higher among all CEOs than the 55% reported by the public at large?

a) Find the standard deviation of the sample proportion based on the null hypothesis.

b) Find the z-statistic.

c) Does the z-statistic seem like a particularly large or small value?

SECTION 13.5

7. For each of the following, write out the null and alternative hypothesis, being sure to state whether it is one-sided or two-sided.

a) A company reports that last year 40% of their reports in accounting were on time. From a random sample this year, they want to know if that proportion has changed.

b) A company wants to know if average click-through rates (in minutes) are shorter than the 5.4 minutes that customers spent on their old website before making a purchase.

c) Last year, 42% of the employees enrolled in at least one wellness class at the company's site. Using a survey, they want to know if a greater percentage is planning to take a wellness class this year.

d) A political candidate wants to know from recent polls if she's going to garner a majority of votes in next week's election.

8. For each of the following, write out the alternative hypothesis, being sure to indicate whether it is one-sided or two-sided.

a) *Consumer Reports* discovered that 20% of a certain computer model had warranty problems over the first three months. From a random sample, the manufacturer wants to know if a new model has improved that rate.

b) The last time a philanthropic agency requested donations, 4.75% of people responded. From a recent pilot mailing, they wonder if that rate has increased.

c) The average age of a customer of a clothing store is 35.2 years. The company wants to know if customers who use their website are younger on average.

d) A student wants to know if other students on her campus prefer Coke or Pepsi.

SECTION 13.6

9. The owner of the store from Chapter 12, Exercise 1, wants to know if the mean age of all customers is 25 years old.

a) What is the null hypothesis?

b) Is the alternative one- or two-sided?

c) What is the value of the test statistic?

d) What is the P-value of the test statistic?

e) What do you conclude at $\alpha = 0.05$?

f) Given the confidence interval you found in Chapter 12, Exercise 7, what might you have concluded? Explain.

10. The analyst in Chapter 12, Exercise 2, wants to know if the mean purchase amount of all transactions is at least $40.

a) What is the null hypothesis?
b) Is the alternative one- or two-sided?
c) What is the value of the test statistic?
d) What is the P-value of the test statistic?
e) What do you conclude at $\alpha = 0.05$?
f) Given the confidence interval you found in Chapter 12, Exercise 8, what might you have concluded? Explain.

SECTION 13.7

11. Which of the following statements are true? If false, explain briefly.

a) Using an alpha level of 0.05, a P-value of 0.04 results in rejecting the null hypothesis.
b) The alpha level depends on the sample size.
c) With an alpha level of 0.01, a P-value of 0.10 results in rejecting the null hypothesis.
d) Using an alpha level of 0.05, a P-value of 0.06 means the null hypothesis is true.

12. Which of the following statements are true? If false, explain briefly.

a) It is better to use an alpha level of 0.05 than an alpha level of 0.01.
b) If we use an alpha level of 0.01, then a P-value of 0.001 is statistically significant.
c) If we use an alpha level of 0.01, then we reject the null hypothesis if the P-value is 0.001.
d) If the P-value is 0.01, we reject the null hypothesis for any alpha level greater than 0.01.

SECTION 13.8

13. For each of the following situations, find the critical value(s) for z or t.

a) $H_0: p = 0.5$ vs. $H_A: p \neq 0.5$ at $\alpha = 0.05$.
b) $H_0: p = 0.4$ vs. $H_A: p > 0.4$ at $\alpha = 0.05$.
c) $H_0: \mu = 10$ vs. $H_A: \mu \neq 10$ at $\alpha = 0.05$; $n = 36$.
d) $H_0: p = 0.5$ vs. $H_A: p > 0.5$ at $\alpha = 0.01$; $n = 345$.
e) $H_0: \mu = 20$ vs. $H_A: \mu < 20$ at $\alpha = 0.01$; $n = 1000$.

14. For each of the following situations, find the critical value for z or t.

a) $H_0: \mu = 105$ vs. $H_A: \mu \neq 105$ at $\alpha = 0.05$; $n = 61$.
b) $H_0: p = 0.05$ vs. $H_A: p > 0.05$ at $\alpha = 0.05$.
c) $H_0: p = 0.6$ vs. $H_A: p \neq 0.6$ at $\alpha = 0.01$.
d) $H_0: p = 0.5$ vs. $H_A: p < 0.5$ at $\alpha = 0.01$; $n = 500$.
e) $H_0: p = 0.2$ vs. $H_A: p < 0.2$ at $\alpha = 0.01$.

SECTION 13.9

15. Suppose that you are testing the hypotheses $H_0: p = 0.20$ vs. $H_A: p \neq 0.20$. A sample of size 250 results in a sample proportion of 0.25.

a) Construct a 95% confidence interval for p.

b) Based on the confidence interval, at $\alpha = .05$ can you reject H_0? Explain.
c) What is the difference between the standard error and standard deviation of the sample proportion?
d) Which is used in computing the confidence interval?

16. Suppose that you are testing the hypotheses $H_0: p = 0.40$ vs. $H_A: p > 0.40$. A sample of size 200 results in a sample proportion of 0.55.

a) Construct a 90% confidence interval for p.
b) Based on the confidence interval, at $\alpha = .05$ can you reject H_0? Explain.
c) What is the difference between the standard error and standard deviation of the sample proportion?
d) Which is used in computing the confidence interval?

17. Suppose that you are testing the hypotheses $H_0: \mu = 16$ vs. $H_A: \mu < 16$. A sample of size 25 results in a sample mean of 16.5 and a standard deviation of 2.0.

a) What is the standard error of the mean?
b) What is the critical value of t^* for a 90% confidence interval?
c) Construct a 90% confidence interval for μ.
d) Based on the confidence interval, at $\alpha = 0.05$ can you reject H_0? Explain.

18. Suppose that you are testing the hypotheses $H_0: \mu = 80$ vs. $H_A: \mu \neq 80$. A sample of size 61 results in a sample mean of 75 and a standard deviation of 1.5.

a) What is the standard error of the mean?
b) What is the critical value of t^* for a 95% confidence interval?
c) Construct a 95% confidence interval for μ.
d) Based on the confidence interval, at $\alpha = 0.05$ can you reject H_0? Explain.

SECTION 13.10

19. For each of the following situations, state whether a Type I, a Type II, or neither error has been made. Explain briefly.

a) A bank wants to know if the enrollment on their website is above 30% based on a small sample of customers. They test $H_0: p = 0.3$ vs. $H_A: p > 0.3$ and reject the null hypothesis. Later they find out that actually 28% of all customers enrolled.
b) A student tests 100 students to determine whether other students on her campus prefer Coke or Pepsi and finds no evidence that preference for Coke is not 0.5. Later, a marketing company tests all students on campus and finds no difference.
c) A human resource analyst wants to know if the applicants this year score, on average, higher on their placement exam than the 52.5 points the candidates averaged last year. She samples 50 recent tests and finds the average to be 54.1 points. She fails to reject the null hypothesis that the mean is 52.5 points. At the end of the year, they find that the candidates this year had a mean of 55.3 points.

d) A pharmaceutical company tests whether a drug lifts the headache relief rate from the 25% achieved by the placebo. They fail to reject the null hypothesis because the P-value is 0.465. Further testing shows that the drug actually relieves headaches in 38% of people.

20. For each of the following situations, state whether a Type I, a Type II, or neither error has been made.

a) A test of $H_0: \mu = 25$ vs. $H_A: \mu > 25$ rejects the null hypothesis. Later it is discovered that $\mu = 24.9$.

b) A test of $H_0: p = 0.8$ vs. $H_A: p < 0.8$ fails to reject the null hypothesis. Later it is discovered that $p = 0.9$.

c) A test of $H_0: p = 0.5$ vs. $H_A: p \neq 0.5$ rejects the null hypothesis. Later is it discovered that $p = 0.65$.

d) A test of $H_0: p = 0.7$ vs. $H_A: p < 0.7$ fails to reject the null hypothesis. Later is it discovered that $p = 0.6$.

CHAPTER EXERCISES

21. Hypotheses. Write the null and alternative hypotheses to test each of the following situations.

a) An online clothing company is concerned about the timeliness of their deliveries. The VP of Operations and Marketing recently stated that she wanted the percentage of products delivered on time to be greater than 90%, and she wants to know if the company has succeeded.

b) A realty company recently announced that the proportion of houses taking more than three months to sell is now greater than 50%.

c) A financial firm's accounting reports that after improvements in their system, they now have an error rate below 2%.

22. More hypotheses. Write the null and alternative hypotheses to test each of the following situations.

a) A business magazine article reports that, in 1990, 35% of CEOs had an MBA degree. Has the percentage changed?

b) Recently, 20% of cars of a certain model have needed costly transmission work after being driven between 50,000 and 100,000 miles. The car manufacturer hopes that the redesign of a transmission component has solved this problem.

c) A market researcher for a cola company decides to field test a new flavor soft drink, planning to market it only if he is sure that over 60% of the people like the flavor.

23. Deliveries. The clothing company in Exercise 21a looks at a sample of delivery reports. They test the hypothesis that 90% of the deliveries are on time against the alternative that greater than 90% are on time and find a P-value of 0.22. Which of these conclusions is appropriate?

a) There's a 22% chance that 90% of the deliveries are on time.

b) There's a 78% chance that 90% of the deliveries are on time.

c) There's a 22% chance that the sample they drew shows the correct percentage of on-time deliveries

d) There's a 22% chance that natural sampling variation could produce a sample with an observed proportion of on-time deliveries such as the one they obtained if, in fact, 90% of deliveries are on time.

24. House sales. The realty company in Exercise 21b looks at a recent sample of houses that have sold. On testing the null hypothesis that 50% of the houses take more than three months to sell against the hypothesis that more than 50% of the houses take more than three months to sell, they find a P-value of 0.034. Which of these conclusions is appropriate?

a) There's a 3.4% chance that 50% of the houses take more than three months to sell.

b) If 50% of the houses take more than three months to sell, there's a 3.4% chance that a random sample would produce a sample proportion as high as the one they obtained.

c) There's a 3.4% chance that the null hypothesis is correct.

d) There's a 96.6% chance that 50% of the houses take more than three months to sell.

25. P-value. Have harsher penalties and ad campaigns increased seat-belt use among drivers and passengers? Observations of commuter traffic have failed to find evidence of a significant change compared with three years ago. Explain what the study's P-value of 0.17 means in this context.

26. Another P-value. A company developing scanners to search for hidden weapons at airports has concluded that a new device is significantly better than the current scanner. The company made this decision based on a P-value of 0.03. Explain the meaning of the P-value in this context.

27. Ad campaign. An information technology analyst believes that they are losing customers on their website who find the checkout and purchase system too complicated. She adds a one-click feature to the website, to make it easier but finds that only about 10% of the customers are using it. She decides to launch an ad awareness campaign to tell customers about the new feature in the hope of increasing the percentage. She doesn't see much of a difference, so she hires a consultant to help her. The consultant selects a random sample of recent purchases, tests the hypothesis that the ads produced no change against the alternative that the percent who use the one-click feature is now greater than 10%, and finds a P-value of 0.22. Which conclusion is appropriate? Explain.

a) There's a 22% chance that the ads worked.

b) There's a 78% chance that the ads worked.

c) There's a 22% chance that the null hypothesis is true.

d) There's a 22% chance that natural sampling variation could produce poll results like these if the use of the one-click feature has increased.

e) There's a 22% chance that natural sampling variation could produce poll results like these if there's really no change in website use.

28. Mutual funds. A mutual fund manager claims that at least 70% of the stocks she selects will increase in price over the next year. We examined a sample of 200 of her selections over the past three years. Our P-value turns out to be 0.03. Test an appropriate hypothesis. Which conclusion is appropriate? Explain.

a) There's a 3% chance that the fund manager is correct.
b) There's a 97% chance that the fund manager is correct.
c) There's a 3% chance that a random sample could produce the results we observed, so it's reasonable to conclude that the fund manager is correct.
d) There's a 3% chance that a random sample could produce the results we observed if $p = .7$, so it's reasonable to conclude that the fund manager is not correct.
e) There's a 3% chance that the null hypothesis is correct.

29. Product effectiveness. A pharmaceutical company's old antacid formula provided relief for 70% of the people who used it. The company tests a new formula to see if it is better and gets a P-value of 0.27. Is it reasonable to conclude that the new formula and the old one are equally effective? Explain.

30. Car sales. A German automobile company is counting on selling more cars to the younger market segment—drivers under the age of 20. The company's market researchers survey to investigate whether or not the proportion of today's high school seniors who own their own cars is higher than it was a decade ago. They find a P-value of 0.017. Is it reasonable to conclude that more high school seniors have cars? Explain.

31. False claims? A candy company claims that in a large bag of holiday M&M's® half the candies are red and half the candies are green. You pick candies at random from a bag and discover that of the first 20 you eat, 12 are red.

a) If it were true that half are red and half are green, what is the probability you would have found that at least 12 out of 20 were red?
b) Do you think that half of the M&M's® candies in the bag are really red? Explain.

32. Scratch off. A retail company offers a "scratch off" promotion. Upon entering the store, you are given a card. When you pay, you may scratch off the coating. The company advertises that half the cards are winners and have immediate cash-back savings of $5 (the others offer $1 off any future purchase of coffee in the cafe). You aren't sure the percentage is really 50% winners.

a) The first time you shop there, you get the coffee coupon. You try again and again get the coffee coupon. Do two failures in a row convince you that the true fraction of winners isn't 50%? Explain.
b) You try a third time. You get coffee again! What's the probability of not getting a cash savings three times in a row if half the cards really do offer cash savings?

c) Would three losses in a row convince you that the store is cheating?
d) How many times in a row would you have to get the coffee coupon instead of cash savings to be pretty sure that the company isn't living up to its advertised percentage of winners? Justify your answer by calculating a probability and explaining what it means.

33. E-commerce. A market researcher at a major clothing company that has traditionally relied on catalog mail-order sales decides to investigate whether the amount of online sales has changed. She compares the mean monthly online sales of the past several months with a historical figure for mean monthly sales for online purchases. She gets a P-value of 0.01. Explain in this context what the 1% represents.

34. Performance standards. The United States Golf Association (USGA) sets performance standards for golf balls. For example, the initial velocity of the ball may not exceed 250 feet per second when measured by an apparatus approved by the USGA. Suppose a manufacturer introduces a new kind of ball and provides a randomly selected sample of balls for testing. Based on the mean speed in the sample, the USGA comes up with a P-value of 0.34. Explain in this context what the 34% represents.

35. Social Security payments. The average monthly Social Security benefit for widows and widowers in 2005 was $967 (*Statistical Abstract of the United States*, U.S. Census Bureau). Payments vary from region to region. In Texas, the 2005 average monthly benefit equaled $940. A social welfare advocate from a rural Texas county believed the 2005 average Social Security benefit for widows and widowers differed significantly from the overall Texas average. To test this assumption, the advocate randomly sampled 100 widow/widower monthly benefit payments and found the sample mean = $915 with a standard deviation of $90.

a) Find and interpret a 95% confidence interval estimate for the sample mean 2005 Social Security widow/widower benefit in the rural Texas county.
b) In a hypothesis test performed to determine whether the county's average payment was different, the test was rejected with a P-value = .007 (using the same sample results shown above and a level of significance = .05). Explain how the confidence interval constructed in part a is consistent with the hypothesis test results. Your discussion should include level of confidence, the interval bounds, the P-value, and the hypothesis test decision.

36. Social Security payments, part 2. In a neighboring county, a newspaper wrote that widow/widower Social Security benefits were significantly lower in the county than in all the other counties of Texas. The newspaper reported that the average 2005 monthly benefit for widows and widowers was $900, based on a random sample of 100 with standard deviation of $90.

a) Find and interpret a 95% confidence interval estimate for the sample mean Social Security widow/widower benefit in this county.

b) Is this county's average widow/widower benefit different from the Texas average of $940?

37. Spike poll. In August 2004, *Time* magazine reported the results of a random telephone poll commissioned by the Spike network. Of the 1302 men who responded, only 39 said that their most important measure of success was their work.

a) Estimate the percentage of all American males who measure success primarily from their work. Use a 98% confidence interval. Don't forget to check the conditions first.

b) Some believe that few contemporary men judge their success primarily by their work. Suppose we wished to conduct a hypothesis test to see if the fraction has fallen below the 5% mark. What does your confidence interval indicate? Explain.

c) What is the significance level of this test? Explain.

38. Stocks. A young investor in the stock market is concerned that investing in the stock market is actually gambling, since the chance of the stock market going up on any given day is 50%. She decides to track her favorite stock for 250 days and finds that on 140 days the stock was "up."

a) Find a 95% confidence interval for the proportion of days the stock is "up." Don't forget to check the conditions first.

b) Does your confidence interval provide any evidence that the market is not random? Explain.

c) What is the significance level of this test? Explain.

39. Economy. In 2008, a Gallup Poll asked 2336 U.S. adults, aged 18 or over, how they rated economic conditions. In a poll conducted from January 27 through February 1, 2008, 24% rated the economy as Excellent/Good. A recent media outlet claimed that the percentage of Americans who felt the economy was in Excellent/Good shape was, in fact, 28%. Does the Gallup Poll support this claim?

a) Test the appropriate hypothesis. Find a 95% confidence interval for the sample proportion of U.S. adults who rated the economy as Excellent/Good. Check conditions.

b) Does your confidence interval provide evidence to support the claim?

c) What is the significance level of the test in b? Explain.

40. Economy, part 2. The same Gallup Poll data from Exercise 39 also reported that 33% of those surveyed rated the economy as Poor. The same media outlet claimed the true proportion to be 30%. Does the Gallup Poll support this claim?

a) Test the appropriate hypothesis. Find a 95% confidence interval for the sample proportion of U.S. adults who rated the economy as Poor. Check conditions.

b) Does your confidence interval provide evidence to support the claim.

c) What is the significance level of the test in b? Explain.

41. Convenient alpha. An enthusiastic junior executive has run a test of his new marketing program. He reports that it resulted in a "significant" increase in sales. A footnote on his report explains that he used an alpha level of 7.2% for his test. Presumably, he performed a hypothesis test against the null hypothesis of no change in sales.

a) If instead he had used an alpha level of 5%, is it more or less likely that he would have rejected his null hypothesis? Explain.

b) If he chose the alpha level 7.2% so that he could claim statistical significance, explain why this is not an ethical use of statistics.

42. Safety. The manufacturer of a new sleeping pill suspects that it may increase the risk of sleepwalking, which could be dangerous. A test of the drug fails to reject the null hypothesis of increased sleepwalking when tested at alpha = .01.

a) If the test had been performed at alpha = .05, would the test have been more or less likely to reject the null hypothesis of no increase in sleepwalking?

b) Which alpha level do you think the company should use? Why?

43. Product testing. Since many people have trouble programming their VCRs, an electronics company has developed what it hopes will be easier instructions. The goal is to have at least 96% of customers succeed at being able to program their VCRs. The company tests the new system on 200 people, 188 of whom were successful. Is this strong evidence that the new system fails to meet the company's goal? A student's test of this hypothesis is shown here. How many mistakes can you find?

$H_0: \hat{p} = 0.96$
$H_A: \hat{p} \neq 0.96$
SRS, $0.96(200) > 10$
$\frac{188}{200} = 0.94; SD(\hat{p}) = \sqrt{\frac{(0.94)(0.06)}{200}} = 0.017$
$z = \frac{0.96 - 0.94}{0.017} = 1.18$
$P = P(z > 1.18) = 0.12$

There is strong evidence that the new system does not work.

44. Marketing. In November 2001, the *Ag Globe Trotter* newsletter reported that 90% of adults drink milk. A regional farmers organization planning a new marketing campaign across its multicounty area polls a random sample of 750 adults living there. In this sample, 657 people said that they drink milk. Do these responses provide strong evidence that the 90% figure is not accurate for this region? Correct the mistakes you find in the following student's attempt to test an appropriate hypothesis.

$H_0: \hat{p} = 0.9$

$H_A: \hat{p} < 0.9$

SRS, $750 > 10$

$\dfrac{657}{750} = 0.876; SD(\hat{p}) = \sqrt{\dfrac{(0.88)(0.12)}{750}} = 0.012$

$z = \dfrac{0.876 - 0.94}{0.012} = -2$

$P = P(z > -2) = 0.977$

There is more than a 97% chance that the stated percentage is correct for this region.

45. E-commerce, part 2. The average age of online consumers a few years ago was 23.3 years. As older individuals gain confidence with the Internet, it is believed that the average age has increased. We would like to test this belief.

a) Write appropriate hypotheses.
b) We plan to test the null hypothesis by selecting a random sample of 40 individuals who have made an online purchase during 2007. Do you think the necessary assumptions for inference are satisfied? Explain.
c) The online shoppers in our sample had an average age of 24.2 years, with a standard deviation of 5.3 years. What's the P-value for this result?
d) Explain (in context) what this P-value means.
e) What's your conclusion?

46. Fuel economy. A company with a large fleet of cars hopes to keep gasoline costs down and sets a goal of attaining a fleet average of at least 26 miles per gallon. To see if the goal is being met, they check the gasoline usage for 50 company trips chosen at random, finding a mean of 25.02 mpg and a standard deviation of 4.83 mpg. Is this strong evidence that they have failed to attain their fuel economy goal?

a) Write appropriate hypotheses.
b) Are the necessary assumptions to perform inference satisfied?
c) Test the hypothesis and find the P-value.
d) Explain what the P-value means in this context.
e) State an appropriate conclusion.

47. Pricing for competitiveness. SLIX wax is developing a new high performance fluorocarbon wax for cross country ski racing designed to be used under a wide variety of conditions. In order to justify the price marketing wants, the wax needs to be very fast. Specifically, the mean time to finish their standard test course should be less than 55 seconds for the former Olympic champion who is now their consultant. To test it, the consultant will ski the course 8 times.

a) The champion's times are 56.3, 65.9, 50.5, 52.4, 46.5, 57.8, 52.2, and 43.2 seconds to complete the test course. Should they market the wax? Explain.
b) Suppose they decide not to market the wax after the test, but it turns out that the wax really does lower the champion's

average time to less than 55 seconds. What kind of error have they made? Explain the impact to the company of such an error.

48. Popcorn. Pop's Popcorn, Inc. needs to determine the optimum power and time settings for their new licorice-flavored microwave popcorn. They want to find a combination of power and time that delivers high-quality popcorn with less than 10% of the kernels left unpopped, on average—a value that their market research says is demanded by their customers. Their research department experiments with several settings and determines that power 9 at 4 minutes is optimum. Their tests confirm that this setting meets the less than 10% requirement. They change the instructions on the box and promote a new money back guarantee of less than 10% unpopped kernels.

a) If, in fact, the setting results in more than 10% kernels unpopped, what kind of error have they made? What will the consequence be for the company?
b) To reduce the risk of making an error, the president (Pop himself) tells them to test 8 more bags of popcorn (selected at random) at the specified setting. They find the following percentage of unpopped kernels: 7, 13.2, 10, 6, 7.8, 2.8, 2.2, 5.2. Does this provide evidence that the setting meets their goal of less than than 10% unpopped? Explain.

49. Environment. In the 1980s, it was generally believed that congenital abnormalities affected about 5% of the nation's children. Some people believe that the increase in the number of chemicals in the environment has led to an increase in the incidence of abnormalities. A recent study examined 384 children and found that 46 of them showed signs of an abnormality. Is this strong evidence that the risk has increased? (We consider a P-value of around 5% to represent reasonable evidence.)

a) Write appropriate hypotheses.
b) Check the necessary assumptions.
c) Perform the mechanics of the test. What is the P-value?
d) Explain carefully what the P-value means in this context.
e) What's your conclusion?
f) Do environmental chemicals cause congenital abnormalities?

50. Billing company. A billing company that collects bills for doctors' offices in the area is concerned that the percentage of bills being paid by Medicare has risen. Historically, that percentage has been 31%. An examination of 8368 recent bills reveals that 32% of these bills are being paid by Medicare. Is this evidence of a change in the percent of bills being paid by Medicare?

a) Write appropriate hypotheses.
b) Check the assumptions and conditions.
c) Perform the test and find the P-value.
d) State your conclusion.
e) Do you think this difference is meaningful? Explain.

51. Education. The National Center for Education Statistics monitors many aspects of elementary and secondary education nationwide. Their 1996 numbers are often used as a baseline to assess changes. In 1996, 34% of students had not been absent from school even once during the previous month. In the 2000 survey, responses from 8302 students showed that this figure had slipped to 33%. Officials would be concerned if student attendance were declining. Do these figures give evidence of a decrease in student attendance?

a) Write appropriate hypotheses.
b) Check the assumptions and conditions.
c) Perform the test and find the P-value.
d) State your conclusion.
e) Do you think this difference is meaningful? Explain.

52. Consumer confidence. At various times in 2007, when asked if economic conditions were getter better or worse, consistently more than 20% of U.S. adults said better. On January 19–20, 2008, when Gallup polled 2590 U.S. adults, only 13% said that conditions were getting better. Do these responses give evidence that consumer confidence has decreased from the 2007 level?

a) Write appropriate hypotheses.
b) Check the assumptions and conditions.
c) Perform the test and find the P-value.
d) State your conclusion.
e) Do you think this difference is meaningful? Explain.

53. Retirement. A survey of 1000 workers indicated that approximately 520 have invested in an individual retirement account. National data suggests that 44% of workers invest in individual retirement accounts.

a) Create a 95% confidence interval for the proportion of workers who have invested in individual retirement accounts based on the survey.
b) Does this provide evidence of a change in behavior among workers? Using your confidence interval, test an appropriate hypothesis and state your conclusion.

54. Customer satisfaction. A company hopes to improve customer satisfaction, setting as a goal no more than 5% negative comments. A random survey of 350 customers found only 10 with complaints.

a) Create a 95% confidence interval for the true level of dissatisfaction among customers.
b) Does this provide evidence that the company has reached its goal? Using your confidence interval, test an appropriate hypothesis and state your conclusion.

55. Maintenance costs. A limousine company is concerned with increasing costs of maintaining their fleet of 150 cars. After testing, the company found that the emissions systems of 7 out of the 22 cars they tested failed to meet pollution control guidelines. They had forecasted costs assuming that a total of 30 cars would need updating to meet the latest guidelines. Is this strong evidence that more than 20% of the fleet might be out of compliance? Test an appropriate hypothesis and state your conclusion. Be sure the appropriate assumptions and conditions are satisfied before you proceed.

56. Damaged goods. An appliance manufacturer stockpiles washers and dryers in a large warehouse for shipment to retail stores. Sometimes in handling them the appliances get damaged. Even though the damage may be minor, the company must sell those machines at drastically reduced prices. The company goal is to keep the proportion of damaged machines below 2%. One day an inspector randomly checks 60 washers and finds that 5 of them have scratches or dents. Is this strong evidence that the warehouse is failing to meet the company goal? Test an appropriate hypothesis and state your conclusion. Be sure the appropriate assumptions and conditions are satisfied before you proceed.

57. WebZine. A magazine called *WebZine* is considering the launch of an online edition. The magazine plans to go ahead only if it's convinced that more than 25% of current readers would subscribe. The magazine contacts a simple random sample of 500 current subscribers, and 137 of those surveyed expressed interest. What should the magazine do? Test an appropriate hypothesis and state your conclusion. Be sure the appropriate assumptions and conditions are satisfied before you proceed.

58. Truth in advertising. A garden center wants to store leftover packets of vegetable seeds for sale the following spring, but the center is concerned that the seeds may not germinate at the same rate a year later. The manager finds a packet of last year's green bean seeds and plants them as a test. Although the packet claims a germination rate of 92%, only 171 of 200 test seeds sprout. Is this evidence that the seeds have lost viability during a year in storage? Test an appropriate hypothesis and state your conclusion. Be sure the appropriate assumptions and conditions are satisfied before you proceed.

59. Women executives. A company is criticized because only 13 of 43 people in executive-level positions are women. The company explains that although this proportion is lower than it might wish, it's not surprising given that only 40% of their employees are women. What do you think? Test an appropriate hypothesis and state your conclusion. Be sure the appropriate assumptions and conditions are satisfied before you proceed.

60. Jury. Census data for a certain county shows that 19% of the adult residents are Hispanic. Suppose 72 people are called for jury duty, and only 9 of them are Hispanic. Does this apparent underrepresentation of Hispanics call into question the fairness of the jury selection system? Explain.

61. Nonprofit. A nonprofit company concerned with the school dropout rates in the United States has designed a tutoring program aimed at students between 16 to 18 years old. The National Center for Education Statistics reported that the high school dropout rate in the United States for the year 2000 was 10.9%. One school district, who adopted the use of the nonprofit's tutoring program and whose dropout rate has always been very close to the national average, reported in 2004 that 175 of their 1782 students dropped out. Is their experience evidence that the tutoring program has been effective? Explain.

62. Real estate. A national real estate magazine advertised that 15% of first-time home buyers had a family income below $40,000. A national real estate firm believes this percentage is too low and samples 100 of its records. The firm finds that 25 of its first-time home buyers did have a family income below $40,000. Does the sample suggest that the proportion of first-time home buyers with an income less than $40,000 is more than 15%? Comment and write up your own conclusions based on an appropriate confidence interval as well as a hypothesis test. Include any assumptions you made about the data.

63. Public relations. According to the U.S. Department of Transportation (DOT), passengers filed more complaints about airline service in 2007 than in 2006. One airline's public relations department says that their airline rarely loses luggage. Furthermore, it claims that when it does, 90% of the time the bags are recovered and delivered within 24 hours. A consumer group surveys a large number of air travelers and finds that 103 of 122 people who lost luggage were reunited with their missing items within 24 hours. Does this cast doubt on the airline's claim? Explain.

64. TV ads. A start-up company is about to market a new computer printer. It decides to gamble by running commercials during the Super Bowl. The company hopes that name recognition will be worth the high cost of the ads. The goal of the company is that over 40% of the public recognize its brand name and associate it with computer equipment. The day after the game, a pollster contacts 420 randomly chosen adults and finds that 181 of them know that this company manufactures printers. Would you recommend that the company continue to advertise during the Super Bowl? Explain.

65. Business ethics. One study reports that 30% of newly hired MBAs are confronted with unethical business practices during their first year of employment. One business school dean wondered if her MBA graduates had similar experiences. She surveyed recent graduates from her school's MBA program to find that 27% of the 120 graduates from the previous year claim to have encountered unethical business practices in the workplace. Can she conclude that her graduates' experiences are different?

66. Stocks, part 2. A young investor believes that he can beat the market by picking stocks that will increase in value. Assume that on average 50% of the stocks selected by a portfolio manager will increase over 12 months. Of the 25 stocks that the young investor bought over the last 12 months, 14 have increased. Can he claim that he is better at predicting increases than the typical portfolio manager?

67. U.S. politics. The national elections in 2008 apparently drew more interest and debate among voters than prior U.S. elections. A national sample of 2020 U.S. adults, aged 18 and older, surveyed over the telephone (using both landlines and cell phones) between January 30 and February 2 in 2008 by Gallup revealed that 71% reported that they had given "quite a lot" of thought to the upcoming election for president. Is there any evidence that the percentage has changed from the historically reported benchmark of 58% during the same timeframe in 2004?

a) Find the z-score of the observed proportion.
b) Compare the z-score to the critical value for a 0.1% significance level using a two-sided alternative.
c) Explain your conclusion.

68. iPod reliability. MacInTouch reported that several versions of the iPod reported failure rates of 20% or more. From a customer survey, the color iPod, first released in 2004, showed 64 failures out of 517. Is there any evidence that the failure rate for this model may be lower than the 20% rate of previous models?

a) Find the z-score of the observed proportion.
b) Compare the z-score to the critical value for a 0.1% significance level using a one-sided alternative.
c) Explain your conclusion.

69. Testing cars. A clean air standard requires that vehicle exhaust emissions not exceed specified limits for various pollutants. Many states require that cars be tested annually to be sure they meet these standards. Suppose state regulators double-check a random sample of cars that a suspect repair shop has certified as okay. They will revoke the shop's license if they find significant evidence that the shop is certifying vehicles that do not meet standards.

a) In this context, what is a Type I error?
b) In this context, what is a Type II error?
c) Which type of error would the shop's owner consider more serious?
d) Which type of error might environmentalists consider more serious?

70. Quality control. Production managers on an assembly line must monitor the output to be sure that the level of defective products remains small. They periodically inspect a random sample of the items produced. If they find a significant increase in the proportion of items that must be rejected, they will halt the assembly process until the problem can be identified and repaired.

a) Write null and alternative hypotheses for this problem.
b) What is the Type I and Type II error in this context?
c) Which type of error would the factory owner consider more serious?
d) Which type of error might customers consider more serious?

71. Testing cars, again. As in Exercise 69, state regulators are checking up on repair shops to see if they are certifying vehicles that do not meet pollution standards.

a) In this context, what is meant by the power of the test the regulators are conducting?
b) Will the power be greater if they test 20 or 40 cars? Why?
c) Will the power be greater if they use a 5% or a 10% level of significance? Why?
d) Will the power be greater if the repair shop's inspectors are only a little out of compliance or a lot? Why?

72. Quality control, part 2. Consider again the task of the quality control inspectors in Exercise 70.

a) In this context, what is meant by the power of the test the inspectors conduct?
b) They are currently testing 5 items each hour. Someone has proposed they test 10 items each hour instead. What are the advantages and disadvantages of such a change?
c) Their test currently uses a 5% level of significance. What are the advantages and disadvantages of changing to a significance level of 1%?
d) Suppose that as a day passes one of the machines on the assembly line produces more and more items that are defective. How will this affect the power of the test?

73. Statistics software. A Statistics professor has observed that for several years about 13% of the students who initially enroll in his Introductory Statistics course withdraw before the end of the semester. A salesperson suggests that he try a statistics software package that gets students more involved with computers, predicting that it will cut the dropout rate. The software is expensive, and the salesperson offers to let the professor use it for a semester to see if the dropout rate goes down significantly. The professor will have to pay for the software only if he chooses to continue using it.

a) Is this a one-tailed or two-tailed test? Explain.
b) Write the null and alternative hypotheses.
c) In this context, explain what would happen if the professor makes a Type I error.
d) In this context, explain what would happen if the professor makes a Type II error.
e) What is meant by the power of this test?

74. Radio ads. A company is willing to renew its advertising contract with a local radio station only if the station can prove that more than 20% of the residents of the city have heard the ad and recognize the company's product. The radio station conducts a random phone survey of 400 people.

a) What are the hypotheses?
b) The station plans to conduct this test using a 10% level of significance, but the company wants the significance level lowered to 5%. Why?
c) What is meant by the power of this test?
d) For which level of significance will the power of this test be higher? Why?
e) They finally agree to use $\alpha = 0.05$, but the company proposes that the station call 600 people instead of the 400 initially proposed. Will that make the risk of Type II error higher or lower? Explain.

75. Statistics software, part 2. Initially, 203 students signed up for the Statistics course in Exercise 73. They used the software suggested by the salesperson, and only 11 dropped out of the course.

a) Should the professor spend the money for this software? Support your recommendation with an appropriate test.
b) Explain what your P-value means in this context.

76. Radio ads, part 2. The company in Exercise 74 contacts 600 people selected at random, and 133 can remember the ad.

a) Should the company renew the contract? Support your recommendation with an appropriate test.
b) Explain carefully what your P-value means in this context.

77. Investment. Investment style plays a role in constructing a mutual fund. Many individual stocks can be grouped into two distinct groups: Growth and Value. A Growth stock is one with high earning potential and often pays little or no dividends to shareholders. Conversely, Value stocks are commonly viewed as steady or more conservative stocks with a lower earning potential. A family is trying to decide what type of funds to invest in. An independent advisor claims that Value Mutual Funds provided an annualized return of greater than 8% over a recent 5-year period. Below are the summary statistics for the 5-year return for a random sample of such Value funds.

Variable	N	Mean	SE Mean	StDev
5 yr Return	35	8.418	0.493	2.916

Minimum	Q1	Median	Q3	Maximum
2.190	6.040	7.980	10.840	14.320

Test the hypothesis that the mean 5-year return for value funds is greater than 8%, assuming a significance level of 5%. What does this evidence say about the portfolio managers' claim that the annualized 5-year return was greater than 8%? State your conclusion.

78. Manufacturing. A tire manufacturer is considering a newly designed tread pattern for its all-weather tires. Tests have indicated that these tires will provide better gas mileage and longer tread life. The last remaining test is for braking effectiveness. The company hopes the tire will allow a car traveling at 60 mph to come to a complete stop within an average of 125 feet after the brakes are applied. They will adopt the new tread pattern unless there is strong evidence that the tires do not meet this objective. The distances (in feet) for 10 stops on a test track were 129, 128, 130, 132, 135, 123, 102, 125, 128, and 130. Should the company adopt the new tread pattern? Test an appropriate hypothesis and state your conclusion. Explain how you dealt with the outlier and why you made the recommendation you did.

79. Collections. Credit card companies lose money on cardholders who fail to pay their minimum payments. They use a variety of methods to encourage their delinquent cardholders to pay their credit card balances, such as letters, phone calls and eventually the hiring of a collection agency. To justify the cost of using the collection agency, the agency must collect an average of at least $200 per customer. After a trial period during which the agency attempted to collect from a random sample of 100 delinquent cardholders, the 90% confidence interval on the mean collected amount was reported as ($190.25, $250.75). Given this, what recommendation(s) would you make to the credit card company about using the collection agency?

80. Free gift. A philanthropic organization sends out "free gifts" to people on their mailing list in the hope that the receiver will respond by sending back a donation. Typical gifts are mailing labels, greeting cards, or post cards. They want to test out a new gift that costs $0.50 per item to produce and mail. They mail it to a "small" sample of 2000 customers and find a 90% confidence interval of the mean donation to be ($0.489, $0.879). As a consultant, what recommendation(s) would you make to the organization about using this gift?

81. Collections, part 2. The owner of the collection agency in Exercise 79 is quite certain that they can collect more than $200 per customer on average. He urges that the credit card company run a larger trial. Do you think a larger trial might help the company make a better decision? Explain.

82. Free gift, part 2. The philanthropic organization of Exercise 80 decided to go ahead with the new gift. In mailings to 98,000 prospects, the new mailing yielded an average of $0.78. If they had decided based on their initial trial *not* to use this gift, what kind of error would they have made? What aspects of their initial trial might have suggested to you (as their consultant) that a larger trial would be worthwhile?

Just Checking Answers

1 You can't conclude that the null hypothesis is true. You can conclude only that the experiment was unable to reject the null hypothesis. They were unable, on the basis of 12 patients, to show that aspirin was effective.

2 The null hypothesis is $H_0: p = 0.75$.

3 With a P-value of 0.0001, this is very strong evidence against the null hypothesis. We can reject H_0 and conclude that the improved version of the drug gives relief to a higher proportion of patients.

4 The parameter of interest is the proportion, p, of all delinquent customers who will pay their bills. $H_0: p = 0.30$ and $H_A: p > 0.30$.

5 At $\alpha = 0.05$, you can't reject the null hypothesis because 0.30 is contained in the 90% confidence interval—it's plausible that sending the DVDs is no more effective than sending letters.

6 The confidence interval is from 29% to 45%. The DVD strategy is more expensive and may not be worth it. We can't distinguish the success rate from 30% given the results of this experiment, but 45% would represent a large improvement. The bank should consider another trial, increasing the sample size to get a narrower confidence interval.

7 The confidence interval suggests that at 95% confidence, even in the worst case, the arrangement will generate a mean profit of at least $37.95 − $35.00 = $2.95 per customer. On the other hand, it may generate a profit as high as $55.05 − $35.00 = $20.05 per customer.

8 A Type I error would mean deciding that the DVD success rate is higher than 30%, when it isn't. The bank would adopt a more expensive method for collecting payments that's no better than its original, less expensive strategy.

9 A Type II error would mean deciding that there's not enough evidence to say the DVD strategy works when in fact it does. The bank would fail to discover an effective method for increasing revenue from delinquent accounts.

10 Higher; the larger the effect size, the greater the power. It's easier to detect an improvement to a 60% success rate than to a 32% rate.

11 No error has been made. The confidence interval from the test market suggested that the mean revenue was above the $35 cost per customer, and in fact that's what happened.

Comparing Two Groups

14

Visa Global Organization

Today, more than one billion people and 24 million merchants use the Visa card in 170 countries worldwide. But back in the early 1950s when the idea of cashless transactions first took hold, only Diners Club and some retailers, notably oil companies, issued charge cards. The vast majority of purchases were made by cash or personal check. Bank of America pioneered its BankAmericard program in Fresno, California, in 1958, and American Express issued the first plastic card in 1959. The idea of a credit "card" really gained momentum a decade later when a group of banks formed a joint venture to create a centralized system of payment. National BankAmericard, Inc. (NBI) took ownership of the credit card system in 1970 and for simplicity and marketability changed its name to Visa in 1976. (The name Visa is pronounced nearly the same way in every language.) That year, Visa processed 679,000 transactions—a volume that is processed on average every four minutes today. As technology changed, so did the credit card industry. By 1986, cardholders were able to use their Visa cards to get cash from ATMs.

Now a global organization, Visa is divided into several regional entities including: Visa Asia Pacific, Visa Canada, Visa Europe, and Visa USA. The Visa system is currently able to handle a load of about 6800 transactions a second. That was nearly surpassed on December 23, 2005, during the height of the Christmas season, when Visa processed an average of nearly 6400 transaction messages every second.[1]

Champion vs. Challenger

In the credit card industry, among others, it is common to run experiments testing a new idea (offer, promotion, incentive, etc.) against the current, or "business as usual" practice. The new idea is called the *challenger* and the previously tested offer is the *champion*. Much of the direct mail that you see consists of marketing experiments whose goal is to determine if the challenger's performance warrants dethroning the champion and using the challenger going forward.

For credit card promotions, the typical measure used to judge success of the challenger offer is the amount charged on the card, from which the credit card issuer derives revenue. Because customers vary so much in the amount they charge on their cards, it is more efficient to compare the amount charged after the offer is received to the amount charged before the offer is received. This change in amount charged is often referred to as the *spend lift*. The word *lift* connotes the optimistic sentiment that the offer will increase the amount a customer charges, although it can be and often is negative.

The credit card business can be extremely profitable. In 2003, card issuers worldwide made $2.5 billion per month before taxes. The typical American household has eight cards with a total of $7500 debt on them, usually at rates much higher than conventional loans and mortgages. Not surprisingly, the credit card business is also intensely competitive. Rival banks and lending agencies are constantly trying to create new products and offers to win new customers, keep current customers, and provide incentives for current customers to charge more on their cards.

Are some credit card promotions more effective than others? For example, do customers spend more using their credit card if they know they will be given "double miles" or "double coupons" toward flights, hotel stays, or store purchases? To answer questions such as this, credit card issuers often perform experiments on a sample of customers, making them an *offer* of an incentive, while other customers receive no offer. Promotions cost the company money, so the company needs to estimate the size of any increased revenue to judge whether it is sufficient to cover their expenses. By comparing the performance of the two offers on the sample, they can decide whether the new offer would provide enough potential profit if they were to "roll it out" and offer it to their entire customer base.

Experiments that compare two groups are common throughout both science and industry. Other applications include comparing the effects of a new drug with the traditional therapy, the fuel efficiency of two car engine designs, or the sales of new products on two different customer segments. Usually the experiment is carried out on a subset of the population, often a much smaller subset. Using statistics, we can make statements about whether the means of the two groups differ in the population at large, and how large that difference might be.

14.1   Comparing Two Means

The natural display for comparing the means of two groups is side-by-side boxplots (see Figure 14.1). For the credit card promotion, the company judges performance by comparing the *mean* spend lift (the change in spending from before receiving the promotion to after receiving it) for the two samples. If the difference in spend

---

[1]Source: Visa International Service Association, www.corporate.visa.com.

Figure 14.1 Side-by-side boxplots show a small increase in spending for the group that received the promotion.

lift between the group that received the promotion and the group that didn't is high enough, this will be viewed as evidence that the promotion worked. Looking at the two boxplots, it's not obvious that there's much of a difference. Can we conclude that the slight increase seen for those who received the promotion is more than just random fluctuation? We'll need statistical inference.

For two groups, the statistic of interest is the difference in the observed means of the offer and no offer groups: $\bar{y}_{Offer} - \bar{y}_{No\ Offer}$. We've offered the promotion to a random sample of cardholders and used another sample of cardholders, who got no special offer, as a control group. We know what happened in our samples, but what we'd really like to know is the difference of the means in the population at large:

$$\mu_{Offer} - \mu_{No\ Offer}$$

We compare two means in much the same way as we compared a single mean to a hypothesized value. But now the population model parameter of interest is the *difference* between the means. In our example, it's the true difference between the mean spend lift for customers offered the promotion and for customers for whom no offer was made. We estimate the difference with $\bar{y}_{Offer} - \bar{y}_{No\ Offer}$. How can we tell if a difference we observe in the sample means indicates a real difference in the underlying population means? We'll need to know the sampling distribution model and standard deviation of the difference. Once we know those, we can build a confidence interval and test a hypothesis just as we did for a single mean.

We have data on 500 randomly selected customers who were offered the promotion and another randomly selected 500 who were not. It's easy to find the mean and standard deviation of the spend lift for each of these groups. From these, we can find the standard deviations of the means, but that's not what we want. We need the standard deviation of the *difference* in their means. For that, we can use a simple rule: *If the sample means come from independent samples, the variance of their sum or difference is the sum of their variances.*

> ## Variances Add for Sums *and* Differences
>
> At first, it may seem that this can't be true for differences as well as for sums. Here's some intuition about why variation increases even when we subtract two random quantities. Grab a full box of cereal. The label claims that it contains 16 ounces of cereal. We know that's not exact. There's a random quantity of cereal in the box with a mean (presumably) of 16 ounces and some variation from box to box. Now pour a 2-ounce serving of cereal into a bowl. Of course, your serving isn't exactly 2 ounces. There's some variation there, too. How much cereal would you guess was left in the box? Can you guess as accurately as you could for the full box? The mean should be 14 ounces. But does the amount left in the box have *less* variation than it did before you poured your serving? Almost certainly not! *After* you pour your bowl, the amount of cereal in the box is still a random quantity (with a smaller mean than before), but you've made it *more variable* because of the uncertainty in the amount you poured. However, notice that we don't add the *standard deviations* of these two random quantities. As we'll see, it's the *variance* of the amount of cereal left in the box that's the sum of the two variances.

As long as the two groups are independent, we find the standard deviation of the *difference* between the two sample means by adding their variances and then taking the square root:

$$SD(\bar{y}_1 - \bar{y}_2) = \sqrt{Var(\bar{y}_1) + Var(\bar{y}_2)}$$
$$= \sqrt{\left(\frac{\sigma_1}{\sqrt{n_1}}\right)^2 + \left(\frac{\sigma_2}{\sqrt{n_2}}\right)^2}$$
$$= \sqrt{\frac{\sigma_1^2}{n_1} + \frac{\sigma_2^2}{n_2}}$$

An Easier Rule?

The formula for the degrees of freedom of the sampling distribution of the difference between two means is complicated. So some books teach an easier rule: The number of degrees of freedom is always at *least* the smaller of $n_1 - 1$ and $n_2 - 1$ and at most $n_1 + n_2 - 2$. The problem is that if you need to use this rule, you'll have to be conservative and use the lower value. And *that* approximation can be a poor choice because it can give less than *half* the degrees of freedom you're entitled to from the correct formula. So, we don't recommend using it unless you have lots of data.

Of course, usually we don't know the true standard deviations of the two groups, σ_1 and σ_2, so we substitute the estimates, s_1 and s_2, and find a *standard error*:

$$SE(\bar{y}_1 - \bar{y}_2) = \sqrt{\frac{s_1^2}{n_1} + \frac{s_2^2}{n_2}}.$$

Just as we did for one mean, we'll use the standard error to see how big the difference really is. You shouldn't be surprised that, just as for a single mean, the ratio of the difference in the means to the standard error of that difference has a sampling model that follows a Student's t distribution.

What else do we need? Only the degrees of freedom for the Student's t-model. Unfortunately, *that* formula isn't as simple as $n - 1$. The problem is that the sampling model isn't *really* Student's t, but only something close. The reason is that we estimated two different variances (s_1^2 and s_2^2) and they may be different. That extra variability makes the distribution even more variable than the Student's t for either of the means. But by using a special, adjusted degrees of freedom value, we can find a Student's t-model that is so close to the right sampling distribution model that nobody can tell the difference. The adjustment formula is straightforward but doesn't help our understanding much, so we leave it to the computer or calculator. (If you are curious and really want to see the formula, look in the footnote.[2])

> ## A Sampling Distribution for the Difference Between Two Means
>
> When the conditions are met (see Section 14.3), the standardized sample difference between the means of two independent groups,
>
> $$t = \frac{(\bar{y}_1 - \bar{y}_2) - (\mu_1 - \mu_2)}{SE(\bar{y}_1 - \bar{y}_2)},$$
>
> can be modeled by a Student's t-model with a number of degrees of freedom found with a special formula. We estimate the standard error with
>
> $$SE(\bar{y}_1 - \bar{y}_2) = \sqrt{\frac{s_1^2}{n_1} + \frac{s_2^2}{n_2}}.$$

---

[2]The result is due to Satterthwaite and Welch.

Satterthwaite, F. E. (1946). "An Approximate Distribution of Estimates of Variance Components," *Biometrics Bulletin* 2: 110–114.

Welch, B. L. (1947). "The Generalization of Student's Problem when Several Different Population Variances are Involved," *Biometrika* 34: 28–35.

$$df = \frac{\left(\dfrac{s_1^2}{n_1} + \dfrac{s_2^2}{n_2}\right)^2}{\dfrac{1}{n_1 - 1}\left(\dfrac{s_1^2}{n_1}\right)^2 + \dfrac{1}{n_2 - 1}\left(\dfrac{s_2^2}{n_2}\right)^2}$$

This approximation formula usually doesn't even give a whole number. If you are using a table, you'll need a whole number, so round down to be safe. If you are using technology, the approximation formulas that computers and calculators use for the Student's t-distribution can deal with fractional degrees of freedom.

For Example	Sampling distribution of the difference of two means

The owner of a large car dealership wants to understand the negotiation process for buying a new car. Cars are given a "sticker price," but that a potential buyer may negotiate a better price is well known. He wonders if there is a difference in how men and women negotiate and who, if either, obtains the larger discount.

The owner takes a random sample of 100 customers from the last six months' sales and finds that 54 were men and 46 were women. On average the 54 men received a discount of $962.96 with a standard deviation of $458.95; the 46 women received an average discount of $1262.61 with a standard deviation of $399.70.

Question: What is the mean difference of the discounts received by men and women? What is its standard error? If there is no difference between them, does this seem like an unusually large value?

Answer: The mean difference is $1262.61 - $962.96 = $263.65. The women received, on average, a discount that was larger by $263.65. The standard error is

$$SE(\bar{y}_{Women} - \bar{y}_{Men}) = \sqrt{\frac{s^2_{Women}}{n_{Women}} + \frac{s^2_{Men}}{n_{Men}}} = \sqrt{\frac{(399.70)^2}{46} + \frac{(458.95)^2}{54}} = \$85.87$$

So, the difference is $263.65/85.87 = 3.07 standard errors away from 0. That sounds like a reasonably large number of standard errors for a Student's t statistic with at least 45 degrees of freedom. (The approximation formula gives 97.94 degrees of freedom.)

14.2 The Two-Sample *t*-Test

Notation Alert!

Δ_0 (pronounced "delta naught") isn't so standard that you can assume everyone will understand it. We use it because it's the capital Greek letter "D" for "difference."

The Two-Sample *t*-Methods

Two-sample *t*-methods assume that the two groups are independent. This is a crucial assumption. If it's not met, it is not safe to use these methods. One common way for groups to fail independence is when each observation in one group is related to one (and only one) observation in the other group. For example, if we test the *same* subjects before and after an event, or if we measure a variable on both husbands and wives. In that case, the observations are said to be **paired** and you'll need to use the paired *t*-methods discussed in section 14.7.

Sometimes you may see the two-sample *t*-methods referred to as the two-sample **independent** *t*-methods for emphasis. In this book, however, when we say two-sample, we'll always assume that the groups are independent.

Now we've got everything we need to construct the hypothesis test, and you already know how to do it. It's the same idea we used when testing one mean against a hypothesized value. Here, we start by hypothesizing a value for the true difference of the means. We'll call that hypothesized difference Δ_0. (It's so common for that hypothesized difference to be zero that we often just assume $\Delta_0 = 0$.) We then take the ratio of the difference in the means from our samples to its standard error and compare that ratio to a critical value from a Student's *t*-model. The test is called the **two-sample *t*-test**.

Two-sample *t*-test

When the appropriate assumptions and conditions are met, we test the hypothesis:

$$H_0: \mu_1 - \mu_2 = \Delta_0$$

where the hypothesized difference Δ_0 is almost always 0. We use the statistic:

$$t = \frac{(\bar{y}_1 - \bar{y}_2) - \Delta_0}{SE(\bar{y}_1 - \bar{y}_2)}.$$

The standard error of $\bar{y}_1 - \bar{y}_2$ is:

$$SE(\bar{y}_1 - \bar{y}_2) = \sqrt{\frac{s^2_1}{n_1} + \frac{s^2_2}{n_2}}.$$

When the null hypothesis is true, the statistic can be closely modeled by a Student's *t*-model with a number of degrees of freedom given by a special formula. We use that model to compare our *t* ratio with a critical value for *t* or to obtain a P-value.

For Example   The *t*-test for the difference of two means

Question: We saw (on page 403) that the difference between the average discount obtained by men and women appeared to be large if we assume that there is no true difference. Test the hypothesis, find the P-value, and state your conclusions.

Answer: The null hypothesis is: $H_0: \mu_{Women} - \mu_{Men} = 0$ vs. $H_A: \mu_{Women} - \mu_{Men} \neq 0$. The difference $\bar{y}_{Women} - \bar{y}_{Men}$ is \$263.65 with a standard error of \$84.26. The *t*-statistic is the difference divided by the standard error:

$$t = \frac{\bar{y}_{Women} - \bar{y}_{Men}}{SE(\bar{y}_{Women} - \bar{y}_{Men})} = \frac{263.65}{85.87} = 3.07.$$ The approximation formula gives 97.94 degrees of freedom (which is close to the maximum possible of $n_1 + n_2 - 2 = 98$). The P-value (from technology) for $t = 3.07$ with 97.94 df is 0.0028. We reject the null hypothesis. There is strong evidence to suggest that the difference in mean discount received by men and women is not 0.

14.3   Assumptions and Conditions

Before we can perform a two-sample *t*-test, we have to check the assumptions and conditions.

Independence Assumption

The data in each group must be drawn independently and at random from each group's own homogeneous population or generated by a randomized comparative experiment. We can't expect that the data, taken as one big group, come from a homogeneous population because that's what we're trying to test. But without randomization of some sort, there are no sampling distribution models and no inference. We should think about whether the independence assumption is reasonable. We can also check two conditions:

Randomization Condition: Were the data collected with suitable randomization? For surveys, are they a representative random sample? For experiments, was the experiment randomized?

10% Condition: We usually don't check this condition for differences of means. We'll check it only if we have a very small population or an extremely large sample. We needn't worry about it at all for randomized experiments.

Normal Population Assumption

As we did before with Student's *t*-models, we need the assumption that the underlying populations are *each* Normally distributed. So we check one condition.

Nearly Normal Condition: We must check this for *both* groups; a violation by either one violates the condition. As we saw for single sample means, the Normality Assumption matters most when sample sizes are small. When either group is small ($n < 15$), you should not use these methods if the histogram or Normal probability plot shows skewness. For *n*'s closer to 40, a mildly skewed histogram is OK, but you should remark on any outliers you find and not work with severely skewed data. When both groups are bigger than that, the Central Limit Theorem starts to work unless the data are severely skewed or there are extreme outliers, so the Nearly Normal Condition for the data matters less. Even in large samples, however, you should still be on the lookout for outliers, extreme skewness, and multiple modes.

Independent Groups Assumption

To use the two-sample *t* methods, the two groups we are comparing must be independent of each other. In fact, the test is sometimes called the two *independent samples t*-test. No statistical test can verify that the groups are independent. You have to think about how the data were collected. The assumption would be violated, for example, if one group were comprised of husbands and the other group, their wives. Whatever we measure on one might naturally be related to the other. Similarly, if we compared subjects' performances before some treatment with their performances afterward, we'd expect a relationship of each "before" measurement with its corresponding "after" measurement. Measurements taken for two groups over time when the observations are taken at the same time may be related—especially if they share, for example, the chance they were influenced by the overall economy or world events. In cases such as these, where the observational units in the two groups are related or matched, *the two-sample methods of this chapter can't be applied*. When this happens, we need a different procedure, discussed in section 14.7.

Guided Example   **Credit Card Promotions and Spending**

Our preliminary market research has suggested that a new incentive may increase customer spending. However, before we invest in this promotion on the entire population of cardholders, let's test a hypothesis on a sample. To judge whether the incentive works, we will examine the change in spending (called the *spend lift*) over a six-month period. We will see whether the *spend lift* for the group that received the offer was greater than the *spend lift* for the group that received no offer. If we observe differences, how will we know whether these differences are important (or real) enough to justify our costs?

PLAN

Setup State what we want to know.

Identify the *parameter* we wish to estimate. Here our parameter is the difference in the means, not the individual group means.

Identify the *population(s)* about which we wish to make statements.

Identify the variables and context.

We want to know if cardholders who are offered a promotion spend more on their credit card. We have the spend lift (in \$) for a random sample of 500 cardholders who were offered the promotion and for a random sample of 500 customers who were not. H_0: The mean spend lift for the group who received the offer is the same as for the group who did not:

$H_0: \mu_{Offer} = \mu_{No\ Offer}$

H_A: The mean spend lift for the group who received the offer is higher:

$H_A: \mu_{Offer} > \mu_{No\ Offer}$

(continued)

Make a graph to compare the two groups and check the distribution of each group. For completeness, we should report any outliers. If any outliers are extreme enough, we should consider performing the test both with and without the outliers and reporting the difference.

The boxplots and histograms show the distribution of both groups. It looks like the distribution for each group is unimodal and fairly symmetric.

The boxplots indicate several outliers in each group, but we have no reason to delete them, and their impact is minimal.

Model Check the assumptions and conditions.

For large samples like these with quantitative data, we often don't worry about the 10% Condition.

✓ **Independence Assumption.** We have no reason to believe that the spending behavior of one customer would influence the spending behavior of another customer in the same group. The data report the "spend lift" for each customer for the same time period.

✓ **Randomization Condition.** The customers who were offered the promotion were selected at random.

✓ **Nearly Normal Condition.** The samples are large, so we are not overly concerned with this condition, and the boxplots and histograms show unimodal, symmetric distributions for both groups.

✓ **Independent Groups Assumption.** Customers were assigned to groups at random. There's no reason to think that those in one group can affect the spending behavior of those in the other group.

Under these conditions, it's appropriate to use a Student's t-model.

State the sampling distribution model for the statistic. Here the degrees of freedom will come from the approximation formula in footnote 2.

Specify your method.

We will use a two-sample t-test.

DO

Mechanics List the summary statistics. Be sure to include the units along with the statistics. Use meaningful subscripts to identify the groups.

Use the sample standard deviations to find the standard error of the sampling distribution.

The best alternative is to let the computer use the approximation formula for the degrees of freedom and find the P-value.

We know $n_{No\ Offer} = 500$ and $n_{Offer} = 500$.

From technology, we find:

$$\bar{y}_{No\ Offer} = \$7.69 \qquad \bar{y}_{Offer} = \$127.61$$
$$s_{No\ Offer} = \$611.62 \quad s_{Offer} = \$566.05$$

The observed difference in the two means is:

$$\bar{y}_{Offer} - \bar{y}_{No\ Offer} = \$127.61 - \$7.69 = \$119.92$$

The groups are independent, so:

$$SE(\bar{y}_{Offer} - \bar{y}_{No\ Offer}) = \sqrt{\frac{(611.62)^2}{500} + \frac{(566.05)^2}{500}}$$
$$= \$37.27$$

The observed t-value is:

$$t = 119.92/37.27 = 3.218$$

with 992.0 df (from technology).

(To use critical values, we could find that the one-sided 0.01 critical value for a t with 992.0 df is $t^* = 2.33$.

Our observed t-value is larger than this, so we could reject the null hypothesis at the 0.01 level.)

Using software to obtain the P-value, we get:

```
Promotional Group   N     Mean    StDev
No                  500    7.69   611.62
Yes                 500  127.61   566.05

Difference = mu(1) − mu(0)
Estimate for difference: 119.9231
t = 3.2178, df = 992.007
One-sided P-value = 0.0006669
```

REPORT

Conclusion Interpret the test results in the proper context.

MEMO

Re: Credit Card Promotion

Our analysis of the credit card promotion experiment found that customers offered the promotion spent more than those not offered the promotion. The difference was statistically significant, with a P-value < 0.001. So we conclude that this promotion will increase spending. The difference in spend lift averaged $119.92, but our analyses so far have not determined how much income this will generate for the company and thus whether the estimated increase in spending is worth the cost of the offer.

Just Checking

Many office "coffee stations" collect voluntary payments for the food consumed. Researchers at the University of Newcastle upon Tyne performed an experiment to see whether the image of eyes watching would change employee behavior.[3] Each week on the cupboard behind the "honesty box," they alternated pictures of eyes looking at the viewer with pictures of flowers. They measured the consumption of milk to approximate the amount of food consumed and recorded the contributions (in £) each week per liter of milk. The table summarizes their results.

	Eyes	Flowers
n (# weeks)	5	5
$\bar{y}$	0.417 £/liter	0.151 £/liter
s	0.1811	0.067

1 What null hypothesis were the researchers testing?
2 Check the assumptions and conditions needed to test whether there really is a difference in behavior due to the difference in pictures.

For Example   Checking assumptions and conditions for a two-sample *t*-test

Question: In the example on page 404, we rejected the null hypothesis that the mean discount received by men and women is the same. Here are the histograms of the discounts for both women and men. Check the assumptions and conditions and state any concerns that you might have about the conclusion we reached.

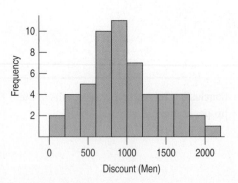

Answer: We were told that the sample was random, so the Randomization Condition is satisfied. There is no reason to think that the men's and women's responses are related (as they might be if they were husband and wife pairs) so the Independent Group Assumption is plausible. Because both groups have more than 40 observations, the discounts can be mildly skewed, which is the case. There are no obvious outliers (there is a small gap in the women's distribution, but the observations are not far from the center), so all the assumptions and conditions seem to be satisfied. We have no real concerns about the conclusion we reached that the mean difference is not 0.

---

[3]Melissa Bateson, Daniel Nettle, and Gilbert Roberts, "Cues of Being Watched Enhance Cooperation in a Real-World Setting," *Biol. Lett.* Doi:10.1098/rsbl.2006.0509.

14.4 A Confidence Interval for the Difference Between Two Means

We rejected the null hypothesis that customers' mean spending would not change when offered a promotion. Because the company took a random sample of customers for each group, and our P-value was convincingly small, we concluded this difference is not zero for the population. Does this mean that we should offer the promotion to all customers?

A hypothesis test really says nothing about the size of the difference. All it says is that the observed difference is large enough that we can be confident it isn't zero. That's what the term "statistically significant" means. It doesn't say that the difference is important, financially significant, or interesting. Rejecting a null hypothesis simply says that a value as extreme as the observed statistic is unlikely to have been observed if the null hypothesis were true.

So, what recommendations can we make to the company? Almost every business decision will depend on looking at a range of likely scenarios—precisely the kind of information a confidence interval gives. We construct the confidence interval for the difference in means in the usual way, starting with our observed statistic, in this case $(\bar{y}_1 - \bar{y}_2)$. We then add and subtract a multiple of the standard error $SE(\bar{y}_1 - \bar{y}_2)$ where the multiple is based on the Student's t distribution with the same df formula we saw before.

Confidence Interval for the Difference Between Two Means

When the conditions are met, we are ready to find a **two-sample t-interval** for the difference between means of two independent groups, $\mu_1 - \mu_2$. The confidence interval is:

$$(\bar{y}_1 - \bar{y}_2) \pm t^*_{df} \times SE(\bar{y}_1 - \bar{y}_2),$$

where the standard error of the difference of the means is:

$$SE(\bar{y}_1 - \bar{y}_2) = \sqrt{\frac{s_1^2}{n_1} + \frac{s_2^2}{n_2}}.$$

The critical value t^*_{df} depends on the particular confidence level, and on the number of degrees of freedom.

Guided Example Confidence Interval for Credit Card Spending

We rejected the null hypothesis that the mean spending in the two groups was equal. But, to find out whether we should consider offering the promotion nationwide, we need to estimate the magnitude of the spend lift.

PLAN

Setup State what we want to know.

Identify the *parameter* we wish to estimate. Here our parameter is the difference in the means, not the individual group means.

We want to find a 95% confidence interval for the mean difference in spending between those who are offered a promotion and those who aren't.

(continued)

Identify the *population(s)* about which we wish to make statements.

Identify the variables and context.

Specify the method.

We looked at the boxplots and histograms of the groups and checked the conditions before. The same assumptions and conditions are appropriate here, so we can proceed directly to the confidence interval.

We will use a two-sample *t*-interval.

DO

Mechanics Construct the confidence interval. Be sure to include the units along with the statistics. Use meaningful subscripts to identify the groups.

Use the sample standard deviations to find the standard error of the sampling distribution.

In our previous analysis, we found:

$$\bar{y}_{No\ Offer} = \$7.69 \qquad \bar{y}_{Offer} = \$127.61$$
$$s_{No\ Offer} = \$611.62 \qquad s_{Offer} = \$566.05$$

The observed difference in the two means is:

$$\bar{y}_{Offer} - \bar{y}_{No\ Offer} = \$127.61 - \$7.69 = \$119.92$$

and the standard error is:

$$SE(\bar{y}_{Offer} - \bar{y}_{No\ Offer}) = \$37.27$$

The best alternative is to let the computer use the approximation formula for the degrees of freedom and find the confidence interval.

From technology, the df is 992.007, and the 0.025 critical value for *t* with 992.007 df is 1.96. So the 95% confidence interval is:

$$\$119.92 \pm 1.96(\$37.27) = (\$46.87, \$192.97)$$

Ordinarily, we rely on technology for the calculations. In our hand calculations, we rounded values at intermediate steps to show the steps more clearly. The computer keeps full precision and is the one you should report. The difference between the hand and computer calculations is about $0.08.

Using software to obtain these computations, we get:

```
95 percent confidence interval:
 (46.78784, 193.05837)
sample means:
No Offer    Offer
7.690882    127.613987
```

REPORT

Conclusion Interpret the test results in the proper context.

MEMO

Re: Credit Card Promotion Experiment

In our experiment, the promotion resulted in an increased spend lift of $119.92 on average. Further analysis gives a 95% confidence interval of ($46.79, $193.06). In other words, we expect with 95% confidence that under similar conditions, the mean spend lift that we achieve when we roll out the offer to all similar customers will be in this interval. We recommend that the company consider whether the values in this interval will justify the cost of the promotion program.

For Example   A confidence interval for the difference between two means

Question: We've concluded on page 408 that, on average, women receive a larger discount than men at this car dealership. How big is the difference, on average? Find a 95% confidence interval for the difference.

Answer: We've seen that the difference from our sample is $263.65 with a standard error of $85.84 and that it has 97.94 degrees of freedom. The 95% critical value for a t with 97.94 degrees of freedom is 1.984.

$$\bar{y}_{Women} - \bar{y}_{Men} \pm t^*_{97.97}\, SE(\bar{y}_{Women} - \bar{y}_{Men}) = 263.65 \pm 1.984 \times 85.87 = (\$93.28, \$434.02)$$

We are 95% confident that, on average, women received a discount that is between $93.28 and $434.02 larger than the men at this dealership.

14.5   The Pooled *t*-Test

If you bought a used camera in good condition from a friend, would you pay the same as you would if you bought the same item from a stranger? A researcher at Cornell University[4] wanted to know how friendship might affect simple sales such as this. She randomly divided subjects into two groups and gave each group descriptions of items they might want to buy. One group was told to imagine buying from a friend whom they expected to see again. The other group was told to imagine buying from a stranger.

Here are the prices they offered to pay for a used camera in good condition.

Price Offered for a Used Camera ($)	
Buying from a Friend	Buying from a Stranger
275	260
300	250
260	175
300	130
255	200
275	225
290	240
300	

WHO	University students
WHAT	Prices offered to pay for a used camera ($)
WHEN	1990s
WHERE	Cornell University
WHY	To study the effects of friendship on transactions

The researcher who designed the friendship study was interested in testing the impact of friendship on negotiations. Previous theories had doubted that friendship had a measurable effect on pricing, but she hoped to find such an effect. The usual null hypothesis is that there's no difference in means and that's what we'll use for the camera purchase prices.

When we performed the *t*-test earlier in the chapter, we used an approximation formula that adjusts the degrees of freedom to a lower value. When $n_1 + n_2$ is only 15, as it is here, we don't really want to lose any degrees of freedom. Because this is an experiment, we might be willing to make another assumption. The null hypothesis says that whether you buy from a friend or a stranger should have no effect on the mean amount you're willing to pay for a camera. If it has no effect on the means, should it affect the variance of the transactions?

---

[4]J. J. Halpern (1997). "The Transaction Index: A Method for Standardizing Comparisons of Transaction Characteristics Across Different Contexts," *Group Decision and Negotiation*, 6, no. 6: 557–572.

If we're willing to *assume* that the variances of the groups are equal (at least when the null hypothesis is true), then we can save some degrees of freedom. To do that, we have to *pool* the two variances that we estimate from the groups into one common, or *pooled*, estimate of the variance:

$$s^2_{pooled} = \frac{(n_1 - 1)s_1^2 + (n_2 - 1)s_2^2}{(n_1 - 1) + (n_2 - 1)}.$$

(If the two sample sizes are equal, this is just the average of the two variances.)

Now we just substitute this pooled variance in place of each of the variances in the standard error formula. Remember, the standard error formula for the difference of two independent means is:

$$SE(\bar{y}_1 - \bar{y}_2) = \sqrt{\frac{s_1^2}{n_1} + \frac{s_2^2}{n_2}}.$$

We substitute the common pooled variance for each of the two variances in this formula, making the pooled standard error formula simpler:

$$SE_{pooled}(\bar{y}_1 - \bar{y}_2) = \sqrt{\frac{s^2_{pooled}}{n_1} + \frac{s^2_{pooled}}{n_2}} = s_{pooled}\sqrt{\frac{1}{n_1} + \frac{1}{n_2}}.$$

The formula for degrees of freedom for the Student's *t*-model is simpler, too. It was so complicated for the two-sample *t* that we stuck it in a footnote. Now it's just df = $(n_1 - 1) + (n_2 - 1)$.

Substitute the pooled-*t* estimate of the standard error and its degrees of freedom into the steps of the confidence interval or hypothesis test and you'll be using pooled-*t* methods. Of course, if you decide to use a pooled-*t* method, you must defend your assumption that the variances of the two groups are equal.

To use the pooled-*t* methods, you'll need to add the **Equal Variance Assumption** that the variances of the two populations from which the samples have been drawn are equal. That is, $\sigma_1^2 = \sigma_2^2$. (Of course, we can think about the standard deviations being equal instead.)

Pooled *t*-Test and Confidence Interval for the Difference Between Means

The conditions for the **pooled *t*-test** for the difference between the means of two independent groups (commonly called a pooled *t*-test) are the same as for the two-sample *t*-test with the additional assumption that the variances of the two groups are the same. We test the hypothesis:

$$H_0: \mu_1 - \mu_2 = \Delta_0,$$

where the hypothesized difference Δ_0 is almost always 0, using the statistic

$$t = \frac{(\bar{y}_1 - \bar{y}_2) - \Delta_0}{SE_{pooled}(\bar{y}_1 - \bar{y}_2)}.$$

The standard error of $\bar{y}_1 - \bar{y}_2$ is:

$$SE_{pooled}(\bar{y}_1 - \bar{y}_2) = s_{pooled}\sqrt{\frac{1}{n_1} + \frac{1}{n_2}},$$

where the pooled variance is:

$$s^2_{pooled} = \frac{(n_1 - 1)s_1^2 + (n_2 - 1)s_2^2}{(n_1 - 1) + (n_2 - 1)}.$$

When the conditions are met and the null hypothesis is true, we can model this statistic's sampling distribution with a Student's *t*-model with $(n_1 - 1) + (n_2 - 1)$ degrees of freedom. We use that model to obtain a P-value for a test or a margin of error for a confidence interval.

The corresponding **pooled-*t* confidence interval** is:

$$(\bar{y}_1 - \bar{y}_2) \pm t^*_{df} \times SE_{pooled}(\bar{y}_1 - \bar{y}_2),$$

where the critical value *t** depends on the confidence level and is found with $(n_1 - 1) + (n_2 - 1)$ degrees of freedom.

Guided Example   Role of Friendship in Negotiations

The usual null hypothesis in a pooled *t*-test is that there's no difference in means and that's what we'll use for the camera purchase prices.

PLAN

Setup State what we want to know.

Identify the *parameter* we wish to estimate. Here our parameter is the difference in the means, not the individual group means. Identify the variables and context.

Hypotheses State the null and alternative hypotheses.

The research claim is that friendship changes what people are willing to pay.[5] The natural null hypothesis is that friendship makes no difference.

We didn't start with any knowledge of whether friendship might increase or decrease the price, so we choose a two-sided alternative.

Make a graph. Boxplots are the display of choice for comparing groups. We'll also want to check the distribution of each group. Histograms may do a better job.

REALITY CHECK Looks like the prices are higher if you buy from a friend. The two ranges barely overlap, so we'll be pretty surprised if we don't reject the null hypothesis.

We want to know whether people are likely to offer a different amount for a used camera when buying from a friend than when buying from a stranger. We wonder whether the difference between mean amounts is zero. We have bid prices from 8 subjects buying from a friend and 7 subjects buying from a stranger, found in a randomized experiment.

H_0: The difference in mean price offered to friends and the mean price offered to strangers is zero:

$$\mu_F - \mu_S = 0,$$

H_A: The difference in mean prices is not zero:

$$\mu_F - \mu_S \neq 0.$$

(continued)

---

[5]This claim is a good example of what is called a "research hypothesis" in many social sciences. The only way to check it is to deny that it's true and see where the resulting null hypothesis leads us.

Model Think about the assumptions and check the conditions. (Because this is a randomized experiment, we haven't sampled at all, so the 10% Condition doesn't apply.)

✓ **Independence Assumption.** There is no reason to think that the behavior of one subject influenced the behavior of another.

✓ **Randomization Condition.** The experiment was randomized. Subjects were assigned to treatment groups at random.

✓ **Independent Groups Assumption.** Randomizing the experiment gives independent groups.

✓ **Nearly Normal Condition.** Histograms of the two sets of prices show no evidence of skewness or extreme outliers.

Buy from Friend

Buy from Stranger

State the sampling distribution model.

Specify the method.

Because this is a randomized experiment with a null hypothesis of no difference in means, we can make the Equal Variance Assumption. If, as we are assuming from the null hypothesis, the treatment doesn't change the means, then it is reasonable to assume that it also doesn't change the variances. Under these assumptions and conditions, we can use a Student's t-model to perform a pooled t-test.

DO

Mechanics List the summary statistics. Be sure to use proper notation.

From the data:

$$n_F = 8 \qquad n_S = 7$$
$$\bar{y}_F = \$281.88 \quad \bar{y}_S = \$211.43$$
$$s_F = \$18.31 \quad s_S = \$46.43$$

Use the null model to find the P-value. First determine the standard error of the difference between sample means.

The pooled variance estimate is:

$$s_p^2 = \frac{(n_F - 1)s_F^2 + (n_S - 1)s_S^2}{n_F + n_S - 2}$$

$$= \frac{(8 - 1)(18.31)^2 + (7 - 1)(46.43)^2}{8 + 7 - 2}$$

$$= 1175.48$$

The standard error of the difference becomes:

$$SE_{pooled}(\bar{y}_F - \bar{y}_S) = \sqrt{\frac{s_p^2}{n_F} + \frac{s_p^2}{n_S}}$$

$$= 17.744$$

Make a graph. Sketch the *t*-model centered at the hypothesized difference of zero. Because this is a two-tailed test, shade the region to the right of the observed difference and the corresponding region in the other tail.

Find the *t*-value.

The observed difference in means is:

$$(\bar{y}_F - \bar{y}_S) = 281.88 - 211.43 = \$70.45$$

which results in a t-ratio

$$t = \frac{(\bar{y}_F - \bar{y}_S) - (0)}{SE_{pooled}(\bar{y}_F - \bar{y}_S)} = \frac{70.45}{17.744} = 3.97$$

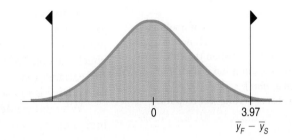

A statistics program can find the P-value.

The computer output for a pooled *t*-test appears here.

```
Pooled T Test for friend vs. stranger
            N    Mean    StDev    SE Mean
Friend      8    281.9   18.3     6.5
Stranger    7    211.4   46.4     18

t = 3.9699, df = 13, P-value = 0.001600
Alternative hypothesis: true difference in
means is not equal to 0

95 percent confidence interval:
(32.11047, 108.78238)
```

REPORT

Conclusion Link the P-value to your decision about the null hypothesis and state the conclusion in context.

Be cautious about generalizing to items whose prices are outside the range of those in this study. The confidence interval can reveal more detailed information about the size of the difference. In the original article (referenced in footnote 4 in this chapter), the researcher tested several items and proposed a model relating the size of the difference to the price of the items.

MEMO

Re: Role of friendship in negotiations.

Results of a small experiment show that people are likely to offer a different amount for a used camera when bargaining with a friend than when bargaining with a stranger. The difference in mean offers was statistically significant (P = .0016).

The confidence interval suggests that people tend to offer more to a friend than they would to a stranger. For the camera, the 95% confidence interval for the mean difference in price was $32.11 to $108.78, but we suspect that the actual difference may vary with the price of the item purchased.

Just Checking

Recall the experiment to see whether pictures of eyes would improve compliance in voluntary contributions at an office coffee station.

3 What alternative hypothesis would you test?

4 The P-value of the test was less than 0.05. State a brief conclusion.

When Should You Use the Pooled *t*-Test?

When the variances of the two groups are in fact equal, the two methods give pretty much the same result. Pooled methods have a small advantage (slightly narrower confidence intervals, slightly more powerful tests) mostly because they usually have a few more degrees of freedom, but the advantage is slight. When the variances are *not* equal, the pooled methods are just not valid and can give poor results. You have to use the two-sample methods instead.

As the sample sizes get bigger, the advantages that come from a few more degrees of freedom make less and less difference. So the advantage (such as it is) of the pooled method is greatest when the samples are small—just when it's hardest to check the conditions. And the difference in the degrees of freedom is greatest when the variances are not equal—just when you can't use the pooled method anyway. Our advice is to use the two-sample *t* methods to compare means.

Why did we devote a whole section to a method that we don't recommend using? That's a good question. The answer is that pooled methods are actually very important in Statistics, especially in the case of designed experiments, where we start by assigning subjects to treatments at random. We know that at the start of the experiment each treatment group is a random sample from the same population,[6] so each treatment group begins with the same population variance. In this case, assuming the variances are equal after we apply the treatment is the same as assuming that the treatment doesn't change the variance. When we test whether the true means are equal, we may be willing to go a bit farther and say that the treatments made no difference *at all*. That's what we did in the friendship and bargaining experiment. Then it's not much of a stretch to assume that the variances have remained equal.

The other reason to discuss the pooled *t*-test is historical. Until recently, many software packages offered the pooled *t*-test as the default for comparing means of two groups and required you to specifically request the *two-sample t-test* (or sometimes the misleadingly named "unequal variance *t*-test") as an option. That's changing, but be careful when using software to specify the right test.

There is also a hypothesis test that you could use to test the assumption of equal variances. However, it is sensitive to failures of the assumptions and works poorly for small sample sizes—just the situation in which we might care about a difference in the methods. When the choice between two-sample *t* and pooled-*t* methods makes a difference (that is, when the sample sizes are small), the test for whether the variances are equal hardly works at all.

Even though pooled methods are important in Statistics, the ones for comparing two means have good alternatives that don't require the extra assumption. The two-sample methods apply to more situations and are safer to use.

> Because the advantages of pooling are small, and you are allowed to pool only rarely (when the equal variances assumption is met), our advice is: *don't*.
> It's never wrong *not* to pool.

For Example   The pooled *t*-test

Question: Would the dealership owner of the example on page 403 have reached a different conclusion had he used the pooled *t*-test for the difference in mean discounts?

Answer: The difference in the pooled *t*-test is that it assumes that the variances of the two groups are equal, and pools those two estimates to find the standard error of the difference. The pooled estimate of the common standard deviation is:

$$s_{pooled} = \sqrt{\frac{(n_{Women} - 1)s^2_{Women} + (n_{Men} - 1)s^2_{Men}}{n_{Women} + n_{Men} - 2}} = \sqrt{\frac{(46 - 1)(399.70)^2 + (54 - 1)(458.95)^2}{46 + 54 - 2}}$$

$$= 432.75$$

[6]That is, the population of experimental subjects. Remember that to be valid, experiments do not need a representative sample drawn from a population because we are not trying to estimate a population model parameter.

We use that to find the *SE* of the difference:

$$SE_{pooled}(\bar{y}_{Women} - \bar{y}_{Men}) = s_{pooled}\sqrt{\frac{1}{n_{Women}} + \frac{1}{n_{Men}}} = 432.75\sqrt{\frac{1}{46} + \frac{1}{54}} = \$86.83$$

(Without pooling, our estimate was \$85.87.) This pooled *t* has $46 + 54 - 2 = 98$ degrees of freedom.

$$t_{98} = \frac{\bar{y}_{Women} - \bar{y}_{Men}}{SE_{pooled}(\bar{y}_{Women} - \bar{y}_{Men})} = \frac{263.65}{86.83} = 3.04$$

The P-value for a *t* of 3.04 with 98 degrees of freedom is (from technology) 0.0030. The two-sample *t*-test value was 0.0028. There is no practical difference between these, and we reach the same conclusion that the difference is not 0. The assumption of equal variances did not affect the conclusion.

*14.6 Tukey's Quick Test

If you think that the *t*-test is a lot of work for what seemed like an easy comparison, you're not alone. The famous statistician John Tukey[7] was once challenged to come up with a simpler alternative to the two-sample *t*-test that, like the 68–95–99.7 Rule, had critical values that could be remembered easily. The test he came up with asks you only to count and to remember three numbers: 7, 10, and 13.

When you first looked at the boxplots of the friendship data, you might have noticed that they didn't overlap very much. That's the basis for Tukey's test. To use Tukey's test, one group must have the highest value, and the other must have the lowest. We just count how many values in the high group are higher than *all* the values of the lower group. Add to this the number of values in the low group that are lower than *all* the values of the higher group. (You can count ties as $\frac{1}{2}$.) If the total of these exceedances is 7 or more, we can reject the null hypothesis of equal means at $\alpha = 0.05$. The "critical values" of 10 and 13 correspond to α's of 0.01 and 0.001.

Let's try it. The "Friend" group has the highest value (\$300), and the "Stranger" group has the lowest value (\$130). Six of the values in the Friend group are higher than the highest value of the Stranger group (\$260), and one is a tie. Six of the Stranger values are lower than the lowest value for the Friend group. That's a total of $12\frac{1}{2}$ exceedances. That's more than 10, but less than 13. So the P-value is between 0.01 and 0.001—just what we found with the pooled *t*.

This is a remarkably good test. The only assumption it requires is that the two samples be independent. It's so simple to do that there's no reason not to do one to check your two-sample *t* results. If they disagree, check the assumptions. Tukey's quick test, however, is not as widely known or accepted as the two-sample *t*-test, so you still need to know and use the two-sample *t*.

For Example Tukey's Quick Test

Here are the discounts (in order) received by the men (see page 403):

130, 158, 303, 340, 353, 390, 415, 423, 536, 566, 588 ... 1606, 1616, 1658, 1763, 1840, 1881, 2030

Here are the discounts received by the women:

503, 526, 574, 579, 603, 630, 794, 831 ... 1727, 1742, 1748, 2142, 2192

(continued)

[7]Famous seems like an understatement for John Tukey. The *New York Times* called Tukey "one of the most influential statisticians" of the 20th century. The *Times* also noted that he is credited with inventing the term "bit" (in its computer sense) and with the first printed use of the word "software." Tukey also invented both the stem-and-leaf display and the boxplot.

Question: Is Tukey's Quick Test applicable? If so, find the number of exceedances and report the P-value and conclusion. How does the test compare to the two-sample *t*-test?

Answer: Tukey's Quick Test is applicable because the groups are independent; also, one group (Men) has the smallest value ($130) and the other group (Women) has the largest value ($2192).

There were 8 discounts received by men that were smaller than all the women's discounts and 2 discounts received by women that were larger than all the men's discounts for a total of 10 exceedances. That corresponds to a P-value of about 0.01 (but larger than 0.001). This gives strong evidence to reject the null hypothesis that the two distributions are the same. We conclude that the mean difference is not 0. The two-sample *t*-test gave a P-value of 0.0028, which is in line with this test.

14.7   Paired Data

The two-sample *t*-test depends crucially on the assumption that the cases in the two groups are independent of each other. When is that assumption violated? Most commonly, it's when we have data on the *same* cases in two different circumstances. For example, we might want to compare the same customers' spending at our website last January to this January, or we might have each participant in a focus group rate two different product designs. Data such as these are called **paired.** When pairs arise from an experiment, we compare measurements before and after a treatment, and the pairing is a type of *blocking*. When they arise from an observational study, it is called *matching*.

When the data are paired, you *should not* use the two-sample (or pooled two-sample) method. You must decide, from the way the data were collected whether the data are paired. Be careful. There is no statistical test to determine whether the data are paired. You must decide whether the data are paired from understanding how they were collected and what they mean (check the W's).

Once we recognize that our data are matched pairs, it makes sense to concentrate on the *difference* between the two measurements for each individual. That is, we look at the collection of pairwise differences in the measured variable. For example, if studying customer spending, we would analyze the *difference* between this January's and last January's spending for each customer. Because it is the *differences* we care about, we can treat them as if there was a single variable of interest holding those differences. With only one variable to consider, we can use a simple one-sample *t*-test. A **paired *t*-test** is just a one-sample *t*-test for the mean of the pairwise differences. The sample size is the number of pairs.

Because the paired *t*-test is mechanically just the same as a one-sample *t*-test of the pairwise differences for each case, the assumptions and conditions for those differences are exactly the same as the ones we used for the one-sample *t*-test. However, because the paired *t*-test is used so often, it's a good idea to go over how the assumptions and conditions apply to this new setting.

> **Are the Data Paired?**
>
> To decide whether your data are paired you'll need to understand the business context of the data and how the data were collected. Pairing is not a choice, but a fact about the data. If the data are paired, you can't treat the groups as independent. But to be paired, the data must be related between the two groups in a way that you can justify. That justification may involve knowledge about the data not present in the variables themselves.

Paired Data Assumption

The data must actually be paired. You can't just decide to pair data from independent groups. When you have two groups with the same number of observations, it may be tempting to match them up, but that's not valid. You can't pair data just because they "seem to go together." To use paired methods you must determine from knowing how the data were collected whether the two groups are paired or independent. Usually the context will make it clear.

Be sure to recognize paired data when you have it. Remember, two-sample *t*-methods aren't valid unless the groups are independent, and paired groups aren't independent.

Independence Assumption

For these methods, it's the *differences* that must be independent of each other. This is just the one-sample *t*-test assumption of independence, now applied to the differences. As always, randomization helps to ensure independence.

Randomization Condition. Randomness can arise in many ways. The *pairs* may be a random sample. For example, we may be comparing opinions of husbands and wives from a random selection of couples. In an experiment, the order of the two treatments may be randomly assigned, or the treatments may be randomly assigned to one member of each pair. In a before-and-after study, like this one, we may believe that the observed differences are a representative sample from a population of interest. If we have any doubts, we'll need to include a control group to be able to draw conclusions. What we want to know usually focuses our attention on where the randomness should be.

10% Condition. When we sample from a finite population, we should be careful not to sample more than 10% of that population. Sampling too large a fraction of the population calls the independence assumption into question. As with other quantitative data situations, we don't usually explicitly check the 10% condition, but be sure to think about it.

Normal Population Assumption

We need to assume that the population of *differences* follows a Normal model. We don't need to check the data in each of the two individual groups. In fact, the data from each group can be quite skewed, but the differences can still be unimodal and symmetric.

Nearly Normal Condition. This condition can be checked with a histogram of the differences. As with the one-sample *t*-methods, this assumption matters less as we have more pairs to consider. You may be pleasantly surprised when you check this condition. Even if your original measurements are skewed or bimodal, the *differences* may be nearly Normal. After all, the individual who was way out in the tail on an initial measurement is likely to still be out there on the second one, giving a perfectly ordinary difference.

For Example   Paired data

Question: The owner of the car dealership (see page 403) is curious about the maximum discount his salesmen are willing to give to customers. In particular, two of his salespeople, Frank and Ray, seem to have very different ideas about how much discount to allow. To test his suspicion, he selects 30 cars from the lot and asks each to say how much discount they would allow a customer. If we want to test whether the mean discount given by them is the same, what test would he use? Explain.

Answer: A paired *t*-test. The responses (maximum discount in $) are not independent between the two salespeople because they are evaluating the same 30 cars.

14.8   The Paired *t*-Test

The paired *t*-test is mechanically a one-sample *t*-test. We treat the differences as our variable. We simply compare the mean difference to its standard error. If the *t*-statistic is large enough, we reject the null hypothesis.

Paired *t*-test

When the conditions are met, we are ready to test whether the mean paired difference is significantly different from a hypothesized value (called Δ_0). We test the hypothesis:

$$H_0: \mu_d = \Delta_0,$$

where the d's are the pairwise differences and Δ_0 is almost always 0.

We use the statistic:

$$t = \frac{\bar{d} - \Delta_0}{SE(\bar{d})},$$

where $\bar{d}$ is the mean of the pairwise differences, n is the number of *pairs*, and

$$SE(\bar{d}) = \frac{s_d}{\sqrt{n}},$$

where s_d is the standard deviation of the pairwise differences.

When the conditions are met and the null hypothesis is true, the sampling distribution of this statistic is a Student's *t*-model with $n - 1$ degrees of freedom and we use that model to obtain the P-value.

Similarly, we can construct a confidence interval for the true difference. As in a one-sample *t*-interval, we center our estimate at the mean difference in our data. The margin of error on either side is the standard error multiplied by a critical *t*-value (based on our confidence level and the number of pairs we have).

Paired *t*-interval

When the conditions are met, we are ready to find the confidence interval for the mean of the paired differences. The confidence interval is:

$$\bar{d} \pm t^*_{n-1} \times SE(\bar{d}),$$

where the standard error of the mean difference is $SE(\bar{d}) = \frac{s_d}{\sqrt{n}}$.

The critical value t^* from the Student's *t*-model depends on the particular confidence level that you specify and on the degrees of freedom, $n - 1$, which is based on the number of pairs, n.

Just Checking

Think about each of the following situations. Would you use a two-sample t or paired t-method (or neither)? Why?

5 Random samples of 50 men and 50 women are surveyed on the amount they invest on average in the stock market on an annual basis. We want to estimate any gender difference in how much they invest.

6 Random samples of students were surveyed on their perception of ethical and community service issues both in their first year and fourth year at a university. The university wants to know whether their required programs in ethical decision-making and service learning change student perceptions.

7 A random sample of work groups within a company was identified. Within each work group, one male and one female worker were selected at random. Each was asked to rate the secretarial support that their workgroup received. When rating the same support staff, do men and women rate them equally on average?

8 A total of 50 companies are surveyed about business practices. They are categorized by industry, and we wish to investigate differences across industries.

9 These same 50 companies are surveyed again one year later to see if their perceptions, business practices, and R&D investment have changed.

Guided Example   Seasonal Spending

Economists and credit card banks know that people tend to spend more near the holidays in December. In fact, sales in the few days after Thanksgiving (the fourth Thursday of the month in the United States) provide an indication of the strength of the holiday season sales and an early look at the strength of the economy in general. After the holidays, spending decreases substantially. Because credit card banks receive a percentage of each transaction, they need to forecast how much the average spending will increase or decrease from month to month. How much less do people tend to spend in January than December? For any particular segment of cardholders, a credit card bank could select two random samples—one for each month—and simply compare the average amount spent in January with that in December. A more sensible approach might be to select a single random sample and compare the spending between the two months for *each cardholder*. Designing the study in this way and examining the paired differences gives a more precise estimate of the actual change in spending.

Here we have a sample of 911 cardholders from a particular market segment and the amount they charged on their credit card in both December 2004 and January 2005. We can test whether the mean difference in spending is 0 by using a paired *t*-test and create a **paired *t*-confidence interval** to estimate the true mean difference in spending between the two months.

WHO	Cardholders in a particular market segment of a major credit card issuer
WHAT	Amount charged on their credit card in December and January
WHEN	2004–2005
WHERE	United States
WHY	To estimate the amount of decrease in spending one could expect after the holiday shopping season

PLAN

Setup State what we want to know.

Identify the *parameter* we wish to estimate and the sample size.

We want to know how much we can expect credit card charges to change, on average, from December to January for this market segment. We have the total amount charged in December 2004 and January 2005 for $n = 911$ cardholders in this segment. We want to test whether the mean spending is the same for customers in the two months and find a confidence interval for the true mean difference in charges between these two months for all cardholders in this segment. Because we know that people tend to spend more in December, we will look at the difference: *December spend − January spend* and use a one-sided test. A positive difference will mean a decrease in spending.

Hypotheses State the null and alternative hypotheses.

H_O: Mean spending was the same in December and January; the mean difference is zero: $\mu_d = 0$.

H_A: Mean spending was greater in December than January; the mean difference was greater than zero: $\mu_d > 0$.

Model Check the conditions.

State why the data are paired. Simply having the same number of individuals in each group or displaying them in side-by-side columns, doesn't make them paired.

✓ **Paired Data Assumption:** The data are paired because they are measurements on the same cardholders in two different months.

✓ **Independence Assumption:** The behavior of any individual is independent of the behavior of the others, so the differences are mutually independent.

(continued)

Think about what we hope to learn and where the randomization comes from.

Make a picture of the differences. Don't plot separate distributions of the two groups—that would entirely miss the pairing. For paired data, it's the Normality of the differences that we care about. Treat those paired differences as you would a single variable, and check the Nearly Normal condition.

✓ **Randomization Condition:** This was a random sample from a large market segment.

✓ **Nearly Normal Condition:** The distribution of the differences is unimodal and symmetric. Although the tails of the distribution are long, the distributions are symmetric. (This is typical of the behavior of credit card spending.) There are no isolated cases that would unduly dominate the mean difference, and our sample size is large so the Central Limit Theorem will protect us.

Specify the sampling distribution model.

Choose the method.

The conditions are met, so we'll use a Student's *t*-model with $(n - 1) = 910$ degrees of freedom, perform a paired *t*-test, and find a paired *t*-confidence interval.

DO

Mechanics *n* is the number of *pairs*, in this case, the number of cardholders.

$\bar{d}$ is the mean difference.

s_d is the standard deviation of the differences.

Make a picture. Sketch a *t*-model centered at the observed mean of 788.18.

The computer output tells us:

$$n = 911 \text{ pairs}$$
$$\bar{d} = \$788.18$$
$$s_d = \$3,740.22$$

Find the standard error and the *t*-score of the observed mean difference. There is nothing new in the mechanics of the paired *t*-methods. These are the mechanics of the *t*-test and *t*-interval for a mean applied to the differences.

We estimate the standard error of $\bar{d}$ using:

$$SE(\bar{d}) = \frac{s_d}{\sqrt{n}} = \frac{3740.22}{\sqrt{911}} = \$123.919$$

We find $t = \dfrac{\bar{d} - 0}{SE(\bar{d})} = \dfrac{788.180}{123.919} = 6.36$

A *t*-statistic with 910 degrees of freedom and a value of 6.36 has a one-sided P-value <0.001.

The critical value is:

$$t^*_{910} = 1.96$$

The margin of error, $ME = t^*_{910} \times SE(\bar{d})$

$$= 1.96 \times 123.919 = 242.88$$

So a 95% CI is $\bar{d} \pm ME = (\$545.30, \$1031.06)$.

REPORT

Conclusion Link the results of the confidence interval to the context of the problem.

MEMO

Re: Credit Card Expenditure Changes

In the sample of cardholders studied, the change in expenditures between December and January averaged $788.18, which means that, on average, cardholders spend $788.18 less in January than the month before. There is strong evidence to suggest that the mean difference is not zero and that customers really do spend more in December than January. Although we didn't measure the change for all cardholders in the segment, we can be 95% confident that the true mean decrease in spending from December to January was between $545.30 and $1031.06.

For Example   The paired *t*-test

Here are some summary statistics from the study undertaken by the car dealership owner (see page 419):

$$\bar{y}_{Frank} = \$414.48 \qquad \bar{y}_{Ray} = \$478.88$$
$$SD_{Frank} = \$87.33 \qquad SD_{Ray} = \$175.12$$
$$\bar{y}_{Diff} = \$64.40 \qquad SD_{Diff} = \$146.74$$

Question: Test the hypothesis that the mean maximum discount that Frank and Ray would give is the same. Give a 95% confidence interval for the mean difference.

Answer: We use a paired *t*-test because Frank and Ray were asked to give opinions about the same 30 cars.

$$t_{n-1} = \frac{\bar{d}}{SE(\bar{d})}$$

$$SE(\bar{d}) = \frac{s_d}{\sqrt{n}} = \frac{146.74}{\sqrt{30}} = \$26.79$$

$$t_{29} = \frac{64.40}{26.79} = 2.404$$

(continued)

which has a (two-sided) P-value of 0.0228. We reject the null hypothesis that the mean difference is 0 and conclude that there is strong evidence to suggest that they are not the same.

A 95% confidence interval:

$$t^*_{29} = 2.045 \text{ at 95\% confidence}$$

$$\bar{d} \pm t^*_{29} \times SE(\bar{d}) = 64.40 \pm 2.045 \times 26.79$$

$$= (\$9.61, \$119.19)$$

shows that Ray gives, on average, somewhere between \$9.61 and \$119.19 more for his maximum discount than Frank does.

What Can Go Wrong?

- **Watch out for paired data when using the two-sample *t*-test.** The Independent Groups Assumption deserves special attention. Some researcher designs *deliberately* violate the Independent Groups Assumption. For example, suppose you wanted to test a diet program. You select 10 people at random to take part in your diet. You measure their weights at the beginning of the diet and after 10 weeks of the diet. So, you have two columns of weights, one for *before* and one for *after*. But the data are related; each "after" weight goes naturally with the "before" weight for the *same* person. If the samples are *not* independent, you can't use two-sample methods. This is probably the main thing that can go wrong when using two-sample methods.

- **Don't use individual confidence intervals for each group to test the difference between their means.** If you make 95% confidence intervals for the means of each group separately and you find that the intervals don't overlap, you can reject the hypothesis that the means are equal (at the corresponding α level). But, if the intervals do overlap, that doesn't mean that you *can't* reject the null hypothesis. The margin of error for the difference between the means is smaller than the sum of the individual confidence interval margins of error. Comparing the individual confidence intervals is like adding the standard deviations. But we know that it's the variances that we add, and when we do it right, we actually get a more powerful test. So, don't test the difference between group means by looking at separate confidence intervals. Always make a two-sample *t*-interval or perform a two-sample *t*-test.

- **Look at the plots.** The usual (by now) cautions about checking for outliers and non-Normal distributions apply. The simple defense is to make and examine plots (side-by-side boxplots and histograms for two independent samples, histograms of the differences for paired data). You may be surprised how often this simple step saves you from the wrong or even absurd conclusions that can be generated by a single undetected outlier. You don't want to conclude that two methods have very different means just because one observation is atypical.

- **Don't use a paired *t*-method when the samples aren't paired.** When two groups don't have the same number of values, it's easy to see that they can't be paired. But just because two groups have the same number of observations doesn't mean they can be paired, even if they are shown side by side in a table. We might have 25 men and 25 women in our study, but they might be completely independent of one another. If they were siblings or spouses, we might consider them paired. Remember that you cannot *choose* which method to use based on your preferences. Only if the data are from an experiment or study in which observations were paired, can you use a paired method.

- **Don't forget to look for outliers when using paired methods.** For two-sample *t*-methods watch out for outliers in *either* group. For paired *t*-methods, the outliers we care about now are in the differences. A subject who is extraordinary both before and after a treatment may still have a perfectly typical difference. But one outlying difference can completely distort your conclusions. Be sure to plot the differences (even if you also plot the data) when using paired methods.

Do What We Say, Not What We Do

Precision machines used in industry often have a bewildering number of parameters that have to be set, so experiments are performed in an attempt to try to find the best settings. Such was the case for a hole-punching machine used by a well-known computer manufacturer to make printed circuit boards. The data were analyzed by one of the authors, but because he was in a hurry, he didn't look at the boxplots first and just performed *t*-tests on the experimental factors. When he found extremely small P-values even for factors that made no sense, he plotted the data. Sure enough, there was one observation 1,000,000 times bigger than the others. It turns out that it had been recorded in microns (millionths of an inch), while all the rest were in inches.

Ethics in Action

Advocacy groups for equity and diversity in the workplace often cite middle managers as an obstacle to many organizations' efforts to be more inclusive. In response to this concern, Michael Schrute, the CEO for a large manufacturing company, asked Albert Fredericks, VP of Human Resources, to look into the possibility of instituting some type of diversity training for the company's middle managers. One option under consideration was an online education program that focused on cultural diversity, gender equity, and disability awareness. Although cost-effective, Albert suspected that an online program would not be as effective as traditional training for middle managers. In order to evaluate the online program under consideration, 20 middle managers were selected to participate. Before beginning, they were given a test to assess their knowledge of and sensitivity to various diversity and equity issues. Out of a possible perfect score of 100, the test average

was 63.65. Each of the 20 managers then completed the 6-week online diversity education program and was retested. The average on the test after completing the online program was 69.15. Although the group achieved a higher mean test score after completing the program, the two-sample *t*-test revealed that this average test score was not significantly higher than the average prior to completing the online program ($t = -0.94$, P-value = 0.176). Albert was not surprised and began to explore more traditional diversity educational programs.

ETHICAL ISSUE *The pretest and posttest design violates independence and therefore the two-sample t-test is not appropriate (related to Item A, ASA Ethical Guidelines).*

ETHICAL SOLUTION *Use the correct test. The two-sample t-test is not appropriate for these data. Use a paired t-test instead. Using the correct test shows that the online diversity education program was effective.*

What Have We Learned?

Learning Objectives

■ Know how to test whether the difference in the means of two independent groups is equal to some hypothesized value.

- The two-sample *t*-test is appropriate for independent groups. It uses a special formula for degrees of freedom.
- The Assumptions and Conditions are the same as for one-sample inferences for means with the addition of assuming that the groups are independent of each other.
- The most common null hypothesis is that the means are equal.

■ Be able to construct and interpret a confidence interval for the difference between the means of two independent groups.

- The confidence interval inverts the *t*-test in the natural way.

■ Know how and when to use pooled *t* inference methods.

- There is an additional assumption that the variances of the two groups are equal.
- This may be a plausible assumption in a randomized experiment.

■ Recognize when you have paired or matched samples and use an appropriate inference method.

- Paired *t* methods are the same as one-sample *t* methods applied to the pairwise differences.
- If data are paired they cannot be independent, so two-sample *t* and pooled-*t* methods would not be applicable.

Terms

Paired data　Data are paired when the observations are collected in pairs or the observations in one group are naturally related to observations in the other. The simplest form of pairing is to measure each subject twice—often before and after a treatment is applied. Pairing in experiments is a form of blocking and arises in other contexts. Pairing in observational and survey data is a form of matching.

Paired *t*-test　A hypothesis test for the mean of the pairwise differences of two groups. It tests the null hypothesis $H_0: \mu_d = \Delta_0$, where the hypothesized difference is almost always 0, using the statistic $t = \dfrac{\bar{d} - \Delta_0}{SE(\bar{d})}$ with $n-1$ degrees of freedom, where $SE(\bar{d}) = \dfrac{s_d}{\sqrt{n}}$ and n is the number of pairs.

Paired *t*-confidence interval　A confidence interval for the mean of the pairwise differences between paired groups found as $\bar{d} \pm t^*_{n-1} \times SE(\bar{d})$, where $SE(\bar{d}) = \dfrac{s_d}{\sqrt{n}}$ and n is the number of pairs.

*Pooled *t*-interval　A confidence interval for the difference in the means of two independent groups used when we are willing and able to make the additional assumption that the variances of the groups are equal. It is found as:

$$(\bar{y}_1 - \bar{y}_2) \pm t^*_{df} \times SE_{pooled}(\bar{y}_1 - \bar{y}_2),$$

$$\text{where } SE_{pooled}(\bar{y}_1 - \bar{y}_2) = s_{pooled}\sqrt{\frac{1}{n_1} + \frac{1}{n_2}},$$

and the pooled variance is

$$s^2_{pooled} = \frac{(n_1 - 1)s_1^2 + (n_2 - 1)s_2^2}{(n_1 - 1) + (n_2 - 1)}.$$

The number of degrees of freedom is $(n_1 - 1) + (n_2 - 1)$.

*Pooled *t*-test*

A hypothesis test for the difference in the means of two independent groups when we are willing and able to assume that the variances of the groups are equal. It tests the null hypothesis

$$H_0: \mu_1 - \mu_2 = \Delta_0,$$

where the hypothesized difference Δ_0 is almost always 0, using the statistic

$$t_{df} = \frac{(\bar{y}_1 - \bar{y}_2) - \Delta_0}{SE_{pooled}(\bar{y}_1 - \bar{y}_2)},$$

where the pooled standard error is defined as for the pooled interval and the degrees of freedom is $(n_1 - 1) + (n_2 - 1)$.

*Pooling

Data from two or more populations may sometimes be combined, or *pooled*, to estimate a statistic (typically a pooled variance) when we are willing to assume that the estimated value is the same in both populations. The resulting larger sample size may lead to an estimate with lower sample variance. However, pooled estimates are appropriate only when the required assumptions are true.

Two-sample *t*-interval

A confidence interval for the difference in the means of two independent groups found as

$$(\bar{y}_1 - \bar{y}_2) \pm t^*_{df} \times SE(\bar{y}_1 - \bar{y}_2), \text{ where}$$

$$SE(\bar{y}_1 - \bar{y}_2) = \sqrt{\frac{s_1^2}{n_1} + \frac{s_2^2}{n_2}}$$

and the number of degrees of freedom is given by the approximation formula in footnote 2 of this chapter, or with technology.

Two-sample *t*-test

A hypothesis test for the difference in the means of two independent groups. It tests the null hypothesis

$$H_0: \mu_1 - \mu_2 = \Delta_0,$$

where the hypothesized difference Δ_0 is almost always 0, using the statistic

$$t_{df} = \frac{(\bar{y}_1 - \bar{y}_2) - \Delta_0}{SE(\bar{y}_1 - \bar{y}_2)},$$

with the number of degrees of freedom given by the approximation formula in footnote 2 of this chapter, or with technology.

Technology Help: Two-Sample Methods

Here's some typical computer package output with comments:

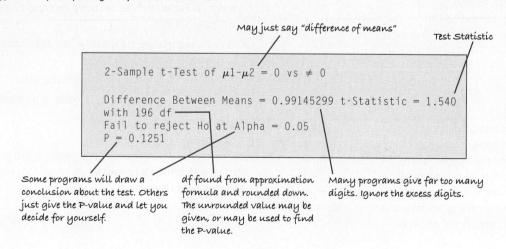

May just say "difference of means"

Test Statistic

```
2-Sample t-Test of μ1-μ2 = 0 vs ≠ 0

Difference Between Means = 0.99145299  t-Statistic = 1.540
with 196 df
Fail to reject Ho at Alpha = 0.05
P = 0.1251
```

Some programs will draw a conclusion about the test. Others just give the P-value and let you decide for yourself.

df found from approximation formula and rounded down. The unrounded value may be given, or may be used to find the P-value.

Many programs give far too many digits. Ignore the excess digits.

(continued)

Most statistics packages compute the test statistic for you and report a P-value corresponding to that statistic. And statistics packages make it easy to examine the boxplots of the two groups, so you have no excuse for skipping the important check of the Nearly Normal Condition.

Some statistics software automatically tries to test whether the variances of the two groups are equal. Some automatically offer both the two-sample-*t* and pooled-*t* results. Ignore the test for the variances; it has little power in any situation in which its results could matter. If the pooled and two-sample methods differ in any important way, you should stick with the two-sample method. Most likely, the Equal Variance Assumption needed for the pooled method has failed.

The degrees of freedom approximation usually gives a fractional value. Most packages seem to round the approximate value down to the next smallest integer (although they may actually compute the P-value with the fractional value, gaining a tiny amount of power).

There are two ways to organize data when we want to compare two independent groups. The first, called **unstacked data**, lists the data in two columns, one for each group. Each list can be thought of as a variable. In this method, the variables in the credit card example would be "Offer" and "No Offer." Graphing calculators usually prefer this form, and some computer programs can use it as well.

The alternative way to organize the data is as **stacked data**. What is the response variable for the credit card experiment? It's the "Spend Lift"—the amount by which customers increased their spending. But the values of this variable in the unstacked lists are in both columns, and actually there's an experiment factor here, too—namely, whether the customer was offered the promotion or not. So we could put the data into two different columns, one with the "Spend Lift"s in it and one with a "Yes" for those who were offered the promotion and a "No" for those who weren't. The stacked data would look like this:

Spend Lift	Offer
969.74	Yes
915.04	Yes
197.57	No
77.31	No
196.27	Yes
. . .	. . .

This way of organizing the data makes sense as well. Now the factor and the response variables are clearly visible. You'll have to see which method your program requires. Some packages even allow you to structure the data either way.

The commands to do inference for two independent groups on common statistics technology are not always found in obvious places. Here are some starting guidelines.

EXCEL   XLSTAT

From the Data Tab, Analysis Group,

- Choose **Data Analysis**.
- Choose the appropriate *t*-Test from the menu.
- Alternatively (if the Data Analysis Tool Pack is not installed), in the Formulas Tab, choose More functions > Statistical > TTEST (T.TEST in Excel 2010), and specify Type = 3 in the resulting dialog.
- Fill in the cell ranges for the two groups, the hypothesized difference, and the alpha level.

Comments

Excel expects the two groups to be in separate cell ranges. Notice that, contrary to Excel's wording, we do not need to assume that the variances are *not* equal to perform a two-sample *t*; we simply choose not to assume that they *are* equal.

JMP

From the **Analyze** menu,

- Select **Fit y by x**.
- Select variables: a **Y, Response** variable that holds the data and an **X, Factor** variable that holds the group names. JMP will make a dotplot.
- Click the **red triangle** in the dotplot title, and choose **Unequal variances**.

    The *t*-test is at the bottom of the resulting table.

- Find the P-value from the Prob7F section of the table (they are the same).

Comments

JMP expects data in one variable and category names in the other. Don't be misled: There is no need for the variances to be unequal to use two-sample *t* methods.

MINITAB

From the **Stat** menu,

- Choose the **Basic Statistics** submenu.
- From that menu, choose **2-sample t....** and select data in "one column" or "two columns" depending on whether the data are stacked or unstacked. You may also enter summarized data.
- Then fill in the dialog.

Comments

The **Graphs** button enables you to create boxplots of your two samples and the **options** button allows you to conduct either a one-sided or two-sided test.

SPSS

From the **Analyze** menu,

- Choose the **Compare Means** submenu.
- From that, choose the **Independent-Samples t-test** command. Specify the data variable and "group variable."
- Then type in the labels used in the group variable. SPSS offers both the two-sample and pooled-*t* results in the same table.

Comments

SPSS expects the data in one variable and group names in the other. If there are more than two group names in the group variable, only the two that are named in the dialog box will be compared.

Technology Help: Paired *t*

Most statistics programs can compute paired-*t* analyses. Some may want you to find the differences yourself and use the one-sample *t*-methods. Those that perform the entire procedure will need to know the two variables to compare. The computer, of course, cannot verify that the variables are naturally paired. Most programs will check whether the two variables have the same number of observations, but some stop there, and that can cause trouble.

Most programs will automatically omit any pair that is missing a value for either variable. You must look carefully to see whether that has happened.

As we've seen with other inference results, some packages pack a lot of information into a simple table, but you must locate what you want for yourself. Here's a generic example with comments.

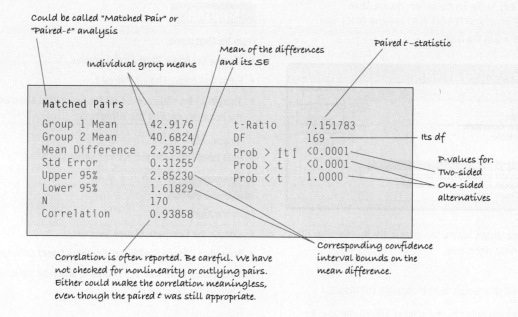

Could be called "Matched Pair" or "Paired-*t*" analysis

Individual group means

Mean of the differences and its SE

Paired *t*-statistic

```
Matched Pairs

Group 1 Mean     42.9176        t-Ratio    7.151783
Group 2 Mean     40.6824        DF         169
Mean Difference   2.23529       Prob > |t|  <0.0001
Std Error         0.31255       Prob > t    <0.0001
Upper 95%         2.85230       Prob < t    1.0000
Lower 95%         1.61829
N               170
Correlation       0.93858
```

Its df

P-values for:
Two-sided
One-sided
alternatives

Correlation is often reported. Be careful. We have not checked for nonlinearity or outlying pairs. Either could make the correlation meaningless, even though the paired *t* was still appropriate.

Corresponding confidence interval bounds on the mean difference.

(continued)

Other packages try to be more descriptive. It may be easier to find the results, but you may get less information from the output table.

Groups may have missing values. Only cases with both values present are used in a paired-t analysis. You may not learn that from some packages.

SD (differences)

Even simple tables can have superfluous numbers such as these.

$SE(\bar{d})$

$\bar{d}$

CI corresponds to specified α.

Some packages let you specify the alternative and report only results for that alternative.

t-statistic and its P-value (You may need to calculate $n_d - 1$ for yourself to get the df.)

Computers make it easy to examine the boxplots of the two groups and the histogram of the differences—both important steps. Some programs offer a scatterplot of the two variables. That can be helpful. In terms of the scatterplot, a paired *t*-test is about whether the points tend to be above or below the 45° line $y = x$. (Note that pairing says nothing about whether the scatterplot should be straight. That doesn't matter for our *t*-methods.)

EXCEL   XLSTAT

- Select **Data Analysis** from the **Analysis** Group on the **Data** Tab.
- From the **Data Analysis** menu, choose **t-test: paired two sample for Means**. Alternatively (if the Data Analysis Tool Pack is not installed), in the Formulas Tab, choose More functions>Statistical>TTEST (T.TEST in Excel 2010), and specify Type = 3 in the dialog.

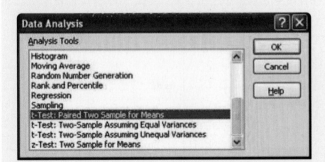

- Fill in the cell ranges for the two groups, the hypothesized difference, and the alpha level.

Comments
Excel expects the two groups to be in separate cell ranges.

Warning: Do not compute this test in Excel without checking for missing values. If there are any missing values (empty cells), Excel will usually give a wrong answer. Excel compacts each list, pushing

values up to cover the missing cells, and then checks only that it has the same number of values in each list. The result is mismatched pairs and an entirely wrong analysis. To avoid this error, delete any pair in which one of the values is missing.

JMP

From the **Analyze** menu,
- Select **Matched Pairs**.
- Specify the columns holding the two groups in the **Y Paired Response** Dialog.
- Click **OK**.

MINITAB

From the **Stat** menu,
- Choose the **Basic Statistics** submenu.
- From that menu, choose **Paired t**…
- Then fill in the dialog boxes for the two paired samples, or fill in the summary data for the differences.

Comments
Minitab takes "First sample" minus "Second sample."

SPSS

From the **Analyze** menu,
- Choose the **Compare Means** submenu.
- From that, choose the **Paired-Samples t-test** command.
- Select pairs of variables to compare, and click the arrow to add them to the selection box.

Comments
You can compare several pairs of variables at once. Options include the choice to exclude cases missing in any pair from all tests.

Brief CASE

Real Estate

In Chapter 6 we examined the regression of the sales price of a home on its size and saw that larger homes generally fetch a higher price. How much can we learn about a house from the fact that it has a fireplace or more than the average number of bedrooms? Data for a random sample of 1047 homes from the upstate New York area can be found in the file **Saratoga_Real_Estate**. There are 6 quantitative variables: *Price($)*, *Living Area(sq.ft)*, *Bathrooms(#)*, *Bedrooms(#)*, *Lot Size(Acres)*, and *Age(years)*, and one categorical variable, *Fireplace?(1 = Yes; 0 = No)* denoting whether the house has at least one fireplace. We can use *t*-methods to see, for example, whether homes with fireplaces sell for more, on average, and by how much. For the quantitative variables, create new categorical variables by splitting them at the median or some other splitting point of your choice, and compare home prices above and below this value. For example, the median number of *Bedrooms* of these homes is 2. You might compare the prices of homes with 1 or 2 bedrooms to those with more than 2. Write up a short report summarizing the differences in mean price based on the categorical variables that you created.

Consumer Spending Patterns (Data Analysis)

You are on the financial planning team for monitoring a high spending segment of a credit card. You know that customers tend to spend more during December before the holidays, but you're not sure about the pattern of spending in the months after the holidays. Look at the data set **Consumer_spending_post_holiday**. It contains the monthly credit card spending of 1200 customers during the months December, January, February, March, and April. Report on the spending differences between the months. If you had failed to realize that these are paired data, what difference would that have made in your reported confidence intervals and tests?

Exercises

SECTION 14.1

1. A developer wants to know if the houses in two different neighborhoods were built at roughly the same time. She takes a random sample of six houses from each neighborhood and finds their ages from local records. The table shows the data for each sample (in years).

Neighborhood 1	Neighborhood 2
57	50
60	43
46	35
62	53
67	46
56	55

a) Find the sample means for each neighborhood.
b) Find the estimated difference of the mean ages of the two neighborhoods.
c) Find the sample variances for each neighborhood.
d) Find the sample standard deviations for each neighborhood.
e) Find the standard error of the difference of the two sample means.

2. A market analyst wants to know if the new website he designed is showing increased page views per visit. A customer is randomly sent to one of two different websites, offering the same products, but with different designs.

Here are the page views from five randomly chosen customers from each website:

Website A	Website B
9	10
4	13
14	2
7	3
2	7

a) Find the sample mean page views for each website.
b) Find the estimated difference of the sample mean page views of the two websites.
c) Find the sample variances for each website.
d) Find the sample standard deviations for each website.
e) Find the standard error of the difference of the sample means.

3. The developer in Exercise 1 hires an assistant to collect a random sample of houses from each neighborhood and finds that the summary statistics for the two neighborhoods look as follows:

Neighborhood 1	Neighborhood 2
$n_1 = 30$	$n_2 = 35$
$\bar{y}_1 = 57.2$ yrs	$\bar{y}_2 = 47.6$ yrs
$s_1 = 7.51$ yrs	$s_2 = 7.85$ yrs

a) Find the estimated mean age difference between the two neighborhoods.
b) Find the standard error of the estimated mean difference.
c) Calculate the t-statistic for the observed difference in mean ages assuming that the true mean difference is 0.

4. Not happy with the previous results, the analyst in Exercise 2 takes a much larger random sample of customers from each website and records their page views. Here are the data:

Website 1	Website 2
$n_1 = 80$	$n_2 = 95$
$\bar{y}_1 = 7.7$ pages	$\bar{y}_2 = 7.3$ pages
$s_1 = 4.6$ pages	$s_2 = 4.3$ pages

a) Find the estimated mean difference in page visits between the two websites.
b) Find the standard error of the estimated mean difference.
c) Calculate the t-statistic for the observed difference in mean page visits assuming that the true mean difference is 0.

SECTION 14.2

5. For the data in Exercise 1, we want to test the null hypothesis that the mean age of houses in the two neighborhoods is the same. Assume that the data come from a population that is Normally distributed.

a) Using the values you found in Exercise 1, find the value of the t-statistic for the difference in mean ages for the null hypothesis.
b) Calculate the degrees of freedom from the formula in the footnote of page 402.
c) Calculate the degrees of freedom using the rule that df = $\min(n_1 - 1, n_2 - 1)$.
d) Find the P-value using the degrees of freedom from part b. (You can either round the number of df and use a table or use technology or a website).
e) Find the P-value using the degrees of freedom from part d.
f) What do you conclude at $\alpha = 0.05$?

6. For the data in Exercise 2, we want to test the null hypothesis that the mean number of page visits is the same for the two websites. Assume that the data come from a population that is Normally distributed.

a) Using the values you found in Exercise 2, find the value of the t-statistic for the difference in mean ages for the null hypothesis.
b) Calculate the degrees of freedom from the formula in the footnote of page 402.
c) Calculate the degrees of freedom using the rule that df = $\min(n_1 - 1, n_2 - 1)$.
d) Find the P-value using the degrees of freedom from part b. (You can either round the number of df and use a table or use technology or a website).
e) Find the P-value using the degrees of freedom from part c.
f) What do you conclude at $\alpha = 0.05$?

7. Using the data in Exercise 3, test the hypothesis that the mean ages of houses in the two neighborhoods is the same. You may assume that the ages of houses in each neighborhood follow a Normal distribution.

a) Calculate the P-value of the statistic knowing that the approximation formula gives 53.4 df (you will need to round or to use technology).
b) Calculate the P-value of the statistic using the rule that df is at least $\min(n_1 - 1, n_2 - 1)$.
c) What do you conclude at $\alpha = 0.05$?

8. Using the data in Exercise 4, test the hypothesis that the mean number of page views from the two websites is the same. You may assume that the number of page views from each website follow a Normal distribution.

a) Calculate the P-value of the statistic knowing that the approximation formula gives 148.0 df.
b) Calculate the P-value of the statistic using the rule that df is at least $\min(n_1 - 1, n_2 - 1)$.
c) What do you conclude at $\alpha = 0.05$?

SECTION 14.4

9. Using the data in Exercise 1, and assuming that the data come from a distribution that is Normally distributed,

a) Find a 95% confidence interval for the mean difference in ages of houses in the two neighborhoods.
b) Is 0 within the confidence interval?
c) What does it say about the null hypothesis that the mean difference is 0?

10. Using the data in Exercise 2, and assuming that the data come from a distribution that is Normally distributed,

a) Find a 95% confidence interval for the mean difference in page views from the two websites.
b) Is 0 within the confidence interval?
c) What does it say about the null hypothesis that the mean difference is 0?

11. Using the summary statistics in Exercise 3, and assuming that the data come from a distribution that is Normally distributed,

a) Find a 95% confidence interval for the mean difference in ages of houses in the two neighborhoods using the df given in Exercise 7.
b) Why is the confidence interval narrower than the one you found in Exercise 9?
c) Is 0 within the confidence interval?
d) What does it say about the null hypothesis that the mean difference is 0?

12. Using the summary statistics in Exercise 4, and assuming that the data come from a distribution that is Normally distributed,

a) Find a 95% confidence interval for the mean difference in page views from the two websites.
b) Why is the confidence interval narrower than the one you found in Exercise 10?
c) Is 0 within the confidence interval?
d) What does it say about the null hypothesis that the mean difference is 0?

SECTION 14.5

13. For the data in Exercise 1,

a) Test the null hypothesis at $\alpha = 0.05$ using the pooled t-test. (Show the t-statistic, P-value, and conclusion.)
b) Find a 95% confidence interval using the pooled degrees of freedom.
c) Are your answers different from what you previously found in Exercise 9? Explain briefly why or why not.

14. For the data in Exercise 2,

a) Test the null hypothesis $\alpha = 0.05$ using the pooled t-test. (Show the t-statistic, P-value, and conclusion.)
b) Find a 95% confidence interval using the pooled degrees of freedom.

c) Are your answers different from what you previously found in Exercise 10? Explain briefly why or why not.

15. For the data in Exercise 3,

a) Test the null hypothesis $\alpha = 0.05$ using the pooled t-test. (Show the t-statistic, P-value, and conclusion.)
b) Find a 95% confidence interval using the pooled degrees of freedom.
c) Are your answers different from what you previously found in Exercise 11? Explain briefly why or why not.

16. For the data in Exercise 4,

a) Test the null hypothesis $\alpha = 0.05$ using the pooled t-test. (Show the t-statistic, P-value, and conclusion.)
b) Find a 95% confidence interval using the pooled degrees of freedom.
c) Are your answers different from what you previously found in Exercise 12? Explain briefly why or why not.

SECTION 14.6

17. Is Tukey's test applicable to the data from Exercise 1? Why or why not? If it is, perform Tukey's test and state your conclusion.

18. Is Tukey's test applicable to the data from Exercise 2? Why or why not? If it is, perform Tukey's test and state your conclusion.

SECTION 14.7

19. For each of the following scenarios, say whether the data should be treated as independent or paired samples. Explain briefly. If paired, explain what the pairing involves.

a) An efficiency expert claims that a new ergonomic desk chair makes typing at a computer terminal easier and faster. To test it, 15 volunteers are selected. Using both the new chair and their old chair, each volunteer types a randomly selected passage for 2 minutes and the number of correct words typed is recorded.
b) A developer wants to know if the houses in two different neighborhoods have the same mean price. She selects 10 houses from each neighborhood at random and tests the null hypothesis that the means are equal.
c) A manager wants to know if the mean productivity of two workers is the same. For a random selection of 30 hours in the past month, he compares the number of items produced by each worker in that hour.

20. For each of the following scenarios, say whether the data should be treated as independent or paired samples. Explain briefly. If paired, explain what the pairing involves.

a) An efficiency expert claims that a new ergonomic desk chair makes typing at a computer terminal easier and faster.

To test it, 30 volunteers are selected. Half of the volunteers will use the new chair and half will use their old chairs. Each volunteer types a randomly selected passage for 2 minutes and the number of correct words typed is recorded.

b) A real estate agent wants to know how much extra a fireplace adds to the price of a house. She selects 25 city blocks. In each block she randomly chooses a house with a fireplace and one without and records the assessment value.

c) A manager wants to know if the mean productivity of two workers is the same. For each worker he randomly selects 30 hours in the past month and compares the number of items produced.

SECTION 14.8

21. A supermarket chain wants to know if their "buy one, get one free" campaign increases customer traffic enough to justify the cost of the program. For each of 10 stores they select two days at random to run the test. For one of those days (selected by a coin flip), the program will be in effect. They want to test the hypothesis that there is no mean difference in traffic against the alternative that the program increases the mean traffic. Here are the results in number of customer visits to the 10 stores:

Store #	With Program	Without Program
1	140	136
2	233	235
3	110	108
4	42	35
5	332	328
6	135	135
7	151	144
8	33	39
9	178	170
10	147	141

a) Are the data paired? Explain.
b) Compute the mean difference.
c) Compute the standard deviation of the differences.
d) Compute the standard error of the mean difference.
e) Find the value of the *t*-statistic.
f) How many degrees of freedom does the *t*-statistic have?
g) Is the alternative one- or two-sided? Explain.
h) What is the P-value associated with this *t*-statistic? (Assume that the other assumptions and conditions for inference are met.)
i) At $\alpha = 0.05$, what do you conclude?

22. A city wants to know if a new advertising campaign to make citizens aware of the dangers of driving after drinking has been effective. They count the number of drivers who have been stopped with more alcohol in their systems than the law allows for each day of the week in the week before

and the week a month after the campaign starts. Here are the results:

Day of week	Before	After
M	5	2
T	4	0
W	2	2
Th	4	1
F	6	8
S	14	7
Su	6	7

a) Are the data paired? Explain.
b) Compute the mean difference.
c) Compute the standard deviation of the differences.
d) Compute the standard error of the mean difference.
e) Find the value of the *t*-statistic.
f) How many degrees of freedom does the *t*-statistic have?
g) Is the alternative one- or two-sided? Explain.
h) What is the P-value associated with this *t*-statistic? (Assume that the other assumptions and conditions for inference are met.)
i) At $\alpha = 0.05$, what do you conclude?

23. In order to judge whether the program is successful, the manager of the supermarket chain in Exercise 21 wants to know the plausible range of values for the mean increase in customers using the program. Construct a 90% confidence interval.

24. A new operating system is installed in every workstation at a large company. The claim of the operating system manufacturer is that the time to shut down and turn on the machine will be much faster. To test it an employee selects 36 machines and tests the combined shut down and restart time of each machine before and after the new operating system has been installed. The mean and standard deviation of the differences (before – after) is 23.5 seconds with a standard deviation of 40 seconds.

a) What is the standard error of the mean difference?
b) How many degrees of freedom does the *t*-statistic have?
c) What is the 90% confidence interval for the mean difference?
d) What do you conclude at $\alpha = 0.05$?

CHAPTER EXERCISES

25. Hot dogs and calories. Consumers increasingly make food purchases based on nutrition values. In the July 2007 issue, *Consumer Reports* examined the calorie content of two kinds of hot dogs: meat (usually a mixture of pork, turkey, and chicken) and all beef. The researchers purchased samples of several different brands. The meat hot dogs averaged 111.7 calories, compared to 135.4 for the beef hot dogs. A test of the null hypothesis that there's no difference in mean calorie content yields a P-value of 0.124. What would you conclude?

26. Hot dogs and sodium. The *Consumer Reports* article described in Exercise 25 also listed the sodium content (in mg) for the various hot dogs tested. A test of the null hypothesis that beef hot dogs and meat hot dogs don't differ in the mean amounts of sodium yields a P-value of 0.110. What would you conclude?

27. Learning math. The Core Plus Mathematics Project (CPMP) is an innovative approach to teaching mathematics that engages students in group investigations and mathematical modeling. After field tests in 36 high schools over a three-year period, researchers compared the performances of CPMP students with those taught using a traditional curriculum. In one test, students had to solve applied algebra problems using calculators. Scores for 320 CPMP students were compared with those of a control group of 273 students in a traditional math program. Computer software was used to create a confidence interval for the difference in mean scores (*Journal for Research in Mathematics Education*, 31, no. 3, 2000).

```
Conf. level: 95%
Variable: μ(CPMP) − μ(Ctrl)
Interval: (5.573, 11.427)
```

a) What is the margin of error for this confidence interval?
b) If we had created a 98% confidence interval, would the margin of error be larger or smaller?
c) Explain what the calculated interval means in this context.
d) Does this result suggest that students who learn mathematics with CPMP will have significantly higher mean scores in applied algebra than those in traditional programs? Explain.

28. Sales performance. A chain that specializes in healthy and organic food would like to compare the sales performance of two of its primary stores in the state of Massachusetts. These stores are both in urban, residential areas with similar demographics. A comparison of the weekly sales randomly sampled over a period of nearly two years for these two stores yields the following information:

```
Store     N Mean   StDev Minimum Median Maximum
Store #1 9 242170 23937  211225 232901  292381
Store #2 9 235338 29690  187475 232070  287838
```

a) Create a 95% confidence interval for the difference in the mean store weekly sales.
b) Interpret your interval in context.
c) Does it appear that one store sells more on average than the other store?
d) What is the margin of error for this interval?
e) Would you expect a 99% confidence interval to be wider or narrower? Explain.
f) If you computed a 99% confidence interval, would your conclusion in part c change? Explain.

29. CPMP, again. During the study described in Exercise 27, students in both CPMP and traditional classes took another algebra test that did not allow them to use calculators. The table shows the results. Are the mean scores of the two groups significantly different? Assume that the assumptions for inference are satisfied.

Math Program	n	Mean	SD
CPMP	312	29.0	18.8
Traditional	265	38.4	16.2

a) Write an appropriate hypothesis.
b) Here is computer output for this hypothesis test. Explain what the P-value means in this context.

```
2-Sample t-Test of μ1 − μ2 ≠ 0
t-Statistic = −6.451 w/574.8761 df
P < 0.0001
```

c) State a conclusion about the CPMP program.

30. IT training costs. An accounting firm is trying to decide between IT training conducted in-house and the use of third party consultants. To get some preliminary cost data, each type of training was implemented at two of the firm's offices located in different cities. The table below shows the average annual training cost per employee at each location. Are the mean costs significantly different? Assume that the assumptions for inference are satisfied.

IT Training	n	Mean	SD
In-House	210	$490.00	$32.00
Consultants	180	$500.00	$48.00

a) Write the appropriate hypotheses.
b) Below is computer output for this hypothesis test. Explain what the P-value means in this context.

```
2-Sample t-Test of μ₁ − μ₂ ≠ 0
t-Statistic = −2.38 w/303df
P = .018
```

c) State a conclusion about IT training costs.

31. CPMP and word problems. The study of the new CPMP mathematics methodology described in Exercise 27 also tested students' abilities to solve word problems. This table shows how the CPMP and traditional groups performed. What do you conclude? (Assume that the assumptions for inference are met.)

Math Program	n	Mean	SD
CPMP	320	57.4	32.1
Traditional	273	53.9	28.5

32. Statistical training. The accounting firm described in Exercise 30 is interested in providing opportunities for its auditors to gain more expertise in statistical sampling methods. They wish to compare traditional classroom instruction with online self-paced tutorials. Auditors were assigned at random to one type of instruction, and the auditors were then given an exam. The table shows how the two groups performed. What do you conclude? (Assume the assumptions for inference are met.)

Program	n	Mean	SD
Traditional	296	74.5	11.2
Online	275	72.9	12.3

33. Trucking company. A trucking company would like to compare two different routes for efficiency. Truckers are randomly assigned to two different routes. Twenty truckers following Route A report an average of 40 minutes, with a standard deviation of 3 minutes. Twenty truckers following Route B report an average of 43 minutes, with a standard deviation of 2 minutes. Histograms of travel times for the routes are roughly symmetric and show no outliers.

a) Find a 95% confidence interval for the difference in average time for the two routes.
b) Will the company save time by always driving one of the routes? Explain.

34. Change in sales. Suppose the specialty food chain from Exercise 28 wants to now compare the change in sales across different regions. An examination of the difference in sales over a 37-week period in a recent year for 8 stores in the state of Massachusetts compared to 12 stores in nearby states reveals the following descriptive statistics for relative increase in sales. (If these means are multiplied by 100, they show % increase in sales.)

```
State    N     Mean      StDev

MA       8     0.0738    0.0666
Other    12    0.0559    0.0503
```

a) Find the 90% confidence interval for the difference in relative increase in sales over this time period.
b) Is there a significant difference in increase in sales between these two groups of stores? Explain.
c) What would you like to see to check the conditions?

T 35. Cereal company. A food company is concerned about recent criticism of the sugar content of their children's cereals. The data show the sugar content (as a percentage of weight) of several national brands of children's and adults' cereals.

> **Children's cereals:** 40.3, 55, 45.7, 43.3, 50.3, 45.9, 53.5, 43, 44.2, 44, 47.4, 44, 33.6, 55.1, 48.8, 50.4, 37.8, 60.3, 46.6

> **Adults' cereals:** 20, 30.2, 2.2, 7.5, 4.4, 22.2, 16.6, 14.5, 21.4, 3.3, 6.6, 7.8, 10.6, 16.2, 14.5, 4.1, 15.8, 4.1, 2.4, 3.5, 8.5, 10, 1, 4.4, 1.3, 8.1, 4.7, 18.4

a) Write the null and alternative hypotheses.
b) Check the conditions.
c) Find the 95% confidence interval for the difference in means.
d) Is there a significant difference in mean sugar content between these two types of cereals? Explain.

T 36. Foreclosure rates. According to recent reports, home foreclosures were up 47% in March 2008 compared to the previous year (realestate.msn.com; April 2008). The data show home foreclosure rates (as % change from the previous year) for a sample of cities in two regions of the U.S., the Northeast and the Southwest.

> **Northeast:** 2.99, −2.36, 3.03, 1.01, 5.77, 9.95, −3.52, 7.16, −3.34, 4.75, 5.25, 6.21, 1.67, −2.45, −0.55, 3.45, 4.50, 1.87, −2.15, −0.75

> **Southwest:** 10.15, 23.05, 18.95, 21.16, 17.45, 12.67, 13.75, 29.42, 11.45, 16.77, 12.67, 13.69, 25.81, 21.16, 19.67, 11.88, 13.67, 18.00, 12.88

a) Write the null and alternative hypotheses.
b) Check the conditions.
c) Test the hypothesis and find the P-value.
d) Is there a significant difference in the mean home foreclosure rates between these two regions of the U.S.? Explain.

37. Investment. Investment style plays a role in constructing a mutual fund. Each individual stock is grouped into two distinct groups: "Growth" and "Value." A Growth stock is one with high earning potential and often pays little or no dividends to shareholders. Conversely, Value stocks are commonly viewed as steady, or more conservative, with a lower earning potential. You are trying to decide what type of funds to invest in. Because you are saving toward your retirement, if you invest in a Value fund, you hope that the fund remains conservative. We would call such a fund "consistent." If the fund did not remain consistent and became higher risk, that could impact your retirement savings. The funds in this data set have been identified as either being "style consistent" or "style drifter." Portfolio managers wonder whether consistency provides the optimal chance for successful retirement, so they believe that style consistent funds outperform style drifters. Out of a sample of 140 funds, 66 were identified as style consistent, while 74 were identified as style drifters. Their statistics for their average return over 5 years are:

Type	N	Mean	StDev	Minimum	Q1	Q3	Maximum 5-yr Return
Consistent	66	9.382	2.675	1.750	7.675	11.110	15.920
Drifter	74	8.563	3.719	−0.870	5.928	11.288	17.870

a) Write the null and alternative hypotheses.
b) Find the 95% confidence interval of the difference in mean return between style consistent and style drifter funds.
c) Is there a significant difference in 5-year return for these two types of funds? Explain.

38. Technology adoption. The Pew Internet & American Life Project (www.pewinternet.org/) conducts surveys to gauge how the Internet and technology impact daily life of individuals, families, and communities. In a recent survey Pew asked respondents if they thought that computers and technology give people more or less control over their lives. Companies that are involved in innovative technologies use the survey results to better understand their target market. One might suspect that younger and older respondents might differ in their opinions of whether computers and technology give them more control over their lives. A subset of the data from this survey (*February–March 2007 Tracking Data Set*) shows the mean ages of two groups of respondents, those who reported that they believed that computers and technology give them "more" control and those that reported "less" control.

Group	N	Mean	StDev	Min	Q1	Med	Q3	Max
More	74	54.42	19.65	18	41.5	53.5	68.5	99.0
Less	29	54.34	18.57	20	41.0	58.0	70.0	84.0

a) Write the null and alternative hypotheses.
b) Find the 95% confidence interval for the difference in mean age between the two groups of respondents.
c) Is there a significant difference in the mean ages between these two groups? Explain.

T 39. Product testing. A company is producing and marketing new reading activities for elementary school children that it believes will improve reading comprehension scores. A researcher randomly assigns third graders to an eight-week program in which some will use these activities and others will experience traditional teaching methods. At the end of the experiment, both groups take a reading comprehension exam. Their scores are shown in the back-to-back stem-and-leaf display. Do these results suggest that the new activities are better? Test an appropriate hypothesis and state your conclusion.

New Activities		Control
	1	07
4	2	068
3	3	377
96333	4	12222238
9876432	5	355
721	6	02
1	7	
	8	5

T 40. Product placement. The owner of a small organic food store was concerned about her sales of a specialty yogurt manufactured in Greece. As a result of increasing fuel costs, she recently had to increase its price. To help boost sales, she decided to place the product on a different shelf (near eye level for most consumers) and in a location near other popular international products. She kept track of sales (number of containers sold per week) for six months after she made the change. These values are shown below, along with the sales numbers for the six months prior to making the change, in stem-and-leaf displays.

After Change				Before Change
3	2		2	0
3	9		2	899
4	23		3	224
4	589		3	7789
5	0012		4	0000223
5	55558		4	5567
6	00123		5	0
6	67		5	6
7	0			

Do these results suggest that sales are better after the change in product placement? Test an appropriate hypothesis and state your conclusion. Be sure to check assumptions and conditions.

T 41. Acid streams. Researchers collected samples of water from streams in the Adirondack Mountains to investigate the effects of acid rain. They measured the pH (acidity) of the water and classified the streams with respect to the kind of substrate (type of rock over which they flow). A lower pH means the water is more acidic. Here is a plot of the pH of the streams by substrate (limestone, mixed, or shale):

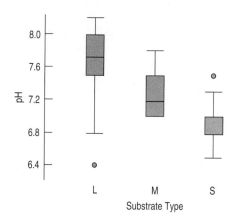

Here are selected parts of a software analysis comparing the pH of streams with limestone and shale substrates:

```
2-Sample t-Test of μ1 − μ2 = 0
Difference Between Means = 0.735
t-Statistic = 16.30 w/133 df
P ≤ 0.0001
```

a) State the null and alternative hypotheses for this test.
b) From the information you have, do the assumptions and conditions appear to be met?
c) What conclusion would you draw?

T 42. Hurricanes. It has been suggested that global warming may increase the frequency of hurricanes. The data show the number of hurricanes recorded annually before and after 1970. Has the frequency of hurricanes increased since 1970?

Before (1944–1969)	After (1970–2000)
3, 3, 1, 2, 4, 3, 8, 5, 3, 4, 2,	2, 1, 0, 1, 2, 3, 2, 1, 2, 2, 2,
6, 2, 2, 5, 2, 2, 7, 1, 2, 6, 1,	3, 1, 1, 1, 3, 0, 1, 3, 2, 1, 2,
3, 1, 0, 5	1, 1, 0, 5, 6, 1, 3, 5, 3, 4, 2, 3, 6, 7, 2

a) Write the null and alternative hypotheses.
b) Are the conditions for hypothesis testing satisfied?
c) If so, test the hypothesis.

T 43. Ginko test. A pharmaceutical company is producing and marketing a ginkgo biloba supplement to enhance memory. In an experiment to test the product, subjects were assigned randomly to take ginkgo biloba supplements or a placebo. Their memory was tested to see whether it improved. Here are boxplots comparing the two groups and some computer output from a two-sample *t*-test computed for the data.

```
2-Sample t-Test of μ_G − μ_p > 0
Difference Between Means = −0.9914
t-Statistic = −1.540 w/196 df
P = 0.9374
```

a) Explain in this context what the P-value means.
b) State your conclusion about the effectiveness of ginkgo biloba.
c) Proponents of ginkgo biloba continue to insist that it works. What type of error do they claim your conclusion makes?

T 44. Designated hitter 2009. American League baseball teams play their games with the designated hitter rule, meaning that pitchers do not bat. The league believes that replacing the pitcher, traditionally a weak hitter, with another player in the batting order produces more runs and

generates more interest among fans. The data provided in the file on the CD include the average numbers of runs scored per game (*Runs per game*) by American League and National League teams for the 2009 season (http://espn.go.com/mlb/stats/team/_/stat/batting/year/2009).

a) Create an appropriate display of these data. What do you see?
b) With a 95% confidence interval, estimate the mean number of runs scored by American League teams.
c) With a 95% confidence interval, estimate the mean number of runs scored by National League teams.
d) Explain why you should not use two separate confidence intervals to decide whether the two leagues differ in average number of runs scored.

45. Productivity. A factory hiring people to work on an assembly line gives job applicants a test of manual agility. This test counts how many strangely shaped pegs the applicant can fit into matching holes in a one-minute period. The table summarizes the data by gender of the job applicant. Assume that all conditions necessary for inference are met.

	Male	Female
Number of subjects	50	50
Pegs placed:		
Mean	19.39	17.91
SD	2.52	3.39

a) Find 95% confidence intervals for the average number of pegs that males and females can each place.
b) Those intervals overlap. What does this suggest about any gender-based difference in manual agility?
c) Find a 95% confidence interval for the difference in the mean number of pegs that could be placed by men and women.
d) What does this interval suggest about any gender-based difference in manual agility?
e) The two results seem contradictory. Which method is correct: doing two-sample inference, or doing one-sample inference twice?
f) Why don't the results agree?

46. Online shopping. Online shopping statistics are routinely reported by www.shop.org. Of interest to many online retailers are gender-based differences in shopping preferences and behaviors. The average monthly online expenditures are reported for males and females:

		Male	Female
GROUP	*n*	45	45
	Mean	$352	$310
	StDev	$95	$80

a) Find 95% confidence intervals for the average monthly online expenditures for males and females.
b) These intervals overlap. What does this suggest about any gender-based difference in monthly online expenditures?
c) Find a 95% confidence interval for the *difference* in average monthly online expenditures between males and females.
d) The two results seem contradictory. Which method is correct: doing two-sample inference, or doing one-sample inference twice?

T 47. Designated hitter 2009, part 2. Do the data in Exercise 44 suggest that the American League's designated hitter rule may lead to more runs per game scored?

a) Write the null and alternative hypotheses.
b) Find a 95% confidence interval for the difference in mean, Runs per game, and interpret your interval.
c) Test the hypothesis stated above in part a and find the P-value.
d) Interpret the P-value and state your conclusion. Does the test suggest that the American League scores more runs on average?

48. Online shopping, again. In 2004, it was reported that the average male spends more money shopping online per month than the average female, $204 compared to $186 (www.shop.org; accessed April 2008). Do the data reported in Exercise 46 indicate that this is still true?

a) Write the null and alternative hypotheses.
b) Test the hypothesis stated in part a and find the P-value.
c) Interpret the P-value and state your conclusion. Does the test suggest that males continue to spend more online on average than females?

T 49. Water hardness. In an investigation of environmental causes of disease, data were collected on the annual mortality rate (deaths per 100,000) for males in 61 large towns in England and Wales. In addition, the water hardness was recorded as the calcium concentration (parts per million, ppm) in the drinking water. The data set also notes for each town whether it was south or north of Derby. Is there a significant difference in mortality rates in the two regions? Here are the summary statistics.

```
Summary of:          mortality
For categories in:   Derby
```

Group	Count	Mean	Median	StdDev
North	34	1631.59	1631	138.470
South	27	1388.85	1369	151.114

a) Test appropriate hypotheses and state your conclusion.
b) The boxplots of the two distributions show an outlier among the data north of Derby. What effect might that have had on your test?

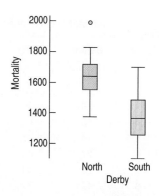

50. Sustainable stocks. The earnings per share ratio (EPS) is one of several important indicators of a company's profitability. There are several categories of "sustainable" stocks including natural foods/health and green energy/bio fuels. Below are earnings per share for a sample of stocks from both of these categories (Yahoo Financial, April 6, 2008). Is there a significant difference in earnings per share values for these two groups of sustainable stocks?

Group	Count	Mean	Median	StDev
Foods/Health	15	0.862	1.140	0.745
Energy/Fuel	16	−0.320	−0.545	0.918

a) Test appropriate hypotheses and state your conclusion.
b) Based upon the boxplots of the two distributions shown below, what might you suspect about your test? Explain.

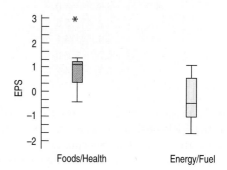

T 51. Job satisfaction. A company institutes an exercise break for its workers to see if this will improve job satisfaction, as measured by a questionnaire that assesses workers' satisfaction.

Worker Number	Job Satisfaction Index	
	Before	**After**
1	34	33
2	28	36
3	29	50
4	45	41
5	26	37
6	27	41
7	24	39
8	15	21
9	15	20
10	27	37

a) Identify the procedure you would use to assess the effectiveness of the exercise program and check to see if the conditions allow for the use of that procedure.
b) Test an appropriate hypothesis and state your conclusion.

52. ERP effectiveness. When implementing a packaged Enterprise Resource Planning (ERP) system, many companies report that the module they first install is Financial Accounting. Among the measures used to gauge the effectiveness of their ERP system implementation is acceleration of the financial close process. Below is a sample of 8 companies that report their average time (in weeks) to financial close before and after the implementation of their ERP system.

Company	Before	After
1	6.5	4.2
2	7.0	5.9
3	8.0	8.0
4	4.5	4.0
5	5.2	3.8
6	4.9	4.1
7	5.2	6.0
8	6.5	4.2

a) Identify the procedure you would use to assess the effectiveness of the ERP system and check to see if the conditions allow for the use of that procedure.
b) Test an appropriate hypothesis and state your conclusion.

53. Delivery time. A small appliance company is interested in comparing delivery times of their product during two months. They are concerned that the summer slow-downs in August cause delivery times to lag during this month. Given the following delivery times (in days) of their appliances to the customer for a random sample of 6 orders each month, test if delivery times differ across these two months.

June	54	49	68	66	62	62
August	50	65	74	64	68	72

54. Branding. In June 2002, the *Journal of Applied Psychology* reported on a study that examined whether the content of TV shows influenced the ability of viewers to recall brand names of items featured in the commercials. The researchers randomly assigned volunteers to watch one of three programs, each containing the same nine commercials. One of the programs had violent content, another sexual content, and the third neutral content. After the shows ended, the subjects were asked to recall the brands of products that were advertised.

		Program Type		
		Violent	**Sexual**	**Neutral**
Brands Recalled	*n*	108	108	108
	Mean	2.08	1.71	3.17
	SD	1.87	1.76	1.77

a) Do these results indicate that viewer memory for ads may differ depending on program content? Test the hypothesis that there is no difference in ad memory between programs with sexual content and those with violent content. State your conclusion.
b) Is there evidence that viewer memory for ads may differ between programs with sexual content and those with neutral content? Test an appropriate hypothesis and state your conclusion.

55. Ad campaign. You are a consultant to the marketing department of a business preparing to launch an ad campaign for a new product. The company can afford to run ads during one TV show, and has decided not to sponsor a show with sexual content. You read the study described in Exercise 54 and then use a computer to create a confidence interval for the difference in mean number of brand names remembered between the groups watching violent shows and those watching neutral shows.

```
Two-Sample t
  95% CI for μ_viol − μ_neut: (−1.578, −0.602)
```

a) At the meeting of the marketing staff, you have to explain what this output means. What will you say?
b) What advice would you give the company about the upcoming ad campaign?

56. Branding, part 2. In the study described in Exercise 54, the researchers also contacted the subjects again, 24 hours later, and asked them to recall the brands advertised. Results for the number of brands recalled are summarized in the table.

	Program Type		
	Violent	**Sexual**	**Neutral**
No. of subjects	101	106	103
Mean	3.02	2.72	4.65
SD	1.61	1.85	1.62

a) Is there a significant difference in viewers' abilities to remember brands advertised in shows with violent vs. neutral content?

b) Find a 95% confidence interval for the difference in mean number of brand names remembered between the groups watching shows with sexual content and those watching neutral shows. Interpret your interval in this context.

57. Ad recall. In Exercises 54 and 56, we see the number of advertised brand names people recalled immediately after watching TV shows and 24 hours later. Strangely enough, it appears that they remembered more about the ads the next day. Should we conclude this is true in general about people's memory of TV ads?

a) Suppose one analyst conducts a two-sample hypothesis test to see if memory of brands advertised during violent TV shows is higher 24 hours later. The P-value is 0.00013. What might she conclude?

b) Explain why her procedure was inappropriate. Which of the assumptions for inference was violated?

c) How might the design of this experiment have tainted these results?

d) Suggest a design that could compare immediate brand name recall with recall one day later.

58. Hybrid SUVs. The Chevy Tahoe Hybrid got a lot of attention in 2008. It is a relatively high-priced hybrid SUV that makes use of the latest technologies for fuel efficiency. One of the more popular hybrid SUVs on the market is the modestly priced Ford Escape Hybrid. A consumer group was interested in comparing the gas mileage of these two models. In order to do so, each vehicle was driven on the same 10 routes that combined both highway and city streets. The results showed that the mean mileage for the Chevy Tahoe was 29 mpg and for the Ford Escape it was 31 mpg. The standard deviations were 3.2 mpg and 2.5 mpg, respectively.

a) An analyst for the consumer group computed the two-sample *t* 95% confidence interval for the difference between the two means as $(-.71, 4.71)$. What conclusion would he reach based on this analysis?

b) Why is this procedure inappropriate? What assumption is violated?

c) In what way do you think this may have impacted the results?

59. Science scores. Newspaper headlines recently announced a decline in science scores among high school seniors. In 2000, 15,109 seniors tested by the National Assessment in Education Program (NAEP) scored a mean of 147 points. Four years earlier, 7537 seniors had averaged 150 points. The standard error of the difference in the mean scores for the two groups was 1.22.

a) Have the science scores declined significantly? Cite appropriate statistical evidence to support your conclusion.

b) The sample size in 2000 was almost double that in 1996. Does this make the results more convincing or less? Explain.

60. Credit card debt. The average household credit card debt has been reported to be between $8000 and $10,000. Often of interest is the average credit card debt carried by college students. In 2008, the average credit card debt for college students was reported to be $2200 based on 12,500 responses. A year earlier it was reported to be $2190 based on survey of 8200 college students. The standard error of the difference in mean credit card balances was $1.75.

a) Has the average credit card balance carried by college students increased significantly? Cite appropriate statistical evidence to support your conclusion.

b) Is this a meaningful difference to the typical student? Is it meaningful to a credit card company?

c) The sample size in 2008 is one and a half times that in 2007. Does this make the results more or less convincing? Explain.

61. The Internet. The NAEP report described in Exercise 59 compared science scores for students who had home Internet access with the scores of those who did not, as shown in the graph. They report that the differences are statistically significant.

a) Explain what "statistically significant" means in this context.

b) If their conclusion is incorrect, which type of error did the researchers commit?

c) Does this prove that using the Internet at home can improve a student's performance in science?

d) What companies might be interested in this information?

62. Credit card debt public or private. The average credit card debt carried by college students was compared at public versus private universities. It was reported that a significant difference existed between the two types of institutions and that students at private universities carried higher credit card debt.

a) Explain what "statistically significant" means in this context.

b) If this conclusion is incorrect, which type of error was committed?

c) Does this prove that students who choose to attend public institutions will carry lower credit card debt?

63. Pizza sales. A national food product company believes that it sells more frozen pizza during the winter months than during the summer months. Average weekly sales for a sample of stores in the Baltimore area over a three-year period provided the following data for sales volume (in pounds) during the two seasons.

Season	N	Mean	StDev	Minimum	Maximum
Winter	38	31234	13500	15312	73841
Summer	40	22475	8442	12743	54706

a) How much difference is there between the mean amount of this brand of frozen pizza sold (in pounds) between the two seasons? (Assume that this time frame represents typical sales in the Baltimore area.)
b) Construct and interpret a 95% confidence interval for the difference between weekly sales during the winter and summer months.
c) Suggest factors that might have influenced the sales of the frozen pizza during the winter months.

64. More pizza sales. Here's some additional information about the pizza sales data presented in Exercise 63. It is generally thought that sales spike during the weeks leading up to AFC and NFC football championship games, as well as leading up to the Super Bowl at the end of January each year. If we omit those 6 weeks of sales from this three-year period of weekly sales, the summary statistics look like this. Do sales appear to be higher during the winter months after omitting those weeks most influenced by football championship games?

Season	N	Mean	StDev	Minimum	Maximum
Winter	32	28995	9913	15312	48354
Summer	40	22475	8442	12743	54706

a) Write the null and alternative hypotheses.
b) Test the null hypotheses and state your conclusion.
c) Suggest additional factors that may influence pizza sales not accounted for in this exercise.

T 65. Olympic heats. In Olympic running events, preliminary heats are determined by random draw, so we should expect the ability level of runners in the various heats to be about the same, on average. Here are the times (in seconds) for the 400-m women's run in the 2004 Olympics in Athens for preliminary heats 2 and 5. Is there any evidence that the mean time to finish is different for randomized heats? Explain. Be sure to include a discussion of assumptions and conditions for your analysis.

Country	Name	Heat	Time
USA	HENNAGAN Monique	2	51.02
BUL	DIMITROVA Mariyana	2	51.29
CHA	NADJINA Kaltouma	2	51.50
JAM	DAVY Nadia	2	52.04
BRA	ALMIRAO Maria Laura	2	52.10
FIN	MYKKANEN Kirsi	2	52.53
CHN	BO Fanfang	2	56.01
BAH	WILLIAMS-DARLING Tonique	5	51.20
BLR	USOVICH Svetlana	5	51.37
UKR	YEFREMOVA Antonina	5	51.53
CMR	NGUIMGO Mireille	5	51.90
JAM	BECKFORD Allison	5	52.85
TOG	THIEBAUD-KANGNI Sandrine	5	52.87
SRI	DHARSHA K V Damayanthi	5	54.58

T 66. Swimming heats. In Exercise 65 we looked at the times in two different heats for the 400-m women's run from the 2004 Olympics. Unlike track events, swimming heats are *not* determined at random. Instead, swimmers are seeded so that better swimmers are placed in later heats. Here are the times (in seconds) for the women's 400-m freestyle from heats 2 and 5. Do these results suggest that the mean times of seeded heats are not equal? Explain. Include a discussion of assumptions and conditions for your analysis.

Country	Name	Heat	Time
ARG	BIAGIOLI Cecilia Elizabeth	2	256.42
SLO	CARMAN Anja	2	257.79
CHI	KOBRICH Kristel	2	258.68
MKD	STOJANOVSKA Vesna	2	259.39
JAM	ATKINSON Janelle	2	260.00
NZL	LINTON Rebecca	2	261.58
KOR	HA Eun-Ju	2	261.65
UKR	BERESNYEVA Olga	2	266.30
FRA	MANAUDOU Laure	5	246.76
JPN	YAMADA Sachiko	5	249.10
ROM	PADURARU Simona	5	250.39
GER	STOCKBAUER Hannah	5	250.46
AUS	GRAHAM Elka	5	251.67
CHN	PANG Jiaying	5	251.81
CAN	REIMER Brittany	5	252.33
BRA	FERREIRA Monique	5	253.75

67. Tee tests. Does it matter what kind of tee a golfer places the ball on? The company that manufactures "Stinger" tees claims that the thinner shaft and smaller head will lessen resistance and drag, reducing spin and allowing the ball to travel farther. In August 2003, Golf Laboratories, Inc. compared the distance traveled by golf balls hit off regular wooden tees to those hit off Stinger tees. All the balls were struck by the same golf club using a robotic device set to swing the club head at approximately 95 miles per hour. Summary statistics from the test are shown in the table. Assume that 6 balls were hit off each tee and that the data were suitable for inference.

		Total Distance (yards)	Ball Velocity (mph)	Club Velocity (mph)
Regular tee	Avg.	227.17	127.00	96.17
	SD	2.14	0.89	0.41
Stinger tee	Avg.	241.00	128.83	96.17
	SD	2.76	0.41	0.52

Is there evidence that balls hit off the Stinger tees would have a higher initial velocity?

68. Tee tests, part 2. Given the test results on golf tees described in Exercise 67, is there evidence that balls hit off Stinger tees would travel farther? Again assume that 6 balls were hit off each tee and that the data were suitable for inference.

69. Marketing slogan. A company is considering marketing their classical music as "music to study by." Is this a valid slogan? In a study conducted by some Statistics students, 62 people were randomly assigned to listen to rap music, music by Mozart, or no music while attempting to memorize objects pictured on a page. They were then asked to list all the objects they could remember. Here are summary statistics for each group.

	Rap	Mozart	No Music
Count	29	20	13
Mean	10.72	10.00	12.77
SD	3.99	3.19	4.73

a) Does it appear that it is better to study while listening to Mozart than to rap music? Test an appropriate hypothesis and state your conclusion.
b) Create a 90% confidence interval for the mean difference in memory score between students who study to Mozart and those who listen to no music at all. Interpret your interval.

70. Marketing slogan, part 2. Using the results of the experiment described in Exercise 69, does it matter whether one listens to rap music while studying, or is it better to study without music at all?

a) Test an appropriate hypothesis and state your conclusion.
b) If you concluded there is a difference, estimate the size of that difference with a 90% confidence interval and explain what your interval means.

71. Mutual funds returns. You have heard that if you leave your money in mutual funds for a longer period of time, you will see a greater return. So you would like to compare the 3-year and 5-year returns of a random sample of mutual funds to see if indeed, your return is expected to be greater if you leave your money in the funds for 5 years.

a) Using the data provided, check the conditions for this test.
b) Write the null and alternative hypotheses for this test.
c) Test the hypothesis and find the P-value if appropriate.
d) Find a 95% confidence interval for the mean difference.

72. Mutual funds returns, part 2. An investor now tells you that if you leave your money in as long as 10 years, you will see an even greater return, so you would like to compare the 5-year and 10-year returns of a random sample of mutual funds to see if your return is expected to be greater if you leave your money in the funds for 10 years.

a) Using the data provided, check the conditions for this test.
b) Write the null and alternative hypotheses for this test.
c) Test the hypothesis and find the P-value if appropriate.
d) Find a 95% confidence interval for the mean difference.

73. Real estate, two towns. Residents of neighboring towns in a state in the United States have an ongoing disagreement over who lays claim to the higher average price of a single-family home. Since you live in one of these towns, you decide to obtain a random sample of homes listed for sale with a major local realtor to investigate if there is actually any difference in the average home price.

a) Using the data provided on the CD, check the conditions for this test.
b) Write the null and alternative hypotheses for this test.
c) Test the hypothesis and find the P-value.
d) What is your conclusion?

74. Real estate, two towns, bigger sample. Residents of one of the towns discussed in Exercise 73 claim that since their town is much smaller, the sample size should be increased. Instead of random sampling 30 homes, you decide to sample 42 homes from the database to test the difference in the mean price of single-family homes in these two towns.

a) Using the data provided on the CD, check the conditions for this test.
b) Write the null and alternative hypotheses for this test.
c) Test the hypothesis and find the P-value.
d) What is your conclusion? Did the sample size make a difference?

75. Designated hitter, part 3. For the same reasons identified in Exercise 44, a friend of yours claims that the average number of home runs hit per game is higher in the American

League than in the National League. Using the same 2009 data as in Exercises 44 and 47, you decide to test your friend's theory.

a) Using the data provided on the CD, check the conditions for this test.
b) Write the null and alternative hypotheses for this test.
c) Test the hypothesis and find the P-value.
d) What is your conclusion?

76. Statistics journals. When a professional statistician has information to share with colleagues, he or she will submit an article to one of several Statistics journals for publication. This can be a lengthy process; typically, the article must be circulated for "peer review" and perhaps edited before being accepted for publication. Then the article must wait in line with other articles before actually appearing in print. In the Winter 1998 issue of *Chance* magazine, Eric Bradlow and Howard Wainer reported on this delay for several journals between 1990 and 1994. For 288 articles published in *The American Statistician*, the mean length of time between initial submission and publication was 21 months, with a standard deviation of 8 months. For 209 *Applied Statistics* articles, the mean time to publication was 31 months, with a standard deviation of 12 months. Create and interpret a 90% confidence interval for the difference in mean delay, and comment on the assumptions that underlie your analysis.

77. Labor force. Values for the labor force participation rate (proportion) of women (LFPR) are published by the U.S. Bureau of Labor Statistics. We are interested in whether there was a difference between female participation in 1968 and 1972, a time of rapid change for women. We check LFPR values for 19 randomly selected cities for 1968 and 1972. Here is software output for two possible tests.

```
Paired t-Test of μ(1 − 2)
Test Ho: μ(1972−1968) = 0 vs Ha: μ(1972−1968) ≠ 0
Mean of Paired Differences = 0.0337
t-Statistic = 2.458 w/18 df
p = 0.0244
2-Sample t-Test of μ1−μ2
Ho: μ1 − μ2 = 0 Ha: μ1 − μ2 ≠ 0
Test Ho: μ(1972) − μ(1968) = 0 vs
Ha: μ(1972) − μ(1968) ≠ 0
Difference Between Means = 0.0337
t-Statistic = 1.496 w/35 df
p = 0.1434
```

a) Which of these tests is appropriate for these data? Explain.
b) Using the test you selected, state your conclusion.

78. Cloud seeding. It has long been a dream of farmers to summon rain when it is needed for their crops. Crop losses to drought have significant economic impact. One possibility is cloud seeding in which chemicals are dropped into clouds in

an attempt to induce rain. Simpson, Alsen, and Eden (*Technometrics*, 1975) report the results of trials in which clouds were seeded and the amount of rainfall recorded. The authors report on 26 seeded (Group 2) and 26 unseeded (Group 1) clouds. Each group has been sorted in order of the amount of rainfall, largest amount first. Here are two possible tests to study the question of whether cloud seeding works.

```
Paired t-Test of μ(1 − 2)
Mean of Paired Differences = −277.4
t-Statistic = −3.641 w/25 df p = 0.0012
2-Sample t-test of μ1 − μ2
Difference Between Means = −277.4
t-Statistic = −1.998 w/33 df p = 0.0538
```

a) Which of these tests is appropriate for these data? Explain.
b) Using the test you selected, state your conclusion.

79. Online insurance. After seeing countless commercials claiming one can get cheaper car insurance from an online company, a local insurance agent was concerned that he might lose some customers. To investigate, he randomly selected profiles (type of car, coverage, driving record, etc.) for 10 of his clients and checked online price quotes for their policies. The comparisons are shown in the table. His statistical software produced the following summaries (where *PriceDiff = Local − Online*):

Variable	Count	Mean	StdDev
Local	10	799.200	229.281
Online	10	753.300	256.267
PriceDiff	10	45.9000	175.663

Local	Online	PriceDiff
568	391	177
872	602	270
451	488	−37
1229	903	326
605	677	−72
1021	1270	−249
783	703	80
844	789	55
907	1008	−101
712	702	10

At first, the insurance agent wondered whether there was some kind of mistake in this output. He thought the Pythagorean Theorem of Statistics should work for finding the standard deviation of the price differences—in other words, that $SD(Local − Online) = \sqrt{SD^2(Local) + SD^2(Online)}$. But when he checked, he found that $\sqrt{(229.281)^2 + (256.267)^2} = 343.864$, not 175.663 as given by the software. Tell him where his mistake is.

80. Windy. Alternative sources of energy are of increasing interest throughout the energy industry. Wind energy has great potential. But appropriate sites must be found for the turbines. To select the site for an electricity-generating wind turbine, wind speeds were recorded at several potential sites every 6 hours for a year. Two sites not far from each other looked good. Each had a mean wind speed high enough to qualify, but we should choose the site with a higher average daily wind speed. Because the sites are near each other and the wind speeds were recorded at the same times, we should view the speeds as paired. Here are the summaries of the speeds (in miles per hour):

Variable	Count	Mean	StdDev
site2	1114	7.452	3.586
site4	1114	7.248	3.421
site2 – site4	1114	0.204	2.551

Is there a mistake in this output? Why doesn't the Pythagorean Theorem of Statistics work here? In other words, shouldn't $SD(site2 - site4) = \sqrt{SD^2(site2) + SD^2(site4)}$? But $\sqrt{(3.586)^2 + (3.421)^2} = 4.956$, not 2.551 as given by the software. Explain why this happened.

81. Online insurance, part 2. In Exercise 79, we saw summary statistics for 10 drivers' car insurance premiums quoted by a local agent and an online company. Here are displays for each company's quotes and for the difference (*Local – Online*):

a) Which of the summaries would help you decide whether the online company offers cheaper insurance? Why?
b) The standard deviation of *PriceDiff* is quite a bit smaller than the standard deviation of prices quoted by either the local or online companies. Discuss why.
c) Using the information you have, discuss the assumptions and conditions for inference with these data.

82. Windy, part 2. In Exercise 80, we saw summary statistics for wind speeds at two sites near each other, both being considered as locations for an electricity-generating wind turbine. The data, recorded every 6 hours for a year, showed each of the sites had a mean wind speed high enough to qualify, but how can we tell which site is best? Here are some displays:

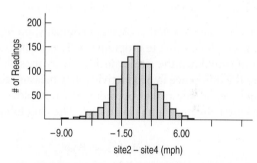

a) The boxplots show outliers for each site, yet the histogram shows none. Discuss why.
b) Which of the summaries would you use to select between these sites? Why?
c) Using the information you have, discuss the assumptions and conditions for paired *t* inference for these data. (*Hint:* Think hard about the independence assumption in particular.)

83. Online insurance, part 3. Exercises 79 and 81 give summaries and displays for car insurance premiums quoted by a local agent and an online company. Test an appropriate hypothesis to see if there is evidence that drivers might save money by switching to the online company.

84. Windy, part 3. Exercises 80 and 82 give summaries and displays for two potential sites for a wind turbine. Test an appropriate hypothesis to see if there is evidence that either of these sites has a higher average wind speed.

85. Employee athletes. An ergonomics consultant is engaged by a large consumer products company to see what they can do to increase productivity. The consultant recommends an "employee athlete" program, encouraging every employee to devote 5 minutes an hour to physical activity. The company worries that the gains in productivity will be offset by the loss in time on the job. They'd like to know if the program increases or decreases productivity. To measure it, they monitor a random sample of 145 employees who word process, measuring their hourly key strokes both before and after the program is instituted. Here are the data:

Keystrokes per Hour

	Before	After	Difference (After − Before)
Mean	1534.2	1556.9	22.7
SD	168.5	149.5	113.6
N	145	145	145

a) What are the null and alternative hypotheses?
b) What can you conclude? Explain.
c) Give a 95% confidence interval for the mean change in productivity (as measured by keystrokes per hour).

86. Employee athletes, part 2. A small company, on hearing about the employee athlete program (see Exercise 85) at the large company down the street, decides to try it as well. To measure the difference in productivity, they measure the average number of keystrokes per hour of 23 employees before and after the program is instituted. The data follow:

Keystrokes per Hour

	Before	After	Difference (After − Before)
Mean	1497.3	1544.8	47.5
SD	155.4	136.7	122.8
N	23	23	23

a) Is there evidence to suggest that the program increases productivity?
b) Give a 95% confidence interval for the mean change in productivity (as measured by keystrokes per hour).
c) Given this information and the results of Exercise 85, what recommendations would you make to the company about the effectiveness of the program?

87. Exercise equipment. A leading manufacturer of exercise equipment wanted to collect data on the effectiveness of their equipment. An August 2001 article in the journal *Medicine and Science in Sports and Exercise* compared how long it would take men and women to burn 200 calories during light or heavy workouts on various kinds of exercise

equipment. The results summarized in the following table are the average times for a group of physically active young men and women whose performances were measured on a representative sample of exercise equipment:

AVERAGE MINUTES TO BURN 200 CALORIES				
	Hard Exertion		Light Exertion	
Machine Type	Men	Women	Men	Women
Treadmill	12	17	14	22
X-C skier	12	16	16	23
Stair climber	13	18	20	37
Rowing machine	14	16	21	25
Exercise rider	22	24	27	36
Exercise bike	16	20	29	44

a) On average, how many minutes longer than a man must a woman exercise at a light exertion rate in order to burn 200 calories? Find a 95% confidence interval.
b) Estimate the average number of minutes longer a woman must work out at light exertion than at heavy exertion to get the same benefit. Find a 95% confidence interval.
c) These data are actually averages rather than individual times. How might this affect the margins of error in these confidence intervals?

T 88. Market value. Real estate agents want to set correctly the price of a house that's about to go on the real estate market. They must choose a price that strikes a balance between one that is so high that the house takes too long to sell and one that's so low that not enough value will go to the homeowner. One appraisal method is the "Comparative Market Analysis" approach by which the market value of a house is based on recent sales of similar homes in the neighborhood. Because no two houses are exactly the same, appraisers have to adjust comparable homes for such features as extra square footage, bedrooms, fireplaces, upgrading, parking facilities, swimming pool, lot size, location, and so on. The appraised market values and the selling prices of 45 homes from the same region are found on the CD.

a) Test the hypothesis that on average, the market value and the sale price of homes from this region are the same.
b) Find a 95% confidence interval for the mean difference.
c) Explain your findings in a sentence or two in context.

T 89. Stopping distance. In an experiment on braking performance, a tire manufacturer measured the stopping distance for one of its tire models. On a test track, a car made repeated stops from 60 miles per hour. Twenty tests were run, 10 each on both dry and wet pavement, with results shown in the following table. (Note that actual *braking distance*, which takes into account the driver's reaction time, is much longer, typically nearly 300 feet at 60 mph!)

a) Find a 95% confidence interval for the mean dry pavement stopping distance. Be sure to check the appropriate assumptions and conditions, and explain what your interval means.

b) Find a 95% confidence interval for the mean increase in stopping distance on wet pavement. Be sure to check the appropriate assumptions and conditions, and explain what your interval means.

Stopping Distance (ft)	
Dry Pavement	**Wet Pavement**
145	211
152	191
141	220
143	207
131	198
148	208
126	206
140	177
135	183
133	223

90. Stopping distances, again. For another test of the tires in Exercise 89, the company tried them on 10 different cars, recording the stopping distance for each car on both wet and dry pavement. Results are shown in the following table.

	Stopping Distance (ft)	
Car #	**Dry Pavement**	**Wet Pavement**
1	150	201
2	147	220
3	136	192
4	134	146
5	130	182
6	134	173
7	134	202
8	128	180
9	136	192
10	158	206

a) Find a 95% confidence interval for the mean dry pavement stopping distance. Be sure to check the appropriate assumptions and conditions, and explain what your interval means.

b) Find a 95% confidence interval for the mean increase in stopping distance on wet pavement. Be sure to check the appropriate assumptions and conditions, and explain what your interval means.

91. Airfares. In recent years, the airline industry has been severely criticized for a variety of service-related issues including poor on-time performance, canceled flights, and lost luggage. Some believe airline service is declining while the price of airline fares is increasing. A sample of 10 third quarter changes in airfares is shown below.

Origin	Third Quarter 2006	Third Quarter 2007	Percent Change from 3rd Qtr 2006
Cincinnati, OH	511.11	575.67	12.6
Salt Lake City, UT	319.29	344.48	7.9
Dallas Love, TX	185.12	198.74	7.4
New York JFK, NY	324.75	345.97	6.5
Hartford, CT	341.05	363.17	6.5
Charleston, SC	475.10	367.08	−22.7
Columbus, OH	322.60	277.24	−14.1
Kona, HI	206.50	180.40	−12.6
Memphis, TN	418.70	382.29	−8.7
Greensboro/High Point, NC	411.95	377.41	−8.4

Source: Bureau of Transportation Statistics. Top Five Third Quarter U.S. Domestic Average Itinerary Fare Increases and Decreases, 3rd Qtr 2006–3rd Qtr 2007—Top 100 Airports Based on 2006 U.S. Originating Domestic Passengers. Fares based on 2006 U.S. domestic itinerary fares, round-trip or one-way for which no return is purchased. Averages do not include frequent flyer fares. www.bts.gov/press_releases/2008

a) Does the percent change in airfare from the third quarter 2006 column represent paired data? Why or why not?

b) Was there an actual change, on average, in airline fares between the two quarters? Perform the test on both the actual and percentage differences. Discuss the results of the test and explain how you chose between the fares and the percent differences as the data to test.

92. Grocery prices. WinCo Foods, a large discount grocery retailer in the western United States promotes itself as the lowest priced grocery retailer. In newspaper ads printed and distributed during January 2008, WinCo Foods published a price comparison for products between WinCo and several competing grocery retailers. One of the retailers compared against WinCo was Wal-Mart, also known as a low price competitor. WinCo selected a variety of products, listed the price of the product charges at each retailer, and showed the sales receipt to prove the prices at WinCo were the lowest in the area. A sample of the product and their price comparison at both WinCo and Wal-Mart are shown in the following table:

Item	WinCo Price	Wal-Mart Price
Bananas (lb)	0.42	0.56
Red Onions (lb)	0.58	0.98
Mini Peeled Carrots (1 lb bag)	0.98	1.48
Roma Tomatoes (lb)	0.98	2.67
Deli Tater Wedges (lb)	1.18	1.78
Beef Cube Steak (lb)	3.83	4.118
Beef Top Round London Broil (lb)	3.48	4.12
Pillsbury Devils Food Cake Mix (18.25 oz)	0.88	0.88
Lipton Rice and Sauce Mix (5.6 oz)	0.88	1.06
Sierra Nevada Pale Ale (12 - 12 oz bottles)	12.68	12.84
GM Cheerios Oat Clusters (11.3 oz)	1.98	2.74
Charmin Bathroom Tissue (12 roll)	5.98	7.48
Bumble Bee Pink Salmon (14.75 oz)	1.58	1.98
Pace Thick & Chunky Salsa, Mild (24 oz)	2.28	2.78
Nalley Chili, Regular w/Beans (15 oz)	0.78	0.78
Challenge Butter (lb quarters)	2.18	2.58
Kraft American Singles (12 oz)	2.27	2.27
Yuban Coffee FAC (36 oz)	5.98	7.56
Totino's Pizza Rolls, Pepperoni (19.8 oz)	2.38	2.42
Rosarita Refried Beans, Original (16 oz)	0.68	0.73
Barilla Spaghetti (16 oz)	0.78	1.23
Sun-Maid Mini Raisins (14 – .5 oz)	1.18	1.36
Jif Peanut Butter, Creamy (28 oz)	2.54	2.72
Dole Fruit Bowl, Mixed Fruit (4 – 4 oz)	1.68	1.98
Progresso Chicken Noodle Soup (19 oz)	1.28	1.38
Precious Mozzarella Ball, Part Skim (16 oz)	3.28	4.23
Mrs. Cubbison Seasoned Croutons (6 oz)	0.88	1.12
Kellogg's Raisin Bran (20 oz)	1.98	2.50
Campbell's Soup at Hand, Cream of Tomato (10.75 oz)	1.18	1.26

a) Do the prices listed indicate that, on average, prices at WinCo are lower than prices at Wal-Mart?

b) At the bottom of the price list, the following statement appears: "Though this list is not intended to represent a typical weekly grocery order or a random list of grocery items, WinCo continues to be the area's low price leader." Why do you think WinCo added this statement?

c) What other comments could be made about the statistical validity of the test on price comparisons given in the ad?

Just Checking Answers

1. $H_0: \mu_{eyes} - \mu_{flowers} = 0$

2. ✓ **Independence Assumption:** The amount paid by one person should be independent of the amount paid by others.

   ✓ **Randomization Condition:** This study was observational. Treatments alternated a week at a time and were applied to the same group of office workers.

   ✓ **Nearly Normal Condition:** We don't have the data to check, but it seems unlikely there would be outliers in either group.

   ✓ **Independent Groups Assumptions:** The same workers were recorded each week, but week-to-week independence is plausible.

3. $H_A: \mu_{eyes} - \mu_{flowers} \neq 0$. An argument could be made for a one-sided test because the research hypothesis was that eyes would improve honest compliance.

4. Office workers' compliance in leaving money to pay for food at an office coffee station was different when a picture of eyes was placed behind the "honesty box" than when the picture was one of flowers.

5. These are independent groups sampled at random, so use a two-sample t confidence interval to estimate the size of the difference.

6. If the same random sample of students was sampled both in the first year and again in the fourth year of their university experience, then this would be a paired t-test.

7. A male and female are selected from each work group. The question calls for a paired t-test.

8. Since the sample of companies is different in each of the industries, this would be a two-sample test.

9. Since the same 50 companies are surveyed twice to examine a change in variables over time, this would be a paired t-test.

Inference for Counts: Chi-Square Tests

SAC Capital

Hedge funds, like mutual funds and pension funds, pool investors' money in an attempt to make profits. Unlike these other funds, however, hedge funds are not required to register with the U.S. Securities and Exchange Commission (SEC) because they issue securities in "private offerings" only to "qualified investors" (investors with either $1 million in assets or annual income of at least $200,000).

Hedge funds don't necessarily "hedge" their investments against market moves. But typically these funds use multiple, often complex, strategies to exploit inefficiencies in the market. For these reasons, hedge fund managers have the reputation for being obsessive traders.

One of the most successful hedge funds is SAC Capital, which was founded by Steven (Stevie) A. Cohen in 1992 with nine employees and $25 million in assets under management (AUM). SAC Capital returned annual gains of 40% or more through much of the 1990s and is now reported to have more than 800 employees and nearly

$14 billion in assets under management. According to *Forbes*, Cohen's $6.4 billion fortune ranks him as the 36th wealthiest American.

Cohen, a legendary figure on Wall Street, is known for taking advantage of any information he can find and for turning that information into profit. SAC Capital is one of the most active trading organizations in the world. According to *Business Week* (7/21/2003), Cohen's firm "routinely accounts for as much as 3% of the NYSE's average daily trading, plus up to 1% of the NASDAQ's—a total of at least 20 million shares a day."

In a business as competitive as hedge fund management, information is gold. Being the first to have information and knowing how to act on it can mean the difference between success and failure. Hedge fund managers look for small advantages everywhere, hoping to exploit inefficiencies in the market and to turn those inefficiencies into profit.

Wall Street has plenty of "wisdom" about market patterns. For example, investors are advised to watch for "calendar effects," certain times of year or days of the week that are particularly good or bad: "As goes January, so goes the year" and "Sell in May and go away." Some analysts claim that the "bad period" for holding stocks is from the sixth trading day of June to the fifth-to-last trading day of October. Of course, there is also Mark Twain's advice:

> October. This is one of the peculiarly dangerous months to speculate in stocks. The others are July, January, September, April, November, May, March, June, December, August, and February.
>
> —Pudd'nhead Wilson's Calendar

One common claim is that stocks show a weekly pattern. For example, some argue that there is a *weekend effect* in which stock returns on Mondays are often lower than those of the immediately preceding Friday. Are patterns such as this real? We have the data, so we can check. Between October 1, 1928 and June 6, 2007, there were 19,755 trading sessions. Let's first see how many trading days fell on each day of the week. It's not exactly 20% for each day because of holidays. The distribution of days is shown in Table 15.1.

Day of Week	Count	% of days
Monday	3820	19.3369%
Tuesday	4002	20.2582
Wednesday	4024	20.3695
Thursday	3963	20.0607
Friday	3946	19.9747

Table 15.1   The distribution of days of the week among the 19,755 trading days from October 1, 1928 to June 6, 2007. We expect about 20% to fall in each day, with minor variations due to holidays and other events.

Of these 19,755 trading sessions, 10,272, or about 52% of the days, saw a gain in the Dow Jones Industrial Average (DJIA). To test for a pattern, we need a model. The model comes from the supposition that any day is as likely to show a gain as any other. In any sample of positive or "up" days, we should expect to see the same distribution of days as in Table 15.1—in other words, about 19.34% of "up" days would be Mondays, 20.26% would be Tuesdays, and so on. Here is the distribution of days in one such random sample of 1000 "up" days.

Day of Week	Count	% of days in the sample of "up" days
Monday	192	19.2%
Tuesday	189	18.9
Wednesday	202	20.2
Thursday	199	19.9
Friday	218	21.8

Table 15.2   The distribution of days of the week for a sample of 1000 "up" trading days selected at random from October 1, 1928 to June 6, 2007. If there is no pattern, we would expect the proportions here to match fairly closely the proportions observed among all trading days in Table 15.1.

Of course, we expect some variation. We wouldn't expect the proportions of days in the two tables to match exactly. In our sample, the percentage of Mondays in Table 15.2 is slightly lower than in Table 15.1, and the proportion of Fridays is a little higher. Are these deviations enough for us to declare that there is a recognizable pattern?

15.1   Goodness-of-Fit Tests

To address this question, we test the table's **goodness-of-fit**, where *fit* refers to the null model proposed. Here, the null model is that there is no pattern, that the distribution of *up* days should be the same as the distribution of trading days overall. (If there were no holidays or other closings, that would just be 20% for each day of the week.)

Assumptions and Conditions

Data for a goodness-of-fit test are organized in tables, and the assumptions and conditions reflect that. Rather than having an observation for each individual, we typically work with summary counts in categories. Here, the individuals are trading days, but rather than list all 1000 trading days in the sample, we have totals for each weekday.

Counted Data Condition. The data must be counts for the categories of a categorical variable. This might seem a silly condition to check. But many kinds of values can be assigned to categories, and it is unfortunately common to find the methods of this chapter applied incorrectly (even by business professionals) to proportions or quantities just because they happen to be organized in a two-way table. So check to be sure that you really have counts.

Independence Assumption

Independence Assumption. The counts in the cells should be independent of each other. You should think about whether that's reasonable. If the data are a random sample you can simply check the randomization condition.

Randomization Condition. The individuals counted in the table should be a random sample from some population. We need this condition if we want to generalize our conclusions to that population. We took a random sample of 1000 trading days on which the DJIA rose. That lets us assume that the market's performance on any one day is independent of performance on another. If we had selected 1000 consecutive trading days, there would be a risk that market performance on one day could affect performance on the next, or that an external event could affect performance for several consecutive days.

Sample Size Assumption

Sample Size Assumption. We must have enough data for the methods to work. We usually just check the following condition:

Expected Cell Frequency Condition. We should expect to see at least 5 individuals in each cell. The expected cell frequency condition should remind you of—and is, in fact, quite similar to—the condition that np and nq be at least 10 when we test proportions.

Chi-Square Model

We have observed a count in each category (weekday). We can compute the number of up days we'd *expect* to see for each weekday if the null model were true. For the trading days example, the expected counts come from the null hypothesis that the up days are distributed among weekdays just as trading days are. Of course, we could imagine almost any kind of model and base a null hypothesis on that model.

To decide whether the null model is plausible, we look at the differences between the expected values from the model and the counts we observe. We wonder: Are these differences so large that they call the model into question, or could they have arisen from natural sampling variability? We denote the *differences* between these observed and expected counts, $(Obs - Exp)$. As we did with variance, we square them. That gives us positive values and focuses attention on any cells with large differences. Because the differences between observed and expected counts generally get larger the more data we have, we also need to get an idea of the *relative* sizes of the differences. To do that, we divide each squared difference by the expected count for that cell.

The test statistic, called the **chi-square (or chi-squared) statistic**, is found by adding up the sum of the squares of the deviations between the observed and expected counts divided by the expected counts:

$$\chi^2 = \sum_{all\ cells} \frac{(Obs - Exp)^2}{Exp}.$$

The chi-square statistic is denoted χ^2, where χ is the Greek letter chi (pronounced kī). The resulting family of sampling distribution models is called the **chi-square models**.

The members of this family of models differ in the number of degrees of freedom. The number of degrees of freedom for a goodness-of-fit test is $k - 1$, where k is the number of cells—in this example, 5 weekdays.

We will use the chi-square statistic only for testing hypotheses, not for constructing confidence intervals. A small chi-square statistic means that our model fits the data well, so a small value gives us no reason to doubt the null hypothesis. If the observed counts don't match the expected counts, the statistic will be large. If the calculated statistic value is large enough, we'll reject the null hypothesis. So the chi-square test is always one-sided. What could be simpler? Let's see how it works.

Expected Cell Frequencies

Companies often want to assess the relative successes of their products in different regions. However, a company whose sales regions had 100, 200, 300, and 400 representatives might not expect equal sales in all regions. They might expect observed sales to be proportional to the size of the sales force. The null hypothesis in that case would be that the proportions of sales were 1/10, 2/10, 3/10, and 4/10, respectively. With 500 total sales, their expected counts would be 50, 100, 150, and 200.

Notation Alert! —————

We compare the counts *observed* in each cell with the counts we *expect* to find. The usual notation uses *Obs* and *Exp* as we've used here. The expected counts are found from the null model.

Notation Alert! —————

The only use of the Greek letter χ in Statistics is to represent the chi-square statistic and the associated sampling distribution. This violates the general rule that Greek letters represent population parameters. Here we are using a Greek letter simply to name a family of distribution models and a statistic.

For Example   Goodness of fit test

Atara manages 8 call center operators at a telecommunications company. To develop new business, she gives each operator a list of randomly selected phone numbers of rival phone company customers. She also provides the operators with a script that tries to convince the customers to switch providers. Atara notices that some operators have found more than twice as many new customers as others, so she suspects that some of the operators are performing better than others.

The 120 new customer acquisitions are distributed as follows:

Operator	1	2	3	4	5	6	7	8
New customers	11	17	9	12	19	18	13	21

Question: Is there evidence to suggest that some of the operators are more successful than others?

Answer: Atara has randomized the potential new customers to the operators so the Randomization Condition is satisfied. The data are counts and there are at least 5 in each cell, so we can apply a chi-square goodness-of-fit test to the null hypothesis that the operator performance is uniform and that each of the operators will convince the same number of customers. Specifically we expect each operator to have converted 1/8 of the 120 customers that switched providers.

Operator	1	2	3	4	5	6	7	8
Observed	11	17	9	12	19	18	13	21
Expected	15	15	15	15	15	15	15	15
Observed-Expected	−4	2	−6	−3	4	3	−2	6
(Obs-Exp)^2	16	4	36	9	16	9	4	36
(Obs-Exp)^2/Exp	16/15 = 1.07	4/15 = 0.27	36/15 = 2.40	9/15 = 0.60	16/15 = 1.07	9/15 = 0.60	4/15 = 0.27	36/15 = 2.40

$$\sum \frac{(Obs - Exp)^2}{Exp} = 1.07 + 0.27 + 2.40 + \cdots + 2.40 = 8.67$$

The number of degrees of freedom is $k - 1 = 7$.

$$P(\chi^2_7 > 8.67) = 0.2772.$$

8.67 is not a surprising value for a Chi-square statistic with 7 degrees of freedom. So, we fail to reject the null hypothesis that the operators actually find new customers at different rates.

By Hand

The chi-square calculation

Here are the steps to calculate the chi-square statistic:

1. **Find the expected values.** These come from the null hypothesis model. Every null model gives a hypothesized proportion for each cell. The expected value is the product of the total number of observations times this proportion. (The result need not be an integer.)
2. **Compute the residuals.** Once you have expected values for each cell, find the residuals, $Obs - Exp$.
3. **Square the residuals.** $(Obs - Exp)^2$
4. **Compute the components.** Find $\dfrac{(Obs - Exp)^2}{Exp}$ for each cell.

(continued)

5. **Find the sum of the components.** That's the chi-square statistic,

$$\chi^2 = \sum_{all\ cells} \frac{(Obs - Exp)^2}{Exp}.$$

6. **Find the degrees of freedom.** It's equal to the number of cells minus one.

7. **Test the hypothesis.** Large chi-square values mean lots of deviation from the hypothesized model, so they give small P-values. Look up the critical value from a table of chi-square values such as Table X in Appendix D, or use technology to find the P-value directly.

The steps of the chi-square calculations are often laid out in tables. Use one row for each category, and columns for observed counts, expected counts, residuals, squared residuals, and the contributions to the chi-square total:

	A	B Observed	C Expected	D Residual = (Obs - Exp)	E (Obs-Exp)^2	F Component = (Obs - Exp)^2 / Exp
2	Monday	192	193.369	-1.369	1.879	0.0097
3	Tuesday	189	202.582	-13.582	184.461	0.9105
4	Wednesday	202	203.695	-1.695	2.874	0.0141
5	Thursday	199	200.607	-1.607	2.584	0.0129
6	Friday	218	199.747	18.253	333.176	1.668

Table 15.3 Calculations for the chi-square statistic in the trading days example, can be performed conveniently in **Excel**. Set up the calculation in the first row and Fill Down, then find the sum of the rightmost column. The CHIDIST function looks up the chi square total to find the P-value.

Guided Example   Stock Market Patterns

We have counts of the "up" days for each day of the week. The economic theory we want to investigate is whether there is a pattern in "up" days. So, our null hypothesis is that across all days in which the DJIA rose, the days of the week are distributed as they are across all trading days. (As we saw, the trading days are not quite *evenly* distributed because of holidays, so we use the *trading days* percentages as the null model.) We refer to this as *uniform*, accounting for holidays. The alternative hypothesis is that the observed percentages are *not* uniform. The test statistic looks at how closely the observed data match this idealized situation.

PLAN **Setup** State what you want to know.

Identify the variables and context.

Hypotheses State the null and alternative hypotheses. For χ^2 tests, it's usually easier to state the hypotheses in words than in symbols.

We want to know whether the distribution for "up" days differs from the null model (the trading days distribution). We have the number of times each weekday appeared among a random sample of 1000 "up" days.

H_0: The days of the work week are distributed among the up days as they are among all trading days.

H_A: The trading days model does not fit the up days distribution.

	Model Think about the assumptions and check the conditions.	✓ **Counted Data Condition** We have counts of the days of the week for all trading days and for the "up" days.
		✓ **Independence Assumption** We have no reason to expect that one day's performance will affect another's, but to be safe we've taken a random sample of days. The randomization should make them far enough apart to alleviate any concerns about dependence.
		✓ **Randomization Condition** We have a random sample of 1000 days from the time period.
		✓ **Expected Cell Frequency Condition** All the expected cell frequencies are much larger than 5.
	Specify the sampling distribution model. Name the test you will use.	The conditions are satisfied, so we'll use a χ^2 model with $5 - 1 = 4$ degrees of freedom and do a **chi-square goodness-of-fit test.**

DO | **Mechanics** To find the expected number of days, we take the fraction of each weekday from all days and multiply by the number of "up" days.

The expected values are:

> Monday: 193.369
> Tuesday: 202.582
> Wednesday: 203.695
> Thursday: 200.607
> Friday: 199.747

For example, there were 3820 Mondays out of 19,755 trading days.

So, we'd expect there would be $1000 \times 3820/19{,}755$ or 193.369 Mondays among the 1000 "up" days.

Each cell contributes a value equal to $\dfrac{(Obs - Exp)^2}{Exp}$ to the chi-square sum.

And we observe:

> Monday: 192
> Tuesday: 189
> Wednesday: 202
> Thursday: 199
> Friday: 218

Add up these components. If you do it by hand, it can be helpful to arrange the calculation in a table or spreadsheet.

$$\chi^2 = \frac{(192 - 193.369)^2}{193.369} + \cdots + \frac{(218 - 199.747)^2}{199.747}$$
$$= 2.615$$

The P-value is the probability in the upper tail of the χ^2 model. It can be found using software or a table (see Table X in Appendix D).

Using Table X in Appendix D, we find that for a significance level of 5% and 4 degrees of freedom, we'd need a value of 9.488 or more to have a P-value less than .05. Our value of 2.615 is less than that.

Large χ^2 statistic values correspond to small P-values, which would lead us to reject the null hypothesis, but the value here is not particularly large.

Using a computer to generate the P-value, we find:

$$\text{P-value} = P(\chi^2_4 > 2.615) = 0.624$$

(continued)

REPORT   Conclusion Link the P-value to your decision. Be sure to say more than a fact about
the distribution of counts. State your conclusion in terms of what the data mean.

MEMO

Re: Stock Market Patterns

Our investigation of whether there are day-of-the-week
patterns in the behavior of the DJIA in which one day or
another is more likely to be an "up" day found no evidence
of such a pattern. Our statistical test indicated that
a pattern such as the one found in our sample of trading
days would happen by chance about 62% of the time.

We conclude that there is, unfortunately, no evidence of a
pattern that could be used to guide investment in the
market. We were unable to detect a "weekend" or other
day-of-the-week effect in the market.

15.2   Interpreting Chi-Square Values

When we calculated χ^2 for the trading days example, we got 2.615. That value was
not large for 4 degrees of freedom, so we were unable to reject the null hypothesis.
In general, what *is* big for a χ^2 statistic?

Think about how χ^2 is calculated. In every cell any deviation from the expected
count contributes to the sum. Large deviations generally contribute more, but if
there are a lot of cells, even small deviations can add up, making the χ^2 value larger.
So the more cells there are, the higher the value of χ^2 has to be before it becomes
significant. For χ^2, the decision about how big is big depends on the number of
degrees of freedom.

Unlike the Normal and t families, χ^2 models are skewed. Curves in the χ^2 family change both shape and center as the number of degrees of freedom grows. For
example, Figure 15.1 shows the χ^2 curves for 5 and for 9 degrees of freedom.

| Figure 15.1   The χ^2 curves for 5 and 9 degrees of freedom.

Notice that the value $\chi^2 = 10$ might seem somewhat extreme when there are
5 degrees of freedom, but appears to be rather ordinary for 9 degrees of freedom.
Here are two simple facts to help you think about χ^2 models:

- The mode is at $\chi^2 = $ df $- 2$. (Look at the curves; their peaks are at 3 and 7.)
- The expected value (mean) of a χ^2 model is its number of degrees of freedom.
  That's a bit to the right of the mode—as we would expect for a skewed distribution.

Goodness-of-fit tests are often performed by people who have a theory of what the
proportions *should* be in each category and who believe their theory to be true. In
some cases, unlike our market example, there isn't an obvious null hypothesis against
which to test the proposed model. So, unfortunately, in those cases, the only null
hypothesis available is that the proposed theory is true. And as we know, the hypothesis testing procedure allows us only to reject the null or fail to reject it. We can
never confirm that a theory is in fact true; we can never confirm the null hypothesis.

At best, we can point out that the data are consistent with the proposed theory. But this doesn't prove the theory. The data *could* be consistent with the model even if the theory were wrong. In that case, we fail to reject the null hypothesis but can't conclude anything for sure about whether the theory is true.

> ## Why Can't We Prove the Null?
>
> A student claims that it really makes no difference to your starting salary how well you do in your Statistics class. He surveys recent graduates, categorizes them according to whether they earned an A, B, or C in Statistics, and according to whether their starting salary is above or below the median for their class. He calculates the proportion above the median salary for each grade. His null model is that in each grade category, 50% of students are above the median. With 40 respondents, he gets a P-value of .07 and declares that Statistics grades don't matter. But then more questionnaires are returned, and he finds that with a sample size of 70, his P-value is .04. Can he ignore the second batch of data? Of course not. If he could do that, he could claim almost any null model was true just by having too little data to refute it.

15.3 Examining the Residuals

Chi-square tests are always one-sided. The chi-square statistic is always positive, and a large value provides evidence against the null hypothesis (because it shows that the fit to the model is *not* good), while small values provide little evidence that the model doesn't fit. In another sense, however, chi-square tests are really many-sided; a large statistic doesn't tell us *how* the null model doesn't fit. In our market theory example, if we had rejected the uniform model, we wouldn't have known *how* it failed. Was it because there were not enough Mondays represented, or was it that all five days showed some deviation from the uniform?

When we reject a null hypothesis in a goodness-of-fit test, we can examine the residuals in each cell to learn more. In fact, whenever we reject a null hypothesis, it's a good idea to examine the residuals. (We don't need to do that when we fail to reject because when the χ^2 value is small, all of its components must have been small.) Because we want to compare residuals for cells that may have very different counts, we standardize the residuals. We know the mean residual is zero,[1] but we need to know each residual's standard deviation. When we tested proportions, we saw a link between the expected proportion and its standard deviation. For counts, there's a similar link. To standardize a cell's residual, we divide by the square root of its expected value[2]:

$$\frac{(Obs - Exp)}{\sqrt{Exp}}.$$

Notice that these **standardized residuals** are the square roots of the components we calculated for each cell, with the plus (+) or the minus (−) sign indicating whether we observed more or fewer cases than we expected.

The standardized residuals give us a chance to think about the underlying patterns and to consider how the distribution differs from the model. Now that we've divided each residual by its standard deviation, they are z-scores. If the null hypothesis was true, we could even use the 68–95–99.7 Rule to judge how extraordinary the large ones are.

---

[1] Residual = observed − expected. Because the total of the expected values is the same as the observed total, the residuals must sum to zero.

[2] It can be shown mathematically that the square root of the expected value estimates the appropriate standard deviation.

Here are the standardized residuals for the trading days data:

Standardized Residual $= \dfrac{(Obs - Exp)}{\sqrt{Exp}}$

	Standardized Residual $= \dfrac{(Obs - Exp)}{\sqrt{Exp}}$
Monday	−0.0984
Tuesday	−0.9542
Wednesday	−0.1188
Thursday	−0.1135
Friday	1.292

Table 15.4   Standardized residuals.

None of these values is remarkable. The largest, Friday, at 1.292, is not impressive when viewed as a z-score. The deviations are in the direction suggested by the "weekend effect," but they aren't quite large enough for us to conclude that they are real.

For Example   Examining residuals from a chi-square test

Question: In the call center example (see page 453), examine the residuals to see if any operators stand out as having especially strong or weak performance.

Answer: Because we failed to reject the null hypothesis, we don't expect any of the standardized residuals to be large, but we will examine them nonetheless.

The standardized residuals are the square roots of the components (from the bottom row of the table in the Example on page 453).

Standardized Residuals	−1.03	0.52	−1.55	−0.77	1.03	0.77	−0.52	1.55

As we expected, none of the residuals are large. Even though Atara notices that some of the operators enrolled more than twice the number of new customers as others, the variation is typical (within two standard deviations) of what we would expect if all their performances were, in fact, equal.

15.4   The Chi-Square Test of Homogeneity

Skin care products are big business. According to the American Academy of Dermatology, "the average adult uses at least seven different products each day," including moisturizers, skin cleansers, and hair cosmetics.[3] Growth in the skin care market in China during 2006 was 15%, fueled, in part, by massive economic growth. But not all cultures and markets are the same. Global companies must understand cultural differences in the importance of various skin care products in order to compete effectively.

The GfK Roper Reports® Worldwide Survey, which we first saw in Chapter 3, asked 30,000 consumers in 23 countries about their attitudes on health, beauty, and other personal values. One question participants were asked was how important is "Seeking the utmost attractive appearance" to you? Responses were a scale with 1 = Not at all important and 7 = Extremely important. Is agreement with this

---

[3]www.aad.org/public/Publications/pamphlets/Cosmetics.htm.

	Respondents in the GfK Roper Reports Worldwide Survey
WHO	Respondents in the GfK Roper Reports Worldwide Survey
WHAT	Responses to questions relating to perceptions of food and health
WHEN	Fall 2005; published in 2006
WHERE	Worldwide
HOW	Data collected by GfK Roper Consulting using a multistage design
WHY	To understand cultural differences in the perception of the food and beauty products we buy and how they affect our health

question the same across the five countries for which we have data (China, France, India, U.K., and U.S.)? Here is a contingency table with the counts.

			Country			
Appearance		China	France	India	U.K.	U.S.	Total
7—Extremely important	197	274	642	210	197	**1520**
6	257	405	304	252	203	**1421**
5	315	364	196	348	250	**1473**
4—Average importance	480	326	263	486	478	**2033**
3	98	82	41	125	100	**446**
2	63	46	36	70	58	**273**
1—Not at all important	92	38	53	62	29	**274**
Total	**1502**	**1535**	**1535**	**1553**	**1315**	**7440**

| Table 15.5   Responses to how important is "Seeking the utmost attractive appearance."

We can compare the countries more easily by examining the column percentages.

			Country			
Appearance		China	France	India	U.K.	U.S.	Row %
7—Extremely important	13.12%	17.85	41.82	13.52	14.98	**20.43%**
6	17.11	26.38	19.80	16.23	15.44	**19.10**
5	20.97	23.71	12.77	22.41	19.01	**19.80**
4—Average importance	31.96	21.24	17.13	31.29	36.35	**27.33**
3	6.52	5.34	2.67	8.05	7.60	**5.99**
2	4.19	3.00	2.35	4.51	4.41	**3.67**
1—Not at all important	6.13	2.48	3.45	3.99	2.21	**3.68**
Total	**100%**	**100**	**100**	**100**	**100**	**100%**

| Table 15.6   Responses as a percentage of respondents by country.

The stacked bar chart of the responses by country shows the patterns more vividly:

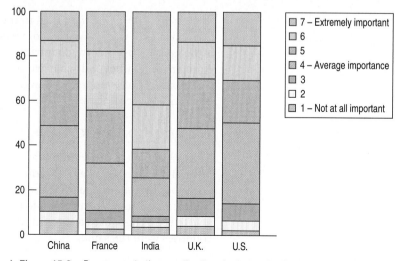

Figure 15.2   Responses to the question how important is "Seeking the utmost attractive appearance" by country. India stands out for the proportion of respondents who said Important or Extremely important.

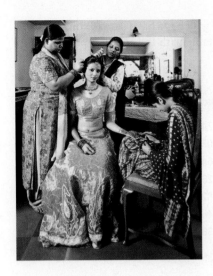

It seems that India stands out from the other countries. There is a much larger proportion of respondents from India who responded *Extremely Important*. But are the observed differences in the percentages real or just natural sampling variation? Our null hypothesis is that the proportions choosing each alternative are the same for each country. To test that hypothesis, we use a **chi-square test of homogeneity**. This is just another chi-square test. It turns out that the mechanics of the test of this hypothesis are nearly identical to the chi-square goodness-of-fit test we just saw in Section 15.1. The difference is that the goodness-of-fit test compared our observed counts to the expected counts from a *given* model. The test of homogeneity, by contrast, has a null hypothesis that the distributions are the same for all the groups. The test examines the differences between the observed counts and what we'd expect under that assumption of homogeneity.

For example, 20.43% (the row %) of *all* 7440 respondents said that looking good was extremely important to them. If the distributions were homogeneous across the five countries (as the null hypothesis asserts), then that proportion should be the same for all five countries. So 20.43% of the 1315 U.S. respondents, or 268.66, would have said that looking good was extremely important. That's the number we'd *expect* under the null hypothesis.

Working in this way, we (or, more likely, the computer) can fill in expected values for each cell. The following table shows these expected values for each response and each country.

		Country					
		China	France	India	U.K.	U.S.	Total
Appearance	7—Extremely important	306.86	313.60	313.60	317.28	268.66	**1520**
	6	286.87	293.18	293.18	296.61	251.16	**1421**
	5	297.37	303.91	303.91	307.47	260.35	**1473**
	4—Average importance	410.43	419.44	419.44	424.36	359.33	**2033**
	3	90.04	92.02	92.02	93.10	78.83	**446**
	2	55.11	56.32	56.32	56.99	48.25	**273**
	1—Not at all important	55.32	56.53	56.53	57.19	48.43	**274**
	Total	**1502**	**1535**	**1535**	**1553**	**1315**	**7440**

Table 15.7 Expected values for the responses. Because these are theoretical values, they don't have to be integers.

The term *homogeneity* refers to the hypothesis that things are the same. Here, we ask whether the distribution of responses about the importance of looking good is the same across the five countries. The chi-square test looks for differences large enough to step beyond what we might expect from random sample-to-sample variation. It can reveal a large deviation in a single category or small but persistent differences over all the categories—or anything in between.

Assumptions and Conditions

The assumptions and conditions are the same as for the chi-square test for goodness-of-fit. The **Counted Data Condition** says that these data must be counts. You can never perform a chi-square test on a quantitative variable. For example, if Roper had recorded how much respondents spent on skin care products, you wouldn't be able to use a chi-square test to determine whether the mean expenditures in the five countries were the same.[4]

---

[4]To do that, you'd use a method called Analysis of Variance (see Chapter 21).

Independence Assumption. So that we can generalize, we need the counts to be independent of each other. We can check the **Randomization Condition**. Here, we have random samples, so we *can* assume that the observations are independent and draw a conclusion comparing the populations from which the samples were taken.

We must be sure we have enough data for this method to work. The **Sample Size Assumption** can be checked with the **Expected Cell Frequency Condition**, which says that the expected count in each cell must be at least 5. Here, our samples are certainly large enough.

Following the pattern of the goodness-of-fit test, we compute the component for each cell of the table:

$$\textbf{Component} = \frac{(\textbf{Obs} - \textbf{Exp})^2}{\textbf{Exp}}.$$

Summing these components across all cells gives the chi-square value:

$$\chi^2 = \sum_{all\ cells} \frac{(Obs - Exp)^2}{Exp}.$$

The degrees of freedom are different than they were for the goodness-of-fit test. For a test of homogeneity, there are $(R - 1) \times (C - 1)$ degrees of freedom, where R is the number of rows and C is the number of columns.

In our example, we have $6 \times 4 = 24$ degrees of freedom. We'll need the degrees of freedom to find a P-value for the chi-square statistic.

By Hand

How to find expected values

In a contingency table, to test for homogeneity, we need to find the expected values when the null hypothesis is true. To find the expected value for row i and column j, we take:

$$Exp_{ij} = \frac{Total_{Row\ i} \times Total_{Col\ j}}{Table\ Total}$$

Here's an example:

Suppose we ask 100 people, 40 men and 60 women, to name their magazine preference: *Sports Illustrated, Cosmopolitan,* or *The Economist,* with the following result, shown in **Excel**:

	A	B	C	D	E
1	Actual	SI	Cosmo	Economist	Total
2	Men	25	5	10	40
3	Women	10	45	5	60
4	Total	35	50	15	100

Then, for example, the expected value under homogeneity for *Men* who prefer *The Economist* would be:

$$Exp_{13} = \frac{40 \times 15}{100} = 6.$$

Performing similar calculations for all cells gives the expected values:

6	Expected	SI	Cosmo	Economist	Total
7	Men	14	20	6	40
8	Women	21	30	9	60
9	Total	35	50	15	100

Guided Example   Attitudes on Appearance

How we think about our appearance, in part, depends on our culture. To help providers of beauty products with global markets, we want to examine whether the responses to the question "How important is seeking the utmost attractive appearance to you?" varied in the five markets of China, France, India, the U.K., and the U.S. We will use the data from the GfK Roper Reports Worldwide Survey.

PLAN	**Setup** State what you want to know. Identify the variables and context.	We want to know whether the distribution of responses to how important is "Seeking the utmost attractive appearance" is the same for the five countries for which we have data: China, France, India, U.K., and U.S.
	Hypotheses State the null and alternative hypotheses.	H_O: The responses are homogeneous (have the same distribution for all five countries). H_A: The responses are not homogeneous.
	Model Think about the assumptions and check the conditions.	We have counts of the number of respondents in each country who choose each response. ✓ **Counted Data Condition** The data are counts of the number of people choosing each possible response. ✓ **Randomization Condition** The data were obtained from a random sample by a professional global marketing company. ✓ **Expected Cell Frequency Condition** The expected values in each cell are all at least 5.
	State the sampling distribution model. Name the test you will use.	The conditions seem to be met, so we can use a χ^2 model with $(7 - 1) \times (5 - 1) = 24$ degrees of freedom and use a **chi-square test of homogeneity**.
DO	**Mechanics** Show the expected counts for each cell of the data table. You could make separate tables for the observed and expected counts or put both counts in each cell. A segmented bar chart is often a good way to display the data.	The observed and expected counts are in Tables 15.5 and 15.7. The bar graph shows the column percentages:

Use software to calculate χ^2 and the associated P-value.	$\chi^2 = 802.64$
Here, the calculated value of the χ^2 statistic is extremely high, so the P-value is quite small.	P-value $= P(\chi^2_{24} > 802.64) < 0.001$, so we reject the null hypothesis.

REPORT   **Conclusion** State your conclusion in the context of the data. Discuss whether the distributions for the groups appear to be different. For a small table, examine the residuals.

MEMO

Re: Importance of Appearance

Our analysis of the Roper data shows large differences across countries in the distribution of how important respondents say it is for them to look attractive. Marketers of cosmetics are advised to take notice of these differences, especially when selling products to India.

If you find that simply rejecting the hypothesis of homogeneity is a bit unsatisfying, you're in good company. It's hardly a shock that responses to this question differ from country to country especially with samples sizes this large. What we'd really like to know is where the differences were and how big they were. The test for homogeneity doesn't answer these interesting questions, but it does provide some evidence that can help us. A look at the standardized residuals can help identify cells that don't match the homogeneity pattern.

For Example   Testing homogeneity

Question: Although annual inflation in the United States has been low for several years, many Americans fear that inflation may return. In May 2010, a Gallup poll asked 1020 adults nationwide, "Are you very concerned, somewhat concerned, or not at all concerned that inflation will climb?" Does the distribution of responses appear to be the same for Conservatives as Liberals?

Ideology	Very Concerned	Somewhat Concerned	Not at all Concerned	Total
Conservative	232	83	25	340
Liberal	143	126	71	340
Total	**375 (55.15%)**	**209 (30.74%)**	**96 (14.12%)**	**680**

Answer: This is a test of homogeneity, testing whether the distribution of responses is the same for the two ideological groups. The data are counts, the Gallup poll selected adults randomly (stratified by ideology), and all expected cell frequencies are much greater than 5 (see table below).

There are $(3 - 1) \times (2 - 1)$ or 2 degrees of freedom.

If the distributions were the same, we would expect each cell to have expected values that are 55.15%, 30.74% and 14.12% of the row totals for Very Concerned, Somewhat Concerned and Not at all Concerned respectively. These values can be computed explicitly from:

$$Exp_{ij} = \frac{TotalRow_i \times TotalCol_j}{Table\ Total}$$

So, in the first cell (Conservative, Very Concerned):

$$Exp_{11} = \frac{TotalRow_1 \times TotalCol_1}{Table\ Total} = \frac{340 \times 375}{680} = 187.5$$

Expected counts for all cells are:

Expected Numbers	Very Concerned	Somewhat Concerned	Not at all Concerned
Conservative	187.5	104.5	48.0
Liberal	187.5	104.5	48.0

The components $\dfrac{(Obs - Exp)^2}{Exp}$ are:

Components	Very Concerned	Somewhat Concerned	Not at all Concerned
Conservative	10.56	4.42	11.02
Liberal	10.56	4.42	11.02

Summing these gives $\chi^2 = 10.56 + 4.42 + \cdots + 11.02 = 52.01$, which, with 2 df, has a P-value of < 0.0001.

We, therefore, reject the hypothesis that the distribution of responses is the same for Conservatives and Liberals.

15.5   Comparing Two Proportions

Many employers require a high school diploma. In October 2000, U.S. Department of Commerce researchers contacted more than 25,000 24-year-old Americans to see if they had finished high school and found that 84.9% of the 12,460 men and 88.1% of the 12,678 women reported having high school diplomas. Should we conclude that girls are more likely than boys to complete high school?

The U.S. Department of Commerce gives percentages, but it's easy to find the counts and put them in a table. It looks like this:

	Men	Women	Total
HS diploma	10,579	11,169	**21,748**
No diploma	1,881	1,509	**3,390**
Total	**12,460**	**12,678**	**25,138**

Table 15.8   Numbers of men and women who had earned high school diploma or not, by 2000, in a sample of 25,138 24-year-old Americans.

Overall, $\dfrac{21,748}{25,138} = 86.5144\%$ of the sample had received high school diplomas. So, under the homogeneity assumption, we would expect the same percentage of the 12,460 men (or $0.865144 \times 12,460 = 10,779.7$ men) to have diplomas. Completing the table, the expected counts look like this:

	Men	Women	Total
HS diploma	10,779.7	10,968.3	**21,748**
No diploma	1,680.3	1,709.7	**3,390**
Total	**12,460**	**12,678**	**25,138**

Table 15.9   The expected values.

The chi-square statistic with $(2 - 1) \times (2 - 1) = 1$ df is:

$$\chi_1^2 = \frac{(10579 - 10779.7)^2}{10779.7} + \frac{(11169 - 10968.3)^2}{10968.3} + \frac{(1881 - 1680.3)^2}{1680.3}$$
$$+ \frac{(1509 - 1709.7)^2}{1709.7} = 54.941$$

This has a P-value < 0.001, so we reject the null hypothesis and conclude that the distribution of receiving high school diplomas is different for men and women.

A chi-square test on a 2×2 table, which has only 1 df, is equivalent to testing whether two proportions (in this case, the proportions of men and women with diplomas) are equal. There is an equivalent way of testing the equality of two proportions that uses a z-statistic, and it gives exactly the same P-value. You may encounter the z-test for two proportions, so remember that it's the same as the chi-square test on the equivalent 2×2 table.

Even though the z-test and the chi-square test are equivalent for testing whether two proportions are the same, the z-test can also give a confidence interval. This is crucial here because we rejected the null hypothesis with a large sample size. The confidence interval can tell us how large the difference may be.

Confidence Interval for the Difference of Two Proportions

As we saw, 88.1% of the women and 84.9% of the men surveyed had earned high school diplomas in the United States by the year 2000. That's a difference of 3.2%. If we knew the standard error of that quantity, we could use a z-statistic to construct a confidence interval for the true difference in the population. It's not hard to find the standard error. All we need is the formula[5]:

$$SE(\hat{p}_1 - \hat{p}_2) = \sqrt{\frac{\hat{p}_1 \hat{q}_1}{n_1} + \frac{\hat{p}_2 \hat{q}_2}{n_2}}$$

The confidence interval has the same form as the confidence interval for a single proportion, with this new standard error:

$$(\hat{p}_1 - \hat{p}_2) \pm z^* SE(\hat{p}_1 - \hat{p}_2).$$

Confidence interval for the difference of two proportions

When the conditions are met, we can find the confidence interval for the difference of two proportions, $p_1 - p_2$. The confidence interval is

$$(\hat{p}_1 - \hat{p}_2) \pm z^* SE(\hat{p}_1 - \hat{p}_2),$$

where we find the standard error of the difference as

$$SE(\hat{p}_1 - \hat{p}_2) = \sqrt{\frac{\hat{p}_1 \hat{q}_1}{n_1} + \frac{\hat{p}_2 \hat{q}_2}{n_2}}$$

from the observed proportions.

The critical value z^* depends on the particular confidence level that you specify.

---

[5]The standard error of the difference is found from the general fact that the variance of a difference of two independent quantities is the *sum* of their variances. See Chapter 8 for details.

For high school graduation, a 95% confidence interval for the true difference between women's and men's rates is:

$$(0.881 - 0.849) \pm 1.96 \times \sqrt{\frac{(0.881)(0.119)}{12678} + \frac{(0.849)(0.151)}{12460}}$$

$$= (0.0236, 0.0404), \text{ or } 2.36\% \text{ to } 4.04\%.$$

We can be 95% confident that women's rates of having a HS diploma by 2000 were 2.36 to 4.04% higher than men's. With a sample size this large, we can be quite confident that the difference isn't zero. But is it a difference that matters? That, of course, depends on the *reason* we are asking the question. The confidence interval shows us the effect size—or at least, the interval of plausible values for the effect size. If we are considering changing hiring or recruitment policies, this difference may be too small to warrant much of an adjustment even though the difference is statistically "significant." Be sure to consider the effect size if you plan to make a business decision based on rejecting a null hypothesis using chi-square methods.

For Example   A confidence interval for the difference of proportions

Question: In the Gallup poll on inflation (see page 463), 68.2% (232 of 340) of those identifying themselves as Conservative were very concerned about the rise of inflation, but only 42.1% (143 of 340) of Liberals responded the same way. That's a difference of 26.1% in this sample of 680 adults. Find a 95% confidence interval for the true difference.

Answer: The confidence interval can be found from:

$$(\hat{p}_C - \hat{p}_L) \pm z^* SE(\hat{p}_C - \hat{p}_L) \text{ where } SE(\hat{p}_C - \hat{p}_L) = \sqrt{\frac{\hat{p}_C \hat{q}_C}{n_C} + \frac{\hat{p}_L \hat{q}_L}{n_L}} = \sqrt{\frac{(0.682)(0.338)}{340} + \frac{(0.421)(0.579)}{340}} = 0.037.$$

Since we know the 95% confidence critical value for z is 1.96, we have:

$$0.261 \pm 1.96(0.037) = (0.188, 0.334).$$

In other words, we are 95% confident that the proportion of Conservatives who are very concerned by inflation is between 18.8% and 33.4% higher than the same proportion of Liberals.

15.6   Chi-Square Test of Independence

We saw that the importance people place on their personal appearance varies a great deal from one country to another, a fact that might be crucial for the marketing department of a global cosmetics company. Suppose the marketing department wants to know whether the age of the person matters as well. That might affect the kind of media channels they use to advertise their products. Do older people feel as strongly as younger people that personal appearance is important?

		13–19	20–29	30–39	40–49	50–59	60+	Total
	7—Extremely important	396	337	300	252	142	93	1520
	6	325	326	307	254	123	86	1421
	5	318	312	317	270	150	106	1473
Appearance	4—Average importance	397	376	403	423	224	210	2033
	3	83	83	88	93	54	45	446
	2	37	43	53	58	37	45	273
	1—Not at all important	40	37	53	56	36	52	274
	Total	1596	1514	1521	1406	766	637	7440

(Age across top columns)

Table 15.10   Responses to the question about personal appearance by age group.

When we examined the five countries, we thought of the countries as five different groups, rather than as levels of a variable. But here, we can (and probably should) think of *Age* as a second variable whose value has been measured for each respondent along with his or her response to the appearance question. Asking whether the distribution of responses changes with *Age* now raises the question of whether the variables personal *Appearance* and *Age* are independent.

Whenever we have two variables in a contingency table like this, the natural test is a **chi-square test of independence**. Mechanically, this chi-square test is identical to a test of homogeneity. The difference between the two tests is in how we think of the data and, thus, what conclusion we draw.

Here we ask whether the response to the personal appearance question is independent of age. Remember, that for any two events, **A** and **B**, to be independent, the probability of event **A** given that event **B** occurred must be the same as the probability of event **A**. Here, this means the probability that a randomly selected respondent thinks personal appearance is extremely important is the same for all age groups. That would show that the response to the personal *Appearance* question is independent of that respondent's *Age*. Of course, from a table based on data, the probabilities will never be exactly the same. But to tell whether they are different enough, we use a chi-square test of independence.

Now we have two categorical variables measured on a single population. For the homogeneity test, we had a single categorical variable measured independently on two or more populations. Now we ask a different question: "Are the variables independent?" rather than "Are the groups homogeneous?" These are subtle differences, but they are important when we draw conclusions.

> **Homogeneity or Independence?**
> The only difference between the test for homogeneity and the test for independence is in the decision you need to make.

Assumptions and Conditions

Of course, we still need counts and enough data so that the expected counts are at least five in each cell.

If we're interested in the independence of variables, we usually want to generalize from the data to some population. In that case, we'll need to check that the data are a representative random sample from that population.

Guided Example   Personal Appearance and Age

We previously looked at whether responses to the question "How important is seeking the utmost attractive appearance to you?" varied in the five markets of China, France, India, the U.K., and the U.S., and we reported on the cultural differences that we saw. Now we want to help marketers discover whether a person's age influences how they respond to the same question. We have the values of *Age* in six age categories. Rather than six different groups, we can view *Age* as a variable, and ask whether the variables *Age* and *Appearance* are independent.

PLAN   **Setup** State what you want to know.

Identify the variables and context.

We want to know whether the categorical variables personal *Appearance* and *Age* are statistically independent. We have a contingency table of 7440 respondents from a sample of five countries.

(continued)

Hypotheses State the null and alternative hypotheses.

We perform a test of independence when we suspect the variables may not be independent. We are making the claim that knowing the respondents' A age will change the distribution of their response to the question about personal *Appearance*, and testing the null hypothesis that it is *not* true.

H_0: *Personal Appearance and Age are independent.*[6]

H_A: *Personal Appearance and Age are not independent.*

Model Check the conditions.

✓ **Counted Data Condition** We have counts of individuals categorized on two categorical variables.

✓ **Randomization Condition** These data are from a randomized survey conducted in 30 countries. We have data from five of them. Although they are not an SRS, the samples within each country were selected to avoid biases.

✓ **Expected Cell Frequency Condition** The expected values are all much larger than 5.

This table shows the expected counts below for each cell. The expected counts are calculated exactly as they were for a test of homogeneity; in the first cell, for example, we expect $\frac{1520}{7440} = 20.43\%$ of 1596 which is 326.065.

Expected Values

Appearance		13–19	20–29	30–39	40–49	50–59	60+
	7—Extremely important	326.065	309.312	310.742	287.247	156.495	130.140
	6	304.827	289.166	290.503	268.538	146.302	121.664
	5	315.982	299.748	301.133	278.365	151.656	126.116
	4—Average importance	436.111	413.705	415.617	384.193	209.312	174.062
	3	95.674	90.759	91.178	84.284	45.919	38.186
	2	58.563	55.554	55.811	51.591	28.107	23.374
	1—Not at all important	58.777	55.758	56.015	51.780	28.210	23.459

The stacked bar graph shows that the response seems to be dependent on *Age*. Older people tend to think personal appearance is less important than younger people.

---

[6]As in other chi-square tests, the hypotheses are usually expressed in words, without parameters. The hypothesis of independence itself tells us how to find expected values for each cell of the contingency table. That's all we need.

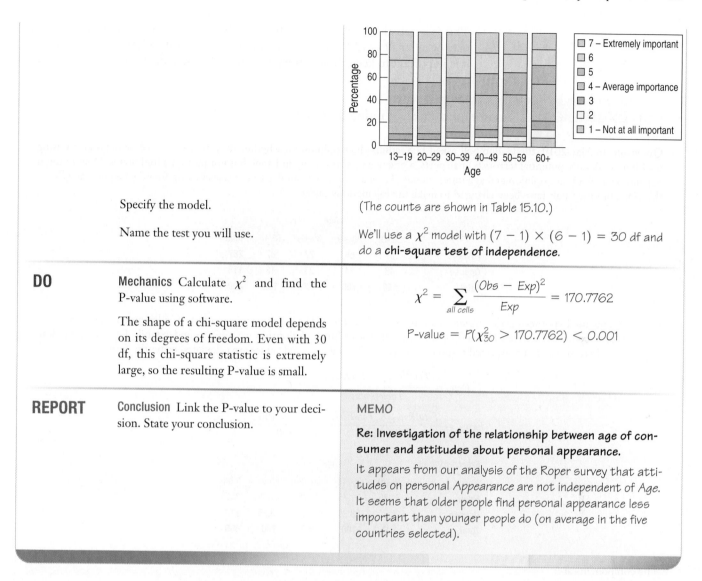

Specify the model.

(The counts are shown in Table 15.10.)

Name the test you will use.

We'll use a χ^2 model with $(7 - 1) \times (6 - 1) = 30$ df and do a **chi-square test of independence**.

DO

Mechanics Calculate χ^2 and find the P-value using software.

The shape of a chi-square model depends on its degrees of freedom. Even with 30 df, this chi-square statistic is extremely large, so the resulting P-value is small.

$$\chi^2 = \sum_{all\ cells} \frac{(Obs - Exp)^2}{Exp} = 170.7762$$

$$\text{P-value} = P(\chi^2_{30} > 170.7762) < 0.001$$

REPORT

Conclusion Link the P-value to your decision. State your conclusion.

MEMO

Re: Investigation of the relationship between age of consumer and attitudes about personal appearance.

It appears from our analysis of the Roper survey that attitudes on personal *Appearance* are not independent of *Age*. It seems that older people find personal appearance less important than younger people do (on average in the five countries selected).

We rejected the null hypothesis of independence between *Age* and attitudes about personal *Appearance*. With a sample size this large, we can detect very small deviations from independence, so it's almost guaranteed that the chi-square test will reject the null hypothesis. Examining the residuals can help you see the cells that deviate farthest from independence. To make a meaningful business decision, you'll have to look at effect sizes as well as the P-value. We should also look at each country's data individually since country to country differences could affect marketing decisions.

Suppose the company was specifically interested in deciding how to split advertising resources between the teen market and the 30–39-year-old market. How much of a difference is there between the proportions of those in each age group that rated personal *Appearance* as very important (responding either 6 or 7)?

For that we'll need to construct a confidence interval on the difference. From Table 15.10, we find that the percentages of those answering 6 and 7 are 45.17% and 39.91% for the teen and 30–39-year-old groups, respectively. The 95% confidence interval is:

$$(\hat{p}_1 - \hat{p}_2) \pm z^* SE(\hat{p}_1 - \hat{p}_2)$$

$$= (0.4517 - 0.3991) \pm 1.96 \times \sqrt{\frac{(0.4517)(0.5483)}{1596} + \frac{(0.3991)(0.6009)}{1521}}$$

$$= (0.018, 0.087), \text{ or } (1.8\% \text{ to } 8.7\%)$$

This is a statistically significant difference, but now we can see that the difference may be as small as 1.8%. When deciding how to allocate advertising expenditures, it is important to keep these estimates of the effect size in mind.

For Example   A chi-square test of independence

Question: In May 2010, the Gallup poll asked U.S. adults their opinion on whether they are in favor of or opposed to using profiling to identify potential terrorists at airports, a practice used routinely in Israel, but not in the United States. Does opinion depend on age? Or are opinion and age independent? Here are numbers similar to the ones Gallup found (the percentages are the same, but the totals have been changed to make the calculations easier).

	Age				
	18–29	30–49	50–64	65 +	Total
Favor	57	66	77	87	**287**
Oppose	43	34	23	13	**113**
Total	**100**	**100**	**100**	**100**	**400**

Answer: The null hypothesis is that *Opinion* and *Age* are independent. We can view this as a test of independence as opposed to a test of homogeneity if we view *Age* and *Opinion* are variables whose relationship we want to understand. This was a random sample and there are at least 5 expected responses in every cell. The expected values are calculated using the formula:

$$Exp_{ij} = \frac{TotalRow_i \times TotalCol_j}{Table\ Total} \Rightarrow$$

$$Exp_{11} = \frac{TotalRow_1 \times TotalCol_1}{Table\ Total} = \frac{287 \times 100}{400} = 71.75$$

Expected Values	Age				
	18–29	30–49	50–64	65 +	Total
Favor	71.75	71.75	71.75	71.75	**287**
Oppose	28.25	28.25	28.25	28.25	**113**
Total	**100**	**100**	**100**	**100**	**400**

The components are:

Components	Age			
	18–29	30–49	50–64	65 +
Favor	3.03	0.46	0.38	3.24
Oppose	7.70	1.17	0.98	8.23

There are $(r - 1) \times (c - 1) = 1 \times 3 = 3$ degrees of freedom. Summing all the components gives:

$$\chi_3^2 = 3.03 + 0.46 + \cdots + 8.23 = 25.20,$$

which has a P-value < 0.0001.

Thus, we reject the null hypothesis and conclude that *Age* and *Opinion* about Profiling are not independent. Looking at the residuals,

Residuals	Age			
	18–29	30–49	50–64	65 +
Favor	−1.74	−0.68	0.62	1.80
Oppose	2.78	1.08	−0.99	−2.87

we see a pattern. These two variable fail to be independent because increasing age is associated with more favorable attitudes toward profiling.

Bar charts arranged in *Age* order make the pattern clear:

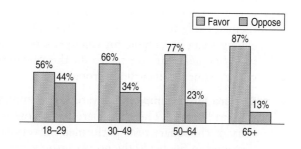

Just Checking

Which of the three chi-square tests would you use in each of the following situations—goodness-of-fit, homogeneity, or independence?

1 A restaurant manager wonders whether customers who dine on Friday nights have the same preferences among the chef's four special entrées as those who dine on Saturday nights. One weekend he has the wait staff record which entrées were ordered each night. Assuming these customers to be typical of all weekend diners, he'll compare the distributions of meals chosen Friday and Saturday.

2 Company policy calls for parking spaces to be assigned to everyone at random, but you suspect that may not be so.

There are three lots of equal size: lot A, next to the building; lot B, a bit farther away; and lot C on the other side of the highway. You gather data about employees at middle management level and above to see how many were assigned parking in each lot.

3 Is a student's social life affected by where the student lives? A campus survey asked a random sample of students whether they lived in a dormitory, in off-campus housing, or at home and whether they had been out on a date 0, 1–2, 3–4, or 5 or more times in the past two weeks.

Chi-square tests and causation

Chi-square tests are common. Tests for independence are especially widespread. Unfortunately, many people interpret a small P-value as proof of causation. We know better. Just as correlation between quantitative variables does not demonstrate causation, a failure of independence between two categorical variables does not show a cause-and-effect relationship between them, nor should we say that one variable *depends* on the other.

The chi-square test for independence treats the two variables symmetrically. There is no way to differentiate the direction of any possible causation from one variable to the other. While we can see that attitudes on personal *Appearance* and *Age* are related, we can't say that getting older *causes* you to change attitudes. And certainly it's not correct to say that changing attitudes on personal appearance makes you older.

Of course, there's never any way to eliminate the possibility that a lurking variable is responsible for the observed lack of independence. In some sense, a failure of independence between two categorical variables is less impressive than a strong, consistent association between quantitative variables. Two categorical variables can fail the test of independence in many ways, including ways that show no consistent pattern of failure. Examination of the chi-square standardized residuals can help you think about the underlying patterns.

What Can Go Wrong?

- **Don't use chi-square methods unless you have counts.** All three of the chi-square tests apply only to counts. Other kinds of data can be arrayed in two-way tables. Just because numbers are in a two-way table doesn't make them suitable for chi-square analysis. Data reported as proportions or percentages can be suitable for chi-square procedures, *but only after they are converted to counts.* If you try to do the calculations without first finding the counts, your results will be wrong.

- **Beware large samples.** Beware *large* samples? That's not the advice you're used to hearing. The chi-square tests, however, are unusual. You should be wary of chi-square tests performed on very large samples. No hypothesized distribution fits perfectly, no two groups are exactly homogeneous, and two variables are rarely perfectly independent. The degrees of freedom for chi-square tests don't grow with the sample size. With a sufficiently large sample size, a chi-square test can always reject the null hypothesis. But we have no measure of how far the data are from the null model. There are no confidence intervals to help us judge the effect size except in the case of two proportions.

- **Don't say that one variable "depends" on the other just because they're not independent.** "Depend" can suggest a model or a pattern, but variables can fail to be independent in many different ways. When variables fail the test for independence, it may be better to say they are "associated."

Ethics in Action

eliberately Different specializes in unique accessories for the home such as hand-painted switch plates and hand-embroidered linens, offered through a catalog and a website. Its customers tend to be women, generally older, with relatively high household incomes. Although the number of customer visits to the site has remained the same, management noticed that the proportion of customers visiting the site who make a purchase has been declining. Megan Cally, the product manager for Deliberately Different, was in charge of working with the market research firm hired to examine this problem. In her first meeting with Jason Esgro, the firm's consultant, she directed the conversation toward website design. Jason mentioned several reasons for consumers abandoning online purchases, the two most common being concerns about transaction security and unanticipated shipping/handling charges. Because Deliberately Different's shipping charges are reasonable, Megan asked him to look further into the issue of security concerns. They developed a survey that randomly sampled customers who had visited the website. They contacted these customers by e-mail and asked them to respond to a brief survey, offering the chance of winning a prize, which would be awarded at random among the respondents. A total of 2450 responses were received. The analysis of the responses included chi-square tests for independence,

checking to see if responses on the security question were independent of gender and income category. Both tests were significant, rejecting the null hypothesis of independence. Megan reported to management that concerns about online transaction security were dependent on gender, and income, so Deliberately Different began to explore ways in which they could assure their older female customers that transactions on the website are indeed secure. As product manager, Megan was relieved that the decline in purchases was not related to product offerings.

ETHICAL ISSUE *The chance of rejecting the null hypothesis in a chi-square test for independence increases with sample size. Here the sample size is very large. In addition, it is misleading to state that concerns about security depend on gender, age, and income. Furthermore, patterns of association were not examined (for instance, with varying age categories). Finally, as product manager, Megan intentionally steered attention away from examining the product offerings, which could be a factor in declining purchases. Instead she reported to management that they have pinpointed the problem without noting that they had not explored other potential factors (related to Items A and H, ASA Ethical Guidelines).*

ETHICAL SOLUTION *Interpret results correctly, cautioning about the large sample size and looking for any patterns of association, realizing that there is no way to estimate the effect size.*

What Have We Learned?

Learning Objectives

- Recognize when a chi-square test of goodness of fit, homogeneity, or independence is appropriate.
- For each test, find the expected cell frequencies.
- For each test, check the assumptions and corresponding conditions and know how to complete the test.
  - Counted data condition.
  - Independence assumption; randomization makes independence more plausible.
  - Sample size assumption with the expected cell frequency condition; expect at least 5 observations in each cell.
- Interpret a chi-square test.
  - Even though we might believe the model, we cannot prove that the data fit the model with a chi-square test because that would mean confirming the null hypothesis.
- Examine the standardized residuals to understand what cells were responsible for rejecting a null hypothesis.
- Compare two proportions.
- State the null hypothesis for a test of independence and understand how that is different from the null hypothesis for a test of homogeneity.
  - Both are computed the same way. You may not find both offered by your technology. You can use either one as long as you interpret your result correctly.

Terms

Chi-square models
Chi-square models are skewed to the right. They are parameterized by their degrees of freedom and become less skewed with increasing degrees of freedom.

Chi-square (or chi-squared) statistic
The chi-square statistic is found by summing the chi-square components. Chi-square tests can be used to test goodness-of-fit, homogeneity, or independence.

Chi-square goodness-of-fit test
A test of whether the distribution of counts in one categorical variable matches the distribution predicted by a model. A chi-square test of goodness-of-fit finds

$$\chi^2 = \sum_{all\ cells} \frac{(Obs - Exp)^2}{Exp},$$

where the expected counts come from the predicting model. It finds a P-value from a chi-square model with $n - 1$ degrees of freedom, where n is the number of categories in the categorical variable.

Chi-square test of homogeneity
A test comparing the distribution of counts for two or more groups on the same categorical variable. A chi-square test of homogeneity finds

$$\chi^2 = \sum_{all\ cells} \frac{(Obs - Exp)^2}{Exp},$$

where the expected counts are based on the overall frequencies, adjusted for the totals in each group. We find a P-value from a chi-square distribution with $(R - 1) \times (C - 1)$ degrees of freedom, where R gives the number of categories (rows) and C gives the number of independent groups (columns).

Chi-square test of independence
A test of whether two categorical variables are independent. It examines the distribution of counts for one group of individuals classified according to both variables. A chi-square test of *independence* uses the same calculation as a test of homogeneity. We find a P-value from a chi-square distribution with $(R - 1) \times (C - 1)$ degrees of freedom, where R gives the number of categories in one variable and C gives the number of categories in the other.

(continued)

Standardized residual

In each cell of a two-way table, a standardized residual is the square root of the chi-square component for that cell with the sign of the *Observed − Expected* difference:

$$\frac{(Obs - Exp)}{\sqrt{Exp}}$$

When we reject a chi-square test, an examination of the standardized residuals can sometimes reveal more about how the data deviate from the null model.

Technology Help: Chi-Square

Most statistics packages associate chi-square tests with contingency tables. Often chi-square is available as an option only when you make a contingency table. This organization can make it hard to locate the chi-square test and may confuse the three different roles that the chi-square test can take. In particular, chi-square tests for goodness-of-fit may be hard to find or missing entirely. Chi-square tests for homogeneity are computationally the same as chi-square tests for independence, so you may have to perform the mechanics as if they were tests of independence and interpret them afterward as tests of homogeneity.

Most statistics packages work with data on individuals rather than with the summary counts. If the only information you have is the table of counts, you may find it more difficult to get a statistics package to compute chi-square. Some packages offer a way to reconstruct the data from the summary counts so that they can then be passed back through the chi-square calculation, finding the cell counts again. Many packages offer chi-square standardized residuals (although they may be called something else).

EXCEL   XLSTAT

Excel offers the function **CHITEST (actual_range, expected_range),** which computes a chi-square P-value for independence. Both ranges are of the form UpperLeftCell: LowerRightCell, specifying two rectangular tables. The two tables must be of the same size and shape. The function is called CHISQ.TEST in Excel 2010.

Comments
In order to use this function, you will have to compute the expected values table, typically using the column total divided by the sample size method.

JMP

From the **Analyze** menu,

- Select **Fit Y by X**.

- Choose one variable as the Y, response variable, and the other as the X, factor variable. Both selected variables must be Nominal or Ordinal.

- JMP will make a plot and a contingency table. Below the contingency table, **JMP** offers a **Tests** panel. In that panel, the Chi Square for independence is called a **Pearson ChiSquare.** The table also offers the P-value.

- Click on the contingency Table title bar to drop down a menu that offers to include a **Deviation** and Cell **Chi square** in each cell of the table.

Comments
JMP will choose a chi-square analysis for a **Fit Y by X** if both variables are nominal or ordinal (marked with an N or O), but not otherwise. Be sure the variables have the right type. Deviations are the observed—expected differences in counts. Cell chi-squares are

the squares of the standardized residuals. Refer to the deviations for the sign of the difference. Look under **Distributions** in the **Analyze** menu to find a chi-square test for goodness-of-fit.

From the **Start** menu,

- Choose the **Tables** submenu.
- From that menu, choose **Chi Square Test**. . . .
- In the dialog, identify the columns that make up the table. Minitab will display the table and print the chi-square value and its P-value.

Comments

Alternatively, select the **Cross Tabulation** . . . command to see more options for the table, including expected counts and standardized residuals.

From the **Analyze** menu,

- Choose the **Descriptive Statistics** submenu.
- From that submenu, choose **Crosstabs**. . . .
- In the Crosstabs dialog, assign the row and column variables from the variable list. Both variables must be categorical.
- Click the **Cells** button to specify that standardized residuals should be displayed.
- Click the **Statistics** button to specify a chi-square test.

Comments

SPSS offers only variables that it knows to be categorical in the variable list for the Crosstabs dialog. If the variables you want are missing, check that they have the right type.

Brief CASE

Health Insurance

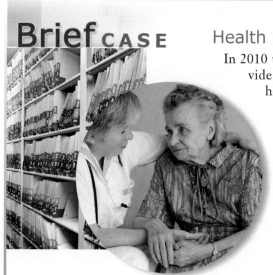

In 2010 the U.S. Congress passed the historic health care reform bill that will provide some type of coverage for the 32 million Americans currently without health care insurance. Just how widespread was the lack of medical coverage? The media claims that the segments of the population most at risk are women, children, the elderly and the poor. The tables give the number of uninsured (in thousands) by sex, by age and by household income in 2008.[7] Using the appropriate summary statistics, graphical displays, statistical tests, and confidence intervals, investigate the accuracy of the media's statement using these data. Be sure to discuss your assumptions, methods, results, and conclusions. (Note: some rows and colums may not add exactly to totals due to rounding.)

	Sex		
	Male	Female	**Total**
Uninsured	25,208	21,131	**46,340**
Insured	122,886	132,257	**255,143**
Total	**148,094**	**153,388**	**301,483**

	Age			
	0-17	18-64	65+	**Total**
Uninsured	7,348	38,345	646	**46,340**
Insured	67,161	150,841	37,142	**255,143**
Total	**74,510**	**189,185**	**37,788**	**301,483**

(continued)

[7]Source: U.S. Census Bureau, Current Population Survey, Annual Social and Economic Supplement, 2009. http://www.census.gov/hhes/www/cpstables/032009/health/h01_001.htm.

		Household Income			
	<$25,000	$25,000–$49,999	$50,000–$74,999	$75,000+	Total
Uninsured	13,673	14,908	8,034	9,725	46,340
Insured	42,142	54,712	49,491	108,798	255,143
Total	55,814	69,621	57,525	118,523	301,483

Loyalty Program

A marketing executive tested two incentives to see what percentage of customers would enroll in a new web-based loyalty program. The customers were asked to log on to their accounts on the Web and provide some demographic and spending information. As an incentive, they were offered either Nothing (No Offer), Free flight insurance on their next flight (Free Insurance), or a free companion Airline ticket (Free Flight). The customers were segmented according to their past year's spending patterns as spending primarily in one of five areas: *Travel, Entertainment, Dining, Household,* or *Balanced.* The executive wanted to know whether the incentives resulted in different enrollment rates (*Response*). Specifically, she wanted to know how much higher the enrollment rate for the free flight was compared to the free insurance. She also wanted to see whether *Spending Pattern* was associated with *Response*. Using the data **Loyalty_Program**, write up a report for the marketing executive using appropriate graphics, summary statistics, statistical tests, and confidence intervals.

Exercises

SECTION 15.1

1. If there is no seasonal effect on human births, we would expect equal numbers of children to be born in each season (winter, spring, summer and fall). A student takes a census of her statistics class and finds that of the 120 students in the class, 25 were born in winter, 35 in spring, 32 in summer and 28 in fall. She wonders if the excess in the spring is an indication that births are not uniform throughout the year.

a) What is the expected number of births in each season if there is no "seasonal effect" on births?
b) Compute the χ^2 statistic.
c) How many degrees of freedom does the χ^2 statistic have?

2. At a major credit card bank, the percentages of people who historically apply for the Silver, Gold and Platinum cards are 60%, 30% and 10% respectively. In a recent sample of customers responding to a promotion, of 200 customers, 110 applied for Silver, 55 for Gold and 35 for Platinum. Is there evidence to suggest that the percentages for this promotion may be different from the historical proportions?

a) What is the expected number of customers applying for each type of card in this sample if the historical proportions are still true?
b) Compute the χ^2 statistic.
c) How many degrees of freedom does the χ^2 statistic have?

SECTION 15.2

3. For the births in Exercise 1,

a) If there is no seasonal effect, about how big, on average, would you expect the χ^2 statistic to be (what is the mean of the χ^2 distribution)?
b) Does the statistic you computed in Exercise 1 seem large in comparison to this mean? Explain briefly.
c) What does that say about the null hypothesis?
d) Find the $\alpha = 0.05$ critical value for the χ^2 distribution with the appropriate number of df.
e) Using the critical value, what do you conclude about the null hypothesis at $\alpha = 0.05$?

4. For the customers in Exercise 2,

a) If the customers apply for the three cards according to the historical proportions, about how big, on average, would you expect the χ^2 statistic to be (what is the mean of the χ^2 distribution)?

b) Does the statistic you computed in Exercise 2 seem large in comparison to this mean? Explain briefly.

c) What does that say about the null hypothesis?

d) Find the $\alpha = 0.05$ critical value for the χ^2 distribution with the appropriate number of df.

e) Using the critical value, what do you conclude about the null hypothesis at $\alpha = 0.05$?

SECTION 15.3

5. For the data in Exercise 1,

a) Compute the standardized residual for each season.

b) Are any of these particularly large? (Compared to what?)

c) Why should you have anticipated the answer to part b?

6. For the data in Exercise 2,

a) Compute the standardized residual for each type of card.

b) Are any of these particularly large? (Compared to what?)

c) What does the answer to part b say about this new group of customers?

SECTION 15.4

7. An analyst at a local bank wonders if the age distribution of customers coming for service at his branch in town is the same as at the branch located near the mall. He selects 100 transactions at random from each branch and researches the age information for the associated customer. Here are the data:

| | Age | | | |
	Less than 30	30–55	56 or older	Total
In-town branch	20	40	40	100
Mall branch	30	50	20	100
Total	100	100	200	200

a) What is the null hypothesis?

b) What type of test is this?

c) What are the expected numbers for each cell if the null hypothesis is true?

d) Find the χ^2 statistic.

e) How many degrees of freedom does it have?

f) Find the critical value at $\alpha = 0.05$.

g) What do you conclude?

8. A market researcher working for the bank in Exercise 2 wants to know if the distribution of applications by card is the same for the past three mailings. She takes a random

sample of 200 from each mailing and counts the number applying for Silver Gold and Platinum. The data follow:

| | Type of Card | | | |
	Silver	Gold	Platinum	Total
Mailing 1	120	50	30	200
Mailing 2	115	50	35	200
Mailing 3	105	55	40	200
Total	340	155	105	600

a) What is the null hypothesis?

b) What type of test is this?

c) What are the expected numbers for each cell if the null hypothesis is true?

d) Find the χ^2 statistic.

e) How many degrees of freedom does it have?

f) Find the critical value at $\alpha = 0.05$.

g) What do you conclude?

SECTION 15.5

9. Marketers want to know about the differences between men's and women's use of the Internet. A Pew Research poll in April 2009 from a random sample of US adults found that 2393 of 3037 men use the Internet, at least occasionally, while 2378 of 3166 women did.

a) Find the proportions of men and women who said they use the Internet at least occasionally.

b) What is the difference in proportions?

c) What is the standard error of the difference?

d) Find a 95% confidence interval for the difference between percentages of usage by men and women nationwide.

10. From the same poll as in Exercise 9, marketers want to know about the differences in use of the Internet by income. Of 2741 who reported earning less than $50,000 a year, 1813 said they used the Internet. Of the 2353 people who reported earnings of $50,000 a year or more, 2201 said they used the Internet.

a) Find the proportions of those earning at least $50,000 a year and those earning less than $50,000 a year who said they use the Internet.

b) What is the difference in proportions?

c) What is the standard error of the difference?

d) Find a 95% confidence interval for the difference in the proportion of people who use the Internet between the two income groups.

SECTION 15.6

11. The same poll as in Exercise 9 also asked the questions "Did you use the Internet yesterday?" and "Are you White, Black, or Hispanic/Other?" Is the response to the question about the Internet independent of race? The data follow:

	Did You Use the Internet Yesterday?	
	Yes	No
White	2546	856
Black	314	146
Hispanic/Other	431	174

a) Under the null hypothesis, what are the expected values?
b) Compute the χ^2 statistic.
c) How many degrees of freedom does it have?
d) Find the critical value for $\alpha = 0.05$.
e) What do you conclude?

12. The same poll as in Exercise 9 also asked the questions "Did you use the Internet yesterday?" and "What is your educational level?" Is the response to the question about the internet independent of educational level? The data follow:

	Did You Use the Internet Yesterday?	
	Yes	No
Less Than High School	209	131
High School	932	550
Some College	958	346
College Grad	1447	247

a) Under the null hypothesis, what are the expected values?
b) Compute the χ^2 statistic.
c) How many degrees of freedom does it have?
d) Find the critical value for $\alpha = 0.01$.
e) What do you conclude?

CHAPTER EXERCISES

13. Concepts. For each of the following situations, state whether you'd use a chi-square goodness-of-fit test, chi-square test of homogeneity, chi-square test of independence, or some other statistical test.

a) A brokerage firm wants to see whether the type of account a customer has (Silver, Gold, or Platinum) affects the type of trades that customer makes (in person, by phone, or on the Internet). It collects a random sample of trades made for its customers over the past year and performs a test.
b) That brokerage firm also wants to know if the type of account affects the size of the account (in dollars). It performs a test to see if the mean size of the account is the same for the three account types.

c) The academic research office at a large community college wants to see whether the distribution of courses chosen (Humanities, Social Science, or Science) is different for its residential and nonresidential students. It assembles last semester's data and performs a test.

14. Concepts, part 2. For each of the following situations, state whether you'd use a chi-square goodness-of-fit test, a chi-square test of homogeneity, a chi-square test of independence, or some other statistical test.

a) Is the quality of a car affected by what day it was built? A car manufacturer examines a random sample of the warranty claims filed over the past two years to test whether defects are randomly distributed across days of the work week.
b) A researcher for the American Booksellers Association wants to know if retail sales/sq. ft. is related to serving coffee or snacks on the premises. She examines a database of 10,000 independently owned bookstores testing whether retail sales (dollars/sq. ft.) is related to whether or not the store has a coffee bar.
c) A researcher wants to find out whether education level (some high school, high school graduate, college graduate, advanced degree) is related to the type of transaction most likely to be conducted using the Internet (shopping, banking, travel reservations, auctions). He surveys 500 randomly chosen adults and performs a test.

15. Dice. After getting trounced by your little brother in a children's game, you suspect that the die he gave you is unfair. To check, you roll it 60 times, recording the number of times each face appears. Do these results cast doubt on the die's fairness?

Face	Count
1	11
2	7
3	9
4	15
5	12
6	6

a) If the die is fair, how many times would you expect each face to show?
b) To see if these results are unusual, will you test goodness-of-fit, homogeneity, or independence?
c) State your hypotheses.
d) Check the conditions.
e) How many degrees of freedom are there?
f) Find χ^2 and the P-value.
g) State your conclusion.

16. Quality control. Mars, Inc. says that the colors of its M&M's® candies are 14% yellow, 13% red, 20% orange,

24% blue, 16% green and 13% brown. (www.mms.com/us/about/products/milkchocolate). On his way home from work the day he was writing these exercises, one of the authors bought a bag of plain M&M's. He got 29 yellow, 23 red, 12 orange, 14 blue, 8 green, and 20 brown. Is this sample consistent with the company's advertised proportions? Test an appropriate hypothesis and state your conclusion.

a) If the M&M's are packaged in the advertised proportions, how many of each color should the author have expected in his bag of M&M's?
b) To see if his bag was unusual, should he test goodness-of-fit, homogeneity, or independence?
c) State the hypotheses.
d) Check the conditions.
e) How many degrees of freedom are there?
f) Find χ^2 and the P-value.
g) State a conclusion.

17. Quality control, part 2. A company advertises that its premium mixture of nuts contains 10% Brazil nuts, 20% cashews, 20% almonds, 10% hazelnuts, and that the rest are peanuts. You buy a large can and separate the various kinds of nuts. Upon weighing them, you find there are 112 grams of Brazil nuts, 183 grams of cashews, 207 grams of almonds, 71 grams of hazelnuts, and 446 grams of peanuts. You wonder whether your mix is significantly different from what the company advertises.

a) Explain why the chi-square goodness-of-fit test is not an appropriate way to find out.
b) What might you do instead of weighing the nuts in order to use a χ^2 test?

18. Sales rep travel. A sales representative who is on the road visiting clients thinks that, on average, he drives the same distance each day of the week. He keeps track of his mileage for several weeks and discovers that he averages 122 miles on Mondays, 203 miles on Tuesdays, 176 miles on Wednesdays, 181 miles on Thursdays, and 108 miles on Fridays. He wonders if this evidence contradicts his belief in a uniform distribution of miles across the days of the week. Is it appropriate to test his hypothesis using the chi-square goodness-of-fit test? Explain.

19. Maryland lottery. For a lottery to be successful, the public must have confidence in its fairness. One of the lotteries in Maryland is Pick-3 Lottery, where 3 random digits are drawn each day.[8] A fair game depends on every value (0 to 9) being equally likely at each of the three positions. If not, then someone detecting a pattern could take advantage of that and beat the lottery. To investigate the randomness, we'll look at data collected over a recent

32-week period. Although the winning numbers look like three-digit numbers, in fact, each digit is a randomly drawn numeral. We have 654 random digits in all. Are each of the digits from 0 to 9 equally likely? Here is a table of the frequencies.

Group	Count	%
0	62	9.480
1	55	8.410
2	66	10.092
3	64	9.786
4	75	11.468
5	57	8.716
6	71	10.856
7	74	11.315
8	69	10.550
9	61	9.327

a) Select the appropriate procedure.
b) Check the assumptions.
c) State the hypotheses.
d) Test an appropriate hypothesis and state your results.
e) Interpret the meaning of the results and state a conclusion.

20. Employment discrimination? Census data for New York City indicate that 29.2% of the under-18 population is white, 28.2% black, 31.5% Latino, 9.1% Asian, and 2% are of other ethnicities. The New York Civil Liberties Union points out that of 26,181 police officers, 64.8% are white, 14.5% black, 19.1% Hispanic, and 1.4% Asian. Do the police officers reflect the ethnic composition of the city's youth?

a) Select the appropriate procedure.
b) Check the assumptions.
c) State the hypotheses.
d) Test an appropriate hypothesis and state your results.
e) Interpret the meaning of the results and state a conclusion.

T 21. Titanic. Here is a table showing who survived the sinking of the *Titanic* based on whether they were crew members or passengers booked in first-, second-, or third-class staterooms.

	Crew	First	Second	Third	Total
Alive	212	202	118	178	**710**
Dead	673	123	167	528	**1491**
Total	**885**	**325**	**285**	**706**	**2201**

[8]Source: Maryland State Lottery Agency, www.mdlottery.com.

a) If we draw an individual at random from this table, what's the probability that we will draw a member of the crew?
b) What's the probability of randomly selecting a third-class passenger who survived?
c) What's the probability of a randomly selected passenger surviving, given that the passenger was in a first-class state-room?
d) If someone's chances of surviving were the same regardless of their status on the ship, how many members of the crew would you expect to have lived?
e) State the null and alternative hypotheses we would test here (and the name of the test).
f) Give the degrees of freedom for the test.
g) The chi-square value for the table is 187.8, and the corresponding P-value is barely greater than 0. State your conclusions about the hypotheses.

22. Promotion discrimination? The table shows the rank attained by male and female officers in the New York City Police Department (NYPD). Do these data indicate that men and women are equitably represented at all levels of the department? (All possible ranks in the NYPD are shown.)

Rank	Male	Female
Officer	21,900	4281
Detective	4058	806
Sergeant	3898	415
Lieutenant	1333	89
Captain	359	12
Higher ranks	218	10

a) What's the probability that a person selected at random from the NYPD is a female?
b) What's the probability that a person selected at random from the NYPD is a detective?
c) Assuming no bias in promotions, how many female detectives would you expect the NYPD to have?
d) To see if there is evidence of differences in ranks attained by males and females, will you test goodness-of-fit, homogeneity, or independence?
e) State the hypotheses.
f) Test the conditions.
g) How many degrees of freedom are there?
h) Find the chi-square value and the associated P-value.
i) State your conclusion.
j) If you concluded that the distributions are not the same, analyze the differences using the standardized residuals of your calculations.

23. Birth order and college choice. Students in an Introductory Statistics class at a large university were classified by birth order and by the college they attend.

Birth Order (1 = oldest or only child)

College	1	2	3	4 or more	Total
Arts and Sciences	34	14	6	3	57
Agriculture	52	27	5	9	93
Social Science	15	17	8	3	43
Professional	13	11	1	6	31
Total	114	69	20	21	224

Expected Values
Birth Order (1 = oldest or only child)

College	1	2	3	4 or more
Arts and Sciences	29.0089	17.5580	5.0893	5.3438
Agriculture	47.3304	28.6473	8.3036	8.7188
Social Science	21.8839	13.2455	3.8393	4.0313
Professional	15.7768	9.5491	2.7679	2.9063

a) What kind of chi-square test is appropriate—goodness-of-fit, homogeneity, or independence?
b) State your hypotheses.
c) State and check the conditions.
d) How many degrees of freedom are there?
e) The calculation yields $\chi^2 = 17.78$, with P = 0.0378. State your conclusion.
f) Examine and comment on the standardized residuals. Do they challenge your conclusion? Explain.

Standardized Residuals
Birth Order (1 = oldest or only child)

College	1	2	3	4 or more
Arts and Sciences	0.92667	−0.84913	0.40370	−1.01388
Agriculture	0.67876	−0.30778	−1.14640	0.09525
Social Science	−1.47155	1.03160	2.12350	−0.51362
Professional	−0.69909	0.46952	−1.06261	1.81476

24. Automobile manufacturers. *Consumer Reports* uses surveys given to subscribers of its magazine and website (www.ConsumerReports.org) to measure reliability in automobiles. This annual survey asks about problems that consumers have had with their cars, vans, SUVs, or trucks during the previous 12 months. Each analysis is based on the number of problems per 100 vehicles.

Origin of Manufacturer

	Asia	Europe	U.S.	Total
No Problems	88	79	83	250
Problems	12	21	17	50
Total	100	100	100	300

Expected Values

	Asia	Europe	U.S.
No Problems	83.33	83.33	83.33
Problems	16.67	16.67	16.67

a) State your hypotheses.
b) State and check the conditions.
c) How many degrees of freedom are there?
d) The calculation yields $\chi^2 = 2.928$, with $P = 0.231$. State your conclusion.
e) Would you expect that a larger sample might find statistical significance? Explain.

T 25. Cranberry juice. It's common folk wisdom that cranberries can help prevent urinary tract infections in women. A leading producer of cranberry juice would like to use this information in their next ad campaign, so they need evidence of this claim. In 2001, the *British Medical Journal* reported the results of a Finnish study in which three groups of 50 women were monitored for these infections over 6 months. One group drank cranberry juice daily, another group drank a lactobacillus drink, and the third group drank neither of those beverages, serving as a control group. In the control group, 18 women developed at least one infection compared with 20 of those who consumed the lactobacillus drink and only 8 of those who drank cranberry juice. Does this study provide supporting evidence for the value of cranberry juice in warding off urinary tract infections in women?

a) Select the appropriate procedure.
b) Check the assumptions.
c) State the hypotheses.
d) Test an appropriate hypothesis and state your results.
e) Interpret the meaning of the results and state a conclusion.
f) If you concluded that the groups are not the same, analyze the differences using the standardized residuals of your calculations.

T 26. Car company. A European manufacturer of automobiles claims that their cars are preferred by the younger generation and would like to target university students in their next ad campaign. Suppose we test their claim with our own survey. A random survey of autos parked in the student lot and the staff lot at a large university classified the brands by country of origin, as seen in the following table. Are there differences in the national origins of cars driven by students and staff?

Driver

Origin	Student	Staff
American	107	105
European	33	12
Asian	55	47

a) Is this a test of independence or homogeneity?
b) Write appropriate hypotheses.
c) Check the necessary assumptions and conditions.
d) Find the P-value of your test.
e) State your conclusion and analysis.

T 27. Market segmentation. The Chicago Female Fashion Study[9] surveyed customers to determine characteristics of the "frequent" shoppers at different department stores in the Chicago area. Suppose you are a marketing manager at one of the department stores. You would like to know if a customer's shopping frequency and her age are related. Here are the data:

Shopping Frequency	Age 18–24	25–44	45–54	55 or over
Never/Hardly Ever	32	171	45	24
1–2 times/yr	18	134	40	37
3–4 times/yr	21	109	48	27
≥ 5 times/yr	39	134	71	50

Standardized Residuals Age

Shopping Frequency	18–24	25–44	45–54	55 or over
Never/Hardly Ever	0.3803	1.7974	−1.4080	−2.2094
1–2 times/yr	−1.4326	0.7595	−0.9826	0.9602
3–4 times/yr	−0.3264	−0.3151	0.9556	−0.2425
≥ 5 times/yr	1.1711	−2.1360	1.4235	1.4802

---

[9] Original *Market Segmentation Exercise* prepared by K. Matsuno, D. Kopcso, and D. Tigert, Babson College in 1997 (Babson Case Series #133-C97A-U).

a) Is this a test of homogeneity or independence?
b) Write an appropriate hypothesis.
c) Are the conditions for inference satisfied?
d) The calculation yields $\chi^2 = 26.084$, P-value = 0.002. State your conclusion.
e) Given the standardized residuals in the table, state a complete conclusion.

28. Seafood company. A large company in the northeastern United States that buys fish from local fishermen and distributes them to major companies and restaurants is considering launching a new ad campaign on the health benefits of fish. As evidence, they would like to cite the following study. Medical researchers followed 6272 Swedish men for 30 years to see if there was any association between the amount of fish in their diet and prostate cancer ("Fatty Fish Consumption and Risk of Prostate Cancer," *Lancet*, June 2001).

	Prostate Cancer	
Fish Consumption	No	Yes
Never/seldom	110	14
Small part of diet	2420	201
Moderate part	2769	209
Large part	507	42

a) Is this a survey, a retrospective study, a prospective study, or an experiment? Explain.
b) Is this a test of homogeneity or independence?
c) Do you see evidence of an association between the amount of fish in a man's diet and his risk of developing prostate cancer?
d) Does this study prove that eating fish does not prevent prostate cancer? Explain.

29. Shopping. A survey of 430 randomly chosen adults finds that 47 of 222 men and 37 of 208 women had purchased books online.

a) Is there evidence that the sex of the person and whether they buy books online are associated?
b) If your conclusion in fact proves to be wrong, did you make a Type I or Type II error?
c) Give a 95% confidence interval for the difference in proportions of buying online for men and women.

30. Information technology. A recent report suggests that Chief Information Officers (CIO's) who report directly to Chief Financial Officers (CFO's) rather than Chief Executive Officers (CEO's) are more likely to have IT agendas that deal with cost cutting and compliance (SearchCIO.com, March 14, 2006). In a random sample of 535 companies, it was found that CIO's reported directly to CFO's in 173 out of 335 service firms and in 95 out of 200 manufacturing companies.

a) Is there evidence that type of business (service versus manufacturing) and whether or not the CIO reports directly to the CFO are associated?
b) If your conclusion proves to be wrong, did you make a Type I or Type II error?
c) Give a 95% confidence interval for the difference in proportions of companies in which the CIO reports directly to the CFO between service and manufacturing firms.

31. Fast food. GfK Roper Consulting gathers information on consumer preferences around the world to help companies monitor attitudes about health, food, and health care products. They asked people in many different cultures how they felt about the following statement: *I try to avoid eating fast foods.*

In a random sample of 800 respondents, 411 people were 35 years old or younger, and, of those, 197 agreed (completely or somewhat) with the statement. Of the 389 people over 35 years old, 246 people agreed with the statement.

a) Is there evidence that the percentage of people avoiding fast food is different in the two age groups?
b) Give a 90% confidence interval for the difference in proportions.

32. Computer gaming. In order to effectively market electronic games, a manager wanted to know what age group of boys played more. A survey in 2006 found that 154 of 223 boys aged 12–14 said they "played computer or console games like Xbox or PlayStation . . . or games online." Of 248 boys aged 15–17, 154 also said they played these games.

a) Is there evidence that the percentage of boys who play these types of games is different in the two age groups?
b) Give a 90% confidence interval for the difference in proportions.

33. Foreclosure rates. The two states with the highest home foreclosure rates in March 2008 were Nevada and Colorado (realestate.msn.com, April 2008). In the second quarter of 2008, there were 8 foreclosures in a random sample of 1098 homes in Nevada, and 6 in a sample of 1460 homes in Colorado.

a) Is there evidence that the percentage of foreclosures is different in the two states?
b) Give a 90% confidence interval for the difference in proportions.

34. Labor force. Immigration reform has focused on dividing illegal immigrants into two groups: long-term and short-term. According to a recent report, short-term unauthorized workers make up nearly 6% of the U.S. labor force in construction (Pew Hispanic Center Fact Sheet, April 13, 2006). The regions of the country with the lowest percentage of unauthorized short-term immigrant construction workers are the Northeast and the Midwest. In a random sample of 958 construction workers from the Northeast, 66 are illegal short-term immigrants. In the Midwest, 42 out of a sample of 1070 are illegal short-term immigrants.

a) Is there evidence that the percentage of construction workers who are illegal short-term immigrants differs in the two regions?

b) Give a 90% confidence interval for the difference in proportions.

T 35. Market segmentation, part 2. The survey described in Exercise 27 also investigated the customers' marital status. Using the same definitions for *Shopping Frequency* as in Exercise 27, the calculations yielded the following table. Test an appropriate hypothesis for the relationship between marital status and the frequency of shopping at the same department store as in Exercise 27, and state your conclusions.

	Counts			
	Single	**Widowed**	**Married**	**Total**
Never/Hardly Ever	105	5	162	**272**
1–2 times/yr	53	15	161	**229**
3–4 times/yr	57	8	140	**205**
≥ 5 times/yr	72	15	207	**294**
Total	**287**	**43**	**670**	**1000**

36. Investment options. The economic slowdown in early 2008 and the possibility of future inflation prompted a full service brokerage firm to gauge the level of interest in inflation-beating investment options among its clients. It surveyed a random sample of 1200 clients asking them to indicate the likelihood that they would add inflation-linked annuities and bonds to their portfolios within the next year. The table below shows the distribution of responses by the investors' tolerance for risk. Test an appropriate hypothesis for the relationship between risk tolerance and the likelihood of investing in inflation linked options.

	Risk Tolerance			
	Averse	**Neutral**	**Seeking**	**Total**
Certain Will Invest	191	93	40	**324**
Likely to Invest	82	106	123	**311**
Not Likely to Invest	64	110	101	**275**
Certain Will not Invest	63	91	136	**290**
Total	**400**	**400**	**400**	**1200**

Likelihood of Investing in Inflation-Linked Options

37. Accounting. The Sarbanes Oxley (SOX) Act was passed in 2002 as a result of corporate scandals and in an attempt to regain public trust in accounting and reporting practices. Two random samples of 1015 executives were surveyed and asked their opinion about accounting practices in both 2000 and in 2006. The table below summarizes all 2030 responses to the question, "Which of the following do you consider most critical to establishing ethical and legal accounting and reporting practices?" Did the distribution of responses change from 2000 to 2006?

		2000	**2006**
Responses	**Training**	142	131
	IT Security	274	244
	Audit Trails	152	173
	IT Policies	396	416
	No Opinion	51	51

a) Select the appropriate procedure.
b) Check the assumptions.
c) State the hypotheses.
d) Test an appropriate hypothesis and state your results.
e) Interpret the meaning of the results and state a conclusion.

38. Entrepreneurial executives. A leading CEO mentoring organization offers a program for chief executives, presidents, and business owners with a focus on developing entrepreneurial skills. Women and men executives that recently completed the program rated its value. Are perceptions of the program's value the same for men and women?

		Men	**Women**
Perceived Value	**Excellent**	3	9
	Good	11	12
	Average	14	8
	Marginal	9	2
	Poor	3	1

a) Will you test goodness-of-fit, homogeneity, or independence?
b) Write appropriate hypotheses.
c) Find the expected counts for each cell, and explain why the chi-square procedures are not appropriate for this table.

T 39. Market segmentation, part 3. The survey described in Exercise 27 also investigated the customers' emphasis on *Quality* by asking them the question: "For the same amount of money, I will generally buy one good item rather than several of lower price and quality." Using the same definitions for *Shopping Frequency* as in Exercise 27, the calculations yielded the following table. Test an appropriate hypothesis for the relationship between a customer's emphasis on *Quality* and the *Shopping Frequency* at this department store.

a) Select the appropriate procedure.
b) Check the assumptions.
c) State the hypotheses.
d) Test an appropriate hypothesis and state your results.
e) Interpret the meaning of the results and state a conclusion.

Counts			
		Moderately	
	Disagree	**Disagree/Agree**	**Agree**	**Total**
Never/Hardly Ever	15	97	160	272
1–2 times/yr	28	107	94	229
3–4 times/yr	30	90	85	205
≥ 5 times/yr	35	140	119	294
Total	108	434	458	1000

40. Online shopping. A recent report concludes that while Internet users like the convenience of online shopping, they do have concerns about privacy and security (*Online Shopping*, Washington, DC, Pew Internet & American Life Project, February 2008). Respondents were asked to indicate their level of agreement with the statement "I don't like giving my credit card number or personal information online." The table gives a subset of responses. Test an appropriate hypothesis for the relationship between age and level of concern about privacy and security online.

	Strongly Agree	Agree	Disagree	Strongly Disagree	Total
Ages 18–29	127	147	138	10	422
Ages 30–49	141	129	78	55	403
Ages 50–64	178	102	64	51	395
Ages 65 +	180	132	54	14	380
Total	626	510	334	130	1600

(Age Category)

a) Select the appropriate procedure.
b) Check the assumptions.
c) State the hypotheses.
d) Test an appropriate hypothesis and state your results.
e) Interpret the meaning of the results and state a conclusion.

41. Entrepreneurial executives again. In some situations where the expected counts are too small, as in Exercise 38, we can complete an analysis anyway. We can often proceed after combining cells in some way that makes sense and also produces a table in which the conditions are satisfied. Here is a new table displaying the same data, but combining "Marginal" and "Poor" into a new category called "Below Average."

Perceived Value	Men	Women
Excellent	3	9
Good	11	12
Average	14	8
Below Average	12	3

a) Find the expected counts for each cell in this new table, and explain why a chi-square procedure is now appropriate.
b) With this change in the table, what has happened to the number of degrees of freedom?
c) Test your hypothesis about the two groups and state an appropriate conclusion.

42. Small business. The director of a small business development center located in a mid-sized city is reviewing data about its clients. In particular, she is interested in examining if the distribution of business owners across the various stages of the business life cycle is the same for white-owned and Hispanic-owned businesses. The data are shown below.

Stage in Business	White-Owned	Hispanic-Owned
Planning	11	9
Starting	14	11
Managing	20	2
Getting Out	15	1

a) Will you test goodness-of-fit, homogeneity, or independence?
b) Write the appropriate hypotheses.
c) Find the expected counts for each cell and explain why chi-square procedures are not appropriate for this table.
d) Create a new table by combining categories so that a chi-square procedure can be used.
e) With this change in the table, what has happened to the number of degrees of freedom?
f) Test your hypothesis about the two groups and state an appropriate conclusion.

43. Racial steering. A subtle form of racial discrimination in housing is "racial steering." Racial steering occurs when real estate agents show prospective buyers only homes in neighborhoods already dominated by that family's race. This violates the Fair Housing Act of 1968. According to an article in *Chance* magazine (Vol. 14, no. 2, 2001), tenants at a large apartment complex recently filed a lawsuit alleging racial steering. The complex is divided into two parts: Section A and Section B. The plaintiffs claimed that white potential renters were steered to Section A, while African-Americans were steered to Section B. The following table displays the data that were presented in court to show the locations of recently rented apartments. Do you think there is evidence of racial steering?

New Renters			
	White	Black	Total
Section A	87	8	95
Section B	83	34	117
Total	170	42	212

44. Titanic, again. Newspaper headlines at the time and traditional wisdom in the succeeding decades have held that women and children escaped the *Titanic* in greater proportion than men. Here's a table with the relevant data. Do you think that survival was independent of whether the person was male or female? Defend your conclusion.

	Female	Male	Total
Alive	343	367	710
Dead	127	1364	1491
Total	470	1731	2201

45. Racial steering, revisited. Find a 95% confidence interval for the difference in the proportions of Black renters in the two sections for the data in Exercise 43.

46. Titanic, one more time. Find a 95% confidence interval for the difference in the proportion of women who survived and the proportion of men who survived for the data in Exercise 44.

47. Industry sector and outsourcing. Many companies have chosen to outsource segments of their business to external providers in order to cut costs and improve quality and/or efficiencies. Common business segments that are outsourced include Information Technology (IT) and Human Resources (HR). The data below show the types of outsourcing decisions made (no outsourcing, IT only, HR only, both IT and HR) by a sample of companies from various industry sectors.

		No Outsourcing	IT Only	HR Only	Both IT and HR
Industry Sector	Healthcare	810	6429	4725	1127
	Financial	263	1598	549	117
	Industrial Goods	1031	1269	412	99
	Consumer Goods	66	341	305	197

Do these data highlight significant differences in outsourcing by industry sector?

a) Select the appropriate procedure.
b) Check the assumptions.
c) State the hypotheses.
d) Test an appropriate hypothesis and state your results.
e) Interpret the meaning of the results and state a conclusion.

48. Industry sector and outsourcing, part 2. Consider only the companies that have outsourced their IT and HR business segments. Do these data suggest significant differences between companies in the financial and industrial goods sectors with regard to their outsourcing decisions?

		IT Only	HR Only	Both IT and HR
Industry Sector	Financial	1598	549	117
	Industrial Goods	1269	412	99

a) Select the appropriate procedure.
b) Check the assumptions.
c) State the hypotheses.
d) Test an appropriate hypothesis and state your results.
e) Interpret the meaning of the results and state the conclusion.

	Employee Job Satisfaction			
	Very Satisfied	Satisfied	Somewhat Satisfied	Not Satisfied
Exploitative Authoritarian	27	82	43	48
Benevolent Authoritarian	50	19	56	75
Laissez Faire	52	88	26	34
Consultative	71	83	20	26
Participative	101	59	20	20

(Row labels grouped under "Management Styles")

49. Management styles. Use the survey results in the table at the top of the page to investigate differences in employee job satisfaction among organizations in the United States with different management styles.

a) Select the appropriate procedure.
b) Check the assumptions.
c) State the hypotheses.
d) Test an appropriate hypothesis and state your results.
e) Interpret the meaning of the results and state a conclusion.

50. Ranking companies. Every year Fortune Magazine lists the 100 best companies to work for, based on criteria such as pay, benefits, turnover rate, and diversity. In 2008, the top three were Google, Quicken Loans, and Wegmans Food Markets (*Fortune*, February 4, 2008). Of the best 100 companies to work for, 33 experienced double digit job growth (10%–68%), 49 experienced single digit job growth (1%–9%), and 18 experienced no growth or a decline. A closer examination of the top 30 showed that 15 had job growth in the double digits, 11 in the single digits, and only 4 had no growth or a decline. Is there anything unusual about job growth among the 30 top companies?

a) Select the appropriate procedure.
b) Check the assumptions.
c) State the hypotheses.
d) Test an appropriate hypothesis and state your results.
e) Interpret the meaning of the results and state a conclusion.

51. Businesses and blogs. The Pew Internet & American Life Project routinely conducts surveys to gauge the impact of the Internet and technology on daily life. A recent survey asked respondents if they read online journals or blogs, an Internet activity of potential interest to many businesses. A subset of the data from this survey *(February–March 2007 Tracking Data Set)* shows responses to this question. Test whether reading online journals or blogs is independent of generation.

	Read online journal or blog			
	Yes, Yesterday	Yes, but not Yesterday	No	Total
Gen-Y (18–30)	29	35	62	126
Gen X (31–42)	12	34	137	183
Trailing Boomers (43–52)	15	34	132	181
Leading Boomers (53–61)	7	22	83	112
Matures (62+)	6	21	111	138
Total	69	146	525	740

(Row labels grouped under "Generation")

52. Businesses and blogs again. The Pew Internet & American Life Project survey described in Exercise 51 also asked respondents if they ever created or worked on their own online journal or blog. Again, a subset of the data from this survey *(February–March 2007 Tracking Data Set)* shows responses to this question. Test whether creating online journals or blogs is independent of generation.

	Create online journal or blog . . .			
	Yes/ Yesterday	Yes/Not Yesterday	No	Total
Gen Y (18–30)	18	24	85	127
Gen X (31–42)	6	15	162	183
Boomers (43–61)	5	15	273	293
Matures (62+)	3	3	132	138
Total	32	57	652	741

(Row labels grouped under "Generation")

53. Information systems. In a recent study of enterprise resource planning (ERP) system effectiveness, researchers asked companies about how they assessed the success of their ERP systems. Out of 335 manufacturing companies surveyed, they found that 201 used return on investment (ROI), 100 used reductions in inventory levels, 28 used improved data quality, and 6 used on-time delivery. In a survey of 200 service firms, 40 used ROI, 40 used inventory levels, 100 used improved data quality, and 20 used on-time delivery. Is there evidence that the measures used to assess ERP system effectiveness differ between service and manufacturing firms? Perform the appropriate test and state your conclusion.

	GDP		
	Top 40%	**Bottom 60%**	**Total**
West (Far West, Southwest, and Rocky Mtn.)	5	10	15
Midwest (Great Lakes and Plains States)	5	7	12
Southeast	5	7	12
Northeast (Mideast and New England States)	5	7	12
Total	20	31	51

	GDP % Change		
	Top 40%	**Bottom 60%**	**Total**
West (Far West, Southwest, and Rocky Mtn.)	13	2	15
Midwest (Great Lakes and Plains States)	2	10	12
Southeast	4	8	12
Northeast (Mideast and New England States)	2	10	12
Total	21	30	51

54. U.S. Gross Domestic Product. The U.S. Bureau of Economic Analysis provides information on the Gross Domestic Product (GDP) in the United States by state (www.bea.gov). The Bureau recently released figures that showed the real GDP by state for 2007. Using the data in the table at the top of the page, examine if GDP and *Region* of the country are independent. (Alaska and Hawaii are part of the West Region. D.C is included in the Mideast Region.)

55. Economic growth. The U.S. Bureau of Economic Analysis also provides information on the growth of the U.S. economy (www.bea.gov). The Bureau recently released figures that they claimed showed a growth spurt in the western region of the United States. Using the table and map below, determine if the percent change in real GDP by state for 2005–2006 was independent of region of the country. (Alaska and Hawaii are part of the West Region. D.C is included in the Mideast Region.)

56. Economic growth, revisited. The U.S. Bureau of Economic Analysis provides information on the GDP in the United States by metropolitan area (www.bea.gov). The Bureau recently released figures that showed the percent change in real GDP by metropolitan area for 2004–2005. Using the data in the following table, examine if there is independence of the growth in metropolitan GDP and region of the country. (Alaska and Hawaii are part of the West Region. Some of the metropolitan areas may have been combined for this analysis.)

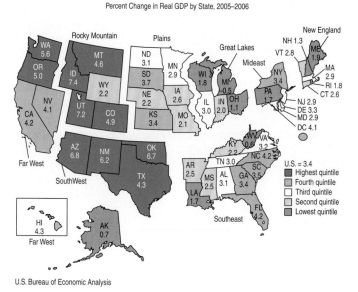

Percent Change in Real GDP by State, 2005–2006

U.S. Bureau of Economic Analysis

	GDP Growth		
	Top two quintiles (top 40%)	Bottom three quintiles (bottom 60%)	Total
West (Far West, Southwest, and Rocky Mtn.)	62	46	108
Midwest (Great Lakes and Plains States)	9	87	96
Southeast	38	58	96
Northeast (Mideast and New England States)	12	36	48
Total	121	227	348

Percent Change in Real GDP by Metropolitan Area, 2004–2005

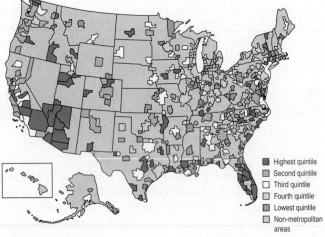

- Highest quintile
- Second quintile
- Third quintile
- Fourth quintile
- Lowest quintile
- Non-metropolitan areas

U.S. Bureau of Economic Analysis

Just Checking Answers

1 This is a test of homogeneity. The clue is that the question asks whether the distributions are alike.

2 This is a test of goodness-of-fit. We want to test the model of equal assignment to all lots against what actually happened.

3 This is a test of independence. We have responses on two variables for the same individuals.

C A S E
Study

Investment Strategy Segmentation

In the aftermath of the financial crisis of 2008, brokerage firms struggled to get individual investors back into the stock market. To gain competitive advantage, market analysts in nearly every industry segment their customers in order to place advertisements where they can have the most impact. The brokerage business is no different. Because different groups of people invest differently, customizing advertising to the needs of these groups leads to more efficient advertising placement and response. Brokerage firms get their information about the investment practices of individuals from a variety of sources, among which the U.S. Census Bureau figures prominently.

The U.S. Census Bureau, with the help of the Bureau of Labor Statistics (BLS) and the Internal Revenue Service (IRS), monitors the incomes and expenditures of Americans. A random sample of Americans are surveyed periodically about their investment practices. In the file CSIII.txt you'll find a random sample of 1000 people from the 48,842 records found in the file Census Income Data set on the University of California at Irvine machine learning repository. This subset was sampled from those that showed some level of investment in the stock market as evidenced by claiming *Capital Gains ($)*, *Capital Losses ($)*, or *Dividends ($)*. Included as well are the demographic variables for these people: *Age (years)*, *Sex (male/female)*, *Union Member (Yes/No)*, *Citizenship (several categories)*, *College (No College/Some)*, *Married (Married/Single)*, *Filer Status (Joint/Single)*.

In order to support her segmentation efforts, a market analyst at an online brokerage firm wants to study differences in investment behaviors. If she can find meaningful differences in the types and amounts of investing that various groups engage in, she can use that information to inform the advertising and marketing departments in their strategies to attract new investors. Using the techniques of Part III, including confidence intervals and hypothesis tests, what differences in investment behaviors can you find among the various demographic groups?

Some specific questions to consider:

1. Do men and women invest similarly? Construct confidence intervals for the differences in mean *Capital Gains*, *Capital Losses* and *Dividends*.

    Be sure to make a suitable display to check assumptions and conditions. If you find outlier(s), consider the analysis with and without the outlier(s).

2. Make a suitable display to compare investment results for the various levels of *Citizenship*.

3. Do those who file singly have the same investment results as those who file jointly? Select, perform, and interpret an appropriate test.

4. Compare differences in investment results for the other demographic variables, being careful to check assumptions and conditions.

    Once again, make suitable displays. Are there outliers to be concerned with? Discuss.

Summarize your findings and conclusions about the investment practices of various groups. Write a short report in order to help the market analyst.

Inference for Regression

Nambé Mills

Nambé (nam ′bei) Mills, Inc. was founded in 1951 near the tiny village of Nambé Pueblo, about 10 miles north of Santa Fe, New Mexico. Known for its elegant, functional cooking and tableware, Nambé Mills now sells its products in luxury stores throughout the world. Many of its products are made from an eight-metal alloy created at the Los Alamos National Laboratory (where the atomic bomb was developed during World War II) and now used exclusively by Nambé Mills. The alloy has the luster of silver and the solidity of iron, but its main component is aluminum. In fact, it does not contain silver, lead, or pewter (a tin and copper alloy), and it does not tarnish. Because it's a trade secret, Nambé Mills does not divulge the rest of the formula. Up to 15 craftsmen may be involved in the production process of an item, which includes molding, pouring, grinding, polishing, and buffing.

B ecause Nambé Mills's metal products are sand-cast, they must go through a lengthy production process. To rationalize its production schedule, management examined the total polishing times of 59 tableware items. Here's a scatterplot showing the retail price of the items and the amount of time (in minutes) spent in the polishing phase (Figure 16.1).

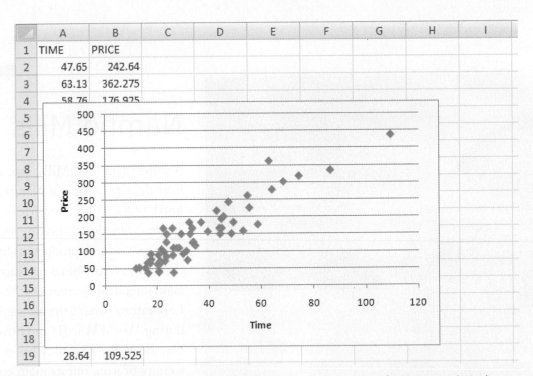

Figure 16.1   A scatterplot of *Price* ($) against polishing *Time* (minutes) for Nambé tableware products shows that the items that take longer to polish have a higher price, on average. Plot created in **Excel**.

Back in Chapter 6 we modeled relationships like this by fitting a straight line. The equation of the least squares line for these data is:

$$\widehat{Price} = -4.871 + 4.200 \times Time$$

The slope says that, on average, the price increases by $4.20 for every extra minute of polishing time.

How useful is this model? When we fit linear models before, we used them to describe the relationship between the variables, and we interpreted the slope and intercept as descriptions of the data. Now we'd like to know what the regression model can tell us beyond the sample we used to generate this regression. To do that, we'll want to make confidence intervals and test hypotheses about the slope and intercept of the regression line.

16.1   The Population and the Sample

Our data are a sample of 59 items. If we take another sample, we hope the regression line will be similar to the one we found here, but we know it won't be exactly the same. Observations vary from sample to sample. But we can imagine a true line that summarizes the relationship between *Price* and *Time*. Following our usual conventions, we write the idealized line with Greek letters and consider the coefficients (slope and intercept) to be parameters: β_0 is the intercept, and β_1 is the slope. Corresponding to our fitted line of $\hat{y} = b_0 + b_1 x$, we write $\mu_y = \beta_0 + \beta_1 x$. We write μ_y instead of y because the regression line assumes that the *means* of the y values for each value of x fall exactly on the line. We can picture the relationship as in Figure 16.2. The means fall exactly on the line (for our idealized model), and the y values at each x are distributed around them.

Figure 16.2 There's a distribution of *Prices* for each value of polishing *Time*. The regression model assumes that the means line up perfectly like this.

Notation Alert!

We use lowercase Greek betas (β) to denote the coefficients in the regression model. We estimate them with the b's in the fitted regression equation. We used β earlier for the probability of a Type II error, but β here is not related to its earlier use.

Now, if only we had all the values in the population, we could find the slope and intercept of this *idealized regression line* explicitly by using least squares.

Of course, not all the individual y's are at these means. In fact, the line will miss most—and usually all—of the plotted points. Some y's lie above the line and some below the line, so like all models, this one makes errors. If we want to account for each individual value of y in our model, we have to include these errors, which we denote by ε:

$$y = \beta_0 + \beta_1 x + \varepsilon.$$

This equation has an ε to soak up the deviation at each point, so the model gives a value of y for each value of x.

We estimate the β's by finding a regression line, $\hat{y} = b_0 + b_1 x$, as we did in Chapter 6. The residuals, $e = y - \hat{y}$, are the sample-based versions of the errors, ε. We'll use them to help us assess the regression model.

We know that least squares regression will give us reasonable estimates of the parameters of this model from a random sample of data. We also know that our estimates won't be equal to the parameters in the idealized or "true" model. Our challenge is to account for the uncertainty in our estimates by making confidence intervals as we've done for means and proportions. For that, we need to make some assumptions about the model and the errors.

For Example   National survey

The Cornell National Social Survey (CNSS) is a national annual survey of a random sample of 1000 adults in the continental United States. Researchers in various disciplines can submit questions for the survey. According to the Cornell Survey Research Institute's site, https://sri.cornell.edu/sri/cnss.cfm:

> The survey is administered by telephone, using a random digit dial (RDD) method. The survey contains a small core of demographic questions, with all remaining questions being generated by Cornell researchers. The interviews average no more than 20 minutes in length and are conducted by the Survey Research Institute.

(continued)

The data are made available to the public and can be used by researchers and business professionals.

A chain store that sells bulk foods is planning a marketing campaign. Using data from the CNSS they construct variables that report for each respondent the weekly amount spent by their household on food (both groceries and eating out). A regression of this variable against household income (in $1000) results in the model:

$$\widehat{Food\$/wk} = 101.33 + 0.77\ HHincome$$

Question: Write an interpretation of this model. How should the store managers interpret the relationship between spending on food and household income?

Answer: The average amount spent on food can be estimated as $101.33 plus $0.77 for each $1000 of household income. Managers might use this to target advertising to different communities based on census data about median incomes in those communities.

16.2   Assumptions and Conditions

Back in Chapter 6 when we fit lines to data, we needed both the **Linearity** and the **Equal Variance Assumptions**, and we checked the corresponding conditions. Now, when we want to make inferences about the coefficients of the line, we'll have to assume more, so we'll add more conditions.

We also need to be careful about the order in which we check conditions. So we number the assumptions, and check conditions for each in order: (1) Linearity Assumption, (2) Independence Assumption, (3) Equal Variance Assumption, and (4) Normal Population Assumption.

1. Linearity Assumption

If the true relationship of two quantitative variables is far from linear and we use a straight line to fit the data, our entire analysis will be useless, so we always check linearity first (and we check the **Quantitative Variable Condition** for both variables as well).

The **Linearity Condition** is satisfied if a scatterplot looks straight. It's generally not a good idea to draw a line through the scatterplot when checking. That can fool your eye into seeing the plot as straighter than it really is. Recall the errors, or residuals, we computed in Chapter 6 for each observation. Sometimes it's easier to see violations of this condition by looking at a scatterplot of the residuals against x or against the predicted values, $\hat{y}$. That plot should have no pattern if the condition is satisfied.

If the scatterplot is straight enough, we can go on to some assumptions about the errors. If not, we stop here, or consider transforming the variables to make the scatterplot more linear.[1]

2. Independence Assumption

The errors in the true underlying regression model (the ε's) must be independent of each other. As usual, there's no way to be sure that the Independence Assumption is true.

---

[1]We've seen that re-expressing variables can help us understand relationships in several other contexts back in Chapters 5 and 6. Chapter 17 discusses transformations for regression more thoroughly.

When we care about inference for the regression parameters, it's often because we think our regression model might apply to a larger population. In such cases, we can check the **Randomization Condition** that the individuals are a random sample from that population.

We can also check displays of the regression residuals for evidence of patterns, trends, or clumping, any of which would suggest a failure of independence. In the special case when we have a time series, a common violation of the Independence Assumption is for consecutive errors to be correlated with each other (autocorrelated). The error our model makes today may be similar to the one it made yesterday. We can check this violation by plotting the residuals against time, which is usually the *x*-variable for a time series anyway, and looking for patterns.

3. Equal Variance Assumption

The variability of *y* should be about the same for all values of *x*. In Chapter 6, we looked at the standard deviation of the residuals (s_e) to measure the size of the scatter. Now we'll need this standard deviation to build confidence intervals and test hypotheses. The standard deviation of the residuals is the building block for the standard errors of all the regression parameters. But it only makes sense if the scatter of the residuals is the same everywhere. In effect, the standard deviation of the residuals "pools" information across all of the individual distributions of *y* at each *x*-value, and pooled estimates are appropriate only when they combine information for groups with the same variance.

We check the **Equal Spread Condition** with a scatterplot of the residuals against *x* or *ŷ*. Make sure the spread is nearly constant across the plot. Be alert for a "fan" shape or other tendency for the variation to grow or shrink in one part of the scatterplot.

If the plot is straight enough, the data are independent, and the spread is about the same everywhere, we can move on to the final assumption and its associated condition.

4. Normal Population Assumption

We assume the errors around the idealized regression line at each value of *x* follow a Normal model. We need this assumption so that we can use a Student's *t*-model for inference.

As with other times when we've used Student's *t*, we'll settle for the residuals satisfying the **Nearly Normal Condition**.[2] As noted before, the Normality Assumption becomes less important as the sample size grows because the model is about means and the Central Limit Theorem takes over. A histogram of the residuals is one way to check whether they are nearly Normal. Alternatively, we can look at a **Normal probability plot** of the residuals (see Figure 16.3). Normal probability plots expose deviations from the Normal model more efficiently than histograms. Recall from Section 8.4 that when the distribution of the data is Normal, the Normal probability plot will look like a diagonal straight line. Deviations from a straight line indicate that the distribution is not Normal.

Figure 16.3 A Normal probability plot graphs values (here the residuals from the Nambé mills regression) against the values that would be expected in a sample from a standard Normal containing the same number of values.

---

[2]*This* is why we check the conditions in order. We check that the residuals are independent and that the variation is the same for all *x*'s before we can lump all the residuals together to check the Normal Condition.

Another common failure of Normality is the presence of an outlier. So, we still check the **Outlier Condition** to ensure that no point is exerting too much influence on the fitted model. We can check for outliers with a histogram of the residuals or by checking whether the edge values of a normal probability plot line up with the rest of the values.

Summary of Assumptions and Conditions

If all four assumptions were true, the idealized regression model would look like Figure 16.4.

Figure 16.4   The regression model has a distribution of *y*-values for each *x*-value. These distributions follow a Normal model with means lined up along the line and the same standard deviations.

At each value of *x*, there is a distribution of *y*-values that follows a Normal model, and each of these Normal models is centered on the line and has the same standard deviation. Of course, we don't expect the assumptions to be exactly true. As George Box said, "all models are wrong." But the linear model is often close enough to be useful.

In regression, there's a little catch. The best way to check many of the conditions is with the residuals, but we get the residuals only *after* we compute the regression. Before we compute the regression, however, we should check at least one of the conditions.

So we work in this order:

1. **Make a scatterplot of the data** to check the Linearity Condition (and always check that the variables are quantitative as well). (This checks the **Linearity Assumption**.)

2. If the data are straight enough, **fit a regression and find the residuals, *e*, and predicted values, $\hat{y}$.**

3. If you know when the measurements were made, **plot the residuals against time** to check for evidence of patterns that suggest they may not be independent (**Independence Assumption**).

4. **Make a scatterplot of the residuals against *x* or the predicted values.** This plot should have no pattern. Check in particular for any bend (which would suggest that the data weren't that straight after all), for any thickening (or thinning), and, of course, for any unusual observations. (If you discover any errors, correct them or omit those points, and go back to step 1. Otherwise, consider performing two regressions—one with and one without the unusual observations.) (**Equal Variance Assumption**)

5. If the scatterplots look OK, then **make a histogram and Normal probability plot of the residuals** to check the **Nearly Normal** and **Outlier Conditions** (**Normal Population Assumption**).

"Truth will emerge more readily from error than from confusion."

—FRANCIS BACON
(1561–1626)

For Example   Checking assumptions and conditions

Here are some displays related to the regression of the percentage of income spent on food against household income (see page 493). Use them and the other information provided to check the following regression assumptions and conditions. For each, indicate whether you think it is satisfied and why:

Question: Do the assumptions and conditions for regression inference appear to be met?

Answer:

- **Linearity:** The scatterplot of the data is not strongly linear, but there is no evidence of nonlinearity.
- **Independence:** The data are from a national random sample. Respondents are almost surely independent of one another.
- **Equal Variance:** The data may be less variable at the low end (which would be understandable because households with lower income would be constrained from spending too much).
- **Normal Population:** The normal probability plot shows some bend and the histogram is unimodal, but has a longer upper tail. However, the sample size is large, so we probably can proceed with inference.

16.3   The Standard Error of the Slope

There's only one regression model for the population. Sample regressions try to estimate the parameters, β_0 and β_1. We expect the estimated slope for any sample, b_1, to be close to—but not actually equal to—the model slope, β_1. If we could see the collection of slopes from many samples (imagined or real) we would see a distribution of values around the true slope. That's the sampling distribution of the slope.

What is the standard deviation of this distribution? What aspects of the data affect how much the slopes vary from sample to sample?

- **Spread around the line.** Figure 16.5 shows samples from two populations. Which underlying population would give rise to the more consistent slopes?

Figure 16.5   Which of these scatterplots would give the more consistent regression slope estimate if we were to sample repeatedly from its underlying population?

Less scatter around the line means the slope will be more consistent from sample to sample. Recall that we measure the spread around the line with the **residual standard deviation:**

$$s_e = \sqrt{\dfrac{\sum (y - \hat{y})^2}{n - 2}}.$$

The less scatter around the line, the smaller the residual standard deviation and the stronger the relationship between x and y.

- **Spread of the x's:** Here are samples from two more populations (Figure 16.6). Which of these would yield more consistent slopes?

Figure 16.6   Which of these scatterplots would give the more consistent regression slope estimate if we were to sample repeatedly from the underlying population?

A plot like the one on the right has a broader range of x-values, so it gives a more stable base for the slope. We might expect the slopes of samples from situations like that to vary less from sample to sample. A large standard deviation of x, s_x, as in the figure on the right, provides a more stable regression.

- **Sample size.** What about the two scatterplots in Figure 16.7?

Figure 16.7   Which of these scatterplots would give the more consistent regression slope estimate if we were to sample repeatedly from the underlying population?

It shouldn't shock you that a larger sample size (scatterplot on the right), n, gives more consistent estimates from sample to sample.

Let's summarize what we've seen in these three figures:

> ## The standard error of the regression slope
>
> Three aspects of the scatterplot that affect the standard error of the regression slope are:
>
> - Spread around the line: s_e
> - Spread of x values: s_x
> - Sample size: n
>
> These are in fact the *only* things that affect the standard error of the slope. The formula for the standard error of the slope is:
>
> $$SE(b_1) = \frac{s_e}{s_x\sqrt{n-1}}.$$

The error standard deviation, s_e, is in the *numerator*, since a larger spread around the line *increases* the slope's standard error. On the other hand, the *denominator* has both a sample size term $(\sqrt{n-1})$ and s_x because increasing either of these *decreases* the slope's standard error.

For Example | Standard error

Here is more of the regression output from a regression of the amount spent on food *vs* household income (see page 497):

```
Dependent variable is: Food $/wk
R squared = 15.1%
s = 91.10 with 905 − 2 = 903 degrees of freedom
```

Variable	Coefficient	SE(Coeff)	t-ratio	P-value
Intercept	101.334	5.812	89.1	≤0.0001
HHincome	0.769860	0.0608	12.2	≤0.0001

Question: What is the standard error of the slope and how is it computed? Locate the values in the computer output that are used to calculate the standard error and show how they work together to do that. You'll also need to know that $s_{\text{HHincome}} = 49.91$.

Answer: $s_e = 91.10$

$s_x = 49.91$

$n = 905$

$$SE(b_1) = \frac{91.10}{49.91\sqrt{905-1}} = 0.0608^{\,3}$$

16.4 A Test for the Regression Slope

We know b_1 varies from sample to sample. As you'd expect, its sampling distribution model is centered at β_1, the slope of the idealized regression line. Now we can estimate its standard deviation with $SE(b_1)$. What about its shape? Here the Central Limit Theorem and Gosset come to the rescue again. When we standardize the slopes by subtracting the model mean and dividing by their standard error, we get a Student's t-model, this time with $n - 2$ degrees of freedom.

---

[3] 0.0607 with these values from the regression table because they have been rounded.

> ### The sampling distribution for the regression slope
> When the conditions are met, the standardized estimated regression slope,
>
> $$t = \frac{b_1 - \beta_1}{SE(b_1)},$$
>
> follows a Student's t-model with $n - 2$ degrees of freedom. We calculate the standard
>
> error as $SE(b_1) = \dfrac{s_e}{s_x \sqrt{n-1}}$, where $s_e = \sqrt{\dfrac{\sum(y - \hat{y})^2}{n-2}}$, n is the number of data values,
>
> and s_x is the standard deviation of the x-values.

The same reasoning applies for the intercept. We write:

$$\frac{b_0 - \beta_0}{SE(b_0)} \sim t_{n-2}.$$

We could use this statistic to construct confidence intervals and test hypotheses, but often the value of the intercept isn't interesting. Most hypothesis tests and confidence intervals for regression are about the slope. But in case you really want to see the formula for the standard error of the intercept, we've parked it in a footnote.[4]

Now that we have the standard error of the slope and its sampling distribution, we can test a hypothesis about it and make confidence intervals. The usual null hypothesis about the slope is that it's equal to 0. Why? Well, a slope of zero would say that y doesn't tend to change linearly when x changes—in other words, that there is no linear association between the two variables. If the slope were zero, there wouldn't be much left of our regression equation.

A null hypothesis of a zero slope questions the entire claim of a linear relationship between the two variables, and often that's just what we want to know. In fact, every software package or calculator that does regression simply assumes that you want to test the null hypothesis that the slope is really zero.

> ### The t-test for the regression slope
> When the assumptions and conditions are met, we can test the hypothesis $H_0: \beta_1 = 0$ vs. $H_A: \beta_1 \neq 0$ (or a one-sided alternative hypothesis) using the standardized estimated regression slope,
>
> $$t = \frac{b_1 - \beta_1}{SE(b_1)},$$
>
> which follows a Student's t-model with $n - 2$ degrees of freedom. We can use the t-model to find the P-value of the test.

This is just like every other t-test we've seen: a difference between the statistic and its hypothesized value divided by its standard error. This test is the *t-test* that the regression slope is 0, usually referred to as the **t-test for the regression slope**.

Another use of these values might be to make a confidence interval for the slope. We can build a confidence interval in the usual way, as an estimate plus or minus a margin of error. As always, the margin of error is just the product of the standard error and a critical value.

---

[4]$SE(b_0) = s_e \sqrt{\dfrac{1}{n} + \dfrac{\bar{x}^2}{\sum(x - \bar{x})^2}}$

> ## The confidence interval for the regression slope
>
> When the assumptions and conditions are met, we can find a confidence interval for β_1 from
>
> $$b_1 \pm t^*_{n-2} \times SE(b_1),$$
>
> where the critical value t^* depends on the confidence level and has $n - 2$ degrees of freedom.

Guided Example   Nambé Mills

Now that we have a method to draw inferences from our regression equation, let's try it out on the Nambé Mills data.

The slope of the regression gives the impact of *Time* on *Price*. Let's test the hypothesis that the slope is different from zero.

PLAN

Setup State the objectives.

Identify the parameter you wish to estimate. Here our parameter is the slope.

Identify the variables and their context.

Hypotheses Write the null and alternative hypotheses.

Model Check the assumptions and conditions.

We want to test the theory that the price of a particular item at Nambé Mills is related to the time it takes to polish it. We have data for 59 items sold by Nambé Mills. The slope of this relationship will indicate the impact of *Time* on *Price*. Our null hypothesis will be that the slope of the regression is 0.

H_0: The *Price* of an item is not related to the polishing *Time*: $\beta_1 = 0$.
H_A: The *Price* is, in fact, related to the *Time*: $\beta_1 \neq 0$.

✓ **Linearity Condition:** There is no obvious curve in the scatterplot of y versus x.

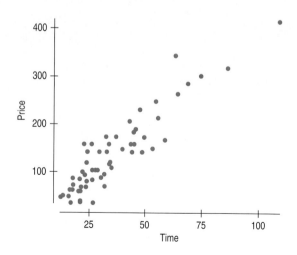

✓ **Independence Assumption:** These data are on 59 different items manufactured by the company. There is no reason to suggest that the error in price of one item should be influenced by another.

(continued)

Make graphs. Because our scatterplot of *y* versus *x* seems straight enough, we can find the least squares regression and plot the residuals.

Usually, we check for suggestions that the Independence Assumption fails by plotting the residuals against time. Patterns or trends in that plot raise our suspicions.

✓ **Randomization Condition:** The data are *not* a random sample, but we assume they are representative of the prices and polishing times of Nambé Mills items.

✓ **Equal Spread Condition:** The plot of residuals against the predicted values shows no obvious patterns. The spread is about the same for all predicted values, and the scatter appears random.

✓ **Nearly Normal Condition:** A histogram of the residuals is unimodal and symmetric, and the normal probability plot is reasonably straight.

State the sampling distribution model.

Under these conditions, the sampling distribution of the regression slope can be modeled by a Student's *t*-model with $(n - 2) = 59 - 2 = 57$ degrees of freedom, so we'll proceed with a regression slope *t*-test.

DO | **Mechanics** The regression equation can be found from the formulas in Chapter 6, but regressions are almost always found from a computer program or calculator.

Here's the computer output for this regression.

Variable	Coefficient	SE(coeff)	t-ratio	P-value
Intercept	−4.871	9.654	−0.50	0.6159
Time	4.200	0.2385	17.61	<0.0001

$s = 32.54$ R-Sq = 84.5%

The P-values given in the regression output table are from the Student's *t*-distribution on $(n - 2) = 57$ degrees of freedom. They are appropriate for two-sided alternatives.

The P-value <0.0001 means that the association we see in the data is unlikely to have occurred by chance. Therefore, we reject the null hypothesis and conclude that there is strong evidence that the *Price* is linearly related to the polishing *Time*.

Create a confidence interval for the true slope. To obtain the *t*-value for 57 degrees of freedom, use the *t*-table at the back of your textbook. The estimated slope and SE for the slope are obtained from the regression output.

A 95% confidence interval for β_1 is:

$$b_1 \pm t^*_{n-2} \times SE(b_1) = (3.722, 4.678) \text{ \$/minute}$$

Interpret the interval.

Simply rejecting the standard null hypothesis doesn't guarantee that the size of the effect is large enough to be important.

I am 95% confident that the price increases, on average, between $3.72 and $4.68 for each additional minute of polishing time. (Technically: I am 95% confident that the interval from $3.72 to $4.68 per minute captures the true rate at which the *Price* increases with polishing *Time*.)

REPORT | **Conclusion** State the conclusion in the proper context.

MEMO

Re: Nambé Mills Pricing

We investigated the relationship between polishing time and pricing of 59 Nambé Mills items. The regression analysis showed that, on average, the price increased $4.20 for every additional minute of polishing time.

Assuming that these items are representative, we are 95% confident that the actual price of a metal item manufactured by Nambé Mills increases, on average, between $3.72 and $4.68 for each additional minute of polishing work required.

For Example   Testing the slope

Question: Using the computer output from the example on page 499, state and test the standard null hypothesis for the slope. Tell what the hypothesis means in terms of the variables as well as in symbols, and explain what your conclusion means.

Answer: The null hypothesis is that there is no linear relationship between the amount spent on food and household income in U.S. families:

$$H_0: \beta_1 = 0$$
$$H_A: \beta_1 \neq 0$$

With a P-value less than 0.001, we can be quite confident in rejecting the null hypothesis. We conclude that there is indeed a linear relationship between the amount spent on food and household income in U.S. families.

16.5　A Hypothesis Test for Correlation

We just tested whether the slope, β_1, was 0. To test it, we estimated the slope from the data and then, using its standard error and the t-distribution, measured how far the slope was from 0: $t = \dfrac{b_1 - 0}{SE(b_1)}$. What if we wanted to test whether the *correlation* between x and y is 0? We write ρ for the parameter (true population value) of the correlation, so we're testing $H_0: \rho = 0$. Remember that the regression slope estimate is $b_1 = r\dfrac{s_y}{s_x}$. The same is true for the parameter versions of these statistics: $\beta_1 = \rho\dfrac{\sigma_y}{\sigma_x}$. That means that if the slope is really 0, then the correlation has to be 0, too. So if we test $H_0: \beta_1 = 0$, that's really the same as testing $H_0: \rho = 0$. Sometimes a researcher, however, might want to test correlation without fitting a regression, so you'll see the test of correlation as a separate test (it's also slightly more general), but the results are mathematically the same even though the form looks a little different. Here's the **t-test for the correlation coefficient.**

> ### The t-test for the correlation coefficient
>
> When the conditions are met, we can test the hypothesis $H_0: \rho = 0$ vs. $H_A: \rho \neq 0$ using the test statistic:
>
> $$t = r\sqrt{\frac{n - 2}{1 - r^2}},$$
>
> which follows a Student's t-model with $n - 2$ degrees of freedom. We can use the t-model to find the P-value of the test.

For Example　　Testing correlation

Question: In the Cornell National Social Survey (see page 493), the correlation between the *Age* of respondents and their *Expenditures on Electronics* is -0.142. Does this indicate a statistically significant association?

Answer:

$$t = -0.142\sqrt{\frac{1000 - 2}{1 - (-0.142)^2}} = -4.53,$$

which is a large enough value to be statistically significant. We can reject the null hypothesis that the true correlation, ρ, is zero.

Just Checking

General economic theory suggests that as unemployment rises and jobs become harder to find, more students will enroll in universities. Researchers analyzed college enrollment at the University of New Mexico and unemployment data in New Mexico to determine whether or not there is any statistical relationship between the two variables. The data were collected by the University of New Mexico over a period of 29 years, starting with 1961 and ending with 1989. The variable *Enrollment* is in number of students and the variable *Unemp* is a percentage. Here is some regression output for these data.

Predictor	Coeff	SE(Coeff)	t-ratio	P-value
Intercept	3957	4000	0.99	0.331
Unemp	1133.8	513.1	2.21	0.036

s = 3049.50　R-Sq = 15.3%

1 What would you like to see before proceeding with inference on this regression? Why?

2 Assuming the assumptions and conditions for regression are met, find the 95% confidence interval for the slope.

3 Clearly state the null and alternative hypothesis for the slope. Interpret the P-value.

4 Is there a strong relationship between enrollment and unemployment?

5 Interpret the value of R-Sq in the output.

6 The correlation between enrollment and unemployment for this sample is 0.391, which gives a *t* value of 2.21 with 27 degrees of freedom and a two-sided P-value of 0.036. What does this say about the true correlation between enrollment and unemployment? Does this give you any new information?

16.6   Standard Errors for Predicted Values

We've seen how to construct the confidence interval for a slope or intercept, but we're often interested in prediction. We know how to compute predicted values of *y* for any value of *x*. We first did that in Chapter 6. This predicted value would be our best estimate, but it's still just an informed guess. Now, however, we have standard errors. We can use those SE's to construct confidence intervals for the predictions and to report our uncertainty honestly.

Suppose we want to predict the *Price* of an item that takes 40 minutes of *Time* to polish. From our model of Nambé Mills items, we can use polishing *Time* to get a reasonable estimate. We know that items with 40 minutes of polishing time may not all have the same price. It makes sense to compute a confidence interval. But we must first be clear about what we are asking. There are two different questions we could ask:

Do we want to estimate the *mean Price* for items that have a polishing *Time* of 40 minutes?
or,
Do we want to estimate the *Price* for a *particular* item whose polishing *Time* is 40 minutes?

What's the difference between the two questions? If we were the manufacturer, we might be more naturally interested in the *mean Price* of all items that take a certain *Time* to polish to better understand how the manufacturing process affects pricing. On the other hand, if we're a customer interested in purchasing an item, we might be more interested in estimating an *individual* item's *Price*. Both questions are interesting. The predicted *Price* value is the same for both, but one question leads to a much more precise interval than the other. If your intuition says that it's easier to be more precise about the mean than about the individuals, you're on the right track. Because individual items vary much more than means, we can predict the *mean Price* for all items with a lot more precision than we can predict the *Price* of a particular item with the same polishing *Time*.

Let's start by predicting the *Price* for a new *Time*, one that was not necessarily part of the original data set (although, of course, it could be a value in the data). To emphasize this, we'll call this *x*-value "*x* sub new" and write it x_ν.[5] As an example, we'll take x_ν to be 40 minutes. The regression equation predicts *Price* by $\hat{y}_\nu = b_0 + b_1 x_\nu$. Now that we have the predicted value, we can construct intervals around this number. Both intervals take the form:

$$\hat{y}_\nu \pm t^*_{n-2} \times SE.$$

---

[5]The Greek letter ν is called "nu."

Even the t^* value is the same for both. It's the critical value (from Table T or technology) for $n - 2$ degrees of freedom and the specified confidence level. The difference between the two intervals is in the standard errors.

> ### The confidence interval for the mean response
> When the conditions are met, we find the confidence interval for the mean response value μ_ν at a value x_ν as
>
> $$\hat{y}_\nu \pm t^*_{n-2} \times SE,$$
>
> where the standard error is
>
> $$SE(\hat{\mu}_\nu) = \sqrt{SE^2(b_1) \times (x_\nu - \bar{x})^2 + \frac{s_e^2}{n}}.$$

The details behind the standard error can be found in the Math Box on page 508, but the ideas behind the interval are best understood by looking at an example. Figure 16.8 shows the confidence interval for the true mean of y for each x. In this plot, the intervals for the mean *Prices* at each value of *Time* are shown together as confidence bands. Notice that the bands get wider as we attempt to predict values that lie farther away from the mean *Time* (35.82 minutes). (That's the $(x_\nu - \bar{x})^2$ term in the SE formula.) As we move away from the mean x value, there is more uncertainty associated with our prediction. We can see, for example, that a 95% confidence interval for the mean *Price* of items that take 40 minutes to polish would extend from about $150 to $170. (It's actually $153.90 to $172.34.) The interval is much wider for the mean *Price* of items that take 100 minutes to polish.

Figure 16.8   The confidence intervals for the mean *Price* at a given polishing *Time* are shown as the green dotted lines. Near the mean *Time* (35.8 minutes) our confidence interval for the mean *Price* is much narrower than for values far from the mean, like 100 minutes.

Like all confidence intervals, the width of these confidence intervals varies with the sample size. A sample larger than 59 items would result in narrower intervals. A regression on 10,000 items would have much narrower bands. The last factor affecting our confidence intervals is the spread of the data around the line.

If there is more spread around the line, predictions are less certain, and the confidence interval bands are wider.

From Figure 16.8, it's easy to see that most *points* don't fall within the confidence interval bands—and we shouldn't expect them to. These bands show confidence intervals for the *mean*. An even larger sample would have given even narrower bands. Then we'd expect an even smaller percentage of the points to fall within them.

If we want an interval likely to cover an individual price, we need to use a wider one, called a **prediction interval**. Figure 16.9 shows these prediction intervals for the Nambé Mills data. Prediction intervals are based on the same quantities as the confidence intervals, but in order to cover a percentage of all the future predictions, they include an extra term for the spread around the line. As we can see in Figure 16.9, these bands also widen as we move from the mean of x, but it's less obvious because the extra width across the entire range of x makes the change harder to see.

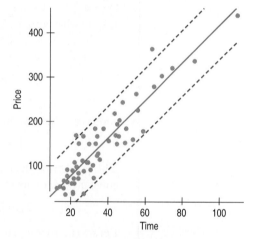

Figure 16.9   Prediction intervals (in red) estimate the intervals that contain 95% of the distribution of the y values that might be observed at a given value of x. If the assumptions and conditions hold, then there's about a 95% chance that a particular y-value at x_ν will be covered by the interval.

The standard errors for prediction depend on the same kinds of things as the coefficients' standard errors. If there is more spread around the line, we'll be less certain when we try to predict the response. Of course, if we're less certain of the slope, we'll be less certain of our prediction. If we have more data, our estimate will be more precise. And there's one more piece. If we're farther from the center of our data, our prediction will be less precise. It's a lot easier to predict a data point near the middle of the data set than to extrapolate far from the center.

The prediction interval for an individual response

When the conditions are met, we can find the prediction interval for all values of y at a value x_ν as

$$\hat{y}_\nu \pm t^*_{n-2} \times SE,$$

where the standard error is

$$SE(\hat{y}_\nu) = \sqrt{SE^2(b_1) \times (x_\nu - \bar{x})^2 + \frac{s_e^2}{n} + s_e^2}.$$

The critical value t^* depends on the confidence level that you specify.

Remember to keep the distinction between the two kinds of intervals when looking at computer output. The narrower ones are confidence intervals for the *mean* and the wider ones are prediction intervals for *individual* response values.

Math Box

The Standard Errors of the Predicted Values

Some insight into the differences between the two intervals can be gained by looking at the formulas for their standard errors and how they're derived.

To predict a *y*-value for a new value of *x*, x_ν, we'd have

$\hat{y}_\nu = b_0 + b_1 x_\nu$, which, because $b_0 = \bar{y} - b_1\bar{x}$, can be written $\hat{y}_\nu = b_1(x_\nu - \bar{x}) + \bar{y}$.

We use $\hat{y}_\nu$ in two ways. First, we use it to estimate the mean value of all the *y*'s at x_ν, in which case we *call* it $\hat{\mu}_\nu$.

To create a confidence interval for the mean value, we need to measure the variability in this prediction

$$Var(\hat{\mu}_\nu) = Var(b_1(x_\nu - \bar{x}) + \bar{y}).$$

We now call on the Pythagorean Theorem of Statistics: The slope, b_1, and mean, $\bar{y}$, are independent, so their variances add.

$$Var(\hat{\mu}_\nu) = Var(b_1(x_\nu - \bar{x})) + Var(\bar{y}).$$

The horizontal distance from our specific *x*-value to the mean, $x_\nu - \bar{x}$, is a constant, so it comes "out" of the variance.

$$Var(\hat{\mu}_\nu) = (Var(b_1))(x_\nu - \bar{x})^2 + Var(\bar{y}).$$

Let's write that equation in terms of standard deviations.

$$SD(\hat{\mu}_\nu) = \sqrt{(SD^2(b_1))(x_\nu - \bar{x})^2 + SD^2(\bar{y})}.$$

Because we'll need to estimate these standard deviations using sample statistics, we're really dealing with standard errors.

$$SE(\hat{\mu}_\nu) = \sqrt{(SE^2(b_1))(x_\nu - \bar{x})^2 + SE^2(\bar{y})}.$$

We know that the standard deviation of a mean, $\bar{y}$, is $\frac{\sigma}{\sqrt{n}}$. Here we'll estimate σ using s_e, which describes the variability in how far the line we drew through our sample mean may lie above or below the true mean.

$$SE(\hat{\mu}_\nu) = \sqrt{(SE^2(b_1))(x_\nu - \bar{x})^2 + \left(\frac{s_e}{\sqrt{n}}\right)^2}.$$

$$= \sqrt{(SE^2(b_1))(x_\nu - \bar{x})^2 + \frac{s_e^2}{n}}.$$

And there it is—the standard error we need to create a confidence interval for a mean response.[6]

---

[6]You may see the standard error expressions written in other, equivalent ways. The most common alternatives are

$$SE(\hat{\mu}_\nu) = s_e\sqrt{\frac{1}{n} + \frac{(x_\nu - \bar{x})^2}{\sum(x - \bar{x})^2}} \quad \text{and} \quad SE(\hat{y}_\nu) = s_e\sqrt{1 + \frac{1}{n} + \frac{(x_\nu - \bar{x})^2}{\sum(x - \bar{x})^2}}.$$

When we try to predict an *individual* response, we also must worry about how far the true point may lie above or below the regression line. We represent that uncertainty by adding another term, e, to the original equation to get:

$$y = \hat{\mu}_\nu + e = b_1(x_\nu - \bar{x}) + \bar{y} + e.$$

To make a long story short (and the equation just a bit longer), that additional term simply adds one more standard error to the sum of the variances.

$$SE(\hat{y}_\nu) = \sqrt{(SE^2(b_1))(x_\nu - \bar{x})^2 + \frac{s_e^2}{n} + s_e^2}.$$

We've written the predicted value as $\hat{y}_\nu$ instead of $\hat{\mu}_\nu$, this time not because it's a different value, but to emphasize that we're *using* it to predict an individual now and not the mean of all y values at x_ν.

16.7 Using Confidence and Prediction Intervals

Now that we have standard errors, we can ask how well our analysis can predict the mean price for objects that take 25 minutes to polish. The regression output table provides most of the numbers we need.

```
Variable  Coefficient  SE(coeff)  t-ratio  P-value
Intercept   -4.871      9.654     -0.50    0.6159
Time         4.200      0.2385    17.61   <0.0001

s = 32.54  R-Sq = 84.5%
```

The regression model gives a predicted value at $x_\nu = 25$ minutes of:

$$-4.871 + 4.200(25) = \$100.13$$

Using this, we'll first find the 95% confidence interval for the mean *Price* for all objects whose polishing *Time* is 25 minutes. We find the standard error from the formula using the values in the regression output.

$$SE(\hat{\mu}_\nu) = \sqrt{(SE^2(b_1))(x_\nu - \bar{x})^2 + \left(\frac{s_e}{\sqrt{n}}\right)^2}$$

$$= \sqrt{(0.2385)^2(25 - 35.82)^2 + \left(\frac{32.54}{\sqrt{59}}\right)^2} = \$4.96$$

The t^* value that excludes 2.5% in either tail with $59 - 2 = 57$ df is (according to the tables) 2.002.

Putting it all together, we find the margin of error as:

$$ME = 2.002(4.96) = \$9.93$$

So, we are 95% confident that the interval

$$\$100.13 \pm 9.93 = (\$90.20, \$110.06)$$

includes the true mean *Price* of objects whose *Time* is 25 minutes.

Suppose, however, that instead of the mean price, we want to know how much an item that needs 25 minutes of polishing will cost. The confidence interval we just found is too narrow. It may contain the mean price, but it's unlikely to cover many of the individual price values. To make a prediction interval for an *individual* item's price with a polishing time of 25 minutes, we need the larger standard error formula to account for the greater variability. Using the formula

$$SE(\hat{y}_\nu) = \sqrt{(SE^2(b_1))(x_\nu - \bar{x})^2 + \frac{s_e^2}{n} + s_e^2} = \$32.92,$$

we find the ME to be

$$ME = t^*SE(\hat{y}_\nu) = 2.002 \times 32.92 = \$65.91,$$

and so the prediction interval is

$$\hat{y} \pm ME = 100.13 \pm 65.91 = (\$34.22, \$166.04).$$

Notice how much wider this interval is than the 95% confidence interval. Most of the time we will use a software package to compute and display these intervals. Most packages generate displays that show the regression line along with both the 95% confidence and prediction intervals (combining what we've shown in Figure 16.8 and 16.9). This makes it easier to see how much wider the prediction intervals are than the corresponding confidence intervals (see Figure 16.10).

Figure 16.10   A scatterplot of *Price* versus *Time* with a least squares regression line. The inner lines (green) near the regression line show the 95% confidence intervals, and the outer lines (red) show the 95% prediction intervals. Most of the points are contained within the prediction intervals (as they should be), but not within the confidence interval for the means.

For Example   Predictions

Questions: Working with the regression model we found in earlier examples (see page 493), corporate management wants to predict the amount spent on food.

1. State and test the standard null hypothesis (at $\alpha = 0.05$) for the regression slope in the regression output.

2. Construct the corresponding 95% confidence interval for the slope.

3. Construct a 95% confidence interval for the predicted mean value of the amount spent on food by a household with an annual income of $100,000. You'll need to know also that the Mean household income is $81,000.

4. Construct a 95% prediction interval for the amount spent on food by the Smith household, which has an annual income of $100,000.

Answers:

1. $H_0: \beta_{HHIncome} = 0$
   $H_A: \beta_{HHIncome} \neq 0$ From the regression table, P < 0.0001, so we can easily reject the null hypothesis.

2. $0.769860 - 1.96 \times 0.0608 < \beta_{HHIncome} < 0.769860 + 1.96 \times 0.0608$

3. $\$871.194 \pm 1.96 \times 19.2$

4. $\$871.194 \pm 1.96 \times 93.1$

What Can Go Wrong?

In this chapter we've added inference to the regression explorations that we did in Chapter 6. Everything covered in that chapter that could go wrong with regression can still go wrong.

With inference, we've put numbers on our estimates and predictions, but these numbers are only as good as the model. Here are the main things to watch out for:

- **Don't fit a linear regression to data that aren't straight.** This is the most fundamental assumption. If the relationship between x and y isn't approximately linear, there's no sense in fitting a straight line to it.

- **Watch out for changing spread.** The common part of confidence and prediction intervals is the estimate of the error standard deviation, the spread around the line. If it changes with x, the estimate won't make sense. Imagine making a prediction interval for these data:

When x is small, we can predict y precisely, but as x gets larger, it's much harder to pin y down. Unfortunately, if the spread changes, the single value of s_e won't pick that up. The prediction interval will use the average spread around the line, with the result that we'll be too pessimistic about our precision for low x-values and too optimistic for high x-values. A re-expression of y (as we'll see in Chapter 17) is often a good fix for changing spread.

- **Watch out for non-Normal errors.** When we make a prediction interval for an individual y-value, the Central Limit Theorem can't come to our rescue. For us to believe the prediction interval, the errors must follow the Normal model. Check the histogram and Normal probability plot of the residuals to see if this assumption looks reasonable.

- **Watch out for extrapolation.** It's tempting to think that because we have prediction *intervals*, they'll take care of all of our uncertainty so we don't have to worry about extrapolating. Wrong. The interval is only as good as the model. The uncertainty our intervals predict is only correct if our model is true. There's no way to adjust for wrong models. That's why it's always dangerous to predict for x-values that lie far from the center of the data.

(continued)

• **Watch out for high-influence points and unusual observations.** We should always be on the lookout for a few points that have undue influence on our estimates, and regression analysis is certainly no exception.

• **Watch out for one-tailed tests.** Because tests of hypotheses about regression coefficients are usually two-tailed, software packages report two-tailed P-values. If you are using that type of software to conduct a one-tailed test about the slope, you'll need to divide the reported P-value by two.

Ethics in Action

The need for senior care businesses that offer companionship and non-medical home services is increasing as the U.S. population continues to age. One such franchise, Independent Senior Care, tries to set itself apart from its competitors by offering an additional service to prospective franchisees. In addition to standard information packets that provide tools, training, and mentorship opportunities, Independent Senior Care has an analyst on staff, Allen Ackman, to help prospective franchisees evaluate the feasibility of opening an elder care business in their area. Allen was contacted recently by Kyle Sennefeld, a recent business school graduate with a minor in gerontology, who is interested in starting a senior care franchise in northeastern Pennsylvania. Allen decides to use a regression model that relates annual profit to the number of residents over the age of 65 that live within a 100-mile radius of a franchise location. Even though the R^2 for this model is small, the variable is statistically significant, and the model is easy to explain to prospective franchisees. Allen sends Kyle a report that estimates the annual profit at Kyle's proposed location. Kyle was excited to see that opening an Independent Senior Care franchise in northeastern Pennsylvania would be a good business decision.

ETHICAL ISSUE *The regression model has a small R^2, so its predictive ability is questionable. Related to Items A and B, ASA Ethical Guidelines.*

ETHICAL SOLUTION *Disclose the value of R^2 along with the prediction results and disclose if the regression is being used to extrapolate outside the range of x values. Allen should provide a prediction interval as well as an estimate of the profit. Because Kyle will be assessing his franchise's chances for profit from this interval, Allen should make sure it is a prediction interval and not a confidence interval for the mean profit at all similar locations.*

What Have We Learned?

Learning Objectives

■ Apply your understanding of inference for means using Student's *t* to inference about regression coefficients.

■ Know the Assumptions and Conditions for inference about regression coefficients and how to check them, in this order:

• **Linearity Assumption,** checked with the Linearity Condition by examining a scatterplot of *y* vs. *x* or of the residuals plotted against the predicted values.

• **Independence Assumption,** which can't be checked, but is more plausible if the data were collected with appropriate randomization—the Randomization Condition.

- **Equal Variance Assumption**, which requires that the spread around the regression model be the same everywhere. We check it with the Equal Spread Condition, assessed with a scatterplot of the residuals versus the predicted values.

- **Normal Population Assumption**, which is required to use Student's t-models unless the sample size is large. Check it with the Nearly Normal Condition by making a histogram or normal probability plot of the residuals.

■ Know the components of the standard error of the slope coefficient:

- The standard deviation of the residuals, $s_e = \sqrt{\dfrac{\sum(y - \hat{y})^2}{n - 2}}$

- The standard deviation of x, $s_x = \sqrt{\dfrac{(x - \bar{x})^2}{n - 1}}$

- The sample size, n

■ Be able to find and interpret the standard error of the slope.

- $SE(b_1) = \dfrac{s_e}{s_x\sqrt{n - 1}}$

- The standard error of the slope is the estimated standard deviation of the sampling distribution of the slope.

■ State and test the standard null hypothesis on the slope.

- $H_0: \beta_1 = 0$. This would mean that x and y are not linearly related.

- We test this null hypothesis using the t-statistic $t = \dfrac{b_1 - 0}{SE(b_1)}$

■ Know how to use a t-test to test whether the true correlation is zero.

- $t = r\sqrt{\dfrac{n - 2}{1 - r^2}}$

■ Construct and interpret a confidence interval for the predicted mean value corresponding to a specified value, x_ν.

- $\hat{y}_\nu \pm t^*_{n-2} \times SE$, where $SE(\hat{\mu}_\nu) = \sqrt{SE^2(b_1) + (x_\nu - \bar{x})^2 + \dfrac{s_e^2}{n}}$

■ Construct and interpret a confidence interval for an individual predicted value corresponding to a specified value, x_ν.

- $\hat{y}_\nu \pm t^*_{n-2} \times SE$, where $SE(\hat{y}_\nu) = \sqrt{SE^2(b_1) + (x_\nu - \bar{x})^2 + \dfrac{s_e^2}{n} + s_e^2}$

Terms

Confidence interval for the regression slope

When the assumptions are satisfied, we can find a confidence interval for the slope parameter from $b_1 \pm t^*_{n-2} \times SE(b_1)$. The critical value, t^*_{n-2}, depends on the confidence interval specified and on the Student's t-model with $n - 2$ degrees of freedom.

Confidence interval for the mean response

Different samples will give different estimates of the regression model and, so, different predicted values for the same value of x. We find a confidence interval for the mean of these predicted values at a specified x-value, x_ν, as

$$\hat{y}_\nu \pm t^*_{n-2} \times SE(\hat{\mu}_\nu),$$

where

$$SE(\hat{\mu}_\nu) = \sqrt{SE^2(b_1) \times (x_\nu - \bar{x})^2 + \frac{s_e^2}{n}}.$$

The critical value, t^*_{n-2}, depends on the specified confidence level and the Student's t-model with $n - 2$ degrees of freedom.

(continued)

Residual standard deviation	The measure, denoted s_e, of the spread of the data around the regression line:

$$s_e = \sqrt{\frac{\sum (y - \hat{y})^2}{n - 2}} = \sqrt{\frac{\sum e^2}{n - 2}}.$$

Prediction interval for an individual response	An interval that captures most of the individual values. Confidence intervals specify likely values for parameters. Prediction intervals do the same for individual observations. The prediction interval takes the form

$$\hat{y}_{\nu} \pm t^*_{n-2} \times SE(\hat{y}_{\nu}),$$

where

$$SE(\hat{y}_{\nu}) = \sqrt{SE^2(b_1) \times (x_{\nu} - \bar{x})^2 + \frac{s_e^2}{n} + s_e^2}.$$

The critical value, t^*_{n-2}, depends on the specified confidence level and the Student's t-model with $n - 2$ degrees of freedom. The extra s_e^2 in $SE(\hat{y}_{\nu})$ makes the interval wider than the corresponding confidence interval for the mean.

t-test for the regression slope	The usual null hypothesis is that the true value of the slope is zero. The alternative is that it is not. A slope of zero indicates a complete lack of linear relationship between y and x. To test $H_0: \beta_1 = 0$ we find

$$t = \frac{b_1 - 0}{SE(b_1)},$$

where $SE(b_1) = \dfrac{s_e}{s_x \sqrt{n-1}}$, $s_e = \sqrt{\dfrac{\sum (y - \hat{y})^2}{n - 2}}$, n is the number of cases, and s_x is the standard deviation of the x-values. We find the P-value from the Student's t-model with $n - 2$ degrees of freedom.

t-test for the correlation coefficient	When the conditions are met, we can test the hypothesis $H_0: \rho = 0$ vs. $H_A: \rho \neq 0$ using the test statistic

$$t = r \sqrt{\frac{n - 2}{1 - r^2}},$$

which follows a Student's t-model with $n - 2$ degrees of freedom. We can use the t-model to find the P-value of the test.

Technology Help: Regression Analysis

All statistics packages make a table of results for a regression. These tables differ slightly from one package to another, but all are essentially the same. We've seen some examples of such tables already.

All packages offer analyses of the residuals. With some, you must request plots of the residuals as you request the regression. Others let you find the regression first and then analyze the residuals afterward. Either way, your analysis is not complete if you don't check the residuals with a histogram or Normal probability plot and a scatterplot of the residuals against x or the predicted values.

You should, of course, always look at the scatterplot of your two variables before computing a regression.

Regressions are almost always found with a computer or calculator. The calculations are too long to do conveniently by hand for data sets of any reasonable size. No matter how the regression is computed, the results are usually presented in a table that has a standard form. Here's a portion of a typical regression results table, along with annotations showing where the numbers come from.

The regression table gives the coefficients (once you find them in the middle of all this other information). This regression predicts *Price* from *Time*. The regression equation is

$$\widehat{Price} = -4.871 + 4.200 \ Time$$

and the R^2 for the regression is 84.5%.

The column of *t*-ratios gives the test statistics for the respective null hypotheses that the true values of the coefficients are zero. The corresponding P-values are also usually reported.

EXCEL   XLSTAT

- In Excel 2007 and later, select **Data Analysis** from the **Analysis Group** on the Data Tab.
- Select Regression from the **Analysis Tools** list.
- Click the **OK** button.

- Enter the data range holding the Y-variable in the box labeled "Y-range."
- Enter the range of cells holding the X-variable in the box labeled "X-range."
- Select the **New Worksheet Ply** option.
- Select **Residuals, Residual Plots, and Line Fit Plots** options. Click the **OK** button.

  By right clicking on Line Fit Plot data, you can delete the predicted values and/or add the least squares line, which Excel calls a Trendline.

Comments

The *Y* and *X* ranges do not need to be in the same rows of the spreadsheet, although they must cover the same number of cells. But it is a good idea to arrange your data in parallel columns as in a data table.

Although the dialog offers a Normal probability plot of the residuals, the data analysis add-in does not make a correct probability plot, so don't use this option.

JMP

- From the **Analyze** menu, select **Fit Y by X**.
- Select variables: a Y, Response variable, and an X, Factor variable. Both must be continuous (quantitative).
- JMP makes a scatterplot.
- Click on the red triangle beside the heading labeled **Bivariate Fit. . .** and choose **Fit Line**. JMP draws the least squares regression line on the scatterplot and displays the results of the regression in tables below the plot.
- The portion of the table labeled "Parameter Estimates" gives the coefficients and their standard errors, *t*-ratios, and P-values.

Comments

JMP chooses a regression analysis when both variables are "Continuous." If you get a different analysis, check the variable types.

(continued)

The Parameter table does not include the residual standard deviation s_e. You can find that as Root Mean Square Error in the Summary of Fit panel of the output.

Comments

You can also start by choosing a Fitted Line plot from the **Regression** submenu to see the scatterplot first—usually good practice.

MINITAB

- Choose **Regression** from the **Stat** menu.
- Choose **Regression. . .** from the **Regression** submenu.
- In the Regression dialog, assign the Y-variable to the Response box and assign the X-variable to the Predictors box.
- Click the **Graphs** button.
- In the Regression-Graphs dialog, select **Standardized residuals**, and check **Normal plot of residuals, Residuals versus fits** and **Residuals versus order**.
- Click the **OK** button to return to the Regression dialog.
- Click the **OK** button to compute the regression.

SPSS

- Choose **Regression** from the **Analyze** menu.
- Choose **Linear** from the **Regression** submenu.
- In the Linear Regression dialog that appears, select the Y-variable and move it to the dependent target. Then move the X-variable to the independent target.
- Click the **Plots** button.
- In the Linear Regression Plots dialog, choose to plot the *SRESIDs against the *ZPRED values.
- Click the **Continue** button to return to the Linear Regression dialog.
- Click the **OK** button to compute the regression.

Brief CASE

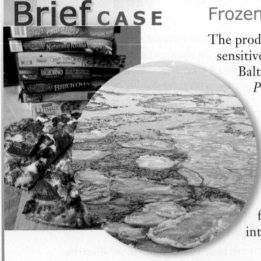

Frozen Pizza

The product manager at a subsidiary of Kraft Foods, Inc. is interested in learning how sensitive sales are to changes in the unit price of a frozen pizza in Dallas, Denver, Baltimore, and Chicago. The product manager has been provided data on both *Price* and *Sales* volume every fourth week over a period of nearly four years for the four cities (**Frozen_Pizza**).

Examine the relationship between *Price* and *Sales* for each city. Be sure to discuss the nature and validity of this relationship. Is it linear? Is it negative? Is it significant? Are the conditions of regression met? Some individuals in the product manager's division suspect that frozen pizza sales are more sensitive to price in some cities than in others. Is there any evidence to suggest that? Write up a short report on what you find. Include 95% confidence intervals for the mean *Sales* if the *Price* is $2.50 and discuss how that interval changes if the *Price* is $3.50.

Global Warming?

Every spring, Nenana, Alaska, hosts a contest in which participants try to guess the exact minute that a wooden tripod placed on the frozen Tanana River will fall through the breaking ice. The contest started in 1917 as a diversion for railroad engineers, with a jackpot of $800 for the closest guess. It has grown into an event in which hundreds of thousands of entrants enter their guesses on the Internet and vie for more than $300,000.

Because so much money and interest depends on the time of the ice breakup, it has been recorded to the nearest minute with great accuracy ever since 1917 (**Nenana**).

And because a standard measure of breakup has been used throughout this time, the data are consistent. An article in *Science* ("Climate Change in Nontraditional Data Sets," *Science* 294, October 2001) used the data to investigate global warming. Researchers are interested in the following questions. What is the rate of change in the date of breakup over time (if any)? If the ice is breaking up earlier, what is your conclusion? Does this necessarily suggest global warming? What could be other reasons for this trend? What is the predicted breakup date for the year 2015? (Be sure to include an appropriate prediction or confidence interval.) Write up a short report with your answers.

Exercises

SECTION 16.1

1. You are considering opening an Estonian cuisine restaurant and wonder how much people in your target community spend on eating out relative to their incomes. You commission a telephone survey and build the following regression model for individual respondents:

$$\widehat{EatOut\$/wk} = 17.28 + 0.354\,HHIncome$$

where household income is in $1000's.

Write an interpretation of this model.

2. For the model found in Exercise 1, predict how much a person with a household income of $75,000 would be expected to spend eating out in a week.

SECTION 16.2

3. A website that rents movies online recorded the age and the number of movies rented during the past month for some of their customers. Here are their data:

Age	Rentals
35	9
40	8
50	4
65	3
40	10
30	12

Make a scatterplot for these data. What does it tell you about the relationship between these two variables? From what you know, which of the following conditions can you check from the scatterplot? Are they satisfied?

a) Linearity
b) Independence
c) Equal spread
d) Normal population

4. Here's a scatterplot of the % of income spent on food versus Household income for respondents to the Cornell National Social Survey:

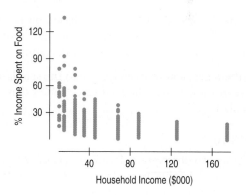

For each of the regression assumptions, state whether it is satisfied, not satisfied, or can't be determined from this plot.

a) Linearity
b) Independence
c) Equal spread
d) Normal population

SECTION 16.3

5. For the data of Exercise 3, the best fitting regression line has $b_0 = 18.9$ and $b_1 = -0.260$.

a) Use the estimated regression equation to predict *Rentals* for all six values of *Age*.
b) Find the residuals e_i.
c) Calculate the residual standard deviation, s_e.

6. A training center, wishing to demonstrate the effectiveness of their methods, tests some of their clients after different numbers of days of training, recording their scores on a sample test. Their data are:

Training Days	Correct Responses
1	4
4	6
8	7
10	9
12	10

The regression model they calculate is

$$\widehat{correct\ responses} = 3.525 + 0.525\ Training\ days.$$

a) Use the model to predict the correct responses for each number of training days.
b) Find the residuals, e_i.
c) Calculate the residual standard deviation, s_e.

7. For the regression of Exercise 5, based on the data of exercise 3, find the standard error of the regression slope. Show all three values that go into the calculation.

8. For the regression of Exercise 6, find the standard error of the regression slope. Show all three values that go into the calculation.

SECTION 16.4

9. A data set of 5 observations for *Concession Sales per person* ($) at a theater and *Minutes before the movie begins* results in the following estimated regression model:

$$\widehat{Sales} = 4.3 + 0.265\ Minutes.$$

The standard error of the regression slope is 0.0454.

a) Compute the value of the *t*-statistic to test if there is a significant relationship between *Sales* and *Minutes*.
b) What are the degrees of freedom associated with the *t*-statistic?
c) What is the P-value associated with the *t*-statistic?
d) At $\alpha = .05$, can you reject the standard null hypothesis for the slope? Explain.

10. A soap manufacturer tested a standard bar of soap to see how long it would last. A test subject showered with the soap each day for 15 days and recorded the *Weight*

(in grams) of the soap after the shower. The resulting regression computer output looks, in part, like this:

```
Dependent variable is: Weight
R squared = 99.5%
s = 2.949
```

Variable	Coefficient	SE(Coeff)	t-ratio	P-value
Intercept	123.141	1.382	89.1	≤0.0001
Day	−5.57476	0.1068	−52.2	≤0.0001

Find the following facts in this output, or determine them from what you know.

a) The standard deviation of the residuals
b) The slope of the regression line
c) The standard error of b_1
d) The P-value appropriate for testing $H_0: \beta_1 = 0$ versus $H_A: \beta_1 \neq 0$
e) Is the null hypothesis rejected at $\alpha = 0.05$?

SECTION 16.5

11. A study of 98 investment advisors asked them to predict the value of a "penny" stock a week ahead. The correlation between the values predicted and the actual value was 0.20. Was this evidence that they could predict penny stock performance?

a) State the null and alternative hypotheses about the correlation.
b) Calculate the test statistic.
c) Perform the test at $\alpha = 0.05$.

12. For a physical exam to qualify for company-provided insurance, 14 employees had their blood pressure measured. Their Systolic and Diastolic BP are given here;

SBP	DBP
140	90
208	114
140	90
164	104
142	88
124	82
120	80
142	72
160	90
142	98
225	125
140	90
220	114
150	85

A regression, found with technology, reports an R^2 of 81.3%.

a) Make a scatterplot of the data. Is it appropriate to interpret correlation?

b) State the null and alternative hypotheses about the correlation.

c) Calculate the test statistic. (*Hint*: Remember the relationship between correlation and R².)

d) Perform the test at $\alpha = 0.01$ and state your conclusion.

SECTION 16.6

13. Here are data from a small bookstore.

Number of Salespeople Working	Sales (in $1000's)
2	10
3	11
7	13
9	14
10	18
10	20
12	20
15	22
16	22
20	26
$\bar{x} = 10.4$	$\bar{y} = 17.6$
$SD(x) = 5.64$	$SD(y) = 5.34$

The regression line is:

$$\widehat{Sales} = 8.10 + 0.914\ Number\ of\ Salespeople\ Working.$$

The assumptions and conditions for regression are met, and from technology we learn that

$$SE(b_1) = 0.0873\ \ s_e = 1.477$$

a) Find the predicted *Sales* on a day with 12 employees working.

b) Find a 95% confidence interval for the mean *Sales* on days that have 12 employees working.

c) Find the 95% prediction interval for *Sales* on a day with 12 employees working.

14. A study of external disk drives finds the following:

Capacity (in GB)	Price (in $)
80	29.95
120	35.00
250	49.95
320	69.95
1000	99.00
2000	205.00
4000	449.00
$\bar{x} = 1110.00$	$\bar{y} = 133.98$
$SD(x) = 1446.89$	$SD(y) = 151.26$

The least squares line was found to be: $\widehat{Price} = 18.64 + 0.104\ Capacity$ with $s_e = 17.95$ and $SE(b_1) = 0.0051$.

a) Find the predicted *Price* of a 1000 GB hard drive.

b) Find a 95% confidence interval for the mean *Price* of all 1000 GB disk drives.

c) Find the 95% prediction interval for the *Price* of a 1000 GB hard drive.

15. In Exercise 1 we saw a regression based on a survey of households in a community served by a potential Estonian restaurant that might open. The regression model,

$$\widehat{EatOut\$/wk} = 17.28 + 0.354\ HHIncome$$

related the amount respondents said they spent individually to eat out each week to their household income in $1000's.

a) A 95% prediction interval for a customer with a household income of $80,000 is ($35.60, $55.60). Explain to the restaurant owner how she should interpret this interval.

b) A 95% confidence interval for the mean amount spent weekly to eat out by people with a household income of $80,000 is ($40.60, $50.60). Explain to the restaurant owner how to interpret this interval.

c) Now explain to her why these intervals are different.

16. In Exercise 9 we saw a regression to predict the sales per person at a movie theater in terms of the time (in minutes) before the show. The model was:

$$\widehat{Sales} = 4.3 + 0.265\ Minutes.$$

a) A 90% prediction interval for a concessions customer 10 minutes before the movie starts is ($4.60, $9.30). Explain how to interpret this interval.

b) A 90% confidence interval for the mean of sales per person 10 minutes before the movie starts is ($6.65, $7.25). Explain how to interpret this interval.

c) Which interval is of particular interest to the concessions manager? Which one is of particular interest to you, the moviegoer?

SECTION 16.7

17. Teenagers were stopped randomly at a mall and asked how many Jonas Brothers songs they could name. Here are the data and a regression model.

$$\widehat{Songs} = 19.3 - 0.795\ Age\ \text{with}\ s_e = 1.3\ \text{and}\ SE(b_1) = .182$$

Age	Songs
12	10
13	8
17	7
19	5
20	2

a) What is the predicted value of *Songs* when *Age* = 16?
b) Develop a 95% interval for the predicted mean value of *Songs* when *Age* = 16.
c) Develop a 95% interval for the predicted individual value of *Songs* when *Age* = 16.
d) Which interval is wider? Why?

18. For the blood pressure data of Exercise 12, a regression, found with technology, reports

```
Dependent variable is: SBP(mm Hg)
R squared = 81.3%
s = 15.44 with 14 − 2 = 12 degrees of freedom
```

Variable	Coefficient	SE(Coeff)	t-ratio	P-value
Intercept	−37.7147	27.48	−1.37	0.1951
DBP	2.07640	0.2877	7.22	≤0.0001

a) Find a 95% prediction interval for the Systolic Blood Pressure of someone who has a Diastolic BP of 100 mm Hg.
b) Find a 95% confidence interval for the mean SBP of people with a DBP of 100 mm Hg.

CHAPTER EXERCISES

19. Online shopping. Several studies have found that the frequency with which shoppers browse Internet retailers is related to the frequency with which they actually purchase products and/or services online. Here are data showing the age of respondents and their answer to the question "how many minutes do you browse online retailers per week?"

Age	Browsing Time (min/wk)
22	492
50	186
44	180
32	384
55	120
60	120
38	276
22	480
21	510
45	252
52	126
33	360
19	570
17	588
21	498

a) Make a scatterplot for these data.
b) Do you think a linear model is appropriate? Explain.
c) Find the equation of the regression line.
d) Check the residuals to see if the conditions for inference are met.

20. Climate change. The Earth's climate is getting warmer. Most climate scientists agree that one important cause of the warming is the increase in atmospheric levels of carbon dioxide (CO_2), a greenhouse gas. Here is part of a regression analysis of the mean annual CO_2 concentration in the atmosphere, measured in parts per thousand (ppt), at the top of Mauna Loa in Hawaii and the mean annual air temperature over both land and sea across the globe, in degrees Celsius. The scatterplots and residuals plots indicated that the data were appropriate for inference and the response variable is *Temp*.

Variable	Coeff	SE(Coeff)
Intercept	10.707	0.4810
CO_2	0.010	0.0013

```
R squared = 67.8%
s = 0.0985 with 29 − 2 = 27 degrees of freedom
```

a) Write the equation of the regression line.
b) Find the value of the correlation and test whether the true correlation is zero. Is there evidence of an association between CO_2 level and global temperature?
c) Find the *t*-value and P-value for the slope. Is there evidence of an association between CO_2 level and global temperature? What do you know from the slope and *t*-test that you might not have known from testing the correlation?
d) Do you think predictions made by this regression will be very accurate? Explain.

21. Movie budgets. How does the cost of a movie depend on its length? Data on the cost (millions of dollars) and the running time (minutes) for major release films of 2005 are summarized in these plots and computer output:

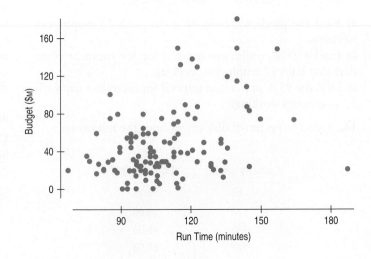

```
Dependent variable is: Budget($M)
R squared = 15.4%
s = 32.95 with 120 − 2 = 118 degrees of freedom
```

Variable	Coefficient	SE(Coeff)	t-ratio	P-value
Intercept	−31.39	17.12	−1.83	0.0693
Run time	0.71	0.15	4.64	≤0.0001

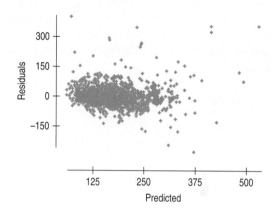

```
Dependent variable is: Price
R squared = 59.5%
s = 53.79 with 1064 − 2 = 1062 degrees of freedom
```

Variable	Coefficient	SE(Coeff)	t-ratio	P-value
Intercept	−3.1169	4.688	−0.665	0.5063
Size	94.4539	2.393	39.465	≤0.0001

a) Explain in words and numbers what the regression says.
b) The intercept is negative. Discuss its value, taking note of the P-value.
c) The output reports *s* = 32.95. Explain what that means in this context.
d) What's the value of the standard error of the slope of the regression line?
e) Explain what that means in this context.

T 22. House prices. How does the price of a house depend on its size? Data from Saratoga, New York, on 1064 randomly selected houses that had been sold include data on price ($1000's) and size (1000's ft²), producing the following graphs and computer output:

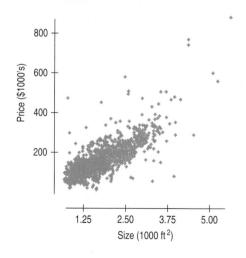

a) Explain in words and numbers what the regression says.
b) The intercept is negative. Discuss its value, taking note of its P-value.
c) The output reports *s* = 53.79. Explain what that means in this context.
d) What's the value of the standard error of the slope of the regression line?
e) Explain what that means in this context.

T 23. Movie budgets, part 2. Exercise 21 shows computer output examining the association between the length of a movie and its cost.

a) Check the assumptions and conditions for inference.
b) Find a 95% confidence interval for the slope and interpret it.

T 24. House prices, part 2. Exercise 22 shows computer output examining the association between the sizes of houses and their sale prices.

a) Check the assumptions and conditions for inference.
b) Find a 95% confidence interval for the slope and interpret it.

25. Water hardness. In an investigation of environmental causes of disease, data were collected on the annual mortality rate (deaths per 100,000) for males in 61 large towns in England and Wales. In addition, the water hardness was recorded as the calcium concentration (parts per million, or ppm) in the drinking water. Here are the scatterplot and regression analysis of the relationship between mortality and calcium concentration, where the dependent variable is *Mortality*.

Variable	Coeff	SE(Coeff)
Intercept	1676.36	29.30
Calcium	−3.226	0.485

R squared = 42.9%
s = 143.0 with 61 − 2 = 59 degrees of freedom

a) Is there an association between the hardness of the water and the mortality rate? Write the appropriate hypothesis.
b) Assuming the assumptions for regression inference are met, what do you conclude?
c) Create a 95% confidence interval for the slope of the true line relating calcium concentration and mortality.
d) Interpret your interval in context.

26. Mutual fund returns. In March 2002, *Consumer Reports* listed the rate of return for several large cap mutual funds over the previous 3-year and 5-year periods. (Here, "large cap" refers to companies worth over $10 billion.) It's common for advertisements to carry the disclaimer that "past returns may not be indicative of future performance." Do these data indicate that there was an association between 3-year and 5-year rates of return?

Fund Name	Annualized Returns (%)	
	3-year	5-year
Ameristock	7.9	17.1
Clipper	14.1	18.2
Credit Suisse Strategic Value	5.5	11.5
Dodge & Cox Stock	15.2	15.7
Excelsior Value	13.1	16.4
Harbor Large Cap Value	6.3	11.5
ICAP Discretionary Equity	6.6	11.4
ICAP Equity	7.6	12.4
Neuberger Berman Focus	9.8	13.2
PBHG Large Cap Value	10.7	18.1
Pelican	7.7	12.1
Price Equity Income	6.1	10.9
USAA Cornerstone Strategy	2.5	4.9
Vanguard Equity Income	3.5	11.3
Vanguard Windsor	11.0	11.0

27. Youth unemployment. The United Nations has developed a set of millennium goals for countries, and the United Nations Statistics Division (UNSD) maintains databases to measure economic progress toward these goals (unstats.un.org/unsd/mdg). Data extracted from this source are on the disk. One measure that is tracked is the youth (ages 15–24) unemployment rate in different countries. Is the unemployment rate for the male youth related to the unemployment rate for the female youth?

a) Find a regression model predicting *Male Rate* from the *Female Rate* in 2005 for the sample of 57 countries provided by UNSD.
b) Examine the residuals to determine if a linear regression is appropriate.
c) Test an appropriate hypothesis to determine if the association is significant.
d) What percentage of the variability in the *Male Rate* is accounted for by the regression model?

28. Male unemployment. Using the unemployment data provided by the United Nations, investigate the association between the male unemployment rate in 2004 and 2005 for a sample of 52 countries. (The sample is smaller than in Exercise 27, since not all countries reported rates in both years.)

a) Find a regression model predicting the *2005-Male* rate from the *2004-Male* rate.
b) Examine the residuals to determine if a linear regression is appropriate.
c) Test an appropriate hypothesis to determine if the association is significant.
d) What percentage of the variability in the *2005-Male* rate is accounted for by the *regression model*?

T 29. Used cars. Classified ads in a newspaper offered several used Toyota Corollas for sale. Listed below are the ages of the cars and the advertised prices.

Age (yr)	Prices Advertised ($)
1	13990
1	13495
3	12999
4	9500
4	10495
5	8995
5	9495
6	6999
7	6950
7	7850
8	6999
8	5995
10	4950
10	4495
13	2850

a) Make a scatterplot for these data.
b) Do you think a linear model is appropriate? Explain.
c) Find the equation of the regression line.
d) Check the residuals to see if the conditions for inference are met.

T 30. Property assessments. The following software results provide information about the size (in square feet) of 18 homes in Ithaca, New York, and the city's assessed value of those homes, where the response variable is *Assessment*.

Predictor	Coeff	SE(Coeff)	t-ratio	P-value
Intercept	37108.85	8664.33	4.28	0.0006
Size	11.90	4.29	2.77	0.0136

s=4682.10   R-Sq=32.5%

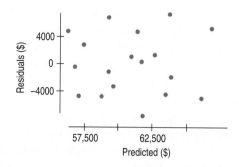

a) Explain why inference for linear regression is appropriate with these data.
b) Is there a significant linear association between the *Size* of a home and its *Assessment*? Test an appropriate hypothesis and state your conclusion.
c) What percentage of the variability in assessed value is accounted for by this regression?
d) Give a 90% confidence interval for the slope of the true regression line, and explain its meaning in the proper context.
e) From this analysis, can we conclude that adding a room to your house will increase its assessed value? Why or why not?
f) The owner of a home measuring 2100 square feet files an appeal, claiming that the $70,200 assessed value is too high. Do you agree? Explain your reasoning.

T 31. Used cars, part 2. Based on the analysis of used car prices you did for Exercise 29, if appropriate, create a 95% confidence interval for the slope of the regression line and explain what your interval means in context.

T 32. Assets and sales. A business analyst is looking at a company's assets and sales to determine the relationship (if any) between the two measures. She has data (in $million) from a random sample of 79 Fortune 500 companies, and obtained the linear regression below:

The regression equation is Assets = 1867.4 + 0.975 Sales

Predictor	Coeff	SE(Coeff)	t-ratio	P-value
Constant	1867.4	804.5	2.32	0.0230
Sales	0.975	0.099	9.84	≤0.0001

s = 6132.59 R-Sq = 55.7% R-Sq(adj) = 55.1%

Use the data provided to find a 95% confidence interval, if appropriate, for the slope of the regression line and interpret your interval in context.

T 33. Fuel economy. A consumer organization has reported test data for 50 car models. We will examine the association between the weight of the car (in thousands of pounds) and the fuel efficiency (in miles per gallon). Use the data provided on the disk to answer the following questions, where the response variable is *Fuel Efficiency* (mpg).

a) Create the scatterplot and obtain the regression equation.
b) Are the assumptions for regression satisfied?
c) Write the appropriate hypotheses for the slope.
d) Test the hypotheses and state your conclusion.

34. Auto Batteries. In October 2002, *Consumer Reports* listed the price (in dollars) and power (in cold cranking amps) of auto batteries. We want to know if more expensive batteries are generally better in terms of starting power. Here are the regression and residual output, where the response variable is *Power*.

```
Dependent variable is: Power
R squared = 25.2%
s = 116.0 with 33 − 2 = 31 degrees of freedom
```

Variable	Coefficient	SE(Coeff)	t-ratio	P-value
Intercept	384.594	93.55	4.11	0.0003
Price	4.146	1.282	3.23	0.0029

a) How many batteries were tested?
b) Are the conditions for inference satisfied? Explain.

c) Is there evidence of a linear association between the price and cranking power of auto batteries? Test an appropriate hypothesis and state your conclusion.
d) Is the association strong? Explain.
e) What is the equation of the regression line?
f) Create a 90% confidence interval for the slope of the true line.
g) Interpret your interval in this context.

T 35. SAT scores. How strong was the association between student scores on the Math and Verbal sections of the old SAT? Scores on this exam ranged from 200 to 800 and were widely used by college admissions offices. Here are summary statistics, regression analysis, and plots of the scores for a graduating class of 162 students at Ithaca High School, where the response variable is *Math Score*.

Predictor	Coeff	SE(Coeff)	t-ratio	P-value
Intercept	209.55	34.35	6.10	<0.0001
Verbal	0.675	0.057	11.88	<0.0001

s = 71.75   R-Sq = 46.9%

a) Is there evidence of a linear association between *Math* and *Verbal* scores? Write an appropriate hypothesis.
b) Discuss the assumptions for inference.
c) Test your hypothesis and state an appropriate conclusion.

36. Productivity. How strong is the association between labor productivity and labor costs? Data from the Bureau of Labor Statistics, specifically 2006 index values (computed using 1997 as the base year) for labor productivity and unit labor costs across 53 industries, are used to examine this relationship (ftp://ftp.bls.gov; accessed May 2008). Here are the results of a regression analysis where the response variable is *Labor Productivity*.

Predictor	Coeff	SE(Coeff)	t-ratio	P-value
Intercept	212.67	12.18	17.46	<0.0001
Unit Labor Cost	−0.768	0.111	−6.92	<0.0001

s = 14.39   R-Sq = 48.4%

a) Is there evidence of a linear association between *Labor Productivity* and Unit Labor Costs? Write appropriate hypotheses.
b) Test your null hypothesis and state an appropriate conclusion (assume that assumptions and conditions are met).

37. Football salaries. Football owners are constantly in competition for good players. The more wins, the more likely that the team will provide good business returns for the owners. Of course, the resources that each of the 32 teams has in the National Football League (NFL) vary. Does the size of the payroll matter? Here is a scatterplot and regression showing the association between team salaries in the NFL in 2006 and number of wins.

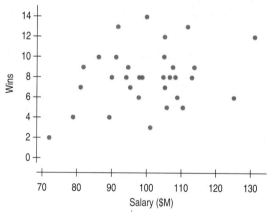

Predictor	Coeff	SE(Coeff)	t-ratio	P-value
Intercept	1.783	3.964	0.45	0.6560
Salary ($M)	0.062	0.039	1.58	0.1244

s = 2.82   R-Sq = 7.7%

a) State the hypotheses about the slope.
b) Perform the hypothesis test and state your conclusion in context.

38. Female president. The Gallup organization has, over six decades, periodically asked the following question:

If your party nominated a generally well-qualified person for president who happened to be a woman, would you vote for that person?

We wonder if the proportion of the public who have "no opinion" on this issue has changed over the years. Here is a regression for the proportion of those respondents whose response to this question about voting for a woman president was "no opinion." Assume that the conditions for inference are satisfied and that the response variable is *No Opinion*.

Predictor	Coeff	SE(Coeff)	t-ratio	P-value
Intercept	7.693	2.445	3.15	0.0071
Year	−0.042	0.035	−1.21	0.2460

s = 2.28 R-Sq = 9.5%

a) State the hypotheses about the slope (both numerically and in words) that describes how voters' thoughts have changed about voting for a woman.
b) Assuming that the assumptions for inference are satisfied, perform the hypothesis test and state your conclusion.
c) Examine the scatterplot corresponding to the regression for No Opinion. How does it change your opinion of the trend in "no opinion" responses? Do you think the true slope is negative as shown in the regression output?

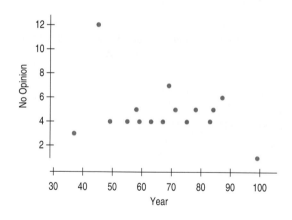

39. Fuel economy, part 2. Consider again the data in Exercise 33 about the gas mileage and weights of cars.

a) Create a 95% confidence interval for the slope of the regression line.
b) Explain in this context what your confidence interval means.

T 40. SAT scores, part 2. Consider the high school SAT scores data from Exercise 35.

a) Find a 90% confidence interval for the slope of the true line describing the association between Math and Verbal scores.

b) Explain in this context what your confidence interval means.

T 41. Wilshire 5000. It is common economic theory that the money flowing into and out of mutual funds (fund flows) is related to the performance of the stock market. Another way of stating this is that investors are more likely to place money into mutual funds when the market is performing well. One way to measure market performance is the Wilshire 5000 Total Market Return (%), which is a value-weighted return. (The return of each stock in the index is weighted by its percent of market value for all stocks.) Here are the scatterplot and regression analysis. The data are recorded monthly. The response variable is *Fund Flows* ($ million) and the explanatory variable is *Market Return* (%), using data from January 1990 to October 2002.

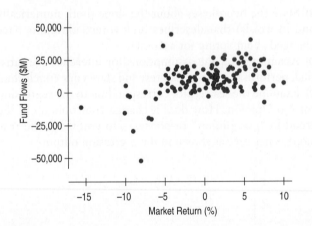

```
Dependent variable is: Fund Flows
R squared = 17.9%
s = 10999 with 154 - 2 = 152 degrees of freedom
```

Variable	Coeff	SE(Coeff)	t-ratio	P-value
Intercept	9599.10	896.9	10.7	≤0.0001
Market Return	1156.40	201.1	5.75	≤0.0001

a) State the null and alternative hypotheses under investigation.

b) Assuming that the assumptions for regression inference are reasonable, test the hypothesis.

c) State your conclusion.

42. Marketing managers. Are wages for various marketing managerial positions related? One way to determine this is to examine the relationship between the mean hourly wages for two managerial occupations in marketing: sales managers and advertising managers. The average hourly wage for both occupations is reported for all U.S. states and territories by the U.S. Bureau of Labor Statistics (data.bls.gov/oes; Occupational Employment Statistics; accessed May 2008). Here are the regression analysis results.

Predictor	Coeff	SE(Coeff)	t-ratio	P-value
Constant	10.317	4.382	2.35	0.0227
Sales Mgr Avg Hourly Wage	0.56349	0.09786	5.76	<0.0001

a) State the null and alternative hypothesis under investigation.

b) Assuming that the assumptions for regression inference are reasonable, test the null hypothesis.

c) State your conclusion.

T 43. Cost index. Recall the *Worldwide Cost of Living Survey* from Chapter 6 that determined the cost of living in the most expensive cities in the world as an index. This index scales New York City as 100 and expresses the cost of living in other cities as a percentage of the New York cost. For example, in 2007 the cost of living index in Tokyo was 122.1, which means that it was 22% higher than New York. The output shows the regression of the 2007 on the 2006 index for the most expensive cities in 2007, where *Index 2007* is the response variable.

Predictor	Coeff	SE(Coeff)	t-ratio	P-value
Intercept	12.02	12.25	0.98	0.3446
Index 2006	0.943	0.115	8.17	<0.0001

s = 4.45 R-Sq = 83.7%

a) State the hypotheses about the slope (both numerically and in words).

b) Perform the hypothesis test and state your conclusion in context.

c) Explain what the *R*-squared in this regression means.

d) Do these results indicate that, in general, cities with a higher index in 2006 had a higher index in 2007? Explain.

44. Job growth. *Fortune Magazine* publishes the top 100 companies to work for every year. Among the information listed is the percentage growth in jobs at each company. The output below shows the regression of the 2008 job growth (%) on the 2006 job growth for a sample of 29 companies. Note that *Job Growth 2008* is the response variable. (money.cnn.com/magazines/fortune/bestcompanies/full_list; accessed May, 2008).

```
Dependent variable is: Job Growth 2008
R squared = 25.6%  R squared (adjusted) = 22.9%
s = 6.129 with 29 - 2 = 27 degrees of freedom

Variable    Coefficient  SE(Coeff)  t-ratio  P-value

Intercept   2.993        1.441      2.08     0.0475
Job Growth
   2006     0.399        0.131      3.05     0.0051
```

a) State the hypotheses about the slope (both numerically and in words).
b) Assuming that the assumptions for inference are satisfied, perform the hypothesis test and state your conclusion in context.
c) Explain what the *R*-squared in this regression means.
d) Do these results indicate that, in general, companies with a higher job growth in 2006 had a higher job growth in 2008? Explain.

45. Oil prices. The Organization of Petroleum Exporting Countries (OPEC) is a cartel, so it artificially sets prices. But are prices related to production? Using the data provided on the disk for crude oil prices ($/barrel) and oil production (thousand barrels per day) between 2001 and 2007, answer the following questions. Use the *Crude price* as the response variable and *Production* as the predictor variable.

a) Examine a scatterplot for the two variables and test the conditions for regression.
b) Do you think there is a linear association between oil prices and production? Explain.

46. Attendance. Traditionally, athletic teams that perform better grow their fan base and generate greater attendance at games or matches. This should hold true regardless of the sport—whether it's soccer, football, or American baseball. Using the data provided for the number of wins and attendance in 2006 for the 14 American League baseball teams, answer the following questions. Use *Home Attendance* as the dependent variable and *Wins* as the explanatory variable.

a) Examine a scatterplot for the two variables and test the conditions for regression.
b) Do you think there is a linear association between *Home Attendance* and *Wins*? Explain.

47. Printers. In March 2002, *Consumer Reports* reviewed several models of inkjet printers. The following table shows the speed of the printer (in pages per minute) and the cost per page printed. Is there evidence of an association between *Speed* and *Cost*? Test an appropriate hypothesis and state your conclusion.

Speed (ppm)	Cost (cents/page)
4.6	12.0
5.5	8.5
4.5	6.2
3.8	3.4
4.6	2.6
3.7	4.0
4.7	5.8
4.7	8.1
4.0	9.4
3.1	14.9
1.9	2.6
2.2	4.3
1.8	4.6
2.0	14.8
2.0	4.4

48. Little league product testing. Ads for a Little League instructional video claimed that the techniques would improve the performances of Little League pitchers. To test this claim, 20 Little Leaguers threw 50 pitches each, and we recorded the number of strikes. After the players participated in the training program, we repeated the test. The following table shows the number of strikes each player threw before and after the training. A test of paired differences failed to show that this training was effective in improving a player's ability to throw strikes. Is there any evidence that the *Effectiveness (After − Before)* of the video depends on the player's *Initial Ability (Before)* to throw strikes? Test an appropriate hypothesis and state your conclusion. Propose an explanation for what you find.

Before	After	Before	After
28	35	33	33
29	36	33	35
30	32	34	32
32	28	34	30
32	30	34	33
32	31	35	34
32	32	36	37
32	34	36	33
32	35	37	35
33	36	37	32

Number of Strikes (out of 50)

49. Fuel economy, part 3. Consider again the data in Exercise 33 about the fuel economy and weights of cars.

a) Create a 95% confidence interval for the average fuel efficiency among cars weighing 2500 pounds, and explain what your interval means.

b) Create a 95% prediction interval for the gas mileage you might get driving your new 3450-pound SUV, and explain what that interval means.

50. SAT scores, part 3. Consider the high school SAT scores data from Exercise 35 once more. The mean Verbal score was 596.30.

a) Find a 90% confidence interval for the mean SAT Math score for all students with an SAT Verbal score of 500.

b) Find a 90% prediction interval for the Math score of the senior class president, if you know she scored 710 on the Verbal section.

51. Wilshire 5000, part 2. Using the same mutual fund flow data provided in Exercise 41, answer the following questions.

a) Find the 95% prediction interval for month that reports a market return of 8%.

b) Do you think predictions made by this regression will be very accurate? Explain.

52. Assets and sales, part 2. A business analyst was interested in the relationship between a company's assets and its sales. She collected data (in millions of dollars) from a random sample of 79 Fortune 500 companies and created the following regression analysis. Economists commonly take the logarithm of these variables to make the relationship more nearly linear, and she did too. (These are base 10 logs.) The dependent variable is *LogSales*. The assumptions for regression inference appeared to be satisfied.

```
Dependent variable is: LogSales
R squared = 33.9%
s = 0.4278 with 79 - 2 = 77 degrees of freedom
```

Variable	Coefficient	SE(Coeff)	t-ratio	P-value
Intercept	1.303	0.3211	4.06	0.0001
LogAssets	0.578	0.0919	6.28	≤ 0.0001

a) Is there a significant linear association between *LogAssets* and *LogSales*? Find the *t*-value and P-value to test an appropriate hypothesis and state your conclusion in context.

b) Do you think that a company's assets serve as a useful predictor of their sales?

53. All the efficiency money can buy. A sample of 84 model-2004 cars from an online information service was examined to see how fuel efficiency (as highway mpg) relates to the cost (Manufacturer's Suggested Retail Price in dollars) of cars. Here are displays and computer output:

```
Dependent variable is: Highway MPG
R squared = 30.1%
s = 5.298 with 84 - 2 = 82 degrees of freedom
```

Variable	Coefficient	SE(Coeff)	t-ratio	P-value
Constant	33.06	1.299	25.5	≤0.0001
MSRP	−2.165e-4	3.639e-5	−5.95	≤0.0001

a) State what you want to know, identify the variables, and give the appropriate hypotheses.

b) Check the assumptions and conditions.

c) If the conditions are met, complete the analysis.

54. Energy use. Based on data collected from the United Nations Millennium Indicators Database related to measuring the goal of *ensuring environmental sustainability*, investigate the association between energy use (kg oil equivalent per $1000 GDP) in 1990 and 2004 for a sample of 96 countries (unstats.un.org/unsd/mi/mi_goals.asp; accessed May 2008).

a) Find a regression model showing the relationship between *2004 Energy Use* (reponse variable) and *1990 Energy Use* (predictor variable).

b) Examine the residuals to determine if a linear regression is appropriate.

c) Test an appropriate hypothesis to determine if the association is significant.

d) What percentage of the variability in *2004 Energy Use* is explained by *1990 Energy Use*?

55. Youth unemployment, part 2. Refer to the United Nations data referenced in Exercise 27. Here is a scatterplot showing the regression line, 95% confidence interval, and

95% prediction interval, using 2005 youth unemployment data for a sample of 57 nations. The response variable is the *Male Rate*, and the predictor variable is the *Female Rate*.

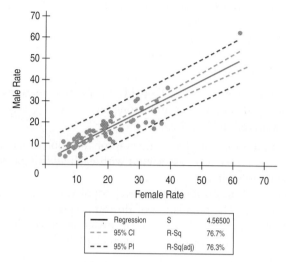

a) Explain the meaning of the 95% prediction interval in this context.
b) Explain the meaning of the 95% confidence interval in this context.
c) Identify the unusual observation, and discuss its potential impact on the regression.

56. Male unemployment, part 2. Refer to the United Nations data referenced in Exercise 28. Here is a scatterplot showing the regression line, 95% confidence interval, and 95% prediction interval, using 2005 and 2004 male unemployment data for a sample of 52 nations. The response variable is the *2005-Male Rate*, and the predictor variable is the *2004-Male Rate*.

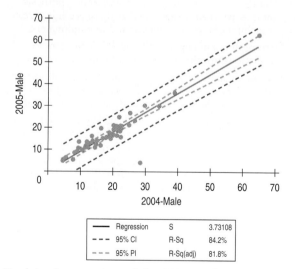

a) Explain the meaning of the 95% prediction interval in this context.
b) Explain the meaning of the 95% confidence interval in this context.
c) Identify the unusual observation, and discuss its potential impact on the regression.

57. Energy use, part 2. Examine the regression and scatterplot showing the regression line, 95% confidence interval, and 95% prediction interval using *1990* and *2004 energy use* (kg oil equivalent per $1000 GDP) for a sample of 96 countries. The response variable is *2004 Energy Use*.

R Squared	0.870303	
R Squared Adj	0.868923	
S	43.6401	

Term	Coeff	SE(Coeff)	t-ratio	P-value
Intercept	23.979345	9.148568	2.62	0.0102
Use 1990	0.8073999	0.032148	25.12	<.0001

a) Explain the meaning of the 95% prediction interval in this context.
b) Explain the meaning of the 95% confidence interval in this context.

58. Global reach. The Internet has revolutionized business and offers unprecedented opportunities for globalization. However, the ability to access the Internet varies greatly among different regions of the world. One of the variables the United Nations collects data on each year is *Personal Computers per 100 Population* (unstats.un.org/unsd/cdb/cdb_help/cdb_quick_start.asp) for various countries. Below is a scatterplot showing the regression line, 95% confidence interval, and 95% prediction interval using 2000 and 2004 computer adoption (personal computers per 100 population) for a sample of 85 countries. The response variable is *PC/100 2004*.

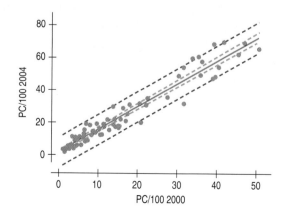

a) Working with the data set on the DVD with the same name as this exercise, find a regression model showing the relationship between personal computer adoption in 2004 *PC/100 2000* (the response variable) and personal computer adoption in 2000 *PC/100 2000* (the predictor variable).
b) Explain the meaning of the 95% prediction interval in this context.
c) Explain the meaning of the 95% confidence interval in this context.

T 59. Seasonal spending. Spending on credit cards decreases after the Christmas spending season (as measured by amount charged on a credit card in December). The data set on the DVD with the same name as this exercise contains the monthly credit card charges of a random sample of 99 cardholders.

a) Build a regression model to predict January spending from December's spending.
b) How much, on average, will cardholders who charged $2000 in December charge in January?
c) Give a 95% confidence interval for the average January charges of cardholders who charged $2000 in December.
d) From part c, give a 95% confidence interval for the average decrease in the charges of cardholders who charged $2000 in December.
e) What reservations, if any, do you have about the confidence intervals you made in parts c and d?

T 60. Seasonal spending, part 2. Financial analysts know that January credit card charges will generally be much lower than those of the month before. What about the difference between January and the next month? Does the trend continue? The data set on the DVD contains the monthly credit card charges of a random sample of 99 cardholders.

a) Build a regression model to predict February charges from January charges.
b) How much, on average, will cardholders who charged $2000 in January charge in February?
c) Give a 95% confidence interval for the average February charges of cardholders who charged $2000 in January.
d) From part c, give a 95% confidence interval for the average decrease in the charges of cardholders who charged $2000 in January.
e) What reservations, if any, do you have about the confidence intervals you made in parts c and d?

61. Environment. The Environmental Protection Agency is examining the relationship between the ozone level (in parts per million) and the population (in millions) of U.S. cities. Part of the regression analysis is shown, where the dependent variable is *Ozone*.

```
Dependent variable is: Ozone
R squared = 84.4%
s = 5.454 with 16 − 2 = 14 df
```

Variable	Coefficient	SE(Coeff)
Intercept	18.892	2.395
Pop	6.650	1.910

a) We suspect that the greater the population of a city, the higher its ozone level. Is the relationship significant? Assuming the conditions for inference are satisfied, find the *t*-value and P-value to test the appropriate hypothesis. State your conclusion in context.
b) Do you think that the population of a city is a useful predictor of ozone level? Use the values of both R^2 and *s* in your explanation.

62. Environment, part 2. Consider again the relationship between the population and ozone level of U.S. cities that you analyzed in Exercise 61.

a) Give a 90% confidence interval for the approximate increase in *Ozone* level associated with each additional million city inhabitants.
b) For the cities studied, the mean population was 1.7 million people. The population of Boston is approximately 0.6 million people. Predict the mean ozone level for cities of that size with an interval in which you have 90% confidence.

Just Checking Answers

1 I would need to see a scatterplot to see if the linearity assumption is reasonable, to make sure that there are no outliers, and a residual plot to check the equal spread condition. I'd also like to see a histogram or Normal probability plot of the residuals to make sure the nearly normal condition is satisfied. Finally, I'd like to see the residuals plotted in time to see if the residuals appear independent. Without verifying these conditions, I wouldn't know whether my analysis is valid.

2 The 95% CI for the slope is $1133.8 \pm 2.052 \ (513.1)$, or (80.9, 2186.7).

3 H_0: The slope $\beta_1 = 0$. H_A: The slope $\beta_1 \neq 0$. Since the P-value = 0.036, we reject the null hypothesis (at $\alpha = 0.05$) and conclude that there is a linear relationship between enrollment and unemployment.

4 Strength is a judgment call, but I'd be hesitant to call a relationship with an R^2 value of only 15% strong.

5 Approximately 15% of the variation in enrollment at the University of New Mexico is accounted for by variation in the unemployment rate in New Mexico.

6 The test says that we can reject the hypothesis that the correlation is 0 (at $\alpha = 0.05$) and conclude that there is a linear relationship between enrollment and unemployment. This is exactly what the test of the slope in part 3 told us. The correlation is 0, if and only if the slope is 0. There is no new information here.

Understanding Residuals

Kellogg's

John Harvey Kellogg was a physician who ran the Battle Creek Sanitarium. He was an advocate of vegetarian diets and of living without caffeine, alcohol, or sex. He was an early advocate of peanut butter, patenting one of the first devices for making it. But he is best remembered for his work with his brother Will Keith Kellogg. Together, in 1897, they founded the Sanitas Food Company to manufacture whole grain cereals.

At the start of the 20th century, breakfast was typically a large, high-fat meal of eggs and meat for the wealthy and a less nutritious meal of porridge or gruel for the poor. The Kellogg brothers introduced toasted corn flakes as a healthy and affordable alternative. But in 1906, they argued when Will wanted to add sugar to the recipe— an idea that horrified John Harvey. Will founded the Battle Creek Toasted Corn Flake Company, which eventually became the Kellogg Company, using its founder's "W. K. Kellogg" signature as a logo—a marketing concept that survives in the script "Kellogg's" on their boxes to this day.

True to its roots in healthy nutrition, Kellogg's brands include Kashi® health foods, Morningstar Farms® vegetarian foods, and Nutri-Grain® cereal bars. (However, they also make Cocoa Krispies®, Froot Loops®, Cheez-Its®, and Keebler® Cookies.) In 1923, Kellogg hired the first dietitian to work in the food industry, and in the 1930s, Kellogg was the first company to print nutrition information on their boxes. The W. K. Kellogg Institute for Food and Nutrition Research, a world-class research facility, opened in 1997. The company continues to advocate for healthy nutrition, offering advice and education on their website and partnering with organizations such as the American Heart Association.

Regression may be the most widely used Statistics method. It is used every day throughout the world to help make good business decisions. The applications of linear regression are limitless. It can be used to predict customer loyalty, staffing needs at hospitals, sales of automobiles, and almost anything that can be quantified. Because regression is so widely used, it's also widely abused and misinterpreted. This chapter presents examples of regressions in which things are not quite as simple as they may seem at first and shows how you can still use regression to discover what the data have to say.

The residuals from the fitting of a linear regression hold a remarkable amount of information about the model. Remember that a residual is the difference between the actual data and the value we predict for it: $e = y - \hat{y}$. Residuals can help tell you how well the model is performing and provide clues for fixing it if it's not working as well as it could. In this chapter, we'll show a variety of ways in which detecting patterns in the residuals can help you improve the model. Examining residuals can reveal more about the data than was apparent at first, or even second, glance. That's why no regression analysis is ever complete without a display of the residuals and a thorough examination of what they have to say.

17.1   Examining Residuals for Groups

WHO	Breakfast cereals
WHAT	Nutrition information
WHEN	Not stated
WHERE	U.S.

It seems that ever since the Kellogg brothers fought over sugar in breakfast cereals, it has been a concern. Using data from those nutrition labels introduced by Kellogg, we can examine the relationship between the calories in a serving and the amount of sugar (in grams). Figure 17.1 appears to satisfy the conditions for regression; the relationship is linear with no outliers.

The least squares regression model,

$$\widehat{calories} = 89.5 + 2.50 \, sugar$$

has an R^2 of 32%. Figure 17.2 shows the residuals.

Figure 17.1   *Calories* versus *Sugar* content (grams) per serving of breakfast cereal.

Figure 17.2   Residuals for the regression plotted against predicted *Calories*.

At first glance, Figure 17.2 seems to have no particular structure, and as you may remember from Chapter 6, that's exactly what we hope to see. But let's check a histogram of the residuals.

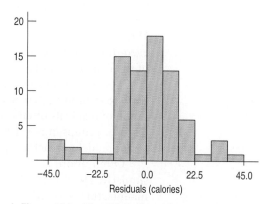

Figure 17.3   The distribution of the regression residuals shows modes above and below the central large mode. These may be worth a second look.

How would you describe the shape of this histogram? It looks like there might be small modes on either side of the central body of the data. A few cereals stand out with larger negative residuals—that is, fewer calories than we might have predicted. And a few stand out with larger positive residuals. Of course, the sample size here is not very large. We can't say for sure that there are three modes, but it's worth a closer look.

Figure 17.4   A scatterplot of the residuals vs. predicted values for the cereal regression. The green x's are cereals whose calorie content is higher than the linear model predicts. The red −'s show cereals with fewer calories than the model predicts. Is there something special about these cereals?

Let's look more carefully at the residuals. Figure 17.4 repeats the scatterplot of Figure 17.3, but with the points in those modes marked. Now we can see that those two groups stand away from the central pattern in the scatterplot. Doing a little more work and examining the data set, we find that the high-residual cereals (green x's) are *Just Right Fruit & Nut; Muesli Raisins, Dates & Almonds; Peaches & Pecans; Mueslix Crispy Blend;* and *Nutri-Grain Almond Raisin.* Do these cereals have something in common? These high-calorie cereals all market themselves as "healthy." This might be surprising, but in fact, "healthy" cereals often contain more fat. They often contain nuts and oil which are "natural" and don't necessarily contain sugar, but are higher in fat than grain and sugar. So, they may have more calories than we might expect from looking at their sugar content alone.

The low-residual (red –'s) cereals are *Puffed Rice*, *Puffed Wheat*, three bran cereals, and *Golden Crisps*. These cereals have fewer calories than we would expect based on their sugar content. We might not have grouped these cereals together before. What they have in common is a low calorie count *relative to their sugar content*— even though their sugar contents are quite different. (They are low calorie because of their shape and structure.)

These observations may not lead us to question the overall linear model, but they do help us understand that other factors may be part of the story. An exploration of residuals often leads us to discover more about individual cases. When we discover groups in our data, we may decide to analyze them separately, using a different model for each group.

Often, more research can help us discover why certain cases tend to behave similarly. Here, certain cereals group together in the residual plot because cereal manufacturers aim cereals at different segments of the market. A common technique used to attract different customers is to place different types of cereals on certain shelves. Cereals for kids tend to be on the "kid's shelf," at their eye level. Toddlers aren't likely to grab a box from this shelf and beg, "Mom, can we please get this *All-Bran with Extra Fiber*?"

How can we take this extra information into account in our analysis? Figure 17.5 shows a scatterplot of *Calories* and *Sugar*, colored according to the shelf on which the cereals were found, with a separate regression line fit for each shelf. Now we can see that the top shelf is unlike the bottom two shelves. We might want to report two regressions, one for the top shelf and one for the bottom two shelves.[1]

Figure 17.5 *Calories* and *Sugars* colored according to the shelf on which the cereal was found in a supermarket, with regression lines fit for each shelf individually. Do these data appear homogeneous? That is, do the cereals seem to all be from the same population of cereals? Or are there kinds of cereals that we might want to consider separately?

---

[1]Another alternative is to fit a multiple regression model by adding variables (called dummy or indicator variables) that distinguish the groups. This method will be discussed in Chapter 19.

For Example   Diamond prices

The price of a jewelry diamond depends on the 4 C's: Carat weight, Cut, Clarity, and Color. Diamond colors are assigned letters from D, for colorless stones, through K, for increasingly more yellow stones. A K-color stone is only faintly yellow and still considered jewelry quality. We've collected a sample of diamond prices from a website. Here is a scatterplot of *Price* vs *Carat Weight* with points colored according to *Color*.

Question: How does the relationship of *Price* and *Carat Weight* change with *Color*?

Answer: For a given *Carat Weight*, diamonds with better (earlier letter) colors are generally worth more. This is not true for every diamond, but it is generally true. There may be greater variability in the middle of the color range with H-color diamonds showing more variability that either D- or K-color stones.

17.2   Extrapolation and Prediction

When the Data Are Years

We usually don't enter them as four-digit numbers. Here, we used 0 for 1970, 10 for 1980, and so on. It's common to assign 0 to the date of the first observation in our data set if we are working with a time series. Another option is to enter two digits for the year, using 88 for 1988, for instance. Rescaling years like this often makes calculations easier and equations simpler. But be careful; if 1988 is 88, then 2004 is 104 (not 4).

Linear models give a predicted value for each case in the data. Put a new *x*-value into the equation, and it gives a predicted value, $\hat{y}$, to go with it. But when the new *x*-value lies far from the data we used to build the regression, how trustworthy is the prediction?

The simple answer is that the farther the new *x*-value is from $\bar{x}$, the center of the *x*-values, the less trust we should place in the predicted value. Once we venture into new *x* territory, such a prediction is called an **extrapolation**. Extrapolations are dangerous because they require the additional—and questionable—assumption that nothing about the relationship between *x* and *y* changes, even at extreme values of *x* and beyond. Extrapolations can get us into deep trouble, especially if we try to predict far into the future.

As a cautionary example, let's examine oil prices from 1972 to 1981.[2] In the mid 1970s, in the midst of an energy crisis, oil prices surged, and long lines at gas stations were common. In 1970, the price of oil was about $3 a barrel. A few years later, it had surged to $15. In 1975, a survey of 15 top econometric forecasting models (built by groups that included Nobel prize-winning economists) found predictions for 1985 oil prices that ranged from $50 to $200 a barrel—which would be $181 to $726(!) dollars a barrel in 2005 dollars. (We shall use constant 2005 dollars from now on, to factor out inflation.) How close were these forecasts? Let's look at Figure 17.6.

The regression model for the *Price* of oil against *Time* (Years since 1970) for these data is

$$\widehat{Price} = -0.85 + 7.39\ Time,$$

which says that prices increased, on average, $7.39 per year, or nearly $75 in 10 years. If they continued to increase linearly, it would have been easy to predict oil

---

[2]We will discuss special models for fitting data when *x* is time in Chapter 20, but simple regression models are often used. Even when using more sophisticated methods, the dangers of extrapolation don't disappear.

Figure 17.6   The price of oil per barrel in constant (2005) dollars from 1971 to 1982 shows a linear trend increasing at about $7 a year.

prices. And indeed, many forecasters made that assumption. So, how well did they do? Well, in the period from 1982 to 1998, oil prices didn't exactly continue that steady increase. In fact, they went down so much that by 1998, prices (adjusted for inflation) were the lowest they'd been since before World War II (Figure 17.7).

Figure 17.7   Time series plot of price of oil in constant (2005) dollars shows a fairly constant decrease over time.

For example, the average price of oil in 1985 turned out to be less than $30 per barrel—not quite the $100 predicted by the model. Extrapolating out beyond the original data by just four years produced some vastly inaccurate forecasts. While the time series plot in Figure 17.7 shows a fairly steady decline, this pattern clearly didn't continue (or oil would be free by now).

In the 1990s, the U.S. government decided to include scenarios in their forecasts. The result was that the Energy Information Administration (EIA) offered *two* 20-year forecasts for oil prices after 1998 in their Annual Energy Outlook (AEO). Both of these scenarios, however, called for relatively modest increases in oil prices (Figure 17.8).

Figure 17.8   This graph, adapted from one by the Energy Information Administration, shows oil prices from 1970 to 1998 with two sets of forecasts for the period 1999 to 2020.

So, how accurate have these forecasts been? Let's compare these predictions to the actual prices in constant (2005) dollars (Figure 17.9).

Figure 17.9 Here are the same EIA forecasts as in Figure 17.8, together with the actual prices from 1981 to 2007. Neither forecast predicted the sharp run-up in the first decade of the 21st century.

The experts seem to have missed the sharp run-up in oil prices in the first decade of the 21st century. Where do you think oil prices will go in the *next* decade? Your guess may be as good as anyone's. Clearly, these forecasts did not take into account many of the unforeseen global and economic events that occurred since 2000. Providing accurate long-term forecasts is extremely difficult.

Extrapolation far from the data is dangerous. Linear models are based on the *x*-values of the data at hand and cannot be trusted beyond that span. Some phenomena do exhibit a kind of inertia that allows us to guess that the currently observed systematic behavior will continue outside this range. When *x* is time, you should be especially wary. Such regularity can't be counted on in phenomena such as stock prices, sales figures, hurricane tracks, or public opinion.

Extrapolating from current trends is a mistake made not only by regression beginners or the naïve. Professional forecasters are prone to the same mistakes, and sometimes the errors are striking. However, because the temptation to predict the future is so strong, our more realistic advice is this:

If you extrapolate far into the future, be prepared for the actual values to be (possibly quite) different from your predictions.

"Prediction is difficult, especially about the future."

—NIELS BOHR, DANISH PHYSICIST

For Example   A big diamond

The Cullinan I diamond (the "Star of Africa") owned by the Queen of England is, at 530.20 carats, the largest D-color cut diamond in the world. It is mounted at the head of the scepter with the cross in the British crown jewels.

It has an estimated value of about $400 million. When we restrict our analysis of the diamonds we saw on page 535 to D-color diamonds, the regression model of *Price* on *Carat Weight* is

$$\widehat{Price} = -2129.14 + 8081.12 CaratWt$$

Question: What does this model predict for the *Price* of the Cullinan I diamond? Is the estimate of $400 million consistent with this model? Do you think this model should be relied on to establish prices for large diamonds?

Answer: The predicted price from the model is $4,282,482. That's only about 1/100th the price placed on the diamond itself. The rarity of such large diamonds is likely to make them far more valuable than a model based on much smaller stones could predict. Extrapolation that far from the data on which a model is based is not likely to work well.

17.3   Unusual and Extraordinary Observations

Your credit card company makes money each time you use your card. To encourage you to use your card, the card issuer may offer you an incentive such as airline miles, rebates, or gifts.[3] Of course, this is profitable to the company only if the increased use brings in enough revenue to offset the cost of the incentives. New ideas for offers (referred to as "creatives") are typically tested on a sample of cardholders before they are rolled out to the entire segment or population, a process referred to as a "campaign." Typically, the new offer (the "challenger") is tested against a control group who may be offered nothing or the current best offer ("the champion").

One campaign offered one of the highest-performing market segments an incentive for three months: one redeemable anytime air mile for each dollar spent. They hoped that the cardholders would increase their spending enough to pay for the campaign, but they feared that some cardholders would move spending forward into the incentive period, with a resulting drop in spending afterward.

For this particular segment, the typical cardholder charged about $1700 a month. During the campaign period, the group averaged around $1919.61 a month, a difference that was both statistically and financially significant. But analysts were suprised to see that the increase in spending continued well beyond the offer period. To investigate it, they made a scatterplot like the one shown in Figure 17.10.

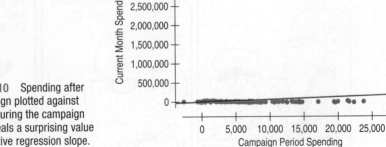

Figure 17.10 Spending after the campaign plotted against spending during the campaign period reveals a surprising value and a positive regression slope.

---

[3]There are websites dedicated to finding credit card "deals." Search "credit card rewards."

The outlying point, at the top of the graph, represents a cardholder who charged nearly $3 million in the month after the free miles period ended. Remarkably, the point was verified to be a real purchase! Nevertheless, this cardholder is clearly not typical of the rest of the segment. To answer the company's question, we need to examine the plot without the outlying point (Figure 17.11).

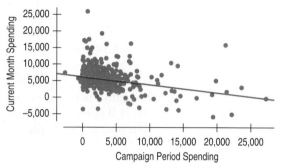

Figure 17.11 A plot of current spending against the spending during the campaign period, with the outlier set aside. Now the slope is negative, and significantly so.

The plot does show that those with the largest charges during the campaign spent less in the month after the campaign. Just one outlier was capable of changing the slope's direction from strongly negative to strongly positive. On the basis of this finding, the analysts decided to focus only on people whose spending during *both* periods was less than $10,000 a month, figuring that if someone decides to spend more than $10,000 on their credit card, their primary motivation is probably not the airline miles incentive.

Outliers, Leverage, and Influence

By providing a simple description of how data behave, models help us see when and how data values are unusual. In regression, a point can stand out in two ways. A case can have a large residual, as our $3 million spender certainly did. Because they are not like the other cases, points with large residuals always deserve special attention and are called **outliers**.

A data point can also be unusual if its *x*-value is far from the mean of the *x*-values. Such a point is said to have high **leverage**. The physical image of a lever is exactly right. The least squares line must pass through $(\bar{x}, \bar{y})$, so you can picture that point as the fulcrum of the lever. Just as sitting farther from the center of a seesaw gives you more leverage, points with values far from $\bar{x}$ pull more strongly on the regression line.

A point with high leverage has the potential to change the regression line but it doesn't always use that potential. If the point lines up with the pattern of the other points, it doesn't change our estimate of the line. By sitting so far from $\bar{x}$ though, it may appear to strengthen the relationship, inflating the correlation and R^2.

How can you tell if a high-leverage point changes the model? Just fit the linear model twice, both with and without the point in question. We say that a point is **influential** if omitting it from the analysis gives a very different model (as the high spender did for the credit card spending in Figure 17.10).[4]

"Give me a place to stand and I will move the Earth."

—ARCHIMEDES
(287–211 BCE)

Influence and Leverage

Influence depends on both the leverage and the residual; a case with high leverage whose *y*-value sits right on the line fit to the rest of the data is not influential. A case with low leverage but a very large residual can be influential. The only way to be sure is to fit the regression with and without the potential influential point.

---

[4]Some textbooks use the term *influential point* for any observation that influences the slope, intercept, or R^2. We'll reserve the term for points that influence the slope.

Unusual points in a regression often tell us more about the data and the model than any other cases. Whenever you have—or suspect that you have—influential points, you should fit the linear model to the other cases alone and then compare the two regression models to understand how they differ. A model dominated by a single point is unlikely to be useful for understanding the rest of the cases. The best way to understand unusual points is against the background of the model established by the other data values. Don't give in to the temptation to delete points simply because they don't fit the line. That can give a false picture of how well the model fits the data. But often the best way to identify interesting cases and subgroups is to note that they are influential and to find out what makes them special.

Not all points with large influence have large residuals. Sometimes, their influence pulls the regression line so close that it makes the residual deceptively small. Influential points like that can have a shocking effect on the regression. Figure 17.12 shows IQ plotted against shoe size from a fanciful study of intelligence and foot size. The outlier is Bozo the clown, known for his large feet and hailed as a comic genius.

Figure 17.12   Bozo the clown's extraordinarily large shoes give his data point high-leverage in the regression of: $\widehat{IQ} = 93.3 + 2.08\ Shoe\ Size$, even though the R^2 is 25%. Wherever Bozo's IQ happens to be, the regression line will follow.

Although this is a silly example, it illustrates an important and common potential problem. Almost all of the variance accounted for ($R^2 = 25\%$) is due to *one* point, namely, Bozo. Without Bozo, there is little correlation between shoe size and IQ. If we run the regression after omitting Bozo, we get an R^2 of only 0.7%—a weak linear relationship (as one might expect). One single point exhibits a great influence on the regression analysis.

Just Checking

Each of these scatterplots shows an unusual point. For each, tell whether the point is a high-leverage point, would have a large residual, and/or is influential.

What should you do with a high-leverage point? Sometimes these values are important (they may be customers with extremely high incomes or employees with unusually long service to the company), and they may say more about the relationship between y and x than any of the other data values. However, at other times, high-leverage points are values that really don't belong with the rest of the data. Such points should probably be omitted, and a linear model found without them for comparison. When in doubt, it's usually best to fit regressions both with and without the points and compare the two models.

- **Warning:** Influential points can hide in plots of residuals. Points with high-leverage pull the line close to them, so they often have small residuals. You'll see influential points more easily in scatterplots of the original data, and you'll see their effects by finding a regression model with and without the points.

For Example   Colored diamonds

Some true diamonds are colored from inclusions or structural differences. Can the prices of these diamonds be predicted with the models we make for traditional clear diamonds?

One online site lists a black diamond that weighs 1.44 carats for $3050. Black diamonds are true diamonds but may have formed from impacts of comets, asteroids, or meteors rather than from geological pressures deep underground. They are probably best compared to the most-colored K-quality diamonds. Here is a scatterplot of *Price* vs *Carat Weight* for K-color diamonds. The black diamond (shown with a red x) has both high leverage and large residual.

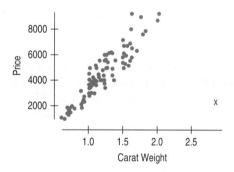

The same site lists a blue diamond that weighs 0.82 carats for $16,500. Blue diamonds are a luxury item, best compared to the best D-color clear diamonds. Here is a scatterplot of *Price* vs *Carat Weight* for D-color diamonds with the blue diamond shown as a red x.

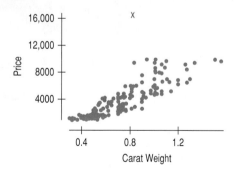

Question: What would be the effect of including these exotic diamonds on regression lines fit to each of these scatterplots?

Answer: The black diamond has high leverage because it is heavier than any of the other diamonds considered. It also has a large negative residual because it is much less expensive than a regression for K-color diamonds would predict. As a result, it will pull the regression line down, making the slope smaller.

The blue diamond has a large positive residual because it is much more expensive even than the regression on perfect D-color diamonds would predict. However, because its carat weight is in the middle of the range of weights considered for the regression, it has little leverage. Including this case will increase the intercept slightly, but won't change the slope very much.

17.4   Working with Summary Values

Scatterplots of statistics summarized over groups tend to show less variability than we would see if we measured the same variables on individuals. This is because the summary statistics themselves vary less than the data on the individuals.

Wind power is getting increasing attention as an alternative, carbon-free method of generating electricity. Of course, there must be enough wind to make it cost-effective. In a study to find a site for a wind generator, wind speeds were collected four times a day (at 6:00 a.m., noon, 6:00 p.m., and midnight) for a year at several possible sites. Figure 17.13 plots the wind speeds for two of these sites. The correlation is 0.736.

Figure 17.13   The wind speed at sites 2 and 4 are correlated.

What would happen to the correlation if we used only one measurement per day? If, instead of plotting four data points for each day, we record an average speed for each day, the resulting scatterplot shows less variation, as Figure 17.14 shows. The correlation for these values increases to 0.844.

Figure 17.14   Daily average wind speeds show less variation.

Let's average over an even longer time period. Figure 17.15 shows *monthly* averages for the year (plotted on the same scale). Now the correlation is 0.942.

What these scatterplots show is that summary statistics exhibit less scatter than the data on individuals on which they're based and can give us a false impression of how well a line summarizes the data. There's no simple correction for this phenomenon. If we're given summary data, we usually can't get the original values back. You should be a bit suspicious of conclusions based on regressions of summary data. They may look better than they really are.

Another way to reduce the number of points in a data set is to select or sample points rather than average them. This can be especially important with data, such as

Figure 17.15 Monthly averages are even less variable.

the wind speeds, that are measured over time. For example, if instead of finding the *mean* for each day, we select just one of the four daily measurements—say the one made at noon on each day. We would have just as many points as in Figure 17.14, but the correlation is 0.730—essentially the same as for the full data. Figure 17.16 shows the relationship.

Figure 17.16 Selecting only the noon measurements doesn't reduce the variation. Compare this scatterplot to Figures 17.13 and 17.14.

17.5 Autocorrelation

Time series data that are collected at regular time points often have the property that points near each other in time will be related. When values at time t are correlated with values at time t-1 we say the values are **autocorrelated** in the first order. If values are correlated with values two time periods back, we say second-order autocorrelation is present, and so on.

A regression model applied to autocorrelated data will have errors that are not independent, and that violates an assumption for regression. The statistical tests and confidence intervals for the slope depend on independence, and its violation can render these tests and intervals invalid. Fortunately, there is a statistic, called the Durbin-Watson statistic, that can detect first-order autocorrelation from the residuals of a regression analysis.

The Brief Case of Chapter 16 provided data on *Price* and *Sales* volume of frozen pizza for several cities for one week each month. We actually have the data for *every* week in the same three-year period. Here is the regression of *Sales* volume on *Price* for Dallas.

Why Is Autocorrelation a Problem?

When data are highly correlated over time, each data point is similar to those arround it. So each data point provides less additional information than if the points had been independent. All regression inference is based on independent errors so we need to check for autocorrelation.

	Coeff	SE(Coeff)	t-value	P-value
Intercept	139547	11302	12.347	<0.0001
Price	−33527	4308	−7.783	<0.0001

A plot of the residuals against predicted values shows nothing particularly unusual.

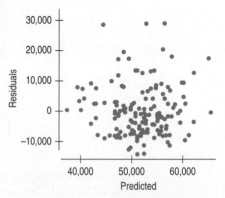

Figure 17.17   A scatterplot of the residuals vs. predicted values for the 156 weeks of pizza sales data reveals no obvious patterns.

Figure 17.18   A scatterplot of the residuals against weeks for the 156 weeks of pizza sales data seems to show some trend.

But because these data points are consecutive weekly data, we should investigate the residuals *vs.* time. Here we've plotted the *Residuals* against *Week*, consecutively from week 1 to week 156.

It may not be obvious that there's a pattern here. Autocorrelation can be difficult to see in residuals. It does seem, however, that there is a tendency in Figure 17.18 for the residuals to be related to nearby points. Notice the overall positive trend. We shouldn't see such a trend in residuals that are independent of each other.

The **Durbin-Watson** statistic estimates the autocorrelation by summing squares of consecutive differences and comparing the sum with its expected value under the null hypothesis of no autocorrelation. The Durbin-Watson statistic is computed as follows:

$$D = \frac{\sum_{t=2}^{n}(e_t - e_{t-1})^2}{\sum_{t=1}^{n}e_t^2}$$

where e_t is the residual at time t. The statistic always falls in the interval from 0 to 4. When the null hypothesis of no autocorrelation is true, the value of D should be 2. Values of D below 2 are evidence of positive autocorrelation, while values of D above 2 indicate possible negative autocorrelation. Positive autocorrelation is more common than negative autocorrelation. How far below or above 2 does D need to be to show "strong" or significant autocorrelation? It may be surprising, but the answer to this question depends only on the sample size, n, and the number of predictors in the regression model, k, which for simple regression is equal to 1.

A standard Durbin-Watson table (see Table D in Appendix D) shows the sample size down the left-hand column, so that each row corresponds to a different sample size n, with the number of predictors k across the top. For each k there are two columns: d_L and d_U. (The significance level of the table is also shown at the top of the page.) The test has several possible outcomes:

If $D < d_L$ (lower critical value), then there is evidence of positive autocorrelation.

If $d_L < D < d_U$, then the test is inconclusive.

If $D > d_U$ (upper critical value), then there is no evidence of positive autocorrelation.

Why 0 and 4?

Notice that if the adjacent residuals are equal (perfectly correlated), then the numerator and the value of D equals 0. If, on the other hand, the residuals are equal but have opposite signs (perfectly negatively correlated), then each difference is twice the residual. Then after squaring, the numerator will be 2^2, or four times the denominator.

To test negative autocorrelation, we use the same values of d_L and d_U, but we subtract them from 4:

> If $D > 4 - d_L$ (lower critical value), then there is evidence of negative autocorrelation.
>
> If $4 - d_L < D < 4 - d_U$, then the test is inconclusive.
>
> If $D < 4 - d_U$ (upper critical value), then there is no evidence of negative autocorrelation.

We usually rely on technology to compute the statistic. For the pizza example, we have $n = 156$ weeks and one predictor (*Price*), so $k = 1$. The value of D is $D = 0.8812$. Using the table in Appendix D, we find the largest value of n listed is $n = 100$, and at $\alpha = 0.05$, $d_L = 1.65$. Because our value is less than that, we conclude that there is evidence of positive autocorrelation. (A software package would find the P-value to be < 0.0001.) We conclude that the residuals are *not* independent but that residuals from one week have a positive correlation with the residuals from the preceding week. The standard errors and test for the slope are not valid since we don't have independence.

For Example   Gemstone imports

The U.S. Geological Survey reports facts about a range of mineral and material commodities (http://minerals.usgs.gov/ds/2005/140/) including gemstones. From their report, we find that imports of gemstones increased between 1990 and 2008:

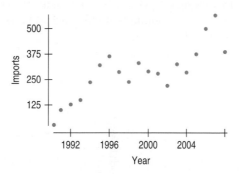

A regression gives the following:

```
Dependent variable is: Imports
R squared = 66.1%  R squared (adjusted) = 64.1%
s = 78.20 with 19 - 2 = 17 degrees of freedom
```

Variable	Coefficient	SE(Coeff)	t-ratio	P-value
Intercept	-37412.4	6548	-5.71	≤0.0001
Year	18.8581	3.275	5.76	≤0.0001

The Durbin-Watson statistic is $D = 1.077$.

Question: Why should we check the Durbin-Watson statistic for this analysis? What does the statistic say about these data?

Answer: The data are a time series. For time series there is always a possibility of autocorrelation, which we test for with the Durbin-Watson statistic. A value of 1.077 for a regression with $n = 19$ and one predictor is below the lower limit (d_L) for significance, which we can find from Table D in Appendix D to be 1.18. So we conclude that there is evidence of a positive autocorrelation in these data. We should not rely on the standard errors calculated for this regression.

Time series methods (see Chapter 20) attempt to deal with the problem of autocorrelation by modeling the errors. Another solution is to find a predictor variable that accounts for some of the autocorrelation and removes the dependence in the residuals (see Chapter 19). A simple solution that often works is to sample from the time series so that the values are more distant in time and thus less likely to be correlated. If we take every fourth week of data (as we did in Chapter 16), starting at week 4, from the Dallas pizza data, our regression becomes:

	Coeff	SE(Coeff)	t-ratio	P-value
Intercept	148350	22266	6.663	8.01e-08
Price	−36762	8583	−4.283	0.000126

Now, $D = 1.617$. With $n = 39$, the upper critical value d_U is 1.54. Since our new value of D is larger than that, we see no evidence of autocorrelation. Output from technology shows the P-value:

```
Durbin-Watson test
D = 1.6165,  P = 0.098740
```

We should feel more comfortable basing our confidence and prediction intervals on this model.

17.6   Transforming (Re-expressing) Data
Linearity

Increasing gas prices and concern for the environment have lead to increased attention to automobile fuel efficiency. The most important factor in fuel efficiency is the weight of the car.

Figure 17.19   *Fuel Efficiency* (mpg) *vs. Weight* (thousands of pounds) shows a strong, apparently linear, negative trend.

The relationship is strong ($R^2 = 81.6\%$), clearly negative, and apparently linear. The regression equation

$$\widehat{Fuel\ Efficiency} = 48.7 - 8.4\ Weight$$

says that fuel efficiency drops by 8.4 mpg per 1000 pounds, starting from a value of 48.7 mpg. We check the **Linearity Condition** by plotting the residuals versus either the x variable or the predicted values.

The scatterplot of the residuals against *Weight* (Figure 17.20) holds a surprise. Residual plots should have no pattern, but this one has a bend. Look back at the original scatterplot. The scatter of points isn't really straight. There's a slight bend to the plot, but the bend is much easier to see in the residuals.

Figure 17.20   Plotting residuals against weight reveals a bend. The bend can be seen if you look carefully at the original scatterplot, but here it's easier to see.

When the relationship isn't straight, we shouldn't fit a regression or summarize the strength of the association with correlation. But often we can make the relationship straighter. All we have to do is **re-express (or transform)** one or both of the variables with a simple function. In this case, there's a natural function. In the U.S., automobile fuel efficiency is measured in miles per gallon. But throughout the rest of the world, things are different. Not only do other countries use metric measures, and thus kilometers and liters, but they measure fuel efficiency in liters per 100 kilometers. That's the *reciprocal* of miles per gallon (times a scale constant). That is, the gas amount (gallons or liters) is in the numerator, and the distance (miles or kilometers) is now in the denominator.

There's no reason to prefer one form or the other, so let's try the (negative) reciprocal form.

Figure 17.21   The reciprocal of *Fuel Efficiency vs. Weight* is straighter.

The residuals look better as well.

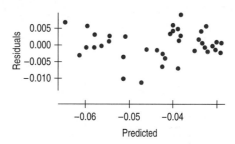

Figure 17.22   Residuals from the regression of *Fuel Efficiency (−1/mpg)* on *Weight* show less bend.

There's a clear improvement using the reciprocal, so we should use the reciprocal as the response in our regression model.

Are we allowed to re-express fuel efficiency as its reciprocal? For these data, it seems like a good idea. After all, we chose a form of the variable used by most people in the world. The general idea of transforming data to improve and simplify its structure extends beyond straightening scatterplots. In fact, you use re-expressions in everyday life. How fast can you go on a bicycle? If you measure your speed, you probably do it in distance per time (miles per hour or kilometers per hour). In 2005, during a 12-mile-long time trial in the Tour de France, Dave Zabriskie averaged nearly 35 mph (54.7 kph), beating Lance Armstrong by 2 seconds. You probably realize that's a tough act to follow. It's fast. You can tell that at a glance because you have no trouble thinking in terms of distance covered per time.

If you averaged 12.5 mph (20.1 kph) for a mile run, would that be fast? Would it be fast for a 100-meter dash? Even if you run the mile often, you probably have to stop and calculate. Although we measure speed of bicycles in distance per time, we don't usually measure running speed that way. Instead, we re-express it as the *reciprocal*—time per distance (minutes per mile, seconds per 100 meters, etc.). Running a mile in under 5 minutes (12 mph) is fast. A mile at 16 mph would be close to the world record (that's a 3-minute 45-second mile).

The point is that there is no single natural way to measure speed. In some cases, we use distance traveled per time, and in other cases, we use the reciprocal. It's just because we're used to thinking that way in each case, not because one way is correct. It's important to realize that the way these quantities are measured is not sacred. It's usually just convenience or custom. When we re-express a quantity to make it satisfy certain conditions, we may either leave it in those new units when we explain the analysis to others or convert it back to the original units.

Goals of Re-expression

We re-express data for several reasons. Each of these goals helps make the data more suitable for analysis by our methods. We'll illustrate each goal by looking at data about large companies.

Goal 1 *Make the distribution of a variable (as seen in its histogram, for example) more symmetric.* It's easier to summarize the center of a symmetric distribution, and for nearly symmetric distributions, we can use the mean and standard deviation. If the distribution is unimodal, then the resulting distribution may be closer to the Normal model, allowing us to use the 68–95–99.7 Rule.

Here are the *Assets* of 77 large companies as reported by *Forbes*.

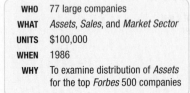

WHO	77 large companies
WHAT	*Assets, Sales,* and *Market Sector*
UNITS	$100,000
WHEN	1986
WHY	To examine distribution of *Assets* for the top *Forbes* 500 companies

Figure 17.23 The distribution of the *Assets* of large companies is skewed to the right. Data on wealth often look like this.

The skewed distribution is made much more symmetric by taking logs.

Figure 17.24   Taking logs makes the distribution more symmetric.

Goal 2 *Make the spread of several groups (as seen in side-by-side boxplots) more alike,* even if their centers differ. Groups that share a common spread are easier to compare. We'll see methods later in the book that can be applied only to groups with a common standard deviation. We saw an example of re-expression for comparing groups with boxplots in Chapter 5.

Here are the *Assets* of these companies by *Market Sector.*

Figure 17.25   Assets of large companies by *Market Sector.* It's hard to compare centers or spreads, and there seem to be a number of high outliers.

Taking logs makes the individual boxplots more symmetric and gives them spreads that are more nearly equal.

Figure 17.26   After re-expressing using logs, it's much easier to compare across *Market Sectors.* The boxplots are more symmetric, most have similar spreads, and the companies that seemed to be outliers before are no longer extraordinary. Two new outliers have appeared in the finance sector. They are the only companies in that sector that are not banks.

This makes it easier to compare *Assets* across *Market Sectors*. It can also reveal problems in the data. Some companies that looked like outliers on the high end turned out to be more typical. But two companies in the Finance sector now stick out. They are not banks. Unlike the rest of the companies in that sector, they may have been placed in the wrong sector, but we couldn't see that in the original data.

Goal 3 *Make the form of a scatterplot more nearly linear:* Linear scatterplots are easier to describe. The value of re-expressing data to straighten a relationship is that we can fit a linear model once the relationship is straight.

Here are *Assets* plotted against the logarithm of *Sales*.

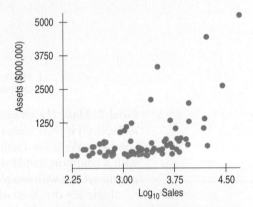

Figure 17.27 *Assets* vs. log *Sales* shows a positive association (bigger sales goes with bigger *Assets*) with a bent shape.

Note that the plot of *Assets* versus log *Sales* shows that the points go from tightly bunched at the left to widely scattered at the right—a "fan" shape. The plot's shape is bent. Taking logs makes the relationship much more linear (see Figure 17.28). If we re-express the company *Assets* using logarithms, we get a graph that shows a more linear association. Also note that the "fan" shape has disappeared and the variability at each value of *x* is about the same.

Figure 17.28 Log *Assets* vs. log *Sales* shows a positive linear association.

Goal 4 *Make the scatter in a scatterplot or residual plot spread out evenly rather than following a fan shape.* Having an even scatter is a condition of many methods of Statistics, as we'll see in later chapters. This goal is closely related to Goal 2, but it often comes along with Goal 3. Indeed, a glance back at the scatterplot (Figure 17.27) shows that the plot for *Assets* is much more spread out on the right than on the left, while the plot for log *Assets* (Figure 17.28) has roughly the same variation in log *Assets* for any *x*-value.

For Example   Re-expressing diamond prices

When we first looked at diamond prices (on page 535), we noted that we needed to model prices for different color diamonds separately. In other examples, we restricted attention to either the D- or K-color diamonds at either end of the color spectrum. Here is a scatterplot of *Price* vs *Carat Weight* for the E- and F-color stones:

Question: Which assumptions for regression are violated in this plot?

Answer: The relationship is not straight. It also spreads out more on the right than on the left.

17.7   The Ladder of Powers

We've seen that taking logs or reciprocals can improve an analysis of relationships. Other transformations can be helpful too, but how do we know which re-expression to use? We could use trial and error to choose a re-expression, but there's an easier way. We can choose our re-expressions from a family of simple mathematical expressions that move data toward our goals in a consistent way. This family includes the most common ways to re-express data. More important, the members of the family line up in order, so that the farther you move away from the original data (the "1" position), the greater the effect on the data. This fact lets you search systematically for a transformation that works—stepping a bit farther from "1" or taking a step back toward "1" as you see the results.

Where to start? It turns out that certain kinds of data are more likely to be helped by particular re-expressions. Knowing that gives you a good place to start your search for a mathematical expression. We call this collection of re-expressions the **Ladder of Powers.** The following table shows some of the most useful powers with each one specified as a single value.

Power	Name	Comment
2	The square of the data values, y^2	Try this for unimodal distributions that are skewed to the left.
1	The raw data—no change at all. This is "home base." The farther you step from here up or down the ladder, the greater the effect.	Data that can take on both positive and negative values with no bounds are less likely to benefit from re-expression.
1/2	The square root of the data values $\sqrt{y}$	Counts often benefit from a square root re-expression. For counted data, start here.
"0"	Although mathematicians define the "0-th" power differently, for us the place is held by the logarithm.	Measurements that cannot be negative, and especially values that grow by percentage increases such as salaries or populations, often benefit from a log re-expression. When in doubt, start here. If your data have zeros, try adding a small constant to all values before finding the logs.

(continued)

Power	Name	Comment
−1/2	The (negative) reciprocal square root $-1/\sqrt{y}$	An uncommon re-expression, but sometimes useful. Changing the sign to take the *negative* of the reciprocal square root preserves the direction of relationships, which can be a bit simpler.
−1	The (negative) reciprocal, $-1/y$	Ratios of two quantities (miles per hour, for example) often benefit from a reciprocal. (you have about a 50-50 chance that the original ratio was taken in the "wrong" order for simple statistical analysis and would benefit from re-expression.) Often, the reciprocal will have simple units (hours per mile). Change the sign if you want to preserve the direction of relationships. If your data have zeros, try adding a small constant to all values before finding the reciprocal.

The Ladder of Powers orders the *effects* that the re-expressions have on data. If you try, say, taking the square roots of all the values in a variable and it helps, but not enough, then moving farther down the ladder to the logarithm or reciprocal root will have a similar effect on your data, but even stronger. If you go too far, you can always back up. But don't forget—when you take a negative power, the *direction* of the relationship will change. You can always change the sign of the response if you want to keep the same direction.

For Example   Finding a re-expression for diamond prices

We have seen (on page 551) that the relationship between *Price* and *Carat Weight* is not linear and spreads out toward the right side of the plot. We can re-express *Price*, hoping to improve these problems. Here are three possible re-expressions of *Price*: square root, $\log_{10}$, and (negative) reciprocal.

Question: Which of these does the best job of making the relationship linear? Does the plot indicate that we have satisfied the regression assumptions?

Answer: The relationship of $\log_{10}$ *Price* to *Carat Weight* is much more linear and the spread has evened out across the plot. A regression of $\log_{10}$ *Price* on *Carat Weight* would be appropriate.

Guided Example

What determines the value of commercial real estate? In Chapter 6 we used a simple linear relationship with an R^2 of 62.4% to see that living area was a good predictor of the value of a family home. Does the same relationship hold for commercial property? Are other factors important?

We have available data on 286 commercial properties in and around a major U.S. city.

PLAN

Plan State your objective. Identify the quantitative variables you wish to examine. Define each variable. (State the W's.)

Our objective is to investigate the association between the *Price* of a commercial property ($M) and its *Size* (square ft).

We have data for 286 commercial properties in and around a major U.S. city. We will build and diagnose a regression model.

Plot the data.

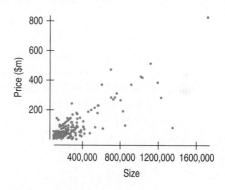

Check the conditions for regression.

✓ **Quantitative Variable Condition:** Both variables are quantitative.

✓ **Linearity Condition:** The scatterplot is straight enough.

✓ **Outlier Condition:** There are no obvious outliers.

✓ **Equal Spread Condition:** The plot fans out to the right. That makes us suspect skewness, so we'll look at a histogram:

REALITY CHECK It isn't surprising to find that *Price* is skewed to the high end. Variables like this that cannot get too small but have no real limit on the upper end are often skewed like this.

The skewness of *Price* suggests a re-expression to make the distribution more nearly symmetric and to even out the spread in the scatterplot.

(continued)

DO

Log₁₀ is easier to interpret than the natural log. The $\log_{10}$ of 1,000,000 is 6; the $\log_{10}$ of 10,000,000 is 7. So these prices are mostly between \$1M and \$100M, with some near \$1B ($\log_{10} = 9$).

We will investigate re-expressions of the variables.

A histogram of $\text{Log}_{10}(\textit{Price})$ is much more symmetric:

(Rather than find the log of *Price* in \$M, we used the original *Price* in \$ to aid interpretation.)

The scatterplot of $\text{Log}_{10}(\textit{Price})$ vs. *Size* is not straight:

It seems natural to try a square root re-expression for a variable like *Size* that is measured in squared units (ft²):

This is an improvement, but still not straight. We'll try the next two steps on the Ladder of Powers: log and $-1/\sqrt{y}$:

We change the sign when we take the reciprocal to preserve the direction—here a positive relationship.

The log and (negative) reciprocal root look good. We'll pick the reciprocal root but could just as well have used the log.

A regression fit to this relationship looks like this:

```
Dependent variable is: Log₁₀(Price)
R² = 45.7%   R²(adjusted) = 45.5%
s = 0.3484 with 286 − 2 = 284 df
```

Variable	Coeff	SE(Coeff)	t-ratio	P-value
Intercept	8.842	0.0857	103	≤0.0001
$-1/\sqrt{Size}$	561.375	36.34	15.4	≤0.0001

As often happens when we re-express data for one reason (in this example, to improve linearity and make the variables more nearly symmetric), we also improve other aspects of the data. Except for these outliers, the residuals have a more symmetric distribution and are more nearly Normal than they would have been for the original variables. When we test coefficients, one condition is that the residuals be nearly Normal, so this can be important.

A plot of the residuals shows some outliers:

Without knowing more, we can't be sure why these properties appear to be so inexpensive. Perhaps they are an opportunity for a developer.

One suggestive observation is that two of them are among the oldest of the properties in our data having been built

(continued)

before 1890. But two other properties are still older, and one of our outliers isn't old at all, so that's not a full explanation:

If we set aside the three outliers, the R^2 increases to 59.2%

REPORT

Conclusion Describe the steps taken and what you learned about the data.

MEMO

Our regression analysis of commercial properties showed that the relationship between price and size was not linear—unlike what we found previously for residential properties. Re-expressing the variables, we found that the $\log_{10}$ of *Price* can be predicted from $1/\sqrt{Size}$ with an R^2 of 59.2%. That's not as strong a relationship as we found for residential properties.

We also found three outliers, which we set aside for further investigation. They may present investment opportunities because they appear to be unexpectedly inexpensive.

Just Checking

4 You want to model the relationship between prices for various items in Paris and Hong Kong. The scatterplot of Hong Kong prices vs. Paris prices shows a generally straight pattern with a small amount of scatter. What re-expression (if any) of the Hong Kong prices might you start with?

5 You want to model the population growth of the United States over the past 200 years with a percentage growth that's nearly constant. The scatterplot shows a strongly upwardly curved pattern. What re-expression (if any) of the population might you start with?

What Can Go Wrong?

This entire chapter has warned about things that can go wrong in a regression analysis. So let's just recap. When you make a linear model:

- **Make sure the relationship is straight enough to fit a regression model.** Check the Linearity Condition on the scatterplot of y against x and always examine the residuals for evidence that the Linearity Assumption has failed. It's often easier to see deviations from a straight line in the residuals plot than in the scatterplot of the original data. Pay special attention to the most extreme residuals because they may have something to add to the story told by the linear model.

- **Be on guard for different groups.** Check for evidence that the data consist of separate subsets. If you find subsets that behave differently, consider fitting a different linear model to each subset.

- **Beware of extrapolating.** Beware of extrapolation beyond the x-values that were used to fit the model. Although it's common to use linear models to extrapolate, be cautious.

- **Beware of extrapolating far into the future.** Be especially cautious about extrapolating far into the future with linear models. A linear model assumes that changes over time will continue forever at the same rate you've observed in the past. Predicting the future is particularly tempting and particularly dangerous.

- **Look for unusual points.** Unusual points always deserve attention and may well reveal more about your data than the rest of the points combined. Always look for them and try to understand why they stand apart. Making a scatterplot of the data is a good way to reveal high-leverage and influential points. A scatterplot of the residuals against the predicted values is a good tool for finding points with large residuals.

- **Beware of high-leverage points, especially of those that are influential.** Influential points can alter the regression model a great deal. The resulting model may say more about one or two points than about the overall relationship.

- **Consider setting aside outliers and re-running the regression.** To see the impact of outliers on a regression, try running two regressions, one with and one without the extraordinary points, and then discuss the differences.

- **Treat unusual points honestly.** If you remove enough carefully selected points, you will eventually get a regression with a high R^2. But it won't get you very far. Some data are not simple enough for a linear model to fit very well. When that happens, report the failure and stop.

- **Be alert for autocorrelation.** Data measured over time may fail the Independence Assumption. A Durbin-Watson test can check for that.

- **Watch out when dealing with data that are summaries.** Be cautious in working with data values that are themselves summaries, such as means or medians. Such statistics are less variable than the data on which they are based, so they tend to inflate the impression of the strength of a relationship.

- **Re-express your data when necessary.** When the data don't have the right form for the model you are fitting, your analysis can't be valid. Be alert for opportunities to re-express data to achieve simpler forms.

Ethics in Action

Certain types of fish and shellfish, such as swordfish and shark, are known to contain high levels of mercury. The FDA and EPA publish guidelines on the recommended consumption per week for various types of fish to protect public health, particularly the health of children and pregnant women. There has been a recent resurgence of media attention on the potentially high levels of mercury in tuna. A few years ago the state of California tried unsuccessfully to include mercury warnings on canned tuna labels. Then the *New York Times* (January 23, 2008) published a study that showed the levels of mercury in high-grade sushi tuna are above FDA recommended cutoffs. This prompted James Halibut, director for the U.S. Pacific Coast Tuna Association, to check on the progress being made by researcher Gary Waters. Gary, a biologist at a regional university, is examining factors that affect mercury levels in tuna. His work is being funded, in part, by a grant from the association. Since most of the tuna caught off the U.S. Pacific Coast is relatively small in size, James told Gary that he is particularly interested in the relationship between the weight of tuna and mercury levels. Gary had found a significant linear relationship between mercury levels and tuna weight, but it was not very strong. Moreover, when he analyzed the residuals from the regression with tuna weight, he discovered an influential outlier. In fact, Gary had performed the regression analysis twice, including and then excluding this point. Omitting this point further reduced the strength of the relationship. Feeling a bit pressured by James, Gary decided to only discuss with him the regression results including this point.

ETHICAL ISSUE *An influential outlier needs to be examined further; including it to make a linear relationship appear stronger than it would otherwise be is unethical. Related to Item H, ASA Ethical Guidelines.*

ETHICAL SOLUTION *The results, even though not as favorable as expected by the funding agency, need to be discussed honestly. Gary must disclose the outlier and its effects on the relationship. James should not place any pressure on the researcher.*

What Have We Learned?

Learning Objectives

- Be skeptical of regression models. Always plot and examine the residuals for unexpected behavior. Be alert to a variety of possible violations of the standard regression assumptions and know what to do when you find them.
- Be alert for subgroups in the data.
  - Often these will turn out to be separate groups that should not be analyzed together in a single analysis.
  - Often identifying subgroups can help us understand what is going on in the data.
- Be especially cautious about extrapolating beyond the data.
- Look out for unusual and extraordinary observations.
  - Cases that are extreme in x have high leverage and can affect a regression model strongly.
  - Cases that are extreme in y have large residuals.
  - Cases that have both high leverage and large residuals are influential. Setting them aside will change the regression model, so you should consider whether their influence on the model is appropriate or desirable.
- Notice when you are working with summary values.
  - Summaries vary less than the data they summarize, so they may give the impression of greater certainty than your model deserves.
- Diagnose and treat nonlinearity.
  - If a scatterplot of y vs. x isn't straight, a linear regression model isn't appropriate.
  - Re-expressing one or both variables can often improve the straightness of the relationship.
  - The powers, roots, and the logarithm provide an ordered collection of re-expressions so you can search up and down the "ladder of powers" to find an appropriate one.

Terms

Autocorrelation   The correlation of a variable with itself, offset by one or more positions. Autocorrelation is especially important for variables measured over time.

Durbin-Watson Statistic   A statistic that tests the strength of autocorrelation in a variable. It is commonly applied to check whether regression residuals are not mutually independent for data measured over time.

Extrapolation   Although linear models provide an easy way to predict values of y for a given value of x, it is unsafe to predict for values of x far from the ones used to find the linear model equation. Be cautious when extrapolating.

Influential   If omitting a point from the data changes the regression model substantially, that point is considered influential.

Leverage   Data points whose x-values are far from the mean of x are said to exert leverage on a linear model. High-leverage points pull the line close to them, so they can have a large effect on the line, sometimes completely determining the slope and intercept. Points with high enough leverage can have deceptively small residuals.

Outlier   Any data point that stands away from the regression line by having a large residual is called an outlier.

Transformation (or Re-expression)   A function—typically a simple power or root—applied to the values of a quantitative variable to make its distribution more symmetric and/or to simplify its relationship with other variables.

Technology Help: Examining Residuals

Most statistics technology offers simple ways to check whether your data satisfy the conditions for regression and to re-express the data when that is called for. We have already seen that these programs can make a simple scatterplot. They can also help us check the conditions by plotting residuals. Most statistics packages offer a way to re-express and compute with variables. Some packages permit you to specify the power of a re-expression with a slider or other moveable control, possibly while watching the consequences of the re-expression on a plot or analysis. This is an effective way to find a good re-expression.

EXCEL   ✗LSTAT

The Data Analysis add-in for Excel includes a Regression command. The dialog box it shows offers to make plots of residuals.

Excel is an excellent place to re-express data. Just use Excel's built-in functions as you would for any calculation. Changing a value in the original column will change the re-expressed value.

	A	B	C
1	Weight	City MPG	-1/MPG
2	3252	22	=-1/B2
3	3219	20	

Comments

Do not use the Normal probability plot offered in the regression dialog. It is not what it claims to be and is wrong.

JMP

- From the **Analyze** menu, choose **Fit Y by X**.
- Select **Fit Line**.
- Under Linear Fit, select **Plot Residuals**. You can also choose to **Save Residuals**.
- Subsequently, from the **Distribution** menu, choose **Normal quantile plot** or **histogram** for the residuals.

To re-express a variable in JMP,

- Double-click to the right of the last column of data to create a new column.
- Name the new column and select it.
- Choose **Formula** from the **Cols** menu.
- In the Formula dialog, choose the transformation and variable that you wish to assign to the new column.
- Click the **OK** button.

JMP places the re-expressed data in the new column.

Comments

The log and square root re-expressions are found in the **Transcendental** menu of functions in the formula dialog.

MINITAB

From the **Stat** menu, choose **Regression**.

- From the Regression submenu, select **Regression**.
- In the Regression dialog, enter the response variable name in the "Response" box and the predictor variable name in the "Predictor" box.
- To specify saved results, in the Regression dialog, click **Storage**.
- Check "Residuals" and "Fits." Click **OK**.
- To specify displays, in the Regression dialog, click **Graphs**.
- Under "Residual Plots," select "Individual plots" and check "Residuals versus fits." Click **OK**.
- Now back in the Regression dialog, click **OK**. Minitab computes the regression and the requested saved values and graphs.

To re-express a variable in MINITAB,

- Choose **Calculator** from the **Calc** menu.
- In the Calculator dialog, specify a name for the new re-expressed variable.
- Use the **Functions List**, the calculator buttons, and the **Variables list** box to build the expression. Click **OK**.

(continued)

- From the **Analyze** menu, choose **Regression**.
- From the Regression submenu, choose **Linear**. After assigning variables to their roles in the regression, click the "**Plots . . .**" button.
- In the Plots dialog, you can specify a Normal probability plot of residuals and scatterplots of various versions of standardized residuals and predicted values.

To re-express a variable in SPSS, choose **Compute** from the **Transform** menu.

- Enter a name in the Target Variable field.
- Use the calculator and Function List to build the expression.
- Move a variable to be re-expressed from the source list to the Numeric Expression field. Click the **OK** button.

Comments

A plot of ***ZRESID** against ***PRED** will look most like the residual plots we've discussed. SPSS standardizes the residuals by dividing by their standard deviation. (There's no need to subtract their mean; it must be zero.) The standardization doesn't affect the scatterplot.

Brief CASE

Gross Domestic Product

The Gross Domestic Product (GDP) per capita is a widely used measure of a country's (or state's) economy. It is defined as the total market value of all goods and services produced within a country (or state) in a specified period of time. The most common computation of GDP includes five items: consumption, gross investment, government spending, exports, and imports (which negatively impact the total). The Census Bureau reports the GDP for each state in the United States quarterly. The government also reports annual personal income totals (seasonally adjusted in $millions) by state and each state's population. Let's examine how personal income is related to GDP at the state level. Use the data in the file **GDP** to investigate the relationship between GDP and personal income.

Find a model to predict *Personal Income* from *GDP*. Write a short report detailing what you find. Be sure to include appropriate plots, look for influential points, and consider transforming either or both variables.

Repeat the analysis after dividing both variables by state population in 2005 to create per capita versions of the variables. Discuss which regression you think best helps to describe how personal income and GDP are related. Be sure to examine the residuals and discuss the regression assumptions.

Energy Sources

Renewable sources of energy are of growing importance in the economy. The U.S. government (www.stat-usa.gov) reports the amount of renewable energy generated (in thousands of kilowatt-hours) in each of the states for each of several renewable sources. Some states do not have reports, so there are missing values in the data. The data for 2004 are in the file **Alternative_Energy**.

Consider the relationship of hydroelectric power with energy from wind. Find a model for this relationship. Be sure to deal with any extraordinary or influential points. You may want to transform one or both variables. Discuss your residual analysis and the regression assumptions.

Now graph the relationship between (re-expressed) hydroelectric power and wind-generated power. Locate subgroups of states within this plot and discuss how they differ. Do you think a single model for the relationship of these renewable sources is appropriate? Summarize your conclusions in a report.

Exercises

SECTION 17.1

1. An analysis of spending by a sample of credit card bank cardholders shows that spending by cardholders in January (*Jan*) is related to their spending in December (*Dec*):

The assumptions and conditions of the linear regression seemed to be satisfied and an analyst was about to predict January spending using the model shown above:

$$\widehat{Jan} = \$612.07 + 0.403\ Dec$$

Another analyst worried that different types of cardholders might behave differently. She examined the spending patterns of the cardholders and placed them into five market *Segments*. When she plotted the data using different colors and symbols for the five different segments, she found the following:

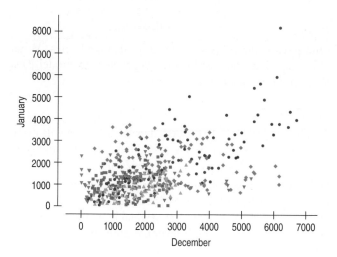

Look at this plot carefully and discuss why she might be worried about the predictions from the model $\widehat{Jan} = \$612.07 + 0.403\ Dec$.

2. A concert production company examined its records. The manager made the following scatterplot. The company places concerts in two venues, a smaller, more intimate theater (plotted with blue circles) and a larger auditorium-style venue.

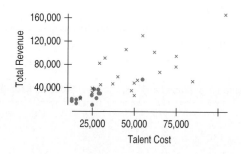

a) Describe the relationship between *Talent Cost* and *Total Revenue*. (Remember: direction, form, strength, outliers.)
b) How are the results for the two venues similar?
c) How are they different?

3. The analyst in Exercise 1 tried fitting the regression line to each market segment separately and found the following:

What does this say about her concern in Exercise 1? Was she justified in worrying that the overall model $\widehat{Jan} = \$612.07 + 0.403\ Dec$ might not accurately summarize the relationship? Explain briefly.

4. The concert production company of Exercise 2 made a second scatterplot, this time relating *Total Revenue* to *Ticket Sales*.

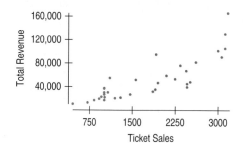

a) Describe the relationship between *Ticket Sales* and *Total Revenue*.
b) How are the results for the two venues similar?
c) How are they different?

SECTION 17.2

5. Here are the data from the small bookstore we saw in Chapter 16 Exercise 13.

Number of Sales People Working	Sales (in $1000)
2	10
3	11
7	13
9	14
10	18
10	20
12	20
15	22
16	22
20	26
$\bar{x} = 10.4$	$\bar{y} = 17.6$
$SD(x) = 5.64$	$SD(y) = 5.34$

The regression line is:

$$\widehat{Sales} = 8.10 + 0.913 \; Number\ of\ Sales\ People\ Working$$

and we can assume that the assumptions and conditions for regression are met. Calculations with technology find that

$$s_e = 1.477.$$

a) Find the predicted sales on a day with 500 employees working.
b) Find a 95% prediction interval for the sales on a day with 500 employees working.
c) Are these predictions likely to be useful? Explain.

6. Here are prices for the external disk drives we saw in Chapter 16 Exercise 14:

Capacity (in GB)	Price (in $)
80	29.95
120	35.00
250	49.95
320	69.95
1000	99.00
2000	205.00
4000	449.00
$\bar{x} = 1110.00$	$\bar{y} = 133.98$
$SD(x) = 1446.89$	$SD(y) = 151.26$

The least squares line is $\widehat{Price} = 18.62 + 0.104 \; Capacity$.

The assumptions and conditions for regression are met.

$$SE(b_1) = 0.0051$$

a) Disk drives keep growing in capacity. Some tech experts now talk about *Petabyte* (PB = 1000 TB = 1,000,000 GB) drives. What does this model predict that a Petabyte-capacity drive will cost?
b) Find a 95% prediction interval for the price of a 1 PB drive.
c) Are these predictions likely to be useful? Explain.

7. A regression of *Total Revenue* on *Ticket Sales* by the concert production company of Exercises 2 and 4 finds the model

$$\widehat{Revenue} = -14{,}228 + 36.87 \; TicketSales$$

a) Management is considering adding a stadium-style venue that would seat 10,000. What does this model predict that revenue would be if the new venue were to sell out?
b) Why would it be a poor business decision to assume that this model accurately predicts revenue for this situation?

8. The production company of Exercise 7 offers advanced sales to "Frequent Buyers" through its Web site. Here's a relevant scatterplot:

One performer refused to permit advanced sales. What effect has that point had on the regression to model *Total Revenue* from *Advanced Sales*?

SECTION 17.3

9. The bookstore in Exercise 5 decides to have a gala event in an attempt to drum up business. They hire 100 employees for the day and bring in a total of $42,000.

a) Find the regression line predicting *Sales* from *Number of people working* with the new point added.
b) What has changed from the original regression equation?
c) Is the new point a high leverage point or an influential point?
d) Does the new point have a large residual? Explain.

10. The data for hard drives in Exercise 6 omitted a 200 GB drive that sold for $299.00.

a) Find the regression line predicting *Price* from *Capacity* with this hard drive added.
b) What has changed from the original regression equation?

c) Is the new point a high leverage point or an influential point?

d) Does the new point have a large residual? Explain.

SECTION 17.5

11. A beverage company specializing in sales of champagne reports four years of quarterly sales as follows (in millions of $):

Quarter	Sales ($M)
1	12
2	41
3	15
4	48
5	25
6	55
7	23
8	69
9	51
10	80
11	54
12	87
13	64
14	94
15	62
16	108

The regression equation is *Predicted Sales* = 14.15 + 4.87 *Quarter*.

a) Find the residuals.

b) Plot the residuals against *Quarter*. Comment.

c) Compute the Durbin-Watson statistic.

d) At $\alpha = 0.05$, what are the values of d_L and d_U?

e) Is there evidence of positive autocorrelation? Explain.

f) Is there evidence of negative autocorrelation? Explain.

12. A company fits a regression to predict monthly *Orders* over a period of 48 months. The Durbin-Watson statistic on the residuals is 0.875.

a) At $\alpha = 0.01$, what are the values of d_L and d_U?

b) Is there evidence of positive autocorrelation? Explain.

c) Is there evidence of negative autocorrelation? Explain.

13. The manager of the concert production company considered in earlier exercises considers the regression of *Total Revenue* on *Ticket Sales* (see Exercise 4) and computes the Durbin-Watson statistic, obtaining a value of 0.51.

a) Consult Table D in Appendix D using $k = 1$ and $n = 45$ and complete the test at $\alpha = 0.05$.

b) Specify what this statistic tests and what the test says about these data.

14. The regression of *Total Revenue* on *Total Expenses* for the concerts of Exercises 13 gives the following model:

```
Dependent variable is: Total Revenue
R squared = 56.4%   R squared (adjusted) = 55.0%
s = 24269 with 34 − 2 = 32 degrees of freedom
```

Variable	Coefficient	SE(Coeff)	t-ratio	P-value
Intercept	−3688.54	9453	−0.390	0.6990
Total Expenses	0.731250	0.1137	6.43	≤0.0001

a) The Durbin-Watson statistic for this analysis is 0.73. Consult Table D in Appendix D and complete the test at $\alpha = 0.05$.

b) What do you conclude from this test?

SECTION 17.6

15. A scatterplot of *Salary* against *Years Experience* for some employees, and the scatterplot of residuals against predicted *Salary* from the regression line are shown in the figures. On the basis of these plots, would you recommend a re-expression of either *Salary* or *Years Experience*? Explain.

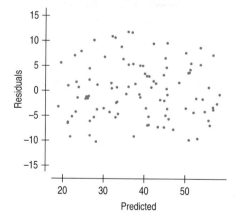

16. A study of homes looking at the relationship between *Age* of a home and *Price* produced the following scatterplot. A regression was fit to the data as shown below.

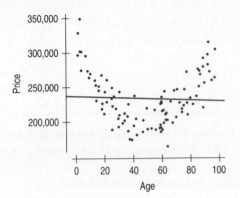

On the basis of this plot, would you advise using this regression? Explain.

17. A small company has developed an improved process for making solar panels. The company needs to set its prices and wants those prices to reflect the efficiencies of producing larger batches. The data show the following:

#Units	Cost per Unit
10	7.389
20	6.049
30	4.953
40	4.055
50	3.320
60	2.718
70	2.459
80	2.225
90	2.013
100	1.822
120	1.648
140	1.491
160	1.349
180	1.221
200	1.105

A scatterplot of *Cost per Unit* vs. *#Units* looks like this:

What should be done to make the relationship more nearly linear?

18. One possible model for the manufacturing process of Exercise 17 is the following:

```
Dependent variable is: Log(Cost per unit)
R squared = 90.1%   R squared (adjusted) = 89.3%
s = 0.0841 with 15 − 2 = 13 degrees of freedom
```

Variable	Coefficient	SE(Coeff)	t-ratio	P-value
Intercept	0.7618	0.0405	18.8	≤0.0001
#Units	−0.0041	0.0004	−10.9	≤0.0001

Using this model, predict the cost per unit in a batch of 300.

SECTION 17.7

19. A quickly growing company shows the following scatterplot of customers vs. time (in months).

a) What re-expression might you suggest for the number of customers?
b) What power in the ladder of powers does that correspond to?

20. For the regression in Exercise 19:

A student is tries taking the reciprocal of customers and produces the plot shown below:

a) Would you recommend this transformation? Why or why not?
b) What would you suggest?

CHAPTER EXERCISES

T 21. Marriage age 2007. Weddings are one of the fastest growing businesses; about $40 billion is spent on weddings in the United States each year. But demographics may be changing, and this could affect wedding retailers' marketing plans. Is there evidence that the age at which women get married has changed over the past 100 years? The scatterplot shows the trend in age at first marriage for American women. (www.census.gov)

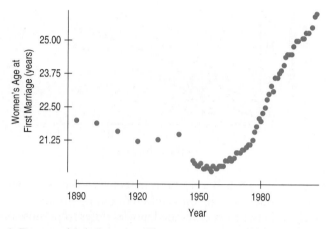

a) Do you think there is a clear pattern? Describe the trend.
b) Is the association strong?
c) Is the correlation high? Explain.
d) Do you think a linear model is appropriate for these data? Explain.

22. Smoking 2004. Even with campaigns to reduce smoking, Americans still consume more than four packs of cigarettes per month per adult. (ssdc.ucsd.edu/tobacco/sales/) The Centers for Disease Control and Prevention track cigarette smoking in the United States. How has the percentage of people who smoke changed since the danger became clear during the last half of the 20th century? The scatterplot shows percentages of smokers among men 18–24 years of age, as estimated by surveys, from 1965 through 2004. (www.cdc.gov/nchs/)

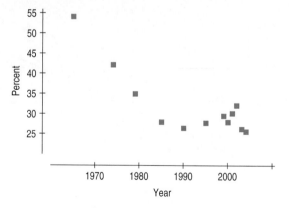

a) Do you think there is a clear pattern? Describe the trend.
b) Is the association strong?
c) Is a linear model appropriate for these data? Explain.

23. Human Development Index. The United Nations Development Programme (UNDP) collects data in the developing world to help countries solve global and national development challenges. In the UNDP annual Human Development Report, you can find data on over 100 variables for each of 177 countries worldwide. One summary measure used by the agency is the Human Development Index (HDI), which attempts to summarize in a single number the progress in health, education, and economics of a country. In 2006, the HDI was as high as 0.965 for Norway and as low as 0.331 for Niger. The gross domestic product per capita (GDPPC), by contrast, is often used to summarize the *overall* economic strength of a country. Is the HDI related to the GDPPC? Here is a scatterplot of *HDI* against *GDPPC*.

a) Explain why fitting a linear model to these data would be misleading.
b) If you fit a linear model to the data, what do you think a scatterplot of residuals versus predicted *HDI* will look like?
c) There is an outlier (Luxembourg) with a *GDPPC* of around $70,000. Will setting this point aside improve the model substantially? Explain.

24. HDI, part 2. The United Nations Development Programme (UNDP) uses the Human Development Index (HDI) in an attempt to summarize in one number the progress in health, education, and economics of a country. The number of cell phone subscribers per 1000 people is positively associated with economic progress in a country. Can the number of cell phone subscribers be used to predict the HDI? Here is a scatterplot of *HDI* against cell phone subscribers:

a) Explain why fitting a linear model to these data might be misleading.

b) If you fit a linear model to the data, what do you think a scatterplot of residuals versus predicted *HDI* will look like?

25. Good model? In justifying his choice of a model, a consultant says "I know this is the correct model because $R^2 = 99.4\%$."

a) Is this reasoning correct? Explain.

b) Does this model allow the consultant to make accurate predictions? Explain.

26. Bad model? An intern who has created a linear model is disappointed to find that her R^2 value is a very low 13%.

a) Does this mean that a linear model is not appropriate? Explain.

b) Does this model allow the intern to make accurate predictions? Explain.

27. Movie dramas. Here's a scatterplot of the production budgets (in millions of dollars) vs. the running time (in minutes) for major release movies in 2005. Dramas are plotted in red and all other genres are plotted in blue. A separate least squares regression line has been fitted to each group. For the following questions, just examine the plot:

a) What are the units for the slopes of these lines?

b) In what way are dramas and other movies similar with respect to this relationship?

c) In what way are dramas different from other genres of movies with respect to this relationship?

28. Movie ratings. Does the cost of making a movie depend on its audience? Here's a scatterplot of the same data we examined in Exercise 27. Movies with an R rating are colored blue, those with a PG-13 rating are purple, and those with a PG rating are green. Regression lines have been

found for each group. (The red points are G-rated, but there were too few to fit a line reliably.)

a) In what ways is the relationship between run times and budgets similar for the three different ratings groups?

b) How do the costs of R-rated movies differ from those of PG-13 and PG rated movies? Discuss both the slopes and the intercepts.

c) The film *King Kong*, with a run time of 187 minutes, is the purple point sitting at the lower right. If it were omitted from this analysis, how might that change your conclusions about PG-13 movies?

29. Oakland passengers. Much attention has been paid to the challenges faced by the airline industry. Patterns in customer demand are an important variable to watch. The scatterplot below, created in Excel, shows the number of passengers departing from Oakland (CA) airport month by month since the start of 1997. Time is shown as years since 1990, with fractional years used to represent each month. (Thus, June of 1997 is 7.5—halfway through the 7th year after 1990.) www.oaklandairport.com

Here's a regression and the residuals, created in Excel, plotted against *Years since 1990*:

SUMMARY OUTPUT	
Dependent Variable is: Passengers	

Regression Statistics	
R Square	0.711
Standard Error	104330

	Coefficients
Intercept	282584
Years since 1990	59704.4

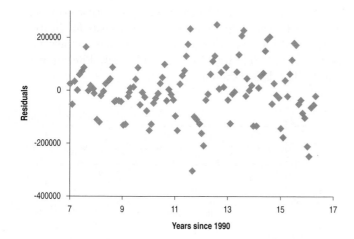

a) Interpret the slope and intercept of the regression model.
b) What does the value of R^2 say about how successful the model is?
c) Interpret s_e in this context.
d) Compute the Durbin-Watson statistic and comment.
e) Would you use this model to predict the numbers of passengers in 2010 (*YearsSince1990* = 20)? Explain.
f) There's a point near the middle of this time span with a large negative residual. Can you explain this outlier?

30. Tracking hurricanes 2009. Like many businesses, The National Hurricane Center also participates in a program to improve the quality of data and predictions by government agencies. They report their errors in predicting the path of hurricanes. The following scatterplot shows the trend in 48-hour tracking errors since 1970. (www.nhc.noaa.gov)

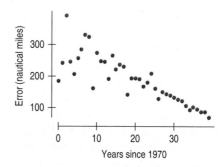

```
Dependent variable is: Error
R-squared = 71.0%  s = 40.68

Variable      Coeff
Intercept     293.917
Years-1970    −5.37882
```

a) Interpret the slope and intercept of the regression model.
b) Interpret s_e in this context.
c) If the Center had a goal of achieving an average tracking error of 65 nautical miles by 2015, will they make it? Explain.
d) Use technology to compute the Durbin-Watson statistic and comment.
e) What cautions would you state about your conclusion?

31. Unusual points. Each of the four scatterplots a–d that follow shows a cluster of points and one "stray" point. For each, answer questions 1–4:

1) In what way is the point unusual? Does it have high leverage, a large residual, or both?
2) Do you think that point is an influential point?
3) If that point were removed from the data, would the correlation become stronger or weaker? Explain.
4) If that point were removed from the data, would the slope of the regression line increase, decrease, or remain the same? Explain.

32. More unusual points. Each of the following scatterplots a–d shows a cluster of points and one "stray" point. For each, answer questions 1–4:

1) In what way is the point unusual? Does it have high leverage, a large residual, or both?
2) Do you think that point is an influential point?
3) If that point were removed from the data, would the correlation become stronger or weaker? Explain.
4) If that point were removed from the data, would the slope of the regression line increase, decrease, or remain the same? Explain.

33. The extra point. The scatterplot shows five blue data points at the left. Not surprisingly, the correlation for these points is $r = 0$. Suppose *one* additional data point is added at one of the five positions suggested below in green. Match each point (a–e) with the correct new correlation from the list given.

1) −0.90
2) −0.40
3) 0.00
4) 0.05
5) 0.75

34. The extra point, part 2. The original five points in Exercise 33 produce a regression line with slope 0. Match each of the green points (a–e) with the slope of the line after that one point is added:

1) −0.45
2) −0.30
3) 0.00
4) 0.05
5) 0.85

35. What's the cause? A researcher gathering data for a pharmaceutical firm measures blood pressure and the percentage of body fat for several adult males and finds a strong positive association. Describe three different possible cause-and-effect relationships that might be present.

36. What's the effect? Published reports about violence in computer games have become a concern to developers and distributors of these games. One firm commissioned a study of violent behavior in elementary-school children. The researcher asked the children's parents how much time each child spent playing computer games and had their teachers rate each child's level of aggressiveness when playing with other children. The researcher found a moderately strong positive correlation between computer game time and aggressiveness score. But does this mean that playing computer games increases aggression in children? Describe three different possible cause-and-effect explanations for this relationship.

37. Heating cost. Small businesses must track every expense. A flower shop owner tracked her costs for heating and related it to the average daily Fahrenheit temperature, finding the model $\widehat{Cost} = 133 - 2.13\,Temp$. The residuals plot for her data is shown.

a) Interpret the slope of the line in this context.
b) Interpret the *y*-intercept of the line in this context.
c) During months when the temperature stays around freezing, would you expect cost predictions based on this model to be accurate, too low, or too high? Explain.
d) What heating cost does the model predict for a month that averages 10°?
e) During one of the months on which the model was based, the temperature did average 10°. What were the actual heating costs for that month?
f) Do you think the home owner should use this model? Explain.
g) Would this model be more successful if the temperature were expressed in degrees Celsius? Explain.

38. Fuel economy. How does the speed at which a car drives affect fuel economy? Owners of a taxi fleet, watching their bottom line sink beneath fuel costs, hired a research firm to tell them the optimal speed for their taxis to drive. Researchers drove a compact car for 200 miles at speeds ranging from 35 to 75 miles per hour. From their data, they created the model $\widehat{Fuel\ Efficiency} = 32 - 0.1\ Speed$ and created this residual plot:

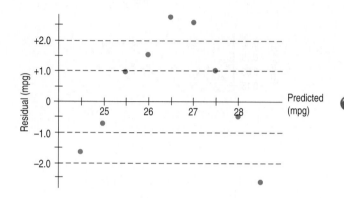

a) Interpret the slope of this line in context.
b) Explain why it's silly to attach any meaning to the *y*-intercept.
c) When this model predicts high *Fuel Efficiency*, what can you say about those predictions?
d) What *Fuel Efficiency* does the model predict when the car is driven at 50 mph?
e) What was the actual *Fuel Efficiency* when the car was driven at 45 mph?
f) Do you think there appears to be a strong association between *Speed* and *Fuel Efficiency*? Explain.
g) Do you think this is the appropriate model for that association? Explain.

T 39. Interest rates. Here's a plot showing the federal rate on 3-month Treasury bills from 1950 to 1980, and a regression model fit to the relationship between the *Rate* (in %) and *Years since 1950*. (www.gpoaccess.gov/eop/)

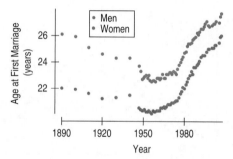

```
Dependent variable is: Rate
R-squared = 77.4%  s = 1.239

Variable        Coeff

Intercept       0.640282
Year-1950       0.247637
```

a) What is the correlation between *Rate* and *Year*?
b) Interpret the slope and intercept.
c) What does this model predict for the interest rate in the year 2000?
d) Compute the Durbin-Watson statistic and comment.
e) Would you expect this prediction to have been accurate? Explain.

T 40. Ages of couples 2007. In Exercise 21 we looked at the age at which women married as one of the variables considered by those selling wedding services. Another variable of concern is the *difference* in age of the two partners. The graph shows the ages of both men and women at first marriage. (www.census.gov)

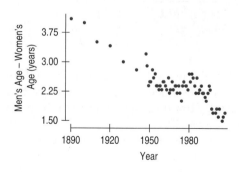

Clearly, the pattern for men is similar to the pattern for women. But are the two lines getting closer together?

Here is a timeplot showing the *difference* in average age (men's age – women's age) at first marriage, the regression analysis, and the associated residuals plot.

```
Dependent variable is: Age Difference
R-squared = 77.4% s = 0.2334

Variable      Coeff
Constant      36.2639
Year          -0.017181
```

a) What is the correlation between *Age Difference* and *Year*?
b) Interpret the slope of this line.
c) Predict the average age difference in 2015.
d) Compute the Durbin-Watson statistic and comment.
e) Describe reasons why you might not place much faith in that prediction.

41. Interest rates, part 2. In Exercise 39 you investigated the federal rate on 3-month Treasury bills between 1950 and 1980. The scatterplot below shows that the trend changed dramatically after 1980, so we've built a new regression model that includes only the data since 1980 (from $x = 30$ and on in the plot below).

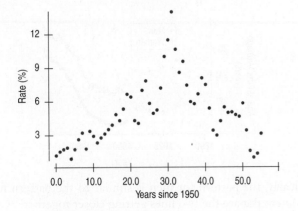

```
Dependent variable is: Rate
R-squared = 74.5% s = 1.630

Variable      Coeff
Intercept     21.0688
Year-50       -0.356578
```

a) How does this model compare to the one in Exercise 39?
b) What does this model estimate the interest rate to have been in 2000? How does this compare to the rate you predicted in Exercise 39?
c) Do you trust this newer predicted value? Explain.
d) Given these two models, what would you predict the interest rate on 3-month Treasury bills will be in 2020?

42. Ages of couples 2007, part 2. Has the trend of decreasing difference in age at first marriage seen in Exercise 40 gotten stronger recently? Here are the scatterplot and residual plot for the data from 1975 through 2007, along with a regression for just those years.

```
Dependent variable is: Men - Women
R-squared = 74.1%  s = 0.1824

Variable      Coeff
Intercept     65.3103
Year          -0.031725
```

a) Why is R^2 higher for the first model (in Exercise 40)?
b) Is this linear model appropriate for the post-1975 data? Explain.
c) What does the slope say about marriage ages since 1975?
d) Explain why it's not safe to interpret the *y*-intercept.

43. Colorblind. Although some women are colorblind, this condition is found primarily in men. An advertisement for socks marked so they were easy for someone who was colorblind to match started out "There's a strong correlation between sex and colorblindness." Explain in statistics terms why this isn't a correct statement (whether or not it might be a good ad).

44. New homes. A real estate agent collects data to develop a model that will use the *Size* of a new home (in square feet) to predict its *Sale Price* (in thousands of dollars). Which of these is most likely to be the slope of the regression line: 0.008, 0.08, 0.8, or 8? Explain.

45. Residuals. Suppose you have fit a linear model to some data and now take a look at the residuals. For each of the following possible residuals plots, tell whether you would try a re-expression and, if so, why.

46. Residuals, part 2. Suppose you have fit a linear model to some data and now take a look at the residuals. For each of the following possible residuals plots, tell whether you would try a re-expression and, if so, why.

a)   b)   c)

🔵 **47.** Oakland passengers, part 2. In Exercise 29, we created a linear model describing the trend in the number of passengers departing from Oakland (CA) airport each month since the start of 1997. Here's the residual plot, but with lines added to show the order of the values in time:

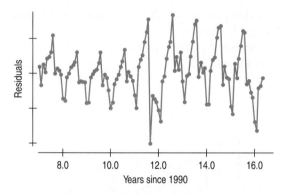

a) Can you account for the pattern shown here?
b) Would a re-expression help us deal with this pattern? Explain.

🔵 **48.** Home Depot sales. The home retail industry has experienced relatively consistent annual growth over the past few decades. Here is a scatterplot of the *Net Sales* ($B) of *The Home Depot* from 1995 through 2004, along with a regression and a time series plot of the residuals.

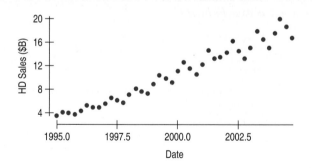

```
Dependent variable is: HDSales
R-squared = 95.5%
s = 1.044 with 40 − 2 = 38 degrees of freedom
```

Variable	Coeff	SE(Coeff)	t-ratio	P-value
Constant	−3234.87	114.4	−28.3	≤0.0001
Date	1.62283	0.0572	28.4	≤0.0001

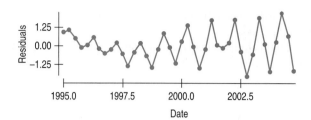

a) What does the R^2 value in the regression mean?
b) What features of the residuals should be noted with regards to this regression?
c) What features of the residuals might be dealt with by a re-expression? Which ones would not be helped by a re-expression?
d) Can you use the regression model to help in your understanding of the growth of this market?

49. Models. For each of the models listed below, predict y when $x = 2$.

a) $\hat{y} = 1.2 + 0.8x$
b) $\ln \hat{y} = 1.2 + 0.8x$
c) $\sqrt{\hat{y}} = 1.2 + 0.8x$
d) $\dfrac{1}{\hat{y}} = 1.2 + 0.8x$
e) $\hat{y} = 1.2x^{0.8}$

50. More models. For each of the models listed below, predict y when $x = 2$.

a) $\hat{y} = 1.2 + 0.8 \log x$
b) $\log \hat{y} = 1.2 + 0.8x$
c) $\hat{y} = 1.2 + 0.8\sqrt{x}$
d) $\hat{y} = 1.2(0.8^x)$
e) $\hat{y} = 0.8x^2 + 1.2x + 1$

51. Models, again. Find the predicted value of y, using each model for $x = 10$.

a) $\hat{y} = 2 + 0.8 \ln x$
b) $\log \hat{y} = 5 - 0.23x$
c) $\dfrac{1}{\sqrt{\hat{y}}} = 17.1 - 1.66x$

52. Models, last time. Find the predicted value of y, using each model when $x = 4$.

a) $\hat{y} = 10 + \sqrt{x}$
b) $\dfrac{1}{y} = 14.5 - 3.45x$
c) $\sqrt{y} = 3.0 + 0.5x$

🔵 **53.** Lobster industry 2008. According to the Maine Department of Marine Resources, in 2008 more than 69,700,687 pounds of lobster were landed in Maine—a catch worth more than $244,195,000. The lobster fishing industry is carefully controlled and licensed, and facts about it have been recorded for more than a century, so it is an important industry that we can examine in detail. We'll look at annual data (available at www.maine.gov/dmr) from 1950 through 2008.

The value of the annual lobster catch has grown. Here's a scatterplot of the value in millions of dollars over time:

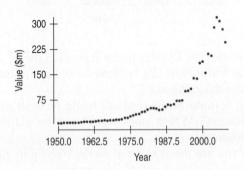

a) Which regression assumptions and conditions appear to be violated according to this plot?

Here's a scatterplot of the *log* of the value:

b) Discuss the same assumptions as in part a. Does taking logs make these data suitable for regression?

After performing a regression on the log values, we obtain the following plot of residuals:

c) Discuss what this plot shows. Would a different transformation be likely to do better than the *log*? Explain.

T 54. Lobster traps. Lobster are caught in traps, which are baited and left in the open ocean. Licenses to fish for lobster are limited, there is a small additional fee for each

trap in use, and there are limits on the numbers of traps that can be placed in each of seven fishing zones. But those limits have changed over time. Here's a scatterplot of the number of traps per licensed lobster fisher over time:

a) Does this plot satisfy the regression assumptions and conditions? Explain.

A regression of *Traps/Fisher* vs. *Year* yields the following plot of residuals:

b) What can you see in the plots of residuals that may not have been clear in the original scatterplot of the data?

T 55. Lobster value. Here's a regression model relating the *logValue* of the annual Maine lobster catch to the number of licensed lobster *Fishers*:

```
Dependent variable is: LogValue
R-squared = 18.9%   R-squared (adjusted) = 17.3%
s = 0.4306 with 56 − 2 = 54 degrees of freedom
```

Variable	Coeff	SE(Coeff)	t-ratio	P-value
Intercept	6.43619	0.3138	20.5	≤0.0001
Fishers	1.56021e-4	0.0000	3.54	0.0008

a) The number of licensed lobster fishers has fluctuated over the years between roughly 5000 and 10,000. Recently the number has been just over 7000. But licenses are in demand (and tightly restricted). What does this model predict the value of the catch would be in a year if there were 10,000 licensed fishers? (Take care to interpret the coefficient correctly and to compute the inverse of the log transformation.)

b) Interpret the slope coefficient. Do more fishers cause a higher valued harvest? Suggest alternative explanations.

T 56. Lobster price. Of course, what matters most to the individual entrepreneur—the licensed commercial lobster fisher—is the price of lobster. Here's an analysis relating that price ($/lb) to the number of traps (millions):

```
Dependent variable is: Price/lb
R-squared = 93.7%   R-squared (adjusted) = 93.6%
s = 0.2890 with 56 − 2 = 54 degrees of freedom
```

Variable	Coeff	SE(Coeff)	t-ratio	P-value
Intercept	−0.276454	0.0812	−3.40	0.0013
Traps(M)	1.25210	0.0441	28.4	≤0.0001

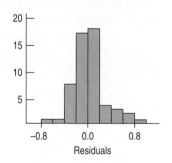

a) Are the assumptions and conditions for regression inference satisfied?
b) What does the coefficient of *Traps* mean in this model? Does it predict that licensing more traps would cause an increase in the price of lobster? Suggest some alternative explanations.

57. GDP. The scatterplot shows the gross domestic product (GDP) of the United States in billions of dollars plotted against years since 1950.

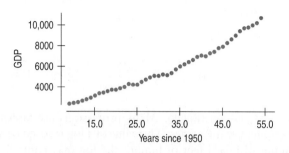

A linear model fit to the relationship looks like this:

```
Dependent variable is: GDP
R-squared = 97.2%  s = 406.6
```

Variable	Coeff
Intercept	240.171
Year−1950	177.889

a) Does the value 97.1% suggest that this is a good model? Explain.
b) Here's a scatterplot of the residuals. Now do you think this is a good model for these data? Explain?

58. Better GDP model? Consider again the post-1950 trend in U.S. GDP we examined in Exercise 57. Here are a regression and residual plot when we use the log of GDP in the model. Is this a better model for GDP? Explain.

```
Dependent variable is: LogGDP
R-squared = 99.4% s = 0.0150
```

Variable	Coeff
Intercept	3.29092
Year−1950	0.013881

59. Logs (not logarithms). Many professions use tables to determine key quantities. The value of a log is based on the number of board feet of lumber the log may contain. (A board foot is the equivalent of a piece of wood 1 inch thick, 12 inches wide, and 1 foot long. For example, a 2″ × 4″ piece that is 12 feet long contains 8 board feet.) To estimate the amount of lumber in a log, buyers measure the diameter inside the bark at the smaller end. Then they look in a table based on the Doyle Log Scale. The table below shows the estimates for logs 16 feet long.

Diameter of Log	8″	12″	16″	20″	24″	28″
Board Feet	16	64	144	256	400	576

a) What transformation of *Board Feet* makes this relationship linear?
b) How much lumber would you estimate that a log 10 inches in diameter contains?
c) What does this model suggest about logs 36 inches in diameter?

T 60. Life expectancy. Life insurance rates are based on life expectancy values compiled for large demographic groups. But with improvements in medical care and nutrition, life expectancies have been changing. Here is a table from the National Vital Statistics Report that gives the Life Expectancy for white males in the United States every decade during the last century (1 = 1900 to 1910, 2 = 1911 to 1920, etc.). Consider a linear model to predict future increases in life expectancy. Would re-expressing either variable make a better model?

Decade	1	2	3	4	5	6	7	8	9	10
Life exp.	48.6	54.4	59.7	62.1	66.5	67.4	68.0	70.7	72.7	74.9

T 61. OECD GDP. The Organization for Economic Cooperation and Development (OECD) is an organization comprised of thirty countries. To belong, a country must support the principles of representative democracy and a free market economy. How have these countries grown in the decade from 1988 to 1998–2000? Here are the GDP per capita for 24 of the OECD members (in year 2000 dollars). (http://www.sba.gov/advo/research/rs264tot.pdf)

Country	1988 GDP/Capita	1998–2000 GDP/Capita
Australia	18,558	23,713
Austria	25,626	31,192
Belgium	24,204	30,506
Canada	19,349	22,605
Denmark	31,517	38,136
Finland	25,682	31,246
France	24,663	29,744
Germany	27,196	32,256
Greece	10,606	13,181
Ireland	13,050	27,282
Italy	17,339	20,710
Japan	36,301	44,154
Korea	7038	12,844
Mexico	3024	3685
Netherlands	23,159	30,720
New Zealand	15,480	17,979
Norway	28,241	37,934
Portugal	8935	12,756
Spain	12,879	17,197
Sweden	26,634	30,873
Switzerland	43,375	46,330
Turkey	2457	2947
United Kingdom	17,676	22,153
United States	25,324	31,296

Make a model of 1998–2000 GDP/Capita in terms of 1988 GDP/Capita. Plot the residuals and discuss any concerns you may have.

T 62. Orange production. Orange growers know that the larger an orange the higher the price it will bring. But as the number of oranges on a tree increases, the fruit tends to be smaller. Here's a table of that relationship. Create a model for this relationship, and express any concerns you may have.

Number of Oranges/Tree	Average Weight/Fruit (lb)
50	0.60
100	0.58
150	0.56
200	0.55
250	0.53
300	0.52
350	0.50
400	0.49
450	0.48
500	0.46
600	0.44
700	0.42
800	0.40
900	0.38

63. Human Development Index, part 3. In Exercise 23 we saw that the United Nations Development Programme (UNDP) uses the Human Development Index (HDI) in an attempt to summarize the progress in health, education, and economics of a country with one number. The gross domestic product per capita (GDPPC) attempts to summarize the wealth produced by a country in one number. Here is a plot of *GDPPC* against *HDI* for 172 countries throughout the world:

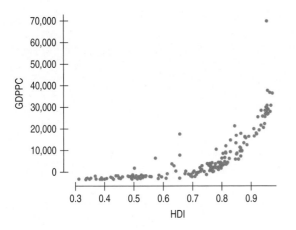

GDPPC is measured in dollars. Incomes and other economic measures tend to be highly right skewed. Taking logs often makes the distribution more unimodal and symmetric. Compare the histogram of *GDPPC* to the histogram of *log(GDPPC)*.

64. Human Development Index, part 4. In Exercise 63 we examined the relationship between *log(GDPPC)* and *HDI* for 172 countries. The number of cell phone subscribers (per 1000 people) is also positively associated with economic progress in a country. Here's a scatterplot of *CellPhones* (subscribers per 1000 people) against *HDI* for 154 countries:

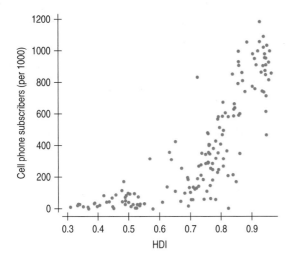

a) *CellPhones* is a count of subscribers (per 1000). What re-expression is often useful for counts? Examine the histogram of *CellPhones* and the histogram of *Cellphones* using the re-expression you suggested. Comment.
b) Use the re-expression in a) for the scatterplot against *HDI*. Comment.
c) Why might you be skeptical of using this relationship to predict the number of cell phone users based on the *HDI*?

65. Lobster fishers. How has the number of licensed lobster fishers changed? Here's a plot of the number of *Fishers vs. Year*:

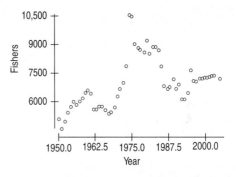

This plot isn't straight. Would a transformation help? If so, which one? If not, why not?

66. Lobster price. How has the price of a lobster changed? Here's a plot tracking the price of lobster per pound in constant year 2000 dollars.

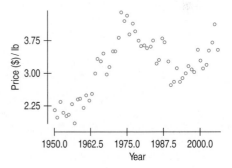

This plot is not straight. Would a transformation help? If so, which one? If not, why not?

Just Checking Answers

1 Not high-leverage, not influential, large residual
2 High-leverage, not influential, small residual
3 High-leverage, influential, not large residual
4 None
5 Logarithm

Multiple Regression

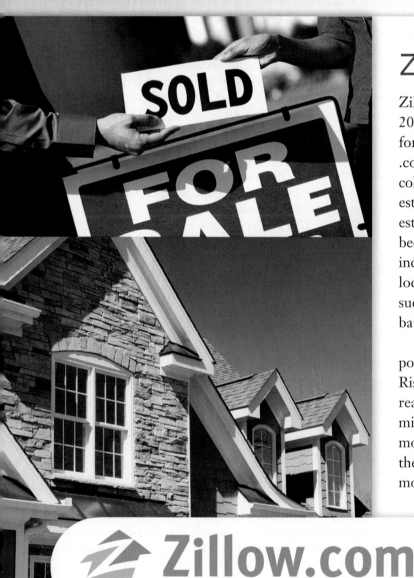

Zillow.com

Zillow.com is a real estate research site, founded in 2005 by Rich Barton and Lloyd Frink. Both are former Microsoft executives and founders of Expedia.com, the Internet-based travel agency. Zillow collects publicly available data and provides an estimate (called a Zestimate®) of a home's worth. The estimate is based on a model of the data that Zillow has been able to collect on a variety of predictor variables, including the past history of the home's sales, the location of the home, and characteristics of the house such as its size and number of bedrooms and bathrooms.

The site is enormously popular among both potential buyers and sellers of homes. According to Rismedia.com, Zillow is one of the most-visited U.S. real estate sites on the Web, with approximately 5 million unique users each month. These users include more than one-third of all mortgage professionals in the U.S.—or approximately 125,000—in any given month. Additionally, 90% of Zillow users are homeowners, and two-thirds are either buying and selling now, or plan to in the near future.

ow exactly does Zillow figure the worth of a house? According to the Zillow.com site, "We compute this figure by taking zillions of data points—much of this data is public—and entering them into a formula. This formula is built using what our statisticians call 'a proprietary algorithm'—big words for 'secret formula.' When our statisticians developed the model to determine home values, they explored how homes in certain areas were similar (i.e., number of bedrooms and baths, and a myriad of other details) and then looked at the relationships between actual sale prices and those home details." These relationships form a pattern, and they use that pattern to develop a model to estimate a market value for a home. In other words, the Zillow statisticians use a model, most likely a regression model, to predict home value from the characteristics of the house. We've seen how to predict a response variable based on a single predictor. That's been useful, but the types of business decisions we'll want to make are often too complex for simple regression.[1] In this chapter, we'll expand the power of the regression model to take into account many predictor variables into what's called a multiple regression model. With our understanding of simple regression as a base, getting to multiple regression isn't a big step, but it's an important and worthwhile one. Multiple regression is probably the most powerful and widely used statistical tool today.

As anyone who's ever looked at house prices knows, house prices depend on the local market. To control for that, we will restrict our attention to a single market. We have a random sample of 1057 home sales from the public records of sales in upstate New York, in the region around the city of Saratoga Springs. The first thing often mentioned in describing a house for sale is the number of bedrooms. Let's start with just one predictor variable. Can we use *Bedrooms* to predict home *Price*?

WHO	Houses
WHAT	Sale price (2002 dollars) and other facts about the houses
WHEN	2002–2003
WHERE	Upstate New York near Saratoga Springs
WHY	To understand what influences housing prices and how to predict them

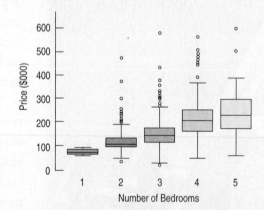

Figure 18.1   Side-by-side boxplots of *Price* against *Bedrooms* show that price increases, on average, with more bedrooms.

The number of *Bedrooms* is a quantitative variable, but it holds only a few values (from 1 to 5 in this data set). So a scatterplot may not be the best way to examine the relationship between *Bedrooms* and *Price*. In fact, at each value for *Bedrooms* there is a whole distribution of prices. Side-by-side boxplots of *Price* against *Bedrooms* (Figure 18.1) show a general increase in price with more bedrooms, and an approximately linear growth.

Figure 18.1 also shows a clearly increasing spread from left to right, violating the Equal Spread Condition, and that's a possible sign of trouble. For now, we'll proceed cautiously. We'll fit the regression model, but we will be cautious about using inference methods for the model. Later we'll add more variables to increase the power and usefulness of the model.

---

[1]When we need to note the difference, a regression with a single predictor is called a **simple regression**.

The output from a linear regression model of *Price* on *Bedrooms* shows:

SUMMARY OUTPUT				
Response Variable is Price				
Regression Statistics				
R Square	0.214			
Standard Error	68432.21			
Observations	1057			
	Coefficients	*Standard Error*	*t Stat*	*P-value*
Intercept	14349.48	9297.69	1.54	0.1230
Bedrooms	48218.91	2843.88	16.96	0.0000

Table 18.1 Linear regression in Excel of *Price* on *Bedrooms*.

Apparently, just knowing the number of bedrooms gives us some useful information about the sale price. The model tells us that, on average, we'd expect the price to increase by almost $50,000 for each additional bedroom in the house, as we can see from the slope value of $48,218.91:

$$\widehat{Price} = 14349.48 + 48218.91 \times Bedrooms.$$

Even though the model does tell us something, notice that the R^2 for this regression is only 21.4%. This means that the variation in the number of bedrooms accounts for about 21.4% of the variation in house prices. Perhaps some of the other facts about these houses can account for portions of the remaining variation.

The standard deviation of the residuals, $s = 68,432$, tells us that the model only does a modestly good job of accounting for the price of a home. Approximating with the 68–95–99.7 Rule, we'd guess that only about 68% of home prices predicted by this model would be within $68,432 of the actual price. That's not likely to be close enough to be useful for a home buyer.

18.1 The Multiple Regression Model

For simple regression, we wrote the predicted values in terms of one predictor variable:

$$\hat{y} = b_0 + b_1 x.$$

To include more predictors in the model, we simply write the same regression model with more predictor variables. The resulting **multiple regression** looks like this:

$$\hat{y} = b_0 + b_1 x_1 + b_2 x_2 + \ldots + b_k x_k$$

where b_0 is still the intercept and each b_k is the estimated coefficient of its corresponding predictor x_k. Although the model doesn't look much more complicated than a simple regression, it isn't practical to determine a multiple regression by hand. This is a job for a statistics program on a computer. Remember that for simple regression, we found the coefficients for the model using the least squares solution, the one whose coefficients made the sum of the squared residuals as small as possible. For multiple regression, a statistics package does the same thing and can find the coefficients of the least squares model easily.

If you know how to find the regression of *Price* on *Bedrooms* using a statistics package, you can probably just add another variable to the list of predictors in your program to compute a multiple regression. A multiple regression of *Price* on the two variables *Bedrooms* and *Living Area* generates a multiple regression table like this one.

```
Response variable: Price

R² = 57.8%
s = 50142.4 with 1057 - 3 = 1054 degrees of freedom

Variable        Coeff       SE(Coeff)   t-ratio   P-value
Intercept       20986.09    6816.3      3.08      0.0021
Bedrooms        -7483.10    2783.5      -2.69     0.0073
Living Area     93.84       3.11        30.18     ≤0.0001
```

Table 18.2   Multiple regression output for the linear model predicting *Price* from *Bedrooms* and *Living Area*.

You should recognize most of the numbers in this table, and most of them mean what you expect them to. The value of R^2 for a regression on two variables gives the fraction of the variability of *Price* accounted for by both predictor variables together. With *Bedrooms* alone predicting *Price*, the R^2 value was 21.4%, but this model accounts for 57.8% of the variability in *Price* and the standard deviation of the residuals is now $50,142.40. We shouldn't be surprised that the variability explained by the model has gone up. It was for this reason—the hope of accounting for some of that leftover variability—that we introduced a second predictor. We also shouldn't be surprised that the size of the house, as measured by *Living Area*, also contributes to a good prediction of house prices. Collecting the coefficients of the multiple regression of *Price* on *Bedrooms* and *Living Area* from Table 18.2, we can write the estimated regression as:

$$\widehat{Price} = 20,986.09 - 7,483.10 Bedrooms + 93.84 Living\ Area.$$

As before, we define the residuals as:

$$e = y - \hat{y}.$$

The standard deviation of the residuals is still denoted as s (or also sometimes as s_e as in simple regression—for the same reason—to distinguish it from the standard deviation of y). The degrees of freedom calculation comes right from our definition. The degrees of freedom is the number of observations ($n = 1057$) minus one for each coefficient estimated:

$$df = n - k - 1,$$

where k is the number of predictor variables and n is the number of cases. For this model, we subtract 3 (the two coefficients and the intercept). To find the standard deviation of the residuals, we use that number of degrees of freedom in the denominator:

$$s_e = \sqrt{\frac{\sum(y - \hat{y})^2}{n - k - 1}}.$$

For each predictor, the regression output shows a coefficient, its standard error, a *t*-ratio, and the corresponding P-value. As with simple regression, the *t*-ratio measures how many standard errors the coefficient is away from 0. Using a Student's *t*-model, we can find its P-value and use that to test the null hypothesis that the true value of the coefficient is 0.

What's different? With so much of the multiple regression looking just like simple regression, why devote an entire chapter to the subject?

There are several answers to this question. First, and most important, is that the meaning of the coefficients in the regression model has changed in a subtle, but important, way. Because that change is not obvious, multiple regression coefficients are often misinterpreted. We'll show some examples to explain this change in meaning.

Second, multiple regression is an extraordinarily versatile model, underlying many widely used statistics methods. A sound understanding of the multiple regression model will help you to understand these other applications as well.

Third, multiple regression offers you a first glimpse into statistical models that use more than two quantitative variables. The real world is complex. Simple models of the kind we've shown so far are a great start, but they're not detailed enough to be useful for understanding, predicting, and making business decisions in many real-world situations. Models that use several variables can be a big step toward realistic and useful modeling of complex phenomena and relationships.

For Example   The multiple regression model

A large clothing store has recently sent out a special catalog of Fall clothes, and Leiram, its marketing analyst, wants to find out which customers responded to it and which ones bought the most. She plans to conduct an RFM (Recency, Frequency, Monetary) analysis. The RFM method is based on the principle that 80% of your business comes from the best 20% of your customers, and states that the following attributes will be useful predictors of who the best customers will be:

- How *recently* the customer has purchased (Recency)
- How *frequently* the customer shops (Frequency)
- How *much* the customer spends (Monetary)

For each customer, Leiram has information for the past 5 years on: *Date of Last Purchase*, *Number of Purchases* and *Total Amount Spent*. In addition, she has demographic information including *Age*, *Marital Status*, *Sex*, *Income*, and *Number of Children*. She chooses a random sample of 149 customers who bought something from the catalog and who have purchased at least 3 times in the past 5 years (*Number of Purchases*). She wants to model how much they bought from the new catalog (*Respond Amount*).

Leiram fits the following multiple regression model to the response variable *Respond Amount*:

```
Response Variable: Respond Amount
R² = 91.48% Adjusted R² = 91.31%
s = 18.183 with 149 − 4 = 145 degrees of freedom
```

Variable	Coeff	SE(Coeff)	t-ratio	P-value
Intercept	111.01	6.459	17.187	<0.0001
Income	0.00091	0.00012	7.643	<0.0001
Total Amount Spent	0.154	0.00852	18.042	<0.0001
Number of Purchases	−14.81	0.716	−20.695	<0.0001

Questions: How much of the variation in *Respond Amount* is explained by this model? What does the term $s = 18.183$ mean? Which variables seem important in the model?

Answers: This model has an R^2 of 91.48% which means that the model has explained 91.48% of the variation in *Respond Amount* using the three predictors: *Income*, *Total Amount Spent*, and *Number of Purchases*. The term $s = 18.183$ means that the standard deviation of the residuals is about $18.18. Using the 68–95–99.7 Rule we know that most prediction errors will be no larger than about $36.36. All terms seem important in this model since all three have very large t-ratios and correspondingly small P-values.

18.2   Interpreting Multiple Regression Coefficients

It makes sense that both the number of bedrooms and the size of the living area would influence the price of a house. We'd expect both variables to have a positive effect on price—houses with more bedrooms typically sell for more money, as do larger houses. But look at the coefficient for *Bedrooms* in the multiple regression equation. It's negative: -7483.09. How can it be that the coefficient of *Bedrooms* in the multiple regression is negative? And not just slightly negative, its t-ratio is large enough for us to be quite confident that the true value is really negative. Yet from Table 18.1, we saw the coefficient was equally clearly positive when *Bedrooms* was the sole predictor in the model (see Figure 18.2).

Figure 18.2 The slope of *Bedrooms* is positive. For each additional bedroom, we would predict an additional $48,000 in the price of a house from the simple regression model of Table 18.1.

Figure 18.3 For the 96 houses with *Living Area* between 2500 and 3000 square feet, the slope of *Price* on *Bedrooms* is negative. For each additional bedroom, restricting data to homes of this size, we would predict that the house's *Price* was about $17,800 lower.

The explanation of this apparent paradox is that in a multiple regression, coefficients have a more subtle meaning. Each coefficient takes into account the other predictor(s) in the model.

Think about a group of houses of about the same size. For the *same size* living area, a house with more bedrooms is likely to have smaller rooms. That might actually make it *less* valuable. To see this in the data, let's look at a group of similarly sized homes from 2500 to 3000 square feet of living area, and examine the relationship between *Bedrooms* and *Price* just for houses in this size range (see Figure 18.3).

For houses with between 2500 and 3000 square feet of living area, it appears that homes with *fewer* bedrooms have a higher price, on average, than those with more bedrooms. When we think about houses in terms of *both* variables, we can see that this makes sense. A 2500-square-foot house with five bedrooms would have either relatively small, cramped bedrooms or not much common living space. The same size house with only three bedrooms could have larger, more

Just Checking

Body fat percentage is an important health indicator, but it is difficult to measure accurately. One way to do so is via dual energy X-ray absorpitometry (DEXA) at a cost of several hundred dollars per image. Insurance companies want to know if body fat percentage can be estimated from easier to measure characteristics such as *Height* for 250 adult men and *Weight*. A scatterplot of *Percent Body Fat* against *Height* for 250 adult men shows no pattern, and the correlation is −0.03 which is not statistically significant. A multiple regression using *Height (inches)*, *Age (years)*, and *Weight (pounds)* finds the following model:

	Coeff	SE(Coeff)	t-ratio	P-value
Intercept	57.27217	10.39897	5.507	<0.0001
Height	−1.27416	0.15801	−8.064	<0.0001
Weight	0.25366	0.01483	17.110	<0.0001
Age	0.13732	0.02806	4.895	<0.0001

$s = 5.382$ on 246 degrees of freedom
Multiple R-squared: 0.584,
F-statistic: 115.1 on 3 and 246 DF, P-value: <0.0001

1 Interpret the R^2 of this regression model.
2 Interpret the coefficient of *Age*.
3 How can the coefficient of *Height* have such a small P-value in the multiple regression when the correlation between *Height* and *Percent Body Fat* was not statistically distinguishable from zero?

attractive bedrooms and still have adequate common living space. What the coefficient of *Bedrooms* is saying in the multiple regression is that, after accounting for living area, houses with more bedrooms tend to sell for a *lower* price. In other words, what we saw by *restricting* our attention to homes of a certain size and seeing that additional bedrooms had a negative impact on price was generally true across all sizes. What seems confusing at first is that without taking *Living Area* into account, *Price* tends to go *up* with more bedrooms. But that's because *Living Area* and *Bedrooms* are also related. Multiple regression coefficients must always be interpreted in terms of the other predictors in the model. That can make their interpretation more subtle, more complex, and more challenging than when we had only one predictor. This is also what makes multiple regression so versatile and effective. The interpretations are more sophisticated and more appropriate.

There's a second common pitfall in interpreting coefficients. Be careful not to interpret the coefficients causally. For example, this analysis cannot tell a homeowner how much the price of his home will change if he combines two of his four bedrooms into a new master bedroom. And it can't be used to predict whether adding a 100-square-foot child's bedroom onto the house would increase or decrease its value. The model simply reports the relationship between the number of *Bedrooms* and *Living Area* and *Price* for existing houses. As always with regression, we should be careful not to assume causation between the predictor variables and the response.

For Example   Interpreting multiple regression coefficients

Question: In the regression model of *Respond Amount*, interpret the intercept and the regression coefficients of the three predictors.

Answer: The model says that from a base of $111.01 of spending, customers (who have purchased at least 3 times in the last 12 months), on average, spent $0.91 for every $1000 of *Income* (after accounting for *Number of Purchases* and *Total Amount Spent*), $0.154 for every dollar they have spent in the past 5 years (after accounting for *Income* and *Number of Purchases*), but $14.81 less for every additional purchase that they've made in the last 5 years (after accounting for *Income* and *Total Amount Spent*). It is important to note especially that the coefficient in *Number of Purchases* is negative, but only after accounting for both *Income* and *Total Amount Spent*.

18.3 Assumptions and Conditions for the Multiple Regression Model

We can write the multiple regression model like this, numbering the predictors arbitrarily (the order doesn't matter), writing betas for the model coefficients (which we will estimate from the data), and including the errors in the model:

$$y = \beta_0 + \beta_1 x_1 + \beta_2 x_2 + \ldots + \beta_k x_k + \varepsilon.$$

The assumptions and conditions for the multiple regression model are nearly the same as for simple regression, but with more variables in the model, we'll have to make a few changes, as described in the following sections.

Linearity Assumption

We are fitting a linear model.[2] For that to be the right kind of model for this analysis, we need to verify an underlying linear relationship. But now we're thinking about

---

[2] By *linear* we mean that each *x* appears simply multiplied by its coefficient and added to the model, and that no *x* appears in an exponent or some other more complicated function. That ensures that as we move along any *x*-variable, our prediction for y will change at a constant rate (given by the coefficient) if nothing else changes.

several predictors. To confirm that the assumption is reasonable, we'll check the Linearity Condition for *each* of the predictors.

Linearity Condition. Scatterplots of *y* against each of the predictors are reasonably straight. The scatterplots need not show a strong (or any) slope; we just check to make sure that there isn't a bend or other nonlinearity. For the real estate data, the scatterplot is linear in both *Bedrooms* and *Living Area* as we saw in Chapter 16.

As in simple regression, it's a good idea to check the residual plot for any violations of the linearity condition. We can fit the regression and plot the residuals against the predicted values (Figure 18.4), checking to make sure we don't find patterns—especially bends or other nonlinearities.

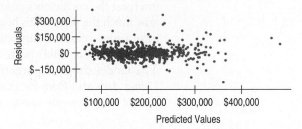

Figure 18.4 A scatterplot of residuals against the predicted values shows no obvious pattern.

Independence Assumption

As with simple regression, the errors in the true underlying regression model must be independent of each other. As usual, there's no way to be sure that the Independence Assumption is true, but we should think about how the data were collected to see if that assumption is reasonable. We should check the randomization condition as well.

Randomization Condition. Ideally, the data should arise from a random sample or randomized experiment. Randomization assures us that the data are representative of some identifiable population. If you can't identify the population, you can interpret the regression model as a description of the data you have, but you can't interpret the hypothesis tests at all because such tests are about a regression model for a specific population. Regression methods are often applied to data that were not collected with randomization. Regression models fit to such data may still do a good job of modeling the data at hand, but without some reason to believe that the data are representative of a particular population, you should be reluctant to believe that the model generalizes to other situations.

We also check the regression residuals for evidence of patterns, trends, or clumping, any of which would suggest a failure of independence. In the special case when one of the *x*-variables is related to time (or *is* itself *Time*), be sure that the residuals do not have a pattern when plotted against that variable. In addition to checking the plot of residuals against the predicted values, we recommend that you check the individual plots of the residuals against each of the explanatory, or *x*, variables in the model. These individual plots can yield important information on necessary transformations, or re-expressions, for the predictor variables.

The real estate data were sampled from a larger set of public records for sales during a limited period of time. The error we make in predicting one is not likely to be related to the error we make in predicting another.

Equal Variance Assumption

The variability of the errors should be about the same for all values of *each* predictor. To see whether this assumption is valid, we look at scatterplots and check the Equal Spread Condition.

Equal Spread Condition. The same scatterplot of residuals against the predicted values (Figure 18.4) is a good check of the consistency of the spread. We saw what appeared to be a violation of the equal spread condition when *Price* was plotted against *Bedrooms* (Figure 18.2). But here in the multiple regression, the problem has dissipated when we look at the residuals. Apparently, much of the tendency of houses with more bedrooms to have greater variability in prices was accounted for in the model when we included *Living Area* as a predictor.

If residual plots show no pattern, if the errors are plausibly independent, and if the plots of residuals against each *x*-variable don't show changes in variation, we can feel good about interpreting the regression model. Before we test hypotheses, however, we must check one final assumption: the normality assumption.

Normality Assumption

We assume that the errors around the idealized regression model at any specified values of the *x*-variables follow a Normal model. We need this assumption so that we can use a Student's *t*-model for inference. As with other times when we've used Student's *t*, we'll settle for the residuals satisfying the Nearly Normal Condition. As with means, the assumption is less important as the sample size grows. For large data sets, we have to be careful only of extreme skewness or large outliers. Our inference methods will work well even when the residuals are moderately skewed, if the sample size is large. If the distribution of residuals is unimodal and symmetric, there is little to worry about.[3]

Nearly Normal Condition. Because we have only one set of residuals, this is the same set of conditions we had for simple regression. Look at a histogram or Normal probability plot of the residuals.

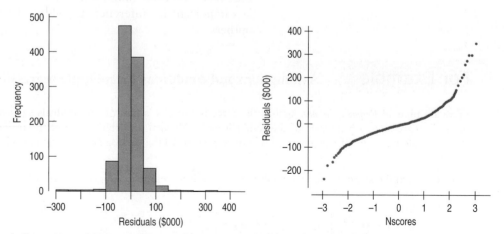

Figure 18.5 A histogram of the residuals shows a unimodal, symmetric distribution, but the tails seem a bit longer than one would expect from a Normal model. The Normal probability plot confirms that.

---

[3]The only time we need strict adherence to the Normality Assumption of the errors is when finding prediction intervals for individuals in multiple regression. Because they are based on individual Normal probabilities and not the Central Limit Theorem, the errors must closely follow a Normal model.

The histogram of residuals in the real estate example certainly looks unimodal and symmetric. The Normal probability plot has some bend on both sides, which indicates that the residuals in the tails straggle away from the center more than Normally distributed data would. However, as we have said before, the Normality Assumption becomes less important as the sample size grows, and here we have no skewness and more than 1000 cases. (The Central Limit Theorem helps our confidence intervals and tests based on the *t*-statistic to be valid when we have large samples.)

Let's summarize all the checks of conditions that we've made and the order in which we've made them.

1. Check the Linearity Condition with scatterplots of the *y*-variable against each *x*-variable.

2. If the scatterplots are straight enough, fit a multiple regression model to the data. (Otherwise, either stop or consider re-expressing an *x*-variable or the *y*-variable.)

3. Find the residuals and predicted values.

4. Make a scatterplot of the residuals against the predicted values (and ideally against each predictor variable separately). These plots should look patternless. Check, in particular, for any bend (which would suggest that the data weren't all that straight after all) and for any changes in variation. If there's a bend, consider re-expressing the *y* and/or the *x* variables. If the variation in the plot grows from one side to the other, consider re-expressing the *y*-variable. If you re-express a variable, start the model fitting over.

5. Think about how the data were collected. Was suitable randomization used? Are the data representative of some identifiable population? If the data are measured over time, check for evidence of patterns that might suggest they're not independent by plotting the residuals against time to look for patterns.

6. If the conditions check out this far, feel free to interpret the regression model and use it for prediction.

7. Make a histogram and Normal probability plot of the residuals to check the Nearly Normal Condition. If the sample size is large, the Normality is less important for inference, but always be on the lookout for skewness or outliers.

For Example   Assumptions and conditions for multiple regression

Here are plots of *Respond Amount* against the three predictors, a plot of the residuals against the predicted values from the multiple regression, a histogram of the residuals and a Normal probability plot of the residuals. Recall that the data come from a random sample of customers who both responded to the catalog and who had purchased at least 3 times in the past 5 years.

Question: Do the assumptions and conditions for multiple regression appear to have been met?

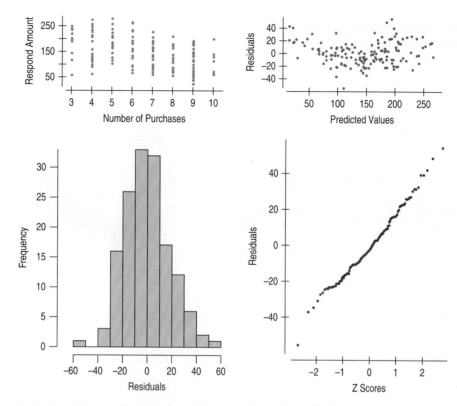

Answer: Because the sample is random, the randomization condition is satisfied and we assume the responses are independent. The plots of *Respond Amount* against each predictor are reasonably linear with the possible exception of *Number of Purchases*. There may also be some curvature and increasing spread in the residual plot. The histogram of residuals is unimodal and symmetric with no extreme outliers. The Normal probability plot shows that the distribution is fairly Normally distributed. The conditions are not completely satisfied and we should proceed somewhat cautiously especially with regard to *Number of Purchases*.

Guided Example   Housing Prices

Zillow.com attracts millions of users each month who are interested in finding out how much their house is worth. Let's see how well a multiple regression model can do. The variables available include:

Price   The price of the house as sold in 2002

Living Area The size of the living area of the house in square feet

Bedrooms   The number of bedrooms

Bathrooms   The number of bathrooms (a half bath is a toilet and sink only)

Age   Age of the house in years

Fireplaces   Number of fireplaces in the house

PLAN	**Setup** State the objective of the study. Identify the variables.	We want to build a model to predict house prices for a region in upstate New York. We have data on *Price* ($), *Living Area* (sq ft), *Bedrooms* (#), *Bathrooms* (#), *Fireplaces* (#), and *Age* (in years).

(continued)

Model Think about the assumptions and check the conditions.

Linearity Condition

To fit a regression model, we first require linearity. Scatterplots (or side-by-side boxplots) of *Price* against all potential predictor variables are shown.

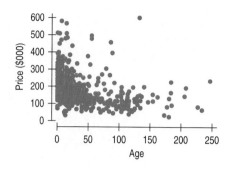

Remarks

There are a few anomalies in the plots that deserve discussion. The plot of *Price* against *Bathrooms* shows a positive relationship, but it is not quite linear. There seem to be two slopes, one from 1–2 bathrooms and then a steeper one from 2–4. For now, we'll proceed cautiously, realizing that any slope we find will average these two. The plot of *Price* against *Fireplaces* shows an outlier—an expensive home with four fireplaces. We tried setting this home aside and running the regression without it, but its influence on the coefficients was not large, so we decided to include it in the model. The plot of *Price* against *Age* shows that there may be some curvature. We may want to consider re-expressing *Age* to improve the linearity of the relationship.

✓ **Independence Assumption.** We can regard the house prices as being independent of one another. The error we'll make in predicting one isn't likely to be related to the error we'll make in predicting another.

✓ **Randomization Condition.** These 1057 houses are a random sample of a much larger set.

✓ **Equal Spread Condition.** A scatterplot of residuals vs. predicted values shows no evidence of changing spread. There is a group of homes whose residuals are larger (both negative and positive) than the vast majority. This is also seen in the long tails of the histogram of residuals.

We need the Nearly Normal Condition only if we want to do inference and the sample size is not large. If the sample size is large, as it is here, we need the distribution to be Normal only if we plan to produce prediction intervals.

✓ **Nearly Normal Condition, Outlier Condition.** The histogram of residuals is unimodal and symmetric, but long tailed. The Normal probability plot supports that.

(continued)

Under these conditions, we can proceed with caution to a multiple regression analysis. We will return to some of our concerns about curvature in some plots in the discussion.

DO	**Mechanics** We always fit multiple regression models with computer software. An output table like this one isn't exactly what any of the major packages produce, but it is enough like all of them to look familiar.	Here is some computer output for the multiple regression, using all five predictors.

	Coeff	SE(Coeff)	t-ratio	P-value
Intercept	15712.702	7311.427	2.149	0.03186
Living Area	73.446	4.009	18.321	<0.0001
Bedrooms	−6361.311	2749.503	−2.314	0.02088
Bathrooms	19236.678	3669.080	5.243	<0.0001
Fireplaces	9162.791	3194.233	2.869	0.00421
Age	−142.740	48.276	−2.957	0.00318

```
Residual standard error: 48615.95 on 1051 degrees of freedom
Multiple R-squared: 0.6049,
F-statistic: 321.8 on 5 and 1051 DF, P-value: <0.0001
```

The estimated equation is:

$$\widehat{Price} = 15{,}712.70 + 73.45 Living\ Area -$$
$$6361.31 Bedrooms + 19{,}236.68 Bathrooms +$$
$$9162.79 Fireplaces - 142.74 Age$$

All of the P-values are small which indicates that even with five predictors in the model, all are contributing. The R^2 value of 60.49% indicates that more than 60% of the overall variation in house prices has been accounted for by this model. The residual standard error of $48,620 gives us a rough indication that we can predict the price of a typical home to within about $2 \times \$48{,}620 = \$97{,}240$. That seems too large to be useful, but the model does give some idea of the price of a home.

REPORT	**Summary and Conclusions** Summarize your results and state any limitations of your model in the context of your original objectives.	**MEMO** **Re: Regression Analysis of Home Price Predictions** A regression model of *Price* on *Living Area*, *Bedrooms*, *Bathrooms*, *Fireplaces*, and *Age* accounts for 60.5% of the variation in the price of homes in upstate New York. The residual standard deviation is $48,620. A statistical test of each coefficient shows that each one is almost certainly not zero, so each of these variables appears to be a contributor to the price of a house.

This model reflects the common wisdom in real estate about the importance of various aspects of a home. An important variable not included is the location, which every real estate agent knows is crucial to pricing a house. This is ameliorated by the fact that all these houses are in the same general area. However, knowing more specific information about where they are located would almost certainly help the model. The price found from this model is to be used as a starting point for comparing a home with comparable homes in the area.

The model may be improved by re-expressing one or more of the predictors, especially *Age* and *Bathrooms*. We recommend caution in interpreting the coefficients across the entire range of these predictors.

18.4 Testing the Multiple Regression Model

There are several hypothesis tests in the multiple regression output, but all of them talk about the same thing. Each is concerned with whether the underlying model parameters (the slopes and intercept) are actually zero. The first of these hypotheses is one we skipped over for simple regression (for reasons that will be clear in a minute).

Now that we have more than one predictor, there's an overall test we should perform before we consider inference for the coefficients. We ask the global question: Is this multiple regression model any good at all? If home prices were set randomly or based on other factors than those we have as predictors, then the best estimate would just be the mean price.

To address the overall question, we'll test the null hypothesis that all the slope coefficients are zero:

$$H_0: \beta_1 = \ldots = \beta_k = 0 \text{ vs. } H_A: \text{at least one } \beta \neq 0.$$

We can test this hypothesis with an **F-test**. (It's the generalization of the *t*-test to more than one predictor.) The sampling distribution of the statistic is labeled with the letter F (in honor of Sir Ronald Fisher). The F-distribution has two degrees of freedom, k, the number of predictors, and $n - k - 1$. In our Guided Example, we have $k = 5$ predictors and $n = 1057$ homes, which means that the F-value of 321.8 has 5 and $1057 - 5 - 1 = 1051$ degrees of freedom. The regression output (page 590) shows that it has a P-value < 0.0001. The null hypothesis is that all the coefficients are 0, in which case the regression model predicts no better than the mean. The alternative is that at least one coefficient is non-zero. The test is one-sided—bigger F-values mean smaller P-values. If the null hypothesis were true, the F-statistic would have a value near 1. The F-statistic here is quite large, so we can easily reject the null hypothesis and conclude that the multiple regression model for predicting house prices with these five variables is better than just using the mean.[4]

Once we check the F-test and reject its null hypothesis—and, if we are being careful, *only* if we reject that hypothesis—we can move on to checking the test statistics for the individual coefficients. Those tests look like what we did for the slope of a simple regression in Chapter 16. For each coefficient, we test the null hypothesis that the slope is zero against the (two-sided) alternative that it isn't zero. The

F-test for Simple Regression?

Why didn't we check the F-test for simple regression? In fact, we did. When you do a simple regression with statistics software, you'll see the F-statistic in the output. But for simple regression, it gives the same information as the *t*-test for the slope. It tests the null hypothesis that the slope coefficient is zero, and we already test that with the *t*-statistic for the slope. In fact, the square of that *t*-statistic is equal to the F-statistic for the simple regression, so it really is the identical test.

---

[4]To know how big the F-value has to be, we need a table of F-values. There are F tables in the back of the book, and most regression tables include a P-value for the F-statistic.

regression table gives a standard error for each coefficient and the ratio of the estimated coefficient to its standard error. If the assumptions and conditions are met (and now we need the Nearly Normal Condition or a large sample), these ratios follow a Student's *t*-distribution:

$$t_{n-k-1} = \frac{b_j - 0}{SE(b_j)}.$$

Where did the degrees of freedom $n - k - 1$ come from? We have a rule of thumb that works here. The degrees of freedom value is the number of data values minus the number of estimated coefficients, minus 1 for the intercept. For the house price regression on five predictors, that's $n - 5 - 1$. Almost every regression report includes both the *t*-statistics and their corresponding P-values.

We can build a confidence interval in the usual way, with an estimate plus or minus a margin of error. As always, the margin of error is the product of the standard error and a critical value. Here the critical value comes from the *t*-distribution on $n - k - 1$ degrees of freedom, and the standard errors are in the regression table. So a confidence interval for each slope β_j is:

$$b_j \pm t^*_{n-k-1} \times SE(b_j).$$

The tricky parts of these tests are that the standard errors of the coefficients now require harder calculations (so we leave it to technology), and the meaning of a coefficient, as we have seen, depends on all the other predictors in the multiple regression model.

That last point is important. If we fail to reject the null hypothesis for a multiple regression coefficient, it does *not* mean that the corresponding predictor variable has no linear relationship to y. It means that the corresponding predictor contributes nothing to modeling y *after allowing for all the other predictors*.

The multiple regression model looks so simple and straightforward. It *looks* like each β_j tells us the effect of its associated predictor, x_j, on the response variable, y. But that is not true. This is, without a doubt, the most common error that people make with multiple regression. In fact:

- The coefficient β_j in a multiple regression can be quite different from zero even when it is possible there is no simple linear relationship between y and x_j.

- It is even possible that the multiple regression slope changes sign when a new variable enters the regression. We saw this for the *Price* on *Bedrooms* real estate example when *Living Area* was added to the regression.

So we'll say it once more: The coefficient of x_j in a multiple regression depends as much on the *other* predictors as it does on x_j. Failing to interpret coefficients properly is the most common error in working with regression models.

For Example   *Testing a multiple regression model*

Leiram tries another model, adding the variable *Age* to see if that improves the model:

```
Response Variable: Respond Amount
R² = 91.50%; Adjusted R² = 91.23%
s = 18.179 with 149 - 5 = 144 degrees of freedom
```

Variable	Coeff	SE(Coeff)	t-ratio	P-value
Intercept	114.91	9.244	12.439	<0.0001
Income	0.00091	0.00012	7.619	<0.0001
Total Amount Spent	0.154	0.00855	18.007	<0.0001
Number of Purchases	−14.79	0.719	−20.570	<0.0001
Age	−0.1264	0.2144	−0.5898	0.5563

Question: Has the variable *Age* improved the model? Would you leave the term in? Comment.

Answer: Of course, we would like to see the residual plots, but given this output it appears that although the R^2 value has increased from 0.9148 to 0.9150, the *t*-ratio for *Age* is only -0.5898 with a P-value of 0.5563. This indicates that there is no evidence to suggest that the slope for *Age* is different from 0. We cannot reject that null hypothesis. There is no reason to leave *Age* in this model.

18.5 Adjusted R^2, and the F-statistic

In Chapter 16, for simple linear regression, we interpreted R^2 as the variation in y accounted for by the model. The same interpretation holds for multiple regression, where now the model contains more than one predictor variable. The R^2 value tells us how much (as a fraction or percentage) of the variation in y is accounted for by the model with all the predictor variables included.

There are some relationships among the standard error of the residuals, s_e, the F-ratio, and R^2 that are useful for understanding how to assess the value of the multiple regression model. To start, we can write the standard error of the residuals as:

$$s_e = \sqrt{\frac{SSE}{n - k - 1}},$$

where $SSE = \sum e^2$ is called the **Sum of Squared Residuals** (or errors—the E is for error). As we know, a larger SSE (and thus s_e) means that the residuals are more variable and that our predictions will be correspondingly less precise.

We can look at the total variation of the response variable, y, which is called the **Total Sum of Squares** and is denoted SST: $SST = \sum (y - \bar{y})^2$. For any regression model, we have no control over SST, but we'd like SSE to be as small as we can make it by finding predictor variables that account for as much of that variation as possible. In fact, we can write an equation that relates the total variation SST to SSE:

$$SST = SSR + SSE,$$

where $SSR = \sum (\hat{y} - \bar{y})^2$ is called the **Regression Sum of Squares** because it comes from the predictor variables and tells us how much of the total variation in the response is due to the regression model. For a model to account for a large portion of the variability in y, we need SSR to be large and SSE to be small. In fact, R^2 is just the ratio of SSR to SST:

$$R^2 = \frac{SSR}{SST} = 1 - \frac{SSE}{SST}.$$

When the SSE is nearly 0, the R^2 value will be close to 1.

In Chapter 16, we saw that for the relationship between two quantitative variables, testing the standard null hypothesis about the correlation coefficient, $H_0: \rho = 0$, was equivalent to testing the standard null hypothesis about the slope, $H_0: \beta_1 = 0$. A similar result holds here for multiple regression. Testing the overall hypothesis tested by the F-statistic, $H_0: \beta_1 = \beta_2 = \ldots = \beta_k = 0$, is equivalent to testing whether the true multiple regression R^2 is zero. In fact, the F-statistic for testing that all the slopes are zero can be found as:

$$F = \frac{R^2/k}{(1 - R^2)/(n - k - 1)} = \frac{\dfrac{SSR}{SST}\dfrac{1}{k}}{\dfrac{SSE}{SST}\dfrac{1}{n - k - 1}} = \frac{SSR/k}{SSE/(n - k - 1)} = \frac{MSR}{MSE}.$$

Mean Squares

Whenever a sum of squares is divided by its degrees of freedom, the result is called a mean square. For example, the Mean Square for Error, which you may see written as MSE, is found as $SSE/(n - k - 1)$. It estimates the variance of the errors.

Similarly $SST/(n - 1)$ divides the total sum of square by *its* degrees of freedom. That is sometimes called the Mean Square for Total and denoted MST. We've see this one before; the MST is just the variance of y.

And SSR/k is the Mean Square for Regression.

In other words, using an F-test to see whether any of the true coefficients is different from 0 is equivalent to testing whether the R^2 value is different from zero. A rejection of either hypothesis says that at least one of the predictors accounts for enough variation in y to distinguish it from noise. Unfortunately, the test doesn't say which slope is responsible. You need to look at individual t-tests on the slopes to determine that. Because removing one predictor variable from the regression equation can change any number of slope coefficients, it is not straightforward to determine the right subset of predictors to use. We'll return to that problem when we discuss model selection in Chapter 19.

R^2 and Adjusted R^2

Adding a predictor variable to a multiple regression equation does not always increase the amount of variation accounted for by the model, but it can never reduce it. Adding new predictor variables will always keep the R^2 value the same or increase it. It can never decrease it. But, even if the R^2 value grows, that doesn't mean that the resulting model is a better model or that it has greater predictive ability. If we have a model with k predictors (all of which have statistically significant coefficients at some α level) and want to see if including a new variable, x_{k+1}, is warranted, we could fit the model with all $k+1$ variables and simply test the slope of the added variable with a t-test of the slope.

This method can test whether the last added variable adds significantly to the model, but choosing the "best" subset of predictors is not necessarily straightforward. We'll spend time discussing strategies for model selection in Chapter 19. The trade-off between a small (parsimonious) model and one that fits the data well is one of the great challenges of any serious model-building effort. Various statistics have been proposed to provide guidance for this search, and one of the most common is called adjusted R^2. **Adjusted R^2** imposes a "penalty" for each new term that's added to the model in an attempt to make models of different sizes (numbers of predictors) comparable. It differs from R^2 because it can shrink when a predictor is added to the regression model or grow when a predictor is removed if the predictor in question doesn't contribute usefully to the model. In fact, it can even be negative.

For a multiple regression with k predictor variables and n cases, it is defined as

$$R^2_{adj} = 1 - (1 - R^2)\frac{n-1}{n-k-1} = 1 - \frac{SSE/(n-k-1)}{SST/(n-1)}.$$

In the Guided Example, we saw that the regression of *Price* on *Bedrooms*, *Bathrooms*, *Living Area*, *Fireplaces*, and *Age* resulted in an R^2 of 0.6049. All the coefficients had P-values well below 0.05. The adjusted R^2 value for this model is 0.6030. If we add the variable *Lot Size* to the model, we get the following regression model:

	Coeff	SE(Coeff)	t-ratio	P-value
Intercept	15360.011	7334.804	2.094	0.03649
Living Area	73.388	4.043	18.154	<0.00001
Bedrooms	-6096.387	2757.736	-2.211	0.02728
Bathrooms	18824.069	3676.582	5.120	<0.00001
Fireplaces	9226.356	3191.788	2.891	0.00392
Age	-152.615	48.224	-3.165	0.00160
Lot Size	847.764	1989.112	0.426	0.67005

Residual standard error: 48440 on 1041 degrees of freedom
Multiple R-squared: 0.6081, Adjusted R-squared: 0.6059
F-statistic: 269.3 on 6 and 1041 DF, P-value: <0.0001

The most striking feature of this regression table as compared to the one in the Guided Example on page 590, is that although most of the coefficients have changed very little, the coefficient of *Lot Size* is far from significant, with a *P*-value of 0.670. Yet, the adjusted R^2 value is actually higher than for the previous model. This is why we warn against putting too much faith in this statistic. Especially for large samples, the adjusted R^2 does not always adjust downward enough to make sensible model choices. The other problem with comparing these two models is that 9 homes had missing values for *Lot Size*, which means that we're not comparing the models on exactly the same data set. When we matched the two models on the smaller data set, the adjusted R^2 value actually did "make the right decision" but just barely—0.6059 versus 0.6060 for the model without *Lot Size*. One might expect a larger difference considering we added a variable whose *t*-ratio is much less than 1.

The lesson to be learned here is that there is no "correct" set of predictors to use for any real business decision problem, and finding a reasonable model is a process that takes a combination of science, art, business knowledge, and common sense. Look at the adjusted R^2 value for any multiple regression model you fit, but be sure to think about all the other reasons for including or not including any given predictor variable. We will have much more to say about this important subject in Chapter 19.

For Example R^2 and adjusted R^2

Question: The model for *Respond Amount* that included *Age* as a predictor (see For Example on page 592) had an adjusted R^2 value of 0.9123 (or 91.23%) The original model (see For Example on page 581) had an adjusted R^2 value of 0.9131. Explain how that is consistent with the decision we made to drop *Age* from the model.

Answer: The *t*-ratio indicated that the slope for *Age* was not significantly different from 0, and thus that we should drop *Age* from the model. Although the R^2 value for the model that included *Age* was slightly higher, the adjusted R^2 value was slightly lower. Adjusted R^2 adds a penalty for each added term that allows us to compare models of different sizes. Because the adjusted R^2 value for the smaller model (without *Age*) is higher than the larger model (with *Age*), it indicates that the inclusion of *Age* is not justified, which is consistent with the recommendation from the *t*-ratio.

---

*18.6   The Logistic Regression Model

Business decisions often depend on whether something will happen or not. Will my customers leave my wireless service at the end of their subscription? How likely is it that my customers will respond to the offer I just mailed? The response variable is either Yes or No—a dichotomous response. By definition, that's a categorical response, so we can't use linear regression methods to predict it.

If we coded this categorical response variable by giving the Yes values the value 1 and the No values the value 0, we could "pretend" that it's quantitative and try to fit a linear model. Let's imagine that we're trying to model whether someone will respond to an offer based on how much they spent in the last year at our company. The regression of *Purchase (1 = Yes: 0 = No)* on *Spending* might look like Figure 18.6.

The model shows that as past spending increases, something having to do with responding to the offer increases. Look at *Spending* near $500. Nearly all the customers who spend that much responded to the offer, while for those customers who spent less than $100 last year, almost none responded. The line seems to show the proportion of those responding for different values of *Spending*. Looking at the plot, what would you predict the proportion of responders to be for those that spent about $225? You might say that the proportion was about 0.50.

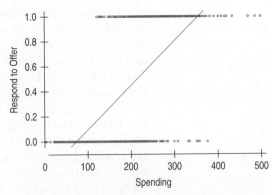

Figure 18.6   A linear model of Respond to Offer shows that as Spending increases, the proportion of customers responding to the offer increases as well. Unfortunately, the line predicts values outside the interval (0,1), but those are the only sensible values for proportions or probabilities.

What's wrong with this interpretation? We know proportions (or probabilities) must lie between 0 and 1. But the linear model has no such restrictions. The line crosses 1.0 at about $350 in *Spending* and goes below 0 at about $75 in *Spending*. However, by transforming the probability, we can make the model behave better and get more sensible predictions for all values of *Spending*. We could just cut off all values greater than 1 at 1 and all values less than 0 at 0. That would work fine in the middle, but we know that things like customer behavior don't change that abruptly. So we'd prefer a model that curves at the ends to approach 0 and 1 gently. That's likely to be a better model for what really happens. A simple function is all we need. There are several that can do the job, but one common model is the logistic regression model. Here is a plot of the logistic regression model predictions for the same data.

Figure 18.7   A logistic model of *Respond* to *Spending* predicts values between 0 and 1 for the proportion of customers who *Respond* for all values of *Spending*.

For many values of *Spending*, especially those near the mean value of *Spending*, the predictions of the probability of responding is similar, but the logistic transformation approaches the limits at 0 and 1 smoothly. A logistic regression is an example of a nonlinear regression model. The computer finds the coefficients by solving a system of equations that in some ways mimic the least squares calculations we did for linear regression. The **logistic regression** model looks like this:

$$\ln\left(\frac{p}{1-p}\right) = \beta_0 + \beta_1 x_1 + \cdots + \beta_k x_k.$$

where ln( ) denotes the natural logarithm.

In other words, it's not the probabilities themselves that are modeled by a (multiple) regression model, but a transformation (the logistic function) of the probabilities. When the probabilities are transformed back and plotted, we get an S-shaped curve (when plotted against each predictor), as seen in Figure 18.7.

As with multiple regression, computer output for a logistic regression model provides estimates for each coefficient, standard errors, and a test for whether the coefficient is zero. Unlike multiple regression, the coefficients for each variable are tested with a Chi-square statistic[5] rather than a t-statistic, and the overall R^2 is no longer available. (Some computer programs use a z-statistic to test individual coefficients. The tests are equivalent.) Because the probabilities are not a linear function of the predictors, it is more difficult to interpret the coefficients. The fitted logistic regression equation can be written:

$$\ln\left(\frac{\hat{p}}{1 - \hat{p}}\right) = b_0 + b_1 x_1 + \dots + b_k x_k.$$

Some researchers try to interpret the logistic regression equation directly by realizing that the expression on the left of the logistic regression equation, $\ln\left(\frac{\hat{p}}{1 - \hat{p}}\right)$, can be thought of as the predicted log (base e) odds of the outcome. This is sometimes denoted logit($\hat{p}$). The higher the log odds, the higher the probability. Negative log odds indicate that the probability is less than 0.5 because then $\frac{p}{1 - p}$ will be less than 1 and thus have a negative logarithm. Positive log odds indicate a probability greater than 0.5. If the coefficient of a particular predictor is positive, it means that higher values of it are associated with higher log odds and thus a higher probability of the response. So, the *direction* is interpretable in the same way as a multiple regression coefficient. But an increase of one unit of the predictor *increases* the predicted *log odds* by an amount equal to the coefficient of that predictor, after allowing for the effects of all the other predictors. It does not increase the predicted *probability* by that amount.

The transformation back to probabilities is straightforward, but nonlinear. Once we have fit the logistic regression equation

$$\ln\left(\frac{\hat{p}}{1 - \hat{p}}\right) = \text{logit}\left(\hat{p}\right) = b_0 + b_1 x_1 + \dots + b_k x_k,$$

we can find the individual probability estimates from the equation:

$$\hat{p} = \frac{1}{1 + e^{-(b_0 + b_1 x_1 + \dots + b_k x_k)}} = \frac{e^{(b_0 + b_1 x_1 + \dots + b_k x_k)}}{1 + e^{(b_0 + b_1 x_1 + \dots + b_k x_k)}}.$$

The assumptions and conditions for fitting a logistic regression are similar to multiple regression. We still need the independence assumption and randomization condition. We no longer need the linearity or equal spread condition. However, it's a good idea to plot the response variable against each predictor variable to make sure that there are no outliers that could unduly influence the model. A customer who both spent $10,000 and responded to the offer could possibly change the shape of the curve shown in Figure 18.7. However, residual analysis for logistic regression models is beyond the scope of this book.

---

[5]Yes, the same sampling distribution model as we used in Chapter 15.

For Example   Logistic regression

Leiram wants to build a model to predict *whether* a customer will respond to the new catalog offer. The variable *Respond* indicates 1 for responders and 0 for nonresponders. After trying several models, she settles on the following logistic regression model:

Response Variable: Respond

Variable	Coeff	SE(Coeff)	z-value	P-value
Intercept	−3.298	0.4054	−8.136	<.0001
Age	0.08365	0.012226	6.836	<.0001
Days Since Last Purchase	0.00364	0.000831	4.378	<.0001

Questions: Interpret the model. Would you keep both predictor variables in the model? Why? Who would be more likely to respond, a 20-year-old customer who has recently purchased another item, or a 50-year-old customer who has not purchased in a year?

Answers: The model says:

$$\text{logit}(\hat{p}) = -3.298 + 0.08365 \, Age + 0.00364 \, Days \, Since \, Last \, Purchase,$$

which means that the probability of responding goes up with both *Age* and *Days Since Last Purchase*. Both terms have large z-values and correspondingly low P-values, which indicate that they contribute substantially to the model. We should keep both terms. Because both terms are positively associated with increasing (log) odds of purchasing, both an older person and one who has *not* purchased recently are more likely to respond. A 50-year-old who has not purchased recently would be more likely to respond than the 20-year-old who has.

Guided Example   Time on Market

A real estate agent used information on 1115 houses, such as we used in our multiple regression Guided Example. She wants to predict whether each house sold in the first 3 months it was on the market based on other variables. The variables available include:

Sold	1 = Yes—the house sold within the first 3 months it was listed; 0 = No, it did not sell within 3 months.	*Bedrooms*	The number of bedrooms
		Bathrooms	The number of bathrooms (a half bath is a toilet and sink only)
Price	The price of the house as sold in 2002.	*Age*	Age of the house in years
Living Area	The size of the living area of the house in square feet	*Fireplaces*	Number of fireplaces in the house

| **PLAN** | **Setup** State the objective of the study. Identify the variables.

Model Think about the assumptions and check the conditions. | We want to build a model to predict whether a house will sell within the first 3 months it's on the market based on *Price* (\$), *Living Area* (sq ft), *Bedrooms* (#), *Bathrooms* (#), *Fireplaces* (#), and *Age* (years). Notice that now *Price* is a predictor variable in the regression.

Outlier Condition

To fit a logistic regression model, we check that there are no outliers in the predictors that may unduly influence the model.

Here are scatterplots of *Sold* (1 = Yes; 0 = No) against each predictor. (Plots of *Sold* against the variables *Bathrooms*, *Bedrooms*, and *Fireplaces* are uninformative because these predictors are discrete.) |

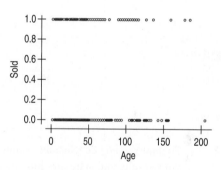

✓ **Outlier Condition.** There do not seem to be any outliers in the predictors. Of course, there can't be any in the response variable; it is only 0 or 1.

✓ **Independence Assumption.** We can regard the house prices as being independent of one another since they come from a fairly large geographic area.

✓ **Randomization Condition.** These 1115 houses are a random sample from a larger collection of houses.

We can fit a logistic regression predicting *Sold* from the six predictor variables.

DO

Mechanics We always fit logistic regression models with computer software. An output table like this one isn't exactly what any of the major packages produce, but it is enough like all of them to look familiar.

Here is the computer output for the logistic regression using all six predictors.

	Coeff	SE(Coeff)	z-value	P-value
Intercept	−3.222e + 00	3.826e − 01	−8.422	<0.0001
Living Area	−1.444e − 03	2.518e − 04	−5.734	<0.0001
Age	4.900e − 03	2.823e − 03	1.736	0.082609
Price	1.693e − 05	1.444e − 06	11.719	<0.0001
Bedrooms	4.805e − 01	1.366 e − 01	3.517	0.000436
Bathrooms	−1.813e − 01	1.829e − 01	−0.991	0.321493
Fireplaces	−1.253e − 01	1.633 e − 01	−0.767	0.442885

(continued)

Strategies for model selection will be discussed in Chapter 19.

The estimated equation is:

$$\text{logit}(\hat{p}) = -3.22 - 0.00144 Living\ Area$$
$$+ 0.0049 Age + 0.00000169 Price$$
$$+ 0.481 Bedrooms - 0.181 Bathrooms$$
$$- 0.125 Fireplaces$$

Three of the P-values are quite small, two are large (*Bathrooms* and *Fireplaces*), and one is marginal (*Age*). After examining several alternatives, we chose the following model.

	Coeff	SE(Coeff)	z-value	P-value
Intercept	-3.351e+00	3.601e-01	-9.305	<0.0001
Living Area	-1.574e-03	2.342e-04	-6.719	<0.0001
Age	6.106e-03	2.668e-03	2.289	0.022102
Bedrooms	4.631e-01	1.354e-01	3.421	0.000623
Price	1.672e-05	1.428e-06	11.704	<0.0001

The estimated logit equation is:

$$\text{logit}(\hat{p}) = -3.351 - 0.00157 Living\ Area$$
$$+ 0.00611 Age + 0.00000167 Price$$
$$+ 0.463 Bedrooms$$

While interpretation is difficult, it appears that for a house of a given size (and age and bedrooms), higher-priced homes may have a higher chance of selling within 3 months. For a house of given size and price and age, having more bedrooms may be associated with a greater chance of selling within 3 months.

REPORT

Summary and Conclusions Summarize your results and state any limitations of your model in the context of your original objectives.

MEMO

Re: Logistic Regression Analysis of Selling

A logistic regression model of *Sold* on various predictors was fit, and a model based on *Living Area*, *Bedrooms*, *Price*, and *Age* found that these four predictors were statistically significant in predicting the probability that a house will sell within 3 months. More thorough analysis to understand the meaning of the coefficients is needed, but each of these variables appears to be an important predictor of whether the house will sell quickly.

However, knowing more specific information about other characteristics of a house and where it is located would almost certainly help the model.

What Can Go Wrong?

Interpreting Coefficients

- **Don't claim to "hold everything else constant" for a single individual.** It's often meaningless to say that a regression coefficient says what we expect to happen if all variables but one were held constant for an individual and the predictor in question changed. While it's mathematically correct, it often just doesn't make any sense. For example, in a regression of salary on years of experience, years of education, and age, subjects can't gain a year of experience or get another year of education without getting a year older. Instead, we *can* think about all those who satisfy given criteria on some predictors and ask about the conditional relationship between y and one x for those individuals.

- **Don't interpret regression causally.** Regressions are usually applied to observational data. Without deliberately assigned treatments, randomization, and control, we can't draw conclusions about causes and effects. We can never be certain that there are no variables lurking in the background, causing everything we've seen. Don't interpret b_1, the coefficient of x_1 in the multiple regression, by saying: "If we were to change an individual's x_1 by 1 unit (holding the other x's constant), it would change his y by b_1 units." We have no way of knowing what applying a change to an individual would do.

- **Be cautious about interpreting a regression model as predictive.** Yes, we do call the x's predictors, and you can certainly plug in values for each of the x's and find a corresponding *predicted value*, $\hat{y}$. But the term "prediction" suggests extrapolation into the future or beyond the data, and we know that we can get into trouble when we use models to estimate $\hat{y}$ values for x's not in the range of the data. Be careful not to extrapolate very far from the span of your data. In simple regression, it was easy to tell when you extrapolated. With many predictor variables, it's often harder to know when you are outside the bounds of your original data.[6] We usually think of fitting models to the data more as modeling than as prediction, so that's often a more appropriate term.

- **Be careful when interpreting the signs of coefficients in a multiple regression.** Sometimes our primary interest in a predictor is whether it has a positive or negative association with y. As we have seen, though, the sign of the coefficient also depends on the other predictors in the model. Don't look at the sign in isolation and conclude that "the direction of the relationship is positive (or negative)." Just like the value of the coefficient, the sign is about the relationship after allowing for the linear effects of the other predictors. The sign of a variable can change depending on which other predictors are in or out of the model. For example, in the regression model for house prices, we saw the coefficient of *Bedrooms* change sign when *Living Area* was added to the model as a predictor. It isn't correct to say either that houses with more bedrooms sell for more on average or that they sell for less. The truth is more subtle and requires that we understand the multiple regression model.

- **If a coefficient's *t*-statistic is not significant, don't interpret it at all.** You can't be sure that the value of the corresponding parameter in the underlying regression model isn't really zero.

---

[6]With several predictors we can wander beyond the data because of the *combination* of values even when individual values are not extraordinary. For example, houses with 1 bathroom and houses with 5 bedrooms can both be found in the real estate records, but a single house with 5 bedrooms and only 1 bathroom would be quite unusual. The model we found is not appropriate for predicting the price of such an extraordinary house.

What Else Can Go Wrong?

- **Don't fit a linear regression to data that aren't straight.** This is the most fundamental regression assumption. If the relationship between the *x*'s and *y* isn't approximately linear, there's no sense in fitting a linear model to it. What we mean by "linear" is a model of the form we have been writing for the regression. When we have two predictors, this is the equation of a plane, which is linear in the sense of being flat in all directions. With more predictors, the geometry is harder to visualize, but the simple structure of the model is consistent; the predicted values change consistently with equal size changes in any predictor.

  Usually we're satisfied when plots of *y* against each of the *x*'s are straight enough. We'll also check a scatterplot of the residuals against the predicted values for signs of nonlinearity.

- **Watch out for changing variance in the residuals.** The estimate of the error standard deviation shows up in all the inference formulas. But that estimate assumes that the error standard deviation is the same throughout the extent of the *x*'s so that we can combine all the residuals when we estimate it. If s_e changes with any *x*, these estimates won't make sense. The most common check is a plot of the residuals against the predicted values. If plots of residuals against several of the predictors all show a thickening and especially if they also show a bend, then consider re-expressing *y*.

- **Make sure the errors are nearly Normal.** All of our inferences require that, unless the sample size is large, the true errors be modeled well by a Normal model. Check the histogram and Normal probability plot of the residuals to see whether this assumption looks reasonable.

- **Watch out for high-influence points and outliers.** We always have to be on the lookout for a few points that have undue influence on our model, and regression is certainly no exception. Chapter 19 discusses this issue in greater depth.

Ethics in Action

Alpine Medical Systems, Inc. is a large provider of medical equipment and supplies to hospitals, doctors, clinics, and other health care professionals. Alpine's VP of Marketing and Sales, Kenneth Jadik, asked one of the company's analysts, Nicole Haly, to develop a model that could be used to predict the performance of the company's sales force. Based on data collected over the past year, as well as records kept by Human Resources, she considered five potential independent variables: (1) gender, (2) starting base salary, (3) years of sales experience, (4) personality test score, and (5) high school grade point average. The dependent variable (sales performance) is measured as the sales dollars generated per quarter. In discussing the results with Nicole, Kenneth asks to see the full regression model with all five independent variables included. Kenneth notes that a *t*-test for the coefficient of gender shows no significant effect on sales performance and recommends that it be eliminated from the model. Nicole reminds him of the company's history of offering lower starting base salaries to women, recently corrected under court order. If instead, starting base salary is removed from the model, gender is statistically significant, and its coefficient indicates that women on the sales force outperform men (taking into account the other variables). Kenneth argues that because gender is not significant when all predictors are included, it is the variable that should be omitted.

ETHICAL ISSUE *The choice of predictors for the regression model is politically motivated. Because gender and base salary are related, it is impossible to separate their effects on sales performance, and inappropriate to conclude that one or the other is irrelevant. Related to Item A, ASA Ethical Guidelines.*

ETHICAL SOLUTION *The situation is more complex than a single model can explain. Both the model with gender but not base salary and the one with base salary but not gender should be reported. Then the discussion of these models should point out that the two variables are related because of previous company policy and note that the conclusion that those with lower base salary have better sales and the conclusion that women tend to have better sales performance are equivalent as far as these data are concerned.*

What Have We Learned?

Learning Objectives

- Know how to perform a multiple regression, using the technology of your choice.

  - Technologies differ, but most produce similar-looking tables to hold the regression results. Know how to find the values you need in the output generated by the technology you are using.

- Understand how to interpret a multiple regression model.

  - The meaning of a multiple regression coefficient depends on the other variables in the model. In particular, it is the relationship of y to the associated x after removing the linear effects of the other x's.

- Be sure to check the Assumptions and Conditions before interpreting a multiple regression model.

  - The **Linearity Assumption** asserts that the form of the multiple regression model is appropriate. We check it by examining scatterplots. If the plots appear to be linear, we can fit a multiple regression model.

  - The **Independence Assumption** requires that the errors made by the model in fitting the data be mutually independent. Data that arise from random samples or randomized experiments usually satisfy this assumption.

  - The **Equal Variance Assumption** states that the variability around the multiple regression model should be the same everywhere. We usually check the **Equal Spread Condition** by plotting the residuals against the predicted values. This assumption is needed so that we can pool the residuals to estimate their standard deviation, which we will need for inferences about the regression coefficients.

  - The **Normality Assumption** says that the model's errors should follow a Normal model. We check the **Nearly Normal Condition** with a histogram or normal probability plot of the residuals. We need this assumption to use Student's t models for inference, but for larger sample sizes, it is less important.

- Know how to state and test hypotheses about the multiple regression coefficients.

  - The standard hypothesis test for each coefficient is

    $$H_0: \beta_j = 0 \text{ vs.}$$
    $$H_A: \beta_j \neq 0$$

  - We test these hypotheses by referring the test statistic

    $$\frac{b_j - 0}{SE(b_j)}$$

    to the Student's t distribution on $n - k - 1$ degrees of freedom, where k is the number of coefficients estimated in the multiple regression.

- Interpret other associated statistics generated by a multiple regression

  - R^2 is the fraction of the variation in y accounted for by the multiple regression model.

  - Adjusted R^2 attempts to adjust for the number of coefficients estimated.

(continued)

- The F-statistic tests the overall hypothesis that the regression model is of no more value than simply modeling y with its mean.

- The standard deviation of the residuals,

$$s_e = \sqrt{\frac{\sum e^2}{n - k - 1}}$$

provides an idea of how precisely the regression model fits the data.

Terms

Adjusted R^2 An adjustment to the R^2 statistic that attempts to allow for the number of predictors in the model. It is sometimes used when comparing regression models with different numbers of predictors:

$$R_{adj}^2 = 1 - (1 - R^2)\frac{n - 1}{n - k - 1} = 1 - \frac{SSE/(n - k - 1)}{SST/(n - 1)}.$$

F-test The F-test is used to test the null hypothesis that the overall regression is no improvement over just modeling y with its mean:

$$H_0: \beta_1 = \ldots = \beta_k = 0. \text{ vs } H_A: \text{at least one } \beta \neq 0$$

If this null hypothesis is not rejected, then you should not proceed to test the individual coefficients.

*Logistic regression A regression model that models a binary (0/1) response variable based on quantitative predictor variables.

Multiple regression A linear regression with two or more predictors whose coefficients are found by least squares. When the distinction is needed, a least squares linear regression with a single predictor is called a *simple regression*. The multiple regression model is:
$y = \beta_0 + \beta_1 x_1 + \ldots + \beta_k x_k + \varepsilon.$

Regression Sum of Squares, SSR A measure of the total variation in the response variable due to the model. $SSR = \sum (\hat{y} - \bar{y})^2.$

Sum of Squared Errors or Residuals, SSE A measure of the variation in the residuals. $SSE = \sum (y - \hat{y})^2.$

Total Sum of Squares, SST A measure of the variation in the response variable. $SST = \sum (y - \bar{y})^2.$ Note that $\frac{SST}{n - 1} = Var(y).$

t-ratios for the coefficients The t-ratios for the coefficients can be used to test the null hypotheses that the true value of each coefficient is zero against the alternative that it is not. The t distribution is also used in the construction of confidence intervals for each slope coefficient.

Technology Help: Regression Analysis

All statistics packages make a table of results for a regression. The table for multiple regression looks very similar to the table for simple regression. You'll want to look at the Analysis of Variance (ANOVA) table, and you'll see information for each of the coefficients.

Most packages offer to plot residuals against predicted values. Some will also plot residuals against the x's. With some packages, you must request plots of the residuals when you request the regression. Others let you find the regression first and then analyze the residuals afterward. Either way, your analysis is not complete if you don't check the residuals with a histogram or Normal probability plot and a scatterplot of the residuals against the x's or the predicted values.

One good way to check assumptions before embarking on a multiple regression analysis is with a scatterplot matrix. This is sometimes abbreviated SPLOM (or Matrix Plot) in commands.

Multiple regressions are always found with a computer or programmable calculator. Before computers were available, a full multiple regression analysis could take months or even years of work.

EXCEL XLSTAT>

- In Excel 2007 or Excel 2010, select **Data Analysis** from the **Analysis Group** on the Data Tab.
- Select **Regression** from the **Analysis Tools** list.
- Click the **OK** button.
- Enter the data range holding the Y-variable in the box labeled "Y-range."
- Enter the range of cells holding the X-variables in the box labeled "X-range."
- Select the **New Worksheet Ply** option.
- Select **Residuals** options. Click the **OK** button.

Comments

The Y and X ranges do not need to be in the same rows of the spreadsheet, although they must cover the same number of cells. But it is a good idea to arrange your data in parallel columns as in a data table. The X-variables must be in adjacent columns. No cells in the data range may hold nonnumeric values or be left blank.

Although the dialog offers a Normal probability plot of the residuals, the data analysis add-in does not make a correct probability plot, so don't use this option.

JMP

- From the **Analyze** menu, select **Fit Model**.
- Specify the response, Y. Assign the predictors, X, in the **Construct Model Effects** dialog box.
- Click on **Run Model**.

Comments

JMP chooses a regression analysis when the response variable is "Continuous." The predictors can be any combination of quantitative or categorical. If you get a different analysis, check the variable types.

MINITAB

- Choose **Regression** from the **Stat** menu.
- Choose **Regression . . .** from the **Regression** submenu.
- In the Regression dialog, assign the Y-variable to the Response box and assign the X-variables to the Predictors box.
- Click the **Graphs** button.
- In the Regression-Graphs dialog, select **Standardized residuals**, and check **Normal plot of residuals** and **Residuals versus fits**.
- Click the **OK** button to return to the Regression dialog.
- Click the **OK** button to compute the regression.

SPSS

- Choose **Regression** from the **Analyze** menu.
- Choose **Linear** from the **Regression** submenu.
- When the Linear Regression dialog appears, select the Y-variable and move it to the dependent target. Then move the X-variables to the independent target.
- Click the **Plots** button.
- In the Linear Regression Plots dialog, choose to plot the *SRESIDs against the *ZPRED values.
- Click the **Continue** button to return to the Linear Regression dialog.
- Click the **OK** button to compute the regression.

Brief CASE

Golf Success

Professional sports, like many other professions, require a variety of skills for success. That makes it difficult to evaluate and predict success. Fortunately, sports provide examples we can use to learn about modeling success because of the vast amount of data which are available. Here's an example.

What makes a golfer successful? The game of golf requires many skills. Putting well or hitting long drives will not, by themselves, lead to success. Success in golf requires a combination of skills. That makes multiple regression a good candidate for modeling golf achievement.

A number of Internet sites post statistics for the current PGA players. We have data for 204 top players of 2006 in the file **Golfers**.

All of these players earned money on the tour, but they didn't all play the same number of events. And the distribution of earnings is quite skewed. (Tiger Woods earned $662,000 per event. In second place, Jim Furyk earned only $300,000 per event. Median earnings per event were $36,600.) So it's a good idea to take logs of Earnings/Event as the response variable.

The variables in the data file include:

Log$/E   The logarithm of earnings per event

GIR   Greens in Regulation. Percentage of holes played in which the ball is on the green with two or more strokes left for par.

Putts   Average number of putts per hole in which the green was reached in regulation.

Save%   Each time a golfer hits a bunker by the side of a green but needs only one or two additional shots to reach the hole, he is credited with a save. This is the percentage of opportunities for saves that are realized.

DDist   Average Drive Distance (yards). Measured as averages over pairs of drives in opposite directions (to account for wind).

DAcc   Drive Accuracy. Percent of drives landing on the fairway.

Investigate these data. Find a regression model to predict golfers' success (measured in log earnings per event). Write a report presenting your model including an assessment of its limitations. Note: Although you may consider several intermediate models, a good report is about the model you think best, not necessarily about all the models you tried along the way while searching for it.

Exercises

SECTION 18.1

1. A house in the upstate New York area from which the chapter data was drawn has 2 bedrooms and 1000 square feet of living area. Using the multiple regression model found in the chapter,

$$\widehat{Price} = 20{,}986.09 - 7483.10\,Bedrooms + 93.84\,Living\ Area.$$

a) Find the price that this model estimates.
b) The house just sold for $135,000. Find the residual corresponding to this house.
c) What does that residual say about this transaction?

2. A candy maker surveyed chocolate bars available in a local supermarket and found the following least squares regression model:

$$\widehat{Calories} = 28.4 + 11.37\,Fat(g) + 2.91\,Sugar(g).$$

a) The hand-crafted chocolate she makes has 15g of fat and 20g of sugar. How many calories does the model predict for a serving?
b) In fact, a laboratory test shows that her candy has 227 calories per serving. Find the residual corresponding to this candy. (Be sure to include the units.)
c) What does that residual say about her candy?

SECTION 18.2

3. What can predict how much a motion picture will make? We have data on a number of recent releases that includes the *USGross* (in $), the *Budget* ($), the *Run Time* (minutes), and the average number of *Stars* awarded by reviewers. The first several entries in the data table look like this:

Movie	USGross ($M)	Budget ($M)	Run Time (minutes)	Stars
White Noise	56.094360	30	101	2
Coach Carter	67.264877	45	136	3
Elektra	24.409722	65	100	2
Racing Stripes	49.772522	30	110	3
Assault on Precinct 13	20.040895	30	109	3
Are We There Yet?	82.674398	20	94	2
Alone in the Dark	5.178569	20	96	1.5
Indigo	51.100486	25	105	3.5

We want a regression model to predict *USGross*. Parts of the regression output computed in Excel look like this:

```
Dependent variable is: USGross($)
R squared = 47.4%   R squared (adjusted) = 46.0%
s = 46.41 with 120 − 4 = 116 degrees of freedom

Variable    Coefficient   SE(Coeff)   t-ratio   P-value
Intercept   −22.9898      25.70       −0.895    0.3729
Budget($)     1.13442      0.1297      8.75     ≤0.0001
Stars        24.9724       5.884       4.24     ≤0.0001
Run Time     −0.403296     0.2513     −1.60     0.1113
```

a) Write the multiple regression equation
b) What is the interpretation of the coefficient of *Budget* in this regression model?

4. A middle manager at an entertainment company, upon seeing this analysis, concludes that the longer you make a movie, the less money it will make. He argues that his company's films should all be cut by 30 minutes to improve their gross. Explain the flaw in his interpretation of this model.

SECTION 18.3

5. For the movies examined in Exercise 4, here is a scatterplot of *USGross* vs. *Budget*:

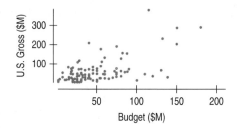

What (if anything) does this scatterplot tell us about the following Assumptions and Conditions for the regression?
a) Linearity condition
b) Equal Spread condition
c) Normality assumption

6. For the movies regression, here is a histogram of the residuals. What does it tell us about these Assumptions and Conditions?

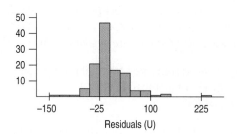

a) Linearity condition
b) Nearly Normal condition
c) Equal Spread condition

SECTION 18.4

7. Regression output for the movies again:

a) What is the null hypothesis tested for the coefficient of *Stars* in this table?
b) What is the *t*-statistic corresponding to this test?
c) What is the P-value corresponding to this *t*-statistic?
d) Complete the hypothesis test. Do you reject the null hypothesis?

8. a) What is the null hypothesis tested for the coefficient of *Run Time* in the regression of Exercise 3?
b) What is the *t*-statistic corresponding to this test?
c) Why is this *t*-statistic negative?
d) What is the P-value corresponding to this *t*-statistic?
e) Complete the hypothesis test. Do you reject the null hypothesis?

SECTION 18.5

9. In the regression model of Exercise 3,

a) What is the R^2 for this regression? What does it mean?
b) Why is the "Adjusted R Square" in the table different from the "R Square"?

10. Here is another part of the regression output for the movies in Exercise 3:

```
Source       Sum of Squares  df   Mean Square  F-ratio
Regression   224995           3   74998.4       34.8
Residual     249799         116   2153.44
```

a) Using the values from the table, show how the value of R^2 could be computed. Don't try to do the calculation, just show what is computed.
b) What is the F-statistic value for this regression?
c) What null hypothesis can you test with it?
d) Would you reject that null hypothesis?

CHAPTER EXERCISES

The next 12 exercises consist of two sets of 6 (one even-numbered, one odd-numbered). Each set guides you through a multiple regression analysis. We suggest that you do all 6 exercises in a set. Remember that the answers to the odd-numbered exercises can be found in the back of the book.

T **11.** Police salaries. Is the amount of violent crime related to what police officers are paid? The U.S. Bureau of Labor Statistics publishes data on occupational employment and wage estimates (www.bls.gov/oes/). Here are data on those states for which 2006 data were available. The variables are:

Violent Crime (crimes per 100,000 population)
Police Officer Wage (mean $/hr)
Graduation Rate (%)

One natural question to ask of these data is how police officer wages are related to violent crime across these states.

First, here are plots and background information.

	Correlations		
	Violent Crime	Graduation Rate	Police Officer Wage
Violent Crime	1.000		
Graduation Rate	−0.682	1.000	
Police Officer Wage	0.103	0.213	1.000

a) Name and check (to the extent possible) the regression assumptions and their corresponding conditions.
b) If we found a regression to predict *Violent Crime* just from *Police Officer Wage*, what would the R^2 of that regression be?

T **12.** Ticket prices. On a typical night in New York, about 25,000 people attend a Broadway show, paying an average price of more than $75 per ticket. *Variety* (www.variety.com), a news weekly that reports on the entertainment industry, publishes statistics about the Broadway show business. The data file on the disk holds data about shows on Broadway for most weeks of 2006–2008. (A few weeks are missing data.) The following variables are available for each week:

Receipts ($ million)
Paid Attendance (thousands)
 # Shows
Average Ticket Price ($)

Viewing this as a business, we'd like to model *Receipts* in terms of the other variables.

First, here are plots and background information.

	Correlations			
	Receipts	Paid Attendance	# Shows	Average Ticket Price
Receipts ($M)	1.000			
Paid Attendance	0.961	1.000		
# Shows	0.745	0.640	1.000	
Average Ticket Price	0.258	0.331	−0.160	1.000

a) Name and check (to the extent possible) the regression assumptions and their corresponding conditions.

b) If we found a regression to predict *Receipts* only from *Paid Attendance*, what would the R^2 of that regression be?

13. Police salaries, part 2. Here's a multiple regression model for the variables considered in Exercise 11.

```
Dependent variable is: Violent Crime
R squared = 53.0%   R squared (adjusted) = 50.5%
s = 129.6 with 37 degrees of freedom
```

Source	Sum of Squares	df	Mean Square	F-ratio
Regression	701648	2	350824	20.9
Residual	621060	37	16785.4	

Variable	Coeff	SE(Coeff)	t-ratio	P-value
Intercept	1390.83	185.9	7.48	<0.0001
Police Officer Wage	9.33	4.125	2.26	0.0297
Graduation Rate	−16.64	2.600	−6.40	<0.0001

a) Write the regression model.

b) What does the coefficient of *Police Officer Wage* mean in the context of this regression model?

c) In a state in which the average police officer wage is $20/hour and the high school graduation rate is 70%, what does this model estimate the violent crime rate would be?

d) Is this likely to be a good prediction? Why do you think that?

14. Ticket prices, part 2. Here's a multiple regression model for the variables considered in Exercise 12:[7]

```
Dependent variable is: Receipts($M)
R squared = 99.9%   R squared (adjusted) = 99.9%
s = 0.0931 with 74 degrees of freedom
```

Source	Sum of Squares	df	Mean Square	F-ratio
Regression	484.789	3	161.596	18634
Residual	0.641736	74	0.008672	

Variable	Coeff	SE(Coeff)	t-ratio	P-value
Intercept	−18.320	0.3127	−58.6	<0.0001
Paid Attendance	0.076	0.0006	126.7	<0.0001
# Shows	0.0070	0.0044	1.59	0.116
Average Ticket Price	0.24	0.0039	61.5	<0.0001

a) Write the regression model.

b) What does the coefficient of *Paid Attendance* mean in this regression? Does that make sense?

c) In a week in which the paid attendance was 200,000 customers attending 30 shows at an average ticket price of $70, what would you estimate the receipts would be?

---

[7]Some values are rounded to simplify the exercises. If you recompute the analysis with your statistics software you may see slightly different numbers.

d) Is this likely to be a good prediction? Why do you think that?

15. Police salaries, part 3. Using the regression table in Exercise 13, answer the following questions.

a) How was the *t*-ratio of 2.26 found for *Police Officer Wage*? (Show what is computed using numbers from the table.)

b) How many states are used in this model. How do you know?

c) The *t*-ratio for *Graduation Rate* is negative. What does that mean?

16. Ticket prices, part 3. Using the regression table in Exercise 14, answer the following questions.

a) How was the *t*-ratio of 126.7 found for *Paid Attendance*? (Show what is computed using numbers found in the table.)

b) How many weeks are included in this regression? How can you tell?

c) The *t*-ratio for the intercept is negative. What does that mean?

17. Police salaries, part 4. Consider the coefficient of *Police Officer Wage* in the regression table of Exercise 13.

a) State the standard null and alternative hypotheses for the true coefficient of *Police Officer Wage*.

b) Test the null hypothesis (at $\alpha = 0.05$) and state your conclusion.

c) A state senate aide challenges your conclusion. She points out that we can see from the scatterplot and correlation (see Exercise 11) that there is almost no linear relationship between police officer wages and violent crime. Therefore, she claims, your conclusion in part a must be mistaken. Explain to her why this is not a contradiction.

18. Ticket prices, part 4. Consider the coefficient of *# Shows* in the regression table of Exercise 14.

a) State the standard null and alternative hypotheses for the true coefficient of *# Shows*.

b) Test the null hypothesis (at $\alpha = 0.05$) and state your conclusion.

c) A Broadway investor challenges your analysis. He points out that the scatterplot of *Receipts* vs. *# Shows* in Exercise 12 shows a strong linear relationship and claims that your result in part a can't be correct. Explain to him why this is not a contradiction.

19. Police salaries, part 5. The Senate aide in Exercise 17 now accepts your analysis but claims that it demonstrates that if the state pays police more, it will actually *increase* the rate of violent crime. Explain why this interpretation is not a valid use of this regression model. Offer some alternative explanations.

20. Ticket prices, part 5. The investor in Exercise 18 now accepts your analysis but claims that it demonstrates that it doesn't matter how many shows are playing on Broadway; receipts will be essentially the same. Explain why this interpretation is not a valid use of this regression model. Be specific.

T 21. Police salaries, part 6. Here are some plots of residuals for the regression of Exercise 13.

Which of the regression conditions can you check with these plots?

Do you find that those conditions are met?

T 22. Ticket prices, part 6. Here are some plots of residuals for the regression of Exercise 14.

Which of the regression conditions can you check with these plots?

Do you find that those conditions are met?

23. Real estate prices. A regression was performed to predict selling price of houses based on *Price* in dollars, *Area* in square feet, *Lotsize* in square feet, and *Age* in years. The R^2 is 92%. The equation from this regression is given here.

$$Price = 169,328 + 35.3\ Area + 0.718\ Lotsize - 6543\ Age$$

One of the following interpretations is correct. Which is it? Explain what's wrong with the others.

a) Each year a house ages, it is worth $6543 less.
b) Every extra square foot of area is associated with an additional $35.50 in average price, for houses with a given lot size and age.
c) Every additional dollar in price means lot size increases 0.718 square feet.
d) This model fits 92% of the data points exactly.

24. Wine prices. Many factors affect the price of wine, including such qualitative characteristics as the variety of grape, location of winery, and label. Researchers developed a regression model considering two quantitative variables: the tasting score of the wine and the age of the wine (in years) when released to market. They found the following regression equation, with an R^2 of 65%, to predict the price (in dollars) of a bottle of wine.

$$Price = 6.25 + 1.22\ Tasting\ Score + 0.55\ Age$$

One of the following interpretations is correct. Which is it? Explain what's wrong with the others.

a) Each year a bottle of wine ages, its price increases about $.55.
b) This model fits 65% of the points exactly.
c) For a unit increase in tasting score, the price of a bottle of wine increases about $1.22.
d) After allowing for the age of a bottle of wine, a wine with a one unit higher tasting score can be expected to cost about $1.22 more.

25. Appliance sales. A household appliance manufacturer wants to analyze the relationship between total sales and the company's three primary means of advertising (television, magazines, and radio). All values were in millions of dollars. They found the following regression equation.

$$Sales = 250 + 6.75\ TV + 3.5\ Radio + 2.3\ Magazine$$

One of the following interpretations is correct. Which is it? Explain what's wrong with the others.

a) If they did no advertising, their income would be $250 million.
b) Every million dollars spent on radio makes sales increase $3.5 million, all other things being equal.
c) Every million dollars spent on magazines increases TV spending $2.3 million.
d) Sales increase on average about $6.75 million for each million spent on TV, after allowing for the effects of the other kinds of advertising.

26. Wine prices, part 2. Here are some more interpretations of the regression model to predict the price of wine developed in Exercise 24. One of these interpretations is correct. Which is it? Explain what is wrong with the others.

a) The minimum price for a bottle of wine that has not aged is $6.25.
b) The price for a bottle of wine increases on average about $.55 for each year it ages, after allowing for the effects of tasting score.
c) Each year a bottle of wine ages, its tasting score increases by 1.22.
d) Each dollar increase in the price of wine increases its tasting score by 1.22.

27. Cost of pollution. What is the financial impact of pollution abatement on small firms? The U.S. government's Small Business Administration studied this and reported the following model.

$$Pollution\ abatement/employee = -2.494 - 0.431$$
$$\ln(Number\ of\ Employees) + 0.698\ \ln(Sales)$$

Pollution abatement is in dollars per employee (Crain, M. W., *The Impact of Regulatory Costs on Small Firms*, available at www.sba.gov/idc/groups/public/documents/sba_homepage/rs264tot.pdf).

a) The coefficient of ln(*Number of Employees*) is negative. What does that mean in the context of this model? What does it mean that the coefficient of ln(*Sales*) is positive?
b) The model uses the (natural) logarithms of the two predictors. What does the use of this transformation say about their effects on pollution abatement costs?

28. OECD economic regulations. A study by the U.S. Small Business Administration modeled the GDP per capita of 24 of the countries in the Organization for Economic Cooperation and Development (OECD) (Crain, M. W., *The Impact of Regulatory Costs on Small Firms*, available at www.sba.gov/idc/groups/public/documents/sba_homepage/rs264tot.pdf). One analysis estimated the effect on GDP of economic regulations, using an index of the degree of OECD economic regulation and other variables. They found the following regression model.

$$GDP/Capita(1998-2002)$$
$$= 10487 - 1343\ OECD\ Economic\ Regulation$$
$$Index + 1.078\ GDP/Capita(1988) - 69.99\ Ethno\text{-}linguistic$$
$$Diversity\ Index + 44.71\ Trade\ as\ share\ of\ GDP\ (1998-2002) -$$
$$58.4\ Primary\ Education(\%Eligible\ Population)$$

All *t*-statistics on the individual coefficients have P-values < 0.05, except the coefficient of *Primary Education*.

a) Does the coefficient of the OECD Economic Regulation Index indicate that more regulation leads to lower GDP/Capita? Explain.

b) The F-statistic for this model is 129.61 (5, 17 df). What do you conclude about the model?
c) If *GDP/Capita(1988)* is removed as a predictor, then the F-statistic drops to 0.694 and none of the *t*-statistics is significant (all P-values > 0.22). Reconsider your interpretation in part a.

29. Home prices. Many variables have an impact on determining the price of a house. A few of these are size of the house (square feet), lot size, and number of bathrooms. Information for a random sample of homes for sale in the Statesboro, Georgia, area was obtained from the Internet. Regression output modeling the asking price with square footage and number of bathrooms gave the following result.

```
Dependent Variable is: Asking Price
s = 67013   R-Sq = 71.1%   R-Sq(adj) = 64.6%
Predictor   Coeff      SE(Coeff)   t-ratio   P-value
Intercept   -152037    85619       -1.78     0.110
Baths       9530       40826        0.23     0.821
Area        139.87     46.67        3.00     0.015
Analysis of Variance
Source      DF    SS            MS          F       P-value
Regression  2     99303550067   49651775033 11.06   0.004
Residual    9     40416679100   4490742122
Total       11    1.39720E + 11
```

a) Write the regression equation.
b) How much of the variation in home asking prices is accounted for by the model?
c) Explain in context what the coefficient of Area means.
d) The owner of a construction firm, upon seeing this model, objects because the model says that the number of bathrooms has no effect on the price of the home. He says that when *he* adds another bathroom, it increases the value. Is it true that the number of bathrooms is unrelated to house price? (*Hint:* Do you think bigger houses have more bathrooms?)

30. Home prices, part 2. Here are some diagnostic plots for the home prices data from Exercise 29. These were generated by a computer package and may look different from the plots generated by the packages you use. (In particular, note that the axes of the Normal probability plot are swapped relative to the plots we've made in the text. We only care about the pattern of this plot, so it shouldn't affect your interpretation.) Examine these plots and discuss whether the assumptions and conditions for the multiple regression seem reasonable.

Normal Probability Plot

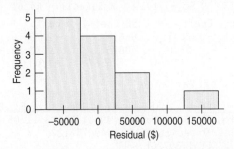

31. Secretary performance. The AFL-CIO has undertaken a study of 30 secretaries' yearly salaries (in thousands of dollars). The organization wants to predict salaries from several other variables. The variables to be considered potential predictors of salary are:

X1 = months of service

X2 = years of education

X3 = score on standardized test

X4 = words per minute (wpm) typing speed

X5 = ability to take dictation in words per minute

A multiple regression model with all five variables was run on a computer package, resulting in the following output.

Variable	Coeff	Std. Error	t-value
Intercept	9.788	0.377	25.960
X1	0.110	0.019	5.178
X2	0.053	0.038	1.369
X3	0.071	0.064	1.119
X4	0.004	0.0307	0.013
X5	0.065	0.038	1.734

s = 0.430 R-sq = 0.863

Assume that the residual plots show no violations of the conditions for using a linear regression model.

a) What is the regression equation?

b) From this model, what is the predicted salary (in thousands of dollars) of a secretary with 10 years (120 months)

of experience, 9th grade education (9 years of education), 50 on the standardized test, 60 wpm typing speed, and the ability to take 30 wpm dictation?

c) Test whether the coefficient for words per minute of typing speed (X4) is significantly different from zero at $\alpha = 0.05$.

d) How might this model be improved?

e) A correlation of age with salary finds $r = 0.682$, and the scatterplot shows a moderately strong positive linear association. However, if X6 = Age is added to the multiple regression, the estimated coefficient of age turns out to be $b_6 = -0.154$. Explain some possible causes for this apparent change of direction in the relationship between age and salary.

T 32. Wal-Mart revenue. Here's a regression of monthly revenue of Wal-Mart Corp, relating that revenue to the Total U.S. Retail Sales, the Personal Consumption Index, and the Consumer Price Index.

```
Dependent variable is: Wal-Mart_Revenue
R squared = 66.7%   R squared (adjusted) = 63.8%
s = 2.327 with 39 - 4 = 35 degrees of freedom
```

Source	Sum of Squares	df	Mean Square	F-ratio
Regression	378.749	3	126.250	23.3
Residual	189.474	35	5.41354	

Variable	Coeff	SE(Coeff)	t-ratio	P-value
Intercept	87.0089	33.60	2.59	0.0139
Retail Sales	0.000103	0.000015	6.67	< 0.0001
Persnl Consmp	0.00001108	0.000004	2.52	0.0165
CPI	−0.344795	0.1203	−2.87	0.0070

a) Write the regression model.

b) Interpret the coefficient of the Consumer Price Index (CPI). Does it surprise you that the sign of this coefficient is negative? Explain.

c) Test the standard null hypothesis for the coefficient of CPI and state your conclusions.

T 33. Wilshire 5000. In Chapter 16, Exercise 41 considered the relationship between the Wilshire 5000 Total Market Return and the amount of money flowing into and out of mutual funds (fund flows) monthly from January 1990 through October of 2002. The data file included data on the unemployment rate in each of the months. The original model looked like this.

```
Dependent variable is: Fund_Flows
R squared = 17.9%   R squared (adjusted) = 17.3%
s = 10999 with 154 − 2 = 152 degrees of freedom

Source       Sum of Squares   df   Mean Square   F-ratio
Regression   4002044231        1   4002044231    33.1
Residual     18389954453     152    120986542

Variable    Coeff    SE(Coeff)   t-ratio   P-value
Intercept   9599.10    896.9      10.7     <0.0001
Wilshire    1156.40    201.1       5.75    <0.0001
```

Adding the *Unemployment Rate* to the model yields:

```
Dependent variable is: Fund_Flows
R squared = 28.8%   R squared (adjusted) = 27.8%
s = 10276 with 154 − 3 = 151 degrees of freedom

Source       Sum of Squares   df   Mean Square   F-ratio
Regression   6446800389        2   3223400194    30.5
Residual     15945198295     151    105597340

Variable      Coeff      SE(Coeff)   t-ratio   P-value
Intercept     30212.2     4365        6.92     ≤0.0001
Wilshire       1183.29     187.9       6.30    <0.0001
Unemployment  −3719.55     773.0      −4.81    <0.0001
```

a) Interpret the coefficient of the *Unemployment Rate*.
b) The *t*-ratio for the *Unemployment Rate* is negative. Explain why.
c) State and complete the standard hypothesis test for the *Unemployment Rate*.

T **34.** Lobster industry 2008, revisited. In Chapter 17, Exercise 53 predicted the annual value of the Maine lobster industry catch from the number of licensed lobser fishers. The lobster industry is an important one in Maine, with annual landings worth about $300,000,000 and employment consequences that extend throughout the state's economy. We saw in Chapter 17 that it was best to transform Value by logarithms. Here's a more sophisticated multiple regression to predict the logValue from other variables published by the Maine Department of Marine Resources (maine.gov/dmr). The predictors are number of *Traps* (millions), number of licensed *Fishers*, and *Water Temperature* (°F).

```
Dependent variable is: LogValue
R squared = 97.4%   R squared (adjusted) = 97.3%
s = 0.0782 with 56 − 4 = 52 degrees of freedom

Source       Sum of Squares   df   Mean Square   F-ratio
Regression   12.0200           3   4.00667       656
Residual      0.317762        52   0.006111

Variable     Coeff       SE(Coeff)   t-ratio   P-value
Intercept     1.70207    0.3706       4.59    <0.0001
Traps(M)      0.575390   0.0152      37.8     <0.0001
Fishers      −0.0000532  0.00001     −5.43    <0.0001
Water Temp   −0.015185   0.0074      −2.05     0.0457
```

a) Write the regression model.
b) Are the assumptions and conditions met?
c) Interpret the coefficient of *Fishers*. Would you expect that restricting the number of lobstering licenses to even fewer fishers would increase the value of the harvest?
d) State and test the standard null hypothesis for the coefficient of *Water Temperature*. Scientists claim that this is an important predictor of the harvest. Do you agree?

T **35.** Lobster industry 2008, part 2. In Chapter 17, Exercise 54 predicted the price ($/lb) of lobster harvested in the Maine lobster fishing industry. Here's a multiple regression to predict the *Price* from the number of *Traps* (millions), the number of *Fishers*, and the *Catch/Trap* (metric tonnes).

```
Dependent variable is: Price/lb
R squared = 94.7%   R squared (adjusted) = 94.4%
s = 0.2703 with 56 − 4 = 52 degrees of freedom

Source       Sum of Squares   df   Mean Square   F-ratio
Regression   68.0259           3   22.6753       310
Residual      3.80006         52    0.073078

Variable     Coefficient   SE(Coeff)   t-ratio   P-value
Intercept     0.565367     0.3770       1.50      0.1397
Traps(M)      1.30744      0.0558      23.4      ≤0.0001
Fishers      −0.00012      0.0000      −2.94      0.0049
Catch/Trap   −9.97788     11.22       −0.889     0.3781
```

a) Write the regression model.
b) Are the assumptions and conditions met?
c) State and test the standard null hypothesis for the coefficient of *Catch/Trap*. Use the standard α-level of .05 and state your conclusion.
d) Does the coefficient of *Catch/Trap* mean that when the catch per trap declines the price will increase?
e) This model has an adjusted R^2 of 94.7%. The previous model of Chapter 17, Exercise 34, had an adjusted R^2 of 93.6%. What does adjusted R^2 say about the two models?

T 36. Price of beef. How is the price of beef related to other factors? The data in the following table give information on the price of beef (*PBE*) and the possible explanatory variables: consumption of beef per capita (*CBE*), retail food price index (*PFO*), food consumption per capita index (*CFO*), and an index of real disposable income per capita (*RDINC*) for the years 1925 to 1941 in the United States.

Year	PBE	CBE	PFO	CFO	RDINC
1925	59.7	58.6	65.8	90.9	68.5
1926	59.7	59.4	68	92.1	69.6
1927	63	53.7	65.5	90.9	70.2
1928	71	48.1	64.8	90.9	71.9
1929	71	49	65.6	91.1	75.2
1930	74.2	48.2	62.4	90.7	68.3
1931	72.1	47.9	51.4	90	64
1932	79	46	42.8	87.8	53.9
1933	73.1	50.8	41.6	88	53.2
1934	70.2	55.2	46.4	89.1	58
1935	82.2	52.2	49.7	87.3	63.2
1936	68.4	57.3	50.1	90.5	70.5
1937	73	54.4	52.1	90.4	72.5
1938	70.2	53.6	48.4	90.6	67.8
1939	67.8	53.9	47.1	93.8	73.2
1940	63.4	54.2	47.8	95.5	77.6
1941	56	60	52.2	97.5	89.5

a) Use computer software to find the regression equation for predicting the price of beef based on all of the given explanatory variables. What is the regression equation?

b) Produce the appropriate residual plots to check the assumptions. Is this inference appropriate for this model? Explain.
c) How much variation in the price of beef can be explained by this model?
d) Consider the coefficient of beef consumption per capita (*CBE*). Does it say that that price of beef goes up when people eat less beef? Explain.

T 37. Wal-Mart revenue, part 2. Wal-Mart is the second largest retailer in the world. The data file on the disk holds monthly data on Wal-Mart's revenue, along with several possibly related economic variables.

a) Using computer software, find the regression equation predicting Wal-Mart revenues from the *Retail Index*, the Consumer Price index (*CPI*), and *Personal Consumption*.
b) Does it seem that Wal-Mart's revenue is closely related to the general state of the economy?

T 38. Wal-Mart revenue, part 3. Consider the model you fit in Exercise 37 to predict Wal-Mart's revenue from the Retail Index, CPI, and Personal Consumption index.

a) Plot the residuals against the predicted values and comment on what you see.
b) Identify and remove the four cases corresponding to December revenue and find the regression with December results removed.
c) Does it seem that Wal-Mart's revenue is closely related to the general state of the economy?

39.* Right-to-work laws. Are state right-to-work laws related to the percent of public sector employees (*Public*) in unions and the percent of private sector employees (*Private*) in unions? (Each predictor variable has a value between 0 and 100.) This data set looks at these percentages for the states in the United States in 1982. The dependent variable is whether the state had a right-to-work law or not (1 = Yes, 0 = No). The computer output for the logistic regression is given here. (Source: N. M. Meltz, "Interstate and Interprovincial Differences in Union Density," *Industrial Relations*, 28:2 (Spring 1989), 142–158 by way of DASL.)

```
Logistic Regression Table
```

Predictor	Coeff	SE(Coeff)	z	P
Intercept	6.19951	1.78724	3.47	0.001
publ	−0.106155	0.0474897	−2.24	0.025
pvt	−0.222957	0.0811253	−2.75	0.006

a) Write out the estimated regression equation.
b) What is the predicted log odds (logit) of the probability that a state has a right-to-work law if 20% of the public and 10% of the private sector employees are in unions?
c) What is the predicted probability of that state?
d) What is the probability log odds (logit) that a state has a right-to-work law if 40% of the public and 20% of the private sector employees are in unions?
e) What is the associated predicted probability?

40.* Cost of higher education. Are there fundamental differences between liberal arts colleges and universities? In this case, we have information on the top 25 liberal arts colleges and the top 25 universities in the United States. We will consider the type of school as our response variable and will use the percent of students who were in the top 10% of their high school class and the amount of money spent per student by the college or university as our explanatory variables. The output from this logistic regression is given here.

Logistic Regression Table

Predictor	Coeff	SE(Coeff)	z	P
Intercept	−13.1461	3.98629	−3.30	0.001
Top 10%	0.0845469	0.0396345	2.13	0.033
$/Student	0.0002594	0.0000860	3.02	0.003

a) Write out the estimated regression equation.
b) Is percent of students in the top 10% of their high school class statistically significant in predicting whether or not the school is a university? Explain.
c) Is the amount of money spent per student statistically significant in predicting whether or not the school is a university? Explain.

T 41. Motorcycles. More than one million motorcycles are sold annually (www.webbikeworld.com). Off-road motorcycles (often called "dirt bikes") are a market segment (about 18%) that is highly specialized and offers great variation in features. This makes it a good segment to study to learn about which features account for the cost (manufacturer's suggested retail price, MSRP) of a dirt bike. Researchers collected data on 2005 model dirt bikes (lib.stat.cmu.edu/datasets/dirtbike_aug.csv). Their original goal was to study market differentiation among brands (*The Dirt on Bikes: An Illustration of CART Models for Brand Differentiation*, Jiang Lu, Joseph B. Kadane, and Peter Boatwright, server1.tepper.cmu.edu/gsiadoc/WP/2006-E57.pdf), but we can use these to predict msrp from other variables.

Here are scatterplots of three potential predictors, *Wheelbase (in)*, *Displacement (cu in)*, and *Bore (in)*.

Comment on the appropriateness of using these variables as predictors on the basis of the scatterplots.

T 42. Motorcycles, part 2. In Exercise 41, we saw data on off-road motorcycles and examined scatterplots. Review those scatterplots. Here's a regression of *MSRP* on both *Displacement* and *Bore*. Both of the predictors are measures of the size of the engine. The displacement is the total volume of air and fuel mixture that an engine can draw in during one cycle. The bore is the diameter of the cylinders.

Dependent variable is: MSRP
R squared = 77.0%   R squared (adjusted) = 76.5%
s = 979.8 with 98 − 3 = 95 degrees of freedom

Variable	Coeff	SE(Coeff)	t-ratio	P-value
Intercept	318.352	1002	0.318	0.7515
Bore	41.1650	25.37	1.62	0.1080
Displacement	6.57069	3.232	2.03	0.0449

a) State and test the standard null hypothesis for the coefficient of *Bore*.
b) Both of these predictors seem to be linearly related to *MSRP*. Explain what your result in part a means.

T 43. Motorcycles, part 3. Here's another model for the *MSRP* of off-road motorcycles.

Dependent variable is: MSRP
R squared = 90.9%   R squared (adjusted) = 90.6%
s = 617.8 with 95 − 4 = 91 degrees of freedom

Source	Sum of Squares	df	Mean Square	F-ratio
Regression	346795061	3	115598354	303
Residual	34733372	91	381685	

Variable	Coeff	SE(Coeff)	t-ratio	P-value
Intercept	−2682.38	371.9	−7.21	<0.0001
Bore	86.5217	5.450	15.9	<0.0001
Clearance	237.731	30.94	7.68	<0.0001
Engine strokes	−455.897	89.88	−5.07	<0.0001

a) Would this be a good model to use to predict the price of an off-road motorcycle if you knew its bore, clearance, and engine strokes? Explain.

b) The Suzuki DR650SE had an *MSRP* of $4999 and a 4-stroke engine, with a bore of 100 inches. Can you use this model to estimate its *Clearance*? Explain.

44. Demographics. Following is a data set on various measures of the 50 United States. The *Murder* rate is per 100,000, *HS Graduation* rate is in %, *Income* is per capita income in dollars, *Illiteracy* rate is per 1000, and *Life Expectancy* is in years. Find a regression model for *Life Expectancy* with three predictor variables by trying all four of the possible models.

a) Which model appears to do the best?

b) Would you leave all three predictors in this model?

c) Does this model mean that by changing the levels of the predictors in this equation, we could affect life expectancy in that state? Explain.

d) Be sure to check the conditions for multiple regression. What do you conclude?

45. Burger King nutrition. Like many fast-food restaurant chains, Burger King (BK) provides data on the nutrition content of its menu items on its website. Here's a multiple regression predicting calories for Burger King foods from *Protein* content (g), *Total Fat* (g), *Carbohydrate* (g), and *Sodium* (mg) per serving.

```
Dependent variable is: Calories
R-squared = 100.0%   R-squared [adjusted] = 100.0%
s = 3.140 with 31 - 5 = 26 degrees of freedom
```

Source	Sum of Squares	df	Mean Square	F-ratio
Regression	1419311	4	354828	35994
Residual	256.307	26	9.85796	

Variable	Coeff	SE(Coeff)	t-ratio	P-value
Intercept	6.53412	2.425	2.69	0.0122
Protein	3.83855	0.0859	44.7	<0.0001
Total fat	9.14121	0.0779	117	<0.0001
Carbs	3.94033	0.0338	117	<0.0001
Na/Serv.	−0.69155	0.2970	−2.33	0.0279

a) Do you think this model would do a good job of predicting calories for a new BK menu item? Why or why not?

b) The mean of *Calories* is 455.5 with a standard deviation of 217.5. Discuss what the value of s in the regression means about how well the model fits the data.

c) Does the R^2 value of 100.0% mean that the residuals are all actually equal to zero?

Building Multiple Regression Models

Bolliger and Mabillard

The story is told that when John Wardley, the award-winning concept designer of theme parks and roller coasters, including *Nitro* and *Oblivion*, was about to test the roller coaster *Nemesis* for the first time, he asked Walter Bolliger, the president of the coaster's manufacturer, B&M, "What if the coaster stalls? How will we get the trains back to the station?" Bolliger replied, "Our coasters never stall. They always work perfectly the first time." And, of course, it did work perfectly.

Roller coaster connoisseurs know that Bolliger & Mabillard Consulting Engineers, Inc. (B&M) is responsible for some of the most innovative roller coasters in the business. The company was founded in the late 1980s when Walter Bolliger and Claude Mabillard left Intamin AG, where they had designed the company's first stand-up coaster. B&M has built its reputation on innovation. They developed the first "inverted" roller coaster in which the train runs under the track with the seats attached to the wheel carriage and pioneered "diving machines," which feature a vertical drop, introduced first with *Oblivion*.

B&M coasters are famous among enthusiasts for particularly smooth rides and for their reliability, easy maintainability, and excellent safety record. Unlike some other manufacturers, B&M does not use powered launches, preferring, as do many coaster connoisseurs, gravity-powered coasters. B&M is an international leader in the roller coaster design field, having designed 24 of the top 50 steel roller coasters on the 2009 Golden Ticket Awards list and 4 of the top 10.

WHO	Roller coasters
WHAT	See Table 19.1 for the variables and their units.
WHERE	Worldwide
WHEN	All were in operation in 2003.
WHY	To understand characteristics that affect speed and duration

Theme parks are big business. In the United States alone, there are nearly 500 theme and amusement parks that generate over $10 billion a year in revenue. The U.S. industry is fairly mature, but parks in the rest of the world are still growing. Europe now generates more than $1 billion a year from their theme parks, and Asia's industry is growing fast. Although theme parks have started to diversify to include water parks and zoos, rides are still the main attraction at most parks, and at the center of the rides is the roller coaster. Engineers and designers compete to make them bigger and faster. For a two-minute ride on the fastest and best roller coasters, fans will wait for hours.

Can we learn what makes a roller coaster fast? What are the most important design considerations in getting the fastest coaster? Here are data on some of the fastest roller coasters in the world today:

Name	Park	Country	Type	Duration (sec)	Speed (mph)	Height (ft)	Drop (ft)	Length (ft)	Inversion?
New Mexico Rattler	Cliff's Amusement Park	USA	Wooden	75	47	80	75	2750	No
Fujiyama	Fuji-Q Highlands	Japan	Steel	216	80.8	259.2	229.7	6708.67	No
Goliath	Six Flags Magic Mountain	USA	Steel	180	85	235	255	4500	No
Great American Scream Machine	Six Flags Great Adventure	USA	Steel	140	68	173	155	3800	Yes
Hangman	Wild Adventures	USA	Steel	125	55	115	95	2170	Yes
Hayabusa	Tokyo SummerLand	Japan	Steel	108	60.3	137.8	124.67	2559.1	No
Hercules	Dorney Park	USA	Wooden	135	65	95	151	4000	No
Hurricane	Myrtle Beach Pavilion	USA	Wooden	120	55	101.5	100	3800	No

A small selection of coasters from the larger data set available on the CD.

| Table 19.1   Facts about some roller coasters. Source: The Roller coaster database at www.rcdb.com.

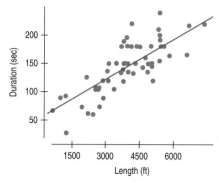

Figure 19.1 The relationship between *Duration* and *Length* looks strong and positive. On average, *Duration* increases linearly with *Length*.

- *Type* indicates what kind of track the roller coaster has. The possible values are "wooden" and "steel." (The frame usually is of the same construction as the track, but doesn't have to be.)
- *Duration* is the duration of the ride in seconds.
- *Speed* is top speed in miles per hour.
- *Height* is maximum height above ground level in feet.
- *Drop* is greatest drop in feet.
- *Length* is total length of the track in feet.
- *Inversions* reports whether riders are turned upside down during the ride. It has the values "yes" or "no."

Customers not only want the ride to be fast; they also want it to last. It makes sense that the longer the track is, the longer the ride will last. Let's have a look at *Duration* and *Length* to see what the relationship is.

```
Dependent variable is:  Duration
R-squared = 62.0%   R-squared (adjusted) = 61.4%
s = 27.23 with 63 - 2 = 61 degrees of freedom
```

Source	Sum of Squares	DF	Mean Square	F-ratio
Regression	73901.7	1	73901.7	99.6
Residual	45243.7	61	741.700	

Variable	Coefficient	SE(Coeff)	t-ratio	P-value
Intercept	53.9348	9.488	5.68	<0.0001
Length	0.0231	0.0023	9.98	<0.0001

| Table 19.2   The regression of *Duration* on *Length* looks strong, and the conditions seem to be met.

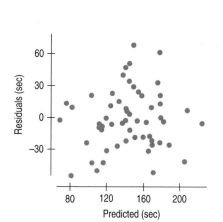

Figure 19.2 Residuals from the regression in Table 19.2

As the scatterplots of the variables (Figure 19.1) and residuals (Figure 19.2) show, the regression conditions seem to be met, and the regression makes sense; we'd expect longer tracks to give longer rides. Starting from the intercept at about 53.9 seconds, the duration of the ride increases, on average, by 0.0231 seconds per foot of track—or 23.1 seconds more for each 1000 additional feet of track.

19.1   Indicator (or Dummy) Variables

Of course, there's more to these data. One interesting variable might not be one you'd naturally think of. Many modern coasters have "inversions" where riders get turned upside down, with loops, corkscrews, or other devices. These inversions add excitement, but they must be carefully engineered, and that enforces some speed limits on that portion of the ride. Riders like speed, but they also like inversions which affect both the speed and the duration of the ride. We'd like to add the information of whether the roller coaster has an inversion to our model. Until now, all our predictor variables have been quantitative. Whether or not a roller coaster has an inversion is a categorical variable ("yes" or "no"). Let's see how to introduce the categorical variable *Inversions* as a predictor in our regression model. Figure 19.3 shows the same scatterplot of *Duration* against *Length*, but now with the roller coasters that have inversions shown as red x's and those without shown as blue dots. There's a separate regression line for each type of roller coaster.

Figure 19.3   The two lines fit to coasters with (red x's) and without (blue dots) inversions are roughly parallel.

It's easy to see that, for a given length, the roller coasters with inversions take a bit longer and that for each type of roller coaster, the slopes of the relationship between duration and length are not quite equal but are similar. If we split the data into two groups—coasters without inversions and those with inversions—and compute the regression for each group, the output looks like this:

```
Dependent variable is: Duration
Cases selected according to: No inversions
R-squared = 69.4%   R-squared (adjusted) = 68.5%
s = 25.12 with 38 − 2 = 36 degrees of freedom
```

Variable	Coefficient	SE(Coeff)	t-ratio	P-value
Intercept	25.9961	14.10	1.84	0.0734
Length	0.0274	0.003	9.03	<0.0001

```
Dependent variable is: Duration
Cases selected according to: Inversions
R-squared = 70.5%  R-squared (adjusted) = 69.2%
s = 23.20 with 25 − 2 = 23 degrees of freedom
```

Variable	Coefficient	SE(Coeff)	t-ratio	P-value
Intercept	47.6454	12.50	3.81	0.0009
Length	0.0299	0.004	7.41	<0.0001

Table 19.3   The regressions computed separately for the two types of roller coasters show similar slopes but different intercepts.

As the scatterplot showed, the slopes are very similar, but the intercepts are different. When we have a situation like this with roughly parallel regressions for each group,[1] there's an easy way to add the group information to a single regression model. We create a new variable that *indicates* what type of roller coaster we have, giving it the value 1 for roller coasters that have inversions and the value 0 for those that don't. (We could have reversed the coding; it's an arbitrary choice.[2]) Such variables are called **indicator variables** or *indicators* because they indicate which category each case is in. They are also often called **dummy variables**. When we add our new indicator, *Inversions*, to the regression model as a second variable, the multiple regression model looks like this.

> *Inversions* = 1 if a coaster has inversions.
> *Inversions* = 0 if not.

```
Dependent variable is: Duration
R-squared = 70.4%   R-squared (adjusted) = 69.4%
s = 24.24 with 63 - 3 = 60 degrees of freedom

Variable      Coefficient   SE(Coeff)   t-ratio   P-value
Intercept     22.3909       11.39       1.97      0.0539
Length         0.0282        0.0024     11.7      <0.0001
Inversions    30.0824        7.290       4.13     <0.0001
```

| **Table 19.4** The regression model with a dummy, or indicator, variable for Inversions.

This looks like a better model than the simple regression for all the data. The R^2 is larger, the *t*-ratios of both coefficients are large, and now we can understand the effect of inversions with a single model without having to compare two regressions. (The residuals look reasonable as well.) But what does the coefficient for *Inversions* mean? Let's see how an indicator variable works when we calculate predicted values for two of the roller coasters listed in Table 19.1.

Name	Park	Country	Type	Duration	Speed	Height	Drop	Length	Inversion?
Hangman	Wild Adventures	USA	Steel	125	55	115	95	2170	Yes
Hayabusa	Tokyo SummerLand	Japan	Steel	108	60.3	137.8	124.67	2559.1	No

Ignoring the variable *Inversions* for the moment, the model (in Table 19.4) says that for all coasters, the predicted *Duration* is:

$$22.39 + 0.0282 \, Length + 30.08 \, Inversions.$$

Now remember that for this indicator variable, the value 1 means that a coaster has an inversion, while a 0 means it doesn't. For *Hayabusa*, with no inversion, the value of *Inversions* is 0, so the coefficient of *Inversions* doesn't affect the prediction at all. With a length of 2259.1 feet, we predict its duration as:[3]

$$22.39 + 0.0282 \, (2559.1) + 30.08 \times 0 = 94.56 \text{ seconds,}$$

which is close to its actual duration of 108 seconds. The *Hangman* (with a length of 2170 feet) has an inversion, and so the model predicts an "additional" 30.0824 seconds for its duration:

$$22.39 + 0.0282 \, (2170.0) + 30.08 \times 1 = 113.66 \text{ seconds.}$$

That compares well with the actual duration of 125 seconds.

---

[1]The fact that the individual regression lines are nearly parallel is a part of the Linearity Condition. You should check that the lines are nearly parallel before using this method or read on to see what to do if they are not parallel enough.

[2]Some implementations of indicator variables use 1 and −1 for the levels of the categories.

[3]We round coefficient values when we write the model but calculate with the full precision, rounding at the end of the calculation.

Notice how the indicator works in the model. When there is an inversion (as in *Hangman*), the value 1 for the indicator causes the amount of the indicator's coefficient, 30.08, to be *added* to the prediction. When there is no inversion (as in *Hayabusa*), the indicator is 0, so nothing is added. Looking back at the scatterplot, we can see that this is exactly what we need. The difference between the two lines is a vertical shift of about 30 seconds. This may seem a bit confusing at first because we usually think of the coefficients in a multiple regression as slopes. For indicator variables, however, they act differently. They're vertical shifts that keep the slopes for the other variables apart.

An indicator variable that is 0 or 1 can only shift the line up and down. It can't change the slope, so it works only when we have lines with the same slope and different intercepts.

Indicators for Three or More Categories

Business and economic variables such as *Month* or *Socioeconomic Class* may have several levels, not just two. You can construct indicators for a categorical variable with several levels by constructing a separate indicator for each of these levels. There's just one thing to keep in mind. If a variable has k levels, you can create only $k - 1$ indicators. You have to choose one of the k categories as a "baseline" and *leave out* its indicator. Then the coefficients of the other indicators can be interpreted as the amount by which their categories differ from the baseline, after allowing for the linear effects of the other variables in the model.

For the two-category variable *Inversions*, we used "no inversion" as the baseline, and coasters with an inversion got a 1. We needed only one variable for two levels. If we wished to represent *Month* with indicators, we would need 11 of them. We might, for example, define *January* as the baseline and make indicators for *February*, *March*, . . . , *November*, and *December*. Each of these indicators would be 0 for all cases except for the ones that had that value for the variable *Month*.

Why couldn't we use a single variable with "1" for *January*, "2" for *February*, and so on? That would require the pretty strict assumption that the responses to the months are linear and equally spaced—that is, that the change in our response variable from January to February is the same in both direction and amount as the change from July to August. That's a pretty severe restriction and usually isn't true. Using 11 indicators releases the model from that restriction even though it adds complexity to the model.

Once you've created multiple indicator variables (up to $k - 1$) for a categorical variable with k levels, it often helps to combine levels with similar characteristics and with similar relationships with the response. This can help keep the number of variables in a multiple regression from exploding.

For Example   Indicator variables

In Chapter 17 we looked at data on the prices of diamonds (see page 535). Although, in general, higher carat weight means higher cost, *Carat Weight* is not the only factor that determines a diamond's price. A typical diamond is pale yellow. The less color a diamond has, the higher its color grade and—generally—its price. We want to build a model that incorporates both *Carat Weight* and *Color* to predict a diamond's price. The data here are a collection of 749 diamonds: *Weight* is between 0.3 and 1.4 carats; color grades are D (highest possible), G (medium), and K (fairly low). As before, we'll use *Log₁₀Price* as the response to make the relationship more linear. Here are scatterplots of *Price* and *Log₁₀Price* versus *Carat Weight* for these diamonds:

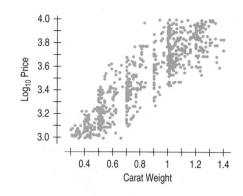

A linear model for $Log_{10}Price$ on *Carat Weight* alone finds:

```
Response Variable: Log₁₀Price
R² = 77.39%   Adjusted R² = 77.36%
s = 0.1364 with 749 - 2 = 747 degrees of freedom

Variable       Coeff      SE(Coeff)    t-ratio    P-value
Intercept      2.76325    0.01532      180.42     <0.0001
Carat.Weight   0.90020    0.01780       50.56     <0.0001
```

Here is a scatterplot of $Log_{10}Price$ vs. *Carat Weight* showing the three different color grades (D, G, and K):

To account for the differences, two indicator variables were created:

$$Color\ D = 1 \text{ if } Color = D \text{ and } 0 \text{ otherwise}$$
$$Color\ G = 1 \text{ if } Color = G \text{ and } 0 \text{ otherwise}$$

A multiple regression of $Log_{10}Price$ was run with three predictors: *Carat Weight*, *Color D*, and *Color G*

Here is the output:

```
Response Variable: Log₁₀Price
R² = 85.66%   Adjusted R² = 85.61%
s = 0.1088 with 749 - 4 = 745 degrees of freedom

Variable       Coeff      SE(Coeff)    t-ratio    P-value
Intercept      2.43978    0.01981      123.14     <0.0001
Carat.Weight   1.02865    0.01589       64.75     <0.0001
ColorD         0.29405    0.01424       20.65     <0.0001
ColorG         0.21883    0.01278       17.12     <0.0001
```

Questions: For the three colors of diamonds, what are the models that predict $Log_{10}Price$? Does the addition of the indicator variables for the colors seem like a good idea? Why are there only two indicator variables?

Answers: The indicator variable adds amounts to the intercept value for the two color levels, D and G. Since diamonds of color K have both dummy variables set to 0, the equation for diamonds with *Color K* is simply:

$$\widehat{Log_{10}Price} = 2.440 + 1.029\ Carat\ Weight$$

(continued)

For diamonds with *Color G*, we add 0.219 to the intercept, so:

$$\widehat{Log_{10}Price} = (2.440 + 0.219) + 1.029\ Carat\ Weight = 2.659 + 1.029\ Carat\ Weight$$

Similarly for diamonds with *Color D*, we add 0.294 to the intercept, so:

$$\widehat{Log_{10}Price} = (2.440 + 0.294) + 1.029\ Carat\ Weight = 2.734 + 1.029\ Carat\ Weight$$

Both indicator variables are highly significant, which means that the differences in intercept are large enough to justify the addition of the variables.

With 3 levels of color, we need only 2 indicator variables. (In general *k* levels require *k* − 1 indicators.)

19.2   Adjusting for Different Slopes— Interaction Terms

WHO	Burger King menu items
WHAT	Calories, Carbohydrates, Protein, Type of item
WHEN	Current
WHERE	Posted in stores and online
WHY	Informing customers

Even consumers of fast food are increasingly concerned with nutrition. So, like many restaurant chains, Burger King publishes the nutrition details of its menu items on its website (www.bk.com). Many customers count calories or carbohydrates. Of course, these are likely to be related to each other. We can examine that relationship in Burger King foods by looking at a scatterplot.

Figure 19.4   *Calories* of Burger King foods plotted against *Carbs* seems to fan out.

Burger King

James McLamore and David Edgerton, graduates of the Cornell University School of Hotel Administration, opened their first Burger King (BK) restaurant in 1954. McLamore had been impressed with the assembly-line–based production innovated by Dick and Mac MacDonald. BK grew in the Miami area and then expanded nationally through franchising. Pillsbury acquired BK in 1967 and expanded it internationally. After a number of owners, BK was purchased by private investors and taken public. The initial public offering (IPO) in 2006 generated $425 million in revenue—the largest IPO of a U.S.-based restaurant chain on record.

It's not surprising to see that an increase in *Carbs* is associated with more *Calories*, but the plot seems to thicken as we move from left to right. Could there be something else going on?

We can divide Burger King foods into two groups, coloring those with meat (including chicken and fish) with red x's and those without meat with blue dots. Looking at the regressions for each group, we see a different picture:

Figure 19.5   Plotting the meat-based and non-meat items separately, we see two distinct linear patterns.

Clearly, meat-based items contribute more calories from their carbohydrate content than do other Burger King foods. But unlike Figure 19.3, when the lines were parallel, we can't account for the kind of difference we see here by just including a dummy variable in a regression. It isn't just the intercepts of the lines that are different; they have entirely different slopes.

We'll start, as before, by constructing the indicator for the two groups, *Meat*, which is 1 for foods that contain meat and 0 for the others. The variable *Meat* can adjust the intercepts of the two lines. To adjust the slopes, we have to construct another variable—the *product* of *Meat* and the predictor variable *Carbs*. The coefficient of this **interaction term** in a multiple regression gives an adjustment to the slope for the cases in the indicated group. The resulting variable *Carbs*Meat* has the value of *Carbs* for foods containing meat (those coded 1 in the *Meat* indicator) and the value 0 for the others. By including this interaction variable in the model, we can adjust the slope of the line fit to the meat-containing foods. Here's the resulting analysis:

```
Dependent variable is: Calories
R-squared = 78.1%   R-squared (adjusted) = 75.7%
s = 106.0 with 32 - 4 = 28 degrees of freedom
```

Source	Sum of Squares	DF	Mean Square	F-ratio
Regression	1119979	3	373326	33.2
Residual	314843	28	11244.4	

Variable	Coefficient	SE(Coeff)	t-ratio	P-value
Intercept	137.395	58.72	2.34	0.0267
Carbs(g)	3.93317	1.113	3.53	0.0014
Meat	−26.1567	98.48	−0.266	0.7925
Carbs*Meat	7.87530	2.179	3.61	0.0012

| **Table 19.5** The regression model with both an indicator variable and an interaction term.

What does the coefficient for the indicator *Meat* do? It provides a different intercept to separate the meat and non-meat items at the origin (where *Carbs* = 0). Each group has its own slope, but the two lines nearly meet at the origin, so there seems to be no need for an additional intercept adjustment. The difference of 26.16 calories is small. That's why the coefficient for the indicator variable *Meat* has a small t-ratio (−0.266).

By contrast, the coefficient of the interaction term, *Carbs*Meat*, says that the slope relating calories to carbohydrates is steeper by 7.88 calories per carbohydrate gram for meat-containing foods than for meat-free foods. Its small P-value suggests that this difference is real. Overall, the regression model predicts calories to be:

$$137.40 + 3.93 \text{ } Carbs - 26.16 \text{ } Meat + 7.88 \text{ } Carbs*Meat.$$

Let's see how these adjustments work. A BK Whopper has 53 grams of carbohydrates and is a meat dish. The model predicts its *Calories* as:

$$137.40 + 3.93 \times 53 - 26.16 \times 1 + 7.88 \times 53 \times 1 = 737.2 \text{ calories,}$$

not far from the measured calorie count of 680. By contrast, the Veggie Burger, with 43 grams of carbohydrates, has value 0 for *Meat* and so has a value of 0 for *Carbs*Meat* as well. Those indicators contribute nothing to its predicted calories:

$$137.40 + 3.93 \times 43 - 26.16 \times 0 + 7.88 \times 0 \times 43 = 306.4 \text{ calories,}$$

close to the 330 measured officially.

For Example   Adding interaction terms

After adding the indicator variables for *Color*, which showed different intercepts for the three different color levels, an analyst wonders if the slopes for the three color levels might be different as well, so she adds two more predictors, *ColorD*Carat Weight* and *ColorG*Carat Weight* to the model (see page 622). The regression output shows:

```
Response Variable: Log₁₀Price
R² = 85.77%   Adjusted R² = 85.67%
s = 0.1085 with 749 − 6 = 743 degrees of freedom
```

Variable	Coeff	SE(Coeff)	t-ratio	P-value
Intercept	2.54151	0.06000	42.361	<0.0001
Carat.Weight	0.92968	0.05734	16.212	<0.0001
ColorD	0.16466	0.06302	2.613	0.00916
ColorG	0.12656	0.06287	2.013	0.04448
Carat.Weight*ColorD	0.14076	0.06356	2.215	0.02709
Carat.Weight*ColorG	0.08809	0.06095	1.445	0.14877

Question: Based on this, what model might you use to predict *Log₁₀Price*?

Answer: The only significant interaction term is the one for *Color D* by *Carat Weight*, so we probably don't need different slopes for levels *G* and *K*. We should refit the regression without the interaction term *ColorG*Carat Weight*. When we do, we find that the term for *Carat Weight*ColorD* is marginally significant. The decision whether to include it is a judgment call. To make a final decision, we should also consider residual analysis and diagnostics. For the sake of simplicity we will proceed with the simpler model with only the indicator variables and not the interaction (the model fit in the previous example).

19.3   Multiple Regression Diagnostics

We often use regression analyses to make important business decisions. By working with the data and creating models, we can learn a great deal about the relationships among variables. As we saw with simple regression, sometimes we can learn as much from the cases that *don't* fit the model as from the bulk of cases that do. Extraordinary cases often tell us more just by the ways in which they fail to conform and the reasons we can discover for those deviations. If a case doesn't conform to the others, we should identify it and, if possible, understand why it is different. In simple regression, a case can be extraordinary by standing away from the model in the *y* direction or by having unusual values in an *x*-variable. In multiple regression, it can also be extraordinary by having an unusual *combination* of values in the *x*-variables. Just as in simple regression, large deviations in the *y* direction show up in the residuals as outliers. Deviations in the *x*'s show up as *leverage*.

Leverage

In a regression of a single predictor and a response, it's easy to see if a value has high leverage, because it's far from the mean of the *x* values in a scatterplot. In a multiple regression with *k* predictor variables, things are more complicated. A point might actually not be far from any of the *x* means and yet still exert large leverage because it has an unusual *combination* of predictor values. Even a graphics program designed to display points in high dimensional spaces may not make it obvious. Fortunately, we can calculate leverage values for each case. These are standard for most multiple regression programs.

The **leverage** of a case is easy to understand. For any case, add (or subtract) 1 to its *y*-value. Recompute the regression, and see how much the *predicted* value of the case changes. The amount of the change is the leverage. It can never be greater than 1 or less than 0. A point with zero leverage has no effect at all on the regression slope, although it does participate in the calculations of R^2, s, and the F- and t-statistics. (The leverage of the ith point in a data set is often denoted by h_i.)

A point with high leverage may not actually influence the regression coefficients if it follows the pattern of the model set by the other points, but it's worth examining simply because of its *potential* to do so. Looking at leverage values can be an effective way to discover cases that are extraordinary on a combination of x-variables. In business, such cases often deserve special attention.

There are no tests for whether the leverage of a case is too large. The average leverage value among all cases in a regression is $1/n$, but that doesn't give us much of a guide. Some packages use rules of thumb to indicate high leverage values,[4] but another common approach is to just make a histogram of the leverages. Any case whose leverage stands out in a histogram of leverages probably deserves special attention. You may decide to leave the case in the regression or to see how the regression model changes when you delete the case, but you should be aware of its potential to influence the regression.

We've already seen that the *Duration* of a roller coaster ride depends linearly on its *Length*. But even more than a long ride, roller coaster customers like speed. So, rather than predict the duration of a roller coaster ride, let's build a model for how fast it travels. A multiple regression with two predictors shows that both the total *Height* and the *Drop* (the maximum distance from the top to the bottom of the largest drop in the ride) are important factors:

Standard Error

In some software regression tables, the standard deviation of the residuals, which we denote s_e, is called a "standard error." In this book, "standard error" refers to the estimated standard deviation of a statistic, and we denote it SE (*statistic*). The standard deviation of the residuals is not a standard error. Together with the notational similarity of s_e and SE, that can be confusing. Don't be confused.

	A	B	C	D	E	F
1	SUMMARY OUTPUT					
2						
3	*Regression Statistics*					
4	Multiple R	0.925				
5	R Square	0.855				
6	Adjusted R Square	0.851				
7	Standard Error	4.608				
8	Observations	76				
9						
10	ANOVA					
11		*df*	*SS*	*MS*	*F*	*Significance F*
12	Regression	2	9140.889	4570.445	215.240	2.44884E-31
13	Residual	73	1550.098	21.234		
14	Total	75	10690.988			
15						
16		*Coefficients*	*Standard Error*	*t Stat*	*P-value*	
17	Intercept	37.01333	1.47723	25.056	1.3777E-37	
18	Height	0.06581	0.01911	3.444	0.00095345	
19	Drop	0.12540	0.01888	6.643	4.75347E-09	

Table 19.6 **Excel** regression output for the regression of *Speed* on *Height* and *Drop*. Both coefficients are highly significant.

[4]One common rule for determining when a leverage is large is to indicate any leverage value greater than $3(k + 1)/n$, where k is the number of predictors.

The regression certainly seems reasonable. The R^2 value is high, and the residual plot looks patternless:

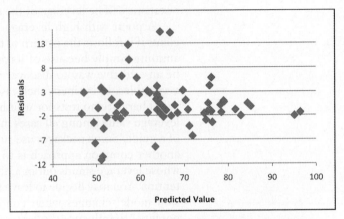

Figure 19.6 An **Excel** scatterplot of residuals against predicted values shows nothing unusual for the regression of *Speed* on *Height* and *Drop*. Note that, although it is easy to make this plot in Excel, it is not one of the residual plots offered by default with Excel's regression.

A histogram of the leverage values, however, shows something interesting:

Figure 19.7 The distribution of leverage values shows a few high values and one extraordinarily high-leverage point. The leverages were calculated in a statistics package. Excel does not offer leverages in its regression, but many add-ins do.

The case with very high leverage is a coaster called *Oblivion*, a steel roller coaster in England that opened as the world's first "vertical drop coaster" in 1998. What's unusual about *Oblivion* is that its *Height* is only about 65 feet above ground (placing it below the median), and yet it drops 180 feet to achieve a top speed of 68 mph. The unique feature of *Oblivion* is that it plunges *underground* nearly 120 feet.

Leverage points can affect not only the coefficients of the model, but also our choice of whether to include a predictor in a regression model as well. The more complex the regression model, the more important it is to look at high-leverage values and their effects.

Residuals and Standardized Residuals

Residuals are not all alike. Consider a point with leverage 1.0. That's the highest a leverage can be, and it means that the line follows the point perfectly. So, a point like that must have a zero residual. And since we know the residual exactly, that residual has zero standard deviation. This tendency is true in general: The larger the leverage, the smaller the standard deviation of its residual. When we want to compare values that have differing standard deviations, it's a good idea to standardize them. We can do that with the regression residuals, dividing each one by an estimate of its own standard deviation. When we do that, the resulting values follow a Student's *t*-distribution. In fact, these standardized residuals are called **Studentized residuals**.[5] It's a good idea to examine the Studentized residuals (rather than the simple residuals) to check the Nearly Normal Condition and the Equal Spread Condition. Any Studentized residual that stands out from the others deserves your attention.

It may occur to you that we've always plotted the *unstandardized* residuals when we made regression models. We treated them as if they all had the same standard deviation when we checked the Nearly Normal Condition. It turns out that this was a simplification. It didn't matter much for simple regression, but for multiple regression models, it's a better idea to use the Studentized residuals when checking the Nearly Normal Condition and when making scatterplots of residuals against predicted values.

Influence Measures

A case that has *both* high leverage and a large Studentized residual is likely to have changed the regression model substantially all by itself. Such a case is said to be **influential**. An influential case cries out for special attention because removing it is likely to give a very different regression model. The surest way to tell whether a case is influential is to try leaving it out[6] and see how much the regression model changes. You should call a case "influential" if omitting it changes the regression model by enough to matter for *your* purposes.

To identify possibly influential cases, check the leverage and Studentized residuals. Two statistics that combine leverage and Studentized residuals into a single measure of influence, **Cook's Distance** (Cook's D) and DFFITs, are offered by many statistics programs. If either of these measures is unusually large for a case, that case should be checked as a possible influential point. Cook's D is found from the leverage, the residual, the number of predictors, and the residual standard error:

$$D_i = \frac{e_i^2}{ks_e^2}\left[\frac{h_i}{(1-h_i)^2}\right].$$

A histogram of the Cook's Distances from the model in Table 19.6 shows a few influential values:

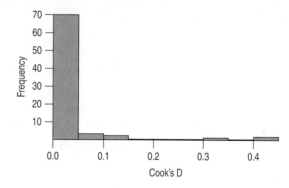

---

[5]There's more than one way to Studentize residuals according to how you estimate the standard deviation of the residuals. You may find statistics packages referring to *externally Studentized residuals* and *internally Studentized residuals*. It is the *externally Studentized* version that follows a *t*-distribution, so those are the ones we recommend.

[6]Or, equivalently, include an indicator variable that selects only for that case. See the discussion in the next section.

Here are coasters with the four highest values of Cook's D.

Name	Type	Duration	Speed	Height	Drop	Length	Inversions	Cook's D
HyperSonic XLC	Steel	NA	80	165	133	1560	0	0.1037782
Oblivion	Steel	NA	68	65	180	1222	0	0.1124114
Volcano, The Blast Coaster	Steel	NA	70	155	80	2757	1	0.3080218
Xcelerator	Steel	62	82	205	130	2202	0	0.4319336

| Table 19.7   Coasters with high Cook's D

In addition to the *Oblivion*, Cook's Distance singles out three other coasters: *HyperSonic XLC*; *Volcano, The Blast Coaster*; and *Xcelerator*. A little research finds that these three coasters *are* different as well. Our model found that *Height* and *Drop* significantly influence a coaster's *Speed*. But these three coasters have something extra—a hydraulic catapult that accelerates the coasters more than gravity alone. In fact, the *Xcelerator* reaches 82 mph in 2.3 seconds, using only 157 feet of track to launch it. Removing these three accelerator coasters from the model has a striking effect.

```
Variable     Coeff        SE(Coeff)     t-ratio     P-value
Intercept    36.47453     1.06456       34.262      <0.0001
Drop          0.17519     0.01493       11.731      <0.0001
Height        0.01600     0.01507        1.062       0.292

s = 3.307 on 70 degrees of freedom
R-squared = 92.5%   Adjusted R-squared = 92.3%
F = 429.5 on 2 and 70 DF, P-value: <0.0001
```

| Table 19.8   Removing the three blast coasters has made *Height* no longer important to the model. A simple regression model, such as in Table 19.9, may be a more effective summary than the model with two predictor variables.

The *Height* of the coaster is no longer a statistically significant predictor, so we might choose to omit that variable.

```
Variable     Coeff         SE(Coeff)     t-ratio     P-value
Intercept    36.743925     1.034798      35.51       <0.0001
Drop          0.189474     0.006475      29.26       <0.0001

s = 3.31 on 71 degrees of freedom
R-squared = 92.3%   Adjusted R-squared = 92.2%
F = 856.3 on 1 and 71 DF, P-value: <0.0001
```

| Table 19.9   A simple linear regression model without the three blast coasters and with *Height* deleted.

Indicators for Influence

One good way to examine the effect of an extraordinary case on a regression is to construct a special indicator variable that is zero for all cases *except* the one we want to isolate. Including such an indicator in the regression model has the same effect as removing the case from the data, but it has two special advantages. First, it makes it clear to anyone looking at the regression model that we have treated that case specially. Second, the *t*-statistic for the indicator variable's coefficient can be used as a test of whether the case is influential. If the P-value is small, then that case really didn't fit well with the rest of the data. Typically, we name such an indicator with the identifier of the case we want to remove. Here's the last roller coaster model in which we have removed the influence of the three blast coasters by constructing indicators for them instead of by removing them from the data. Notice that the coefficient for drop is just the same as the ones we found by omitting the cases.

```
Dependent variable is: Speed
R-squared = 92.7%   R-squared (adjusted) = 92.3%
s = 3.310 with 76 − 5 = 71 degrees of freedom
```

Variable	Coeff	SE(Coeff)	t-ratio	P-value
Intercept	36.7439	1.035	35.5	<0.0001
Drop	0.189474	0.0065	29.3	<0.0001
Xcelerator	20.6244	3.334	6.19	<0.0001
HyperSonic	18.0560	3.334	5.42	<0.0001
Volcano	18.0981	3.361	5.38	<0.0001

Table 19.10 The P-values for the three indicator variables confirm that each of these roller coasters doesn't fit with the others.

Diagnosis Wrapup

What have we learned from diagnosing the regression? We've discovered four roller coasters that may be strongly influencing the model. And for each of them, we've been able to understand why and how it differed from the others. The oddness of *Oblivion* plunging into a hole in the ground may cause us to value *Drop* as a predictor of *Speed* more than *Height*. The three influential cases with high Cook's D values turned out to be different from the other roller coasters because they are "blast coasters" that don't rely only on gravity for their acceleration. Although we can't count on always discovering why influential cases are special, diagnosing influential cases raises the question of what about them might be different and can help us understand our model better.

When a regression analysis has cases that have both high leverage and large Studentized residuals, it would be irresponsible to report only the regression on all the data. You should also compute and discuss the regression found with such cases removed, and discuss the extraordinary cases individually if they offer additional insight. If your interest is to understand the world, the extraordinary cases may tell you more than the rest of the model. If your only interest is in the model (for example, because you hope to use it for prediction), then you'll want to be certain that the model wasn't determined by only a few influential cases, but instead was built on the broader base of the body of your data.

For Example   Regression diagnostics

Two other measures of quality of a diamond (see page 622) are its *Table* size and *Depth*; both are expressed as percentages. *Table* size is the ratio of the flat top part (the table) diameter to the diameter of the stone. *Depth* is the ratio of the distance from table to bottom (culet) to the diameter. The output of the regression of *Log₁₀Price* on *Carat Weight, Color D, Color G, Table,* and *Depth* shows:

```
Response Variable: Log₁₀Price
R² = 77.93%   Adjusted R² = 77.84%
s = 0.1350 with 749 − 4 = 745 degrees of freedom
```

Variable	Coeff	SE(Coeff)	t-ratio	P-value
Intercept	3.464326	0.322336	10.748	<0.0001
Carat.Size	1.031627	0.015799	65.295	<0.0001
ColorD	0.291119	0.014214	20.481	<0.0001
ColorG	0.213842	0.012760	16.759	<0.0001
Table	−0.008497	0.002345	−3.623	0.000311
Depth	−0.008638	0.003765	−2.294	0.022043

(continued)

Influence measures can be adversely affected by indicator variables, so Studentized residuals, Cook's Distance, and leverage were all calculated on this model without the indicator variables for *Color*.

Here are the histograms and boxplots for these three measures. Several points have unusually high values on one or more measures.

The analyst identified 5 diamonds that had high leverage (1) or Cook's Distance (4). Two of these also had large studentized residuals. Indicator variables were created for these 5 points and a regression was run with these new variables:

Variable	Coeff	SE(Coeff)	t-ratio	P-value
Intercept	4.5658	0.3972	11.49	<0.0001
Carat Size	0.9125	0.0177	51.54	<0.0001
Depth	−0.0190	0.0047	−4.04	<0.0001
Table	−0.0111	0.0029	−3.84	0.0001
Diamond 1	0.0478	0.1343	0.36	0.7221
Diamond 2	0.0837	0.1344	0.62	0.5336
Diamond 3	−0.1332	0.1361	−0.98	0.3283
Diamond 4	−0.3205	0.1346	−2.38	0.0176
Diamond 5	−0.4284	0.1346	−3.18	0.0015

Question: What would you do next?

Answer: The indicator variables indicated that of the 5 suspected high influence points, only Diamonds 4 and 5 actually influence the regression. Run the regression again without these two points to see what coefficients change. If there is anything identifiably different about these two diamonds, set them aside or create a new variable for the newly identified characteristic (if possible) that distinguishes these diamonds.

19.4 Building Regression Models

When many possible predictors are available, we will naturally want to select only a few of them for a regression model. But which ones? The first and most important thing to realize is that often there is no such thing as the "best" regression model. In fact, no regression model is "right." Often, several alternative models may be useful or insightful. The "best" for one purpose may not be best for another, and the one with the highest R^2 may not be best for many purposes.

Multiple regressions are subtle. The coefficients often don't mean what they may appear to mean at first. The choice of which predictors to use determines almost everything about the regression. Predictors interact with each other, which complicates interpretation and understanding. So it is usually best to build a parsimonious model, using as few predictors as you can. On the other hand, we don't want to leave out predictors that are theoretically or practically important. Making this trade-off is the heart of the challenge of selecting a good model.[7]

---

[7]This trade-off is sometimes referred to as Occam's Razor after the medieval philosopher William of Occam.

The best regression models, in addition to satisfying the assumptions and conditions of multiple regression, have:

- Relatively few predictors, to keep the model simple.
- A relatively high R^2, indicating that much of the variability in y is accounted for by the regression model.
- A relatively small value of s_e, the standard deviation of the residuals, indicating that the magnitude of the errors is small.
- Relatively small P-values for the F- and t-statistics, showing that the overall model is better than a simple summary with the mean and that the individual coefficients are reliably different from zero.
- No cases with extraordinarily high leverage that might dominate and alter the model.
- No cases with extraordinarily large residuals, and Studentized residuals that appear to be nearly Normal. Outliers can alter the model and certainly weaken the power of any test statistics, and the Nearly Normal Condition is required for inference.
- Predictors that are reliably measured and relatively unrelated to each other.

The term "relatively" in this list is meant to suggest that you should favor models with these attributes over others that satisfy them less, but of course, there are many trade-offs and no absolute rules. In addition to favoring predictors that can be measured reliably, you may want to favor those that are less expensive to measure, especially if your model is intended for prediction with values not yet measured.

It should be clear from this discussion that the selection of a regression model calls for judgment. This is yet another of those decisions in Statistics that just can't be made automatically. Indeed, it is one that we shouldn't want to make automatically; there are so many aspects of what makes a model useful that human judgment is necessary to make a final choice. Nevertheless, there are tools that can help by identifying potentially interesting models.

Best Subsets and Stepwise Regression

How can we find the best multiple regression model? The list of desirable features we just looked at should make it clear that there is no simple definition of the "best" model. The choice of a multiple regression model always requires judgment to choose among potential models. Sometimes it can help to look at models that are "good" in some arbitrary sense to understand some possibilities, but such models should never be accepted blindly.

If we choose a single criterion such as finding a model with the highest adjusted R^2, then, for modest size data sets and a modest number of potential predictors, it is actually possible for computers to search through *all* possible models. The method is called a **Best Subsets Regression**. Often the computer reports a collection of "best" models: the best with three predictors, the best with four, and so on.[8] Of course, as you add predictors, the R^2 can never decrease, but the improvement may not justify the added complexity of the additional predictors. One criterion that might help is to use adjusted R^2. Best subsets regression programs usually offer a choice of criteria, and of course, different criteria usually lead to different "best" models.

Although best subsets programs are quite clever about computing far fewer than all the possible alternative models, they can become overwhelmed by more than a

---

[8]Best subsets regressions don't actually compute every regression. Instead, they cleverly exclude models they know to be worse than some they've already examined. Even so, there are limits to the size of the data and number of variables they can deal with comfortably.

few dozen possible predictors or very many cases. So, unfortunately, they aren't useful in many data mining applications. (We'll discuss those more in Chapter 25.)

Another alternative is to have the computer build a regression "stepwise." In a **stepwise regression,** at each step, a predictor is either added to or removed from the model. The predictor chosen to add is the one whose addition increases the adjusted R^2 the most (or similarly improves some other measure such as an F-statistic). The predictor chosen to remove is the one whose removal reduces the adjusted R^2 least (or similarly loses the least on some other measure). The hope is that by following this path, the computer can settle on a good model. The model will gain or lose a predictor only if that change in the model makes a big enough change in the performance measure. The changes stop when no more changes pass the criterion.

Best subsets and stepwise methods offer both a final model and information about the paths they followed. The intermediate stage models can raise interesting questions about the data and suggest relationships that you might not have thought about. Some programs offer the chance for you to make choices as the process progresses. By interacting with the process at each decision stage, you can exclude a variable that you judge inappropriate for the model (even if including it would help the statistic being optimized) or include a variable that wasn't the top choice at the next step, if you think it is important for your purposes. Don't let a variable that doesn't make sense enter the model just because it has a high correlation, but at the same time, don't exclude a predictor just because you didn't initially think it was important. (That would be a good way to make sure that you never learn anything new.) Finding the balance between these two choices underlies the art of successful model building and makes it challenging.

Unlike best subset methods, stepwise methods can work even when the number of potential predictors is so large that you can't examine them individually. In such cases, using a stepwise method can help you identify potentially interesting predictors, especially when you use it sequentially.

Both methods are powerful. But as with many powerful tools, they require care when you use them. You should be aware of what the automated methods *fail* to do: They don't check the Assumptions and Conditions. Some, such as the independence of the cases, you can check before performing the analyses. Others, such as the Linearity Condition and concerns over outliers and influential cases, must be checked for each model. There's a risk that automated methods will be influenced by nonlinearities, by outliers, by high leverage points, by clusters, and by the need for constructed dummy variables to account for subgroups.[9] And these influences affect not just the coefficients in the final model, but the *selection* of the predictors themselves. If there is a case that is influential for even one possible multiple regression model, a best subsets search is guaranteed to consider that model (because it considers *all* possible models) and have its decision influenced by that one case.

•   **Choosing the Wrong "Best" Model.** Here's a simple example of how stepwise and best subsets regressions can go astray. We might want to find a regression to model *Horsepower* in a sample of cars from the car's engine size (*Displacement*) and its *Weight*. The simple correlations are as follows:

	HP	Disp	Wt
Horsepower	1.000		
Displacement	0.872	1.000	
Weight	0.917	0.951	1.000

---

[9]This risk grows dramatically with larger and more complex data sets—just the kind of data for which these methods can be most helpful.

Because *Weight* has a slightly higher correlation with *Horsepower,* stepwise regression will choose it first. Then, because *Weight* and engine size (*Displacement*) are so highly correlated, once *Weight* is in the model, *Displacement* won't be added to the model. And a best subsets regression will prefer the regression on *Weight* because it has a higher R^2 and adjusted R^2. But *Weight* is, at best, a lurking variable leading to both the need for more horsepower and a larger engine. Don't try to tell an engineer that the best way to increase horsepower is to add weight to the car and that the engine size isn't important! From an engineering standpoint, *Displacement* is a far more appropriate predictor of *Horsepower,* but stepwise regression can't find that model.

Challenges in Building Regression Models

The data set used to construct the regression in the Guided Example of Chapter 18 originally contained more than 100 variables on more than 10,000 houses in the upstate New York area. Part of the challenge in constructing models is simply preparing the data for analysis. A simple scatterplot can often reveal a data value mistakenly coded, but with hundreds of potential variables, the task of checking the data for accuracy, missing values, consistency, and reasonableness can become the major part of the effort. We will return to this issue when we discuss data mining in Chapter 25.

Another challenge in building large models is Type I error. Although we've warned against using 0.05 as an unquestioned guide to statistical significance, we have to start somewhere, and this critical value is often used to test whether a variable can enter (or leave) a regression model. Of course, using 0.05 means that about 1 in 20 times, a variable whose contribution to the model may be negligible will appear as significant. Using something more stringent than 0.05 means that potentially valuable variables may be overlooked. Whenever we use automatic methods (stepwise, best subsets or others), the actual number of different models considered becomes huge, and the probability of a Type I error grows with it. There is no easy remedy for this problem. Building a model that includes predictors that actually contribute to reducing the variation of the response and avoiding predictors that simply add noise to the predictions is the challenge of modern model building. Much current research is devoted to criteria and automatic methods to make this search easier and more reliable, but for the foreseeable future, you will need to use your own judgment and wisdom in addition to your statistical knowledge to build sensible useful regressions.

For Example   Stepwise regression

In addition to the predictor variables from the last example (page 631) we also have information on the quality of the *Cut* (4 grades: *Good, Very Good, Excellent,* and *Ideal*) and the quality of the *Clarity* (7 grades: *VVS2, VVS1,VS2, VS1, SI2, SI1,* and *IF*). A model was fit to predict *Log₁₀Price* from all the predictors we have available: *Carat Weight, Color, Cut, Clarity, Depth,* and *Table* on all 749 diamonds. A backward elimination of terms was performed, and the following model was selected:

```
Response Variable: Log10Price
R² = 94.46%   Adjusted R² = 94.37%
s = 0.06806 with 749 − 13 = 736 degrees of freedom
```

(continued)

Variable	Coeff	SE(Coeff)	t-ratio	P-value
Intercept	2.437903	0.014834	164.349	<0.0001
Carat.Weight	1.200301	0.011365	105.611	<0.0001
ColorD	0.342888	0.009107	37.649	<0.0001
ColorG	0.254024	0.008159	31.135	<0.0001
CutGood	−0.028884	0.011133	−2.594	0.00966
CutVeryGood	−0.015374	0.005503	−2.793	0.00535
CutIdeal	0.012127	0.010022	1.210	0.22662
ClaritySI1	−0.237850	0.011736	−20.267	<0.0001
ClaritySI2	−0.312076	0.012454	−25.058	<0.0001
ClarityVS1	−0.130431	0.012118	−10.764	<0.0001
ClarityVS2	−0.178439	0.011684	−15.273	<0.0001
ClarityVVS1	−0.064518	0.012445	−5.184	<0.0001
ClarityVVS2	−0.076747	0.011878	−6.461	<0.0001

Question: Compare this model with the model based only on *Carat Weight* and *Color* from the first example with indicator variables for *Color*.

Answer: This new model has several advantages over the simpler model. First, the residual standard error is now 0.068 $\log_{10}$Price, a decrease from 0.109 $\log_{10}$Price. (We can transform that back to dollars as $10^{0.068} = \$1.17$—quite a small standard deviation.)

Correspondingly, the R^2 is now 94.46% compared to 85.66%. Nearly all the terms are highly statistically significant. The one exception is the indicator for the *Ideal* level of *Cut*. We could consider omitting this indicator (which would then combine the levels *Excellent* and *Ideal*), but for simplicity we will leave it as is. Before deploying this model, however, we should look at all the assumptions and conditions, which we will do in the final example of this chapter.

Guided Example   Housing Prices

Let's return to the upstate New York data set to predict house prices, this time using a sample of 1734 houses and 16 variables.

The variables available include:

Price	The price of the house as sold in 2002
Lot Size	The size of the land in *acres*
Waterfront	An indicator variable coded as 1 if the property contains waterfront, 0 otherwise
Age	The age of the house in *years*
Land Value	The assessed value of the property without the structures
New Construct	An indicator variable coded as 1 if the house is new construction, 0 otherwise
Central Air	An indicator variable coded as 1 if the house has central air conditioning, 0 otherwise

Fuel Type	A categorical variable describing the main type of fuel used to heat the house: 1 = None; 2 = Gas; 3 = Electric; 4 = Oil; 5 = Wood; 6 = Solar; 7 = Unknown/Other
Heat Type	A categorical variable describing the heating system of the house: 1 = None; 2 = Forced Hot Air; 3 = Hot Water; 4 = Electric
Sewer Type	A categorical variable describing the sewer system of the house: 1 = None/Unknown; 2 = Private (Septic System); 3 = Commercial/Public
Living Area	The size of the living area of the house in *square feet*
Pct College	The percent of the residents of the zip code that attended four-year college (from the U.S. Census Bureau)
Full Baths	The *number* of full bathrooms
Half Baths	The *number* of half bathrooms
Bedrooms	The *number* of bedrooms
Fireplaces	The *number* of fireplaces

PLAN

Setup State the objective of the study.

Identify the variables.

Model Think about the assumptions and check the conditions. A scatterplot matrix is a good way to examine the relationships for the quantitative variables.

We want to build a model to predict house prices for a region in upstate New York. We have data on *Price* ($), and 15 potential predictor variables selected from a much larger list.

✓ **Linearity Condition.** To fit a regression model, we first require linearity. (Scatterplots of *Price* against *Living Area, Age, Bedrooms, Bathrooms,* and *Fireplaces* are similar to the plots shown in the regression of Chapter 18 and are not shown here.)

✓ **Independence Assumption.** We can regard the errors as being independent of one another since the houses themselves were a random sample.

✓ **Randomization Condition.** These 1728 houses are a random sample of a much larger set. That supports the idea that the errors are independent.

To check equal variance and Normality, we usually find a regression and examine the residuals. Linearity is all we need for that.

Remarks

Examination of *Fuel Type* showed that there were only 6 houses that did not have categories 2, 3 or 4.

Because there are 7 levels of Fuel Type, we expect to need 6 indicator variables. We can use *None* as the baseline. *Fuel2* has the value 1 for Gas and 0 for all other types, *Fuel3* has value 1 for Electric and 0 for all other types. As the bar chart shows, only 6 houses have values other than 2, 3, and 4. So we can omit those cases and use just 2 indicators, *Fuel2* and *Fuel3* indicating Gas and Electric respectively, leaving Oil as the baseline. The 6 houses (out of over 1700) set aside have a very small impact on the model.

Two of those had unknown *Heat Type*. We decided to set these 6 houses aside, leaving three categories. So we can use two dummy variables for each. We combined the two *Bathroom* variables into a single variable *Bathrooms* equal to the sum of the full baths plus 0.5*half baths. We now have 17 potential predictor variables.

We started by fitting a model to all of them.

```
Dependent variable is: Price
R-squared = 65.1%   R-squared (adjusted) = 64.8%
s = 58408 with 1728 − 18 = 1710 degrees of freedom
```

Variable	Coeff	SE(Coeff)	t-ratio	P-value
Intercept	18794.0	23333	0.805	0.4207
Lot.Size	7774.34	2246	3.46	0.0006
Waterfront	119046	15577	7.64	≤0.0001
Age	−131.642	58.54	−2.25	0.0246
Land.Value	0.9258	0.048	19.4	≤0.0001
New.Construct	−45234.8	7326	−6.17	≤0.0001
Central.Air	9864.30	3487	2.83	0.0047
Fuel Type[2]	4225.35	5027	0.840	0.4008

(continued)

Variable	Coeff	SE(Coeff)	t-ratio	P-value *(contd.)*
Fuel Type[3]	−8148.11	12906	−0.631	0.5279
Heat Type[2]	−1185.54	12345	−0.096	0.9235
Heat Type[3]	−11974.4	12866	−0.931	0.3521
Sewer Type[2]	4051.84	17110	0.237	0.8128
Sewer Type[3]	5571.89	17165	0.325	0.7455
Living.Area	75.769	4.24	17.9	≤0.0001
Pct.College	−112.405	151.9	0.740	0.4593
Bedrooms	−4963.36	2405	2.06	0.0392
Fireplaces	768.058	2992	0.257	0.7975
Bathrooms	23077.4	3378	6.83	≤0.0001

✓ **Equal Spread Condition.** A scatterplot of the Studentized residuals against predicted values shows no thickening or other patterns. There is a group of homes whose residuals are larger (both negative and positive) than the vast majority, whose Studentized residual values are larger than 3 or 4 in absolute value. We'll revisit them after we've selected our model.

We need the Nearly Normal Condition only if we want to do inference and the sample size is not large. If the sample size is large, we need the distribution to be Normal only if we plan to produce prediction intervals.

✓ **Nearly Normal Condition, Outlier Condition.** The histogram of residuals is unimodal and symmetric, but slightly long tailed.

Under these conditions, we can proceed to search for a suitable multiple regression model using a subset of the predictors. We will return to some of our concerns in the discussion.

DO

Mechanics We first let the stepwise program proceed backward from the full model on all 17 predictors.

Here is the computer output for the multiple regression, starting with all 17 predictor variables and proceeding backward until no more candidates were nominated for exclusion.

```
Dependent variable is: Price
R-squared = 65.1%   R-squared (adjusted) = 64.9%
s = 58345 with 1728 − 12 = 1716 degrees of freedom

Variable         Coeff          SE(Coeff)     t-ratio     P-value
Intercept        9643.14        6546          1.47        0.1409
Lot.Size         7580.42        2049          3.70        0.0002
Waterfront       119372         15365         7.77        ≤0.0001
Age              −139.704       57.18         −2.44       0.0147
Land.Value       0.921838       0.0463        19.9        ≤0.0001
New.Construct    −44172.8       7159          −6.17       ≤0.0001
Central.Air      9501.81        3402          2.79        0.0053
Heat Type[2]     10099.9        4048          2.50        0.0127
Heat Type[3]     −791.243       5215          −0.152      0.8794
Living.Area      75.9000        4.124         18.4        ≤0.0001
Bedrooms         −4843.89       2387          −2.03       0.0426
Bathrooms        23041.0        3333          6.91        ≤0.0001
```

The estimated equation is:

$$\widehat{Price} = 9643.14 + 7580.42\,LotSize$$
$$+ 119{,}372\,Waterfront − 139.70\,Age$$
$$+ 0.922\,LandValue$$
$$− 44{,}172.8\,NewConstuction$$
$$+ 9501.81\,CentralAir$$
$$+ 10099.9\,Heat\;Type2$$
$$− 791.24\,Heat\;Type3$$
$$+ 75.90\,Living\;Area$$
$$− 4843.89\,Bedrooms$$
$$+ 23041\,Bathrooms$$

Nearly all of the P-values are small which indicates that even with 11 predictors in the model, most are contributing. The exception is the indicator for Heat Type 3 which is *Hot Water*. We could consider dropping this indicator and combining those houses with those that have levels 1 (None) and 4 (Electricity). However, for simplicity we have left the model as the stepwise search found it. The R^2 value of 65.1% indicates that more than 65% of the overall variation in house prices has been accounted for by this model and the fact that the adjusted R^2 has actually increased, suggest that we haven't removed any important predictors from the model. The residual standard error of $58,345 gives us a rough indication that we can predict the price of a home to within about 2 × $58,345 = $116,690. If that's close enough to be useful, then our model is potentially useful as a price guide.

(continued)

Remarks

We also tried running the stepwise regression program forward and obtained the same model. There are some houses that have large Studentized residual values and some that have somewhat large leverage, but omitting them from the model did not significantly change the coefficients. We can use this model as a starting basis for pricing homes in the area.

REPORT

Summary and Conclusions Summarize your results and state any limitations of your model in the context of your original objectives.

MEMO

Re: Regression Analysis of Home Price Predictions

A regression model of *Price* on 11 predictors accounts for about 65% of the variation in the price of homes in these data from upstate New York. Tests of each coefficient show that each of these variables appears to be an aspect of the price of a house.

This model reflects the common wisdom in real estate about the importance of various aspects of a home. An important variable not included is the location, which every real estate agent knows is crucial to pricing a house. This is ameliorated by the fact that all these houses are in the same general area. However, knowing more specific information about where they are located would probably improve the model. The price found from this model can be used as a starting point for comparing a home with comparable homes in the area.

As with all multiple regression coefficients, when we interpret the effect of a predictor, we must take account of all the other predictors and be careful not to suggest a causal relationship between the predictor and the response variable. Here are some interesting features of the model. It appears that among houses with the same values of the other variables, those with waterfront access are worth on average about $119,000 more. Among houses with the same values of the other variables, those with more bedrooms have lower sales prices by, on average, $4800 for each bedroom, while among those with the same values of the other variables, those with more bathrooms have higher prices by, on average, $23,000 per bathroom. Not surprisingly, the value of the land is positively associated with the sale price, accounting for, on average, about $0.92 of the sale price for each $1 of assessed land value among houses that are otherwise alike on the other variables.

This model reflects the prices of 1728 homes in a random sample of homes taken in this area.

19.5 Collinearity

Anyone who has shopped for a home knows that houses with more rooms generally cost more than houses with fewer rooms. A simple regression of *Price* on *Rooms* for the New York houses shows:

Variable	Coeff	SE(Coeff)	t-ratio	P-value
Intercept	53015.6	6424.3	8.252	<0.0001
Rooms	22572.9	866.7	26.046	<0.0001

An additional room seems to be "worth" about $22,600 on average to these homes. We can also see that the *Living Area* of a house is an important predictor. Each square foot is worth about $113.12:

Variable	Coeff	SE(Coeff)	t-ratio	P-value
Intercept	13439.394	4992.353	2.692	0.00717
Living Area	113.123	2.682	42.173	<0.0001

Finally, the *Price* increases with the number of *Bedrooms*, with each *Bedroom* worth, on average, $48,218:

Variable	Coeff	SE(Coeff)	t-ratio	P-value
Intercept	59863	8657	6.915	<0.0001
Bedrooms	48218	2656	18.151	<0.0001

But, when we put more than one of these variables into a regression equation simultaneously, things can change. Here's a regression with both *Living Area* and *Bedrooms*:

Variable	Coeff	SE(Coeff)	t-ratio	P-value
Intercept	36667.895	6610.293	5.547	<0.0001
Living Area	125.405	3.527	35.555	<0.0001
Bedrooms	−14196.769	2675.159	−5.307	<0.0001

What the Coefficients Tell Us When Predictors Are Correlated

Sometimes we can understand what the coefficients are telling us even in such paradoxical situations. Here, it seems that a house that allocates more of its living area to bedrooms (and correspondingly less to other areas) will be worth less.

In the second example, we see that more rooms don't make a house worth more if they just carve up the existing living area. The value of more rooms we saw before was probably because houses with more rooms tend to have more living area as well.

Now, it appears that each bedroom is associated with a *lower* sale *Price*.

This type of coefficient change often happens in multiple regression and can seem counterintuitive. When two predictor variables are correlated, their coefficients in a multiple regression (with both of them present) can be quite different from their simple regression slopes. In fact, the coefficient can change from being significantly positive to significantly negative with the inclusion of one correlated predictor, as is the case here with *Bedrooms* and *Living Area*. The problem arises when one of the predictor variables can be predicted well from the others. This phenomenon is called **collinearity**.[10]

Collinearity in the predictors can have other consequences in a multiple regression. If instead of adding *Bedrooms* to the model, we add *Rooms*, we see a different outcome:

Variable	Coeff	SE(Coeff)	t-ratio	P-value
Intercept	11691.586	5521.253	2.118	0.0344
Living Area	110.974	3.948	28.109	<0.0001
Rooms	783.579	1056.568	0.742	0.4584

---

[10] You may see also this problem called "multicollinearity."

The coefficient for *Living Area* has hardly changed at all. It still shows that a house costs about $111 per square foot, but now the coefficient for *Rooms* is indistinguishable from 0. With the addition of *Living Area* to the model the coefficient for *Rooms* changed from having a *t*-statistic over 25, with a very small P-value (in the simple regression) to having a P-value of 0.458. Notice also that the standard errors of the coefficients have increased. The standard error of *Living Area* increased from 2.68 to 3.95. That may not seem like much, but it's an increase of nearly 50%.

This variance inflation of the coefficients is another consequence of collinearity. The stronger the correlation between predictors, the more the variance of their coefficients increases when both are included in the model. Sometimes this effect can change a coefficient from statistically significant to indistinguishable from zero.

Data sets in business often have related predictor variables. General economic variables, such as interest rates, unemployment rates, GDP, and other productivity measures are highly correlated. The choice of which predictors to include in the model can significantly change the coefficients, their standard errors, and their P-values, making both selecting regression models and interpreting them difficult.

How can we detect and deal with collinearity? Let's look at a regression among just the predictor variables. If we regress *Rooms* on *Bedrooms* and *Living Area* we find:

```
Variable      Coeff      SE(Coeff)    t-ratio    P-value
Intercept     0.680      0.141          4.821     <0.0001
Bedrooms      0.948      0.057         16.614     <0.0001
Living Area   0.009      0.000         25.543     <0.0001

s = 1.462 on 1725 degrees of freedom
R-squared = 60.2%   Adjusted R-squared = 60.15%
F = 1304 on 2 and 1725 DF, P-value: <0.0001
```

Look at the R^2 for that regression. What does it tell us? Since R^2 is the fraction of variability accounted for by the regression, in this case, that's the fraction of the variability in *Rooms* accounted for by the other two predictors.

Now we can be precise about collinearity. If that R^2 were 100%, we'd have perfect collinearity. *Rooms* would then be perfectly predictable from the other two predictors and so could tell us nothing new about *Price* because it didn't vary in any way not already accounted for by the predictors already in the model. In fact, we couldn't even perform the calculation. Its coefficient would be indeterminate, and its standard error would be infinite. (Statistics packages usually print warnings when this happens.[11]) Conversely, if the R^2 were 0%, then *Rooms* would bring entirely new information to the model, and we'd have no collinearity at all.

Clearly, there's a range of possible collinearities for each predictor. The statistic that measures the degree of collinearity of the jth predictor with the others is called the **Variance Inflation Factor (VIF)** and is found as:

$$VIF_j = \frac{1}{1 - R_j^2}.$$

The R_j^2 here shows how well the jth predictor can be predicted by the other predictors. The $1 - R_j^2$ term measures what that predictor has left to bring to the regression model. If R_j^2 is high, then not only is that predictor superfluous, but it can damage the regression model. The VIF tells how much the variance of the coefficient has been inflated due to collinearity. The higher the VIF, the higher the

---

[11]Excel does not. It gives 0 as the estimate of most values and a NUM! warning for the standard error of the coefficient.

standard error of its coefficient and the less it can contribute to the regression model. Since R_j^2 can't be less than zero, the minimum value of the VIF is 1.0. The VIF takes into account the influence of all the other predictors—and that's important. You can't judge whether you have a collinearity problem simply by looking at the correlations among the predictors because those only consider each "pair" of predictors.

As a final blow, when a predictor is collinear with the other predictors, it's often difficult to figure out what its coefficient means in the multiple regression. We've blithely talked about "removing the effects of the other predictors," but now when we do that, there may not be much left. What is left is not likely to be about the original predictor, but more about the fractional part of that predictor not associated with the others. In a regression of *Horsepower* on *Weight* and *Engine Size*, once we've removed the effect of *Weight* on *Horsepower*, *Engine Size* doesn't tell us anything *more* about *Horsepower*. That's certainly not the same as saying that *Engine Size* doesn't tell us anything at all about *Horsepower*. It's just that most cars with big engines also weigh a lot. So *Engine Size* may be telling us mostly about sporty cars that have larger engines than expected for their weight.

To summarize, when a predictor is collinear with the other predictors in the model, two things can happen:

1. Its coefficient can be surprising, taking on an unanticipated sign or being unexpectedly large or small.
2. The standard error of its coefficient can be large, leading to a smaller *t*-statistic and correspondingly large P-value.

One telltale sign of collinearity is the paradoxical situation in which the overall *F*-test for the multiple regression model is significant, showing that at least one of the coefficients is significantly different from zero, and yet all of the individual coefficients have small *t*-values, each in effect, denying that *it* is the significant one.

What should you do about a collinear regression model? The simplest cure is to remove some of the predictors. That both simplifies the model and generally improves the *t*-statistics. If several predictors give pretty much the same information, removing some of them won't hurt the model's ability to describe the response variable. Which should you remove? Keep the predictors that are most reliably measured, least expensive to find, or even those that are politically important. Another alternative that may make sense is to construct a new predictor by combining variables. For example, several different measures of a product's durability (perhaps for different parts of it) could be added together to create a single durability measure.

Facts about Collinearity

- The collinearity of any predictor with the others in the model can be measured with its Variance Inflation Factor.
- High collinearity leads to the coefficient being poorly estimated and having a large standard error (and correspondingly low *t*-statistic). The coefficient may seem to be the wrong size or even the wrong sign.
- Consequently, if a multiple regression model has a high R^2 and large F, but the individual *t*-statistics are not significant, you should suspect collinearity.
- Collinearity is measured in terms of the R_j^2 between a predictor and *all* of the other predictors in the model. It is not measured in terms of the correlation between any two predictors. Of course, if two predictors are highly correlated, then the R_j^2 with even more predictors must be at least that large and will usually be even higher.

Just Checking

Just Checking

1 Give two ways that we use histograms to support the construction, inference, and understanding of multiple regression models.

2 Give two ways that we use scatterplots to support the construction, inference, and understanding of multiple regression models.

3 What role does the Normal model play in the construction, inference, and understanding of multiple regression models?

19.6 Quadratic Terms

WHO	Downhill racers in the 2002 Winter Olympics
WHAT	Time, Starting Position
WHEN	2002
WHERE	Salt Lake City

After the women's downhill ski racing event at the 2002 Winter Olympic Games in Salt Lake City, Picabo Street of the U.S. team was disappointed with her 16th place finish after she'd posted the fastest practice time. Changing snow conditions can affect finish times, and in fact, the top seeds can choose their starting positions and try to guess when the conditions will be best. But how much impact was there? On the day of the women's downhill race, it was unusually sunny. Skiers expect conditions to improve and then, as the day wears on, to deteriorate, so they try to pick the optimum time. But their calculations were upset by a two-hour delay. Picabo Street chose to race in 26th position. By then conditions had turned around, and the slopes had begun to deteriorate. Was that the reason for her disappointing finish?

The regression in Table 19.11 seems to support her point. Times did get slower as the day wore on.

```
Dependent variable is: Time

R-squared = 37.9%   R-squared (adjusted) = 36.0%
s = 1.577 with 35 − 2 = 33 degrees of freedom

Variable     Coeff      SE(Coeff)    t-ratio    P-value
Intercept    100.069    0.5597       179        <0.0001
StartOrder   0.108563   0.0242       4.49       <0.0001
```

Table 19.11 Time to ski the women's downhill event at the 2002 Winter Olympics depended on starting position.

But a plot of the residuals warns us that the linearity assumption isn't met.

Figure 19.8 The residuals reveal a bend.

If we plot the data, we can see that re-expression can't help us because as the skiers expected, the times first trended down and then turned around and increased.

Figure 19.9   The original data trend down and then up. That kind of bend can't be improved with re-expression.

How can we use regression here? We can introduce a squared term to the model:

$$\widehat{Time} = b_0 + b_1 startorder + b_2 startorder^2.$$

The fitted function is a *quadratic*, which can follow bends like the one in these data. Table 19.12 has the regression table.

```
Dependent variable is: Time

R-squared = 83.3%   R-squared (adjusted) = 82.3%
s = 0.8300 with 35 - 3 = 32 degrees of freedom
```

Source	Sum of Squares	df	Mean Square	F-ratio
Regression	110.139	2	55.0694	79.9
Residual	22.0439	32	0.688871	

Variable	Coeff	SE(Coeff)	t-ratio	P-value
Intercept	103.547	0.4749	218	<0.0001
StartOrder	−0.367408	0.0525	−6.99	<0.0001
SartOrder^2	0.011592	0.0012	9.34	<0.0001

Table 19.12   A regression model with a quadratic term fits these data better.

This model fits the data better. Adjusted R^2 is 82.3%, up from 36.0% for the linear version. And the residuals look generally unstructured, as Figure 19.10 shows.

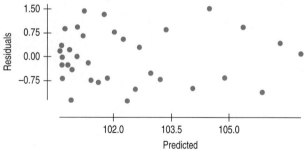

Figure 19.10   The residuals from the quadratic model show no structure.

However, one problem remains. In the new model, the coefficient of *Start Order* has changed from significant and positive to significant and negative. As we've just seen, that's a signal of possible collinearity. Quadratic models often have collinearity problems because for many variables, x and x^2 are highly correlated. In these data, *Start Order* and *Start Order^2* have a correlation of 0.97.

There's a simple fix for this problem. Instead of using *Start Order*^{2}, we can use $(Start\ Order - \overline{Start\ Order})^2$. The form with the mean subtracted has a *zero* correlation with the linear term. Table 19.13 shows the resulting regression.

```
Dependent variable is: Time
R-squared = 83.3%   R-squared (adjusted) = 82.3%
s = 0.8300 with 35 - 3 = 32 degrees of freedom
```

Source	Sum of Squares	df	Mean Square	F-ratio
Regression	110.139	2	55.0694	79.9
Residual	22.0439	32	0.688871	

Variable	Coeff	SE(Coeff)	t-ratio	P-value
Intercept	98.7493	0.3267	302	<0.0001
StartOrder	0.104239	0.0127	8.18	<0.0001
(SO − mean)^2	0.011592	0.0012	9.34	<0.0001

| **Table 19.13**   Using a centered quadratic term alleviates the collinearity.

The predicted values and residuals are the same for these two models,[12] but the coefficients of the second one are easier to interpret.

So, did Picabo Street have a valid complaint? Well, times did increase with start order, but (from the quadratic term), they decreased before they turned around and increased. Picabo's start order of 26 has a predicted time of 101.83 seconds. Her performance at 101.17 was better than predicted, but her residual of −0.66 is not large in magnitude compared to that for some of the other skiers. Skiers who had much later starting positions *were* disadvantaged, but Picabo's start position was only slightly later than the best possible one (about 16th according to this model), and her performance was not extraordinary by Olympic standards.

One final note: Quadratic models can do an excellent job of fitting curved patterns such as this one. But they are particularly dangerous to use for extrapolating beyond the range of the *x*-values. So you should use them with care.

For Example   Quadratic terms

We have fit a model of *Log$_{10}$Price* to *Carat Weight, Color, Cut,* and *Clarity.* (See page 635.) The R^2 is an impressive 94.46%. The 749 diamonds are a random sample of diamonds of three *Color* levels and of *Weight* between 0.3 and 1.5 carats. We transformed *Price* using the log (base 10) to linearize the relationship with *Carat Weight.* However, a plot of residuals vs. predicted values reveals:

There appears to be a curved relationship between residuals and predicted values. What happens if we add (*Carat Weight*)^2 as a predictor?

---

[12]This can be shown algebraically for any quadratic model with a centered squared term.

Here's the regression output:

```
Response Variable: Log₁₀Price
R² = 97.31%   Adjusted R² = 97.27%
s = 0.04741 with 749 - 14 = 735 degrees of freedom
```

Variable	Coeff	SE(Coeff)	t-ratio	P-value
Intercept	2.040547	0.017570	116.138	<0.0001
Carat.Weight	2.365036	0.042400	55.780	<0.0001
Carat.Weight²	−0.699846	0.025028	−27.962	<0.0001
ColorD	0.347479	0.006346	54.755	<0.0001
ColorG	0.252336	0.005683	44.398	<0.0001
CutGood	−0.038896	0.007763	−5.010	<0.0001
CutIdeal	0.016165	0.006982	2.315	0.020877
CutVery Good	−0.014988	0.003834	−3.910	0.000101
ClaritySI1	−0.286641	0.008359	−34.292	<0.0001
ClaritySI2	−0.353371	0.008800	−40.155	<0.0001
ClarityVS1	−0.161288	0.008513	−18.947	<0.0001
ClarityVS2	−0.215559	0.008246	−26.141	<0.0001
ClarityVVS1	−0.077703	0.008682	−8.950	<0.0001
ClarityVVS2	−0.103078	0.008327	−12.378	<0.0001

The plot of residuals against predicted values shows:

A histogram, boxplot, and Normal probability plot of the residuals show:

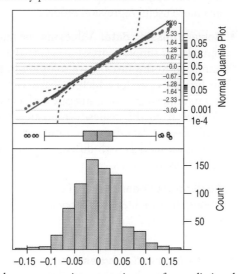

Question: Summarize this regression and comment on its appropriateness for predicting the price of diamonds (of this *Carat Weight*).

Answer: The model for *Log₁₀Price* is based on *Carat Weight*, *Carat Weight²*, *Color (3 levels)*, *Cut (4 levels)*, and *Clarity (7 levels)*. The R^2 for this model is 97.31% (97.27% adjusted), and the residual standard deviation is only 0.047 (in *Log₁₀Price*). Every term included in the model is statistically significant. The assumptions and conditions of multiple regression are all met. The residuals appear to be symmetric and roughly Normal. The inclusion of the squared term for *Carat Weight* has eliminated the pattern in the plot of residuals vs. predicted values. This model seems appropriate to use for other diamonds of this *Carat Weight*.

Regression Roles

We build regression models for a number of reasons. One reason is to model how variables are related to each other in the hope of understanding the relationships. Another is to build a model that might be used to predict values for a response variable when given values for the predictor variables. When we hope to understand, we are often particularly interested in simple, straightforward models in which predictors as unrelated to each other as possible. We are especially happy when the *t*-statistics are large, indicating that the predictors each contribute to the model.

By contrast, when prediction is our goal, we are more likely to care about the overall R^2. Good prediction occurs when much of the variability in y is accounted for by the model. We might be willing to keep variables in our model that have relatively small *t*-statistics simply for the stability that having several predictors can provide. We care less whether the predictors are related to each other because we don't intend to interpret the coefficients anyway, so collinearity is less of a concern.

In both roles, we may include some predictors to "get them out of the way." Regression offers a way to approximately control for factors when we have observational data because each coefficient estimates a relationship *after removing the effects* of the other predictors. Of course, it would be better to control for factors in a randomized experiment, but in the real world of business that's often just not possible.

What Can Go Wrong?

- **Be alert for collinearity when you set aside an influential point.** Removing a high-influence point may surprise you with unexpected collinearity. Alternatively, a single case that is extreme on several predictors can make them appear to be collinear when in fact they would not be if you removed that point. Removing that point may make apparent collinearities disappear (and would probably result in a more useful regression model).

- **Beware missing data.** Values may be missing or unavailable for any case in any variable. In simple regression, when the cases are missing for reasons that are unrelated to the variable we're trying to predict, that's not a problem. We just analyze the cases for which we have data. But when several variables participate in a multiple regression, any case with data missing on any of the variables will be omitted from the analysis. You can unexpectedly find yourself with a much smaller set of data than you started with. Be especially careful, when comparing regression models with different predictors, that the cases participating in the models are the same.

- **Don't forget linearity.** The **Linearity Assumption** requires linear relationships among the variables in a regression model. As you build and compare regression models, be sure to plot the data to check that it is straight. Violations of this assumption make everything else about a regression model invalid.

- **Check for parallel regression lines.** When you introduce an indicator variable for a category, check the underlying assumption that the other coefficients in the model are essentially the same for both groups. If not, consider adding an interaction term.

Ethics in Action

Fred Barolo heads a travel company that offers, among other services, customized travel packages. These packages provide a relatively high profit margin for his company, but Fred worries that a weakened economic outlook will adversely affect this segment of his business. He read in a recent Travel Industry Association report that there is increasing interest among U.S. leisure travelers in trips focused on a unique culinary and wine-related experience. To gain an understanding of travel trends in this niche market, he seeks advice from a market analyst, Smith Nebbiolo, whose firm also handles promotional campaigns. Smith has access to several databases, some through membership with the Travel Industry Association and others through the U.S. Census Bureau and Bureau of Economic Analysis. Smith suggests developing a model to predict demand for culinary wine-related travel, and Fred agrees. As the dependent variable, Smith uses the monthly dollar amount spent on travel packages advertised as culinary wine experiences. He considers a number of monthly economic indicators such as gross domestic product (GDP) and personal consumption expenditures (PCE); variables related to the travel industry (e.g., the American Consumer Satisfaction Index (ACSI) for airlines and hotels, etc.); and factors specific to culinary and wine travel experiences such as advertising expenditure and price. With so many variables, Smith uses an automatic stepwise procedure to select among them. The final model Smith presents to Fred does not include any monthly economic indicators; ACSI for airlines and hotels are included as was advertising expenditure for these types of trips. Fred and Smith discuss how it appears that the economy has little effect on this niche travel market and how Fred should start thinking of how he will promote his new culinary wine-related travel packages.

ETHICAL ISSUE *Although using an automatic stepwise procedure is useful in narrowing down the number of independent variables to consider for the model, usually more thoughtful analysis is required. In this case, many of the potential independent variables are highly correlated with each other. It turns out that the ACSI is a strong predictor of economic indicators such as GDP and PCE. Its presence in the model would preclude the entry of GDP and PCE, but saying that economic factors don't affect the dependent variable is misleading. Further, these data are time-dependent. No variables capturing trend or potential seasonality are considered. Related to Item A, ASA Ethical Guidelines.*

ETHICAL SOLUTION *The interrelationships between the independent variables need to be examined. More expertise is required on the part of the model builder; residuals need to be examined to determine if there are any time-dependent patterns (i.e., seasonality).*

What Have We Learned?

Learning Objectives:

- Use indicator (dummy) variables intelligently.
  - An indicator variable that is 1 for a group and 0 for others is appropriate when the slope for that group is the same as for the others, but the intercept may be different.
  - If the slope for the indicated group is different, then it may be appropriate to include an interaction term in the regression model.
  - When there are three or more categories, use a separate indicator variable for each, but leave one out to avoid collinearity.

- Diagnose multiple regressions to expose any undue influence of individual cases.
  - Leverage measures how far a case is from the mean of all cases when measured on the *x*-variables.
  - The leverage of a case tells how much the predicted value of that case would change if the *y*-value of that case were changed by adding 1 and nothing else in the regression changed.
  - Studentized residuals are residuals divided by their individual standard errors. Externally Studentized residuals follow a *t*-distribution when the regression assumptions are satisfied.
  - A case is influential if it has both sufficient leverage and a large enough residual. Removing an influential case from the data will change the model in ways that matter to your interpretation or intended use. Measures such as *Cook's D* and *DFFITS* combine leverage and Studentized residuals into a single measure of influence.

(continued)

- By assigning an indicator variable to a single influential case, we can remove its influence from the model and test (using the P-value of its coefficient) whether it is in fact influential.

■ **Build multiple regression models when many predictors are available.**

- Seek models with few predictors, a relatively high R^2, a relatively small residual standard deviation, relatively small P-values for the coefficients, no cases that are unduly influential, and predictors that are reliably measured and relatively unrelated to each other.
- Automated methods for seeking regression models include Best Subsets and Stepwise regression. Neither one should be used without carefully diagnosing the resulting model.

■ **Recognize collinearity and deal with it to improve your regression model. Collinearity occurs when one predictor can be well predicted from the others.**

- The R^2 of the regression of one predictor on the others is a suitable measure. Alternatively, the *Variance Inflation Factor*, which is based on this R^2, is often reported by statistics programs.
- Collinearity can have the effect of making the P-values for the coefficients large (not significant) even though the overall regression fits the data well.
- Removing some predictors from the model or making combinations of those that are collinear can reduce this problem.

■ **Consider fitting quadratic terms in your regression model when the residuals show a bend. Re-expressing *y* is another option, unless the bent relationship between *y* and the *x*'s is not monotonic.**

■ **Recognize why a particular regression model is fit.**

- We may want to understand some of the individual coefficients.
- We may simply be interested in prediction and not be concerned about the coefficients themselves.

Terms

Best Subsets Regression	A regression method that checks all possible combinations of the available predictors to identify the combination that optimizes an arbitrary measure of regression success.
Collinearity	When one (or more) of the predictors can be fit closely by a regression on the other predictors, we have collinearity. When collinear predictors are in a regression model, they may have unexpected coefficients and often have inflated standard errors (and correspondingly small *t*-statistics).
Cook's Distance	A measure of the influence of a single case on the coefficients in a multiple regression.
Indicator variable (dummy variable)	A variable constructed to indicate for each case whether it is in a designated group or not. Usually the values are 0 and 1, where 1 indicates group membership.
Influential case	A case is *influential* on a multiple regression model if, when it is omitted, the model changes by enough to matter for your purposes. (There is no specific amount of change defined to declare a case influential.) Cases with high leverage and large Studentized residual are likely to be influential.
Interaction term	A variable constructed by multiplying a predictor variable by an indicator variable. An interaction term adjusts the slope of that predictor for the cases identified by the indicator.
Leverage	A measure of the amount of influence an individual case has on the regression. Moving a case in the *y* direction by 1 unit (while changing nothing else) will move its predicted value by the leverage. The leverage of the ith case is denoted h_i.
Stepwise regression	An automated method of building regression models in which predictors are added to or removed from the model one at a time in an attempt to optimize a measure of the success of the regression. Stepwise methods rarely find the best model and are easily affected by influential cases, but they can be valuable in winnowing down a large collection of candidate predictors.
Studentized residual	When a residual is divided by an independent estimate of its standard deviation, the result is a Studentized residual. The type of Studentized residual that has a *t*-distribution is an *externally Studentized residual*.

Variance inflation factor (VIF)	A measure of the degree to which a predictor in a multiple regression model is collinear with other predictors. It is based on the R^2 of the regression of that predictor on all the other predictors in the model:

$$VIF_j = \frac{1}{1 - R_j^2}$$

Technology Help: Building Multiple Regression Models

Statistics packages differ in how much information they provide to diagnose a multiple regression. Most packages provide leverage values. Many provide far more, including statistics that we have not discussed. But for all, the principle is the same. We hope to discover any cases that don't behave like the others in the context of the regression model and then to understand why they are special.

Many of the ideas in this chapter rely on the concept of examining a regression model and then finding a new one based on your growing understanding of the model and the data. Regression diagnosis is meant to provide steps along that road. A thorough regression analysis may involve finding and diagnosing several models.

EXCEL XLSTAT

Excel does not offer diagnostic statistics with its regression function.

Comments
Although the dialog offers a Normal probability plot of the residuals, the data analysis add-in does not make a correct probability plot, so don't use this option. The "standardized residuals" are just the residuals divided by their standard deviation (with the wrong df), so they too should be ignored.

JMP

- From the **Analyze** menu select **Fit Model**.
- Specify the response, Y. Assign the predictors, X, in the **Construct Model Effects** dialog box.
- Click on **Run Model**.
- Click on the red triangle in the title of the Model output to find a variety of plots and diagnostics available.

Comments
JMP chooses a regression analysis when the response variable is "Continuous."

In JMP, stepwise regression is a *personality* of the Model Fitting platform; it is one of the selections in the Fitting Personality popup menu on the Model Specification dialog. Stepwise provides best subsets with an **All Possible Models** command, accessible from the red triangle drop-down menu on the stepwise control panel after you've computed a stepwise regression analysis.

MINITAB

- Choose **Regression** from the **Stat** menu.
- Choose **Regression. . .** from the **Regression** submenu.

- In the Regression dialog, assign the Y variable to the Response box and assign the X-variables to the Predictors box.
- Click on the **Options** button to obtain the VIF in the regression output.
- In the Regression **Storage** dialog, you can select a variety of diagnostic statistics. They will be stored in the columns of your worksheet.
- Click the **OK** button to return to the Regression dialog.
- To specify displays, click **Graphs**, and check the displays you want.
- Click the **OK** button to return to the Regression dialog.
- Click the **OK** button to compute the regression.

Comments
You will probably want to make displays of the stored diagnostic statistics. Use the usual Minitab methods for creating displays.

Minitab also offers both stepwise and best subsets regression from the **Regression** dialog. Indicate the response variable, the predictors eligible for inclusion, and any predictors that you wish to force into the model.

SPSS

- Choose **Regression** from the **Analyze** menu.
- Choose **Linear** from the **Regression** submenu.
- When the Linear Regression dialog appears, select the Y-variable and move it to the dependent target. Then move the X-variables to the independent target.
- Click the **Save** button.
- In the Linear Regression Save dialog, choose diagnostic statistics. These will be saved in your worksheet along with your data.
- Click the **Continue** button to return to the Linear Regression dialog.
- Click the **OK** button to compute the regression.

Comments
SPSS offers stepwise methods (use the **Method** drop-down menu), but not best subsets (in the student version). Click on the **Statistics** button to find collinearity diagnostics and on the **Save** button for influential point diagnostics. (The residuals SPSS calls "Studentized deleted" are the externally studentized residuals that we've recommended in this chapter.) You may want to plot the saved diagnostics using SPSS's standard graphics methods.

Brief CASE

The Paralyzed Veterans of America (PVA) is a philanthropic organization sanctioned by the U.S. government to represent the interests of those veterans who are disabled. (For more information on the PVA see the opening of Chapter 25.) To generate donations, the PVA sends out greeting cards and mailing address labels periodically with their requests for donations. To increase their efficiency, they would like to be able to model the amount of donations based on past giving and demographic variables on their donors. The data set **PVA** contains data on 3648 donors who gave to a recent solicitation. There are 26 predictor variables and 1 response variable. The response variable (GIFTAMNT) is the amount of money donated by the donor to the last solicitation. Find a model from the 26 predictor variables using any model selection procedure you like to predict this amount. The variables include:

Variables based on the donor's zipcode

MALEVET (% Male veterans)

VIETVETS (% Vietnam Veterans)

WWIIVETS (% WWII veterans)

LOCALGOV (% Employed by local government)

STATEGOV (% Employed by state government)

FEDGOV (% Employed by federal government)

Variables specific to the individual donor

CARDPROM (Number of card promotions received lifetime)

MAXADATE (Date of most recent promotion received in YYMM Year Month format)

NUMPROM (Number of promotions received lifetime)

CARDPRM12 (Number of card promotions received in last 12 months)

NUMPRM12 (Number of promotions received in last 12 months)

NGIFTALL (Number of gifts given lifetime to date)

CARDGIFT (Number of gifts to card promotions given lifetime to date)

MINRAMNT (Amount of smallest gift to date in $)

MINRDATE (Date associated with the smallest gift to date—YYMM format)

MAXRAMNT (Amount of largest gift to date in $)

MAXRDATE (Date associated with the largest gift to date—YYMM format)

LASTGIFT (Amount of most recent gift in $)

AVGGIFT (Average amount of gifts to date in $)

CONTROLN (Control number—unique record identifier)

HPHONE_D (Indicator variable for presence of a published home phone number: 1 = Yes; 0 = No)

CLUSTER2 (Classic Cluster Code—nominal field)

CHILDREN (Number of children living at home)

Response variable

GIFTAMNT (Response variable—amount of last gift in $)

Be sure to include exploratory data analysis and evaluate the relationship among these variables using graphical and correlation analysis to guide you in building your regression models. Write a report summarizing your analysis.

Exercises

SECTION 19.1

1. For each of the following, show how you would code dummy (or indicator) variables to include in a regression model.

a) Company unionization status (Unionized, No Union)
b) Gender (Female, Male)
c) Account Status (Paid on time, Past Due)
d) Political party affiliation (Democrat, Republican, Other)

2. A marketing manager has developed a regression model to predict quarterly sales of his company's down jackets based on price and amount spent on advertising. An intern suggests that he include an indicator (dummy) variable for the Fall quarter.

a) How would you code such a variable? (What values would it have for each quarter?)
b) Why does the intern's suggestion make sense?
c) Do you think a regression with the indicator variable for Fall would model down jacket sales better than one without that predictor?

3. Do movies of different types have different rates of return on their budgets? Here's a scatterplot of *USGross* ($M) vs. *Budget* ($M) for Comedies (red) and Action pictures (blue):

a) What attributes of this plot suggest that it would be appropriate to use an indicator variable in a regression of *USGross* on *Budget*?
b) What would the data values in such an indicator variable be?

4. Here is the regression for Exercise 3 with an indicator variable:

```
Dependent variable is: USGross($M)
R-squared = 33.8%   R-squared (adjusted) = 31.4%
s = 46.50 with 58 − 3 = 55 degrees of freedom

Variable    Coefficient  SE(Coeff)  t-ratio  P-value
Constant    −7.03913     16.20      −0.435   0.6655
Budget($M)   1.00428      0.1895     5.30    ≤0.0001
Comedy      25.4175      13.62       1.87    0.0674
```

a) Write out the regression model.
b) In this regression, the variable *Comedy* is an indicator variable that is 1 for movies that are comedies. How would you interpret the coefficient of *Comedy*?

c) What null hypothesis can we test with the *t*-ratio for *Comedy*?
d) Would you reject that hypothesis at $\alpha = 0.05$?

5. For each of the following, show how you would code dummy (indicator) variables to include in a regression model.

a) Type of residence (Apartment, Condominium, Townhouse, Single family home)
b) Employment status (Full-time, Part-time, Unemployed)

6. A marketing manager has developed a regression model to predict quarterly sales of his company's mid-weight microfiber jackets based on price and amount spent on advertising. An intern suggests that he include indicator (dummy) variables for each quarter.

a) How would you code the variables? (How many dummy variables do you need? What values would they have?)
b) Why does the intern's suggestion make sense?
c) Do you think a regression with the indicator variables would model jacket sales better than one without those predictors?

SECTION 19.2

7. Do movies of different types have different rates of return on their budgets? Here's scatterplot of *USGross* ($M) vs *Budget* ($M) for Comedies (red) and Adventure pictures (green):

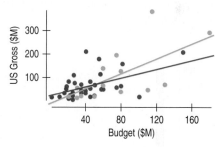

a) What attributes of this plot suggest that it would be appropriate to use both an indicator variable and an interaction term in a regression of *USGross* on *Budget*?
b) How would you code the indicator variable? (Use Adventure as the base level.)
c) How would you construct the interaction term variable?

8. Here is the regression of Exercise 7 with the constructed variables:

```
Dependent variable is: USGross($M)
R-squared = 41.0%   R-squared (adjusted) = 37.4%
s = 58.45 with 53 − 4 = 49 degrees of freedom

Variable    Coefficient  SE(Coeff)  t-ratio  P-value
Intercept   −13.2860     29.20      −0.455   0.6512
Budget($M)    1.58935     0.3457     4.60    ≤0.0001
Budgt*Cmdy   −0.652661    0.4863    −1.34    0.1857
Comedy       34.4795     33.85       1.02    0.3134
```

a) Write out the regression model.

b) In this regression, the variable *Budgt*Cmdy* is an interaction term. How would you interpret the coefficient of *Budgt*Cmdy*?

c) What null hypothesis can we test with the *t*-ratio for *Budgt*Cmdy*?

d) Would you reject that hypothesis at $\alpha = 0.05$?

SECTION 19.3

9. For a regression of *USGross* ($M) for movies on *Budget* ($M), *Run time* (Minutes), and *Stars* (Reviewers' rating average), a histogram of leverages looks like this:

The movie with the highest leverage is *King Kong*, with a leverage of 0.189. If *King Kong* had earned $1M more than it did (and everything else remained the same), how much would the model's prediction of *King Kong*'s U.S. gross revenue change?

10. For the same regression as in Exercise 9, the Cook's Distance values look like this:

The Cook's *D* value for *Star Wars III* is 0.36.

a) If *Star Wars III* were removed from the data and the regression repeated, would you expect a large change in the regression coefficients?

b) Do you think *Star Wars III* should be included in this model if the plan is to use it to understand how these predictors affect movie revenue?

SECTION 19.4

11. An analyst wants to build a regression model to predict spending from the following four predictor variables: *Past Spending*, *Income*, *Net Worth* and *Age*. A correlation matrix of the four predictors shows:

	Income	Net Worth	Age
Past Spending	0.442	0.433	0.446
Income		0.968	0.992
Net Worth			0.976

Why might a stepwise regression search not find the same model as an "all subsets" regression?

12. The analyst in Exercise 11 fits the model with the four predictor variables. The regression output shows:

```
Response Variable: Spending
R² = 84.92%   Adjusted R² = 84.85%
s = 48.45 with 908 − 5 = 903 degrees of freedom

Variable    Coeff         SE(Coeff)    t-ratio   P-value
Intercept   −3.738e + 00  1.564e + 01  −0.239    0.811
Past
  Spending  1.063e − 01   4.203e − 03  25.292    <.0001
Income      1.902e − 03   3.392e − 04  5.606     <.0001
Networth    2.900e − 05   3.815e − 05  0.760     0.447
Age         6.065e − 01   7.631e − 01  0.795     0.427
```

a) How many observations were used in the regression?

b) What might you do next?

c) Is it clear that *Income* is more important to predicting *Spending* than *Networth*? Explain.

SECTION 19.5

13. The analyst from Exercise 11, worried about collinearity, regresses *Age* against *Past Spending*, *Income* and *Networth*. The output shows:

```
Response Variable: Age
R² = 98.75%   Adjusted R² = 98.74%
s = 2.112 with 908 − 4 = 904 degrees of freedom

Variable     Coeff         SE(Coeff)    t-ratio    P-value
(Intercept)  2.000e + 01   1.490e − 01  134.234    <0.0001
Past.
  Spending   3.339e − 04   1.828e − 04  1.826      0.0681
Income       3.811e − 04   7.610e − 06  50.079     <0.0001
Networth     2.420e − 05   1.455e − 06  16.628     <0.0001
```

What is the VIF for *Age*?

14. If the VIF for *Networth* in the regression of Exercise 11 was 20.83, what would the R^2 be from the regression of *Networth* on *Age*, *Income* and *Past Spending*?

SECTION 19.6

15. A collection of houses in a neighborhood of Boston shows the following relationship between *Price* and *Age* of the house:

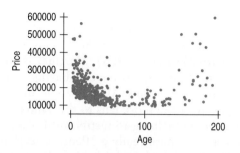

a) Describe the relationship between *Price* and *Age*. Explain what this says in terms of house prices.
b) A linear regression of *Price* on *Age* shows that the slope for *Age* is not statistically significant. Is *Price* unrelated to *Age*? Explain.
c) What might you do instead?

16. A regression model from the collection of houses in Exercise 15 shows the following:

Variable	Coeff	SE(Coeff)	t-ratio	P-value
Intercept	217854.85	4197.417	51.90	<.0001
Age	−1754.254	127.3356	−13.78	<.0001
(Age-38.5122)^2	20.401223	1.327713	15.37	<.0001

a) The slope of *Age* is negative. Does this indicate that older houses cost less, on average? Explain.
b) Why did the model subtract 38.5122 from *Age* in the quadratic term?

CHAPTER EXERCISES

17. Pizza ratings. Manufacturers of frozen foods often reformulate their products to maintain and increase customer satisfaction and sales. So they pay particular attention to evaluations of their products in comparison to their competitors' products. Frozen pizzas are a major sector of the frozen food market, accounting for $2.84 billion in sales in 2007 (www.aibonline.org/resources/statistics/2007pizza.htm). The prestigious Consumer's Union rated frozen pizzas for flavor and quality, assigning an overall score to each brand tested. A regression model to predict the Consumer's Union score from Calories, Type (1 = cheese, 0 = pepperoni), and fat content gives the following result:

```
Dependent variable is: Score
R-squared = 28.7%   R-squared (adjusted) = 20.2%
s = 19.79 with 29 − 4 = 25 degrees of freedom
```

Source	Sum of Squares	df	Mean Square	F-ratio
Regression	3947.34	3	1315.78	3.36
Residual	9791.35	25	391.654	

Variable	Coeff	SE(Coeff)	t-ratio	P-value
Intercept	−148.817	77.99	−1.91	0.0679
Calories	0.743023	0.3066	2.42	0.0229
Type	15.6344	8.103	1.93	0.0651
Fat	−3.89135	2.138	−1.82	0.0807

a) What is the interpretation of the coefficient of *Type* in this regression? According to these results, what type would you expect to sell better—cheese or pepperoni?
b) What displays would you like to see to check assumptions and conditions for this model?

18. Traffic delays. The Texas Transportation Institute (tti.tamu .edu) studies traffic delays. They estimate that in the year 2000, the 75 largest metropolitan areas experienced 3.6 billion vehicle hours of delay, resulting in 5.7 billion gallons of wasted fuel and $67.5 billion in lost productivity. That's about 0.7% of the nation's GDP that year. Data the institute published for the year 2001 include information on the *Total Delay per Person* (hours per year spent delayed by traffic), the *Average Arterial Road Speed* (mph), the *Average Highway Road Speed* (mph), and the *Size* of the city (small, medium, large, very large). The regression model based on these variables looks like this. The variables *Small, Large,* and *Very Large* are indicators constructed to be 1 for cities of the named size and 0 otherwise.

```
Dependent variable is: Delay/person
R-squared = 79.1%   R-squared (adjusted) = 77.4%
s = 6.474 with 68 − 6 = 62 degrees of freedom
```

Source	Sum of Squares	df	Mean Square	F-ratio
Regression	9808.23	5	1961.65	46.8
Residual	2598.64	62	41.9135	

Variable	Coeff	SE(Coeff)	t-ratio	P-value
Intercept	139.104	16.69	8.33	≤0.0001
HiWay MPH	−1.07347	0.2474	−4.34	≤0.0001
Arterial MPH	−2.04836	0.6672	−3.07	0.0032
Small	−3.58970	2.953	−1.22	0.2287
Large	5.00967	2.104	2.38	0.0203
Very Large	3.41058	3.230	1.06	0.2951

a) Why is there no coefficient for *Medium*?
b) Explain how the coefficients of *Small, Large, and Very Large* account for the size of the city in this model.

19. Pizza ratings, part 2. Here's a scatterplot of the residuals against predicted values for the regression model found in Exercise 17.

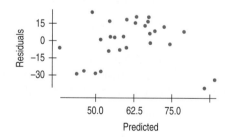

a) The two extraordinary points in the lower right are *Reggio's* and *Michelina's*, two gourmet brands. Interpret these points.
b) Do you think these two pizzas are likely to be influential in the regression. Would setting them aside be likely to change the coefficients? What other statistics might help you decide?

20. Traffic delays, part 2. Here's a scatterplot of the residuals from the regression in Exercise 18 plotted against mean *Highway mph*.

a) The point plotted with an X is Los Angeles. Read the graph and explain what it says about traffic delays in Los Angeles and about the regression model.

b) Is Los Angeles likely to be an influential point in this regression?

21. Wal-Mart revenue. Each week about 100 million customers—nearly one-third of the U.S. population—visit one of Wal-Mart's U.S. stores. How does Wal-Mart's revenue relate to the state of the economy in general? Here's a regression table predicting Wal-Mart's monthly revenue ($Billion) from the end of 2003 through the start of 2007 from the consumer price index (*CPI*), and a scatterplot of the relationship.

```
Dependent variable is: WM_Revenue

R-squared = 11.4%   R-squared (adjusted) = 9.0%
s = 3.689 with 39 − 2 = 37 degrees of freedom
```

Variable	Coeff	SE(Coeff)	t-ratio	P-value
Intercept	−24.4085	19.25	−1.27	0.2127
CPI	0.071792	0.0330	2.18	0.0358

a) The points plotted with X are the four December values. We can construct a variable that is "1" for those four values and "0" otherwise. What is such a variable called?

Here's the resulting regression.

```
Dependent variable is: WM_Revenue

R-squared = 80.3%   R-squared (adjusted) = 79.2%
s = 1.762 with 39 − 3 =  36 degrees of freedom
```

Variable	Coeff	SE(Coeff)	t-ratio	P-value
Intercept	−34.7755	9.238	−3.76	0.0006
CPI	0.087707	0.0158	5.55	≤0.0001
December	10.4905	0.9337	11.2	≤0.0001

b) What is the interpretation of the coefficient of the constructed variable *December*?

c) What additional assumption is required to include the variable *December* in this model? Is there reason to believe that it is satisfied?

22. Baseball attendance. Baseball team owners want to attract fans to games. The New York Mets acquired the pitcher Pedro Martinez in 2005. Martinez is considered one of the best pitchers of his era, having won the Cy Young Award three times. Martinez had his own fans. Possibly, he attracted more fans to the ballpark when he pitched at home, helping to justify his 4-year, $53 million contract. Is there really a "Pedro effect" in attendance? We have data for the Mets home games of the 2005 season. The regression has the following predictors:

Weekend	1 if game is on Saturday or Sunday, 0 otherwise
Yankees	1 if game is against the Yankees (a hometown rivalry), 0 otherwise
Rain Delay	1 if the game was delayed by rain (which might have depressed attendance), 0 otherwise
Opening Day	1 for opening day, 0 for the others
Pedro Start	1 if Pedro was the starting pitcher, 0 otherwise

Here's the regression.

```
Dependent variable is: Attendance

R-squared = 53.9%   R-squared (adjusted) = 50.8%
s = 6998 with 80 − 6 = 74 degrees of freedom
```

Variable	Coeff	SE(Coeff)	t-ratio	P-value
Intercept	28896.9	1161	24.9	≤0.0001
Weekend	9960.50	1620	6.15	≤0.0001
Yankees	15164.3	4218	3.59	0.0006
Rain Delay	−17427.9	7277	−2.39	0.0192
Opening Day	24766.1	7093	3.49	0.0008
Pedro Start	5428.02	2017	2.69	0.0088

a) All of these predictors are of a special kind. What are they called?

b) What is the interpretation of the coefficient for *Pedro Start*?

c) If we're primarily interested in Pedro's effect on attendance, why is it important to have the other variables in the model?

d) Could Pedro's agent claim, based on this regression, that his man attracts more fans to the ballpark? What statistics should he cite?

23. Pizza ratings, part 3. In Exercise 19, we raised questions about two gourmet pizzas. After removing them, the resulting regression looks like this.

```
Dependent variable is: Score

R-squared = 64.4%  R-squared (adjusted) = 59.8%
s = 14.41 with 27 - 4 = 23 degrees of freedom

Source      Sum of Squares  df  Mean Square  F-ratio
Regression  8649.29          3  2883.10       13.9
Residual    4774.56         23   207.590

Variable   Coeff      SE(Coeff)  t-ratio  P-value
Intercept  -363.109   72.15      -5.03    ≤0.0001
Calories     1.56772   0.2824     5.55    ≤0.0001
Type        25.1540    6.214      4.05    0.0005
Fat         -8.82748   1.887     -4.68    0.0001
```

A plot of the residuals against the predicted values for this regression looks like this. It has been colored according to the *Type* of pizza.

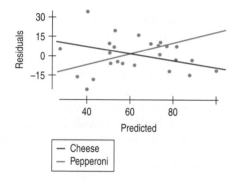

— Cheese
— Pepperoni

a) What does this plot say about how the regression model deals with these two types of pizza?

We constructed another variable consisting of the indicator variable *Type* multiplied by *Calories*. Here's the resulting regression.

```
Dependent variable is: Score

R-squared = 73.1%  R-squared (adjusted) = 68.2%
s = 12.82 with 27 - 5 = 22 degrees of freedom

Source      Sum of Squares  df  Mean Square  F-ratio
Regression  9806.53          4  2451.63       14.9
Residual    3617.32         22   164.424

Variable    Coeff      SE(Coeff)  t-ratio  P-value
Intercept   -464.498   74.73      -6.22    ≤0.0001
Calories      1.92005   0.2842     6.76    ≤0.0001
Type        183.634    59.99       3.06    0.0057
Fat         -10.3847    1.779     -5.84    ≤0.0001
Type*Cals    -0.461496  0.1740    -2.65    0.0145
```

b) Interpret the coefficient of *Type*Cals* in this regression model.
c) Is this a better regression model than the one in Exercises 17 and 19?

T 24. Traffic delays, part 3. Here's a plot of the Studentized residuals from the regression model of Exercise 18 plotted against *ArterialMPH*. The plot is colored according to *City*

Size (Small, Medium, Large, and Very Large), and regression lines are fit for each city size.

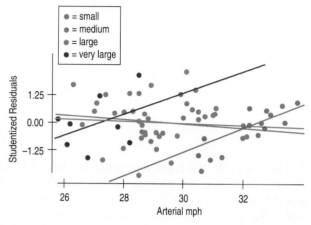

● = small
● = medium
● = large
● = very large

a) The model in Exercise 18 includes indicators for city size. Considering this display, have these indicator variables accomplished what is needed for the regression model? Explain.

Here is another model that adds two new constructed variables to the model in Exercise 18. They are the product of *ArterialMPH* and *Small* and the product of *ArterialMPH* and *VeryLarge*.

```
Dependent variable is: Delay/person
R-squared = 80.7%  R-squared (adjusted) = 78.5%
s = 6.316 with 68 - 8 = 60 degrees of freedom

Source      Sum of Squares  df  Mean Square  F-ratio
Regression  10013.0          7  1430.44       35.9
Residual     2393.82        60    39.8970

Variable     Coeff      SE(Coeff)  t-ratio  P-value
Intercept    153.110    17.42       8.79    ≤0.0001
HiWayMPH      -1.02104   0.2426    -4.21    ≤0.0001
ArterialMPH   -2.60848   0.6967    -3.74    0.0004
Small       -125.979    66.92      -1.88    0.0646
Large          4.89837   2.053      2.39    0.0202
VeryLarge    -89.4993   63.25      -1.41    0.1623
AM*Sml         3.81461   2.077      1.84    0.0712
AM*VLg         3.38139   2.314      1.46    0.1491
```

b) What does the predictor *AM*Sml* (*ArterialMPH* by *Small*) do in this model? Interpret the coefficient.
c) Does this model improve on the model in Exercise 18? Explain.

T 25. Insurance (life expectancy). Insurance companies base their premiums on many factors, but basically all the factors are variables that predict life expectancy. Life expectancy varies from place to place. Here's a regression that models *Life Expectancy* in terms of other demographic variables that we saw in Exercise 44 of Chapter 18. (Refer to that exercise for variable definitions and units.)

Dependent variable is: Life exp

R-squared = 67.0%   R-squared (adjusted) = 64.0%
s = 0.8049 with 50 − 5 = 45 degrees of freedom

Source	Sum of Squares	df	Mean Square	F-ratio
Regression	59.1430	4	14.7858	22.8
Residual	29.1560	45	0.647910	

Variable	Coeff	SE(Coeff)	t-ratio	P-value
Intercept	69.4833	1.325	52.4	≤0.0001
Murder	−0.261940	0.0445	−5.89	≤0.0001
HSGrad	0.046144	0.0218	2.11	0.0403
Income	1.24948e-4	0.0002	0.516	0.6084
Illiteracy	0.276077	0.3105	0.889	0.3787

a) The state with the highest leverage and largest Cook's Distance is Alaska. It is plotted with an x in the residuals plot. Here are a scatterplot of the residuals and a normal probability plot of the leverage values. What evidence do you have from these diagnostic plots that Alaska might be an influential point?

Here's another regression with a dummy variable for Alaska added to the regression model.

Dependent variable is: Life exp

R-squared = 70.8%   R-squared (adjusted) = 67.4%
s = 0.7660 with 50 − 6 = 44 degrees of freedom

Source	Sum of Squares	df	Mean Square	F-ratio
Regression	62.4797	5	12.4959	21.3
Residual	25.8193	44	0.586802	

Variable	Coeff	SE(Coeff)	t-ratio	P-value
Intercept	67.6377	1.480	45.7	≤0.0001
Murder	−0.250395	0.0426	−5.88	≤0.0001
HSGrad	0.055792	0.0212	2.63	0.0116
Illiteracy	0.458607	0.3053	1.50	0.1401
Income	3.68218e-4	0.0003	1.46	0.1511
Alaska	−2.23284	0.9364	−2.38	0.0215

b) What does the coefficient for the dummy variable for Alaska mean? Is there evidence that Alaska is an outlier in this model?

c) Which model would you prefer for understanding or predicting *Life Expectancy*? Explain.

T 26. Cereal nutrition. Breakfast cereal manufacturers publish nutrition information on each box of their product. As we saw in Chapter 17, there is a long history of cereals being associated with nutrition. Here's a regression to predict the number of *Calories* in breakfast cereals from their *Sodium*, *Potassium*, and *Sugar* content, and some diagnostic plots.

Dependent variable is: Calories

R-squared = 38.4%   R-squared (adjusted) = 35.9%
s = 15.60 with 77 − 4 = 73 degrees of freedom

Variable	Coeff	SE(Coeff)	t-ratio	P-value
Intercept	83.0469	5.198	16.0	≤0.0001
Sodium	0.057211	0.0215	2.67	0.0094
Potassium	−0.019328	0.0251	−0.769	0.4441
Sugar	2.38757	0.4066	5.87	≤0.0001

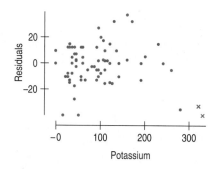

The shaded part of the histogram corresponds to the two cereals plotted with x's in the normal probability plot of the leverages and the residuals plot. These are *All-Bran with Extra Fiber* and *All-Bran*.

a) What do the displays say about the influence of these two cereals on this regression? (The histogram is of the Studentized residuals.)

Here's another regression with dummy variables defined for each of the two bran cereals.

```
Dependent variable is: Calories
R-squared = 50.7%   R-squared (adjusted) = 47.3%
s = 14.15 with 77 − 6 = 71 degrees of freedom

Variable        Coeff      SE(Coeff) t-ratio P-value
Intercept       79.0874    4.839      16.3   ≤0.0001
Sodium          0.068341   0.0198     3.46   0.0009
Potassium       0.043063   0.0272     1.58   0.1177
Sugar           2.03202    0.3795     5.35   ≤0.0001
All-Bran        −50.7963   15.84      −3.21  0.0020
All-Bran Extra  −52.8659   16.03      −3.30  0.0015
```

b) Explain what the coefficients of the bran cereal dummy variables mean.
c) Which regression would you select for understanding the interplay of these nutrition components. Explain. (Note: Both are defensible.)
d) As you can see from the scatterplot, there's another cereal with high potassium. Not too surprisingly, it is *100% Bran*. But it does not have leverage as high as the other two bran cereals. Do you think it should be treated like them (i.e., removed from the model, fit with its own dummy, or left in the model with no special attention, depending on your answer to part c)? Explain.

27. Lobster industry 2008, value diagnosed. In Chapter 18, Exercise 34 (Lobster industry 2008), we constructed the following regression model to predict the log10 of the annual value of Maine's lobster harvest.

```
Dependent variable is: LogValue
R-squared = 97.4%   R-squared (adjusted) = 97.3%
s = 0.0782 with 56 − 4 = 52 degrees of freedom

Variable    Coeff       SE(Coeff) t-ratio P-value
Intercept   7.70207     0.3706     20.8   ≤0.0001
Traps(M)    0.575390    0.0152     37.8   ≤0.0001
Fishers     −5.32221e-5 0.0000     −5.43  ≤0.0001
Water Temp  −0.015185   0.0074     −2.05  0.0457
```

At that time we also examined plots of the residuals which appeared to satisfy the assumptions and conditions for regression inference. Let's look a bit deeper. Here are histograms of the Cook's Distances and Studentized residuals for this model.

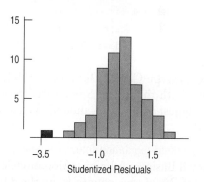

The case with high Cook's Distance is 1994. (You can find a scatterplot of the *logValue* over time in Exercise 53 of Chapter 17.) What does this suggest about this model? What would you recommend?

28. Lobster industry 2008, price diagnosed. In Chapter 18, Exercise 35 (Lobster industry 2008), we constructed the following regression model to predict the *Price* of lobster harvested in Maine's lobster fishing industry.

```
Dependent variable is: Price/lb
R-squared = 94.4%   R-squared (adjusted) = 94.1%
s = 0.2462 with 53 − 4 = 49 degrees of freedom

Variable    Coeff       SE(Coeff) t-ratio P-value
Intercept   0.845123    0.3557     2.38   0.0215
Traps(M)    1.20094     0.0619     19.4   ≤0.0001
Fishers     −1.21820e-4 0.0000     −3.30  0.0018
Catch/Trap  −22.0207    10.96      −2.01  0.0500
```

At that time we also examined plots of the residuals which appeared to satisfy the assumptions and conditions for regression inference. Let's look a bit deeper. Here are histograms of the Cook's Distances and Studentized residuals for this model.

The case with large Cook's Distance is 1994, which was also the year with the lowest Studentized residual. What does this suggest about this model? What would you recommend?

T **29. OECD economic regulation (model building).** A study by the U.S. Small Business Administration modeled the GDP per capita of 24 of the countries in the Organization for Economic Cooperation and Development (OECD) (Crain, M. W., *The Impact of Regulatory Costs on Small Firms*, available at www.sba.gov/idc/groups/public/documents/ sba_homepage/rs264tot.pdf). One analysis estimated the effect on GDP of economic regulations using an index of the degree of OECD economic regulation and other variables. (We considered this model in Exercise 28 of Chapter 18.) They found the following regression model.

```
Dependent variable is: GDP/Capita

R-squared = 97.4%   R-squared (adjusted) = 96.6%
s = 2084 with 24 − 6 = 18 degrees of freedom
```

Source	Sum of Squares	df	Mean Square	F-ratio
Regression	2895078376	5	579015675	133
Residual	78158955	18	4342164	

Variable	Coeff	SE(Coeff)	t-ratio	P-value
Intercept	10487.3	9431	1.11	0.2808
OECD Reg Index	−1343.28	626.4	−2.14	0.0459
Ethnolinguistic Diversity	−69.9875	23.26	−3.01	0.0075
Int'l Trade/GDP	44.7096	14.00	3.19	0.0050
Primary Education(%)	−58.4084	86.11	−0.678	0.5062
1988 GDP/Capita	1.07767	0.0448	24.1	≤0.0001

a) If we remove *Primary Education* from the model, the R^2 decreases to 97.3%, but the adjusted R^2 *increases* to 96.7%. How can that happen? What does it mean? Would you include *Primary Education* in this model?

Here's a part of that regression output:

```
Dependent variable is: GDP/Capita
R-squared = 97.3%   R-squared (adjusted) = 96.7%
s = 2054 with 24 − 5 = 19 degrees of freedom
```

Variable	Coeff	SE(Coeff)	t-ratio	P-value
Intercept	4243.21	2022	2.10	0.0495
OECD Reg Index	−1244.20	600.4	−2.07	0.0521
Ethnolinguistic Diversity	−64.4200	21.45	−3.00	0.0073
Int'l Trade/GDP	40.3905	12.29	3.29	0.0039
1988 GDP/Capita	1.08492	0.0429	25.3	≤0.0001

b) Consider the *t*-statistic for *OECD Regulation* in the reduced model. That was the predictor of interest to this author. Do you agree with his conclusion that OECD regulation reduced GDP/Capita in these countries? Why do you think he chose to include *Primary Education* as a predictor? Explain.

T **30. Dirt bikes.** Off-road motorcycles (often called "dirt bikes") are a segment (about 18%) of the growing motorcycle market. Because dirt bikes offer great variation in features, they are a good market segment to study to learn about which features account for the cost (manufacturer's suggested retail price, *MSRP*) of a bike. Researchers collected data on 2005-model dirt bikes (lib.stat.cmu.edu/ datasets/dirtbike_aug.csv). Their original goal was to study market differentiation among brands. (*The Dirt on Bikes: An Illustration of CART Models for Brand Differentiation*, Jiang Lu, Joseph B. Kadane, and Peter Boatwright). In Chapter 18, Exercises 41, 42, and 43 dealt with these data, but several bikes were removed from those data to simplify the analysis. Now we'll take on the full set.[13]

Here's a regression model and some associated graphs.

```
Dependent variable is: MSRP
R-squared = 91.0%   R-squared (adjusted) = 90.5%
s = 606.4 with 100 − 6 = 94 degrees of freedom
```

Source	Sum of Squares	df	Mean Square	F-ratio
Regression	349911096	5	69982219	190
Residual	34566886	94	367733	

Variable	Coeff	SE(Coeff)	t-ratio	P-value
Intercept	−5514.66	826.2	−6.67	≤0.0001
Bore	83.7950	6.145	13.6	≤0.0001
Clearance	152.617	52.02	2.93	0.0042
Engine Strokes	−315.812	89.83	−3.52	0.0007
Total Weight	−13.8502	3.017	−4.59	≤0.0001
Wheelbase	119.138	34.26	3.48	0.0008

---

[13]Well, in honesty, we've removed one luxury, handmade bike whose MSRP was $19,500 as a clearly identified outlier.

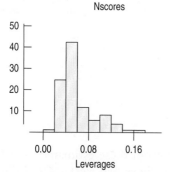

a) List aspects of this regression model that lead to the conclusion that it is likely to be a useful model.
b) What aspects of the displays indicate that the model is a good one?

T 31. Airlines on time. Airlines strive to be on time, in part, because customers can refer to government-published statistics to select flights that are most often on time. We have data for 19 airlines for March 2006 reporting the following variables:

On Time	(number of on-time arrivals)
Cancelled	(number of cancelled flights)
Diverted	(number of diverted flights)
Carrier	(number of delays due to the carrier)
Weather	(number of delays due to weather)
NAS Delay	(Delays due to the National Airspace System (traffic control))
Late Arrival	(number of delays due to late arrival of equipment or crew)

Here's a regression model.

```
Dependent variable is: On Time

R-squared = 93.9%   R-squared (adjusted) = 90.8%
s = 5176 with 19 − 7 = 12 degrees of freedom
```

Source	Sum of Squares	df	Mean Square	F-ratio
Regression	4947151273	6	824525212	30.8
Residual	321546284	12	26795524	

Variable	Coeff	SE(Coeff)	t-ratio	P-value
Intercept	1357.10	2316	0.586	0.5687
Cancelled	−18.5514	6.352	−2.92	0.0128
Diverted	39.5623	95.59	0.414	0.6863
Carrier	10.9620	3.104	3.53	0.0041
Weather	10.4637	9.462	1.11	0.2905
NAS Delay	2.46727	1.091	2.26	0.0431
Late Arrival	4.64874	1.445	3.22	0.0074

a) In view of its P-value, can we interpret the coefficient of *Diverted*?

Here's a scatterplot of *On Time* vs. *Diverted* and another regression. Note that *Diverted* is the response variable in this second regression.

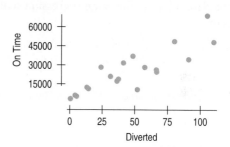

```
Dependent variable is: Diverted
R-squared = 85.6%   R-squared (adjusted) = 80.0%
s = 15.02 with 19 − 6 = 13 degrees of freedom
```

Variable	Coeff	SE(Coeff)	t-ratio	P-value
Intercept	−1.65088	6.703	−0.246	0.8093
Cancelled	−0.016896	0.0178	−0.948	0.3604
Carrier	0.013823	0.0081	1.70	0.1136
Weather	0.050714	0.0236	2.15	0.0509
Late Arrival	6.33917e-3	0.0038	1.67	0.1196
NAS Delay	7.09830e-3	0.0025	2.86	0.0133

b) It seems from the scatterplot that *Diverted* would be a good predictor of *On Time*, but that seems not to be the case. Why do you think the coefficient of *Diverted* is not significant in the first regression?
c) How does the second regression explain this apparent contradiction?
d) Find the value of the Variance Inflation Factor statistic for *Diverted* in the first regression.

T 32. Dirt bikes, part 2. The model in Exercise 30 is missing one predictor that we might have expected to see. *Engine Displacement* is highly correlated ($r = 0.783$) with *MSRP*, but that variable has not entered the model (and, indeed, would have a P-value of 0.54 if it were added to the model). Here is some evidence to explain why that may be. (Hint: Notice that *Displacement* is the response variable in this regression.)

```
Dependent variable is: Displacement
R-squared = 95.9%   R-squared (adjusted) = 95.7%
s = 35.54 with 100 − 6 = 94 degrees of freedom
```

Variable	Coeff	SE(Coeff)	t-ratio	P-value
Intercept	−8.05901	48.42	−0.166	0.8682
Bore	9.10890	0.3601	25.3	≤0.0001
Clearance	3.55912	3.048	1.17	0.2460
Engine Stroke	−27.3943	5.264	−5.20	≤0.0001
Total Weight	1.03749	0.1768	5.87	≤0.0001
Wheelbase	−10.0612	2.008	−5.01	≤0.0001

a) What term describes the reason *Displacement* doesn't contribute to the regression model for *MSRP*?

b) Find the value of the Variance Inflation Factor for *Displacement* in the regression on *MSRP*.

33. Ticket prices, diagnosed. Exercises in Chapter 18 considered weekly receipts from shows on Broadway in New York City. To simplify matters, we omitted a few weeks from those data. Here is the same regression with all the data present.

Dependent variable is: Receipts($M)
R-squared = 93.3%   R-squared (adjusted) = 93.1%
s = 0.9589 with 92 − 4 = 88 degrees of freedom

Source	Sum of Squares	df	Mean Square	F-ratio
Regression	1130.43	3	376.811	410
Residual	80.9067	88	0.919395	

Variable	Coeff	SE(Coeff)	t-ratio	P-value
Intercept	−22.1715	2.221	−9.98	≤0.0001
Paid Attendance	0.087031	0.0046	19.0	≤0.0001
Number of Shows	−0.024934	0.0338	−0.737	0.4632
Average Ticket Price	0.265756	0.0286	9.29	≤0.0001

Here are some diagnostic plots.

a) The two points plotted with X in the residuals vs. predicted plot are the two with the highest Cook's Distances. What does this information say about how those points might be affecting the analysis?

In fact, those points are reports published the last week of November and the first week of December in 2007—a time when the Stagehands Union Local One was on strike, closing down most Broadway shows. It seems that these are not representative weeks for business on Broadway. Removing them results in the following regression.

Dependent variable is: Receipts($M)

R-squared = 98.5%   R-squared (adjusted) = 98.5%
s = 0.3755 with 90 − 4 = 86 degrees of freedom

Source	Sum of Squares	df	Mean Square	F-ratio
Regression	809.793	3	269.931	1915
Residual	12.1250	86	0.140988	

Variable	Coeff	SE(Coeff)	t-ratio	P-value
Intercept	−21.3165	0.8984	−23.7	≤0.0001
Paid Attendance	0.071955	0.0019	37.4	≤0.0001
Number of Shows	0.045793	0.0137	3.35	0.0012
Average Ticket Price	0.274503	0.0115	23.8	≤0.0001

b) What changes in these estimated models support the conclusion that those two weeks were influential points? (*Hint*: The increase in adjusted R^2 isn't one of the reasons.)

c) Which model would be the best to use in analyzing the business of Broadway shows—the one with all the data or the one with the two influential points removed? Explain.

34. OECD and economic regulation model building, part 2. Exercise 29 raised some questions about the regression model built to understand the effect of OECD regulation on GDP/Capita in 24 OECD countries. Let's look more deeply. Here's a histogram of the Cook's Distances for that model.

The country with high Cook's Distance is Ireland.

a) What does the Cook's Distance value suggest about Ireland in this model?

Of the predictors available for this model, by far the best (highest R^2 predictor) is *1988 GDP/Capita*. In a scatterplot of *GDP/Capita* vs. *1988 GDP/Capita*, Ireland stands away from the overall linear trend.

We can define a dummy variable that is "1" for Ireland and "0" for all other countries. The resulting model looks like this.

```
Dependent variable is: GDP/Capita
R-squared = 98.3%   R-squared (adjusted) = 97.7%
s = 1713 with 24 - 7 = 17 degrees of freedom
```

Variable	Coeff	SE(Coeff)	t-ratio	P-value
Intercept	13609.0	7818	1.74	0.0998
1988 GDP/ Capita	1.10397	0.0378	29.2	≤0.0001
OECD Reg Index	−520.181	579.2	−0.898	0.3816
Int'l Trade/ GDP	21.0171	13.81	1.52	0.1463
Ethnolinguistic Diversity	−49.8210	20.19	−2.47	0.0245
Primary Education	−99.3369	72.00	−1.38	0.1856
Ireland	8146.64	2624	3.10	0.0064

b) Explain what the dummy variable for Ireland accomplishes in this model.
c) What do you conclude now about the effects of OECD regulation in these data?

35. Lobster industry 2008, catch model. In Chapter 17, we saw data on the lobster fishing industry in Maine. We have annual data from 1950 through 2005. The annual catch (metric tons) has increased, but is this due to more licensed lobster fishers or to more efficient fishing? Because both fishers and traps are individually licensed, we have detailed data. Here's a scatterplot of *Catch/Fisher* (tonnes) vs. the number of *Traps* (in millions).

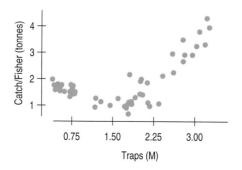

a) Consider the assumptions and conditions for regression. Would re-expressing either variable help? Explain.

Here's a model that includes a quadratic term.

```
Dependent variable is: Catch/Fisher
R-squared = 85.9%   R-squared (adjusted) = 85.3%
s = 0.3340 with 56 - 3 = 53 degrees of freedom
```

Source	Sum of Squares	df	Mean Square	F-ratio
Regression	35.9298	2	17.9649	161
Residual	5.91122	53	0.111765	

Variable	Coeff	SE(Coeff)	t-ratio	P-value
Intercept	2.82022	0.1698	16.8	≤0.0001
Traps	−2.5153	0.2284	−11.0	≤0.0001
Traps^2	0.898526	0.0647	13.9	≤0.0001

b) In 2001, almost 3 million lobster traps were licensed in Maine to 7327 fishers. What would this model estimate as the total catch for that year?
c) The coefficient of *Traps* in this model is negative, but we can see from the scatterplot that the catch has increased substantially as the number of traps increased. What accounts for this apparent anomaly?

36. Lobster industry 2008, water temperature. Like many businesses, the Maine lobster fishing industry is seasonal. But the peak lobstering season has been shifting later in the fall. According to scientists, the biggest factor in the time of the peak is water temperature. The data at www.maine.gov/dmr include water temperature. How has it changed over the past half century? Here's a scatterplot of Water Temperature(°F) vs. Year.

a) Consider the assumptions and conditions for regression. Would re-expressing either variable help? Explain.

A regression model for *Water Temperature* by *Year* looks like this.

```
Dependent variable is: Water Temperature
R-squared = 2.7%   R-squared (adjusted) = 0.9%
s = 1.522 with 56 - 2 = 54 degrees of freedom
```

Variable	Coeff	SE(Coeff)	t-ratio	P-value
Intercept	17.7599	24.88	0.714	0.4784
Year	0.015437	0.0126	1.23	0.2251

b) Do the statistics associated with this model indicate that there is no relationship between *Water Temperature* and *Year*? Explain.

Here's a regression model including a quadratic term.

```
Dependent variable is: Water Temperature
57 total cases of which 1 is missing

R-squared = 60.0%   R-squared (adjusted) = 58.5%
s = 0.9852 with 56 - 3 = 53 degrees of freedom

Variable  Coeff      SE(Coeff) t-ratio P-value
Intercept 19208.2    2204        8.71   ≤0.0001
Year      -19.3947   2.229      -8.70   ≤0.0001
Year²      4.90774e-3 0.0006     8.71   ≤0.0001
```

c) This model fits better, but the coefficient of *Year* is now strongly negative. We can see from the scatterplot that temperatures have not been dropping at 19 degrees per year. Explain this apparent anomaly.

Here's the same regression model[14] but with the quadratic term in the form of $(Year - \overline{Year})^2$.

```
Dependent variable is: Water Temperature
R-squared = 60.0%   R-squared (adjusted) = 58.5%
s = 0.9852 with 56 - 3 = 53 degrees of freedom

Variable       Coeff      SE(Coeff) t-ratio P-value
Intercept      6.77149    16.16     0.419   0.6768
Year           0.020345   0.0082    2.49    0.0159
(Year-mean)²   4.90774e-3 0.0006    8.71   ≤0.0001
```

d) Using the second model, predict what the water temperature will be in 2015, and then explain why you don't trust that prediction.

---

[14]Notice that, even though mathematically the P-value of the coefficient for *Year* should be the same in this table as in the previous regression, it is not. This is because *Year* and *Year²* are so highly correlated (their correlation is 0.99999309) that the collinearity has caused a large enough rounding error to affect the calculation in the previous regression.

Time Series Analysis

CHAPTER

20

Whole Foods Market®

In 1978, twenty-five-year-old John Mackey and twenty-one-year-old Rene Lawson Hardy borrowed $45,000 from family and friends to open the doors of a small natural foods store they named SaferWay in Austin, Texas. Two years later, they joined forces with Clarksville Natural Grocery to open the first Whole Foods Market. With 10,500 square feet of floor space and a staff of 19, their store was large in comparison to other health food stores of the time.

In the mid 1980s, Whole Foods Market began expanding outside of Austin and opened stores elsewhere in the southern United States and California. During the next 15 years, Whole Foods Market grew rapidly, in part due to several mergers and acquisitions. Among the stores they acquired were Wellspring Grocery, Bread & Circus, Mrs. Gooch's Natural Foods, and Fresh Fields Markets. Since 2000, Whole Foods has expanded outside of North America with the purchase of seven Fresh & Wild stores in the United Kingdom. Today, Whole Foods Market, Inc. has 54,000 employees and more than 270 stores. In 2005 they ranked fifth in the Fortune 100 best places to work. In 2007, they merged with Wild Oats of Boulder, Colorado. The firm has continued to grow both by opening new stores and by acquiring related firms.[1]

The decade of the 1990s was a period of growth for most companies, but unlike firms in many other industries, Whole Foods Market continued to grow into the next decade. Here is a time series plot showing the quarterly Sales ($M) plotted by quarter since 1995. If you were asked to summarize the trend in *Sales* over this decade what would you say?

WHAT	Quarterly *Sales*
UNITS	Millions of U.S. dollars
WHEN	1995–2007
WHERE	United States, Canada, and United Kingdom
WHY	To forecast sales for Whole Foods Market

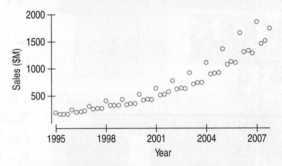

Figure 20.1   Quarterly Sales (in $M) for Whole Foods from 1995 to 2007.

Clearly, sales at Whole Foods Market, Inc. grew between 1995 and 2007, starting at below $250 million and reaching nearly $2 billion per quarter. But we'd like to be able to say something more than that, and ultimately, we'd like to *model* the growth. Time series models are primarily built for predicting into the near future. Some also offer interpretable coefficients.

Corporations often look at time series plots to examine prior patterns in the data and to forecast future values. The executives at Whole Foods Market, Inc. might be interested in understanding patterns in *Sales* to:

• Plan inventory and distribution of goods
• Schedule hiring and staffing
• Understand the impact of seasons, or time of the year, on sales
• Develop advertising campaigns
• Forecast profits and plan corporate strategy

Suppose that you are an analyst at Whole Foods Market, Inc. and you have been asked to forecast *Sales* for the next four quarters. How can you analyze the time series to produce accurate forecasts of *Sales*? How can you measure accuracy and compare your different forecasting models?

---

[1]Source: Official Whole Foods Market website: www.wholefoodsmarket.com.

20.1   What Is a Time Series?

The Whole Foods sales are recorded each financial quarter, and we're interested in the growth of sales over time. Whenever we have data recorded sequentially over time, and we consider *Time* to be an important aspect of the data, we have a **time series**.[2] Most time series are equally spaced at roughly regular intervals, such as monthly, quarterly, or annually.[3] The Whole Foods sales data in Figure 20.1 are a time series measured quarterly. The fiscal year at Whole Foods Market, Inc. starts on or near September 30, so, unlike many companies, the first fiscal quarter reports sales for the end of the calendar year, and includes the holiday season.

Because a time series is likely to be revealed to us sequentially and only up to the present, most time series methods use prior values of a series to predict future values. Some methods describe the overall pattern in the series and predict that it will continue, at least for a short while. Other methods estimate *time series models* that identify structured components like trends and seasonal fluctuation in the series and model those. Some introduce external variables to help predict the response, as in regression, while others simply look at the prior values in the series to try to discern patterns.[4] The objective of most time series analyses is to provide **forecasts** of future values of the time series. Of course, predicting the future is valuable to business decision makers, so time series methods are widely used.

Time series analysis is a rich and sophisticated topic. This chapter can do little more than survey the field and give you a taste of how time series are special. If you want to understand time series in depth, we recommend taking a course devoted to the topic.

20.2   Components of a Time Series

When we examined the distribution of a single variable, we looked at its shape, center, and spread. When we looked at scatterplots of two variables, we asked about the direction, form, and strength. For a time series, we look for the trend, seasonal patterns, and long-term cycles. Some time series exhibit some of these components, some show all of them, and others have no particular large-scale structure at all.

The Trend Component

Look at the time series plot of the Whole Foods *Sales* data in Figure 20.1. What overall pattern are the *Sales* following? Not only have sales at Whole Foods Market been increasing, but they seem to be accelerating. Of course, there are fluctuations around this pattern, but viewed over more than a decade, the overall trend is clear. We'd describe the direction as positive, and the shape as curving upwards (see Figure 20.2). This overall pattern is the **trend component** of the time series. This is often the most interesting aspect of a time series. For example, it is what an investor would want to know about.

Most series have an increasing or decreasing trend with other fluctuations around the trend. Some, however, just fluctuate randomly, much like residual plots from a successful regression. If a series shows no particular trend over time and has a relatively consistent mean, it is said to be **stationary in the mean**.

---

[2]Actually, the methods of this chapter will work for any values that are ordered.

[3]Some series, such as those recording values for trading days or on the first day of each month, are not exactly equally spaced. If there are actual gaps in a time series, researchers use a variety of methods to fill in the missing observations before analyzing the data.

[4]Advanced models, such as dynamic regression models or distributed lag models, are outside the scope of this text.

Figure 20.2 The trend in the Whole Foods data looks nonlinear. In fact, it is increasing and curved.

If the trend grows roughly linearly, we can use the linear regression methods of Chapter 16 to estimate the linear component. We'll do that later in this chapter. (You may wonder if we can use regression for data in which successive errors are not likely to be independent. But if we don't plan to test the coefficients against the standard regression null hypothesis, we don't need any regression assumptions other than linearity.)

The Seasonal Component

Quarterly data often show a seasonal component that repeats every 4 points. The red trace exhibits such a seasonal fluctuation overlaying a linear trend.

Many time series fluctuate regularly. Sales of skis, for example, are always higher in the fall and lower in the spring. Sales of swimsuits peak in the summer. The **seasonal component** of a time series is the part of the variation in a time series that fluctuates in a way that is roughly stable over time with respect to timing, direction, and magnitude. In Figure 20.2, you can easily find that kind of consistent pattern around the general trend. In particular, the first quarter of every fiscal year records more sales than the adjacent quarters, but these fluctuations are relatively small compared to the overall trend. Because seasonal components are usually related to external patterns they are generally stable and predictable. For example, a retail outlet can forecast sales for the next holiday season from information about the previous holiday season and the overall trend. Even though the retail environment may change, they know that this year's holiday season will look more like last year's holiday season than like last April's sales. Typically seasonal components repeat annually, but patterns that repeat more frequently (for example, hourly energy use by a company during a 24-hour period) are still called seasonal components and are modeled with the same methods. A **deseasonalized**, or seasonally adjusted, series is one from which the seasonal component has been removed.

The time between peaks of a seasonal component is referred to as the **period**. The period is independent of the actual timing of peaks and valleys. For example, many retail companies' sales spike during the holiday season in December, and sales of water skis peak in early summer, but both have a seasonal period of one year.

Cyclical Component

Regular cycles in the data with periods longer than one year are referred to as **cyclical components**. Economic and business cycles can sometimes be modeled, but often we do little more than describe them. When a cyclical component can be related to a predictable phenomenon, then it can be modeled based on some regular behavior or by introducing a variable that represents that predictable phenomenon, and adding it to whatever model we are building for the time series.

Irregular Component

We will see a number of ways to model time series. Just as we did with linear models and regression in Chapter 6, in this chapter we'll find that the residuals—the part of the data *not* fit by the model—can be informative. In time series modeling, these residuals are called the **irregular component**. As with regression residuals, it's a good idea to plot the irregular component to look for extraordinary cases or other unexpected patterns. Often our interest in the irregular component is in how variable it is, whether that variability changes over time, and whether there are any outliers or spikes that may deserve special attention. A time series that has a relatively constant variance is said to be **stationary in the variance**.

To summarize, we identify four *components* of a time series:

1. Trend component (T)
2. Seasonal component (S)
3. Cyclical component (C)
4. Irregular component (I)

Table 20.1 provides a summary of the components as applied to the Whole Foods *Sales* data.

Component	Description	Rationale	Period Length
Trend	Positive and nonlinear	Overall increase in sales, with a change in the rate of increase of sales in the last 7 years	
Seasonal	Peaks every first quarter in 1995 through 2007	Larger sales in first quarter; reason unknown at present	4 quarters—yearly
Cyclical	Insufficient data to observe peaks every 2 to 10 years	Could be due to economic cycles and factors, such as inflation, interest rates, and employment that might impact consumer spending	Insufficient data to observe cyclical period
Irregular	Random fluctuations in data	Due to irregular or unpredictable events, such as mergers and acquisitions of other companies, or natural disasters, such as floods	No regular repeating pattern; no period

Table 20.1 Time series components terms applied to the series *Sales* for Whole Foods Market, Inc. from 1995 to 2007.

Even though some time series exhibit only two or one of these components, or even none, an understanding of them can help us structure our understanding of a time series. Just as we look at the direction, form, and strength of a scatterplot, remembering to think about the trend, seasonal, and cyclical parts of a time series can clarify our view of a time series.

Modeling Time Series

I have seen the future and it is just like the present, only longer.

—KEHLOG ALBRAN,
"The Profit"

Methods for forecasting a time series fall into two general classes. Smoothing methods work from the bottom up. They try to "smooth out" the irregular component so any underlying patterns will be easier to see. They have the advantage that they don't assume that there is a trend or seasonal component—and indeed, they'll work even when no seasonal component is present and the trend is complex. For example, we can use smoothing methods on a time series that has only a general cyclical component but no clear trend or seasonal fluctuations to model. The disadvantage of smoothing methods is that they can forecast only the immediate future. Lacking a model of behaviors that can be trusted to continue (such as a seasonal component based on calendar shopping patterns or temperature variation over the year), smoothing models don't have a basis for long-term forecasting. Instead, they rely on the assumption that most time series show patterns that vary more slowly than each successive observation, so the next value in the future will resemble the most recent ones.

When we can discern a trend or both a trend and seasonal component, we'll often prefer regression-based modeling methods. These use the methods of multiple regression we've learned in Chapters 18 and 19 to estimate each component's contribution to the time series and to build a model for the time series. As with any regression-based model, models of this kind can be used to forecast for any value of *Time* and thus can generate forecasts further into the future than one time period. However, as always, we'll need to be cautious with such extrapolations.

The next sections discuss several kinds of smoothing methods, followed by a discussion of regression-based models. Although the smoothing methods don't explicitly use the time series components, it is a good idea to keep them in mind. The regression models explicitly estimate the components as a basis for building the models.

For Example   Truck border crossings in Alaska

The U.S. Department of Transportation Statistics records border crossings into each state on the U.S. border. Commercial traffic, in particular, can be of great interest to businesses that ship to and from Alaska and to economists and investors concerned with those businesses. Here are the border crossings by trucks for Alaska, recorded each month from 1999 through 2009.

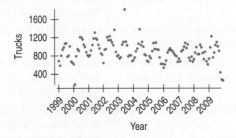

Question: Describe the components of this time series.

Answer: Trend: Generally constant; there may be some slight downward trend from 2002 onward.

Seasonal: Strong seasonal component each 12 months, possibly reflecting the difficulty of driving in Alaska in the winter.

Cyclic: Not enough data to see any long-term cycles; trucking should reflect business cycles to some extent.

Irregular: There is clearly some random fluctuation around the seasonal and trend patterns. There are also a few outlying points that may be worth investigating.

20.3   Smoothing Methods

Most time series contain some random fluctuations that vary up and down rapidly—often for consecutive observations. But, precisely because they are random, these fluctuations provide no help in forecasting. Even if we believe that a time series will continue to fluctuate randomly, we can't predict *how* it will do so. The only aspects of a time series that we have any hope of predicting are those that vary either regularly or slowly. One way to identify these aspects is to smooth away the rapid random fluctuations.[5] To forecast the value of a time series in the future,

---

[5]To an engineer, this would be separating the signal from the rapidly varying noise.

we want to identify the underlying, consistent behavior of the series. In many time series, these slower changes have a kind of inertia. They change and fluctuate, but recent behavior is often a good indication of behavior in the near future. Smoothing methods damp down random fluctuations and try to reveal the underlying behavior so we can use it to forecast values in the immediate future.

The time series for the daily Intel stock price in 2002 (Figure 20.3) shows no regular repeating patterns and no evidence of a regular seasonal effect. But it does show rapid fluctuations and some evidence of longer-term movements.

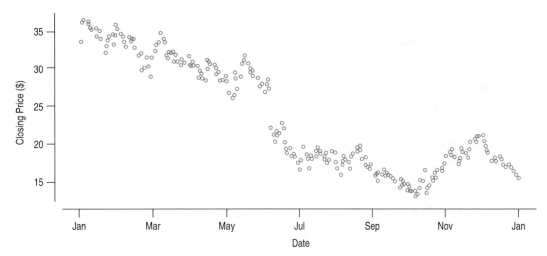

| **Figure 20.3** Daily closing prices for Intel stock in 2002 show no seasonal or other regular patterns.

Methods for smoothing out the apparently random fluctuations generally work by averaging adjacent values in the series. We know from Chapter 12 that means vary less than the underlying data. We can use that principle to find values that are typical of the local value of the series while varying less than the original data.

Simple Moving Average Methods

The most commonly used smoothing method is the method of **moving averages**. A moving average replaces each value in a time series by an average of the adjacent values. The number of values we use to construct the averages is called the *length* of the moving average (*L*). Almost every stock tracking service on the Internet offers a variety of moving averages (often with 50-, 100-, and 200-day lengths) to help track stock movements.

A moving average simply uses the mean of the previous *L* values as the fitted value at each time. Because it focuses on only the recent values, a moving average with a short length can respond to rapid changes in a time series. A moving average with a longer length will respond more slowly. The general form of a moving average is:

$$\widetilde{y}_t = \frac{\sum_{i=t-L+1}^{t} y_i}{L}$$

The length, *L*, of a moving average is a subjective choice, but it must be specified when discussing a moving average. We write MA(*L*) for a moving average of length *L* and use the tilde to denote a moving average calculated from a sequence of data values.

Let's begin by using a moving average of length of 5 on the Intel stock series to illustrate the calculation. The data are in Table 20.2 and the formulas to calculate the smoothed stock price for the fifth and sixth day in the series are:

$$\widetilde{Price_5} = \frac{\sum_1^5 Price_t}{5} = \frac{33.00 + 35.52 + 35.79 + 35.27 + 35.58}{5} = 35.03$$

and

$$\widetilde{Price_6} = \frac{\sum_2^6 Price_t}{5} = \frac{35.52 + 35.79 + 35.27 + 35.58 + 35.36}{5} = 35.50$$

The MA(5) smoothed stock price for each day in the series in 2002 is computed from that day's closing price and the preceding four daily closing prices using similar formulas. If we instead select $L = 15$ for our moving averages, then the calculations will average the 15 previous closing prices (including today's price). Table 20.2 shows the computed values for the two moving averages using $L = 5$ and $L = 15$ and the actual closing price of Intel daily stock for the first 30 days when the market was open in 2002.

There are no moving averages for the first $(L - 1)$ days in the series for each moving average model. What happens to the moving average as the length, L, increases? Of the two smoothed series produced by computing moving averages for the daily Intel stock price using $L = 5$ and $L = 15$, Figure 20.4 shows that the moving average series with the greater length is smoother. That should be what you expected. Of course, a smoother series is not necessarily a better model for the data because it has a hard time following the series when it changes rapidly. Look, for example, at the two series during June of the year. The stock price fell rapidly, but the MA(15) smooth changed too slowly, running above the data for several weeks before it "caught up."

Date	Price	MA(5)	MA(15)
2-Jan-02	$33.00	*	*
3-Jan-02	$35.52	*	*
4-Jan-02	$35.79	*	*
7-Jan-02	$35.27	*	*
8-Jan-02	$35.58	$35.03	*
9-Jan-02	$35.36	$35.50	*
10-Jan-02	$34.65	$35.33	*
11-Jan-02	$34.55	$35.08	*
14-Jan-02	$34.84	$35.00	*
15-Jan-02	$34.68	$34.82	*
16-Jan-02	$33.71	$34.49	*
17-Jan-02	$34.53	$34.46	*
18-Jan-02	$33.48	$34.25	*
22-Jan-02	$31.70	$33.62	*
23-Jan-02	$32.45	$33.17	$34.34
24-Jan-02	$33.20	$33.07	$34.35
25-Jan-02	$33.68	$32.90	$34.23
28-Jan-02	$33.92	$32.99	$34.11
29-Jan-02	$32.68	$33.19	$33.93
30-Jan-02	$33.86	$33.47	$33.82
31-Jan-02	$35.04	$33.84	$33.80
1-Feb-02	$34.67	$34.03	$33.80
4-Feb-02	$33.98	$34.05	$33.76
5-Feb-02	$33.80	$34.27	$33.69
6-Feb-02	$32.92	$34.08	$33.57
7-Feb-02	$32.31	$33.54	$33.48
8-Feb-02	$32.52	$33.11	$33.35
11-Feb-02	$33.57	$33.02	$33.35
12-Feb-02	$32.97	$32.86	$33.44
13-Feb-02	$33.38	$32.95	$33.50

Table 20.2   Moving Averages for $L = 5$, or MA(5), and $L = 15$, or MA(15), for the closing price for Intel stock during the first 30 trading days of 2002.

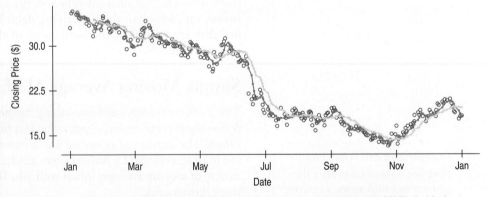

Figure 20.4   A time series plot of the daily Intel stock closing prices in 2002 with MA(5) (in blue) and MA(15) (in yellow) forecasts superimposed. The moving average with the smaller length, $L = 5$, follows the data more closely while the moving average with the longer length is smoother.

Smoothed values and forecast values

To model the smooth pattern that we think might underlay a time series, it makes sense to include the observed value at time t in calculating the smooth value we'll plot at that time point. There's little sense in ignoring information that we have.

When we forecast the next value, that most recent smoothed value is a good choice since it is recent and incorporates information from several recent time periods, damping out the short-term fluctuations.

We use a tilde to denote the smoothed values. We follow the convention from regression of using a hat to denote a predicted value.

To obtain a forecast for a new time point, analysts use the last average in the series:

$$\hat{y}_{t+1} = \tilde{y}_t$$

This is the **simple moving average forecast.** Of course, this method can forecast only one time period into the future, for *Time* = $t + 1$. (You can repeat that value as a forecast beyond period $t + 1$, but unless the time series is essentially unstructured and horizontal, it won't be a very good forecast.) If the length of the moving average is 1 ($L = 1$), then the forecast is simply the next value, which will be the same as the previous one, $\hat{y}_{t+1} = y_t$. As the simplest forecast, this is called the **naïve forecast.**

Often, moving averages are used primarily as summaries of how a time series is changing. The length selected depends on the purpose of the analysis. If the focus is on long-term behavior, a longer moving average is appropriate. But an analyst interested in shorter-term changes would choose a shorter length. Sometimes (as you can see in the sidebar) analysts compare a shorter-length moving average and one with a longer length, hoping to learn something from how they compare.

One potential problem with a moving average is that, as we know from Chapter 5, means can be affected by outliers. An outlier in a time series would be a spike in the series far from the adjacent values. Such a spike will contaminate all of the averages in which it participates, spreading its influence over a number of values.

Weighted Moving Averages

In a simple moving average, we just average the most recent L values. But we can benefit from a more sophisticated averaging scheme. We can assign a *weight* to each value according to how far it is before the current value. The result is a *weighted* average. In a weighted average, each value is multiplied by a *weight* before they are added up, and the total is divided by the sum of the weights:

$$\tilde{y}_t = \frac{\sum w_i y_{t-i}}{\sum w_i}$$

The weights might be specified, or they might be found as part of the smoothing process.

Weighted moving averages form a very general class of smoothers.[6] We will consider two types of weighted moving average smoothers that are commonly used on time series data, exponential smoothers and autoregressive moving averages.

For Example   The value of the euro

The euro is the second-most traded currency (after the dollar), so its value relative to other currencies is of great interest. One euro was equal to one U.S. dollar when the new European currency was launched on January 1, 1999. Since that time the value has fluctuated. On October 26, 2000, one euro cost $0.8252. But on July 15, 2008, it was as high as $1.5990. As the time series plot shows, the fluctuations show little pattern, so smoothing methods may be the best approach to short-term prediction of euro values.

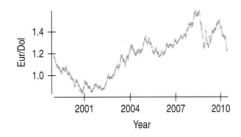

(continued)

[6]In this general form, these smoothers are known as *linear smoothers.* They are important in engineering and finance.

Here are the euro values in dollars for a week in May 2010.

Date	Value ($)
2010-05-27	1.2265
2010-05-28	1.2301
2010-05-29	1.2344
2010-05-30	1.2278
2010-05-31	1.2276
2010-06-01	1.2303
2010-06-02	1.2228

Question: Find a 3-term moving average for these data and predict the value for the next day.

Answer:

Date	Value ($)	MA(3)
2010-05-27	1.2265	*
2010-05-28	1.2301	*
2010-05-29	1.2344	1.2303333
2010-05-30	1.2278	1.2307667
2010-05-31	1.2276	1.2299333
2010-06-01	1.2303	1.2285667
2010-06-02	1.2228	1.2269000

The prediction for 2010-06-03 is $1.227.

Exponential Smoothing Methods

Smoothing methods summarize each value of a time series with an average of recent values. In many time series, recent values of the series are more relevant for modeling than older ones. So a weighted moving average that weights the more recent values more heavily than the older ones makes sense. Exponential smoothing does just that. **Exponential smoothing** is a weighted moving average with weights that decline exponentially into the past. The most recent data are weighted the most and the most distant data are weighted the least. This model is the **single-exponential smoothing (SES) model:**

$$\tilde{y}_t = \alpha y_t + (1 - \alpha)\tilde{y}_{t-1}$$

The choice of the weight α is up to the data analyst, although it is usually restricted to $0 < \alpha < 1$. When $\alpha = 0.50$, the current data point and the entire set of historical data (all points before the current one) are weighted equally. If $\alpha = 0.75$, then historical data are weighted only 25% and the current value has more weight at 75%. If the objective is to produce forecasts that are stable and smoother, then choose a smoothing coefficient closer to zero. If, however, the objective is to react to volatile events rapidly, then choose a smoothing coefficient close to one.[7]

Unlike a simple moving average, exponential smoothing uses *all* previous values, although distant ones typically get very small weight. If we expand the calculation, we can see that the smoothed value at time t is a *weighted* average of the current value and all the previous values, with the weights depending on a smoothing coefficient, α:

$$\tilde{y}_t = \alpha y_t + \alpha(1 - \alpha)y_{t-1} + \alpha(1 - \alpha)^2 y_{t-2} + \alpha(1 - \alpha)^3 y_{t-3} + \cdots.$$

As with the moving average model, we use $\tilde{y}_t$ as our prediction for time $t + 1$.

[7]The initial smoothed value to get the algorithm started is either the initial observed value ($\tilde{y}_1 = y_1$) or the average of some of the initial values. In Minitab, for example, the initial smoothed value is equal to the average of the first six observations.

Figure 20.5 shows the Intel stock prices again, this time with exponentially smoothed values using $\alpha = 0.75$ and $\alpha = 0.10$. You can see that the curve computed using the larger α follows the original series closely. By contrast, the curve computed using the smaller α is smoother, but doesn't follow rapid changes in the series such as the sharp price drop in June.

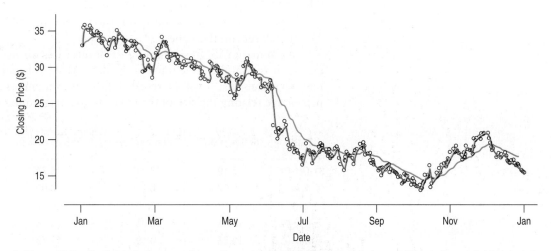

Figure 20.5   The Intel stock prices along with exponential smoothing models ($\alpha = 0.75$ in red, $\alpha = 0.10$ in green). The model with the larger alpha follows the data more closely, and the model with the smaller alpha is smoother.

20.4   Summarizing Forecast Error

Whenever we model a time series, it is natural to ask how closely the model fits the series. A closely related question is how well the model *forecasts* the series. For smoothing models, we use the most recent model value as the forecast for the next time period. When we make a plot of the data and the smoothed series, we usually plot the smoothed value $\tilde{y}_t$ at time t. But if we are interested in the error the model makes when trying to forecast each value, we must compare the data value y_t not to $\tilde{y}_t$, but rather to $\hat{y}_t$, which is actually equal to $\tilde{y}_{t-1}$. We can find a forecast error for each time in the series for which we have such a forecast:

$$e_t = (y_t - \hat{y}_t)$$

When there is a particular forecast of interest, it makes sense to look at the forecast error, e_t. More often, we ask about the overall success of a model at forecasting for a time series. That calls for a summary of the forecast errors and, as often happens in statistics, we have several summaries to choose among.

In Chapter 18, we found the **mean squared error (MSE)** to summarize the magnitude of the errors. We can do the same with the forecast errors:

$$MSE = \frac{1}{n}\Sigma(y_t - \hat{y}_t)^2$$

The MSE penalizes large forecast errors because the errors are squared. It also has the problem that it is not in the same units as the data, but rather in the square of those units. We can address both of these problems by summing the *absolute values* of the errors. That gives the **mean absolute deviation (MAD)**:

$$MAD = \frac{1}{n}\Sigma|y_t - \hat{y}_t|$$

The most common approach to measuring forecast error compares the absolute errors to the magnitude of the estimated quantities and sums up the *proportions* of the values that are in error. Multiplying the proportions by 100 gives the **absolute percentage error (APE)**. If we average the APE for all the forecasts, we have the **mean absolute percentage error (MAPE)**:

$$MAPE = 100 \times \frac{1}{n}\sum \frac{|y_t - \hat{y}_t|}{|y_t|}$$

The MAPE is expressed in percent, so it doesn't depend on the units of the y variable. If you choose to rescale y, both the MSE and MAD will change, but the MAPE will remain the same.

In summary, MSE resembles the error measures we've used for regression models, but it isn't in the same units as the data. MAD is in the same units as the data, but that means that it will be rescaled if the measurements are rescaled. MAPE is a percentage relating the size of the errors to the magnitudes of the data values.

Date	SES α =0.75	Forecast error	SES α = 0.10	Forecast error	Price
31-Oct-02	17.19	0.11	15.74	1.56	$17.30
1-Nov-02	18.02	0.28	16.00	2.30	$18.30
4-Nov-02	18.58	0.19	16.27	2.50	$18.77
5-Nov-02	18.41	−0.06	16.48	1.87	$18.35
6-Nov-02	18.96	0.19	16.75	2.40	$19.15
7-Nov-02	18.57	−0.13	16.92	1.52	$18.44
8-Nov-02	18.26	−0.11	17.04	1.11	$18.15
11-Nov-02	17.57	−0.23	17.07	0.27	$17.34
12-Nov-02	17.73	0.06	17.14	0.65	$17.79
13-Nov-02	18.02	0.10	17.24	0.88	$18.12
14-Nov-02	18.91	0.30	17.44	1.77	$19.21
15-Nov-02	18.83	−0.03	17.57	1.23	$18.80
18-Nov-02	18.62	−0.07	17.67	0.88	$18.55
19-Nov-02	18.27	−0.12	17.72	0.43	$18.15
20-Nov-02	18.93	0.22	17.86	1.29	$19.15
21-Nov-02	19.89	0.32	18.10	2.11	$20.21
22-Nov-02	20.01	0.04	18.29	1.76	$20.05
25-Nov-02	20.36	0.12	18.51	1.97	$20.48
26-Nov-02	20.24	−0.04	18.68	1.52	$20.20
27-Nov-02	20.74	0.16	18.90	2.00	$20.90
29-Nov-02	20.84	0.04	19.10	1.78	$20.88
2-Dec-02	21.00	0.05	19.29	1.76	$21.05
3-Dec-02	20.48	−0.17	19.40	0.91	$20.31
4-Dec-02	19.87	−0.20	19.42	0.25	$19.67
5-Dec-02	19.19	−0.23	19.38	−0.42	$18.96
6-Dec-02	18.83	−0.12	19.31	−0.60	$18.71
9-Dec-02	17.97	−0.29	19.15	−1.47	$17.68
10-Dec-02	18.09	0.04	19.05	−0.92	$18.13
11-Dec-02	18.14	0.02	18.96	−0.80	$18.16
12-Dec-02	18.18	0.01	18.88	−0.69	$18.19
MAD		$0.135		$1.321	
MSE		0.026		2.162	
MAPE		0.714%		6.91%	

Table 20.3   The Intel stock prices along with smoothed values reported as forecasts one period ahead for two models. For this series, the single-exponential smoothing (SES) model with the larger coefficient (α = 0.75) has a lower forecast error.

Table 20.3 shows forecast error for two different single-exponential smoothing (SES) models over the final 30 days of the Intel stock series along with the MAD, MSE, and MAPE computed for the entire time series. The smoothing model that uses a larger smoothing coefficient ($\alpha = 0.75$) more accurately forecasts the daily stock price over this time period.

For Example　Forecast errors

Question: For the euro values in the example on page 673, find the forecasts, the errors, and the measures of forecast error.

Answer:

Date	Value ($)	MA(3)	Forecast	Error
2010-05-27	1.2265	*	*	*
2010-05-28	1.2301	*	*	*
2010-05-29	1.2344	1.2303333	*	*
2010-05-30	1.2278	1.2307667	1.2303333	−0.00253333
2010-05-31	1.2276	1.2299333	1.2307667	−0.00316667
2010-06-01	1.2303	1.2285667	1.2299333	0.00036667
2010-06-02	1.2228	1.2269000	1.2285667	−0.00576667

MSE = 0.00001246

MAD = 0.00295835

MAPE = 0.241%.

20.5　Autoregressive Models

Simple moving averages and exponential smoothing methods are good choices for time series with no regular long-term patterns. But if some patterns are present—even if they don't rise to the level of a well-structured seasonal fluctuation—we may want to choose weights that facilitate modeling that structure—something that exponential smoothing can find difficult to do. Such weights might even be negative. (Imagine a series with values that alternated up and down for successive times. A good weighted average would give a negative weight to the most recent value and a positive weight to the one before that.)

But how can we find appropriate weights and how can we choose among the huge number of possible weights? It turns out that we can use the methods of multiple regression we saw in Chapters 18 and 19, along with the fact that the data come to us in time sequence order, to discover weights for a weighted moving average smoother. We shift the data by a time period or a few time periods. This shift is known as *lagging*, and the resulting variables are called *lagged variables*. For example, Table 20.4 shows the first 15 values of the daily Intel stock prices in 2002 along with the lagged values for lags of one, two, three, and four days.

Date	Price	Price$_{lag1}$	Price$_{lag2}$	Price$_{lag3}$	Price$_{lag4}$
2-Jan-02	$33.00	*	*	*	*
3-Jan-02	$35.52	$33.00	*	*	*
4-Jan-02	$35.79	$35.52	$33.00	*	*
7-Jan-02	$35.27	$35.79	$35.52	$33.00	*
8-Jan-02	$35.58	$35.27	$35.79	$35.52	$33.00
9-Jan-02	$35.36	$35.58	$35.27	$35.79	$35.52
10-Jan-02	$34.65	$35.36	$35.58	$35.27	$35.79
11-Jan-02	$34.55	$34.65	$35.36	$35.58	$35.27
14-Jan-02	$34.84	$34.55	$34.65	$35.36	$35.58
15-Jan-02	$34.68	$34.84	$34.55	$34.65	$35.36
16-Jan-02	$33.71	$34.68	$34.84	$34.55	$34.65
17-Jan-02	$34.53	$33.71	$34.68	$34.84	$34.55
18-Jan-02	$33.48	$34.53	$33.71	$34.68	$34.84
22-Jan-02	$31.70	$33.48	$34.53	$33.71	$34.68
23-Jan-02	$32.45	$31.70	$33.48	$34.53	$33.71

Table 20.4   The lagged values for the first 15 days in the Intel daily stock time series for one, two, three, and four days.

If we fit a regression to predict a time series from its lag1 and lag2 versions,

$$\hat{y} = b_0 + b_1 y_{lag1} + b_2 y_{lag2},$$

each predicted value, $\hat{y}_t$, is just a sum of the two previous values, y_{lag1} and y_{lag2} (plus a constant) weighted by the fitted coefficients b_1 and b_2. That's just a weighted moving average with weights found by the regression.

But wait. Regression methods assume that the errors are mutually independent. And this method works only if recent values can help predict current ones, which makes it likely that recent errors will be like the current one. Isn't this a violation of the regression model? Well, yes and no. The independence assumption is certainly required for inference on the coefficients, for example, to test the standard null hypothesis that the true coefficient is zero. But we're not doing inference here; we're just building a model. And for that purpose, the failure of independence is really more an opportunity than a problem.

In fact, we can specifically account for the association of cases with previous ones. The correlation between a series and a **lagged** version of the same series that is offset by a fixed number of time periods is called **autocorrelation**.[8] Table 20.5 shows some autocorrelations for the Intel series.

	Price	Lag1	Lag2	Lag3	Lag4
Price	1.000				
Lag1	0.992	1.000			
Lag2	0.984	0.992	1.000		
Lag3	0.978	0.984	0.992	1.000	
Lag4	0.973	0.977	0.984	0.992	1.000

Table 20.5   Autocorrelations of the daily Intel closing stock price for the entire year 2002 for lags 1, 2, and 3.

---

[8]Recall that we evaluated the presence of lag-1 autocorrelation using the Durbin-Watson statistic in Chapter 17.

A regression model that is based on an average of prior values in the series weighted according to a regression on lagged versions of the series, is called an **autoregressive model**. A model based on only the first lagged variable is called a first-order autoregressive model, often abbreviated as AR(1).

A pth-order autoregressive model has the form:

$$\hat{y} = b_0 + b_1 y_{lag1} + \cdots + b_p y_{lag\,p}.$$

For the Intel stock price series, we find the coefficients for a fourth-order autoregressive model from a multiple regression of the series on its first four lagged values, as shown in Table 20.6.

```
Dependent variable is: Price
R squared = 98.4%   R squared (adjusted) = 98.4%
s = 0.8788 with 236 − 5 = 231 degrees of freedom
```

Variable	Coefficient	SE(Coeff)	t-ratio	P-value
Intercept	0.195748	0.2025	0.967	0.3348
Lag1	0.958529	0.0652	14.7	≤0.0001
Lag2	−0.046097	0.0906	−0.509	0.6113
Lag3	−0.057307	0.0900	−0.637	0.5249
Lag4	0.132766	0.0638	2.08	0.0385

Table 20.6 A fourth-order autoregressive model for the Intel stock prices. Note that there are 240 values in the series but, because lagged variables have missing values at the beginning of the series, there are only 236 complete cases.

The resulting fourth-order autoregressive model is

$$\hat{y}_t = 0.195748 + 0.958529 y_{lag1} - 0.046097 y_{lag2} - 0.057307 y_{lag3} + 0.132766 y_{lag4}$$

Looking at the coefficients, we can see that the model puts most of its weight on the data value just preceding the one we're estimating (the lag 1 value), some on the lag 4, and relatively little on the other two lags.

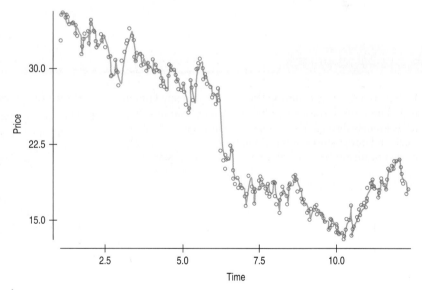

Figure 20.6   The Intel stock price series with an AR(4) model.

We can use an autoregressive model to predict the next value in a time series, but, to predict further into the future, we must substitute predicted values in the lag variables, which of course is likely to make our errors larger.

The forecasts, forecast errors, and APE produced by the two moving average models (MA(5) and MA(15)), the exponential smoothing models ($\alpha = 0.10$) and ($\alpha = 0.75$), and the fourth-order autoregressive model, or AR(4) for the euro prices on page 674, are shown in Table 20.7.

Model	Forecast $(\hat{y}_t)$	Actual Price (y_t)	Forecast Error $(y_t - \hat{y}_t)$	Absolute Percent Error (APE) $100 \times \lvert y_t - \hat{y}_t\rvert / \lvert y_t\rvert$
MA-5	$17.95	$17.58	−$0.37	2.10%
MA-15	$19.40	$17.58	−$1.82	10.35%
SES ($\alpha = 0.10$)	$18.75	$17.58	−$1.17	6.66%
SES ($\alpha = 0.75$)	$17.45	$17.58	$0.13	0.74%
AR(4)	$18.17	$17.58	−$0.59	3.36%

Table 20.7   The forecasts and forecast error for each of the moving average, single exponential smoothing, and autoregressive models for the Intel daily stock price on December 13, 2002.

For Example   An autoregressive model for the euro

Question: Find and interpret an autoregressive model for the euro prices.

Answer:

```
Dependent variable is: Eur/Dol
4188 total cases of which 2
are missing

R squared = 99.9%   R squared
(adjusted) = 99.9%
s = 0.0060 with 4186 − 4 = 4182 degrees of freedom
```

Variable	Coefficient	SE(Coeff)	t-ratio	P-value
Intercept	1.85227e−3	0.0008	2.24	0.0252
Time	3.02310e−7	0.0000	2.05	0.0406
Lag1	1.09975	0.0154	71.5	≤0.0001
Lag2	−0.101845	0.0154	−6.62	≤0.0001

Both the lag1 and lag2 terms are very significant, but the *Time* coefficient is very small with a smaller *t*-statistic. This reflects the fact that the time series fluctuates rapidly over the short term.

Just Checking

J. Crew is a clothing company known for its preppy fashions, including jeans, khakis, and other basic items sold to young professionals through its catalogs, websites, and some 260 retail and outlet stores in the United States. It was recently made prominent because First Lady Michelle Obama shops there. We have their reported quarterly revenue from Q1 2003 through 2007. Here's a time series plot:

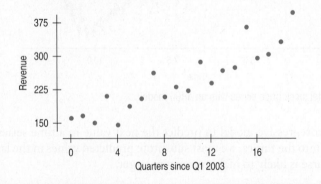

1 Which time series components do you see in this plot?

2 The final four values of this series are

  Q1 2007   297.312
  Q2 2007   304.731
  Q3 2007   332.744
  Q4 2007   399.936

   Without calculating anything, roughly where on the graph would the final value of a 4-point simple moving average be? Do you think it would be a good prediction for Q1 of 2008?

3 If the exponential smooth value for Q3 of 2007 is 319.3, what is the exponential smooth value of Q4 when an α of 0.5 is used?

4 If you wished to fit an autoregressive model, how many terms should you include? Why?

Unlike simple moving average models and single exponential smoothing models, AR models can follow time series that have seasonal fluctuations. The AR method will assign a larger weight to lags that correspond to the period of the fluctuation. For example, AR models will tend to predict quarterly sales that show a seasonal cycle by assigning a large weight to the lag4 version of the series, so sales in the same quarter of the previous year are counted heavily.

Random Walks

All moving average models include the special naïve model that predicts that the next value will be the same as the current one. For an autoregressive model, when the coefficient of y_{lag1} is close to one and the estimated intercept is close to zero, the first-order autoregressive model is approximately

$$\hat{y}_{t+1} = y_t$$

The naïve forecast model is the best we can do when the underlying time series has no other structure. A series with no other structure can be described as just adding a random kick to the previous value:

$$y_{t+1} = y_t + e_t$$

where the "e" are independent random values with some distribution. The "e" are sometimes referred to as "white noise." This kind of time series is called a **random walk** because each new value can be thought of as a random step away from the previous value.

Time series that are modeled by a random walk can have rapid and sudden changes in direction, but they also may have long periods of runs up or down that can be mistaken for cycles.[9] Random walks include series such as the assets of a gambler over time, the location of a molecule in a gas, and the path taken by a foraging animal. The random walk hypothesis in finance predicts that in an efficient market, stock prices should follow a random walk.[10,11] There is a large literature dealing with random walks. They show up in mathematics, economics, finance, and physics, as well as other fields.

Guided Example	**Comparing Time Series Methods**

In Chapter 6 we examined the share price of The Home Depot. Now, let's look at sales. The Home Depot reports sales quaterly, so these are a time series that we can model.

(continued)

[9]This is one reason we recommended that the identification of cycles be based on theory, on established patterns, or on other variables.

[10]In the case of first-order autoregressive models, or in cases where autocorrelation continues to present a problem, one solution is to model the *differenced* time series. For example, in this case, we might model the *change* in Intel daily stock price $(y_t - y_{t-1})$ as the response variable.

[11]Princeton economist Burton Malkiel made the random walk theory of the stock market famous in his book *A Random Walk Down Wall Street: The Time-Tested Strategy for Successful Investing*, first published in 1973. The theory originated in the 1950s with Maurice Kendall, a British statistician.

PLAN

Setup State your objective. Identify the quantitative variables you wish to examine. Report the time frame over which the data have been collected and define each variable.

We want to build time series models for quarterly sales at The Home Depot from 1995 through 2004. We have quarterly sales ($Billion).

Plot Plot the time series and clearly label the axes to identify the scale and units.

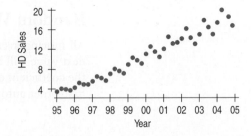

Model Think about the assumptions and check the conditions.

These are quantitative data measured over time at consistent intervals. So it is appropriate to use time series methods.

There is a consistent increasing trend with fluctuations around it that appear to be seasonal because they repeat every four quarters. Some smoothing methods may have difficulty with this kind of series.

DO

Mechanics, Part 1 Try a moving average. For data with a strong seasonal component, such as these, a moving average length that is a multiple of the period is a good idea. But series with a strong trend, such as this one, won't be fit well by an uncentered moving average.

Here is a simple moving average of length 4 as computed by a statistics program:

Variable	Moving Average	Accuracy Measures	
Actual	length 4	MAPE	10.9425
Fits		MAD	1.1999
Forecasts		MSD	2.1150
95.0% PI			

Evaluate how well this method fits the series.

The program reports:

MAD = 1.1999
MSE = 2.115
MAPE = 10.9425

Make a forecast for Q1 2005 ($t = 40$).

The program offers a forecast of 18.2735 $B for the first quarter of 2005.

Mechanics, Part 2 Exponential smoothing can be a good compromise between a simple moving average smoother and a fit with a seasonal component.

In series with a strong trend such as this one, exponential smooths will inevitably lag behind the data.

Let's try an exponential smooth. Now we have to choose a smoothing weight. We'll use $\alpha = 0.5$, which weights the current data value equally with all the rest in the past. Here's the result from a computer-generated smooth:

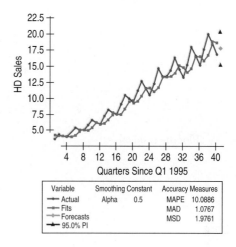

Variable	Smoothing Constant		Accuracy Measures	
━━ Actual	Alpha	0.5	MAPE	10.0886
━━ Fits			MAD	1.0767
━━ Forecasts			MSD	1.9761
━▲━ 95.0% PI				

Evaluate how well this model fits.

The statistics package reports:

MAD = 1.0767
MSE = 1.9761
MAPE = 10.0886

Forecast sales for Q1 2005.

The package generates a prediction of 17.6770 \$B for the first quarter of 2005.

Mechanics, Part 3 An autoregressive model is a multiple regression on the series itself lagged, or offset, by 1, 2, or more time periods. When we know that there is a seasonal component, it is important to include the corresponding lag—here lag4. This model has terms for each lag.

Now let's fit an autoregressive model. Because we know that the seasonal pattern is 4 quarters long, we'll fit four terms, using multiple regression to find the smoothing weights:

```
Dependent variable is: HDSales
R squared = 98.7%   R squared (adjusted) = 98.6%
s = 0.5412 with 36 − 5 = 31 degrees of freedom
```

Variable	Coeff	SE(Coeff)	t-ratio	P-value
Intercept	0.947082	0.2469	3.84	0.0006
Lag1	0.428426	0.1455	2.94	0.0061
Lag2	−0.396168	0.1548	−2.56	0.0156
Lag3	0.353743	0.1612	2.19	0.0358
Lag4	0.646243	0.1552	4.16	0.0002

The AR(4) model is

$$\hat{y}_t = 0.947 + 0.428 y_{t-1} - 0.396 y_{t-2}$$
$$+ 0.354 y_{t-3} + 0.646 y_{t-4}$$

(continued)

Plot the fit.

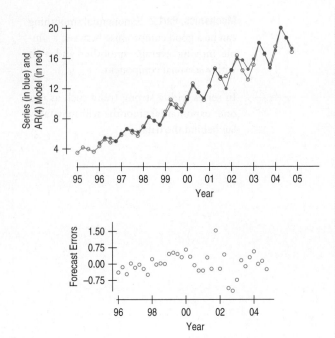

Plot the residuals.

The forecast errors show some disturbance near and just after Q4 2001. We've already noticed this quarter as a high outlier. Notice here how its effects propagate through the fit and residuals. A lack of resistance to outliers is one disadvantage of autoregressive models.

MAD = 0.32667
MSE = 0.227
MAPE = 3.46

Calculate fit measures and a prediction.

A predicted value for the first quarter of 2005 is:

$$0.9470 + 0.4284 \times 16.812 - 0.39617$$
$$\times 18.772 + 0.3537 \times 19.96 + 0.6462$$
$$\times 17.55 = 19.11$$

REPORT

Conclusion Compare the advantages and disadvantages of the time series methods. Be sure to state your interpretations in the proper context.

MEMO

Re: Time Series Analyses of The Home Depot Quarterly Sales

We compared several time series methods to fit quarterly data on sales at The Home Depot for the period from 1995 through 2004. The actual sales in Q1 of 2005 were $19B.

The different methods had differing strengths and weaknesses. The moving average method smooths out most of the seasonal effects and because it lags behind the series, mis-estimates the seasonal effect. But it provides a good description of the trend and a good prediction.

The exponential smoothing method follows the seasonal pattern more closely, but that doesn't improve its prediction.

The autoregressive method works well and makes a good prediction, but it is clearly affected by the single outlying quarter, which contaminates the fit for several quarters afterward.

20.6   Multiple Regression–based Models

We noted earlier that some time series have identifiable components; a trend, a seasonal component, and possibly a cyclical component. Simple moving average models work best on time series that lack any consistent structures like these. Exponential smoothing models don't usually follow a seasonal component well either. Autoregressive moving averages can follow all of these components provided the length of the smoother is at least as long as the period of the seasonal or cyclical component.

When some or all of these components are present, we can gain two distinct advantages by modeling them directly. First, we may be able to forecast beyond the immediate next time period—something the smoothing models can't do easily. Second, we may be able to understand the components themselves and reach a deeper understanding of the time series itself.

Modeling the Trend Component   When a time series has a **linear trend**, the natural thing to do is to model it with a regression. If the trend is linear, a linear regression of y_t on *Time* can model the trend. The residuals would then be a *detrended* version of the time series.

The Home Depot sales data that we examined in the Guided Example seem to have a roughly linear trend. The regression to estimate the trend is in Table 20.8.

```
Dependent variable is: HDSales
R squared = 95.5%   R squared (adjusted) = 95.4%
s = 1.044 with 40 - 2 = 38 degrees of freedom

Variable    Coeff     SE(Coeff)   t-ratio   P-value
Intercept   2.67102   0.3241       8.24     ≤0.0001
Time        0.405707  0.0143      28.4      ≤0.0001
```

Table 20.8   Estimating the trend component in the Home Depot sales data by regression.

One attractive feature of a regression-based model is that the coefficient of *Time* can be interpreted directly as the change in *y* (here, Home Depot Sales) *per* time unit (here, quarters). The trend in Home Depot sales was that they increased by 0.405 billion dollars per quarter.

By contrast, we saw in Figure 20.1 that the Whole Foods sales data *don't* have a linear trend. As we learned in Chapter 17, we can often improve the linearity of a relationship with a re-expression. The re-expression that most often works for time series is the logarithm. That's because many time series grow or shrink exponentially. Typically, the bigger you are, the larger the absolute increment in your profits. Growth by a consistent percentage is *exponential growth*. For example, if sales each year are 5% higher than the previous year, then the overall growth will be exponential. And the logarithm makes exponential growth linear. Figure 20.7

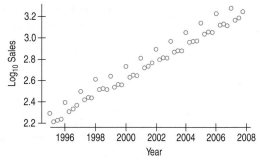

Figure 20.7   The logarithm of Whole Foods quarterly Sales is linear over time.

shows the result of taking logs in the Whole Foods data. The corresponding model
for the trend is

```
Dependent variable is: Log WFSales
R squared = 96.5%   R squared (adjusted) = 96.4%
s = 0.0574 with 52 − 2 = 50 degrees of freedom

Variable   Coeff      SE(Coeff)  t-ratio  P-value
Intercept  −154.663   4.243      −36.4    ≤0.0001
Year       0.078650   0.0021     37.1     ≤0.0001
```

Now the interpretation of the trend coefficient is different. Adding 0.07865 to the
logarithm of sales each year is the same as multiplying sales by $10^{0.07865} \approx 1.20$.
And that's an increase of 20 percent. So we can say that Whole Foods Market's sales
were increasing by 20 percent per year.[12]

Re-expressing the Whole Foods quarterly sales data by logarithms reveals a
second advantage of the re-expression. The seasonal fluctuations evident in
Figure 20.1 grew in magnitude as the sales themselves grew. But in Figure 20.7,
those fluctuations are nearly constant in size. That will make them much easier
to model.

> **Recording *Time***
> The Whole Foods Market Sales are reported quarterly, but the time variable here
> is *Year*. The time variable in a time series often is just a count of the number of
> time periods (weeks, months, quarters, etc.) since the series began. So, the periods
> are often recorded as 1, 2, 3, . . . as we did for the Home Depot sales data. If we
> want to use the actual date as the predictor variable and we have quarterly data, we
> need to express each time period as a fractional year, such as 1995.0, 1995.25,
> 1995.5, ... The only difference this will make is in interpreting the coefficients.
> Moving average methods take no note of how time is recorded, but when you use a
> regression-based model, you must be aware of that so you can interpret the *Time*
> coefficient correctly.
> For example, if we had used a time variable that counted 1, 2, 3, . . . in the
> regression instead of fractional year, the slope coefficient would have been 1/4 of the
> one we found ($0.07865/4 = 0.01966$) and it would have estimated the *quarterly*
> growth in log sales rather than than the annual growth.

Modeling the Seasonal Component   Figure 20.7 shows that the Whole Foods
data have a strong seasonal component. Every fourth quarter there's a spike.
That's not unusual in time series related to retail sales. The simplest version of a
seasonal component is one that adds a different value to the series (in addition to
the trend) for each season. We can see that this is a good description of the
Whole Foods data; the first fiscal quarter is above the overall trend by roughly
the same amount each year. Figure 20.8 shows the pattern.

[12]Of course, if you use natural logarithms rather than base-10 logarithms, the function that returns
to the original units is e^b where b is the Trend coefficient. But the interpretation of the result (and,
allowing for rounding error, the value itself) will be the same.

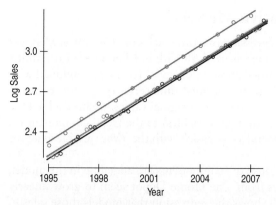

Figure 20.8 The logarithm of Whole Foods quarterly sales. Each quarter is displayed in its own color and has regression line fit to it. The lines differ almost exclusively by a shift up or down.

As we learned in Chapter 19, a pattern such as the one shown in Figure 20.8 can be modeled by introducing an indicator or dummy variable for each season. For example, we can define our indicator variables to be:

$$Q_1 = 1 \text{ in quarter 1 and 0, otherwise}$$
$$Q_2 = 1 \text{ in quarter 2 and 0, otherwise, and}$$
$$Q_3 = 1 \text{ in quarter 3 and 0, otherwise.}$$

Recall from Chapter 19 that for a categorical variable with k levels to enter a regression model we use $k - 1$ indicator variables. We can't use all k because that would create a collinearity. So we leave out one of them. It doesn't really matter which one we choose to leave out, but because the omitted one becomes the standard of comparison, we often choose the one that is the most typical. The intercept coefficient will estimate a level for the period "left out," and the coefficient of each indicator variable estimates the shift up or down in the series relative to that base level. With four quarters, we use three indicator variables. For this example, we'll arbitrarily choose to leave out the indicator for Q4. Then cases in Q4 will have the value zero for all three of our indicator variables (Q_1, Q_2, and Q_3) and the mean adjustment relative to the trend will be estimated by the intercept (b_0). In any other quarter, the adjustment relative to Q4 will be the coefficient for that quarter's indicator variable.

For Example | A regression model for truck border crossings

Question: Find and interpret a multiple regression model for the Alaska truck border crossing data from the example on page 670. (The data are in the file **Trucks**.)

Answer: The model is:

```
Dependent variable is: Trucks
R squared = 54.2%   R squared (adjusted) = 49.6%
s = 156.7 with 132 − 13 = 119 degrees of freedom
```

Variable	Coefficient	SE(Coeff)	t-ratio	P-value
Intercept	27539.6	8719	3.16	0.0020
Decimal Year	−13.3027	4.350	−3.06	0.0028
Jan	−187.342	46.93	−3.99	0.0001
Feb	−160.325	46.93	−3.42	0.0009
Mar	−2.03334	46.93	−0.043	0.9655
May	129.728	46.94	2.76	0.0066
Jun	311.474	46.95	6.63	≤0.0001
Jul	195.491	46.97	4.16	≤0.0001
Aug	125.531	50.71	2.48	0.0147
Sep	23.3453	47.00	0.497	0.6203
Oct	−4.27389	47.03	−0.091	0.9277
Nov	−139.893	47.05	−2.97	0.0036
Dec	−209.693	47.08	−4.45	≤0.0001

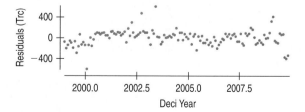

The model finds that the trend is declining over time, but only by about 13 trucks per month per year. (The units of the series are "Trucks per month.") The coefficient of the trend indicates that 13 fewer trucks cross the border each month than in the previous year, after allowing for the (very substantial) seasonal pattern. The model estimates a strong seasonal pattern with crossings low in winter months (November to February) and high in the summer (May to August)—not a surprising pattern for Alaska.

Additive and Multiplicative Models

Adding indicator variables to the regression of a time series on *Time* turns what was a simple one-predictor regression, such as we dealt with in Chapter 6, into a multiple regression, such as those we learned about in Chapter 18. That, combined with the question of whether to work with the original time series or the logarithm of the series, raises a new question. If we model the original values, we have added the seasonal component (in the form of dummy variables) to the trend component (in the form of an intercept coefficient and a regression with the *Time* variable as a predictor). We can write $\hat{y}_t = T + S$.

This is an **additive model** because the components are added up in the model. For example, we've seen that Sales from The Home Depot seem to grow linearly with a seasonal pattern. Table 20.9 shows the regression that models those sales in terms of a trend component and three quarterly dummy variables.

```
Dependent variable is: HDSales
R squared = 98.7%   R squared (adjusted) = 98.5%
s = 0.5915 with 40 − 5 = 35 degrees of freedom

Variable    Coeff      SE(Coeff)   t-ratio   P-value
Intercept   1.38314    0.2534      5.46      ≤0.0001
Time        0.410336   0.0081      50.4      ≤0.0001
Q1          1.22191    0.2657      4.60      ≤0.0001
Q2          2.41907    0.2650      9.13      ≤0.0001
Q3          1.14944    0.2647      4.34      0.0001
```

Table 20.9   A regression to model Home Depot sales with a trend component and three dummy variables representing a seasonal component in an additive model.

The model contains a trend component that predicts growth at about $0.41B per quarter with adjustments for each quarter that are consistent over the entire time period. For example, because Q4 is the quarter without a dummy variable, sales in Q4 are predicted to be on average $1.38 + 0.41\,Time$ billion dollars. The seasonal dummy variable coefficients adjust the predictions for each quarter by adding the value of their coefficients to the intercept. For example, sales in Q1 are predicted to be $1.38 + 0.41\,Time + 1.22 = \$2.60B + 0.41\,Time$ (see Figure 20.9). But you can see from Figure 20.9 that the seasonal fluctuations are small early in the series and grow larger later in the series—a pattern this model doesn't fit.

Figure 20.9   Sales at The Home Depot with predictions from the additive model. The model predicts a consistent seasonal component although the seasonal component of the data varies.

When we examined the Whole Foods sales data, we saw that we could straighten the trend and make the seasonal fluctuations more nearly the same size by finding the *logarithm*. We can still find a multiple regression, but now the response variable is re-expressed by logarithms. The model is in Table 20.10.

```
Dependent variable is: Log WFSales
R squared = 99.6%   R squared (adjusted) = 99.6%
s = 0.0188 with 52 − 5 = 47 degrees of freedom

Variable   Coeff      SE(Coeff)   t-ratio   P-value
Constant   −156.488   1.396       −112      ≤0.0001
Year       0.079541   0.0007      114       ≤0.0001
Q1         0.132511   0.0074      17.9      ≤0.0001
Q2         0.016969   0.0074      2.30      0.0261
Q3         0.012885   0.0074      1.75      0.0874
```

Table 20.10 A regression model for the logarithm of the Whole Foods quarterly Sales data with trend and seasonal components.

Now, however, there's a difference in the model. Because we are modeling the logarithm of sales, when we think in terms of the sales themselves, the model components are multiplied rather than added, so we have a **multiplicative model**,

$$\hat{y} = T \times S.$$

Although we acknowledge that the terms in a multiplicative model are multiplied, we always fit the multiplicative model by taking logs, changing the form to an additive model that can be fit by multiple regression.

As we observed earlier, seasonal fluctuations are often proportional to the overall level of the values in the series; the Q1 lift in sales in a multiplicative model is a *proportion* of overall sales, not a fixed additive increment. Specifically, it is $10^{0.1325} \approx 1.36$—about 36% higher than the sales at that time. Because the sales themselves were growing (at 20% per year), this 36% lift grew as well, in dollar terms. But after taking logs, it is a constant lift, and easier to model.

Look back at the Whole Foods sales in Figure 20.1. You can see this growth in the size of the seasonal component in the plot as well as in the trend. Taking logs not only turns exponential growth into linear growth, but it also tends to stabilize the size of the seasonal fluctuations. In fact, if you look back at the Home Depot data in Figure 20.9, you may notice a similar increase in seasonal fluctuations. Although the trend is already linear and we can fit an additive model to those data, a multiplicative model is likely to work even better.

Cyclical and Irregular Components

Many time series are more complex than the trend and seasonal components can model. The Intel Stock price data of Figure 20.3 is one example. Models of time series components are usually said to include two additional components, a cyclical component and an irregular component. Consistent with their form for the trend and seasonal components, we write for additive models:

$$\hat{y}_t = T + S + C + I,$$

and for multiplicative models:

$$\hat{y}_t = T \times S \times C \times I.$$

The Cyclical Component   Long-term business cycles may influence economic and financial time series. Other time series may be influenced by other long-term fluctuations. Whenever there is a business, economic, or physical cycle whose cause

---

Logarithms Reminder

Logarithms have the special property that they turn multiplication into addition. Before technology, that made them a workhorse tool. To understand why a model for the logarithm of a series is a *multiplicative* model, recall that

If

$$\hat{y} = T \times S$$

then

$$\log(\hat{y}) = \log(T) + \log(S)$$

The terms *exponential model* and *multiplicative model* are equivalent for these models and can be used interchangeably.

is understood and that can be relied on, we should look for an external or **exogenous** variable to model the cycle. The regression models we've been considering can accommodate such additional predictors naturally.

Just calling a long-term fluctuation in the data that isn't fit by the trend or seasonal component a "cyclical component" doesn't add much to our understanding. Cyclical patterns may not be immediately evident in the data, so it is wise to compute and plot the residuals, known for time series models as the irregular component.

Irregular Components   The irregular components are the residuals—what's left over after we fit all the other components. We should examine them to check any assumptions and also to see if there might be other patterns apparent in the residuals that we might model. For multiple regression, most statistics programs plot the residuals against the predicted values, but for time series models, it is essential to plot them against *Time*. Figure 20.10 shows the residuals of the Whole Foods data plotted against year. Two quarters stand out. Both extraordinary quarters are fourth quarters of their years, which suggests that our seasonal model may need to be improved.[13] We also see a possible cyclical pattern with a period of about 4 years. This might be something worth investigating to see if we can add a component to our model.

Figure 20.10   The irregular component, or residuals, of the multiplicative model for the Whole Foods sales. In the beginning years, growth was apparently faster than after 1999 or so. There are also a couple of quarters that deserve attention: Q4 of 2001 and Q4 of 2007 which were underestimated.

For Example   Additive and multiplicative models

Question: Is the model for Alaska truck border crossings in the previous example (see page 687) an additive or a multiplicative model? Describe the components. Which month was left out? Why?

Answer: This is an additive model. There is an overall negative trend of -13.3 truck crossings per month per year. There is a strong seasonal pattern, fit by the indicator variables for month. April was left out. One month had to be left out to avoid collinearity among the indicator variables.[14] There are generally fewer border crossings between October and March and more between April and September—probably due to the difficulty of driving in the winter in Alaska.

The residuals show no cyclic pattern. The irregular component shown in the residuals has a few outlying months that may deserve looking into. (For example, January of 2000 was a month with particularly heavy snows in Alaska.)

---

[13]It's possible to include an indicator (dummy) variable to model a specific event in time. Doing so is known as intervention analysis.

[14]April was selected by technology because it was alphabetically first. That's not a particularly good reason. You should be alert for arbitrary decisions by technology like this.

Continuing our analysis of the J. Crew revenue data form the previous Just Checking (page 680), here is a regression fitting a linear model to predict the *Revenue* ($M):

```
Dependent variable is: Revenue
R squared = 96.1%    R squared (adjusted) = 95.0%
s = 15.83 with 20 - 5 = 15 degrees of freedom
```

Variable	Coeff	SE(Coeff)	t-ratio	P-value
Intercept	193.037	9.87	19.6	≤0.0001
Q since '03	10.2338	0.6255	16.4	≤0.0001
Q1	−64.0281	10.18	−6.29	≤0.0001
Q2	−52.8349	10.09	−5.24	0.0001
Q3	−57.5810	10.03	−5.74	≤0.0001

5 Locate and interpret the trend coefficient.

6 Why is there no term for Q4? Would that term have a positive or negative coefficient?

Car Makers' Boom Years Now Look Like a Bubble

. . . Through most of the 1990s, auto makers sold a little over 15 million cars and light trucks a year in the U.S. market. That changed in the late 1990s: With gasoline prices low and many U.S. consumers feeling flush from the tech-stock boom, auto sales surged. Sales peaked at 17.4 million in 2000 and remained near 17 million for another five years. Heads of General Motors Corp. and Toyota said the U.S. was entering a golden age of the automobile. In 2003, Toyota's head of North American sales predicted the industry would soon be selling 20 million vehicles a year.

They were wrong. Sales started falling in 2006 and this year are expected to be right back where they were in the 1990s, at just over 15 million.[15]

Source: Wall Street Journal, May 20, 2008, page 1.

Forecasting with Regression-based Models

Regression models are easy to use for forecasting because they give us a formula. It is easy to substitute future values for *Time*, for the appropriate dummy variables, and for any exogenous variables and calculate a predicted value. Unlike smoothing models, regression-based models can forecast beyond the next time period. But any kind of forecast is uncertain, and the uncertainty grows the further we extrapolate, so it's wise to limit forecasts to the near term. The forecast error measures we discussed in Section 20.4 apply equally well to regression models and are computed from the residuals (or Irregular component.)

The most reliable part of a regression time series model is the seasonal component, because it is probably driven by fairly regular economic or environmental phenomena. We can reasonably expect those patterns to continue into the future, so we can feel comfortable basing longer term forecasts on them. A company that sees large fourth-quarter holiday season sales can probably count on higher fourth-quarter sales even several years in the future.

The trend component is less reliable. As much as we might like to think that a growing company will continue to grow, it should be clear that no company can grow forever. Exponential growth is even harder to maintain. Changes in the trend can be quite sudden. The business news is filled with stories of companies whose growth "suddenly" stopped or slowed, of products whose sales unexpectedly shot up and then, just as unexpectedly fell back, and of economic predictions by experts that, in retrospect, look ill-informed. While long-term government forecasts can be found for such reliable indicators as the Gross Domestic Product (GDP) and Disposable Income (DI), long-term forecasts should be made with great care. Changes in the economy or the market cannot be anticipated by the trend component, and can change a company's business quite suddenly.

The reliability of cyclical components for forecasting is something you must simply judge for yourself based on your knowledge and understanding of the underlying economic cycles. An empirical cycle that is not understood makes a risky

---

[15]The economic crisis of 2009 led to reorganizations of major auto companies, so any forecast based on previous data became risky.

basis for prediction. A cycle that is understood and is due to underlying phenomena that are stable over time would be a more reliable component of a forecast.

Table 20.11 compares the forecasts from four regression-based models for the Whole Foods Markets quarterly *Sales* to the actual sales in the first (calendar) quarter of 2008. When the forecast error is negative, the model provides an *over*estimate. When the error is positive, the model provides an *under*estimate for the forecast. The first model is the simple linear trend model (additive model). The second model is the additive model with quarterly seasonal components. The third model is the multiplicative trend model, and the fourth model is the multiplicative model with trend and quarterly seasonal components.

Model	Forecast ($M) $\hat{y}_t$	Actual Sales ($M) Q1 2008 ($\hat{y}_t$)	Forecast Error $(y_t - \hat{y}_t)$	Absolute Percent Error $100 \times \dfrac{|y_t - \hat{y}_t|}{y_t}$
Simple Linear Trend Model	1489.717	2457.258	967.541	39.4%
Additive Trend plus Seasonal Component Model	1652.760	2457.258	804.498	32.7%
Multiplicative Trend Model	1930.640	2457.258	526.618	21.4%
Multiplicative Trend plus Seasonal Component Model	2419.637	2457.258	37.621	1.5%

Table 20.11   The forecasts and forecast error for the Whole Foods sales data for the first quarter of 2008 using additive and multiplicative models with and without seasonal components.

From the original scatterplot, we saw that the trend wasn't linear, so it's no surprise that the two additive models grossly underestimate Q1 of 2008. The multiplicative trend model without seasonal components doesn't take into account that Q1 is typically a big quarter for Whole Foods, so it still underestimates it by more than 21%. The multiplicative trend model with seasonal components comes closest, getting within 1.5% of the actual sales.

Guided Example   Comparing Time Series Methods, Part 2

In the first Guided Example, we looked at smoothing models for sales at The Home Depot. But that time series shows a clear linear trend and a seasonal component. So let's try the regression models and see how well they do by comparison.

PLAN

Setup State your objective. Identify the quantitative variables you wish to examine. Report the time frame over which the data have been collected and define each variable.

We want to build regression-based time series models for quarterly sales at The Home Depot from 1995 through 2004. We have quarterly sales ($Billion).

Plot Plot the time series and clearly label the axes to identify the scale and units.	

Model Think about the assumptions and check the conditions.	This is a time series recording quantitative values. There is a consistent, increasing trend and fluctuations around it that may be seasonal. We can start by estimating the trend.

DO

Mechanics, Part 1 Fit an additive trend model using regression.	``` Dependent variable is: HDSales R squared = 95.5% R squared (adjusted) = 95.4% s = 1.044 with 40 − 2 = 38 degrees of freedom ```

```
Variable    Coeff      SE(Coeff)   t-ratio   P-value
Intercept   2.67102    0.3241      8.24      ≤ 0.0001
time        0.405707   0.0143      28.4      ≤ 0.0001
```

The blue series tracks the data. The model is a straight line, shown here in red.

Plot the residuals.	

Evaluate how well this method fits the series.	There's a bend in the irregular component and increasing spread. A seasonal fluctuation is easier to see in the irregular component.
Make a forecast for Q1 of 2005 ($t = 40$).	We can examine various measures of how well this model fits the series. These summaries are mostly useful to compare models, as we will here. MAD = 0.7706 MSE = 1.0356 MAPE = 8.1697 Forecast = 2.671 + 0.4057 × 40 = 18.899($B)

(continued)

Mechanics, Part 2 Fit a multiplicative model with trend and seasonal terms.

We fit a multiplicative model by multiple regression of the log series on both *Time* and dummy variables for three of the four quarters.[16]

```
Dependent variable is: log HDSales
R squared 97.2%    R squared (adjusted) = 96.9%
s = 0.0393 with 40 − 5 = 35 degrees of freedom
```

Variable	Coeff	SE(Coeff)	t-ratio	P-value
Intercept	0.556994	0.0168	33.1	≤ 0.0001
Time	0.018850	0.0005	34.8	≤ 0.0001
Q1	0.049673	0.0177	2.81	0.0080
Q2	0.096707	0.0176	5.49	≤ 0.0001
Q3	0.046880	0.0176	2.67	0.0116

Plot the fit vs. the data (here on the log scale) and the residuals.

Again, the data are the blue series and the model is the more regular red pattern.

This long-term pattern could be a cycle, but we'd need to see it repeat over the next 40 quarters to be sure of that and to be able to fit it. In fact, it probably is not a business cycle.

The Irregular component shows that the growth in Home Depot sales slowed in the middle of this period. There is also one point that stands out above the pattern in the Irregular component. This is Q4 of 2001. Possibly, the events of 9/11/01 encouraged people to fix up their homes.

To make a prediction with a multiplicative model, we use the inverse function of the logarithm.

MAD = 1.077
MSE = 1.0144
MAPE = 12.0386

Forecast = $10^{(0.556994 + 0.018850 \cdot 40 + 0.049673)}$

= 22.94 $B

Note that for comparability, these must be computed on the residuals transformed back to original units: $10^{residuals}$.

Make a forecast for Q1 of 2005.

---

[16]If you don't recall why we must leave out one of the quarters or how dummy variables work, see Chapter 19.

REPORT

Conclusion Compare the advantages and disadvantages of the time series models. Be sure to state your interpretations in the proper context.

MEMO

Re: Time Series Analyses of The Home Depot Quarterly Sales

We compared several time series methods to fit quarterly data on sales at The Home Depot for the period from 1995 through 2004. The actual sales in Q1 of 2005 were $19B, so the best prediction was made by the simplest model—the additive linear trend model.

The different methods had differing strengths and weaknesses. The linear trend model is simple to find and, for these data, was the best predictor. This may be in part because average sales in Q1 are close to the annual average. If we were forecasting either Q2 or Q4, we would expect the multiplicative model with seasonal dummy variables to do better.

The multiplicative component model fit and predicted least well, but by providing a model, it revealed both a slowing of growth in Home Depot sales and an outlying quarter of unexpectedly high sales in Q4 of 2001, possibly related to the events of 9/11/01. Both of these observations may be important in understanding the series.

20.7 Choosing a Time Series Forecasting Method

We've considered several methods for modeling and forecasting time series. How can you choose among them the ones that fit your needs? Your choice depends both on the nature of your data and on what you hope to learn from the data.

Simple moving averages demand the least of the data. You can apply them to almost any time series. However:

- They can forecast well only for the next time period. Longer forecasts just repeat the single forecast value.
- They are sensitive to spikes or outliers in the series, and can smear the shock of a spike across several adjacent time periods.
- They don't do well on series that have a strong trend, tending to lag behind the trend.

Exponential smoothing methods also make few assumptions about the data. They have the advantage of controlling the relative importance of recent values relative to older ones, and do a good job of following the movement of a time series. However:

- They can forecast well only for the next time period. Longer forecasts just repeat the single forecast value.
- They are sensitive to spikes or outliers in the series.
- They don't do well on series that have a strong trend, tending to lag behind the trend.

Autoregressive models use automatically determined weights to allow them to follow time series that may have regular fluctuations, such as a seasonal component, or series with a consistent trend. However:

- They can forecast for a limited span before they must "recycle" forecast values to use them to make more distant forecasts.

- They are sensitive to spikes or outliers in two different ways. Those outliers will influence the regression that determines the smoothing weights. And then again, when the smoothing is done, the effect of spikes in the series can be spread out, contaminating several points.

Regression-based models estimate the trend and seasonal components by regression against *Time* and using dummy variables to represent the seasonal period. They can incorporate exogenous variables to help model business cycles and other phenomena. And, unlike moving average models, they can forecast into the future. However:

- You must decide whether to fit an additive model to the time series (if the trend is linear and the seasonal fluctuations have a consistent size) or to re-express the series by logarithms and fit the resulting multiplicative model.

- Because they are based on regression, these models are sensitive to outliers and failures of linearity. Because they use dummy variables to estimate the seasonal effects, those effects must be consistent in magnitude during the time covered by the data.

- Forecasts depend on the continuation of the trend and seasonal patterns. Although seasonal patterns may be reliable, trends are difficult to forecast and hazardous to assume beyond the near future. Cycles are best forecast when they are based on some identifiable (and predictable) phenomenon.

20.8   Interpreting Time Series Models: The Whole Foods Data Revisited

When you use a time series model based on moving averages, you can (and should) summarize the patterns seen in the series and any patterns noticed in the residuals around the smooth trend. But time series models based on regression encourage us to interpret the coefficients.

Many time series of retail sales have a strong seasonal component. We've seen two in this chapter. But one of those, the quarterly sales at Whole Foods, is problematic. No, there's no problem with the models we've fit. But there is a problem if we interpret them without thinking.

Why should there be a seasonal spike in Whole Foods' sales? Food isn't a seasonal item. To be honest, it took the authors a while before we asked these questions. But that sent us back to the data. It turns out that Whole Foods Market divides its financial year into three quarters of 12 weeks and one of 16 weeks. The spike is entirely due to this bookkeeping anomaly. You can check for yourself; the seasonal peaks are $16/12 = 1.33$ times as big as the other quarters—almost exactly what the multiplicative model estimated them to be.

That doesn't invalidate any of our models. We'd still have to allow for the seasonal peak to model Whole Foods sales. But it is a cautionary tale that warns us not to jump to conclusions when interpreting our models.

What Can Go Wrong?

- **Don't use a linear trend model to describe a nonlinear trend.** Be sure to examine the scatter of the observations around the linear trend line to determine if there is a pattern. A curved pattern may indicate a need to transform the series. Plot the residuals. Variation in the residuals that increases when the central value is higher is a sign of a need to take logarithms. Series that grow or shrink by a constant percentage each time period are exponential and should be transformed before fitting a model.

- **Don't use a trend model for short-term forecasting.** Trend models are most effective for long-term forecasting and are rarely the best models to forecast one or two time periods ahead. Be aware that forecast errors are greater for long-term forecasting.

- **Don't use a moving average or an exponential model for long-term forecasting.** Smoothing models are most effective for short-term forecasting because they require continuous updating as new values in the time series are observed. However, it is less important to transform an exponentially trending series for these models because they base their predictions on recent values.

- **Don't ignore autocorrelation if it is present.** Look at correlations among the lagged versions of the time series. Compute the Durbin-Watson statistic, and make residual plots to identify autocorrelation but don't limit your attention to lag1 autocorrelation (as Durbin-Watson does). If present, then try using lagged variables to model the time series.

Ethics in Action

Kevin Crammer, a broker for a large financial services firm, is getting ready to meet with a new client, Sally. Sally just inherited a rather large sum of money and is looking to invest it long term. She is not an experienced investor, and even though Kevin sent her information about various financial investment options, she is confused and still wants to meet with him to get his advice. Kevin's company has several of its own mutual fund products they like their brokers to push. Kevin selected one of these funds and, based on historical returns, prepared a graph showing how an initial amount of $10,000 invested 20 years would have grown over time. While some cyclical fluctuations tied to economic conditions were noticeable, the underlying trend was upward. Kevin decided to fit a linear trend line to the graph. Checking out the residuals, he did notice a pattern indicating a curvature. Transforming the data by logs yielded a better model. Upon closer examination, he was able to see that the straight line didn't fit as well to the raw data as to the transformed data. Although past pattern is no indication of future behavior, he did use both models to project future values for an initial investment of the size Sally was considering. The linear trend line provided higher values. As he prepared for his meeting with Sally, he thought it would be a good idea to keep the linear trend analysis handy when discussing this investment option.

ETHICAL ISSUE *A linear trend should not be used to model a nonlinear trend. In this case, the linear trend produced forecasts that suited Kevin's purpose and he ignored the observed cyclical variation in the data. Related to Item B, ASA Ethical Guidelines.*

ETHICAL SOLUTION *Present all relevant models. Caution about forecasting future values based on past patterns.*

What Have We Learned?

Learning Objectives

- Be able to recognize when data are in a time series.
  - A *time series* consists of data recorded sequentially over time, usually at equally spaced intervals.
  - Time series analyses often attempt to provide forecasts of future values of the series.

- Recognize the four components of a time series.
  - The Trend component measures the overall tendency of the series to increase or decrease. It is ordinarily estimated as the slope of a regression against *Time*.
  - The Seasonal component measures regular, repeating fluctuations. Often these are due to seasons of the year, but the term applies to any such regular fluctuation. Seasonal components are often estimated by introducing indicator (dummy) variables in a regression model.
  - The Cyclic component accounts for such things as long-term business cycles.
  - The Irregular component is the random fluctuation around the time series model. It corresponds to the residuals from a time series model.

- Use smoothing methods to see past random fluctuations (noise) in a time series to detect an underlying smoother pattern (signal).
  - Simple Moving Average methods average a relatively small number of adjacent values to obtain a smooth value for each time period.
  - Weighted Moving Average methods introduce weights. Because the weights can determine the behavior of the smoother, these are a very general class of time series methods.
  - Exponential smoothing methods are weighted moving averages with weights that decline exponentially into the past.
  - Autoregressive models use regression methods, predicting the time series from versions of the same series offset, or *lagged*, in time. The result is a weighted moving average method in which the weights are estimated from the data.

- Estimate and report forecast error with statistics such as MSE, MAD, and MAPE.

- Use multiple regression methods to model a structured time series, using *Time*, indicator variables for the Seasonal component, and exogenous variables that might account for the Cyclic component.
  - Multiple regression models have the advantage that they can provide forecasts farther into the future.
  - *Additive Models* estimate the components using multiple regression methods to find a model in which the estimates are added to yield the predicted values.
  - Multiplicative models for a time series model the series as a product of its components.
  - Multiplicative models are ordinarily estimated by taking the logarithm of the time series values and then using multiple regression as for additive models.

Terms

Absolute percentage error (APE)

A measure of the error of a forecast:

$$APE = \frac{|y_t - \hat{y}_t|}{|y_t|}$$

Additive model	A model for a time series that models the times series with a sum of all or some of the following terms: a trend, a seasonal pattern, and a cyclical pattern.
Autocorrelation	The correlation between a data sequence, such as a time series, and that same sequence offset (or lagged) by one or more positions. Autocorrelation is one measure of association over time among the individual cases.
Autoregressive model	A pth-order autoregressive model has the form: $$\hat{y} = b_0 + b_1 y_{lag1} + \cdots + b_p y_{lag\,p}.$$
Cyclical component	The part of a model for a time series that describes regular repeating fluctuations with a period of 2 to 10 years.
Deseasonalized	A time series that has had a seasonal component estimated and subtracted.
Exogenous	Variables that are not part of a time series but nevertheless might be helpful in modeling it.
Exponential model	A model for a time series that grows exponentially. It is estimated by transforming the response variable by the logarithm: $$\widehat{log}(y) = b_0 + b_1 Time + other\ terms.$$
Exponential smoothing, Single exponential smoothing (SES model)	An exponential smoother has the form $$\tilde{y}_t = \alpha y_t + (1 - \alpha)\tilde{y}_{t-1}$$ or equivalently, $$\tilde{y}_t = \alpha y_t + \alpha(1 - \alpha)y_{t-1} + \alpha(1 - \alpha)^2 y_{t-2} + \alpha(1 - \alpha)^3 y_{t-3} \ldots$$ The parameter α determines how the smoother behaves. Larger values of α give more weight to recent values of the series. Smaller values give more weight to more distant values.
Forecast	Many analyses of time series attempt to forecast future values. We denote a forecast value $\hat{y}_t$.
Forecast error	The difference between the observed value and the forecasted value for a particular time in a time series: $$e_t = y_t - \hat{y}_t.$$
Irregular component	The part of a time series model that describes random, or unpredictable, behavior; the residuals.
Lagged series	A time series shifted in time by one or more time units.
Linear trend model	A time series model that assumes a constant rate of increase (or decrease) over time: $$\hat{y} = b_0 + b_1 t.$$
Mean absolute deviation (MAD)	A measure of forecast error of the form: $$MAD = \frac{1}{n}\Sigma |y_t - \hat{y}_t|.$$
Mean squared error (MSE)	A measure of forecast error of the form: $$MSE = \frac{1}{n}\Sigma (y_t - \hat{y}_t)^2.$$
Mean absolute percentage error (MAPE)	A measure of forecast error of the form: $$MAPE = 100 \times \frac{1}{n}\Sigma \frac{|y_t - \hat{y}_t|}{|y_t|}.$$
Moving average	An estimate that uses the arithmetic average of the prior L values in a time series to forecast the next value $$\tilde{y}_t = \frac{\displaystyle\sum_{i=t-L+1}^{t} y_i}{L}.$$

(continued)

Multiplicative model	A classical time series model consisting of four components, $$\hat{y} = T \times S \times C \times I,$$ where T is the trend component, S is the seasonal component, C is the cyclic component, and I is the irregular component.
Naïve forecast	Forecasting that the next value in a time series will be equal to the current one: $$\hat{y}_{t+1} = y_t.$$
Period	The time between peaks of a regular oscillation in a time series.
Random walk	A time series that exhibits random periods of upturns and downturns and is best modeled using a naive forecast.
Seasonal component	The part of a model for a time series that fits a regular pattern that has a period of less than or equal to 12 months.
Single exponential smoothing (SES) model	(See Exponential smoothing.)
Simple moving average forecast	A simple moving average averages L consecutive values of a time series, where L is the length of the moving average. When the final L values have been averaged, the resulting smooth value for time t can be used to forecast the series for time $t + 1$. (See Moving average.)
Stationary in the mean	A time series that has a relatively constant mean value over the time frame of the series is said to be stationary in the mean.
Stationary in the variance	A time series that has a relatively constant variance is said to be stationary in the variance. This is equivalent to homoskedasticity. (If a time series is simply said to be stationary, most often what is meant is that it has stationary variance.)
Time series	A time series is data recorded sequentially over time at roughly equally spaced intervals.
Trend component	The part of a model for a time series that fits long-term changes in the mean of the series.

Technology Help

EXCEL   XLSTAT

Excel offers some, but not all, the time series methods in this chapter.

Exponential Smoothing:

- Choose **Tools > Data Analysis > Exponential Smoothing**.
- Enter the input and output range and the smoothing coefficient (Damping factor in Excel).

Moving Averages:

- Choose **Tools > Data Analysis > Moving Average**.
- Enter the input and output range and the Length (Interval in Excel).

JMP

- From the **Analyze** menu choose **Modeling > Time Series.**
- Put the data series in **Y**. The data must be in time order and equally spaced.
- JMP will display a time series plot and report the Autocorrelation.
- Commands for analyzing the series are under the red triangle.

MINITAB

The time series commands are in the **Time Series** submenu of the **Stat** menu. They are generally self-explanatory. Most commands open a dialog in which you specify the series to analyze, specify parameters (e.g., for smoothers), and request predictions.

The commands related to time series are found in the **Time Series** submenu of the **Analyze** menu. Choose **Create Models** for a choice of methods.

To make a time series plot,

- Choose **Analyze > Time Series > Create Models**.
- Specify the variable you wish to plot.
- Click on the **Plots Tab**, and check **Series**.

Trend Analysis:

- Choose **Transform > Date/Time > Assign Periodicity > Days** to create the time variable.
- Then choose **Analyze > Regression > Linear**.
- Enter the variable you wish to analyze as the dependent variable and days as the independent variable.

Moving Averages:

- Choose **Analyze > Time Series > Create Models**.
- Enter the variable you wish to analyze.
- Then select **ARIMA > Criteria**.
- Enter the moving average period length in the criteria box.

Exponential Smoothing:

- Choose **Analyze > Time Series > Create Models**.
- Enter the variable you wish to analyze.
- Then select **Exponential Smoothing > Criteria**.
- Enter the weight (smoothing coefficient) to use.

Lagged Variables:

- Choose **Transform > Create Time Series**.
- Enter the variable for which you wish to create a lag function.
- Select **Lag** and enter the length of the new lagged variable.

Brief CASE

Intel Corporation

Intel Corporation, located in Santa Clara, California, was founded by three engineers in 1968 to develop technology for silicon-based electronic chips. Intel has a reputation as an innovator, and their net revenues more than tripled between 1993 and 2002. Although Intel remains the world's largest maker of chips, the semiconductor industry is impacted by volatility in sales of personal computers and other electronic and computing devices. As PC sales leveled off in the United States, modest semiconductor growth resulted from sales in video games, wireless and smart phones, networking equipment, and handheld devices. In 2002, worldwide semiconductor sales were over $150 billion.

The daily closing prices for Intel stock in the first four months of 2007 are in the file **Intel**. Examine the time series and identify appropriate times series models to forecast the daily stock price for Intel. Download the stock prices for Intel in the month of May 2007. Compare the forecasts for May 2007 obtained using your time series models with the actual closing price using the measures of forecast error discussed in this chapter. Which models are more appropriate and/or more accurate for forecasting Intel daily stock prices in 2007? Explain.

Tiffany & Co.

Tiffany was founded in 1837, when Charles Lewis Tiffany opened his first store in downtown Manhattan. Tiffany retails and distributes a selection of Tiffany & Co. brand jewelry at a range of prices. Today, more than 150 Tiffany & Co. stores sell to customers in U.S. and international markets. In addition to jewelry, it sells Tiffany & Co. brand merchandise in the following categories: timepieces and clocks; sterling silver merchandise; stainless steel flatware; crystal, glassware, china, and other tableware; custom engraved stationery; writing instruments; and fashion accessories. Fragrance

(continued)

products are sold under the trademarks Tiffany, Pure Tiffany, and Tiffany for Men. Tiffany also sells other brands of timepieces and tableware in its U.S. stores.[17]

Tiffany's quarterly sales from 1995 to 2002 are in the file **Tiffany**. Tiffany's quarterly sales (in $million) from 1995 to 2002 are shown here.

Build alternative time series models for Tiffany's sales and forecast future sales. In addition, develop causal, or multiple regression, models to forecast Tiffany's sales by using economic data, such as gross domestic product (GDP), consumer price index (CPI), disposable income, unemployment, and interest rates over this same time period. These economic data are also provided in the data file. Given that Tiffany is known for its high-end quality and price, it was of interest for the corporate executives to see how sensitive Tiffany sales were to economic indicators. Develop a hypothesis before developing your models. Compare the forecasts from your time series models and regression models to quarterly sales in 2003. (These are also included in the data file.) Use alternative measures of forecast error and recommend a model for the executives at Tiffany to use for forecasting their quarterly sales.

Exercises

SECTION 20.1

1. Are the following data time series? If not, explain why.

a) Quarterly earnings of Microsoft Corp.
b) Unemployment in August 2010 by Education level.
c) Time spent in training by workers in NewCo.
d) Numbers of e-mails sent by employees of SynCo each hour in a single day.

2. Are the following data time series? If not, explain why.

a) Reports from the Bureau of Labor Statistics on the number of U.S. adults who are employed full-time in each major sector of the economy.
b) The quarterly Gross Domestic Product (GDP) of France from 1980 to the present.
c) The dates on which a particular employee was absent from work due to illness over the past two years.
d) The number of cases of flu reported by the CDC each week during a flu season.

SECTION 20.2

Here is a table of values from the U.S. Bureau of Labor Statistics (disk file BLS_Output):

Year	Output/hr Labor	Output/unit Capital
1987	64.1	96.4
1988	65.4	99.7
1989	66.1	99.1
1990	67.5	96.1
1991	69.3	92.2
1992	71.9	93.0
1993	73.8	94.3
1994	76.4	97.2
1995	79.8	98.7
1996	82.7	98.0
1997	87.3	100.6
1998	92.0	100.7
1999	96.1	100.4
2000	100.0	100.0
2001	101.6	93.5
2002	108.6	92.3
2003	115.3	93.2
2004	117.9	95.4
2005	123.5	98.9
2006	125.0	100.2

[17]Source: www.shareholder.com/tiffany.

T 3. For the series of Output per hour of labor:

a) Make a time series plot.

b) Describe the Trend component. (Remember: Direction, Form, and Strength)

c) Is there evidence of a Seasonal component?

T 4. For the series of Output per unit of capital:

a) Make a time series plot.

b) Describe the Trend component.

c) Is there evidence of a Seasonal component?

d) Is there evidence of a cyclic component?

SECTION 20.3

Here are data on the monthly price of Delicious apples and gas, which are both components of the Consumer Price Index.[18] The timeplot shows the years 2006–2009 for apples; the data table shows just 2006, for both.

Month	Apples	Gas
Jan	0.963	2.359
Feb	0.977	2.354
Mar	0.935	2.444
Apr	0.958	2.801
May	1.021	2.993
Jun	1.053	2.963
Jul	1.146	3.046
Aug	1.235	3.033
Sep	1.256	2.637
Oct	1.138	2.319
Nov	1.089	2.287
Dec	1.027	2.380

[18]The Consumer Price Index (CPI) represents changes in prices of all goods and services purchased for consumption by urban households. User fees (such as water and sewer service) and sales and excise taxes paid by the consumer are also included. Income taxes and investment items (such as stocks, bonds, and life insurance) are not included. Most of the specific CPI indexes have a 1982–1984 reference base. That is, the average index level (representing the average price level) is set with the 36-month period covering the years 1982, 1983, and 1984 equal to 100 and then measures changes in relation to that figure. An index of 110, for example, means there has been a 10-percent increase in price since the reference period. See www.bls.gov/cpi for more information.

T 5. For the Apple prices:

a) Find a 2-point moving average of the first year (2006).

b) Use it to predict the value for January 2007.

T 6. For the Gas prices:

a) Find a 2-point moving average of the first year.

b) Use it to predict the value for January 2007.

SECTION 20.4

T 7. For the Apple prices smoothed in Exercise 5, the actual value for January 2007 was 1.034. Find the absolute percentage error of your forecast.

T 8. For the Gas prices of Exercise 6, the actual value for January 2007 was 2.321. Find the absolute percentage error of your forecast.

SECTION 20.5

T 9. For the Apple prices of Exercise 5, find the lag1 version of the prices.

T 10. For the Gas prices of Exercise 6, find the lag2 version of the prices.

T 11. A second-order autoregressive model for the apple prices (for all 4 years of data) is

```
Dependent variable is: Apples
R squared = 78.1%    R squared (adjusted) = 71.9%
s = 0.0574 with 10 − 3 = 7 degrees of freedom

Variable        Coefficient  SE(Coeff)  t-ratio  P-value
Intercept        0.327717    0.1881      1.74    0.1250
Lag1<Apples>     1.32814     0.3114      4.27    0.0037
Lag2<Apples>    −0.634127    0.2959     −2.14    0.0693
```

Using the values from the table, what is the predicted value for January 2007 (the value just past those given in the table)?

T 12. A second-order autoregressive model for the gas prices is:

```
Dependent variable is: Gas
R squared = 82.2%    R squared (adjusted) = 77.1%
s = 0.1498 with 10 − 3 = 7 degrees of freedom

Variable        Coefficient  SE(Coeff)  t-ratio  P-value
Intercept        1.28207     0.4644      2.76    0.0281
Lag1<Gas>        1.31432     0.2383      5.51    0.0009
Lag2<Gas>       −0.788250    0.2457     −3.21    0.0149
```

Using values from the table, what is the predicted value for January 2007 (the value just past those given in the table)?

SECTION 20.6

T 13. An Additive regression model for the Apple prices is:

```
Dependent variable is: Apples
R squared = 51.9%    R squared (adjusted) = 34.9%
s = 0.1108 with 47 - 13 = 34 degrees of freedom
```

Variable	Coefficient	SE(Coeff)	t-ratio	P-value
Intercept	-114.825	29.46	-3.90	0.0004
Time	0.057746	0.0147	3.94	0.0004
Jan	-0.001438	0.0783	-0.018	0.9855
Feb	-0.007062	0.0783	-0.090	0.9287
Mar	0.003626	0.0784	0.046	0.9634
Apr	0.015064	0.0784	0.192	0.8488
may	0.058751	0.0785	0.749	0.4592
Jun	0.115689	0.0786	1.47	0.1501
Jul	0.173627	0.0787	2.21	0.0342
Aug	0.169815	0.0788	2.16	0.0383
Sep	0.049253	0.0789	0.624	0.5368
Oct	0.006191	0.0791	0.078	0.9381
Nov	-0.014249	0.0847	-0.168	0.8675

a) What is the name for the kind of variable called *Jan* in this model?
b) Why is there no predictor variable for December?
c) Is there evidence of a cyclic component in these data?

T 14. An additive model for the Gas prices is:

```
Dependent variable is: Gas
R squared = 28.6%    R squared (adjusted) = 3.3%
s = 0.5524 with 47 - 13 = 34 degrees of freedom
```

Variable	Coefficient	SE(Coeff)	t-ratio	P-value
Intercept	66.3101	146.9	0.451	0.6546
Time	-0.031816	0.0732	-0.435	0.6664
Jan	-0.036401	0.3907	-0.093	0.9263
Feb	0.162901	0.3907	0.417	0.6793
Mar	0.395053	0.3908	1.01	0.3192
Apr	0.644704	0.3910	1.65	0.1084
may	0.788105	0.3914	2.01	0.0520
Jun	0.774757	0.3918	1.98	0.0562
Jul	0.673908	0.3923	1.72	0.0949
Aug	0.544059	0.3929	1.38	0.1752
Sep	0.333961	0.3937	0.848	0.4022
Oct	0.167112	0.3945	0.424	0.6745
Nov	-0.029645	0.4226	-0.070	0.9445

a) What is the value predicted by this model for January 2010 (Time = 2010)?
b) Is there evidence of a cyclic component for these data?

CHAPTER EXERCISES

15. Concepts.

a) Which will be smoother, a 50-day or a 200-day moving average?
b) Which will be smoother, a single exponential smoothing (SES) model using $\alpha = 0.10$ or a model using $\alpha = 0.80$?
c) What is the difference in how historical data are used when the smoothing coefficient in a single exponential smoothing (SES) model is raised from 0.10 to 0.80?

16. Concepts, again. We are trying to forecast monthly sales for a company that sells ski equipment and clothing. Assume that the company's sales peak each December and that the monthly sales have been growing at the rate of 1% each month. Answer the following questions.

a) Based on the description of these data, what time series components can you identify?
b) If you identified a seasonal component, what is the period?
c) If you use seasonal dummy variables, specify the dummy variables you would use.
d) After examining the residuals and using the information provided, you decide to transform the sales data. What transformation are you likely to suggest? Why?

17. More concepts. For each of the following time series, suggest an appropriate model:

a) Weekly stock prices that reveal erratic periods of up and down swings.
b) Annual sales that reveal a consistent percentage annual increase.
c) Quarterly sales for a bicycle shop that reveal a seasonal pattern where sales peak in Q2 of each year.

18. Final concepts. For each of the following time series, suggest an appropriate model:

a) Daily stock prices that reveal erratic periods of up and down swings.
b) Monthly sales that reveal a consistent percentage increase from month to month.
c) Quarterly sales for a woman's clothing company that reveal an annual peak each December.

19. Liquid assets. The Bank of New York Company was founded by Alexander Hamilton in 1784 and was a major commercial bank until its merger with the Mellon Financial Corporation in 2007. Their year-end financial reports for the final five years of independent operation give the following values for their liquid assets (Source: *The Financial Times*).

Year	Liquid Assets ($M)
2002	18,546
2003	22,364
2004	22,413
2005	19,881
2006	26,670

a) Use a 3-year moving average to predict what liquid assets would have been in 2007.

b) Predict the value for 2007 using a single exponential smooth with smoothing parameter $\alpha = 0.2$.

20. Baking profits. Sara Lee Corp., maker of food, beverage, and household products, is known especially for its baked products, marketed under its corporate name. For the five years ending July 1 of each year from 2002 to 2006, their bakery division reported the following profits.

Fiscal Year	Profits ($M)
2002	97
2003	98
2004	156
2005	−4
2006	−197

a) Use a 4-year moving average to predict profits for 2007.
b) Predict the profits for 2007 using a single exponential smooth with smoothing parameter $\alpha = 0.5$.
c) Think about the exponential smoother. If the parameter were 0.8 would you expect the prediction to be higher or lower? What if it were 0.2? Explain.

21. Banana prices. The price of bananas fluctuates on the world market. Here are the prices ($/tonne) for the years 2000–2004. (Source: *Holy See Country Review*, 2008).

2000	2001	2002	2003	2004
422.27	584.70	527.61	375.19	524.84

a) Find a 3-year moving average prediction for the price in 2005.
b) Find a prediction for 2005 with an exponential smoothing model with $\alpha = 0.4$.
c) The actual price of bananas in 2005 was 577 $/tonne (you can find current prices at www.imf.org/external/np/res/commod/table3.pdf.) Compute the absolute percentage error for each prediction.

22. Target earnings. Target Corp. operates "big box" stores that sell everyday essentials and fashionable differentiated merchandise. It also operates an online business at target.com. Target's reported gross earnings per share for the years 2003–2006 are given here.

2003	2004	2005	2006
$1.82	2.02	2.17	2.73

a) Find a prediction for 2007 based on a 3-year moving average and one for a 4-year moving average.

b) Find a prediction for 2007 based on an exponential smoothing model with $\alpha = 0.8$.
c) Earnings per share in 2007 were, in fact, $3.18. Compute the absolute percentage error for each prediction.

T 23. Toyota stock prices. The following time series graph shows daily closing stock prices (adjusted for splits and dividends) for Toyota Motor Manufacturing from April 1, 2008, through July 3, 2008 (Source: Yahoo! Finance).

a) Which time series components seem to be present?

The method of moving averages was applied to these data. Here are time series graphs showing moving average results using two different lengths.

b) In which application is a larger length used?

24. Exponential smoothing. The following time series graph shows daily closing stock prices (adjusted for splits and dividends) for Google, Inc. from February 1, 2008, through June 30, 2008 (Source: Yahoo! Finance).

a) Which time series components are evident?

Single exponential smoothing (SES) models were found for these data. Examine the time series graphs on the next page showing two different smoothing coefficients values ($\alpha = 0.2$ and $\alpha = 0.8$).

b) In which graph is a larger value of α used?

25. Men's weekly earnings. This graph shows the quarterly median weekly earnings from the first quarter of 2000 through the fourth quarter of 2007 for men, 25 years of age or older, in the United States (www.bls.gov).

Here are time series plots showing a 2-quarter moving average and an 8-quarter moving average.

Identify which plot is the 2-quarter moving average and which is the 8-quarter moving average. Explain why the better-fitting model fits better.

26. Women's weekly earnings. The following graph shows the quarterly median weekly earnings for U.S. women 25 years of age or older (www.bls.gov). Data are provided from the first quarter of 2000 through the fourth quarter of 2007.

Here is single exponential smoothing model to these data using $\alpha = 0.2$.

Variable	Smoothing Constant	Accuracy Measures	
Actual	Alpha 0.2	MAPE	3.212
Fits		MAD	18.972
Forecasts		MSD	405.401
95.0% PI			

Here is the exponential smoothing model using $\alpha = 0.8$:

Variable	Smoothing Constant	Accuracy Measures	
Actual	Alpha 0.8	MAPE	1.2297
Fits		MAD	7.1648
Forecasts		MSD	73.0814
95.0% PI			

Why does the exponential smoother with the higher coefficient fit the series better? What about this series is important for this result? What does this suggest about using exponential smoothers on such time series?

27. Autoregressive model. Suppose an autoregressive model is used for data in which quarterly sales in 2005 were: 1.9, 1.7, 2.2, and 2.3 ($Billion).

a) If a first-order autoregressive model is developed with estimated parameters of $b_0 = 0.100$ and $b_1 = 1.12$, compute the forecast for Q4 of 2006.
b) Compare this forecast to the actual value ($2.9B) by computing the absolute percentage error (APE). Did you over-forecast or under-forecast?
c) Assuming these quarterly sales have a seasonal component of length 4, use the following model to compute a forecast for Q4 of 2006: $y_t = 0.410 + 1.35\,y_{t-4}$. Compare the APE for this forecast to that in question a. Compare the appropriateness of the different models.

28. Another autoregressive model. Suppose an autoregressive model is used to model sales for a company that peaks twice per year (in June and December).

a) What lagged variables would you try in a regression to forecast sales? Explain.
b) How would you determine which of your lagged variables should remain in the model? Explain.

T 29. Coffee prices. Coffee is the world's second largest legal export commodity (after oil) and is the second largest source of foreign exchange for developing nations. The United States consumes about one-fifth of the world's coffee. The International Coffee Organization (ICO) computes a coffee price index using Colombian, Brazilian, and a mixture of other coffee data. Data are provided for the ICO price index (in $US) from January 2001 to December 2005.

Here's an autoregressive model for the composite prices.

```
Dependent variable is: Composite
R squared = 94.8%   R squared (adjusted) = 94.6%
s = 3.892 with 57 - 3 = 54 degrees of freedom
```

Variable	Coeff	SE(Coeff)	t-ratio	P-value
Intercept	1.59175	1.941	0.820	0.4157
lag1	1.22856	0.1327	9.26	≤0.0001
lag2	-0.247256	0.1350	-1.83	0.0725

a) Here are the last several values of the series: 88.48, 85.31, 78.79, 82.55, and 85.93. What price does this model predict for the next value in the series?
b) The next value in the series was, in fact, 86.85. Compute the APE.
c) Find a prediction based on a 2-point moving average. How does it compare with the AR model?

T 30. Gas prices. We have data on the weekly average retail price (cents per gallon) of regular gas nationwide from 2002 through the early part of 2007. Here's a time series plot.

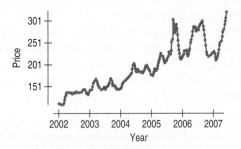

Here are a 3-point moving average and a 3-term autoregressive model fit to these data.

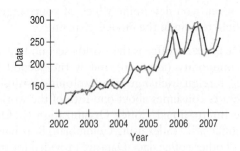

```
Dependent variable is: Price
R squared = 92.5%    R squared (adjusted) = 92.2%
s = 14.60 with 63 - 4 = 59 degrees of freedom

Variable   Coeff       SE(Coeff)  t-ratio  P-value
Intercept   7.93591    7.504       1.06    0.2946
Price1      1.47044    0.1250     11.8    ≤0.0001
Price2     -0.896360   0.2003     -4.47   ≤0.0001
Price3      0.400086   0.1270      3.15    0.0026
```

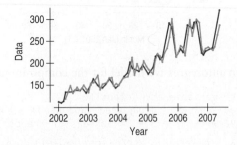

Which model is the better fit for the data?

31. Gallup poll. The Gallup organization periodically asks the following question:

If your party nominated a generally well-qualified person for president who happened to be a woman, would you vote for that person?

Here is a time series plot of the percentage answering "yes" versus the year of the (20th) century. The least squares trend line is given by: $\hat{y}_t = 5.58 + 0.999\,Year$, where *Year* = 37, 45, . . . 99 to represent the years during which the survey was given.

a) The R^2 for this trend line is 94%. A student decided to use this linear model to obtain a forecast for the percent who will respond "yes" in 2012. What value should the student use for *Year*?
b) Find the predicted value for the year 2012. Is it realistic?

T 32. Consumer price index. The most common use of the CPI is as an economic indicator to forecast inflation and evaluate the effectiveness of government policies. Following is the time series plot for the monthly CPI (not seasonally adjusted) from January 2001 to March 2007. The linear trend line is: CPI = $173 + 0.414\,t$, where $t = 0, 1, \ldots 74$ to represent the months in the series.

a) What does the intercept 173 represent in this trend line? What does the slope represent?
b) The R^2 for this trend line is 97%. Use this linear model to obtain a forecast for the CPI in April 2007.
c) The actual CPI in April 2007 was 206.686. Compute the absolute percent error for your forecast.

T 33. Retail gas, part 2. In Exercise 30 we looked at the weekly average retail price (cents per gallon) of regular gas nationwide from 2002 through the early part of 2007. Here's a time series plot:

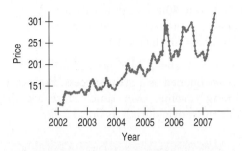

a) What components can you see in this plot?

Here's a linear trend model fit to these data.

```
Dependent variable is: Price
R squared = 82.3%    R squared (adjusted) = 82.2%
s = 21.88 with 282 - 2 = 280 degrees of freedom

Variable   Coeff      SE(Coeff)  t-ratio  P-value
Intercept  -60025.3   1668       -36.0    ≤0.0001
Year        30.0388   0.8322      36.1    ≤0.0001
```

b) Interpret the trend coefficient in this model.
c) How would you interpret the intercept coefficient?

Here's a time series plot of the residuals.

d) Comment on any patterns you see.

Ⓣ **34.** Banks annual interest rates. Average annual interest rates (banks prime lending) in the United States from 1980 through 2005 are shown in the following time series graph (unstats.un.org), created in **Excel**.

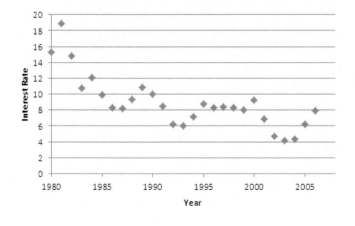

a) What components do you see in this series?

Here's a linear trend model fit to these data.

```
Dependent variable is: Interest Rate
R squared = 59.8%    R squared (adjusted) = 58.1%
s = 2.161 with 27 - 2 = 25 degrees of freedom

Variable   Coeff       SE(Coeff)  t-ratio  P-value
Intercept  657.353     106.4       6.18    ≤0.0001
Year        -0.325342  0.0534     -6.09    ≤0.0001
```

b) Interpret the trend coefficient mean in this model.
c) Predict the interest rate for 2005. Do you trust the prediction? Why or why not?

Here's a time series plot of the residuals from this model, created in **Excel**.

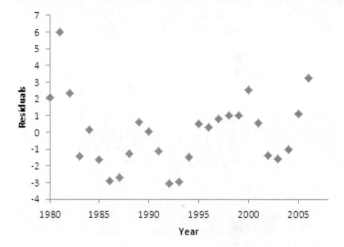

d) Discuss any patterns you see in this plot.
e) Would an exponential model be likely to do better for these data? Explain.

35. Seasonal model. Use the following model to forecast quarterly sales ($Million) for a company (where time is rescaled to begin at zero and Q_2, Q_3, and Q_4 are dummy variables for the indicated quarters), and answer the following questions.

$$\hat{y} = 1.1 + 0.2\,t - 0.1\,Q_2 - 0.5\,Q_3 + 0.5\,Q_4$$

a) For the first quarter of the time series, what are the sales?
b) What is the quarter that on average has the lowest level of sales over the time frame of the series?
c) What is the quarter that on average has the highest level of sales over the time frame of the series?
d) Interpret the coefficient of the dummy variable named Q_4.

36. Another seasonal model. Use the following model to forecast quarterly sales ($000) for a start-up (where time is rescaled to begin at zero and Q_2, Q_3, and Q_4 are dummy variables for the indicated quarters.), and answer the following questions.

$$\hat{y} = 15.1 + 10.5\,t - 5.0\,Q_2 - 7.2\,Q_3 + 7.5\,Q_4$$

a) For the first quarter of the time series, what are the sales?

b) What is the quarter that on average has the lowest level of sales over the time frame of the series?

c) What is the quarter that on average has the highest level of sales over the time frame of the series?

d) Interpret the coefficient of the dummy variable named Q_4.

T 37. Wal-Mart revenue. Wal-Mart has grown rapidly in recent years. Here is the monthly revenue ($Billion) for Wal-Mart from November 2003 to January 2007.

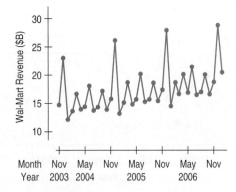

a) What components of a time series do you see in this timeplot?

Here's a regression model fit using dummy variables for months and a *Time* variable that counts from 1 for the first data value in the series.

```
Dependent variable is: WM Rev
R squared = 94.3%    R squared (adjusted) = 91.6%
s = 1.121 with 39 − 13 = 26 degrees of freedom

Variable   Coeff      SE(Coeff)  t-ratio  P-value
Intercept  12.0322    0.6562     18.3     ≤0.0001
Time       0.145241   0.0163     8.93     ≤0.0001
Feb        1.46096    0.8599     1.70     0.1013
Mar        2.84671    0.8585     3.32     0.0027
Apr        1.67981    0.8574     1.96     0.0609
May        0.870232   0.8567     1.02     0.3191
Jun        4.99999    0.8562     5.84     ≤0.0001
Jul        0.106417   0.8560     0.124    0.9020
Aug        0.434176   0.8562     0.507    0.6164
Sep        3.25327    0.8567     3.80     0.0008
Oct        −0.219640  0.8574     −0.256   0.7998
Nov        1.87023    0.7932     2.36     0.0262
Dec        11.5625    0.7927     14.6     ≤0.0001
```

b) Interpret the coefficient of *Time*.

c) Interpret the coefficient of *Dec*.

d) What revenue would you predict for Wal-Mart in February 2007 (the 40th month in this series)?

e) What does it mean that the coefficient for *Oct* is the only negative coefficient in the model?

T 38. Harry Potter revenue. The movie *Harry Potter and the Sorcerer's Stone* opened as a great success. But every movie sees declining revenue over time. Here are the daily revenues for the movie during its first 17 days.

Day	Date	Earnings ($M)
Friday	11/16/01	35
Saturday	11/17/01	30
Sunday	11/18/01	25
Monday	11/19/01	9
Tuesday	11/20/01	10
Wednesday	11/21/01	10
Thursday	11/22/01	10
Friday	11/23/01	22
Saturday	11/24/01	23
Sunday	11/25/01	13
Monday	11/26/01	3
Tuesday	11/27/01	3
Wednesday	11/28/01	3
Thursday	11/29/01	2
Friday	11/30/01	9
Saturday	12/1/01	9
Sunday	12/2/01	6

a) Without plotting the data, what components can you see in this series? Be specific.

For some series, a "seasonal" effect repeats weekly rather than annually. Here's a regression model fit to these data with dummy variables for days of the week. (*Day#* counts days starting at 1.)

```
Dependent variable is: Earnings
R squared = 96.9%    R squared (adjusted) = 94.6%
s = 2.365 with 17 − 8 = 9 degrees of freedom

Variable    Coeff      SE(Coeff)  t-ratio  P-value
Intercept   21.0000    2.090      10.0     ≤0.0001
Day#        −1.42857   0.1194     −12.0    ≤0.0001
Friday      12.4286    2.179      5.70     0.0003
Saturday    12.5238    2.166      5.78     0.0003
Sunday      7.95238    2.160      3.68     0.0051
Monday      −4.28571   2.392      −1.79    0.1068
Tuesday     −2.35714   2.377      −0.992   0.3473
Wednesday   −0.928571  2.368      −0.392   0.7041
```

b) Interpret the coefficient of *Day#*.
c) Interpret the coefficient of *Saturday* in this model.
d) Predict what earnings probably were for Monday 12/3/01. What does this say about the model?
e) What probably happened to earnings after the initial 17 days?

T 39. Hawaii tourism. Much of the public and private industry in Hawaii depends on tourism. The following time series plot shows the total number of visitors to Hawaii by air from the rest of the United States per month from January 2002 to December 2006.

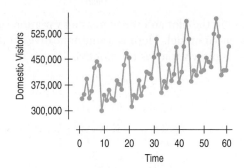

Here is a regression model fit to these data with dummy variables for months and a *Time* variable that starts at 0 and counts months:

```
Dependent variable is: Domestic Visitors
R squared = 95.5%    R squared (adjusted) = 94.4%
s = 14336 with 60 − 13 = 47 degrees of freedom

Variable    Coeff       SE(Coeff)   t-ratio   P-value
Intercept   312,312     6925        45.1      ≤0.0001
Time        2016.84     109.1       18.5      ≤0.0001
Feb         8924.36     9068        0.984     0.3301
Mar         58,693.7    9070        6.47      ≤0.0001
Apr         20,035.3    9073        2.21      0.0321
May         19,501.0    9078        2.15      0.0369
Jun         90,440.8    9084        9.96      ≤0.0001
Jul         132,893     9091        14.6      ≤0.0001
Aug         96,037.3    9099        10.6      ≤0.0001
Sep         −27,919.7   9109        −3.07     0.0036
Oct         1244.42     9120        0.136     0.8921
Nov         −12,181.4   9133        −1.33     0.1887
Dec         39,201.9    9146        4.29      ≤0.0001
```

a) Interpret the P-value for the trend coefficient.
b) You are planning to visit Hawaii and hope to avoid the crowds. When should you go? (Name the two best months.)
c) You are planning to advertise your product (a new sunscreen/massage oil combination) to tourists visiting Hawaii. What two months would be the best times to advertise?
d) How many tourists would you predict for Hawaii in April 2007 (month 63 of this series)?

T 40. Hawaii tourism, part 2. In Exercise 39 we examined domestic tourists who visit Hawaii. Now, let's consider international tourism. Here's a time series plot of international visitors.

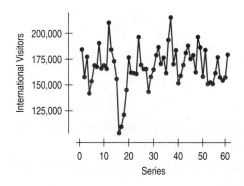

Here's the linear trend model with dummy variables for months.

```
Dependent variable is: International
R squared =62.9%    R squared (adjusted) = 53.4%
s = 13684 with 60 − 13 = 47 degrees of freedom

Variable    Coeff       SE(Coeff)   t-ratio   P-value
Intercept   185963      6610        28.1      ≤0.0001
Time        73.1937     104.1       0.703     0.4854
Feb         −23134.8    8655        −2.67     0.0103
Mar         −14961.8    8657        −1.73     0.0905
Apr         −50026.4    8660        −5.78     ≤0.0001
May         −41912.4    8664        −4.84     ≤0.0001
Jun         −33370.4    8670        −3.85     0.0004
Jul         −21521.2    8677        −2.48     0.0168
Aug         −5021.56    8685        −0.578    0.5659
Sep         −22294.3    8694        −2.56     0.0136
Oct         −20751.5    8705        −2.38     0.0212
Nov         −27338.9    8717        −3.14     0.0029
Dec         6075.27     8730        0.696     0.4899
```

a) Interpret the P-value for the *Time* coefficient.
b) The R^2 for this model is lower than for the model fit to domestic visitors in Exercise 39. Does that mean that an exponential trend model would do better?
c) International tourists often visit Hawaii in January. How can you tell that from this model?
d) Even though the R^2 for this model is lower than the corresponding R^2 for the model fit in Exercise 39 to domestic tourist visits, you might feel more comfortable predicting the number of international visitors for April 2007 with this model than you did predicting the number of domestic visitors with the previous model. Explain why.

T 41. Oakland passengers. The Port of Oakland airport reports the number of passengers passing through each month. At first glance, this is just simple growth, but by recognizing the series as a time series, we may learn more. The following plot was created in **Excel**.

The increasing spread of the data suggests that a multiplicative model may fit better.

Here's an exponential trend model with seasonal (monthly) dummy variables.

```
Dependent variable is: LogPassengers
R squared = 92.4%    R squared (adjusted) = 91.9%
s = 0.0358 with 194 − 13 = 181 degrees of freedom
```

Variable	Coeff	SE(Coeff)	t-ratio	P-value
Intercept	5.64949	0.0101	558	≤0.0001
Years since '90	0.024929	0.0006	45.2	≤0.0001
Feb	−0.013884	0.0127	−1.10	0.2747
Mar	0.048429	0.0127	3.82	0.0002
Apr	0.049028	0.0125	3.93	0.0001
May	0.066227	0.0125	5.30	≤0.0001
Jun	0.082862	0.0127	6.54	≤0.0001
Jul	0.095295	0.0127	7.52	≤0.0001
Aug	0.121779	0.0127	9.61	≤0.0001
Sep	0.029382	0.0127	2.32	0.0216
Oct	0.049596	0.0127	3.91	0.0001
Nov	0.039649	0.0127	3.13	0.0020
Dec	0.043545	0.0127	3.44	0.0007

a) Interpret the slope.
b) Interpret the intercept.
c) Which months have the lowest traffic at Oakland airport? (Hint: Consider all 12 months.)

Here's a plot of the residuals from the model fit to the Oakland airport passengers, created in **Excel**:

d) What components would you now say are in this series?

T 42. Monthly gas prices. In Exercise 30 we saw weekly data on the U.S. average regular retail gasoline price. Here are the data extracted for one week of each month from January 2002 to May 2007.

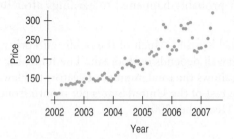

The bend in the plot and the increasing variation suggest a multiplicative model. Here is a multiplicative model with seasonal dummy variables for months.

```
Dependent variable is: Log price
R squared = 92.0%    R squared (adjusted) = 90.2%
s = 0.0377 with 66 − 13 = 53 degrees of freedom
```

Variable	Coeff	SE(Coeff)	t-ratio	P-value
Intercept	2.04566	0.0164	125	≤ 0.0001
Year since 2002	0.071754	0.0030	24.2	≤ 0.0001
Feb	0.002682	0.0210	0.128	0.8988
Mar	0.030088	0.0210	1.43	0.1574
Apr	0.053455	0.0210	2.55	0.0138
May	0.058316	0.0210	2.78	0.0075
Jun	0.044912	0.0221	2.03	0.0471
Jul	0.055008	0.0221	2.49	0.0159
Aug	0.061186	0.0221	2.77	0.0077
Sep	0.057329	0.0221	2.60	0.0122
Oct	0.039903	0.0221	1.81	0.0764
Nov	0.012609	0.0221	0.571	0.5704
Dec	−0.085243	0.0221	−0.386	0.7011

a) Interpret the slope.
b) Interpret the intercept.
c) In what month of the year are gas prices highest?

43. Oakland outlier. The plot of residuals in Exercise 41 shows an outlier that wasn't as evident in the data. The outlier is September 2001. Clearly, this wasn't a typical month for air travel. Here are three models fit to this series, a single exponential smooth, a 12-point moving average, and the fitted values from the seasonal regression model of Exercise 41. Discuss how each deals with the outlier.

The remaining exercises require the use of statistics software. Statistics packages vary in their capabilities and in the default decisions that some make. As a result, depending upon which package you choose, your answers may differ from those in the back of the book.

T 44. **E-commerce.** Quarterly e-commerce retail sales (in millions of dollars) in the United States are provided. (Source: www.census.gov.) Use this time series to answer the following questions.

a) Fit a linear trend model to this series but do not use the last two quarters (Q4 2007 and Q1 2008).
b) Fit an exponential trend model to this series but do not use the last two quarters.
c) Use both models to forecast the quarterly values for Q4 2007 and Q1 2008. Which model produces better forecasts?

d) What other time series components (besides trend) are likely present in this series?

T 45. **E-commerce, part 2.**

a) Fit a linear trend model with dummy variables for the seasonal effect to the e-commerce data in Exercise 44.
b) Fit an exponential trend (multiplicative) model with dummy variables to these data.
c) Which model fits better?

T 46. **J. Crew revenue.** J. Crew launched its first catalog in 1983, and the first J. Crew store opened in 1989 at the South Street Seaport in New York City. Since then, J. Crew has expanded to 198 retail and 65 outlet stores across the country. Quarterly *Revenues* ($Million) for J. Crew follow; use the data provided to answer the questions.

a) Identify the time series components you see in the graph. If you see seasonality, what is the period?
b) Use your answers to part a to develop an autoregressive model.
c) Examine the residuals over time to evaluate the Independence Assumption.

T 47. **Monthly gas prices, part 3.** Using the data from Exercise 42, develop and compare the following models.

a) Fit an appropriate autoregressive model by testing for the significance of each autoregressive term.
b) Obtain a forecast for the same week you did in Exercise 42 (May 28, 2007).
c) Compare your forecast to the actual value (by computing APE) and to the forecasts obtained in Exercise 42.
d) Recommend an appropriate model for forecasting this time series.

T 48. **Oil prices.** Recall the crude oil price data from Chapter 16. A time series plot of monthly crude oil price ($/barrel) from January 2001 to March 2007 is shown here.

Using these data,

a) Fit an appropriate autoregressive model by testing for the significance of each autoregressive term.

b) Obtain a forecast for March 2007.

49. Oil prices, again. Return to the oil price data of Exercise 48.

a) Find a linear model for this series.

b) Find an exponential (multiplicative) model for this series.

c) Use these methods to forecast the crude oil price for March 2007.

d) The actual price for March 2007 was $58.70. Compute measures of forecast error (e.g., MAD and MAPE) and compare the accuracy of the forecasts for the models of Exercises 48 and 49.

50. U.S. unemployment rate. Following is the time series plot for the monthly U.S. *Unemployment rate* (%) from January 1997 to March 2007. These data have been seasonally adjusted (meaning that the seasonal component has already been removed).

a) What time series components do you observe in this series?

b) Develop a 6-month and 12-month moving average model for this series.

c) Fit a single exponential smoothing model to this series.

d) Use these models to forecast the *Unemployment rate* for March 2007.

e) Compute measures of forecast error (e.g., MAD and MAPE) and compare the accuracy of the forecasts for these two models.

51. Mutual funds. The money that is invested in mutual funds is often referred to as "Fund Flows." The following time series plot shows the monthly *Fund Flows* ($M) from January 1990 to October 2002.

a) What time series components do you observe in this series?

b) Develop a 6-month and 12-month moving average model for this series.

c) Fit a single exponential smoothing model to the series.

d) Use these models to forecast the *Fund Flows* for October 2002.

e) Compute measures of forecast error (e.g., MAD and MAPE) and compare the accuracy of the forecasts for these two models.

52. U.S. unemployment rate, part 2. Using the data from Exercise 50, develop and compare the following models.

a) Fit an appropriate autoregressive model by testing for the significance of each autoregressive term (assume $\alpha = 0.05$).

b) Obtain a forecast for March 2007.

c) Compare your forecast for March to the actual value (by computing APE) and to the forecasts obtained in Exercise 50.

d) Recommend an appropriate model for forecasting this time series.

53. Hotel occupancy rates. The monthly occupancy rates for hotels in Honolulu, Hawaii, from January 2000 to December 2004 are available for the same time period as we examined in Exercises 39 and 40.

Use the data to answer the following questions.

a) Identify the time series components you see in the graph. If you see seasonality, what is the period? Use your answers to develop an autoregressive model.

b) Examine the residuals over time to evaluate the Independence Assumption.

c) Obtain a forecast for January 2005.

54. OPEC production. Recall the data for OPEC oil production from Chapter 16, Exercise 45. A time series plot of monthly OPEC production (thousand barrels/day) is shown here.

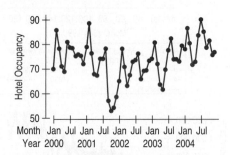

a) What time series components do you observe in this series?
b) Develop a 6-month and 12-month moving average model for this series.
c) Fit a single exponential smoothing model to these series.
d) Use these models to forecast the OPEC oil production for March 2007.
e) Compute measures of forecast error (e.g., MAD and MAPE) and compare the accuracy of the forecasts for these two models.

55. OPEC production, part 2. Using the data from Exercise 54, develop and compare the following models.

a) Fit an appropriate autoregressive model by testing for the significance of each autoregressive term.
b) Obtain a forecast for March 2007.
c) Compare your forecast for March to the actual value (by computing APE) and to the forecasts obtained in Exercise 54.
d) Recommend an appropriate model for forecasting this time series.

56. Interest rates, part 2. Consider the average annual interest rates we saw in Exercise 34. That exercise showed a linear model fit to these data.

a) Fit a linear trend model to this series but do not use the values for 2005 and 2006.
b) Fit an exponential trend model to this series but do not use the values for 2005 and 2006.
c) Use both models to forecast the average interest rate for

2005 and 2006. Which model produces better forecasts?
d) What other time series components (besides trend) are likely present in this series?

Just Checking Answers

1 Trend and seasonal.
2 The value is the mean of the final 4 values, about 333. It might be a bit low as a prediction.
3 $(319.3 + 399.9)/2 = 359.6$.
4 Four terms because there is a strong seasonal component with period 4.
5 10.2338. Revenue grew at about 10.2 million dollars per quarter.
6 Dummy variables require that we leave one out. Q4 coefficient will be positive.

CASE
Study

Health Care Costs

Heart disease is the leading cause of death for both men and women. According to the Centers for Disease Control and Prevention, about 935,000 heart attacks occur in the United States each year.[1] The cost of cardiovascular diseases in the United States, including health care expenditures and lost productivity from deaths and disability, was estimated to be more than $503 billion in 2010.

Many businesses have seen their cost for health insurance grow beyond bounds. Negotiations with insurance companies can depend, in part, on the demographics of a company's workforce. To understand the relationship between costs and other variables companies must understand how factors such as age and sex can help predict health costs. Because diseases and therapies vary greatly, it is best to build a separate model for each disease group. And, of course, heart disease is near the top of the list as an important condition to predict.

Models for health care costs can help in predicting and controlling costs. The data file **Heart_Attack_Costs** holds records for 12,844 patients admitted to hospitals in the Northeast United States with a diagnosis of acute myocardial infarction (heart attack; AMI). The data are available from Medicare records for public use. These data provide a good example of the kinds of models we can build to understand and predict health care expenses.

The variables available in this dataset are:

Charges: the total hospital charges in dollars.

Age: in years

Sex: coded M for males F for females

Diagnosis: the International Classification of Diseases, 10th revision, Clinical Modification code. These codes tell which part of the heart was affected.

DRG: the Diagnosis Related Group, which groups together patients with similar management. In this data set there are three different DRGs.

121: patients with AMIs and cardiovascular complications who did not die.

122: patients with AMIs without cardiovascular complications who did not die.

123: patients with AMIs who died.

LOS: hospital length of stay in days.

Died: 1 for patients who died in hospital and a 0 otherwise.

Build a model to predict *Charges*. Here are some things to consider in building the model:

1. Should *Charges* be re-expressed? Examine displays and find a suitable re-expression.
2. Should *LOS* be re-expressed? Are there unusual features of the distribution that may deserve special attention?
3. Some of these records may hold errors or otherwise be extraordinary. Make appropriate displays and set aside records that may not be reliable.
4. How do men and women differ? Consider, for example, the distribution of *Age* for male and female patients. Should you introduce *Sex* as a variable in your model? How will you do that?
5. Can you use *DRG* in your regression model as it is presented here, or should it be recoded in some way? *Diagnosis* raises similar questions. If our purpose is to predict costs based on demographic information, should these variables be predictors in our model?
6. Ultimately, how well can you predict *Charges*? Consider not just the R^2, but also the s_e. (Be sure to transform the s_e back to dollars if you are working with a transformed version of *Cost*.) Discuss how well your model predicts health care costs due to heart attack.

---

[1]http://www.cdc.gov/chronicdisease/resources/publications/AAG/dhdsp.htm

Design and Analysis of Experiments and Observational Studies

Capital One

Not everyone graduates first in their class at a prestigious business school. But even doing that won't guarantee that the first company you start will become a Fortune 500 company within a decade. Richard Fairbank managed to do both. When he graduated from Stanford Business School in 1981, he wanted to start his own company, but, as he said in an interview with the *Stanford Business Magazine*, he had no experience, no money, and no business ideas. So he went to work for a consulting firm. Wanting to be on his own, he left in 1987 and landed a contract to study the operations of a large credit card bank in New York. It was then that he realized that the secret lay in data. He and his partner, Nigel Morris, asked themselves, "Why not use the mountains of data that credit cards produce to design cards with prices and terms to satisfy different customers?" But they had a hard time selling this idea to the large credit card issuers. At the time all cards carried the same interest rate—19.8% with a $20 annual fee, and almost half of the population didn't qualify for a card. And credit issuers were naturally resistant to new ideas.

Finally, Fairbank and Morris signed on with Signet, a regional bank that hoped to expand its modest credit card operation.

Using demographic and financial data about Signet's customers, they designed and tested combinations of card features that allowed them to offer credit to customers who previously didn't qualify. Signet's credit card business grew and, by 1994, was spun off as Capital One with a market capitalization of $1.1B. By 2000, Capital One was the ninth largest issuer of credit cards with $29.5B in cardholder balances.

Fairbank also introduced "scientific testing." Capital One designs experiments to gather data about customers. For example, customers who hear about a better deal than the one their current card offers may phone, threatening to switch to another bank unless they get a better deal. To help identify which potential card-hoppers were serious, Fairbank designed an experiment. When a card-hopper called, the customer service agent's computer randomly ordered one of three actions: match the claimed offer, split the difference in rates or fees, or just say no. In that way the company could gather data on who switched, who stayed, and how they behaved. Now when a potential card-hopper phones, the computer can give the operator a script specifying the terms to offer—or instruct the operator to bid the card-hopper a pleasant good-bye.

Fairbank attributes the phenomenal success of Capital One to their use of such experiments. According to Fairbank, "Anyone in the company can propose a test and, if the results are promising, Capital One will rush the new product or approach into use immediately." Why does this work for Capital One? Because, as Fairbank says, "We don't hesitate because our testing has already told us what will work."

In 2002, Capital One won the Wharton Infosys Business Transformation Award, which recognizes enterprises that have transformed their businesses by leveraging information technology.

21.1   Observational Studies

Fairbank started by analyzing the data that had already been collected by the credit card company. These data weren't from designed studies of customers. He simply *observed* the behavior of customers from the data that were already there. Such studies are called **observational studies.** Many companies collect data from customers with "frequent shopper" cards, which allow the companies to record each purchase. A company might study that data to identify associations between customer behavior and demographic information. For example, customers with pets might tend to spend more. The company can't conclude that owning a pet *causes* these customers to spend. People who have pets may also have higher incomes on average or be more likely to own their own homes. Nevertheless, the company may decide to make special offers targeted at pet owners.

Observational studies are used widely in public health and marketing because they can reveal trends and relationships. Observational studies that study an outcome in the present by delving into historical records are called **retrospective studies.** When Fairbank looked at the accumulated experiences of Signet bank's credit card customers, he started with information about which customers earned the bank the most money and sought facts about these customers that could identify others like them, so he was performing a retrospective study. Retrospective studies can often generate testable hypotheses because they identify interesting relationships although they can't demonstrate a causal link.

When it is practical, a somewhat better approach is to observe individuals over time, recording the variables of interest and seeing how things turn out. For example, if we thought pet ownership might be a way to identify profitable customers, we might start by selecting a random sample of new customers and ask whether they have a pet. We could then track their performance and compare those who own pets to those who don't. Identifying subjects in advance and collecting data as events unfold would make this a **prospective study.** Prospective studies are often used in public health, where by following smokers or runners over a period of time we may find that one group or the other develops emphysema or arthritic knees (as you might expect), or dental cavities (which you might not anticipate).

Although an observational study may identify important variables related to the outcome we are interested in, there is no guarantee that it will find the right or the most important related variables. People who own pets may differ from the other customers in ways that we failed to observe. It may be this difference— whether we know what it is or not—rather than owning a pet in itself that leads pet owners to be more profitable customers. It's just not possible for observational studies, whether prospective or retrospective, to demonstrate a causal relationship. That's why we need experiments.

For Example   Observational studies

Amtrak launched its high-speed train, the Acela, in December 2000. Not only is it the only high-speed line in the United States, but it currently is the only Amtrak line to operate at a profit. The Acela line generates about one quarter of Amtrak's entire revenue.[1] The Acela is typically used by business professionals because of its fast travel times, high fares, business class seats, and free Wi-Fi. As a new member of the marketing department for the Acela, you want to boost young ridership of the Acela. You examine a sample of last year's customers for whom you have demographic information and find that only 5% of last year's riders were 21 years old or younger, but of those, 90% used the Internet while on board as opposed to 37% of riders older than 21 years.

Question: What kind of study is this? Can you conclude that Internet use is a factor in deciding to take the Acela?

Answer: This is a retrospective observational study. Although we can compare rates of Internet use between those older and younger than 21 years, we cannot come to any conclusions about why they chose to ride the Acela.

Just Checking

In early 2007, a larger-than-usual number of cats and dogs developed kidney failure; many died. Initially, researchers didn't know why, so they used an observational study to investigate.

1 Suppose that, as a researcher for a pet food manufacturer, you are called on to plan a study seeking the cause of this problem. Specify how you might proceed. Would your study be prospective or retrospective?

---

[1]Amtrak News Release ATK–09–074, October 2009.

21.2   Randomized, Comparative Experiments

Experiments are the only way to show cause-and-effect relationships convincingly, so they are a critical tool for understanding what products and ideas will work in the marketplace. An **experiment** is a study in which the experimenter *manipulates* attributes of what is being studied and observes the consequences. Usually, the attributes, called **factors**, are manipulated by being set to particular **levels** and then allocated or assigned to individuals. An experimenter identifies at least one factor to manipulate and at least one response variable to measure. Often the observed **response** is a quantitative measurement such as the amount of a product sold. However, responses can be categorical ("customer purchased"/ "customer didn't purchase"). The combination of factor levels assigned to a subject is called that subject's **treatment.**

The individuals on whom or which we experiment are known by a variety of terms. Humans who are experimented on are commonly called **subjects** or **participants**. Other individuals (rats, products, fiscal quarters, company divisions) are commonly referred to by the more generic term **experimental units.**

You've been the subject of marketing experiments. Every credit card offer you receive is actually a combination of various factors that specify your "treatment," the specific offer you get. For example, the factors might be *Annual Fee, Interest Rate,* and *Communication Channel* (e-mail, direct mail, phone, etc.). The particular treatment you receive might be a combination of *no Annual Fee* and a *moderate Interest Rate* with the offer being sent by *e-mail*. Other customers receive different treatments. The response might be categorical (do you accept the offer of that card?) or quantitative (how much do you spend with that card during the first three months you have it?).

Two key features distinguish an experiment from other types of investigations. First, the experimenter actively and deliberately manipulates the factors to specify the treatment. Second, the experiment assigns the subjects to those treatments at *random.* The importance of **random assignment** may not be immediately obvious. Experts, such as business executives and physicians, may think that they know how different subjects will respond to various treatments. In particular, marketing executives may want to send what they consider the best offer to the their best customers, but this makes fair comparisons of treatments impossible and invalidates the inference from the test. Without random assignment, we can't perform the hypothesis tests that allow us to conclude that differences among the treatments were responsible for any differences we observed in the responses. By using random assignment to ensure that the groups receiving different treatments are comparable, the experimenter can be sure that these differences are *due* to the differences in treatments. There are many stories of experts who were certain they knew the effect of a treatment and were proven wrong by a properly designed study. In business, it is important to get the facts rather than to just rely on what you may think you know from experience.

For Example   A marketing experiment

After finding out that most young riders of the Acela use the Internet while on board (see page 719), you decide to perform an experiment to see how to encourage more young people to take the Acela. After purchasing a mailing list of 16,000 college students, you decide to randomly send 1/4 a coupon worth 10% off their next Acela ride (*Coupon*), 1/4 a 5000 mile Amtrak mile bonus card (*Card*), and 1/4 a free Netflix download during their next Acela trip (*Movie*). The remaining 4000 students will receive no offer (*No Offer*). You plan to monitor the four groups to see which group travels most during the 12 months after sending the offer.

Question: What kind of study is this? What are the factors and levels? What are the subjects? What is the response variable?

Answer: This is an experiment because the factor (type of offer) has been manipulated. The four levels are *Coupon, Card, Movie,* and *No Offer.* The subjects are 16,000 college students. Each of four different offers will be distributed randomly to 1/4 of the college students. The response variable is *Miles Traveled* during the next 12 months on the Acela.

21.3   The Four Principles of Experimental Design

There are four **principles of experimental design.**

1. **Control.** We control sources of variation other than the factors we are testing by making conditions as similar as possible for all treatment groups. In a test of a new credit card, all alternative offers are sent to customers at the same time and in the same manner. Otherwise, if gas prices soar, the stock market drops, or interest rates spike dramatically during the study, those events could influence customers' responses, making it difficult to assess the effects of the treatments. So an experimenter tries to make any other variables that are not manipulated as alike as possible. Controlling extraneous sources of variation reduces the variability of the responses, making it easier to discern differences among the treatment groups.

   There is a second meaning of control in experiments. A bank testing the new creative idea of offering a card with special discounts on chocolate to attract more customers will want to compare its performance against one of their standard cards. Such a baseline measurement is called a control treatment, and the group that receives it is called the **control group.**

2. **Randomize.** In any true experiment, subjects are assigned treatments at random. Randomization allows us to equalize the effects of unknown or uncontrollable sources of variation. Although randomization can't eliminate the effects of these sources, it spreads them out across the treatment levels so that we can see past them. Randomization also makes it possible to use the powerful methods of inference to draw conclusions from your study. Randomization protects us even from effects we didn't know about. Perhaps women are more likely to respond to the chocolate benefit card. We don't need to test equal numbers of men and women—our mailing list may not have that information. But if we randomize, that tendency won't contaminate our results. There's an adage that says "Control what you can, and randomize the rest."

3. **Replicate.** Replication shows up in different ways in experiments. Because we need to estimate the variability of our measurements, we must make more than one observation at each level of each factor. Sometimes that just means making repeated observations. But, as we'll see later, some experiments combine two or more factors in ways that may permit a single observation for each *treatment*—that is, each combination of factor levels. When such an experiment is repeated in its entirety, it is said to be *replicated*. Repeated observations at each treatment are called **replicates.** If the number of replicates is the same for each treatment combination, we say that the experiment is **balanced.**

   A second kind of replication is to repeat the entire experiment for a different group of subjects, under different circumstances, or at a different time. Experiments do not require, and often can't obtain, representative random samples from an identified population. Experiments study the consequences of different levels of their factors. They rely on the random assignment of treatments to the subjects to generate the sampling distributions and to control for other possibly contaminating variables. When we detect a significant difference in response among treatment groups, we can conclude that it is due to the difference in treatments. However, we should take care in generalizing that result too broadly if we've only studied a specialized population. A special offer of accelerated checkout lanes for regular customers may attract more business in December, but it may not be effective in July. Replication in a variety of circumstances can increase our confidence that our results apply to other situations and populations.

4. **Blocking.** Sometimes we can identify a factor not under our control whose effect we don't care about, but which we suspect might have an effect either on

our response variable or on the ways in which the factors we are studying affect that response. Perhaps men and women will respond differently to our chocolate offer. Or maybe customers with young children at home behave differently than those without. Platinum card members may be tempted by a premium offer much more than standard card members. Factors like these can account for some of the variation in our observed responses because subjects at different levels respond differently. But we can't *assign* them at random to subjects. So we deal with them by grouping, or **blocking**, our subjects together and, in effect, analyzing the experiment separately for each block. Such factors are called **blocking factors**, and their levels are called **blocks.** Blocking in an experiment is like stratifying in a survey design. Blocking reduces variation by comparing subjects within these more homogenous groups. That makes it easier to discern any differences in response due to the factors of interest. In addition, we may want to study the effect of the blocking factor itself. Blocking is an important compromise between randomization and control. However, unlike the first three principles, blocking is not required in all experiments.

Just Checking

Following concerns over the contamination of its pet foods by melamine, which had been linked to kidney failure, a manufacturer now claims its products are safe. You are called on to design the study to demonstrate the safety of the new formulation.

2 Identify the treatment and response.

3 How would you implement control, randomization, and replication?

For Example   Experimental design principles

Question: Explain how the four principles of experimental design are used in the Acela experiment described in the previous section (see page 720).

Answer:
Control: It is impossible to control other factors that may influence a person's decision to use the Acela. However, a control group—one that receives no offer—will be used to compare to the other three treatment levels.

Randomization: Although we can't control the other factors (besides *Offer*) that may influence a person's decision to use the Acela, by randomizing which students receive which offer, we hope that the influences of all those other factors will average out, enabling us to see the effect of the four treatments.

Replication: We will send each type of offer to 4000 students. We hope that the response is high enough that we will be able to see differences in *Miles Traveled* among the groups. This experiment is balanced since the number of subjects is the same for all four treatments.

Blocking: We have not blocked the experiment. Possible blocking factors might include demographic variables such as the region of the student's home or college, their sex, or their parent's income.

21.4   Experimental Designs
Completely Randomized Designs

When each of the possible treatments is assigned to at least one subject at random, the design is called a **completely randomized design**. This design is the simplest and easiest to analyze of all experimental designs. A diagram of the procedure can help in thinking about experiments. In this experiment, the subjects are assigned at random to the two treatments.

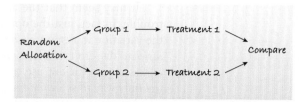

Figure 21.1 The simplest randomized design has two groups randomly assigned two different treatments.

Randomized Block Designs

When one of the factors is a blocking factor, complete randomization isn't possible. We can't randomly assign factors based on people's behavior, age, sex, and other attributes. But we may want to block by these factors in order to reduce variability and to understand their effect on the response. When we have a blocking factor, we randomize the subject to the treatments *within each block*. This is called a **randomized block design.** In the following experiment, a marketer wanted to know the effect of two types of offers in each of two segments: a high spending group and a low spending group. The marketer selected 12,000 customers *from each group* at random and then randomly assigned the three treatments to the 12,000 customers *in each group* so that 4000 customers in each segment received each of the three treatments. A display makes the process clearer.

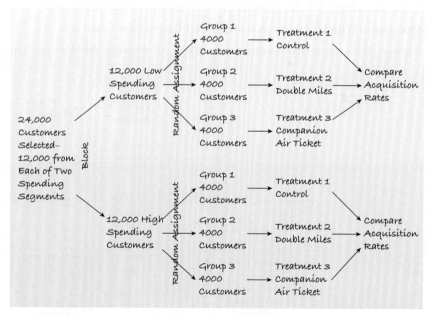

Figure 21.2 This example of a randomized block design shows that customers are randomized to treatments within each segment, or block.

Factorial Designs

An experiment with more than one manipulated factor is called a **factorial design.** A full factorial design contains treatments that represent all possible combinations of all levels of all factors. That can be a lot of treatments. With only three factors, one at 3 levels, one at 4, and one at 5, there would be $3 \times 4 \times 5 = 60$ different treatment combinations. So researchers typically limit the number of levels to just a few.

It may seem that the added complexity of multiple factors is not worth the trouble. In fact, just the opposite is true. First, if each factor accounts for some of the variation in responses, having the important factors in the experiment makes it *easier* to discern the effects of each. Testing multiple factors in a single experiment makes more efficient use of the available subjects. And testing factors together is the only way to see what happens at *combinations* of the levels.

An experiment to test the effectiveness of offering a $50 coupon for free gas may find that the coupon increases customer spending by 1%. Another experiment finds that lowering the interest rate increases spending by 2%. But unless some customers were offered *both* the $50 free gas coupon *and* the lower interest rate, the analyst can't learn whether offering both together would lead to still greater spending or less.

When the combination of two factors has a different effect than you would expect by adding the effects of the two factors together, that phenomenon is called an **interaction.** If the experiment does not contain both factors, it is impossible to see interactions. That can be a major omission because such effects can have the most important and surprising consequences in business.

For Example   Designing an experiment

Continuing the example from page 722, you are considering splitting up the students into two groups before mailing the offers: those who live or go to school in the Northeast corridor, where the Acela operates, and those who don't. Using home and school zip codes, you split the original 12,000 students into those groups and find that they split 4000 in the Northeast corridor and 8000 outside. You plan to randomize the treatments within those two groups and you'll monitor them to see if this factor, *NE corridor*, affects their *Miles Traveled* as well as the type of offer they receive.

Questions: What kind of design would this be? Diagram the experiment.

Answer: It is a randomized block experiment with *NE corridor* as the blocking factor.

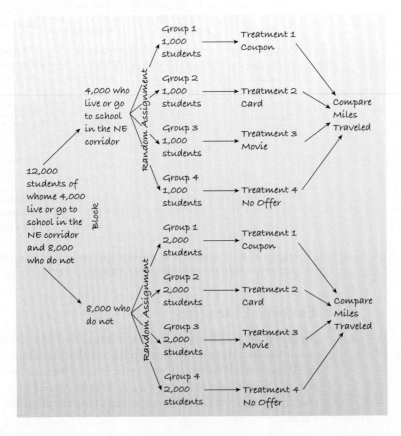

Guided Example   Designing a Direct Mail Experiment

At a major credit card bank, management has been pleased with the success of a recent campaign to cross-sell Silver card customers with the new SkyWest Gold card. But you, as a marketing analyst, think that the revenue of the card can be increased by adding three months of double miles on SkyWest to the offer, and you think the additional gain in charges will offset the cost of the double miles. You want to design a marketing experiment to find out what the difference will be in revenue if you offer the double miles. You've also been thinking about offering a new version of the miles called "use anywhere miles," which can be transferred to other airlines, so you want to test that version as well.

You also know that customers receive so many offers that they tend to disregard most of their direct mail. So, you'd like to see what happens if you send the offer in a shiny gold envelope with the SkyWest logo prominently displayed on the front. How can we design an experiment to see whether either of these factors has an effect on charges?

PLAN		
	State the problem.	We want to study two factors to see their effect on the revenue generated for a new credit card offer.
	Response Specify the response variable.	Revenue is a percentage of the amount charged to the card by the cardholder. To measure the success, we will use the monthly charges of customers who receive the various offers. We will use the three months after the offer is sent out as the collection period and use the total amount charged per customer during this period as the response.
	Factors Identify the factors you plan to test. **Levels** Specify the levels of the factors you will use.	We will offer customers three levels of the factor **miles** for the SkyWest Gold card: regular (no additional) miles, double miles, and double "use anywhere miles." Customers will receive the offer in the standard envelope or the new SkyWest logo envelope (factor **envelope**).
	Experimental Design Observe the principles of design: **Control** any sources of variability you know of and can control. **Randomly** assign experimental units to treatments to equalize the effects of unknown or uncontrollable sources of variation. **Replicate** results by placing more than one customer (usually many) in each treatment group.	We will send out all the offers to customers at the same time (in mid September) and evaluate the response as total charges in the period October through December. A total of 30,000 current Silver card customers will be randomly selected from our customer records to receive one of the six offers. ✓ Regular miles with standard envelope ✓ Double miles with standard envelope ✓ Double "anywhere miles" with standard envelope ✓ Regular miles with Logo envelope ✓ Double miles with Logo envelope ✓ Double "anywhere miles" with Logo envelope

(continued)

Make a Picture A diagram of your design can help you think about it.

Specify any other details about the experiment. You must give enough details so that another experimenter could exactly replicate your experiment.

It's generally better to include details that might seem irrelevant because they may turn out to make a difference.

Specify how to measure the response.

On January 15, we will examine the total card charges for each customer for the period October 1 through December 31.

DO

Once you collect the data, you'll need to display them (if appropriate) and compare the results for the treatment groups. (Methods of analysis for factorial designs will be covered later in the chapter.)

REPORT

To answer the initial question, we ask whether the differences we observe in the means (or proportions) of the groups are meaningful.

Because this is a randomized experiment, we can attribute significant differences to the treatments. To do this properly, we'll need methods from the analysis of factorial designs covered later in the chapter.

MEMO

Re: Test Mailing for Creative Offer and Envelope

The mailing for testing the Double Miles and Logo envelope ideas went out on September 17. On January 15, once we have total charges for everyone in the treatment groups, I would like to call the team back together to analyze the results to see:

✓ Whether offering Double Miles is worth the cost of the miles

✓ Whether the "use anywhere miles" are worth the cost

✓ Whether the Logo envelope increased spending enough to justify the added expense

21.5   Issues in Experimental Design
Blinding and Placebos

Humans are notoriously susceptible to errors in judgment—all of us. When we know what treatment is assigned, it's difficult not to let that knowledge influence our response or our assessment of the response, even when we try to be careful.

Suppose you were trying to sell your new brand of cola to be stocked in a school's vending machines. You might hope to convince the committee designated to make the choice that students prefer your less expensive cola, or at least that they can't taste the difference. You could set up an experiment to see which of the three competing brands students prefer (or whether they can tell the difference at all). But people have brand loyalties. If they know which brand they are tasting, it might influence their rating. To avoid this bias, it would be better to disguise the brands as much as possible. This strategy is called **blinding** the participants to the treatment. Even professional taste testers in food industry experiments are blinded to the treatment to reduce any prior feelings that might influence their judgment.

But it isn't just the subjects who should be blind. Experimenters themselves often subconsciously behave in ways that favor what they believe. It wouldn't be appropriate for you to run the study yourself if you have an interest in the outcome. People are so good at picking up subtle cues about treatments that the best (in fact, the only) defense against such biases in experiments on human subjects is to keep anyone who could affect the outcome or the measurement of the response from knowing which subjects have been assigned to which treatments. So, not only should your cola-tasting subjects be blinded, but also you, as the experimenter, shouldn't know which drink is which—at least until you're ready to analyze the results.

There are two main classes of individuals who can affect the outcome of the experiment:

- Those who could influence the results (the subjects, treatment administrators, or technicians)
- Those who evaluate the results (judges, experimenters, etc.)

When all the individuals in either one of these classes are blinded, an experiment is said to be **single-blind**. When everyone in both classes is blinded, we call the experiment **double-blind**. Double-blinding is the gold standard for any experiment involving both human subjects and human judgment about the response.

Often simply applying *any* treatment can induce an improvement. Every parent knows the medicinal value of a kiss to make a toddler's scrape or bump stop hurting. Some of the improvement seen with a treatment—even an effective treatment—can be due simply to the act of treating. To separate these two effects, we can sometimes use a control treatment that mimics the treatment itself. A "fake" treatment that looks just like the treatments being tested is called a **placebo**. Placebos are the best way to blind subjects so they don't know whether they have received the treatment or not. One common version of a placebo in drug testing is a "sugar pill." Especially when psychological attitude can affect the results, control group subjects treated with a placebo may show an improvement.

The fact is that subjects treated with a placebo sometimes improve. It's not unusual for 20% or more of subjects given a placebo to report reduction in pain, improved movement, or greater alertness or even to demonstrate improved health or performance. This **placebo effect** highlights both the importance of effective blinding and the importance of comparing treatments with a control. Placebo controls are so effective that you should use them as an essential tool for blinding whenever possible.

Just Checking

The pet food manufacturer we've been following hires you to perform the experiment to test whether their new formulation is safe and nutritious for cats and dogs.

4 How would you establish a control group?

5 Would you use blinding? How? (Can or should you use double-blinding?)

6 Both cats and dogs are to be tested. Should you block? Explain.

The best experiments are usually:

- Randomized
- Double-blind
- Comparative
- Placebo-controlled

Confounding and Lurking Variables

A credit card bank wanted to test the sensitivity of the market to two factors: the annual fee charged for a card and the annual percentage rate charged. The bank selected 100,000 people at random from a mailing list and sent out 50,000 offers with a low rate and no fee and 50,000 offers with a higher rate and a $50 annual fee. They discovered that people preferred the low-rate, no-fee card. No surprise. In fact, customers signed up for that card at over twice the rate as the other offer. But the question the bank really wanted to answer was: "How much of the change was due to the rate, and how much was due to the fee?" Unfortunately, there's simply no way to separate out the two effects with that experimental design.

If the bank had followed a factorial design in the two factors and sent out all four possible different treatments—low rate with no fee; low rate with $50 fee; high rate with no fee, and high rate with $50 fee—each to 25,000 people, it could have learned about both factors and could have also learned about the interaction between rate and fee. But we can't tease apart these two effects because the people who were offered the low rate were also offered the no-fee card. When the levels of one factor are associated with the levels of another factor, we say that the two factors are **confounded**.

Confounding can also arise in well-designed experiments. If some other variable not under the experimenter's control but associated with a factor has an effect on the response variable, it can be difficult to know which variable is really responsible for the effect. A shock to the economic or political situation that occurs during a marketing experiment can overwhelm the effects of the factors being tested. Randomization will usually take care of confounding by distributing uncontrolled factors over the treatments at random. But be sure to watch out for potential confounding effects even in a well-designed experiment.

Confounding may remind you of the problem of lurking variables that we discussed in Chapter 6. Confounding variables and lurking variables are alike in that they interfere with our ability to interpret our analyses simply. Each can mislead us, but they are not the same. A lurking variable is associated with two variables in such a way that it creates an apparent, possibly causal relationship between them. By contrast, confounding arises when a variable associated with a factor has an effect on the response variable, making it impossible to separate the effect of the factor from the effect of the confounder. Both confounding and lurking variables are outside influences that make it harder to understand the relationship we are modeling.

21.6   Analyzing a Design in One Factor— The One-Way Analysis of Variance

The most common experimental design used in business is the single factor experiment with two levels. Often these are known as champion/challenger designs because typically they're used to test a new idea (the challenger) against the current version (the champion). In this case, the customers offered the champion are the control group, and the customers offered the challenger (a special deal, a new offer, a new service, etc.) are the test group. As long as the customers are randomly assigned to the two groups, we already know how to analyze data from experiments like these. When the response is quantitative, we can test whether the means are equal with a two-sample *t*-test, and if the response is 0-1 (yes/no), we would test whether the two proportions are equal using a two proportion *z*-test.

But those methods can compare only two groups. What happens when we introduce a third level into our single factor experiment? Suppose an associate in a percussion music supply company, *Tom's Tom-Toms*, wants to test ways to increase the amount purchased from the catalog the company sends out every three months. He decides on three treatments: a coupon for free drum sticks with any purchase, a free practice pad, and a $50 discount on any purchase. The response will be the dollar amount of sales per customer. He decides to keep some customers as a control group by sending them the catalog without any special offer. The experiment is a single factor design with four levels: no coupon, coupon for free drum sticks, coupon for the practice pad, and $50 coupon. He assigns the same number of customers to each treatment randomly.

Now the hypothesis to test isn't quite the same as when we tested the difference in means between two independent groups. To test whether all *k* means are equal, the hypothesis becomes:

$$H_0: \mu_1 = \mu_2 = \ldots = \mu_k$$
$$H_A: \textit{at least one mean is different}$$

The test statistic compares the variance of the means to what we'd expect that variance to be based on the variance of the individual responses. Figure 21.3 illustrates the concept. The differences among the means are the same for the two sets of boxplots, but it's easier to see that they are different when the underlying variability is smaller.

Figure 21.3   The means of the four groups in the left display are the same as the means of the four groups in the right display, but the differences are much easier to see in the display on the right because the variation within each group is less.

Why is it easier to see that the means[2] of the groups in the display on the right are different and much harder to see it in the one on the left? It is easier because we naturally compare the differences *between* the group means to the variation *within* each group. In the picture on the right, there is much less variation within each group so the differences among the group means are evident.

This is exactly what the test statistic does. It's the ratio of the variation among the group means to the variation within the groups. When the numerator is large enough, we can be confident that the differences among the group means are greater than we'd expect by chance, and reject the null hypothesis that they are equal. The test statistic is called the *F*-statistic in honor of Sir Ronald Fisher, who derived the sampling distribution for this statistic. The *F*-statistic showed up in multiple regression (Chapter 18) to test the null hypothesis that all slopes were zero. Here, it tests the null hypothesis that the means of all the groups are equal.

The ***F*-statistic** compares two quantities that measure variation, called *mean squares*. The numerator measures the variation *between* the groups (treatments) and is called the **Mean Square due to Treatments (MST)**. The denominator measures the variation *within* the groups, and is called the **Mean Square due to Error (MSE)**. The *F*-statistic is their ratio:

$$F_{k-1, \, N-k} = \frac{MST}{MSE}$$

We reject the null hypothesis that the means are equal if the *F*–statistic is too big. The critical value for deciding whether *F* is too big depends both on its degrees of freedom and the α-level you choose. Every ***F*-distribution** has two degrees of freedom, corresponding to the degrees of freedom for the mean square in the numerator and for the mean square (usually the MSE) in the denominator. Here, the MST has $k - 1$ degrees of freedom because there are k groups. The MSE has $N - k$ degrees of freedom where N is the total number of observations. Rather than compare the *F*-statistic to a specified critical value, we could find the P-value of this statistic and reject the null hypothesis when that value is small.

This analysis is called an **Analysis of Variance (ANOVA)**, but the hypothesis is actually about *means*. The null hypothesis is that the means are all equal. The collection of statistics—the sums of squares, mean squares, *F*-statistic, and P-value—are usually presented in a table, called the **ANOVA table,** like this one:

Source	DF	Sum of Squares	Mean Square	*F*-Ratio	Prob > *F*
Treatment (Between)	$k - 1$	SST	MST	MST/MSE	P-value
Error (Within)	$N - k$	SSE	MSE		
Total	$N - 1$	SSTotal			

Table 21.1   An ANOVA table displays the treatment and error sums of squares, mean squares, *F*-ratio, and P-value.

- **How does the Analysis of Variance work?** When looking at side-by-side boxplots to see whether we think there are real differences between treatment means, we naturally compare the variation *between* the groups to the variation

---

[2]Of course the boxplots show medians at their centers, and we're trying to find differences among means. But for roughly symmetric distributions like these, the means and medians are very close.

within the groups. The variation between the groups indicates how large an effect the treatments have. The variation within the groups shows the underlying variability. To model those variations, the one-way ANOVA decomposes the data into several parts: the grand average, the treatment effects, and the residuals.

$$y_{ij} = \bar{\bar{y}} + (\bar{y}_i - \bar{\bar{y}}) + (y_{ij} - \bar{y}_i).$$

We can write this as we did for regression as

data = predicted + residual.

To estimate the variation *between* the groups we look at how much their means vary. The SST (sometimes called the *between* sum of squares) captures it like this:

$$SST = \sum_{i=1}^{k} n_i(\bar{y}_i - \bar{\bar{y}})^2$$

where $\bar{y}_i$ is the mean of group i, n_i is the number of observations in group i and $\bar{\bar{y}}$ is the overall mean of all observations.

We compare the *SST* to how much variation there is *within* each group. The SSE captures that like this:

$$SSE = \sum_{i=1}^{k} (n_i - 1)s_i^2$$

where s_i^2 is the sample variance of group i.

To turn these estimates of variation into variances, we divide each sum of squares by its associated degrees of freedom:

$$MST = \frac{SST}{k - 1}$$

$$MSE = \frac{SSE}{N - k}$$

Remarkably (and this is Fisher's real contribution), these two variances estimate the *same* variance when the null hypothesis is true. When it is false (and the group means differ), the MST gets larger.

The *F*-statistic tests the null hypothesis by taking the ratio of these two mean squares:

$$F_{k-1,\ N-k} = \frac{MST}{MSE},$$ and rejecting the hypothesis if the ratio is too large.

The critical value and P-value depend on the two degrees of freedom $k - 1$ and $N - k$.

Let's look at an example. For the summer catalog of the percussion supply company *Tom's Tom-Toms*, 4000 customers were selected at random to receive one of four offers[3]: *No Coupon*, *Free Sticks* with purchase, *Free Pad* with purchase, or $50 off next purchase. All the catalogs were sent out on March 15 and sales data for the month following the mailing were recorded.

---

[3]Realistically, companies often select equal (and relatively small) sizes for the treatment groups and consider all other customers as the control. To make the analysis easier, we'll assume that this experiment just considered 4000 "control" customers. Adding more controls wouldn't increase the power very much.

The first step is to plot the data. Here are boxplots of the spending of the four groups for the month after the mailing:

Figure 21.4   Boxplots of the spending of the four groups show that the coupons seem to have stimulated spending.

Here are summary statistics for the four groups:

SUMMARY			
Groups	*Median*	*Mean*	*SD*
No Coupon	$0.00	$216.68	$390.58
Free Sticks	$233.00	$385.87	$331.10
Free Pad	$157.50	$339.54	$364.17
Fifty Dollars	$232.00	$399.95	$337.07

The ANOVA table (from **Excel**) shows the components of the calculation of the *F*-test.

ANOVA					
Source of Variation	*SS*	*df*	*MS*	*F*	*P-value*
Between Groups	20825966	3	6941988.66	54.6169	<0.0001
Within Groups	507905263	3996	127103.42		
Total	528731229	3999			

Table 21.2   The ANOVA table (from **Excel**) shows that the *F*-statistic has a very small P-value, so we can reject the null hypothesis that the means of the four treatments are equal.

The very small P-value is an indication that the differences we saw in the boxplots are not due to chance, so we reject the null hypothesis of equal means and conclude that the four means are not equal.

For Example   Analyzing a one-way design

You decide to implement the simple one factor completely randomized design sending out four offers (*Coupon, Card, Movie, or No Offer*) to 4000 students each (see page 720). A year later you collect the results and find the following table of means and standard deviations:

Level	Number	Mean	Std Dev	Std Err Mean	Lower 95%	Upper 95%
Coupon	4,000	15.17	72.30	1.14	12.93	17.41
Card	4,000	11.53	62.62	0.99	9.59	13.47
Movie	4,000	13.29	66.51	1.05	11.22	15.35
No Offer	4,000	9.03	50.99	0.81	7.45	10.61

An ANOVA table shows:

Source	DF	Sum of Squares	Mean Square	F Ratio	Prob > F
Offer	3	81,922.26	27,307.42	6.75	0.0002
Error	15,996	64,669,900.04	4,042.88		
Total	15,999	64,751,822.29			

Question: What conclusions can you draw from these data?

Answer: From the ANOVA table, it looks like the null hypothesis that all the means are equal is strongly rejected. The average number of miles traveled seems to have increased about 2.5 miles for students receiving the *Card*, about 4.5 miles for students receiving the free *Movie*, and about 6 miles for those students receiving the *Coupon*.

21.7   Assumptions and Conditions for ANOVA

Whenever we compute P-values and make inferences about a hypothesis, we need to make assumptions and check conditions to see if the assumptions are reasonable. The ANOVA is no exception. Because it's an extension of the two-sample *t*-test, many of the same assumptions apply.

Independence Assumption

The groups must be independent of each other. No test can verify this assumption. You have to think about how the data were collected. The individual observations must be independent as well.

We check the **Randomization Condition**. Did the experimental design incorporate suitable randomization? We were told that the customers were assigned to each treatment group at random.

Equal Variance Assumption

ANOVA assumes that the true variances of the treatment groups are equal. We can check the corresponding **Similar Variance Condition** in various ways:

- Look at side-by-side boxplots of the groups to see whether they have roughly the same spread. It can be easier to compare spreads across groups when they have the same center, so consider making side-by-side boxplots of the residuals.

If the groups have differing spreads, it can make the pooled variance—the MSE—larger, reducing the F-statistic value and making it less likely that we can reject the null hypothesis. So the ANOVA will usually fail on the "safe side," rejecting H_0 less often than it should. Because of this, we usually require the spreads to be quite different from each other before we become concerned about the condition failing. If you've rejected the null hypothesis, this is especially true.

- Look at the original boxplots of the response values again. In general, do the spreads seem to change systematically with the centers? One common pattern is for the boxes with bigger centers to have bigger spreads. This kind of systematic trend in the variances is more of a problem than random differences in spread among the groups and should not be ignored. Fortunately, such systematic violations are often helped by re-expressing the data. If, in addition to spreads that grow with the centers, the boxplots are skewed with the longer tail stretching off to the high end, then the data are pleading for a re-expression. Try taking logs of the dependent variable for a start. You'll likely end up with a much cleaner analysis.

- Look at the residuals plotted against the predicted values. Often, larger predicted values lead to larger magnitude residuals. This is another sign that the condition is violated. If the residual plot shows more spread on one side or the other, it's usually a good idea to consider re-expressing the response variable. Such a systematic change in the spread is a more serious violation of the equal variance assumption than slight variations of the spreads across groups.

Figure 21.5 A plot of the residuals against the predicted values from the ANOVA shows no sign of unequal spread.

Normal Population Assumption

Like Student's t-tests, the F-test requires that the underlying errors follow a Normal model. As before when we faced this assumption, we'll check a corresponding **Nearly Normal Condition**.

Technically, we need to assume that the Normal model is reasonable for the populations underlying each treatment group. We can (and should) look at the side-by-side boxplots for indications of skewness. Certainly, if they are all (or mostly) skewed in the same direction, the Nearly Normal Condition fails (and re-expression is likely to help). However, in many business applications, sample sizes are quite large, and when that is true, the Central Limit Theorem implies that the sampling distribution of the means may be nearly Normal in spite of skewness. Fortunately, the F-test is conservative. That means that if you see a small P-value it's probably safe to reject the null hypothesis for large samples even when the data are nonnormal.

Check Normality with a histogram or a Normal probability plot of all the residuals together. Because we really care about the Normal model within each group, the Normal Population Assumption is violated if there are outliers in any of the groups. Check for outliers in the boxplots of the values for each treatment.

The Normal Probability plot for the *Tom's Tom-Toms* residuals holds a surprise.

Figure 21.6   A normal probability plot shows that the residuals from the ANOVA of the *Tom's Tom-Toms* data are clearly not normal.

Investigating further with a histogram, we see the problem.

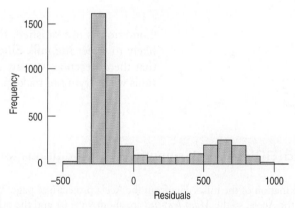

| Figure 21.7   A histogram of the residuals reveals bimodality.

The histogram shows clear bimodality of the residuals. If we look back to histograms of the spending of each group, we can see that the boxplots failed to reveal the bimodal nature of the spending.

The manager of the company wasn't surprised to hear that the spending is bimodal. In fact, he said, "We typically have customers who either order a complete new drum set, or who buy accessories. And, of course, we have a large group of customers who choose not to purchase anything during a given quarter."

These data (and the residuals) clearly violate the Nearly Normal Condition. Does that mean that we can't say anything about the null hypothesis? No. Fortunately, the sample sizes are large, and there are no individual outliers that have undue influence on the means. With sample sizes this large, we can appeal to the Central Limit Theorem and still make inferences about the means. In particular, we are safe in rejecting the null hypothesis. When the Nearly Normal

Figure 21.8 The spending appears to be bimodal for all the treatment groups. There is one mode near $1000 and another larger mode between $0 and $200 for each group.

Condition is not satisfied, the *F*-test will tend to fail on the safe side and be less likely to reject the null. Since we have a very small P-value, we can be fairly sure that the differences we saw were real. On the other hand, we should be very cautious when trying to make predictions about individuals rather than means.

For Example   Assumptions and conditions for ANOVA

Closer examination of the miles data from the Acela project (see page 720) shows that only about 5% of the students overall actually took the Acela, so the *Miles Traveled* are about 95% 0's and the other values are highly skewed to the right.

Question: Are the assumptions and conditions for ANOVA satisfied?

Answer: The responses are independent since the offer was randomized to the students on the mailing list. The distributions of *Miles Traveled* by *Offer* are highly right skewed. Most of the entries are zeros. This could present a problem, but because the sample size is so large (4000 per group), the inference is valid (a simulation shows that the averages of 4000 are Normally distributed). Although the distributions are right skewed, there are no extreme outliers that are influencing the group means. The variances in the four groups also appear to be similar. Thus, the assumptions and conditions appear to be met. (An alternative analysis might be to focus on the *Miles Traveled* only of those who actually took the Acela. The conclusion of the ANOVA would remain the same).

Just Checking

Your experiment to test the new pet food formulation has been completed. One hypothesis you have tested is whether the new formulation is different in nutritional value (measured by having veterinarians evaluate the test animals) from a standard food known to be safe and nutritious. The ANOVA has an *F*-statistic of 1.2, which (for the degrees of freedom in your experiment) has a P-value of 0.87. Now you need to make a report to the company.

7 Write a brief report. Can you conclude that the new formulation is safe and nutritious?

*21.8   Multiple Comparisons

Simply rejecting the null hypothesis is almost never the end of an Analysis of Variance. Knowing that the means are different leads to the question of which ones are different and by how much. Tom, the owner of *Tom's Tom-Toms*, would hardly be satisfied with a consultant's report that told him that the offers generated different amounts of spending, but failed to indicate which offers did better and by how much.

We'd like to know more, but the *F*-statistic doesn't offer that information. What can we do? If we can't reject the null hypothesis, there's no point in further testing. But if we can reject the null hypothesis, we can do more. In particular, we can test whether any pairs or combinations of group means differ. For example, we might want to compare treatments against a control or against the current standard treatment.

We could do *t*-tests for any pair of treatment means that we wish to compare. But each test would have some risk of a Type I error. As we do more and more tests, the risk that we'll make a Type I error grows. If we do enough tests, we're almost sure to reject one of the null hypotheses by mistake—and we'll never know which one.

There is a solution to this problem. In fact, there are several solutions. As a class, they are called methods for **multiple comparisons**. All multiple comparisons methods require that we first reject the overall null hypothesis with the ANOVA's *F*-test. Once we've rejected the overall null, we can think about comparing several—or even all—pairs of group means.

One such method is called the **Bonferroni method**. This method adjusts the tests and confidence intervals to allow for making many comparisons. The result is a wider margin of error (called the **minimum significant difference, or MSD**) found by replacing the critical *t*-value t^* with a slightly larger number. That makes the confidence intervals wider for each pairwise difference and the corresponding Type I error rates lower for each test, and it keeps the overall Type I error rate at or below α.

The Bonferroni method distributes the error rate equally among the confidence intervals. It divides the error rate among J confidence intervals, finding each interval at confidence level $1 - \dfrac{\alpha}{J}$ instead of the original $1 - \alpha$. To signal this adjustment, we label the critical value t^{**} rather than t^*. For example, to make the

Carlo Bonferroni (1892–1960) was a mathematician who taught in Florence. He wrote two papers in 1935 and 1936 setting forth the mathematics behind the method that bears his name.

six confidence intervals comparing all possible pairs of offers at our overall α risk of 5%, instead of making six 95% confidence intervals, we'd use

$$1 - \frac{0.05}{6} = 1 - .0083 = .9917$$

instead of 0.95. So we'd use a critical t^{**} value of 2.64 instead of 1.96. The ME would then become:

$$ME = 2.642 \times 356.52\sqrt{\frac{1}{1000} + \frac{1}{1000}} = 42.12$$

This change doesn't affect our decision that each offer increases the mean sales compared to the *No Coupon* group, but it does adjust the comparison of average sales for the *Free Sticks* offer and the *Free Pad* offer. With a margin of error of $42.12, the difference between average sales for those two offers is now $(385.87 - 339.54) \pm 42.12 = (\$4.21, \$88.45)$.

The confidence interval says that the *Free Sticks* offer generated between $4.21 and $88.45 more sales per customer on average than the *Free Pad* offer. In order to make a valid business decision, the company should now calculate their expected *profit* based on the confidence interval. Suppose they make 8% profit on sales. Then, multiplying the confidence interval

$$0.08 \times (\$4.21, \$88.45) = (\$0.34, \$7.08)$$

we find that the *Free Sticks* generate between $0.34 and $7.08 profit per customer on average. So, if the *Free Sticks* cost $1.00 more than the pads, the confidence interval for profit would be:

$$(\$0.34 - \$1.00, \$7.08 - \$1.00) = (-\$0.66, \$6.08)$$

There is a possibility that the *Free Sticks* may actually be a less profitable offer. The company may decide to take the risk or to try another test with a larger sample size to get a more precise confidence interval.

Many statistics packages assume that you'd like to compare all pairs of means. Some will display the result of these comparisons in a table such as the one to the left. This table indicates that the top two are indistinguishable, that all are distinguishable from *No Coupon*, and that *Free Pad* is also distinguishable from the other three.

The subject of multiple comparisons is a rich one because of the many ways in which groups might be compared. Most statistics packages offer a choice of several methods. When your analysis calls for comparing all possible pairs, consider a multiple comparisons adjustment such as the Bonferroni method. If one of the groups is a control group, and you wish to compare all the other groups to it, there are specific methods (such as Dunnett's methods) available. Whenever you look at differences after rejecting the null hypothesis of equal means, you should consider using a multiple comparisons method that attempts to preserve the *overall* α risk.

Fifty Dollars	$399.95	A		
Free Sticks	$385.87	A		
Free Pad	$339.54		B	
No Coupon	$216.68			C

Table 21.3 The output shows that the two top-performing offers are indistinguishable in terms of mean spending, but that the Free Pad is distinguishable from both those two and from *No Coupon*.

For Example   Multiple comparisons

You perform a multiple comparison analysis using the Bonferroni correction on the Acela data (see page 733) and the output looks like:

Offer			Mean
Coupon	A		15.17
Movie	A		13.29
Card	A	B	11.53
No Offer		B	9.03

Question: What can you conclude?

Answer: From the original ANOVA we concluded that the means were not all equal. Now it appears that we can say that the mean *Miles Traveled is* greater for those receiving the *Coupon* or the *Movie* than those receiving *No Offer*, but we cannot distinguish the mean *Miles Traveled* between those receiving the *Card* or *No Offer*.

A further analysis only of those who actually took the Acela during the 12 months shows a slightly different story:

Coupon	A			300.37
Card	A	B		279.50
Movie		B		256.74
No Offer			C	206.40

Here we see that all offers can be distinguished from the *No Offer* group and that the *Coupon* group performed better than the group receiving the free *Movie*. Of those using the Acela, the *Coupon* resulted in nearly 100 more miles traveled on average during the year of those taking the Acela at least once.

21.9   ANOVA on Observational Data

So far we've applied ANOVA only to data from designed experiments. That application is appropriate for several reasons. The primary one is that randomized comparative experiments are specifically designed to compare the results for different treatments. The overall null hypothesis, and the subsequent tests on pairs of treatments in ANOVA, address such comparisons directly. In addition, the **Equal Variance Assumption** (which we need for all of the ANOVA analyses) is often plausible in a randomized experiment because when we randomly assign subjects to treatments, all the treatment groups start out with the same underlying variance of the experimental units.

Sometimes, though, we just can't perform an experiment. When ANOVA is used to test equality of group means from observational data, there's no *a priori* reason to think the group variances might be equal at all. Even if the null hypothesis of equal means were true, the groups might easily have different variances. But you can use ANOVA on observational data if the side-by-side boxplots of responses for each group show roughly equal spreads and symmetric, outlier-free distributions.

Observational data tend to be messier than experimental data. They are much more likely to be unbalanced. If you aren't assigning subjects to treatment groups, it's harder to guarantee the same number of subjects in each group. And because you are not controlling conditions as you would in an experiment, things tend to be, well, less controlled. The only way we know to avoid the effects of possible lurking variables is with control and randomized assignment to treatment groups, and for observational data, we have neither.

ANOVA is often applied to observational data when an experiment would be impossible or unethical. (We can't randomly break some subjects' legs, but we *can* compare pain perception among those with broken legs, those with sprained ankles, and those with stubbed toes by collecting data on subjects who have already suffered those injuries.) In such data, subjects are already in groups, but not by random assignment.

Be careful; if you have not assigned subjects to treatments randomly, you can't draw *causal* conclusions even when the *F*-test is significant. You have no way to control for lurking variables or confounding, so you can't be sure whether any differences you see among groups are due to the grouping variable or to some other unobserved variable that may be related to the grouping variable.

Balance
Recall that a design is called *balanced* if it has an equal number of observations for each treatment level.

Because observational studies often are intended to estimate parameters, there is a temptation to use pooled confidence intervals for the group means for this purpose. Although these confidence intervals are statistically correct, be sure to think carefully about the population that the inference is about. The relatively few subjects that you happen to have in a group may not be a simple random sample of any interesting population, so their "true" mean may have only limited meaning.

21.10   Analysis of Multifactor Designs

In our direct mail example, we looked at two factors: *Miles* and *Envelope*. *Miles* had three levels: *Regular Miles*, *Double Miles*, and *Double Anywhere Miles*. The factor *Envelope* had two levels: *Standard* and new *Logo*. The three levels of *Miles* and the two levels of *Envelope* resulted in six treatment groups. Because this was a completely randomized design, the 30,000 customers were selected at random, and 5000 were assigned at random to each treatment.

Three months after the offer was mailed out, the total charges on the card were recorded for each of the 30,000 cardholders in the experiment. Here are boxplots of the six treatment groups' responses, plotted against each factor.

Figure 21.9   Boxplots of *Total Charge* by each factor. It is difficult to see the effects of the factors for two reasons. First, the other factor hasn't been accounted for, and second, the effects are small compared to the overall variation in charges.

If you look closely, you may be able to discern a very slight increase in the *Total Charges* for some levels of the factors, but it's very difficult to see. There are two reasons for this. First, the variation due to each factor gets in the way of seeing the effect of the other factor. For example, each customer in the boxplot for the *Logo Envelope* got one of three different offers. If those offers had an effect on spending, then that increased the variation within the *Logo* treatment group. Second, as is typical in a marketing experiment of this kind, the effects are very small compared to the variability in people's spending. That's why companies use such a large sample size.

The analysis of variance for two factors removes the effects of each factor from consideration of the other. It can also model whether the factors interact, increasing or decreasing the effect. In our example, it will separate out the effect of changing the levels of *Miles* and the effect of changing the levels of *Envelope*. It will also test whether the effect of the *Envelope* is the same for the three different *Miles* levels. If the effect is different, that's called an interaction effect between the two factors.

The details of the calculations for the two-way ANOVA with interaction are less important than understanding the summary, the model, and the assumptions and conditions under which it's appropriate to use the model. For a one-way ANOVA, we calculated three sums of squares (SS): the Total SS, the Treatment SS, and the Error SS. For this model, we'll calculate five: the Total SS, the SS due to Factor A, the SS due to Factor B, the SS due to the interaction, and the Error SS.

Let's suppose we have a levels of factor A, b levels of factor B, and r replicates at each treatment combination. In our case, $a = 2, b = 3, r = 5000$, and $a \times b \times r = N$ is 30,000. Then the ANOVA table will look like this.

Source	DF	Sum of Squares	Mean Square	F-Ratio	Prob > F
Factor A	$a - 1$	SSA	MSA	MSA/MSE	P-value
Factor B	$b - 1$	SSB	MSB	MSB/MSE	P-value
Interaction	$(a - 1) \times (b - 1)$	SSAB	MSAB	MSAB/MSE	P-value
Error	$ab(r - 1)$	SSE	MSE		
Total (Corrected)	$N - 1$	SSTotal			

Table 21.4 An ANOVA table for a replicated two-factor design with a row for each factor's sum of squares, interaction sum of square, error, and total.

There are now three null hypotheses—one that asserts that the means of the levels of factor A are equal, one that asserts that the means of the levels of factor B are all equal, and one that asserts that the effects of factor A are *constant* across the levels of factor B (or vice versa). Each P-value is used to test the corresponding hypothesis.

Here is the ANOVA table for the marketing experiment.

Source	DF	Sum of Squares	Mean Square	F-Ratio	Prob > F
Miles	2	201,150,000	100,575,000	66.20	<.0001
Envelope	1	203,090,000	203,090,000	133.68	<.0001
Miles $\times$ Envelope	2	1,505,200	752,600	0.50	0.61
Error	29,994	45,568,000,000	1,519,237		

Table 21.5 The ANOVA table for the marketing experiment. Both the effect of *Miles* and *Envelope* are highly significant, but the interaction term is not.

From the ANOVA table, we can see that both the *Miles* and the *Envelope* effects are highly significant, but that the interaction term is not. An **interaction plot,** a plot of means for each treatment group, is essential for sorting out what these P-values mean.

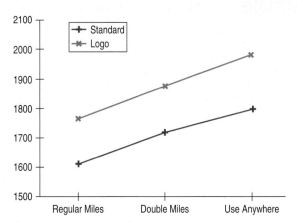

Figure 21.10 An interaction plot of the *Miles* and *Envelope* effects. The parallel lines show that the effects of the three *Miles* offers are roughly the same over the two different *Envelopes* and therefore that the interaction effect is small.

The interaction plot shows the mean *Charges* at all six treatment groups. The levels of one of the factors, in this case *Miles*, are shown on the *x*-axis, and the mean *Charges* of the groups for each *Envelope* level are shown at each *Miles* level. The means of each level of *Envelope* are connected for ease of understanding. Notice that the effect of *Double Miles* over *Regular Miles* is about the same for both the *Standard* and *Logo Envelopes*. And the same is true for the *Use Anywhere* miles. This indicates that the effect of *Miles* is constant for the two different *Envelopes*. The lines are parallel, which indicates that there is no interaction effect.

We reject the null hypothesis that the mean *Charges* at the three different levels of *Miles* are equal (with P-values < 0.0001), and also we reject that the mean *Charges* for *Standard* and *Logo* are the same (with P-value < 0.0001). We have no evidence, however, to suggest that there is an interaction between the factors.

After rejecting the null hypotheses, we can create a confidence interval for any particular treatment mean or perform a hypothesis test for the difference between any two means. If we want to do several tests or confidence intervals, we will need to use a multiple comparisons method that adjusts the size of the confidence interval or the level of the test to keep the *overall* Type I error rate at the level we desire.

When the interaction term is not significant, we can talk about the overall effect of either factor. Because the effect of *Envelope* is roughly the same for all three *Miles* offers (as we know by virtue of not rejecting the hypothesis that the interaction effect is zero), we can calculate and interpret an overall *Envelope* effect. The means of the two *Envelope* levels are:

Logo $1871.75   *Standard* $1707.19

and so the *Logo* envelope generated a difference in average charge of $1871.75 − $1707.19 = $164.56. A confidence interval for this difference is ($136.66, $192.45), which the analysts can use to decide whether the added cost of the *Logo* envelope is worth the expense.

But when an interaction term *is* significant, we must be very careful not to talk about the effect of a factor, *on average*, because the effect of one factor *depends* on the level of the other factor. In that case, we always have to talk about the factor effect at a specific level of the other factor, as we'll see in the next example.

For Example   Multifactor designs

Question: Suppose that you had run the randomized block design from Section 21.4 (see page 724). You would have had two levels of the (blocking) factor *NE Corridor* (*NE* or *not*) and the same four levels of *Offer* (*Coupon, Card, Movie,* and *No Offer*). The ANOVA shows a significant interaction effect between *NE Corridor* and *Offer*. Explain what that means. An analysis of the two groups (*NE* and *not*) separately shows that for the *NE* group, the P-value for testing the four offers is <0.0001 but for the *not* NE group, the P-value is 0.2354. Is this consistent with a significant interaction effect? What would you tell the marketing group?

Answer: A significant interaction effect implies that the effect of one factor is not the same for the levels of another. Thus, it is saying that the effect of the four offers is not the same for those living in the NE corridor as it is for those who do not. The separate analysis explains this further. For those who do not live in the NE, the offers do not significantly change the average number of miles they travel on the Acela. However, for those who live or go to school in the NE, the offers have an effect. This could impact where Amtrak decides to advertise the offers, or to whom they decide to send them.

Math Box

- **How does Two-Way Analysis of Variance work?** In Two-Way ANOVA, we have two factors. Each treatment consists of a level of each of the factors, so we write the individual responses as y_{ij}, to indicate the ith level of the first factor and the jth level of the second factor. The more general formulas are no more informative; just more complex. We will start with an unreplicated design with one observation in each treatment group (although in practice, we'd always recommend replication if possible). We'll call the factors A and B, each with a and b levels, respectively. Then the total number of observations is $n = a \times b$.

  For the first factor (factor A), the *Treatment Sum of Squares*, SSA, is the same as we calculated for one-way ANOVA:

$$SSA = \sum_{i=1}^{a} b(\bar{y}_i - \bar{\bar{y}})^2 = \sum_{j=1}^{b}\sum_{i=1}^{a}(\bar{y}_i - \bar{\bar{y}})^2$$

where b is the number of levels of factor B, $\bar{y}_i$ is the mean of all subjects assigned level i of factor A (regardless of which level of factor B they were assigned), and $\bar{\bar{y}}$ is the *overall* mean of all observations. The mean square for treatment A (MSA) is

$$MSA = \frac{SSA}{a-1}.$$

The treatment sum of squares for the second factor (B) is computed in the same way, but of course the treatment means are now the means for each level of this second factor:

$$SSB = \sum_{j=1}^{b} a(\bar{y}_j - \bar{\bar{y}})^2 = \sum_{i=1}^{a}\sum_{j=1}^{b}(\bar{y}_j - \bar{\bar{y}})^2, \text{ and}$$

$$MSB = \frac{SSB}{b-1}$$

where a is the number of levels of factor A, and $\bar{\bar{y}}$, as before, is the overall mean of all observations. $\bar{y}_j$ is the mean of all subjects assigned the jth level of factor B.

The SSE can be found by subtraction:

$$SSE = SSTotal - (SSA + SSB)$$

where

$$SSTotal = \sum_{i=1}^{a}\sum_{j=1}^{b}(y_{ij} - \bar{\bar{y}})^2.$$

The mean square for error is $MSE = \dfrac{SSE}{N-(a+b-1)}$, where $N = a \times b$.

There are now two F-statistics, the ratio of each of the treatment mean squares to the MSE, which are associated with each null hypothesis.

To test whether the means of all the levels of factor A are equal, we would find a P-value for $F_{a-1, N-(a+b-1)} = \dfrac{MSA}{MSE}$. For factor B, we would find a P-value for $F_{b-1, N-(a-b+1)} = \dfrac{MSB}{MSE}$.

If the experiment is replicated (say r times), we can also estimate the interaction between the two factors, and test whether it is zero. The sums of squares for each factor have to be multiplied by r. Alternatively they can be written as a (triple) sum where the sum is now over factor A, factor B, and the replications:

$$SSA = \sum_{k=1}^{r}\sum_{j=1}^{b}\sum_{i=1}^{a}(\bar{y}_i - \bar{\bar{y}})^2 \text{ and } SSB = \sum_{k=1}^{r}\sum_{i=1}^{a}\sum_{j=1}^{b}(\bar{y}_j - \bar{\bar{y}})^2.$$

(continued)

We find the sum of squares for the interaction effect AB as:

$$SSAB = \sum_{k=1}^{r} \sum_{j=1}^{b} \sum_{i=1}^{a} (\bar{y}_{ij} - \bar{y}_i - \bar{y}_j - \bar{\bar{y}})^2, \text{ and}$$

$$MSAB = \frac{SSAB}{(a-1)(b-1)}.$$

The SSE is the sum of the squared residuals:

$$SSE = \sum_{k=1}^{r} \sum_{j=1}^{b} \sum_{i=1}^{a} (y_{ijk} - \bar{y}_{ij})^2 \text{ and } MSE = \frac{SSE}{ab(r-1)}.$$

There are now three *F*-statistics associated with the three hypotheses (factor A, factor B, and the interaction). They are the ratios of each of these mean squares with the MSE:

$$F_{a-1,\,ab(r-1)} = \frac{MSA}{MSE}, \; F_{b-1,\,ab(r-1)} = \frac{MSB}{MSE}, \text{ and } F_{(a-1)(b-1),\,ab(r-1)} = \frac{MSAB}{MSE}.$$

Note that $N = r \times a \times b$ is the total number of observations in the experiment.

Guided Example   A Follow-up Experiment

After analyzing the data, the bank decided to go with the *Logo* envelope, but a marketing specialist thought that more *Miles* might increase spending even more. A new test was designed to test both the type of *Miles* and the amount. Again, total *Charge* in three months is the response.

PLAN State the problem.	We want to study the two factors *Miles* and *Amount* to see their effect on the revenue generated for a new credit card offer.
Response Specify the response variable.	To measure the success, we will use the monthly charges of customers who receive the various offers. We will use the three months after the offer is sent out as the collection period and the total amount charged per customer during this period as the response.
Factors Identify the factors you plan to test. **Levels** Specify the levels of the factors you will use.	We will offer each customer one of the two levels of the factor *Miles* for the SkyWest Gold card: SkyWest miles or Use Anywhere Miles. Customers are offered three levels of *Miles*: Regular Miles, Double Miles, and Triple Miles.
	We will send out all the offers to customers at the same time (in mid March) and evaluate the response as total charges in the period April through June.

Experimental Design Specify the design.

A total of 60,000 current Gold card customers will be randomly selected from our customer records to receive one of the six offers.

✓ Regular SkyWest miles

✓ Double SkyWest miles

✓ Triple SkyWest miles

✓ Regular *Use Anywhere Miles*

✓ Double *Use Anywhere Miles*

✓ Triple *Use Anywhere Miles*

Make a Picture A diagram of your design can help you think about it. We could also draw this diagram like the one on page 726 with 6 treatment groups, but now we are thinking of the design as having two distinct factors that we wish to evaluate individually, so this form gives the right impression.

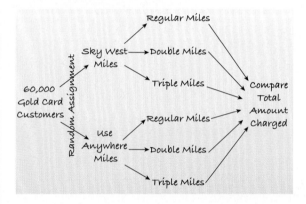

On June 15, we will examine the total card charges for each customer for the period April 1 through June 30.

We want to measure the effect of the two types of *Miles* and the three award *Amounts*.

Specify any other experimental details. You must give enough details so that another experimenter could exactly replicate your experiment.

It's generally better to include details that might seem irrelevant than to leave out matters that could turn out to make a difference.

Specify how to measure the response and your hypotheses.

The three null hypotheses are:

H_O: The mean charges for *Sky West Miles* and *Use Anywhere Miles* are the same (the means for *Miles* are equal).

H_O: The mean charges for *Regular Miles*, *Double Miles*, and *Triple Miles* are the same (the means for *Amount* are equal).

H_O: The effect of *Miles* is the same for all levels of *Amount* (and vice-versa) (no interaction effect).

The alternative for the first hypothesis is that the mean *Charges* for the two levels of *Miles* are different.

The alternative for the second hypothesis is that at least one of the mean charges for the three levels of *Amount* is different.

The alternative for the third hypothesis is that there is an interaction effect.

(continued)

Plot Examine the boxplots and interaction plots.

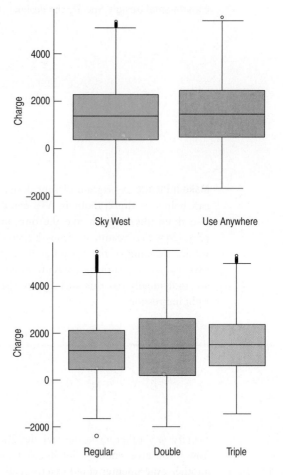

Boxplots by each factor show that there may be a slight increase in charges due to the *Use Anywhere* miles and the *Amount* of miles offered, but the differences are hard to see because of the intrinsic variation in *Charges*.

There are some outliers apparent in the boxplots, but none exerts a large influence on its group mean, so we will leave them in.

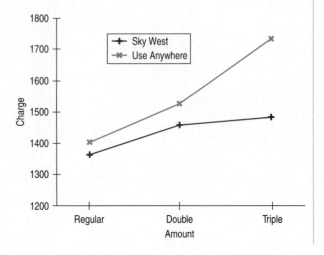

The interaction plot shows that offering *Triple* miles may have a much larger effect for *Use Anywhere* miles than for *Sky West* miles.

Assumptions and Conditions Think about the assumptions and check the conditions.

✓ **Independence Assumption, Randomization Condition.** The experiment was randomized to current cardholders.

✓ **Similar Variance Condition.** The boxplots show that the variances across all groups are similar. (We can recheck with a residual plot after fitting the ANOVA model.)

✓ **Outlier Condition.** There are some outliers, but none appear to be exerting undue influence on the group means.

DO

Show the ANOVA table.

Source	Df	SS	MS	*F*-Ratio	P-Value
Miles	1	103,576,768	103,576,768	61.6216	<0.0001
Amount	2	253,958,660.1	126,979,330	75.5447	<0.0001
Miles × Amount	2	64,760,963.01	32,380,481.51	19.2643	<0.0001
Error	29,994	50,415,417,459	1,680,850		
Total	29,999	50,837,713,850			

Check the remaining conditions on the residuals.

✓ **Nearly Normal Condition.** A histogram of the residuals shows that they are reasonably unimodal and symmetric.

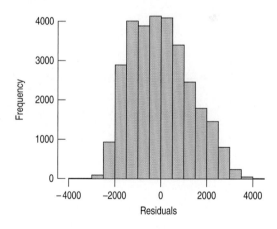

Under these conditions, it is appropriate to interpret the F-ratios and their P-values.

Discuss the results of the ANOVA table.

The F-ratios are all large, and the P-values are all very small, so we reject all three null hypotheses. Because the interaction effect is significant, we cannot talk about the overall effect of the amount of miles but must make the discussion specific to the type of *Miles* offered.

(continued)

Show a table of means, possibly with confidence intervals or tests from an appropriate Multiple Comparisons method.

Level					Mean
Use Anywhere, Triple	A				1732.21
Use Anywhere, Double		B			1526.93
Sky West, Triple		B			1484.34
Sky West, Double		B	C		1460.20
Use Anywhere, Regular			C	D	1401.89
Sky West, Regular				D	1363.94

REPORT

To answer the initial question, we ask whether the differences we observe in the means of the groups are meaningful.

Because this is a randomized experiment, we can attribute significant differences to the treatments.

MEMO

Re: Test Mailing for Creative Offer and Envelope

The mailing for testing the Triple Miles initiative went out in March, and results on charges from April through June were available in early July. We found that *Use Anywhere* miles performed better than the standard *Sky West* miles, but that the amount they increased charges depended on the amount offered.

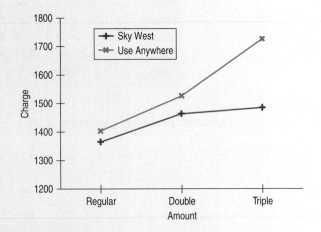

As we can see, *Triple Miles* for the *Sky West* miles didn't increase *Charge* significantly and is probably not worth the added expense. However, *Triple Miles* for the *Use Anywhere* miles generated an average $205 more in average *Charge* (with a confidence interval from $131 to $279). Even at the low end of this interval, we feel that the added revenue of the Triple Miles justifies their cost.

Be sure to make recommendations based on the context of your business decision.

In summary, we recommend offering Triple Miles for the Use Anywhere miles offers but would keep the Double miles offer for the Sky West miles.

What Can Go Wrong?

- **Don't give up just because you can't run an experiment.** Sometimes we can't run an experiment because we can't identify or control the factors. Sometimes it would simply be unethical to run the experiment. (Consider randomly assigning employees to two environments—one where workers were exposed to massive amounts of cigarette smoke and one a smoke-free environment—to see differences in health and productivity.) If we can't perform an experiment, often an observational study is a good choice.

- **Beware of confounding.** Use randomization whenever possible to ensure that the factors not in your experiment are not confounded with your treatment levels. Be alert to confounding that cannot be avoided, and report it along with your results.

- **Bad things can happen even to good experiments.** Protect yourself by recording additional information. An experiment in which the air-conditioning failed for two weeks, affecting the results, was saved by recording the temperature (although that was not originally one of the factors) and estimating the effect the higher temperature had on the response.[4] It's generally good practice to collect as much information as possible about your experimental units and the circumstances of the experiment. For example, in the direct mail experiment, it would be wise to record details of the general economy and any global events (such as a sharp downturn in the stock market) that might affect customer behavior.

- **Don't spend your entire budget on the first run.** Just as it's a good idea to pretest a survey, it's always wise to try a small pilot experiment before running the full-scale experiment. You may learn, for example, how to choose factor levels more effectively, about effects you forgot to control, and about unanticipated confounding.

- **Watch out for outliers.** One outlier in a group can change both the mean and the spread of that group. It will also inflate the Error Mean Square, which can influence the F-test. The good news is that ANOVA fails on the safe side by losing power when there are outliers. That is, you are less likely to reject the overall null hypothesis if you have (and leave) outliers in your data, so they are not likely to cause you to make a Type I error.

- **Watch out for changing variances.** The conclusions of the ANOVA depend crucially on the assumptions of independence and constant variance and (somewhat less seriously as the number of observations in each group increases) on Normality. If the conditions on the residuals are violated, it may be necessary to re-express the response variable to approximate these conditions more closely. ANOVA benefits so greatly from a judiciously chosen re-expression that the choice of a re-expression might be considered a standard part of the analysis.

- **Be wary of drawing conclusions about causality from observational studies.** ANOVA is often applied to data from randomized experiments for which causal conclusions are appropriate. If the data are not from a designed experiment, however, the Analysis of Variance provides no more evidence for causality than any other method we have studied. Don't get into the habit of assuming that ANOVA results have causal interpretations.

(continued)

---

[4]R. D. DeVeaux and M. Szelewski, "Optimizing Automatic Splitless Injection Parameters for Gas Chromatographic Environmental Analysis," *Journal of Chromatographic Science* 27, no. 9 (1989): 513–518.

- **Be wary of generalizing to situations other than the one at hand.** Think hard about how the data were generated to understand the breadth of conclusions you are entitled to draw.

- **Watch for multiple comparisons.** When rejecting the null hypothesis, you can conclude that the means are not *all* equal. But you can't start comparing every pair of treatments in your study with a *t*-test. You'll run the risk of inflating your Type I error rate. Use a multiple comparisons method when you want to test many pairs.

- **Be sure to fit an interaction term when it exists.** When the design is replicated, it is always a good idea to fit an interaction term. If it turns out not to be statistically significant, you can then fit a simpler two-factor main effects model instead.

- **When the interaction effect is significant, don't interpret the main effects.** Main effects can be very misleading in the presence of interaction terms. Look at this interaction plot:

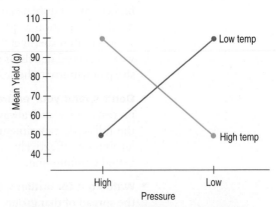

Figure 21.11   An interaction plot of *Yield by Temperature* and *Pressure*. The main effects are misleading. There is no (main) effect of *Pressure* because the average *Yield* at the two pressures is the same. That doesn't mean that *Pressure* has no effect on the *Yield*. In the presence of an interaction effect, be careful when interpreting the main effects.

The experiment was run at two temperatures and two pressure levels. High amounts of material were produced at high pressure with high temperature and at low pressure with low temperature. What's the effect of *Temperature*? Of *Pressure*? Both main effects are 0, but it would be silly (and wrong) to say that neither *Temperature* nor *Pressure* was important. The real story is in the interaction.

Ethics in Action

Professors at many state universities belong to a faculty union. The unionized faculty in one state's university system are preparing for contract negotiations. Cheryl McCrady, recently elected union president at one of the state university campuses, has long been concerned about the salary differential between male and female faculty. As union president, she now has access to faculty salary information, and she decides to run some analyses. After consulting with a few colleagues who regularly use statistics, she settles on using analysis of variance to determine if differences in salary can be attributed to gender accounting for faculty rank (assistant professor, associate professor, and full professor). She's not surprised by the results. While there is no significant interaction effect of gender and rank, she does find that both gender and rank are significant factors in explaining salary differences. Given that discrimination based on gender is a serious issue, she is wondering how she should proceed.

ETHICAL ISSUE *This is an observational study lacking the control of an experimental study. Confounding variables are likely to exist, but are not discussed. For instance, lower paid disciplines (e.g., Education) tend to have more female faculty than higher paid disciplines (e.g., Business). Related to Item A, ASA Guidelines. She should also check for outliers. Special cases, such as a star football coach or Nobel prize winner, may command unusually large salaries but not be relevant to the pay of ordinary faculty members.*

ETHICAL SOLUTION *Make all caveats explicit. This is a complex issue that should not be treated simply.*

What Have We Learned?

Learning Objectives

■ Recognize observational studies.
- A retrospective study looks at an outcome in the present and looks for facts in the past that relate to it.
- A prospective study selects subjects and follows them as events unfold.

■ Know the elements of a designed randomized experiment.
- *Experimental units* (sometimes called *subjects* or *participants*) are assigned at random to *treatments*.
  - The experimenter manipulates *factors*, setting them to specified *levels* to establish the treatments.
- A quantitative *response variable* is measured or observed for each experimental unit.
- We can attribute differences in the response to the differences among the treatments.

■ State and apply the Four Principles of Experimental Design.
- *Control* sources of variation other than the factors being tested. Make the conditions as similar as possible for all treatment groups except for differences among the treatments.
- *Randomize* the assignment of subjects to treatments. *Balance* the design by assigning the same number of subjects to each treatment.
- *Replicate* the experiment on more than one subject.
- *Block* the experiment by grouping together subjects who are similar in important ways that you cannot control.

■ Work with *Blinding* and *Control groups*.
- A *single-blind* study is one in which either all those who can affect the results or all those who evaluate the results are kept ignorant of which subjects receive which treatments.

- A *double-blind* study is one in which both those classes of actors are ignorant of the treatment assignment.
- A *control group* is assigned to a null treatment or to the best available alternative treatment.
  - Control subjects are often administered a *placebo* or null treatment that mimics the treatment being studied but is known to be inactive.

■ Understand how to use Analysis of Variance (ANOVA) to analyze designed experiments.
- ANOVA tables follow a standard layout; be acquainted with it.
- The *F*-statistic is the test statistic used in ANOVA. *F*-statistics test hypotheses about the equality of the means of two or more groups.

Terms

Analysis of Variance (ANOVA)	An analysis method for testing equality of means across treatment groups.
ANOVA table	The ANOVA table is convenient for showing the degrees of freedom, treatment mean square, error mean square, their ratio, *F*-statistic, and its P-value. There are usually other quantities of lesser interest included as well.
Blind, Blinding	Any individual associated with an experiment who is not aware of how subjects have been allocated to treatment groups is said to be blinded.
Blocking, Blocking Factor	When groups of experimental units are similar, it is often a good idea to gather them together into the same level of a factor. The factor is called a blocking factor and its levels are called blocks. By blocking we isolate the variability attributable to the differences between the blocks so that we can see the differences in the means due to the treatments more clearly.
Bonferroni method	One of many methods for adjusting the margin of error to control the overall risk of making a Type I error when testing many pairwise differences between group means.
Confounded	When a factor is associated with another factor in such a way that their effects cannot be separated, we say that these two factors are confounded.
Control	When we limit the levels of a factor not explicitly part of the experiment design, we have controlled that factor. (By contrast, the factors we are testing are said to be *manipulated*.)
Control group	The experimental units assigned to a baseline treatment level, typically either the default treatment, which is well understood, or a null, placebo treatment. Their responses provide a basis for comparison.
Designs	• **Randomized block design:** The randomization occurs only within blocks. • **Completely randomized design:** All experimental units have an equal chance of receiving any treatment. • **Factorial design:** Includes more than one factor in the same design and includes every combination of all the levels of each factor.
Double-blind, Single-blind	There are two classes of individuals who can affect the outcome of an experiment: those who could *influence the results* (subjects, treatment administrators, or technicians) those who *evaluate the results* (judges, treating physicians, etc.) When every individual in *either* of these classes is blinded, an experiment is said to be single-blind. When everyone in *both* classes is blinded, we call the experiment double-blind.
Experiment	An experiment *manipulates* factor levels to create treatments, *randomly assigns* subjects to these treatment levels, and then *compares* the responses of the subject groups across treatment levels.
Experimental units	Individuals on whom an experiment is performed. Usually called subjects or participants when they are human.

F-distribution	The *F*-distribution is the sampling distribution of the *F*-statistic when the null hypothesis that the treatment means are equal is true. The *F* is the ratio of two estimates of variance (mean squares), which are equal when the null hypothesis is true. It has two degrees of freedom parameters, corresponding to the degrees of freedom for the mean squares in the numerator and denominator respectively.
F-statistic	The *F*-statistic for one-way ANOVA is the ratio MST/MSE. When the *F*-statistic is sufficiently large, we reject the null hypothesis that the group means are equal.
F-test	The *F*-test tests the null hypothesis that all the group means are equal against the one-sided alternative that they are not all equal. We reject the hypothesis of equal means if the *F*-statistic exceeds the critical value from the *F*-distribution corresponding to the specified significance level and degrees of freedom.
Factor	A variable whose levels are controlled by the experimenter. Experiments attempt to discover the effects that differences in factor levels may have on the responses of the experimental units.
Interaction	When the effects of the levels of one factor change depending on the level of the other factor, the two factors are said to interact. When interaction terms are present, it is misleading to talk about the main effect of one factor because how large it is *depends* on the level of the other factor.
Interaction plot	A plot that shows the means at each treatment combination, highlighting the factor effects and their behavior at all the combinations.
Level	The specific values that the experimenter chooses for a factor are called the levels of the factor.
Mean Square	A sum of squares divided by its associated degrees of freedom. • **Mean Square due to Error (MSE)** The estimate of the error variance obtained by pooling the variance of each treatment group. The square root of the MSE is the estimate of the error standard deviation, s_p. • **Mean Square due to Treatment (MST)** The estimate of the error variance under the null hypothesis that the treatment means are all equal. If the null hypothesis is not true, the expected value of the MST will be larger than the error variance.
Multiple comparisons	If we reject the null hypothesis of equal means, we often then want to investigate further and compare pairs of treatment group means to see if the corresponding population means differ. If we want to test several such pairs, we must adjust for performing several tests to keep the overall risk of a Type I error from growing too large. Such adjustments are called methods for multiple comparisons.
Observational study	A study based on data in which no manipulation of factors has been employed.
Placebo	A treatment that mimics the treatment to be studied, designed so that all groups think they are receiving the same treatment. Many subjects respond to such a treatment (a response known as a *placebo effect*). Only by comparing with a placebo can we be sure that the observed effect of a treatment is not due simply to the placebo effect.
Placebo effect	The tendency of many human subjects (often 20% or more of experiment subjects) to show a response even when administered a placebo.
Principles of experimental design	• **Control** aspects of the experiment that we know may have an effect on the response, but that are not the factors being studied. • **Randomize** subjects to treatments to even out effects that we cannot control. • **Replicate** over as many subjects as possible. Results for a single subject are just anecdotes. • **Block** to reduce the effects of identifiable attributes of the subjects that cannot be controlled.

Prospective study	An observational study in which subjects are followed to observe future outcomes. Because no treatments are deliberately applied, a prospective study is not an experiment. Nevertheless, prospective studies typically focus on estimating differences among groups that might appear as the groups are followed during the course of the study.
Random assignment	To be valid, an experiment must assign experimental units to treatment groups at random. This is called random assignment.
Response	A variable whose values are compared across different treatments. In a randomized experiment, large response differences can be attributed to the effect of differences in treatment level.
Retrospective study	An observational study in which subjects are selected and then their previous conditions or behaviors are determined. Because retrospective studies are not based on random samples, they usually focus on estimating differences between groups or associations between variables.
Subjects or Participants	When the experimental units are people, they are usually referred to as Subjects or Participants.
Treatment	The process, intervention, or other controlled circumstance applied to randomly assigned experimental units. Treatments are the different levels of a single factor or are made up of combinations of levels of two or more factors.

Technology Help: Analysis of Variance

Most analyses of variance are performed with computers, and all statistics packages present the results in an ANOVA table much like the ones in the chapter. Technology also makes it easy to examine the side-by-side boxplots and check the residuals for violations of the assumptions and conditions. Statistics packages offer different choices among possible multiple comparisons methods. This is a specialized area. Get advice or read further if you need to choose a multiple comparisons method. There are two ways to organize data recorded for several groups. We can put all the response values in a single variable and use a second, "factor," variable to hold the group identities. This is sometimes called *stacked format.* The alternative is an unstacked format, placing the data for each group in its own column or variable. Then the variable identities become the group identifiers. Stacked format is necessary for experiments with more than one factor. Each factor's levels are named in a variable. Some packages can work with either format for simple one-factor designs, and some use one format for some things and the other for others. (Be careful, for example, when you make side-by-side boxplots; be sure to give the appropriate version of that command to correspond to the structure of your data.) Most packages offer to save residuals and predicted values and make them available for further tests of conditions. In some packages, you may have to request them specifically.

Some statistics packages have different commands for models with one factor and those with two or more factors. You must be alert to these differences when analyzing a two-factor ANOVA. It's not unusual to find ANOVA models in several different places in the same package. (Look for terms like "Linear Models.")

To compute a single-factor ANOVA:

- From the tools menu (or the Data Ribbon in Office 2007or 2010), select **Data Analysis**.
- Select **Anova Single Factor** from the list of analysis tools.
- Click the **OK** button.

- Enter the data range in the box provided.
- Check the **Labels in First Row** box, if applicable.

- Enter an alpha level for the *F*-test in the box provided.
- Click the **OK** button.

Comments

The data range should include two or more columns of data to compare. Unlike statistics packages, Excel expects each column of the data to represent a different level of the factor. However, it offers no way to label these levels. The columns need not have the same number of data values, but the selected cells must make up a rectangle large enough to hold the column with the most data values.

The Excel Data Analysis Add-in offers a two-way ANOVA "with and without replication." That command requires the data to be in a special format and cannot deal with unbalanced (i.e., unequal counts in treatment groups) data. See the Excel help files for specific instructions.

JMP

To compute a one-way ANOVA:

- From the **Analyze** menu, select **Fit Y by X**.
- Select variables: a quantitative Y, Response variable, and a categorical X, Factor variable.
- JMP opens the **Oneway** window.
- Click on the red triangle beside the heading, select **Display Options**, and choose **Boxplots**.
- From the same menu, choose the **Means/ANOVA t-test** command.
- JMP opens the one-way ANOVA output.

To compute a two-way ANOVA:

- From the **Analyze** menu, select **Fit Model**.
- Select variables and **Add** them to the **Construct Model Effects** box.
- To specify an interaction, select both factors and press the **Cross** button.
- Click **Run Model**.
- JMP opens a **Fit Least Squares** window.
- Click on the red triangle beside each effect to see the means plots for that factor. For the interaction term, this is the interaction plot.
- Consult JMP documentation for information about other features.

Comments

JMP expects data in "stacked" format with one continuous response and two nominal factor variables.

MINITAB

- Choose **ANOVA** from the Stat menu.
- Choose **One-way...** or **Two-way...** from the **ANOVA** submenu.
- In the dialog, assign a quantitative Y variable to the Response box and assign the categorical X factor(s) to the Factor box.
- In a two-way ANOVA, specify interactions.
- Check the **Store Residuals** check box.
- Click the **Graphs** button.
- In the ANOVA-Graphs dialog, select **Standardized residuals**, and check **Normal plot of residuals** and **Residuals versus fits**.
- Click the **OK** button to return to the ANOVA dialog.
- Click the **OK** button to compute the ANOVA.

Comments

If your data are in unstacked format, with separate columns for each treatment level, Minitab can compute a one-way ANOVA directly. Choose **One-way (unstacked)** from the **ANOVA** submenu. For two-way ANOVA, you must use the stacked format.

SPSS

To compute a one-way ANOVA:

- Choose **Compare Means** from the Analyze menu.
- Choose **One-way ANOVA** from the **Compare Means** submenu.
- In the One-Way ANOVA dialog, select the Y-variable and move it to the dependent target. Then move the X-variable to the independent target.
- Click the **OK** button.

To compute a two-way ANOVA:

- Choose **Analyze > General Linear Model > Univariate**.
- Assign the response variable to the **Dependent Variable** box.
- Assign the two factors to the **Fixed Factor(s)** box. This will fit the model with interactions by default.
- To omit interactions, click on **Model**. Select **Custom**. Highlight the factors. Select **Main Effects** under the **Build Terms** arrow and click the arrow.
- Click **Continue** and **OK** to compute the model.

Comments

SPSS expects data in stacked format. The **Contrasts** and **Post Hoc** buttons offer ways to test contrasts and perform multiple comparisons. See your SPSS manual for details.

Brief CASE

Design, carry out, and analyze your own multifactor experiment. The experiment doesn't have to involve human subjects. In fact, an experiment designed to find the best settings for microwave popcorn, the best paper airplane design, or the optimal weight and placement of coins on a toy car to make it travel farthest and fastest down an incline are all fine ideas. Be sure to define your response variable of interest before you start the experiment and detail how you'll perform the experiment, specifically including the elements you control, how you use randomization, and how many times you replicate the experiment. Analyze the results of your experiment and write up your analysis and conclusions including any recommendations for further testing.

Exercises

SECTION 21.1

1. For the following observational studies, indicate whether they are prospective or retrospective.

a) A company looked at a sample of returned registration cards to estimate the income level of households that purchased their product.

b) A retail outlet encouraged customers to join their "frequent buyers" program and studied whether those who joined were more likely to make use of discount coupons than those who were not members.

2. For the following observational studies, indicate whether they are prospective or retrospective studies.

a) An airline was concerned that new security measures might discourage air travelers. A year after the new security restrictions were put into place, the airlines compared the miles traveled by their frequent fliers before and after the change.

b) Does giving children a flu shot protect parents? Researchers questioned a random sample of families at the end of a flu season. They asked whether the children had been immunized, whether the parents had received flu shots, and who in the family had contracted the flu.

SECTION 21.2

3. For the following experiment, identify the experimental units, the treatments, the response, and the random assignment.

A commercial food lab compared recipes for chocolate chip cookies. They baked cookies with different kinds of chips (milk chocolate, dark chocolate, and semi-sweet). All other ingredients and amounts were the same. Ten trained tasters rated the cookies on a scale of 1 to 10. The cookies were presented to the tasters in a random order.

4. For the following experiment, identify the experimental units, the treatments, the response, and the random assignment.

An investment club decided to compare investment strategies. Starting with nine equal investment amounts, three invested in the "dogs of the Dow"—stocks in the Dow Industrial average that had been underperforming relative to the rest of the Dow average. The relative amounts to invest in each of the stocks were chosen randomly and differently for each fund. Three funds invested following the advice of a TV investment show host, again choosing the specific stocks and allocations randomly for the three funds. And three funds invested by throwing darts at a page from the *Wall Street Journal* that listed stocks on the NYSE, and invested in each of the stocks hit by a dart, throwing a different set of darts for each of the three funds. At the end of six months the funds were compared.

SECTION 21.3

5. For the cookie recipe experiment of Exercise 3, identify how Control, Randomization, and Replication were used.

6. For the investment experiment of Exercise 4, identify how Control, Randomization, and Replication were used.

SECTION 21.4

7. An Internet sale site randomly sent customers to one of three versions of its welcome page. It recorded how long each visitor stayed in the site.

Here is a diagram of that experiment. Fill in the parts of the experiment.

8. An Internet company was concerned that customers who came directly to their site (by typing their URL into a browser) might respond differently than those referred to the site from other sites (such as search engines). They decided to block according to how the customer arrived at their site.

Here is a diagram of that experiment. Fill in the parts.

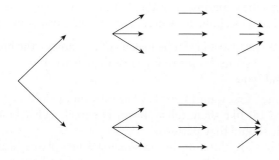

SECTION 21.5

9. For the following experiment, indicate whether it was single-blind, double-blind, or not blinded at all. Explain your reasoning.

Makers of a new frozen entrée arranged for it to be served to randomly selected customers at a restaurant in place of the equivalent entrée ordinarily prepared in the kitchen. After their meal, the customers were asked about the quality of the food.

10. For the following experiment, indicate whether it was single-blind, double-blind, or not blinded at all. Explain your reasoning.

Does a "stop smoking" program work better if it costs more? Smokers responding to an advertisement offering to help them stop smoking were randomly offered a program costing $100 or the same program costing $250. The offer was made individually to each client by presenting a sealed envelope so the clerk providing the offer did not know the details of the offer. At the end of the program (a course and films along with diet and smoking cessation aids), clients were followed for six months to see if they had indeed quit smoking.

SECTION 21.6

11. In a completely randomized design, ten subjects were assigned to each of four treatments of a factor. Below is the partially completed ANOVA table.

Source	DF	Sum of Squares	Mean Square	F-Ratio	P-Value
Treatment (Between)		856.07			
Error (Within)					
Total		1177.97			

a) What are the degrees of freedom for treatment, error, and total?
b) What is SSE?
c) What is MST?
d) What is MSE?

12. Refer to Exercise 11.

a) State the null and alternative hypotheses.
b) Calculate the *F*-ratio.
c) What is the P-value?
d) State your conclusion at $\alpha = .05$.

SECTION 21.10

13. In the experiment described in Exercise 3, in fact the study also compared the use of butter or margarine in the recipes. The design was balanced, with each combination of chip type and oil type tested.

a) What were the factors and factor levels?
b) What were the treatments?
c) If an interaction was found to be significant, what would that mean?

14. The investment club described in Exercise 4 decided to repeat their experiment in a different way. Three members of the club took responsibility for one of each of the three investment "strategies," making the final choices and allocations of investment dollars. For this new experiment:

a) What were the subjects?
b) What were the factors and factor levels?
c) What were the treatments?

CHAPTER EXERCISES

15. Laundry detergents. A consumer group wants to test the efficacy of a new laundry detergent. They take 16 pieces of white cloth and stain each with the same amount of grease. They decide to try it using both hot and cold water settings and at both short and long washing times. Half of the 16 pieces will get the new detergent, and half will get a standard detergent. They'll compare the shirts by using an optical scanner to measure whiteness.

a) What are the factors they are testing?
b) Identify all the factor levels.
c) What is/are the response(s)?

16. Sales scripts. An outdoor products company wants to test a new website design where customers can get information about their favorite outdoor activity. They randomly send half of the customers coming to the website to the new design. They want to see whether the Web visitors spend more time at the site and whether they make a purchase.

a) What are the factors they are testing?
b) Identify all the factor levels.
c) What is/are the response(s)?

17. Laundry detergents, part 2. One member of the consumer group in Exercise 15 is concerned that the experiment will take too long and makes some suggestions to shorten it. Comment briefly on each idea.

a) Cut the runs to 8 by testing only the new detergent. Compare the results to results on the standard detergent published by the manufacturer.
b) Cut the runs to 8 by testing only in hot water.
c) Keep the number of runs at 16, but save time by running all the standard detergent runs first to avoid swapping detergents back and forth.

18. Swimsuits. A swimsuit manufacturer wants to test the speed of its newly designed $550 suit. They design an experiment by having 6 randomly selected Olympic swimmers swim as fast as they can with their old swimsuit first and then swim the same event again with the new, expensive swim suit. They'll use the difference in times as the response variable. Criticize the experiment and point out some of the problems with generalizing the results.

19. Mozart. Will listening to a Mozart piano sonata make you smarter? In a 1995 study, Rauscher, Shaw, and Ky reported that when students were given a spatial reasoning section of a standard IQ test, those who listened to Mozart for 10 minutes improved their scores more than those who simply sat quietly.

a) These researchers said the differences were statistically significant. Explain what that means in this context.
b) Steele, Bass, and Crook tried to replicate the original study. The subjects were 125 college students who participated in the experiment for course credit. Subjects first took the test. Then they were assigned to one of three groups: listening to a Mozart piano sonata, listening to music by Philip Glass, and sitting for 10 minutes in silence. Three days after the treatments, they were retested. Draw a diagram displaying the design of this experiment.
c) The boxplots show the differences in score before and after treatment for the three groups. Did the Mozart group show improvement?
d) Do you think the results prove that listening to Mozart is beneficial? Explain.

20. More Mozart. An advertisement selling specially designed CDs of Mozart's music specifically because they will "strengthen your mind, heal your body, and unlock your creative spirit" claims that "in Japan, a brewery actually reported that their best sake is made when Mozart is played near the yeast." Suppose you wished to design an experiment to test whether this is true. Assume you have the full cooperation of the sake brewery. Specify how you would design the experiment. Indicate factors and response and how they would be measured, controlled, or randomized.

21. Cereal marketing. The makers of Frumpies, "the breakfast of rug rats," want to improve their marketing, so they consult you.

a) They first want to know what fraction of children, ages 10 to 13, like their celery-flavored cereal. What kind of study should they perform?
b) They are thinking of introducing a new flavor, maple-marshmallow Frumpies and want to know whether children will prefer the new flavor to the old one. Design a completely randomized experiment to investigate this question.
c) They suspect that children who regularly watch the Saturday morning cartoon show starring Frump, the flying teenage warrior rabbit who eats Frumpies in every episode, may respond differently to the new flavor. How would you take that into account in your design?

22. Wine marketing. A 2001 Danish study published in the *Archives of Internal Medicine* casts significant doubt on suggestions that adults who drink wine have higher levels of "good" cholesterol and fewer heart attacks. These researchers followed a group of individuals born at a Copenhagen hospital between 1959 and 1961 for 40 years. Their study found that in this group the adults who drank wine were richer and better educated than those who did not.

a) What kind of study was this?
b) It is generally true that people with high levels of education and high socioeconomic status are healthier than

others. How does this call into question the supposed health benefits of wine?

c) Can studies such as these prove causation (that wine helps prevent heart attacks, that drinking wine makes one richer, that being rich helps prevent heart attacks, etc.)? Explain.

23. SAT prep courses. Can special study courses actually help raise SAT scores? One organization says that the 30 students they tutored achieved an average gain of 60 points when they retook the test.

a) Explain why this does not necessarily prove that the special course caused the scores to go up.

b) Propose a design for an experiment that could test the effectiveness of the tutorial course.

c) Suppose you suspect that the tutorial course might be more helpful for students whose initial scores were particularly low. How would this affect your proposed design?

24. Safety switch. An industrial machine requires an emergency shutoff switch that must be designed so that it can be easily operated with either hand. Design an experiment to find out whether workers will be able to deactivate the machine as quickly with their left hands as with their right hands. Be sure to explain the role of randomization in your design.

T 25. Cars (fuel efficiency). These boxplots show the relationship between the number of cylinders in a car's engine and its fuel economy from a study conducted by a major car manufacturer.

a) What are the null and alternative hypotheses? Talk about cars and fuel efficiency, not symbols.

b) Do the conditions for an ANOVA seem to be met here? Why or why not?

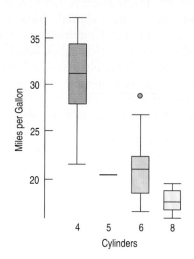

T 26. Wine production. The boxplots display case prices (in dollars) of wine produced by wineries along three of the Finger Lakes in upstate New York.

a) What are the null and alternative hypotheses? Talk about prices and location, not symbols.

b) Do the conditions for an ANOVA seem to be met here? Why or why not?

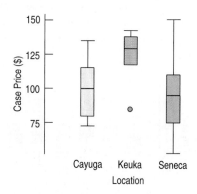

27. Cell phone adoption. Cell phone adoption rates are available for various countries in the United Nations Database (unstats.un.org). Countries were randomly selected from three regions (Africa, Asia, and Europe), and cell phone adoption (per 100 inhabitants) rates retrieved. The boxplots display the data.

a) What are the null and alternative hypotheses (in words, not symbols)?

b) Are the conditions for ANOVA met? Why or why not?

T 28. Marketing managers' salaries. A sample of eight states was selected randomly from each of three regions in the United States (Northeast, Southeast, and West). Mean annual salaries for marketing managers were retrieved from the U.S. Bureau of Labor Statistics (data.bls.gov/oes). The boxplots display the data.

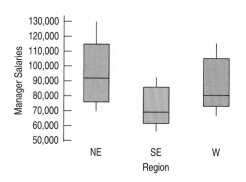

a) What are the null and alternative hypotheses (in words, not symbols)?

b) Are the conditions for ANOVA met? Why or why not?

29. Bank tellers. A bank is studying the average time that it takes 6 of its tellers to serve a customer. Customers line up in the queue and are served by the next available teller. Here is a boxplot of the times it took to serve the last 140 customers.

Source	DF	Sum of Squares	Mean Square	F-ratio	P-value
Teller	5	3315.32	663.064	1.508	0.1914
Error	134	58919.1	439.695		
Total	139	62234.4			

a) What are the null and alternative hypotheses?

b) What do you conclude?

c) Would it be appropriate to run a multiple comparisons test (for example, a Bonferroni test) to see which tellers differ from each other? Explain.

30. Hearing. Vendors of hearing aids test them by having patients listen to lists of words and repeat what they hear. The word lists are supposed to be equally difficult to hear accurately. But the challenge of hearing aids is perception when there is background noise. A researcher investigated four different word lists used in hearing assessment (Loven, 1981). She wanted to know whether the lists were equally difficult to understand in the presence of a noisy background. To find out, she tested 24 subjects with normal hearing and measured the number of words perceived correctly in the presence of background noise. Here are the boxplots of the four lists.

Source	DF	Sum of Squares	Mean Square	F-ratio	P-value
List	3	920.4583	306.819	4.9192	0.0033
Error	92	5738.1667	62.371		
Total	95	6658.6250			

a) What are the null and alternative hypotheses?

b) What do you conclude?

c) Would it be appropriate to run a multiple comparisons test (for example, a Bonferroni test) to see which lists differ from each other in terms of mean percent correct? Explain.

31. E-security. A report released by the Pew Internet & American Life Project entitled *The Internet & Consumer Choice* focused on current online issues. Respondents were asked to indicate their level of agreement (1 = strongly agree to 4 = strongly disagree) with a variety of statements including "I don't like giving my credit card number or personal information online." A part of the data set was used to determine whether the type of community in which the individual resides (Urban, Suburban, or Rural) affected responses. Here are the results in the form of a partially completed Analysis of Variance table.

Source	DF	Sum of Squares	Mean Square	F-ratio	P-value
Community	2	6.615			
Error	183	96.998			
Total	185	103.613			

a) Is this an experimental or observational study? Explain.

b) Is this a prospective or retrospective study? Explain.

c) State the null and alternative hypothesis.

d) Calculate the *F*-statistic.

e) The P-value for this statistic turns out to be 0.002. State the conclusion. Can a causal link be established? Explain.

32. Internet usage. Internet usage rates are available for various countries in the United Nations Common Database (unstats. un.org). Countries were randomly selected from three regions (Africa, Asia, and Europe), and Internet usage (per 100 inhabitants) data from 2005 were retrieved. The data were analyzed to determine if Internet usage rates were the same across regions. The partially completed Analysis of Variance table is shown here.

Source	DF	Sum of Squares	Mean Square	F-ratio	P-value
Region	2	21607			
Error	93	20712			
Total	95	42319			

a) Is this an experimental or observational study? Explain.

b) Is this a prospective or retrospective study? Explain.

c) State the null and alternative hypotheses.

d) Calculate the *F*-statistic.

e) The P-value for this statistic turns out to be <0.001. State the conclusion. Can a causal link be established? Explain.

33. Colorful experiment? In a recent article published in *Quality Progress*, each student in a statistics class had a randomly assigned bag of peanut M&M's® and counted the number of each color (*Blue, Red, Orange, Green, Brown, Yellow*). The bags were all the same size (1.74 ounces). The investigators claimed to use a randomized block design, with *Bag* as the blocking factor. They counted the number of candies of each color in each bag. Their results are reproduced here (Lin, T., and Sanders, M.S., "A Sweet Way to Learn DOE," *Quality Progress*, Feb. 2006, p. 88).

Source	Degrees of Freedom	Sum of Squares	Mean Square	F-ratio	P-value
Bag	13	4.726	0.364	0.10	1.000
Color	5	350.679	70.136	18.72	<0.001
Error	65	243.488	3.746		
Total	83	598.893			

a) Was this an observational or experimental study?
b) What was the treatment? What factors were manipulated?
c) What was the response variable?

34. Six Sigma training. A large financial institution is interested in training its college educated workforce in Six Sigma principles and methods. One part of the training involves basic statistical concepts and tools. Management is considering three approaches: *online, traditional classroom*, and *hybrid (a mix of both)*. Prior to launching the program throughout the entire organization, they decided to pilot test the three approaches. Because they believed that educational background may affect the results, they selected 3 employees from each of 10 different college major programs of study (*liberal arts, accounting, economics, management, marketing, finance, information systems, computer science, operations, other*), and randomly assigned each to one of the three approaches. At the end of training, each participant took an exam. The results are shown here.

Source	Degrees of Freedom	Sum of Squares	Mean Square	F-ratio	P-value
Major	9	2239.47	248.830	21.69	<0.001
Training	2	171.47	85.735	7.47	0.004
Error	18	206.53	11.474		
Total	29	2617.47			

a) Was this an observational study or an experiment?
b) What was the purpose of using Major as a blocking factor?
c) Given the results, was it necessary to use Major as a blocking factor? Explain.
d) State the conclusion from this analysis.

35. E-trust. Online retailers want customers to trust their websites and want to alleviate any concerns potential customers may have about privacy and security. In a study investigating the factors that affect e-trust, participants were randomly assigned to carry out online transactions on fictitious retailers' websites. The sites were configured in one of three ways: (1) *with a third-party assurance seal (e.g., BBBOnLine) displayed*; (2) *a self-proclaimed assurance displayed*; or (3) *no assurance*. In addition, participants made a transaction involving one of three products (*book, camera, or insurance*). These products represent varying degrees of risk. After completing the transaction, they rated how "trustworthy" the website was on a scale of 1 (not at all) to 10 (extremely trustworthy).

a) Is this an experiment or an observational study? Explain.
b) What is the response variable?
c) How many factors are involved?
d) How many treatments are involved?
e) State the hypotheses (in words, not symbols).

36. Injection molding. In order to improve the quality of molded parts, companies often test different levels of parameter settings in order to find the best combinations. Injection molding machines typically have many adjustable parameters. One company used three different mold temperatures (25, 35, and 45 degrees Celsius) and four different cooling times (10, 15, 20, and 25 minutes) to examine how they affect the tensile strength of the resulting molded parts. Five parts were randomly sampled and measured from each treatment combination.

a) Is this an experiment or an observational study? Explain.
b) What is the response variable?
c) What are the factors?
d) How many treatments are involved?
e) State the hypotheses (in words, not symbols).

37. Stock returns. Companies that are ISO 9000 certified have met standards that ensure they have a quality management system committed to continuous improvement. Going through the certification process generally involves a substantial investment that includes the hiring of external auditors. A group of such auditors, wishing to "prove" that ISO 9000 certification pays off, randomly selected a sample of small and large companies with and without ISO 9000 certification. Size was based on the number of employees. They computed the % change in closing stock price from August 2006 to August 2007. The two-way ANOVA results are presented here (data obtained from *Yahoo! Finance*).

Source	DF	Sum of Squares	Mean Square	F-ratio	P-value
ISO 9000	1	2654.4	2654.41	5.78	0.022
Size	1	0.2	0.18	0.004	0.984
Interaction	1	1505.5	1505.49	3.28	0.079
Error	36	16545.9	459.61		
Total	39	20705.9			

a) Is this an experiment or an observational study?
b) State the hypotheses.
c) Given the small P-value associated with the ISO 9000 factor and that the mean annual return for the companies

with ISO 9000 is 30.7% compared to 14.4% for those without, the auditors state that achieving ISO 9000 certification results in higher stock prices. Do you agree with their statement? Explain.

38. Company bonuses. After complaints about gender discrimination regarding bonus incentive pay, a large multinational firm collected data on bonuses awarded during the previous year (% of base pay). Human Resources (HR) randomly sampled male and female managers from three different levels: *senior, middle,* and *supervisory.* The two-way ANOVA results are presented here.

Source	DF	Sum of Squares	Mean Square	F-ratio	P-value
Gender	1	32.033	32.033	9.76	0.005
Level	2	466.200	233.100	70.99	0.000
Interaction	2	20.467	10.233	3.12	0.063
Error	24	78.800	3.283		
Total	29	597.500			

a) Is this an experiment or an observational study?
b) State the hypotheses.
c) Given the small P-value associated with the gender and that the mean annual bonus percent for females is 12.5% compared to 14.5% for males, HR concludes that gender discrimination exists. Do you agree? Explain.

39. Managers' hourly wages. What affects marketing managers' hourly wages? In order to find out, mean hourly wages were retrieved from the U.S. Bureau of Labor Statistics for two managerial occupations in marketing (*Sales managers, Advertising managers*) for a random sample of states from three regions (*Midwest, Southeast, West*) (www .bls.gov/data/#wages). Here are boxplots showing mean hourly wages for the two marketing occupations and the three regions as well as the results for a two-way ANOVA.

Source	DF	Sum of Squares	Mean Square	F-ratio	P-value
Manager Type	1	1325.93	1325.93	31.84	0.000
Region	2	153.55	76.78	1.84	0.176
Interaction	2	32.74	16.37	0.39	0.678
Error	30	1249.32	41.64		
Total	35	2761.55			

a) Is this an experiment or an observational study? Explain.
b) Are the conditions for two-way ANOVA met?
c) If so, perform the hypothesis tests and state your conclusions in terms of hourly wages, occupational type, and region.
d) Is it appropriate to interpret the main effects in this case? Explain.

40. Concrete testing. A company that specializes in developing concrete for construction strives to continually improve the properties of its materials. In order to increase the compressive strength of one of its new formulations, they varied the amount of alkali content (*low, medium, high*). Since the type of sand used may also affect the strength of concrete, they used three different types of sand (Types I, II, III). Four samples were randomly selected from each treatment combination to be tested. The boxplots show the test results on compressive strength (in psi) for the three levels of alkali content and three types of sand. Two-way ANOVA results are also given.

Source	DF	Sum of Squares	Mean Square	F-ratio	P-value
Alkali Content	2	4016600	2008300	46.38	0.000
Sand Type	2	1547817	773908	17.87	0.000
Interaction	4	177533	44383	1.02	0.412
Error	27	1169250	43306		
Total	35	6911200			

a) Is this an experiment or an observational study? Explain.
b) Are the conditions for two-way ANOVA met?
c) If so, perform the hypothesis tests and state your conclusions in terms of compressive strength, alkali content, and sand type.
d) Is it appropriate to interpret the main effects in this case? Explain.

T 41. Production problems. A manufacturing company that makes dental drills was experiencing problems with a specific part on the production line. Management suspected a machining problem that resulted in the length of the part to vary outside of target specification. Two factors were examined: the machine setting (at three levels) and the shift (morning, afternoon, and night). New hires were typically scheduled for night shift, and management believed that their relative inexperience may also be contributing to the variation. Three parts were randomly selected and measured from each treatment combination. The deviation from specified size was measured in microns. The data and two-way ANOVA results are shown.

Size Error	Machine Setting	Shift
1.1	1	Morning
3.6	2	Morning
3.3	3	Morning
2.1	1	Morning
0.9	2	Morning
2.6	3	Morning
0.6	1	Morning
2.3	2	Morning
3.2	3	Morning
2	1	Afternoon
2.4	2	Afternoon
5	3	Afternoon
1.8	1	Afternoon
4.3	2	Afternoon
3.2	3	Afternoon
2.5	1	Afternoon
5	2	Afternoon
2.3	3	Afternoon
3.8	1	Night
5.5	2	Night
5	3	Night
2.9	1	Night
6.7	2	Night
5.8	3	Night
2.8	1	Night
3	2	Night
5.3	3	Night

Source	DF	Sum of Squares	Mean Square	F-ratio	P-value
MachSet	2	17.1119	8.55593	7.3971	0.0045
Shift	2	24.9607	12.4804	10.790	0.0008
Interaction	4	1.4970	0.374259	0.32357	0.8585
Error	18	20.8200	1.15667		
Total	26	64.3896			

a) Is this an experiment or an observational study? Explain.
b) What is the response variable?
c) How many treatments are involved?
d) Based on the two-way ANOVA results, management concluded that shift has a significant impact on the length of the part and that consequently operator inexperience is the root cause of the part problems. Do you agree with this conclusion? Explain.

T 42. Process improvements. One way to improve a process is to eliminate non–value-added activities (e.g., extra movements) and wasted effort (e.g., looking for materials). A consultant was hired to improve the efficiency in a large shop floor operation. She tested three different workspace designs and two different storage/retrieval systems. She measured process flow time for three randomly selected operations through each of the combinations of workspace design and storage/retrieval systems. The data and two-way ANOVA results are shown here.

Workspace Design	Storage System	Flow Time (Days)
1	1	4.5
2	1	3.3
3	1	3.4
1	1	4.0
2	1	3.0
3	1	2.9
1	1	4.2
2	1	3.0
3	1	3.2
1	1	4.5
2	1	3.5
3	1	3.2
1	1	3.8
2	1	2.8
3	1	3.0
1	2	3.0
2	2	3.8
3	2	3.6
1	2	2.8
2	2	4.0
3	2	3.5
1	2	3.0
2	2	3.5
3	2	3.8
1	2	4.0
2	2	4.2
3	2	4.2
1	2	3.0
2	2	3.6
3	2	3.8

Source	DF	Sum of Squares	Mean Square	F-ratio	P-value
Workspace					
Design	2	0.30867	0.15433	1.56	0.230
Storage					
System	1	0.07500	0.07500	0.76	0.392
Interaction	2	4.87800	2.43900	24.72	<0.001
Error	24	2.36800	0.09867		
Total	29	7.62967			

a) Is this an experiment or observational study? Explain.
b) What is the response variable?
c) How many treatments are involved?
d) Based on the two-way ANOVA results, management concludes that neither the workspace design nor the storage/retrieval system impacts process flow time (and that the consultant wasn't worth the money). Do you agree with this conclusion? Explain.

43. Yogurt research. An experiment to determine the effect of several methods of preparing cultures for use in commercial yogurt was conducted by a food science research group. Three batches of yogurt were prepared using each of three methods: traditional, ultrafiltration, and reverse osmosis. A trained expert then tasted each of the 9 samples, presented in random order, and judged them on a scale from 1 to 10. A partially complete Analysis of Variance table of the data follows.

Source	Sum of Squares	Degrees of Freedom	Mean Square	F-ratio
Treatment	17.300			
Residual	0.460			
Total	17.769			

a) Calculate the mean square of the treatments and the mean square of the error.
b) Form the F-statistic by dividing the two mean squares.
c) The P-value of this F-statistic turns out to be 0.000017. What does this say about the null hypothesis of equal means?
d) What assumptions have you made in order to answer part c?
e) What would you like to see in order to justify the conclusions of the F-test?
f) What is the average size of the error standard deviation in the judge's assessment?

44. Smokestack scrubbers. Particulate matter is a serious form of air pollution often arising from industrial production. One way to reduce the pollution is to put a filter, or scrubber, at the end of the smokestack to trap the particulates. An experiment to determine which smokestack scrubber design is best was run by placing four scrubbers of different designs on an industrial stack in random order. Each scrubber was tested 5 times. For each run, the same material was produced, and the particulate emissions coming out of the scrubber were measured (in parts per billion).

A partially complete Analysis of Variance table of the data is shown here.

Source	Sum of Squares	Degrees of Freedom	Mean Square	F-ratio
Treatment	81.2			
Residual	30.8			
Total	112.0			

a) Calculate the mean square of the treatments and the mean square of the error.
b) Form the F-statistic by dividing the two mean squares.
c) The P-value of this F-statistic turns out to be 0.00000949. What does this say about the null hypothesis of equal means?
d) What assumptions have you made in order to answer part c?
e) What would you like to see in order to justify the conclusions of the F-test?
f) What is the average size of the error standard deviation in particulate emissions?

45. Cereal shelf placement. Supermarkets often place similar types of cereal on the same supermarket shelf. The shelf placement for 77 cereals was recorded as their sugar content. Does sugar content vary by shelf? Here's a boxplot and an ANOVA table.

Source	DF	Sum of Squares	Mean Square	F-ratio	P-value
Shelf	2	248.4079	124.204	7.3345	0.0012
Error	74	1253.1246	16.934		
Total	76	1501.5325			

Level	n	Mean	StdDev
1	20	4.80000	4.57223
2	21	9.61905	4.12888
3	36	6.52778	3.83582

a) What kind of design or study is this?
b) What are the null and alternative hypotheses?
c) What does the ANOVA table say about the null hypothesis? (Be sure to report this in terms of sugar content and shelf placement.)

Dependent Variable: SUGARS (Exercise 45)

Bonferroni	(I) SHELF	(J) SHELF	Mean Difference (I-J)	Std. Error	P-value	95% Confidence Interval	
						Lower Bound	Upper Bound
	1	2	−4.819	1.2857	0.001	−7.969	−1.670
		3	−1.728	1.1476	0.409	−4.539	1.084
	2	1	4.819	1.2857	0.001	1.670	7.969
		3	3.091	1.1299	0.023	0.323	5.859
	3	1	1.728	1.1476	0.409	−1.084	4.539
		2	−3.091	1.1299	0.023	−5.859	−0.323

d) Can we conclude that cereals on shelf 2 have a different mean sugar content than cereals on shelf 3? Can we conclude that cereals on shelf 2 have a different mean sugar content than cereals on shelf 1? What can we conclude?

e) To check for significant differences between the shelf means, we can use a Bonferroni test, whose results are shown here. For each pair of shelves, the difference is shown along with its standard error and significance level. What does it say about the questions in part d?

46. Cereal shelf placement, part 2. We also have data on the protein content on the 77 cereals in Exercise 45. Does protein content vary by shelf? Here's a boxplot and an ANOVA table.

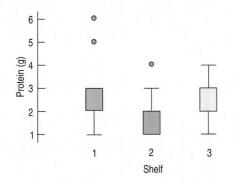

Source	DF	Sum of Squares	Mean Square	F-ratio	P-value
Shelf	2	12.4258	6.2129	5.8445	0.0044
Error	74	78.6650	1.0630		
Total	76	91.0909			

Level	n	Mean	StdDev
1	20	2.65000	1.46089
2	21	1.90476	0.99523
3	36	2.86111	0.72320

a) What kind of design or study is this?

b) What are the null and alternative hypotheses?

c) What does the ANOVA table say about the null hypothesis? (Be sure to report this in terms of protein content and shelf placement.)

d) Can we conclude that cereals on shelf 2 have a lower mean protein content than cereals on shelf 3? Can we conclude that cereals on shelf 2 have a lower mean protein content than cereals on shelf 1? What can we conclude?

e) To check for significant differences between the shelf means, we can use a Bonferroni test, whose results are shown here. For each pair of shelves, the difference is shown along with its standard error and significance level. What does it say about the questions in part d?

Dependent Variable: PROTEIN
Bonferroni

(I) SHELF	(J) SHELF	Mean Difference (I-J)	Std. Error	P-value	95% Confidence Interval	
					Lower Bound	Upper Bound
1	2	0.75	0.322	0.070	−0.04	1.53
	3	−0.21	0.288	1.000	−0.92	0.49
2	1	−0.75	0.322	0.070	−1.53	0.04
	3	−0.96	0.283	0.004	−1.65	−0.26
3	1	0.21	0.288	1.000	−0.49	0.92
	2	0.96	0.283	0.004	0.26	1.65

47. Automotive safety. The National Highway Transportation Safety Administration runs crash tests in which stock automobiles are crashed into a wall at 35 mph with dummies in both the passenger and the driver's seats. The THOR Alpha crash dummy is capable of recording 134 channels of data on the impact of the crash at various sites on the dummy. In this test, 335 cars are crashed. The response variable is a measure of head injury. Researchers want to know whether the seat the dummy is sitting in affects head injury severity, as well as whether the type of car affects severity. Here are boxplots for the 2 different *Seats* (*driver, passenger*) and the 6 different *Size* classifications (*compact, light, medium, mini, pickup, van*).

An interaction plot shows:

A scatterplot of residuals vs. predicted values shows:

The ANOVA table follows:

Source	DF	Sum of Squares	Mean Square	F-ratio	P-value
Seat	1	0.88713	0.88713	25.501	<0.0001
Size	5	1.49253	0.29851	8.581	<0.0001
Seat × Size	5	0.07224	0.01445	0.415	0.838
Error	282	9.8101	0.03479		
Total	293	12.3853			

a) State the null hypotheses about the main effects (in words, not symbols).
b) Are the conditions for two-way ANOVA met?
c) If so, perform the hypothesis tests and state your conclusion. Be sure to state it in terms of head injury severity, seats, and vehicle types.

48. Gas additives. An experiment to test a new gasoline additive, *Gasplus*, was performed on three different cars: a sports car, a minivan, and a hybrid. Each car was tested with both *Gasplus* and regular gas on 10 different occasions, and their gas mileage was recorded. Here are the boxplots.

A two-way ANOVA with interaction model was run, and the following ANOVA table resulted.

Source	DF	Sum of Squares	Mean Square	F-ratio	P-value
Type	2	23175.4	11587.7	2712.2	<0.0001
Additive	1	92.1568	92.1568	21.57	<0.0001
Interaction	2	51.8976	25.9488	6.0736	0.0042
Error	54	230.711	4.27242		
Total	59	23550.2			

A plot of the residuals vs. predicted values showed:

What conclusions about the additive and car types do you draw? Do you see any potential problems with the analysis?

Just Checking Answers

1   Gather reports from veterinarians and pet hospitals. Look into the histories of sick animals. This would be a retrospective observational study.

2   Treatment: Feed the new food and a previously tested, known to be safe, food to pets.

   Response: Judge the health of the animals, possibly by having a veterinarian examine them before and after the feeding trials.

3   Control by choosing similar animals. Perhaps choose just one breed and test animals of the same age and health. Treat them otherwise the same in terms of exercise, attention, and so on.

   Randomize by assigning animals to treatments at random.

   Replicate by having more than one animal fed each formulation.

4   A control group could be fed a standard laboratory food, if we have one known to be safe. Otherwise we could prepare a special food in our test kitchens to be certain of its safety.

5   The veterinarian evaluating the animals should be blind to the treatments. For double-blinding, all technicians handling the animals should also be blinded. That would require making the control food look as much like the test food as possible.

6   Yes. Test dogs and cats separately.

7   No. We have failed to reject the hypothesis of a difference, but that's all we can conclude. There is insufficient evidence to discern any difference. But that should be sufficient for the company's purposes.

Quality Control

Sony

In 1945, at the end of World War II, Masaru Ibuka started a radio repair shop in a bombed-out building in Tokyo. The next year, Akio Morita joined him and together they founded a company called Tokyo Tsushin Kogyo K.K. (or Tokyo Communication Engineering Company), which built Japan's first tape recorder. In the early 1950s, Ibuka traveled to the United States and convinced Bell Labs to license its transistor technology to his Japanese company, which then released Japan's first commercially produced transistor radio. The next May, the company released a new radio with a slim design and improved sound quality. In 1957, it came out with the TR-63—the smallest transistor radio in commercial production. By all accounts, it was a worldwide commercial success.

With increasing international sales, Ibuka and Morita knew they needed a simpler name for their company. Unfortunately, the name Totsuko didn't work well in English, and their initials, TKK, were already in use by a railroad company, so Ibuka and Morita worked to find a word that didn't exist in any language but would be easy to remember. Finally, they combined the Latin word for sound, *sonus*, with the American word *sonny*, which was a popular word meaning boy or young man in the 1950s. From this Sony was born in 1958.

Sony is now one of the world's most recognized brands with sales of nearly $90 billion for the fiscal year ending March 31, 2008. It employs more than 170,000 people worldwide.

あなたに本当に
「ちょうどいい」
クルマ。
いままで
ありましたか？

Unfortunately for the young Sony Corporation, in the 1950s, the phrase "Made in Japan" was synonymous with cheap and shoddily made. Japan's manufacturing capabilities had been destroyed in the latter part of World War II. To rebuild its economy as quickly as possible, Japan started producing and exporting large quantities of inexpensive goods, including electronic products. A rumor often heard in the United States after the war was that Japan had renamed one of its towns Usa so that it could stamp "Made in Usa" on its products. Although this urban legend was false (in fact the town of Usa on the island of Kyushu existed long before WWII) the shoddy quality of the Japanese goods gave credence to the story and kept it alive for years.

The United States sent General Douglas MacArthur to oversee the rebuilding of Japan after the war ended. Realizing the importance of reversing Japan's poor quality image, he recruited two key players, W. Edwards Deming and Joseph Juran, who had both been active in the field of quality control in the United States. In addition, he enlisted the help of Homer Sarasohn to help rebuild the communication infrastructure of Japan.

Just after World War II, Japanese manufacturing plants were designed to produce radios using vacuum tube technology. In early trial runs, Sarasohn was shocked to find that yields were commonly less than 10%. In other words, 90% of the vacuum tubes produced were defective. Even more shocking was that the Japanese were not surprised and had little understanding of quality; other than the Zero fighter (known as the best carrier-based fighter jet in the world at that time), their engineering designs were often flawed with low production yields. Sarasohn decided to work more closely with the plant managers on improving quality at their plants.

One day, Sarasohn made a visit to the Tokyo Communication Engineering Company, a new start-up factory that was building radio equipment. The equipment owed to Sarasohn was overdue and when he arrived at the factory, he found chaos. When the owners (Ibuka and Morita) heard of the surprise inspection, they raced to see Sarasohn in Tokyo who gave them one more chance to prove that they could learn something about quality and apply it to their factory. Luckily Ibuka and Morita learned their lesson, and this young company was the company that eventually changed its name to the Sony Corporation.

These quality control efforts in Japan proved enormously fruitful. In the 30 years after WWII, Japan became a world leader in manufacturing, and today Japanese products consistently rank among the world's highest quality. The United States, the world leader in quality in the 1950s, saw its market share in electronics and automobiles shrink at the expense of Japan in this same time period.

Today, manufacturing has shifted from Japan, Germany, and the United States to China. In 2006, China replaced the United States as the world's largest exporter. Unfortunately, quality control issues often plague Chinese manufacturing. News stories about baby formula and pet food tainted with melamine, toys painted with lead paint, and tires that fail at high speed alarm consumers worldwide. The head of China's top quality-control agency quit in the midst of the scandal about tainted milk in 2008. Quality control continues to be of vital importance to worldwide manufacturing.

22.1   A Short History of Quality Control

The desire to make quality products is part of human nature. Several millennia ago, the builders of the Great Pyramid of Giza had to produce blocks of incredible uniformity to build a pyramid whose four sides were perpendicular to within 3.5 arc seconds. During the Middle Ages in Europe, guilds were formed, in part to enforce quality control on craftsmen but also to punish those who tried to pass off inferior goods. The manufacture of money has long relied on quality control to guarantee that the same amount of precious metal is used in each coin. In seventeenth century England, Samuel Pepys wrote in his famous diary of his inspection of quality at The Royal Mint.

Walter A. Shewhart (1891–1967) was an early pioneer of quality methods. Before his arrival at the Western Electric Company, quality control consisted of inspecting products and removing defective items. Shewhart introduced statistical methods including the quality control chart which bears his name.

Because he helped Japanese industrial leaders rebuild after World War II and to shed the image of Japanese products as cheap and shoddy, W. Edwards Deming (1900–1993) is known as the father of the Japanese post-war industrial revival. He was also regarded by many as the leading quality guru in the United States until his death in 1993.

Statistical methods for quality control were pioneered in the 1920s at the Bell Telephone Laboratories in the United States and at roughly the same time at the General Electric Company in the United Kingdom. Quality control today is a tool in the larger effort of continual process improvement in both the manufacturing and service industries. In this view, work is a process on which data can be gathered and studied. This study can then lead to improvement in the process, which ideally will have an impact on both customer satisfaction and the company's bottom line. The process improvement strategy includes understanding the process, collecting the data, assessing the process, eliminating causes of variation, studying the cause and effect of relationships within the process, and implementing changes. In this chapter we will study some tools used in the assessment of the process data to investigate whether the process has changed, either in its mean or in its variation.

Walter A. Shewhart was a statistician and engineer working at Bell Laboratories in the early part of the twentieth century, responsible for improving the quality of the manufactured telephones. As early as 1924, he introduced a chart in an internal Bell Labs memo to graphically display variation in a process over time. The kind of chart he used, a time plot of a measure of quality, became known as a **control chart**, or sometimes a **Shewhart chart**, in his honor. In the same memo, he defined the fundamental rule of **statistical process control (SPC)** as:

"Variation from common-cause systems should be left to chance, but special causes of variation should be identified and eliminated."

Shewhart's insight was that the variation we see in manufacturing arises from two distinct sources. Some of the random fluctuations are *intrinsic* to the process or **common-cause variation**. Others, which are not, are called **special-cause variation**. When special-cause variation is present, the process is said to be **out of control**. The short-term focus of quality control is to detect and correct the special causes of variation. The long-term focus of continual quality improvement is to reduce the amount of variation from the common causes. Shewhart suggested a system of analysis based on the scientific method specifically to address the special-cause issue, but which has also been used as the basis for continual quality improvement.

The term control chart is somewhat of a misnomer. Certainly one can't control a process simply by graphing measurements of it over time. The term quality control became popular in the latter part of the twentieth century and control charts were an important tool in the quality control effort. More recent views of process improvement view the process as evolving toward higher and higher quality (and less variation) and so the term quality "control" is almost passé. The American Society for Quality Control (ASQC) changed its name in the 1980s to the American Society for Quality (ASQ) when its members realized that their methods could be part of a more general movement toward excellence in the organization.

In 1950, when W. Edwards Deming replaced Sarasohn in Japan, he taught the Shewhart principles and process to Japanese managers. The process became codified into four distinct stages: Plan, Do, Check, and Act—known as the PDCA cycle. In the *Plan* stage, the manager identifies and describes the process to be improved and suggests improvements. In the *Do* stage, the manager implements on a pilot level the improvement strategies. In the *Check* stage, the manager examines variation to see if the improvement strategies have been successful. Finally, in the *Act* stage, depending on the results of the Check stage, the manager implements the improvement strategies on a large scale. These stages are completed continuously and repeatedly, so that they form a cycle, or some say a spiral, of quality improvement.[1]

---

[1]Some organizations use a modification of this known as the Define-Measure-Analyze-Improve-Control (DMAIC) cycle.

| Figure 22.1   The Shewhart PDCA Cycle.

Control and Specification Limits

Shewhart's idea for monitoring the variability in a process was based on probability theory. He knew that if the common-cause variation in a process was Normally distributed, there would be only a small probability that an individual value could be farther than certain limits from the mean. To monitor the process, Shewhart advocated plotting measurements of the process over time, looking for evidence that the process might change and slip "out of control." He called such a plot a control chart, although they are often referred to as Shewhart charts. Typical examples of measurements for manufactured parts are characteristics such as thickness or size, weight, color, or electrical properties, just to name a few possibilities. Before starting to monitor the process, however, the process must first be stable, or "in control"—that is, both the mean and the standard deviation must be constant. Then, if an observation deviates far enough from the mean, it could be taken as a signal that either the process mean or standard deviation has changed.

Here is a control chart for 15 observations from an automobile manufacturing plant. The measurements are the diameters of a bolt used to fasten the car door to the chassis. The desired mean of the bolts' diameters is 23 mm:

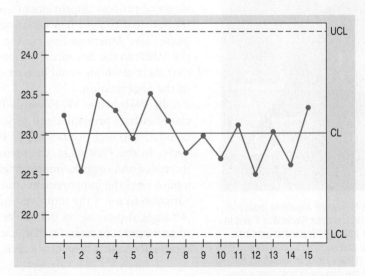

Figure 22.2   A control chart for 15 diameters of bolts with a desired mean of 23 mm.

The main idea of using a control chart to monitor a process is that if the process stays in control, observations will stay within a certain distance of the mean (at least with very high probability). However, if the process mean or standard deviation changes, then the observations will start to drift, and, by monitoring the process, we will be able to discover that change. The question then becomes how far the observations must drift before we decide that the process has changed. So, for every process, **control limits** need to be established to decide when to take action. When points fall outside of the control limits, this provides a signal that the process may have changed and some action is needed. We will discuss how to define the control limits and the various rules for deciding when action is needed in the rest of this chapter.

Using a control chart to monitor a process can be viewed as a graphical hypothesis test. The null hypothesis is that the process has not changed (both the mean and variance are constant). Various rules are used to decide when a signal has occurred (for example, an observation that falls more than 3σ from the mean is a common rule). When such a signal occurs, we reject the null hypothesis and conclude that a change has taken place. Of course, like all hypothesis tests, both type I and type II errors are always possible. In this context, a type I error is a **false alarm**, requiring an unnecessary adjustment to the process. A type II error is a failure to see a change in the process. The trade-off between the two types of errors and their relative costs in particular settings give rise to the many different rules for deciding when to reject the null hypothesis and declare a change in the process. Of course, rules that make it easier to reject the hypothesis will result in more false alarms and rules that make it harder will allow processes to drift farther (on average) before a signal is produced. Rules are often judged by how long they run, on average, without a false alarm when the process is in control. This is called the in-control **average run length** or **ARL**.

Control limits are based on the data themselves, and for that reason it is imperative that the process be in control before starting the monitoring process. Sometimes, however, quality concerns impose *external* limits on the process as well. For example, the diameters of the bolts must fall between certain limits or they just won't work. Limits such as these are **specification limits** because they are based on external specifications and not on how well the parts are actually produced. These upper and lower specification limits are abbreviated as USL and LSL respectively. To understand how well the process is behaving compared to the specification limits, studies known as **process capability studies** are undertaken. Quality control practitioners use these studies to understand the relationships between the specification limits and the actual process parameters (mean and variance). Several different **process capability indices** measuring different aspects of the relationship are commonly used.

Process capability studies can be performed only when the process is in control. In other words, these studies involve only the common-cause variation and not special-cause variation. As long as the observations stay between the LSL and USL, the parts will meet specification. Ideally we would want the upper and lower specification limits to be as far from the process mean as is economically feasible. That width can be measured by USL – LSL. If the process can be centered at the desired mean, then by the 68–95–99.7 Rule, we know that observations farther than 3σ from the mean will be rare. The ratio $C_p = \dfrac{USL - LSL}{6\sigma}$ measures the ratio of the entire specification range (both sides of the mean) to 6 times the standard deviation of the actual process variation. A process with a C_p value of 1.0 would therefore just satisfy the 3σ on each side of the mean criteria and translates to about one observation in about 370 falling outside the specification limits, which is considered too often for most modern manufacturing and service industries. Usually it is recommended that C_p be at least 1.25, and sometimes as high as 2.0, but the actual choice depends on the economics of the situation and the type of industry. The larger this C_p value, the smaller the chance of an observation falling outside the specification limits. A large value of C_p indicates a process with a large potential to produce parts within the specification limits.

In any real process, no matter how stable, the mean will drift slightly even when the process is in control. Because C_p does not explicitly use the location of the mean it tends to overestimate the potential of the process when the mean is not centered and can be misleading. A measure known as C_{pk} is often used instead. When the process mean and standard deviation are known, C_{pk} is defined as:

$$C_{pk} = \min\left[\frac{USL - \mu}{3\sigma}, \frac{\mu - LSL}{3\sigma}\right]$$

Using a period during which the process is in control to estimate the process parameters gives rise to the estimated value of C_{pk}.

$$\hat{C}_{pk} = \min\left[\frac{USL - \hat{\mu}}{3\hat{\sigma}}, \frac{\hat{\mu} - LSL}{3\hat{\sigma}}\right]$$ where the hats indicate that the parameters

are estimated from the process data. Most studies recommend that the estimates of process capability be based on at least 50 observations. As with C_p the choice of how large a value of C_{pk} is acceptable or desired depends on the situation and the associated costs but values below 1.0 are rarely acceptable. Some industries (such as the automotive manufacturing industry) have specific guidelines that they publish. It is also important to remember that these indices rely heavily on the assumption that all observations are Normally distributed.

For Example   Capability indices

Fudge requires fairly strict adherence to a sequence of boiling sugar in milk until it reaches a certain consistency and then beating the mixture while it cools so that it acquires a smooth, creamy texture. A small candy company about 100 miles north of New York City ships fudge worldwide. If its fudge mix has a stable consistency, workers can control the weight of the fudge by simply timing how long they pour the warm fudge into the box prior to shipping. They would like the labeled 18 oz. box of fudge to contain no more than 18.5 oz. of fudge and no less than 17.9 oz. From the last 100 batches, they estimate the standard deviation of the weights to be 0.1 oz. with a mean of 18.1 oz.

Question: Perform a process capability analysis on the fudge shop and comment on your findings.

Answer: Our estimate of C_p is $\hat{C}_p = \dfrac{USL - LSL}{6\hat{\sigma}} = \dfrac{18.5 - 17.9}{6(0.1)} = \dfrac{0.6}{0.6} = 1.0$. Our estimate of C_{pk} is $\hat{C}_{pk} =$

$\min\left[\dfrac{USL - \hat{\mu}}{3\hat{\sigma}}, \dfrac{\hat{\mu} - LSL}{3\hat{\sigma}}\right] = \min\left[\dfrac{18.5 - 18.1}{3(0.1)}, \dfrac{18.1 - 17.9}{3(0.1)}\right] = \min\left[\dfrac{0.4}{0.3}, \dfrac{0.2}{0.3}\right] = 0.67$. The C_p value shows that if the

weights are Normally distributed and the mean is centered, about 1 in 370 boxes will exceed the limits. This is a low C_p value, but it may be acceptable in this industry. However, the mean is not centered between the two limits and the C_{pk} value of 0.67 is too low. The candy company should work to center the mean and/or lower the standard deviation of the weights.

22.2   Control Charts for Individual Observations (Run Charts)

The Sony Corporation of the 1950s used transistors and tubes for their electronic equipment, but nearly all their products today rely on computer chips. And almost all computer chips in use today are built on silicon wafers. Silicon is a natural semiconductor—it can be used both as a conductor of electricity and an insulator to prevent conduction. It is also relatively inexpensive and plentiful. The basic source of silicon is sand, which can be purified to produce chips that are 99.9999% pure silicon. Although there are many different kinds of chips (RAM, processors, etc.), they are all built from the basic silicon wafer. Making a wafer involves a number of steps, all of which must be carried out as consistently as possible to ensure the chip will be able to perform reliably, regardless of its final function. During the manufacturing process a variety of performance characteristics must be checked—physical measurements such as thickness, and chemical properties involving purity and electrical properties such as conductance. All of these have specifications depending on the type of chip being produced. Because billions of chips are manufactured worldwide every year, maintaining quality control on chip manufacturing is big business.

Silicon wafers are produced in varying thicknesses, depending on their diameter. For photovoltaic use, a typical wafer is about 100–200 mm square with a thickness of $200 - 300 \ \mu$m. The mean target thickness of the particular wafer whose production is studied here is $300 \ \mu$m. One way to ensure quality is to keep the thickness as consistent as possible. We will use 250 past measurements of the process which has been judged to be in control to establish control limits. Here are the summary statistics and a histogram:

```
maximum     392.423
quartile    325.067
median      298.521
quartile    277.030
minimum     221.838
Mean        301.111
Std Dev      32.990
```

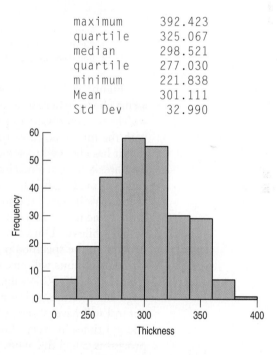

Figure 22.3   A histogram of the thickness shows a nearly Normal distribution centered at 300 μm.

Because the process has been judged to be in control, we can use the statistics from these data to develop control limits. We will base the mean at 300 μm, the desired thickness. The standard deviation of these measurements is 32.99. Using 33 μm as the standard deviation, (σ), the 68–95–99.7 Rule states that, as long as the process remains stable, only about 0.3% or 3 of 1000 observations will fall outside of 3σ from the mean. Using 3σ the lower and upper control limits would be

$$\mu - 3\sigma = 300 - 99 = 201 \text{ and } \mu + 3\sigma = 300 + 99 = 399 \ \mu\text{m}.$$

Using these control limits, a control chart can be monitored to see whether the process stays in control. Here are the subsequent 50 thicknesses of the wafer manufacturing process:

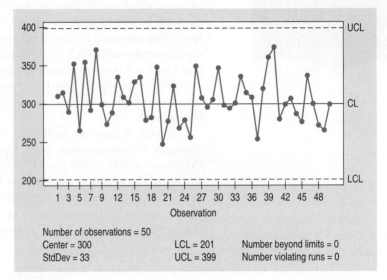

Number of observations = 50
Center = 300
StdDev = 33

LCL = 201
UCL = 399

Number beyond limits = 0
Number violating runs = 0

Figure 22.4 A control chart of individual observations shows no evidence that the process has slipped out of control. The 3σ lower (LCL) and upper (UCL) control limits are shown as dotted lines. The center line is denoted CL.

Figure 22.4 shows a control chart for individual observations (sometimes called a **run chart**), where a measurement is made on every product being produced. As we check each observation we are really performing a graphical hypothesis test, with the null hypothesis that the mean is 300 μm, and the (two-sided) alternative that it has changed to something else. The change may be abrupt, caused perhaps by a breakdown of a machine or the introduction of an impurity, in which case it is called a **level shift**. Or, the change might be gradual, in which case it is called a **trend**. We'll reject the null hypothesis if any point lies outside the 3σ control limits. The use of 3σ control limits as a standard is fairly widespread in the United States, while the United Kingdom and Europe tend to use other multiples of σ that depend on the specified probability of a Type I error, or false alarm signal. Using 3σ control limits will give rise to a false alarm rate of about 0.003 (using $1 - 0.997$ from the 68–95–99.7 Rule), or more precisely, 0.0027. That's the probability of any *single* observation giving rise to a false alarm. Even a process that is perfectly in control will have observations farther than 3σ from the mean, eventually (in fact, about 3 times in every 1000 observations). The average time between alarms for a process is called the **average run length** or **ARL**. The average time between *false* alarms for a process in control is called the **in-control ARL**. We want as large an in-control ARL as possible, but at the same time, we'd like the ARL for a process that is out-of-control to be short. The in-control ARL can be calculated by taking the reciprocal of the false alarm probability. For the 3σ rule, the in-control ARL is:

$$ARL_0 = \frac{1}{p} = \frac{1}{0.0027} = 370.37.$$

On average, we would expect to see a false alarm about every 370 observations for a process that is in control.

If the process mean shifts we'd like the control chart to show an alarm. If the chart shows the process staying in control, that's a Type II error. As in any hypothesis testing situation, the probability of the Type II error depends on how many observations have occurred since the process went out of control and how large the shift is (the effect size). Balancing Type I and Type II errors is an economic decision and depends on the control limits and the rule for deciding when a process has changed. The ability of any chart to detect shifts can be described by its **operating-characteristic (OC) curve** or equivalently by comparing the in-control and out-of-control ARLs. We describe shifts in mean in terms of standard deviations. If the process shifts only one standard deviation, the probability of noticing the shift on the next observation will be very small. If the shift were quite large, say, 5σ, we should be surprised if we didn't see it immediately. Here is an OC curve for the run chart just described:

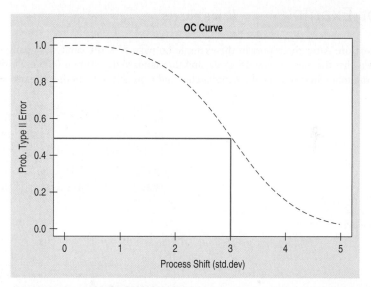

Figure 22.5   The operating-characteristic curve for a run chart. For a shift of 3σ the probability of not detecting it on the next observation is about 0.50.

Even though the probability of detecting a shift of 3σ is only about 0.5 on the *next* observation, the probability of a Type II error after two observations is about $0.5 \times 0.5 = 0.25$, and the probability of failing to detect the shift after 5 observations is $(0.5)^5 = 0.031$. More observations, of course, help guard against making errors.

In addition to the simple 3σ rule there are many other rules commonly used in industry with different OC curves, and different ARLs. The Western Electric Handbook (WECO, 1956) suggests several rules for deciding whether a process is out-of-control:

1. One point outside of the 3σ control limits on either side
2. Two out of three consecutive points beyond the 2σ "warning limits"
3. Four out of five consecutive points beyond 1σ from the mean
4. Eight (or seven) points in a row on the same side of the mean
5. Six points in a row increasing (or decreasing)
6. Fourteen points in a row alternatively increasing and decreasing

These rules are commonly used in practice, and some software (notably Excel and Minitab) implement several other rules as well. You may see 8 or even 10 rules used by software to detect a possible shift in the process. Deciding which combination of rules to use involves a trade-off between false alarms and failure to detect process changes. If the simple 3σ rule is the only one used, the in-control ARL before a false alarm, as we have seen, is about 370. If all 6 WECO rules are implemented, the ARL drops to about 92 observations between false alarms, but decreases the out-of-control ARL as well. Using more rules only decreases both ARLs even more. Each user has to decide whether the ability to detect shifts faster is worth the price of more frequent false alarms. It is also important to note that these tests rely crucially on the assumption that the observations themselves are well modeled by a Normal distribution. In the next section, we will plot *averages* of several observations and the Central Limit Theorem will help somewhat with this restrictive assumption.

For Example   Runs charts

Question: After the report in the example on page 774, the candy company makes some changes to the pouring process and thinks that the mean is now 18.15 oz. and the standard deviation is 0.08 oz. Using these parameter values, construct a runs chart for the following weights of recent batches and comment on whether the process is in control.

18.03	18.00	18.12	17.81	18.23
18.11	18.23	18.19	18.00	18.05
18.12	18.05	18.14	18.16	18.11
18.19	18.26	18.11	18.15	18.07
18.04	18.14	18.00	18.21	18.12

Answer:

It appears that the process is not yet in control. The weight of batch 16 is too low and, in fact, it is outside both the 3σ and the lower specification limit of 17.90.

22.3   Control Charts for Measurements: $\overline{X}$ and R Charts

Although the run chart of the last section is the easiest to implement and to understand, it is not usually practical to measure *every* item coming down a production line, or every phone call coming into a call center, especially if the testing is destructive. Nor are all observations Normally distributed. As an alternative, a sample is taken at periodic intervals and, typically, the *mean* of that sample is plotted. A chart of these sample means is called an $\overline{X}$ **chart** (pronounced X bar chart and often written as x-bar or xbar chart).

For detecting a shift in the mean, the $\overline{X}$ chart has several advantages over the run chart. A practical advantage is that it tests less frequently. Even more important, the averages from a sample of size n are more nearly Normal than individual observations, so that the basic assumption of the control chart that gives us the probabilities of a false alarm and other characterisitcs is more likely to be met. Because the sample sizes used in an $\overline{X}$ chart are usually small, however, it still relies somewhat on the underlying measurements being approximately Normal. Very skewed distributions will make the $\overline{X}$ chart inaccurate for detecting out-of-control signals.

When the standard deviation is assumed to be known (perhaps based on a large amount of past data), the implementation of the $\overline{X}$ chart is essentially the same as the individual control chart. Because a sample mean has a standard deviation of $\dfrac{\sigma}{\sqrt{n}}$ rather than σ, the 3σ control limits for the $\overline{X}$ chart become:

$$ LCL = \mu - 3\frac{\sigma}{\sqrt{n}} $$

$$ UCL = \mu + 3\frac{\sigma}{\sqrt{n}} $$

For the wafer data, if we continue to use $\mu = 300$, and $\sigma = 33$, and take a 5-item sample at regular intervals, the control limits become:

$$ LLC = \mu - 3\frac{\sigma}{\sqrt{n}} = 300 - 3\left(\frac{33}{\sqrt{5}}\right) = 255.7 \ \mu\text{m} $$

$$ ULC = \mu - 3\frac{\sigma}{\sqrt{n}} = 300 + 3\left(\frac{33}{\sqrt{5}}\right) = 344.3 \ \mu\text{m} $$

Here is an $\overline{X}$ chart for 25 samples of size 5 from the wafer production process:

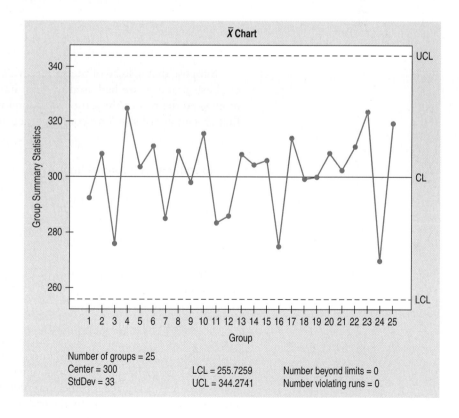

Figure 22.6 An $\overline{X}$ chart for 25 samples of size 5. The process appears to be in control because all sample means are within the control limits. This particular chart use two of the Western Electric rules and reports *Number beyond limits* (rule #1) and *Number violating runs* (variant of rule #4—7 on same side of mean).

Up to now, we have assumed that the past data are numerous, so the mean and standard deviation of the process are known. In practice, we may not have this information, and we will have to estimate both μ and σ. These estimates should be based on as much data as possible—20 to 25 samples at a minimum—and, of course, the process must be in control. If we have k samples of size n, then we can use the average of the averages, called $\overline{\overline{X}}$ (x double bar), to get our estimate of μ:

$$\overline{\overline{X}} = \frac{\bar{x}_1 + \cdots + \bar{x}_k}{k}.$$

If no external center were desired, $\overline{\overline{X}}$ would be the center of our control chart.

To find the control limits, we still need an estimate of the standard deviation, σ. This is typically done by technology, or from the standard deviation of a large in-control sample, but can also be found by using the ranges of the samples rather than their standard deviations. Recall that the range of a sample is defined as the maximum minus the minimum:

$$R = max - min.$$

For Normally distributed data when no other reliable estimate of the standard deviation is available, we can use a well known relationship between the range of a sample and its standard deviation. The mean of the ratio $\dfrac{R}{\sigma}$ is a constant denoted d_2, which depends on the sample size (and is given for various sample sizes in Table R in Appendix D). Using this constant, an estimate of σ is found from $\hat{\sigma} = \dfrac{\overline{R}}{d_2}$ where

$$\overline{R} = \frac{R_1 + \cdots + R_k}{k}.$$

The limits of the control chart then become:

$$LCL = \overline{\overline{X}} - 3\frac{\hat{\sigma}}{\sqrt{n}} = \overline{\overline{X}} - 3\frac{\overline{R}}{d_2\sqrt{n}}$$

$$UCL = \overline{\overline{X}} + 3\frac{\hat{\sigma}}{\sqrt{n}} = \overline{\overline{X}} + 3\frac{\overline{R}}{d_2\sqrt{n}}$$

Suppose that instead of the 250 individual measurements we used previously to establish μ and σ, we had used the first half of these as a calibration set and then monitored the rest of the samples. In total we have 50 samples of size 5. From the first 25 samples of size 5 we compute their means and ranges and find:

$$\overline{\overline{X}} = 295.27; \overline{R} = 81.32.$$

To compute the control limits, we have $n = 5$, and $d_2 = 2.326$, and so:

$$LCL = 295.27 - 3\left(\frac{81.32}{2.326\sqrt{5}}\right) = 248.36$$

$$UCL = 295.27 + 3\left(\frac{81.32}{2.326\sqrt{5}}\right) = 342.18$$

These are not very far from the control limits we found by assuming that the mean and standard deviation were known. But often, control limits have to be adjusted as new information about the process performance becomes available. Quality control charting involves learning about a process as well as monitoring it. A more modern approach to process monitoring uses control charts to monitor the *inputs* to the process instead of just the measurements (outputs). In this way of thinking, if the critical inputs to the process are in control, the resulting outputs will be as well.

As we saw with run charts, the power for detecting a shift depends on the size of the process shift. For $\overline{X}$ charts it also depends on the sample size. To choose an appropriate sample size, quality control monitors examine the operating characteristic curves such as the one shown in Figure 22.7 based on the wafer data. Both the process shift and sample size can be used to find the probability of a Type II error for the next sample.

Figure 22.7   Operating characteristic curves for the wafer data $\overline{X}$ chart. For any proposed value of the process shift (in standard deviations), operators can examine different sample sizes to determine a good sample size with a reasonable probability of a Type II error. For example, for a 1.5σ shift, a sample size of more than 10 would be necessary to produce a probability of a Type II error below 0.10.

In general, we care about both the level of the process and consistency in variation. Although an $\overline{X}$ chart will detect increases in the standard deviation, for more information, the ranges themselves are plotted in what is known as an **R chart**. The procedure for setting up the R chart is very similar to the $\overline{X}$ chart. The only new element is finding the standard deviation of the range. Fortunately, for Normally distributed data, the relationship between σ and R is well-known. There is a new parameter d_3, that depends on the sample size as did d_2. If we assume σ is known, we can find the standard deviation of the range directly by: $\sigma_R = d_3\sigma$. If we need to estimate σ, we can use $\hat{\sigma} = \dfrac{\overline{R}}{d_2}$ as before:

$$\hat{\sigma}_R = d_3\frac{\overline{R}}{d_2}$$

We center the range chart at $\overline{R}$, so the control limits:

$$LCL = \overline{R} - 3\sigma_R$$
$$UCL = \overline{R} + 3\sigma_R$$

become:

$$LCL = \overline{R} - 3d_3\frac{\overline{R}}{d_2}$$

$$UCL = \overline{R} + 3d_3\frac{\overline{R}}{d_2}$$

(If we are given both the mean and standard deviation from the process history and wish to establish limits for R, we can alternatively use $R = d_2\sigma$ as the center line.)

Using our 25 samples of size 5 with $\overline{R} = 81.32$, and finding d_3 for $n = 5$ to be 0.864, we have: $\hat{\sigma}_R = d_3\dfrac{\overline{R}}{d_2} = 0.864\dfrac{81.32}{2.326} = 30.21$. (Had we assumed $\sigma = 33$ we would have $\sigma_R = 0.864(33) = 28.51$, not far from the value we obtained.) This gives us lower and upper control limits for the range of:

$$LCL = \overline{R} - 3d_3\frac{\overline{R}}{d_2} = 81.32 - 3(30.21) = -9.31$$

$$UCL = \overline{R} + 3d_3\frac{\overline{R}}{d_2} = 81.32 + 3(30.21) = 171.95$$

Because a range can never be negative, we set the lower limit at 0.

When not using external values for the process mean and standard deviation, these formulas can (and in industry usually are) simplified by noticing that we can collect the constants and rewrite the limits more simply as:

$$LCL = \overline{R} - 3d_3\frac{\overline{R}}{d_2} = \overline{R}\left(1 - 3\frac{d_3}{d_2}\right) = D_3\overline{R}$$

$$UCL = \overline{R} - 3d_3\frac{\overline{R}}{d_2} = \overline{R}\left(1 + 3\frac{d_3}{d_2}\right) = D_4\overline{R}$$

So now the upper and lower limits for the $\overline{R}$ chart are just multiples of $\overline{R}$ (and are found in Table R). For small sample sizes, $D_3 = 0$ which automatically sets the lower range limit 0. (These simplications should not be used if you want to base the limits on an *external* value of the process standard deviation or mean).

For $n = 5, D_3 = 0$ and $D_4 = 2.114$ so $LCL = 0$ and $UCL = 2.114(81.32) = 171.91$, nearly the same (accounting for rounding) as we found before.

For R charts, the first four WECO rules are often used, although some adaptations for the fact that the chart is generally asymmetric exist as well. The chart below uses rules 1 and 4 (with seven) as it did for the $\overline{X}$ chart. Monitoring the next 25 samples, we see that the range stayed in control:

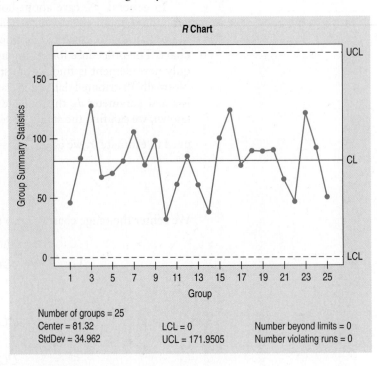

Figure 22.8 An *R* chart for 25 samples of size 5. The process appears to be in control in terms of variation because all sample ranges are within the control limits.

An alternative to the *R* chart is the **S chart**, which uses the standard deviations of samples rather than their ranges. The most common use of $\overline{X}$ and *S* charts occurs when the sample sizes are moderately large (where the range loses statistical efficiency—around $n = 10$ or 12) or when the sample size varies and the ranges of different samples are not comparable. The formulas for finding the control limits for both $\overline{X}$ and *S* are complicated and are usually left to technology to calculate. We refer the interested reader to any number of excellent books on quality control.

In this section we have seen how to establish the most common control limits for $\overline{X}$ and *R* charts. $\overline{X}$ and *R* charts are the most commonly used charts for measuring quantitative values and are used together to monitor both changes in the mean and the spread of a process. However, it should be pointed out that the more rules one uses for detecting signals, the smaller the in-control ARL will be. So, using more of the Western Electric rules or using $\overline{X}$ and *R* charts together will decrease the time to detect a true signal, but will increase the number of false alarms as well. Important alternatives to the $\overline{X}$ charts have been developed, among which the two most widely used are the cumulative sum (CUSUM) and exponentially weighted moving average (EWMA) charts.

The CUSUM chart plots the cumulative sums of all the deviations of the values from the target mean. If the process stays in control, that sum should stay near 0 (although its variance will increase). If the process drifts, the sum will start to accumulate the differences (on one side of the target mean or the other). Because the variance of a sum increases even when the process is in control, the control limits are not lines equally spaced from the mean, but a V-shaped mask that moves as the number of observations increases. The EWMA chart, like the exponential smoothing method of Chapter 20, weights recent observations more heavily so that past observations have decreasing importance as time rolls on. Both of these charts have been shown to provide advantages over $\overline{X}$ charts for many real world situations.

Finally, it is important to point out that not all process changes are undesirable. In fact, in modern quality control, much emphasis is placed on reducing process variation. A shift indicating that the range (or standard deviation) has decreased would almost certainly be viewed as beneficial.

For Example $\overline{X}$ and *R* charts

The fudge makers (see page 774) make more changes to the process and wonder if they have been successful in getting the process under control. They still desire a mean of 18.1 oz. and a process standard deviation of 0.08 oz. Here are the average weights and ranges of 25 batches of size 5:

Batch	Mean	Range	Batch	Mean	Range
1	18.104	0.13	14	18.076	0.20
2	18.014	0.09	15	18.152	0.15
3	18.138	0.14	16	18.114	0.31
4	18.108	0.31	17	18.180	0.36
5	18.118	0.12	18	18.086	0.26
6	18.112	0.18	19	18.104	0.40
7	18.216	0.15	20	18.074	0.20
8	18.332	0.62	21	18.122	0.24
9	18.434	0.19	22	18.146	0.12
10	18.146	0.29	23	18.104	0.36
11	18.032	0.22	24	18.136	0.16
12	18.102	0.39	25	18.104	0.08
13	18.112	0.21			

(continued)

Questions: Using $\overline{X}$ and R charts with the desired mean of 18.15 oz. and the desired standard deviation of 0.08 oz., does the process seem to be in control now? If not, when did the process seem to slip out of control?

Answers: For the $\overline{X}$ chart, we use 18.15 as the center line and the desired process standard deviation of 0.08. The control limits are found at

$$\mu \pm 3\frac{\sigma}{\sqrt{5}} = 18.15 \pm 3\frac{0.08}{\sqrt{5}} = 18.15 \pm 0.107 = 18.043 \; and \; 18.257.$$

Here is the $\overline{X}$ chart of the 25 batches:

Number of groups = 25
Center = 18.15   LCL = 18.04267   Number beyond limits = 4
StdDev = 0.08   UCL = 18.25733   Number violating runs = 2

For the ranges, we use the desired process mean and standard deviation as well. So the center is

$$center = d_2\sigma = 2.326(0.08) = 0.186$$
$$UCL = center + 3(d_3\sigma) = 0.186 + 3(0.864 \times 0.08) = 0.393$$
$$LCL = center - 3(d_3\sigma) = 0.186 - 3(0.864 \times 0.08) = -0.021 \Rightarrow 0$$

Here is the R chart:

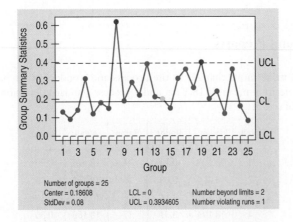

Number of groups = 25
Center = 0.18608   LCL = 0   Number beyond limits = 2
StdDev = 0.08   UCL = 0.3934605   Number violating runs = 1

The process is out of control both in mean and variation. Batches 2 and 11 were too low, and 8 and 9 were too high. Batches 8 and 19 had ranges too high and by Batch 14, there were six ranges on the same side of the center line. The candy company clearly needs to study the process to bring it into control.

22.4   Actions for Out-of-Control Processes

Once a process has been determined to be "out of control," it does not necessarily mean that something is "wrong" with the process. It simply means that a signal has occurred indicating a possible change in the system. An action should be taken to understand the source of the special-cause variation. If the change is determined not to be beneficial, however, those in charge of the process must act quickly to resolve the problem. In the case of a detrimental change, an out-of-control signal may necessitate stopping production of the process until the cause has been determined and a correction put into place. This is the "Act" part of the Shewhart Cycle and a crucial step for process management.

For an established manufacturing process there may be an **out-of-control action plan (OCAP)** in place based on past experience. An operator seeing an out-of-control signal would consult the OCAP and take the steps outlined in it and then continue monitoring the process to see if the process regained an in-control state.

If no such plan exists, as would be the case for a new process, an investigation of possible causes is undertaken. The full details of such an investigation are beyond the scope of this text, but the basic steps involve:

- Forming a team of all those involved in the process to discuss and enumerate the possible causes of the shift. A powerful tool in this effort is the **cause-and-effect diagram** (sometimes called an Ishikawa or fishbone diagram) that helps organize the possible causes and helps prioritize them for further analysis. (See Figure 22.9.)

- If no simple cause can be determined it may be necessary to commence a series of experiments using the most probable causes as the factors. There are efficient designs that can accommodate as many as a dozen or more possible factors in the initial investigation of possible causes.

- Once the most likely cause has been identified, and the optimal value for that factor has been determined, the process inputs are changed in an attempt to drive the process back into control.

- The monitoring process is restarted in the hope that the problem has been resolved, and the Shewhart cycle starts again.

Understanding and controlling processes is a large area of activity in both the manufacturing and service industries involving the efforts of many people in the organization. The American Society for Quality (www.asq.org) provides a wealth of information on the subject.

Figure 22.9 Cause-and-effect diagram for a production problem involving faulty welding between a detector and its base in a gas chromatograph. After a control chart showed the process was out of control (see Figure 22.11), a team was assembled and together they listed 5 potential areas to investigate for process improvement: Materials, Personnel, Welding, Welding Methods, and the Machines involved. For each area, three to four potential factors were listed. From here, the team designed a series of experiments in the factors deemed most likely to affect weld breakage. The factors were manipulated and the resulting weld breakage was measured. Eventually it was determined that the alloy used by one of the suppliers had changed. When the company insisted that the supplier use the alloy that historically had been used for the weld, the problem disappeared.

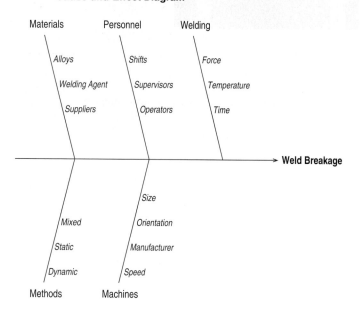

Guided Example

KomTek Technologies produces hip prosthetics in the United States with more than 60% of the U.S. market share. Founded in 1984, KomTek Technologies specializes in the production of nickel- and cobalt-based super alloys—specifically in the production of prosthetic devices. This manufacturing firm sells hip joints to five biomedical companies and these account for 30% to 40% of the firm's annual revenue. This firm uses a cobalt chromium alloy to forge the joints, which are heated in a furnace and then beaten or hammered into shape. Because prosthetic hips are costly to manufacture, the firm produces them only when a customer places an order. The average size of an order is typically between 200 and 600 pieces.[2]

Each part of the manufacturing process is monitored and controlled and the corporation has implemented a rigorous program to assist with quality control. Toward the end of the manufacturing process, dimples need to be hammered into each hip joint to enhance the fit. Occasionally, this hammering process causes the center line to be off-center. The tip length of this dimple is a crucial measurement to monitor.

A major customer that accounts for nearly 40% to 50% of the total hip joint sales at the firm recently placed an order for 1200 prosthetic hips—a larger than usual order. Following the manufacturing and shipment of these hips for this customer, the team at the firm responsible for quality control noticed that their overall cost for filling this order was high. The team members realized that the order was larger than usual, but they need to figure how and why the cost was so high. There was some concern by management that quality control monitoring may not have been in place for the latter part of this order. In particular, there was some suspicion that the last 230 or so hip joints had not been subjected to rigorous quality control.

Guided Example	KomTek Technologies

We want to establish control limits for $\overline{X}$ and R charts for the *Tip Length* and investigate the recent production run to understand why the cost of recent orders has been high.

PLAN   Setup State the objectives.

Identify the type of data you are examining.

We want to examine the variation in the Tip Length of the distal hip for the latter part of a recent large customer order to look for possible quality control issues. The tip length spec is 0.25 cm so we will use this as the center line. To start the monitoring process, we have data from 50 samples of size 5 from previous orders, when the process was judged to be in control (the calibration period), to use to establish our control limits.

---

[2]This example is based on a case study written by J. Hunt, E. Landry, and J. Rao as part of the Babson College case series. Copyright © by Babson College, 1998. The data and setting used in this example are based on the actual case study, but the data have been modified and the conclusions are fictitious.

DO

Mechanics We can compute the limits using the specs provided by the company.

For the calibration period the overall mean is $\bar{\bar{X}} = 0.2507$ inch. We will use 0.25, the specified length, as the center line. A histogram of these 250 observations shows a unimodal and symmetric distribution:

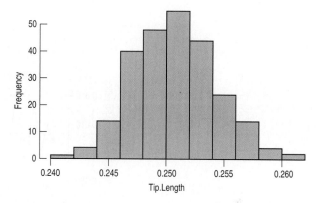

The average range for these 50 samples is $\bar{R} = 0.00847$ cm. Using the range to estimate sigma, we have:

$$\hat{\sigma} = \frac{R}{d_2} = \frac{0.00847}{2.326} = 0.0036 \text{ cm}$$

Our initial control limits for $\bar{X}$ are therefore:

$$LCL = \mu - 3\frac{\hat{\sigma}}{\sqrt{n}} = 0.25 - 3\left(\frac{0.0036}{\sqrt{5}}\right)$$
$$= 0.2452 \text{ cm}$$

$$UCL = \mu + 3\frac{\hat{\sigma}}{\sqrt{n}} = 0.25 + 3\left(\frac{0.0036}{\sqrt{5}}\right)$$
$$= 0.2548 \text{ cm}$$

We use software to create the $\bar{X}$ control chart.

The $\bar{X}$ chart for the calibration data is shown here for completeness (although, of course, we would not expect to see problems here because these are the data on which the limits are based):

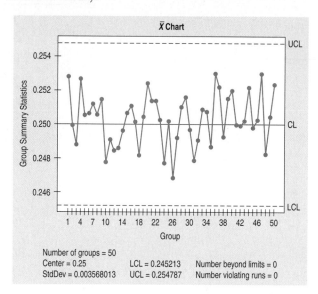

(continued)

We use technology to create the *R* chart on the calibration data.

Similarly, we show the R chart for the calibration period. For this period $\overline{R} = 0.00847$. To establish control limits for the R chart we use

$$LCL = D_3\overline{R} = 0(0.00847) = 0$$
$$UCL = D_4\overline{R} = 2.114(0.00847) = 0.0179$$

as the control limits.

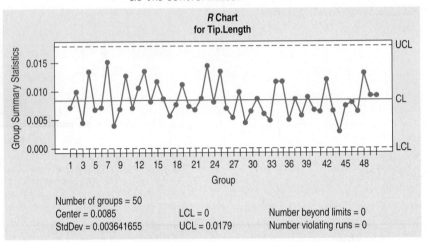

R Chart for Tip.Length

Number of groups = 50
Center = 0.0085   LCL = 0   Number beyond limits = 0
StdDev = 0.003641655   UCL = 0.0179   Number violating runs = 0

Examine the $\overline{X}$ and *R* charts for the recent order.

Using these limits, we now create $\overline{X}$ and R charts for 100 hip joints in the latter part of the recent order, starting with unit 970 (20 samples of size 5). The $\overline{X}$ chart shows several problems:

$\overline{X}$ Chart for Recent.Order

Number of groups = 20
Center = 0.25   LCL = 0.2452   Number beyond limits = 5
StdDev = 0.006079889   UCL = 0.2548   Number violating runs = 0

There are 5 samples whose average tip length falls below the LCL. In addition, there were 9 samples in a row below the center line starting with sample 3. This should have been a signal that the process mean had shifted and some intervention should have taken place.

The R chart shows similar problems:

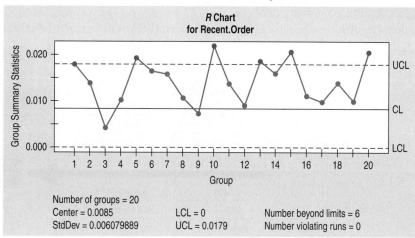

R Chart for Recent.Order

Number of groups = 20
Center = 0.0085
StdDev = 0.006079889
LCL = 0
UCL = 0.0179
Number beyond limits = 6
Number violating runs = 0

There are 6 samples whose range is too large (outside of the UCL). Starting with the first sample, this should have been a sign that something had changed with the process.

REPORT

Conclusion State the conclusion in the context of the problem.

Follow up

MEMO

Re: Patterns of Tip Length

We investigated the variation in the length of the distal tip for this unusually large customer order. It is clear that quality control for the latter part of this order was lax. Several samples from units 970 to 1070 had average tip lengths that were too short. In addition, the volatility in the process also seems to have increased. The quality team notes that the length of the distal tip must be monitored using sampling during the process and the equipment should be checked, recalibrated, and/or changed after every violation of the limits in either the $\bar{X}$ or R charts.

Following this incident, the team was assembled to try to diagnose the problem. None of the inputs to the machine that produces the part seemed to have changed. However, using a fishbone diagram, many possible causes of the problem were identified. After intensive discussion, four factors deemed most likely to affect tip length were selected for further study. An experiment testing these was designed and run. The analysis of the data showed that the tensile strength needed to be increased. A subsequent run of 50 tips showed that the problem had been resolved by this adjustment.

For Example   Actions for out-of-control processes

The fudge company (see page 774) wants to bring its process under control and assembles a team of bakers, process engineers, and marketing staff. The team discusses possible factors influencing the weight and groups them into four main categories:

Raw materials (sugar, butter, milk, flavorings)
Oven (oven temperature, pan temperature, time)

(continued)

Process (timing of stir, amount of stir, pouring of fudge)
Container (pan material, pan size)

Question: Organize these possible factors into a fishbone (Ishikawa) diagram to help them design experiments using these factors.

Answer:

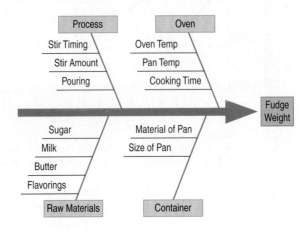

22.5   Control Charts for Attributes: *p* Charts and *c* Charts

We've seen several examples of quantitative measurement that are used to assess quality. For manufacturing processes these might include physical characteristics such as the thickness or diameter of a silicon wafer. In service industries, we might want to measure the length of time a customer has to wait to talk to a representative at a call center or how long it has been since they used one of our branches. Quantitative measurements are common, but we may want to monitor the quality of a process that has only categorical measurements such as color, or product type. But the most common type of categorical measurements used in quality control are **dichotomous**—or yes/no—**variables**. Examples of these include whether a part is defective or nondefective or whether the customer has rated a phone call as satisfactory or not. In such cases, we can monitor the *proportion* of defective items in the sample. If it is determined that the proportion of defectives has increased, the special-cause variation will be investigated and the process will be adjusted to reduce variation. The control chart of proportions is called a ***p* chart**.

Recall that, if we assume the samples are independent, the standard deviation of a sample proportion $\hat{p}$ (writing q for $(1 - p)$) is:

$$SD(\hat{p}) = \sqrt{\frac{pq}{n}}.$$

If we know the desired proportion p, we can use this to establish 3σ control limits for the sample proportion:

$$LCL = p - 3\sqrt{\frac{pq}{n}}$$

$$UCL = p + 3\sqrt{\frac{pq}{n}}$$

In practice, however, we usually use the data from a period believed to be in control to estimate p and then use this value to monitor future data. As we did for $\overline{X}$ charts, we start by taking k samples of size n as our calibration data and base our control limits on the results. We'll call the average proportion of defective items from our k samples $\bar{p} = \dfrac{\hat{p}_1 + \ldots + \hat{p}_k}{k}$. Recall that when we didn't know p, we used the sample proportion to find the standard error:

$$SE(\hat{p}) = \sqrt{\frac{\hat{p}\hat{q}}{n}}.$$

We can use the same formula to estimate the standard deviation and find the control limits by using $\bar{p}$ in place of $\hat{p}$:

$$LCL = \bar{p} - 3\sqrt{\frac{\overline{pq}}{n}}$$

$$UCL = \bar{p} + 3\sqrt{\frac{\overline{pq}}{n}}$$

As with R charts, if the lower limit is negative, we set it to 0.

The usual point outside of the 3σ rule indicates a process that may be out of control. As with $\overline{X}$ charts there are a variety of alternative rules (WECO). Minitab, for example, uses the following four:

1. One point outside of the 3σ lines
2. Nine points in a row outside of 1σ on the same side
3. Six increasing or decreasing points in a row
4. Fourteen points in a row that alternate up and down

For example, suppose that a wafer is defective if its film thickness is either too thin ($<250\ \mu$m) or too thick ($>350\ \mu$m). If we use 25 samples of size 50 as our calibration set we find $\bar{p} = 0.1544$ and so

$$LCL = \bar{p} - 3\sqrt{\frac{\overline{pq}}{n}} = 0.1544 - 3\sqrt{\frac{0.1544 \times 0.8456}{50}} = 0.001$$

$$UCL = \bar{p} + 3\sqrt{\frac{\overline{pq}}{n}} = 0.1544 + 3\sqrt{\frac{0.1544 \times 0.8456}{50}} = 0.307$$

The p chart for the next 25 samples shows no evidence of a change in the proportion:

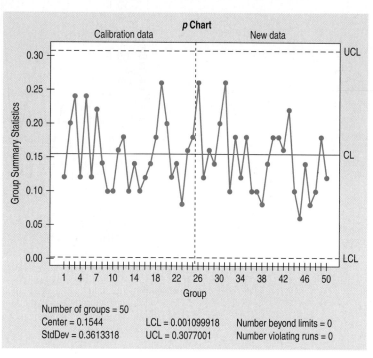

Figure 22.10   A p chart based on 25 samples of size 50 with an average proportion defective of 0.1544. The process appears to be in control in terms of variation because all sample proportions for the 25 samples after the initial calibration are within the control limits.

Number of groups = 50
Center = 0.1544
StdDev = 0.3613318
LCL = 0.001099918
UCL = 0.3077001
Number beyond limits = 0
Number violating runs = 0

(Some computer programs have the ability to accommodate variable sample sizes for p charts, even though the control limits will vary with the sample size. If the sample sizes don't vary widely, it is also possible to use the average sample size and keep the control limits constant.)

Another variation on this type of monitoring focuses on the number of defective items in each sample rather than the proportion. This process is tracked in a **c chart**. The c chart should be used when the occurrence of a defective is *rare*. Then, the probabilities can be based on the Poisson model rather than the Binomial, which can lead to better ability to detect shifts. Because the standard deviation of a Poisson variable with mean μ is $\sqrt{\mu}$, we find the control limits easily. If the standard number of defects c is known we could simply use the 3σ control limits:

$$LCL = c - 3\sqrt{c}$$
$$UCL = c + 3\sqrt{c}$$

When no standard is given, we find the average number of defectives, $\bar{c}$, from k samples of size n and use that in the control limits centered at $\bar{c}$:

$$LCL = \bar{c} - 3\sqrt{\bar{c}}$$
$$UCL = \bar{c} + 3\sqrt{\bar{c}}$$

A more serious type of defect in wafer manufacturing is the failure of a die, the integrated circuit that eventually becomes chips. If the number of failed dies per wafer is too great, the wafer is scrapped. If the number of scrapped wafers becomes too large, the process may need to be recalibrated or examined for other problems.

Suppose that for a period in which the process was in control, the average number of scrapped wafers was 1.44 in 1000. Then, in subsequent samples of size 500 we would expect 0.72 defects per sample. The control limits then become:

$$LCL = \bar{c} - 3\sqrt{\bar{c}} = 0.72 + 3\sqrt{0.72} = -1.826$$
$$UCL = \bar{c} + 3\sqrt{\bar{c}} = 0.72 + 3\sqrt{0.72} = 3.266$$

Number of groups = 25
Center = 0.72
StdDev = 0.7483315

LCL = 0
UCL = 3.265584

Number beyond limits = 0
Number violating runs = 0

Figure 22.11   A *c* chart based on the same 25 samples of size 50 with an average number of defectives of 0.72. The process appears to be in control in terms of variation because all sample counts of defectives for the 25 samples after the initial calibration are within the control limits. Notice that the bottom control limit has been set to 0 because 3 standard deviations below 0.72 was negative.

For Example   **Control Charts for Attributes**

The fudge company (see page 774) solicits customer feedback by surveys and phone interviews. One of the questions asks how satisfied customers were with the fudge on a scale from 1 to 5: 1 = Very Dissatisfied; 2 = Dissatisfied; 3 = Neither Satisfied nor Dissatisfied; 4 = Satisfied; 5 = Very Satisfied. They want 80% or more of their customers to be "Very Satisfied." Here are the percentages saying "Very Satisfied" from the last 15 surveys of 100 people.

Number responding "5" out of 100: 78 78 88 77 76 79 85 75 81 88 86 74 76 81 75

Question: Is the process "in control"?

Answer: Using a p chart with center line $p = 0.8$ the control lines are:

$$UCL = p + 3\sqrt{\frac{pq}{n}} = 0.8 + 3\sqrt{\frac{(.8)(.2)}{100}} = 0.92$$

$$LCL = p - 3\sqrt{\frac{pq}{n}} = 0.8 - 3\sqrt{\frac{(.8)(.2)}{100}} = 0.68$$

From the p chart, it appears that the process is in control.

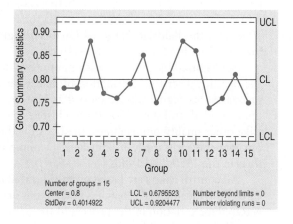

Number of groups = 15
Center = 0.8   LCL = 0.6795523   Number beyond limits = 0
StdDev = 0.4014922   UCL = 0.9204477   Number violating runs = 0

22.6   Philosophies of Quality Control

As U.S. industry saw its global market share slip in the 1970s and 1980s, and worldwide customers began to rely on Japanese and German products for high quality, a renaissance of interest in quality improvement sprang up all over the United States. During this time, many similar quality philosophies, acronyms, and gurus competed for the attention of top management. Most of them share three common ideas from statistical thinking as outlined by the Statistics Division of the American Society for Quality (ASQ):

• All work occurs in a system of interconnected processes,

• Variation exists in all processes, and

• Understanding and reducing variation are keys to success.[3]

Although the United States and Britain were the birthplace of modern quality control, the efforts of the quality movement in Japan caught the attention of U.S. industry in the decades after WWII. Deming is often credited for much of the success of Japan's manufacturing in the latter part of the twentieth century.

---

[3]Glossary of Statistical Terms, Quality Press 1996.

Deming ambitiously shared his ideas of concentrating on the customer and cutting costs by concentrating on quality in all aspects of the company in an effort to reduce waste and raise efficiency. Deming also suggested that the Japanese use market research—a new field in the United States and virtually unknown at the time in Japan—to help satisfy the needs of the customers. At the core of market research was the concept of random sampling to understand customers' desires and needs. Deming also conveyed the idea that managers need to take responsibility for quality control. He taught the Shewhart cycle and explained the importance of managers "acting" on the analytical information provided by the quality control tools. Deming also explained the importance of viewing the supplier as a partner and of building cooperative relationships, so that the supplier, producer, and customer all form a network. The teachings of Deming's philosophy can be summarized as **total quality management (TQM)**, although the term was not used until 1974 by one of the Deming Prize recipients.[4] TQM extends the theory of quality control in that it encompasses the total organization, uses quality control methods, and focuses on the leadership provided by management to ensure long-term success, by focusing on customer satisfaction. It is a broad approach that seeks to improve all areas of the business to better satisfy the customer. The philosophy of TQM plays a central role in the Criteria for Performance Excellence, published in 1988, and the Baldrige National Quality Program (BNQP) that provides benchmarks for quality for organizations today.

Six Sigma® is another business improvement approach that focuses more on the identification and elimination of defects or special-cause variation in business processes. Six Sigma was developed at Motorola during the 1980s but was picked up, expanded, and spread to other industries by General Electric and Allied Signal. Like other quality philosophies, the basis for the Six Sigma approach and philosophy was the early work of Shewhart, Deming, and others who furthered their efforts in the 1970s and 1980s.

All control charts are based on the idea that the process should be monitored to catch individual items or batches that are out of control. The core idea of Six Sigma is to reduce variation in the process to the point that inspection is no longer necessary. If the standard deviation can be made small enough so that the specification limits defining what it means for a part to be out of control are 4½ standard deviations from the mean, then the probability of an out-of-control observation is only about 6 in 1,000,000. In real-life processes, the mean of the process can shift somewhat as well. To account for that the developers of Six Sigma added an allowance of a 1½ standard deviation (sigma) shift. In this case, to achieve the 6 in 1,000,000 probability, the standard deviation has to be so small that 6 standard deviations are still within the control limit. This corresponds to a C_{pk} value of 2.0. At this point, it is no longer necessary (in theory) to monitor the process with control charts for inspection because the probability of being out of control is so small. Although this was the original idea of Six Sigma, it has become a much broader philosophy and approach to quality as practiced by large firms such as the General Electric Company. There is a hierarchy of experts (known as Black Belts) who, after being trained in the Six Sigma philosophy, act as internal quality consultants on a wide variety of quality issues.

---

[4] The American Society for Quality (ASQ) maintains that the term *TQM* was initially used by the U.S. Naval Air Systems Command in 1984 to describe the management approach to quality used in Japan because they did not like the term Total Quality Control.

Ethics in Action

Trax, one of the leading manufacturers of bicycles in the United States, is interested in expanding its line to include high-end road bikes. One of the features of high-end bikes, and one that avid cyclists pay top dollar for, is the use of super-light carbon fiber frames. Trax is currently in the process of selecting a supplier for these frames. Potential suppliers were asked to provide reports that included statistical process control (SPC) and capability data on their carbon fiber frame production processes.

Trax had specified the weight for these frames to be between 2 and 3 pounds with a target mean of ±2.5 pounds. John Dickinson, Vice President of Operations, was in charge of summarizing the information contained in these reports to present at an upcoming senior management meeting at which a final decision on a supplier would be made. As he reviewed the reports, he noticed that one supplier, DBS, could provide the frames at a much lower cost than its competitors. Eager to save the company money, he carefully considered this supplier's SPC and process capability data. John was pleased with their $\overline{X}$ and R charts for weight that indicated the process was stable (in statistical control). He was also pleased to be able to verify from these charts that their reported process standard deviation is only 0.1 pounds and that the C_p index of 1.67 indicates that they would be able to easily meet Trax's weight requirement. While he did notice that the process mean was a bit off target (at 2.8 rather than 2.5 pounds), he decided that the significant cost savings that Trax would realize by choosing this supplier outweighed any potential negative consequences of a shifted process mean. John decided that he would present only cost data and the C_p index for the suppliers under consideration, and conclude his presentation at the senior management meeting with a recommendation that Trax choose DBS as its supplier of carbon frames. He believed that the team would be quite impressed with his cost-saving recommendation.

ETHICAL ISSUE *John is misrepresenting the process capability of DBS. Although the C_p index is greater than one, because the process is not centered on target John should be reporting the C_{pk} index, which in this case would be 0.67. DBS's process will not be able to meet the upper specification limit and some of the carbon fiber frames it supplies to Trax will be over 3 pounds (related to Item C, ASA Ethical Guidelines).*

ETHICAL SOLUTION *John should report all relevant SPC and process capability data, most importantly the C_{pk} index for any process that is not centered on target. He should not let his bias toward wanting to save money, and impress his colleagues, affect his judgment. Carbon fiber frames above the weight limit will adversely affect Trax's bottom line.*

What Have We Learned?

Learning Objectives

■ Understand the principles of quality control.

- A process is in control if the aspects being followed stay within the control limits established for them.

- When a process drifts out of control, it is important to recognize the fact so that adjustments can be made promptly. But we hope to avoid making too many false alarms, so a balance must be found.

■ Use run charts to track individual observations.

- A level shift flags a sudden change. A trend is a more gradual change.

- The ability of a chart to detect shifts is described by its operating characteristics (OC) curve.

- The Western Electric Handbook specified several rules for deciding when a process is out of control.

■ Use $\overline{X}$, R, and S charts to track a process by plotting the means, and ranges or standard deviations of small samples drawn from the process.

- The control limits are often based on estimated standard deviations and means.

- The S chart, which uses the standard deviations of the samples rather than their ranges is more appropriate than the R chart used when the samples are moderately large.

(continued)

- Understand the basic steps in creating an out-of-control action plan (OCAP).
  - A cause-and-effect diagram helps organize possible causes and prioritize them for further analysis.
  - It may be necessary to perform experiments to determine the cause of the problem.
  - Once a cause is identified, process inputs can be adjusted to drive the process back into control.
  - The monitoring process can then be restarted.

- Use *p* charts and *c* charts to track processes that have categorical measurements.
  - A *p* chart tracks proportions using control limits based on the standard deviation of a proportion.
  - A *c* chart tracks the number of defective items in a process.

- Understand the basic philosophies of quality control, such as Total Quality Management (TQM) and Six Sigma®, and how they are applied to improve business performance and quality.

Terms

Average run length (ARL)	The in-control ARL is the average time between false alarms, calculated by taking the reciprocal of the false alarm probability. The out-of-control ARL is the average number of runs until a signal is detected. It depends on the state of the process while out of control.
c chart	A control chart that displays the counts, or number, of occurrences of defectives (or non-defectives if rare) in samples taken from a process over time.
C_p	A process capability index that measures the ratio of the specification limit range to 6 times the standard deviation of the process without taking into consideration possible drift of the process: $$C_p = \frac{USL - LSL}{6\sigma}$$
C_{pk}	A process capability index that measures what the process is capable of producing, estimated by: $$\hat{C}_{pk} = \min\left[\frac{USL - \hat{\mu}}{3\hat{\sigma}}, \frac{\hat{\mu} - LSL}{3\hat{\sigma}}\right]$$
Common-cause variation	Variation characterized by apparent random fluctuations around an average level within specified limits called control limits.
Control chart or Shewhart chart	A graph that displays variation by plotting individual measurements or sample statistics taken from a process over time, sometimes referred to as a Shewhart chart.
Control limits	Horizontal lines drawn on a control chart. Various rules are used to determine when a process is out of control depending on the specific control chart.
Dichotomous variable	A categorical variable that has only two possible outcomes such as success or failure. Also called a yes/no variable.
False alarm	A Type I error indicating that the process is out of control, when in actuality, it was due only to common-cause variation.
Level shift	A sudden and sustained shift in the mean of the measurements being monitored.
Operating-characteristic (OC) curve	A chart that shows the probability of failing to see a process shift (a Type II error) plotted against the size of the shift (in standard deviation units).

Out of control	The presence of special-cause variation in a process.
Out-of-control Action Plan (OCAP)	A set of instructions to follow for well-established processes indicating what action to take following an out-of-control signal.
p chart	A control chart that displays the proportion of defectives (or nondefectives) in samples taken from a process over time.
Process Capability Study	A study that compares the variability of the process to the specification limits imposed on the process by external concerns. Two of the most common process capability indices are C_p and C_{pk}.
R chart	A control chart that plots the ranges of samples taken from a process over time.
Run chart	A control chart that displays individual measurements from a process taken over time.
S chart	A control chart that plots the standard deviations of samples drawn from a process. It may be preferred to the *R* chart when the samples are relatively large (10 or more).
Six Sigma®	A quality control philosophy based on reducing variation to the point where inspection sampling is no longer necessary. The philosophy's name comes from the fact that if specification limits (plus possible drift) are 6 standard deviations from the control center line, the chances of an out-of-control part are negligible.
Special-cause variation	Causes of variation that are not intrinsic to the process and thus whose presence is important to detect.
Specification limits	Limits above (and/or below) which a measurement should not lie, imposed by external considerations.
Statistical process control (SPC)	The examination of variation in a process over time to identify sources of variation.
Total quality management	Extends the theory of quality control to include the total organization and the leadership of management.
Trend	The gradual shift of a process mean.
$\overline{X}$ or *x*-bar chart	Control chart that displays the averages of samples taken from a process over time.

Technology Help: Quality Control Charts

EXCEL   XLSTAT⁷

Excel does not offer quality control charts. A number of third-party add-ins and templates are available.

JMP

From the **Graph** menu select the **Control Chart** submenu. There you will find a variety of control charts.

MINITAB

From the **Stat** menu, select the **Control Charts** submenu. There you will find an additional level of submenus offering Variables Charts for

Subgroups, Variables Charts for Individuals, Attributes charts, and Time Weighted charts. Standard control chart types are found within these submenus.

SPSS

From the **Analyze** menu, select the **Quality Control** submenu. Selecting **Control Charts** from that submenu brings up a dialog box from which you can select the chart type and specify other details.

Brief CASE

Laptop Touchpad Quality

Consumers always seem to want thinner and lighter electronic products, and computer laptops are no exception. One of the challenges of producing extremely thin and light laptops is the manufacturing of the touch pads. In particular, touch pads are difficult to align. A common complaint is that one corner is higher or lower than another, resulting in what is known as "tilting." Tilting makes navigation difficult, so touch pad manufacturers have strict limits on the amount of tilting that they will allow in a pad.

A company producing touch pads will want to ensure that the pads are flat to within a certain acceptable tolerance. One possible measure is the total deviation from flatness as measured by the maximum difference in height between any two corners (in mm). In the data file **flat.control** you will find this flatness measure for 250 touch screens produced during a period that was judged to be in control.

Using the mean and standard deviation from this control period, examine the next 150 samples of 5 pads (in the data file **flat.prod**). Using various quality control rules and appropriate graphs, do you think the process is still in control? Explain.

Instead of samples of size 5 suppose they had used a run chart on the individual pads. Would your answer change? Explain.

Suppose the engineers establish 14 mm as the maximum flatness deviation they will tolerate and want to use that as an upper specification limit. Does your answer change now? Explain.

A qualitative measure is clarity of the pad. Batches of 100 pads are produced and defects are counted where defect is defined as a pad that does not pass the clarity test. From a period that was deemed to be in control, the mean number of defects per 100 was 0.5. Using this mean, determine whether the process stayed in control for the next 50 batches (clarity.prod).

Exercises

SECTION 22.1

1. A producer of beverage containers wants to ensure that a liquid at 90°C will lose no more than 4°C after 30 minutes. Containers are selected at random and subjected to testing. Historical data shows the standard deviation to be 0.2°C. The quality control team has set 3.5 and 4.5°C as the specification limits. Here are the data from the last 12 containers tested:

Container #	Temperature Loss (°C)	Container #	Temperature Loss (°C)
1	3.89	7	4.22
2	3.71	8	4.02
3	4.14	9	4.48
4	4.12	10	3.93
5	3.74	11	3.87
6	3.99	12	3.94

a) Make a control chart of these data using the specification limits given.
b) Does the process seem to be in control?
c) What is the value of C_p for the process?
d) Given this value of C_p, what recommendations might you make to the quality control team?

2. A computer screen manufacturing process must produce screens of uniform size. In particular, for a new tablet, the screen should be 10.1 inches long. The engineers have set 10.16 and 10.04 as the upper and lower specifications on screen length. If the actually process standard deviation is 0.01 inches,

a) What is the value of C_p for the process?
b) What might you recommend to the engineers given this value of C_p?

SECTION 22.2

3. For the data in Exercise 1, instead of the specification limits,

a) Use the historical standard deviation to set up 1, 2 and 3σ limits and draw the control chart.

b) Are any points outside the 3σ limits on either side?

c) Are two of three consecutive points outside of 2σ on either side?

d) Are four out of five consecutive points outside of 1σ on either side?

e) Are there eight points in a row on the same side of the mean?

f) Are any six points in a row increasing (or decreasing)?

g) Are any fourteen points in a row alternatively increasing and decreasing?

h) Based on your answers to parts b–g, what do you conclude?

4. For the data in Exercise 1, suppose the historical standard deviation had been 0.1°C instead of 0.2°C. How would that change your answers to the questions in 3?

SECTION 22.3

5. A process to make dental floss produces spools with a desired mean length of 91.4 m (100 yds). The historical standard deviation is 1 cm (0.01 m). If samples of size 5 are taken and measured,

a) What is the standard deviation of the sample mean?

b) How long (or short) would a sample mean have to be to fall outside the 3σ limit?

c) How would the answers to parts a and b change if the sample were of size 10?

6. As in Exercise 5, a process to make dental floss produces spools with a desired mean length of 91.4 m (100 yds). The historical standard deviation is 1 cm (0.01 m). If samples of size 5 are taken and measured,

a) What is the center line for the *R* chart?

b) What is the standard deviation of the sample range?

c) How large (or small) would the sample range have to be to fall outside the 3σ limit?

SECTION 22.5

7. One of the ways in which the dental floss discussed in Exercise 5 can fail is if it snaps when pulled with a certain tension. Historically, only 2% of the tested floss snaps. Suppose boxes of size 50 are selected at random to be tested.

a) What is the standard deviation for the sample proportion from a box of 50?

b) What are the upper and lower 3σ control limits?

c) How many packages in a box of 50 would have to fail for a process to be declared out of control using the 3σ control limits?

8. Rather than use the proportion of packages that contain defective floss, another quality control inspector simply counts the number of packages. If historically, only 1 package out of 50 is defective,

a) What is the standard deviation of the number of defective packages?

b) What are the upper and lower 3σ control limits on the number of defectives?

c) How does the upper limit compare to the upper limit in Exercise 7, part c?

CHAPTER EXERCISES

9. Super Glue®. Cyanoacrylates, the generic name for several compounds with strong adhesive properties, were invented during WWII during experiments to make a special extra-clear plastic suitable for gun sights. They didn't work for gun sights, however, because they stuck to everything. They were first commercialized by the Eastman Kodak company in the 1950s and are now sold under the names Super Glue and Krazy® Glue. They are often sold in small quantities, typically 2-oz. tubes. The manufacturers want to ensure that the same quantity of glue goes into each tube so they monitored the process from the past three weeks (when there were no production problems) and calculate the mean and standard deviation weight from 50 tubes. If they use those quantities to calculate future limits for quality control monitoring, are they constructing control or specification limits? Explain.

10. MLB®. Major League Baseball (MLB) requires that a ball used in a major league game weigh between 5 and 5.25 oz. (www.MLB.com). Rawlings® is the official supplier of baseballs for MLB. In their production process, do they treat 5 and 5.25 oz. as control or specification limits? Explain.

11. Pole vault. Pole vault technology has evolved considerably since the times of the ancient Greeks and even more recently as new materials have found their way into pole manufacturing. There are currently no regulations on the composition of poles although for safety reasons, there are restrictions on the amount of bend depending on the vaulter's weight. On the other hand, safety considerations have led to the U.S. National High School Federation (NHSF) to require that high school vault landing pads must be no less than 6.0-m wide and 6.15-m long. Should the manufacturers of landing pads treat these as specification or control limits? Explain.

12. Milk production. Whether pasteurized or raw, both the European Economic Council (EEC) and the U.S. Food and Drug Administration (USFDA) require that milk sold to consumers contain no more than 10 coliform bacteria per mL of milk (www.cdfa.ca.gov). Should dairies take this as a control or specification limit when applying quality control to milk production? Explain.

13. Observatories. The two Gemini observatories (www .gemini.edu) were built and are operated by a partnership of seven countries including the United States, United Kingdom, Canada, Chile, Australia, Brazil, and Argentina. Astronomers in each of these countries are able to apply for time on the telescopes for scientific research. Gemini sets strict technical requirements on many aspects of the filters used on the telescope including a strict requirement that they be extremely flat—their total deviation from absolute flatness must be no more than 30 nm as measured by the root mean square deviation. How should filter manufacturers view this requirement, as a control or specification limit? Explain.

14. DVD production. Plastic covers for DVDs must be manufactured with enough precision so that a DVD can properly fit into the cover. Specifically, the diameter of the central plastic piece over which the DVD is placed must not vary more than a certain amount. A manufacturer has been producing DVD covers for several months and has had no complaints from its customers. It decides to use the past 2 months of data to estimate the mean and standard deviation of the process. Are the limits for quality monitoring based on these quantities control or specification limits? Explain.

15. Roulette. An observer at the roulette wheel in Monte Carlo notices that there seems to be no pattern to the numbers that come up during an entire evening. Is this an example of special- or common-cause variation? Explain.

16. Pennsylvania lottery. In 1980, millions of viewers watched as the number 666 was drawn for the Pick 3 Pennsylvania lottery. Although as likely as any other number, lottery authorities and local bookmakers became suspicious when they noticed that a large number of tickets were purchased for the eight numbers 444, 446, 464, 466, 644, 646, 664, and 666. Soon, a handful of players came forward to claim approximately $1.8 million of the then-record $3.5 million payout. Officials eventually discovered that 8 of the 10 balls were weighted, giving them almost no chance to be selected in the drawing. Is this an example of special- or common-cause variation? Explain.

17. Out-of-control rules. Most control charts use three sigmas (standard deviations) to set the upper and lower control limits. For a run chart, what is the probability that an observation would fall beyond 3σ from the mean based on random chance variation alone? What assumptions do you have to make to answer this?

18. Out-of-control rules, part 2. Instead of one observation outside of 3σ from the mean as in Exercise 17, suppose a quality control engineer decides to use only the run chart rule that the system is out of control if 8 observations in a row lie on the same side of the mean. What is the probability that

would happen based on random chance variation alone? What assumptions do you have to make to answer this?

T 19. Basketballs. A company that makes basketballs has the motto: "Our basketballs are ready to play." Therefore, it is important to the company that the basketballs are inflated with the proper amount of air when shipped. Most basketballs are inflated to 7 to 9 pounds per square inch. From a set of calibration data, the following statistics were measured:

$$\overline{\overline{X}} = 8.05$$
$$\overline{R} = 0.65$$

Recently the company selected a random basketball from its production line at four different time periods over its normal production day. The results were recorded and listed in the following table.

	Shift			
	8:00–10:00 AM	10:00 AM	Noon–2:00 PM	2:00–4:00
Day 1	8.05	8.49	7.59	7.90
Day 2	8.16	7.62	7.71	7.90
Day 3	8.30	8.07	8.14	8.12
Day 4	7.55	8.82	8.12	8.18
Day 5	8.18	7.76	7.70	7.65
Day 6	7.88	7.90	7.99	7.62
Day 7	8.01	8.02	7.32	8.10
Day 8	8.32	7.89	8.62	7.89
Day 9	7.94	7.75	7.58	7.74
Day 10	7.38	7.47	7.93	7.73
Day 11	7.44	7.84	7.87	8.33
Day 12	8.30	8.04	7.94	8.09
Day 13	8.46	8.20	7.93	7.89
Day 14	8.10	8.38	8.08	7.80
Day 15	7.70	7.68	8.25	8.40
Day 16	8.32	7.98	7.65	8.32
Day 17	8.13	7.96	8.21	8.51
Day 18	7.69	8.07	7.55	8.26
Day 19	7.86	8.55	8.05	7.63
Day 20	7.97	8.08	7.89	7.97
Day 21	8.73	7.84	7.94	8.07
Day 22	8.14	7.56	7.97	7.86
Day 23	7.64	7.89	7.47	7.66
Day 24	8.33	7.87	7.89	7.62
Day 25	8.00	7.61	7.65	7.39
Day 26	7.31	8.66	7.71	7.29
Day 27	7.58	7.95	8.13	8.52
Day 28	7.91	8.53	7.86	7.96
Day 29	8.34	8.10	8.45	8.16
Day 30	8.04	8.25	7.13	8.12

a) Create an $\overline{X}$ chart based on the calibration data statistics for these 30 samples.
b) Create an R chart based on the calibration data statistics for these 30 samples.
c) Assuming that the process was in control during the calibration period, is the company's process for filling its basketball with the proper amount of air in control?

T 20. Rulers. A company that traditionally made rulers and yardsticks is setting up a manufacturing process to make metersticks. Obviously, accuracy is important for their new product. The company runs a 24-hour production process. Using a calibration set of data they found:

$$\overline{\overline{X}} = 1.000$$
$$\overline{R} = 0.0016$$

They took a sample of 3 metersticks each hour and recorded the results in the following table.

	Sample 1	Sample 2	Sample 3
Hour 1	0.9983	1.0007	1.0010
Hour 2	1.0005	1.0011	1.0014
Hour 3	1.0002	0.9995	0.9997
Hour 4	1.0004	1.0016	1.0007
Hour 5	1.0014	0.9994	1.0008
Hour 6	0.9990	0.9996	0.9984
Hour 7	1.0014	1.0010	1.0017
Hour 8	1.0017	1.0002	1.0002
Hour 9	1.0003	1.0008	0.9993
Hour 10	0.9998	1.0002	0.9992
Hour 11	0.9985	1.0002	1.0006
Hour 12	1.0003	1.0012	0.9994
Hour 13	0.9998	1.0013	0.9995
Hour 14	1.0000	0.9994	0.9992
Hour 15	1.0005	0.9999	1.0016
Hour 16	1.0019	1.0018	1.0010
Hour 17	0.9993	0.9994	0.9993
Hour 18	1.0020	1.0000	0.9978
Hour 19	1.0007	1.0005	0.9991
Hour 20	1.0009	1.0005	0.9999
Hour 21	1.0002	0.9998	1.0021
Hour 22	1.0019	0.9986	0.9996
Hour 23	1.0011	1.0003	0.9993
Hour 24	0.9993	1.0010	1.0001

a) Create an $\overline{X}$ chart based on the calibration data statistics for these 24 hourly samples.
b) Create an R chart based on the calibration data statistics for these 24 hourly samples.

c) Assuming that the process was in control during the calibration period, is the company's process for making accurate meter sticks out of control?

T 21. Graphite production. A graphite manufacturer makes long rolls of flexible graphite to be used to seal components in combustion engines. The specifications state that the mean strength should be 21.2 ounces per square yard with a standard deviation of 0.29. Further specifications state that no roll should have strength less than 20.2 or more than 22.2 ounces per square yard. If there is a defect in terms of the strength of the graphite rolls, the seal will not hold. After the roll is created, a beta scanner takes readings of the basis weight in ounces per square yard. The data is separated into 3 lanes with 20 scans in each lane. A sample consists of one roll from each lane. The results from 20 samples follow:

Sample	Lane 1	Lane 2	Lane 3
1	20.65	21.01	20.76
2	21.27	21.53	21.53
3	21.69	21.53	21.14
4	21.06	20.38	20.62
5	21.18	21.29	20.95
6	21.14	21.01	21.37
7	20.89	20.63	21.24
8	21.73	21.39	20.87
9	21.55	21.34	21.17
10	21.72	21.97	21.84
11	21.26	20.77	21.24
12	21.05	20.70	20.87
13	22.23	22.09	21.62
14	21.27	21.71	21.89
15	21.60	21.03	20.68
16	21.37	21.29	21.13
17	21.19	21.80	21.36
18	21.56	21.64	21.67
19	21.12	20.56	20.57
20	20.88	20.57	20.37

a) Using the given mean and standard deviation, create an $\overline{X}$ chart for these samples.
b) Using the given mean and standard deviation, create an R chart for these samples.
c) Is the company's process for making flexible graphite sheets out of control?
d) Do all the rolls meet the specification limits?

Sample	Lane 1	Lane 2	Lane 3	Lane 4	Lane 5	Lane 6	Lane 7	Lane 8	Lane 9	Lane 10
1	21.56	21.59	21.26	21.16	21.01	21.10	21.22	20.95	20.78	20.66
2	21.42	21.54	21.43	21.44	20.99	21.10	20.78	21.17	21.21	21.10
3	21.48	21.23	21.19	21.34	21.63	21.47	21.18	21.05	21.21	21.22
4	21.14	21.26	21.57	21.34	21.36	21.37	21.42	21.48	21.14	21.02
5	21.53	21.26	21.43	21.26	20.95	20.86	20.99	20.90	20.69	20.80
6	21.40	21.26	21.19	21.38	21.57	21.30	20.97	20.72	20.78	20.70
7	21.84	21.51	21.71	21.34	21.38	21.32	21.32	21.05	20.57	20.27
8	21.83	21.83	21.56	21.17	20.74	20.83	21.00	21.35	21.54	21.41
9	21.60	21.57	21.44	20.99	20.60	20.83	20.85	20.90	21.14	21.36
10	21.84	21.93	21.99	21.97	21.57	21.56	21.17	20.89	21.20	21.28
11	21.96	21.50	21.48	21.37	21.57	21.68	21.82	21.65	21.20	21.19
12	21.27	21.53	21.31	21.22	21.33	21.31	21.46	21.68	21.69	21.52
13	22.03	21.90	22.02	21.46	21.34	21.35	21.44	21.84	21.55	21.63
14	21.30	21.36	21.28	21.29	21.00	20.82	20.55	20.48	20.46	20.68
15	21.65	21.18	20.77	20.98	20.92	20.99	21.27	21.56	21.72	21.67
16	21.46	21.52	21.47	21.45	21.31	21.76	21.92	21.34	21.11	21.19
17	21.72	21.69	21.77	21.44	21.13	21.12	20.91	20.60	20.69	21.07
18	21.67	22.02	21.78	21.34	21.00	20.80	20.30	20.54	20.59	21.16
19	21.44	21.28	21.15	20.97	20.85	21.17	21.87	21.67	21.59	21.20
20	21.38	21.32	20.91	21.10	21.17	21.58	21.53	21.54	21.12	20.69
21	21.23	21.30	21.57	21.42	21.52	21.42	20.90	20.85	20.60	20.71
22	21.65	21.33	21.44	21.27	21.06	20.80	20.82	20.84	20.98	21.22
23	21.59	21.58	21.13	21.34	21.57	21.47	21.55	21.83	21.56	21.02
24	20.92	20.74	20.30	20.07	20.48	20.39	20.81	20.90	21.18	21.18
25	21.02	21.03	21.00	21.04	21.15	21.10	21.24	21.09	21.25	20.86

T 22. Graphite production, 10 lanes. The same graphite manufacturer from Exercise 21 uses an alternative process to make rolls. For this process the data are separated into 10 lanes—a sample consists of one roll from each lane. The results of 25 samples are shown in the table at the top of the page.

a) Using the same mean and standard deviation as given in Exercise 21, create an $\overline{X}$ chart for these samples.
b) Using the mean and standard deviation as in Exercise 21, create an R chart for these samples.
c) Is the company's process for making flexible graphite sheets in control?
d) Are all of these rolls within the specification limits?

T 23. Defect monitoring. The following data are from a production process that makes 250 units each hour. The data were collected over a normal 12-hour shift one day. The data are shown in the table at the right.

Using the first 10 hours as the calibration period,

a) Calculate the center line of the p chart.
b) Calculate the upper control limit of the p chart.
c) Calculate the lower control limit of the p chart.
d) Create a p chart for the samples from the second set of 10 hours.
e) Is this process out of control?

Hour	No. of Defects	Sample Size
1	17	250
2	15	250
3	12	250
4	20	250
5	13	250
6	12	250
7	19	250
8	14	250
9	14	250
10	18	250
11	17	250
12	16	250
13	19	250
14	22	250
15	18	250
16	13	250
17	23	250
18	17	250
19	22	250
20	19	250

T 24. Defect monitoring, second product. The following data are from a production process that makes approximately 250 units each hour. The data were collected over a normal 12-hour shift one day. Historical data shows the proportion of defects to be 6.21%. Either use technology to accommodate the different sample sizes, or use 250 as an approximation for all of them.

Hour	No. of Defects	Sample Size
1	13	241
2	18	265
3	17	256
4	14	249
5	20	221
6	19	276
7	12	244
8	15	261
9	12	259
10	17	243
11	14	237
12	16	257
13	18	252
14	19	260
15	17	245
16	20	255
17	19	262
18	14	244
19	22	270
20	18	254

Assuming the process was previously in control,

a) Calculate the center line.
b) Create a *p* chart for these samples.
c) Is this process out of control?

T 25. Computer chip manufacturing. MediaChip manufactures computer chips specifically for MP3 players. The process uses sophisticated lasers to imprint several hundred chips on a single silicon "wafer." While the process is completely automated and set in a clean room environment, a batch of chips can contain imperfections that render the chips useless. This results in significant cost overruns. MediaChip would like to use a control chart to monitor their manufacturing process. Presently, the production level is set to 250 wafers per day. The average number of defects that can be tolerated is 1.7 per day. The number of defective wafers was tracked and recorded in the following table.

Day	Defects	Day	Defects
1	1	11	3
2	1	12	3
3	2	13	2
4	0	14	2
5	2	15	1
6	2	16	2
7	2	17	1
8	3	18	2
9	2	19	2
10	2	20	1

a) Create a *c* chart.
b) Is MediaChip's manufacturing process in control?

T 26. Faulty pixels. In LCD screens, every attempt is made to make sure that all pixels illuminate when they receive a signal, but it is impossible to ensure that every pixel in a large screen will actually work. A screen is determined to be defective if more than 2 pixels fail to light. Every week 500 screens are randomly selected and a quality control team member counts the number of defective screens. Out of 500 screens, management would like an average of no more than 5 defective screens. The results for the last 50 weeks are shown in the following table.

Week	Defective Screens	Week	Defective Screens	Week	Defective Screens
1	5	18	4	35	6
2	6	19	4	36	6
3	4	20	4	37	7
4	5	21	3	38	3
5	1	22	3	39	8
6	7	23	5	40	5
7	1	24	3	41	7
8	4	25	7	42	5
9	3	26	10	43	5
10	4	27	5	44	5
11	4	28	11	45	5
12	6	29	2	46	8
13	4	30	3	47	3
14	3	31	2	48	6
15	4	32	5	49	4
16	1	33	11	50	4
17	4	34	9		

a) Create a *c* chart.
b) Is the number of defective screens out of control? Comment.

27. Patient complaints. The following data were collected from the number of patient complaints from a small medical service facility over a two-week period. The past year the average number of complaints was 4.57. The management team would like to use that as a baseline for monitoring.

Day	1	2	3	4	5	6	7	8	9	10	11	12	13	14
Complaints	3	6	1	2	12	5	7	0	4	6	3	5	2	8

Based on last year, using a *c* chart:

a) What is the center line?
b) Calculate the upper control limit.
c) Calculate the lower control limit.
d) Create the *c* chart for this two-week period.
e) Is this process out of control?

28. Patient forms. The medical facility in Exercise 27 also keeps track of the number of errors found when transcribing information from patient forms. Last year the average number of errors was 1.26 per day which the facility found acceptable. Here are the past two weeks' data:

Day	1	2	3	4	5	6	7	8	9	10	11	12	13	14
Errors	1	2	1	2	1	2	5	3	3	2	0	3	5	5

Based on last year, using a *c* chart:

a) What is the center line?
b) Calculate the upper control limit.
c) Calculate the lower control limit.
d) Create the *c* chart for this two-week period.
e) Is this process out of control?

T 29. Baseball circumferences. Baseballs used in the major leagues in the United States must adhere to strict standards. One such standard is that the circumference must be between 9 and 9.25 inches. One could look at this standard as a target circumference of 9.125 inches with a tolerance of $+/-0.125$ inches. These data (shown at the bottom of the page) were taken from a random sample of 50 baseballs from a large production batch. The actual circumference and deviation from the target circumference are listed.

Using the specification limits as the control limits,

a) Create a run chart for the baseballs' circumferences.
b) Is the process for making baseballs in terms of their circumference in control?
c) How many baseballs produced were out of spec?
d) When should the quality team have investigated the production process?

Sample	Circumference	Deviation	Sample	Circumference	Deviation
1	8.998	−0.127	26	9.121	−0.004
2	9.181	0.056	27	9.111	−0.014
3	9.147	0.022	28	9.130	0.005
4	9.182	0.057	29	9.116	−0.009
5	9.087	−0.038	30	9.035	−0.090
6	9.154	0.029	31	9.188	0.063
7	9.108	−0.017	32	9.092	−0.033
8	9.109	−0.016	33	9.078	−0.047
9	9.141	0.016	34	9.101	−0.024
10	9.093	−0.032	35	9.113	−0.012
11	9.142	0.017	36	9.087	−0.038
12	9.140	0.015	37	9.054	−0.071
13	9.125	0.000	38	9.177	0.052
14	9.175	0.050	39	9.058	−0.067
15	8.985	−0.140	40	9.179	0.054
16	9.119	−0.006	41	9.147	0.022
17	9.140	0.015	42	9.123	−0.002
18	9.109	−0.016	43	9.097	−0.028
19	9.190	0.065	44	9.108	−0.017
20	9.087	−0.038	45	9.214	0.089
21	9.046	−0.079	46	9.140	0.015
22	9.108	−0.017	47	9.151	0.026
23	9.188	0.063	48	9.041	−0.084
24	9.125	0.000	49	9.061	−0.064
25	9.114	−0.011	50	9.162	0.037

T **30.** Baseball weights. Baseballs used in the major leagues in the United States must adhere to strict standards. One such standard is that the weight must be between 5 and 5.25 ounces. One could look at this standard as a target weight of 5.125 ounces with a tolerance of $+/-0.125$ ounces. The following data were taken from a random sample of 50 baseballs from a large production batch. The actual weight and deviation from the target weight are listed.

Using the specification limits as the control limits,

a) Create a run chart for the baseballs' weights.
b) Is the process for making baseballs in control?
c) How many baseballs produced were out of spec?
d) When should the quality team have investigated the production process?

Sample	Weight (oz.)	Deviation	Sample	Weight (oz.)	Deviation
1	5.060	−0.065	26	5.073	−0.052
2	5.014	−0.111	27	5.133	0.008
3	5.118	−0.007	28	5.075	−0.050
4	5.098	−0.027	29	4.971	−0.154
5	5.069	−0.056	30	5.121	−0.004
6	5.159	0.034	31	5.013	−0.112
7	5.099	−0.026	32	5.181	0.056
8	5.031	−0.094	33	5.151	0.026
9	5.067	−0.058	34	5.124	−0.001
10	5.122	−0.003	35	5.044	−0.081
11	5.243	0.118	36	5.054	−0.071
12	5.149	0.024	37	5.020	−0.105
13	5.124	−0.001	38	5.063	−0.062
14	5.077	−0.048	39	5.163	0.038
15	5.178	0.053	40	5.122	−0.003
16	5.165	0.040	41	5.127	0.002
17	5.156	0.031	42	5.016	−0.109
18	5.121	−0.004	43	5.132	0.007
19	5.056	−0.069	44	5.091	−0.034
20	5.127	0.002	45	5.165	0.040
21	5.100	−0.025	46	5.097	−0.028
22	5.057	−0.068	47	5.121	−0.004
23	5.173	0.048	48	5.165	0.040
24	5.160	0.035	49	5.185	0.060
25	5.126	0.001	50	5.173	0.048

Nonparametric Methods

i4cp

When Rensis Likert ("Lick-urt") (1903–1981) was studying for his PhD in Psychology, he developed a scale for recording opinions with 5 or 7 ordered categories, now known as a *Likert scale*, and demonstrated its value. Likert scales are now widely used, especially for measurements that are ordered information with no quantitative values or units.

Likert was a founder of the Institute for Social Research (ISR) at the University of Michigan, and directed the Institute from its founding in 1946 until 1970. The Institute specializes in research on organizations. While he was at ISR, Likert helped found the Human Resource Institute (HRI) to work with companies such as GE, GM, and IBM to develop human resource planning. The Institute developed metrics for measuring the value of human capital and return on investment in human capital—work called Human Resource Accounting.

In 2006, HRI was acquired by the Institute for Corporate Productivity (i4cp), currently the world's largest vendor-free network of corporations focused on improving workforce productivity. i4cp maintains a staff of more than 20 full-time analysts and researchers and runs an Executives-in-Residence program. i4cp performs and publishes surveys of organizations focused on Human Resource information.

Questions that survey opinion often present respondents with a choice of five or seven ordered alternatives. They'll ask something like "Thinking about your recent car rental experience, would you say that you were

Very satisfied
Satisfied
Neither satisfied nor dissatisfied
Dissatisfied
or
Very dissatisfied"

Data recorded on a *Likert scale* such as this can be quite useful, but it has one major disadvantage. It isn't quantitative. Specifically, although we can be reasonably sure that a response of "Satisfied" is more favorable than a response of "Neither," we can't tell whether the difference between "Very satisfied" and "Satisfied" is the same, in any sense, as the difference between "Dissatisfied" and "Very dissatisfied."

If, for example, we hope to anticipate which customers are likely to return or identify which of our locations lead to less customer satisfaction, we should be very cautious if we want to find a difference in means or compare boxplots.

In fact, we can be reasonably certain that there will be an occasional customer whose dissatisfaction with his experience will be quite extreme and (we hope) some whose joy with their rental experience exceeds all bounds. Our questionnaire can't record such extreme responses.

Another method, which can give a bit more discrimination, would be to ask customers to assign grades—A, A−, B+, and so on—to their rental experience, specifying that the grades can run from A+ (for extreme satisfaction) to F (for extreme dissatisfaction).

23.1   Ranks

Suppose we have such grades for 6 customers from one location and 5 from another location rating their car rental experience. They might look like this:

Airport	Main Street
C+	B
B−	C
A−	D−
D+	B+
A+	C+
B+	

Table 23.1   Customer grades for two car rental sites.

We can get a clearer understanding of our customers' opinions if we *rank* these values. To rank values, all we need to do is put them in order. It doesn't really matter whether we assign rank "1" to the lowest value and count up or assign rank "1" to the highest value and count down—but we should make clear which we choose. Of course, we can always rank any quantitative values, but ranks have the additional advantage that we can assign them even if we have no more than a clear ordering, as we do here.

Airport	Rank	Main Street	Rank
C+	4.5	B	7
B−	6	C	3
A−	10	D−	1
D+	2	B+	8.5
A+	11	C+	4.5
B+	8.5		

Table 23.2 Customer grades with their ranks.

We can assign rank 1 to the lowest grade—here a D−. If there are any ties, we simply give all the values that are tied the *average* rank they are at. In this example, there are two C+ grades and two B+ grades, so those get ranks of 4.5 (averaging 4 and 5) and 8.5, respectively.

To assign these ranks, all we did was to put all the grades in order:

D−	D+	C	C+	C+	B−	B	B+	B+	A−	A+
1	2	3	4.5	4.5	6	7	8.5	8.5	10	11

Table 23.3 All the grades in order. Red indicates the Main Street location.

Because they are "grades," they have no numeric value, but they do have a clear order. The ranks in Table 23.3 order all the grades from the worst (a D− assigned rank 1) to the best (an A+ assigned rank 11). Of course, we could have just as easily assigned the ranks in the opposite order; for our purposes it makes no difference.

Such a ranking can be used whenever we have data that can be placed in order. Of course, that includes data that have numeric values as well as data that tell us no more than the order.

XLSTAT 23.2

The Wilcoxon Rank-Sum/Mann-Whitney Statistic

> Frank Wilcoxon (1892–1965) was a statistician and chemist who worked primarily in industry. Over his career, he published more than 70 papers. Perhaps the best known is a paper from 1945 in which he proposed the two tests bearing his name that are discussed in this chapter.

The manager of the company wants to know whether satisfaction levels differ between these two locations. In formal terms, she wants to test a hypothesis:

H_0: The locations have the same typical satisfaction.

vs.

H_A: The locations differ in typical satisfaction.

How can we perform such a test? The key is that we'll give up an aspect of hypothesis testing that we've maintained for all the inference methods we've seen so far. In all those other methods, a null hypothesis specified a parameter and a null value for that parameter. The parameter was a number that specified something about the distribution of values in a population. Thus, for example, we might hypothesize a value for a mean or for a difference between two means. The parameters would be the underlying population mean or the means of the two populations from which two samples had been drawn.

But with only grades, no parameters are available. Indeed, it might be hard to say just exactly what it would mean for customers to be more satisfied at one location compared to the other. If *all* the grades for one location were better than the grades for the other, we'd be pretty sure which was the better one. That would show up

Wilcoxon

Frank Wilcoxon first proposed this test to compare equal-size groups. H. B. Mann and D. R. Whitney extended it to work with groups of unequal size. Because it is more general, the Mann-Whitney name for the test is the one you are most likely to find in statistics software.

clearly when we ranked the grades together. And if the grades alternated from one location to the other when we put them all in order, we'd probably conclude that there wasn't much difference.

So we need a statistic that measures where the data stand between the extremes of clearly favoring one site and showing no difference between them.

We'll start by computing the sum of the ranks (called the *rank sums*) in each group. In general, we can call them T_1 and T_2, but it helps to name them. In the example, $T_{\text{Airport}} = 42$ and $T_{\text{Main Street}} = 24$.

The sum of all of the ranks for 11 customers, 1 to 11, is $1 + 2 + 3 + \cdots + 11 = 66$. There is a formula for this: The sum of the first n integers 1, 2, up to n is equal to $n(n + 1)/2$. So, $T_{\text{Airport}} + T_{\text{Main Street}} = 66$ and, in general, if you know one rank sum, the other is completely determined.

We don't need to compare rank sums with each other. The rank sum for one of the groups tells the whole story. So we can pick one rank sum and ask whether this is unusually big or small. The conventional choice is to pick the rank sum that has the smaller sample. That's the Main Street location in this example.

What do we know about this rank sum? It could be as small as $1 + 2 + 3 + 4 + 5 = 15$ or as big as $11 + 10 + 9 + 8 + 7 = 45$. You might guess that its expectation (under H_0) would be right in the middle of these two values—that is, equal to 30—and you'd be right. Under the null hypothesis that the two groups are drawn from the same population, the rankings in either sample are a random selection from all of the rankings, so each sample should, on average, receive its fair share of the ranks. The Main Street sample has 5 of the 11 grades, so its expected rank sum is 5/11 of 66, the sum of all the ranks. That's 30.

So we know that the rank sum of 24, which we have for the Main Street location, is a bit low. But to test the hypothesis we must determine whether it is unusually low. That is, if H_0 were true, how likely would it be to find a rank sum of 24 or less for a comparison like this? That probability would be a P-value because it gives the probability of observing the statistic value we have observed—or something more extreme—if the null hypothesis were true.

When we needed a P-value before, we relied on the Central Limit Theorem or a similar result to find the sampling distribution of our statistic when the underlying parameter took on the value in the null hypothesis. But now we have no parameter and no distribution. Methods that work without estimating a parameter are called **nonparametric**. Those that don't rely on a sampling distribution are called **distribution-free**.

But the lack of a sampling distribution is no problem. When dealing with ranks rather than with quantitative values, there are only so many possibilities. In particular, we can consider *all possible* rankings. Under the null hypothesis, each of them is equally likely. So we can calculate what fraction of them would give rank sums less than 24. Fortunately, the "we" here is the computer, which does the work for us. The results are in Table W1 in Appendix D. The "W" is because this test is called the **Wilcoxon rank-sum** test or the **Mann-Whitney** test. The table gives critical values for the rank sum of the sample with fewer observations. The rank sum for the smaller sample here is 24. From the table, for samples of size 5 and 6, the closest value to 24 is 20.4 and the probability beyond that is 0.05. The manager has chosen a two-sided alternative hypothesis, so we'll double that value and conclude that the probability of a difference this great is more than 0.10. We wouldn't reject the null hypothesis, so we can't conclude that there's a significant difference in customer satisfaction between these two sites.

Using the Central Limit Theorem for Larger Groups

Table W1 only goes as far as groups of 10. What can we do for larger groups? The Central Limit Theorem comes to the rescue. We know from the CLT that means tend to be normally distributed as the sample size grows, and that this happens more quickly for well-behaved, unimodal population distributions with no skewness or outliers. If a mean has a Normal sampling distribution, then so does the sum of the values before dividing by n. (After all, n is a constant and can't change the distribution.) So when we add up the ranks, that rank sum will tend toward a Normal sampling distribution as the number of ranks in the sum grows. And we know that the ranks can't be badly distributed. After all, under the null hypothesis, they are Uniformly distributed, so there are no outliers or long tails. A sum of even 10 ranks has a sampling distribution that is close enough to Normal for our purposes. All we need to know is the mean and standard deviation, and we can find those easily:

For groups 1 and 2, with corresponding rank sums, T_1 and T_2,

$$E(T_i) = n_i(n_1 + n_2 + 1)/2 \quad \text{and} \quad Var(T_i) = n_1 n_2 (n_1 + n_2 + 1)/12 \quad [i = 1 \text{ or } 2]$$

So we can find a P-value directly from the Normal tables (or with the 68–95–99.7 Rule.)

When to Use the Wilcoxon/Mann-Whitney Test

The Wilcoxon test applies to situations, such as the customer-provided grades in our example, in which we can't use means. (The Mann-Whitney version is the version that is appropriate when the groups differ in size.) Data consisting only of grades or ordered categories is one example. We might also choose to replace numeric values with their ranks, for example, because we feared that there may be extreme outliers in the data.

These tests make no assumptions about the distribution of the data or the errors. That frees us from checking the corresponding assumptions. But what do we pay for these advantages? Statisticians evaluate statistical methods in terms of their *power*; the ability of the test to reject a false null hypothesis. When it is possible to perform both a parametric and a non-parametric test (because we have quantitative measurements) reducing the data to ranks and using the Wilcoxon/Mann-Whitney test will have about 95% of the power of a corresponding two-sample *t*-test. And, as we have seen, often outliers are interesting in their own right. An analysis that simply ignores them may miss an important fact or instance.

Assumptions and Conditions

Nonparametric methods require no assumptions about distributions. However, we still have some assumptions about the structure of the data:

Independence Assumption. The data values are mutually independent. Appropriate randomization in data collection (checked with the **Randomization Condition**) is one way to accomplish this.

Independent Groups Assumption. The two groups must be independent of each other.

The hypotheses may seem kind of vague. But it's a feature of the method that it doesn't test properties of any population parameter. That's what makes it a *nonparametric* method.

Wilcoxon rank-sum/Mann-Whitney test

When the Assumptions are satisfied, we can test:

H_0: the two samples are from the same distribution

vs.

H_A: one distribution is shifted in location higher or lower than the other.

Calculate $T =$ the rank sum for the smaller sample. Then, estimate the P-value using tables of critical values or the normal approximation, if either sample size is off the table.

For Example   Comparing two groups

Question: In Section 17.7 we saw data on commercial properties in a large U.S. city. One variable available for those properties is a "quality" ranking into one of 6 categories, with 1 as the "best" and 6 as the "worst." A realtor wanted to know how that quality index related to the "extras" available such as food service on site, an atrium, or a fitness center. He grouped the properties into two categories: those that had no more than one of these extras and those that had more than one. How can he use a nonparametric method to compare the quality rankings of the two groups?

Answer: To compare the quality rankings for these two groups he used the Mann-Whitney extension of the Wilcoxon rank-sum test. (Most statistics software packages offer Mann-Whitney rather than Wilcoxon because it doesn't require equal-size groups and is therefore more general.) The Minitab results are typical of those from a statistics package:

```
W = 21642.5
Test of Extras = NoExtras vs Extras not = NoExtras is significant at 0.0008
The test is significant at 0.0003 (adjusted for ties)
```

The realtor concluded that these extras might be an effective quick method of judging the quality of commercial properties in this city.

Guided Example   **Buying from a Friend**

 A researcher was interested in how friendship might affect bargaining and negotiation. She set up an experiment to test whether there is an effect. Subjects were randomly assigned to imagine purchasing a camera from a friend, and others were assigned to imagine purchasing the same camera from a stranger. They were asked to make a single bid. The table shows their bids. The two groups were separated and were blind to the purpose of the study. The question of interest is whether there is a difference between what people tend to offer to a friend and what they would offer to a stranger.

Price Offered for a Used Camera ($)	
Buying from a Friend	**Buying from a Stranger**
275	260
300	250
260	175
300	130
255	200
275	225
290	240
300	

PLAN	**Setup** State the objectives.	We want to test whether people tend to offer a different amount to a friend than they would offer to a stranger when making a purchase offer.
	Check the assumptions and conditions.	The subjects in this experiment were randomly assigned individually to either imagine dealing with a friend or with a stranger. That randomization allows us to assume both independence among the subjects and between the two groups. That's the only assumption we'll need to perform a nonparametric test.
		The null hypothesis is that the two groups are the same—that there's no difference in offering price between those negotiating with a friend and those negotiating with a stranger.

DO	Mechanics Rank the data from the two groups together.								We start by ranking all the bids as a single group. We give average ranks to ties:						

Data	130	175	200	225	240	250	255	260	260	275	275	290	300	300	300
Rank	1	2	3	4	5	6	7	8.5	8.5	10.5	10.5	12	14	14	14
Group	S	S	S	S	S	S	F	S	F	F	F	F	F	F	F

Now we add up the ranks of the smaller (Strangers) group:
$1 + 2 + 3 + 4 + 5 + 6 + 8.5 = 29.5$

From Table W1, the critical value for a two-sided test at $\alpha = 0.05$ and groups of 7 and 8 is 38. Our observed rank sum is less than this so we can reject the null hypothesis and conclude that there is, in fact, a difference in the two groups.

REPORT Conclusion State the conclusion in the context of the problem.

MEMO

Re: Friendship in Negotiations

We investigated whether friendship has an influence on what negotiators are willing to bid for a mid-priced object—in our experiment, a camera. We analyzed data from a randomized single-blind experiment using a nonparametric Wilcoxon rank sum test and found that there is a statistically significant difference. Negotiators appear to bid more when buying from a friend than when buying from a stranger.

23.3 Kruskal-Wallace Test

Suppose now that we have more than two independent groups to compare. We need a generalization of the Wilcoxon test. For example, we may have customer grades for *three* car rental locations. Now the null hypothesis is that all three sites are rated equally. The alternative hypothesis is that they aren't—but without specifying exactly how they might differ. These hypotheses are similar to those in the one-way Analysis of Variance (see Chapter 21).

Site 1	Site 2	Site 3
D	C	C
D	D	A
C	B	C
D	B	B
C	C	A
C	D	C
C	C	D
C	C	B
B	B	D
C	C	A

The manager wants to know of any differences among these locations. The **Kruskal-Wallace** test provides a nonparametric method. First, we rank all of the data values together, assigning average ranks to tied values. We then sum up these ranks for each group and calculate

$$H = \frac{12}{N(N+1)} \sum \frac{T_i^2}{n_i} - 3(N+1)$$

where the T_i are the rank sums for each group and N is the *total* number of values.

The test says that we can reject the null hypothesis H_0: all populations are identical

$$\text{if } H > \chi^2_{k-1,\alpha}$$

where χ^2 denotes the chi-square distribution, and k is the number of groups we are comparing. The chi-square distribution is a large-sample approximation (like the use of Normal for the Wilcoxon rank-sum test). For very small samples, you may want to consult a table of critical values.[1]

For Example   Comparing several groups

The realtor of the example on page 812 catalogued four features of interest to his health-oriented clientele (green construction, atrium, food service on site, and fitness center) in the commercial properties. Properties had between none and four of the features.

Question: Did the overall quality measure provided for these properties reflect these differences?

Answer: The quality measures are just ordered categories with no units, so a Kruskal-Wallace test is appropriate. The Minitab results are:

```
Features   N   Median  Ave Rank     Z
0         69   4.000    176.9      3.86
1         70   3.000    145.0      0.17
2         91   3.000    136.7     -0.95
3         50   3.000    117.2     -2.48
4          6   2.000     64.2     -2.37
Overall  286            143.5
```

```
H = 22.50  DF = 4  P = 0.000
H = 25.94  DF = 4  P = 0.000 (adjusted for ties)
```

The realtor notes from the detailed output that having none of these features corresponds to much lower quality (recall that quality = 1 is top so a high number rank is poor quality), but that the difference between 3 and 4 features is slight.

Guided Example   Customer Satisfaction

A car rental company wants to compare customer satisfaction at three of its rental sites. They have a simple survey that asks customers to grade them on an A to F scale when they return their cars. Are there differences among the three sites?

---

[1]You can find these on the Internet. For example, at http://www.unity.edu/FacultyPages/woods/CVKW.pdf

PLAN	Setup State the objectives of the study.

The objective is to compare customer satisfaction for all three sites.

We'll test the following hypothesis:

H_O: The centers of all groups are the same.

H_A: At least one of the centers is different from the others.

We have three groups. It is reasonable to assume that the customers at the three sites are mutually independent. That's all we need to perform a Kruskal-Wallace test.

DO Rank all the values together, assigning average ranks to values that are tied, just as we did before.

Order	Data	Rank
1	A	2.0
2	A	2.0
3	A	2.0
4	B	6.5
5	B	6.5
6	B	6.5
7	B	6.5
8	B	6.5
9	B	6.5
10	C	16.5
11	C	16.5
12	C	16.5
13	C	16.5
14	C	16.5
15	C	16.5
16	C	16.5
17	C	16.5
18	C	16.5
19	C	16.5
20	C	16.5
21	C	16.5
22	C	16.5
23	C	16.5
24	D	27.0
25	D	27.0
26	D	27.0
27	D	27.0
28	D	27.0
29	D	27.0
30	D	27.0

(continued)

Sum the ranks within each group

Site 1	Rank	Site 2	Rank	Site 3	Rank
D	27.0	C	16.5	C	16.5
D	27.0	D	27.0	A	2.0
C	16.5	B	6.5	C	16.5
D	27.0	B	6.5	B	6.5
C	16.5	C	16.5	A	2.0
C	16.5	D	27.0	C	16.5
C	16.5	C	16.5	D	27.0
C	16.5	C	16.5	B	6.5
B	6.5	B	6.5	D	27.0
C	16.5	C	16.5	A	2.0
	186.5		156.0		122.5

Calculate the corresponding test statistic H:

$$H = \frac{12}{N(N+1)} \sum \frac{T_i^2}{n_i} - 3(N+1)$$

Where N is the total number of observations from all samples, n_i are the individual sample sizes, and T_i is the sum of the ranks in the ith group.

The Kruskal-Wallis test rejects the null hypothesis when the value of H is large, yielding a small P-value.

$$H = \frac{12}{N(N+1)} \left(\frac{T_1^2}{n_i} + \frac{T_2^2}{n_2} + \frac{T_3^2}{n_3} \right) - 3(N+1)$$

$$= \frac{12}{30(30+1)} \left(\frac{186.5^2}{10} + \frac{156^2}{10} + \frac{122.5^2}{10} \right)$$

$$-3(30+1)$$

$$= 2.64$$

The critical value is approximately 5.99, so we can't reject the null hypothesis.

REPORT

MEMO

An examination of customer ratings from three rental sites did not reveal any significant difference in customer satisfaction.

23.4   Paired Data: The Wilcoxon Signed-Rank Test

Companies that devise tests for students (e.g., for college applications) and for professional certification have been looking to online testing as a way to increase efficiency, save money, and provide results to test takers more rapidly. But there is a natural concern about whether someone taking a test at a computer screen will achieve the same result as he or she would taking the test with pencil and paper.

A testing company devised an experiment to investigate the problem. Researchers randomly assigned subjects to take one of two tests (form A or form B) either electronically or with pencil and paper. Subjects then took the *other* test using the other method. The two forms had been designed to be equivalent in difficulty, but nevertheless, that equivalence was checked as part of the experiment. Our concern is whether subjects did equally well with each testing method.

We have two test scores for each subject. Clearly these are paired because they are two scores from the same subject. How can we tell whether there is a difference in methods?

Subject	Online	Paper	Difference
1	13	14	−1
2	13	10	3
3	8	16	−8
4	14	15	−1
5	16	17	−1
6	11	14	−3
7	12	9	3
8	12	12	0
9	16	16	0
10	14	7	7
11	13	8	5
12	13	11	2
13	17	15	2
14	13	11	2
15	14	13	1
16	9	9	0
17	9	15	−6
18	15	14	1
19	12	16	−4
20	10	8	2

Table 23.4 Test scores with two testing methods

The tests we've examined so far require that the groups being compared be mutually independent. What can we do when they are naturally paired? We saw paired data in Chapter 14, where we used methods based on Student's *t*. A nonparametric method suitable for paired data is the **Wilcoxon signed-rank test**.

First we find the pairwise difference for each subject. Then set aside any ties (zero differences) and, using the new count of cases, rank the absolute values of the differences. Under the null hypothesis that the two testing methods are the same, we'd expect these ranks to appear about equally between the groups. Sum up the ranks for the positive differences (T^+) and for all the negative differences (T^-). We can concentrate on the smaller of these two totals, and call it T. This is our test statistic. A table of critical values for the statistic will tell us if it is sufficiently small to reject the null hypothesis.

For our example, $T^+ = 3 + 3 + 7.5 + 7.5 + 7.5 + 7.5 + 11 + 11 + 14 + 16 = 88$ and $T^- = 3 + 3 + 3 + 11 + 13 + 15 + 17 = 65$ so our test statistic is $T = 65$. Consulting Table W2 at the end of the chapter, we see that critical values for $n = 17$ are all substantially less than 65, so we can't reject the null hypothesis.

Assumptions and Conditions: As with our previous consideration of paired data, the primary assumption is that the data are appropriately paired. You can't just match up values arbitrarily. On the other hand, if the data are paired, you should not use methods that assume independent groups. There's no condition to check here; you just have to think it through.

Because this is a nonparametric test, the only other assumption is that the differences are mutually independent.

For Example   **Paired data**

The Lake County, Illinois, Health Department experimented with a flexible four-day work week. For a year, the department recorded the mileage driven by 11 field workers on an ordinary five-day work week. Then it changed to a flexible four-day work week and recorded mileage for another year.[2] Here are the data:

Name	5-Day Mileage	4-Day Mileage
Jeff	2798	2914
Betty	7724	6112
Roger	7505	6177
Tom	838	1102
Aimee	4592	3281
Greg	8107	4997
Larry G.	1228	1695
Tad	8718	6606
Larry M.	1097	1063
Leslie	8089	6392
Lee	3807	3362

Question: Do flexible schedules reduce the demand for resources?

Answer: These data are clearly paired. And we have no particular reason to assume Normal distributions. We can use a Wilcoxon signed-rank test to compare the mileages.

The results are:

H_0: Median(1 − 2) = 0 H_A: Median(1 − 2) ≠ 0
Test H_0: Median(5-day mileage − 4-day mileage) = 0 vs H_A: Median(5-day mileage − 4-day mileage) ≠ 0

```
                Rank Totals Cases  Mean Rank
Positive Ranks 56          8      7
Negative Ranks 10          3      3.333
Ties           •           0      •
Total          66          11     6
P = 0.0420
```

So based on this test we can conclude that there is a difference in mileage.

Guided Example   **Do Hormones Boost Profits?**

"Hormones may boost profits" read the headline of science reporter Anne McIlroy's article in the Toronto *Globe & Mail* on April 15, 2008. Is this true? To examine the question, Canadian researcher John Coates took saliva samples in the morning, twice a day for eight days, from 17 men working on a London mid-size trading floor (trading a wide range of assets, with largest exposure to German interest rate futures), in June 2005, and classified each trader according to whether his testosterone level was high or low on that day (compared with the trader's median over the period). High testosterone days differed from trader to trader, and high days differed from low days on average by 25% in testosterone level. He also recorded the profits or losses (P&L) in pounds sterling of each trader during 11 A.M.–4 P.M. daily. (Coates JM, Herbert J. *PNAS* 2008;105:6167–6172.)

[2]Charles S. Catlin, "Four-day Work Week Imporves Environment," *Journal of Environmental Health*, Denver, 59:7.

| | Setup State the objectives. | We want to perform a test to determine if there is evidence of a testosterone effect on traders' daily profits. For each trader, we have recorded a profit (or loss) average for his high-testosterone days (testosterone level above his own median) during this 8-day period in June 2005, and another profit average for his low-testosterone days during this 8-day period. |

PLAN

Setup State the objectives.

Check the assumptions and conditions.

We want to perform a test to determine if there is evidence of a testosterone effect on traders' daily profits. For each trader, we have recorded a profit (or loss) average for his high-testosterone days (testosterone level above his own median) during this 8-day period in June 2005, and another profit average for his low-testosterone days during this 8-day period.

Hypotheses State the null and alternative hypotheses.

H_0: There is no difference in the distribution of profits between high testosterone days and low testosterone days.

H_A: Profits tend to be either generally higher or generally lower on a trader's high testosterone days than on his low testosterone days.

Model Think about the assumptions and check any conditions.

The data are paired. The researcher took an average on both high and low testosterone days for the same trader, so each trader gives us paired responses.

It is reasonable to assume that the traders are independent of each other, although this is not a random sample of traders.

Coates provided the following data from his study. We've added the pairwise differences.

Trader	P&L Low 11 A.M. testosterone	P&L High 11 A.M. testosterone	Difference
1	127	−63	−190
2	113	725	612
3	−2	−5	−3
4	−292	1780	2072
5	−98	262	360
6	−1308	1401	2709
7	88	2092	2004
8	361	887	526
9	−700	2500	3200
10	143	82	−61
11	1448	1547	99
12	1625	7000	5375
13	−127	38	165
14	−2066	1941	4007
15	183	268	85
16	483	513	30
17	−115	−4	111

Select a method

The differences include some extraordinarily large values. A nonparametric method would be more appropriate for these data. We'll use a Wilcoxon signed-rank test.

DO

Mechanics Compute the difference for each trader (in the previous table). Throw out any zero differences.

n is the number of pairs = # of traders

(continued)

List the differences in order of magnitude. Sum the ranks for either the positive or negative differences. Clearly here the latter will be smaller, so we compute that rank sum.

Here are the differences, in order by magnitude, with ranks shown underneath.

−3	30	−61	85	99	111	165	−190	360
1	2	3	4	5	6	7	8	9

526	612	2004	2072	2712	3200	4007	5375
10	11	12	13	14	15	16	17

Determine if T differs significantly from its expectation under the null, by either using the table of exact critical values or the Normal approximation (if appropriate).

The sum of the ranks for negative differences is $1 + 3 + 8 = 12$. According to the table of critical values, we see that the P-value corresponding to this sum is <0.02, because $P \leq 0.002$ for a rank sum of 14. The other rank sum must be $n(n + 1)/2 - 12 = 153 - 12 = 141$.

We can reject the null hypothesis.

REPORT

Conclusion State the conclusion in the context of the problem.

MEMO
Re: Hormones and Trading
We investigated whether testosterone levels have a measurable effect on the success of traders. We found that there was a statistically significant difference: when traders' testosterone levels were higher, they were more successful.

23.5 *Friedman Test for a Randomized Block Design

Chapter 21 discussed experimental design and the Analysis of Variance. There is a nonparametric alternative for ANOVA as well. Suppose that we have a randomized block design, with more than two treatments to compare. (If we had only two, we'd use methods for paired data.) We need a generalization of the Wilcoxon Signed Rank.

Micky Parker, the owner of *Mick's Comics*, a store that deals in collector's comic books and memorabilia, wondered what method to use to attract customers into his store. He decided to compare three methods:

- Coupons distributed on the street by a model dressed as a super hero
- A sale announced by a large sign in the window
- Music from movies based on comics played loud enough to be heard in the street

Micky reasons that customer traffic is different in the morning, in the afternoon, and in the evening, so he decides to block on time of day. On three weekdays, he randomly decides which method to use for each time of day. After three days, his customer counts look like this:

		Treatment		
		Coupons	Sale	Music
Block	Morning	9	12	2
	Afternoon	5	10	4
	Evening	7	11	1

The method we'll use for these data is the Friedman test. Again, we need to convert to ranks, but now we *rank separately within each block*, yielding the following ranks (in parentheses):

		Treatment		
		Coupons	**Sale**	**Music**
Block	**Morning**	9(2)	12(3)	2(1)
	Afternoon	5(2)	10(3)	4(1)
	Evening	7(2)	11(3)	1(1)
	Rank Sums	6	9	3

We wish to test for significant variation among the treatment rank sums. The Friedman statistic is:

$$F = \frac{12}{bk(k+1)} \sum T_i^2 - 3b(k+1)$$

where T_i are the rank sums for the various treatments (6,9,3 in our example), b is the number of blocks, and k is the number of treatments.

We reject H_0: all treatments are identical, at approximate significance level α if:

$$F > \chi^2_{k-1, \alpha}$$

This is an approximate result. There are tables of exact critical values available for small sample sizes, but this approximation works well for most cases.

23.6 Kendall's Tau: Measuring Monotonicity

In the chapters on correlation and regression, we used a linear model to describe the relationship between two quantitative variables. Sometimes, however, we are primarily concerned with the *direction* of the association, but not particularly worried about whether the form of the relationship is linear. In other situations, we may have data measured on a Likert scale or other data that gives only information about order and not about quantitative value.

A restaurant owner might wonder whether adding more salt really does make fried potatoes taste better. It may be impossible to measure "how much better," but it is certainly possible to ask customers which of several plates of fries they like best and least.

A consistent trend is called *monotone*. Of course, real data are rarely so simple as to line up in a perfect monotone trend. But we can measure how close our data are to overall monotonicity. The statistic we use is **Kendall's τ** (that's a Greek tau).

Kendall's tau measures monotonicity directly. For each pair of points in a scatterplot, it records only whether the slope of a line between those two points is positive, negative, or zero. (If the points have the same *x*-value, the slope between them is ignored.) In a monotone plot, these pairwise slopes would all be positive or all be negative. In a non-monotone plot, the counts of positive and negative slopes will tend to balance out. Tau (often written with the Greek letter τ) is the difference between the number of positive slopes and the number of negative slopes divided by the total number of slopes between pairs.

Obviously, tau can take values between -1.0 and $+1.0$. If every pairwise slope is negative, then tau will equal -1.0, and the plot will show a single stream of points trending consistently downward. If every pairwise slope is positive, then tau will equal $+1.0$, and the plot will show a single stream of points trending consistently upward. In a generally horizontal plot or one that goes up and then down, the counts of positive and negative slopes will balance out and tau will be near zero. For the example in Figure 23.1 the value of tau is -0.787.

Kendall

Sir Maurice Kendall (1907–1983) was a prominent statistician, writer, and teacher. He made significant contributions to the study of randomness (developing tests for randomness that are still used), the theory of statistics, and rank-based statistics. He also served as director of the World Fertility Study, for which the United Nations awarded him their peace medal.

Figure 23.1 For each pair of points, Kendall's tau records whether the slope between them is positive (red), negative (blue), or zero.

A Monotonous Relationship?
You probably think of monotonous as *boring*. A monotone relationship shares the same root, but it's only the *sameness* that's important. A monotone relationship is one that increases or decreases consistently (but it doesn't have to be boring).

Kendall's tau is easy to understand, but is not the kind of statistic that we calculate by hand. Usually, we let the computer do that work. You can get an idea of why by considering how to find Kendall's tau by hand. First, for every *pair* of points record whether it is *concordant*—that is, with a positive slope between the two points, or *discordant*—having a negative slope. If n_c is the number of concordant points, and n_d the number of discordant points, then

$$\tau = \frac{n_c - n_d}{\frac{1}{2}n(n - 1)}$$

This is a simple formula, so why don't we typically find tau by hand? The denominator, $\frac{1}{2}n(n - 1)$, is the number of pairs to check. That number can get very large for even modest size data sets.

For Example   Estimating association for ordered categories

In Chapters 17 (page 535 and later) and 19 (page 622 and later) we looked at data on diamond prices and quality. The methods of those chapters could only deal with color as a category. But diamond colors are ordered with D as the best and later letters labeling lower-quality colors, so it makes sense to ask whether prices change consistently with color quality. We can rank the colors, assigning rank 1 to color D, 2 to color E, and so on. Of course, these ranks have no measurement units, so a least squares regression would make no sense. But it is appropriate to use Kendall's tau to assess the association between *Price* and *Color rank*.

Question: What is Kendall's tau for these data, and what does it mean?

Answer: Kendall's tau for these data is 0.175. It seems that lower-quality (later letters; higher rank) colors are associated with *higher* prices!

Question: That doesn't make sense. How can we explain this apparent anomaly?

Answer: These diamonds are not all the same *Carat weight*. Perhaps larger diamonds of the best color are harder to find. If we restrict our analysis to diamonds that are 1.0 carat in weight, we find that Kendall's tau for the selected diamonds is −0.463. That indicates a strong association with "better" colors associated with higher prices.

23.7 Spearman's Rho

Figure 23.2 Spearman's rho finds the correlation of the *ranks* of the two variables.

One of the problems we have seen with the correlation coefficient is that it is very sensitive to violations of the Linearity Condition. Both outliers and bends in the data make it impossible to interpret correlation. **Spearman's rho** (often denoted by the Greek letter ρ) can deal with both of these problems. Rho replaces the original data values with their ranks within each variable. That is, it replaces the lowest value in x by the number 1, the next lowest value by 2, and so on, until the highest value is assigned the value n. The same ranking method is applied to the y-variable. A scatterplot of the y-ranks against the x-ranks shows the same general trend as the original data, going up whenever the data went up and down whenever the data went down. If the original scatterplot shows a consistent but bent trend, however, the scatterplot of the ranks is likely to be more nearly linear. And if either variable has an extreme outlying value, the ranking process keeps it from being extreme by just counting it as the highest or lowest value, ignoring how extreme it really is.

Spearman's rho is the correlation of the two rank variables. Because this is a correlation coefficient, it must be between −1.0 and 1.0.

Charles Edward Spearman, FRS (1863–1945) was an English psychologist known for work in statistics, as a pioneer of factor analysis, and for Spearman's rank correlation coefficient. He also did fundamental work on measure of human intelligence.

Both Kendall's tau and Spearman's rho have advantages over the correlation coefficient r. For one, they can be used even when we know only the ranks. Also, they measure the consistency of the trend between the variables without insisting that the trend be linear.

They have the added advantage that they are not much affected by outliers. Spearman's rho limits the outlier to the value of its rank, and Kendall's tau cares only about the sign of each slope between points, not how steep it might be. Neither tau nor rho is changed at all by re-expressing variables by functions that don't alter the order of values (that is, by just the sort of functions we might use to straighten a scatterplot). But, unlike the correlation coefficient, neither statistic can be used as a base for more advanced or complex methods, so they tend to be specialized methods used when we care primarily about consistent trend between two variables.

For Example   Estimating association when there are outliers

How are the prices of commercial properties related to the number of stories high they are? A scatterplot for the commercial properties in one city that are built with masonry construction looks like this:

Question: The linearity assumption is clearly not satisfied by these data, and there is one building of more than 20 stories with a low price that is likely to be influential. What is an appropriate way to summarize the relationship between *Price* and *Stories*?

Answer: The standard (Pearson) correlation for these data is 0.280, but it wouldn't be right to interpret it. The high leverage case on the right and the two particularly expensive buildings at the top of the plot are both likely to influence the value of the correlation. Both violate the linearity assumption. Spearman's rho is more appropriate because it is based on the ranks so it doesn't require an assumption of linearity and can deal with outliers in either direction. Spearman's rho for these properties is 0.467. That's probably a better assessment of association.

23.8   When Should You Use Nonparametric Methods?

Nonparametric methods such as those discussed in this chapter (there are others) are particularly valuable when your data contain only information about order. But they can also be valuable for quantitative data—especially when the quantitative variables violate one or more of our assumptions and conditions.

Translating quantitative values to ranks protects us from the influence of outliers (which simply become the "biggest one" without being extreme), from the effects of two or more modes in the data (because the ranks ignore modes), and from skewed distributions and nonlinear relationships. They "charge" for this service by being a bit less powerful than corresponding parametric methods when the assumptions and conditions of those methods *are* met. But when those assumptions are violated, nonparametric methods can be not only more powerful, but also more appropriate. It is unwise to use methods when their underlying assumptions are not met.

On the other hand, the choice of nonparametric methods should not be made to ignore or hide features of the data. For example, although nonparametric methods can protect us from the influence of an outlier, we know already that outliers often hold valuable information themselves. You should certainly examine any outliers even if you elect to then use a nonparametric method. Similarly, nonparametric methods can protect us from the effects of a bimodal variable (where parametric methods assumed a nearly Normal distribution). But if the data are bimodal, they may be trying to tell us that we have two fundamentally different sub-groups in the data. It would be wise not to ignore that information.

Although nonparametric methods are appropriate for skewed data and nonlinear relationships, it may be more effective to re-express one or more variables to improve symmetry and straighten relationships. (See Sections 5.11 and 6.11.) The resulting models using the re-expressed variables may be simpler, more useful, and have wider applicability.

Of course, statistics software makes it easy to try two or more approaches and compare the results. That is never a bad approach.

What Can Go Wrong?

- **Don't forget to check independence.** Nonparametric methods are free of many of the assumptions and conditions we check for other methods, but they are not entirely free of assumptions. The assumption that the cases are independent is especially important when P-values and critical values are determined by probability calculations based on the assumption that all orderings are equally likely.

- **Don't degrade your data unnecessarily.** If you have quantitative data that satisfy the assumptions and conditions of statistics methods such as *t*-tests, correlation, regression, and ANOVA, you are generally better off using those methods rather than the corresponding nonparametric methods. Working with the ranks rather than measurements inevitably throws away some information in return for protection from outliers and violations of other assumptions.

Ethics in Action

Green Edibles, a large manufacturing company specializing in the production of frozen and canned organic vegetables, has successfully implemented six sigma practices to increase efficiencies, eliminate waste, and improve its manufacturing processes. Some members of the senior management team are proposing that the company use the same practices to improve quality in their administrative and nonmanufacturing processes. Since a sizeable investment is required to train employees in quality improvement practices, the team decided to implement a pilot training program that involves only a subset of the company's administrative departments. Patrick Day, Vice President of Human Resources (HR), was placed in charge of implementing the pilot program and evaluating its effectiveness. His first task was to select ten administrative departments to participate in the pilot program; he chose not to include the HR department as it was currently understaffed. Each of the participating departments identified a process to target for improvement and gathered quality-related data on the process (e.g., number of data entry errors) both before and after its employees were trained in quality improvement practices. Since the sample size was small (only ten departments), the data were analyzed using the Wilcoxon rank-sum test. The results revealed no significant improvement in administrative processes after training compared to before; consequently Patrick reported to the senior management team that there was no evidence to suggest that the pilot program was effective in improving processes in administrative and non-manufacturing departments. When the team subsequently decided not to invest in such a company-wide quality improvement training program, Patrick was somewhat relieved. After all, releasing some of his HR employees to attend training sessions would only make his current staffing situation even worse.

ETHICAL ISSUE *The data are paired, not independent. Assuming independent groups when the data are paired increases the chances of not rejecting the null hypothesis when it should be rejected (related to Item C, ASA Ethical Guidelines).*

ETHICAL SOLUTION *Although nonparametric methods have relatively few assumptions and conditions, it is still important to check the assumptions for each test. The Wilcoxon rank-sum test makes two assumptions: (1) Independence and (2) Independent Groups. These data are paired, so these assumptions are violated. The Wilcoxon signed-rank test should have been used. Also, Patrick should not have let his concerns about the current staffing situation in his department affect his judgment.*

What Have We Learned?

Learning Objectives

- Recognize when nonparametric methods may be appropriate.
  - Data recorded as ranks without units can usually be best analyzed with nonparametric methods.
  - Data with unusual distributions or for which we can't reasonably make the assumptions about the distribution as required by parametric methods call for distribution-free (nonparametric) methods.
  - Situations in which the question being asked is best answered by a nonparametric method.

- The Wilcoxon rank-sum statistic, also called the Mann-Whitney statistic, tests whether two independent groups have the same or different typical ranks for some variable.
  - The test requires only that the individuals and the two groups be independent.

- The Kruskal-Wallace test compares three or more groups.

- The Wilcoxon signed-rank test is appropriate for paired data.

- Kendall's tau measures the degree on monotonicity in the relationship between two variables.
  - It can be applied to two quantitative variables if the question of interest is about monotonicity—and especially if we suspect or know that the linearity assumption required by regression and Pearson correlation is not satisfied.

(continued)

- It can be applied to rank data, where it is not appropriate to apply least squares methods at all.

■ **Spearman's rho measures the correlation between the ranks of two variables.**

- It can be used even when the relationship is not linear or when outliers are present in either variable.
- It can be used when one or both variables consist of ranks.

Terms	
Distribution-free methods	Statistical tests that are not based on the sampling distribution of parameters. See Nonparametric methods.
Kendall's tau	A statistic that measures how close the relationship between two variables is to being monotone
Kruskal-Wallis test	A statistical test that uses the sums of the ranks for the groups that are being compared
Mann-Whitney test	See Wilcoxon rank-sum test.
Nonparametric methods	Statistical tests that investigate characteristics of populations without using specific parameters; these tests use the sign or the rank of the data rather than the specific numerical value of the variable, and they can also be used with ordinal data (categorical data with a logical order such as survey results based on a scale)
Spearman's rho	A correlation computed on the ranks of data values rather than their numeric values
Wilcoxon rank-sum test	(also referred to as **Mann-Whitney**) A statistical method used to test for the differences between population centers: whether the median of the difference between two variables is less than, equal to, or greater than zero
Wilcoxon signed-rank test	A statistical test used to test the difference between two related measurements or samples

Brief CASE

Real Estate Reconsidered

In Chapters 6 and 14 we considered what factors might contribute to the price of a home. In Chapter 6 we looked at living area and in Chapter 14 we considered such features as whether the home had a fireplace or more than 2 bedrooms. In those initial analyses, we used correlation and regression and Student's *t*-based methods. However, a careful look at the data reveals that the assumptions required by those methods are dubious. The data file is **Real_Estate**. Return to those questions, this time using appropriate nonparametric methods. Specifically, consider:

Does the price for homes with a fireplace differ significantly from the price for homes without fireplaces? Make an appropriate display and discuss whether a nonparametric method is preferable to the methods of Chapter 14. Select an appropriate nonparametric test and compare the results.

Does the price for homes depend on the number of bedrooms? Code the bedrooms as 1, 2, 3, or 4 or more. Make an appropriate display and discuss whether nonparametric methods are preferable to an ANOVA for comparing central values of these groups. Perform an appropriate nonparametric test and report on your findings.

Does the price of a home depend on its living area? Make a scatterplot and discuss whether a nonparametric method would be more appropriate than Pearson correlation for discussing this question. Select an appropriate method, and discuss your conclusions.

Exercises

SECTION 23.1

1. Which of the following variables are ranks? For those that are not ranks, give the units.

a) Student ratings of a course on a 5-point Likert scale.

b) Students' letter grades in that course.

c) Students' point scores on the final exam.

2. Which of the following variables are ranks? For those that are not ranks, give the units.

a) U.S. change by size: dime, penny, nickel, quarter, half dollar, dollar.

b) U.S. change by value: penny, nickel, dime, quarter, half dollar, dollar.

c) The change in your pocket by year minted.

SECTION 23.2

3. For which one of the following situations would a Wilcoxon rank-sum (or Mann-Whitney) test be appropriate?

a) Comparing the ratings of a new product on a 5-point Likert scale by a panel of consumers to their ratings of a competitor's product on the same scale.

b) Comparing the sweetness of a diet drink (rated from 1 to 10) as judged by a panel of teenage tasters and a panel of adult tasters.

c) Judging whether the height of the water at high tide at a marina increases more or less consistently as the phase of the moon changes from new to full.

d) Testing whether the distributions of the ratings of a new distance learning course were the same as rated (on the usual 5-point Likert scale) by Freshmen, Sophomores, Juniors, and Seniors who had taken the course.

e) Estimating the association between the value of homes and their age in a community known to have a number of very expensive mansions.

4. For which one of the following situations would a Wilcoxon rank-sum (or Mann-Whitney) test be appropriate?

a) The Mohs scale rates the hardness of minerals. If one mineral can scratch another, it is judged to be harder. (Diamond, the hardest mineral, is a 10.) Is hardness related to the weight of a 1 × 1 × 1 cm cube of the mineral?

b) Each member of a group of professional chefs rates the quality of a basket of organic vegetables on a 5-point scale. The same chefs then rate the quality of a basket of non-organic vegetables on the same scale. Are the organic veggies rated higher?

c) During its initial release, is there an association between the number of downloads of a new smartphone game and the number of hours since it has been released?

d) The Richter scale measures the damage potential of earthquakes. Some scientists think it is best used for judging the relative size of earthquakes and not their absolute size. Are earthquakes around the Pacific "ring of fire" rim stronger in general than those around the Atlantic? Compare earthquakes over a recent 5-year period.

e) A dog's intelligence can be judged roughly by how many times a new command must be repeated until it is learned. A researcher trained several dogs of each of 4 different breeds, teaching them 3 new commands. Are the breeds equally "intelligent"?

SECTION 23.3

5. For which one of the following situations would a Kruskal-Wallace test be appropriate?

a) Comparing the ratings of a new product on a 5-point Likert scale by a panel of consumers to their ratings of a competitor's product on the same scale.

b) Comparing the sweetness of a diet drink (rated from 1 to 10) as judged by a panel of teenage tasters and a panel of adult tasters.

c) Judging whether the height of the water at high tide at a marina increases more or less consistently as the phase of the moon changes from new to full.

d) Testing whether the distributions of the ratings of a new distance learning course were the same as rated (on the usual 5-point Likert scale) by Freshmen, Sophomores, Juniors, and Seniors who had taken the course.

e) Estimating the association between the value of homes and their age in a community known to have a number of very expensive mansions.

6. For which one of the following situations would a Kruskal-Wallace test be appropriate?

a) The Mohs scale rates the hardness of minerals. If one mineral can scratch another, it is judged to be harder. (Diamond, the hardest mineral, is a 10.) Is hardness related to the weight of a 1 × 1 × 1 cm cube of the mineral?

b) Each member of a group of professional chefs rates the quality of a basket of organic vegetables on a 5-point scale. The same chefs then rate the quality of a basket of non-organic vegetables on the same scale. Are the organic veggies rated higher?

c) During its initial release, is there an association between the number of downloads of a new smartphone game and the number of hours since it has been released?

d) The Richter scale measures the damage potential of earthquakes. Some scientists think it is best used for judging the relative size of earthquakes and not their absolute size. Are earthquakes around the Pacific "ring of fire" rim stronger in general than those around the Atlantic? Compare earthquakes over a recent 5-year period.

e) A dog's intelligence can be judged roughly by how many times a new command must be repeated until it is learned. A researcher trained several dogs of each of 4 different breeds, teaching them 3 new commands. Are the breeds equally "intelligent?"

SECTION 23.4

7. For which one of the following situations would a Wilcoxon signed-rank test be appropriate?

a) Comparing the ratings of a new product on a 5-point Likert scale by a panel of consumers to their ratings of a competitor's product on the same scale.

b) Comparing the sweetness of a diet drink (rated from 1 to 10) as judged by a panel of teenage tasters and a panel of adult tasters.

c) Judging whether the height of the water at high tide at a marina increases more or less consistently as the phase of the moon changes from new to full.

d) Testing whether the distributions of the ratings of a new distance learning course were the same as rated (on the usual 5-point Likert scale) by Freshmen, Sophomores, Juniors, and Seniors who had taken the course.

e) Estimating the association between the value of homes and their age in a community known to have a number of very expensive mansions.

8. For which one of the following situations would a Wilcoxon signed-rank test be appropriate?

a) The Mohs scale rates the hardness of minerals. If one mineral can scratch another, it is judged to be harder. (Diamond, the hardest mineral, is a 10.) Is hardness related to the weight of a $1 \times 1 \times 1$ cm cube of the mineral?

b) Each member of a group of professional chefs rates the quality of a basket of organic vegetables on a 5-point scale. The same chefs then rate the quality of a basket of non-organic vegetables on the same scale. Are the organic veggies rated higher?

c) During its initial release, is there an association between the number of downloads of a new smartphone game and the number of hours since it has been released?

d) The Richter scale measures the damage potential of earthquakes. Some scientists think it is best used for judging the relative size of earthquakes and not their absolute size. Are earthquakes around the Pacific "ring of fire" rim stronger in general than those around the Atlantic? Compare earthquakes over a recent 5-year period.

e) A dog's intelligence can be judged roughly by how many times a new command must be repeated until it is learned. A researcher trained several dogs of each of 4 different breeds, teaching them 3 new commands. Are the breeds equally "intelligent?"

SECTION 23.6

9. For which one of the following situations would Kendall's tau be appropriate?

a) Comparing the ratings of a new product on a 5-point Likert scale by a panel of consumers to their ratings of a competitor's product on the same scale.

b) Comparing the sweetness of a diet drink (rated from 1 to 10) as judged by a panel of teenage tasters and a panel of adult tasters.

c) Judging whether the height of the water at high tide at a marina increases more or less consistently as the phase of the moon changes from new to full.

d) Testing whether the distributions of the ratings of a new distance learning course were the same as rated (on the usual 5-point Likert scale) by Freshmen, Sophomores, Juniors, and Seniors who had taken the course.

e) Estimating the association between the value of homes and their age in a community known to have a number of very expensive mansions.

10. For which one of the following situations would Kendall's tau be appropriate?

a) The Mohs scale rates the hardness of minerals. If one mineral can scratch another, it is judged to be harder. (Diamond, the hardest mineral, is a 10.) Is hardness related to the weight of a $1 \times 1 \times 1$ cm cube of the mineral?

b) Each member of a group of professional chefs rates the quality of a basket of organic vegetables on a 5-point scale. The same chefs then rate the quality of a basket of non-organic vegetables on the same scale. Are the organic veggies rated higher?

c) During its initial release, is there an association between the number of downloads of a new smartphone game and the number of hours since it has been released?

d) The Richter scale measures the damage potential of earthquakes. Some scientists think it is best used for judging the relative size of earthquakes and not their absolute size. Are earthquakes around the Pacific "ring of fire" rim stronger in general than those around the Atlantic? Compare earthquakes over a recent 5 year period.

e) A dog's intelligence can be judged roughly by how many times a new command must be repeated until it is learned. A researcher trained several dogs of each of 4 different breeds, teaching them 3 new commands. Are the breeds equally "intelligent?"

SECTION 23.7

11. For which one of the following situations would Spearman's rho be appropriate?

a) Comparing the ratings of a new product on a 5-point Likert scale by a panel of consumers to their ratings of a competitor's product on the same scale.

b) Comparing the sweetness of a diet drink (rated from 1 to 10) as judged by a panel of teenage tasters and a panel of adult tasters.

c) Judging whether the height of the water at high tide at a marina increases more or less consistently as the phase of the moon changes from new to full.

d) Testing whether the distributions of the ratings of a new distance learning course were the same as rated (on the usual 5-point Likert scale) by Freshmen, Sophomores, Juniors, and Seniors who had taken the course.

e) Estimating the association between the value of homes and their age in a community known to have a number of very expensive mansions.

12. For which one of the following situations would Spearman's rho be appropriate?

a) The Mohs scale rates the hardness of minerals. If one mineral can scratch another, it is judged to be harder. (Diamond, the hardest mineral, is a 10.) Is hardness related to the weight of a $1 \times 1 \times 1$ cm cube of the mineral?

b) Each member of a group of professional chefs rates the quality of a basket of organic vegetables on a 5-point scale. The same chefs then rate the quality of a basket of non-organic vegetables on the same scale. Are the organic veggies rated higher?

c) During its initial release, is there an association between the number of downloads of a new smartphone game and the number of hours since it has been released?

d) The Richter scale measures the damage potential of earthquakes. Some scientists think it is best used for judging the relative size of earthquakes and not their absolute size. Are earthquakes around the Pacific "ring of fire" rim stronger in general than those around the Atlantic? Compare earthquakes over a recent 5 year period.

e) A dog's intelligence can be judged roughly by how many times a new command must be repeated until it is learned. A researcher trained several dogs of each of 4 different breeds, teaching them 3 new commands. Are the breeds equally "intelligent?"

CHAPTER EXERCISES

T **13.** Product testing. A company is producing and marketing new reading activities for elementary school children that it believes will improve reading comprehension scores. A researcher randomly assigns third graders to an eight-week program in which some will use these activities and others will experience traditional teaching methods. At the end of the experiment, both groups take a reading comprehension exam. Their scores are shown in the back-to-back stem-and-leaf display.

a) Do these results suggest that the new activities are better? Test an appropriate hypothesis with a Wilcoxon (or Mann-Whitney) rank-sum test and state your conclusion.

b) How would you re-design this study? What nonparametric method would be appropriate then?

New Activities		Control
	1	07
4	2	068
3	3	377
96333	4	12222238
9876432	5	355
721	6	02
1	7	
	8	5

14. Product placement. The owner of a small organic food store was concerned about her sales of a specialty yogurt manufactured in Greece. As a result of increasing fuel costs, she recently had to increase its price. To help boost sales, she decided to place the product on a different shelf (near eye level for most consumers) and in a location near other popular international products. She kept track of sales (number of containers sold per week) for six months after she made the change. The values are shown below, along with the sales numbers for the six months prior to making the change, in stem-and-leaf displays.

After Change	
3	2
3	9
4	23
4	589
5	0012
5	55558
6	00123
6	67
7	0

Before Change	
2	0
2	899
3	224
3	7789
4	0000223
4	5567
5	0
5	6

Do these results suggest that sales are better after the change in product placement? Test an appropriate hypothesis and state your conclusion. Be sure to check assumptions and conditions.

T **15.** Cereal sugar. A food company is concerned about recent criticism of the sugar content of their children's cereals. The data show the sugar content (as a percentage of weight) of several national brands of children's and adults' cereals. Are children's cereals really different from adults' cereals in sugar content?

Children's cereals: 40.3, 55, 45.7, 43.3, 50.3, 45.9, 53.5, 43, 44.2, 44, 47.4, 44, 33.6, 55.1, 48.8, 50.4, 37.8, 60.3, 46.6

Adults' cereals: 20, 30.2, 2.2, 7.5, 4.4, 22.2, 16.6, 14.5, 21.4, 3.3, 6.6, 7.8, 10.6, 16.2, 14.5, 4.1, 15.8, 4.1, 2.4, 3.5, 8.5, 10, 1, 4.4, 1.3, 8.1, 4.7, 18.4

a) Write the null and alternative hypotheses.

b) Check the conditions for a Wilcoxon rank-sum (or Mann-Whitney) test.

c) Complete the test and report your conclusion.
d) Make histograms of each group. Discuss whether you would choose a nonparametric test for these data or a two-sample *t*-test.

T 16. Foreclosure rates. Home foreclosures were up 47% in March 2008 compared to the previous year (realestate.msn.com; April 2008). The data show home foreclosure rates (as % change from the previous year) for a sample of cities in two regions of the United States, the Northeast and the Southwest.

Northeast: 2.99, −2.36, 3.03, 1.01, 5.77, 9.95, −3.52, 7.16, −3.34, 4.75, 5.25, 6.21, 1.67, −2.45, −0.55, 3.45, 4.50, 1.87, −2.15, −0.75

Southwest: 10.15, 23.05, 18.95, 21.16, 17.45, 12.67, 13.75, 29.42, 11.45, 16.77, 12.67, 13.69, 25.81, 21.16, 19.67, 11.88, 13.67, 18.00, 12.88

a) Write the null and alternative hypotheses.
b) Check the conditions for a Wilcoxon rank-sum (or Mann-Whitney) test.
c) Complete the test and report your conclusion.
d) Make histograms of each group. Discuss whether you would choose a nonparametric test for these data or a two-sample *t*-test (as in Chapter 14 Exercise 36.)

T 17. Trophy sizes. The table gives the names, dates first awarded, and heights (inches) of major trophies and awards in sports and entertainment. The America's Cup, awarded for sailing, claims to be the oldest such award. The first five were created in the 19th century (which ended in 1900, not in 1899). The others were created after a gap of 26 years. When the creation of new awards was begun again, were the trophies the same size? Compare the two groups with an appropriate nonparametric measure. Complete the test at $\alpha = 0.10$.

Award	Year	Size
America's Cup	1851	27
Kentucky Derby	1875	22
Wimbledon Men's Singles	1887	18.5
Stanley Cup	1892	35.2
Davis Cup	1900	42
Ryder Cup	1927	17
Oscar	1929	13.5
Jules Rimet Trophy (Word Cup Soccer)	1929	14
Heisman Trophy	1935	15
Primetime Emmy	1949	15.5
Vince Lombardi Trophy	1967	22
Commissioner's Trophy (World Series)	1967	24
Larry O'Brien NBA Trophy	1977	24
Webb Ellis Cup (Rugby)	1987	15

T 18. Baseball 2006. American League baseball teams play their games with the designated hitter rule, meaning that pitchers do not bat. The league believes that replacing the pitcher, traditionally a weak hitter, with another player in the batting order produces more runs and generates more interest among fans. Below are the average numbers of runs scored (per game by both teams) in American League and National League stadiums for the 2006 season.

American		National	
11.4	9.9	10.5	9.5
10.5	9.7	10.3	9.4
10.4	9.1	10.0	9.1
10.3	9.0	10.0	9.0
10.2	9.0	9.7	9.0
10.0	8.9	9.7	8.9
9.9	8.8	9.6	8.9
		9.5	7.9

Use a Mann-Whitney method to test whether there is really a difference in number of runs scored.

T 19. Freshman 15. Cornell Professor of Nutrition David Levitsky recruited students from two large sections of an introductory health course to test the validity of the "Freshman 15" theory that students gain 15 pounds their first year. Although they were volunteers, they appeared to match the rest of the freshman class in terms of demographic variables such as sex and ethnicity. The students were weighed during the first week of the semester, then again 12 weeks later.

Subject Number	Initial Weight	Terminal Weight	Subject Number	Initial Weight	Terminal Weight
1	171	168	15	115	118
2	110	111	16	110	113
3	134	136	17	142	146
4	115	119	18	127	127
5	150	155	19	102	105
6	104	106	20	125	125
7	142	148	21	157	158
8	120	124	22	119	126
9	144	148	23	113	114
10	156	154	24	120	128
11	114	114	25	135	139
12	121	123	26	148	150
13	122	126	27	110	112
14	120	115	28	160	163

Subject Number	Initial Weight	Terminal Weight	Subject Number	Initial Weight	Terminal Weight
29	220	224	49	104	105
30	132	133	50	111	112
31	145	147	51	160	162
32	141	141	52	134	134
33	158	160	53	151	151
34	135	134	54	127	130
35	148	150	55	106	108
36	164	165	56	185	188
37	137	138	57	125	128
38	198	201	58	125	126
39	122	124	59	155	158
40	146	146	60	118	120
41	150	151	61	149	150
42	187	192	62	149	149
43	94	96	63	122	121
44	105	105	64	155	158
45	127	130	65	160	161
46	142	144	66	115	119
47	140	143	67	167	170
48	107	107	68	131	131

a) Use a Wilcoxon signed-rank test to test whether these students' weights changed. State your hypotheses, check conditions, and complete the test (with technology).

b) Would you prefer the nonparametric test, or the paired *t*-test asked for in Chapter 14? Explain.

T 20. Job satisfaction. You first read about this exercise break plan in Chapter 14. A company institutes an exercise break for its workers to see whether this will improve job satisfaction, as measured by a questionnaire that assesses workers' satisfaction. Scores for 10 randomly selected workers before and after the implementation of the exercise program are shown in the following table.

Worker Number	Job Satisfaction Index	
	Before	After
1	34	33
2	28	36
3	29	50
4	45	41
5	26	37
6	27	41
7	24	39
8	15	21
9	15	20
10	27	37

a) Identify a nonparametric procedure you could use to assess the effectiveness of the exercise program and check to see whether the conditions allow for the use of that procedure.

b) Test an appropriate hypothesis and state your conclusion.

21. ERP effectiveness. When implementing a packaged Enterprise Resource Planning (ERP) system, many companies report that the module they first install is Financial Accounting. Among the measures used to gauge the effectiveness of their ERP system implementation is acceleration of the financial close process. Below is a sample of 8 companies that report their average time (in weeks) to financial close before and after the implementation of their ERP system. In Chapter 14, we saw these data:

Company	Before	After
1	6.5	4.2
2	7.0	5.9
3	8.0	8.0
4	4.5	4.0
5	5.2	3.8
6	4.9	4.1
7	5.2	6.0
8	6.5	4.2

Use a Wilcoxon signed-rank test to see whether the ERP system seems to be effective.

22. Online insurance. A local insurance agent was concerned that he might lose customers to online discount insurance sites. To investigate, he randomly selected profiles (type of car, coverage, driving record, etc.) for 10 of his clients and checked online price quotes for their policies. The comparisons are shown in the table. We first saw these data in Chapter 14.

Local	Online	PriceDiff
568	391	177
872	602	270
451	488	−37
1229	903	326
605	677	−72
1021	1270	−249
783	703	80
844	789	55
907	1008	−101
712	702	10

Identify an appropriate nonparametric method to compare the two sets of prices and complete the test.

T 23. Friday the 13th, traffic. In 1993 the *British Medical Journal* published an article titled, "Is Friday the 13th Bad for Your Health?" Researchers in Britain examined how Friday the 13th affects human behavior. One question was whether people tend to stay at home more on Friday the 13th—an issue of great concern, for example, to restaurants and theaters. The data in the table are the number of cars passing Junctions 9 and 10 on the M25 motorway for consecutive Fridays (the 6th and 13th) for five different periods.

Year	Month	6th	13th
1990	July	134,012	132,908
1991	September	133,732	131,843
1991	December	121,139	118,723
1992	March	124,631	120,249
1992	November	117,584	117,263

Identify and use an appropriate nonparametric test to judge whether there is a significant difference in traffic between consecutive Fridays when one is the 13th.

T 24. Friday the 13th, accidents. The researchers in Exercise 23 also examined the number of people admitted to emergency rooms for vehicular accidents on 12 Friday evenings (6 each on the 6th and 13th).

Year	Month	6th	13th
1989	October	9	13
1990	July	6	12
1991	September	11	14
1991	December	11	10
1992	March	3	4
1992	November	5	12

Is Friday the 13th "unlucky"? Hospital emergency rooms need to plan for any increased "business" if they can. Use an appropriate nonparametric test to decide whether there is really a difference between consecutive Fridays when one of them is a 13th.

T 25. Cuckoos and quality control. Cuckoos lay their eggs in the nests of other (host) birds. The eggs are then adopted and hatched by the host birds. But the potential host birds lay eggs of different sizes. Does the cuckoo change the size of her eggs for different foster species? The numbers in the table are lengths (in mm) of cuckoo eggs found in nests of three different species of other birds. The data are drawn from the work of O. M. Latter in 1902 and were used in a fundamental textbook on statistical quality control by L. H. C. Tippett (1902–1985), one of the pioneers in that field.

Cuckoo Egg Length (MM)		
Foster Parent Species		
Sparrow	**Robin**	**wagtail**
20.85	21.05	21.05
21.65	21.85	21.85
22.05	22.05	21.85
22.85	22.05	21.85
23.05	22.05	22.05
23.05	22.25	22.45
23.05	22.45	22.65
23.05	22.45	23.05
23.45	22.65	23.05
23.85	23.05	23.45
23.85	23.05	24.05
24.05	23.05	24.05
25.05	23.05	24.05
	23.25	24.85
	23.85	

Use a Kruskal-Wallace test to decide whether cuckoos really do adjust the size of their eggs.

26. Analgesics. A pharmaceutical company tested three formulations of a pain relief medicine for migraine headache sufferers. For the experiment, 27 volunteers were selected and 9 were randomly assigned to one of three drug formulations. The subjects were instructed to take the drug during their next migraine headache episode and to report their pain on a scale of 1 = no pain to 10 = extreme pain, 30 minutes after taking the drug.

Drug	Pain	Drug	Pain	Drug	Pain
A	4	B	6	C	6
A	5	B	8	C	7
A	4	B	4	C	6
A	3	B	5	C	6
A	2	B	4	C	7
A	4	B	6	C	5
A	3	B	5	C	6
A	4	B	8	C	5
A	4	B	6	C	5

a) Use the Kruskal-Wallace test to judge whether there is a difference among the analgesics.
b) What aspect of these data may make the Kruskal-Wallace test more appropriate for these data than the ANOVA methods of Chapter 21?

T 27. Crowdedness. In a *Chance* magazine article (Summer 2005), Danielle Vasilescu and Howard Wainer used data from the United Nations Center for Human Settlements to investigate aspects of living conditions for several countries. Among the variables they looked at were the country's per capita gross domestic product (*GDP*, in $) and *Crowdedness*, defined as the average number of persons per room living in homes there. This scatterplot displays these data for 56 countries:

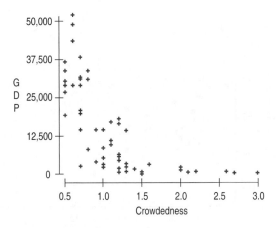

a) The value of Kendall's tau is -0.626. What does this tell us about the association?
b) The value of Spearman's rho is -0.804. What does this tell us about the association?
c) What about these data would make the nonparametric measures appropriate choices?

T 28. Basebal salaries 2009. Baseball is big business, and the top salaries have become quite large. The highest salaries (in millions of dollars per season) for some notable players since 1980 are given in the following table:

Player	Year	Salary ($M)
Nolan Ryan	1980	1.0
George Foster	1982	2.0
Kirby Puckett	1990	3.0
Jose Canseco	1990	4.7
Roger Clemens	1991	5.3
Ken Griffey, Jr.	1986	8.5
Albert Belle	1997	11.0
Pedro Martinez	1998	12.5
Mike Piazza	1999	12.5
Mo Vaughn	1999	13.3
Kevin Brown	1999	15.0
Carlos Delgado	2001	17.0
Alex Rodriguez	2001	22.0
Manny Ramirez	2004	22.5
Alex Rodriguez	2005	26.0
Derek Jeter	2006	20.6
Jason Giambi	2007	23.4
Alex Rodriguez	2009	33.0

a) Kendall's tau for these data is 0.887. What does that say about these data?
b) Based on that result, would you predict that salaries will continue to rise? Explain.

T 29. The Home Depot. Here are data that relate *Sales* at The Home Depot and *Housing Starts*, for quarters between 1995 and 2004.

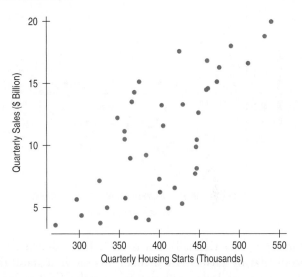

a) Would Kendall's τ or Pearson correlation be the more appropriate summary? Explain.
b) What could we learn from Kendall's tau that we wouldn't learn from the Pearson correlation?

T 30. Carbon footprint. In Chapter 6, Exercise 41, we considered the relationship between the carbon footprint (tons of CO_2 per year) vs. the new Environmental Protection Agency (EPA) highway mileage for 82 family sedans as reported by the U.S. government (www.fueleconomy.gov/feg/byclass.htm). The car with the highest highway mpg and lowest carbon footprint is the Toyota Prius.

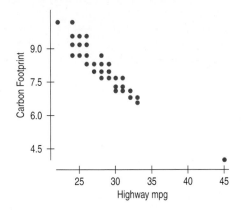

a) Would Spearman's rho be an appropriate summary of the strength of the association shown here?
b) Find rho for the full data and for the relationship with the Prius removed. Comment.

T **31. Income and housing.** The Office of Federal Housing Enterprise Oversight (www.ofheo.gov) collects data on various aspects of housing costs around the United States. Here is a scatterplot of the *Housing Cost Index* versus the *Median Family Income* for each of the 50 states. The correlation is 0.65.

For these data Kendall's tau is 0.51. Does that provide proof that by raising the median income in a state, the Housing Cost Index will rise as a result? Explain what Kendall's tau says and does not say.

T **32. Interest rates and mortgages.** Since 1980, average mortgage interest rates have fluctuated from a low of under 6% to a high of over 14%. Is there a relationship between the amount of money people borrow and the interest rate that's offered? Here is a scatterplot of *Total Mortgages* in the United States (in millions of 2005 dollars) versus *Interest Rate* at various times over the past 26 years. The correlation is −0.84.

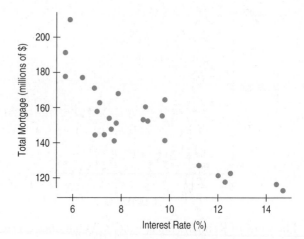

For these data Kendall's tau is −0.61. Does that provide proof that if mortgage rates are lowered, people will take out more mortgages? Explain what Kendall's tau says and does not say.

Decision Making and Risk

Data Description, inc.

Small businesses are 99.7 percent of all U.S. firms. They generate more than half of the private non-farm gross domestic product and create 60 to 80 percent of net new jobs (www.sba.gov). Here's the story of one such firm.

Data Description was founded in 1985 to develop, market, and support the Data Desk® statistics program.[1] Personal computers were becoming more widely available, and graphical desktop interfaces offered new and exciting opportunities for data analysis and graphics. Data Desk was introduced in 1986 for the newly released Macintosh computer and released for the PC in 1997, with expanded capabilities for analysis, and the ability to work efficiently with large (several million case) data sets. Data Description then launched *ActivStats*, a multimedia education product on DVD, which pioneered multimedia teaching in statistics and led to the development of MediaDX, a development platform that deploys a full array of media, narration, animation, video, sound, and interactive tools for teaching.

Data Description employs both a local staff and programmers who work remotely from other

---

[1]Because Data Description was founded by Paul Velleman, one of our co-authors, we have been able to gain access to business decisions that would ordinarily not be publicly available.

parts of the country and "telecommute." In these ways, it is typical of many technologically centered small businesses relying on the Internet for development, marketing, and customer support and competing with larger firms by staying fast and flexible.

Company president John Sammis notes that the Internet has enabled small businesses such as Data Description to compete with larger international firms and that decisions about how best to use the Internet are key to competing successfully. By its twentieth anniversary, Data Description (www.datadesk.com), based in Ithaca, New York, was providing software and services to education and business clients throughout the world.

We make decisions every day, often without knowing how they will turn out. Most of our daily decisions have no serious consequences. But the decisions a business makes determine the success or failure of the venture. The consequences of business decisions can often be quantified in monetary terms, but the costs and benefits themselves often depend on events beyond the control or foresight of decisionmakers. Decisions must be made nevertheless. How can people make intelligent decisions when they don't know for certain what the future holds?

One decision facing Data Description (as with many high-tech companies) is how best to provide technical support for its customers. In the past, Data Description has relied on its thorough documentation, a built-in help system, and free telephone support. But as its user base has grown, telephone-based customer support has become a significant expense. Data Description's president, John Sammis, must decide whether to invest in developing an online FAQ (Frequently Asked Questions) help system as the first line of defense for customer support, to continue free telephone support through the help desk, or to hire and train additional telephone support staff and charge customers for it. For a small business, decisions such as these carry significant costs, but making the right decision can be the difference between keeping customers happy and losing them to larger competitors.

24.1 Actions, States of Nature, and Outcomes

We refer to the alternative choices we can make for a decision as **actions**. Actions are mutually exclusive; if you choose one course of action, you haven't chosen the others. We'll call Data Descriptions' choices:

- Online FAQ
- Free Help Desk
- Pay Help Desk

The facts about the world that affect the consequences of each action are called **states of nature**, or sometimes just **states**. We often simplify the states to make it easier to understand the decision alternatives. We know, for example, that the economy isn't just "good" or "bad," and the number of customers can be any value in some reasonable range. But we'll simplify by grouping the states of nature into a few categories and treating them as discrete possibilities.

For Data Description, the *states of nature* concern the type of question coming into the Help Desk. Management classifies them into two basic types:

- Simple
  or
- Complex

A simple question is one that can be answered with an online FAQ system alone. A complex question requires either human technical support by telephone or backup e-mail support for FAQ users.

Each *action* has consequences that depend on the *state of nature* that actually occurs. These consequences are called the **outcomes** or **payoffs** (because usually in business the consequence can be measured in money).

To make informed decisions, Data Description president John Sammis must estimate the costs and benefits of each action under each state of nature. According to his estimates:

- The FAQ system can answer a simple question for about $3.50. A complex question can't be answered by the FAQ system and needs additional resources. Unfortunately the back-up system is slow and the marketing manager estimates that this would cost about $15 in goodwill and future business. Adding that to the FAQ system cost gives a total cost of $18.50 for a complex question.
- A live telephone support person costs about $10 per question whether it is simple or complex.
- Charging the customer $3 for the telephone support could defray some of the cost, but it might upset a customer with only a simple question. That resulting ill will will cost the company an estimated $15 for each simple question. On the other hand, customers with a complex question might be happy to pay for the telephone support, so the net cost for them would be $7.

For Example   Actions, states of nature, and outcomes

Pat needs to purchase airplane tickets to fly home at the end of the term. The last exam schedule slot is Friday at 3 p.m. Exams are scheduled centrally at Pat's school, and there is no alternative time to take exams. Pat simply must be there. A ticket to fly out on Thursday costs $750. An advanced-purchase ticket for Saturday morning costs $900, but, the end of term being a busy time, Pat doubts this option will be available for many more days. If the exam is scheduled on Thursday or Friday, Pat would have to sacrifice the ticket and purchase a new one for $1200.

Questions:

What are the actions in this decision process?

What are the states of nature?

What are the outcomes?

Answers:

The actions are purchasing a ticket for Thursday or Saturday.

The states of nature are the exam being scheduled before Thursday or afterwards.

The outcomes are $750 for a Thursday ticket and early exam, $900 for a Saturday ticket, and $1950 for a Thursday ticket but late exam.

24.2   Payoff Tables and Decision Trees

We can summarize the actions, states of nature, and corresponding payoffs in a **payoff table**. Table 24.1 shows the payoff table for Data Description's decision about customer support. Customer support is an expense, so all the entries in the payoff table are costs, which Data Description hopes to keep to a minimum. The table shows the costs for a single tech support request.

	State of Nature	
	Simple Question	**Complex Question**
FAQ	$3.50	$18.50
Free Help Desk	$10.00	$10.00
Pay Help Desk	$15.00	$7.00

Table 24.1   A payoff table showing costs of actions for two states of nature. Note that these are costs and not profits. As "payoffs," they might be written as negative values, but it is simpler to just remember that we want to *minimize* costs. Payoff tables of profits look to *maximize* values.

The payoff table is easy to read, but it doesn't display the sequential nature of the decision process. Another way to display this information that shows the dynamics of the decision-making process is with a **decision tree**. The decision tree mimics the actual decision making process by showing actions first followed by the possible states of nature and finally the outcomes that would result from each combination (Figure 24.1).

Figure 24.1   A decision tree for the customer support decision.

The decision process moves from left to right through the decision tree. The decisionmaker chooses an action (indicated by a square node), determining which branch of the tree to follow. The circle nodes indicate branches due to the state of nature, which lie outside the control of the decisionmaker. The tree can expand to show decisions with any (reasonable) number of actions and states of nature.

For Example Payoff tables

Question: Pat must decide which ticket to purchase (see page 837). What is Pat's payoff table?

Answer:

		Exam date	
		Wed or before	After Wed
Ticket	Thursday	$750	$1950
	Sat	$900	$900

24.3 Minimizing Loss and Maximizing Gain

A conservative approach to decision-making would look at the worst possible loss and try to minimize it. Here are the outcomes for each decision and the greatest cost for each.

		State of Nature		
		Simple Question	Complex Question	Greatest Cost
Action	FAQ	$3.50	$18.50	$18.50
	Free Help Desk	$10.00	$10.00	$10.00
	Pay Help Desk	$15.00	$7.00	$15.00

| Table 24.2 The greatest cost for each decision.

The decision to go with a FAQ could cost the company as much as $18.50 when complex calls come in, and the decision for a pay help desk could cost up to $15.00. The safe choice—the one that minimizes the maximum cost—is the free help desk. This is known as the **minimax** choice. (If we are maximizing the minimum return, this can be called a maximin strategy.)

Of course, one could take the other extreme position and try for the maximum gain (or, equivalently here, the minimum cost). In this table, that would call for committing to a FAQ system (and hoping that the calls were simple). Such a choice is known as the **maximax** (when maximizing return) or the **minimin** (when minimizing cost) choice.

Choosing actions based on worst and best case scenarios rarely leads to the best business decision. Instead, successful small business executives rely on knowing their market so they can make less absolute decisions. A more realistic modeling approach takes into account how frequently the decisionmaker expects to experience each of the states of nature, and finds the optimum under that model.

For Example Minimizing loss and maximizing gain

Questions: Recall Pat's payoff table for purchasing an airplane ticket from the example at the top of this page. What is Pat's minimax choice?

What is Pat's minimin choice?

Answers: Pat's minimax choice, minimizing the worst cost, is to purchase a Saturday ticket.

Pat's minimin choice, minimizing the minimum cost, is to purchase a Thursday ticket.

24.4   The Expected Value of an Action

How can Data Description choose an action to maximize profit (or minimize loss)? They can be pretty sure that they'll get a mix of both simple and complex questions. Decisionmakers often estimate the *probability* of a state of nature based on their understanding of their business. Such probabilities can be *subjective* probabilities or can be grounded in data, and viewed as the *frequentist* probabilities we've used earlier in this book. In either case they express the expert's opinion of and belief in the relative likelihood of the states. We'll write $P(s_j)$ for the probability of the jth state of nature.

As with all probabilities, we need to check that the probabilities are *legitimate*. If there are N states of nature, then we require that:

$$P(s_j) \geq 0 \text{ for all } j,$$

$$\sum_{j=1}^{N} P(s_j) = 1.$$

If the probabilities are legitimate, then we can find the **expected value (EV)** of action a_i in essentially the same way as we found the expected value of a discrete random variable in Chapter 8:[2]

$$EV(a_i) = \sum_{j=1}^{N} o_{ij} P(s_j),$$

where there are N possible states of nature and o_{ij} is the outcome or payoff of action i when the state of nature is s_j.

At Data Description, about 40% of the questions are simple, so managers assigned a probability of 0.4 that a support question will be simple and a probability of 0.6 that it will be complex. To compute the expected value, we start by placing the probabilities on the decision tree in the appropriate place. With these values in place, we can find the expected value of each of the actions. The probabilities are associated with each state of nature and are the same regardless of the action. We place these values on the branches associated with each state of nature, repeated for each one of the possible actions (the circle nodes).

> **How Much Risk Are You Willing to Take?**
>
> There are three general types of decisionmakers:
> - a risk-averse individual will sacrifice EV for lower variation
> - a risk-neutral individual will try to maximize EV
> - a risk-tolerant individual may be willing to absorb greater variation in return for the chance of a higher payoff

Action a_i		State s_j	Outcome o_{ij}	$P(s_j) \times o_{ij}$	$EV(a_i)$
FAQ	0.4	Simple	$3.50	0.4 × 3.50 = 1.40	
					1.40 + 11.10 = $12.50
	0.6	Complex	$18.50	0.6 × 18.50 = 11.10	
Free Help	0.4	Simple	$10.00	0.4 × 10.00 = 4.00	
					4.00 + 6.00 = $10.00
	0.6	Complex	$10.00	0.6 × 10.00 = 6.00	
Pay Help	0.4	Simple	$15.00	0.4 × 15.00 = 6.00	
					6.00 + 4.20 = $10.20
	0.6	Complex	$7.00	0.6 × 7.00 = 4.20	

Figure 24.2   Calculating the expected value of each action using a decision tree.

For each combination of outcome and probability, we calculate its contribution to the expected value $P(s_j) \times o_{ij}$. Because only one state of nature can occur for each call, the states of natures are disjoint. We can therefore find the expected value

---

[2]In Chapter 8 we computed the EV for one action, such as investing in a CD. This chapter extends this concept by computing and comparing EVs for different actions.

for each action by summing the contributions over all the states of nature, in this case Simple and Complex (see Figure 24.2).

From the expected values in Figure 24.2, it appears that expanding free telephone support may be the best action because it costs Data Description the *least*. The expected value of that action is a cost of $10.00.

For Example   Expected value of purchasing a ticket

Pat guesses that the course in question has an exam scheduled on the last 2 days of exam period only about 20% of the time (see page 837).

Question: What is the expected value of purchasing a Thursday ticket?

Answer: $0.8 \times 750 + 0.2 \times 1950 = \$990$

24.5   Expected Value with Perfect Information

Unfortunately, we can't predict the state of nature with certainty. But it can be informative to consider what it costs us to live with the uncertainty. If we knew the true state of nature—rather than just the probable state of nature—how much might that be worth to us? To help quantify that, let's consider a single customer support question. Knowing whether it was simple or complex would enable us to take the optimal action in each case. For a simple question we'd offer online FAQ support and the "payoff" would be $3.50. For a complex question we'd provide paid telephone support costing the company $7.00.

Using the probabilities of the two states of nature, we can calculate the expected value of this optimal strategy as:

$$0.6 \times \$7.00 + 0.4 \times \$3.50 = \$5.60$$

This is called the **expected value with perfect information** and is sometimes denoted *EVwPI*.

By contrast, the expected value of the optimal strategy that we calculated without knowing the state of nature was $10.00. The absolute value of the difference between these two quantities is called the **expected value of perfect information**.

$$EVPI = |EVwPI - EV|$$

In our example, this difference is $|\$5.60 - \$10.00| = \$4.40$. (Notice the absolute value in that calculation. Information should increase profit or, as in our example, reduce costs. Either way, the *value* of information is positive.) Our lack of perfect knowledge about the state of nature costs us $4.40 per call. The expected value of information, EVPI, gives the maximum amount the company might be willing to pay for perfect information about the outcome.

For Example   EVwPI and EVPI

Questions: Recall from the example on page 837 that Pat is purchasing an airplane ticket; the payoff table is in the example on page 839. What is Pat's EVwPI? What is the EVPI?

Answer: With perfect information, Pat would purchase the early ticket for $750 80% of the time and the late ticket for $900 20% of the time for an EVwPI of $0.8 \times 750 + 0.2 \times 900 = \$780$.

The EVPI is then $\$990 - \$780 = \$210$.

24.6   Decisions Made with Sample Information

Usually the costs and profits—the payoffs—of alternative actions under various states of nature can be estimated fairly well, because they depend on business-related events and actions that are well understood. By contrast, the probabilities assigned to the states of nature may be based only on expert judgment. These probabilities are sometimes called **prior probabilities** because they are determined before obtaining any additional information on the states of nature. A survey of customers, however, or a planned experiment might provide information that could bring the probabilities closer to reality and make for a more informed decision. Revised probabilities are also called **posterior probabilities**. Should the decisionmaker first gather data to help estimate the probabilities of the states of nature? That's a decision in itself. Surveys and experiments cost money, and the resulting information may not be worth the expense. We can incorporate the decision to gather data into the overall decision process. First, it is worthwhile to ask whether useful information is even possible. If the states of nature in the decision process are about the future health of the economy or the future value of the Dow Jones Industrial Average, it is not likely that we can learn very much new. But market research surveys of customers and trial marketing incentives can often provide useful data.

To improve their estimates of the probabilities making up the mix of simple and complex questions, Data Description could collect data about their customers. Would the added expense of that work be worthwhile? It depends on whether the information about customers helps to predict the likelihood of each type of question. Suppose they could classify their users into those who are *technically sophisticated* users and those who are *beginners*. That information might be useful because for *technically sophisticated* users, Data Description thinks that 75% of the questions will be complex but for *beginners*, that percentage is only 40%. Knowing how to classify their customers will help them make the right decision.

These probabilities are *conditional* probabilities of the kind we first saw in Chapter 7. We write:

$$P(\text{Complex} \mid \text{Sophisticated}) = 0.75; \ P(\text{Simple} \mid \text{Sophisticated}) = 0.25$$
$$P(\text{Complex} \mid \text{Beginner}) = 0.40; \ P(\text{Simple} \mid \text{Beginner}) = 0.60$$

In each case, we know the missing probability from the requirement that the probabilities sum to one.

Now our decision tree must include both the decision of whether or not to conduct the study and the possible outcomes of the study (if it is conducted), each with its appropriate conditional probability. Figure 24.3 shows the results. The costs in Figure 24.3 are the same ones we've used all along from Table 24.1. The expected values when there is no study are those worked out previously in Figure 24.2.

This decision process has two sets of decision nodes. The first decision is whether to conduct the study. The second is what action to take given the information from the study (if available). The outcome of the study is a chance node because we don't know *a priori* what the study will tell us.

To determine a decision strategy we compute the expected payoff at the end of each branch of the tree following the same methods we used before and now using the estimated probabilities that might come from the study. In this example, if we perform the study and it shows that sophisticated users are more common, we'll choose a pay help desk at a cost of $9.00. If we perform the study but it shows that beginners are the norm, we'll choose the FAQ system, at a cost of $9.50. If we don't perform the study at all, our optimal choice was the free phone help at a cost of $10.00. In this example, both alternatives following the study are less costly than the nonstudy choice, so it appears that any information the study may provide will save the company money. To be more precise, however, the company can go further and

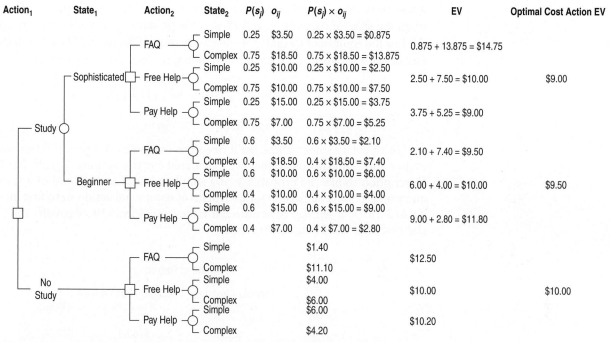

Figure 24.3 A decision tree including a customer history study. For each decision node under each circumstance, the optimal (in this example, lowest cost) alternative is indicated in color, and the corresponding consequence is assigned as the expected value for that alternative EV's for the No Study branch are from Figure 24.2.

estimate the likelihood of the study's outcome. This will give the company an estimate of the **expected value with sample information** and help them estimate the potential worth of the study.

For example, if they think there's a probability of 0.8 that the study will show that most users are now technically sophisticated (and, correspondingly, a probability of 0.2 that it would find that beginners are the norm), then the expected value of the decision with that sample information would be:

$$EVwSI = 0.8 \times \$9.00 + 0.2 \times \$9.50 = \$9.10$$

The absolute value of the difference between this payoff and the payoff without the study is the **expected value of sample information**.

$$EVSI = |EVwSI - EV| = |\$9.10 - \$10.00| = \$0.90$$

An analysis like this can help us understand the value of sample information. It tells us that a study of this sort is likely to be worth about $0.90 per technical support question in its ability to help the company choose the optimal action.

24.7　Estimating Variation

In Chapter 5, we said that means should always be reported with an associated standard deviation, but thus far we've found expected values (which are, after all, means) without finding the corresponding standard deviations. The expected values tell the decisionmaker which action is likely to be worth the most. But to assess the *risk* of the decision, a decisionmaker must understand whether the likely outcome of the decision will always be close to that expected value or whether it might be quite different. We assess that risk by finding the standard deviation of the outcome.

Recall that the calculation of the expected value of an action is:

$$EV(a_i) = \sum_{j=1}^{N} o_{ij} P(s_j),$$

where there are N possible states of nature, o_{ij} is the outcome or payoff of action i when the state of nature is s_j, and $P(s_j)$ is the probability of the jth state of nature. We saw the equivalent formula for expected value in Chapter 8. The corresponding formulas for variance and standard deviation (also seen in Chapter 8) are:

$$Var(a_i) = \sum_{j=1}^{N}(o_{ij} - EV(a_i))^2 P(s_j),$$
$$\text{and } SD(a_i) = \sqrt{Var(a_i)}.$$

To calculate some values, we'll need the outcomes and probabilities we saw earlier and the expected values we calculated for the actions. Recall that Data Description thinks the probability of a simple question is 0.40 and of a complex question is 0.60. The payoffs for each state of nature and action were first given in Table 24.1. Now we also have expected values found from those payoffs and probabilities. Table 24.3 summarizes them.

	State of Nature		
	Simple Question $P = 0.40$	Complex Question $P = 0.60$	$EV(a_i)$
FAQ	$3.50	$18.50	$12.50
Free Help Desk	$10.00	$10.00	$10.00
Pay Help Desk	$15.00	$7.00	$10.20

Table 24.3   The payoff table for the customer support decision along with the expected values computed earlier. Recall that these are costs and thus could be written as negative values. Of course, for finding a standard deviation, writing these as negative numbers would give the same result.

Now, for example, the standard deviation of the outcome for the FAQ action is

$$SD(FAQ) = \sqrt{(3.50 - 12.50)^2 \times 0.40 + (18.50 - 12.50)^2 \times 0.60} = \$7.35$$

and the standard deviation of the Pay Help Desk option is

$$SD(Pay\,Help) = \sqrt{(15.00 - 10.20)^2 \times 0.4 + (7.00 - 10.20)^2 \times 0.6} = \$3.92$$

The standard deviation of the Free Help Desk option is zero. (Can you see why?)

In the customer support example, the choice to rely on an online FAQ system has higher variability. That may make it a less desirable alternative even in situations where its expected value seems to be optimal.

One way to combine the expected value and standard deviation is to find their ratio, the **coefficient of variation**.

$$CV(a_i) = \frac{SD(a_i)}{EV(a_i)}.$$

For the customer support example, the CVs are 0.588 for online FAQ, 0 for Free Help, and 0.384 for Pay Help. Actions with smaller CVs are generally seen as less risky, and may be preferred for that reason by some people. You can see why Data Description may prefer Free Help even though it may be more expensive; the company can predict its cost regardless of the state of nature.

One action may dominate another if its worst outcome is still better than the best outcome for the other. In such cases, the better action should be chosen regardless of how much variation it may have. **Dominance** is defined as the situation in which one decision alternative is never an optimal action regardless of the state of nature. A dominated action can be eliminated as a variable option or strategy.

Some discussions of risk prefer to concentrate on the expected value rather than the standard deviation, so they look at the reciprocal of the CV and give it the name **return to risk ratio**

$$RRR(a_i) = \frac{1}{CV(a_i)} = \frac{EV(a_i)}{SD(a_i)}.$$

The RRR can't be computed for actions with zero standard deviation, because they have no risk. Generally, actions with higher RRR are favored because they are less risky for their expected return. The units of RRR are dollars returned per dollar at risk.

For Example   Estimating variation

Pat must decide about purchasing a ticket (see the example on page 837 for background).

Question: Find the standard deviation of Pat's outcome. What is the RRR of purchasing a Thursday ticket?

Answer: For Thursday ticket: $\sqrt{(750 - 990)^2 \times 0.8 + (1950 - 990)^2 \times 0.2} = 480$

For Saturday ticket, SD $= 0$

$$RRR(\text{Thursday}) = \frac{EV\ (\text{Thursday})}{SD\ (\text{Thursday})} = \frac{990}{480} = 2.06$$

24.8   Sensitivity

One weakness of the methods we've discussed here is that they require that you estimate—subjectively or by research—probabilities of states of nature. You may wish to assess the *sensitivity* of your conclusions about the expected value and standard deviation of alternative actions to the probabilities in your model. One way to do this is to re-compute those values with slightly (or even substantially) different probabilities. If the decision advised by the model doesn't change, you can be more confident in it. But if small changes in the probabilities result in large differences in the estimated value of the action alternatives, you should take great care and not rely too heavily on your decision analysis.[3]

For example, if the probability of a complex question were 0.8 rather than 0.6, would that change the decision? Table 24.4 shows the calculations.

		State of Nature		
		Simple Question $P = 0.20$	Complex Question $P = 0.80$	$EV(a_i)$
Action	FAQ	$3.50	$18.50	$15.50
	Free Help Desk	$10.00	$10.00	$10.00
	Pay Help Desk	$15.00	$7.00	$8.60

Table 24.4   Alternative calculations (compare with Table 24.3) with different probabilities for the states of nature.

---

[3]If you completed the Brief Case at the end of Chapter 8, you have already conducted a small sensitivity analysis.

It seems that the best decision would change if the probability of a complex question were as large as 0.8. The decisionmakers at Data Description should consider how sensitive these expected values are to their probabilities by finding the SD and EV for a plausible range of probabilities and then think about how certain they are about those estimated probabilities.

For Example   Assessing sensitivity

Pat is making a decision about whether to purchase a cheaper ticket with a risk of having to purchase another if there's a conflict with the final exam (see the example on page 837).

Questions: How sensitive is Pat's decision? Suppose, for example, that the probability of a late exam were only 0.1. How would that change the expected values?

Answer: The EV of the Thursday ticket would now be $870, less than the EV of the Saturday ticket (which would still be $900).

24.9   Simulation

Another alternative to assessing the sensitivity of a decision model to the choice of probabilities is to simulate the model for a variety of plausible probability values. Rather than specify single probabilities, you can specify a distribution of plausible values. You can then run a *simulation* in which the computer draws states of nature at random according to your distribution of probabilities and evaluates the consequences. This approach can deal with much more complex decision models than those we've discussed here. The result of the simulation is not a single decision but a distribution of outcomes. Such a distribution may be a more appropriate description for a decisionmaker than any single expected value.

Programs such as ©*Risk* (www.palisade.com/) and Crystal Ball (www.oracle .com/appserver/business-intelligence/crystalball/index.html) provide ways for you to specify actions, states of nature, and the associated probabilities and outcomes and then use simulation methods to generate distributions of outcomes.

Guided Example   Insurance Services

InterCon Travel Health is a Toronto-based company that provides services to health insurers of foreign tourists who travel to the United States and Canada.[4] The primary focus of InterCon is to act as an interface between local health providers and the overseas insurance company who insured the tourist should the traveler find himself or herself in need of medical attention. The customer base of sick or injured tourists is potentially profitable because most tourists who fall ill require only minor treatment and rarely come back for expensive follow-up visits. So InterCon can pass savings along to the foreign insurer by facilitating claim management and payment and can collect processing fees from these insurance companies in return.

---

[4]This example is based on the case by G. Truman, D. Pachamanova, and M. Goldstein entitled "InterCon Travel Health Case Study" in *Journal of the Academy of Business Education*, Vol. 8, Summer 2007, pp. 17–32.

Currently, InterCon charges a processing fee of 9.5%, collected partly from the medical providers and partly from the foreign insurers. It has experienced an average annual growth of 3%. However, to help with rising costs, they may consider increasing the processing fee from 9.5% to 10.5%. Although this would generate additional revenue, they would also incur the risk of losing contracts with insurers and health care providers. Table 24.5 gives the company's estimates of the impact of a change in rates on the annual growth in claims depending on the strength of foreign tourism

	State of Nature	
Action	**Weak Tourism** $P = 0.40$	**Strong Tourism** $P = 0.60$
9.5% Fee	$54.07M	$56.23M
10.5% Fee	$52.97M	$54.56M

Table 24.5 InterCon's estimates of the impact on the annual growth in claims for two different actions (changes in rate) under two possible states of nature (strong or weak tourism).

What would be the best choice for the company?

PLAN	Setup State the objective of the study.	We want to evaluate the two alternative actions the company is considering.
	Identify the variables.	We have estimates of the outcomes and the probabilities of two states of nature: weak or strong tourism.
	Model Think about the assumptions and check the conditions.	The only condition to check is that the probabilities are legitimate; $0.40 + 0.60 = 1.0$, and 0.4 and 0.6 are both between 0 and 1.
DO	Mechanics The calculations of expected value and standard deviation are straightforward.	

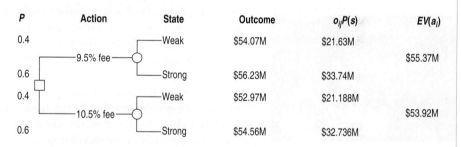

P	Action	State	Outcome	$o_{ij}P(s)$	$EV(a_i)$
0.4		Weak	$54.07M	$21.63M	
	9.5% fee				$55.37M
0.6		Strong	$56.23M	$33.74M	
0.4		Weak	$52.97M	$21.188M	
	10.5% fee				$53.92M
0.6		Strong	$54.56M	$32.736M	

It can help to diagram the decision process with a tree diagram.

For example:

$$0.4 \times \$54.07M = \$21.63M$$
$$\$21.63M + \$33.74M = \$55.37M$$

Standard deviations:

For 9.5% fee:

$$(o_{ij} - EV(a_i)) \qquad (o_{ij} - EV(a_i))^2 P(s_j)$$
$$(54.07 - 55.37) = -1.30 \quad (-1.30)^2 \times 0.4 = 0.676$$
$$(56.23 - 55.37) = 0.86 \quad (0.86^2) \times 0.6 = 0.444$$
$$SD = \sqrt{0.676 + 0.444}$$
$$= \$1.06M$$

(continued)

For 10.5% fee:

SD = \$0.779M

RRR (9.5% fee) = 55.37/1.06 = 52.24

RRR (10.5% fee) = 53.92/0.779 = 69.22

REPORT

Summary and Conclusions Summarize your results and state any limitations of your model in the context of your original objectives.

MEMO

Re: Recommendations for Processing Fee Rate

The expected value of the alternative with the lower processing fee rate is slightly higher (\$55.37 million compared to \$53.92 million). However, because it has a smaller standard deviation, the return to risk ratio (RRR) at the higher processing fee rate of 10.5% is 69.22; while at the lower rate, it is only 52.24.

Because one action has the better expected value but the other seems less risky, the company should determine the impact of the downside of their decision and evaluate the probability estimates for the strength of the foreign tourism industry.

24.10   Probability Trees

Some decisions involve more subtle evaluation of probabilities. Given the probabilities of various states of nature, an analyst can use trees similar to the ones presented earlier to calculate the likelihood of complex combinations of events. This can enable the analyst to compare several possible scenarios. Here's a manufacturing example.

Personal electronic devices, such as personal digital assistants (PDAs) and MP3 players, are getting smaller all the time. Manufacturing components for these devices is a challenge, and at the same time, consumers are demanding more and more functionality and increasing sturdiness. Microscopic and even submicroscopic flaws can develop during their fabrication that can blank out pixels on the screens or cause intermittent performance failures. Defects will always occur, so the quality engineer in charge of the production process must monitor the number of defects and take action if the process seems out of control.

Let's suppose that the engineer is called down to the production line because the number of defects has crossed a threshold and the process has been declared to be out of control. She must decide between two possible actions. She knows that a small adjustment to the robots that assemble the components can fix a variety of problems, but for more complex problems, the entire production line needs to be shut down in order to pinpoint the problem. The adjustment requires that production be stopped for about an hour. But shutting down the line takes at least an entire shift (8 hours). Naturally, her boss would prefer that she make the simple adjustment. But without knowing the source or severity of the problem, she can't be sure whether that will be successful.

If the engineer wants to predict whether the smaller adjustment will work, she can use a probability tree to help make the decision. This tree is much like the decision tree that we saw before, except that now each node corresponds to a state of nature that occurs randomly, whereas in the decision tree, some of the nodes depicted decisions made to select an action. Based on her experience, the engineer thinks that there are three possible problems: (1) the motherboards could have faulty connections, (2) the memory could be the source of the faulty connections, or (3) some of the cases may simply be seating incorrectly in the assembly line. She knows from past experience how often these types of problem crop up and how likely it is that just making an adjustment will fix each type of problem. *Motherboard* problems are rare (10%), *memory* problems have been showing up about 30% of the time, and *case* alignment issues occur most often (60%). We can put those probabilities on the first set of branches.

Figure 24.4 Possible problems and their probabilities.

Notice that we've covered all the possibilities, and so the probabilities sum to one. To this diagram we can now add the *conditional* probabilities that a minor adjustment will fix each type of problem. Most likely the engineer will rely on her experience or assemble a team to help determine these probabilities. For example, the engineer knows that motherboard connection problems are not likely to be fixed with a simple adjustment: $P(\text{Fix}|\text{Motherboard}) = 0.10$. After some discussion, she and her team determine that $P(\text{Fix}|\text{Memory}) = 0.50$ and $P(\text{Fix}|\text{Case alignment}) = 0.80$. At the end of each branch representing the problem type, we draw two possibilities (*Fix* or *Not Fixed*) and write the conditional probabilities on the branches.

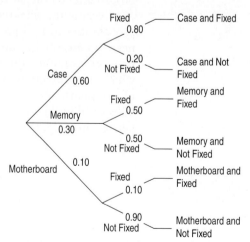

Figure 24.5 Extending the tree diagram, we can show both the problem class and the outcome probabilities. The outcome (Fixed or Not fixed) probabilities are conditional on the problem type, and they change depending on which branch we follow.

At the end of each second branch, we write the *joint event* corresponding to the combination of the two branches. For example, the top branch is the combination of the problem being Case alignment, and the outcome of the small adjustment is that the problem is now Fixed. For each of the joint events, we can use the general multiplication rule to calculate their joint probability. For example:

$$P(Case \text{ and } Fixed) = P(Case) \times P(Fixed|Case)$$
$$= 0.60 \times 0.80 = 0.48$$

We write this probability next to the corresponding event. Doing this for all branch combinations gives Figure 24.6.

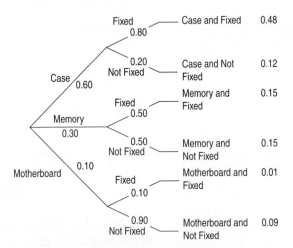

Figure 24.6   We can find the probabilities of compound events by multiplying the probabilities along the branch of the tree that leads to the event, just the way the General Multiplication Rule specifies.

All the outcomes at the far right are disjoint because at every node, all the choices are disjoint alternatives. And those alternatives are *all* the possibilities, so the probabilities on the far right must add up to one.

Because the final outcomes are disjoint, we can add up any combination of probabilities to find probabilities for compound events. In particular, the engineer can answer her question: What's the probability that the problem will be fixed by a simple adjustment? She finds all the outcomes on the far right in which the problem was fixed. There are three (one corresponding to each type of problem), and she adds their probabilities: 0.48 + 0.15 + 0.01 = 0.64. So 64% of all problems are fixed by the simple adjustment. The other 36% require a major investigation.

For Example   Probability trees

A recent Maryland highway safety study found that in 77% of all accidents the driver was wearing a seat belt. Accident reports indicated that 92% of those drivers escaped serious injury (defined as hospitalization or death), but only 63% of the non-belted drivers were so fortunate.

Question: Draw the probability tree.

Answer: Let B = the driver was wearing a seat belt, and NB = no belt.

Let I = serious injury or death, and OK = not seriously injured.

I know P(B) = 0.77, so P(NB) = 1 − 0.77 = 0.23.

Also, P(OK|B) = 0.92, so P(I|B) = 0.08

and P(OK|NB) = 0.63, so P(I|NB) = 0.37

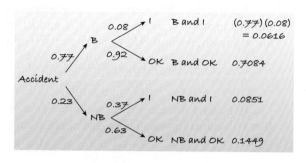

*24.11   Reversing the Conditioning: Bayes's Rule

The engineer in our story decided to try the simple adjustment and, fortunately, it worked. Now she needs to report to the quality engineer on the next shift what she thinks the problem was. Was it more likely to be a case alignment problem or a motherboard problem? We know the probabilities of those problems beforehand, but they change now that we have more information. What are the likelihoods that each of the possible problems was, in fact, the one that occurred?

Unfortunately, we can't read those probabilities from the tree in Figure 24.6. For example, the tree gives us P(*Fixed* and *Case*) = 0.48, but we want P(*Case*|*Fixed*). We know P(*Fixed*|*Case*) = 0.80, but that's not the same thing. It isn't valid to reverse the order of conditioning in a conditional probability statement. To "turn" the probability around, we need to go back to the definition of conditional probability.

$$P(Case \mid Fixed) = \frac{P(Case \text{ and } Fixed)}{P(Fixed)}$$

We can read the probability in the numerator from the tree, and we've already calculated the probability in the denominator by adding all the probabilities on the final branches that correspond to the event *Fixed*. Putting those values in the formula, the engineer finds:

$$P(Case \mid Fixed) = \frac{0.48}{0.48 + 0.15 + 0.01} = 0.75$$

She knew that 60% of all problems were due to case alignment, but now that she knows the problem has been fixed, she knows more. Given the additional information that a simple adjustment was able to fix the problem, she now can increase the probability that the problem was case alignment to 0.75.

It's usually easiest to solve problems like this by reading the appropriate probabilities from the tree. However, we can write a general formula for finding the reverse conditional probability. To understand it, let's review our example again. Let $A_1 = \{Case\}$, $A_2 = \{Memory\}$, and $A_3 = \{Motherboard\}$ represent the three types of problems. Let $B = \{Fixed\}$, meaning that the simple adjustment fixed the problem.

We know $P(B|A_1) = 0.80$, $P(B|A_2) = 0.50$, and $P(B|A_3) = 0.10$. We want to find the reverse probabilities, $P(A_i|B)$, for the three possible problem types. From the definition of conditional probability, we know (for any of the three types of problems):

$$P(A_i|B) = \frac{P(A_i \text{ and } B)}{P(B)}$$

We still don't know either of these quantities, but we use the definition of conditional probability again to find $P(A_i \text{ and } B) = P(B|A_i)P(A_i)$, both of which we know. Finally, we find $P(B)$ by adding up the probabilities of the three events.

$$P(B) = P(A_1 \text{ and } B) + P(A_2 \text{ and } B) + P(A_3 \text{ and } B) =$$
$$P(A_1|B)P(A_1) + P(A_2|B)P(A_2) + P(A_3|B)P(A_3)$$

In general, we can write this for n events A_i that are mutually exclusive (each pair is disjoint) and exhaustive (their union is the whole space). Then:

$$P(A_i|B) = \frac{P(B|A_i)P(A_i)}{\sum_j P(B|A_j)P(A_j)}$$

This formula is known as Bayes's rule, after the Reverend Thomas Bayes (1702–1761), even though historians doubt that it was actually Bayes who first came up with the reverse conditioning probability. When you need to find reverse conditional probabilities, we recommend drawing a tree and finding the appropriate probabilities as we did at the beginning of the section, but the formula gives the general rule.

For Example   Reversing the probability (Bayes's Rule) *

Use the probability tree (see the example on page 850) and Bayes's Rule to show how to answer the following.

Question: If a driver has been seriously injured in an accident, what is the probability that he or she was wearing a seat belt?

Answer: Reading the probabilities from the probability tree of the previous example,

$$P(NB|I) = \frac{P(NB \text{ and } I)}{P(I)} = \frac{0.0851}{0.0616 + 0.0851} = 0.58$$

Even though only 23% of drivers weren't wearing seat belts, they accounted for 58% of all the deaths and serious injuries.

Working with the original probabilities (and those we could find with just the Complement rule in the previous example) and using Bayes's Rule:

$$P(NB|I) = \frac{P(I|NB)P(NB)}{P(I|NB)P(NB) + P(I|B)P(B)}$$
$$= \frac{0.37 \times 0.23}{0.37 \times 0.23 + 0.08 \times 0.77} = 0.58$$

24.12   More Complex Decisions

The methods of this chapter can be extended to decisions that are more complex and that may have second or third stages conditional on the outcomes of the first decisions. For example, customer support questions might be handled best by an initial online FAQ system that then offers customers with complex questions a choice of free e-mail support or paid telephone support. To work out the expected values for these choices, we'd need to estimate the probabilities of a customer with a complex question choosing free e-mail or paid telephone support—something with which the company has no prior experience.

Or the probabilities may be changed by a change in circumstances. When the company releases an upgraded version, technical support calls increase, but there's a higher probability of simple calls. As the customer base learns the new version, their questions generally become more sophisticated.

What Can Go Wrong?

- **Decision trees do not tell the decisionmaker the correct decision.** Models can be useful, but they are based on assumptions. Decision trees model the expected outcome and risk for each decision. The best decision depends on the level of risk the decisionmaker is willing to accept, as well as the state of nature that occurs.

- **The computation of the expected values and standard deviations are sensitive to the probabilities.** The probabilities required by the calculations of expected value and standard deviation may be reliably based on past experience or data, or they may just represent the subjective judgment of the decisionmaker. Don't trust them blindly. Using a decision model and ignoring the sensitivity of the model to different probabilities can lead to poor decisions and misleading conclusions. One way to address the problem is to solve the model with a range of plausible probabilities to assess how sensitive the conclusions are to the choice of probabilities. Another way is to use software that simulates the results for randomly generated probabilities from an appropriate distribution.

- **Be sure that the probabilities are legitimate and that the action choices are disjoint.** The calculations in this chapter depend upon both of these being true.

Ethics in Action

Nelson Greene develops real estate in the southwestern United States. His next planned project is to build a housing development on a parcel of land in New Mexico, not far from Santa Fe. However, with the recent decline in housing prices, particularly in areas of the southwest like Las Vegas, Nelson is reconsidering his decision. He set up a meeting with some key individuals, including Tom Barnick, a representative from the township supervisors. Tom and the township council were positive about the proposed development since it would provide increased tax revenue. So Tom wanted to influence Nelson to go forward with the development. Tom had some analysis prepared, which he presented at the meeting. Based on several assumptions, including the probabilities associated with different housing market conditions in New Mexico (improving, worsening, or staying the same), the analysis showed a positive expected payoff for going forward with the project. Tom mentioned that the probabilities were obtained from a small group of housing market experts, although they had difficulty reaching consensus. Nelson noticed that not all the payoffs associated with going forward with the project were positive for each possible market condition, but the positive expected payoff made him view the project more favorably.

ETHICAL ISSUE *The expected payoff is a long run average and not an actual outcome for a decision. Although the expected payoff was positive, there was no mention of the expected payoff for not going forward with the development.*

ETHICAL SOLUTION *If the experts couldn't agree on the probabilities, the sensitivity analysis should be done for a range of probabilities. Related to Item A, ASA Ethical Guidelines.*

What Have We Learned?

Learning Objectives

■ Assess costs and benefits of various actions given possible states of nature.
  • Use judgment or data.

■ Estimate the likelihood of the possible states of nature.
  • Use data, if they are available.
  • Confirm that probabilities are legitimate.

■ Evaluate likely outcomes (costs and benefits) of decisions when faced with uncertainty about the state of nature.
  • Make a decision tree and find the expected value of actions.
  • Evaluate the relative risk of alternative decisions.

Terms

Action	Alternative choices that one can make for a decision.
Coefficient of variation	The coefficient of variation (CV) of each decision shows the "$ at risk" for each "$ returned": $$CV(a_i) = SD(a_i)/EV(a_i).$$
Decision tree	A decision tree organizes the actions, states of nature, and outcomes to display the sequence of the elements in a decision.
Dominance	An alternative dominates when it pays off at least as much as (or costs less than) the alternative it dominates.
Expected value (EV)	The expected value of an action under uncertain states of nature is: $$EV(a_i) = \sum_{j=1}^{N} o_{ij}P(s_j)$$ where there are N possible states of nature and o_{ij} is the outcome or payoff of action i when the state of nature is s_j.
Expected value of perfect information (EVPI)	The difference between the expected value with perfect information (EVwPI) and the expected value (without perfect information) of the strategy (EV): $$EVPI = |EVwPI - EV|$$
Expected value with perfect information (EVwPI)	The expected value of a hypothetical strategy assuming that probabilities of the states of nature are known.
Expected value of sample information (EVSI)	The difference between the expected value with sample information (EVwSI) and the expected value (without sample information) of the strategy (EV): $$EVSI = |EVwSI - EV|$$
Expected value with sample information (EVwSI)	The expected value of a hypothetical strategy assuming that probabilities of the states of nature are estimated.
Maximax choice (or strategy)	The action that maximizes the maximum possible return across all states of nature in a payoff table.
Minimax choice (or strategy)	The action that minimizes the maximum possible cost across all states of nature in a payoff table.
Minimin choice (or strategy)	The action that minimizes the minimum possible cost across all states of nature in a payoff table.
Outcome (or payoff)	The consequence of an action combined with a state of nature, usually expressed in monetary units (positive or negative).
Payoff table	A table containing the actions, states of nature, and the corresponding payoffs for each combination of the two.

Probability tree	A probability tree displays conditional probabilities.
Return to risk ratio	The return to risk ratio (RRR) measures the expected return of an action relative to the risk, represented by its standard deviation:

$$RRR(a_i) = 1/CV = EV(a_i)/SD(a_i).$$

Standard deviation of an action, Variance of an action	The standard deviation of an action measures the variability of possible outcomes, calculated in terms of their estimated probabilities of occurring. The standard deviation is the square root of the variance:

$$Var(a_i) = \sum_{j=1}^{N} (o_{ij} - EV(a_i))^2 \, P(s_j)$$

$$SD(a_i) = \sqrt{Var(a_i)}$$

States of nature	The facts about the world that affect the consequences of an action, usually summarized into a few possible values.

Brief CASE

Texaco-Pennzoil

"Oil is big business." A classic example of this is the Texaco-Pennzoil court case, which appeared in the book *Making Hard Decisions*[5] and in a subsequent case study by T. Reilly and N. Sharpe (2001). In 1984, a merger was hammered out between two oil giants, Pennzoil and Getty Oil. Before the specifics had been agreed to in a written and binding form, another oil giant—Texaco—offered Getty Oil more money. Ultimately, Getty sold out to Texaco.

Pennzoil immediately sued Texaco for illegal interference, and in late 1985 was awarded $11.1 billion—an enormous award at the time. (A subsequent appeal reduced the award to $10.3 billion.) The CEO of Texaco threatened to fight the judgment all the way to the U.S. Supreme Court, citing improper negotiations held between Pennzoil and Getty. Concerned about bankruptcy if forced to pay the required sum of money, Texaco offered Pennzoil $2 billion to settle the case. Pennzoil considered the offer, analyzed the alternatives, and decided that a settlement price closer to $5 billion would be more reasonable.

The CEO of Pennzoil had a decision to make. He could make the low-risk decision of accepting the $2 billion offer, or he could decide to make the counteroffer of $5 billion. If Pennzoil countered with $5 billion, what are the possible outcomes? First, Texaco could accept the offer. Second, Texaco could refuse to negotiate and demand settlement in the courts. Assume that the courts could order one of the following:

- Texaco must pay Pennzoil $10.3 billion.
- Texaco must pay Pennzoil's figure of $5 billion.
- Texaco wins and pays Pennzoil nothing.

The award associated with each outcome—whether ordered by the court or agreed upon by the two parties—is what we will consider to be the "payoff" for Pennzoil. To simplify Pennzoil's decision process, we make a few assumptions. First, we assume that Pennzoil's objective is to maximize the amount of the settlement. Second, the likelihood of each of the outcomes in this high-profile case is based on similar cases. We will

---

[5]Clemen, R. T. and Reilly, T. (2001), *Making Hard Decisions*, New York: Brooks/Cole

assume that there is an even chance (50%) that Texaco will refuse the counteroffer and go to court. According to a *Fortune* article,[6] the CEO of Pennzoil believed that should the offer be refused, Texaco had a chance to win the case with appeals, which would leave Pennzoil with high legal fees and no payoff. Based on prior similar court cases and expert opinion, assume that there is also a 50% probability that the court will order a compromise and require Texaco to pay Pennzoil the suggested price of $5 billion. What are the remaining options for the court? Assume that the probabilities of the other two alternatives—Pennzoil receiving the original total award ($10.3 billion) or Pennzoil getting nothing—are almost equal, with the likelihood of the original verdict being upheld slightly greater (30%) than the likelihood of reversing the decision (20%).

Evaluate the expected payoff and risk of each decision for Pennzoil.

Insurance Services, Revisited

InterCon Travel Health is a Toronto-based company that provides services to health insurers of foreign tourists who travel to the United States and Canada.[7] As described in the Guided Example in this chapter, the primary focus of InterCon is to act as an interface between local health providers and the overseas insurance company who insured the tourist should the traveler find himself or herself in need of medical attention. The customer base of sick or injured tourists is potentially profitable because most tourists who fall ill require only minor treatment and rarely come back for expensive follow-up visits. Therefore, InterCon can pass savings along to the foreign insurer by facilitating claim management and payment and can collect processing fees from these insurance companies in return.

Currently, this Canadian company charges a processing fee of 9.5% (collected partly from the medical providers and partly from the foreign insurers) and has experienced an average annual growth rate of 3%. However, a backlog of claims and a cyclical pattern to the claims (due to high tourist seasons) has caused a delay in filing claims with the foreign insurers, resulting in a number of noncollectible claims. In addition, while the growing company has been trying to keep costs to a minimum, they are considering adding a new information technology system to help them streamline the process. Most of the company's revenue comes from claims for inpatient hospital stays, which average approximately $10,000. These claims represent 20% of InterCon's claims, but over 80% of its claim revenue. The remaining claims arise from clinic and emergency room visits. Currently, the company has about 20,000 annual claims, so the associated revenue is estimated to be as follows.

Claim Class	# Claims by Class	% of Total	Average Claim Amount	Claim Revenue ($M)	% of Total	Fee Rate	Fees ($M)
Class A	4000	20%	$10000	$40.0	85%	9.5%	$3.8
Class B	6000	30%	$1000	$6.0	13%	9.5%	$0.57
Class C	10000	50%	$100	$1.0	2%	9.5%	$0.095
Total (Annual)	**20000**			**$47.0**			**$4.465**

To help with rising costs, this firm may consider raising the processing fee as high as 11.0%. While this would generate additional revenue, the firm would also incur the risk of losing contracts with insurers and health care providers. Here are estimates of

---

[6]*Fortune*, May 11, 1987, pp. 50–58.

[7]This case is based on the case by G. Truman, D. Pachamanova, and M. Goldstein entitled "InterCon Travel Health Case Study" in *Journal of the Academy of Business Education*, Vol. 8, Summer 2007, pp. 17–32.

the impact of a change in rates on the annual growth in claims depending on the strength of foreign tourism (reverse of the strength of the U.S. economy) for the 9.5% processing fee rate. Find the payoffs (total revenue) for the 11.0% fee rate based on the preceding table and the growth rates provided.

Foreign Tourism	Probability	Processing Fee Rate at 9.5% Growth Rate	Processing Fee Rate at 9.5% Payoff ($)	Processing Fee Rate at 11.0% Growth Rate	Processing Fee Rate at 11.0% Payoff ($)
Weak	0.40	3%	54069129	1%	??
Strong	0.60	4%	56231894	3%	??

What is the expected revenue for the respective fee alternatives based on the growth rate and total income from claims and fees? Suppose the manager gets new information and revises the probabilities for the strength of tourism to be even at 50-50. What are the new values for expected revenue, CV, and return to risk ratio?

Exercises

SECTION 24.1

1. Which of the following are Actions and which are States of Nature?

a) Whether Unemployment is above or below 10%.
b) Whether a new product should be brought to market before or after the beginning of the fiscal year.
c) Whether to tell employees to stay home because of a blizzard.

2. Which of the following are Actions and which are States of Nature?

a) Whether to invest in solar energy companies.
b) Whether your chief competitor decides to launch a new advertising campaign.
c) Whether the decision you made to out-source manufacturing last year turned out to be the cost-saver you thought it would be.

SECTION 24.2

3. You are called on to decide how your company should produce its new cell phone screen defroster (for use by skiers and others spending time out of doors in the cold.) You develop the following cost matrix ($000's):

Production	State of Nature (State of the Economy) Recession	Stable	Expansion
Out-source	220	350	300
In-house	150	240	390

Draw the corresponding decision tree.

4. Here is a decision tree for the profits (in $000's) you project for your sales of the cell phone screen defroster, depending on the weather this coming winter and your choice of advertising method:

Action	State	Outcome
Point of Sale	Warm	550
	Moderate	600
	Cold	710
Magazine	Warm	410
	Moderate	590
	Cold	800
Internet	Warm	750
	Moderate	710
	Cold	590

Write out the corresponding profit matrix.

SECTION 24.3

5. For the cost matrix of Exercise 3,

a) What is the minimax choice?
b) What is the minimin choice?

6. For the decision tree of Exercise 4,

a) What is the maximin choice?
b) What is the maximax choice?

SECTION 24.4

7. For the cost matrix of Exercise 3

a) Suppose $P(Recession) = .2$, $P(Stable) = .2$, and $P(Expansion) = .6$. What is the expected value of each action?

b) What is the best choice using the expected-value approach?

8. For the decision tree of Exercise 4

a) Suppose $P(Warm) = .5$, $P(Moderate) = .3$, and $P(Cold) = .2$. What is the expected value of each action?

b) What is the best choice using the expected-value approach?

SECTION 24.7

9. For the probabilities of Exercise 7 and the cost matrix of Exercise 3, using the expected values you found in Exercise 7, compute the standard deviation of values associated with each action and the corresponding coefficient of variation.

10. For the probabilities of Exercise 8 and the decision tree of Exercise 4, using the expected values found in Exercise 8, compute the standard deviations of the values associated with each action and the corresponding coefficient of variation.

SECTION 24.10

11. According to U.S. Census data, 68% of the civilian U.S. labor force self-identifies as White, 11% as Black, 13% as Hispanic or Latino, and the remaining 8% as other ethnic origin. Among Whites in the labor force 54% are Male and 46% Female. Among Blacks, 52% are Male and 48% Female, and among Hispanic/Latinos, 58% are Male and 42% are Female. Breakdowns by gender are not available for other categories.

a) Here is a probability tree for these relationships. Fill in the probabilities as given:

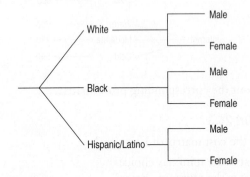

b) What is the probability that a randomly selected worker is a Female Hispanic/Latino?

c) For a randomly selected worker, what is $P(Female \mid White)$?

12. Lie detectors are controversial instruments, barred from use as evidence in many courts. Nonetheless, many employers use lie detector screening as part of their hiring process. There has been some research, but no agreement, about the reliability of polygraph tests. Based on this research, suppose that a polygraph can detect 65% of lies, but incorrectly identifies 15% of true statements as lies.

The company gives everyone a polygraph test, asking "Have you ever stolen anything from your place of work?" Naturally, all the applicants answer "No," but the company has evidence to suggest that 5% of the applicants are lying. When the polygraph indicates that the applicant is lying, that person is ineligible for a job.

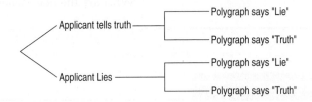

a) Here is the outline of a probability tree for this situation. Fill in the probabilities

b) What is the probability that a random job applicant will be an honest worker who is cleared by the polygraph?

*SECTION 24.11

13. Using the probability tree you made in Exercise 11, if a randomly selected worker is Female, what is the probability that she is White?

14. Using the probability tree you made in Exercise 12, what is the probability that a job applicant who is rejected by the polygraph was actually trustworthy?

CHAPTER EXERCISES

15. Flight decision. You are planning a trip home at the end of the semester and need to make plane reservations soon. However, you've just had a preliminary interview with a consulting firm that seemed to go very well. There's a chance that they will want you to stay for a few days at the end of the semester for a round of interviews at their offices, which will mean you'll have to change the date of the flight if you make the reservation now. Suppose that you can purchase a changeable fare for $750 or a nonrefundable fare for $650 for which a change costs $150. Construct a payoff table for this set of actions using total cost as the "payoff."

16. Product introduction. A small company has the technology to develop a new personal data assistant (PDA), but it worries about sales in the crowded market. They estimate that it will cost $600,000 to develop, launch, and market the product. Analysts have produced revenue estimates for three scenarios: If sales are high, they will sell $1.2M worth of the phones; if sales are moderate, they will sell $800,000 worth; and if sales are low, they will sell only $300,000 worth. Construct a payoff table for this set of actions using net profit as the "payoff." Don't forget the possible action of doing nothing.

17. Advertising strategies. After a series of extensive meetings, several of the key decisionmakers for a small marketing firm have produced the following payoff table (expected profit per customer) for various advertising strategies and two possible states of the economy.

		Consumer Confidence	
		Rising	Falling
Action	Prime Time Ad	$20.00	$2.00
	Targeted Web Marketing	$12.00	$10.00
	Direct Mail Piece	$10.00	$15.00

Construct a decision tree for this payoff table.

18. Energy investment. An investment bank is thinking of investing in a start-up alternative energy company. They can become a major investor for $6M, a moderate investor for $3M, or a small investor for $1.5M. The worth of their investment in 12 months will depend on how the price of oil behaves between then and now. A financial analyst produces the following payoff table with the net worth of their investment (predicted worth − initial investment) as the payoff.

		Price of Oil		
		Substantially Higher	About the Same	Substantially Lower
Action	Major investment	$5,000,000	$3,000,000	−$2,000,000
	Moderate investment	$2,500,000	$1,500,000	−$1,000,000
	Small investment	$1,000,000	$500,000	−$100,000

Construct a decision tree for this payoff table.

19. Flight decision tree. Construct a decision tree for the payoff table in Exercise 15.

20. Product introduction tree. Construct a decision tree for the payoff table in Exercise 16.

21. Flight decision expected value. If you think the probability of being called for an interview is 0.30, calculate the expected value of each action in Exercise 15. Which is the better action in this case?

22. Product introduction expected value. An analyst for the company in Exercise 16 thinks the probabilities of high, moderate, and low sales are 0.2, 0.5, and 0.3, respectively. In this case calculate the expected value of each action. Which is the best action in this case?

23. Flight decision, part 4. For the decision of Exercise 15, you've just learned that you are on the short list and now estimate the chance that you'll be called for an interview is 0.70. Does that change your choice of actions?

24. Product introduction, part 4. For the product launch decision of Exercise 16, the economy isn't looking that good. Your very cautious boss says that he thinks there's a 60% chance of low sales, and a 30% chance of moderate sales. Which course should the company follow?

25. Advertising strategies decisions. For the payoff table in Exercise 17, find the action with the highest expected value.

a) If forecasters think the probability of rising consumer confidence is 0.70, what is its expected value?
b) What action would have the highest expected value if they think the probability of rising consumer confidence is only 0.40?

26. Energy investment decisions.

a) For the payoff table in Exercise 18, find the investment strategy under the assumption that the probability that the price of oil goes substantially higher is 0.4 and that the probability that it goes substantially lower is 0.2.
b) What if those two probabilities are reversed?

27. Advertising strategies EVPI.

a) For the advertising strategies of Exercise 25 and using the probability of 0.70 for rising consumer confidence, what is the Expected Value of Perfect Information (EVPI)?
b) What is the EVPI if the probability of rising consumer confidence is only 0.40?

28. Energy investment EVPI. For the energy investment of Exercise 26 and using both of the probabilities considered in that exercise, find the Expected Value of Perfect Information.

29. Advertising strategies with information. The company from Exercises 17, 25, and 27 has the option of hiring an economics consulting firm to predict consumer confidence. The company has already considered that the probability of rising consumer confidence could be as high as 0.70 or as low as 0.40. They could ask the consultants for their choice between those two probabilities, or they could just pick a probability in the middle, such as .50, and choose a strategy based on that.

a) Draw the decision tree including the decision to hire the consultants.
b) Would the consultants' information be useful? Explain.
c) The company thinks there's an equal chance of either of the consulting alternatives being what the consultants report. What's the value to the company (per customer) of the extra information?

30. Energy investment with information. The company in Exercises 18, 26, and 28 could send a team to Saudi Arabia to obtain additional information about the probabilities that oil will increase or decrease in price. They hope that the fact-finding trip would choose between the two alternatives considered in Exercise 26, or they could just estimate that the probabilities are equal.

a) Make a decision tree for these decisions.
b) Should the company send the fact-finding trip? Explain.
c) The company's experts estimate that if they send the fact-finding mission, there's a 70% chance that they'll conclude there's a 0.4 probability of higher oil prices. What would the value of the additional information be to the company?

31. Investing in equipment. *KickGrass Lawncare* is a service that cares for lawns in a large, affluent community. Shawn Overgrowth, the owner, is considering the purchase of new zero-turn riding lawn tractors, which would allow him to expand his business. The tractors cost $6300 each, and he would purchase two of them. Another alternative is to purchase three additional mowers of the current type to add to his current equipment. Those would cost $475 apiece. Or he could face the coming gardening season with his existing equipment. Shawn estimates that in a good growing season, the tractors would allow him to expand his business by $40,000. But if the summer is hot and dry (so lawns don't grow) or cold and wet (ditto), he'd only be able to add about $15,000 in contracts. If he purchases the mowers, he could expand his business by $10,000 in a good year or by just $5,000 in a bad one. And if he spends nothing, he won't expand his business. In a bad year, his income would contract by about $1,000.

Construct a payoff table and tree diagram for Shawn's decision. Don't forget to include his expenses in the calculations.

32. Market segmentation. Demand and price are related; raising prices typically lowers demand. Many companies understand that if they can *segment* their market and offer different prices to different segments, they can often capture more revenue. *Aaron'sAir* is a small commuter airline.

They typically charge $150 for a one-way flight between a resort island they serve and the mainland. In times of low, medium, and high demand, Aaron (the owner and pilot) estimates that he'll sell 100, 200, or 500 seats per week, respectively. He is considering offering two different fares based on whether his customers stay over a Saturday night on the island. He thinks that business travelers coming to the island for conferences and retreats would typically not stay, but vacationers would. He expects that the low fare will attract additional customers. However, he anticipates that some of his regular customers will also pay less. The two fares would be $90 and $210. Aaron estimates that in times of low demand he'd sell 30 high-fare and 80 low-fare tickets—revenue of $30 \times \$210 + 80 \times \$90 = \$13,500$. In times of medium demand, he estimates 110 high-fare and 250 low-fare tickets, for an estimated revenue of $45,600. And in times of high demand, he expects 500 low-fare customers and 250 high-fare customers, yielding $97,500.

Make a payoff table and decision tree for this decision.

33. Investing in equipment, maxes and mins. Shawn Overgrowth, whom we met in Exercise 31, is an entrepreneur who is optimistic about the growing season. What choice should he make to maximize his return? His assistant, Lance Broadleaf, is very conservative, and argues that *KickGrass* should minimize their potential downside results. Which alternative decision does he argue for?

34. Market segmentation minimax. Aaron, who we met in Exercise 32, tends to be optimistic about business conditions. What is his maximax strategy that would maximize his results?

35. "Weather" or not to invest in equipment. Shawn Overgrowth, from Exercise 31, estimates that the probability of a good growing season is 0.70. Based on that:

a) Find the EV for his actions.
b) Find the standard deviations.
c) Compute the RRRs. Which action is preferred based on the RRRs?

36. Market segmentation and chance. *Aaron'sAir* (see Exercises 32 and 34) estimates that high-demand periods (which depend on the weather and on bookings for conferences) occur with probability .3 and medium demand periods occur with probability .5. The rest are low-demand periods.

a) What's the expected value of each of Aaron's alternative actions?
b) What are the standard deviations for each action?
c) What are the RRRs? Based on the RRRs, what action is best?

37. Equipment and data. Shawn, of Exercises 31, 33, and 35 could obtain long-range predictions of the growing conditions for next summer. He thinks that those might show a probability of good growing conditions as low as 50% or as high as 80%. If he doesn't obtain those predictions, he'll make a decision based on his previous estimates (see Exercise 35).

a) Draw the decision tree.

b) If Shawn thinks there's a 60% chance the long-range predictions will predict a 50% chance of good conditions, find the corresponding EVwSI.

c) Should Shawn purchase the long-range predictions?

38. Segments and surveys. *Aaron'sAir* (see Exercises 32, 34, and 36), could purchase a market survey from a firm that has advised the island tourist and conference bureau. He thinks their projections would help him determine whether the probability of high demand might be as high as .5 or as low as .2, with the corresponding probabilities for medium demand being .3 and .4. If he doesn't purchase the market survey, he'll make a decision based on his previous best estimates (see Exercise 36).

a) Draw the decision tree.

b) Aaron thinks the market survey is likely to be optimistic. He'd estimate a 65% probability that it would predict the higher (.5) probability of high demand. What would be the EVwSI?

c) If the consultant's report costs $2,000, should Aaron pay for it?

39. Investment strategy. An investor is considering adding a stock to her portfolio. Assuming she buys 100 shares, here is an estimated payoff table for the alternative stocks if she holds onto them for six months. The value of the stock depends on whether or not an acquisition is approved for one of the companies, since the companies are actually competitors. She estimates the probability of an acquisition to be 0.3.

	Acquisition?	
	Yes (0.3)	No (0.7)
Stock A	$5000	−$1000
Stock B	−$500	$3500

a) Compute the EV for each alternative decision.

b) Compute the SD for each decision.

c) Compute the CV and RRR for each decision.

d) Which stock would you choose and why?

40. Mutual fund investing. An investor is considering how to invest her money. She has two options—either a domestic mutual fund that only invests in blue chip stocks or an international aggressive mutual fund that invests in young technical firms. The payoff (profit) after one year for these investments depends on the state of the economy.

	Economy	
	Improves (0.5)	Declines (0.5)
Domestic Mutual Fund	$1500	$1000
International Mutual Fund	$3500	−$1000

a) Compute the EV for each alternative decision.

b) Compute the SD for each decision.

c) Compute the CV and RRR for each decision.

d) Which mutual fund would you invest in and why?

41. Bicycle sales. A bicycle shop owner is deciding which products to stock. His distributor will give him a deal if he buys more of the same kind of bike. The payoff table shows monthly sales for a high-end bike (selling at $950) or a moderately priced bike (selling at $500). Based on past experience, the shop owner makes the following assumption about the demand for the high-end bike. Demand will be low, moderate, or high with probabilities 0.3, 0.5, and 0.2, respectively. He also assumes that if demand is low for the high-end bike, it will be higher for the moderately-priced bike.

	Demand for High-End Bike		
	Low Demand (0.3)	Moderate Demand (0.5)	High Demand (0.2)
High-End Bike	$1900	$4750	$7600
Moderately-Priced Bike	$4000	$2500	$1000

a) Compute the EV for each alternative product (decision).

b) Compute the SD for each decision.

c) Compute the CV and RRR for each decision.

d) Which bike would you stock and why?

42. Bicycle sales, part 2. The bike shop owner has now done a bit more research and believes that the demand for high-end bikes has shifted, so that now the low demand is 50% likely and the high demand is only 10% likely. How does this change your responses to Exercise 41? Find the new RRR. Does your recommendation to the shop owner change?

43. Website experiment. Summit Projects provides marketing services and website management for many companies that specialize in outdoor products and services (www .summitprojects.com). To understand customer Web behavior, the company experiments with different offers and website design. The results of such experiments can help to maximize the probability that customers purchase products during a visit to a website. Possible actions by the website include offering the customer an instant discount, offering the customer free shipping, or doing nothing. A recent experiment found that customers make purchases 6% of the time when offered the instant discount, 5% when offered free shipping, and 2% when no special offer was given. Suppose 20% of the customers are offered the discount and an additional 30% are offered free shipping.

a) Construct a probability tree for this experiment.
b) What percent of customers who visit the site made a purchase?
c) Given that a customer made a purchase, what is the probability that they were offered free shipping?

44. Website experiment, part 2. The company in Exercise 43 performed another experiment in which they tested three website designs to see which one would lead to the highest probability of purchase. The first (design A) used enhanced product information, the second (design B) used extensive iconography, and the third (design C) allowed the customer to submit their own product ratings. After 6 weeks of testing, the designs delivered probabilities of purchase of 4.5%, 5.2%, and 3.8%, respectively. Equal numbers of customers were sent randomly to each website design.

a) Construct a probability tree for this experiment.
b) What percent of customers who visited the site made a purchase?
c) What is the probability that a randomly selected customer was sent to design C?
d) Given that a customer made a purchase, what is the probability that the customer had been sent to design C?

45. Contract bidding. As manager for a construction firm, you are in charge of bidding on two large contracts. You believe the probability you get contract #1 is 0.8. If you get contract #1, the probability you also get contract #2 will be 0.2, and if you do not get #1, the probability you get #2 will be 0.4.

a) Sketch the probability tree.
b) What is the probability you will get both contracts?
c) Your competitor hears that you got the second contract but hears nothing about the first contract. Given that you got the second contract, what is the probability that you also got the first contract?

46. Extended warranties. A company that manufactures and sells consumer video cameras sells two versions of their popular hard disk camera, a basic camera for $750 and a deluxe version for $1250. About 75% of customers select the basic camera. Of those, 60% purchase the extended warranty for an additional $200. Of the people who buy the deluxe version, 90% purchase the extended warranty.

a) Sketch the probability tree for total purchases.
b) What is the percentage of customers who buy an extended warranty?

c) What is the expected revenue of the company from a camera purchase (including warranty if applicable)?
d) Given that a customer purchases an extended warranty, what is the probability that he or she bought the deluxe version?

47. Computer reliability. Laptop computers have been growing in popularity according to a study by Current Analysis Inc. Laptops now represent more than half the computer sales in the United States. A campus bookstore sells both types and in the last semester sold 56% laptops and 44% desktops. Reliability rates for the two types of machines are quite different, however. In the first year, 5% of desktops require service, while 15% of laptops have problems requiring service.

a) Sketch a probability tree for this situation.
b) What percentage of computers sold by the bookstore last semester required service?
c) Given that a computer required service, what is the probability that it was a laptop?

48. Titanic survival. Of the 2201 passengers on the RMS *Titanic*, only 711 survived. The practice of "women and children first" was first used to describe the chivalrous actions of the sailors during the sinking of the HMS *Birkenhead* in 1852, but became popular after the sinking of the *Titanic*, during which 53% of the children and 73% of the women survived, but only 21% of the men survived. Part of the protocol stated that passengers enter lifeboats by ticket class as well. Here is a table showing survival by ticket class.

| | | Class | | | | |
		First	Second	Third	Crew	Total
Survival	Alive	203	118	178	212	**711**
		28.6%	16.6%	25.0%	29.8%	**100%**
	Dead	122	167	528	673	1490
		8.2%	11.2%	35.4%	45.2%	**100%**

a) Find the conditional probability of survival for each type of ticket.
b) Draw a probability tree for this situation.
c) Given that a passenger survived, what is the probability they had a first-class ticket?

Introduction to Data Mining

CHAPTER

25

Paralyzed Veterans of America

The Paralyzed Veterans of America (PVA) is a philanthropic and service organization chartered by the U.S. government to serve the needs of U.S. veterans who suffer from spinal cord injury or disease. Since 1946, PVA has raised money to support a variety of activities, including advocacy for veterans' health care, research and education in spinal cord injury and disease, and support for veterans' benefits and rights.

PVA raises money in a variety of ways, but the majority of their fund-raising comes from direct mail campaigns, in which they send free address labels or greeting cards to potential donors on their mailing list and request a donation for receiving these gifts.

PVA sends out regular solicitations to a list of more than four million donors. In 2006, PVA received about $86 million in donations, but the effort cost them over $34 million in postage, administrative, and gift expenses. Most of that money was spent on mailings to people who never responded. In fact, in any one solicitation, groups like PVA are lucky to get responses from more than a few percent of the people they contact. Response rates for commercial companies such as large credit card banks are so low that they often measure responses not in percentage points, but in basis

863

points—hundredths of a percent. If the PVA could avoid mailing to just half of the people who won't respond, they could save $17 million a year and produce only half as much wasted paper. Can statistical methods help them decide who should get what mail?

25.1   Direct Marketing

Rather than advertise in various media, companies sometimes try to appeal directly to the consumer. The appeal is usually in the form of an offer to buy something, to enroll in a special program, or, in the case of a philanthropic group, to donate money or time. You've undoubtedly seen these efforts in the form of mail, e-mail, or phone calls. Direct mail, often called "junk mail," generates about four million tons of paper waste per year. Direct e-mail, often called "spam," accounts for 95% of all e-mails according to The European Network and Information Security Agency (ENISA).

Companies use direct mail and e-mail because, compared to other options, it is inexpensive and effective. To make it more effective, companies want to identify people who are most likely to respond. In other words, the companies would like to *target* their promotions and offers.

To help them decide how likely a customer might be to respond to a particular offer, a company might build a model to estimate the probability. The data used to build the model is usually a combination of data on customers that the company has collected and other data that it has bought. **Data mining** is the name for a process that uses a variety of data analysis tools to discover patterns and relationships in data to help build useful models and make predictions. A company's goal with data mining might be to direct their mailings to people more likely to respond to the so-licitations. For example, they might look at past customer behavior and other de-mographic information and build a model (or models) to predict who is most likely to respond. Many of the modeling techniques that we've covered in this book—especially multiple regression and logistic regression—are used in data mining. But because data mining has benefited from work in machine learning, computer science, and artificial intelligence, as well as statistics, it has a much richer set of tools than those we've discussed in this book.

25.2   The Data

Companies and philanthropic organizations collect an incredible amount of infor-mation about their customers. You've contributed to this data gathering yourself, perhaps without even knowing it. Every time you purchase something with a credit card, use a loyalty card at a supermarket, order by phone or on the Internet, or call an 800 number, your transaction is recorded. For a credit card company, this **transactional data** might contain hundreds of entries a year for each customer. For a philanthropic organization, the data might also include a history of when solicita-tions were sent, whether the donor responded to each solicitation, and how much they contributed. For a credit card bank, the transactional data would include every purchase the customer made. Every time a transaction is attempted, information is sent back to the transactional database to check that the card is valid, has not been reported stolen, and that the credit limit of the customer has not been exceeded—that's why there's a small delay before your transaction is approved every time you charge something. Although companies collect transactional data to facilitate the transactions themselves, they recognize that these data hold a wealth of information about their business. The challenge is to find ways to extract—to mine—that information.

In addition to transactional data, companies often have separate databases containing information about customers and about products (inventory, price, and shipping costs, for example). The databases can be linked to each other in a relational database. The properties of relational databases and some simple examples were discussed in Chapter 2.

The variables in the customer database are of two types: individual and regional. The individual variables are typically first gathered when the customer opens an account, registers on a website, or fills out a warranty card and are specific to that customer. They might include **demographic variables** such as a customer's age, income, and number of children. The company can add to these variables others that arise from the customer's interactions with the company, including some that may summarize variables in the transactional database. For example, the total amount spent each month might be a variable in the customer database that is updated from the customer's individual purchases, stored in the transactional database. The company might also purchase additional demographic data. Some demographic data are based on zip codes and can be obtained from agencies like the U.S. Census Bureau. These data can provide information on the average income, education, home value, and ethnic composition of the neighborhood in which a customer lives, but are not customer-specific.

Customer-specific data can also be purchased from a variety of commercial organizations. For example, a credit card company may want to send out an offer of free flight insurance to customers who travel frequently. To help know which customers those are, the company may buy information about customers' magazine subscriptions to identify customers who subscribe to travel or leisure magazines. The sharing and selling of individual information is controversial and raises privacy concerns, especially when the purchased data involve health records and personal information. In fact, concerns over the sharing of health data in the United States led to a set of strict guidelines known as HIPAA (Health Insurance Portability and Accountability Act) regulations. Restrictions on the collection and sharing of information about customers vary widely from country to country.

The PVA customer database[1] is a typical mix of variables. It contains 481 variables on each donor. There are 479 potential predictor variables and two responses: *TARGET_B*, a 0/1 variable which indicates whether the donor contributed to the most recent campaign and *TARGET_D* which gives the dollar amount of the contribution. Table 25.1 shows the first 18 records for a subset of the 481 variables found in the PVA data. We can guess what some of the variables mean from their names, but others are more mysterious.

Information about the variables, including their definitions, how they are collected, the date of collection, etc. is collectively called **metadata**. The variables in the PVA data set presented below are typical of the types of data found in the customer records of many companies. We shouldn't be surprised to find out that *AGE* is the donor's age measured in years and that *ZIP* is the donor's postal zip code. But without the metadata, it would be hard to know that *TCODE* is the code used before the title on the address label (0 = blank; 1 = Mr.; 2 = Mrs.; 28 = Ms., and so on) or that *RFA_2A* is a summary of past giving.

[1]The PVA made some of their data available for the 1998 Knowledge Discovery and Data Mining (KDD) contest. The object of the contest was to build a model to predict which donors should receive the next solicitation based on demographic and past giving information. The results were presented at the KDD conference (www.kdnuggets.com). The variables discussed in this chapter are the ones that PVA made available.

ODATEDW	OSOURCE	TCODE	STATE	ZIP	DOB	RFA_2A	AGE	OWN	INC	SEX	WEALTH	AVGGIFT	TARGET_B	TARGET_D
9401	L16	2	GA	30738	6501	F	33	U	5	F	2	11.66667	0	0
9001	L01	1	MI	49028	2201	F	76	H	1	M	2	8.777778	0	0
8601	DNA	1	TN	37079	0	E		U	1	M		8.619048	1	10
8601	AMB	1	WI	53719	3902	G	59			M		16.27273	0	0
8601	EPL	2	TX	79925	1705	E	81	H	2	F	6	10.15789	0	0
8701	LIS	1	IN	46771	0	F				M		8.871333	0	0
9201	GRI	1	IL	60016	1807	F	79	H	4	M	6	13.8	0	0
9401	HOS	0	KS	67218	5001	G	48	U	7	F	7	18.33333	0	0
8901	DUR	0	MI	48304	1402	F	84	H	7	M	9	12.90909	0	0
8601	AMB	0	FL	34746	1412	F	83	H	2	F	3	9.090909	0	0
9501	CWR	2	LA	70582	0	D		U	5	F		5.8	0	0
9501	ARG	0	MI	48312-	4401	E	54			F		8	0	0
8601	ASC	0	TX	75644	2401	G	74	U	7	F		13.20833	0	0
9501	DNA	28	CA	90059	2001	E	78	H	1	F		10	0	0
9201	SYN	0	FL	33167	1906	F	79	U	2	M	3	10.09091	0	0
9401	MBC	2	MO	63084	3201	F	66	H	5	F		10	0	0
9401	HHH	28	WI	54235	0	F		H	7	F		20	0	0
9101	L02	28	AL	36108	4006	F	58	H	5	F		10.66667	0	0

Table 25.1   Part of the customer records from the PVA data set. Shown here are 15 of the 481 variables and 18 of the nearly 100,000 customer records used in the 1998 KDD (Knowledge Discovery and Data Mining) data mining competition.[2] The actual PVA customer database contains several million customer records.

About 10% of the PVA variables describe past giving behavior collected by the PVA itself. More than half of the variables are regional data based on zip codes, most likely purchased from the Census Bureau. The rest of the variables are donor-specific, either gathered by the PVA or purchased from other organizations.

Sometimes disparate databases are gathered or merged in a central repository called a **data warehouse**. Maintaining the data warehouse is a huge job, and companies spend millions of dollars each year for the software, hardware, and technical personnel to do just that. Once a company has invested in creating and maintaining a data warehouse, it's only natural that they would want to get as much value as they can from it. For example, at PVA, analysts might use the data to build models to predict who will respond to a direct mail campaign; a credit card company, similarly, might want to predict who is most likely to accept an offer for a new credit card or services.

25.3   The Goals of Data Mining

The purpose of data mining is to extract useful information hidden in these large databases. With a database as large as a typical data warehouse, that search can be like looking for a needle in a haystack. How can analysts hope to find what they're looking for? They may start their search by using a sequence of queries to deduce facts about customer behavior, asking specific questions based on the data. Guided by their knowledge of the specific business, they may try a query-driven approach, asking a series of specific questions to deduce patterns. Such an approach is **online analytical processing** or **OLAP**. Analysts in sales, marketing, budgeting, inventory,

---

[2]The KDD cup is the leading Data Mining competition in the world, and is organized by the SIGKDD interest group of the ACM (Association for Computing Machinery).

and finance often use OLAP to answer specific questions involving many variables. An OLAP question for the PVA might be: "How many customers under the age of 65 with incomes between $40,000 and $60,000 in the Western region who have not donated in the previous two years gave more than $25 to the most recent solicitation?" Although OLAP is efficient in answering such multivariable queries, it is question-specific. OLAP produces answers for specific queries, typically as tables, but does not build a predictive model, and so it is not appropriate to generalize using OLAP.

In contrast to an OLAP query, the outcome of a data mining analysis is a **predictive model**—a model that uses predictor variables to predict a response. For a quantitative response variable (as in linear regression), the model will predict the *value* of the response, whereas for a categorical response, the model estimates the probability that the response variable takes on a certain value (as in logistic regression). Both linear and logistic multiple regressions are common tools of a data miner. Like all statistical models and unlike query-based methods such as OLAP, data mining generalizes to other similar situations through its predictive model. For example, an analyst using an OLAP query might find that customers in a certain age group responded to a recent product promotion. But without building a model, the analyst can't understand the relationship between the customer's age and the success of the promotion, and thus may be unable to predict how the product will do with a larger set of customers. The goal of a data mining project is to increase business understanding and knowledge by building a model to answer a specific set of questions raised at the beginning of the project.

Data mining is similar to traditional statistical analysis in that it involves exploratory data analysis and modeling. However, there are several aspects of data mining that distinguish it from more traditional statistical analysis. Although there is no consensus on exactly what data mining is and how it differs from statistics, some of the most important differences include:

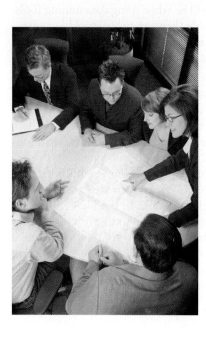

- **The size of the databases.** Although no particular size is required for an analysis to be considered data mining, an analysis involving only a few hundred cases or only a handful of variables would not usually be considered data mining.

- **The exploratory nature of data mining.** Unlike a statistical analysis that might test a hypothesis or produce a confidence interval, the outcome of a data mining effort is typically a model used for prediction. Usually, the data miner is not interested in the values of the parameters of a specific model or in testing hypotheses.

- **The data are "happenstance."** In contrast to data arising from a designed experiment or survey, the typical data on which data mining is performed have not been collected in a systematic way. Thus, warnings not to confuse association with causation are especially pertinent for data mining. Moreover, the sheer number of variables involved makes any search for relationships among variables prone to Type I errors.

- **The results of a data mining effort are "actionable."** For business applications, there should be a consensus of what the problem of interest is and how the model will help solve the problem. There should be an action plan in place for a variety of possible outcomes of the model. Exploring large databases out of curiosity or just to see what they contain is not likely to be productive.

- **The modeling choices are automatic.** Typically the data miner will try several different types of models to see what each has to say, but doesn't want to spend a lot of time choosing which variables to include in the model or making the kinds of choices one might make in a more traditional statistical analysis. Unlike the analyst who wants to understand the terms in a stepwise regression, the data miner is more concerned with the predictive ability of the model. As long as the resulting model can help make decisions about who should receive the next offer, who is best suited for an online coupon, or who is most likely to switch cable providers next month, the data miner is likely to be satisfied.

25.4   Data Mining Myths

Data mining software usually contains a variety of exploratory and model-building tools and a graphical user interface designed to guide the user through the data mining process. Some people buy data mining software hoping that the tools will find information in their databases and write reports that spew knowledge with little or no effort or input from the user. Software vendors often capitalize on this hope by exaggerating the capabilities and the automatic nature of data mining to increase sales of their software. Data mining can often *assist* the analyst to find meaningful patterns and help predict future customer behavior, but the more the analyst knows about his or her business, the more likely he or she will be successful using data mining. Data mining is not a magic wand that can overcome poor data quality or collection. A product may have tools for detecting outliers and may be able to impute (assign a value for) missing data values, but all the issues you learned about good statistical analysis are still relevant for data mining.

Here are some of the more common myths about what data mining can do.

- **Myth 1:** Find answers to unasked questions

  Even though data mining can build a model to help answer a specific question, it does not have the ability to answer questions that haven't been asked. In fact, formulating a precise question is a key first step in any data mining project.

- **Myth 2:** Automatically monitor a database for interesting patterns

  Data mining techniques build predictive models and can answer queries, but do not find interesting patterns on their own any more than regression models do.

- **Myth 3:** Eliminate the need to understand the business

  In fact, the more an analyst understands his or her business, the more effective the data mining effort will be.

- **Myth 4:** Eliminate the need to collect good data

  Good data is as important for a data mining model as for any other statistical model that you've encountered. While some data mining software contains tools to help with missing data and data transformations, there is no substitute for quality data.

- **Myth 5:** Eliminate the need for good data analysis skills

  The better the data analysis skills of the miner—the skills you've learned in every chapter of this book—the better the analysis will be when using data mining tools. Data mining tools are more powerful and flexible than a statistical tool like regression but are similar in the way they work and in how they're implemented.

25.5   Successful Data Mining

The size of a typical data warehouse makes any analysis challenging. The ability to store data is growing faster than the ability to use it effectively. Commercial data warehouses often contain terabytes (TB)—more than 1,000,000,000,000 (1 trillion) bytes—of data (one TB is equivalent to about 260,000 digitized songs), and warehouses containing petabytes (PB—one PB = 1000 TB) are now common. The tracking database of the United Parcel Service (UPS) is estimated to be on the order of 16 TB, or roughly the digital size of all the books in the U.S. Library of Congress. According to *Wired* magazine, about 20 TB of photos are uploaded to Facebook every month. All the U.S. Census data from 1790 to 2000 would take about 600 TB. However, it's estimated that the servers at *Google* process a petabyte of data every 72 minutes.[3] Data miners hope to uncover some important strategic information lying hidden within these massive collections of data.

---

[3] *Wired*, Issue 16.07, June 2008.

To have a successful data mining outcome, the first step is to have a well-defined business problem. With 500 variables, there are over 100,000 possible two-way relationships between pairs of variables. The number of pairs that will be related just by chance is likely to be large. And it's human nature to find many of these relationships interesting and even to posit plausible reasons why two variables might be associated. Some of these variables may appear to provide a useful predictive model, when in fact they do not. A well-defined business objective can help you avoid going down a lot of blind paths.

As in painting a house, much of the effort of a data mining project is in the preparation, cleaning, and checking of the data. It is estimated that for any data mining project, 65% to 90% of the time spent is spent in such **data preparation**. Data preparation involves investigating missing values, correcting wrong and inconsistent entries, reconciling data definitions, and possibly creating new variables from the original ones. The data may need to be extracted from several databases and combined. Errors must be corrected or eliminated, and outliers must be identified.

A successful data mining effort often requires a substantial amount of time devoted to basic examination, cleaning, and preparation of the data before any modeling is performed. It will have a clear objective, agreed to by a team of people who will share the work and the responsibility of the data mining. It will have an action plan once the results of the data mining are known, whether those results are what the team expected or desired. Finally, the data mining should be accompanied by as much knowledge about both the data and the business question as possible. Blind searches for patterns in large databases are rarely fruitful and will waste valuable analytic resources.

For Example   Data preparation

Data mining analyses often deal with large numbers of variables. For the PVA data, there are 479 potential predictors for modeling whether a donor will donate to the next campaign. Each of these variables may present issues that should be dealt with before proceeding to a full analysis. But usually a smaller number are selected for examination. Here's one example.

Question: Looking just at the variable *AGE*, what corrections or adjustments would be appropriate so that it can be a more effective predictor in a regression-based model?

Answer: Let's look at a histogram and boxplot for the variable *AGE* for all 94,649 records of the PVA data set.

	Quantiles	
100.0%	maximum	98.000
99.5%		95.000
97.5%		90.000
90.0%		83.000
75.0%	quartile	75.000
50.0%	median	62.000
25.0%	quartile	48.000
10.0%		39.000
2.5%		31.000
0.5%		26.000
0.0%	minimum	1.000

Mean	61.598019
Std Dev	16.666865
Std Err Mean	0.0624749
upper 95% Mean	61.720469
lower 95% Mean	61.475568
N	71170

| Figure 25.1   Graphical output and statistical summaries of the variable *AGE* help to identify questionable values.

(continued)

There is a group of cases with ages below 20 that should immediately draw our attention. Are people that young likely to be donors? A closer examination reveals that there are even a handful of cases whose ages are below 15, of whom 17 are younger than 5 years old and 9 are 1 year old. There are also 23,479 missing values for the variable *AGE*. Our domain knowledge (what we know about the world and likely donors to such an organization) tells us that values below a certain age (probably below 19, but certainly below 10) or above a certain age (100? 110?) are almost certainly wrong, so we can see whether any of these values can be corrected from other records, or we can remove them from the analysis.[4]

It is harder to know whether the data in other variables are correct. And, while checking one or two variables to see if their values make sense is practical, examining bar charts, histograms, and boxplots for 479 variables would be a daunting task.

25.6   Data Mining Problems

Data mining can address different types of problems, some of which we have already encountered. When the goal is to predict a quantitative response variable, the problem is generically called a **regression problem,** regardless of whether linear regression is used (or even considered) as one of the models. When the response variable is categorical, the problem is referred to as a **classification problem** because the model will either guess the most likely category for each or assign a probability to each class for that code. For example, for the classification problem of predicting *whether or not* a particular donor will give to the next campaign, a model will produce either the most probable class (*donate or not*) or the probabilities of each of these. Predicting the *amount* of money a donor will donate is a regression problem.

Both of the previous problems are referred to as **supervised problems**. In a supervised setting, we are given a set of data for which we know the response. That is, for the PVA data, we *know* for at least one group of donors the values of the responses *TARGET-B* and *TARGET-D*. We know whether they gave to the last campaign and how much they donated. The data miner would construct a model based on a portion of the original data, called the **training set**. In order to assess how well the model will work in the future on data that the model hasn't encountered, the modeler uses the original data set that was withheld from the model building and then tests the predictions of the model on these data. This second data set is called the **test set**.

By contrast, there are problems for which there is no particular response variable. In these **unsupervised problems**, the goal may be to build clusters of cases with similar attributes. For example, a company may want to cluster their customers into groups with similar buying behaviors and tastes. Such analyses resemble segmentation analyses performed by marketing analysts. In this case, there is no response variable. All the predictors (or a subset of them) are used to build clusters based on an index that measures the similarity between the customers. Many different algorithms are available to find clusters.

25.7   Data Mining Algorithms

The methods used for the problems discussed in the previous section are often referred to as **algorithms**, a term that describes a sequence of steps with a specific purpose. You may hear a method (even one like linear regression) referred to as a model, an algorithm, a tool, or generically as a method. The terms seem to be used interchangeably. This section touches on only a few of the models used in data

---

[4]Further analysis revealed that age was calculated from date of birth by a computer algorithm, but only the last two digits of the date of birth year were used. So, some people (perhaps no longer living) who were in the data base had ages 100 years too young.

mining. Some of the most common methods used for prediction are decision trees and neural networks (discussed below), support vector machines, belief nets, the regression methods discussed in Chapters 18 and 19, and random forests. It's a lively research field and new algorithms are appearing all the time.

Tree Models

Chapter 24 discussed decision trees in which the analyst examined a small number of actions and possible states of nature in a sequence to identify the action with the largest payoff. In data mining, the term **decision tree** is used to describe a *predictive model.* Data mining decision trees are superficially similar in form to the decision trees we saw in Chapter 24, but that's where the similarity ends. These tree models use *data* to select predictor variables that give predictions of the response. The models are entirely driven by the data with no input from the user.

The way a tree model works is fairly simple. To illustrate the process, imagine that we want to predict whether someone will default on a mortgage. To build the tree, we use past data from a group of customers on whom we have similar information and for whom we know whether they defaulted or not. For this simplified example, let's assume we'll predict it only from the following variables:

- Age (years)
- Household Income ($)
- Years on the Job (years)
- Debt ($)
- Homeowner (Yes/No)

The tree model tries to find predictors that can distinguish those people who will default on their mortgages from those who won't. To do that, it first examines *every* potential predictor variable and every possible way of *splitting* that variable into two groups. For example, it looks at *Age*, and for every value of *Age*, it calculates the default rate of those *above* that value and *below* that value. Keeping track of the difference in default rates for the two groups defined by every possible predictor variable split at every possible way, it chooses that pair of predictors and *split points* that produce the greatest difference in default rates.[5] The tree algorithm decides which variables to split and where to split them by trying (essentially) all combinations of variables and split points until it finds the best splits. For a categorical predictor, the model considers every possible way of putting the categories into two groups. There are several criteria used by different algorithms to define "best," but they all have in common some attempt to find a way of splitting the cases so that the resulting two groups will have the largest possible difference in default rates. After the algorithm finds the first split, it continues searching again on the two resulting groups, finding the next best variable and split point (possibly using the same variable as the previous split again). It continues in this way until one of several criteria is met (for example, number of customers too small, not a large enough difference in rates is found) and then stops at what are called **terminal nodes,** where the model produces a prediction. The predictions at the terminal nodes are simply the average (if the response is quantitative) or the proportions of each category (for a classification problem) of the cases at that node.

The tree for our hypothetical mortgage example is shown in Figure 25.2. To understand the tree, start at the top and imagine being presented with a new customer. The first question the tree asks is "Is the *Household Income* more than $40,000?" If the answer is yes, move down to the right. For these cases, *Debt* is the

---

[5]This is a slight simplification. There are actually several possible variants on the criteria for splitting. Readers who want to know the details can read more advanced books on data mining or decision trees.

next variable split, this time at $10,000. If this customer's *Debt* exceeded $10,000 move down and to the right again. The tree estimates that customers like this (both *Household Income* > $40,000 and *Debt* > $10,000) defaulted at a rate of 5% (0.05). Just to the left are customers whose *Household Income* was > $40,000 but whose *Debt* was ≤ $10,000. They defaulted at a rate of 1%. For customers with incomes less than $40,000 (the left branch of the first split), the next variable split was not *Debt*, but *Job*. For those who have been at their job more than 5 years, the default rate was 6%, but was 11% for those with less time at their present job. The modeler may, at this point, label the outcomes as *risk* categories, for example by calling 1% *Very low*, 5 and 6% *Moderate*, and 11% *High*.

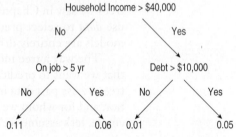

Figure 25.2   Part of a tree model on a hypothetical examination of mortgage default. The tree selects *Household Income* as the most important variable to split on and selects $40,000 as the cut point. For those customers whose income is more than $40,000, *Debt* is the next most important variable, while for those whose income is less than $40,000, how long they've held their current job is more important.

Tree models are very easy to implement and, in principle, are easy to interpret. They have the advantage of showing their logic clearly and are easy to explain to people who don't have a deep background in statistics. Even when they aren't used as a final model in a data mining project, they can be very useful for selecting a smaller subset of variables on which to do further analysis.

Unlike many other data mining algorithms, a tree can handle a great number of potential predictor variables, both categorical and quantitative. This makes it a great model for starting a data mining project. The modeler can then choose whether to use this model as a final model for predicting the response, or to use the variables suggested by the tree as inputs to other models that are more computationally intensive and less able to handle very large numbers of predictors.

For Example   Decision tree

Decision trees are calculated by data mining software. Several different algorithms may be used, but the important thing to know is how to understand what the resulting decision tree says about the data. Figure 25.3 shows a tree model found for the PVA data using *TARGET_B* as a response variable. *TARGET_B* is "YES" if the donor contributed to the most recent solicitation and "NO" if not.

Question: What can we learn about the PVA data from this decision tree?

Answer: The goal of the decision tree is to characterize those potential donors who responded to the most recent solicitation with a donation. The PVA tree model starts at the root node with all 94,649 customers (listed under Count in the top box of the figure).

The first branching of the tree is according to the variable *RFA_4*, a variable that summarizes past giving. The algorithm identifies this as the best single predictor of *TARGET_B*. So, the first split occurs by splitting the levels of that variable into two groups, with 29,032 customers split to the left and the remaining 65,617 customers to the right. You can tell that because *RFA_4* is listed first in both boxes of the first split. The variable *RFA_4* comprises codes that contain information on how recently the last gift solicitation was received, how frequently the donor has given, and how large the last gift was. (Examples of these codes are A3C, S4B, etc., seen in parentheses after the variable name in the output display). The percentages next to the levels 0 and 1 indicate

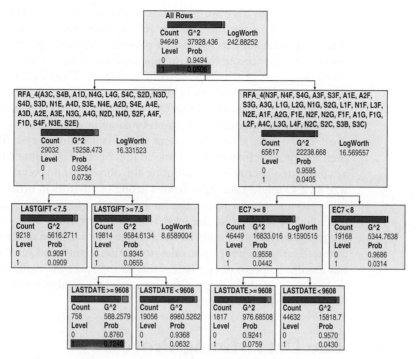

| Figure 25.3   Part of a tree model run on the PVA data using JMP® software.

the proportion of those who donated or not (*TARGET_B 1 or 0*, respectively) in each node. So although only 7.36% of those in the left node contributed, that was almost twice the 4.05% in the right node who contributed.

The tree continues to split each subpopulation. Notice that the variable chosen to split the left branch of the tree (*LASTGIFT*) is not the same as the variable chosen to split the right branch (*EC7*). The terminal nodes are found in the bottom row. The terminal node at the bottom left of the figure shows that 12.4% of the 758 potential donors in this small subgroup donated, a substantial improvement over the average of all potential donors (5.06%) shown in the root node at the top. Even at this stage of the analysis, if the PVA directed their solicitations to names on their list that satisfied the conditions found in the path through the tree to this node, they might reduce their costs and increase their yield substantially.

It's interesting to see that of the 479 potential predictor variables, 3 of the top 4 variables selected by the decision tree algorithm involve past giving history rather than demographics. *RFA_4* is a summary of past giving, *LASTGIFT* measures the amount of the last gift, and *LASTDATE* reports how recently the donor made the last gift. Only *EC7* is a demographic variable, measuring the percentage of residents in the donor's zip code that have at least a bachelor's degree. By letting the tree grow further, we could grow the list of potential predictors to 20 or 30, a much larger number than shown here, but still a manageable subset of the original 479 variables.

Neural Networks

Another popular data mining tool is the multilayer perceptron or (artificial) **neural network**. This algorithm sounds more impressive than it really is. Although it was inspired by models that tried to mimic the function of the brain, it's really just an automatic, flexible, nonlinear regression tool. That is, it models a single response variable with a function of some number of predictor variables but, unlike multiple regression, it constructs a more complex function for the relationship. This can have the advantage of fitting the data better but has the disadvantage that complex functions are harder to interpret and understand. Indeed, modelers often don't even look at the models themselves. There are certainly things to be learned by studying how neural networks build models, but, for our purposes, we'll view them as "black box" models—models that don't produce an equation or graphical display that we can examine, but simply predict the response without much information about *how* it's doing it.

RFA_2A	2.31%
RFA_2	1.64%
RFA_2F	1.27%
LASTDATE	1.18%
INCOME	1.03%
PEPSTRFL	1.00%
ADATE_4	0.83%
ADATE_3	0.74%
RDATE_7	0.73%
LIFESRC	0.73%
HVP1	0.69%
RDATE_6	0.61%
DMA	0.55%
GENDER	0.54%
RDATE_3	0.51%

Table 25.2   The top fifteen variables ranked in importance by the neural network node in Clementine®. The importance percentages indicate relative importance only and are not meant to be interpreted in an absolute sense.

Even though they are black boxes, neural network algorithms do leave some clues. A listing of the most important variables is a common output feature, and some neural network software even provide plots of the predicted relationships between the response and the most important variables as measured by how much the predictions change when the variable is deleted. Table 25.2 shows a list of the top fifteen variables in the PVA problem ranked by "importance" from a neural network output in a data mining software package called Clementine®.

Many other algorithms are used by data miners, and new ones are being developed all the time. Typically, a data miner will build several different models and then test them before selecting one to use, or the data miner might decide to form a "committee" of models, combining the output from several models, much in the way a CEO might take input from a group of advisors in order to make a decision. For a classification problem, the final prediction for a case might be the class that's predicted most often by the models in the committee. For a regression problem, the prediction might simply be the average prediction of all the models in the committee. Averaging models provides protection against choosing the "wrong" model, but usually at a cost of not being able to interpret the resulting predictions. The best ways to combine many different models is an active and exciting area of current research in data mining.

25.8   The Data Mining Process

Data mining projects require a number of different skills. For that reason, a data mining project should be a team effort. No one person is likely to have the business knowledge, computer and database management skills, software expertise, and statistical training needed for all the steps in the process. Because data mining projects tend to be complex, it's useful to map out the steps for a successful project. A group of data mining experts have shared their combined expertise in a project called the Cross Industry Standard Process for Data Mining (CRISP-DM). A CRISP-DM schematic of the data mining cycle appears in Figure 25.4.

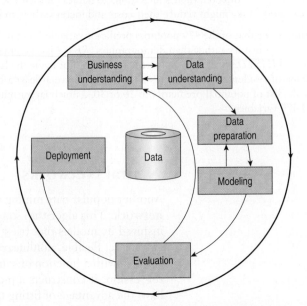

Figure 25.4   A picture of the CRISP-DM data mining process.

In this schematic, the process starts with the *Business understanding* phase. This is where the problem to be addressed is carefully articulated. It is best to have a specific problem before starting. A goal of understanding how to best manage customers sounds good but is not precise enough for a data mining project. A better, more specific question might be to understand which customers are most likely to switch cell phone providers in the next three months. It's important to have all the members of the data mining team involved at this stage, and the team should be representative of all parts of the business that may be affected by the resulting business decisions. If key constituents aren't represented, the model might not answer the right question. It can't be overemphasized how important it is to have consensus on a precise, correctly formulated problem to be addressed before continuing with the data mining effort.

The phase of *Data understanding* is central to the entire data mining project. If you want to know which customers are likely to switch providers, you have to understand your data and have data that can support such an exploration. For example, you'll need a sample that contains both customers who have recently switched and those who have been loyal customers for some time. There should also be variables in the database that could reasonably explain or predict the behavior. The choice of those variables has to rely on the business knowledge of the team. At the beginning of the project, it's best to include all the variables that might be useful, but keep in mind that having *too* many variables can make the model selection phase more difficult. It is crucial to understand the data warehouse, what it contains, and what its limitations are at this stage.

Once the variables have been selected and the response variable(s) agreed upon, it's time to begin the *Data preparation* phase for modeling. As mentioned earlier, this can be a time-consuming part of the process and is likely to be a team effort. Investigating missing values, correcting wrong and inconsistent entries, reconciling data definitions, and merging data sources are all challenging issues. Some of these can be handled automatically, while others require painstaking detailed analysis. The team will have to decide how much effort is reasonable to make the data set as complete and reliable as possible, given the time and resource constraints of the project.

Once the data have been prepared, the analysts begin the *Modeling* phase by exploring and developing models. If the number of variables is very large, a preliminary model (such as a tree) might be considered in order to narrow the candidate predictor variables down to a reasonable number. If the number of predictors is small enough, modelers can use traditional graphical analysis (histograms, bar charts) for each variable to start and then investigate the relationship between each predictor and the response with bivariate graphs (scatterplots, boxplots, or segmented bar charts, depending on the variable types). The more knowledge of the data and variables that goes into the model, the higher the chance of success for the project.

The analysts should now have several models that fit the response variable with differing levels of accuracy in the training set. Once the analyst has several models that seem reasonable (based on domain knowledge and performance on the training set), the *Evaluation* phase can begin. Looking at the structure of the models and deciding what predictors are important for each model should give the analysts information about how the predictors can predict the response variable of interest. In the *Evaluation* phase, the candidate models are tested against the test set, and various criteria are used to judge the models. For example, for a regression problem with a quantitative response, the sum of squares of the residuals from predicting the response on the test set might be compared. For a (two-level) classification problem, the two types of errors that arise (predicting YES when the true response is NO and vice versa) are weighed. Different costs on the two types of errors might be warranted, and the total cost of misclassification should reflect that. At this point, the business question that motivated the project should be revisited. Does the model help to answer the question? If not, it may be necessary to go back to one of the previous steps to investigate why that happened.

If the model (or average of several models) seems to give insight into the business problem, then it's time for the *Deployment* phase. Usually, that means using the model to predict an outcome on a larger data set than the original, or on more recent data than were used to build the model. The reason that so many phases of the CRISP-DM diagram have arrows both to and from them is that this process is not a straightforward movement through the phases—it is an iterative and interconnected process. Knowledge gained at any phase may trigger a reexamination of an earlier phase. Even the "final" phase, deployment, is not really final. In many business situations, the environment changes rapidly, so models can become stale quickly. Although a data mining project is complex and involves the efforts of many people, mapping out the different phases can help assure that the project is as successful as possible.

25.9   Summary

There are many similarities between the modeling process of data mining and the basic approach to modeling that you've been learning throughout this book. What makes data mining different really has to do with the large number of algorithms and types of models available to the data miner and with the size and complexity of the data sets. But most of what makes a data mining project successful would make any statistical analysis successful. The same principles of understanding and exploring variables and their relationships are key to both processes. A famous statistician, Jerry Friedman, was once asked if there was a difference between statistics and data mining. Before answering, he asked if the questioner wanted the long answer or the short answer. Because the response was "the short answer," Jerry said simply, "No." We never got to hear the long answer, but we suspect it may have contained some of the differences discussed in this chapter.

Having a good set of statistical and data analysis tools is a great beginning to becoming a successful data miner. Being willing to learn new techniques, whether they come from statistics, computer science, machine learning, or other disciplines, is essential. Learning to work with other people whose skills complement yours will not only make the task more pleasant, but also is key to a successful data mining project. The need to understand the information contained in large databases will increase in the years ahead, so it will continue to be important to know about this rapidly growing field.

What Can Go Wrong?

- **Be sure that the question to be answered is specific.** Make sure that the business question to be addressed is specific enough so that a model can help to answer it. A goal as vague as "improving the business" is not likely to lead to a successful data mining project.

- **Be sure that the data have the potential to answer the question.** Check the variables to see whether a model can reasonably be built to predict the response. For example, if you want to know which type of customers are going to a particular Web page, make sure that data are being collected that link the Web page to the customer visiting the site.

- **Beware of overfitting to the data.** Because data mining tools are powerful and because the data sets used to train them are usually large, it is easy to think you're fitting the data well. Make sure you validate the model on a test set— a data set not used to fit the model.

- **Make sure that the data are ready to use in the data mining model.** Typically, data warehouses contain data from several different sources. It's important to confirm that variables with the same name actually measure the same thing in two different databases. Missing values, incorrect entries, and different time scales are all challenges to be overcome before the data can be used in the model building phase.

- **Don't try it alone.** Data mining projects require a variety of skills and a lot of work. Assembling the right team of people to carry out the effort is crucial.

Ethics in Action

With U.S. consumers becoming more environmentally conscious, there has been an explosion of eco-friendly products on the market. One notable entry has been the gas-electric hybrid car. A large nonprofit environmental group would like to target customers likely to purchase hybrid cars in the future with a message expressing the urgency to do so sooner than later. They understand that direct mailings are very effective in this regard, but staying true to their environmental concerns, they want to avoid doing a nontargeted mass mailing. The executive team met to discuss the possibility of using data mining to help identify their target audience. The initial discussion revolved around data sources. Although they have several databases on demographics and transactional information for consumers who have purchased green products and donated to organizations that promote sustainability, someone suggested that they get data on political party affiliation. After all, there was a Green party, and Democrats tend to be more concerned about environmental issues than Republicans. Another member of the team was genuinely surprised that this was even possible. She wondered how ethical it is to use information about individuals that they may assume is being kept confidential.

ETHICAL ISSUE *Should information about individuals be collected without their knowledge? Data mining raises general concerns about privacy and confidentiality. Related to Item D, ASA Ethical Guidelines.*

ETHICAL SOLUTION *Data should never be used without the consent of the person to whom the data refer unless care has been taken to disguise the identity or otherwise ensure that the identity of the person cannot be inferred from the data themselves.*

What Have We Learned?

Learning Objectives

- Understand the uses and value of data mining in business.
- Recognize data mining approaches.

    - OLAP approaches answer specific questions about data in a database that may involve many variables.
    - Predictive model approaches build complex models that attempt to predict a response.

- Be aware that data preparation is essential to sound data mining.

    - Data must be cleaned of errors and impossible or implausible values.
    - Missing values may be a serious issue.
    - Data preparation may take the majority of time and effort devoted to a data mining project.

- Be able to articulate the goal of a data mining project and understand that data mining efforts differ according to the kind of dependent variable.

    - When the goal is to predict a quantitative response, the problem is called a regression problem.
    - When the goal is to classify individuals according to a categorical variable, the problem is called a classification problem.

- Be aware of the variety of models and algorithms that are used to build predictive models.

    - Tree models build interpretable models that select predictor variables and split points to best predict the response in either a regression or classification problem.

(continued)

- Neural networks build black box models that use all the predictor variables for both regression and classification problems.
- Many other data mining algorithms are used as well, and new competitors are continually being developed.

■ **Understand that successful data mining is a team effort that requires a variety of skills.**

- Defining the problem and objective of interest is crucial for success.
- The CRISP data mining process provides a good outline for a data mining project.

Terms

Algorithm	A set of instructions used for calculation and data processing. The algorithm specifies how the model is built from the data.
Classification problem	A prediction problem that involves a categorical response variable. (See also regression problem.)
Data mining	A process that uses a variety of data analysis tools to discover patterns and relationships in data that are useful for making predictions.
Data preparation	The process of cleaning data and checking its accuracy prior to modeling. Data preparation includes investigating missing values, correcting wrong and inconsistent entries, and reconciling data definitions.
Data warehouse	A digital repository for several large databases.
Decision tree (data mining version)	A model that predicts either a categorical or quantitative response in which the branches represent variable splits and the terminal nodes provide the predicted value.
Demographic variable	A variable containing information about a customer's personal characteristics or the characteristics of the region in which the customer lives. Commonly used demographics include age, income, race, and education.
Metadata	Information about the data including when and where the data were collected.
Neural network	A model based on analogies with human brain processing that uses combinations of predictor variables and nonlinear regression to predict either a categorical or quantitative response variable.
Online analytical processing (OLAP)	An approach for providing answers to queries that typically involve many variables simultaneously.
Predictive model	A model that provides predictions for the response variable.
Regression problem	A prediction problem that has a quantitative response variable. (See classification problem.)
Supervised problem	A classification or regression problem in which the analyst is given a set of data for which the response is known to use to build a model.
Terminal node	The final leaves of a decision tree where the predictions for the response variable are found.
Test set	The data set used in a supervised classification or regression problem *not used* to build the predictive model. The test set is withheld from the model building stage of the process, and the model's predictions on the test set are used to evaluate the model performance.
Training set	The data set used in a supervised classification or regression problem to build the predictive model.
Transactional data	Data describing an event involving a transaction, usually an exchange of money for goods or services.
Unsupervised problem	A problem, unlike classification or regression problems, where there is no response variable. The goal of an unsupervised problem is typically to put similar cases together into homogeneous groups or clusters.

CASE Study

Marketing Experiment

Many credit cards offer "free" miles to entice customers to increase the use of their cards. These miles can be exchanged for flights or hotel rooms, depending on the type of card, and the miles are very popular among cardholders. The miles are not free to the credit card company, however, which must pay the airline or hotel chain for the services provided to the customer in exchange for the miles.

To help evaluate how much revenue a new campaign that offers double miles might generate on a standard miles card, a marketing analyst ran an experiment on 4800 cardholders, half of whom were randomly chosen to receive the new card with the double-miles offer (*Miles* = Double Miles) and half of whom were using the standard card (*Miles* = Standard Miles).

The analyst had also read some research that claimed that making cardholders enroll on the Internet increases their awareness of the card and will "engage" the customer more, leading to increased use of the card. So, she gave half of each group the requirement to enroll the card on the Internet before the miles would be earned (*Enroll* = Yes or No).

Finally, cardholders are segmented into three types of users: those who use the card mainly for retail purchases (*Cardholder Segment* = Retail), those who use the card mainly for travel and leisure purchases (*Cardholder Segment* = Travel & Leisure), and those who the company is unable to segment (*Cardholder Segment* = Unspecified).

To measure effectiveness of the factors, the marketing analyst measures the increase in spending after three months of the program (*Spend Lift*).

The data set **Market_Experiment** contains the data for the 4800 cardholders in the experiment. The variables are:

> *Miles:* coded Double Miles or Standard Miles
> *Enroll:* coded Yes if enrollment is required; No otherwise
> *Cardholder Segment:* coded Retail, Travel & Leisure, or Unspecified
> *Spend Lift* ($): the amount of increased or decreased use of the card after three months of the program

The analyst wants to know if any of the factors are important for understanding card use. Build a model to predict *Spend Lift* using the three predictor variables. Here are some things to consider:

1. Examine displays to see if the conditions of your model are satisfied.

2. Are any of the interactions of the variables significant?

3. If the three-way interaction is important, considering running separate models for the three levels of *Cardholder Segment*. Use interaction plots for each segment to understand whether the two-way interaction between *Miles* and *Enroll* changes for the three segments.

4. Keep in mind that the credit card company gains about $0.02 profit on every dollar of increased spending. Assume that adding double miles costs $0.005 for each mile awarded, but Internet enrollment is essentially free.

5. Write a report giving recommendations on whether the double-miles program is successful. Be sure to include confidence intervals for the effects and the profit to the company for both the *Miles* and *Enroll* factor effect. What effect does the cardholder segment have on your recommendations?

Answers

CHAPTER 2

SECTION EXERCISE ANSWERS

1. a) Each row represents a different house. It is a case.
   b) There are 7 variables including the house identifier.

3. a) House _ID is an identifier (categorical, not ordinal).
   Neighborhood is categorical (nominal).
   Mail_ZIP is categorical (nominal – ordinal in a sense, but only on a national level).
   YR_BUILT is quantitative (units – year).
   FULL_MARKET_VALUE is quantitative (units – dollars).
   SFLA is quantitative (units – square feet).
   b) These data are cross-sectional. All variables were measured at about the same time.

5. It is not clear if the data were obtained from a survey. They are certainly not from an experiment. Most likely they are just a collection of recent sales. We don't know if those sales are representative of all sales, so we should be cautious in drawing conclusions from these data about the housing market in general.

CHAPTER EXERCISE ANSWERS

7. Answers will vary.

9. *Who*—months; *What*—money spent on advertising ($ thousand) and sales ($ million); *When*—monthly from 2004–2006; *Where*—United States; *Why*—To compare money spent on advertising to sales; *How*—not specified; *Variables*—There are two quantitative variables. The data form a time series.

11. *Who*—quarters; *What*—sales ($ million), unemployment rate (%), and inflation rate (%); *When*—quarterly from 2004–2006; *Where*—United States; *Why*—to relate sales to unemployment and inflation rates; *How*—not specified; *Variables*—There are 3 quantitative variables: sales ($ million), unemployment rate (%), and inflation rate (%). The data form a time series.

13. *Who*—MBA applicants; *What*—sex, age, whether or not they accepted, whether or not they attended, and the reasons; *When*—not specified; *Where*—United States; *Why*—The researchers wanted to investigate any patterns in female acceptance and attendance; *How*—Data obtained internally from admissions office; *Variables*—There are 5 variables. Sex, whether or not they accepted, whether or not they attended, and the reasons are all categorical variables; only age (years) of applicant is quantitative. The data are a cross section.

15. *Who*—experiment volunteers; *What*—herbal cold remedy or sugar solution, and cold severity; *When*—not specified; *Where*—major pharmaceutical firm; *Why*—Scientists were testing the efficacy of a herbal compound on the severity of the common cold; *How*—The scientists set up an experiment; *Variables*—There are 2 variables. Type of treatment (herbal or sugar solution) is categorical, and severity rating is quantitative; *Concerns*—The severity of a cold seems subjective and difficult to quantify. Also, the scientists may feel pressure to report negative findings about the herbal product. The data are cross sectional and from a designed experiment.

17. *Who*—vineyards; *What*—size of vineyard (in acres), number of years in existence, state, varieties of grapes grown, average case price (in dollars), gross sales (probably in dollars), and percent profit; *When*—not specified; *Where*—not specified; *Why*—Business analysts hoped to provide information that would be helpful to producers of U.S. wines; *How*—not specified; *Variables*—There are 5 quantitative variables and 2 categorical variables. Size of vineyard, number of years in existence, average case price, gross sales, and percent profit are quantitative variables. State and variety of grapes grown are categorical variables. The data are a cross section collected in a designed survey.

19. *Who*—every model of automobile in the United States; *What*—vehicle manufacturer, vehicle type, weight (probably in pounds), horsepower (in horsepower), and gas mileage (in miles per gallon) for city and highway driving; *When*—This information is collected currently; *Where*—United States; *Why*—The Environmental Protection Agency uses the information to track fuel economy of vehicles; *How*—The data is collected from the manufacturer of each model; *Variables*—There are 6 variables. City mileage, highway mileage, weight, and horsepower are quantitative variables. Manufacturer and type of car are categorical variables. The data are a cross section.

21. *Who*—states in the United States; *What*—state name, whether or not the state sponsors a lottery, the number of numbers in the lottery, the number of matches required to win, and the probability of holding a winning ticket; *When*—1998; *Where*—United States; *Why*—It is likely that this study was performed in order to compare the chances of winning the lottery in each state; *How*—Although not specified, the researchers probably simply gathered data from a number of different sources, such as state lottery website and publications; *Variables*—There are 5 variables. State name and whether or not the state sponsors a lottery are categorical variables, and number, matches, and probability of winning are quantitative variables. The data are a cross section.

23. *Who*—students in an MBA statistics class; *What*—total personal investment in stock market ($), number of different stocks held, total invested in mutual funds ($), and name of each mutual fund; *When*—not specified; *Where*—United States; *Why*—The information was collected for use in classroom illustrations; *How*—An online survey was conducted. Presumably, participation was required for all members of the class; *Variables*—There are 4 variables. Name of mutual fund is a categorical variable. Number of stocks held, total amount invested in market ($), and in mutual funds ($) are quantitative variables. The data are a cross section.

25. *Who*—Indy 500 races; *What*—year, winner, car, time (hours), speed (mph) and car #. *When*—1911–2009; *Where*—Indianapolis, Indiana; *Why*—It is interesting to examine the trends in Indy 500 races; *How*—Official statistics are kept for the race every year; *Variables*—There are 6 variables. Winner, car, and car # are categorical variables. Year, time, and speed are quantitative variables. The quantitative variables can be viewed as time series.

27. Each row should be a single mortgage loan. Columns hold the borrower name (which identifies the rows) and amount.

29. Each row is a week. Columns hold week number (to identify the row), sales prediction, sales, and difference.

31. Cross-sectional.

33. Time series.

CHAPTER 3

SECTION EXERCISE ANSWERS

1. a) False. Sampling error cannot be avoided, even with unbiased samples.
   b) True.
   c) True.
   d) False. Randomization will match the characteristics in a way that is unbiased. We can't possibly think of all the characteristics that might be important or match our sample to the population on all of them.

3. a) Professional food preparers in the United States.
   b) *Chef's Collaborative* membership listing.
   c) Proportion who believe that food safety has improved.
   d) Simple random sample.

5. a) No. It would be nearly impossible to get exactly 500 males and 500 females by random chance.
   b) A stratified sample, stratified by whether the respondent is male or female.

7. A systematic sample.

9. a) Population—Human resources directors of Fortune 500 companies.
   b) Parameter—Proportion who don't feel surveys intruded on their work day.
   c) Sampling Frame—List of HR directors at Fortune 500 companies.
   d) Sample—23% who responded.
   e) Method—Questionnaire mailed to all (nonrandom).
   f) Bias—Hard to generalize because who responds is related to the question itself.

11. a) Professional food preparers in the United States.
   b) The members who attended the recent symposium.
   c) The sampling frame is not necessarily representative of the entire group of food preparers. Those who attended the symposium may have different opinions from those who didn't. His sample isn't random and may be biased toward those most interested in the topic. Finally, the script is biased and may lead to an estimate of a higher proportion who think food safety has improved than is true within the population.

13. a) Answers will vary. Question 1 seems appropriate. Question 2 predisposes the participant to agree that $50 is a reasonable price, and does not seem appropriate.
   b) Question 1 is the more neutrally worded. Question 2 is biased in its wording.

15. a) True.
   b) True.
   c) False. Measurement error refers to inaccurate responses. Sampling error refers to sample-to-sample variability.
   d) True.

17. a) This is a multistage design, with a cluster sample at the first stage and a simple random sample for each cluster.
   b) If any of the three churches you pick at random is not representative of all churches then you'll introduce sampling error by the choice of that church.

CHAPTER EXERCISE ANSWERS

19. a) Voluntary response.
   b) We have no confidence at all in estimates from such studies.

21. a) The population of interest is all adults in the United States aged 18 and older.

   b) The sampling frame is U.S. adults with landline telephones.
   c) Some members of the population (e.g., many college students) don't have landline phones, which could create a bias.

23. a) Population—Consumers Union subscribers.
   b) Parameter—Proportion who have used and benefited from alternative medicine.
   c) Sampling Frame—All Consumers Union subscribers.
   d) Sample—Those who responded.
   e) Method—Census of subscribers (nonrandom).
   f) Bias—Voluntary response: Those who respond may have strong feelings one way or another.

25. a) Population—Adults.
   b) Parameter—Proportion who think drinking and driving is a serious problem.
   c) Sampling Frame—Bar patrons.
   d) Sample—Every 10th person leaving the bar.
   e) Method—Systematic sampling.
   f) Bias—Those interviewed had just left a bar. They probably think drinking and driving is less of a problem than do adults in general.

27. a) Population—Soil around a former waste dump.
   b) Parameter—Concentrations of toxic chemicals.
   c) Sampling Frame—Accessible soil around the dump.
   d) Sample—16 soil samples.
   e) Method—Not clear.
   f) Bias—Don't know if soil samples were randomly chosen. If not, may be biased toward more or less polluted soil.

29. a) Population—Snack food bags.
   b) Parameter—Weight of bags, proportion passing inspection.
   c) Sampling Frame—All bags produced each day.
   d) Sample—10 randomly selected cases, 1 bag from each case for inspection.
   e) Method—Multistage sampling.
   f) Bias—Should be unbiased.

31. Bias. Only people watching the news will respond, and their preference may differ from that of other voters. The sampling method may systematically produce samples that don't represent the population of interest.

33. a) Voluntary response. Only those who both see the ad *and* feel strongly enough will respond.
   b) Cluster sampling. One town may not be typical of all.
   c) Attempted census. Will have nonresponse bias.
   d) Stratified sampling with follow-up. Should be unbiased.

35. a) This is a systematic sample.
   b) It is likely to be representative of those waiting for the roller coaster. Indeed, it may do quite well if those at the front of the line respond differently (after their long wait) than those at the back of the line.
   c) The sampling frame is patrons willing to wait for the roller coaster on that day at that time. It should be representative of the people in line, but not of all people at the amusement park.

37. Only those who think it worth the wait are likely to be in line. Those who don't like roller coasters are unlikely to be in the sampling frame, so the poll won't get a fair picture of whether park patrons overall would favor still more roller coasters.

39. a) Biased toward yes because of "pollute." "Should companies be responsible for any costs of environmental cleanup?"
   b) Biased toward no because of "enforce" and "strict." "Should companies have dress codes?"

41. a) Not everyone has an equal chance. People with unlisted numbers, people without phones, and those at work cannot be reached.

b) Generate random numbers and call at random times.

c) Under the original plan, those families in which one person stays home are more likely to be included. Under the second plan, many more are included. People without phones are still excluded.

d) It improves the chance of selected households being included.

e) This takes care of phone numbers. Time of day may be an issue. People without phones are still excluded.

43. a) Answers will vary.

b) The amount of change you typically carry. Parameter is the true mean amount of change. Population is the amount on each day around noon.

c) Population is now the amount of change carried by your friends. The average estimates the mean of these amounts.

d) Possibly for your class. Probably not for larger groups. Your friends are likely to have similar needs for change during the day.

45. a) Assign numbers 001 to 120 to each order. Use random numbers to select 10 transactions to examine.

b) Sample proportionately within each type. (Do a stratified random sample.)

47. a) Select three cases at random; then select one jar randomly from each case.

b) Use random numbers to choose three cases from numbers 61 through 80; then use random numbers between 1 and 12 to select the jar from each case.

c) No. Multistage sampling.

49. a) Depends on the Yellow Page listings used. If from regular (line) listings, this is fair if all doctors are listed. If from ads, probably not, as those doctors may not be typical.

b) Not appropriate. This cluster sample will probably contain listings for only one or two business types.

CHAPTER 4

SECTION EXERCISE ANSWERS

1. a) Frequency table:

None	AA	BA	MA	PhD
164	42	225	52	29

b) Relative frequency table (divide each number by 512 and multiply by 100):

None	AA	BA	MA	PhD
32.03%	8.20%	43.95%	10.16%	5.66%

3. a)

b)

c)

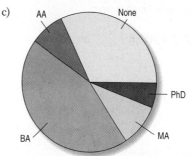

5. a) The vast majority of employees have either no college degree or a bachelor's degree (44% and 32%, respectively). About 10% have master's degrees, 8% have associate's degrees, and nearly 6% have PhDs.

b) I would not be comfortable generalizing this to any other division or company. These data were collected only from my division. Other companies might have vastly different educational distributions.

7. a)

	Totals
<1 year	95
1–5 years	205
more than 5 years	212

b) Yes

None	AA	BA	MA	PhD
164	42	225	52	29

9. a)

(%)	None	AA	BA	MA	PhD
<1 year	6.1	7.1	22.2	38.5	41.4
1–5 years	25.6	21.4	49.8	51.9	51.7
more than 5 years	68.3	71.4	28	9.6	6.9

b) No. The distributions look quite different. More than 2/3 of those with no college degree have been with the company longer than 5 years, but almost none of the PhDs (less than 7%) have been there that long.

c)

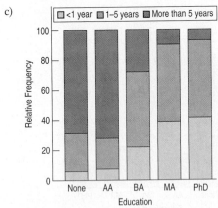

d) It's possible to see it in the table, but the stacked bar chart makes the differences much clearer.

CHAPTER EXERCISE ANSWERS

11. Answers will vary.

13. Answers will vary.

15. a) Yes, the categories divide a whole.
  b) Coca-Cola.

17. a) The pie chart does a better job of showing portions of a whole.
  b) There is no bar for "Other."

19. a) Yes, it is reasonable to assume that heart and respiratory disease caused approximately 38% of U.S. deaths in this year, since there is no possibility for overlap. Each person could only have one cause of death.
  b) Since the percentages listed add up to 73.7%, other causes must account for 26.3% of U.S. deaths.
  c) A bar graph or pie chart would be appropriate if a category for Other with 26.3% were added.

21. WebEx Communications, Inc. has the majority of the market share for web conferencing (58.4%), and Microsoft has approximately a quarter of the market share. There appears to be room for both to grow, because other companies comprise about 15% of market share. A pie chart or bar chart would be appropriate.

23. a) They total more than 100%; overlapping categories.
  b)

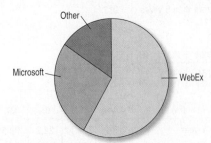

c) No because the percentages do not total 100%.
d) (Answers will vary). More than 50% of business owners say that they have either updated their computers or made other noncomputer capital investments (or both). Smaller percentages of business owners (from 10 to 37%) made changes in either their hiring or salary structure.

25. The bar chart shows that grounding is the most frequent cause of oil spillage for these 312 spills, and allows the reader to rank the other types as well. If being able to differentiate between these close counts is required, use the bar chart. The pie chart is also acceptable as a display, but it's difficult to tell whether, for example, there is a greater percentage of spills caused by grounding or collisions. To showcase the causes of oil spills as a fraction of all 312 spills, use the pie chart.

27. a) 31%
  b) It looks like India's percentage is about 6 times as big, but it's not even twice as big.
  c) Start the percentages at 0% on the vertical axis, not 40%.
  d)

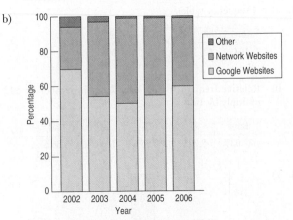

e) The percentage of people who say that wealth is important to them is highest in China and India (around 70%), followed by France (around 60%) and then the U.S. and U.K. where the percentage was only about 45%.

29. a) They must be column percentages because the sums are greater than 100% across the rows and all the columns add to 100%.
  b)

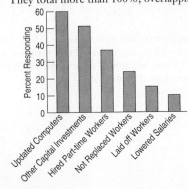

c) The main source of revenue for Google is from their own websites, which in 2002 was 70%, dropped down to 50% in 2004, and by 2006 was back up to 60%. The second largest source is from other network websites. Licensing and other revenue was 6% in 2002, but since 2004 has been only 1%.

31. a) 62.5%
  b) 35%
  c) 15%
  d) 50%

e) 61%

f) 65%

g) There does not appear to be any relationship between the performance of a stock on a single day and its performance over the prior year.

33. a) 10.2%; 13.2%

b) 1.8%; 1.5%

c) Residential sales decreased by 3.3%.

35. a) 11.4% G; 14.3% PG; 48.6% PG-13; 25.7% R

b) 0% G; 0% PG%; 57.9% PG-13; 42.1% R

c)

d) *Genre* and *Rating* are not independent. Thriller/Horror movie are all PG-13 or R and Drama are nearly so. Comedy moves are nearly 40% G and PG and only 10% R. Action/Adventure movies are nearly 15% G and 15% PG.

37. a) 62.7%

b) 62.8%

c) 62.5%

d) 23.9% from Asia, 1.9% Europe, 7.8% Latin America, 3.7% Middle East, and 62.7% North America.

e) The column percentages are given in the table.

| | | MBA Program | | |
		Two-Yr	Evening	Total
Origin	Asia	18.90	31.73	**23.88**
	Europe	3.05	0.00	**1.87**
	Latin America	12.20	0.96	**7.84**
	Middle East	3.05	4.81	**3.73**
	North America	62.80	62.50	**62.69**
	Total	**100.00**	**100.00**	**100.00**

f) No. The distributions appear to be different. For example, the percentage from Latin America among those in Two-Yr. programs is nearly 20% while for those in Evening programs it is less than 1%.

39. a) 6%

b) 5%

c) 3.5%

d) 56.7%

e) 73.8%

f) Here are row percentages:

2004–2008	6%	23%	57%	14%	**100.0%**
1999–2003	6%	19%	53%	22%	**100.0%**

The distributions are quite similar, although there were fewer R rated films in the second half of the period (14% of 20 top films, compared to 22%). The increase was in both PG and PG-13 films.

41. The study by the University of Texas Southwestern Medical Center provides evidence of an association between having a tattoo and contracting hepatitis C. Around 33% of the subjects who were tattooed in a commercial parlor had hepatitis C, compared with 13% of those tattooed elsewhere, and only 3.5% of those with no tattoo. If having a tattoo and having hepatitis C were independent, we would have expected these percentages to be roughly the same.

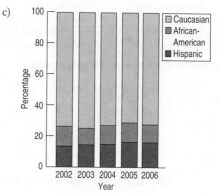

43. a) 8%

b) No, because we're not given counts or totals.

c) 92%

d) There appears to be little, if any, relationship between revenue category and education level of the women CEOs.

45. a) 14.5% Hispanic, 12.5% African-American, and 73.0% Caucasian.

b) For 2006, 15.7% Hispanic, 12.0% African-American, 72.3% Caucasian.

c)

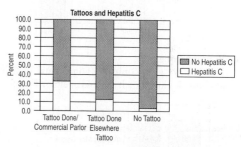

d) The (conditional) distribution of *Ethnicity* is almost the same across the five *Years*, however there seems to be a slight increase in the percentage of Hispanics who go to the movies from 13.1% in 2002 to 15.7% in 2006.

47. a) Row percentages.

b) A slightly higher percentage of urban women's business centers are established (at least 5 years old).

49. a) Row percentages.

b) No. We are given only the conditional distributions. We have no idea how much are sold in either Europe or America.

c)

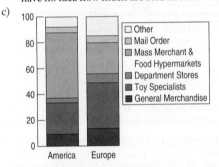

d) In America more than 50% of all toys are sold by large mass merchant discounters and food hypermarkets and only 25% are sold in toy specialty stores. In Europe 36% of all toys are sold in toy specialty stores while a relatively small 24% are sold in the large discount and hypermarket chains.

51. a) The marginal totals have been added to the table:

		Hospital Size		
		Large	Small	Total
Procedure	Major surgery	120 of 800	10 of 50	**130 of 850**
	Minor surgery	0 of 200	20 of 250	**30 of 450**
	Total	**130 of 1000**	**30 of 300**	**160 of 1300**

160 of 1300, or about 12.3% of the patients had a delayed discharge.

b) Major surgery patients were delayed 15.3% of the time. Minor Surgery patients were delayed 6.7% of the time.

c) Large Hospital had a delay rate of 13%. Small Hospital had a delay rate of 10%. The small hospital has the lower overall rate of delayed discharge.

d) Large Hospital: Major Surgery 15% and Minor Surgery 5%. Small Hospital: Major Surgery 20% and Minor Surgery 8%.

Even though the small hospital had the lower overall rate of delayed discharge, the large hospital had a lower rate of delayed discharge for each type of surgery.

e) Yes. While the overall rate of delayed discharge is lower for the small hospital, the large hospital did better with *both* major surgery and minor surgery.

f) The small hospital performs a higher percentage of minor surgeries than major surgeries. 250 of 300 surgeries at the small hospital were minor (83%). Only 200 of the large hospital's 1000 surgeries were minor (20%). Minor surgery had a lower delay rate than major surgery (6.7% to 15.3%), so the small hospital's overall rate was artificially inflated. The larger hospital is the better hospital when comparing discharge delay rates.

53. a) 1284 applicants were admitted out of a total of 3014 applicants. 1284/3014 = 42.6%

		Males Accepted (of applicants)	Females Accepted (of applicants)	Total
Program	1	511 of 825	89 of 108	**600 of 933**
	2	352 of 560	17 of 25	**369 of 585**
	3	137 of 407	132 of 375	**269 of 782**
	4	22 of 373	24 of 341	**46 of 714**
	Total	**1022 of 2165**	**262 of 849**	**1284 of 3014**

b) 1022 of 2165 (47.2%) of males were admitted. 262 of 849 (30.9%) of females were admitted.

c) Since there are four comparisons to make, the table below organizes the percentages of males and females accepted in each program. Females are accepted at a higher rate in every program.

Program	Males	Females
1	61.9%	82.4%
2	62.9%	68.0%
3	33.7%	35.2%
4	5.9%	7.0%

d) The comparison of acceptance rate within each program is most valid. The overall percentage is an unfair average. It fails to take the different numbers of applicants and different acceptance rates of each program. Women tended to apply to the programs in which gaining acceptance was difficult for everyone. This is an example of Simpson's Paradox.

CHAPTER 5

1. a)

b)

c)

d) 1|14
    2|0225699
    3|0002224558
    4|244488

3. a) Unimodal.
   b) Around 35 years old.
   c) Fairly symmetric.
   d) No outliers.

5. a) About the same. The distribution is fairly symmetric.
   b) 31.84 years.
   c) 32 years.

7. a) Q1 26 ; Q3 38 (Answers may vary slightly.)
   b) Q1 26; Q3 38
   c) IQR = 12 years
   d) sd = 9.84 years

9. a) The distribution is skewed to the right. There are a few negative values. The range is about $6000.
   b) The mean will be larger because the distribution is right skewed.
   c) Because of the skewness, the median is a better summary.

11. a)

   b) No.
   c) 32 + 1.5*12 = 50 years old

13. a) Skewed to the right, since the mean is much greater than the median.
   b) Yes, at least one high outlier, since 250 is far greater than Q3 + 1.5 IQRs.

c) We don't know how far the high whisker should go because we don't know the largest value inside the fence.

15. The ages of the women are generally higher than the men by about 10 years. As the boxplot shows, more than 3/4 of the women are older than all the men.

17. Sales in Location #1 were higher than sales in Location #2 in every week. The company might want to compare other stores in locations like these to see if this phenomenon holds true for other locations.

19. a) 11 has a z-score of −2.12; 48 has a z-score of +1.64.
   b) The min, 11, is more extreme.
   c) 61.36 years old.

21. 16, 16, 36, 49, 49, 64, 100
   Skewed to the high end.

23. Answers will vary.

25. This distribution is nearly symmetric and unimodal, centered at around $2500. The range is about $6000. Most of the tuitions lie between $1000 and $4000.

27. The distribution is unimodal and skewed to the right with two high outliers. The median is near 10%.

29. a) Five-Number Summary (Answers may vary depending on software.)

Min.	1st Qu.	Median	3rd Qu.	Max.
−10.820	7.092	11.270	17.330	94.940

b)   Median = 11.275%; IQR = 10.24%

c)

d)   The histogram makes the skewness of the distribution clear.

31.   The stem-and-leaf display shows that many of the acreage values end in 0 or 5. Perhaps this is evidence that they are rounding or estimating the value. There is a high outlier near 240.

```
24 | 0
22 |
20 |
18 |
16 |
14 | 0
12 | 0
10 | 0
 8 | 0
```

```
6 | 920
4 | 553500
2 | 8655210987520
0 | 751000086
```

Key: 8|0 = 80 acres

33.   a)   Wayne Gretzky—Games played per season

```
8 | 000000122
7 | 8899
7 | 0344
6 |
6 | 4
5 |
5 |
4 | 58
4 |
```

Key: 7|8 = 78 games

b)

c)   The distribution of the number of games played per season by Wayne Gretzky is skewed to the low end and has low outliers. The median is 78, and the range is 37 games.

d)   There are two outlier seasons with 45 and 48 games. He may have been injured. The season with 64 games is also separated by a gap.

35.   a)   The median because the distribution is skewed.

b)   Lower, because the distribution is skewed toward the low end.

c)   That display is not a histogram. It's a time series plot using bars to represent each point. The histogram should split up the number of home runs into bins, not display the number of home runs over time.

37.   a)   Descriptive Statistics: Price ($)

Minimum	Q1	Median	Q3	Maximum
2.21	2.51	2.61	2.72	3.05

b)   Range = max − min = 3.05 − 2.21 = $0.84; IQR = Q3 − Q1 = 2.72 − 2.51 = $0.21

c)

d)   Symmetric with one high outlier. The mean is $2.62, with a standard deviation of $0.156.

e)   There is one unusually high price that is greater than $3.00 per frozen pizza.

39.   As we can see from the histogram and boxplot, the distribution of gasoline use is unimodal and skewed to the left with two low outliers, the District of Columbia and New York State. D.C. is a city, and New York may be dominated by New York City. Because of public transportation, gasoline usage is lower per capita in cities. The median usage is 485.7 gal./yr per capita with an IQR of about 81.75 gal./yr (values from different software may vary slightly). The minimum is D.C. with 209.5, and the maximum is Wyoming with 589.18 gal./yr.

Gasoline Usage
(gallons/yr per capita)

41. a) 1611 yards.
    b) Between Quartile 1 = 5585.75 yards, and Quartile 3 = 6131 yards.
    c) The distribution of golf course lengths appears roughly symmetric, so the mean and SD are appropriate.
    d) The distribution of the lengths of all the golf courses in Vermont is roughly unimodal and symmetric. The mean length of the golf courses is approximately 5900 yards and the standard deviation is 386.6 yd.

43. a) A boxplot is shown. A histogram would also be appropriate.

    b) Descriptive Statistics: Sales ($) (Different statistics software may yield different results.)

Variable	N	Mean	SE Mean	StDev	Minimum
Sales ($)	18	107845	11069	46962	62006

Q1	Median	Q3	Maximum
173422.5	95975	112330.0	224504

    The mean sale is $107,845, and the median is $95,975. The mean is higher because the outliers pull it up.
    c) The median because the distribution has outliers.
    d) The standard deviation of the distribution is $46,962, and the IQR is $38,907.50 (answers may vary due to different quartile algorithms).
    e) The IQR because the outliers inflate the standard deviation.
    f) The mean would decrease. The standard deviation would decrease. The median and IQR would be relatively unaffected.

45. A histogram shows that the distribution is unimodal and skewed to the left. There do not appear to be any outliers. The median failure rate for these 17 models is 16.2%. The middle 50% of the models have failure rates between 10.87% and 21.2%. The best rate is 3.17% for the 60GB Video model, and the worst is the 40GB Click Wheel at 29.85%.

47. a) Gas prices increased over the three-year period, and the spread increased as well. The distribution of prices in 2002 was skewed to the left with several low outliers. Since then, the distribution has been increasingly skewed to the right. There is a high outlier in 2004, although it appears to be pretty close to the upper fence.
    b) The distribution of gas prices in 2004 shows the greatest range and the biggest IQR, so the prices varied a great deal.

49. a) Seneca Lake.
    b) Seneca Lake.
    c) Keuka Lake.
    d) Cayuga Lake vineyards and Seneca Lake vineyards have approximately the same average case price, of about $200, while a typical Keuka Lake vineyard has a case price of about $260. Keuka Lake vineyards have consistently high case prices, between $240 and $280, with one low outlier at about $170 per case. Cayuga Lake vineyards have case prices from $140 to $270, and Seneca Lake vineyards have highly variable case prices, from $100 to $300.

51. a) The median speed is the speed at which 50% of the winning horses ran slower. Find 50% on the left, move straight over to the graph and down to a speed of about 36 mph.
    b) Q1 = 34.5 mph, and Q3 = 36.5 mph.
    c) Range = 7 mph
    IRQ = 2 mph.
    d)

    e) The distribution of winning speeds in the Kentucky Derby is skewed to the left. The lowest winning speed is just under 31 mph, and the fastest speed is about 37.5 mph. The median speed is approximately 36 mph, and 75% of winning speeds are above 34.5 mph. Only a few percent of winners have had speeds below 33 mph.

53. a) Class 3.
    b) Class 3.
    c) Class 3 because it is the most highly skewed.
    d) Class 1.
    e) Probably Class 1. But without the actual scores, it is impossible to calculate the exact IQRs.

55. There is an extreme outlier for the slow-speed drilling. One hole was drilled almost an inch away from the center of the target! If that distance is correct, the engineers at the computer production plant should investigate the slow-speed drilling process closely. It may be plagued by extreme, intermittent inaccuracy. The outlier in the slow-speed drilling process is so extreme that no graphical display can display the distribution in a meaningful way while including that outlier. That distance should be removed before looking at a plot of the drilling distances.

With the outlier removed, we can see that the slow drilling process is more accurate. The greatest distance from the target for the slow drilling process, 0.000098 inches, is still more accurate than the smallest distance for the fast drilling process, 0.000100 inches.

57. a) The mean of 54.41 is meaningless. These are categorical values.
   b) Typically, the mean and standard deviation are influenced by outliers and skewness.
   c) No. Summary statistics are only appropriate for quantitative data.

59. Over this 3-month period, International Funds generally outperformed the other two. Almost half of the International Funds outperformed all the funds in the other two categories. U.S. Domestic Large Cap Funds did better than U.S. Domestic Small/Mid Cap Funds in general. Large Cap funds had the least variation of the three types.

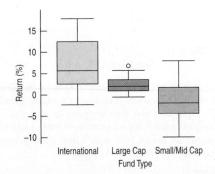

61. a) Even though MLS ID numbers are categorical identifiers, they are assigned sequentially, so this graph has some information. Most of the houses listed long ago have sold and are no longer listed.
   b) A histogram is generally not an appropriate display for categorical data.

63. a)

b) The distribution is fairly uniform. There do not appear to be any decades that would be considered to be outliers.

c)

d) This graph does not support the claim that the number of hurricanes has increased in recent decades.

65. What is the x-axis? If it is time, what are the units? Months? Years? Decades? How is "productivity" measured?

67. The house that sells for \$400,000 has a z-score of $(400000 - 167900)/77158 = 3.01$, but the house with 4000 sq. ft. of living space has a z-score of $(4000 - 1819)/663 = 3.29$. So it's even more unusual.

69. U.S. z-scores are -0.04 and 1.63, total $= 1.59$. Ireland z-scores are 0.25 and 2.77, total 3.02. So Ireland "wins" the consumption battle.

71. a) The histogram shows that the distribution of prices is strongly skewed to the right.

b) Prices were relatively stable until the late 1990s, when they started increasing. Since 2005, the prices have been higher and unstable.
   c) The time series plot because the prices change over time.

73. a) The bimodality of the distribution.
   b) The trend over time.
   c) The time series plot because it reveals much more of the structure of the data.
   d) Unemployment decreased steadily from about 5.6% in 1995 to below 4.0% by 2000, and then increased sharply to between 5.5 and 6.0% from 2002 to 2004.

75. a) The distribution is skewed. That makes it difficult to estimate anything meaningful from the graph.
   b) Transform these data using either square roots or logs.

CHAPTER 6

1. a)

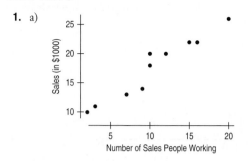

b) Positive.
c) Linear.
d) Strong.
e) No.

3. a) Years of experience.
b) Salary.
c) Salary.

5. a) True.
b) False. It will not change the correlation.
c) False. Correlation has no units.

7. Correlation does not demonstrate causation. The analyst's argument is that sales staff cause sales. However, the data may reflect the store hiring more people as sales increase, so any causation would run the other way.

9. a) False. The line usually touches none of the points. We minimize the sum of the squared errors.
b) True.
c) False. It is the sum of the squares of all the residuals that is minimized.

11. a) $2 \times 0.965 = 1.93$ SDs.
b) $17.6 + 1.93 \times 5.34 = 27.906$ or $27,906.
c) 0.965 SDs below the mean.
d) $12,447.

13. a) $b_1 = 0.914$ if found by hand. $b_1 = 0.913$ if found by technology. (Difference is due to rounding error.)
b) It means that an additional 0.914 ($1000) or $914 of sales is associated with each additional sales person working.
c) $b_0 = 8.10$
d) It would mean that, on average, we expect sales of 8.10 ($1000) or $8100 with 0 sales people working. Doesn't really make sense in this context.
e) $\overline{Sales} = 8.10 + 0.914$ *Number of Sales People Working*.
f) $24.55 ($1000) or $24,550. (24,540 if using the technology solution.)
g) 0.45 ($1000) or $450. ($460 with technology.)
h) Underestimated.

15. The winners may be suffering from regression to the mean. Perhaps they weren't really better than other rookie executives, but just happened to have a lucky year.

17. a) Thousands of dollars.
b) 2.77 (the largest residual in magnitude).
c) 0.07 (the smallest residual in magnitude).

19. $R^2 = 93.12\%$ About 93% of the variance in *Sales* can be accounted for by the regression of *Sales* on *Number of Sales Workers*.

21. 16, 16, 36, 49, 49, 64, 100.
They are skewed to the high end.

23. a) Number of text messages: explanatory; cost: response. To predict cost from number of text messages. Positive direction. Linear shape. Possibly an outlier for contracts with fixed cost for texting.

b) Fuel efficiency: explanatory; sales volume: response. To predict sales from fuel efficiency. There may be no association between mpg and sales volume. Environmentalists hope that a higher mpg will encourage higher sales, which would be a positive association. We have no information about the shape of the relationship.
c) Neither variable is explanatory. Both are responses to the lurking variable of temperature.
d) Price: explanatory variable; demand: response variable. To predict demand from price. Negative direction. Linear shape in a narrow range, but curved over a larger range of prices.

25. a) None
b) 3 and 4
c) 2, 3, and 4
d) 1 and 2
e) 3 and 1

27. a)

Number of Broken Pieces per Batch (24 batches)

b) Unimodal, skewed to the right. The skewness.
c) The positive, somewhat linear relation between batch number and broken pieces.

29. a) 0.006
b) 0.777
c) −0.923
d) −0.487

31. a) *Price*
b) *Sales*
c) Sales decrease by 24,369.49 pounds per dollar.
d) It is just a base value. It means nothing because stores won't set their price to $0.
e) 56,572.32 pounds
f) 3427.69 pounds

33. a) *Salary*
b) *Wins*
c) On average, teams win 0.062 more games per million dollars in salary.
d) Number of wins predicted for a team that spends $0 on salaries. This is not meaningful here.
e) 0.62 games more
f) 4.883 games. Better.
g) 3.117 games

35. 47,084.23 pounds

37. "Packaging" isn't a variable. At best, it is a category. There's no basis for computing a correlation.

39. The model is meaningless because the variable Region is not quantitative. The slope makes no sense because Region has no units. The boxplot comparisons are fine, but the regression is meaningless.

41. a) There is a strong negative linear association between Carbon Footprint and Highway mpg.
b) Quantitative variables, straight enough. The Prius is far from the rest of the data. But it is in line with the linear pattern. It is correct to regard it as an outlier or not; that's a matter of judgment.

c) $r = -0.94$; Removing the Prius reduces the correlation. Data values far from the main body of the data and in line with the linear trend tend to increase correlation and may make it misleading.

43. a) Positive association.

b) Plot is not linear, violating the linearity condition. There may be an outlier at 17 rooms.

45. a) The variables are both quantitative (with units % of GDP), the plot is reasonably straight, there are no outliers, and the spread is roughly constant (although the spread is large).

b) About 21% of the variation in the growth rates of developing countries is accounted for by the growth rates of developed countries.

c) Years 1970–2007

47. a) $\overline{Growth\ (Developing\ Countries)} = 3.46 + 0.433\ Growth\ (Developed\ Countries)$

b) The predicted growth of developing countries in years of 0 growth in developed countries. Yes, this makes sense.

c) On average, GDP in developed countries increased 0.433% for a 1% increase in growth in developed countries.

d) 5.192%

e) More; we would predict 4.61%.

f) 1.48%

49. a) Yes, the scatterplot is straight enough, variables are quantitative and there are no outliers.

b) Teams that score more runs generally have higher attendance.

c) There is a positive association, but correlation doesn't imply causation.

51. a) Positive, roughly straight (slight bending at the upper right), and moderately strong.

b) Yes (although answers may vary). It makes sense that a state's finances would influence both amounts in the same way.

c) $\overline{Public.4yr} = 2826.00 + 1.216\ Public.2yr$

d) Yes

e) A public four-year education costs, on average, $2826 + 1.216 \times$ average tuition at two-year colleges.

f) 44.29% of the variation in the average tuition of four-yr state colleges is accounted for by the regression on two-yr college tuition.

53. a) The predicted value of the money *Flow* if the *Return* was 0%.

b) An increase of 1% in mutual fund return was associated with an increase of $771 million in money flowing into mutual funds.

c) $9747 million

d) −$4747 million; Overestimated

55. a) Model seems appropriate. Residual plot looks fine.

b) Model not appropriate. Relationship is nonlinear.

c) Model not appropriate. Spread is increasing.

57. There are two outliers that inflate the R^2 value and affect the slope and intercept. Without those two points, the R^2 drops from 79% to about 31%. The analyst should set aside those two customers and refit the model.

59. 0.03

61. a) R^2 is an indication of the strength of the model, not the appropriateness of the model.

b) The student should have said, "The model predicts that quarterly sales will be $10 million when $1.5 million is spent on advertising."

63. a) Quantitative variable condition: Both variables are quantatitive (*GPA* and *Starting Salary*).

b) Linearity condition: Examine a scatterplot of *Starting Salary* by *GPA*.

c) Outlier condition: Examine the scatterplot.

d) Equal spread condition: Plot the regression residuals versus predicted values.

65. a)

b) There is a strong, positive, linear association between *Price* and *Year* of used BMW 840s.

c) Yes

d) 0.757

e) 57.4% of the variability in *Price* of a used BMW 840 can be accounted for by the *Year* the car was made.

f) The relationship is not perfect. Other factors, such as options, condition, and mileage, may account for some of the variability in price.

67. a) The association between cost of living in 2007 and 2006 is linear, positive, and strong. The linearity of the scatterplot indicates that the linear model is appropriate.

b) 83.7% of the variability in cost of living in 2007 can be explained by variability in cost of living in 2006.

c) 0.915

d) Moscow had a cost of living of 123.9% of New York's in 2006. According to the model, Moscow is predicted to have a cost of living in 2007 that is about 128.8% of New York's. Moscow actually had a cost of living in 2007 that was 134.4% of New York's, so its residual was about 5.6%.

69. a) 0.578

b) CO_2 levels account for 33.4% of the variation in mean temperature.

c) $\overline{Mean\ Temperature} = 15.3066 + 0.004\ CO_2$

d) The predicted mean temperature has been increasing at an average rate of 0.004 degrees (C)/ppm of CO_2.

e) One *could* say that with no CO_2 in the atmosphere, there would be a temperature of 15.3066 degrees Celsius, but this is extrapolation to a nonsensical point.

f) No

g) Predicted 16.7626 degrees C.

71. a) Spearman's rank correlation would be more appropriate because the relationship is not straight.

b) The relationship with the square root of *Displacement* is more nearly linear and would be more appropriate for fitting a linear model.

CHAPTER 7

SECTION EXERCISE ANSWERS

1. a) Independent (unless a large group of one gender comes to the ATM machine together).

b) Independent. The last digit of one student's SS number provides no information about another.

c) Not independent. How you perform on one test provides information about other tests.

3. a) Won't work, but won't hurt. Each number drawn is equally likely and independent of the others, so this set of numbers is just as likely as any other in the next drawing.

b) Won't work, but won't hurt. Each number drawn is equally likely and independent of the others, so randomly generated numbers are just as likely as any others in the next drawing.

5. a) 0.75
   b) 0.25
   c) $0.75^2 = 0.5625$
   d) $0.75 + 0.75 - (0.5625) = 0.9375$ or $1 - (0.25^2) = 0.9375$

7. a) $760/1200 = 0.63$
   b) Marginal.
   c) $210/1200 = 0.175$
   d) Joint.

9. a) $210/500 = 0.42$
   b) $210/760 = 0.2763$
   c) $440/500 = 0.88$

11. a) $100\% - 30\% = 70\%$
   b) Marginal.
   c)

		Online Banking		
		Yes	No	
Age	Under 50	0.25	0.15	**0.40**
	50 or Older	0.05	0.55	**0.60**
		0.30	**0.70**	**1.00**

   d) $0.25/0.40 = 0.625$
   e) No, because the conditional probability of banking online for those under 50 is 0.625. The probability of banking online is 0.30 which is not the same.

CHAPTER EXERCISE ANSWERS

13. a) Outcomes are equally likely and independent.
   b) This is likely a personal probability expressing his degree of belief that there will be a rate cut.

15. a) There is no such thing as the "law of averages." The overall probability of an airplane crash does not change due to recent crashes.
   b) There is no such thing as the "law of averages." The overall probability of an airplane crash does not change due to a period in which there were no crashes.

17. a) It would be foolish to insure your neighbor's house for $300. Although you would probably simply collect $300, there is a chance you could end up paying much more than $300. That risk is not worth the $300.
   b) The insurance company insures many people. The overwhelming majority of customers pay and never have a claim. The few customers who do have a claim are offset by the many who simply send their premiums without a claim. The relative risk to the insurance company is low.

19. a) Yes.
   b) Yes.
   c) No, probabilities sum to more than 1.
   d) Yes.
   e) No, sum isn't 1 and one value is negative.

21. 0.078

23. The events are disjoint. Use the addition rule.
   a) 0.72
   b) 0.89
   c) 0.28

25. a) 0.5184
   b) 0.0784
   c) 0.4816

27. a) The repair needs for the two cars must be independent of one another.
   b) This may not be reasonable. An owner may treat the two cars similarly, taking good (or poor) care of both. This may decrease (or increase) the likelihood that each needs to be repaired.

29. a) 0.68
   b) 0.32
   c) 0.04

31. a) 0.340
   b) 0.080

33. a) 0.4712
   b) 0.7112
   c) $1 - P(\text{interview}) = 1 - 0.2888 = 0.7112$

35. a) The events are disjoint (an M&M can't be two colors at once), so use the addition rule where applicable.
     i) 0.30
     ii) 0.30
     iii) 0.90
     iv) 0
   b) The events are independent (picking out one M&M doesn't affect the outcome of the next pick), so use the multiplication rule.
     i) 0.027
     ii) 0.128
     iii) 0.512
     iv) 0.271

37. a) Disjoint.
   b) Independent.
   c) No. Once you know that one of a pair of disjoint events has occurred, the other one cannot occur, so its probability has become zero.

39. a) 0.125
   b) 0.125
   c) 0.875
   d) Independence.

41. a) 0.0225
   b) 0.092
   c) 0.00008
   d) 0.556

43. a) Your thinking is correct. There are 47 cards left in the deck, 26 black and only 21 red.
   b) This is not an example of the Law of Large Numbers. The card draws are not independent.

45. a) 0.550
   b) 0.792
   c) 0.424
   d) 0.918

47. a) 0.333
   b) 0.429
   c) 0.667

49. a) 0.11
   b) 0.27
   c) 0.407
   d) 0.344

51. No. 28.8% of men with OK blood pressure have high cholesterol, but 40.7% of men with high blood pressure have high cholesterol.

53. a) 0.086
    b) 0.437
    c) 0.156
    d) 0.174
    e) 0.177
    f) No.

55. a) 0.47
    b) 0.266
    c) Having a garage and a pool are not independent events.
    d) Having a garage and a pool are not disjoint events.

57. a) 96.5%
    b) The probability of U.S. adults having a landline, given that they have a cell phone, is $58.2/(58.2 + 2.8)$ or about 95.4%. About 96.6% of U.S. adults have a landline. It appears that having a cell phone and having a landline are independent, since the probabilities are roughly the same.

59. No. 12.5% of the cars were of European origin, but about 16.9% of the students drive European cars.

61. a) 15.4%
    b) 11.4%
    c) 73.9%
    d) 18.5%

CHAPTER 8

SECTION EXERCISE ANSWERS

1. a) Discrete.
    b) Yes/no.

3. 1.2

5. 0.748

7. a) $19
    b) $7

(continued in next column)

9. a) $\mu = 30$; $\sigma = 6$
    b) $\mu = 26$; $\sigma = 5$
    c) $\mu = 30$; $\sigma = 5.39$
    d) $\mu = -10$; $\sigma = 5.39$

11. a) 110 and 12.
    b) 450 and 40.
    c) 190 and 14.422.
    d) X and Y are independent for the SD calculation, but not necessarily for the sum.

13. Yes.

15. Geometric, $p = 0.26$.

17. x
    a) 0.5488
    b) 0.4512

CHAPTER EXERCISE ANSWERS

19. a) $1, 2, \ldots, n$
    b) Discrete.

21. a) 0,1,2,3,4
    b) Discrete.
    c) No, the outcomes are not equally likely.

23. a) $0.60
    b) $11.59
    c) $-$0.40

25. a) 2.25 lights
    b) 1.26 lights

27. a) No, the probability he wins the second changes depending on whether he won the first.
    b) 0.42
    c) 0.08
    d)

X	0	1	2
$P(X = x)$	0.42	0.50	0.08

    e) $E(X) = 0.66$ tournaments; $\sigma = 0.62$ tournaments

29. a) No, the probability of one battery being dead will depend on the state of the other one since there are only 10 batteries.
    b)

Number good	0	1	2
P(number good)	$\left(\dfrac{3}{10}\right)\left(\dfrac{2}{9}\right) = \dfrac{6}{90}$	$\left(\dfrac{3}{10}\right)\left(\dfrac{7}{9}\right) + \left(\dfrac{7}{10}\right)\left(\dfrac{3}{9}\right) = \dfrac{42}{90}$	$\left(\dfrac{7}{10}\right)\left(\dfrac{6}{9}\right) = \dfrac{42}{90}$

    c) $\mu = 1.4$ batteries
    d) $\sigma = 0.61$ batteries

31. $\mu = E$ (total wait time) $= 74.0$ seconds
$\sigma = SD$ (total wait time) ≈ 20.57 seconds
(Answers to standard deviation may vary slightly due to rounding of the standard deviation of the number of red lights each day.) The standard deviation may be calculated only if the days are independent of each other. This seems reasonable.

33. a) The standard deviation is large because the profits on insurance are highly variable. Although there will be many small gains, there will occasionally be large losses when the insurance company has to pay a claim.
    b) $\mu = E$ (two policies) $= \$300$
$\sigma = SD$(two policies) $\approx \$8,485.28$
    c) $\mu = E$(1000 policies) $= \$150,000$
$\sigma = SD$(1000 policies) $= \$189,736.66$
    d) A natural disaster affecting many policyholders such as a large fire or hurricane.

35. a) B = number basic; D = number deluxe
Net Profit $= 120B + 150D - 200$
    b) $928.00
    c) $187.45
    d) Mean—no; SD—yes (sales are independent).

37. a) $\mu = E$(time) $= 2.6$ days
$\sigma = SD$(time) ≈ 0.707 days
    b) $\mu = E$(combined time) $= 3.7$ days
$\sigma = SD$(combined time) ≈ 0.768 days

39. a) Let X_i = price of ith Hulk figure sold; Y_i = price of ith Iron Man figure sold; Insertion Fee = $0.55; T = Closing Fee $= 0.875(X_1 + X_2 + \cdots + X_{19} + Y_1 + \cdots + Y_{13})$
Net Income $= (X_1 + X_2 + \cdots + X_{19} + Y_1 + \cdots + Y_{13}) - 32(0.55) - 0.0875(X_1 + X_2 + \cdots + X_{19} + Y_1 + \cdots + Y_{13})$
    b) $\mu = E$ (net income) $= \$313.24$
    c) $\sigma = SD$ (net income) $= \$6.05$
    d) Yes, to compute the standard deviation.

41. a) No, these are not Bernoulli trials. The possible outcomes are 1, 2, 3, 4, 5, and 6. There are more than two possible outcomes.

b) Yes, these may be considered Bernoulli trials. There are only two possible outcomes: Type A and not Type A. Assuming the 120 donors are representative of the population, the probability of having Type A blood is 43%. The trials are not independent because the population is finite, but the 120 donors represent less than 10% of all possible donors.

c) No, these are not Bernoulli trials. The probability of choosing a man changes after each promotion and the 10% condition is violated.

d) No, these are not Bernoulli trials. We are sampling without replacement, so the trials are not independent. Samples without replacement may be considered Bernoulli trials if the sample size is less than 10% of the population, but 500 is more than 10% of 3000.

e) Yes, these may be considered Bernoulli trials. There are only two possible outcomes: sealed properly and not sealed properly. The probability that a package is unsealed is constant at about 10%, as long as the packages checked are a representative sample of all.

43. a) 0.0819
b) 0.0064
c) 0.16
d) 0.992

45. $E(X) = 14.28$, so 15 patients

47. a) 0.078 pixels
b) 0.280 pixels
c) 0.375
d) 0.012

49. a) 0.274
b) 0.355
c) 0.043

51. a) 0.090
b) 0.329
c) 0.687

53. a) A uniform; all numbers should be equally likely to be selected.
b) 0.5
c) 0.001

55. a) The Poisson model
b) 0.9502
c) 0.0025

CHAPTER 9

SECTION EXERCISE ANSWERS

1. In economics she scored 1.25 standard deviations above the mean. On the math exam she scored 1.50 standard deviations above the mean, so she did "better" on the math exam.

3. You scored 2.2 standard deviations above the mean.

5. a) According to the 68–95–99.7 Rule, only 5% of the distribution is beyond 2 standard deviations from the mean, so only 2.5% is more than 2 standard deviations above the mean. So less than 3% of the distribution is above a z-score of 2.20. You qualify.

b) You need to assume that the distribution is unimodal and symmetric for the 68–95–99.7 Rule to apply.

7. a)

b) 18.6 to 31.0 mpg
c) 16%
d) 13.5%
e) less than 12.4 mpg

9. a) 6.68%
b) 98.78%
c) 71.63%
d) 61.71%

11. a) 0.842
b) -0.675
c) -1.881
d) -1.645 to 1.645

13. Yes. The histogram is unimodal and symmetric and the Normal probability plot is straight.

15. a) $\mu = E(\text{miles remaining}) = 164$ miles
$\sigma = SD(\text{miles remaining}) \approx 19.799$ miles
b) 0.580

17. a) 0.141 (0.175 with continuity correction)
b) Answers may vary. That's a fairly high proportion, but the decision depends on the relative costs of not selling seats and bumping passengers.

19. a) A uniform; all numbers should be equally likely to be selected.
b) 0.02
c) 0.10

CHAPTER EXERCISE ANSWERS

21. a) 16%
b) 50%
c) 95%
d) 0.15%

23. a) 2.4%
b) 8.0%
c) -8.8%
d) $(-3.2\% < x < 8.0\%)$

25. a) 50%
b) 16%
c) 2.5%
d) More than 1.542 is more unusual.

27. a) $x > 1.492$
b) $x < 1.459$
c) $(1.393 < x < 1.525)$
d) $x < 1.393$

29. a) 21.6% (using technology).
b) 48.9%
c) 59.9%
d) 33.4%

31. a) $x > 9.58\%$
     b) $x < -2.31\%$
     c) $(-0.54\% < x < 5.34\%)$
     d) $x > -2.31\%$

33. a) 0.98%
     b) 15.4%
     c) 7.56%

35. a) 79.58
     b) 18.50
     c) 95.79
     d) −2.79

37. $z_{SAT} = 1.30$; $z_{ACT} = 2$. The ACT score is the better score because it is farther above the mean in standard deviation units than the SAT score.

39. a) To know about their consistency and how long they might last. Standard deviation measures variability, which translates to consistency in everyday use. A type of battery with a small standard deviation would be more likely to have life spans close to their mean life span than a type of battery with a larger standard deviation.
     b) The second company's batteries have a higher mean life span, but a larger standard deviation, so they have more variability. The decision is not clear-cut. The first company's batteries are not likely to fail in less than 21 months, but that wouldn't be surprising for the second company. But the second company's batteries could easily last longer than 39 months—a span very unlikely for the first company.

41. CEOs can have between 0 and maybe 40 (or possibly 50) year's experience. A standard deviation of 1/2 year is impossible because many CEOs would be 10 or 20 SDs away from the mean, whatever it is. An SD of 16 years would mean that 2 SDs on either side of the mean is plus or minus 32, for a range of 64 years. That's too high. So, the SD must be 6 years.

43. a) The Poisson model
     b) The exponential model
     c) 1/3 minutes
     d) 0.0473

45. 0.053

47. 0.025

49. Any Job Satisfaction score more than 2 standard deviations below the mean or less than $100 - 2(12) = 76$ might be considered unusually low. We would expect to find someone with a Job Satisfaction score less than $100 - 3(12) = 64$ very rarely.

51. a) About 16%
     b) One standard deviation below the mean is −1.27 hours, which is impossible.
     c) Because the standard deviation is larger than the mean, the distribution is strongly skewed to the right, not symmetric.

53. a)

     b) 30.85%
     c) 17.00%
     d) IQR = Q3 − Q1 = 32.38
     e) Above 212.87 points.

CHAPTER 10

SECTION EXERCISE ANSWERS

1. a) Normal.
    b) 0.36
    c) They wouldn't change. The shape is still approximately Normal and the mean is still the true proportion.

3. a) 0.0339
    b) 0.5
    c) 0.842 (0.843 using a rounded answer from part a)
    d) 0.01
    e) 0.039

5. Yes. Assuming the survey is random, they should be independent. We don't know the true proportion, so we can't check np and nq, but we have observed 10 successes, which is sufficient.

7.

9. a) Normal.
    b) 215 mg/dl
    c) 4.63 mg/dl
    d) Only c would change, to 3.0 mg/dl.

11. a) 0.0357
     b) 400

13. a) One would expect many small fish and a few large ones.
     b) We don't know the exact distribution, but we know it isn't Normal, so we don't have a model we can use to determine a probability.
     c) Probably not. With a skewed distribution, a sample of size 5 is not large enough to use the Central Limit Theorem.
     d) With a sample of 60, we can use the CLT even for a skewed population. The standard deviation is 0.30.
     e) Yes, 4.5 pounds is more than 2 standard deviations above the mean, so by the 68–95–99.7 rule, it would happen less than 2.5% of the time.

CHAPTER EXERCISE ANSWERS

15. All the histograms are centered near 0.05. As n gets larger, the histograms approach the Normal shape, and the variability in the sample proportions decreases.

17. a)

n	Observed mean	Theoretical mean	Observed st. dev.	Theoretical st. dev.
20	0.0497	0.05	0.0479	0.0487
50	0.0516	0.05	0.0309	0.0308
100	0.0497	0.05	0.0215	0.0218
200	0.0501	0.05	0.0152	0.0154

     b) They are all quite close to what we expect from the theory.
     c) The histogram is unimodal and symmetric for $n = 200$.
     d) The success/failure condition says that np and nq should both be at least 10, which is not satisfied until $n = 200$ for $p = 0.05$. The theory predicted my choice.

19. a) Symmetric, because probabilities of success and failure are equal.
     b) 0.5
     c) 0.125
     d) $np = 8 < 10$

21. a) About 68% should have proportions between 0.4 and 0.6, about 95% between 0.3 and 0.7, and about 99.7% between 0.2 and 0.8.

b) $np = 12.5, nq = 12.5$; both are ≥ 10.

c)

$np = nq = 32$; both are ≥ 10.

d) Becomes narrower (less spread around 0.5).

23. This is a fairly unusual result: about 2.26 SDs above the mean. The probability of that is about 0.012. So, in a class of 100 this is certainly a reasonable possibility.

25. a)

b) Both $np = 56$ and $nq = 24 \geq 10$. Drivers *may* be independent of each other, but if flow of traffic is very fast, they may not be. Or weather conditions may affect all drivers. In these cases they may get more or fewer speeders than they expect.

27. a) Assume that these children are typical of the population. They represent fewer than 10% of all children. We expect 20.4 nearsighted and 149.6 not; both are at least 10.

b)

c) Probably between 12 and 29.

29. a) $\mu = 7\%, \sigma = 1.8\%$

b) Assume that clients pay independently of each other, that we have a random sample of all possible clients, and that these represent less than 10% of all possible clients. $np = 14$ and $nq = 186$ are both at least 10.

c) 0.048

31.

These are not random samples, and not all colleges may be typical (representative). $np = 296, nq = 104$ are large enough.

33. Yes; if their students were typical, a retention rate of $522/603 = 86.6\%$ would be over 7 standard deviations above the expected rate of 74%.

35. 0.212. Reasonable that those polled are independent of each other and represent less than 10% of all potential voters. We assume the sample was selected at random. Success/Failure Condition met: $np = 208, nq = 192$. Both ≥ 10.

37. 0.088 using $N(0.08, 0.022)$ model.

39. Answers will vary. Using $\mu + 3\sigma$ for "very sure," the restaurant should have 89 nonsmoking seats. Assumes customers at any time are independent of each other, a random sample, and represent less than 10% of all potential customers. $np = 72, nq = 48$, so Normal model is reasonable ($\mu = 0.60, \sigma = 0.045$).

41. a) Normal, center at μ, standard deviation $\sigma/\sqrt{n}$.

b) Standard deviation will be smaller. Center will remain the same.

43. a) The histogram is unimodal and slightly skewed to the right, centered at 36 inches with a standard deviation near 4 inches.

b) All the histograms are centered near 36 inches. As n gets larger, the histograms approach the Normal shape and the variability in the sample means decreases. The histograms are fairly normal by the time the sample reaches size 5.

45. a)

n	Observed mean	Theoretical mean	Observed st. dev.	Theoretical st. dev.
2	36.314	36.33	2.855	2.842
5	36.314	36.33	1.805	1.797
10	36.341	36.33	1.276	1.271
20	36.339	36.33	0.895	0.899

b) They are all very close to what we would expect.

c) For samples as small as 5, the sampling distribution of sample means is unimodal and very symmetric.

d) The distribution of the original data is nearly unimodal and symmetric, so it doesn't take a very large sample size for the distribution of sample means to be approximately Normal.

47.

Normal, $\mu = 3.4$, $\sigma = 0.07$. We assume that the students are randomly assigned to the seminars and represent less than 10% of all possible students, and that individual's GPAs are independent of one another.

49. a) The mean of 100 coins varies less from sample to sample than the individual coins.

b) $SD(\bar{y}) = \dfrac{\sigma}{\sqrt{n}} = \dfrac{0.09}{\sqrt{100}} = 0.009$

CHAPTER 11

SECTION EXERCISE ANSWERS

1. a) p is the proportion of all backpacks entering the stadium which contain alcoholic beverages; $\hat{p}$ is the proportion in the sample $\hat{p} = 17/130 = 13.08\%$. Yes. This seems to be a random sample.

b) p is the proportion of all visitors to the website who approve of recent bossnapping. $\hat{p}$ is the proportion in the sample $\hat{p} = 49.2\%$. No. This is a volunteer sample and may be biased.

c) This question is about the mean weight, not a proportion. The methods of this chapter are not appropriate.

3. a) 0.35
b) 0.034
c) (0.282, 0.418) or without rounding (0.283, 0.417)

5. a) False. Doesn't make sense. Workers are not proportions.
b) True.
c) False. Our best guess is 0.48 not 0.95.
d) False. Our best guess is 0.48, but we're not sure that's correct.
e) False. The statement should be about the true proportion, not future samples.

7. a) *Population*—all customers who recently bought new cars; *sample*—167 people surveyed about their experience; p—proportion of all new car buyers who are dissatisfied with the salesperson; $\hat{p}$—proportion of new car buyers surveyed who are dissatisfied with the salesperson (3%); we can't use the methods of Chapter 11 because only 5 people were dissatisfied. Also, sample may not be representative.

b) *Population*—all college students; *sample*—2883 who were asked about their cell phones at the football stadium; p—proportion of all college students with cell phones; $\hat{p}$—proportion of college students at the football stadium with cell phones (77.8%); we should be cautious—students entering the football stadium may not represent all students. Sample may be biased.

9. a) About 2401 (using 1.96 standard errors).
b) About 4148 (using 2.576 standard errors).
c) About 385 (using 1.96 standard errors).

11. a) 141
b) 318
c) 564

CHAPTER EXERCISE ANSWERS

13. He believes the true proportion is within 4% of his estimate, with some (probably 95%) degree of confidence.

15. a) *Population*—all cars in the local area; *sample*—134 cars actually stopped at the checkpoint; p—population proportion of all cars with safety problems; $\hat{p}$—proportion of cars in the sample that actually have safety problems (10.4%); if sample (a cluster sample) is representative, then the methods of this chapter will apply.

b) *Population*—general public; *sample*—602 viewers that logged on to the website; p—population proportion of the general public that think corporate corruption is "worse"; $\hat{p}$—proportion that logged onto the website and voted that corporate corruption is "worse" (81.1%); can't use the methods of this chapter—sample is biased and nonrandom.

17. a) Not correct. This implies certainty.
b) Not correct. Different samples will give different results. Most likely, none of the samples will have *exactly* 88% on-time orders.
c) Not correct. A confidence interval says something about the unknown population proportion, not the sample proportion in different samples.
d) Not correct. In this sample, we *know* that 88% arrived on time.
e) Not correct. The interval is about the parameter, not about the days.

19. a) False.
b) True.
c) True.
d) False.

21. We are 90% confident that between 29.9% and 47.0% of U.S. cars are made in Japan.

23. a) 0.025 or 2.5%
b) The pollsters are 90% confident that the true proportion of adults who do not use e-mail is within 2.5% of the estimated 38%.
c) A 99% confidence interval requires a larger margin of error. In order to increase confidence, the interval must be wider.
d) 0.039 or 3.9%.
e) Smaller margins of error will give us less confidence in the interval.

25. a) (12.7%, 18.6%)
b) We are 95% confident that between 12.7% and 18.6% of all accidents involve teenage drivers.
c) About 95% of all random samples of size 582 will produce intervals that contain the true proportion of accidents involving teenage drivers.
d) Contradicts—the interval is completely below 20%.

27. Probably nothing. Those who bothered to fill out the survey are a voluntary response sample, which may be biased.

29. This was a random sample of less than 10% of all Internet users; there were $703 \times 0.18 = 127$ successes and 576 failures, both at least 10. We are 95% confident that between 15.2% and 20.8% of Internet users have downloaded music from a site that was not authorized. (Answer could be 15.2% to 20.9% if $n = 127$ is used instead of 0.18).

31. a) $385/550 = 0.70$; 70% of U.S. chemical companies in the sample are certified.
b) This was a random sample, but we don't know if it is less than 10% of all U.S. chemical companies; there were $550(0.70) = 385$ successes and 165 failures, both at least 10. We are 95% confident that between 66.2% and 73.8% of the chemical companies in the United States are certified. It appears that the proportion of companies certified in the United States is less than in Canada.

33. a) There may be response bias based on the wording of the question.
b) (45.5%, 51.5%)
c) The margin of error based on the pooled sample is smaller, since the sample size is larger.

35. a) The interval based on the survey conducted by the college Statistics class will have the larger margin of error, since the sample size is smaller.

b) Both samples are random and are probably less than 10% of the city's voters (provided the city has more than 12,000 voters); there were 636 successes and 564 failures for the newspaper, both at least 10; there were 243 successes and 207 failures for the Statistics class, both at least 10; Newspaper poll: (50.2%, 55.8%); Statistics class: (49.4%, 58.6%).

c) The Statistics class should conclude that the outcome is too close to call because 50% is in their interval.

37. a) This was a random sample of less than 10% of all English children; there were $2700(0.20) = 540$ successes and 2160 failures, both at least 10; (18.2%, 21.8%).

b) We are 98% confident that between 18.2% and 21.8% of English children are deficient in vitamin D.

c) About 98% of random samples of size 2700 will produce confidence intervals that contain the true proportion of English children that are deficient in vitamin D.

d) No. The interval says nothing about causation.

39. a) This is not a random sample; even though it's representative we are not sure if we have less than 10% of all WBCs; there are 8 successes which is not greater than 10, so the sample is not large enough.

b) Since the conditions are not met but the sample is representative, we could perform a pseudo observation confidence interval adding two successes and two failures to the data. Then a 90% CI would be (0.251, 0.582).

41. a) This was a random sample of less than 10% of all companies in Vermont; there were 12 successes and 0 failures, which is not greater than 10, so the sample is not large enough.

b) Using the pseudo observation confidence interval, $\tilde{p} = \dfrac{14}{16} = 0.875$; (0.713, 1.037); so we can say that with 95% confidence the true proportion is at least 71.3%.

43. a) This was a random sample of less than 10% of all self-employed taxpayers; there were 20 successes and 206 failures, both at least 10.

b) (5.1%, 12.6%)

c) We are 95% confident that between 5.1% and 12.6% of all self-employed individuals had their tax returns audited in the past year.

d) If we were to select repeated samples of 226 individuals, we'd expect about 95% of the confidence intervals we created to contain the true proportion of all self-employed individuals who were audited.

45. a) This was a random sample of less than 10% of all U.S. adults; there were $703 \times 0.13 = 91$ successes and 612 failures, both at least 10.

b) (10.5%, 15.5%)

47. a) The 95% confidence interval for the true proportion of all 18- to 29-year-olds who believe the U.S. is ready for a woman president will be about twice as wide as the confidence interval for the true proportion of all U.S. adults, since it is based on a sample about one-fourth as large. (Assuming approximately equal proportions)

b) This was a random sample of less than 10% of all U.S. 15- to 29-year-old-adults; there were $250 \times 0.62 = 155$ successes and 95 failures, both at least 10. We are 95% confident that between 56.0% and 68.0% of 18- to 29-year-olds believe the U.S. is ready for a woman president.

49. a) The parameter is the proportion of digital songs in student libraries that are legal. The population is all songs held in digital libraries. The sample size is 117,079 songs, not 168 students.

b) This was a cluster sample (168 clusters of songs in the digital libraries); 117,079 is less than 10% of all digital songs; the number of legal songs and illegal songs in the sample are both much greater than 10.

c) We are 95% confident that between 22.9% and 23.3% of digital songs were legally purchased.

d) The very large sample size has made the confidence interval unreasonably narrow. It is hard to believe that such a narrow interval really captures the parameter of interest. Additionally, these data were collected in a cluster sample of only 168 students. This gives us less certainty about our ability to capture the true parameter.

51. a) This was a random sample of less than 10% of all Internet users; there were $703(0.64) = 450$ successes and 253 failures, both at least 10. We are 90% confident that between 61.0% and 67.0% of Internet users would still buy a CD.

b) In order to cut the margin of error in half, they must sample 4 times as many users; $4 \times 703 = 2812$ users.

53. 1801

55. 384 total, using $p = 0.15$

57. Since $z^* \approx 1.634$, which is close to 1.645, the pollsters were probably using 90% confidence.

59. This was a random sample of less than 10% of all customers; there were 67 successes and 433 failures, both at least 10. From the data set, $\hat{p} = 67/500 = 0.134$. We are 95% confident that the true proportion of customers who spend $1000 per month or more is between 10.4% and 16.4%.

61. a) This was a random sample of less than 10% of all MA males; there were 2662 successes and 398 failures, both at least 10.

b) (0.858, 0.882)

c) We are 95% confident that between 85.8% and 88.2% of MA males have health insurance.

CHAPTER 12

SECTION EXERCISE ANSWERS

1. a) 1.968 years
   b) 0.984 years (half as large)

3. a) 24
   b) 99

5. a) 2.064
   b) 1.984

7. a) (27.78, 35.90) years
   b) 4.06 years
   c) (27.92, 35.76) years. Slightly narrower.

9. Independence: The data were from a random survey and should be independent.

Randomization: the data were selected randomly.

10% Condition: These customers are fewer than 10% of the customer population.

Nearly Normal: The histogram is unimodal and symmetric, which is sufficient.

11. a) Four times as big, or $n = 100$.
   b) 100 times a big, or $n = 2500$.

CHAPTER EXERCISE ANSWERS

13. a) 1.74
   b) 2.37

15. As the variability of a sample increases, the width of a 95% confidence interval increases, assuming that sample size remains the same.

17. a) ($4.382, $4.598)
b) ($4.400, $4.580)
c) ($4.415, $4.565)

19. a) Not correct. A confidence interval is for the mean weight gain of the population of all cows. It says nothing about individual cows.
b) Not correct. A confidence interval is for the mean weight gain of the population of all cows, not individual cows.
c) Not correct. We don't need a confidence interval about the average weight gain for cows in this study. We are certain that the mean weight gain of the cows in this study is 56 pounds.
d) Not correct. This statement implies that the average weight gain varies. It doesn't.
e) Not correct. This statement implies that there is something special about our interval, when this interval is actually one of many that could have been generated, depending on the cows that were chosen for the sample.

21. The assumptions and conditions for a *t*-interval are not met. With a sample size of only 20, the distribution is too skewed. There is also a large outlier that is pulling the mean higher.

23. a) The data are a random sample of all days; the distribution is unimodal and symmetric with no outliers.
b) ($122.20, $129.80)
c) We are 90% confident that the interval $122.20 to $129.80 contains the true mean daily income of the parking garage.
d) 90% of all random samples of size 44 will produce intervals that contain the true mean daily income of the parking garage.
e) $128 is a plausible value.

25. a) We can be more confident that our interval contains the mean parking revenue.
b) Wider (and less precise) interval
c) By collecting a larger sample, they could create a more precise interval without sacrificing confidence.

27. a) $2350 \pm 2.009 (59.51)$ Interval: (2230.4, 2469.6)
b) The assumptions and conditions that must be satisfied are:
1) Independence: probably OK.
2) Nearly Normal condition: can't tell.
3) Sample size of 51 is large enough.
c) We are 95% confident the interval $2230.4 to $2469.6 contains the true mean increase in sales tax revenue.
Examples of what the interval *does not* mean: The mean increase in sales tax revenue is $2350 95% of the time. 95% of all increases in sales tax revenue increases will be between $2230.4 and $2469.6. There's 95% confidence the next small retailer will have an increase in sales tax revenue between $2230.4 and $2469.6.

29. a) Given no time trend, the monthly on-time departure rates should be independent. Though not a random sample, these months should be representative, and they're fewer than 10% of all months. The histogram looks unimodal, but slightly left-skewed; not a concern with this large sample.
b) (80.57%, 81.80%)
c) We can be 90% confident that the interval from 80.57% to 81.80% holds the true mean monthly percentage of on-time flight departures.

31. a) The histogram of the lab fees shows 2 extreme outliers, so with the outliers included, the conditions for inference are violated.

b) With the outliers left in, the 95% confidence interval is (44.9, 81.6) minutes. If we remove the two extreme outliers, it is (54.6, 68.4).
In either case, we would be reluctant to conclude that the mean is above 55 minutes. The sample size is small and the presence of two large outliers advises us to be cautious about conclusions from this sample.

33. a) The assumptions and conditions that must be satisfied are: The data come from a nearly normal distribution.
The air samples were selected randomly, and there is no bias present in the sample.
b) The histogram of air samples is not nearly normal, but the sample size is large, so inference is OK.

35. a) $14.90 - 11.6$ or ± 3.3 miles per hour
b) The sample size for ME $= 2$ should be increased to $1.96 \times 8/2 = 7.84$.
$(7.84)^2 = 61.466 = 62$

37. a) Interval: $653 to $707
b) The confidence interval suggests the mean audit cost is higher than $650.

39. a) The timeplot shows no pattern, so it seems that the measurements are independent. Although this is not a random sample, an entire year is measured, so it is likely that we have representative values. We certainly have fewer than 10% of all possible wind readings. The histogram appears nearly normal.
b) A 95% confidence interval for the true mean speed is (7.795, 8.243) mph. Because there are many plausible values below 8 mph, we can not be confident that the true mean is at least 8 mph. We would not recommend that the turbine be placed here.

CHAPTER 13

1. a) Let $p =$ probability of winning on the slot machine. H_0: $p = 0.01$ vs. H_A: $p \neq 0.01$
b) Let $\mu =$ mean spending per customer this year. H_0: $\mu = \$35.32$ vs. H_A: $\mu \neq \$35.32$
c) Let $p =$ proportion of patients cured by the new drug. H_0: $p = 0.3$ vs. H_A: $p \neq 0.3$
d) Let $p =$ proportion of clients now using the website. H_0: $p = 0.4$ vs. H_A: $p \neq 0.4$

3. a) False. A high P-value shows that the data are consistent with the null hypothesis, but provides no evidence for rejecting the null hypothesis.
b) False. It results in rejecting the null hypothesis, but does not prove that it is false.
c) False. A high P-value shows that the data are consistent with the null hypothesis but does not prove that the null hypothesis is true.
d) False. Whether a P-value provides enough evidence to reject the null hypothesis depends on the risk of a type I error that one is willing to assume (the α level).

5. a) $SE(\hat{p}) = \sqrt{\dfrac{p_0 q_0}{n}} = \sqrt{\dfrac{(0.35)(0.65)}{300}} = 0.028$
b) $z = \dfrac{\hat{p} - p_0}{SD(\hat{p})} = \dfrac{0.46 - 0.35}{0.028} = 3.93$
c) Yes, that's an unusually large *z*-value.

7. a) H_0: $p = 0.40$ vs. H_A: $p \neq 0.40$. Two-sided.
b) H_0: $\mu = 5.4$ vs. H_A: $\mu < 5.4$. One-sided.
c) H_0: $p = 0.42$ vs. H_A: $p > 0.42$. One-sided.
d) H_0: $p = 0.50$ vs. H_A: $p > 0.50$. One-sided.

9. a) $H_0: \mu = 25$
b) Two-sided: $\mu \neq 25$
c) 3.476
d) 0.0020
e) Reject H_0. There is strong evidence that the mean age is not 25.
f) Because the 95% confidence interval does not contain 25, we can reject H_0 at $\alpha = 0.05$.

11. a) True.
b) False. The alpha level is set independently and does not depend on the sample size.
c) False. The P-value would have to be less than 0.01 to reject the null hypothesis.
d) False. It simply means we do not have enough evidence at that alpha level to reject the null hypothesis.

13. a) $z = \pm 1.96$
b) $z = 1.645$
c) $t = \pm 2.03$
d) $z = 2.33$; n is not relevant for critical values for z.
e) $z = -2.33$

15. a) (0.196, 0.304)
b) No, because 0.20 is a plausible value.
c) The SE is based on $\hat{p}$: $SE(\hat{p}) = \sqrt{\dfrac{\hat{p}\hat{q}}{n}} = \sqrt{\dfrac{(0.25)(0.75)}{250}} = 0.0274$. The SD is based on the hypothesized value 0.20, so $SD(\hat{p}) = \sqrt{\dfrac{p_0 q_0}{n}}\sqrt{\dfrac{(0.20)(0.80)}{250}} = 0.0253$.
d) The SE since it is sample based.

17. a) $SE = \dfrac{s}{\sqrt{n}} = \dfrac{2.0}{\sqrt{25}} = 0.4$
b) t^* with 24 df for a 90% confidence interval is 1.711.
c) (15.82, 17.18)
d) No, we fail to reject H_0 (one-sided) at $\alpha = 0.05$ because 16.0 is in the corresponding two-sided 90% confidence interval.

19. a) Type I error. The actual value is not greater than 0.3 but they rejected the null hypothesis.
b) No error. The actual value is 0.50, which was not rejected.
c) Type II error. The actual value was 55.3 points, which is greater than 52.5.
d) Type II error. The null hypothesis was not rejected, but it was false. The true relief rate was greater than 0.25.

CHAPTER EXERCISE ANSWER

21. a) Let p be the percentage of products delivered on time. $H_0: p = 0.90$ vs. $H_A: p > 0.90$
b) Let p be the proportion of houses taking more than 3 months to sell. $H_0: p = 0.50$ vs. $H_A: p > 0.50$
c) Let p be the error rate. $H_0: p = 0.02$ vs. $H_A: p < 0.02$

23. Statement d is correct.

25. If the rate of seat-belt usage after the campaign is the same as the rate of seat-belt usage before the campaign, there is a 17% chance of observing a rate of seat-belt usage after the campaign this large or larger in a sample of the same size by natural sampling variation alone.

27. Statement e is correct.

29. No, we can say only that there is a 27% chance of seeing the observed effectiveness just from natural sampling if $p = 0.7$. There is no *evidence* that the new formula is more effective, but we can't conclude that they are equally effective.

31. a) 0.186 (using the normal model) 0.252 using exact probabilities.
b) It seems reasonable to think there really may have been half of each. We would expect to get 12 or more reds out of 20

more than 15% of the time, so there's no real evidence that the company's claim is not true. The two sided P-value is greater than 0.30.

33. If in fact the mean monthly sales due to online purchases has not changed, then only 1 out of every 100 samples would be expected to have mean sales as different from the historical figure as the mean sales observed in the sample.

35. a) $897.14 to $932.86
We are 95% confident the interval $897.14 to $932.86 contains the true mean Social Security benefit for widows and widowers in the Texas county.
b) With a P-value of .007, the hypothesis test results are significant and we reject the null hypothesis. We conclude the mean benefit payment for the Texas county is different from the $940 for the state. The 95% confidence interval estimate is $897.14 to $932.86. Since the interval does not contain the hypothesized mean of $940, we have evidence that the mean is unlikely to be $940.

37. a) Conditions are satisfied: random sample; less than 10% of population; more than 10 successes and failures; (1.9%, 4.1%).
b) Since 5% is not in the interval, there is strong evidence that fewer than 5% of all men use work as their primary measure of success.
c) $\alpha = 0.01$; it's a lower-tail test based on a 98% confidence interval.

39. a) Conditions are satisfied: random sample; less than 10% of population; more than 10 successes and failures; (0.223, 0.257); we are 95% confident that the true proportion of U.S. adults who rate the economy as Excellent/Good is between 0.223 and 0.257.
b) No. Since 0.28 is not within the interval, there is evidence that the proportion is not 28%.
c) $\alpha = 0.05$; it's a two-tail test based on a 95% confidence interval.

41. a) Less likely
b) Alpha levels must be chosen *before* examining the data. Otherwise the alpha level could always be selected to reject the null hypothesis.

43. 1. Use p, not $\hat{p}$, in hypotheses.
2. The question is about *failing* to meet the goal. H_A should be $p < 0.96$.
3. Did not check $nq = (200)(0.04) = 8$. Since $nq < 10$, the Success/Failure condition is violated. Didn't check the 10% condition.
4. $\hat{p} = \dfrac{188}{200} = 0.94$; $SD(\hat{p}) = \sqrt{\dfrac{pq}{n}} = \sqrt{\dfrac{(0.96)(0.04)}{200}} \approx 0.014$
The student used $\hat{p}$ and $\hat{q}$.
5. z is incorrect; should be $z = \dfrac{0.94 - 0.96}{0.014} \approx -1.43$.
6. $P = P(z < -1.43) = 0.076$
7. There is only weak evidence that the new system has failed to meet the goal.

45. a) $H_0: \mu = 23.3$; $H_A: \mu > 23.3$
b) **Randomization condition:** The 40 online shoppers were selected randomly. **Nearly Normal condition:** We should examine the distribution of the sample to check for serious skewness and outliers, but with a large sample of 40 shoppers, it should be safe to proceed.
c) 0.145
d) If the mean age of shoppers is still 23.3 years, there is a 14.5% chance of getting a sample mean of 24.2 years or older simply from natural sampling variation.
e) There is no evidence to suggest that the mean age of online shoppers has increased from the mean of 23.3 years.

47. a) $H_0: \mu = 55$; $H_A: \mu < 55$; **Independence assumption:** Since the times are not randomly selected, we will assume that the times are independent and representative of all the champion's times. **Nearly Normal condition:** The histogram of the times is unimodal and roughly symmetric; P-value = 0.235; fail to reject H_0. There is insufficient evidence to conclude the mean time is less than 55 seconds. They should not market the new ski wax.

   b) Type II error. They won't market a competitive wax and thus lose the potential profit from having done so.

49. a) Let p = the percentage of children with genetic abnormalities. $H_0: p = 0.05$ vs. $H_A: p > 0.05$

   b) SRS (not clear from information provided); $384 < 10\%$ of all children; $np = (384)(0.05) = 19.2 > 10$ and $nq = (384)(0.95) = 364.8 > 10$.

   c) $z = 6.28$, $P < 0.0001$

   d) If 5% of children have genetic abnormalities, the chance of observing 46 children with genetic abnormalities in a random sample of 384 children is essentially 0.

   e) Reject H_0. There is strong evidence that more than 5% of children have genetic abnormalities.

   f) We don't know that environmental chemicals cause genetic abnormalities, only that the rate is higher now than in the past.

51. a) Let p = the percentage of students in 2000 with perfect attendance the previous month. $H_0: p = 0.34$ vs. $H_A: p < 0.34$

   b) Although not specifically stated, assume that the National Center for Education Statistics used random sampling; $8302 < 10\%$ of all students; $np = (8302)(0.34) = 2822.68 > 10$ and $nq = (8302)(0.66) = 5479.32 > 10$.

   c) $z = -1.92$, $P = 0.027$

   d) Reject H_0 at $\alpha = 0.05$. There is evidence to suggest that the percentage of students with perfect attendance in the previous month has decreased in 2000.

   e) This result is statistically significant at $\alpha = 0.05$, but it's not clear that it has any practical significance since the percentage dropped only from 34% to 33%.

53. a) SRS (not clear from information provided); $1000 < 10\%$ of all workers; $n\hat{p} = 520 > 10$ and $n\hat{q} = 480 > 10$; (0.489, 0.551); we are 95% confident that between 48.9% and 55.1% of workers have invested in individual retirement accounts.

   b) Let p = the percentage of workers who have invested. $H_0: p = 0.44$ vs. $H_A: p \neq 0.44$; since 44% is not in the 95% confidence interval, we reject H_0 at $\alpha = 0.05$. There is strong evidence that the percentage of workers who have invested in individual retirement accounts was not 44%. In fact, our sample indicates an increase in the percentage of adults who invest in individual retirement accounts.

55. Let p = the percentage of cars with faulty emissions. $H_0: p = 0.20$ vs. $H_A: p > 0.20$; two conditions are not satisfied: $22 > 10\%$ of the population of 150 cars and $np = (22)(0.20) = 4.4 < 10$. It's not a good idea to proceed with a hypothesis test.

57. Let p = the percentage of readers interested in an online edition. $H_0: p = 0.25$ vs. $H_A: p > 0.25$; SRS; $500 < 10\%$ of all potential subscribers; $np = (500)(0.25) = 125 > 10$ and $nq = (500)(0.75) = 375 > 10$; $z = 1.24$, $P = 0.1076$. Since the P-value is high, we fail to reject H_0. There is insufficient evidence to suggest that the proportion of interested readers is greater than 25%. The magazine should not publish the online edition.

59. Let p = the proportion of female executives. $H_0: p = 0.40$ vs. $H_A: p < 0.40$; data are for all executives in this company and may not be able to be generalized to all companies; $np = (43)(0.40) = 17.2 > 10$ and $nq = (43)(0.60) = 25.8 > 10$; $z = -1.31$, $P = 0.0955$. Since the P-value is high,

we fail to reject H_0. There is insufficient evidence to suggest proportion of female executives is any different from the overall proportion of 40% female employees at the company.

61. Let p = the proportion of dropouts at this high school. $H_0: p = 0.109$ vs. $H_A: p < 0.109$; assume that the students at this high school are representative of all students nationally; $1792 < 10\%$ of all students nationally; $np = (1782)(0.109) = 194.238 > 10$ and $nq = (1782)(0.891) = 1587.762 > 10$; $z = -1.46$, $P = 0.072$. Since the P-value = 0.072 > 0.05, we fail to reject H_0. There is insufficient evidence of a decrease in dropout rate from 10.9%.

63. Let p = the proportion of lost luggage returned the next day. $H_0: p = 0.90$ vs. $H_A: p < 0.90$; it is reasonable to think that the people surveyed were independent with regard to their luggage woes; although not stated, we will hope that the survey was conducted randomly, or at least that these air travelers are representative of all air travelers for that airline; $122 < 10\%$ of all air travelers on the airline; $np = (122)(0.90) = 109.8 > 10$ and $nq = (122)(0.10) = 12.2 > 10$; $z = -2.05$, $P = 0.0201$. Since the P-value is low, we reject H_0. There is evidence that the proportion of lost luggage returned the next day is lower than the 90% claimed by the airline.

65. H_0: These MBA students are exposed to unethical practices at a similar rate to others in the program ($p = 0.30$).
H_A: These students are exposed to unethical practices at a different rate than other students ($p \neq 0.30$).
There is no reason to believe that students' rates would influence others; the professor considers this class typical of other classes; $120 < 10\%$ of all students in the MBA program; 27% of $120 = 32.4$—use 32 graduates; $np = 36 > 10$ and $nq = 84 > 10$; $z = -0.717$, $P = 0.4733$. Since the P-value is > 0.05, we fail to reject the null hypothesis. There is little evidence that the rate at which these students are exposed to unethical business practices is different from that reported in the study.

67. a) $z = 11.8$

   b) $11.8 > 3.29$, if we assume a two-sided 0.1% significance level.

   c) We conclude that the percent of U.S. adults giving "quite a lot" of thought to the upcoming election is significantly different in 2008 than it was in 2004.

69. a) The regulators decide that the shop is not meeting standards when it actually is.

   b) The regulators certify the shop when it is not meeting the standards.

   c) Type I

   d) Type II

71. a) The probability of detecting that the shop is not meeting standards when they are not.

   b) 40 cars produces higher power because n is larger

   c) 10%; more chance to reject H_0.

   d) A lot; larger problems are easier to detect.

73. a) One-tailed; we are testing to see if a decrease in the dropout rate is associated with the software.

   b) H_0: The dropout rate does not change following the use of the software ($p = 0.13$).
   H_A: The dropout rate decreases following the use of the software ($p < 0.13$).

   c) The professor buys the software but the dropout rate has not actually decreased.

   d) The professor doesn't buy the software and the dropout rate has actually decreased.

   e) The probability of buying the software when the dropout rate has actually decreased.

75. a) H_0: The dropout rate does not change following the use of the software ($p = 0.13$).
H_A: The dropout rate decreases following the use of the software ($p < 0.13$).
One student's decision about dropping out should not influence another's decision; this year's class of 203 students is probably representative of all statistics students; $203 < 10\%$ of all students; $np = (203)(0.13) = 26.39 > 10$ and $nq = (203)(0.87) = 176.61 > 10$; $z = -3.21$, $P = 0.0007$. Since the P-value is very low, we reject H_0. There is strong evidence that the dropout rate has dropped since use of the software program was implemented. As long as the professor feels confident that this class of statistics students is representative of all potential students, then he should buy the program.
b) The chance of observing 11 or fewer dropouts in a class of 203 is only 0.07% if the dropout rate is really 13%.

77. Independence assumption: We assume that these mutual funds were selected at random and that 35 funds are less than 10% of all value funds. **Nearly Normal condition:** A histogram shows a nearly normal distribution.
H_0: $\mu = 8$; H_A: $\mu > 8$; P-value $= 0.201$; fail to reject H_0. There is insufficient evidence that the mean 5-year return is greater than 8% for value funds.

79. Given this confidence interval, we cannot reject the null hypothesis of a mean $200 collection using $\alpha = 0.05$. However, the confidence interval suggests that there may be a large upside potential. The collection agency may be collecting as much as $250 per customer on average, or as little as $190 on average. If the possibility of collecting $250 on average is of interest to them, they may want to collect more data.

81. Yes, there is a large ($50) upside potential. The larger trial will likely narrow the confidence interval and make the decision clearer.

CHAPTER 14

1. a) $\bar{y}_1 = 58.0$ years; $\bar{y}_2 = 47.0$ years
b) 11 years
c) $s_1^2 = 50$; $s_2^2 = 54$
d) $s_1 = 7.07$ years; $s_2 = 7.35$ years
e) 4.16 years

3. a) 9.6 years
b) 1.91 years
c) 5.03

5. a) 2.64
b) 9.985
c) 5
d) 0.0247
e) 0.0459
f) For either method of calculating df, there is reasonably strong evidence to suggest that we reject the null hypothesis and conclude that the mean age of houses is not the same in the two neighborhoods.

7. a) <0.001
b) <0.001
c) In both cases, reject H_0. There is very strong evidence to suggest that there is a difference in the mean ages of houses in the two neighborhoods.

9. a) (1.72, 20.28) years
b) No
c) It suggests that we should reject H_0 since 0 is not a plausible value for the true mean difference.

11. a) (5.77, 13.43) years
b) The sample sizes are larger.

c) No
d) It suggests that we should reject H_0 since 0 is not a plausible value for the true mean difference.

13. a) $t = 2.642$; P-value $= 0.0246$. Reject H_0 and conclude that there is sufficient evidence to suggest that the true means are different.
b) (1.72, 20.28) years
c) Not really. Because the standard deviations are fairly close, the two methods will result in essentially the same confidence intervals and hypothesis tests.

15. a) $t = 5.01$; P-value < 0.001. Reject H_0. There is strong evidence to reject the hypothesis that the means are equal.
b) (5.77, 13.43) years
c) Very close to previous results because the standard deviations are fairly close. In such a case the two methods will result in essentially the same confidence intervals and hypothesis tests.

17. Yes. The groups are independent and one group (Neighborhood 2) has the lowest value, while the other (Neighborhood 1) has the highest. There are 2 observations in Neighborhood 2 lower than all in Neighborhood 1, and 5 in Neighborhood 1 higher than all those in Neighborhood 2, for a total of 7. The difference is significant at the 0.05 level.

19. a) Paired. Each pair consists of a volunteer using each chair.
b) Not paired. The samples are random within each neighborhood.
c) Paired. Each pair consists of an hour in which the productivity of the two workers is compared.

21. a) Yes, each pair is a store in which the customers with and without the program are compared.
b) 3 customers
c) 4.52 customers
d) 1.43 customers
e) 2.098
f) 9
g) One-sided. They want to know if traffic increased.
h) 0.0327
i) We can reject the null hypothesis. There is evidence of an increase in traffic.

23. (0.379, 5.621) customers

25. The P-value is too high to reject H_0 at any reasonable α-level.

27. a) 2.927 points
b) Larger
c) We are 95% confident that the mean score for the CPMP math students will be between 5.573 and 11.427 points higher on this assessment than the mean score of the traditional students.
d) Since the entire interval is above 0, there is strong evidence that students who learn with CPMP will have higher mean scores in applied algebra than those in traditional programs.

29. a) H_0: $\mu_C - \mu_T = 0$; H_A: $\mu_C - \mu_T \neq 0$
b) If the mean scores for the CPMP and traditional students are really equal, there is less than a 1 in 10,000 chance of seeing a difference as large or larger than the observed difference of 9.4 points just from natural sampling variation.
c) There is strong evidence that the CPMP students have a different mean score than the traditional students. The evidence suggests that the CPMP students have a higher mean score.

31. H_0: $\mu_C - \mu_T = 0$; H_A: $\mu_C - \mu_T \neq 0$.
$P = 0.1602$; fail to reject H_0. There is no evidence that the CPMP students have a different mean score on the word problems test than the traditional students.

33. a) (1.36, 4.64); df = 33.1
b) Since the CI does not contain 0, there is evidence that Route A is faster on average.

35. a) $H_0: \mu_C - \mu_A = 0$; $H_A: \mu_C - \mu_A \neq 0$
b) Independent groups assumption: The percentage of sugar in the children's cereals is unrelated to the percentage of sugar in adult cereals. Randomization condition: It is reasonable to assume that the cereals are representative of all children's cereals and adult cereals, in regard to sugar content. Nearly Normal condition: The histogram of adult cereal sugar content is skewed to the right, but the sample sizes are reasonably large. The Central Limit Theorem allows us to proceed.
c) (32.15, 40.82)%
d) Since the 95% confidence interval does not contain 0, we can conclude that the mean sugar content for the two cereals is significantly different at the 5% level of significance.

37. a) $H_0: \mu_C - \mu_D = 0$; $H_A: \mu_C - \mu_D \neq 0$
b) (−0.256, 1.894)
c) Since the confidence interval contains 0, there is insufficient evidence to conclude that the mean return over a 5-year period is different for consistent style funds as opposed to style drifters.

39. $H_0: \mu_N - \mu_C = 0$; $H_A: \mu_N - \mu_C > 0$. Independent groups assumption: Student scores in one group should not have an impact on the scores of students in the other group. Randomization condition: Students were randomly assigned to classes. Nearly Normal condition: The histograms of the scores are unimodal and symmetric. P = 0.017; reject H_0. There is evidence that the students taught using the new activities have a higher mean score on the reading comprehension test than the students taught using traditional methods.

41. a) $H_0: \mu_L - \mu_S = 0$; $H_A: \mu_L - \mu_S \neq 0$
b) Independent groups assumption: pH levels from the two types of streams are independent. Independence assumption: Since we don't know if the streams were chosen randomly, assume that the pH level of one stream does not affect the pH of another stream. This seems reasonable. Nearly Normal condition: The boxplots provided show that the pH levels of the streams may be skewed (since the median is either the upper or lower quartile for the shale streams and the lower whisker of the limestone streams is stretched out), and there are outliers. However, since there are 133 degrees of freedom, we know that the sample sizes are large. It should be safe to proceed.
c) $P \leq 0.0001$; reject H_0. There is strong evidence that the streams with limestone substrates have mean pH levels different from those of streams with shale substrates. The limestone streams are less acidic on average.

43. a) If the mean memory scores for people taking ginkgo biloba and people not taking it are the same, there is a 93.74% chance of seeing a difference in mean memory score this large or larger simply from natural sampling variability.
b) Since the P-value is so high, there is no evidence that the mean memory test score for ginkgo biloba users is higher than the mean memory test score for non-users.
c) Type II

45. a) Males: (18.67, 20.11) pegs; females: (16.95, 18.87) pegs
b) It may appear to suggest that there is no difference in the mean number of pegs placed by males and females, but a two-sample *t*-interval should be constructed to assess the difference in mean number of pegs placed.
c) (0.29, 2.67) pegs

d) We are 95% confident that the mean number of pegs placed by males is between 0.29 and 2.67 pegs higher than the mean number of pegs placed by females.
e) Two-sample *t*-interval.
f) If you attempt to use two confidence intervals to assess a difference in means, you are actually adding standard deviations. But it's the variances that add, not the standard deviations. The two-sample difference of means procedure takes this into account.

47. a) $H_0: \mu_A - \mu_N = 0$; $H_A: \mu_A - \mu_N > 0$
b) We are 95% confident that the mean number of runs scored by American League teams is between 0.064 and 0.708 runs higher than the mean number of runs scored by National League teams.
c) t = 2.47; P = 0.021; reject H_0.
d) There is evidence that the mean number of runs scored per game by the AL is greater than in the NL.

49. a) $H_0: \mu_N - \mu_S = 0$; $H_A: \mu_N - \mu_S \neq 0$. t = 6.47, df = 53.49, P < 0.001
Since the P-value is low, we reject H_0. There is strong evidence that the mean mortality rate is different for towns north and south of Derby. There is evidence that the mortality rate north of Derby is higher.
b) Since there is an outlier in the data north of Derby, the conditions for inference are not satisfied, and it is risky to use the two-sample *t*-test. The outlier should be removed, and the test should be performed again.

51. a) Paired data assumption: The data are before and after job satisfaction rating for the same workers. Randomization condition: The workers were randomly selected to participate. Nearly Normal condition: A histogram of differences between before and after job satisfaction ratings is roughly unimodal and symmetric.
b) $H_0: \mu_d = 0$; $H_A: \mu_d > 0$; t = 3.60; df = 9; P-value = 0.0029; reject H_0. There is evidence that the mean job satisfaction rating has increased since the implementation of the exercise program.

53. Independent groups assumption: Assume that orders in June are independent of orders in August. Independence assumption: Orders were a random sample. Nearly normal condition: Hard to check with small sample, but no outliers. $H_0: \mu_J - \mu_A = 0$; $H_A: \mu_J - \mu_A \neq 0$; t = −1.17; P = 0.274; fail to reject H_0. Thus, although the mean delivery time during August is higher, the difference in delivery time from June is not significant. A larger sample may produce a different result.

55. a) We are 95% confident that the mean number of ads remembered by viewers of shows with violent content will be between 1.6 and 0.6 lower than the mean number of brand names remembered by viewers of shows with neutral content.
b) If they want viewers to remember their brand names, they should consider advertising on shows with neutral content, as opposed to shows with violent content.

57. a) She might attempt to conclude that the mean number of brand names recalled is greater after 24 hours.
b) The groups are not independent. They are the same people, asked at two different time periods.
c) A person with high recall right after the show might tend to have high recall 24 hours later as well. Also, the first interview may have helped the people to remember the brand names for a longer period of time than they would have otherwise.
d) Randomly assign half of the group watching that type of content to be interviewed immediately after watching, and assign the other half to be interviewed 24 hours later.

59. a) Using df ≥ 7536, we are 95% confident that the mean score in 2000 was between 0.61 and 5.39 points lower than the mean score in 1996. Since 0 is not contained in the interval, this provides evidence that the mean score has decreased from 1996 to 2000.

b) Both sample sizes are very large, which will make the standard errors of these samples very small. They are both likely to be very accurate. The difference in sample size shouldn't make you any more certain or any less certain.

61. a) The differences that were observed between the group of students with Internet access and those without were too great to be attributed to natural sampling variation.

b) Type I

c) No. There may be many other factors.

d) It might be used to market computer services to parents.

63. a) 8759 pounds

b) Independent groups assumption: Sales in different seasons should be independent. Randomization condition: Not a random sample of weeks, but it is of stores. Nearly Normal condition: Can't verify, but we will proceed cautiously. We are 95% confident that the interval 3630.54 to 13,887.39 pounds contains the true difference in mean sales between winter and summer.

c) Weather and sporting events may impact pizza sales.

65. H_0: $\mu_2 - \mu_5 = 0$; H_A: $\mu_2 - \mu_5 \neq 0$. Independent groups assumption: The two heats were independent. Randomization condition: Runners were randomly assigned. Nearly Normal condition: Boxplots show an outlier in the distribution of times in heat 2. Perform the test twice, once with the outlier and once without. With outlier: $t = 0.035$; df = 10.82; P = 0.972; fail to reject H_0. Without the outlier in heat 2: $t = -1.141$; df = 8.83; P = 0.287; fail to reject H_0. Regardless of whether the outlier is included or excluded, there is no evidence that the mean time to finish differs between the two heats.

67. H_0: $\mu_S - \mu_R = 0$; H_A: $\mu_S - \mu_R > 0$. Assuming the conditions are satisfied, it is appropriate to model the sampling distribution of the difference in means with a Student's *t*-model, with 7.03 degrees of freedom (from the approximation formula). $t = 4.57$; P = 0.0013; reject H_0. There is strong evidence that the mean ball velocity for Stinger tees is higher than the mean velocity for regular tees.

69. a) H_0: $\mu_M - \mu_R = 0$; H_A: $\mu_M - \mu_R > 0$. Independent groups assumption: The groups are not related in regards to memory score. Randomization condition: Subjects were randomly assigned to groups. Nearly Normal condition: We don't have the actual data. We will assume that the distributions of the populations of memory test scores are Normal. $t = -0.70$; df = 45.88; P = 0.7563; fail to reject H_0. There is no evidence that the mean number of objects remembered by those who listen to Mozart is higher than the mean number of objects remembered by those who listen to rap music.

b) We are 90% confident that the mean number of objects remembered by those who listen to Mozart is between 0.19 and 5.35 objects lower than the mean of those who listened to no music.

71. a) Paired data assumption: These data are paired as they are the 3 and 5 year returns of the same mutual funds. Randomization condition: random sample of funds. Nearly Normal condition: histogram of the differences shows a low outlier. We will perform the analysis with and without the outlier.

b) H_0: $\mu_5 - \mu_3 = 0$; H_A: $\mu_5 - \mu_3 > 0$

c) With the outlier, $t = 15.48$; P-value < 0.0001. Without the outlier $t = 21.31$; P-value < 0.0001. In either case we re-

ject the null hypothesis and conclude that the mean difference of 5 year − 3 year returns is greater than 0.

d) With the outlier: (4.22, 5.50); without the outlier (4.59, 5.56).

73. a) Independent groups assumption: The prices in the two towns are not related. Randomization condition: Each sample was a random sample of prices. Nearly Normal condition: Both histograms are reasonably unimodal and symmetric with no outliers, so will use the two-sample *t*-test.

b) H_0: $\mu_1 = \mu_2$; H_A: $\mu_1 \neq \mu_2$

c) $t = -0.58$; P = 0.567; fail to reject H_0.

d) We conclude that the mean price of homes in these two towns is not significantly different.

75. a) Independent groups assumption: The home runs hit in different leagues are independent. Randomization condition: Not a random sample, but we will assume it's representative. Nearly Normal condition: Both histograms are reasonably symmetric with no outliers. Unimodality is questionable, but we will proceed with caution. We will use the two-sample *t*-test.

b) H_0: $\mu_{AL} - \mu_{NL} = 0$; H_A: $\mu_{AL} - \mu_{NL} > 0$

c) $t = 2.478$; P = 0.0197

d) We can conclude that the mean number of home runs is different in the two leagues. We reject the null hypothesis.

77. a) The paired *t*-test is appropriate. The labor force participation rate for two different years was paired by city.

b) Since the P-value = 0.0244, there is evidence of a difference in the average labor force participation rate for women between 1968 and 1972. The evidence suggests an increase in the participation rate for women.

79. Adding variances requires that the variables be independent. These price quotes are for the same cars, so they are paired. Drivers quoted high insurance premiums by the local company will be likely to get a high rate from the online company, too.

81. a) The histogram—we care about differences in price.

b) Insurance cost is based on risk, so drivers are likely to see similar quotes from each company, making the differences relatively smaller.

c) The price quotes are paired; they were for a random sample of fewer than 10% of the agent's customers; the histogram of differences looks approximately Normal.

83. H_0: $\mu(\text{Local} - \text{Online}) = 0$ vs. H_A: $\mu(\text{Local} - \text{Online}) > 0$; $t = 0.826$ with 9 df. With a P-value of 0.215, we cannot reject the null hypothesis. These data don't provide evidence that online premiums are lower, on average.

85. a) H_0: $\mu_d = 0$ vs. H_A: $\mu_d \neq 0$

b) $t_{144} = \dfrac{22.7}{113.6/\sqrt{145}} = 2.406$; 2-sided P = 0.017

We are able to reject the null hypothesis (with a P-value of 0.017) and conclude that mean number of keystrokes per hour has changed.

c) 95% CI for mean keystrokes per hour
$22.7 \pm t_{0.025,\,144}\,s/\sqrt{n} = (4.05, 41.35)$.

87. a) Paired data assumption: The data are paired by type of exercise machine. Randomization condition: Assume that the men and women participating are representative of all men and women in terms of number of minutes of exercise required to burn 200 calories. Nearly Normal condition: The histogram of differences between women's and men's times is roughly unimodal and symmetric. We are 95% confident that women take an average of 4.8 to 15.2 minutes longer to burn 200 calories than men when exercising at a light exertion rate.

b) Nearly Normal condition: There is no reason to think that this histogram does not represent differences drawn from a Normal population. We are 95% confident that women exercising with light exertion take an average of 4.9 to 20.4 minutes longer to burn 200 calories than women exercising with hard exertion.

c) Since these data are averages, we expect the individual times to be more variable. Our standard error would be larger, resulting in a larger margin of error.

89. a) Randomization condition: These stops are probably representative of all such stops for this type of car, but not for all cars. Nearly Normal condition: A histogram of the stopping distances is roughly unimodal and symmetric. We are 95% confident that the mean dry pavement stopping distance for this type of car is between 133.6 and 145.2 feet.

b) Independent groups assumption: The wet pavement stops and dry pavement stops were made under different conditions and not paired in any way. Randomization condition: These stops are probably representative of all such stops for this type of car, but not for all cars. Nearly Normal condition: The histogram of wet pavement stopping distances is more uniform than unimodal, but no outliers. We are 95% confident that the mean stopping distance on wet pavement is between 51.4 and 74.6 feet longer than the mean stopping distance on dry pavement.

91. a) Yes. The percent change is computed using paired data. Same cities both years.

b) Data are paired. Differences (both actual and percent change) are roughly unimodal and symmetric. Actual differences: $H_0: \mu_d = 0$; $H_A: \mu_d \neq 0$.
Mean difference: $(3Q2007 - 3Q2006) = -10.07$; $t = -0.65$; $P = 0.535$; do not reject H_0. There is no evidence the third quarter 2006 airfares are significantly different from the third quarter 2007 airfares.
T-test on the percent change: $H_0: \mu = 0$; $H_A: \mu \neq 0$
The average percent change in airfare between third quarter 2006 and third quarter 2007 is -2.56%. Airfares were lower by 2.56% in 2007 vs 2006. $t = -0.67$; $P = 0.520$; do not reject H_0. There is no evidence the average percent change in airfare differs between the third quarter 2006 and third quarter 2007.
Both tests provide the same statistical conclusion. However, in general if there were extreme or outlying paired observations that skewed the mean difference, then the mean percent change might be a more appropriate test because it would eliminate the variability due to extreme differences.

CHAPTER 15

1. a) (30, 30, 30, 30), 30 for each season
   b) 1.933
   c) 3

3. a) 3
   b) No, it's smaller than the mean.
   c) It would say that there is no evidence to reject the null hypothesis that births are distributed uniformly across the seasons.
   d) 7.815
   e) Do not reject the null hypothesis. As in part c, there is no evidence to suggest that births are not distributed uniformly across the seasons.

5. a) $(-0.913, 0.913, 0.365, -0.365)$
   b) No, they are quite small for z-values.
   c) Because we did not reject the null hypothesis, we shouldn't expect any of the standardized residuals to be large.

7. a) The age distributions of customers at the two branches are the same.
   b) A chi-square test of homogeneity
   c)

	Age			
	Less than 30	30–55	56 or older	Total
In-town branch	25	45	30	100
Mall branch	25	45	30	100
Total	50	90	60	200

   d) 9.778
   e) 2
   f) 5.991
   g) Reject H_0 and conclude that the age distributions at the two branches are not the same.

9. a) Men 78.79%, women 75.11%
   b) Difference = 0.0368 or 3.68%
   c) SE = 0.0107
   d) (0.0159, 0.0578) or 1.59 to 5.78%

11. a)

	Did You Use the Internet Yesterday?	
	Yes	No
White	2506.38	895.62
Black	338.90	121.10
Hispanic/Other	445.73	159.27

   b) 11.176
   c) 2
   d) 5.991
   e) Reject the null hypothesis and conclude that Race and Internet use are not independent.

13. a) Chi-square test of independence; One sample, two variables. We want to see if the variable *Account type* is independent of the variable *Trade type*.
   b) Some other statistical test; the variable *Account size* is quantitative, not counts.
   c) Chi-square test of homogeneity; we have two samples (residential and nonresidential students) and one variable, *Courses*. We want to see if the distribution of *Courses* is the same for the two groups.

15. a) 10
   b) Goodness-of-fit
   c) H_0: The die is fair. (All faces have $p = 1/6$.)
      H_A: The die is not fair. (Some faces are more or less likely to come up than others.)
   d) Count data; rolls are random and independent of each other; expected frequencies are all greater than 5.
   e) 5
   f) $\chi^2 = 5.600$; $P = 0.3471$
   g) Since $P = 0.3471$ is high, fail to reject H_0. There is not enough evidence to conclude that the die is unfair.

17. a) Weights are quantitative, not counts.
   b) Count the number of each type of nut, assuming the company's percentages are based on counts rather than weights (which is not clear).

19. a) Goodness-of-fit
   b) Count data; assume the lottery mechanism uses randomization and guarantees independence; expected frequencies are all greater than 5.
   c) H_0: Likelihood of drawing each numeral is equal.
      H_A: Likelihood of drawing each numeral is *not* equal.
   d) $\chi^2 = 6.46$; df $= 9$; P $= 0.693$; fail to reject H_0.
   e) The P-value says that if the drawings were in fact fair, an observed chi-square value of 6.46 or higher would occur about 69% of the time. This is not unusual at all, so we won't reject the null hypothesis that the values are uniformly distributed. The variation that we observed seems typical of that expected if the digits were drawn equally likely.

21. a) 40.2%
   b) 8.1%
   c) 62.2%
   d) 285.48
   e) Chi-square test of independence H_0: Survival was independent of status on the ship.
      H_A: Survival was not independent of the status.
   f) 3
   g) We reject the null hypothesis. Survival depended on status. We can see that first-class passengers were more likely to survive than passengers of any other class.

23. a) Independence
   b) H_0: College choice is independent of birth order.
      H_A: There is an association between college choice and birth order.
   c) Count data; not a random sample of students, but assume that it is representative; expected counts are low for both the Social Science and Professional Colleges for both third and fourth or higher birth order. We'll keep an eye on these when we calculate the standardized residuals.
   d) 9
   e) With a P-value this low, we reject the null hypothesis. There is some evidence of an association between birth order and college enrollment.
   f) None are particularly large, but 4 cells have expected counts less than 5, so we should be cautious in making conclusions. Perhaps we should group 3 with 4 or more and redo the analysis. Unfortunately, 3 of the 4 largest standardized residuals are in cells with expected counts less than 5. We should be very wary of drawing conclusions from this test.

25. a) Chi-square test for homogeneity
   b) Count data; assume random assignment to treatments (although not stated); expected counts are all greater than 5.
   c) H_0: The proportion of infection is the same for each group.
      H_A: The proportion of infection is different among the groups.
   d) $\chi^2 = 7.776$; df $= 2$; P $= 0.02$; reject H_0.
   e) Since the P-value is low, we reject the null hypothesis. There is strong evidence of difference in the proportion of urinary tract infections for cranberry juice drinkers, lactobacillus drinkers, and women that drink neither of the two beverages.
   f) The standardized residuals are:

	Cranberry	Lactobacillus	Control
Infection	−1.87276	1.191759	0.681005
No infection	1.245505	−0.79259	−0.45291

The significant difference appears to be primarily due to the success of cranberry juice.

27. a) Independence
   b) H_0: Age is independent of frequency of shopping at this department store.
      H_A: Age is not independent of frequency of shopping at this department store.
   c) Count data; assume survey was conducted randomly (not specifically stated); expected counts are all greater than 5.
   d) Since the P-value is low, we reject the null hypothesis. There is evidence of an association between age and frequency of shopping at this department store.
   e) Given the negative residuals for the low frequency categories among the older women and the positive residuals for the higher frequency categories among the older women, we conclude that older women in this survey shop more frequently at this department store than expected.

29. a) P $= 0.3766$. With a P-value this high, we fail to reject. There is not enough evidence to conclude that either men or women are more likely to make online purchases of books.
   b) Type II.
   c) $(-4.09\%, 10.86\%)$

31. a) P < 0.001. There is strong evidence that the proportions of the two groups are not equal.
   b) $(0.096, 0.210)$ (proportion of old − proportion of young)

33. a) No, the P-value $= 0.281$.
   b) 90% confidence interval is $(-0.0019, 0.0082)$.

35. H_0: Marital status is independent of frequency of shopping.
   H_A: Marital status is not independent of frequency of shopping.
   Count data; assume survey was conducted randomly (not specifically stated); expected counts are all greater than 5.
   $\chi^2 = 23.858$; df $= 6$; P-value $= 0.001$
   Since the P-value is low, reject the null hypothesis. There is strong evidence of an association between marital status and frequency of shopping at this department store. Based on the residuals, married customers shopped at this store more frequently than expected, and more single women shopped never/hardly ever than expected.

37. a) Chi-square test for homogeneity.
   b) Count data; executives were surveyed randomly; expected counts are all greater than 5.
   c) H_0: The distribution of attitudes about critical factors affecting ethical and legal accounting practices was the same in 2000 and 2006.
      H_A: The distribution of attitudes about critical factors affecting ethical and legal accounting practices was not the same in 2000 and 2006.
   d) $\chi^2 = 4.030$; df $= 4$; P $= 0.4019$
   e) Since P-value is high, we fail to reject the null hypothesis. There is no evidence of a change in the distribution of attitudes about factors affecting ethical and legal accounting practices between 2000 and 2006.

39. a) Chi-square test of independence.
   b) Count data; assume survey was conducted randomly (not specifically stated); expected counts are all greater than 5.
   c) H_0: Emphasis on quality is independent of frequency of shopping.
      H_A: Emphasis on quality is not independent of frequency of shopping.
   d) $\chi^2 = 30.007$; df $= 6$; P < 0.001

e) Since the P-value is low, reject the null hypothesis. There is strong evidence of an association between emphasis on quality and frequency of shopping at this department store.

41. a)

	Men	Women
Excellent	6.667	5.333
Good	12.778	10.222
Average	12.222	9.778
Below Average	8.333	6.667

Count data; assume that these executives are representative of all executives that have ever completed the program; expected counts are all greater than 5.

b) Decreased from 4 to 3.

c) $\chi^2 = 9.306$; $P = 0.0255$. Since the P-value is low, we reject the null hypothesis. There is evidence that the distributions of responses about the value of the program for men and women executives are different.

43. H_0: There is no association between race and the section of the apartment complex in which people live.
H_A: There is an association between race and the section of the apartment complex in which people live.
Count data; assume that the recently rented apartments are representative of all apartments in the complex; expected counts are all greater than 5.
$\chi^2 = 14.058$; df $= 1$; $P < 0.001$
Since the P-value is low, we reject the null hypothesis. There is strong evidence of an association between race and the section of the apartment complex in which people live. An examination of the components shows us that whites are more likely to rent in Section A (component $= 6.2215$) and that blacks are more likely to rent in Section B (component $= 5.0517$).

45. $\hat{p}_B - \hat{p}_A = 0.206$
95% CI $= (0.107, 0.306)$

47. a) Chi-square test for independence.
b) Count data; assume that the sample was taken randomly; expected counts are all greater than 5.
c) H_0: Outsourcing is independent of industry sector.
H_A: There is an association between outsourcing and industry sector.
d) $\chi^2 = 2815.968$; df $= 9$; P-value is essentially 0.
e) Since the P-value is so low, we reject the null hypothesis. There is strong evidence of an association between outsourcing and industry sector.

49. a) Chi-square test for homogeneity. (Could be independence if the categories are considered exhaustive.)
b) Count data; assume that the sample was taken randomly; expected counts are all greater than 5.
c) H_0: The distribution of employee job satisfaction level attained is the same for different management styles.
H_A: The distribution of employee job satisfaction level attained is different for different management styles.
d) $\chi^2 = 178.453$; df $= 12$; P-value is essentially 0.
e) Since the P-value is so low, we reject the null hypothesis. There is strong evidence that the distribution of employee job satisfaction level attained is different across management styles. Generally, exploitative authoritarian management is more likely to have lower levels of employee job satisfaction than consultative or participative styles.

51. Assumptions and conditions for test of independence satisfied.
H_0: Reading online journals or blogs is independent of generation. H_A: There is an association between reading online journals or blogs and generation. $\chi^2 = 48.408$; df $= 8$; $P < .001$.

We reject the null hypothesis and conclude that reading online journals or blogs is not independent of generational age.

53. Chi-square test of homogeneity (unless these two types of firms are considered the only two types in which case it's a test of independence).
Count data; assume that the sample was random; expected counts are all greater than 5.
H_0: Systems used have same distribution for both types of industry.
H_A: Distributions of type of system differs in the two industries.
$\chi^2 = 157.256$; df $= 3$; P-value is essentially 0.
Since the P-value is low, we can reject the null hypothesis and conclude that the type of ERP system used differs across industry type. Those in manufacturing appear to use more of the inventory management and ROI systems.

55. Chi-square test of independence.
Count data; we assume that this time period is representative; expected counts are less than 5 in three cells, but very close.
H_0: GDP % change is independent of Region of the United States.
H_A: GDP % change is not independent of Region of the United States.
$\chi^2 = 19.0724$; df $= 3$; $P < 0.001$.
The P-value is low so we reject the null hypothesis and conclude that GDP % change and Region are not independent. The result seems clear enough even though the expected counts of 3 cells were slightly below 5.

CHAPTER 16

1. People in this community spend about $17.28 plus 0.354 for each $1000 of household income eating out each week.

3. a) Linearity: the plot appears linear.

b) Independence: It is likely that the customers are independent. It is likely that errors made by the regression will be independent.
c) Equal spread: the plot appears to have a consistent spread.
d) Normality: We can't tell from what we know.

5. a) 9.8, 8.5, 5.9, 2.0, 8.5, 11.1
b) $-0.8, -0.5, -1.9, 1.0, 1.5, 0.9$
c) 1.46

7. $s_x = 12.5167$, $s_e = 1.46$, $n = 6$, $SE(b_1) = 0.052$

9. a) 5.837
b) 3
c) 0.01
d) Yes, the P-value is smaller than 0.05.

11. a) H_0: $\rho = 0$
H_A: $\rho \neq 0$
b) $t_{96} = 2.0$
c) The t statistic is just barely larger than the critical value. We can reject the null hypothesis.

13. a) 19.07 ($1000)
b) $19.07 \pm 1.12 = (17.94, 20.19)$ in $1000
c) $19.07 \pm 3.59 = (15.48, 22.66)$ in $1000's

15. a) 95% of potential customers in the community from households that have annual incomes of $80,000 can be expected to spend between $35.60 and $55.60 eating out each week. Pricing an individual meal in this range might attract that demographic.
b) We can be 95% confident that the mean amount spent weekly by individuals whose household incomes are $80,000 is between $40.60 and $50.60. The mean is a summary for this demographic.
c) We can estimate means more accurately than we can estimate values for individuals. That's why the confidence interval for the mean is narrower than the corresponding interval for an individual.

17. a) 6.58
b) (4.705, 8.413) songs
c) (2.025, 11.093) songs
d) Prediction interval is wider than the confidence interval because of greater uncertainty when predicting an individual value versus a mean value.

19. a)
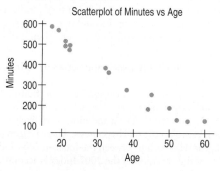

b) This scatterplot appears to have curvature at both ends of the age distribution so a linear regression may not be completely appropriate.
c) The regression equation is $\widehat{Minutes} = 750 - 11.5$.
d) The residual plots are:

The nearly normal condition is satisfied. There may be some curvature to the residual plot.

21. a) $\widehat{Budget} = -31.39 + 0.71\ Run\ Time$. The model suggests that movies cost about $710,000 per minute to make.
b) A negative starting value makes no sense, but the P-value of 0.07 indicates that we can't discern a difference between our estimated value and zero. The statement that a movie of zero length should cost $0 makes sense.
c) Amounts by which movie costs differ from predictions made by this model vary, with a standard deviation of about $33 million.
d) 0.15 $m/min
e) If we constructed other models based on different samples of movies, we'd expect the slopes of the regression lines to vary, with a standard deviation of about $150,000 per minute.

23. a) The scatterplot looks straight enough, the residuals look random and roughly normal, and the residuals don't display any clear change in variability although there may be some increasing spread.
b) I'm 95% confident that the cost of making longer movies increases at a rate of between 0.41 and 1.01 million dollars per minute. (CI is 0.41 to 1.02 using raw data.)

25. a) H_0: There is no linear relationship between calcium concentration in water and mortality rates for males. ($\beta_1 = 0$)
H_A: There is a linear relationship between calcium concentration in water and mortality rates for males. ($\beta_1 \neq 0$)
b) $t = -6.65$, P < 0.0001; reject the null hypothesis. There is strong evidence of a linear relationship between calcium concentration and mortality. Towns with higher calcium concentrations tend to have lower mortality rates.
c) For 95% confidence, use $t^*_{59} \approx 2.001$, or estimate from the table $t^*_{50} \approx 2.009$; $(-4.20, -2.26)$.
d) We are 95% confident that the average mortality rate decreases by between 2.27 and 4.19 deaths per 100,000 for each additional part per million of calcium in drinking water.

27. a) $\widehat{male\ rate} = 2.376 + 0.75564 \times (female\ rate)$.
b) The scatterplot shows a high-leverage point. The residual plot suggests that it may not fit with the other points, so the regression should be run both with and without that point to examine its impact.
c) H_0: There is no linear relationship between male and female unemployment rate. ($\beta_1 = 0$)
H_A: There is a linear relationship between male and female unemployment rate. ($\beta_1 \neq 0$)
$t = 13.46$, df = 55, P < 0.001; reject the null hypothesis. There is strong evidence of a positive linear relationship between the male and female unemployment rate.
d) 76.7%

29. a)
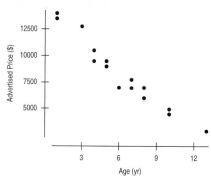

b)   Yes, the plot seems linear.

c) $\overline{Advertised\ Price} = 14286 - 959 \times Age$

d)

The residual plot shows some possible curvature. Inference may not be valid here, but we will proceed (with caution).

31. Based on these data, we are 95% confident that a used car's *Price* decreases between $819.50 and $1098.50 per year.

33. a)

$\widehat{mpg} = 48.7393 - 8.21362 \times (weight)$

b)   Yes, the conditions seem satisfied. Histogram of residuals is unimodal and symmetric; residual plot looks okay, but some "thickening" of the plot with increasing values. There may be one possible outlier.

c) H_0: There is no linear relationship between the weight of a car and its fuel efficiency. ($\beta_1 = 0$)
     H_A: There is a linear relationship between the weight of a car and its mileage. ($\beta_1 \neq 0$)

d) $t = -12.2$, df $= 48$, $P < 0.0001$; reject the null hypothesis. There is strong evidence of a linear relationship between weight of a car and its mileage. Cars that weigh more tend to have lower gas mileage.

35. a) H_0: There is no linear relationship between SAT Verbal and Math scores. ($\beta_1 = 0$)
     H_A: There is a linear relationship between SAT Verbal and Math scores. ($\beta_1 \neq 0$)

b)   Assumptions seem reasonable, since conditions are satisfied. Residual plot shows no patterns (one outlier); histogram is unimodal and roughly symmetric.

c) $t = 11.9$, df $= 160$, $P < 0.0001$; reject the null hypothesis. There is strong evidence of a linear relationship between SAT Verbal and Math scores. Students with higher SAT Verbal scores tend to have higher SAT Math scores.

37. a) H_0: There is no linear relationship between team salaries and number of wins. ($\beta_1 = 0$)
     H_A: There is a linear relationship between team salaries and wins. ($\beta_1 \neq 0$)

b) $t = 1.58$, $P = 0.124$, we fail to reject the null hypothesis. There is no evidence of a linear relationship between team salary and team wins in 2006.

39. a) $(-9.57, -6.86)$ mpg per 1000 pounds

b)   We are 95% confident that the mean mileage of cars decreases by between 6.86 and 9.57 miles per gallon for each additional 1000 pounds of weight.

41. a) H_0: There is no linear relationship between market return and fund flows. ($\beta_1 = 0$)
     H_A: There is a linear relationship between market return and fund flows. ($\beta_1 \neq 0$)

b) $t = 5.75$, $P < 0.001$; reject the null hypothesis. There is strong evidence of a linear relationship between money invested in mutual funds and market performance.

c)   Greater investment in mutual funds tends to be associated with higher market return. I would suggest investigating the unusual observations and finding out when and why they occurred.

43. a) H_0: There is no linear association between the Index in 2006 and 2007. ($\beta_1 = 0$)
     H_A: There is a linear association between the Index in 2006 and 2007. ($\beta_1 \neq 0$)

b) $t = 8.17$; $P < 0.001$. The association is not likely to occur by chance. We reject the null hypothesis. There is strong evidence of a linear relationship between the Index in 2006 and 2007.

c)   83.7% of the variation in the 2007 Index is accounted for by the 2006 Index.

d)   On average, as one increases, so does the other. However, this does not mean if the index is high in one year, it will necessarily be high in the other.

45. a)

The conditions for regression fail, since the scatterplot bends.

b)   No, the relationship does not appear to be linear.

47. H_0: There is no linear relationship between speed and cost of printers. ($\beta_1 = 0$)
H_A: There is a linear relationship between speed and cost of printers. ($\beta_1 \neq 0$)

Assumptions do not seem reasonable. Scatterplot isn't straight. The histogram of the residuals is skewed to the right. The conditions are not met, so inference is not appropriate.

49. a) We are 95% confident that the mean fuel efficiency of cars that weigh 2500 pounds is between 27.34 and 29.07 miles per gallon.
b) We are 95% confident that a car weighing 3450 pounds will have fuel efficiency between 15.44 and 25.37 miles per gallon.

51. a) $(-3145, 40845)$
b) Since the SE for the slope is relatively large and the R^2 is relatively small (18%), predictions using this regression will be imprecise.

53. a) We'd like to know if there is a linear association between *Price* and *Fuel Efficiency* in cars. We have data on 2004 model-year cars giving their highway mpg and retail price. H_0: $\beta_1 = 0$ (no linear relationship between *Price* and *Fuel Efficiency*); H_A: $\beta_1 \neq 0$.
b) The scatterplot fails the Straight Enough Condition. It shows a bend and it has an outlier. There is also some spreading from right to left, which would violate the Plot Thickens Condition. We cannot continue the analysis.
c) The conditions are not met; the regression equation should not be interpreted.

55. a) The 95% prediction interval shows the interval of uncertainty for a single predicted male unemployment rate, given a specific female unemployment rate.
b) The 95% confidence interval shows the interval of uncertainty for the mean male unemployment rate given a sample of female unemployment rates. Because this is an interval for an average, the variation or uncertainty is less, so the interval is narrower.

c) The unusual observation is the former Yugoslav Republic of Macedonia. Besides being an outlier, it is also a potential leverage value because its female rate is so removed from the average female rate for this sample of countries. Without this leverage value, $\widehat{male\ rate} = 4.459 + 0.62951 \times (female\ rate)$. Note that the slope remains significant (P $<$ 0.001), and the R^2 decreases slightly compared to the output in Exercise 27 without this leverage value.

57. a) The 95% prediction interval shows the interval of uncertainty for the predicted energy use in 2004 based on energy use in 1999 for a single country.
b) The 95% confidence interval shows the interval of uncertainty for the mean energy use in 2004 based on the same energy use in 1999 for a sample of countries. Because this is an interval for an average, the variation or uncertainty is less, so the interval is narrower.

59. a) $\widehat{Jan} = 120.73 + 0.6695 \times Dec$. We are told this is an SRS. One cardholders' spending should not affect another's. These are quantitative data with no apparent bend in the scatterplot. The residual plot shows some increased spread for larger values of January charges. A histogram of residuals is unimodal and slightly skewed to the right with several high outliers. We will proceed cautiously.
b) $1519.73
c) ($1330.24, $1709.24)
d) ($290.76, $669.76)
e) The residuals show increasing spread, so the confidence intervals may not be valid. I would be skeptical of interpreting them too literally.

61. a) H_0: There is no linear relationship between population and ozone level. ($\beta_1 = 0$)
H_A: There is a positive linear relationship between population and ozone level. ($\beta_1 > 0$)
$t = 3.48$, P $= 0.0018$; reject the null hypothesis. There is strong evidence of a positive linear relationship between ozone level and population. Cities with larger populations tend to have higher ozone levels.
b) City population is a good predictor of ozone level. Population explains 84% of the variability in ozone level and s is just over 5 parts per million.

CHAPTER 17

1. The different segments are not scattered at random throughout the residual plot. Each segment may have a different relationship.

3. Yes, it is clear that the relationship between January and December spending is not the same for all five segments. Using one overall model to predict January spending would be very misleading.

5. a) 464.8 ($000)
b) $465.1 \pm 98.63 = (366.47, 563.73)$ in $1000
c) No. Extrapolating what their sales might be with 500 employees based on data with between 2 and 20 employees working is likely to be very wrong.

7. a) $354,472
b) An extrapolation this far from the data is unreliable.

9. a) $\widehat{Sales} = 14.32 + 0.297\ Number\ of\ Sales\ People\ Working$
b) The intercept is higher and the slope is much lower.
c) High leverage and influential.
d) Not really. The line is so influenced by the new point that it passes close to it.

11. a)

Quarter	Sales ($M)	Predicted	Residual
1	12	19.02	−7.02
2	41	23.89	17.11
3	15	28.76	−13.76
4	48	33.63	14.37
5	25	38.5	−13.5
6	55	43.37	11.63
7	23	48.24	−25.24
8	69	53.11	15.89
9	51	57.98	−6.98
10	80	62.85	17.15
11	54	67.72	−13.72
12	87	72.59	14.41
13	64	77.46	−13.46
14	94	82.33	11.67
15	62	87.2	−25.2
16	108	92.07	15.93

b)

c) $D = 3.62$
d) 1.08, 1.36
e) No. $D > 1.36$
f) Yes. $D > 4 − 1.08$

13. a) $d_L = 1.48$, $d_U = 1.57$. The observed value is less than d_L so we can reject the null hypothesis of no autocorrelation.
b) The statistic tests for time-related correlation. We can conclude that there is evidence of positive autocorrelation.

15. No. The scatterplot shows a linear relationship and the residual plot has no obvious pattern. No transformation is suggested.

17. A re-expression is called for. Taking the log or reciprocal of *Cost per Unit* would probably improve the linearity.

19. a) Log
b) 0

21. a) The trend appears to be somewhat linear up to about 1940, but from 1940 to about 1970 the trend appears to be nonlinear. From 1975 or so to about 1995, the trend appears to be linear.
b) Relatively strong for certain periods.
c) No, as a whole the graph is clearly nonlinear. Within certain periods (1975 to about 1995) the correlation is high.
d) Overall, no. You could fit a linear model to the period from 1975 to about 1995, but why? You don't need to interpolate, since every year is reported, and extrapolation seems dangerous.

23. a) The relationship is not straight.
b) It will be curved downward.
c) No. The relationship will still be curved.

25. a) No. We need to see the scatterplot first to see if the conditions are satisfied.
b) No, the linear model might not fit the data at all.

27. a) Millions of dollars per minute of run time.
b) Budgets for dramas increase at the same rate per minute as other movies.
c) Regardless of run time, dramas cost less by about $20 million.

29. a) The use of Oakland airport has been growing at about 59,700 passengers/year, starting from about 283,000 in 1990.
b) 71% of the variation in passengers is accounted for by this model.
c) Errors in predictions based on this model have a standard deviation of 104,330 passengers.
d) DW = 0.912; Strong evidence of autocorrelation.
e) No, that would extrapolate too far from the years we've observed, and the series is autocorrelated.
f) The negative residual is September 2001. Air traffic was artificially low following the attacks on 9/11.

31. a) 1) High leverage, small residual.
2) No, not influential for the slope.
3) Correlation would decrease because outlier has large z_x and z_y, increasing correlation.
4) Slope wouldn't change much because the outlier is in line with other points.
b) 1) High leverage, probably small residual.
2) Yes, influential.
3) Correlation would weaken and become less negative because scatter would increase.
4) Slope would increase toward 0, since outlier makes it negative.
c) 1) Some leverage, large residual.
2) Yes, somewhat influential.
3) Correlation would strengthen, since scatter would decrease.
4) Slope would increase slightly.
d) 1) Little leverage, large residual.
2) No, not influential.
3) Correlation would become stronger and become more negative because scatter would decrease.
4) Slope would change very little.

33. 1) e   2) d   3) c   4) b   5) a

35. Perhaps high blood pressure causes high body fat, high body fat causes high blood pressure, or both could be caused by a lurking variable such as a genetic or lifestyle issue.

37. a) *Cost* decreases by $2.13 per degree of average daily *Temp*. So warmer temperatures indicate lower costs.
b) For an avg. monthly temperature of 0°F, the cost is predicted to be $133.
c) Too high; the residuals (observed − predicted) around 32°F are negative, showing that the model overestimates the costs.
d) $111.7
e) About $105.7
f) No, the residuals show a definite curved pattern. The data are probably not linear.
g) No, there would be no difference. The relationship does not depend on the units.

39. a) 0.88
b) Interest rates during this period grew at about 0.25% per year, starting from an interest rate of about 0.64%.
c) Substituting 50 in the model yields a predicted value of about 13%.
d) DW = 0.9527; P-value = 0.002; Strong evidence of autocorrelation.
e) Not really. Extrapolating 20 years beyond the end of these data would be dangerous and unlikely to be accurate. Errors are autocorrelated as well.

41. a) The two models both fit well, but they have very different slopes.

b) This model predicts the interest rate in 2000 to be 3.24%, much lower than the other model predicts.

c) We can trust the new predicted value because it is in the middle of the data used for the regression.

d) The best answer is "I can't predict that."

43. *Sex* and *Colorblindness* are both categorical variables, not quantitative. Correlation is meaningless for them, but we can say that the variables are associated.

45. a) No re-expression needed.
b) Re-express to straighten the relationship.
c) Re-express to equalize spread.

47. a) There's an annual pattern, so the residuals cycle up and down.

b) No, this kind of pattern can't be helped by re-expression.

49. a) 2.8   b) 16.44   c) 7.84   d) 0.36   e) 2.09

51. a) 3.842   b) 501.187   c) 4.0

53. a) Linearity condition violated. Possibly increased variation at higher values. Independence may be in doubt since this is a time series.

b) This plot looks both straighter and more consistent in variation.

c) The residuals show a pattern, but it tracks up and down and could not be improved by an alternative transformation.

55. a) About $100,000,000
b) *logValue* grows at 0.00015 per licensed *Fisher*. Possibly more fishers cause a more valuable harvest, but greater value would likely attract more fishers, so the causation might run the other way. Or improving technology could both lead to more value and attract more fishers.

57. a) Although more than 97% of the variation in GDP can be accounted for by this model, we should examine a scatterplot of the residuals to see if it's appropriate.

b) No. The residuals show clear curvature.

59. a)

$$\widehat{\sqrt{Bdft}} = -4 + diam$$

The model is exact.

b) 36 board feet.
c) 1024 board feet.

61. Dependent variable is: 1998 GDP/Cap
R-squared = 94.2%   R-squared (adjusted) = 93.9%
s = 2803 with 24 − 2 = 22 degrees of freedom

Source	Sum of Squares	df	Mean Square	F-ratio
Regression	2800375675	1	2800375675	356
Residual	172861656	22	7857348	

Variable	Coeff	SE(Coeff)	t-ratio	P-value
Intercept	3295.80	1307	2.52	0.0194
88 GDP CAP	1.090	0.058	18.9	<0.0001

The regression has a high R^2 but the residuals show two influential points, Ireland with the largest positive residual and Switzerland with the largest negative residual. Assuming the data for these countries are correct, they should be set aside from the analysis and further work done to understand why each is unusual. Reports both with and without the outliers should be compared.

63. The histogram is much more unimodal and symmetric after we take logs.

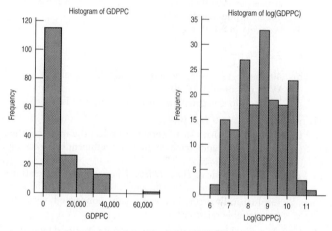

65. The plot cannot be straightened by a transformation because it is not a simple monotonically increasing or decreasing shape, but rather it oscillates up and down.

CHAPTER 18

1. a) $99,859.89
b) $35,140.11
c) The house sold for about $35,100 more than our estimate.

3. a) $\widehat{USGross} = -222.9898 + 1.13442 Budget + 24.9724 Stars - 0.403296 RunTime$

b) After allowing for the effects of *RunTime* and *Stars*, each million dollars spent making a film yields about 1.13 million dollars in gross revenue.

5. a) Linearity: The plot is reasonably linear with no bends.
b) Equal spread: The plot fails the Equal Spread condition. It is much more spread out to the right than on the left.
c) Normality: A scatterplot of two of the variables doesn't tell us anything about the distribution of the residuals.

7. a) $H_0: \beta_{Stars} = 0$
b) $t = 4.24$
c) $P \le 0.0001$
d) Reject the null hypothesis and conclude that the coefficient of *Stars* is not zero.

9. a) $R^2 = 0.474$ or 47.4%

About 47.4% of the variation in *USGross* is accounted for by the least squares regression on *Budget*, *RunTime*, and *Stars*.

b) Adjusted R^2 accounts for the number of predictors, so it differs from R^2, which does not.

11. a) Linearity: The scatterplots appear linear.

Independence: States are not a random sample, but they may be independent of each other.

Equal variance: The scatterplot of *Violent Crime* vs. *Police Officer Wage* may become less spread to the right, or maybe that's just fewer data points.

Normality: To check the Nearly Normal condition, we'll need to look at the residuals; we can't check it with these plots.

b) 1.1%

13. a) $\widehat{Violent\ Crime} = 1390.83 + 9.33\ Police\ Officer\ Wage - 16.64\ Graduation\ Rate$

b) After allowing for the effects of graduation rate (or, alternatively, among states with similar graduation rates), states with higher police officer wages have more crime at the rate of 9.3 crimes per 100,000 for each dollar per hour of average wage.

c) 412.63 crimes per 100,000

d) Not very good; R^2 is only 53%.

15. a) $2.26 = \dfrac{9.33}{4.125}$

b) 40, the degrees of freedom is 37, and that's equal to $n - k - 1$. With two predictors, $40 - 2 - 1 = 37$.

c) The *t*-ratio is negative because the coefficient is negative.

17. a) $H_0: \beta_{Officer} = 0$ $H_A: \beta_{Officer} \neq 0$

b) P = 0.0297; that's small enough to reject the null hypothesis at $\alpha = .05$ and conclude that the coefficient is discernibly different from zero.

c) The coefficient of *Police Officer Wage* reports the relationship *after allowing for the effects* of *Graduation Rate*. The scatterplot and correlation were only concerned with the two-variable relationship.

19. This is a causal interpretation, which is not supported by regression. For example, among states with high graduation rates, it may be that those with higher violent crime rates choose (or are obliged) to spend more to hire police officers, or that states with higher costs of living must pay more to attract qualified police officers but also have higher crime rates.

21. Constant variance condition: met by the residuals vs. predicted plot

Nearly Normal condition: met by the Normal probability plot

23. a) Doesn't mention other predictors; suggests direct relationship

b) Correct

c) Can't predict *x* from *y*

d) Incorrect interpretation of R^2

25. a) Extrapolates far from the data

b) Suggests a perfect relationship

c) Can't predict one explanatory variable from another

d) Correct

27. a) The sign of the coefficient for ln(*number of employees*) is negative. This means that for businesses that have the same amount of sales, those with more employees spend less per employee on pollution abatement on average. The sign of the coefficient for ln(*sales*) is positive. This means that for businesses with the same number of employees, those with larger sales spend more on pollution abatement on average.

b) The logarithms mean that the effects become less severe (in dollar terms) as companies get larger either in sales or in number of employees.

29. a) $\widehat{Price} = -152{,}037 + 9530\ Baths + 139.87\ Area$

b) $R^2 = 71.1\%$

c) For houses with the same number of bathrooms, each square foot of area is associated with an increase of $139.87 in the price of the house, on average.

d) The regression model says that for houses of the same size, those with more bathrooms are not priced higher. It says nothing about what would happen if a bathroom were added to a house. That would be a predictive interpretation, which is not supported by regression.

31. a) The regression equation is: $\widehat{Salary} = 9.788 + 0.110\ Service + 0.053\ Education + 0.071\ Test\ Score + 0.004\ Typing\ wpm + 0.065\ Dictation\ wpm$

b) $Salary = 29.205$ or $29,205

c) The *t*-value is 0.013 with 24 df. P-value = .4949, which is not significant at $\alpha = 0.05$.

d) Take out the explanatory variable for typing speed since it is not significant.

e) *Age* is likely to be collinear with several of the other predictors already in the model. For example, secretaries with longer terms of *Service* will probably also be older.

33. a) After allowing for the effects measured by the Wilshire 5000 index, an increase of one point in the *Unemployment Rate* is associated with a decrease on average of about $3719.55 million in the *Funds flow*.

b) The *t*-ratio divides the coefficient by its standard error. The coefficient here is negative.

c) $H_0: \beta = 0$, $H_A: \beta \neq 0$; P-value < 0.0001 is very small, so reject the null hypothesis.

35. a) $\widehat{Price} = 0.565 + 1.3\ Traps - 0.00012\ Fishers - 9.98\ Catch/Trap$

b) Residuals show no pattern and have equal spread. Histogram of residuals is nearly Normal. We might wonder whether values from year to year are mutually independent.

c) $H_0: \beta_{catch/trap} = 0$, $H_A: \beta_{catch/trap} \neq 0$; the P-value is 0.3781. It appears that *Catch/Trap* does not contribute to the model.

d) Adjusted R^2 says to prefer the more complex model.

37. a)
```
R squared = 66.7%  R squared (adjusted) = 63.8%
s = 2.327 with 39 - 4 = 35 degrees of freedom
```

Variable	Coeff	SE(Coeff)	t-ratio	P-value
Intercept	87.0089	33.60	2.59	0.0139
CPI	−0.344795	0.1203	−2.87	0.0070
Personal Consumption	1.10842e−5	0.0000	2.52	0.0165
Retail Sales	1.03152e−4	0.0000	6.67	≤0.0001

$\widehat{Revenue} = 87.0 - 0.344\ CPI + 0.000011\ Percsonal\ Consumption + 0.0001\ Retail\ Sales$

b) R^2 is 66.7%, and all *t*-ratios are significant. It looks like these variables can account for much of the variation in Wal-Mart revenue.

39. a) $\widehat{Logit(Right\text{-}to\text{-}work)} = 6.19951 - 0.106155\ publ - 0.222957\ pvt$

b) 1.847

c) 0.864

d) −2.506

e) 0.075

41. *Displacement* and *Bore* would be good predictors. Relationship with *Wheelbase* isn't linear.

43. a) Yes, R^2 of 90.9% says that most of the variability of *MSRP* is accounted for by this model.
b) No, a regression model may not be inverted in this way.

45. a) Yes, R^2 is very large.
b) The value of *s*, 3.140 calories, is very small compared with the initial standard deviation of *Calories*. This indicates that the model fits the data quite well, leaving very little variation unaccounted for.
c) A true value of 100% would indicate zero residuals, but with real data such as these, it is likely that the computed value of 100% is rounded up from a slightly lower value.

CHAPTER 19

1. a) Unionized = 1, No Union = 0 (or the reverse)
b) Female = 1, Male = 0 (or the reverse)
c) Paid on time = 1, Past Due = 0 (or the reverse)
d) Two variables. One variable could have Democrat = 1, 0 otherwise and another Republican = 1 and 0 otherwise.

3. a) Parallel or nearly parallel slopes for the two groups with different levels suggest that an indicator would be appropriate.
b) Comedy = 1, Action = 0

5. (One possible answer for each: other equivalent codings are possible.)
a) Three variables: One variable that is 1 for Apartment, 0 for others. One that is 1 for Condo, 0 for others. One that is 1 for Townhouse, 0 for others.
b) Two variables: One that is 1 for Full-time, 0 for others. One that is 1 for Part-time, 0 for others.

7. a) Slopes that are not parallel
b) Comedy = 1, Adventure = 0
c) The interaction term is the product of the indicator variable and the Budget variable.

9. $0.189M. This is the definition of leverage.

11. Because of the high correlation between several of the predictor variables, a stepwise search might find a very different model from the best of the all subsets model.

13. 80.00

15. a) It is curved upward at both extremes of *Age*. It says that the most expensive homes are either quite new or very old.
b) No. Although the relationship is not linear in *Age*, it is probably quadratic.
c) Add a quadratic term in (Age^2).

17. a) Cheese pizzas (*Type* 1) scored about 15.6 points higher, after allowing for the effects of calories and fat. Cheese pizzas can be expected to sell better than pepperoni pizzas based on these results.
b) Scatterplots of the data and of residuals to check linearity and constant variance; a histogram or probability plot of residuals to check normality.

19. a) Both pizzas are predicted to have high scores. But both have scores more than 30 points lower than predicted.
b) Yes, their unusually large predicted values indicate that they must be different from the other pizzas in some ways, and their large (negative) residuals indicate that they would pull on the regression. It might be helpful to see leverage values or Cook's Distance values for these points.

21. a) An indicator or "dummy" variable
b) Sales are about 10.5 billion dollars higher in December, after accounting for the *CPI*.
c) Must assume that the slope is the same for the December points as for the others. That appears to be true in the scatterplot.

23. a) There are different slopes for the two types.
b) This is an interaction term. It says that cheese pizzas (*Type* 1) have scores that grow less rapidly with *Calories* (a slope 0.46 smaller) than do pepperoni pizzas.
c) The adjusted R^2 for this model is higher than the adjusted R^2 for the previous model. Also, the *t*-ratios are larger. Overall, this looks like a more successful regression model.

25. a) Large Cook's Distance suggests that Alaska is influential. Its leverage is high, and its residual is one of the lowest.
b) After allowing for the other predictors, Alaska's life expectancy is 2.23 years lower than the model predicts. The P-value of 0.02 says that Alaska is an outlier because the coefficient of the dummy variable is significantly not zero.
c) The model is better (higher adjusted R^2, smaller standard deviation of residuals) when Alaska is removed from the model with a dummy variable.

27. The *Value* for 1994 is not particularly low, but apparently, judging from its Studentized residual, it was unexpectedly low for the number of traps and fishers and the water temperature. Any point with a large Cook's Distance is potentially influential, so we might want to set 1994 aside to see whether that changes the model.

29. a) R^2 adjusted accounts for the number of predictors. If removing a predictor causes it to increase, then that predictor contributed little to the model. *Primary Education* would best be omitted from the model.
b) Whether the P-value for *OECD Regulation* is technically less than 0.05 or greater than 0.05 depends on including *Primary Education* in the model. We might be suspicious that the author, in his desire to claim that regulation harms *GDP/Capita*, has found an irrelevant predictor that has the effect of nudging the P-value just below .05.

31. a) With a P-value of 0.686, we would not conclude that the true coefficient is not zero. It cannot be interpreted.
b, c) The R^2 of the second regression shows that *Diverted* is linearly related to the other predictors and thus suffers from collinearity.
d) 6.9

33. a) High Cook's Distance values indicate that the points may be influential. The points have both unusually low predicted values and large residuals, so this seems likely.
b) Two of the coefficients have not changed very much, but the coefficient of *#Shows* has changed from negative but not significant to positive and strongly significant. A point is influential if removing it from the model changes any coefficients in important ways.
c) The second model would be better. The first one is influenced by extraordinary events, which are not the kinds of weeks we'd want to model.

35. a) The linearity condition is violated. Because the curve is not monotonic (consistently rising or falling), no re-expression can help.
b) About 24,087 tonnes.
c) *Traps* and *Traps^2* are highly correlated. The collinearity accounts for the change in the coefficient of *Traps*.

CHAPTER 20

1. a) Yes
   b) No, not vs. time
   c) No, for each worker, not over time
   d) Yes

3. a)

   b) Trend: positive, slightly curved, strong.
   c) No evidence of a seasonal component.

5. a)

*
0.97
0.9560
0.9465
0.9895
1.0370
1.0995
1.1905
1.2455
1.1970
1.1135
1.0580

   b) 1.058

7. 2.32% (If your value was negative, you forgot the absolute value.)

9.

*
0.96300000
0.97700000
0.93500000
0.95800000
1.0210000
1.0530000
1.1460000
1.2350000
1.2560000
1.1380000
1.0890000

11. 1.0012

13. a) Dummy or indicator variable.
   b) A set of indicator variables must omit one to avoid collinearity.
   c) No

15. a) The 200-day moving average will be smoother.
   b) The SES model using $\alpha = 0.10$ is smoother.

c) When the α is raised from 0.10 to 0.80, the most recent data point is weighted more heavily, so the model responds more quickly and is not as smooth.

17. a) Random walk, so use AR(1) or naive forecast.
   b) Exponential model
   c) Seasonal dummy variable model, or AR model that uses sales from the same quarter in prior years (lag4).

19. a) 22,988 $M
   b) 21,239 $M

21. a) 475.88 $/tonne
   b) 448.37 $/tonne
   c) 0.1753, 0.2229

23. a) Irregular
   b) Second graph

25. The 2-quarter moving average is shown in A and fits better. This series has a strong, consistent trend component, so a longer-span moving average is dragged down below the series.

27. a) $2.676B
   b) APE = 7.7% (underforecast)
   c) Forecast is $2.975B; APE = 2.6% (overforecast).

29. a) 86.75
   b) 0.115%
   c) 84.24; the AR model prediction is closer.

31. a) 112
   b) The forecast will be over 100%, which is not possible.

33. a) Linear trend and irregular components
   b) Prices have increased at about 30 cents per year.
   c) This intercept has no interpretation because year zero has no meaning for this series.
   d) There is increased variability starting in the middle of 2005.

35. a) $1.1 million
   b) Q3
   c) Q4
   d) Sales on average are $0.5 million greater in Q4 than in Q1.

37. a) A positive trend component and a seasonal component
   b) Wal-Mart revenues have been increasing at about $0.145 billion per month.
   c) Revenues in December tend to be about $11.56 billion more than in January (the base month).
   d) $19.3028 billion
   e) October is the only month in which Wal-Mart revenues are typically lower than they are in January, after allowing for the overall growth trend in revenue.

39. a) The P-value indicates that the time trend is significantly different from zero.
   b) September and November
   c) July and August
   d) 459,408.22

41. a) The log of the number of passengers grew at 0.0249 per month.
   b) In January 1990 there were about $10^{5.64949} = 446,159$ passengers.
   c) January and February; we can tell that January is low because it is the base for the dummy variables and the other coefficients are positive.
   d) Trend, seasonal, and possibly an 8-year cyclic component.

43. The single exponential smooth is pulled down by the outlier. The moving average spreads out the effect of the outlier across several months. The seasonal regression is not noticeably affected. (The September dummy coefficient absorbs the effect and spreads it out over other Septembers, but that isn't evident in the plot.)

45. a)
```
Dependent variable is: E-Commerce Sales (Millions $)
R squared = 96.9%   R squared (adjusted) = 96.4%
s = 1740 with 34 − 5 = 29 degrees of freedom

Variable    Coeff     SE(Coeff)  t-ratio  P-value
Intercept   2022.8    777.4      2.60     0.014
Time        900.138   30.45      29.6     ≤0.0001
Q2          −486.223  845.9      −0.575   0.5699
Q3          −709.611  845.4      0.839    0.4081
Q4          3288.03   820.7      4.01     0.0004
```

Note: Choosing a different quarter as the base quarter will yield a different model, but one with the same R^2.

b)
```
Dependent variable is: logE-commerce Sales
R squared = 98.7%   R squared (adjusted) = 98.5%
s = 0.0297 with 34 − 5 = 29 degrees of freedom

Variable    Coeff        SE(Coeff)  t-ratio  P-value
Intercept   3.73222      0.0136     274      ≤0.0001
Time        0.024502     0.0005     47.1     ≤0.0001
Q2          −2.63785e-3  0.0145     −0.182   0.8566
Q3          −5.43049e-3  0.0145     −0.376   0.7099
Q4          0.072158     0.0140     5.14     ≤0.0001
```

c) The multiplicative model fits better judging from R^2 or adjusted R^2.

47. a) The regression equation is:
U.S. Retail Gas Price (cents/gal) = 0.253 + 1.40 lag1 − 0.400 lag2

b) Forecast is $3.27.

c)

Model	Forecast ($)	Actual	APE (%)
Linear	2.38	3.19	25.4
Exponential	2.43	3.19	23.8
AR(2)	3.27	3.19	2.5

d) The AR(2) model does the best of those compared here.

49. a–c)

Model	Forecast ($)	Actual	APE (%)
Linear	60.95	58.70	3.8
Exponential	65.94	58.70	12.3
AR(1)	53.07	58.70	9.6

d) The linear model seems to fit best.

51. a) Only the irregular component; this is most likely a random walk, which is one reason these models are ineffective.

b) See sample computer output in Solutions.

c) See sample computer output in Solutions.

d) and e)

Model	MAPE	Forecast ($M)	Actual
MA(6)	79%	−12342.3	−7709
MA(12)	81%	−27.67	−7709
SES*	98%	−15886.0	−7709

*Using optimal alpha of 0.47; SES models using different coefficients will produce different values of MAPE.

53. a) Hotel Occupancy = 11.6 + 0.550 lag1 + 0.293 lag6

b) Residuals over time appear random, except for September 2001.

c) 80.04%

55. a) The regression equation is:
OPEC-Prod (Th Bar/Day) = 1148 + 0.962 lag1

b) Forecast is 31245.6.

c)

Model	Forecast (thousand barrels/day)	Actual	APE (%)
MA(6)	31689.8	31286.5	1.29
MA(12)	31937.5	31286.5	2.08
SES	31300.2	31286.5	0.04
AR(1)	31245.6	31286.5	0.13

d) The single exponential smoothing model seems to fit best.

CHAPTER 21

1. a) Retrospective
b) Prospective

3. Experimental units: cookies
Treatments: chocolate chip choice (milk, dark, semi-sweet)
Response: Rating by trained testers
Randomization: random presentation to the tasters

5. Control: All other aspects of the recipes were kept the same
Randomization: Cookies presented to tasters in random order
Replication: 10 tasters did the rating

7.

9. Single-blind. Customers were blind to treatments, but the kitchen staff knew.

11. a) 3, 36, and 39
b) 321.90
c) 285.36
d) 8.94

13. a) Chips (milk, dark, semi), and Oil (butter, margarine)
b) 6 treatments: milk/Butter, dark/butter, semi/butter, milk/margarine, dark/margarine, semi/margarine
c) For example: The relative ratings of the chips were different for cookies baked with butter than for cookies made with margarine.

15. a) Temperature, Washing time, Detergent
b) Temperature (Hot, Cold), Washing time (Short, Long), Detergent (Standard, New).
c) Whiteness as measured by an optical scanner

17. a) Both detergents should be tested under the same conditions to ensure that the results are comparable.
b) It will be impossible to generalize to cold water washing.
c) Treatments should be run in random order to ensure that no other unforeseen influences could affect the response in a systematic way.

19. a) The differences among the Mozart and quiet groups were more than would have been expected from ordinary sampling variation.
b)

    c) The Mozart group seems to have the smallest median difference and thus the *least* improvement, but there does not appear to be a significant difference.

    d) No, the difference does not seem significant compared with the usual variation.

21. a) Observational; randomly select a group of children, ages 10 to 13, have them taste the cereal, and ask if they like the cereal.

    b) Answers may vary. Get volunteers ages 10 to 13. Each volunteer will taste one cereal, randomizing the assignment. Compare the percentage of favorable ratings for each cereal.

    c) Answers may vary. From the volunteers, identify the children who watch Frump and identify the children who do not watch Frump. Use a blocked design to reduce variation in cereal preference that may be associated with watching the Frump cartoon.

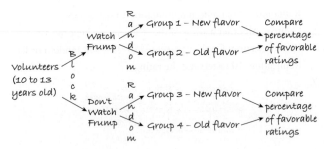

23. a) The students were not randomly assigned. Those who signed up for the prep course may be a special group whose scores would have improved anyway.

    b) Answers may vary. Find a group of volunteers who are willing to participate. Give all volunteers the SAT exam. Randomly assign the subjects to the review or no review group. Give the tutoring to one group. After a reasonable time, retest both groups. Check to see that the tutored group had a significant improvement in scores when compared with the no review group.

    c) After the volunteers have taken the first SAT exam, separate the volunteers into blocks of low, average, and high SAT exam score performance. Now assign half of each block to the review and half to the no review groups. Give the tutoring. Now retest all groups. Compare the differences between treatments for each block.

25. a) The null hypothesis is that the mean *Mileage* for all cylinder levels is the same. The alternative is that not all means are equal.

    b) Randomization not specified, so inference is not clear. The spreads of the four groups look very different. The response variable probably should be re-expressed before proceeding with an Analysis of Variance. Consider omitting the single 5-cylinder car from the study since it is the only one in its group. Boxplots appear symmetric, but other plots should be used to check the Nearly Normal Condition.

27. a) H_0: Mean cell phone adoption rates are equal in the three regions.
H_A: Mean cell phone adoption rates are not all equal in the three regions.

    b) Countries were selected randomly from the three regions, so the randomization condition is met. However, boxplots suggest the spreads may differ, and show several outliers, which may require attention. Other plots should be used to check the Nearly Normal Condition.

29. a) H_0: $\mu_1 = \mu_2 = \mu_3 = \mu_4 = \mu_5 = \mu_6$ (That is, each of the tellers takes, on average, the same amount of time to serve a customer.) vs. the alternative that not all means are equal. Here μ_k refers to the mean *Time* it takes *Teller k* to serve a customer.

    b) With a P-value of 0.19, we can't reject the null hypothesis.

    c) No, because we do not reject the null hypothesis, we cannot perform multiple comparisons.

31. a) Observational; the factor was not deliberately manipulated.

    b) Retrospective; using previously collected data.

    c) H_0: Mean responses to the e-security questions are all equal for the three types of community.
H_A: Mean responses to the e-security questions are not all equal for the three types of community.

    d) $F = 6.24$

    e) With such a small P-value, we can reject the null hypothesis and conclude that mean responses to the e-security question are not all equal for the three types of community. A causal link cannot be established because this is an observational study.

33. a) Observational

    b) There is no treatment. No factors were manipulated.

    c) Numbers of candies of each color.

35. a) Experiment; the factors are deliberately manipulated to specify treatments, and the subjects are assigned to the treatments randomly.

    b) Website trustworthiness measured on a 10-point scale.

    c) There are two factors: (1) website configuration with respect to assurance and (2) type of product purchased.

    d) There are nine treatments.

    e) H_0: The mean *trustworthiness ratings* of websites are the same for those *with a third-party assurance seal*, *a self-proclaimed assurance displayed*, and *no assurance seal*.
H_0: The mean *trustworthiness* of websites is the same for online purchases of *books*, *cameras*, and *insurance*.
H_0: There is no interaction effect. (The effect of each factor is the same at all levels of the other factor.)

37. a) Observational study.

    b) H_0: The mean *percentage changes in stock prices* are the same for companies *with and without ISO 9000 certification*.
H_0: The mean *percentage changes in stock prices* are the same for *large and small* companies.
H_0: There is no interaction effect. (The effect of each factor is the same at all levels of the other factor.)

    c) It is difficult to establish a causal link not only because this is an observational study, but also because of the possibility of confounding or lurking variables. For instance, industry sector may be related to percentage changes in stock prices and confounded with ISO 9000 certification but is unaccounted for in this study.

39. a) Observational study; the factors are not deliberately manipulated to specify treatments.

    b) Boxplots of the hourly wages by both manager type and region indicate that the equal variance assumption is reasonable. Also, the boxplots indicate fairly symmetric distributions. However, other plots should be used to check the Nearly Normal Condition once the model has been fit. The states were randomly selected from the three regions, so the randomization condition is satisfied.

c) Based on the P-values, we can conclude that there is no significant interaction effect. There is no evidence that the mean hourly wages are different across the three regions but are different for sales and advertising managers.

d) Yes, because the interaction effect is not significant.

41. a) Experiment; the machine setting factor was deliberately manipulated. Shift is a blocking factor.

b) Size error of the part is the response variable.

c) There are nine treatments.

d) There is no significant interaction effect so we can conclude that both the Shift and the Machine Setting appear to have a significant effect. However, from this table alone, we cannot determine which of these effects is more important.

43. a) MST = 8.65; MSE = 0.0767

b) $F = 112.78$

c) The data provide very strong evidence that the means are not equal.

d) We have assumed that the experimental runs were performed in random order, that the variances of the treatment groups are similar, and that the residuals are nearly Normal.

e) A boxplot of the *Scores* by *Method*, a plot of residuals vs. predicted values, a Normal probability plot, and a histogram of the residuals.

f) $s_p = \sqrt{0.0767} = 0.277$ points

45. a) An observational study

b) The null hypothesis is that the mean *Sugar Content* is the same for the cereals on each *Shelf*. The alternative is that not all the means are equal.

c) The P-value of 0.0012 provides strong evidence that the mean *Sugar Content* is not the same for each *Shelf*.

d) We cannot conclude that cereals on *Shelf* 2 have a different mean *Sugar Content* than cereals on *Shelf* 3 or that cereals on *Shelf* 2 have a different mean *Sugar Content* than cereals on *Shelf* 1. We can conclude only that the means are not all equal.

e) The Bonferroni test shows that at $\alpha = .05$ the shelves are different, except for Shelf 1 and Shelf 3, which are not discernibly different. Now we can conclude that the mean *Sugar Content* of cereals on *Shelf* 2 is *not* equal to the mean *Sugar Content* on *Shelf* 1 and *Shelf* 3. In other words, we can conclude from this test what we wanted to conclude in part d.

47. a) H_0: Head injuries are, on average, about the same regardless of the size of the car

H_0: Which seat you sit in doesn't affect the degree of head injury.

b) The conditions appear to be met. We assume the data were collected independently, the boxplots show that the variance is roughly constant, and there are no patterns in the scatterplot of residuals vs. predicted values.

c) There is no significant interaction. The P-values for both *Seat* and *Size* are <0.0001. Thus, we reject both the null hypotheses and conclude that both the *Seat* and the *Size* of car affect the severity of head injury.

CHAPTER 22

1. a)

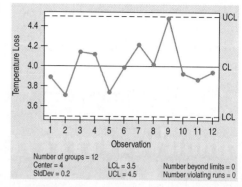

b) Yes, but one container was quite close to the upper limit.

c) 0.833

d) They need to decrease the process standard deviation. There is too great a chance that a container will fall outside the specification limits.

3. a)

b) No.

c) No.

d) No.

e) No.

f) No.

g) No.

h) The process seems to be in control.

5. a) 0.00447 m (or 0.447 cm)

b) Outside (91.387, 91.413) m

c) $SD(\overline{X}) = 0.00316$m; Outside (91.391, 91.409) m

7. a) 0.0198

b) (0, 0.0794)

c) 4 (round up 3.97)

9. These would be control limits because they are based on the process itself and not on external criteria.

11. These are lower specification limits set by the NHSF below which no pad should fall.

13. This is an upper specification limit set by Gemini.

15. This is common-cause variation based on random fluctuations.

17. 0.0027 based on the observations being Normal.

19. a) and b)

$$\overline{\overline{X}} = 8.05; \quad \overline{R} = 0.65 \Rightarrow \hat{\sigma} = \frac{\overline{R}}{d_2} = \frac{0.65}{2.059} = 0.316$$

$$\overline{X} \; chart: \quad LCL: 8.05 - 3\left(\frac{0.316}{\sqrt{4}}\right) = 7.576$$

$$UCL: 8.05 + 3\left(\frac{0.316}{\sqrt{4}}\right) = 8.524$$

$R\ chart$: $LCL = 0(0.65) = 0$
   $UCL = 2.282(0.65) = 1.483$

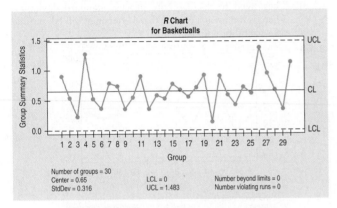

c) Both charts indicate that the process is in control as defined by the calibration period.

21. a) and b)
$\mu = 21.2;\ \sigma = 0.29 \Rightarrow \sigma_R = d_3\sigma = 0.888 \times 0.29 = 0.258$

$\bar{X}\ chart$: $LCL = \mu - 3\dfrac{\sigma}{\sqrt{n}} = 21.2 - 3\dfrac{0.29}{\sqrt{3}} = 20.70$

$UCL = \mu + 3\dfrac{\sigma}{\sqrt{n}} = 21.2 + 3\dfrac{0.29}{\sqrt{3}} = 21.70$

$Center\ line: R = d_2\sigma = 1.693 \times 0.29 = 0.491$

$R\ chart$: $LCL = D_3 R = 0$
   $UCL = D_4 R = 2.574(0.491) = 1.264$

c) The process is out of control. Although the R chart appears to be in control, the $\bar{X}$ chart shows numerous samples outside the 3σ limits.

d)

All but one of the samples meets the specification limits.

23. a) 0.0616
b) 0.1072
c) 0.0160
d) See the following chart.
e) The process seems to be in control.

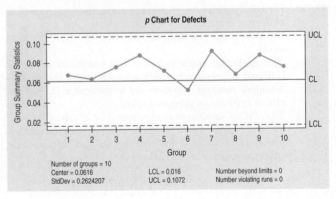

25. a) See the following chart.

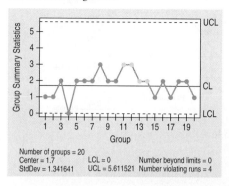

Number of groups = 20
Center = 1.7   LCL = 0   Number beyond limits = 0
StdDev = 1.341641   UCL = 5.611521   Number violating runs = 4

b) Although no day exceeds the upper control limit, there are more than 9 days on the same side of the desired center line. The process appears to be out of control.

27. a) 4.57
b) 10.99
c) 0
d) See the following chart.
e) The process is not in control on day 5.

29. a)

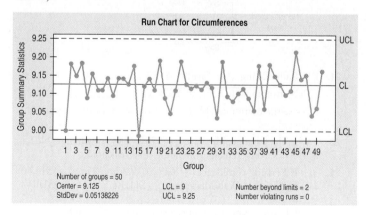

Number of groups = 50
Center = 9.125   LCL = 9   Number beyond limits = 2
StdDev = 0.05138226   UCL = 9.25   Number violating runs = 0

b) No, there are two balls that are too small.
c) Two.
d) Right away. The first ball produced was too small.

CHAPTER 23

1. a) Ranks
b) Ranks
c) Not ranks—units are points earned on the exam.

3. b

5. d

7. a

9. c (possibly e)

11. e (possibly c)

13. a) The assumptions and conditions are met. The experiment was randomized, so the groups are independent and the individual participants are independent. From technology we find that P = 0.01, which is small enough to reject the null hypothesis.
b) It would be better to design this as a before/after study, testing the children at the start of the teaching period, then randomly assigning them to treatment groups, and finally testing them again and comparing their results. An appropriate nonparametric method would be the Wilcoxon signed-rank test.

15. a) H_0: Distributions of sugar content have the same center.
H_A: They have different centers.
b) We don't know whether these cereals are a random sample from any particular population, but they are likely to be mutually independent.
c) P < 0.001 (according to technology). We can reject the null hypothesis and conclude that there is a difference in sugar content.
d) The distribution for Adults' cereals is skewed but the distribution for Children's cereals is symmetric. The skewness is a mild violation of the nearly Normal condition for the *t*-test, but we can't re-express one without re-expressing the other the same way, and that would make the currently symmetric distribution skewed. The nonparametric method is a better choice.

17. Use a Mann-Whitney test

Variable	n	median
size19th	5	27.00
size20th	9	15.50

W = 54.5
Test of medians equal vs medians not equal is significant at 0.0278
Reject the null at $\alpha = 0.10$

19. The subjects are not randomly selected, but are probably representative of the larger population of Freshmen. They are almost surely independent of each other.

H_0: No change over time; the median change is 0.

H_A: Change over time.

P < 0.001.

However, the change is in the unexpected direction. The students actually *lost* weight.

The *paired* *t*-test is probably more appropriate here because all assumptions and conditions it requires are met and it can provide a confidence interval for weight change, which would be a more complete response.

21. $z = 2.032$, $P = 0.042$

Reject the null hypothesis. ERP appears to be effective.

23. Wilcoxon signed-rank test

$P = 0.0625$

Do not reject the null hypothesis.

Wilcoxon Signed-Rank Test: 13thDiff

Test of median = 0.000000 versus median not = 0.000000

	N	N for Test	Wilcoxon Statistic	P	Estimated Median
13thDiff	5	5	15.0	0.059	1889

25.

species	N	Median	Ave Rank	Z
Robin	16	22.55	19.2	−1.46
Sparrow	14	23.05	26.7	1.26
Wagtail	15	23.05	23.7	0.24
Overall	44		23.0	

H = 2.51 DF = 2 P = 0.285
H = 2.55 DF = 2 P = 0.279 (adjusted for ties)

Do not reject the null hypothesis.

27. a) The relationship is strongly monotonic.
b) There is a strong association between GDP and Crowdedness.
c) The association is not linear, so Pearson correlation is not appropriate.

29. a) Because the plot shows some bending, Kendall's tau may be more appropriate.
b) Tau tells us how consistently sales at The Home Depot grow as housing starts grow, but not the amount by which they grow.

31. Kendall's tau tells us that there is a fairly consistent positive association between the housing cost index and median family income. But it does not allow any conclusions about causation.

CHAPTER 24

1. a) State of Nature
b) Action
c) Action

3.

Action	State	Outcome
Out-source	Recession	220
	Stable	350
	Expansion	300
In-House	Recession	150
	Stable	240
	Expansion	390

5. a) Out-source
b) In-house

7. a) EV(Out-source) = 294, EV(In-house) = 312
b) Out-source

9. SD(Out-Source) = 41.76, CV(Out-Source) = 0.14,
SD(In-house) = 99.68, CV(In-house) = 0.32

11. a)

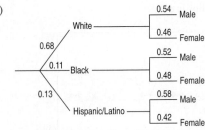

b) 0.0546
c) 0.46

13. 0.744

15.

		Stay for Interview	
		No	Yes
Action	Fully Changeable Fare	$750	$750
	Nonrefundable Fare	$650	$800

17.

19.

21. EV(Fully Changeable Fare) = $750
EV(Nonrefundable Fare) = $695

Choose the nonrefundable fare because we want to minimize the expected cost.

23. EV(Fully Changeable Fare) = $750, EV(not changeable) is now $755, so the changeable fare is a better choice.

25. a) If P(rising confidence) = 0.70, EV(Prime Time) = $14.60. E(Targeted Web Marketing) = $11.40; EV(Direct Mail) = $11.50
Choose Prime Time since expected payoff is highest.
b) If P (rising confidence) = 0.40, EV(Prime Time) = $9.20; E(Targeted Web Marketing) = $10.80; EV(Direct Mail) = $13.
Choose Direct Mail in this case since expected payoff is highest.

27. a) $3.90
b) $4.00

29. a)

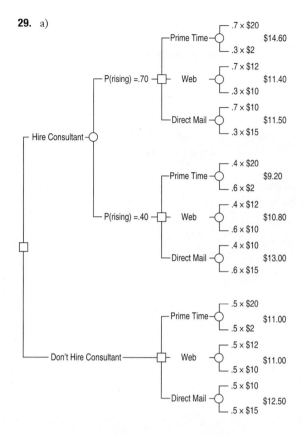

b) Yes, both alternatives with information have a better return.
c) $1.30

31.

		Growing Season	
		Good	Bad
Purchase Decision	2 Tractors	$27,400	$2400
	3 Mowers	$8575	$3575
	No Purchase	$0	−$1000

33. Shawn favors Tractors, with an upside gain of $27,400. Lance prefers the Mowers, whose worst result is $3575.

35. a) and b)

		EV	SD	RRR
Purchase Decision	2 Tractors	$19,900.00	11,456.44	1.737
	3 Mowers	$7075	2291.28	3.088
	No Purchase	−$300.00	458.26	−0.655

c) Purchasing Mowers has the highest RRR, which is preferred.

37. a)

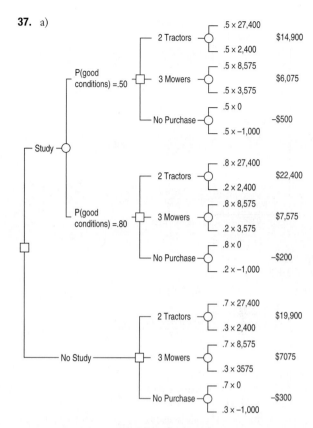

b) $17,900
c) No. In all scenarios, the Tractors investment seems the best course to take.

39. a) EV(Stock A) = $800; EV(Stock B) = $2300
b) SD(Stock A) = $2749.55; SD(Stock B) = $1833.03
c) CV(Stock A) = 3.437; CV(Stock B) = 0.797; RRR (Stock A) = 0.291; RRR(Stock B) = 1.255
d) Stock B; it has a higher expected value and only a slightly higher SD. Its RRR is much higher, which shows it has a better reward to risk ratio.

41. a) EV(High-End Bike) = $4465; EV(Moderately-Priced Bike) = $2650
b) SD(High-End Bike) = $1995.00; SD(Moderately-Priced Bike) = $1050.00
c) CV(High-End Bike) = 0.447; CV(Moderately-Priced Bike) = 0.396; RRR(High-End Bike) = 2.238; RRR (Moderately-Priced Bike) = 2.524
d) Answers may vary. The high-end bike has a much higher EV, but also a higher SD. Its RRR is lower but is very close to the moderately-priced bike. The decision will depend on the store owner's appetite for risk.

43. a)

b)   3.7%
c)   0.405

45. a)

b)   0.16
c)   0.667

47. a)

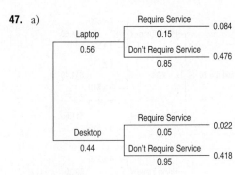

b)   10.6%
c)   0.792

Technology Help: Excel with XLStat

To ensure that XLStat recognizes different symbols for missing data, click **Options** on the right of the XLStat ribbon. Then click the **Missing data** tab and check the buttons next to **Consider empty cells as missing data** and **Consider also the following values as missing data**. Under the latter, select all options. Click **Save**.

CHAPTER 3

To use a specific sampling scheme to sample from data:

- Choose the **XLStat tab** at the top of the Excel ribbon.
- On the far left of the ribbon, click **Preparing data**.
- Choose the **Data sampling** option.

- Enter the cell range of your data.
- Choose the type of sampling from the drop-down list. The sampling types available in XLStat are described in the XLStat help manual.

CHAPTER 4

To create a contingency table from unsummarized data:

- On the XLStat tab, choose **Preparing data**.
- From the menu, choose **Create a contingency table**.
- In the dialog box, enter your data range on the General tab. Your data should be in two columns, one of which is the row variable and the other is the column variable.

- On the Outputs tab, check the box next to **Contingency table** and optionally choose **Percentages/Row** or **Column** to see the conditional distributions.

To make a bar chart or pie chart:

- Choose **Visualizing data**, and then select **Univariate plots**.
- For quantitative data, enter the cell range of your data in the **Quantitative data** field.
- Select the type of chart on the **Charts (1)** tab.

- For qualitative data, enter the cell range of your data in the **Qualitative data** field.
- Select the type of chart on the **Charts (2)** tab.

- To make a segmented or side-by-side bar chart, enter the column variable in the **Qualitative variable** field and the row variable in the **Subsamples** field; choose **Stacked bars** and **Cluster bars** on the **Charts (2)** tab.

CHAPTER 5

To make a histogram:

- Choose the **Visualizing data** from the XLStat ribbon tab, and then choose **Histograms**.

- Enter the cell range of your data in the **Data** field.

Note: You can either choose the number of bins or choose specific bin ranges on the **Options** tab. If you specify your own bins, XLStat requires that the bin data contain the lower bound of the first bin, and then each successive upper bound of the bins, as in the image below.

The Histogram command also reports the summary statistics; specify which statistics are reported on the **Outputs** tab in the dialog box.

Boxplots and stem and leaf plots can be made by choosing **Visualizing data** and then **Univariate plots**.

CHAPTER 6

To analyze regression statistics and plots:

- Select **Modeling data**, and then choose **Linear regression**.
- Enter the cell range of the dependent variable in the **Quantitative** field.
- If the explanatory variable is quantitative, check the quantitative box. Otherwise, be sure to check the qualitative box.
- Enter the cell range of the explanatory variable in the appropriate field.

- To plot the scatterplot and residual plot, in the Linear Regression dialog box, select the **Charts** tab.

- Select **Regression charts** and **Predictions and residuals**. These plots appear at the bottom of the worksheet report that is produced. The regression statistics and linear model are near the top of the report.

CHAPTER 9

XLStat will not calculate Normal distribution probabilities, but it will make Normal probability plots (XLStat calls these Q-Q plots):

- Select **Visualizing data**, and then **Univariate plots**.
- On the General tab, click the Quantitative data box and then select the data on your worksheet.
- Note that under the **Charts (1)** tab **Normal Q-Q plots** is checked.

- Click **OK**.
- If prompted, click **Continue**.

CHAPTER 11

To find a one-proportion z interval:

- Select **Parametric Tests**, and then choose **Tests for one proportion**.
- Under the **General** tab, choose the **Data format** to be either **Frequency** or **Proportion** depending on the form of your data.

- Enter the **frequency** of your variable (or **proportion**) and the **sample size**.
- Enter the **test proportion**.

Under **Data format** choose the appropriate button.

- Under the **Options** tab choose the **Alternative hypothesis** of **Proportion – Test proportion $\neq$ D**. Enter **0** for the **Hypothesized difference (D)**.
- Enter 5 under **Significance Level**. The output will show the 95% confidence interval.

- Choose the radial button for **Sample** under **Variance**.
- Choose the **Wald confidence interval**.

CHAPTER 12

To find a one sample z interval or a one sample t interval:

- Choose **Parametric Tests**, and then **One-sample t-test and z-test**.
- Under the **General** tab, enter your data cell range and choose either **z-test** or **Student's t test**.
- On the **Options** tab, choose the **Alternative hypothesis** of **Mean 1 $\neq$ Theoretical mean**.

- For calculating just a confidence interval, you can leave the **Theoretical mean** field blank. If you are also conducting a hypothesis test, enter in the theoretical mean here.
- Under **Significance Level**, enter in the desired level of significance. The output will yield the $(1 - \alpha)100\%$ confidence level.

CHAPTER 13

To conduct a test for one proportion, a one-mean z test or a one-mean t test:

- Choose **Parametric Tests**, and then **Tests for one proportion** or **One-sample t-test and z-test**.
- Complete the dialog box as you did for a confidence interval. If you are doing a test on a proportion, fill in the field for **Test proportion** with the proportion from your null hypothesis. Under the Options tab, choose **Test proportion** under **Variance**.
- If you are doing a test on a mean, fill in the field for **Theoretical mean** with the population mean from your null hypothesis.

CHAPTER 14

To conduct a *t* test for the difference between two means:

- Choose **Parametric tests**, and then **Two-sample t-test and z-test**.

- If the two samples are in two separate columns in your workbook, choose the option **One column per sample**.

- If your data are *stacked* (one column lists the data, and the other the group identifier) choose the option **One column per variable**.

- If the data are paired, choose the option **Paired samples**.

- Enter the range of cell data.

- On the **Options** tab, for a test that is not pooled uncheck the box next to **Assume equality**.

The output yields the results of a hypothesis test and a confidence interval for the difference between means.

CHAPTER 15

To perform a Chi-Square Goodness-of-Fit Test:

- Choose **Parametric Tests**, and then select **Multinomial goodness of fit test.**

- In one column you should have the observed frequencies of the categories of the variable. Enter this under **Frequencies**.

- In another column you should have either the expected frequencies or the proportions. Enter this under either **Expected frequencies** (or expected proportions) and choose the appropriate **data format**.

- Be sure to check the box that says **Chi-square test**.

- Enter in the desired **Significance level**.

- Select **OK**.

- Select **Continue** (if prompted).

To perform a Chi-Square Test for Homogeneity

- Choose **Correlation/Association tests**, and then select **Tests on contingency tables**.

- If your data already are in a contingency table, choose the **Data format** option **Contingency table**.

- If your data are not in a contingency table, choose **Qualitative variables**.

- Enter the cell range of your data.

- On the **Options** tab, check **Chi-square test**.

- On the **Outputs** tab, choose **Proportions/Row** or **Proportions/Column** if you wish to see the conditional distributions.

CHAPTER 16

To analyze regression statistics and plots:

- Choose the **Modeling data** menu, and then select **Linear regression**.

- On the **Charts** tab, select **Regression charts** and **Predictions and residuals** to plot the scatterplot and residual plot. These plots appear at the bottom of the worksheet report. The regression statistics and linear model appear near the top of the report.

To form prediction intervals:

- Type the *x*-values into a cell in the workbook.

- On the **Prediction** tab, enter that cell in the **Quantitative** field. The prediction interval will be found at the end of the generated report.

CHAPTER 17

To make a scatterplot by groups:

- Choose **Visualizing data**, and then select **Scatter plots**.

- Enter the explanatory, response, and group variable cell ranges.

- In the resulting plot, click anywhere on the plot to access the **Chart Tools Layout** tab.

- Select the **Trendline** button to add a trendline for each group.

- Click on each trendline to change its color in the **Format trendline** menu.

- The Durbin-Watson statistic is calculated in the results for **Linear regression** from the **Modeling data** button. It is reported as **DW** in the Goodness-of-fit statistics section of the report.

CHAPTER 18

Building a multiple regression model in XLStat is the same as building a simple linear one:

- Choose **Modeling data**, and then select **Linear regression**.

- In the **X/Explanatory variable** field, enter the cell range of all columns of explanatory variable data.

Note: The Pearson version of XLStat does not do logistic regression.

CHAPTER 19

XLStat can handle both qualitative (indicator) and quantitative explanatory variables. To build a "best" model:

- Choose **Modeling data**.

- Select all desired explanatory variables in the dialog box for **Linear regression**.

- On the **Options** tab, XLStat gives you options for building a **Best model** using various criteria that you can select, or building a model based on significance in a **stepwise** or **forward** or **backward** direction.

CHAPTER 20

To construct an autoregressive moving average (ARMA(p,q)) model:

- Select **Modeling data**, then select **ARMA**.
- Enter the cell range of the data under **Time series**.
- Note that you can center the data by selecting the appropriate box.
- XLStat requires input of parameters of the model such that $p + q > 0$.

CHAPTER 21

To perform a 1-way ANOVA:

- Select **Modeling data**, and then select **ANOVA**. Note that XLStat requires stacked data.
- Enter the quantitative response variable cell range in the **Y/Dependent variables** field.

- Enter the group variable in the **Qualitative** field under **X/Explanatory variables**.

To do a multiple comparison procedure:

- Select an appropriate procedure on the **Outputs** tab.

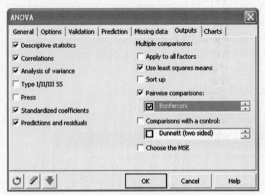

- Select the box for **Pairwise comparisons** and scroll to see several options.

CHAPTER 22

To implement control charts for desired bounds:

- From the XLStat tab, choose the **XLStat-SPC** toolbar.
- Select either **Subgroup charts** (for x-bar and r charts) or **Attribute charts** (for p and c charts).
- Select the desired chart from the **Mode** tab in the dialog box.
- Under the General tab, type the data range into the **Data** field.

- If your data have sampled values in rows for each sample, choose the **Data Format Columns**.
- In the **Options** tab, you can type in your UCL and LCL directly, but if you leave these fields blank, XLStat will compute these for you.

- If you have a specific value for the mean, type it in. Under the **Estimation** tab, choose **R bar** as the **method for sigma**.

CHAPTER 23

To conduct the Wilcoxon Rank Sum/Mann-Whitney test or the Wilcoxon signed-rank test for two samples:

- Choose **Nonparametric tests**, and then select **Comparison of two samples**.
- Choose the **Mann-Whitney test** option on the **General** tab for the Wilcoxon Rank Sum test.

To conduct a Kruskal-Wallace test,

- Choose **Nonparametric tests, and** then select **Comparison of k samples** for the Kruskal-Wallace test.
- If the data are unstacked and the samples are of unequal sizes, choose the option **Remove the observations** from the **Missing data** tab.

To find Kendall and Spearman coefficients:

- Choose **Similarity/Dissimilarity matrices** from the **Describing data** tab.

- Be sure **Quantitative** is selected under Data (for Kendall and Spearman coefficients)
- From the drop-down menu, choose the appropriate procedure.

Chapter 1
Page 1 (newspaper background), Lawrence Manning/ Corbis; (Financial Times screenshot), *Financial Times* screenshot dated April 3, 2008, has been reprinted with permission from the *Financial Times*, London; (man with laptop), Ocean/Corbis; (Wall Street), Ocean/Corbis.

Chapter 2
Page 7 (keyboard button), iStockphoto; (screen capture), Gregory Bajor/iStockphoto; (shopping bag), Feng Yu/iStockphoto; (Amazon.com distribution of *Harry Potter*), Justin Sullivan/Getty Images News; **Page 12** (dial telephone), Shutterstock; **Page 13** (punch card), Pearson; **Page 20** (screen capture), Gregory Bajor/iStockphoto; (credit card), Konstantin Inozemtsev/iStockphoto.

Chapter 3
Page 25 (woman at desk), Digital Vision; (spectators waving American flags), Stockbyte; (Dewey Defeats Truman newspaper), *New York World and Telegraph/* Library of Congress Prints and Photographs Division [LC-USZ62-115068]; **Page 27** (pot of soup), Michael LaMotte/Cole Group/PhotoDisc/Getty Images; **Page 33** (group of similar businesspeople), Digital Vision; **Page 34** (jigsaw puzzle), Shutterstock; **Page 36** (*Calvin and Hobbes* 9.20.1993), *Calvin and Hobbes* Copyright © 1993 Bill Watterson. Distributed by Universal Press Syndicate. Reprinted with permission. All rights reserved; **Page 40** (*The Wizard of Id* 2.23.1991), By permission of Johnny Hart FLP and Creators Syndicate, Inc.; **Page 44** (woman at desk), Digital Vision; (market survey), iStockphoto.

Chapter 4
Page 51 (male jogger), image100/Corbis; (girl walking in water), Mike Watson Images/Corbis; (man/woman on beach), Ben Moon/Keen, Inc.; (blue sandal), Keen, Inc.; **Page 53** (Google Analytics screenshot), Google; **Page 55** (dark blue sandal), Keen, Inc.; **Page 65** (bowl of fruit), Joao Virissimo/Shutterstock; **Page 72** (man/woman on beach), Ben Moon/Keen, Inc.; (couple squatting at river), Tetra Images/Corbis.

Chapter 5
Page 85 (AIG logo), Andrew Scheck/Shutterstock; (stack of coins), Ivan Mateev/iStockphoto; (financial newspaper series), Pali Rao /iStockphoto; (New York city skyline), DigitalVision; **Page 101** (girl with shopping bag), Blend Images/fotolia; **Page 108** (shoppers at toy store), Blend Images/fotolia; **Page 115** (Ithaca Times), collection of the author; **Page 121** (luxury hotel pool), Klaas Lingbeek-van Kranen/iStockphoto; (downtown Honolulu), Aimin Tang/iStockphoto.

Chapter 6
Page 137 (row of appliances), Kostiantyn Postumitenko/ iStockphoto; (garden path), Andrey Volokhatiuk/ iStockphoto; (drill), Roman Chieml/iStockphoto; **Page 141** (Descartes), Portrait of Rene Descartes by Sebastien Bourdon ca. 1640-1670/The Art Archive/ Corbis; **Page 145** (wallet with credit cards), Shutterstock; **Page 148** (child reading), Shutterstock; **Page 154** (shoe size/height correlation), Losevsky Pavel /Shutterstock; (Sir Francis Galton), public domain; **Page 160** (home for sale), Shutterstock; **Page 167** (building fire), Shutterstock; **Page 173** (New York skyline), Klaas Lingbeek-van Kranen/iStockphoto; (financial material), Pali Rao/ iStockphoto.

Chapter 7
Page 189 (crumpled dollar bills), Frank Bean/UpperCut Images/Getty Images; (bills with a pen and a calculator), Sean Russell/fStop/Getty Images; (African American businessman), DreamPictures /Getty Images; (large detached house), Phillip Spears/Photodisc/Getty Images; **Page 192** (Jacob Bernouilli), St. Andrews University/The MacTutor History of Mathematics Archive; **Page 193** (Keno), Jean-Loup Gautreau/AFP/Getty Images; **Page 198** (multicolored M&M candies), Adam Rountree/AP Images; **Page 203** (mother and children with bath tub), LM Productions/Digital Visions/Getty Images; **Page 207** (clothing rack), Martin Mcelligott/iStockphoto; (female shoppers), Jacob Wackerhausen/iStockphoto.

Chapter 8
Page 217 (Met Life building, Walter Gropius), Howard Architectural Models, Inc.; (pens for business men and women), AlbertSmirnov/iStockphoto; (contract series), Pali Rao/iStockphoto; (balancing accounts), Julien Bastide/iStockphoto; **Page 221** (warehouse shipping line), iStockphoto; **Page 227** (*Calvin and Hobbes* 4.13.1993), *Calvin and Hobbes* Copyright © 1993 Bill Watterson. Distributed by Universal Press Syndicate. Reprinted with permission. All rights reserved.; (Bernouille), St. Andrews University/The MacTutor History of Mathematics Archive; **Page 231** (blood donors), Keith Brofsky/Getty Images; **Page 237** (female investor), Novic/Dreamstime; (calculator, stock market data, and pen), iStockphoto.

Chapter 9
Page 245 (company seal), Chris Steer/iStockphoto; (Wall Street sign), Chris Steer/iStockphoto; (stock listings on multiple screens), Dr Rave/iStockphoto; **Page 252** (shopper and cereal), David Buffington/Getty Images; **Page 257** (stacking boxes on conveyor belt), Walter

Hodges/Digital Vision/Getty Images; **Page 269** (U.S. savings bonds), Richard Cano/Stockphoto; (stock market), Frank van den Bergh/iStockphoto.

Chapter 10
Page 277 (MBNA logo), Scott Boehm/Getty Images; (neoclassical columns), Rena Schild/iStockphoto; (stack of credit cards), Ilya Genkin/iStockphoto; **Page 283** (central California neighborhood), Galina Barksaya/Shutterstock; **Page 286** (Pierre-Simon Laplace), Picture Desk/Kobal Collection; **Page 294** (living room), M. Eric Honeycutt/iStockphoto; (front porch), iStockphoto.

Chapter 11
Page 305 (news ticker board in Times Square), iStockphoto; (business graph), José Luis Gutiérrez/iStockphoto; (inauguration of Franklin D. Roosevelt), National Archives/Getty Images; **Page 310** (*Garfield*), Copyright © 2010 Jim Davis/Universal Uclick; **Page 313** (protest at *Sony* France), sipa/Newscom; **Page 316** (pile of junk mail), Photodisc; **Page 322** (watch and money), Clint Hild/iStockphoto; (stock market), iStockphoto.

Chapter 12
Page 331 (Arthur Guinness historical portrait), public domain; (bottle and glass), Roulier/Turiot/photocuisine/Corbis; (Eugene Hackett, brewery worker), Bert Hardy/Hulton Archive/Getty; **Page 333** (William Gossett), University of York Department of Mathematics; **Page 339** (mailboxes), Shutterstock; **Page 340** (car accident), Shutterstock; **Page 348** (for sale sign), Andy Dean/iStockphoto; (San Fernando Valley), Scott Leigh/iStockphoto.

Chapter 13
Page 357 (NYSE floor), Mario Tama/Staff/Getty Images; (market rates), Mike Bentley/iStockphoto; (watch), Alan Crosthwaite/iStockphoto; **Page 361** (jury trial), Alamy; **Page 365** (switching cell phone plans), Digital Vision; **Page 366** (baseball game), Shutterstock; **Page 369** (damaged house), Shutterstock; **Page 371** (Sir Ronald Fisher), Photo Researchers, Inc.; **Page 376** (credit card promotion), Shutterstock; **Page 388** (jet), Mark Evans/iStockphoto; (molten metal), iStockphoto.

Chapter 14
Page 399 (Bank of America), Getty Images; (Visa goes public), Getty Images; (credit card), iStockphoto; **Page 401** (child and cereal), Getty Images; **Page 405** (paying with a credit card), Thinkstock; **Page 408** (eyes, female), iStockphoto; (yellow flowers), Shutterstock; (eyes, male, white), iStockphoto; (purple flowers), Dreamstime; (blue eyes, male), iStockphoto; (orange flowers), Dreamstime; (eyes, male, black), iStockphoto; (white flowers), Shutterstock; (eyes, female, Asian), Shutterstock; (multi-colored flowers), Dreamstime; **Page 409** (stack of

$100 bills), Shutterstock; **Page 411** (photo shoot), Shutterstock; **Page 413** (friendship negotiation), Shutterstock; **Page 421** (holiday shoppers), Blend Image/Getty; **Page 431** (blueprint), Nicholas Belton/iStockphoto; (home at night), M. Eric Honeycutt/iStockphoto.

Chapter 15
Page 449 (investor's crystal ball), iStockphoto; (stock trader), Jonathan Kirn/Stone/Getty Images; (Wall Street sign), Christine Balderas/iStockphoto; **Page 454** (stock market patterns), Shutterstock; **Page 460** (India bride beauty ritual), Stone/Getty Images; **Page 462** (French beauty products), PhotoEdit; **Page 464** (graduating high school girls), Digital Vision; **Page 467** (woman applying makeup), Photographers Direct; **Page 475** (medical records), Mark Kostich/iStockphoto; (elderly care), Alexander Raths/iStockphoto.

Chapter 16
Page 491 (molten metal), iStockphoto; (distressed metal surface), Bill Noll/iStockphoto; (Nambé products), JB Cumberland, Inc., NY; **Page 501** (candles), JB Cumberland, Inc., NY; **Page 504** (older students in university), Digital Vision/Getty Images; **Page 512** (active seniors and caregiver), Photodisc/Getty Images; **Page 516** (frozen pizza), Associated Photo; (melting river ice), Josh Webb/iStockphoto.

Chapter 17
Page 531 (bowl of cereal), Christine Balderas/iStockphoto; (wheat field), Hougaard Malan/iStockphoto; (Kellogg's logo with Tony the Tiger), Lawrence Lucier/Stringer/Getty Images; **Page 537** (gas prices with a humorous slant), Shutterstock; (*Foxtrot* 5.10.2002), Copyright © 2002 Bill Amend. Reprinted with permission from Universal Press Syndicate. All rights reserved; **Page 538** (Star of Africa diamond), Romantic jewels in the regalia, worn in the royal crowns at the coronation of King George VI (1895–1952) and by Queen Elizabeth (1900–2002) on 12th May 1937 (colour litho), English School, (20th century)/Private collection/The Bridgeman Art Library International; **Page 539** (Archimedes), Annenberg Rare Book & Manuscript Library; **Page 542** (wind farm), Michael Betts/Digital Vision/Getty Images; **Page 548** (Tour de France), Bryn Lennon/Getty Images; **Page 560** (cargo ship), Dan Barnes/iStockphoto; (windmills), Nicolas Skaanild/ iStockphoto.

Chapter 18
Page 577 (selling a house), Sean Locke/iStockphoto; (new home), M. Eric Honeycutt/iStockphoto; (Zillow.com logo), Zillow. Zillow logo used with permission from Zillow.com; **Page 581** (house and for-sale sign), Shutterstock; **Page 587** (real estate agent), Rob Marmion/Shutterstock; **Page 596** (customers at appliance store),

Yamini Chao/Digital Vision/Getty Images; **Page 598** (mortgage application), iStockphoto; **Page 606** (golf ball on grass with driver), FreezeFrameStudio/iStockphoto; (golf swing), Andrew Penner/iStockphoto.

Chapter 19
Page 617 (thinking businessman), Suprijono Suharjoto/ iStockphoto; (construction of coaster), Manfred Steinbach/iStockphoto; **Page 618** (roller coaster), Marcio Silva/iStockphoto; **Page 620** (roller coaster), Shutterstock; **Page 624** (cheeseburger), Joao Virissimo/ Shutterstock; **Page 628** (Oblivion drop coaster), Andrew Fox/Alamy; **Page 636** (house for sale), Paul Matthew Photography/Shutterstock; **Page 644** (Picabo Street), Scott Markewitz/Photographer's Choice/Getty Images; **Page 652** (Old Glory), Andrew Rich/iStockphoto; (stack of junk mail and unpaid bills), ShaneKato/iStockphoto.

Chapter 20
Page 665 (Whole Foods Market/Glendale California), Whole Foods Market; (fork with fresh salad), Ranplett/ iStockphoto; (Japanese eggplant), Alison Stieglitz/ iStockphoto; (leeks), Alison Stieglitz/iStockphoto; **Page 672** (stock market), Shutterstock; **Page 681** (seasonal plants), Gina Sanders/Shutterstock; **Page 692** (shopping at housewares store), Katharine Andriotis Photography, LLC/Editorial/Alamy; **Page 696** (grocery shopping), Kenneth V. Pilon/Shutterstock; **Page 701** (diamond necklace), Andrew Manley/iStockphoto; (computer chip), Andrey Volodin/Exclusive to iStockphoto.

Chapter 21
Page 717 (money), Onebluelight/iStockphoto; (business graph), Alex Slobodkin/iStockphoto; (credit card), Anthony Taylor/iStockphoto; **Page 725** (direct mail credit card offers), Michael Blann/Photodisc/Getty Images; **Page 731** (drum kit), Thor Jorgen Udvang/Shutterstock; **Page 737** (Carlo Bonferroni), St. Andrew's University History of Mathematics Archive; **Page 744** (credit card customer), David J. Green - Lifestyle / Alamy; **Page 756** (paper airplane), iStockphoto/Iain Sarjeant; (bowl of popcorn), iStockphoto/Martin Trebbin.

Chapter 22
Page 769 (woman at Sony Playstation display), YoShikazu Tsuno/Getty Images; (Sony AIBO robotic dog), Sony Electronics Inc.; (Sony headquarters), Dorling Kindersley Media Library; (Sony Vaio laptop), Dorling Kindersley Media Library; **Page 771** (Walter Shewhart), Reprinted with permission of Alcatel-Lucent USA Inc.; (W. Edwards Deming), Catherine Karnow/CORBIS; **Page 775** (silicon wafer), Corbis Premium RF/ALAMY; **Page 786** (hip replacement parts), Skyhawk/Shutterstock; **Page 798** (touch pad), Dr Bouz /iStockphoto; (woman working on computers), Getty Images, Inc. AFP.

Chapter 23
Page 807 (Rensis Likert), isr.umich.edu; (flow chart), Getty Images, Inc.- Photodisc./Royalty Free; (puzzle), M Stay/iStockphoto; (team work), shironosov/iStockphoto; **Page 812** (girls with camera), Floresco Productions/ Corbis; **Page 814** (rental car location), Luke MacGregor/ Reuters/Corbis; **Page 818** (lab testing), Olivier Le Queinec/Shutterstock; **Page 823** (Charles Edward Spearman), University College London Dept. of Psychology; **Page 826** (cozy chair by the fire), Diane39/ iStockphoto (suburban home), ziggymaj/iStockphoto.

Chapter 24
Page 835 (various images and logos), Data Description, Inc.; (computer studies), Chris Schmidt/iStockphoto/; **Page 846** (tourists and health coverage), Shutterstock; **Page 848** (PDAs and/or MP3s manufacturing), Photodisc; **Page 855** (gas station), Nick M. Do/iStockphoto; (gavel and scale), iStockphoto/Christine Balderas.

Chapter 25
Page 863 (envelopes), Jarek Szymanski /iStockphoto; (stamp honoring veterans), Clifford Mueller/iStockphoto; (U.S. medals on camouflage), Laura Young/iStockphoto; **Page 867** (businesspeople reviewing data), Getty Images.

Table D

Critical Values d_L and d_U of the Durbin-Watson Statistic (Critical Values are One-Sided)[a]

$\alpha = 0.05$

n	$k=1$ d_L	d_U	$k=2$ d_L	d_U	$k=3$ d_L	d_U	$k=4$ d_L	d_U	$k=5$ d_L	d_U
15	1.08	1.36	.95	1.54	.82	1.75	.69	1.97	.56	2.21
16	1.10	1.37	.98	1.54	.86	1.73	.74	1.93	.62	2.15
17	1.13	1.38	1.02	1.54	.90	1.71	.78	1.90	.67	2.10
18	1.16	1.39	1.05	1.53	.93	1.69	.82	1.87	.71	2.06
19	1.18	1.40	1.08	1.53	.97	1.68	.86	1.85	.75	2.02
20	1.20	1.41	1.10	1.54	1.00	1.68	.90	1.83	.79	1.99
21	1.22	1.42	1.13	1.54	1.03	1.67	.93	1.81	.83	1.96
22	1.24	1.43	1.15	1.54	1.05	1.66	.96	1.80	.86	1.94
23	1.26	1.44	1.17	1.54	1.08	1.66	.99	1.79	.90	1.92
24	1.27	1.45	1.19	1.55	1.10	1.66	1.01	1.78	.93	1.90
25	1.29	1.45	1.21	1.55	1.12	1.66	1.04	1.77	.95	1.89
26	1.30	1.46	1.22	1.55	1.14	1.65	1.06	1.76	.98	1.88
27	1.32	1.47	1.24	1.56	1.16	1.65	1.08	1.76	1.01	1.86
28	1.33	1.48	1.26	1.56	1.18	1.65	1.10	1.75	1.03	1.85
29	1.34	1.48	1.27	1.56	1.20	1.65	1.12	1.74	1.05	1.84
30	1.35	1.49	1.28	1.57	1.21	1.65	1.14	1.74	1.07	1.83
31	1.36	1.50	1.30	1.57	1.23	1.65	1.16	1.74	1.09	1.83
32	1.37	1.50	1.31	1.57	1.24	1.65	1.18	1.73	1.11	1.82
33	1.38	1.51	1.32	1.58	1.26	1.65	1.19	1.73	1.13	1.81
34	1.39	1.51	1.33	1.58	1.27	1.65	1.21	1.73	1.15	1.81
35	1.40	1.52	1.34	1.58	1.28	1.65	1.22	1.73	1.16	1.80
36	1.41	1.52	1.35	1.59	1.29	1.65	1.24	1.73	1.18	1.80
37	1.42	1.53	1.36	1.59	1.31	1.66	1.25	1.72	1.19	1.80
38	1.43	1.54	1.37	1.59	1.32	1.66	1.26	1.72	1.21	1.79
39	1.43	1.54	1.38	1.60	1.33	1.66	1.27	1.72	1.22	1.79
40	1.44	1.54	1.39	1.60	1.34	1.66	1.29	1.72	1.23	1.79
45	1.48	1.57	1.43	1.62	1.38	1.67	1.34	1.72	1.29	1.78
50	1.50	1.59	1.46	1.63	1.42	1.67	1.38	1.72	1.34	1.77
55	1.53	1.60	1.49	1.64	1.45	1.68	1.41	1.72	1.38	1.77
60	1.55	1.62	1.51	1.65	1.48	1.69	1.44	1.73	1.41	1.77
65	1.57	1.63	1.54	1.66	1.50	1.70	1.47	1.73	1.44	1.77
70	1.58	1.64	1.55	1.67	1.52	1.70	1.49	1.74	1.46	1.77
75	1.60	1.65	1.57	1.68	1.54	1.71	1.51	1.74	1.49	1.77
80	1.61	1.66	1.59	1.69	1.56	1.72	1.53	1.74	1.51	1.77
85	1.62	1.67	1.60	1.70	1.57	1.72	1.55	1.75	1.52	1.77
90	1.63	1.68	1.61	1.70	1.59	1.72	1.57	1.75	1.54	1.78
95	1.64	1.69	1.62	1.71	1.60	1.73	1.58	1.75	1.56	1.78
100	1.65	1.69	1.63	1.72	1.61	1.74	1.59	1.76	1.57	1.78

$\alpha = 0.01$

n	$k=1$ d_L	d_U	$k=2$ d_L	d_U	$k=3$ d_L	d_U	$k=4$ d_L	d_U	$k=5$ d_L	d_U
15	.81	1.07	.70	1.25	.59	1.46	.49	1.70	.39	1.96
16	.84	1.09	.74	1.25	.63	1.44	.53	1.66	.44	1.90
17	.87	1.10	.77	1.25	.67	1.43	.57	1.63	.48	1.85
18	.90	1.12	.80	1.26	.71	1.42	.61	1.60	.52	1.80
19	.93	1.13	.83	1.26	.74	1.41	.65	1.58	.56	1.77
20	.95	1.15	.86	1.27	.77	1.41	.68	1.57	.60	1.74
21	.97	1.16	.89	1.27	.80	1.41	.72	1.55	.63	1.71
22	1.00	1.17	.91	1.28	.83	1.40	.75	1.54	.66	1.69
23	1.02	1.19	.94	1.29	.86	1.40	.77	1.53	.70	1.67
24	1.04	1.20	.96	1.30	.88	1.41	.80	1.53	.72	1.66
25	1.05	1.21	.98	1.30	.90	1.41	.83	1.52	.75	1.65
26	1.07	1.22	1.00	1.31	.93	1.41	.85	1.52	.78	1.64
27	1.09	1.23	1.02	1.32	.95	1.41	.88	1.51	.81	1.63
28	1.10	1.24	1.04	1.32	.97	1.41	.90	1.51	.83	1.62
29	1.12	1.25	1.05	1.33	.99	1.42	.92	1.51	.85	1.61
30	1.13	1.26	1.07	1.34	1.01	1.42	.94	1.51	.88	1.61
31	1.15	1.27	1.08	1.34	1.02	1.42	.96	1.51	.90	1.60
32	1.16	1.28	1.10	1.35	1.04	1.43	.98	1.51	.92	1.60
33	1.17	1.29	1.11	1.36	1.05	1.43	1.00	1.51	.94	1.59
34	1.18	1.30	1.13	1.36	1.07	1.43	1.01	1.51	.95	1.59
35	1.19	1.31	1.14	1.37	1.08	1.44	1.03	1.51	.97	1.59
36	1.21	1.32	1.15	1.38	1.10	1.44	1.04	1.51	.99	1.59
37	1.22	1.32	1.16	1.38	1.11	1.45	1.06	1.51	1.00	1.59
38	1.23	1.33	1.18	1.39	1.12	1.45	1.07	1.52	1.02	1.58
39	1.24	1.34	1.19	1.39	1.14	1.45	1.09	1.52	1.03	1.58
40	1.25	1.34	1.20	1.40	1.15	1.46	1.10	1.52	1.05	1.58
45	1.29	1.38	1.24	1.42	1.20	1.48	1.16	1.53	1.11	1.58
50	1.32	1.40	1.28	1.45	1.24	1.49	1.20	1.54	1.16	1.59
55	1.36	1.43	1.32	1.47	1.28	1.51	1.25	1.55	1.21	1.59
60	1.38	1.45	1.35	1.48	1.32	1.52	1.28	1.56	1.25	1.60
65	1.41	1.47	1.38	1.50	1.35	1.53	1.31	1.57	1.28	1.61
70	1.43	1.49	1.40	1.52	1.37	1.55	1.34	1.58	1.31	1.61
75	1.45	1.50	1.42	1.53	1.39	1.56	1.37	1.59	1.34	1.62
80	1.47	1.52	1.44	1.54	1.42	1.57	1.39	1.60	1.36	1.62
85	1.48	1.53	1.46	1.55	1.43	1.58	1.41	1.60	1.39	1.63
90	1.50	1.54	1.47	1.56	1.45	1.59	1.43	1.61	1.41	1.64
95	1.51	1.55	1.49	1.57	1.47	1.60	1.45	1.62	1.42	1.64
100	1.52	1.56	1.50	1.58	1.48	1.60	1.46	1.63	1.44	1.65

[a] n = number of observations; k = number of independent variables.

Source: This table is reproduced from Biometrika, 41 (1951): 173 and 175, with the permission of the Biometrika Trustees.

Table F

$\alpha = .01$

Numerator df

Denominator df	1	2	3	4	5	6	7	8	9	10	11	12	13	14	15	16	17	18	19	20	21	22
1	4052.2	4999.3	5403.5	5624.3	5764.0	5859.0	5928.3	5981.0	6022.4	6055.9	6083.4	6106.7	6125.8	6143.0	6157.0	6170.0	6181.2	6191.4	6200.7	6208.7	6216.1	6223.1
2	98.50	99.00	99.16	99.25	99.30	99.33	99.36	99.38	99.39	99.40	99.41	99.42	99.42	99.43	99.43	99.44	99.44	99.44	99.45	99.45	99.45	99.46
3	34.12	30.82	29.46	28.71	28.24	27.91	27.67	27.49	27.34	27.23	27.13	27.05	26.98	26.92	26.87	26.83	26.79	26.75	26.72	26.69	26.66	26.64
4	21.20	18.00	16.69	15.98	15.52	15.21	14.98	14.80	14.66	14.55	14.45	14.37	14.31	14.25	14.20	14.15	14.11	14.08	14.05	14.02	13.99	13.97
5	16.26	13.27	12.06	11.39	10.97	10.67	10.46	10.29	10.16	10.05	9.96	9.89	9.82	9.77	9.72	9.68	9.64	9.61	9.58	9.55	9.53	9.51
6	13.75	10.92	9.78	9.15	8.75	8.47	8.26	8.10	7.98	7.87	7.79	7.72	7.66	7.60	7.56	7.52	7.48	7.45	7.42	7.40	7.37	7.35
7	12.25	9.55	8.45	7.85	7.46	7.19	6.99	6.84	6.72	6.62	6.54	6.47	6.41	6.36	6.31	6.28	6.24	6.21	6.18	6.16	6.13	6.11
8	11.26	8.65	7.59	7.01	6.63	6.37	6.18	6.03	5.91	5.81	5.73	5.67	5.61	5.56	5.52	5.48	5.44	5.41	5.38	5.36	5.34	5.32
9	10.56	8.02	6.99	6.42	6.06	5.80	5.61	5.47	5.35	5.26	5.18	5.11	5.05	5.01	4.96	4.92	4.89	4.86	4.83	4.81	4.79	4.77
10	10.04	7.56	6.55	5.99	5.64	5.39	5.20	5.06	4.94	4.85	4.77	4.71	4.65	4.60	4.56	4.52	4.49	4.46	4.43	4.41	4.38	4.36
11	9.65	7.21	6.22	5.67	5.32	5.07	4.89	4.74	4.63	4.54	4.46	4.40	4.34	4.29	4.25	4.21	4.18	4.15	4.12	4.10	4.08	4.06
12	9.33	6.93	5.95	5.41	5.06	4.82	4.64	4.50	4.39	4.30	4.22	4.16	4.10	4.05	4.01	3.97	3.94	3.91	3.88	3.86	3.84	3.82
13	9.07	6.70	5.74	5.21	4.86	4.62	4.44	4.30	4.19	4.10	4.02	3.96	3.91	3.86	3.82	3.78	3.75	3.72	3.69	3.66	3.64	3.62
14	8.86	6.51	5.56	5.04	4.69	4.46	4.28	4.14	4.03	3.94	3.86	3.80	3.75	3.70	3.66	3.62	3.59	3.56	3.53	3.51	3.48	3.46
15	8.68	6.36	5.42	4.89	4.56	4.32	4.14	4.00	3.89	3.80	3.73	3.67	3.61	3.56	3.52	3.49	3.45	3.42	3.40	3.37	3.35	3.33
16	8.53	6.23	5.29	4.77	4.44	4.20	4.03	3.89	3.78	3.69	3.62	3.55	3.50	3.45	3.41	3.37	3.34	3.31	3.28	3.26	3.24	3.22
17	8.40	6.11	5.19	4.67	4.34	4.10	3.93	3.79	3.68	3.59	3.52	3.46	3.40	3.35	3.31	3.27	3.24	3.21	3.19	3.16	3.14	3.12
18	8.29	6.01	5.09	4.58	4.25	4.01	3.84	3.71	3.60	3.51	3.43	3.37	3.32	3.27	3.23	3.19	3.16	3.13	3.10	3.08	3.05	3.03
19	8.18	5.93	5.01	4.50	4.17	3.94	3.77	3.63	3.52	3.43	3.36	3.30	3.24	3.19	3.15	3.12	3.08	3.05	3.03	3.00	2.98	2.96
20	8.10	5.85	4.94	4.43	4.10	3.87	3.70	3.56	3.46	3.37	3.29	3.23	3.18	3.13	3.09	3.05	3.02	2.99	2.96	2.94	2.92	2.90
21	8.02	5.78	4.87	4.37	4.04	3.81	3.64	3.51	3.40	3.31	3.24	3.17	3.12	3.07	3.03	2.99	2.96	2.93	2.90	2.88	2.86	2.84
22	7.95	5.72	4.82	4.31	3.99	3.76	3.59	3.45	3.35	3.26	3.18	3.12	3.07	3.02	2.98	2.94	2.91	2.88	2.85	2.83	2.81	2.78
23	7.88	5.66	4.76	4.26	3.94	3.71	3.54	3.41	3.30	3.21	3.14	3.07	3.02	2.97	2.93	2.89	2.86	2.83	2.80	2.78	2.76	2.74
24	7.82	5.61	4.72	4.22	3.90	3.67	3.50	3.36	3.26	3.17	3.09	3.03	2.98	2.93	2.89	2.85	2.82	2.79	2.76	2.74	2.72	2.70
25	7.77	5.57	4.68	4.18	3.85	3.63	3.46	3.32	3.22	3.13	3.06	2.99	2.94	2.89	2.85	2.81	2.78	2.75	2.72	2.70	2.68	2.66
26	7.72	5.53	4.64	4.14	3.82	3.59	3.42	3.29	3.18	3.09	3.02	2.96	2.90	2.86	2.81	2.78	2.75	2.72	2.69	2.66	2.64	2.62
27	7.68	5.49	4.60	4.11	3.78	3.56	3.39	3.26	3.15	3.06	2.99	2.93	2.87	2.82	2.78	2.75	2.71	2.68	2.66	2.63	2.61	2.59
28	7.64	5.45	4.57	4.07	3.75	3.53	3.36	3.23	3.12	3.03	2.96	2.90	2.84	2.79	2.75	2.72	2.68	2.65	2.63	2.60	2.58	2.56
29	7.60	5.42	4.54	4.04	3.73	3.50	3.33	3.20	3.09	3.00	2.93	2.87	2.81	2.77	2.73	2.69	2.66	2.63	2.60	2.57	2.55	2.53
30	7.56	5.39	4.51	4.02	3.70	3.47	3.30	3.17	3.07	2.98	2.91	2.84	2.79	2.74	2.70	2.66	2.63	2.60	2.57	2.55	2.53	2.51
32	7.50	5.34	4.46	3.97	3.65	3.43	3.26	3.13	3.02	2.93	2.86	2.80	2.74	2.70	2.65	2.62	2.58	2.55	2.53	2.50	2.48	2.46
35	7.42	5.27	4.40	3.91	3.59	3.37	3.20	3.07	2.96	2.88	2.80	2.74	2.69	2.64	2.60	2.56	2.53	2.50	2.47	2.44	2.42	2.40
40	7.31	5.18	4.31	3.83	3.51	3.29	3.12	2.99	2.89	2.80	2.73	2.66	2.61	2.56	2.52	2.48	2.45	2.42	2.39	2.37	2.35	2.33
45	7.23	5.11	4.25	3.77	3.45	3.23	3.07	2.94	2.83	2.74	2.67	2.61	2.55	2.51	2.46	2.43	2.39	2.36	2.34	2.31	2.29	2.27
50	7.17	5.06	4.20	3.72	3.41	3.19	3.02	2.89	2.78	2.70	2.63	2.56	2.51	2.46	2.42	2.38	2.35	2.32	2.29	2.27	2.24	2.22
60	7.08	4.98	4.13	3.65	3.34	3.12	2.95	2.82	2.72	2.63	2.56	2.50	2.44	2.39	2.35	2.31	2.28	2.25	2.22	2.20	2.17	2.15
75	6.99	4.90	4.05	3.58	3.27	3.05	2.89	2.76	2.65	2.57	2.49	2.43	2.38	2.33	2.29	2.25	2.22	2.18	2.16	2.13	2.11	2.09
100	6.90	4.82	3.98	3.51	3.21	2.99	2.82	2.69	2.59	2.50	2.43	2.37	2.31	2.27	2.22	2.19	2.15	2.12	2.09	2.07	2.04	2.02
120	6.85	4.79	3.95	3.48	3.17	2.96	2.79	2.66	2.56	2.47	2.40	2.34	2.28	2.23	2.19	2.15	2.12	2.09	2.06	2.03	2.01	1.99
140	6.82	4.76	3.92	3.46	3.15	2.93	2.77	2.64	2.54	2.45	2.38	2.31	2.26	2.21	2.17	2.13	2.10	2.07	2.04	2.01	1.99	1.97
180	6.78	4.73	3.89	3.43	3.12	2.90	2.74	2.61	2.51	2.42	2.35	2.28	2.23	2.18	2.14	2.10	2.07	2.04	2.01	1.98	1.96	1.94
250	6.74	4.69	3.86	3.40	3.09	2.87	2.71	2.58	2.48	2.39	2.32	2.26	2.20	2.15	2.11	2.07	2.04	2.01	1.98	1.95	1.93	1.91
400	6.70	4.66	3.83	3.37	3.06	2.85	2.68	2.56	2.45	2.37	2.29	2.23	2.17	2.13	2.08	2.05	2.01	1.98	1.95	1.92	1.90	1.88
1000	6.66	4.63	3.80	3.34	3.04	2.82	2.66	2.53	2.43	2.34	2.27	2.20	2.15	2.10	2.06	2.02	1.98	1.95	1.92	1.90	1.87	1.85

Table F (cont.)

Numerator df

$\alpha = .01$	23	24	25	26	27	28	29	30	32	35	40	45	50	60	75	100	120	140	180	250	400	1000
1	6228.7	6234.3	6239.9	6244.5	6249.2	6252.9	6257.1	6260.4	6266.9	6275.3	6286.4	6295.7	6302.3	6313.0	6323.7	6333.9	6339.5	6343.2	6347.9	6353.5	6358.1	6362.8
2	99.46	99.46	99.46	99.46	99.46	99.46	99.46	99.47	99.47	99.47	99.48	99.48	99.48	99.48	99.48	99.49	99.49	99.49	99.49	99.50	99.50	99.50
3	26.62	26.60	26.58	26.56	26.55	26.53	26.52	26.50	26.48	26.45	26.41	26.38	26.35	26.32	26.28	26.24	26.22	26.21	26.19	26.17	26.15	26.14
4	13.95	13.93	13.91	13.89	13.88	13.86	13.85	13.84	13.81	13.79	13.75	13.71	13.69	13.65	13.61	13.58	13.56	13.54	13.53	13.51	13.49	13.47
5	9.49	9.47	9.45	9.43	9.42	9.40	9.39	9.38	9.36	9.33	9.29	9.26	9.24	9.20	9.17	9.13	9.11	9.10	9.08	9.06	9.05	9.03
6	7.33	7.31	7.30	7.28	7.27	7.25	7.24	7.23	7.21	7.18	7.14	7.11	7.09	7.06	7.02	6.99	6.97	6.96	6.94	6.92	6.91	6.89
7	6.09	6.07	6.06	6.04	6.03	6.02	6.00	5.99	5.97	5.94	5.91	5.88	5.86	5.82	5.79	5.75	5.74	5.72	5.71	5.69	5.68	5.66
8	5.30	5.28	5.26	5.25	5.23	5.22	5.21	5.20	5.18	5.15	5.12	5.09	5.07	5.03	5.00	4.96	4.95	4.93	4.92	4.90	4.89	4.87
9	4.75	4.73	4.71	4.70	4.68	4.67	4.66	4.65	4.63	4.60	4.57	4.54	4.52	4.48	4.45	4.41	4.40	4.39	4.37	4.35	4.34	4.32
10	4.34	4.33	4.31	4.30	4.28	4.27	4.26	4.25	4.23	4.20	4.17	4.14	4.12	4.08	4.05	4.01	4.00	3.98	3.97	3.95	3.94	3.92
11	4.04	4.02	4.01	3.99	3.98	3.96	3.95	3.94	3.92	3.89	3.86	3.83	3.81	3.78	3.74	3.71	3.69	3.68	3.66	3.64	3.63	3.61
12	3.80	3.78	3.76	3.75	3.74	3.72	3.71	3.70	3.68	3.65	3.62	3.59	3.57	3.54	3.50	3.47	3.45	3.44	3.42	3.40	3.39	3.37
13	3.60	3.59	3.57	3.56	3.54	3.53	3.52	3.51	3.49	3.46	3.43	3.40	3.38	3.34	3.31	3.27	3.25	3.24	3.23	3.21	3.19	3.18
14	3.44	3.43	3.41	3.40	3.38	3.37	3.36	3.35	3.33	3.30	3.27	3.24	3.22	3.18	3.15	3.11	3.09	3.08	3.06	3.05	3.03	3.02
15	3.31	3.29	3.28	3.26	3.25	3.24	3.23	3.21	3.19	3.17	3.13	3.10	3.08	3.05	3.01	2.98	2.96	2.95	2.93	2.91	2.90	2.88
16	3.20	3.18	3.16	3.15	3.14	3.12	3.11	3.10	3.08	3.05	3.02	2.99	2.97	2.93	2.90	2.86	2.84	2.83	2.81	2.80	2.78	2.76
17	3.10	3.08	3.07	3.05	3.04	3.03	3.01	3.00	2.98	2.96	2.92	2.89	2.87	2.83	2.80	2.76	2.75	2.73	2.72	2.70	2.68	2.66
18	3.02	3.00	2.98	2.97	2.95	2.94	2.93	2.92	2.90	2.87	2.84	2.81	2.78	2.75	2.71	2.68	2.66	2.65	2.63	2.61	2.59	2.58
19	2.94	2.92	2.91	2.89	2.88	2.87	2.86	2.84	2.82	2.80	2.76	2.73	2.71	2.67	2.64	2.60	2.58	2.57	2.55	2.54	2.52	2.50
20	2.88	2.86	2.84	2.83	2.81	2.80	2.79	2.78	2.76	2.73	2.69	2.67	2.64	2.61	2.57	2.54	2.52	2.50	2.49	2.47	2.45	2.43
21	2.82	2.80	2.79	2.77	2.76	2.74	2.73	2.72	2.70	2.67	2.64	2.61	2.58	2.55	2.51	2.48	2.46	2.44	2.43	2.41	2.39	2.37
22	2.77	2.75	2.73	2.72	2.70	2.69	2.68	2.67	2.65	2.62	2.58	2.55	2.53	2.50	2.46	2.42	2.40	2.39	2.37	2.35	2.34	2.32
23	2.72	2.70	2.69	2.67	2.66	2.64	2.63	2.62	2.60	2.57	2.54	2.51	2.48	2.45	2.41	2.37	2.35	2.34	2.32	2.30	2.29	2.27
24	2.68	2.66	2.64	2.63	2.61	2.60	2.59	2.58	2.56	2.53	2.49	2.46	2.44	2.40	2.37	2.33	2.31	2.30	2.28	2.26	2.24	2.22
25	2.64	2.62	2.60	2.59	2.58	2.56	2.55	2.54	2.52	2.49	2.45	2.42	2.40	2.36	2.33	2.29	2.27	2.26	2.24	2.22	2.20	2.18
26	2.60	2.58	2.57	2.55	2.54	2.53	2.51	2.50	2.48	2.45	2.42	2.39	2.36	2.33	2.29	2.25	2.23	2.22	2.20	2.18	2.16	2.14
27	2.57	2.55	2.54	2.52	2.51	2.49	2.48	2.47	2.45	2.42	2.38	2.35	2.33	2.29	2.26	2.22	2.20	2.18	2.17	2.15	2.13	2.11
28	2.54	2.52	2.51	2.49	2.48	2.46	2.45	2.44	2.42	2.39	2.35	2.32	2.30	2.26	2.23	2.19	2.17	2.15	2.13	2.11	2.10	2.08
29	2.51	2.49	2.48	2.46	2.45	2.44	2.42	2.41	2.39	2.36	2.33	2.30	2.27	2.23	2.20	2.16	2.14	2.12	2.10	2.08	2.07	2.05
30	2.49	2.47	2.45	2.44	2.42	2.41	2.40	2.39	2.36	2.34	2.30	2.27	2.25	2.21	2.17	2.13	2.11	2.10	2.08	2.06	2.04	2.02
32	2.44	2.42	2.41	2.39	2.38	2.36	2.35	2.34	2.32	2.29	2.25	2.22	2.20	2.16	2.12	2.08	2.06	2.05	2.03	2.01	1.99	1.97
35	2.38	2.36	2.35	2.33	2.32	2.30	2.29	2.28	2.26	2.23	2.19	2.16	2.14	2.10	2.06	2.02	2.00	1.98	1.96	1.94	1.92	1.90
40	2.31	2.29	2.27	2.26	2.24	2.23	2.22	2.20	2.18	2.15	2.11	2.08	2.06	2.02	1.98	1.94	1.92	1.90	1.88	1.86	1.84	1.82
45	2.25	2.23	2.21	2.20	2.18	2.17	2.16	2.14	2.12	2.09	2.05	2.02	2.00	1.96	1.92	1.88	1.85	1.84	1.82	1.79	1.77	1.75
50	2.20	2.18	2.17	2.15	2.14	2.12	2.11	2.10	2.08	2.05	2.01	1.97	1.95	1.91	1.87	1.82	1.80	1.79	1.76	1.74	1.72	1.70
60	2.13	2.12	2.10	2.08	2.07	2.05	2.04	2.03	2.01	1.98	1.94	1.90	1.88	1.84	1.79	1.75	1.73	1.71	1.69	1.66	1.64	1.62
75	2.07	2.05	2.03	2.02	2.00	1.99	1.97	1.96	1.94	1.91	1.87	1.83	1.81	1.76	1.72	1.67	1.65	1.63	1.61	1.58	1.56	1.53
100	2.00	1.98	1.97	1.95	1.93	1.92	1.91	1.89	1.87	1.84	1.80	1.76	1.74	1.69	1.65	1.60	1.57	1.55	1.53	1.50	1.47	1.45
120	1.97	1.95	1.93	1.92	1.90	1.89	1.87	1.86	1.84	1.81	1.76	1.73	1.70	1.66	1.61	1.56	1.53	1.51	1.49	1.46	1.43	1.40
140	1.95	1.93	1.91	1.89	1.88	1.86	1.85	1.84	1.81	1.78	1.74	1.70	1.67	1.63	1.58	1.53	1.50	1.48	1.46	1.43	1.40	1.37
180	1.92	1.90	1.88	1.86	1.85	1.83	1.82	1.81	1.78	1.75	1.71	1.67	1.64	1.60	1.55	1.49	1.47	1.45	1.42	1.39	1.35	1.32
250	1.89	1.87	1.85	1.83	1.82	1.80	1.79	1.77	1.75	1.72	1.67	1.64	1.61	1.56	1.51	1.46	1.43	1.41	1.38	1.34	1.31	1.27
400	1.86	1.84	1.82	1.80	1.79	1.77	1.76	1.75	1.72	1.69	1.64	1.61	1.58	1.53	1.48	1.42	1.39	1.37	1.33	1.30	1.26	1.22
1000	1.83	1.81	1.79	1.77	1.76	1.74	1.73	1.72	1.69	1.66	1.61	1.58	1.54	1.50	1.44	1.38	1.35	1.33	1.29	1.25	1.21	1.16

Denominator df

Table F (cont.)

Numerator df

α = .05	1	2	3	4	5	6	7	8	9	10	11	12	13	14	15	16	17	18	19	20	21	22
1	161.4	199.5	215.7	224.6	230.2	234.0	236.8	238.9	240.5	241.9	243.0	243.9	244.7	245.4	245.9	246.5	246.9	247.3	247.7	248.0	248.3	248.6
2	18.51	19.00	19.16	19.25	19.30	19.33	19.35	19.37	19.38	19.40	19.40	19.41	19.42	19.42	19.43	19.43	19.44	19.44	19.44	19.45	19.45	19.45
3	10.13	9.55	9.28	9.12	9.01	8.94	8.89	8.85	8.81	8.79	8.76	8.74	8.73	8.71	8.70	8.69	8.68	8.67	8.67	8.66	8.65	8.65
4	7.71	6.94	6.59	6.39	6.26	6.16	6.09	6.04	6.00	5.96	5.94	5.91	5.89	5.87	5.86	5.84	5.83	5.82	5.81	5.80	5.79	5.79
5	6.61	5.79	5.41	5.19	5.05	4.95	4.88	4.82	4.77	4.74	4.70	4.68	4.66	4.64	4.62	4.60	4.59	4.58	4.57	4.56	4.55	4.54
6	5.99	5.14	4.76	4.53	4.39	4.28	4.21	4.15	4.10	4.06	4.03	4.00	3.98	3.96	3.94	3.92	3.91	3.90	3.88	3.87	3.86	3.86
7	5.59	4.74	4.35	4.12	3.97	3.87	3.79	3.73	3.68	3.64	3.60	3.57	3.55	3.53	3.51	3.49	3.48	3.47	3.46	3.44	3.43	3.43
8	5.32	4.46	4.07	3.84	3.69	3.58	3.50	3.44	3.39	3.35	3.31	3.28	3.26	3.24	3.22	3.20	3.19	3.17	3.16	3.15	3.14	3.13
9	5.12	4.26	3.86	3.63	3.48	3.37	3.29	3.23	3.18	3.14	3.10	3.07	3.05	3.03	3.01	2.99	2.97	2.96	2.95	2.94	2.93	2.92
10	4.96	4.10	3.71	3.48	3.33	3.22	3.14	3.07	3.02	2.98	2.94	2.91	2.89	2.86	2.85	2.83	2.81	2.80	2.79	2.77	2.76	2.75
11	4.84	3.98	3.59	3.36	3.20	3.09	3.01	2.95	2.90	2.85	2.82	2.79	2.76	2.74	2.72	2.70	2.69	2.67	2.66	2.65	2.64	2.63
12	4.75	3.89	3.49	3.26	3.11	3.00	2.91	2.85	2.80	2.75	2.72	2.69	2.66	2.64	2.62	2.60	2.58	2.57	2.56	2.54	2.53	2.52
13	4.67	3.81	3.41	3.18	3.03	2.92	2.83	2.77	2.71	2.67	2.63	2.60	2.58	2.55	2.53	2.51	2.50	2.48	2.47	2.46	2.45	2.44
14	4.60	3.74	3.34	3.11	2.96	2.85	2.76	2.70	2.65	2.60	2.57	2.53	2.51	2.48	2.46	2.44	2.43	2.41	2.40	2.39	2.38	2.37
15	4.54	3.68	3.29	3.06	2.90	2.79	2.71	2.64	2.59	2.54	2.51	2.48	2.45	2.42	2.40	2.38	2.37	2.35	2.34	2.33	2.32	2.31
16	4.49	3.63	3.24	3.01	2.85	2.74	2.66	2.59	2.54	2.49	2.46	2.42	2.40	2.37	2.35	2.33	2.32	2.30	2.29	2.28	2.26	2.25
17	4.45	3.59	3.20	2.96	2.81	2.70	2.61	2.55	2.49	2.45	2.41	2.38	2.35	2.33	2.31	2.29	2.27	2.26	2.24	2.23	2.22	2.21
18	4.41	3.55	3.16	2.93	2.77	2.66	2.58	2.51	2.46	2.41	2.37	2.34	2.31	2.29	2.27	2.25	2.23	2.22	2.20	2.19	2.18	2.17
19	4.38	3.52	3.13	2.90	2.74	2.63	2.54	2.48	2.42	2.38	2.34	2.31	2.28	2.26	2.23	2.21	2.20	2.18	2.17	2.16	2.14	2.13
20	4.35	3.49	3.10	2.87	2.71	2.60	2.51	2.45	2.39	2.35	2.31	2.28	2.25	2.22	2.20	2.18	2.17	2.15	2.14	2.12	2.11	2.10
21	4.32	3.47	3.07	2.84	2.68	2.57	2.49	2.42	2.37	2.32	2.28	2.25	2.22	2.20	2.18	2.16	2.14	2.12	2.11	2.10	2.08	2.07
22	4.30	3.44	3.05	2.82	2.66	2.55	2.46	2.40	2.34	2.30	2.26	2.23	2.20	2.17	2.15	2.13	2.11	2.10	2.08	2.07	2.06	2.05
23	4.28	3.42	3.03	2.80	2.64	2.53	2.44	2.37	2.32	2.27	2.24	2.20	2.18	2.15	2.13	2.11	2.09	2.08	2.06	2.05	2.04	2.02
24	4.26	3.40	3.01	2.78	2.62	2.51	2.42	2.36	2.30	2.25	2.22	2.18	2.15	2.13	2.11	2.09	2.07	2.05	2.04	2.03	2.01	2.00
25	4.24	3.39	2.99	2.76	2.60	2.49	2.40	2.34	2.28	2.24	2.20	2.16	2.14	2.11	2.09	2.07	2.05	2.04	2.02	2.01	2.00	1.98
26	4.23	3.37	2.98	2.74	2.59	2.47	2.39	2.32	2.27	2.22	2.18	2.15	2.12	2.09	2.07	2.05	2.03	2.02	2.00	1.99	1.98	1.97
27	4.21	3.35	2.96	2.73	2.57	2.46	2.37	2.31	2.25	2.20	2.17	2.13	2.10	2.08	2.06	2.04	2.02	2.00	1.99	1.97	1.96	1.95
28	4.20	3.34	2.95	2.71	2.56	2.45	2.36	2.29	2.24	2.19	2.15	2.12	2.09	2.06	2.04	2.02	2.00	1.99	1.97	1.96	1.95	1.93
29	4.18	3.33	2.93	2.70	2.55	2.43	2.35	2.28	2.22	2.18	2.14	2.10	2.08	2.05	2.03	2.01	1.99	1.97	1.96	1.94	1.93	1.92
30	4.17	3.32	2.92	2.69	2.53	2.42	2.33	2.27	2.21	2.16	2.13	2.09	2.06	2.04	2.01	1.99	1.98	1.96	1.95	1.93	1.92	1.91
32	4.15	3.29	2.90	2.67	2.51	2.40	2.31	2.24	2.19	2.14	2.10	2.07	2.04	2.01	1.99	1.97	1.95	1.94	1.92	1.91	1.90	1.88
35	4.12	3.27	2.87	2.64	2.49	2.37	2.29	2.22	2.16	2.11	2.07	2.04	2.01	1.99	1.96	1.94	1.92	1.91	1.89	1.88	1.87	1.85
40	4.08	3.23	2.84	2.61	2.45	2.34	2.25	2.18	2.12	2.08	2.04	2.00	1.97	1.95	1.92	1.90	1.89	1.87	1.85	1.84	1.83	1.81
45	4.06	3.20	2.81	2.58	2.42	2.31	2.22	2.15	2.10	2.05	2.01	1.97	1.94	1.92	1.89	1.87	1.86	1.84	1.82	1.81	1.80	1.78
50	4.03	3.18	2.79	2.56	2.40	2.29	2.20	2.13	2.07	2.03	1.99	1.95	1.92	1.89	1.87	1.85	1.83	1.81	1.80	1.78	1.77	1.76
60	4.00	3.15	2.76	2.53	2.37	2.25	2.17	2.10	2.04	1.99	1.95	1.92	1.89	1.86	1.84	1.82	1.80	1.78	1.76	1.75	1.73	1.72
75	3.97	3.12	2.73	2.49	2.34	2.22	2.13	2.06	2.01	1.96	1.92	1.88	1.85	1.83	1.80	1.78	1.76	1.74	1.73	1.71	1.70	1.69
100	3.94	3.09	2.70	2.46	2.31	2.19	2.10	2.03	1.97	1.93	1.89	1.85	1.82	1.79	1.77	1.75	1.73	1.71	1.69	1.68	1.66	1.65
120	3.92	3.07	2.68	2.45	2.29	2.18	2.09	2.02	1.96	1.91	1.87	1.83	1.80	1.78	1.75	1.73	1.71	1.69	1.67	1.66	1.64	1.63
140	3.91	3.06	2.67	2.44	2.28	2.16	2.08	2.01	1.95	1.90	1.86	1.82	1.79	1.76	1.74	1.72	1.70	1.68	1.66	1.65	1.63	1.62
180	3.89	3.05	2.65	2.42	2.26	2.15	2.06	1.99	1.93	1.88	1.84	1.81	1.77	1.75	1.72	1.70	1.68	1.66	1.64	1.63	1.61	1.60
250	3.88	3.03	2.64	2.41	2.25	2.13	2.05	1.98	1.92	1.87	1.83	1.79	1.76	1.73	1.71	1.68	1.66	1.65	1.63	1.61	1.60	1.58
400	3.86	3.02	2.63	2.39	2.24	2.12	2.03	1.96	1.90	1.85	1.81	1.78	1.74	1.72	1.69	1.67	1.65	1.63	1.61	1.60	1.58	1.57
1000	3.85	3.00	2.61	2.38	2.22	2.11	2.02	1.95	1.89	1.84	1.80	1.76	1.73	1.70	1.68	1.65	1.63	1.61	1.60	1.58	1.57	1.55

Denominator df

Table F (cont.)

Numerator df

$\alpha = .05$	23	24	25	26	27	28	29	30	32	35	40	45	50	60	75	100	120	140	180	250	400	1000
1	248.8	249.1	249.3	249.5	249.6	249.8	250.0	250.1	250.4	250.7	251.1	251.5	251.8	252.2	252.6	253.0	253.3	253.4	253.6	253.8	254.0	254.2
2	19.45	19.45	19.46	19.46	19.46	19.46	19.46	19.46	19.46	19.47	19.47	19.47	19.48	19.48	19.48	19.49	19.49	19.49	19.49	19.49	19.49	19.49
3	8.64	8.64	8.63	8.63	8.63	8.62	8.62	8.62	8.61	8.60	8.59	8.59	8.58	8.57	8.56	8.55	8.55	8.55	8.54	8.54	8.53	8.53
4	5.78	5.77	5.77	5.76	5.76	5.75	5.75	5.75	5.74	5.73	5.72	5.71	5.70	5.69	5.68	5.66	5.66	5.65	5.65	5.64	5.64	5.63
5	4.53	4.53	4.52	4.52	4.51	4.50	4.50	4.50	4.49	4.48	4.46	4.45	4.44	4.43	4.42	4.41	4.40	4.39	4.39	4.38	4.38	4.37
6	3.85	3.84	3.83	3.83	3.82	3.82	3.81	3.81	3.80	3.79	3.77	3.76	3.75	3.74	3.73	3.71	3.70	3.70	3.69	3.69	3.68	3.67
7	3.42	3.41	3.40	3.40	3.39	3.39	3.38	3.38	3.37	3.36	3.34	3.33	3.32	3.30	3.29	3.27	3.27	3.26	3.25	3.25	3.24	3.23
8	3.12	3.12	3.11	3.10	3.10	3.09	3.08	3.08	3.07	3.06	3.04	3.03	3.02	3.01	2.99	2.97	2.97	2.96	2.95	2.95	2.94	2.93
9	2.91	2.90	2.89	2.89	2.88	2.87	2.87	2.86	2.85	2.84	2.83	2.81	2.80	2.79	2.77	2.76	2.75	2.74	2.73	2.73	2.72	2.71
10	2.75	2.74	2.73	2.72	2.72	2.71	2.70	2.70	2.69	2.68	2.66	2.65	2.64	2.62	2.60	2.59	2.58	2.57	2.57	2.56	2.55	2.54
11	2.62	2.61	2.60	2.59	2.59	2.58	2.58	2.57	2.56	2.55	2.53	2.52	2.51	2.49	2.47	2.46	2.45	2.44	2.43	2.43	2.42	2.41
12	2.51	2.51	2.50	2.49	2.48	2.48	2.47	2.47	2.46	2.44	2.43	2.41	2.40	2.38	2.37	2.35	2.34	2.33	2.33	2.32	2.31	2.30
13	2.43	2.42	2.41	2.41	2.40	2.39	2.39	2.38	2.37	2.36	2.34	2.33	2.31	2.30	2.28	2.26	2.25	2.25	2.24	2.23	2.22	2.21
14	2.36	2.35	2.34	2.33	2.33	2.32	2.31	2.31	2.30	2.28	2.27	2.25	2.24	2.22	2.21	2.19	2.18	2.17	2.16	2.15	2.15	2.14
15	2.30	2.29	2.28	2.27	2.27	2.26	2.25	2.25	2.24	2.22	2.20	2.19	2.18	2.16	2.14	2.12	2.11	2.11	2.10	2.09	2.08	2.07
16	2.24	2.24	2.23	2.22	2.21	2.21	2.20	2.19	2.18	2.17	2.15	2.14	2.12	2.11	2.09	2.07	2.06	2.05	2.04	2.03	2.02	2.02
17	2.20	2.19	2.18	2.17	2.17	2.16	2.15	2.15	2.14	2.12	2.10	2.09	2.08	2.06	2.04	2.02	2.01	2.00	1.99	1.98	1.98	1.97
18	2.16	2.15	2.14	2.13	2.13	2.12	2.11	2.11	2.10	2.08	2.06	2.05	2.04	2.02	2.00	1.98	1.97	1.96	1.95	1.94	1.93	1.92
19	2.12	2.11	2.11	2.10	2.09	2.08	2.08	2.07	2.06	2.05	2.03	2.01	2.00	1.98	1.96	1.94	1.93	1.92	1.91	1.90	1.89	1.88
20	2.09	2.08	2.07	2.07	2.06	2.05	2.05	2.04	2.03	2.01	1.99	1.98	1.97	1.95	1.93	1.91	1.90	1.89	1.88	1.87	1.86	1.85
21	2.06	2.05	2.05	2.04	2.03	2.02	2.02	2.01	2.00	1.98	1.96	1.95	1.94	1.92	1.90	1.88	1.87	1.86	1.85	1.84	1.83	1.82
22	2.04	2.03	2.02	2.01	2.00	2.00	1.99	1.98	1.97	1.96	1.94	1.92	1.91	1.89	1.87	1.85	1.84	1.83	1.82	1.81	1.80	1.79
23	2.01	2.01	2.00	1.99	1.98	1.97	1.97	1.96	1.95	1.93	1.91	1.90	1.88	1.86	1.84	1.82	1.81	1.81	1.79	1.78	1.77	1.76
24	1.99	1.98	1.97	1.97	1.96	1.95	1.95	1.94	1.93	1.91	1.89	1.88	1.86	1.84	1.82	1.80	1.79	1.78	1.77	1.76	1.75	1.74
25	1.97	1.96	1.96	1.95	1.94	1.93	1.93	1.92	1.91	1.89	1.87	1.86	1.84	1.82	1.80	1.78	1.77	1.76	1.75	1.74	1.73	1.72
26	1.96	1.95	1.94	1.93	1.92	1.91	1.91	1.90	1.89	1.87	1.85	1.84	1.82	1.80	1.78	1.76	1.75	1.74	1.73	1.72	1.71	1.70
27	1.94	1.93	1.92	1.91	1.90	1.90	1.89	1.88	1.87	1.86	1.84	1.82	1.81	1.79	1.76	1.74	1.73	1.72	1.71	1.70	1.69	1.68
28	1.92	1.91	1.91	1.90	1.89	1.88	1.88	1.87	1.86	1.84	1.82	1.80	1.79	1.77	1.75	1.73	1.71	1.71	1.69	1.68	1.67	1.66
29	1.91	1.90	1.89	1.88	1.88	1.87	1.86	1.85	1.84	1.83	1.81	1.79	1.77	1.75	1.73	1.71	1.70	1.69	1.68	1.67	1.66	1.65
30	1.90	1.89	1.88	1.87	1.86	1.85	1.85	1.84	1.83	1.81	1.79	1.77	1.76	1.74	1.72	1.70	1.68	1.68	1.66	1.65	1.64	1.63
32	1.87	1.86	1.85	1.85	1.84	1.83	1.82	1.82	1.80	1.79	1.77	1.75	1.74	1.71	1.69	1.67	1.66	1.65	1.64	1.63	1.61	1.60
35	1.84	1.83	1.82	1.82	1.81	1.80	1.79	1.79	1.77	1.76	1.74	1.72	1.70	1.68	1.66	1.63	1.62	1.61	1.60	1.59	1.58	1.57
40	1.80	1.79	1.78	1.77	1.77	1.76	1.75	1.74	1.73	1.72	1.69	1.67	1.66	1.64	1.61	1.59	1.58	1.57	1.55	1.54	1.53	1.52
45	1.77	1.76	1.75	1.74	1.73	1.73	1.72	1.71	1.70	1.68	1.66	1.64	1.63	1.60	1.58	1.55	1.54	1.53	1.52	1.51	1.49	1.48
50	1.75	1.74	1.73	1.72	1.71	1.70	1.69	1.69	1.67	1.66	1.63	1.61	1.60	1.58	1.55	1.52	1.51	1.50	1.49	1.47	1.46	1.45
60	1.71	1.70	1.69	1.68	1.67	1.66	1.65	1.65	1.64	1.62	1.59	1.57	1.56	1.53	1.51	1.48	1.47	1.46	1.44	1.43	1.41	1.40
75	1.67	1.66	1.65	1.64	1.63	1.63	1.62	1.61	1.60	1.58	1.55	1.53	1.52	1.49	1.47	1.44	1.42	1.41	1.40	1.38	1.37	1.35
100	1.64	1.63	1.62	1.61	1.60	1.59	1.58	1.57	1.56	1.54	1.52	1.49	1.48	1.45	1.42	1.39	1.38	1.36	1.35	1.33	1.31	1.30
120	1.62	1.61	1.60	1.59	1.58	1.57	1.56	1.55	1.54	1.52	1.50	1.47	1.46	1.43	1.40	1.37	1.35	1.34	1.32	1.30	1.29	1.27
140	1.61	1.60	1.59	1.57	1.57	1.56	1.55	1.54	1.53	1.51	1.48	1.46	1.44	1.41	1.38	1.35	1.33	1.32	1.30	1.29	1.27	1.25
180	1.59	1.58	1.57	1.56	1.55	1.54	1.53	1.52	1.51	1.49	1.46	1.44	1.42	1.39	1.36	1.33	1.31	1.30	1.28	1.26	1.24	1.22
250	1.57	1.56	1.55	1.54	1.53	1.52	1.51	1.50	1.49	1.47	1.44	1.42	1.40	1.37	1.34	1.31	1.29	1.27	1.25	1.23	1.21	1.18
400	1.56	1.54	1.53	1.52	1.51	1.50	1.50	1.49	1.47	1.45	1.42	1.40	1.38	1.35	1.32	1.28	1.26	1.25	1.23	1.20	1.18	1.15
1000	1.54	1.53	1.52	1.51	1.50	1.49	1.48	1.47	1.46	1.43	1.41	1.38	1.36	1.33	1.30	1.26	1.24	1.22	1.20	1.17	1.14	1.11

Denominator df

Table F (cont.)

Numerator df

α = .1	1	2	3	4	5	6	7	8	9	10	11	12	13	14	15	16	17	18	19	20	21	22
1	39.9	49.5	53.6	55.8	57.2	58.2	58.9	59.4	59.9	60.2	60.5	60.7	60.9	61.1	61.2	61.3	61.5	61.6	61.7	61.7	61.8	61.9
2	8.53	9.00	9.16	9.24	9.29	9.33	9.35	9.37	9.38	9.39	9.40	9.41	9.41	9.42	9.42	9.43	9.43	9.44	9.44	9.44	9.44	9.45
3	5.54	5.46	5.39	5.34	5.31	5.28	5.27	5.25	5.24	5.23	5.22	5.22	5.21	5.20	5.20	5.20	5.19	5.19	5.19	5.18	5.18	5.18
4	4.54	4.32	4.19	4.11	4.05	4.01	3.98	3.95	3.94	3.92	3.91	3.90	3.89	3.88	3.87	3.86	3.86	3.85	3.85	3.84	3.84	3.84
5	4.06	3.78	3.62	3.52	3.45	3.40	3.37	3.34	3.32	3.30	3.28	3.27	3.26	3.25	3.24	3.23	3.22	3.22	3.21	3.21	3.20	3.20
6	3.78	3.46	3.29	3.18	3.11	3.05	3.01	2.98	2.96	2.94	2.92	2.90	2.89	2.88	2.87	2.86	2.85	2.85	2.84	2.84	2.83	2.83
7	3.59	3.26	3.07	2.96	2.88	2.83	2.78	2.75	2.72	2.70	2.68	2.67	2.65	2.64	2.63	2.62	2.61	2.61	2.60	2.59	2.59	2.58
8	3.46	3.11	2.92	2.81	2.73	2.67	2.62	2.59	2.56	2.54	2.52	2.50	2.49	2.48	2.46	2.45	2.45	2.44	2.43	2.42	2.42	2.41
9	3.36	3.01	2.81	2.69	2.61	2.55	2.51	2.47	2.44	2.42	2.40	2.38	2.36	2.35	2.34	2.33	2.32	2.31	2.30	2.30	2.29	2.29
10	3.29	2.92	2.73	2.61	2.52	2.46	2.41	2.38	2.35	2.32	2.30	2.28	2.27	2.26	2.24	2.23	2.22	2.22	2.21	2.20	2.19	2.19
11	3.23	2.86	2.66	2.54	2.45	2.39	2.34	2.30	2.27	2.25	2.23	2.21	2.19	2.18	2.17	2.16	2.15	2.14	2.13	2.12	2.12	2.11
12	3.18	2.81	2.61	2.48	2.39	2.33	2.28	2.24	2.21	2.19	2.17	2.15	2.13	2.12	2.10	2.09	2.08	2.08	2.07	2.06	2.05	2.05
13	3.14	2.76	2.56	2.43	2.35	2.28	2.23	2.20	2.16	2.14	2.12	2.10	2.08	2.07	2.05	2.04	2.03	2.02	2.01	2.01	2.00	1.99
14	3.10	2.73	2.52	2.39	2.31	2.24	2.19	2.15	2.12	2.10	2.07	2.05	2.04	2.02	2.01	2.00	1.99	1.98	1.97	1.96	1.96	1.95
15	3.07	2.70	2.49	2.36	2.27	2.21	2.16	2.12	2.09	2.06	2.04	2.02	2.00	1.99	1.97	1.96	1.95	1.94	1.93	1.92	1.92	1.91
16	3.05	2.67	2.46	2.33	2.24	2.18	2.13	2.09	2.06	2.03	2.01	1.99	1.97	1.95	1.94	1.93	1.92	1.91	1.90	1.89	1.88	1.88
17	3.03	2.64	2.44	2.31	2.22	2.15	2.10	2.06	2.03	2.00	1.98	1.96	1.94	1.93	1.91	1.90	1.89	1.88	1.87	1.86	1.86	1.85
18	3.01	2.62	2.42	2.29	2.20	2.13	2.08	2.04	2.00	1.98	1.95	1.93	1.92	1.90	1.89	1.87	1.86	1.85	1.84	1.84	1.83	1.82
19	2.99	2.61	2.40	2.27	2.18	2.11	2.06	2.02	1.98	1.96	1.93	1.91	1.89	1.88	1.86	1.85	1.84	1.83	1.82	1.81	1.81	1.80
20	2.97	2.59	2.38	2.25	2.16	2.09	2.04	2.00	1.96	1.94	1.91	1.89	1.87	1.86	1.84	1.83	1.82	1.81	1.80	1.79	1.79	1.78
21	2.96	2.57	2.36	2.23	2.14	2.08	2.02	1.98	1.95	1.92	1.90	1.87	1.86	1.84	1.83	1.81	1.80	1.79	1.78	1.78	1.77	1.76
22	2.95	2.56	2.35	2.22	2.13	2.06	2.01	1.97	1.93	1.90	1.88	1.86	1.84	1.83	1.81	1.80	1.79	1.78	1.77	1.76	1.75	1.74
23	2.94	2.55	2.34	2.21	2.11	2.05	1.99	1.95	1.92	1.89	1.87	1.84	1.83	1.81	1.80	1.78	1.77	1.76	1.75	1.74	1.74	1.73
24	2.93	2.54	2.33	2.19	2.10	2.04	1.98	1.94	1.91	1.88	1.85	1.83	1.81	1.80	1.78	1.77	1.76	1.75	1.74	1.73	1.72	1.71
25	2.92	2.53	2.32	2.18	2.09	2.02	1.97	1.93	1.89	1.87	1.84	1.82	1.80	1.79	1.77	1.76	1.75	1.74	1.73	1.72	1.71	1.70
26	2.91	2.52	2.31	2.17	2.08	2.01	1.96	1.92	1.88	1.86	1.83	1.81	1.79	1.77	1.76	1.75	1.73	1.72	1.71	1.71	1.70	1.69
27	2.90	2.51	2.30	2.17	2.07	2.00	1.95	1.91	1.87	1.85	1.82	1.80	1.78	1.76	1.75	1.74	1.72	1.71	1.70	1.70	1.69	1.68
28	2.89	2.50	2.29	2.16	2.06	2.00	1.94	1.90	1.87	1.84	1.81	1.79	1.77	1.75	1.74	1.73	1.71	1.70	1.69	1.69	1.68	1.67
29	2.89	2.50	2.28	2.15	2.06	1.99	1.93	1.89	1.86	1.83	1.80	1.78	1.76	1.75	1.73	1.72	1.71	1.69	1.68	1.68	1.67	1.66
30	2.88	2.49	2.28	2.14	2.05	1.98	1.93	1.88	1.85	1.82	1.79	1.77	1.75	1.74	1.72	1.71	1.70	1.69	1.68	1.67	1.66	1.65
32	2.87	2.48	2.26	2.13	2.04	1.97	1.91	1.87	1.83	1.81	1.78	1.76	1.74	1.72	1.71	1.69	1.68	1.67	1.66	1.65	1.64	1.64
35	2.85	2.46	2.25	2.11	2.02	1.95	1.90	1.85	1.82	1.79	1.76	1.74	1.72	1.70	1.69	1.67	1.66	1.65	1.64	1.63	1.62	1.62
40	2.84	2.44	2.23	2.09	2.00	1.93	1.87	1.83	1.79	1.76	1.74	1.71	1.70	1.68	1.66	1.65	1.64	1.62	1.61	1.61	1.60	1.59
45	2.82	2.42	2.21	2.07	1.98	1.91	1.85	1.81	1.77	1.74	1.72	1.70	1.68	1.66	1.64	1.63	1.62	1.60	1.59	1.58	1.58	1.57
50	2.81	2.41	2.20	2.06	1.97	1.90	1.84	1.80	1.76	1.73	1.70	1.68	1.66	1.64	1.63	1.61	1.60	1.59	1.58	1.57	1.56	1.55
60	2.79	2.39	2.18	2.04	1.95	1.87	1.82	1.77	1.74	1.71	1.68	1.66	1.64	1.62	1.60	1.59	1.58	1.56	1.55	1.54	1.53	1.53
75	2.77	2.37	2.16	2.02	1.93	1.85	1.80	1.75	1.72	1.69	1.66	1.63	1.61	1.60	1.58	1.57	1.55	1.54	1.53	1.52	1.51	1.50
100	2.76	2.36	2.14	2.00	1.91	1.83	1.78	1.73	1.69	1.66	1.64	1.61	1.59	1.57	1.56	1.54	1.53	1.52	1.50	1.49	1.48	1.48
120	2.75	2.35	2.13	1.99	1.90	1.82	1.77	1.72	1.68	1.65	1.63	1.60	1.58	1.56	1.55	1.53	1.52	1.50	1.49	1.48	1.47	1.46
140	2.74	2.34	2.12	1.99	1.89	1.81	1.76	1.71	1.68	1.64	1.62	1.59	1.57	1.55	1.54	1.52	1.51	1.50	1.48	1.47	1.46	1.45
180	2.73	2.33	2.11	1.98	1.88	1.80	1.75	1.70	1.67	1.63	1.61	1.58	1.56	1.54	1.53	1.51	1.50	1.48	1.47	1.46	1.45	1.44
250	2.73	2.32	2.11	1.97	1.87	1.79	1.74	1.69	1.66	1.62	1.60	1.57	1.55	1.53	1.51	1.50	1.49	1.47	1.46	1.45	1.44	1.43
400	2.72	2.32	2.10	1.96	1.86	1.79	1.73	1.69	1.65	1.61	1.59	1.56	1.54	1.52	1.50	1.49	1.47	1.46	1.45	1.44	1.43	1.42
1000	2.71	2.31	2.09	1.95	1.85	1.78	1.72	1.68	1.64	1.61	1.58	1.55	1.53	1.51	1.49	1.48	1.46	1.45	1.44	1.43	1.42	1.41

Denominator df

Table F (cont.)

Numerator df

$\alpha = .1$	23	24	25	26	27	28	29	30	32	35	40	45	50	60	75	100	120	140	180	250	400	1000
1	61.9	62.0	62.1	62.1	62.1	62.2	62.2	62.3	62.3	62.4	62.5	62.6	62.7	62.8	62.9	63.0	63.1	63.1	63.1	63.2	63.2	63.3
2	9.45	9.45	9.45	9.45	9.45	9.46	9.46	9.46	9.46	9.46	9.47	9.47	9.47	9.47	9.48	9.48	9.48	9.48	9.49	9.49	9.49	9.49
3	5.18	5.18	5.17	5.17	5.17	5.17	5.17	5.17	5.17	5.16	5.16	5.16	5.15	5.15	5.15	5.14	5.14	5.14	5.14	5.14	5.14	5.1
4	3.83	3.83	3.83	3.83	3.82	3.82	3.82	3.82	3.81	3.81	3.80	3.80	3.80	3.79	3.78	3.78	3.78	3.77	3.77	3.77	3.77	3.76
5	3.19	3.19	3.19	3.18	3.18	3.18	3.18	3.17	3.17	3.16	3.16	3.15	3.15	3.14	3.13	3.13	3.12	3.12	3.12	3.11	3.11	3.11
6	2.82	2.82	2.81	2.81	2.81	2.81	2.80	2.80	2.80	2.79	2.78	2.77	2.77	2.76	2.75	2.75	2.74	2.74	2.74	2.73	2.73	2.72
7	2.58	2.58	2.57	2.57	2.56	2.56	2.56	2.56	2.55	2.54	2.54	2.53	2.52	2.51	2.51	2.50	2.49	2.49	2.49	2.48	2.48	2.47
8	2.41	2.40	2.40	2.40	2.39	2.39	2.39	2.38	2.38	2.37	2.36	2.35	2.35	2.34	2.33	2.32	2.32	2.31	2.31	2.30	2.30	2.30
9	2.28	2.28	2.27	2.27	2.26	2.26	2.26	2.25	2.25	2.24	2.23	2.22	2.22	2.21	2.20	2.19	2.18	2.18	2.18	2.17	2.17	2.16
10	2.18	2.18	2.17	2.17	2.17	2.16	2.16	2.16	2.15	2.14	2.13	2.12	2.12	2.11	2.10	2.09	2.08	2.08	2.07	2.07	2.06	2.06
11	2.11	2.10	2.10	2.09	2.09	2.08	2.08	2.08	2.07	2.06	2.05	2.04	2.04	2.03	2.02	2.01	2.00	2.00	1.99	1.99	1.98	1.98
12	2.04	2.04	2.03	2.03	2.02	2.02	2.01	2.01	2.01	2.00	1.99	1.98	1.97	1.96	1.95	1.94	1.93	1.93	1.92	1.92	1.91	1.91
13	1.99	1.98	1.98	1.97	1.97	1.96	1.96	1.96	1.95	1.94	1.93	1.92	1.92	1.90	1.89	1.88	1.88	1.87	1.87	1.86	1.86	1.85
14	1.94	1.94	1.93	1.93	1.92	1.92	1.92	1.91	1.91	1.90	1.89	1.88	1.87	1.86	1.85	1.83	1.83	1.82	1.82	1.81	1.81	1.80
15	1.90	1.90	1.89	1.89	1.88	1.88	1.88	1.87	1.87	1.86	1.85	1.84	1.83	1.82	1.80	1.79	1.79	1.78	1.78	1.77	1.76	1.76
16	1.87	1.87	1.86	1.86	1.85	1.85	1.84	1.84	1.83	1.82	1.81	1.80	1.79	1.78	1.77	1.76	1.75	1.75	1.74	1.73	1.73	1.72
17	1.84	1.84	1.83	1.83	1.82	1.82	1.81	1.81	1.80	1.79	1.78	1.77	1.76	1.75	1.74	1.73	1.72	1.71	1.71	1.70	1.70	1.69
18	1.82	1.81	1.80	1.80	1.80	1.79	1.79	1.78	1.78	1.77	1.75	1.74	1.74	1.72	1.71	1.70	1.69	1.69	1.68	1.67	1.67	1.66
19	1.79	1.79	1.78	1.78	1.77	1.77	1.76	1.76	1.75	1.74	1.73	1.72	1.71	1.70	1.69	1.67	1.67	1.66	1.65	1.65	1.64	1.64
20	1.77	1.77	1.76	1.76	1.75	1.75	1.74	1.74	1.73	1.72	1.71	1.70	1.69	1.68	1.66	1.65	1.64	1.64	1.63	1.62	1.62	1.61
21	1.75	1.75	1.74	1.74	1.73	1.73	1.72	1.72	1.71	1.70	1.69	1.68	1.67	1.66	1.64	1.63	1.62	1.62	1.61	1.60	1.60	1.59
22	1.74	1.73	1.73	1.72	1.72	1.71	1.71	1.70	1.69	1.68	1.67	1.66	1.65	1.64	1.63	1.61	1.60	1.60	1.59	1.59	1.58	1.57
23	1.72	1.72	1.71	1.70	1.70	1.69	1.69	1.69	1.68	1.67	1.66	1.64	1.64	1.62	1.61	1.59	1.59	1.58	1.57	1.57	1.56	1.55
24	1.71	1.70	1.70	1.69	1.69	1.69	1.68	1.67	1.66	1.65	1.64	1.63	1.62	1.61	1.59	1.58	1.57	1.57	1.56	1.55	1.54	1.54
25	1.70	1.69	1.68	1.68	1.67	1.67	1.66	1.66	1.65	1.64	1.63	1.62	1.61	1.59	1.58	1.56	1.56	1.55	1.54	1.54	1.53	1.52
26	1.68	1.68	1.67	1.67	1.66	1.66	1.65	1.65	1.64	1.63	1.61	1.60	1.59	1.58	1.57	1.55	1.54	1.54	1.53	1.52	1.52	1.51
27	1.67	1.67	1.66	1.65	1.65	1.64	1.64	1.64	1.63	1.62	1.60	1.59	1.58	1.57	1.55	1.54	1.53	1.53	1.52	1.51	1.50	1.50
28	1.66	1.66	1.65	1.64	1.64	1.63	1.63	1.63	1.62	1.61	1.59	1.58	1.57	1.56	1.54	1.53	1.52	1.51	1.51	1.50	1.49	1.48
29	1.65	1.65	1.64	1.63	1.63	1.62	1.62	1.62	1.61	1.60	1.58	1.57	1.56	1.55	1.53	1.52	1.51	1.50	1.50	1.49	1.48	1.47
30	1.64	1.64	1.63	1.63	1.62	1.62	1.61	1.61	1.60	1.59	1.57	1.56	1.55	1.54	1.52	1.51	1.50	1.49	1.49	1.48	1.47	1.46
32	1.63	1.62	1.62	1.61	1.60	1.60	1.59	1.59	1.58	1.57	1.56	1.54	1.53	1.52	1.50	1.49	1.48	1.47	1.47	1.46	1.45	1.44
35	1.61	1.60	1.60	1.59	1.58	1.58	1.57	1.57	1.56	1.55	1.53	1.52	1.51	1.50	1.48	1.47	1.46	1.45	1.44	1.43	1.43	1.42
40	1.58	1.57	1.57	1.56	1.56	1.55	1.55	1.54	1.53	1.52	1.51	1.49	1.48	1.47	1.45	1.43	1.42	1.42	1.41	1.40	1.39	1.38
45	1.56	1.55	1.55	1.54	1.53	1.53	1.52	1.52	1.51	1.50	1.48	1.47	1.46	1.44	1.43	1.41	1.40	1.40	1.38	1.37	1.37	1.36
50	1.54	1.54	1.53	1.52	1.52	1.51	1.51	1.50	1.49	1.48	1.46	1.45	1.44	1.42	1.41	1.39	1.38	1.37	1.36	1.35	1.34	1.33
60	1.52	1.51	1.50	1.50	1.49	1.49	1.48	1.48	1.47	1.45	1.44	1.42	1.41	1.40	1.38	1.36	1.35	1.34	1.33	1.32	1.31	1.30
75	1.49	1.49	1.48	1.47	1.47	1.46	1.45	1.45	1.44	1.43	1.41	1.40	1.38	1.37	1.35	1.33	1.32	1.31	1.30	1.29	1.27	1.26
100	1.47	1.46	1.45	1.45	1.44	1.43	1.43	1.42	1.41	1.40	1.38	1.37	1.35	1.34	1.32	1.29	1.28	1.27	1.26	1.25	1.24	1.22
120	1.46	1.45	1.44	1.43	1.43	1.42	1.41	1.41	1.40	1.38	1.37	1.35	1.34	1.32	1.30	1.28	1.26	1.26	1.24	1.23	1.22	1.20
140	1.45	1.44	1.43	1.42	1.42	1.41	1.41	1.40	1.39	1.38	1.36	1.34	1.33	1.31	1.29	1.26	1.25	1.24	1.23	1.22	1.20	1.19
180	1.43	1.43	1.42	1.41	1.40	1.40	1.39	1.39	1.38	1.36	1.34	1.33	1.32	1.29	1.27	1.25	1.23	1.22	1.21	1.20	1.18	1.16
250	1.42	1.41	1.41	1.40	1.39	1.39	1.38	1.37	1.36	1.35	1.33	1.31	1.30	1.28	1.26	1.23	1.22	1.21	1.19	1.18	1.16	1.14
400	1.41	1.40	1.39	1.39	1.38	1.37	1.37	1.36	1.35	1.34	1.32	1.30	1.29	1.26	1.24	1.21	1.20	1.19	1.17	1.16	1.14	1.12
1000	1.40	1.39	1.38	1.38	1.37	1.36	1.36	1.35	1.34	1.32	1.30	1.29	1.27	1.25	1.23	1.20	1.18	1.17	1.15	1.13	1.11	1.08

Denominator df

Table R

n	d_2	d_3	D_3	D_4
2	1.128	0.852	0	3.267
3	1.693	0.888	0	2.574
4	2.059	0.880	0	2.282
5	2.326	0.864	0	2.114
6	2.534	0.848	0	2.004
7	2.704	0.833	0.076	1.924
8	2.847	0.820	0.136	1.864
9	2.970	0.808	0.184	1.816
10	3.078	0.797	0.223	1.777
11	3.173	0.787	0.256	1.744
12	3.258	0.779	0.283	1.717
13	3.336	0.771	0.307	1.693
14	3.407	0.763	0.328	1.672
15	3.472	0.756	0.347	1.653
16	3.532	0.750	0.363	1.637
17	3.588	0.744	0.378	1.622
18	3.640	0.738	0.391	1.608
19	3.689	0.734	0.403	1.597
20	3.735	0.728	0.415	1.585
21	3.778	0.724	0.425	1.575
22	3.819	0.721	0.434	1.566
23	3.858	0.716	0.443	1.557
24	3.895	0.711	0.451	1.548
25	3.931	0.709	0.459	1.541

Two-tail probability	0.20	0.10	0.05	0.02	0.01	
One-tail probability	0.10	0.05	0.025	0.01	0.005	
Table T df						df
Values of t_α 1	3.078	6.314	12.706	31.821	63.657	1
2	1.886	2.920	4.303	6.965	9.925	2
3	1.638	2.353	3.182	4.541	5.841	3
4	1.533	2.132	2.776	3.747	4.604	4
5	1.476	2.015	2.571	3.365	4.032	5
6	1.440	1.943	2.447	3.143	3.707	6
7	1.415	1.895	2.365	2.998	3.499	7
8	1.397	1.860	2.306	2.896	3.355	8
9	1.383	1.833	2.262	2.821	3.250	9
10	1.372	1.812	2.228	2.764	3.169	10
11	1.363	1.796	2.201	2.718	3.106	11
12	1.356	1.782	2.179	2.681	3.055	12
13	1.350	1.771	2.160	2.650	3.012	13
14	1.345	1.761	2.145	2.624	2.977	14
15	1.341	1.753	2.131	2.602	2.947	15
16	1.337	1.746	2.120	2.583	2.921	16
17	1.333	1.740	2.110	2.567	2.898	17
18	1.330	1.734	2.101	2.552	2.878	18
19	1.328	1.729	2.093	2.539	2.861	19
20	1.325	1.725	2.086	2.528	2.845	20
21	1.323	1.721	2.080	2.518	2.831	21
22	1.321	1.717	2.074	2.508	2.819	22
23	1.319	1.714	2.069	2.500	2.807	23
24	1.318	1.711	2.064	2.492	2.797	24
25	1.316	1.708	2.060	2.485	2.787	25
26	1.315	1.706	2.056	2.479	2.779	26
27	1.314	1.703	2.052	2.473	2.771	27
28	1.313	1.701	2.048	2.467	2.763	28
29	1.311	1.699	2.045	2.462	2.756	29
30	1.310	1.697	2.042	2.457	2.750	30
32	1.309	1.694	2.037	2.449	2.738	32
35	1.306	1.690	2.030	2.438	2.725	35
40	1.303	1.684	2.021	2.423	2.704	40
45	1.301	1.679	2.014	2.412	2.690	45
50	1.299	1.676	2.009	2.403	2.678	50
60	1.296	1.671	2.000	2.390	2.660	60
75	1.293	1.665	1.992	2.377	2.643	75
100	1.290	1.660	1.984	2.364	2.626	100
120	1.289	1.658	1.980	2.358	2.617	120
140	1.288	1.656	1.977	2.353	2.611	140
180	1.286	1.653	1.973	2.347	2.603	180
250	1.285	1.651	1.969	2.341	2.596	250
400	1.284	1.649	1.966	2.336	2.588	400
1000	1.282	1.646	1.962	2.330	2.581	1000
∞	1.282	1.645	1.960	2.326	2.576	∞
Confidence levels	80%	90%	95%	98%	99%	

Two tails

One tail

Table W1

Critical Values of T_L and T_U for the Wilcoxon Rank Sum Test: Independent Samples

Test statistic is the rank sum associated with the smaller sample (if equal sample sizes, either rank sum can be used). Reject the null hypothesis at the indicated α level if the test statistic falls below the lower bound, T_L, or above the upper bound, T_U. For groups larger than 10, use the Normal approximation given in the text.

a. $\alpha = .025$ one-tailed; $\alpha = .05$ two-tailed

n_2 \ n_1	3		4		5		6		7		8		9		10	
	T_L	T_U	T_L	T_U	T_L	T_U	T_L	T_U	T_L	T_U	T_L	T_U	T_L	T_U	T_L	T_U
3	5	16	6	18	6	21	7	23	7	26	8	28	8	31	9	33
4	6	18	11	25	12	28	12	32	13	35	14	38	15	41	16	44
5	6	21	12	28	18	37	19	41	20	45	21	49	22	53	24	56
6	7	23	12	32	19	41	26	52	28	56	29	61	31	65	32	70
7	7	26	13	35	20	45	28	56	37	68	39	73	41	78	43	83
8	8	28	14	38	21	49	29	61	39	73	49	87	51	93	54	98
9	8	31	15	41	22	53	31	65	41	78	51	93	63	108	66	114
10	9	33	16	44	24	56	32	70	43	83	54	98	66	114	79	131

a. $\alpha = .05$ one-tailed; $\alpha = .10$ two-tailed

n_2 \ n_1	3		4		5		6		7		8		9		10	
	T_L	T_U	T_L	T_U	T_L	T_U	T_L	T_U	T_L	T_U	T_L	T_U	T_L	T_U	T_L	T_U
3	6	15	7	17	7	20	8	22	9	24	9	27	10	29	11	31
4	7	17	12	24	13	27	14	30	15	33	16	36	17	39	18	42
5	7	20	13	27	19	36	20	40	22	43	24	46	25	50	26	54
6	8	22	14	30	20	40	28	50	30	54	32	58	33	63	35	67
7	9	24	15	33	22	43	30	54	39	66	41	71	43	76	46	80
8	9	27	16	36	24	46	32	58	41	71	52	84	54	90	57	95
9	10	29	17	39	25	50	33	63	43	76	54	90	66	105	69	111
10	11	31	18	42	26	54	35	67	46	80	57	95	69	111	83	127

Source: From F. Wilcoxon and R. A. Wilcox, "Some Rapid Approximate Statistical Procedures," 1964. Copyright © 1964 by BASF Corporation.

Table W2

Critical Values of T_0 in the Wilcoxon Paired Difference Signed-Rank Test. Reject the null hypothesis at the indicated α levels if the test statistic is larger than the corresponding critical value.

One-Tailed	Two-Tailed	$n = 5$	$n = 6$	$n = 7$	$n = 8$	$n = 9$	$n = 10$
$\alpha = .05$	$\alpha = .10$	1	2	4	6	8	11
$\alpha = .025$	$\alpha = .05$		1	2	4	6	8
$\alpha = .01$	$\alpha = .02$			0	2	3	5
$\alpha = .005$	$\alpha = .01$				0	2	3
		$n = 11$	$n = 12$	$n = 13$	$n = 14$	$n = 15$	$n = 16$
$\alpha = .05$	$\alpha = .10$	14	17	21	26	30	36
$\alpha = .025$	$\alpha = .05$	11	14	17	21	25	30
$\alpha = .01$	$\alpha = .02$	7	10	13	16	20	24
$\alpha = .005$	$\alpha = .01$	5	7	10	13	16	19
		$n = 17$	$n = 18$	$n = 19$	$n = 20$	$n = 21$	$n = 22$
$\alpha = .05$	$\alpha = .10$	41	47	54	60	68	75
$\alpha = .025$	$\alpha = .05$	35	40	46	52	59	66
$\alpha = .01$	$\alpha = .02$	28	33	38	43	49	56
$\alpha = .005$	$\alpha = .01$	23	28	32	37	43	49
		$n = 23$	$n = 24$	$n = 25$	$n = 26$	$n = 27$	$n = 28$
$\alpha = .05$	$\alpha = .10$	83	92	101	110	120	130
$\alpha = .025$	$\alpha = .05$	73	81	90	98	107	117
$\alpha = .01$	$\alpha = .02$	62	69	77	85	93	102
$\alpha = .005$	$\alpha = .01$	55	61	68	76	84	92
		$n = 29$	$n = 30$	$n = 31$	$n = 32$	$n = 33$	$n = 34$
$\alpha = .05$	$\alpha = .10$	141	152	163	175	188	201
$\alpha = .025$	$\alpha = .05$	127	137	148	159	171	183
$\alpha = .01$	$\alpha = .02$	111	120	130	141	151	162
$\alpha = .005$	$\alpha = .01$	100	109	118	128	138	149
		$n = 35$	$n = 36$	$n = 37$	$n = 38$	$n = 39$	
$\alpha = .05$	$\alpha = .10$	214	228	242	256	271	
$\alpha = .025$	$\alpha = .05$	195	208	222	235	250	
$\alpha = .01$	$\alpha = .02$	174	186	198	211	224	
$\alpha = .005$	$\alpha = .01$	160	171	183	195	208	
		$n = 40$	$n = 41$	$n = 42$	$n = 43$	$n = 44$	$n = 45$
$\alpha = .05$	$\alpha = .10$	287	303	319	336	353	371
$\alpha = .025$	$\alpha = .05$	264	279	295	311	327	344
$\alpha = .01$	$\alpha = .02$	238	252	267	281	297	313
$\alpha = .005$	$\alpha = .01$	221	234	248	262	277	292
		$n = 46$	$n = 47$	$n = 48$	$n = 49$	$n = 50$	
$\alpha = .05$	$\alpha = .10$	389	408	427	446	466	
$\alpha = .025$	$\alpha = .05$	361	379	397	415	434	
$\alpha = .01$	$\alpha = .02$	329	345	362	380	398	
$\alpha = .005$	$\alpha = .01$	307	323	339	356	373	

Source: From F. Wilcoxon and R. A. Wilcox, "Some Rapid Approximate Statistical Procedures," 1964. Copyright © 1964 by BASF Corporation.

Right-tail probability		0.10	0.05	0.025	0.01	0.005
Table X	df					
Values of χ^2_α	1	2.706	3.841	5.024	6.635	7.879
	2	4.605	5.991	7.378	9.210	10.597
	3	6.251	7.815	9.348	11.345	12.838
	4	7.779	9.488	11.143	13.277	14.860
	5	9.236	11.070	12.833	15.086	16.750
	6	10.645	12.592	14.449	16.812	18.548
	7	12.017	14.067	16.013	18.475	20.278
	8	13.362	15.507	17.535	20.090	21.955
	9	14.684	16.919	19.023	21.666	23.589
	10	15.987	18.307	20.483	23.209	25.188
	11	17.275	19.675	21.920	24.725	26.757
	12	18.549	21.026	23.337	26.217	28.300
	13	19.812	22.362	24.736	27.688	29.819
	14	21.064	23.685	26.119	29.141	31.319
	15	22.307	24.996	27.488	30.578	32.801
	16	23.542	26.296	28.845	32.000	34.267
	17	24.769	27.587	30.191	33.409	35.718
	18	25.989	28.869	31.526	34.805	37.156
	19	27.204	30.143	32.852	36.191	38.582
	20	28.412	31.410	34.170	37.566	39.997
	21	29.615	32.671	35.479	38.932	41.401
	22	30.813	33.924	36.781	40.290	42.796
	23	32.007	35.172	38.076	41.638	44.181
	24	33.196	36.415	39.364	42.980	45.559
	25	34.382	37.653	40.647	44.314	46.928
	26	35.563	38.885	41.923	45.642	48.290
	27	36.741	40.113	43.195	46.963	49.645
	28	37.916	41.337	44.461	48.278	50.994
	29	39.087	42.557	45.722	59.588	52.336
	30	40.256	43.773	46.979	50.892	53.672
	40	51.805	55.759	59.342	63.691	66.767
	50	63.167	67.505	71.420	76.154	79.490
	60	74.397	79.082	83.298	88.381	91.955
	70	85.527	90.531	95.023	100.424	104.213
	80	96.578	101.879	106.628	112.328	116.320
	90	107.565	113.145	118.135	124.115	128.296
	100	118.499	124.343	129.563	135.811	140.177

Table Z				SECOND DECIMAL PLACE IN Z							
Areas under the standard Normal curve	*0.09*	*0.08*	*0.07*	*0.06*	*0.05*	*0.04*	*0.03*	*0.02*	*0.01*	*0.00*	*z*
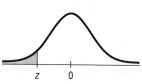										0.0000†	−3.9
	0.0001	0.0001	0.0001	0.0001	0.0001	0.0001	0.0001	0.0001	0.0001	0.0001	−3.8
	0.0001	0.0001	0.0001	0.0001	0.0001	0.0001	0.0001	0.0001	0.0001	0.0001	−3.7
	0.0001	0.0001	0.0001	0.0001	0.0001	0.0001	0.0001	0.0001	0.0002	0.0002	−3.6
	0.0002	0.0002	0.0002	0.0002	0.0002	0.0002	0.0002	0.0002	0.0002	0.0002	−3.5
	0.0002	0.0003	0.0003	0.0003	0.0003	0.0003	0.0003	0.0003	0.0003	0.0003	−3.4
	0.0003	0.0004	0.0004	0.0004	0.0004	0.0004	0.0004	0.0005	0.0005	0.0005	−3.3
	0.0005	0.0005	0.0005	0.0006	0.0006	0.0006	0.0006	0.0006	0.0007	0.0007	−3.2
	0.0007	0.0007	0.0008	0.0008	0.0008	0.0008	0.0009	0.0009	0.0009	0.0010	−3.1
	0.0010	0.0010	0.0011	0.0011	0.0011	0.0012	0.0012	0.0013	0.0013	0.0013	−3.0
	0.0014	0.0014	0.0015	0.0015	0.0016	0.0016	0.0017	0.0018	0.0018	0.0019	−2.9
	0.0019	0.0020	0.0021	0.0021	0.0022	0.0023	0.0023	0.0024	0.0025	0.0026	−2.8
	0.0026	0.0027	0.0028	0.0029	0.0030	0.0031	0.0032	0.0033	0.0034	0.0035	−2.7
	0.0036	0.0037	0.0038	0.0039	0.0040	0.0041	0.0043	0.0044	0.0045	0.0047	−2.6
	0.0048	0.0049	0.0051	0.0052	0.0054	0.0055	0.0057	0.0059	0.0060	0.0062	−2.5
	0.0064	0.0066	0.0068	0.0069	0.0071	0.0073	0.0075	0.0078	0.0080	0.0082	−2.4
	0.0084	0.0087	0.0089	0.0091	0.0094	0.0096	0.0099	0.0102	0.0104	0.0107	−2.3
	0.0110	0.0113	0.0116	0.0119	0.0122	0.0125	0.0129	0.0132	0.0136	0.0139	−2.2
	0.0143	0.0146	0.0150	0.0154	0.0158	0.0162	0.0166	0.0170	0.0174	0.0179	−2.1
	0.0183	0.0188	0.0192	0.0197	0.0202	0.0207	0.0212	0.0217	0.0222	0.0228	−2.0
	0.0233	0.0239	0.0244	0.0250	0.0256	0.0262	0.0268	0.0274	0.0281	0.0287	−1.9
	0.0294	0.0301	0.0307	0.0314	0.0322	0.0329	0.0336	0.0344	0.0351	0.0359	−1.8
	0.0367	0.0375	0.0384	0.0392	0.0401	0.0409	0.0418	0.0427	0.0436	0.0446	−1.7
	0.0455	0.0465	0.0475	0.0485	0.0495	0.0505	0.0516	0.0526	0.0537	0.0548	−1.6
	0.0559	0.0571	0.0582	0.0594	0.0606	0.0618	0.0630	0.0643	0.0655	0.0668	−1.5
	0.0681	0.0694	0.0708	0.0721	0.0735	0.0749	0.0764	0.0778	0.0793	0.0808	−1.4
	0.0823	0.0838	0.0853	0.0869	0.0885	0.0901	0.0918	0.0934	0.0951	0.0968	−1.3
	0.0985	0.1003	0.1020	0.1038	0.1056	0.1075	0.1093	0.1112	0.1131	0.1151	−1.2
	0.1170	0.1190	0.1210	0.1230	0.1251	0.1271	0.1292	0.1314	0.1335	0.1357	−1.1
	0.1379	0.1401	0.1423	0.1446	0.1469	0.1492	0.1515	0.1539	0.1562	0.1587	−1.0
	0.1611	0.1635	0.1660	0.1685	0.1711	0.1736	0.1762	0.1788	0.1814	0.1841	−0.9
	0.1867	0.1894	0.1922	0.1949	0.1977	0.2005	0.2033	0.2061	0.2090	0.2119	−0.8
	0.2148	0.2177	0.2206	0.2236	0.2266	0.2296	0.2327	0.2358	0.2389	0.2420	−0.7
	0.2451	0.2483	0.2514	0.2546	0.2578	0.2611	0.2643	0.2676	0.2709	0.2743	−0.6
	0.2776	0.2810	0.2843	0.2877	0.2912	0.2946	0.2981	0.3015	0.3050	0.3085	−0.5
	0.3121	0.3156	0.3192	0.3228	0.3264	0.3300	0.3336	0.3372	0.3409	0.3446	−0.4
	0.3483	0.3520	0.3557	0.3594	0.3632	0.3669	0.3707	0.3745	0.3783	0.3821	−0.3
	0.3859	0.3897	0.3936	0.3974	0.4013	0.4052	0.4090	0.4129	0.4168	0.4207	−0.2
	0.4247	0.4286	0.4325	0.4364	0.4404	0.4443	0.4483	0.4522	0.4562	0.4602	−0.1
	0.4641	0.4681	0.4721	0.4761	0.4801	0.4840	0.4880	0.4920	0.4960	0.5000	−0.0

† For $z \leq -3.90$ the areas are 0.0000 to four decimal places.

Table Z (cont.)

Areas under the standard Normal curve

SECOND DECIMAL PLACE IN Z										
z	*0.00*	*0.01*	*0.02*	*0.03*	*0.04*	*0.05*	*0.06*	*0.07*	*0.08*	*0.09*
0.0	0.5000	0.5040	0.5080	0.5120	0.5160	0.5199	0.5239	0.5279	0.5319	0.5359
0.1	0.5398	0.5438	0.5478	0.5517	0.5557	0.5596	0.5636	0.5675	0.5714	0.5753
0.2	0.5793	0.5832	0.5871	0.5910	0.5948	0.5987	0.6026	0.6064	0.6103	0.6141
0.3	0.6179	0.6217	0.6255	0.6293	0.6331	0.6368	0.6406	0.6443	0.6480	0.6517
0.4	0.6554	0.6591	0.6628	0.6664	0.6700	0.6736	0.6772	0.6808	0.6844	0.6879
0.5	0.6915	0.6950	0.6985	0.7019	0.7054	0.7088	0.7123	0.7157	0.7190	0.7224
0.6	0.7257	0.7291	0.7324	0.7357	0.7389	0.7422	0.7454	0.7486	0.7517	0.7549
0.7	0.7580	0.7611	0.7642	0.7673	0.7704	0.7734	0.7764	0.7794	0.7823	0.7852
0.8	0.7881	0.7910	0.7939	0.7967	0.7995	0.8023	0.8051	0.8078	0.8106	0.8133
0.9	0.8159	0.8186	0.8212	0.8238	0.8264	0.8289	0.8315	0.8340	0.8365	0.8389
1.0	0.8413	0.8438	0.8461	0.8485	0.8508	0.8531	0.8554	0.8577	0.8599	0.8621
1.1	0.8643	0.8665	0.8686	0.8708	0.8729	0.8749	0.8770	0.8790	0.8810	0.8830
1.2	0.8849	0.8869	0.8888	0.8907	0.8925	0.8944	0.8962	0.8980	0.8997	0.9015
1.3	0.9032	0.9049	0.9066	0.9082	0.9099	0.9115	0.9131	0.9147	0.9162	0.9177
1.4	0.9192	0.9207	0.9222	0.9236	0.9251	0.9265	0.9279	0.9292	0.9306	0.9319
1.5	0.9332	0.9345	0.9357	0.9370	0.9382	0.9394	0.9406	0.9418	0.9429	0.9441
1.6	0.9452	0.9463	0.9474	0.9484	0.9495	0.9505	0.9515	0.9525	0.9535	0.9545
1.7	0.9554	0.9564	0.9573	0.9582	0.9591	0.9599	0.9608	0.9616	0.9625	0.9633
1.8	0.9641	0.9649	0.9656	0.9664	0.9671	0.9678	0.9686	0.9693	0.9699	0.9706
1.9	0.9713	0.9719	0.9726	0.9732	0.9738	0.9744	0.9750	0.9756	0.9761	0.9767
2.0	0.9772	0.9778	0.9783	0.9788	0.9793	0.9798	0.9803	0.9808	0.9812	0.9817
2.1	0.9821	0.9826	0.9830	0.9834	0.9838	0.9842	0.9846	0.9850	0.9854	0.9857
2.2	0.9861	0.9864	0.9868	0.9871	0.9875	0.9878	0.9881	0.9884	0.9887	0.9890
2.3	0.9893	0.9896	0.9898	0.9901	0.9904	0.9906	0.9909	0.9911	0.9913	0.9916
2.4	0.9918	0.9920	0.9922	0.9925	0.9927	0.9929	0.9931	0.9932	0.9934	0.9936
2.5	0.9938	0.9940	0.9941	0.9943	0.9945	0.9946	0.9948	0.9949	0.9951	0.9952
2.6	0.9953	0.9955	0.9956	0.9957	0.9959	0.9960	0.9961	0.9962	0.9963	0.9964
2.7	0.9965	0.9966	0.9967	0.9968	0.9969	0.9970	0.9971	0.9972	0.9973	0.9974
2.8	0.9974	0.9975	0.9976	0.9977	0.9977	0.9978	0.9979	0.9979	0.9980	0.9981
2.9	0.9981	0.9982	0.9982	0.9983	0.9984	0.9984	0.9985	0.9985	0.9986	0.9986
3.0	0.9987	0.9987	0.9987	0.9988	0.9988	0.9989	0.9989	0.9989	0.9990	0.9990
3.1	0.9990	0.9991	0.9991	0.9991	0.9992	0.9992	0.9992	0.9992	0.9993	0.9993
3.2	0.9993	0.9993	0.9994	0.9994	0.9994	0.9994	0.9994	0.9995	0.9995	0.9995
3.3	0.9995	0.9995	0.9995	0.9996	0.9996	0.9996	0.9996	0.9996	0.9996	0.9997
3.4	0.9997	0.9997	0.9997	0.9997	0.9997	0.9997	0.9997	0.9997	0.9997	0.9998
3.5	0.9998	0.9998	0.9998	0.9998	0.9998	0.9998	0.9998	0.9998	0.9998	0.9998
3.6	0.9998	0.9998	0.9999	0.9999	0.9999	0.9999	0.9999	0.9999	0.9999	0.9999
3.7	0.9999	0.9999	0.9999	0.9999	0.9999	0.9999	0.9999	0.9999	0.9999	0.9999
3.8	0.9999	0.9999	0.9999	0.9999	0.9999	0.9999	0.9999	0.9999	0.9999	0.9999
3.9	1.0000[†]									

[†] For $z \geq 3.90$, the areas are 1.0000 to four decimal places.

ROW					TABLE OF RANDOM DIGITS				
1	96299	07196	98642	20639	23185	56282	69929	14125	38872	94168
2	71622	35940	81807	59225	18192	08710	80777	84395	69563	86280
3	03272	41230	81739	74797	70406	18564	69273	72532	78340	36699
4	46376	58596	14365	63685	56555	42974	72944	96463	63533	24152
5	47352	42853	42903	97504	56655	70355	88606	61406	38757	70657
6	20064	04266	74017	79319	70170	96572	08523	56025	89077	57678
7	73184	95907	05179	51002	83374	52297	07769	99792	78365	93487
8	72753	36216	07230	35793	71907	65571	66784	25548	91861	15725
9	03939	30763	06138	80062	02537	23561	93136	61260	77935	93159
10	75998	37203	07959	38264	78120	77525	86481	54986	33042	70648
11	94435	97441	90998	25104	49761	14967	70724	67030	53887	81293
12	04362	40989	69167	38894	00172	02999	97377	33305	60782	29810
13	89059	43528	10547	40115	82234	86902	04121	83889	76208	31076
14	87736	04666	75145	49175	76754	07884	92564	80793	22573	67902
15	76488	88899	15860	07370	13431	84041	69202	18912	83173	11983
16	36460	53772	66634	25045	79007	78518	73580	14191	50353	32064
17	13205	69237	21820	20952	16635	58867	97650	82983	64865	93298
18	51242	12215	90739	36812	00436	31609	80333	96606	30430	31803
19	67819	00354	91439	91073	49258	15992	41277	75111	67496	68430
20	09875	08990	27656	15871	23637	00952	97818	64234	50199	05715
21	18192	95308	72975	01191	29958	09275	89141	19558	50524	32041
22	02763	33701	66188	50226	35813	72951	11638	01876	93664	37001
23	13349	46328	01856	29935	80563	03742	49470	67749	08578	21956
24	69238	92878	80067	80807	45096	22936	64325	19265	37755	69794
25	92207	63527	59398	29818	24789	94309	88380	57000	50171	17891
26	66679	99100	37072	30593	29665	84286	44458	60180	81451	58273
27	31087	42430	60322	34765	15757	53300	97392	98035	05228	68970
28	84432	04916	52949	78533	31666	62350	20584	56367	19701	60584
29	72042	12287	21081	48426	44321	58765	41760	43304	13399	02043
30	94534	73559	82135	70260	87936	85162	11937	18263	54138	69564
31	63971	97198	40974	45301	60177	35604	21580	68107	25184	42810
32	11227	58474	17272	37619	69517	62964	67962	34510	12607	52255
33	28541	02029	08068	96656	17795	21484	57722	76511	27849	61738
34	11282	43632	49531	78981	81980	08530	08629	32279	29478	50228
35	42907	15137	21918	13248	39129	49559	94540	24070	88151	36782
36	47119	76651	21732	32364	58545	50277	57558	30390	18771	72703
37	11232	99884	05087	76839	65142	19994	91397	29350	83852	04905
38	64725	06719	86262	53356	57999	50193	79936	97230	52073	94467
39	77007	26962	55466	12521	48125	12280	54985	26239	76044	54398
40	18375	19310	59796	89832	59417	18553	17238	05474	33259	50595

Selected Formulas

$Range = Max - Min$

$IQR = Q3 - Q1$

Outlier Rule-of-Thumb: $y < Q1 - 1.5 \times IQR$ or $y > Q3 + 1.5 \times IQR$

$$\bar{y} = \frac{\sum y}{n}$$

$$s = \sqrt{\frac{\sum (y - \bar{y})^2}{n - 1}}$$

$$z = \frac{y - \mu}{\sigma} \text{ (model based)} \qquad z = \frac{y - \bar{y}}{s} \text{ (data based)}$$

$$r = \frac{\sum z_x z_y}{n - 1}$$

$$\hat{y} = b_0 + b_1 x \qquad \text{where } b_1 = r\frac{s_y}{s_x} \text{ and } b_0 = \bar{y} - b_1\bar{x}$$

$$P(\mathbf{A}) = 1 - P(\mathbf{A}^C)$$

$$P(\mathbf{A} \text{ or } \mathbf{B}) = P(\mathbf{A}) + P(\mathbf{B}) - P(\mathbf{A} \text{ and } \mathbf{B})$$

$$P(\mathbf{A} \text{ and } \mathbf{B}) = P(\mathbf{A}) \times P(\mathbf{B}|\mathbf{A})$$

$$P(\mathbf{B}|\mathbf{A}) = \frac{P(\mathbf{A} \text{ and } \mathbf{B})}{P(\mathbf{A})}$$

If $\mathbf{A}$ and $\mathbf{B}$ are independent, $P(\mathbf{B}|\mathbf{A}) = P(\mathbf{B})$

$$E(X) = \mu = \sum x \cdot P(x) \qquad\qquad Var(X) = \sigma^2 = \sum (x - \mu)^2 P(x)$$

$$E(X \pm c) = E(X) \pm c \qquad\qquad Var(X \pm c) = Var(X)$$

$$E(aX) = aE(X) \qquad\qquad Var(aX) = a^2 Var(X)$$

$$E(X \pm Y) = E(X) \pm E(Y) \qquad\qquad Var(X \pm Y) = Var(X) + Var(Y)$$

$$\text{if } X \text{ and } Y \text{ are independent}$$

Geometric: $P(x) = q^{x-1}p \qquad \mu = \frac{1}{p} \qquad \sigma = \sqrt{\frac{q}{p^2}}$

Binomial: $P(x) = {}_nC_x p^x q^{n-x} \qquad \mu = np \qquad \sigma = \sqrt{npq}$

$$\hat{p} = \frac{x}{n} \qquad \mu(\hat{p}) = p \qquad SD(\hat{p}) = \sqrt{\frac{pq}{n}}$$

Poisson probability model for successes: Poisson (λ)

λ = mean number of successes.

X = number of successes.

$$P(X = x) = \frac{e^{-\lambda}\lambda^x}{x!}$$

$$\text{Expected value:} \qquad E(X) = \lambda$$

$$\text{Standard deviation:} \qquad SD(X) = \sqrt{\lambda}$$

Sampling distribution of $\bar{y}$:

(CLT) As n grows, the sampling distribution approaches the Normal model with

$$\mu(\bar{y}) = \mu_y \qquad SD(\bar{y}) = \frac{\sigma}{\sqrt{n}}$$

Inference:

Confidence interval for parameter = **statistic ± critical value × SE(statistic)**

$$\text{Test statistic} = \frac{statistic - parameter}{SD(statistic)}$$

Parameter	Statistic	*SD*(statistic)	*SE*(statistic)
p	$\hat{p}$	$\sqrt{\dfrac{pq}{n}}$	$\sqrt{\dfrac{\hat{p}\hat{q}}{n}}$
μ	$\bar{y}$	$\dfrac{\sigma}{\sqrt{n}}$	$\dfrac{s}{\sqrt{n}}$
$\mu_1 - \mu_2$	$\bar{y}_1 - \bar{y}_2$	$\sqrt{\dfrac{\sigma_1^2}{n_1} + \dfrac{\sigma_2^2}{n_2}}$	$\sqrt{\dfrac{s_1^2}{n_1} + \dfrac{s_2^2}{n_2}}$
μ_d	$\bar{d}$	$\dfrac{\sigma_d}{\sqrt{n}}$	$\dfrac{s_d}{\sqrt{n}}$
σ_ε	$s_e = \sqrt{\dfrac{\sum(y - \hat{y})^2}{n - 2}}$	(divide by $n - k - 1$ in multiple regression)	
β_1	b_1	(in simple regression)	$\dfrac{s_e}{s_x\sqrt{n - 1}}$
μ_ν	$\hat{y}_\nu$	(in simple regression)	$\sqrt{SE^2(b_1) \cdot (x_\nu - \bar{x})^2 + \dfrac{s_e^2}{n}}$
y_ν	$\hat{y}_\nu$	(in simple regression)	$\sqrt{SE^2(b_1) \cdot (x_\nu - \bar{x})^2 + \dfrac{s_e^2}{n} + s_e^2}$

Pooling: For testing difference between proportions: $\hat{p}_{pooled} = \dfrac{y_1 + y_2}{n_1 + n_2}$

For testing difference between means: $s_p = \sqrt{\dfrac{(n_1 - 1)s_1^2 + (n_2 - 1)s_2^2}{n_1 + n_2 - 2}}$

Substitute these pooled estimates in the respective SE formulas for both groups when assumptions and conditions are met.

Chi-square: $\chi^2 = \sum \dfrac{(Obs - Exp)^2}{Exp}$

Assumptions for Inference	And the Conditions That Support or Override Them

Proportions (z)

- **One sample**
  1. Individuals are independent.
  2. Sample is sufficiently large.

  1. SRS and $n < 10\%$ of the population.
  2. Successes and failures each ≥ 10.

Means (t)

- **One Sample** (df $= n - 1$)
  1. Individuals are independent.
  2. Population has a Normal model.

  1. SRS and $n < 10\%$ of the population.
  2. Histogram is unimodal and symmetric.*

- **Matched pairs** (df $= n - 1$)
  1. Data are matched.
  2. Individuals are independent.
  3. Population of differences is Normal.

  1. (Think about the design.)
  2. SRS and $n < 10\%$ OR random allocation.
  3. Histogram of differences is unimodal and symmetric.*

- **Two independent samples** (df from technology)
  1. Groups are independent.
  2. Data in each group are independent.
  3. Both populations are Normal.

  1. (Think about the design.)
  2. SRSs and $n < 10\%$ OR random allocation.
  3. Both histograms are unimodal and symmetric.*

Distributions/Association (χ^2)

- **Goodness of fit** (df $=$ # of cells $- 1$; one variable, one sample compared with population model)
  1. Data are counts.
  2. Data in sample are independent.
  3. Sample is sufficiently large.

  1. (Are they?)
  2. SRS and $n < 10\%$ of the population.
  3. All expected counts ≥ 5.

- **Homogeneity** [df $= (r - 1)(c - 1)$; many groups compared on one variable]
  1. Data are counts.
  2. Data in groups are independent.
  3. Groups are sufficiently large.

  1. (Are they?)
  2. SRSs and $n < 10\%$ OR random allocation.
  3. All expected counts ≥ 5.

- **Independence** [df $= (r - 1)(c - 1)$; sample from one population classified on two variables]
  1. Data are counts.
  2. Data are independent.
  3. Sample is sufficiently large.

  1. (Are they?)
  2. SRSs and $n < 10\%$ of the population.
  3. All expected counts ≥ 5.

Regression with k predictors (t, df $= n - k - 1$)

- **Association** of each quantitative predictor with the response variable
  1. Form of relationship is linear.

  2. Errors are independent.
  3. Variability of errors is constant.

  4. Errors follow a Normal model.

  1. Scatterplots of y against each x are straight enough. Scatterplot of residuals against predicted values shows no special structure.
  2. No apparent pattern in plot of residuals against predicted values.
  3. Plot of residuals against predicted values has constant spread, doesn't "thicken."
  4. Histogram of residuals is approximately unimodal and symmetric, or Normal probability plot is reasonably straight.*

Analysis of Variance (F, df dependent on number of factors and number of levels in each)

- **Equality** of the mean response across levels of categorical predictors
  1. Additive Model (if there are 2 factors with no interaction term).
  2. Independent errors.
  3. Equal variance across treatment levels.

  4. Errors follow a Normal model.

  1. Interaction plot shows parallel lines (otherwise include an interaction term if possible).
  2. Randomized experiment or other suitable randomization.
  3. Plot of residuals against predicted values has constant spread. Boxplots (partial boxplots for 2 factors) show similar spreads.
  4. Histogram of residuals is unimodal and symmetric, or Normal probability plot is reasonably straight.

*Less critical as n increases

Quick Guide to Inference

Plan				Do				Report
Inference about?	**One group or two?**	**Procedure**	**Model**	**Parameter**	**Estimate**	**SE**		**Chapter**
Proportions	One sample	1-Proportion z-Interval	z	p	$\hat{p}$	$\sqrt{\dfrac{\hat{p}\hat{q}}{n}}$		10
		1-Proportion z-Test				$\sqrt{\dfrac{p_0 q_0}{n}}$		11
Means	One sample	t-Interval t-Test	t df $= n - 1$	μ	$\bar{y}$	$\dfrac{s}{\sqrt{n}}$		12
	Two independent groups	2-Sample t-Test 2-Sample t-Interval	t df from technology	$\mu_1 - \mu_2$	$\bar{y}_1 - \bar{y}_2$	$\sqrt{\dfrac{s_1^2}{n_1} + \dfrac{s_2^2}{n_2}}$		13
	Matched pairs	Paired t-Test Paired t-Interval	t df $= n - 1$	μ_d	$\bar{d}$	$\dfrac{s_d}{\sqrt{n}}$		14
Distributions (one categorical variable)	One Sample	Goodness-of-Fit	χ^2 df $= cells - 1$					15
	Many independent groups	Homogeneity χ^2 Test	χ^2 df $= (r - 1)(c - 1)$			$\sum \dfrac{(Obs - Exp)^2}{Exp}$		
Independence (two categorical variables)	One sample	Independence χ^2 Test						
Association (two quantitative variables)	One sample	Linear Regression t-Test or Confidence Interval for β	t df $= n - 2$	β_1	b_1	$\dfrac{s_e}{s_x \sqrt{n - 1}}$ (compute with technology)		16
		*Confidence Interval for μ_v		μ_v	$\hat{y}_v$	$\sqrt{SE^2(b_1) \cdot (x_v - \bar{x})^2 + \dfrac{s_e^2}{n}}$		
		*Prediction Interval for y_v		y_v	$\hat{y}_v$	$\sqrt{SE^2(b_1) \cdot (x_v - \bar{x})^2 + \dfrac{s_e^2}{n} + s_e^2}$		
Association (one quantitative variable fit modeled by k quantitative variables)	One sample	Multiple Regression t-test or Confidence interval for each β_j	t $df = n - (k + 1)$	β_j	b_j	(from technology)		17, 18, 19
		F test for regression model	F $df = k$ and $n - (k + 1)$			MST/MSE		18, 19
Association (one quantitative and two or more categorical variables)	Two or more	ANOVA	F $df = k - 1$ and $N - k$			MST/MSE		23

Index

Note: Page numbers in **boldface** indicate chapter-level topics; FE indicates For Example references; n indicates a footnote.

Assumptions for Inference	And the Conditions That Support or Override Them

Proportions (z)

- **One sample**
  1. Individuals are independent.
  2. Sample is sufficiently large.

  1. SRS and $n < 10\%$ of the population.
  2. Successes and failures each ≥ 10.

Means (t)

- **One Sample** (df $= n - 1$)
  1. Individuals are independent.
  2. Population has a Normal model.

  1. SRS and $n < 10\%$ of the population.
  2. Histogram is unimodal and symmetric.*

- **Matched pairs** (df $= n - 1$)
  1. Data are matched.
  2. Individuals are independent.
  3. Population of differences is Normal.

  1. (Think about the design.)
  2. SRS and $n < 10\%$ OR random allocation.
  3. Histogram of differences is unimodal and symmetric.*

- **Two independent samples** (df from technology)
  1. Groups are independent.
  2. Data in each group are independent.
  3. Both populations are Normal.

  1. (Think about the design.)
  2. SRSs and $n < 10\%$ OR random allocation.
  3. Both histograms are unimodal and symmetric.*

Distributions/Association (χ^2)

- **Goodness of fit** (df $=$ # of cells $- 1$; one variable, one sample compared with population model)
  1. Data are counts.
  2. Data in sample are independent.
  3. Sample is sufficiently large.

  1. (Are they?)
  2. SRS and $n < 10\%$ of the population.
  3. All expected counts ≥ 5.

- **Homogeneity** [df $= (r - 1)(c - 1)$; many groups compared on one variable]
  1. Data are counts.
  2. Data in groups are independent.
  3. Groups are sufficiently large.

  1. (Are they?)
  2. SRSs and $n < 10\%$ OR random allocation.
  3. All expected counts ≥ 5.

- **Independence** [df $= (r - 1)(c - 1)$; sample from one population classified on two variables]
  1. Data are counts.
  2. Data are independent.
  3. Sample is sufficiently large.

  1. (Are they?)
  2. SRSs and $n < 10\%$ of the population.
  3. All expected counts ≥ 5.

Regression with k predictors (t, df $= n - k - 1$)

- **Association** of each quantitative predictor with the response variable
  1. Form of relationship is linear.

  2. Errors are independent.
  3. Variability of errors is constant.

  4. Errors follow a Normal model.

  1. Scatterplots of y against each x are straight enough. Scatterplot of residuals against predicted values shows no special structure.
  2. No apparent pattern in plot of residuals against predicted values.
  3. Plot of residuals against predicted values has constant spread, doesn't "thicken."
  4. Histogram of residuals is approximately unimodal and symmetric, or Normal probability plot is reasonably straight.*

Analysis of Variance (F, df dependent on number of factors and number of levels in each)

- **Equality** of the mean response across levels of categorical predictors
  1. Additive Model (if there are 2 factors with no interaction term).
  2. Independent errors.
  3. Equal variance across treatment levels.

  4. Errors follow a Normal model.

  1. Interaction plot shows parallel lines (otherwise include an interaction term if possible).
  2. Randomized experiment or other suitable randomization.
  3. Plot of residuals against predicted values has constant spread. Boxplots (partial boxplots for 2 factors) show similar spreads.
  4. Histogram of residuals is unimodal and symmetric, or Normal probability plot is reasonably straight.

*Less critical as n increases